Techniques of Scientific Computing (Part 2)

Handbook of Numerical Analysis

General Editors:

P.G. Ciarlet

Analyse Numérique, Tour 55–65
Université Pierre et Marie Curie
4 Place Jussieu
75005 PARIS, France

J.L. Lions

Collège de France
Place Marcelin Berthelot
75005 PARIS, France

ELSEVIER
Amsterdam • Lausanne • New York • Oxford • Shannon • Tokyo

Volume V

Techniques of Scientific Computing (Part 2)

1997
ELSEVIER
Amsterdam • Lausanne • New York • Oxford • Shannon • Tokyo

ELSEVIER SCIENCE B.V.
Sara Burgerhartstraat 25
P.O. Box 211, 1000 AE Amsterdam, The Netherlands

For information on published and forthcoming volumes URL = http://www.elsevier.nl/locate/hna

Library of Congress Catalog Card Number: 89-23314

ISBN: 0 444 82278 X

This book is printed on acid-free paper.

Printed in The Netherlands

General Preface

During the past decades, giant needs for ever more sophisticated mathematical models and increasingly complex and extensive computer simulations have arisen. In this fashion, two indissociable activities, *mathematical modeling* and *computer simulation*, have gained a major status in all aspects of science, technology, and industry.

In order that these two sciences be established on the safest possible grounds, mathematical rigor is indispensable. For this reason, two companion sciences, *Numerical Analysis* and *Scientific Software*, have emerged as essential steps for validating the mathematical models and the computer simulations that are based on them.

Numerical Analysis is here understood as the part of *Mathematics* that describes and analyzes all the numerical schemes that are used on computers; its objective consists in obtaining a clear, precise, and faithful, representation of all the "information" contained in a mathematical model; as such, it is the natural extension of more classical tools, such as analytic solutions, special transforms, functional analysis, as well as stability and asymptotic analysis.

The various volumes comprising the *Handbook of Numerical Analysis* will thoroughly cover all the major aspects of Numerical Analysis, by presenting accessible and in-depth surveys, which include the most recent trends.

More precisely, the Handbook will cover the *basic methods of Numerical Analysis*, gathered under the following general headings:

- Solution of Equations in $\mathbb{R}^n$,
- Finite Difference Methods,
- Finite Element Methods,
- Techniques of Scientific Computing,
- Optimization Theory and Systems Science.

It will also cover the *numerical solution of actual problems of contemporary interest in Applied Mathematics*, gathered under the following general headings:

- Numerical Methods for Fluids,
- Numerical Methods for Solids,
- Specific Applications.

"Specific Applications" include: Meteorology, Seismology, Petroleum Mechanics, Celestial Mechanics, etc.

Each heading is covered by several *articles*, each of which being devoted to a specialized, but to some extent "independent", topic. Each article contains a thorough description and a mathematical analysis of the various methods in actual use, whose practical performances may be illustrated by significant numerical examples.

Since the Handbook is basically expository in nature, only the most basic results are usually proved in detail, while less important, or technical, results may be only stated or commented upon (in which case specific references for their proofs are systematically provided). In the same spirit, only a "selective" bibliography is appended whenever the roughest counts indicate that the reference list of an article should comprise several thousand items if it were to be exhaustive.

Volumes are numbered by capital Roman numerals (as Vol. I, Vol. II, etc.), according to their *chronological appearance.*

Since all the articles pertaining to a given *heading* may not be simultaneously available at a given time, a given heading usually appears in more than one volume; for instance, if articles devoted to the heading "Solution of Equations in $\mathbb{R}^n$" appear in Volumes I and III, these volumes will include "Solution of Equations in $\mathbb{R}^n$ (Part 1)" and "Solution of Equations in $\mathbb{R}^n$ (Part 2)" in their respective titles. Naturally, all the headings dealt with within a given volume appear in its title; for instance, the complete title of Volume I is "Finite Difference Methods (Part 1) — Solution of Equations in $\mathbb{R}^n$ (Part 1)".

Each article is subdivided into *sections*, which are numbered consecutively throughout the article by *Arabic numerals*, as Section 1, Section 2, ..., Section 14, etc. Within a given section, *formulas, theorems, remarks, and figures*, have their own independent numberings; for instance, with Section 14, formulas are numbered consecutively as (14.1), (14.2), etc., theorems are numbered consecutively as Theorem 14.1, Theorem 14.2, etc. For the sake of clarity, the article is also subdivided into *chapters*, numbered consecutively throughout the article by *capital Roman numerals*; for instance, Chapter I comprises Sections 1 to 9, Chapter II comprises Sections 10 to 16, etc.

P.G. CIARLET
J.L. LIONS
May 1989

Contents of Volume V

Contents of the Handbook

Techniques of Scientific Computing (Part 2)

Numerical Path Following

Eugene L. Allgower[1] and Kurt Georg[1]

Department of Mathematics
Colorado State University
Ft. Collins, CO 80523, USA

[1]Partially supported by the National Science Foundation via grant # DMS-9104058.

HANDBOOK OF NUMERICAL ANALYSIS, VOL. V
Techniques of Scientific Computing (Part 2)
Edited by P.G. Ciarlet and J.L. Lions

Contents

CHAPTER I

Introduction

Because of their versatility and robustness, numerical continuation or path following methods have been finding ever wider use in scientific applications. Our aim here is to give an introduction to the main ideas of numerical path following and to present some of the recent advances in this subject regarding new adaptations, applications, and analysis of efficiency and complexity. Introductions into aspects of the subject of numerical path following may be found in the books by GARCIA and ZANGWILL [1981], GOULD and TOLLE [1983], KELLER [1987], RHEINBOLDT [1986], SEYDEL [1988], and TODD [1976a]. The philosophy and notation of the present paper will be that of our book ALLGOWER and GEORG [1990], which also contains an extensive bibliography up to 1990. An updating of the literature is given in ALLGOWER and GEORG [1993].

The viewpoint which will be adopted here is that numerical continuation methods are techniques for numerically approximating a solution curve c which is implicitly defined by an underdetermined system of equations. In the literature of numerical analysis, the terms *numerical continuation* and *path following* are used interchangeably.

There are various objectives for which the numerical approximation of c can be used, and depending upon the particular objective, the approximating technique ought to be adapted accordingly. In fact, continuation is a unifying concept under which various numerical methods that may seem to have very little in common can be interpreted and related. For example, simplicial fixed point methods for solving problems in mathematical economics, the generation of bifurcation diagrams of nonlinear eigenvalue problems involving partial differential equations, and the recently developed interior point methods for solving linear programming problems seem to be quite unrelated. Nevertheless, there is some benefit in considering them as specific examples of path following. We personally are struck by the remarkable fact that a technique which was initially developed for overcoming difficulties involved with nonlinear problems now turns out to be extremely useful even for treating various problems which are essentially linear: e.g., linear eigenvalue problems, linear programming, and linear complementarity problems.

For readers who wish to skip directly to topics which are of particular interest to them, we give a brief outline of the contents of the paper.

Chapter II presents the basic ideas of predictor corrector path following methods and some convergence results are given. For theoretical and basic considerations, it is convenient to consider a solution curve parametrized according to (standard) arclength, since this parametrization is globally valid. However, specific implementations practically always deal with approximations of the arclength, the precise arclength is never calculated (except for using initial value solvers on Eq. (1.8) which is not advisable). A typical example is the concept of pseudo arclength which is discussed in Section 2.

In Chapter III predictor steps and steplength strategies are studied. Several alternatives for higher order predictors are explored. Devices for calculating special points on a curve are also discussed, since they can be viewed in the context of steplength adaptation.

In Chapter IV the technical aspects of implementing predictor corrector methods are addressed, e.g., the numerical linear algebra involved in performing the predictor and corrector steps, approximating the derivative terms which arise in the course of numerical path following, update methods, special considerations for large scale problems. In particular, it is pointed out that the use of the pseudo inverse $H'(u)^+$ and use of the bordering algorithm, i.e., the use of $\tilde{H}'(u)^{-1}$ for an augmented sytem $\tilde{H}$, are closely related (also in terms of computational expense).

Chapter V deals with various applications of path following methods. We begin with a brief discussion of homotopy methods for fixed point problems and global Newton methods. Here the relationship can be seen of path following methods to celebrated results of POINCARÉ [1881–1886], KLEIN [1882–1883] and BERNSTEIN [1910]. In fact, the path following methods involving homotopy deformations are essentially constructive implementations of degree arguments. The problem of finding multiple solutions is examined in detail. Of particular interest are recent homotopy methods for finding all solutions of polynomial systems of equations. We survey some path following aspects of nonlinear eigenvalue problems, and address the question of handling bifurcations. Also three rather new developments in path following are discussed: (1) The solution of linear eigenvalue problems via special homotopy approaches, (2) the handling of parametric programming problems by following certain branches of critical points via active set strategies, (3) the path following aspects involved in the interior point methods for solving linear and quadratic programming problems.

Chapter VI presents an introduction to the principles of piecewise linear methods. These methods view path following in a different light: Instead of approximately following a solution curve of a smooth system $H(u) = 0$, they exactly follow the solution curve (i.e., a polygonal path) of a piecewise affine system $\tilde{H}(u) = 0$ which may or may not be related to a smooth problem $H(u) = 0$. Some instances where these methods are useful are discussed, e.g., linear complementarity problems or homotopy methods where predictor corrector methods are not implementable, because of lack of smoothness. We also briefly address the related topic of approximating implicitly defined surfaces. Several error estimate results are given.

The issue of the computational complexity of path following is considered in Chapter VII. This issue is related to the Newton–Kantorovich theory and is currently of considerable interest in the context of interior point methods.

We conclude by listing some available software related to path following and indicate how the reader might access these codes. We make no attempt to compare or evaluate the various codes. Our opinion is that path following codes ought to be adapted to the special purposes for which they are to be used. Although there are some general purpose codes, probably none will slay every dragon and by making appropriate adaptations for the special purpose at hand, improved efficiency can usually be achieved.

Many colleagues helped us with suggestions and references. In particular, we are grateful to Hubert Schwetlick for valuable contributions.

CHAPTER II

Basics

1. Predictor corrector path following

In the context of numerical continuation methods, one considers curves which are implicitly defined by an underdetermined system of equations

$$H(u) = 0 \tag{1.1}$$

where $H : \mathbb{R}^{N+1} \to \mathbb{R}^N$ is a smooth map.

When we say that a map is smooth we shall mean that it has as many continuous derivatives as the context of the subsequent discussion requires. For convenience, the reader may assume smoothness means C^∞.

More generally, if $H : \mathbb{R}^{N+K} \to \mathbb{R}^N$, we would expect the solution set of $H(u) = 0$ to describe a K-dimensional manifold. However, to ensure this, we need a nondegeneracy condition: We call u a *regular point* of H if the Jacobian $H'(u)$ has maximal rank. We call y a *regular value* of H if u is a regular point of H whenever $H(u) = y$. In particular, if $y \notin \operatorname{range} H$, then y is a regular value. If a point or value is not regular, then it is called *singular*.

With these conditions, we may now state the classical implicit function theorem in a global setting, see, e.g., MILNOR [1969]:

THEOREM 1.1 (Implicit function theorem). *Let $H : \mathbb{R}^{N+K} \to \mathbb{R}^N$ be a smooth map such that $0 \in \operatorname{range}(H)$. Then*

$$M = \{x \in \mathbb{R}^{N+K}\colon\ H(x) = 0,\ x \text{ is a regular point of } H\}$$

is a smooth K-dimensional manifold.

Unless we explicitly say otherwise, we now assume hereafter that $H : \mathbb{R}^{N+1} \to \mathbb{R}^N$ is smooth. In particular, if $u_0 \in \mathbb{R}^{N+1}$ is a regular point of H such that $H(u_0) = 0$, it follows from the implicit function theorem that the solution set $H^{-1}(0)$ can be locally parametrized about u_0 with respect to some parameter, say s. We thus obtain the *solution curve* $c(s)$ of the equation $H(u) = 0$.

For convenience of the theoretical discussions, we usually consider parametrization with respect to arclength, though in practical implementations it is often more convenient to use related parametrizations which are easier to incorporate, see in particular

KELLER's [1977] pseudo arclength parametrization. In terms of the intrinsic properties of the curve there is no difference which underlying parametrization is used.

Hence, by a re-parametrization (according to arclength), we obtain a smooth solution curve $c\colon J \to \mathbb{R}^{N+1}$ for some open interval J containing zero such that for all $s \in J$:

$$c(0) = u_0, \tag{1.2}$$

$$H'\big(c(s)\big)\dot{c}(s) = 0, \tag{1.3}$$

$$\big\|\dot{c}(s)\big\| = 1, \tag{1.4}$$

$$\det\begin{pmatrix} H'(c(s)) \\ \dot{c}(s)^* \end{pmatrix} > 0. \tag{1.5}$$

The above conditions uniquely determine the tangent $\dot{c}(s)$ with a specific orientation. Here and in the following, B^* denotes the Hermitian transpose of B, $\|u\|$ the Euclidean norm of u, H' the total derivative (e.g., the Jacobian) of H, and $\dot{c}$ the derivative of c with respect to arclength. Condition (1.4) normalizes the parametrization to arclength. This is only for theoretical convenience, and it is not an intrinsic restriction. Condition (1.5) chooses one of the two possible orientations.

The preceding discussion motivates the following definition: Let A be an $(N, N+1)$-matrix with maximal rank. For the purpose of our exposition, the unique vector $t(A) \in \mathbb{R}^{N+1}$ satisfying the conditions

$$At = 0, \qquad \|t\| = 1, \qquad \det\begin{pmatrix} A \\ t^* \end{pmatrix} > 0, \tag{1.6}$$

will be called the *tangent vector induced by A.*

The vector $\dot{c}(s)$ will hereafter be referred to as the *oriented unit tangent vector to the solution curve c.* At a point $u \in \mathbb{R}^{N+1}$ the matrix

$$\begin{pmatrix} H'(u) \\ t(H'(u))^* \end{pmatrix} \tag{1.7}$$

is called the *augmented Jacobian* of H at u.

Making use of this definition, the above solution curve $c(s)$ is characterized as the solution of the initial value problem

$$\dot{u} = t\big(H'(u)\big), \qquad u(0) = u_0 \tag{1.8}$$

which in this context is sometimes attributed to DAVIDENKO [1953], see also BRANIN [1972]. Note that the domain

$$\big\{u \in \mathbb{R}^{N+1}\colon\ u \text{ is a regular point of } H\big\}$$

is open, and thus the right-hand side of the differential equation is defined on an open domain. The initial value problem (1.8) is not used in efficient path following algorithms, but it serves as a useful device for analyzing the path. The following two statements illustrate this:

THEOREM 1.2. *Let (a, b) be the maximal interval of existence for the solution $c(s)$ of the initial value problem* (1.8). *If a is finite, then $c(s)$ converges to a singular zero point of H as $s \to a$, $s > a$. An analogous statement holds if b is finite.*

PROOF. Since $c(s)$ satisfies the defining initial value problem (1.8), we have

$$(s_1) - c(s_2) = \int_{s_2}^{s_1} t(H'(c(\xi)))\,\mathrm{d}\xi \quad \text{for } s_1, s_2 \in (a, b).$$

Because the integrand has unit norm, it follows that

$$\|c(s_1) - c(s_2)\| \leqslant |s_1 - s_2| \quad \text{for } s_1, s_2 \in (a, b).$$

If $\{s_n\}_{n=1}^{\infty} \subset (a, b)$ is a sequence such that $s_n \to a$ as $n \to \infty$, then the above inequality shows that the sequence $\{c(s_n)\}_{n=1}^{\infty}$ is Cauchy. Hence it converges to a point $\tilde{u}$. By continuity it follows that $H(\tilde{u}) = 0$. We will show by contradiction that $c(a)$ is a singular point of H. Suppose that $\tilde{u}$ is a regular point of H. Then using the initial point $u(0) = \tilde{u}$ in the defining initial value problem (1.8), we obtain a local solution $\tilde{c}(s)$. This solution is unique in some neighborhood U of $\tilde{u}$ (*local uniqueness of solutions*). If $\xi > 0$ is small enough, then $c(a + \xi) \in U$, and hence $c(a + \xi) = \tilde{c}(\xi)$ for $\xi > 0$ holds by the local uniqueness of solutions. Thus it follows that c can be extended beyond a by setting $c(a + \xi) := \tilde{c}(\xi)$ for $\xi \leqslant 0$, contradicting the maximality of the interval (a, b). □

THEOREM 1.3. *Let zero be a regular value of H. Then the solution curve c is defined on the real line and satisfies one of the following two conditions:*

(a) *The curve c is diffeomorphic to a circle. More precisely, there is a period $T > 0$ such that $c(s_1) = c(s_2)$ if and only if $s_1 - s_2$ is an integer multiple of T.*

(b) *The curve c is diffeomorphic to the real line. More precisely, c is injective, and $c(s)$ has no accumulation point for $s \to \pm\infty$.*

PROOF. Since zero is a regular value, no zero point of H is singular, and by the preceding theorem, c is defined on all of $\mathbb{R}$. Furthermore, since the defining differential equation $\dot{u} = t(H'(u))$ is autonomous, the following translation invariance holds: for all $s_0 \in \mathbb{R}$, the curve $s \mapsto c(s_0 + s)$ is also a solution of $\dot{u} = t(H'(u))$. Let us now consider the two possibilities:

(1) The curve c is not injective. We define $T := \min\{s > 0\colon\ c(s) = c(0)\}$. By the local uniqueness of the solutions of initial value problems and by the above mentioned translation invariance, the assertion (a) follows.

(2) The curve c is injective. We show assertion (b) by contradiction. Let us assume without loss of generality that $\tilde{u}$ is an accumulation point of $c(s)$ as $s \to \infty$. By

continuity, $H(\tilde{u}) = 0$. Since $\tilde{u}$ is a regular point of H, we can use the initial point $u(0) = \tilde{u}$ in (1.8) to obtain a solution $\tilde{c}$. By local uniqueness, the two curves c and $\tilde{c}$ must coincide up to a translation, and hence there exists an $s_1 > 0$ such that $c(s_1) = \tilde{u}$. Since $\tilde{u}$ is also an accumulation point of $c(s_1 + s)$ as $s \to \infty$, and since the curve $s \mapsto c(s_1 + s)$ is also a solution curve, the above argument can be repeated to obtain an $s_2 > 0$ such that $c(s_1 + s_2) = \tilde{u}$. This contradicts the injectivity of c. □

Since the solution curve c is characterized by the initial value problem (1.8), it is evident that the numerical methods for solving initial value problems could immediately be used to numerically trace c. However, in general this is not an efficient approach, since it ignores the strong contractive properties which the curve c has relative to corrector steps in view of the fact that it satisfies the equation $H(u) = 0$. In fact, a typical path following method consists of a succession of two different steps:

Predictor step. An approximate step along the curve, usually in the general direction of the tangent of the curve. The initial value problem (1.8) provides motivation for generating predictor steps in the spirit of the techniques of numerical solution of initial value problems.

Corrector steps. One or more iterative steps for solving $H(u) = 0$ (typically of Newton or gradient type) which bring the predicted point back to the curve.

It is usual to call such procedures *predictor corrector* path following methods. However, let us emphasize that this name should not be confused with the predictor corrector multistep methods for initial value problems, since the latter employ correctors which do not converge back to the solution curve.

The following pseudocode shows the basic steps of a generic predictor corrector method, see also Fig. 1.1.

ALGORITHM 1.1 (*Generic predictor corrector method*).

```
input  u_0 ∈ R^{N+1} such that H(u_0) ≈ 0 (initial point)
       h > 0 initial steplength
output u_i, i = 0,1,2,..., approximating the solution curve

for i = 1,2,...
  % predictor step
    predict v_i such that H(v_i) ≈ 0 and ||v_i − u_{i−1}|| ≈ h  ...
      and v_i − u_{i−1} points in the direction of traversing
  % corrector step
    let u_i ∈ R^{N+1} approximately solve ...
      min_w{||v_i − w||: H(w) = 0}
  % steplength adaptation
    choose a new steplength h > 0
end for
```

The predictor corrector type of algorithms for curve following seem to date to HASELGROVE [1961]. In contrast to the present predictor corrector methods, the classical embedding methods assume that the solution path is parametrized with respect

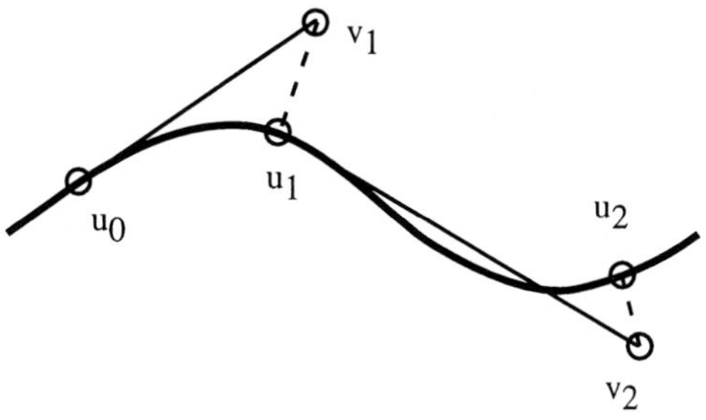

FIG. 1.1. A generic predictor corrector method.

to an explicit parameter which is identified with the last variable in H. Hence, we consider Eq. (1.1) in the form

$$H(x, \lambda) = 0. \tag{1.9}$$

If we assume that the partial derivative $H_x(x, \lambda)$ is nonsingular, then the solution curve can be parametrized in the form $(x(\lambda), \lambda)$. This assumption has the drawback that *folds* are excluded, i.e., points such that $H(x, \lambda) = 0$ and $\det H_x(x, \lambda) = 0$. Such points are sometimes called *turning points* in the literature. However, the assumption has the advantage that the corrector steps can be more easily handled, in particular if H_x is sparse. In some applications it is known a priori that no folds are present, and then the embedding method is applicable. But even in this case, the predictor corrector methods of the type in Algorithm 1.1 are likely to be more efficient since it may be necessary to use very small steps in λ if $\det H_x$ is near zero. For purposes of illustration we present a generic embedding method which is an analogue of the above predictor corrector method:

ALGORITHM 1.2 (*Generic embedding method*).

input $(x_0, \lambda_0) \in \mathbb{R}^{N+1}$ such that $H(x_0, \lambda_0) \approx 0$ (initial point)
    $h > 0$, an initial steplength
output (x_i, λ_i), $i = 0, 1, 2, \ldots$,
    approximating the solution curve

for $i = 1, 2, \ldots$
    $\lambda_i \leftarrow \lambda_{i-1} + h$
    find $x_i \in \mathbb{R}^N$ such that $H(x_i, \lambda_i) \approx 0$
    choose a new steplength $h > 0$
end for

In the preceding algorithm the predictor step is hidden; the predictor point would correspond to the starting point of an iterative method for solving $H(x, \lambda_i) = 0$. The most commonly used starting point is the previous point x_{i-1}.

2. Pseudo arclength algorithms

It is common to blend aspects of the above two Algorithms 1.1, 1.2. A simple example is to use a predictor tangent to the curve $(x(\lambda), \lambda)$ in the embedding Algorithm 1.2. A more sophisticated example is the use of a so-called pseudo arclength parametrization in connection with a bordering algorithm (see Section 13) in the corrector phase of the predictor corrector method, as introduced by KELLER [1977, 1983, 1987]. To avoid dealing with the arclength parameter, one can also adopt a strategy of parameter switching, see, e.g., RHEINBOLDT [1980, 1981]. In this approach, one changes the embedding parameter when nearing a fold point with respect to the current parameter during the traversing of the curve.

Let us give a brief account of these ideas. We assume that 0 is a regular value of H. If $\bar{u}$ is a current solution point, i.e., $H(\bar{u}) = 0$, a (local) *pseudo arclength parameter* s is introduced via an additional equation

$$F(u, s) = 0 \tag{2.1}$$

where $F : \mathbb{R}^{N+1} \times \mathbb{R} \to \mathbb{R}$ is sufficiently smooth and such that $F(\bar{u}, 0) = 0$. We denote by F_u the (row of) partial derivatives of F with respect to u and assume that the matrix

$$\begin{pmatrix} H'(\bar{u}) \\ F_u(\bar{u}, 0) \end{pmatrix}$$

is nonsingular, i.e., $F_u(\bar{u}, 0)\, t(H'(\bar{u})) \neq 0$, or better $F_u(\bar{u}, 0)\, t(H'(\bar{u})) > 0$ if we want to follow the solution curve in the direction of positive orientation. The implicit function theorem guarantees that the solution curve can be parametrized near $\bar{u}$ according to s. Differentiating the system

$$\begin{pmatrix} H(u) \\ F(u, s) \end{pmatrix} = 0$$

with respect to the pseudo arclength parameter s yields the initial value problem

$$\dot{u} = \begin{pmatrix} H'(u) \\ -F_u(u, s)/F_s(u, s) \end{pmatrix}^{-1} \begin{pmatrix} 0 \\ 1 \end{pmatrix}, \qquad u(0) = \bar{u}. \tag{2.2}$$

We compare this characterization of the solution curve with (1.8) which can also be written in the form

$$\dot{u} = t(H'(u)) = \begin{pmatrix} H'(u) \\ t(H'(u))^* \end{pmatrix}^{-1} \begin{pmatrix} 0 \\ 1 \end{pmatrix}, \qquad u(0) = \bar{u}. \tag{2.3}$$

By applying a standard sensitivity analysis to the initial value problems (2.2), (2.3) we obtain the following.

THEOREM 2.1. *Let $c(s)$ be a solution to the initial value problem* (2.3) *and $d(s)$ the solution of* (2.2). *Hence, the two curves differ only in the choice of the parameter, $c(s)$ having exact arclength parameter and $d(s)$ a pseudo arclength parameter. Let*

$$\left\| \lim_{s\to 0+} -F_u\big(d(s),s\big)^*/F_s\big(d(s),s\big) - t\big(H'(\bar{u})\big)\right\| \leqslant \varepsilon.$$

Then

$$\big\|\dot{c}(s) - \dot{d}(s)\big\| \leqslant \mathrm{O}(\varepsilon) + \mathrm{O}(s),$$
$$\big\|c(s) - d(s)\big\| \leqslant \mathrm{O}(\varepsilon s) + \mathrm{O}\big(s^2\big).$$

The above theorem justifies the name "pseudo arclength" for the parameter introduced in (2.1) for the case that ε is small or even zero. In most applications, a convenient choice of the pseudo arclength parameter is easier to handle than the exact arclength parameter and leads to an efficient algorithm. The exact arclength parametrization is mainly useful for theoretical considerations, since it is global and simple. In fact, very few implementations of numerical continuation methods really maintain an arclength parametrization, the typical implementation uses an approximation in one way or another.

The two different parametrizations also suggest different corrector iterations. The pseudo arclength parametrization suggests a Newton iteration of the form

$$u_{n+1} = u_n - \begin{pmatrix} H'(u_n) \\ F_u(u_n, s) \end{pmatrix}^{-1} \begin{pmatrix} H(u_n) \\ F(u_n, s) \end{pmatrix}. \tag{2.4}$$

The exact arclength parametrization suggests the Gauss–Newton iteration

$$u_{n+1} = u_n - H'(u_n)^+ H(u_n), \tag{2.5}$$

see (3.3), (5.2) and Sections 4–5. However, implementations of numerical continuation methods are free to choose any corrector iteration, including the above two conceptually simplest ones. Furthermore, we will see in Sections 12, 13 that the numerical linear algebra and the computational expense involved in a step of one of the two methods (2.4) or (2.5) are very similar.

Let us list a few pseudo arclength choices which are common. A very popular choice is linked to the Euler predictor (5.1) of Section 5.

$$F(u, s) = t\big(H'(\bar{u})\big)^*(u - \bar{u}) - s. \tag{2.6}$$

It follows that

$$-F_u(u,s)^*/F_s(u,s) = t\big(H'(\bar{u})\big)$$

and hence $\varepsilon = 0$ in the above Theorem 2.1.

Another popular choice is a secant predictor, where the tangent $t(H'(\bar{u}))$ in (2.6) which may be computationally expensive to obtain is replaced by the secant $\|\bar{u} - p\|^{-1}(\bar{u} - p)$ which essentially comes for free if we use a previous point p on the solution curve. It is easy to see that $\varepsilon = \mathrm{O}(\|\bar{u} - p\|)$ in this case.

Another choice of a pseudo arclength parameter is linked to the idea of finding a point on the solution curve at Euclidean distance s from $\bar{u}$. This leads to

$$F(u, s) = \|u - \bar{u}\| - s.$$

It follows that

$$-F_u(d(s), s)^*/F_s(d(s), s) = (d(s) - \bar{u})/\|d(s) - \bar{u}\| = t(H'(\bar{u})) + \mathrm{O}(s)$$

and hence again $\varepsilon = 0$. This parametrization has been successfully used by MENZEL and SCHWETLICK [1985].

Finally, the simplest choice is a parametrization according to some chosen co-ordinate (usually the eigenvalue parameter, if possible):

$$F(u, s) = e_i^*(u - \bar{u}) - s$$

where e_i is a unit basis vector, $i \in \{1, \ldots, N + 1\}$. In this case, the iteration (2.4) can be reduced by setting $e_i^* u = e_0^* \bar{u} + s$ and iterating

$$E_i u_{n+1} = E_i u_n - \left(H'(u_n)E_i\right)^{-1} H(u_n)$$

where E_i is obtained from the $(N + 1, N + 1)$ identity matrix by deleting column i. This choice of parametrization (at least for the corrector phase) in connection with a suitable strategy for switching co-ordinates is very popular in structural engineering, see, e.g., RHEINBOLDT [1980, 1981].

Let us conclude this section by giving a pseudo code.

ALGORITHM 2.1 (*Generic pseudo arclength method*).

input $u_0 \in \mathbb{R}^{N+1}$ such that $H(u_0) \approx 0$ (initial point)
    $h > 0$ initial steplength
output u_i, $i = 0, 1, 2, \ldots$, approximating the solution curve

for $i = 1, 2, \ldots$
  % choose a local pseudo arclength parameter
    find F such that $F_u(u_{i-1}, 0)\, t(H'(u_{i-1})) > 0$
  % predictor step
    predict v_i such that $H(v_i) \approx 0$ and $F(v_i, h) \approx 0$
  % corrector step
    let $u_i \in \mathbb{R}^{N+1}$ approximately solve ...
    $H(u) = 0$ and $F(u, h) = 0$ starting an iteration at v_i

    % steplength adaptation
        choose a new steplength $h > 0$
    end for

3. The Moore–Penrose inverse

The corrector step in Algorithm 1.1 tries to approximately solve an underdetermined nonlinear system of equations from a given starting point (predictor point). An analogue of Newton's method for underdetermined nonlinear systems can be formulated by using the Moore–Penrose inverse. This is often called the *Gauss–Newton method*. It has been discussed in several monographs, see, e.g., ORTEGA and RHEINBOLDT [1970] or BEN-ISRAEL and GREVILLE [1974]. More precisely, a Gauss–Newton step approximately solves the minimization problem in Algorithm 1.1. Since this concept will occur on several occasions in the following discussions, we begin with a brief review of the role of the Moore–Penrose inverse in this regard.

Let us consider the minimization problem in Algorithm 1.1. If u_i is a solution, then it satisfies the Lagrangian equations

$$H(u_i) = 0, \qquad u_i - v_i = H'(u_i)^* \lambda \tag{3.1}$$

for some vector of multipliers $\lambda \in \mathbb{R}^N$. The second condition is equivalent to $u_i - v_i \in \operatorname{range}(H'(u_i)^*) = \{t(H'(u_i))\}^\perp$. Thus a necessary condition for u_i to solve the minimization problem is that it satisfies the equations

$$H(u_i) = 0, \qquad t\big(H'(u_i)\big)^* (u_i - v_i) = 0. \tag{3.2}$$

In Newton's method, this nonlinear system is solved approximately via a linearization about v_i. To illustrate this, let us consider the Taylor expansion about v_i:

$$\begin{aligned} H(u_i) &= H(v_i) + H'(v_i)(u_i - v_i) + \mathrm{O}\big(\|u_i - v_i\|^2\big), \\ t\big(H'(u_i)\big)^* (u_i - v_i) &= t\big(H'(v_i)\big)^* (u_i - v_i) + \mathrm{O}\big(\|u_i - v_i\|^2\big). \end{aligned}$$

Hence, the linearization of (3.2) consists of neglecting the higher order terms $\mathrm{O}(\|u_i - v_i\|^2)$. As is usual in Newton's method, we obtain an approximation $\mathcal{N}(v_i)$ to the solution u_i, which has a truncation error of second order. Hence, the Newton point $\mathcal{N}(v_i)$ satisfies the following equations:

$$\begin{aligned} H(v_i) + H'(v_i)\big(\mathcal{N}(v_i) - v_i\big) &= 0, \\ t\big(H'(v_i)\big)^* \big(\mathcal{N}(v_i) - v_i\big) &= 0. \end{aligned}$$

As can be seen from the discussion below, see (3.4), the solution of the latter equations can be written in terms of the Moore–Penrose inverse $H'(v_i)^+$ of $H'(v_i)$:

$$\mathcal{N}(v_i) := v_i - H'(v_i)^+ H(v_i). \tag{3.3}$$

The map $\mathcal{N}$ defined on the regular points of H will for our purposes be called the *Gauss–Newton map*. Note that the Gauss–Newton method is analogous to the classical Newton's method, with the only formal difference being that the Moore–Penrose inverse $H'(v)^+$ replaces the standard inverse.

For completeness, let us review some features of the Moore–Penrose inverse which pertain to our special context. Let A be an $(N, N+1)$ matrix with maximal rank, and let $t(A)$ be its tangent vector, see (1.6). Then it is easy to see that the following statements are equivalent for all $b \in \mathbb{R}^N$ and $x \in \mathbb{R}^{N+1}$:

$$Ax = b \quad \text{and} \quad t(A)^* x = 0, \tag{3.4}$$

$$x = A^+ b \quad \text{where } A^+ := A^*(AA^*)^{-1}, \tag{3.5}$$

$$x \text{ solves the problem } \min_w \{\, \|w\|\colon\ Aw = b\,\}. \tag{3.6}$$

The last problem is one of the standard linear least squares problems. The matrix A^+ is the *Moore–Penrose inverse* of A, see, e.g., GOLUB and VAN LOAN [1989] for a general discussion on how to implement the action of A^+. Note that we are using the Moore–Penrose inverse only for the special case of an $(N, N+1)$ matrix with maximal rank.

Some further properties of the Moore–Penrose inverse are:

$$A^+ A = \mathrm{Id} - t(A)t(A)^* \tag{3.7}$$

(orthogonal projection onto range A^*),

$$AA^+ = \mathrm{Id}, \tag{3.8}$$

$$A^+ = \big(\mathrm{Id} - t(A)t(A)^*\big)B \quad \text{holds for any right inverse } B \text{ of } A. \tag{3.9}$$

4. The Gauss–Newton method

In (3.3) the Gauss–Newton map was introduced as a corrector step. Full theoretical discussions of the Gauss–Newton method can be found in several monographs, see, e.g., ORTEGA and RHEINBOLDT [1970] or BEN-ISRAEL and GREVILLE [1974], see also DEUFLHARD and HEINDL [1979]. Let us discuss some important features of the Gauss–Newton map $v \mapsto \mathcal{N}(v)$. Since our context deals with the curve-following problem, we confine our discussion to the case $H : \mathbb{R}^{N+1} \to \mathbb{R}^N$. Generalizations to the case $H : \mathbb{R}^{N+K} \to \mathbb{R}^N$ for $K > 1$ are straightforward.

THEOREM 4.1. *Let $H : \mathbb{R}^{N+1} \to \mathbb{R}^N$ be a smooth map having zero as a regular value. Then there exists an open neighborhood $U \supset H^{-1}(0)$ such that the following assertions hold:*

(1) *Points in U are regular.*

(2) *U is stable under the Gauss–Newton map, i.e., $\mathcal{N}(U) \subset U$.*

(3) *For $v \in U$ the sequence $\{\mathcal{N}^i(v)\} \subset U$ converges. Let us call the limit $\mathcal{N}^\infty(v)$. Clearly, $\mathcal{N}^\infty(v) \in H^{-1}(0)$.*

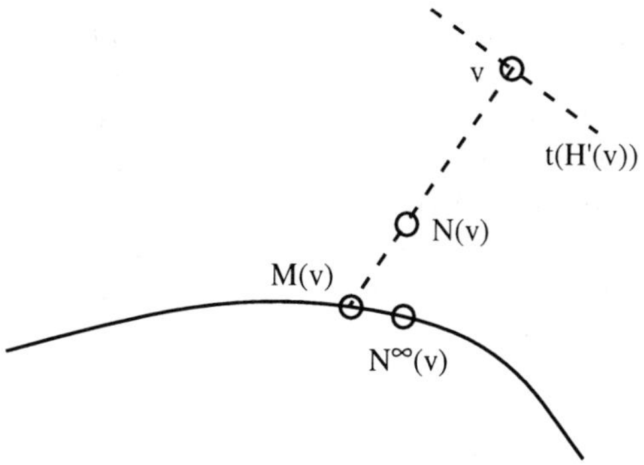

FIG. 4.1. Diagram for the Proof of Theorem 4.1.

(4) *The following quadratic convergence*

$$\|\mathcal{N}(v) - \mathcal{N}^\infty(v)\| \leqslant \mathrm{O}\big(\|v - \mathcal{N}^\infty(v)\|^2\big)$$

holds uniformly for v *in compact subsets of* U.

PROOF. We will only give a brief sketch of the main points. See ALLGOWER and GEORG ([1990], (3.4.1)) for further details.

For $v \in \mathbb{R}^{N+1}$ we define $\mathcal{M}(v)$ as the solution of the minimization problem

$$\min_w \{\|w - v\|:\ H(w) = 0\},$$

see Fig. 4.1. It is clear that we can find an open neighborhood $U_1 \supset H^{-1}(0)$ consisting of regular points such that the map $\mathcal{M}: U_1 \to H^{-1}(0)$ is uniquely defined. Recall, see (3.2) and following, that a Gauss–Newton step was motivated by this minimization. It is hence clear from classical results on Newton's method that

$$\|\mathcal{N}(v) - \mathcal{M}(v)\| \leqslant \mathrm{O}\big(\|v - \mathcal{M}(v)\|^2\big)$$

holds uniformly for v in compact subsets of U_1.

Let us introduce the following notation: $B_\rho(w) := \{u:\ \|u - w\| < \rho\}$. Let us choose for each $w \in H^{-1}(0)$ an $\varepsilon(w) > 0$ such that $B_{\varepsilon(w)}(w) \subset U_1$. From the above estimates it is not hard, but somewhat technical to show that for each $w \in H^{-1}(0)$ there is a $\delta(w) > 0$, $\delta(w) < \varepsilon(w)$, such that $\{\mathcal{N}^i(v)\} \subset B_{\varepsilon(w)}(w)$ and $\mathcal{N}^\infty(v) \subset B_{\varepsilon(w)}(w)$ for $v \in B_{\delta(w)}(w)$.

We now define

$$U_2 := \bigcup_{w \in H^{-1}(0)} B_{\delta(w)}(w)$$

and

$$U := \{u\colon \{\mathcal{N}^i(u)\}_{i=0,1,2,\ldots} \subset U_2\}.$$

Let us show that U is open. Choose $v \in U$. We set $w := \mathcal{N}^\infty(v)$. It is possible to find a ball $B_\rho(w)$ such that $\mathcal{N}^i(B_\rho(w)) \subset B_{\delta(w)}(w)$ for $i = 1, 2, \ldots$. Let $k > 0$ be such that $\mathcal{N}^i(v) \in B_\rho(w)$ for $i \geqslant k$. Then

$$\{u\colon u, \mathcal{N}(u), \ldots, \mathcal{N}^k(u) \in U_2,\ \mathcal{N}^k(u) \in B_\rho(w)\}$$

is an open neighborhood of v contained in U.

The remaining assertions of the theorem are easy to obtain. □

For a sensitivity analysis and for certain applications it is worthwhile to note that the map $v \mapsto \mathcal{N}^\infty(v)$ is smooth. This fact is surprisingly difficult to show:

THEOREM 4.2. *An additional conclusion to Theorem* 4.1 *is that the map* $\mathcal{N}^\infty : U \to H^{-1}(0)$ *is smooth.*

PROOF. We only sketch the main steps of the proof which is due to BEYN [1992]. In view of the fact that $\mathcal{N} : U \to U$ is smooth, we only need to find for each $u_0 \in H^{-1}(0)$ an open neighborhood U_0 of u_0 such that the restriction $\mathcal{N}^\infty : U_0 \to H^{-1}(0)$ is smooth. Furthermore, without loss of generality, we assume $u_0 = 0$. We abbreviate $t := t(H'(0))$.

We next observe that near $u_0 = 0$, the solution curve can be parametrized by $c(h)$ which satisfies the system

$$H(c(h)) = 0, \qquad t^*(ht - c(h)) = 0.$$

For $0 < \rho < 1$ we introduce a Banach space of sequences which converge to zero with at least the geometrical rate ρ:

$$\mathcal{B}_\rho := \Big\{\{v_n\} \subset \mathbb{R}^{N+1}\colon \underbrace{\sup \|\rho^{-n} v_n\|}_{\text{norm!}} < \infty\Big\}.$$

Let us denote $W := (\ker H'(0))^\perp$. For a sufficiently small open neighborhood $\mathcal{V}$ of $\mathcal{B}_\rho \times W \times \mathbb{R}$ the map

$$\Gamma : \mathcal{V} \to \mathcal{B}_\rho \times W$$

is defined by

$$\begin{pmatrix} \{v_n\} \\ w \\ h \end{pmatrix} \mapsto \begin{pmatrix} \{\mathcal{N}(v_n + c(h)) - (v_{n+1} + c(h))\} \\ (\mathrm{Id} - tt^*)v_0 - w \end{pmatrix}.$$

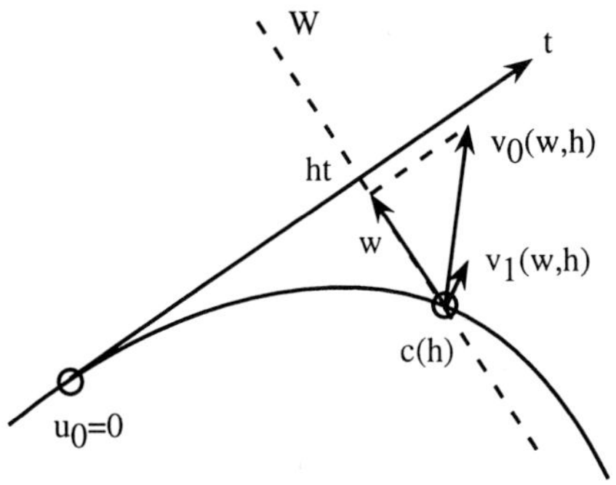

FIG. 4.2. Diagram for the Proof of Theorem 4.2

Note that $\Gamma(0) = 0$.

Next, it is possible to show that the derivative

$$\partial_1 \Gamma : \mathcal{B}_\rho \to \mathcal{B}_\rho \times W$$

is an isomorphism. Hence, by the implicit function theorem, the solution of the equation

$$\Gamma \begin{pmatrix} \{v_n\} \\ w \\ h \end{pmatrix} = 0$$

can be (locally) parametrized by w and h, see Fig. 4.2. From the definition of Γ it is clear that $\{v_n(w, h) + c(h)\}$ is obtained via Gauss–Newton steps, such that the starting point $v_0(w, h) + c(h)$ satisfies the condition $(\mathrm{Id} - tt^*)v_0(w, h) = w$.

Finally, we consider the map

$$\Phi : W \times \mathbb{R} \to \mathbb{R}^{N+1}$$

defined by

$$(w, h) \mapsto c(h) + v_0(w, h).$$

Clearly $\Phi(0) = 0$, and it can be seen that $\Phi'(0)(w, h) = w + ht$ which is nothing else but the natural isomorphism between $W \times \mathbb{R}$ and $\mathbb{R}^{N+1}$. Hence, by the inverse function theorem, Φ can be locally inverted.

Let z with $\|z\|$ sufficiently small be a starting point for a Gauss–Newton iteration. Applying Φ^{-1} we obtain $(w, h) := \Phi^{-1} z$. Hence, we can identify z with the starting point of a Gauss–Newton sequence $\{v_n(w, h) + c(h)\}$ converging to $c(h)$. The smoothness of $\mathcal{N}^\infty$ now becomes evident from the following composition of smooth maps:

$$\mathcal{N}^\infty : z \mapsto (w, h) \mapsto c(h). \qquad \square$$

5. Euler–Newton continuation

A straightforward way of approximating a solution curve is to alternatingly perform a predictor step along the tangent of the curve

$$v = u + ht\big(H'(u)\big), \tag{5.1}$$

and a Gauss–Newton corrector step:

$$u := v - H'(v)^+ H(v). \tag{5.2}$$

The predictor step along the tangent can also be viewed as an Euler step for the differential Eq. (1.8). Here $h > 0$ represents a current stepsize.

The following algorithm sketches this procedure where approximate Euler and Newton steps are used.

ALGORITHM 5.1 (*Euler–Newton method*).

input $u_0 \in \mathbb{R}^{N+1}$ such that $H(u_0) \approx 0$ (initial point)
  $h > 0$ initial steplength
output u_i , $i = 0, 1, 2, \ldots$, approximating the solution curve
for $i = 1, 2, \ldots$
  approximate $A_{i-1} \approx H'(u_{i-1})$
  $v_i \leftarrow u_{i-1} + ht(A_{i-1})$ % *predictor step*
  $u_i \leftarrow v_i - A_{i-1}^+ H(v_i)$ % *corrector step*
  choose a new steplength $h > 0$
end for

Let us first state a convergence result, see ALLGOWER and GEORG ([1990], (5.2.1)), which ensures that the above algorithm safely follows the solution curve under reasonable assumptions.

THEOREM 5.1. *Let $H : \mathbb{R}^{N+1} \to \mathbb{R}^N$ be a smooth map having zero as a regular value and let $H(u_0) = 0$. Denote by $c_h(s)$ the polygonal path, starting at u_0, going through all points u_i generated by Algorithm 5.1 with fixed steplength $h > 0$. Denote by $c(s)$ the corresponding curve in $H^{-1}(0)$ given by the initial value problem (1.8). For definiteness, we assume that $c_h(0) = c(0) = u_0$, and that both curves are parametrized with respect to arclength. If the estimate $\|A_i - H'(u_i)\| = \mathrm{O}(h)$ holds uniformly for* *the approximation in the loop of the algorithm, then, for a given maximal arclength* s_0, *the following quadratic bounds hold uniformly for* $0 \leqslant s \leqslant s_0$:

$$\left\|H\left(c_h(s)\right)\right\| \leqslant O\left(h^2\right), \qquad \left\|c_h(s) - c(s)\right\| \leqslant O\left(h^2\right).$$

PROOF. Let us sketch the main arguments of the proof, see also ALLGOWER and GEORG ([1990], Theorem 5.2.1). Let U be a compact neighborhood of $c([0, s_0])$ which consists only of regular points of H. We define the following constants for U:

$$\alpha := \max\left\{\left\|H'(v)\right\|: \ v \in U\right\}; \tag{5.3}$$

$$\beta := \max\left\{\left\|H'(v)^+\right\|: \ v \in U\right\}; \tag{5.4}$$

$$\gamma := \max\left\{\left\|H''(v)\right\|: \ v \in U\right\}. \tag{5.5}$$

From the estimates below it is evident that Algorithm 5.1 generates only predictor and corrector points in U for sufficiently small steplength h and so long as the maximal arclength s_0 is not exceeded. To make the proof more brief, we use the Landau symbol O when we actually describe global constants in terms of α, β, γ. The readers may convince themselves that in fact all asymptotic estimates are uniform with respect to U.

Let us first show

$$H(u_i) = O\left(h^2\right). \tag{5.6}$$

We proceed by induction. Let us assume that the estimate (5.6) is true for a current corrector point u_i. For the next predictor step

$$v_{i+1} = u_i + ht(A_i)$$

we obtain

$$H(v_{i+1}) = H(u_i) + hH'(u_i)t(A_i) + O\left(h^2\right),$$

and since $A_i t(A_i) = 0$, this implies

$$H(v_{i+1}) = H(u_i) + h\underbrace{\left(H'(u_i) - A_i\right)}_{O(h)} t(A_i) + O\left(h^2\right) = O\left(h^2\right).$$

For the next corrector step

$$u_{i+1} - v_{i+1} - A_i^+ H(v_{i+1})$$

we obtain

$$H(u_{i+1}) = H(v_{i+1}) - H'(v_{i+1})A_i^+ H(v_{i+1}) + \underbrace{O\left(\left\|A_i^+ H(v_{i+1})\right\|^2\right)}_{O(h^4)}. \tag{5.7}$$

Next we observe that

$$A_i - H'(v_{i+1}) = \big[A_i - H'(u_i)\big] + \big[H'(u_i) - H'(v_{i+1})\big]$$

and hence

$$A_i = H'(v_{i+1}) + \mathrm{O}(h).$$

Since the map $A \mapsto A^+$ is smooth, it follows that

$$A_i^+ = H'(v_{i+1})^+ + \mathrm{O}(h).$$

Substituting this into (5.7) yields

$$\begin{aligned} H(u_{i+1}) &= \underbrace{H(v_{i+1}) - H'(v_{i+1})H'(v_{i+1})^+H(v_{i+1})}_{=0} + \mathrm{O}(h^3) + \mathrm{O}(h^4) \\ &= \mathrm{O}(h^3). \end{aligned}$$

The additional order (as compared to (5.6)) serves to keep the above mentioned constants uniform, and assertion (5.6) is shown.

Let us next show that

$$\|u_i - u_{i+1}\| = h\big(1 + \mathrm{O}(h^2)\big). \tag{5.8}$$

In fact, $u_i - v_{i+1} = ht(A_i)$ and $v_{i+1} - u_{i+1} = -A_i^+H(v_{i+1})$ are orthogonal to each other, and hence, combining the two differences yields

$$\|u_i - u_{i+1}\|^2 = h^2 + \mathrm{O}(h^4) = h^2\big(1 + \mathrm{O}(h^2)\big)$$

and thus (5.8) is obtained by taking the square root.

For $\tau \in [0, 1]$, an easy consequence of Taylor's formula is

$$H\big(\tau u_i + (1-\tau)u_{i+1}\big) = \tau H(u_i) + (1-\tau)H(u_{i+1}) + \mathrm{O}\big(\|u_{i+1} - u_i\|^2\big).$$

Together with (5.6) and (5.8), this immediately implies the first assertion of the theorem.

For sufficiently small step sizes h, it is clear that for each u_i there is a unique s_i such that $\|u_i - c(s_i)\| = \min_s \|u_i - c(s)\|$. By the choice of the s_i, it follows that the orthogonality

$$u_i - c(s_i) \perp \dot{c}(s_i) \tag{5.9}$$

holds, see Fig. 5.1. Furthermore, it is not difficult to see that

$$c(s_i) - u_i \;\to\; 0 \quad \text{as } h \to 0. \tag{5.10}$$

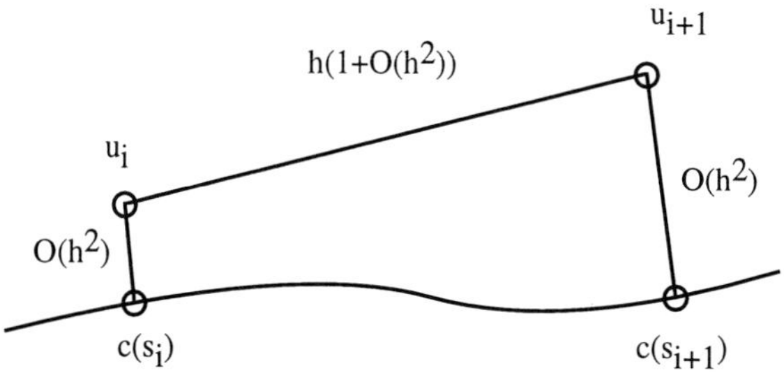

FIG. 5.1. Diagram for the estimates.

Let us next show that in fact

$$\|c(s_i) - u_i\| = O(h^2). \tag{5.11}$$

From Taylor's formula it follows that

$$H(u_i) = H\big(c(s_i)\big) + H'\big(c(s_i)\big)\big(u_i - c(s_i)\big) + O\big(\|u_i - c(s_i)\|^2\big).$$

Now $H(c(s_i)) = 0$, and if we multiply through with $H'(c(s_i))^+$, then (5.9) and (3.7) imply

$$H'\big(c(s_i)\big)^+ H(u_i) = \big(u_i - c(s_i)\big) + O\big(\|u_i - c(s_i)\|^2\big).$$

Now assertion (5.11) follows from (5.6) and (5.10).

Let us define $\Delta s_i := s_{i+1} - s_i$. Differentiating $\|\dot{c}(s)\|^2 \equiv 1$ implies $\dot{c}(s) \perp \ddot{c}(s)$, and Taylor's formula yields

$$\begin{aligned}
\|c(s_{i+1}) - c(s_i)\|^2 &= \left\| \int_0^1 \dot{c}(s_i + \xi \Delta s_i) \Delta s_i \, \mathrm{d}\xi \right\|^2 \\
&= \left\|\dot{c}(s_i)\Delta s_i + \tfrac{1}{2}\ddot{c}(s_i)(\Delta s_i)^2 + O\big((\Delta s_i)^3\big)\right\|^2 \\
&= (\Delta s_i)^2 + O\big((\Delta s_i)^4\big)
\end{aligned}$$

and consequently

$$\|c(s_{i+1}) - c(s_i)\| = \Delta s_i\big(1 + O\big((\Delta s_i)^2\big)\big). \tag{5.12}$$

From (5.8), (5.11) and (5.12) it is straightforward to obtain

$$\Delta s_i\big[1 + O\big((\Delta s_i)^2\big)\big] = h\big[1 + O(h)\big]$$

and consequently

$$\begin{aligned}h &= \Delta s_i\big(1+\mathrm{O}((\Delta s_i))\big),\\ \Delta s_i &= h\big(1+\mathrm{O}(h)\big).\end{aligned}$$

Hence the terms $\mathrm{O}(h)$ and $\mathrm{O}(\Delta s_i)$ can be used interchangeably.

From the orthogonality relation $(u_i - c(s_i)) \perp \dot c(s_i)$ and Taylor's formula we obtain

$$\begin{aligned}\big(u_i - c(s_i)\big)^*\big(c(s_{i+1}) - c(s_i)\big) &= \big(u_i - c(s_i)\big)^*\big(\dot c(s_i)\Delta s_i + \mathrm{O}((\Delta s_i)^2)\big)\\ &= \mathrm{O}\big(h^2\big)\mathrm{O}\big((\Delta s_i)^2\big)\\ &= \mathrm{O}\big(h^4\big).\end{aligned}$$

Therefore

$$\begin{aligned}\|u_{i+1} - u_i\|^2 &= \big\|\big(u_{i+1} - c(s_{i+1})\big) + \big(c(s_{i+1}) - c(s_i)\big) + \big(c(s_i) - u_i\big)\big\|^2\\ &= \big\|c(s_{i+1}) - c(s_i)\big\|^2 + \mathrm{O}\big(h^4\big).\end{aligned}$$

Taking square roots and using (5.12) yields

$$\begin{aligned}\|u_{i+1} - u_i\| &= \big\|c(s_{i+1}) - c(s_i)\big\| + \frac{1}{2}\frac{\mathrm{O}(h^4)}{\|c(s_{i+1}) - c(s_i)\|}\\ &= \big\|c(s_{i+1}) - c(s_i)\big\| + \mathrm{O}\big(h^3\big).\end{aligned} \tag{5.13}$$

If we sum up over the terms $\|u_{i+1} - u_i\|$ (the nodes of c_h), $\|c(s_{i+1} - c(s_i)\|$ and the Δs_i, respectively, then the summation drops the order of h by one degree, and hence the corresponding summations (performed to a certain arclength) differ by $\mathrm{O}(h^2)$:

$$\sum \|u_{i+1} - u_i\| = \sum \big\|c(s_{i+1}) - c(s_i)\big\| + \mathrm{O}\big(h^2\big)$$

by (5.13) and

$$\sum \big\|c(s_{i+1}) - c(s_i)\big\| = \sum \Delta s_i + \mathrm{O}\big(h^2\big)$$

by (5.12). This implies

$$\big\|c_h(s_i) - u_i\big\| = \mathrm{O}\big(h^2\big). \tag{5.14}$$

Now we are in a position to show the second assertion of the theorem: Let $s_i \leqslant s \leqslant s_{i+1}$ and $s = \tau s_i + (1-\tau)s_{i+1}$ for a suitable $\tau \in [0,1]$. Taylor's formula and the estimates (5.14) and (5.11) imply:

$$\begin{aligned}&\big\|c(s) - c_h(s)\big\|\\ &\quad= \big\|c\big(\tau s_i + (1-\tau)s_{i+1}\big) - c_h\big(\tau s_i + (1-\tau)s_{i+1}\big)\big\|\\ &\quad\leqslant \big\|\big[\tau c(s_i) + (1-\tau)c(s_{i+1})\big] - c_h\big(\tau s_i + (1-\tau)s_{i+1}\big)\big\| + \mathrm{O}\big(h^2\big)\end{aligned}$$

$$\leqslant \left\| \left[\tau c(s_i) + (1-\tau)c(s_{i+1})\right] - \left[\tau u_i + (1-\tau)u_{i+1}\right] \right\| + O(h^2)$$
$$\leqslant \tau \left\| c(s_i) - u_i \right\| + (1-\tau)\left\| c(s_{i+1}) - u_{i+1} \right\| + O(h^2)$$
$$\leqslant O(h^2). \qquad \square$$

Similar methods and results hold for pseudo arclength continuation or other forms of continuation methods.

Some major points which remain to be clarified are:

– How do we formulate efficient steplength strategies? See Chapter III for details.

– How do we efficiently handle the numerical linear algebra involved in the calculation of the quantities $t(A)$ and $A^+H(v)$ which are essentially involved in the predictor and corrector steps respectively? See Chapter IV for details.

CHAPTER III

Predictors

The convergence considerations of Theorem 5.1 were carried out under the assumption that the steplength of Algorithm 5.1 was uniformly constant throughout. This assumption is also typical for complexity studies, see Chapter VII. Such an approach is of course not efficient for any practical implementation. An efficient algorithm needs to incorporate an automatic strategy for controlling the steplength. In this respect the predictor corrector methods are similar to the methods for numerically integrating initial value problems in ordinary differential equations. To some extent of course, the steplength strategy depends upon the accuracy with which it is desired to numerically trace a solution curve. Path following methods usually split into two categories:

– Either it is wished to approximate the solution curve with some given accuracy, e.g., for plotting purposes, or if one wants to calculate a contour integral along the curve,

– or the objective is just to safely follow the curve as fast as possible, until a certain point is reached, e.g., a zero point or critical point with respect to some additional functional defined on the curve.

We briefly sketch some ideas which are used to adjust the steplength.

6. Steplength control via error models

One method, due to DEN HEIJER and RHEINBOLDT [1981], is based upon an error model for the corrector iteration. For Gauss–Newton corrector steps, such error models can be obtained by analyzing the Newton–Kantorovich theory. The steplength is controlled by the number of steps which are taken in the corrector iteration until a given stopping criterion is fulfilled.

We sketch a somewhat modified and simplified version of this steplength strategy. Let us assume that u is a point on the solution curve, and consider, for simplicity, an Euler predictor $v_0(h) = u + ht(H'(u))$. Let $v_0(h)$, $v_1(h)$, … , $v_k(h)$ be an iterative corrector process for approximating a nearby point on the curve. Suppose a certain stopping criterion is met after k iterations. The exact nature of the criterion is not important in this context. We assume theoretical convergence to $v_\infty(h)$.

It is assumed that there exists a constant $\gamma > 0$ (which is independent of h) such that the *modified error*

$$\varepsilon_i(h) := \gamma\big\|v_\infty(h) - v_i(h)\big\|$$

satisfies inequalities of the following type

$$\varepsilon_{i+1}(h) \leqslant \psi\big(\varepsilon_i(h)\big),$$

where $\psi : \mathbb{R} \to \mathbb{R}$ is a known monotone function such that $\psi(0) = 0$. For example, if Newton's method is employed, Den Heijer and Rheinboldt suggest two models:

$$\psi(\varepsilon) = \frac{\varepsilon^2}{3 - 2\varepsilon}, \quad 0 \leqslant \varepsilon \leqslant 1, \tag{6.1}$$

$$\psi(\varepsilon) = \frac{\varepsilon + \sqrt{10 - \varepsilon^2}}{5 - \varepsilon^2}\varepsilon^2, \quad 0 \leqslant \varepsilon \leqslant 1. \tag{6.2}$$

We may evaluate a posteriori the quotient

$$\omega(h) := \frac{\|v_k(h) - v_{k-1}(h)\|}{\|v_k(h) - v_0(h)\|} \approx \frac{\|v_\infty(h) - v_{k-1}(h)\|}{\|v_\infty(h) - v_0(h)\|} = \frac{\varepsilon_{k-1}(h)}{\varepsilon_0(h)}.$$

Using the estimate $\varepsilon_{k-1}(h) \leqslant \psi^{k-1}(\varepsilon_0(h))$, we obtain

$$\omega(h) \leqslant \frac{\psi^{k-1}(\varepsilon_0(h))}{\varepsilon_0(h)}.$$

This motivates taking the solution ε of the equation

$$\omega(h) = \frac{\psi^{k-1}(\varepsilon)}{\varepsilon}$$

as an estimate for $\varepsilon_0(h)$.

We now try to choose the steplength $\tilde{h}$ so that the corrector process satisfies the stopping criterion after a chosen number (say $\tilde{k}$) of iterations. Such a steplength leads to the modified error $\varepsilon_0(\tilde{h})$. Hence, we want the modified error $\varepsilon_{\tilde{k}}(\tilde{h})$ after $\tilde{k}$ iterations to be so small that the stopping criterion is satisfied. Using the inequality $\varepsilon_{\tilde{k}}(\tilde{h}) \leqslant \psi^{\tilde{k}}(\varepsilon_0(\tilde{h}))$, we accept the solution ε of the equation

$$\psi^{\tilde{k}}(\varepsilon) = \psi^k\big(\varepsilon_0(h)\big)$$

as an estimate for $\varepsilon_0(\tilde{h})$. Now we use the asymptotic expansion

$$\big\|v_\infty(h) - v_0(h)\big\| = Ch^2 + \mathrm{O}\big(h^3\big)$$

to obtain the approximation

$$\left(\frac{h}{\tilde{h}}\right)^2 \approx \frac{\varepsilon_0(h)}{\varepsilon_0(\tilde{h})},$$

which can be used to determine $\tilde{h}$. This steplength $\tilde{h}$ will now be used in the next predictor step. It is usually safeguarded by some additional considerations such as limiting the steplength to some interval $h_{\min} \leqslant \tilde{h} \leqslant h_{\max}$, or limiting the factor $1/2 \leqslant h/\tilde{h} \leqslant 2$, etc.

7. Steplength control via asymptotic expansion

Another method, based upon asymptotic estimates, is due to GEORG [1983]. The basic idea in this approach is to observe the performance of the corrector procedure in dependence on a given steplength h. This dependence is typically expressed by some asymptotic expansions in h. The aim is then to adapt the steplength in such a way as to strive toward a desired performance.

Let us discuss this approach by means of a simple, but typical example. Suppose that a point u on the solution curve has been approximated. Suppose further that a steplength $h > 0$ and a predictor point

$$v_0(h) = u + ht\big(H'(u)\big) \tag{7.1}$$

are given. Then a Gauss–Newton type iterative corrector process is performed:

$$v_{i+1}(h) = v_i(h) - H'\big(v_0(h)\big)^+ H\big(v_i(h)\big). \tag{7.2}$$

The sequence $\{v_i(h)\}$ converges to the next point $v_\infty(h)$ on the curve.

The steplength strategy is motivated by the following question: Given the observed performance of the corrector process for the steplength h, which steplength $\tilde{h}$ would have been "best" for obtaining $v_\infty(\tilde{h})$ from u? This "ideal" steplength $\tilde{h}$ is determined via asymptotic estimates, and it is then taken as the steplength for the next predictor step. This strategy depends primarily upon two factors: the particular predictor corrector method being utilized, and the criteria used in deciding what performance is considered "best".

Let us illustrate this technique with the case of the *contraction rate* of the corrector process (7.2):

$$\kappa(u,h) := \frac{\|H'(v_0(h))^+ H(v_1(h))\|}{\|H'(v_0(h))^+ H(v_0(h))\|}.$$

Since Newton's method is locally quadratically convergent, it is plain that $\kappa(u,h)$ will decrease (and hence Newton's method will become faster) as h decreases. The following theorem characterizes the asymptotic behavior of $\kappa(u,h)$ with respect to h, see also (6.1.2) in ALLGOWER and GEORG [1990].

THEOREM 7.1. *Suppose that*

$$H''(u)\big[t\big(H'(u)\big),t\big(H'(u)\big)\big] \neq 0.$$

Then

$$\kappa(u,h) = \kappa_2(u)h^2 + O\big(h^3\big)$$

for some constant $\kappa_2(u) \geqslant 0$ *which is independent of* h *and depends smoothly on* u.

Let us first point out that the above assumption implies that the solution curve has nonzero curvature at u. In fact, if $c(s)$ denotes the solution curve (parametrized with respect to arclength), then, by differentiating $H(c(s)) = 0$ twice with respect to s, we obtain

$$H'\big(c(s)\big)\ddot{c}(s) + H''\big(c(s)\big)\big[\dot{c}(s),\dot{c}(s)\big] = 0.$$

Note that $\ddot{c}(s) \perp \dot{c}(s)$, and hence from (3.4) we obtain

$$\ddot{c}(s) = -H'\big(c(s)\big)^+ H''\big(c(s)\big)\big[\dot{c}(s),\dot{c}(s)\big].$$

PROOF. For convenience we hereafter abbreviate $H''(u)[b]^2 := H''(u)[b,b]$. Expanding $H(u+ht(H'(u)))$ about u we obtain

$$\begin{aligned} H\big(v_0(h)\big) &= C_1(u)h^2 + O\big(h^3\big), \quad \text{where} \\ C_1(u) &:= \tfrac{1}{2}H''(u)\big[t\big(H'(u)\big)\big]^2, \end{aligned}$$

since $H(u) = 0$ and $H'(u)t(H'(u)) = 0$. Because the maps $u \mapsto H'(u)$, $u \mapsto H'(u)^+$ and $u \mapsto H''(u)$ are smooth, we have

$$\begin{aligned} H'\big(v_0(h)\big) &= H'(u) + O(h), \\ H'\big(v_0(h)\big)^+ &= H'(u)^+ + O(h), \\ H''\big(v_0(h)\big) &= H''(u) + O(h) \end{aligned}$$

and hence

$$\begin{aligned} H'\big(v_0(h)\big)^+ H\big(v_0(h)\big) &= C_2(u)h^2 + O\big(h^3\big), \quad \text{where} \\ C_2(u) &:= H'(u)^+ C_1(u). \end{aligned}$$

Now we expand $H(v_0(h) - H'(v_0(h))^+ H(v_0(h)))$ about $v_0(h)$:

$$\begin{aligned} H(v_1(h)) &= \underbrace{H(v_0(h)) - H'(v_0(h))H'(v_0(h))^+ H(v_0(h))}_{=0} \\ &\quad + \tfrac{1}{2} H''(v_0(h))\left[H'(v_0(h))^+ H(v_0(h))\right]^2 + O(h^5) \\ &= C_3(u)h^4 + O(h^5), \quad \text{where} \end{aligned}$$

$$C_3(u) := \tfrac{1}{2} H''(u)\left[C_2(u)\right]^2.$$

Furthermore we have

$$\begin{aligned} H'(v_0(h))^+ H(v_1(h)) &= C_4(u)h^4 + O(h^5), \quad \text{where} \\ C_4(u) &:= H'(u)^+ C_3(u). \end{aligned}$$

Finally we obtain

$$\begin{aligned} \kappa(u,h) &= \kappa_2(u)h^2 + O(h^3), \quad \text{where} \\ \kappa_2(u) &:= \frac{\|C_4(u)\|}{\|C_2(u)\|}. \end{aligned} \tag{7.3}$$

Note that the assumption implies that $C_1(u) \neq 0$ and hence $C_2(u) \neq 0$. The smoothness of $\kappa_2(u)$ follows from the smoothness of the vectors $C_2(u)$ and $C_4(u)$. □

In view of asymptotic relation (7.3), the steplength modification $h \to \tilde{h}$ is now easy to explain. Assume that an Euler predictor (7.1) and a Gauss–Newton iteration (7.2) has been performed with steplength h. Then $H'(v_0(h))^+ H(v_0(h))$ and $H'(v_0(h))^+ H(v_1(h))$ will have been calculated and thus $\kappa(u,h)$ can be obtained without any significant additional cost. Now an a posteriori estimate

$$\kappa_2(u) = \frac{\kappa(u,h)}{h^2} + O(h)$$

is available.

In order to have a robust and efficient method we want to continually adapt the steplength h so that a nominal prescribed contraction rate $\tilde{\kappa}$ is maintained. The choice of $\tilde{\kappa}$ will generally depend upon the nature of the problem at hand, and on the desired security with which we want to traverse the curve. That is, the smaller $\tilde{\kappa}$ is chosen, the greater will be the security with which the method will follow the curve. When using the term *securely* or *safely* following the curve we mean that a safeguard prevents the method from jumping to a different part of the curve (at a significantly different arclength value) or to a different connected component of $H^{-1}(0)$. Depending on the structure of the solution manifold $H^{-1}(0)$, this may be an important issue.

Once $\tilde{\kappa}$ has been chosen, we will consider a steplength $\tilde{h}$ to be appropriate if $\kappa(u,\tilde{h}) \approx \tilde{\kappa}$. By using the above equation and neglecting higher order terms we obtain the formula

$$\tilde{h} = h\sqrt{\frac{\tilde{\kappa}}{\kappa(u,h)}}$$

as the steplength for the next predictor step.

In a similar way, other quantities which are important for the performance of the path following method can be taken into account, e.g., the angle of two successive predictor directions, the size of the first Gauss–Newton step (which gives an approximation of the distance of the predictor point to the curve), or the function value $H(v_0(h))$. All these quantities admit asymptotic expansions in h (with varying order). For example, Algorithm (6.1.10) and Program 1 in ALLGOWER and GEORG [1990] incorporate such features in the steplength strategy. Let us summarize the above discussion in a basic Euler–Newton algorithm where the steplength is only monitored via a nominal contraction rate $\tilde{\kappa}$.

ALGORITHM 7.1 (*Basic steplength Euler–Newton method*).

input $u_0 \in \mathbb{R}^{N+1}$ such that $H(u_0) \approx 0$ (initial point)
$\quad 1 > \tilde{\kappa} > 0$ nominal contraction rate
$\quad h > 0$ initial steplength
output u_i, $i = 0, 1, 2, \ldots$, approximating the solution curve
for $i = 1, 2, \ldots$
$\quad$ approximate $A_{i-1} \approx H'(u_{i-1})$
$\quad$ % *predictor step*
$\quad v_{i,0} \leftarrow u_{i-1} + ht(A_{i-1})$
$\quad$ approximate $B_i \approx H'(v_{i,0})$
$\quad$ % *corrector iteration*
$\quad$ for $k = 1, 2, \ldots$ until convergence
$\qquad v_{i,k} \leftarrow v_{i,k-1} - B_i^+ H(v_{i,k-1})$
$\quad$ end for
$\quad u_i \leftarrow v_{i,k}$
$\quad \kappa \leftarrow \dfrac{\|B_i^+ H(v_{i,1})\|}{\|B_i^+ H(v_{i,0})\|}$
$\quad h \leftarrow h\sqrt{\tilde{\kappa}/\kappa}$
end for

KEARFOTT [1989, 1990] proposes interval arithmetic techniques to determine a first order predictor which stresses secure path following.

8. Special points on the curve

One of the main purposes of numerical continuation methods concerns the accurate determination of certain points on the solution curve $c(s)$ which are of special interest. The following are some examples.

(1) In the applications dealing with homotopy methods, the equation $H(x,\lambda)=0$ for $x \in \mathbb{R}^N$ and $\lambda \in \mathbb{R}$ generally has a known starting point (x_0, λ_0). The homotopy path $c(s)$ passes through this point, and we seek a point $(\bar{x}, \bar{\lambda})$ on $c(s)$ such that $H(\bar{x}, \bar{\lambda}) = 0$ for a certain value $\bar{\lambda}$ of the homotopy parameter λ. Examples of applications of homotopy methods are given in Chapter V.

(2) Fold points in $H^{-1}(0)$ may be of interest when the equation represents a branch of solutions for a nonlinear eigenvalue problem involving the eigenvalue parameter λ, see also Section 24. Such points are characterized by the fact that λ has a local extremum on $H^{-1}(0)$. In physics and engineering applications, a fold point can signify a change in the stability of the solutions. A vast literature exists for calculating such points, the following papers are a sample: BOLSTAD and KELLER [1986], CHAN [1984b], FINK and RHEINBOLDT [1986, 1987], MELHEM and RHEINBOLDT [1982], PÖNISCH and SCHWETLICK [1981, 1982], SCHWETLICK [1984a,b], USHIDA and CHUA [1984].

(3) Simple bifurcation points will be discussed in Section 24. There we show how to detect the presence of such points along the curve c. It is of interest to accurately approximate a bifurcation point. They are of great interest since they represent points at which the stability of the solutions changes.

To unify our discussion, let $f : \operatorname{range} c \to \mathbb{R}$ be a smooth functional. There are two general types of special points on the curve c which we shall consider:

Zero points. In this case we seek points $c(s)$ such that $f(c(s)) = 0$. The homotopy method is such a case if we set $f(x,\lambda) := \lambda - \bar{\lambda}$. Simple bifurcation points are another such case if we set, e.g.,

$$f\big(c(s)\big) := \det \begin{pmatrix} H'(c(s)) \\ \dot{c}(s)^* \end{pmatrix}.$$

Extremal points. In this case we seek extremal points (usually maxima or minima) of $f(c(s))$. Fold points are such a case if we set $f(x,\lambda) := \lambda$. Certain regularization methods may also be formulated as determining a fold point on an implicitly defined curve.

We now treat these two general cases in greater detail.

8.1. Calculating zero points

Let $H : \mathbb{R}^{N+1} \to \mathbb{R}^N$ be a smooth map, let $c(s)$ be a smooth solution curve parametrized with respect to arclength (for the sake of convenience), and let $f : \operatorname{range} c \to \mathbb{R}$ be a smooth functional. Suppose that some point $c(s_n)$ has been found which is an approximate zero point of f. For example, it would be reasonable to take $c(s_n)$ as an approximate zero point if a predictor corrector method produced two successive points $c(s_{n-1})$ and $c(s_n)$ such that $f(c(s_{n-1}))f(c(s_n)) < 0$. Then it is reasonable to replace the usual steplength adaptation used to traverse the curve c by a *Newton steplength adaptation* which is motivated by the following one-dimensional Newton method for solving the equation $f(c(s)) = 0$:

$$s_{n+1} = s_n - \frac{f(c(s_n))}{f'(c(s_n))\dot{c}(s_n)} . \tag{8.1}$$

This suggests that we can take the new steplength

$$h := -\frac{f(c(s_n))}{f'(c(s_n))\dot{c}(s_n)} \tag{8.2}$$

at $u := c(s_n)$ in order to obtain a predictor point $v = u + ht(H'(u))$, which should lead to a better approximation of a zero point of f on c.

The following algorithm illustrates for a simple Euler–Newton method how a standard steplength adaptation can be switched to the above Newton steplength adaptation in order to approximate a zero point of f on c while traversing c.

ALGORITHM 8.1 (*Newton steplength adaptation*).

input $u \in \mathbb{R}^{N+1}$ such that $H(u) \approx 0$ (initial point)
    $h > h_{\text{min}} > 0$ (initial, minimal steplength)
output u such that $H(u) = 0$ and $f(u) = 0$

$\nu \leftarrow$ false % *switch*
repeat $v \leftarrow u + ht\big(H'(u)\big)$ % *predictor step*
    repeat $v \leftarrow v - H'(v)^+ H(v)$ % *corrector loop*
    until convergence
    if $f(u)f(v) \leqslant 0$; $\nu \leftarrow$ true; end if
    if $\nu =$ true
        $h \leftarrow -\dfrac{f(v)}{f'(v)\,t\big(H'(v)\big)}$ % *Newton steplength*
    else choose a new $h > 0$ % *standard steplength*
    end if
    $u \leftarrow v$ % *new point along the curve*
until $|h| < h_{\text{min}}$

A sufficient condition for a sequence of points u produced by Algorithm 8.1 to converge to a solution $\bar{u}$ of

$$H(u) = 0, \qquad f(u) = 0,$$

is that the predictor point is sufficiently near $\bar{u}$ and that

$$\det \begin{pmatrix} H'(\bar{u}) \\ f'(\bar{u}) \end{pmatrix} \neq 0. \tag{8.3}$$

Under these assumptions quadratic convergence can be shown.

Algorithm 8.1 requires the quantity

$$\frac{\mathrm{d}}{\mathrm{d}s} f\big(c(s)\big) = f'\big(c(s)\big)\dot{c}(s),$$

and this may be inconvenient to obtain. As an example, we will see in Section 24 that bifurcation points $c(\bar{s})$ are points where

$$f\big(c(s)\big) = \det \begin{pmatrix} H'(c(s)) \\ \dot{c}(s)^* \end{pmatrix} = 0 \tag{8.4}$$

holds. In this case, furnishing $\mathrm{d}f(c(s))/\mathrm{d}s$ would be undesirable, since it would require, at least formally, the calculation of H''. Thus it is reasonable to formulate the secant analogue of (8.1) which leads to the following Newton steplength adaptation in Algorithm 8.1:

$$h := -\frac{f(v)}{f(v) - f(u)}\, h.$$

This reduces the above mentioned quadratic convergence to superlinear convergence.

For the case of calculating a simple bifurcation point, care should be taken since the augmented Jacobian

$$\begin{pmatrix} H'(u) \\ t(H'(u))^* \end{pmatrix}$$

is ill-conditioned near the bifurcation point, and hence the evaluation of (8.4) encounters instabilities. But the above mentioned superlinear convergence of the secant method generally overcomes this difficulty since the instability generally only manifests itself at a predictor point which can already be accepted as an adequate approximation of the bifurcation point.

Obviously, if one zero point $\bar{u}$ of the functional f on the curve c has been approximated, the predictor corrector method can be restarted in order to seek additional zero points.

8.2. Calculating extremal points

The aim now is to give some specific details for calculating an extremal point on the solution curve c for a smooth functional $f: \text{range}\, c \to \mathbb{R}$. Clearly, a necessary condition which must hold at a local extremum $c(\bar{s})$ of f is that the equation

$$f'\big(c(s)\big)\dot{c}(s) = 0 \tag{8.5}$$

holds. Following the same motivation as in the previous section, we can formulate the analogous switchover to a Newton steplength adaptation:

$$h := -\frac{f'(v)\dot{v}}{f'(v)\ddot{v} + f''(v)[\dot{v}, \dot{v}]}, \tag{8.6}$$

where $v = c(s)$, $\dot{v} = \dot{c}(s) = t(H'(v))$, $\ddot{v} = \ddot{c}(s)$, and $v = c(s)$ is the point currently approximated on c in Algorithm 8.1.

The quantities v and $\dot{v}$ are readily obtained. Let us discuss how to calculate an approximation of $\ddot{v}$. By differentiating the equation $H(c(s)) = 0$ we obtain $H'(v)\dot{v} = 0$ and

$$H''(v)[\dot{v}, \dot{v}] + H'(v)\ddot{v} = 0. \tag{8.7}$$

Differentiating $\|\dot{c}(s)\|^2 = 1$ yields the orthogonality

$$\dot{v}^* \perp \ddot{v},$$

and we obtain from (8.7) and (3.4):

$$\ddot{v} = -H'(v)^+ H''(v)[\dot{v}, \dot{v}]. \tag{8.8}$$

To approximate $H''(v)[\dot{v}, \dot{v}]$ we can use the centered difference formula

$$\frac{H(v + \varepsilon\dot{v}) - 2H(v) + H(v - \varepsilon\dot{v})}{\varepsilon^2} = H''(v)[\dot{v}, \dot{v}] + \mathrm{O}(\varepsilon^2), \tag{8.9}$$

see Section 16.2 for more details. Now (8.8), (8.9) provide an approximation of $\ddot{v}$ in the Newton steplength adaptation (8.6). If necessary, higher order formulae and, in particular, an extrapolation method may be used to obtain higher precision approximations of $\ddot{v}$ since the truncation error in (8.9) is expandable in powers of ε^2.

The following example illustrates this approach for the case of calculating a fold point with respect to the last co-ordinate, i.e., $f(x, \lambda) = \lambda$ in the case of a nonlinear eigenvalue problem $H(x, \lambda) = 0$. Then $f(v) = v(N+1)$ (i.e., the last co-ordinate), and (8.6) takes the particular form

$$h := -\frac{\dot{v}[N+1]}{\ddot{v}[N+1]}.$$

A second possible method for calculating a local extremal point of $f(c(s))$ is to use a secant steplength adaptation applied to Eq. (8.5). Analogously to the discussion in the previous section we obtain the corresponding Newton steplength adaptation

$$h := -\frac{f'(v)\dot{v}}{f'(v)\dot{v} - f'(u)\dot{u}}\, h,$$

where v is the latest point and u the previous point on the solution curve. The advantage of using this in Algorithm 9.2.3 is that the need to calculate $\ddot{v}$ is avoided. Under assumptions analogous to (8.3) superlinear convergence of the sequence u generated by the algorithm to a local extremum of $f(c(s))$ can be shown.

An alternative way for choosing the stepsize based on Hermite interpolation of $\dot{\lambda}$ at the latest two points is described in PÖNISCH and SCHWETLICK [1982].

8.3. Calculating fold and bifurcation points

As we already mentioned above, calculating a fold point may be viewed as calculating a special point on the curve which is characterized as an extremal point with respect to a particularly simple functional (e.g., the last co-ordinate or bifurcation parameter). A more efficient approach is to extend the equation $H(u) = 0$ in order to directly characterize such points. A rapidly convergent Newton-like iteration can then be applied to the extended system of equations. A starting value is obtained from the curve following process. We refer to MOORE and SPENCE [1980] as a pioneer of this method, see also PÖNISCH and SCHWETLICK [1981].

The case of bifurcation points is more complex and will be described in more detail in Section 24. In particular, the above mentioned technique of extending the equations $H(u) = 0$ in order to characterize and approximate such points is discussed in Section 24.2.

9. Higher order predictors

The steplength strategies we have discussed up to now have been based upon the Euler predictor, which is only of local order two. This is very often satisfactory since it is usually used in conjunction with rapidly converging correctors such as Newton type correctors. However, for large systems, often less rapidly convergent iterative methods such as Krylov subspace methods are used. Hence, one may expect to obtain improved efficiency by using higher order predictors.

Another instance where it is desirable to use higher order predictors occurs when the solution curve needs to be approximated very well at all points, e.g., for plotting purposes or numerical integration, see Section 30 for the latter. Since in such cases a polynomial approximation or something similar needs to be performed anyway, one may as well expend this numerical effort at the outset by performing higher order predictors.

Variable order predictors and corresponding steplength strategies may be generated in a way reminiscent of the ones used in multistep methods for solving initial value problems, see, e.g., SHAMPINE and GORDON [1975]. Such a method was suggested by

GEORG [1982, 1983]. LUNDBERG and POORE [1991] have made an implementation using variable order Adams–Bashforth predictors. Their numerical results show that there is often a definite benefit to be derived by using higher order predictors.

Let us sketch a general philosophy for monitoring the order and steplength of higher order predictors. Let u_n be a current point on the solution curve c which can be locally parametrized via the parameter s (not necessarily arclength). For simplicity, let us assume that $c(0) = u_n$. Consider a polynomial predictor of the form

$$c(h) \approx p_k(h) = u + \sum_{i=1}^{k} c_i h^i, \tag{9.1}$$

$$c_i \approx \frac{c^{(i)}(0)}{i!}, \tag{9.2}$$

which represents an approximation of the Taylor formula. We see essentially two different ways for obtaining the coefficients c_i: (1) by divided differences or polynomial interpolation making use of previously calculated points on the curve, (2) by successive numerical differentiation at u. The former is less expensive to calculate, but the latter is likely to be more accurate and stable. We elaborate on this in the next sections.

We sketch one possible way of determining the next steplength and the next order in the predictor. Let $\varepsilon > 0$ be a given tolerance. The term $\|c_k\| h^k$ can be viewed as a rough estimate for the truncation error of the predictor $p_{k-1}(h)$. Hence, by solving $\|c_k\| h^k = \varepsilon$ for h, we estimate

$$h_k := \left(\frac{\varepsilon}{\|c_k\|}\right)^{1/k} \tag{9.3}$$

as the steplength for the predictor p_{k-1} in order to remain within the given tolerance. Due to instabilities of various kinds, we anticipate that eventually

$$h_2 < h_3 < \cdots < h_q \geqslant h_{q+1} \tag{9.4}$$

will hold for some q. Hence, the predictor p_{q-1} with steplength h_q is our next choice.

This idea can be implemented and modified in various ways, and needs some stabilizing safeguards, such as setting a maximum increase in steplength and in the order and a maximal steplength and order. The strategy to be developed depends on the objective of the application at hand.

10. Interpolation predictors

Let us consider an example of a polynomial interpolation predictor. Assume that the points $u_0, u_1, \ldots, u_n$ along the solution curve c have already been generated. In certain versions of the continuation method, also the corresponding tangents $t_0 := t(H'(u_0)), \ldots, t_n := t(H'(u_n))$ are available. The idea is to use an interpolating polynomial p_q of degree q (with coefficients in $\mathbb{R}^{N+1}$) satisfying $p_q(0) = u_n$ as a predicting polynomial. We need to express the interpolating polynomial in terms of a suitable parameter ξ. Naturally, the arclength parameter s which we always consider for theoretical discussions would be ideal to use. This is done by LUNDBERG and POORE [1991]. However, for purposes of exposition, we shall avoid the additional complexity of obtaining precise numerical approximations of the arclength s_i such that $c(s_i) = u_i$. We therefore propose to use a local parametrization ξ induced by the current approximate tangent $t \approx t(H'(u_n))$, which does not need to be very accurate. We assume however, that the normalization $\|t\| = 1$ holds. This local parametrization $c(\xi)$ is defined as the locally unique solution of the system

$$\begin{aligned} &H(u) = 0, \\ &t^*(u_n + \xi t - u) = 0, \end{aligned} \tag{10.1}$$

for ξ in some open interval containing zero. It follows immediately that

$$c(\xi_i) = u_i \quad \text{where } \xi_i = t^*(u_i - u_n).$$

Differentiating (10.1) with respect to ξ yields

$$\frac{\mathrm{d}c(\xi)}{\mathrm{d}\xi} = \frac{\dot{c}(s)}{t^*\dot{c}(s)}.$$

If the tangents t_i at the points u_i are available for use, we may form a Hermite interpolating polynomial p_q. Otherwise, a standard interpolating polynomial using Newton's formula is generated.

Hence, the interpolation polynomial is of the form (Newton)

$$\begin{aligned} p_q(h) = {} & c[\xi_n] + c[\xi_n, \xi_{n-1}](h - \xi_n) + c[\xi_n, \xi_{n-1}, \xi_{n-2}](h - \xi_n)(h - \xi_{n-1}) \\ & + \cdots + c[\xi_n, \ldots, \xi_{n-q}](h - \xi_n) \cdots (h - \xi_{n-q+1}), \end{aligned}$$

or (Hermite)

$$\begin{aligned} p_{2i+2}(h) = {} & c[\xi_n] + c[\xi_n, \xi_n](h - \xi_n) + c[\xi_n, \xi_n, \xi_{n-1}](h - \xi_n)^2 + \cdots \\ & + c[\xi_n, \xi_n, \ldots, \xi_{n-i}, \xi_{n-i}, \xi_{n-i-1}](h - \xi_n)^2 \cdots (h - \xi_{n-i})^2, \\ p_{2i+1}(h) = {} & c[\xi_n] + c[\xi_n, \xi_n](h - \xi_n) + c[\xi_n, \xi_n, \xi_{n-1}](h - \xi_n)^2 + \cdots \\ & + c[\xi_n, \xi_n, \ldots, \xi_{n-i}, \xi_{n-i}](h - \xi_n)^2 \cdots (h - \xi_{n-i-1})^2(h - \xi_{n-i}), \end{aligned}$$

where the coefficients are obtained via the formulae

$$\begin{aligned} \xi_i &:= t^*(u_i - u_n), \\ c[\xi_i] &:= u_i, \\ c[\xi_i, \xi_i] &= \frac{\mathrm{d}c(\xi_i)}{\mathrm{d}\xi} := \frac{t_i}{t^*t_i}, \end{aligned}$$

$$c[\xi_i,\dots,\xi_j] := \frac{c[\xi_i,\dots,\xi_{j+1}] - c[\xi_{i-1},\dots,\xi_j]}{\xi_i - \xi_j} \quad \text{for } i > j,\ \xi_i \neq \xi_j.$$

11. Taylor polynomial predictors

Inexpensive higher order predictors are generally based on polynomial interpolation. It is known that such predictors are not very stable, in particular for larger stepsizes. In view of the stability of Newton's method as a corrector, it may be advantageous to also use more stable predictors. MACKENS [1989] has proposed such predictors which are based on Taylor's formula and which are obtained by successive numerical differentiation, see also SCHWETLICK and CLEVE [1987] as a predecessor. However, the gain in stability has to be paid for by additional evaluations of the map H and additional applications of the Moore–Penrose inverse of the Jacobian H' (where it may be assumed that H' has already been decomposed into some factorization).

Let us give a brief account of Mackens' approach. We assume again that $u_n = c(0)$ is a current point on the solution curve, and consider a local parametrization $c(\xi)$ with respect to the current tangent $t := t(H'(u_n))$, i.e., $c(\xi)$ solves

$$H(u) = 0, \qquad t^*\big(u - (u_n + \xi t)\big) = 0.$$

Note that by differentiating with respect to ξ we obtain immediately

$$c'(0) = t \quad \text{and} \quad c^{(k)}(0) \perp t \quad \text{for } k > 1. \tag{11.1}$$

Consider the Taylor expansion

$$c(\xi) = \sum_{i=0}^{k} \frac{1}{i!} c^{(i)}(0)\xi^i + \mathrm{O}\big(\xi^{k+1}\big).$$

In principle, the derivatives of c can be obtained by differentiating the equation $H(c(\xi)) = 0$ repeatedly and evaluating at zero:

$$\begin{aligned} H'(u_n)c'(0) &= \qquad 0 &&=: R_1, \\ H'(u_n)c''(0) &= -H''(u_n)[c'(0), c'(0)] &&=: R_2, \\ &\vdots \\ H'(u_n)c^{(k)}(0) &= \qquad \cdots &&=: R_k. \end{aligned} \tag{11.2}$$

Note that the orthogonality (11.1) and (3.4) implies that

$$c^{(k)}(0) = H'(u_n)^+ R_k \tag{11.3}$$

for $k > 1$.

The crucial observation is to note that the values R_k can also be obtained via

$$\left(\frac{\mathrm{d}}{\mathrm{d}\xi}\right)^{(k)} H\left(\sum_{i=0}^{k-1}\frac{1}{i!}c^{(i)}(0)\xi^i\right)\Bigg|_{\xi=0} = -R_k \tag{11.4}$$

for $k > 1$.

PROOF. Let $p(\xi) := \sum_{i=0}^{k-1}(1/i!)c^{(i)}(0)\xi^i$. On the one hand we have

$$\left(\frac{\mathrm{d}}{\mathrm{d}\xi}\right)^{(k)} H\big(c(\xi)\big) = H'\big(c(\xi)\big)c^{(k)}(\xi) - R_k(\xi) = 0,$$

and on the other hand

$$\left(\frac{\mathrm{d}}{\mathrm{d}\xi}\right)^{(k)} H\big(p(\xi)\big) = H'\big(p(\xi)\big)p^{(k)}(\xi) + S_k(\xi).$$

The expressions $-R_k(\xi)$ and $S_k(\xi)$ contain higher derivatives of H at c and p, respectively, and derivatives of c and p up to order $k-1$. Except for replacing the c's by the p's, the expressions $-R_k(\xi)$ and $S_k(\xi)$ are the same. Since $c^{(i)}(0) = p^{(i)}(0)$ for $i < k$, and since $p^{(k)} = 0$, we obtain $-R_k(0) = S_k(0)$. □

By combining (11.2)–(11.4), it becomes evident that the derivatives of c can be obtained recursively. We summarize the discussion with the following.

ALGORITHM 11.1 (*Taylor coefficients*).

input $u_n \in \mathbb{R}^{N+1}$ such that $H(u_n) = 0$
output $c_k = c^{(k)}(0)$ for $k = 0, 1, 2, \ldots$

$c_0 \leftarrow u_n$
$c_1 \leftarrow t(H'(u_n))$
for $k = 2, 3, \ldots$

$$R_k \leftarrow -\left(\frac{\mathrm{d}}{\mathrm{d}\xi}\right)^{(k)} H\left(\sum_{i=0}^{k-1}\frac{1}{i!}c_i\xi^i\right)\Bigg|_{\xi=0}$$

$c_k \leftarrow H'(u_n)^+ R_k$

end for

The kth derivative employed in the above algorithm for obtaining the coefficient R_k can be numerically approximated by difference formulae (preferably by central difference formulae). The formulae make use of a certain meshsize, say δ. The resulting truncation errors are then propagated to the higher derivatives in the above algorithm. MACKENS [1989] gives a careful discussion of this truncation error propagation.

CHAPTER IV

Special Topics

12. Implementation of the Moore–Penrose inverse

The main ingredients of the predictor corrector method from the linear algebra standpoint consist of obtaining an approximation of the tangent $t(A)$ and either an implementation of an action of the Moore–Penrose inverse A^+ for a given approximation $A \approx H'(u_i)$ of the Jacobian, or an action of

$$\begin{pmatrix} A \\ t^* \end{pmatrix}^{-1} \tag{12.1}$$

where t^* is an appropriate additional row. In this section we concentrate on the Moore–Penrose inverse, and in the next section on the bordering algorithm for an efficient implementation of the action of (12.1).

A straightforward and simple (but not the most efficient) way to handle the numerical linear algebra would be to use a QR factorization:

$$A^* = Q \begin{pmatrix} R \\ 0^* \end{pmatrix}, \tag{12.2}$$

where Q is an $(N+1, N+1)$ orthogonal matrix, and R is a nonsingular (N, N) upper triangular matrix. We assume that A is an $(N,\ N+1)$ matrix with maximal rank. If q denotes the last column of Q, then $t(A) = \sigma q$, where the orientation defined in (1.5) leads to the choice

$$\sigma = \operatorname{sign}(\det Q\ \det R). \tag{12.3}$$

Hence σ is easy to determine. The Moore–Penrose inverse of A can be obtained from the same decomposition in the following way:

$$A^+ = A^*(AA^*)^{-1} = Q \begin{pmatrix} (R^*)^{-1} \\ 0^* \end{pmatrix}. \tag{12.4}$$

Of course, one does not generally calculate A^+, but instead the action of A^+ involved in the above equation, i.e., a backsolving and a matrix multiplication. The matrix Q will probably be available in factored form.

Similar ideas apply if an LU decomposition is employed:

$$PA^* = L\begin{pmatrix} U \\ 0^* \end{pmatrix}, \tag{12.5}$$

where L is a lower triangular $(N+1, N+1)$ matrix, U is an $(N,\ N)$ upper triangular matrix, and P is a permutation matrix corresponding to partial pivoting of the columns of A, which is in general necessary to improve the numerical stability.

Let us first consider the calculation of the tangent vector $t(A)$. From (12.5) it follows that

$$A = (U^*, 0)L^*P. \tag{12.6}$$

Hence, if we set

$$y := P^*(L^*)^{-1}e_{N+1},$$

then it is readily seen from (12.6) that $Ay = 0$. Of course $y \neq 0$, and can be calculated by one backsolving and a permutation of its co-ordinates. Hence $t(A) = \sigma y/\|y\|$, where the sign σ is determined by evaluating the sign of the determinant of

$$\begin{aligned}(A^*, y) &= \left(P^*L\begin{pmatrix} U \\ 0^* \end{pmatrix},\ P^*(L^*)^{-1}e_{N+1}\right) \\ &= P^*L\left(\begin{pmatrix} U \\ 0^* \end{pmatrix},\ L^{-1}(L^*)^{-1}e_{N+1}\right).\end{aligned}$$

The last entry $\|(L^*)^{-1}e_{N+1}\|^2$ of $L^{-1}(L^*)^{-1}e_{N+1}$ must be positive, and hence

$$\sigma = \operatorname{sign}\det(A^*, y) = \operatorname{sign}\det(P)\det(L)\det(U).$$

The right hand side is easily determined.

Let us now turn to the problem of determining the Moore–Penrose inverse. From (12.6) it follows that

$$B := P^*(L^*)^{-1}\begin{pmatrix} (U^*)^{-1} \\ 0^* \end{pmatrix}$$

is a right inverse of A, and hence $A^+ = (\operatorname{Id} - t(A)t(A)^*)B$ by (3.9). Finally, let us note that one action of A^+ amounts to essentially one forward solving with U^*, one back solving with L^*, and one scalar product for the orthogonal projection with $(\operatorname{Id} - t(A)t(A)^*)$.

The above methods are useful for small dense matrices A. However, in many applications of path following methods, the corresponding matrix A is large and sparse, and then the above procedure is inefficient. Among such applications are the approximation of branches of nonlinear eigenvalue problems or the central path methods of linear and nonlinear programming. Let us discuss some ideas which are useful in dealing with such situations.

In many applications of path following, one encounters matrices A with the following structure:

$$A = (A_0, \quad b), \tag{12.7}$$

where equations of the form $A_0 x = y$ permit a fast linear solver. We will refer to one solving of such a system as one *action* of A_0^{-1}. In the path following it is convenient to consider an augmented matrix of the form:

$$\tilde{A} := \begin{pmatrix} A_0 & b \\ c^* & d \end{pmatrix} \tag{12.8}$$

where the additional row $(c^* \ d)$ is typically generated via the last predictor direction. See also the discussion of the bordering algorithm in Section 13.

A standard block elimination

$$\begin{pmatrix} A_0 & b \\ c^* & d \end{pmatrix} \to \begin{pmatrix} A_0 & b \\ 0 & d - c^* A_0^{-1} b \end{pmatrix}$$

may be employed to generate the Schur complement

$$s := d - c^* A_0^{-1} b$$

of A_0 in the augmented matrix $\tilde{A}$. Clearly,

$$\det \tilde{A} = s \det A_0. \tag{12.9}$$

It can be checked that if $\tilde{A}$ is nonsingular, then

$$\tilde{A}^{-1} = \begin{pmatrix} A_0^{-1} + A_0^{-1} b s^{-1} c^* A_0^{-1} & -A_0^{-1} b s^{-1} \\ -s^{-1} c^* A_0^{-1} & s^{-1} \end{pmatrix}. \tag{12.10}$$

As an easy consequence, the tangent $t(A)$ is obtained via

$$t(A) = \sigma y / \|y\|, \tag{12.11}$$

where y denotes the last column of $\tilde{A}^{-1}$. The sign $\sigma \in \{\pm 1\}$ can either be obtained from an angle test with the previous predictor direction or from (12.9). In fact,

$$\begin{aligned}
\det\begin{pmatrix} (A_0, b) \\ y* \end{pmatrix} &= \det\begin{pmatrix} A_0 & b \\ -(A_0^{-1}bs^{-1})^* & s^{-1} \end{pmatrix} \\
&= s^{-1}\det\begin{pmatrix} A_0 & b \\ -b^*(A_0^*)^{-1} & 1 \end{pmatrix} \\
&= s^{-1}\det\begin{pmatrix} A_0 & b \\ 0 & 1 + b^*(A_0^*)^{-1}A_0^{-1}b \end{pmatrix} \\
&= s^{-1}\det(A_0)\left(1 + \left\|A_0^{-1}b\right\|^2\right)
\end{aligned}$$

shows that

$$\sigma = \operatorname{sign}(s \det A_0). \tag{12.12}$$

To obtain the Moore–Penrose inverse, let us first note that the matrix $(\tilde{A}^{-1})_N$ consisting of the first N columns of $\tilde{A}^{-1}$ is a right inverse of A. Therefore (3.9) implies

$$A^+ = \left(I - t(A)t(A)^*\right)\left(\tilde{A}^{-1}\right)_N. \tag{12.13}$$

Note that after an initial computational expense of one action of A_0^{-1}, i.e., $A_0^{-1}b$, and one scalar product (to calculate s), the cost of calculating $t(A)$ amounts to essentially one additional scalar product, the cost of one action of $\tilde{A}^{-1}$ amounts to essentially one additional action of A_0^{-1} and one additional scalar product, and the cost of one action of A^+ amounts to essentially one additional call of the action of A_0^{-1} and two additional scalar products (assuming that $t(A)$ has been calculated).

Among the fast solvers which are of importance here are direct solvers for sparse linear systems, or preconditioned iterative solvers such as conjugate-gradient or other Krylov subspace methods, see, e.g., FREUND, GOLUB and NACHTIGAL [1992].

The above Schur complement construction is also valid if b, c and d are matrices (of appropriate size). This is of interest in parametric optimization, see LUNDBERG and POORE [1993].

If A_0 is symmetric, then an Arnoldi method may be used as a fast linear solver. As a byproduct, the generated Hessenberg matrix may be used to approximate the largest and smallest eigenvalues of A_0. In typical applications, this will often be sufficient to determine the sign of $\det A_0$, see HUITFELDT and RUHE [1990].

13. The bordering algorithm

The popular bordering algorithm of KELLER [1977, 1983], see also CHAN [1984a], CHAN and SAAD [1985], MENZEL and SCHWETLICK [1978, 1985], is related to the ideas of the previous section. These approaches are akin to KELLER's [1977] pseudo arclength method, see Section 2, in which the equation $H(u) = 0$ is extended by an additional condition $F(u, s) = 0$ which models a pseudo arclength parametrization. This viewpoint is often convenient, in particular for structured problems such as those arising from discretizations of nonlinear eigenvalue problems.

Let us give a brief account of the bordering algorithm. We describe the implicit equations in the form $H(u) = H(x, \lambda) = 0$ where $H : \mathbb{R}^N \times \mathbb{R} \to \mathbb{R}^N$ is sufficiently smooth and $\lambda \in \mathbb{R}$ is an exceptional parameter such as a nonlinear eigenvalue parameter in discretizations of boundary value problems. We assume that 0 is a regular value of H. The derivative $A_0 := H_x$ of H with respect to the space variables $x \in \mathbb{R}^N$ usually leads to some sparse matrix, and solving equations with this sparse matrix can typically be accomplished by some special solver such as an iterative method or a multigrid method, see, e.g., CHAN and KELLER [1982], HACKBUSCH [1994], FREUND, GOLUB and NACHTIGAL [1992]. Hence, it is desirable to exploit this special solver in a predictor corrector method.

Introducing the pseudo arclength parameter s via $F(u, s) = 0$, see (2.1), we consider the Newton iteration (2.4) where we need to solve a linear system of equations involving the matrix

$$\tilde{A} := \begin{pmatrix} A_0 & b \\ c^* & d \end{pmatrix} = \begin{pmatrix} H'(u) \\ F_u(u, s) \end{pmatrix} = \begin{pmatrix} H_x(\bar{x}, \bar{\lambda}) & H_\lambda(\bar{x}, \bar{\lambda}) \\ F_x(\bar{x}, \bar{\lambda}, 0) & F_\lambda(\bar{x}, \bar{\lambda}, 0) \end{pmatrix}.$$

We can now follow a discussion very similar to the one starting at (12.8). We refer to the computation of an action of $\tilde{A}^{-1}$ according to (12.10) as the *bordering algorithm*. In particular, we have seen that after an initial computational cost of one action of A_0^{-1} and one scalar product, one action of $\tilde{A}^{-1}$ can be obtained essentially for the cost of one action of A_0^{-1} and one scalar product. It is also pointed out there that the tangent $t(A_0, b)$ can be obtained for essentially one additional scalar product. Hence, solving for a bordered system of equations, i.e., a system involving the matrix $\tilde{A}$, is only slightly cheaper than performing one action of the pseudo inverse $H'(u)^+ = (A_0, b)^+$.

A critical case is the situation where A_0 is nearly singular. One would then expect that the bordering algorithm would fail, since the inversion of A_0 is badly conditioned. However, KELLER [1982, 1987] points out that under certain circumstances the bordering algorithm can still be used. More precisely, let us assume that $\tilde{A}$ is nonsingular (i.e., $\text{cond}(\tilde{A}) = \text{O}(1)$), and

$$PA_0Q = \begin{pmatrix} L & 0 \\ l^* & 1 \end{pmatrix} \begin{pmatrix} U & u \\ 0 & \varepsilon \end{pmatrix}$$

is an LU factorization of A_0 where column and row pivots (indicated by the permutation matrices Q and P) are used in such a way that ε catches the order of singularity of $\tilde{A}$, i.e., $\text{cond}(A_0) = \text{O}(\varepsilon^{-1})$. If this factorization is used to calculate an action of $\tilde{A}^{-1}$ according to (12.10), then even for small ε this bordering algorithm is acceptable, though possibly some cancellation of digits must be expected. Numerical experience confirms this analysis.

One last remark involves the weighting of the space variables x against the eigenvalue parameter λ. If the nonlinear system $H(x, \lambda) = 0$ is obtained via a discretization of some boundary value problem over a domain in $\mathbb{R}^n$ with stepsize h, then the standard Euclidean norm $\|(x, \lambda)\| = \sqrt{\|x\|^2 + \lambda^2}$ and the corresponding scalar product would lead to a very bad scaling problem for λ, e.g., when calculating the tangent or approximating the arclength. A better choice would be to try to approximate the L^2 norm of the underlying function space, e.g.,

$$\|(x, \lambda)\| := \sqrt{h^n \|x\|^2 + \lambda^2},$$

see KELLER ([1987], p. 86).

14. Update methods

In this section we will briefly describe how to incorporate an analogue of Broyden's update method into a predictor corrector algorithm. A more extensive discussion may be found in ALLGOWER and GEORG ([1990], Chapter 7).

Let us first recall the Broyden update method for solving a zero-point problem $F(x) = 0$ where $F : \mathbb{R}^N \to \mathbb{R}^N$ is smooth. For a general reference on update methods we suggest the book of DENNIS and SCHNABEL [1983]. Suppose that $F(\bar{x}) = 0$ with $F'(\bar{x})$ having maximal rank N. It is well-known that Newton's method

$$x_{n+1} = x_n - F'(x_n)^{-1} F(x_n), \quad n = 0, 1, \ldots,$$

is locally quadratically convergent. Even when an adequate starting point x_0 has been chosen, there remains the drawback that after every iteration the Jacobian matrix $F'(x_n)$ needs to be calculated and a new matrix decomposition has to be obtained in order to solve the linear system

$$F'(x_n) s_n = -F(x_n)$$

for $s_n := x_{n+1} - x_n$. On the other hand, if an approximate Jacobian is held fixed, say for example $A := F'(x_0)$, a familiar Newton Chord method is obtained:

$$x_{n+1} = x_n - A^{-1} F(x_n).$$

This method offers the advantage that A stays fixed. Thus, once the matrix decomposition for A has been obtained, further iterations may be cheaply carried out. The drawback of the Newton Chord method is that the local convergence is only linear.

The method of BROYDEN [1965] involves the use of previously calculated data to iteratively improve the quality of the approximation $A \approx F'(x_n)$ via successive rank-one updates. A more general class of update methods, usually called quasi-Newton methods, have since been developed which take into account possible special structure of the Jacobian F' such as positive definiteness, symmetry or sparseness. It is possible to prove local superlinear convergence of a large class of these methods under the standard hypotheses for Newton's method.

For general purpose updates, i.e., when no special structure is present, the so-called "good formula" of Broyden appears to be rank-one update of choice. For this reason we will confine our discussion of update methods for curve-following to an analogue of this formula. Similar but more complicated extensions of the discussion below can be given for the case that special structure is present, see also BOURJI and WALKER [1990], WALKER [1990].

Let us motivate our discussion by reviewing the Broyden update formula for solving $F(x) = 0$ via a Newton-type method. From Taylor's formula, we have

$$F'(x_n)(x_{n+1} - x_n) = F(x_{n+1}) - F(x_n) + \mathrm{O}\big(\|x_{n+1} - x_n\|^2\big). \tag{14.1}$$

By neglecting the higher order term in (14.1), and setting

$$s_n := x_{n+1} - x_n, \qquad y_n = F(x_{n+1}) - F(x_n),$$

we obtain the *secant equation*

$$As_n = y_n, \tag{14.2}$$

which should be satisfied (at least to first order) by an approximate Jacobian $A \approx F'(x_n)$.

When Newton-type steps

$$x_{n+1} = x_n - A_n^{-1} F(x_n)$$

are performed using some approximate Jacobian $A_n \approx F'(x_n)$, it is natural to require that the next approximate Jacobian A_{n+1} should satisfy the secant equation

$$A_{n+1}s_n = y_n, \tag{14.3}$$

since the data s_n and y_n are already available.

Clearly, Eq. (14.3) does not uniquely determine A_{n+1}, since it involves N equations in N^2 unknowns. An additional natural consideration is that if A_n was a good approximation to $F'(x_n)$, then this quality ought to be incorporated in formulating subsequent approximations. This leads to the idea of obtaining A_{n+1} from A_n by the *least change principle*, i.e., among all matrices A satisfying the secant equation, we choose the one with the smallest distance from A_n. Thus we are led to the following definition: We define the updated approximate Jacobian A_{n+1} as the solution of the problem

$$\min_A \big\{\|A - A_n\|_{\mathrm{F}}\colon\ As_n = y_n\big\} \tag{14.4}$$

where the norm $\|\bullet\|_{\mathrm{F}}$ is the *Frobenius norm*:

$$\|A\|_{\mathrm{F}} = \left(\sum_{i,j=1}^{N} \big(A(i,j)\big)^2 \right)^{1/2}.$$

A straightforward calculation shows that the solution to (14.4) is given explicitly by

$$A_{n+1} = A_n + \frac{y_n - A_n s_n}{\|s_n\|^2} s_n^*, \tag{14.5}$$

which is generally referred to as *Broyden's good update formula.*

Let us now describe the analogue of Broyden's method for updating the Jacobian along the solution curve $c(s)$. Suppose we are given two approximate zero-points $u_n, u_{n+1} \in \mathbb{R}^{N+1}$ of H and the corresponding values $H(u_n), H(u_{n+1})$. Motivated by the preceding discussion we again set

$$s_n := u_{n+1} - u_n,$$
$$y_n := H(u_{n+1}) - H(u_n),$$

and we consider the analogous secant equation (14.2) where $A \approx H'(u_n)$ is an approximate Jacobian. The corresponding *Broyden update on the points* u_n, u_{n+1} is again given by (14.5). The following algorithm illustrates how this update formula may be incorporated into an Euler–Newton method. Let us stress that updates are performed for both predictor and corrector steps. Note that, for convenience, the corrector points are now the even numbered u_n, and the predictor points the odd numbered u_n.

ALGORITHM 14.1 (*Euler–Newton method using updates*).

```
input  u_0 ∈ R^{N+1} such that H(u_0) ≈ 0 (initial point)
       h > 0 initial steplength
output u_i, i = 0,2,4,..., approximating the solution curve

approximate A_0 ≈ H'(u_0)
for i = 1,3,5,...
    u_i ← u_{i-1} + h t(A_{i-1})   % predictor step
    update A_i from A_{i-1} on u_i, u_{i-1}   % predictor update
    u_{i+1} ← u_i − A_i^+ H(u_i)   % corrector step
    update A_{i+1} from A_i on u_{i+1}, u_i   % corrector update
    choose a new steplength h > 0
end for
```

Let us make a few remarks concerning the above algorithm. It does not necessarily generate a reliable approximation of the Jacobians $H'(u_i)$, i.e., a relation such as

$$\|A_i - H'(u_i)\| = \mathrm{O}(h),$$

(which we assumed in Theorem 5.1) does not hold in general. The reason behind this is that we are in general not assured that the update data spans the whole space $\mathbb{R}^{N+1}$ sufficiently well. To put this more precisely, let S denote the $(N+1,k)$ matrix (for some fixed $k \geqslant 2N$), whose columns have unit norm and indicate the last k directions in which the update formula was used in Algorithm 14.1. Then the condition

$$\operatorname{cond}(SS^*) < C$$

for some $C > 0$ independent of h and of whichever last k directions S are taken, is sufficient to ensure the convergence of Algorithm 14.1 for h sufficiently small, in the sense of Theorem 5.1. See GEORG ([1982], Chapter 4) for a sketch of a proof. However, it is unrealistic to expect that this condition will in general be satisfied. Let us briefly describe how this difficulty may be circumvented. For details we refer to ALLGOWER and GEORG ([1990], Chapter 7).

Setting

$$t_n := t(A_n),$$
$$w_n := \frac{H(u_{n+1}) - H(u_n) - A_n(u_{n+1} - u_n)}{\|u_{n+1} - u_n\|},$$
$$v_n := \frac{u_{n+1} - u_n}{\|u_{n+1} - u_n\|},$$

we see that the update formula (14.5) may be written as

$$A_{n+1} := A_n + w_n v_n^*.$$

Two types of updates arise, namely predictor updates and corrector updates. For the predictor update we have $u_{n+1} = u_n + ht_n$ and consequently

$$w_n = \frac{H(u_{n+1}) - H(u_n)}{h},$$
$$v_n = t_n.$$

It can be verified that

$$t_n \perp A_n^+ w_n,$$
$$t_{n+1} = \frac{t_n - A_n^+ w_n}{\|t_n - A_n^+ w_n\|},$$
$$A_{n+1}^+ = (\mathrm{Id} - t_{n+1} t_{n+1}^*) A_n^+.$$

From this it is clear that updates based *only* upon predictor steps cannot in general maintain good approximations of the Jacobian H'.

For the corrector update we have $u_{n+1} = u_n - A_n^+ H(u_n)$ and consequently

$$w_n = \frac{H(u_{n+1})}{\|A_n^+ H(u_n)\|}, \tag{14.6}$$
$$v_n = -\frac{A_n^+ H(u_n)}{\|A_n^+ H(u_n)\|}.$$

Setting

$$D_n := 1 + v_n^* A_n^+ w_n,$$

it can be verified in this case that

$$t_{n+1} = \text{sign } D_n t_n,$$
$$A_{n+1}^+ = \left(\text{Id} - \frac{A_n^+ w_n v_n^*}{D_n} \right) A_n^+.$$

From this it is again clear that updates based *only* upon corrector steps cannot maintain a good approximation of the Jacobian H'.

A consequence of (14.6) is that

$$\|A_n^+ w_n\| = \frac{\|A_n^+ H(u_{n+1})\|}{\|A_n^+ H(u_n)\|}. \tag{14.7}$$

The vector $A_n^+ w_n$ arises naturally in the corrector update formula, and its norm gives a reasonable measure for the contraction rate of the corrector step. If this rate is large, then this may be attributed to one of two factors: either the predictor point was too far from the solution curve, or the Jacobian H' is poorly approximated by A_n. Often the first of these two possibilities is easy to check, e.g., by monitoring one of the quantities $\|H(u_n)\|$, $\|A_n^+ H(u_n)\|$. Hence $\|A_n^+ w_n\|$ affords us an empirical measure of the quality of the approximation of the Jacobian.

Let us return to the basic issue of Algorithm 14.1, namely to assure that a good approximation of the Jacobian H' is maintained. To do this, we may be guided by a device proposed by POWELL [1970a,b] in which he suggests monitoring the directions of the differences contributing to the updates. In our context, this amounts to monitoring whether the condition number $\text{cond}(SS^*)$ discussed above is sufficiently small. Instead of following Powell's approach of introducing additional directions, we combine a stepsize adaptation with several tests, the most important being the measuring of the above contraction rate (14.7). If the criteria of the tests are satisfied, the algorithm performs a successive predictor corrector step with increased stepsize. If however, the criteria are not satisfied, the predictor corrector step is repeated with reduced stepsize. In both cases, the predictor and corrector updates which were already performed are *not* rejected. This enables the method to generally update an approximation of the Jacobian in directions which are most needed. The following detailed version of Algorithm 14.1 incorporates these considerations. It is given as a possible implementation of the above discussion.

ALGORITHM 14.2 (*Euler–Newton method monitoring updates*).

input $u_0 \in \mathbb{R}^{N+1}$ such that $H(u_0) \approx 0$ (initial point)
$h > 0$ initial steplength
$\delta_0 > 0$ minimal residual
$\kappa \in (0, 1)$ maximal contraction rate

```
output  u_i, i = 0,2,4,..., approximating the solution curve
approximate A ≈ H'(u_0)
i ← 1
while a stopping criterion is not met
   % predictor step
      u_i ← u_{i-1} + h t(A)
      A ← A + h^{-1}(H(u_i) − H(u_{i-1})) t(A)*
   % perturbation
      if ||A^+ H(u_i)|| ⩽ δ_0
         choose p such that ||A^+(H(u_i) − p)|| = δ_0
      else p ← 0
      end if
   % corrector step
      u_{i+1} ← u_i − A^+(H(u_i) − p)
      A ← A + ||u_{i+1} − u_i||^{-2}(H(u_{i+1}) − p)(u_{i+1} − u_i)*
   % contraction test
      if ||A^+(H(u_i) − p)||^{-1} ||A^+(H(u_{i+1}) − p)|| > κ
         h ← h/2; break loop
      end if
      h ← 2h
      i ← i + 2
end while
```

Let us emphasize that usually an implementation of the above algorithm updates some decomposition of the matrix A_n at each step. For details on such update methods we refer to GOLUB and VAN LOAN ([1989], Section 12.6) or ALLGOWER and GEORG ([1990], Chapter 16). The perturbation term p in the above algorithm serves several purposes. The main purpose is to prevent numerical instability in the corrector update formula due to cancellation of digits. As a general rule for choosing δ_0, one may require that $v - w$ should carry at least half as many significant digits as the computer arithmetic carries. Another purpose of the perturbation p is to safeguard the algorithm against intrinsic instabilities such as those arising from singular points on the curve, e.g., bifurcation points. In fact, the above algorithm will usually bifurcate off at simple bifurcation points.

It is sometimes also useful to incorporate more tests in the step "contraction test" of the above algorithm, e.g., so that a maximal distance to the curve or a maximal angle between successive secants is not exceeded.

Several versions of the above algorithm have been implemented and tested on a variety of nonsparse problems, see, e.g., GEORG [1981a,b, 1982], and they have turned out to be very efficient. It should be noted that an iteration of the corrector step has not been incorporated. This needs only to be done if it is wished to follow the curve closely. However, one additional corrector step could easily be performed at low cost since $A^+(H(w) - p)$ is calculated in the "contraction test".

Let us again emphasize that it is important that although a predictor or corrector point may not be accepted, because it may not satisfy the test criteria, the update information which it contributes to approximating the Jacobian is nevertheless utilized. Finally, let us point out that the essential feature of the above algorithm is that it updates *along* the curve as well as in the corrector steps. This is to be distinguished from Euler–Newton methods where the Jacobian is precisely calculated at the predictor point, and updates are only used to accelerate the Newton-type corrector iteration. In this case, proofs of superlinear convergence have been given in BOURJI and WALKER [1990] and WALKER [1990].

15. Linear solvers exploiting symmetry

This section uses some technical concepts from group theory, and can be skipped by the reader without loss of continuity.

Many problems in science and mathematics may be transformed via a linear or nonlinear change of coordinates into related problems. A tranformation which carries a problem or equation into itself is called a *symmetry* of the equation. Usually the symmetry is derived from underlying geometric symmetries of the domain or body on which the problem is considered. The incorporation of symmetry considerations in the analysis and solution of such problems may result in enormous gains in insight and efficiency.

The numerical treatment of problems such as partial differential equations and integral equations generally involves discretizations. In this setting a general principle may be formulated: if a domain enjoys a symmetry group, and an operator equation which expresses a coordinate-free physical law is to be solved on that domain, and if the discretization is chosen to incorporate or respect the symmetries, then the resulting numerical problem (which is usually a matrix equation) is amenable to a block decomposition which greatly reduces the computational cost of determining the solution.

Symmetry groups have long played a significant role in physics, engineering and other areas of science. As BOSSAVIT [1986] points out, any randomly selected issue of the Journal of Mathematical Physics will generally contain four or five papers with entries "group", "symmetry" or "representation theory" in their list of key words. Also in nonlinear analysis, there is considerable literature involving symmetry aspects, more recently in bifurcation phenomena, see, e.g., GOLUBITSKY and SCHAEFFER [1985], GOLUBITSKY, STEWART and SCHAEFFER [1988], HEALEY [1988a,b, 1989], VANDERBAUWHEDE [1982]. A similar search in any numerical analysis journal will typically turn up no such papers at all. It is surprising to see how little systematic use of symmetry groups has been made in the numerical treatment of operator equations exhibiting geometric symmetries.

A start in this direction was made in FÄSSLER [1976] and the book STIEFEL and FÄSSLER [1979]. In fact, there is a very early forerunner by STIEFEL [1952]. Thereafter several papers began to appear, see, e.g., GUY and MANGEOT [1981], BALLISTI, HAFNER and LEUCHTMAN [1982] and BOSSAVIT [1986, 1993]. Recently symmetry reduction methods for linear systems have been investigated, see ALLGOWER, BÖHMER, GEORG and MIRANDA [1992a], ALLGOWER and FÄSSLER [1993], ALLGOWER, GEORG and MIRANDA [1993a], ALLGOWER, GEORG and WALKER [1993b], GEORG and MI-

RANDA [1992, 1993].

The usefulness of exploiting the actions of symmetry groups in continuation methods is illustrated with the example of bifurcation in Section 24.3, see also Section 25. In this section we want to show how the exploitation of symmetry can significantly reduce the computational cost of solving linear systems of equations (which is one of the basic cost factors in continuation methods).

To briefly illustrate the approach, let us consider an operator equation

$$\mathcal{L}f = g \tag{15.1}$$

where f and g are elements of a Banach space $\mathcal{F}$ of functions defined on a domain $\mathcal{D}$ in $\mathbb{R}^3$. For example, if the operator equation describes a partial differential equation, then $\mathcal{D}$ is a three-dimensional body, and if the operator equation describes a boundary integral formulation, then $\mathcal{D}$ is the surface of a three-dimensional body.

Often, $\mathcal{D}$ exhibits geometric symmetries (e.g., $\mathcal{D}$ may be a cube) which may be described by the action of a certain finite group Γ of orthogonal transformations. For many classical equations, these symmetries are respected by the operator equations. This situation is usually described by saying that the operator $\mathcal{L}$ is *equivariant* with respect to the group action (i.e., commutes with the group actions).

More precisely, if $\gamma \in \Gamma$ is an orthogonal transformation which is an isometry of the body $\mathcal{D}$, then γ acts on the Banach space $\mathcal{F}$ by the formula

$$(\gamma f)(x) = f(\gamma^{-1}x)$$

for $f \in \mathcal{F}$ and $x \in \mathcal{D}$. The equivariance of the operator $\mathcal{L}$ is expressed by requiring that

$$\mathcal{L}\gamma = \gamma\mathcal{L}$$

for all isometries $\gamma \in \Gamma$.

We are now in a position to discuss the numerical solution of the operator equation $\mathcal{L}f = g$ where $\mathcal{L}$ is equivariant with respect to some group Γ of isometries, and in particular, to exploit the symmetry in order to reduce the computational effort as much as possible. It should be stressed that no symmetry assumptions are made concerning either g or f in the equation $\mathcal{L}f = g$.

If the linear operator equation $\mathcal{L}f = g$ is suitably discretized (respecting the action of a symmetry group Γ), then a matrix equation $Lu = v$ results, where L is an equivariant (N, N)-matrix with respect to Γ which now acts as a group of permutations on $\{1, \ldots, N\}$, i.e.,

$$L(\gamma i, \gamma j) = L(i, j) \quad \text{for } \gamma \in \Gamma. \tag{15.2}$$

For example, let us consider a collocation method (e.g., with constant elements) for the operator equation $\mathcal{L}f = g$ based on a choice of basis functions $\{\phi_j\}_{j\in J} \subset \mathcal{F}$ and collocation points $\{s_j\}_{j\in J}$ indexed by the same set J, such that interpolation is possible, i.e., the matrix $\phi_j(s_k)$ has nonzero determinant. Higher order collocation or Galerkin methods can be treated in a very similar way. Following standard techniques we can write the collocation equation as

$$Lu = v,$$

where the entries of the square matrix L are given by

$$L(k,j) := (\mathcal{L}\phi_j)(s_k),$$

the column v is given by $v(k) := g(s_k)$, and the unknown column u represents the coefficients of the approximate solution, i.e.,

$$f \approx \sum_{j\in J} u(j)\phi_j.$$

The standard assumption in collocation methods is that L is a nonsingular matrix.

In order to exploit the symmetry, we need to make the assumption that for all $\gamma \in \Gamma$ we have

$$\{\gamma\phi_j\}_{j\in J} \text{ is a permutation of } \{\phi_j\}_{j\in J},$$

and

$$\{\gamma s_j\}_{j\in J} \text{ is a permutation of } \{s_j\}_{j\in J},$$

which signifies that the chosen configuration for the collocation method respects the symmetry structure. This will usually be easy to satisfy in practical cases. For simplicity, let us assume that the induced permutations of the index set J are the same for the two actions. These conditions lead immediately to the equivariance of the system matrix L as described in (15.2).

Let us look at the simplest case, i.e., Γ is a cyclic group of order N and $\gamma : i \mapsto i+1$ is the generating element of Γ. Here the indices i of the matrix entries are viewed modulo N. It can be easily seen that the matrix L has the circulant form (illustrated for $N = 5$)

$$L = \begin{pmatrix} a_1 & a_5 & a_4 & a_3 & a_2 \\ a_2 & a_1 & a_5 & a_4 & a_3 \\ a_3 & a_2 & a_1 & a_5 & a_4 \\ a_4 & a_3 & a_2 & a_1 & a_5 \\ a_5 & a_4 & a_3 & a_2 & a_1 \end{pmatrix}.$$

Circulant matrices have been studied and used extensively, see, e.g., DAVIS [1979] and VAN LOAN [1992]. The symmetry reduction method for solving the equation $Lu = v$

is well known in this case and consists of an application of the finite Fourier transform (illustrated for $N = 5$)

$$F = \frac{1}{\sqrt{N}} \begin{pmatrix} 1 & 1 & 1 & 1 & 1 \\ 1 & \omega & \omega^2 & \omega^3 & \omega^4 \\ 1 & \omega^2 & \omega^4 & \omega^6 & \omega^8 \\ 1 & \omega^3 & \omega^6 & \omega^9 & \omega^{12} \\ 1 & \omega^4 & \omega^8 & \omega^{12} & \omega^{16} \end{pmatrix}, \tag{15.3}$$

where ω is a primitive Nth root of unity, and the exponents could (of course) be modified modulo N. It is seen that F is unitary. Hence we consider the transformed equation

$$\hat{L}\hat{u} = \hat{v} \quad \text{where } \hat{L} := FLF^*, \ \hat{u} := Fu, \ \hat{v} := Fv.$$

If we denote the first column of L by a, then it is seen that $\hat{L} = \sqrt{N}\,\operatorname{diag}(\hat{a})$, i.e., the finite Fourier transform reduces the problem of solving $Lu = v$ to the (trivial) solving of N scalar equations $\sqrt{N}\,\hat{a}(i)\hat{u}(i) = \hat{v}(i)$ and three finite Fourier transforms, namely $a \mapsto \hat{a}$, $v \mapsto \hat{v}$ and $\hat{u} \mapsto u$. Systems involving block circulant matrices are handled analogously, e.g., the column a now becomes a column of block matrices.

Interestingly enough, the above finite Fourier transform method has not been generalized to arbitrary equivariant matrices until very recently, i.e., TAUSCH [1993] has interpreted the symmetry reduction approach in GEORG and MIRANDA [1993] in this way. Let us give a brief sketch of this generalization for the case of block structures.

Hence, we assume that the matrix L splits into (k, k)-blocks $L(i, j)$, $i, j = 1, \ldots, N$. The equivariance (15.2) is now written for these blocks. Let us further assume (for simplicity) that the actions of the permutations $\gamma \in \Gamma$ are fixed point free, i.e.,

$$\gamma j = j \tag{15.4}$$

for some $\gamma \in \Gamma$ and some $j \in \{1, \ldots, N\}$ implies $\gamma = 1$, the identity element of Γ. If we take the above described block structure as large as possible (possibly rearranging the ordering), it is clear that the order of Γ equals N, hence, for the purposes of this exposition, we index L via Γ. The equivariance now leads to

$$L(\alpha, \beta) = L\big(\beta^{-1}\alpha, 1\big) =: a\big(\beta^{-1}\alpha\big) \quad \text{where } \alpha, \beta \in \Gamma.$$

The matrix L is now determined by its first block column a.

To describe the decomposition of L via equivariance, we recall the concept of the irreducible representations of a group. Let $U(i)$ denote the group of unitary (i, i)-matrices. An irreducible representation (of dimension i) $\rho : \Gamma \to U(i)$ is a group homomorphism without a proper invariant subspace, i.e., there is no proper linear subspace $V \subset \mathbb{C}^i$, $\dim V \notin \{0, i\}$, such that $\rho(\Gamma)V = V$ for all $\gamma \in \Gamma$. Two such representations are equivalent if they result from each other via a unitary similarity transformation. It is well known, see, e.g., SERRE [1977], that any finite group admits a complete list $\mathcal{R}(\Gamma)$ of irreducible representations which are mutually nonequivalent. Furthermore,

$$\sum_{\rho\in\mathcal{R}(\Gamma)} \dim\rho^2 = \text{order}(\Gamma) = N. \tag{15.5}$$

Let us now construct the analogue F of the finite Fourier transform. A typical block row of F has the elements

$$\sqrt{\frac{\dim\rho}{N}}\,\big(\rho(\gamma)(i,j)\,I_k\big)_{\gamma\in\Gamma}$$

where i, j are some indices in $\{1,\ldots,\dim\rho\}$ and I_k is the (k,k) identity matrix. By (15.5), F is a (kn, kn)-matrix, and the well-known orthogonality relations in group representation theory, see, e.g., SERRE [1977], imply that F is unitary.

Now we can repeat for equivariant matrices the program described above for circulant matrices:

$$\hat{L}\hat{u} = \hat{v} \quad \text{where } \hat{L} := FLF^*,\ \hat{u} := Fu,\ \hat{v} := Fv.$$

However, $\hat{L}$ is not a simple block diagonal matrix generated by $\hat{a}$. To describe its structure, let us index the block entries of $\hat{L}$ by $\rho(i,j)$ where $\rho\in\mathcal{R}(\Gamma)$ and $i,j\in\{1,\ldots,\dim\rho\}$. This is possible by (15.5). Then

$$\hat{L}\big(\rho(i,j),\rho'(i',j')\big) = \delta_{\rho\rho'}\,\delta_{ii'}\,\hat{a}\big(\rho(j',j)\big).$$

Hence, a typical block diagonal entry of $\hat{L}$ referring to ρ consists of $\dim\rho$ identical blocks each of size $k\dim\rho$.

By applying this "generalized" finite Fourier transform, we thus obtain a subdivision of the linear problem $Lu = v$ into much smaller subproblems.

As an illustration, let us consider the most simple example of an equivariant matrix with respect to a nonabelian group: let

$$D_3 = \{1, R, R^2, F, FR, FR^2\}$$

be the symmetry group of an equilateral triangle. It is generated, e.g., by a $2\pi/3$ rotation R and a reflection F across the y-axis, see Fig. 15.1. It has 2 irreducible representations of dimension one and 1 irreducible representation of dimension two.

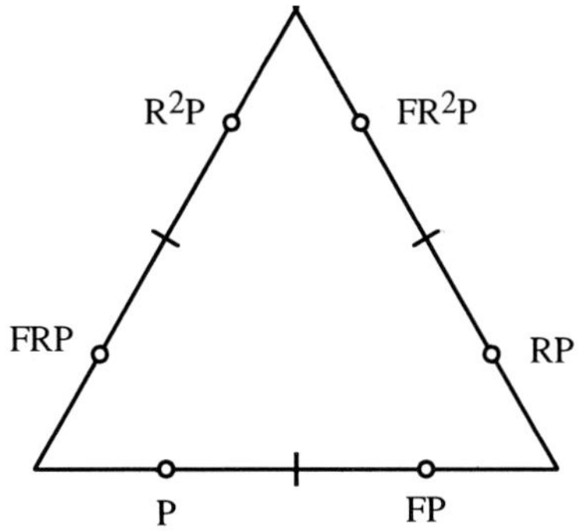

FIG. 15.1. The group D_3.

The multiplication table of the group (the elements ordered as above) has the following form:

	γ_1	γ_2	γ_3	γ_4	γ_5	γ_6
γ_1	γ_1	γ_2	γ_3	γ_4	γ_5	γ_6
γ_2	γ_2	γ_3	γ_1	γ_6	γ_4	γ_5
γ_3	γ_3	γ_1	γ_2	γ_5	γ_6	γ_4
γ_4	γ_4	γ_5	γ_6	γ_1	γ_2	γ_3
γ_5	γ_5	γ_6	γ_4	γ_3	γ_1	γ_2
γ_6	γ_6	γ_4	γ_5	γ_2	γ_3	γ_1

A complete list of irreducible representations is captured in the following generalized Fourier transform:

$$F = \begin{pmatrix} \frac{\sqrt{6}}{6} & \frac{\sqrt{6}}{6} & \frac{\sqrt{6}}{6} & \frac{\sqrt{6}}{6} & \frac{\sqrt{6}}{6} & \frac{\sqrt{6}}{6} \\ \frac{\sqrt{6}}{6} & \frac{\sqrt{6}}{6} & \frac{\sqrt{6}}{6} & -\frac{\sqrt{6}}{6} & -\frac{\sqrt{6}}{6} & -\frac{\sqrt{6}}{6} \\ \frac{\sqrt{3}}{3} & -\frac{\sqrt{3}}{6} & -\frac{\sqrt{3}}{6} & -\frac{\sqrt{3}}{3} & \frac{\sqrt{3}}{6} & \frac{\sqrt{3}}{6} \\ 0 & -1/2 & 1/2 & 0 & 1/2 & -1/2 \\ 0 & 1/2 & -1/2 & 0 & 1/2 & -1/2 \\ \frac{\sqrt{3}}{3} & -\frac{\sqrt{3}}{6} & -\frac{\sqrt{3}}{6} & \frac{\sqrt{3}}{3} & -\frac{\sqrt{3}}{6} & -\frac{\sqrt{3}}{6} \end{pmatrix}.$$

Then the equivariant matrix

$$L = \begin{pmatrix} 1 & 3 & 4 & 6 & 7 & 9 \\ 4 & 1 & 3 & 7 & 9 & 6 \\ 3 & 4 & 1 & 9 & 6 & 7 \\ 6 & 7 & 9 & 1 & 3 & 4 \\ 7 & 9 & 6 & 4 & 1 & 3 \\ 9 & 6 & 7 & 3 & 4 & 1 \end{pmatrix}$$

is transformed into the matrix

$$\hat{L} = \begin{pmatrix} 30 & 0 & 0 & 0 & 0 & 0 \\ 0 & -14 & 0 & 0 & 0 & 0 \\ 0 & 0 & -1/2 & -\frac{\sqrt{3}}{2} & 0 & 0 \\ 0 & 0 & -\frac{3\sqrt{3}}{2} & -9/2 & 0 & 0 \\ 0 & 0 & 0 & 0 & -1/2 & -\frac{\sqrt{3}}{2} \\ 0 & 0 & 0 & 0 & -\frac{3\sqrt{3}}{2} & -9/2 \end{pmatrix}.$$

Let us give an idea of the computational savings the method provides for the simple case that all linear problems are solved via a direct solver. Not exploiting the symmetry results in CN^3k^3 flops. On the other hand, using the symmetry reduction method leads to an overhead of 3 applications of the generalized finite Fourier transform, i.e., $3N^2k^2$ flops, and additional flops for solving the various subproblems, i.e.,

$$Ck^3 \sum_{\rho \in \mathcal{R}(\Gamma)} (\dim \rho)^3.$$

Neglecting the overhead gives a reduction factor of

$$\frac{1}{N^3} \sum_{\rho \in \mathcal{R}(\Gamma)} (\dim \rho)^3.$$

For example, if the full symmetry group of the three-dimensional cube is considered, then the reduction factor equals approximately 0.00116.

Further advantages are that the reduced problems remain sparse if the original problem is sparse. Furthermore, the diagonal blocks have at most the condition number of the large problem, and often the condition numbers are significantly reduced.

Another advantage of the symmetry reduction method is that it greatly facilitates the study of bifurcation phenomena, since the bifurcation structures of the reduced problems are much simpler, owing again to the fact that the multiple eigenvalues of the large problem have become distributed over the various subproblems. See Section 24.3 for more details.

16. Derivatives

An integral part of continuation methods involves the calculation of the Jacobian of a vector valued function, and occasionally also higher derivatives. Let us sketch in this section some ideas for implementing such derivatives.

Of course, the update methods described in Section 14 can be viewed as a method to numerically approximate the Jacobian H'. Since this is explicitly incorporated into a path following algorithm, we have treated these ideas separately in a special section.

A very special case of numerical differentiation (of the implicitly defined solution curve) is discussed in Section 11. Let us just note in passing that automatic differentiation (which will be discussed next) does not handle implicit function definitions.

An obvious way to obtain derivatives is to resort to a symbolic software such as MAPLE or MATHEMATICA. However, especially for higher dimensions and more complicated equations $H(u) = 0$, these methods become cumbersome, and the exact analytic expressions for the derivatives may become unmanagable.

16.1. Automatic differentiation

Recently, a new topic in computer science has evolved dealing with automatic differentiation, see, e.g., the surveys of GRIEWANK [1989] and IRI and KUBOTA [1987] or the proceedings edited by GRIEWANK and CORLISS [1991]. This is a technique which is reasonable to consider integrating into a continuation code. In fact, this has been done for HOMPACK in connection with systems of equations arising in the analysis of integrated circuits, see MELVILLE, TRAJKOVIC, FANG and WATSON [1993b] and MELVILLE, MOINIAN, FELDMANN and WATSON [1993a]. The authors report that simple timing studies show that for each Jacobian matrix evaluated at a point of the solution curve, considerably more time is spent in HOMPACK performing linear algebra computations on the matrix than is spent in constructing the derivatives.

Let us sketch the basic idea of automatic differentiation. Assume that a function $f(x_1, \dots, x_n)$ can be expressed by means of a code of the following form:

ALGORITHM 16.1 (*Function*).

input $x_1, \dots, x_n$
output $y = f(x_1, \dots, x_n)$
for $i = n+1, \dots, m$
  $x_i \leftarrow f_i(\{x_j\}_{j \in I_i})$
end for
$y \leftarrow x_m$

The variables $x_{n+1}, \dots, x_{m-1}$ store intermediate results and can be discarded when they are no longer needed. The functions f_i depend explicitly on the quantities $\{x_j\}_{j \in I_i}$, which have already been computed, hence $j < i$ if $j \in I_i$. Moreover, the f_i are assumed to represent basic computational steps in the sense that f_i is one of the basic operations $\{+, -, *, /\}$ possibly with one constant operand as, e.g., $x_1 + 3$, or else an elementary function such as $\{\sin, \cos, \exp, \log, \dots\}$. Then the partial derivatives $\partial_j f_i = (\partial / \partial x_j) f_i$ are available.

For example, consider $y = f(x_1, x_2) = (x_1 + 2x_2)\sin(x_1x_2)$ which gives

$$x_3 \leftarrow 2x_2$$

$$x_4 \leftarrow x_1 + x_3$$

$$x_5 \leftarrow x_1x_2$$

$$x_6 \leftarrow \sin x_5$$

$$x_7 \leftarrow x_4x_6$$

A straightforward way of obtaining the gradient of f would be to differentiate all the ingredients of Algorithm 16.1.

ALGORITHM 16.2 (*Function and gradient, forward mode*).

input $x_1, \ldots, x_n$
output $y = f(x_1, \ldots, x_n)$, $g = \nabla f(x_1, \ldots, x_n)$
for $i = 1, \ldots, n$
  $\nabla x_i \leftarrow e_i$ % *initialization with unit base vectors*
end for
for $i = n+1, \ldots, m$
  $x_i \leftarrow f_i(\{x_j\}_{j\in I_i})$
  $\nabla x_i \leftarrow \sum_{j\in I_i} \partial_j f_i(\{x_j\}_{j\in I_i})\nabla x_j$
end for
$y \leftarrow x_m$, $g \leftarrow \nabla x_m$

This is the so-called forward mode of automatic differentiation. It has the advantage of being easy to implement, but it increases the computational effort of evaluating the gradient compared to evaluating the original function (Algorithm 16.1) by a factor $O(n)$.

For our example we would obtain the following sequence:

$$\nabla x_3 \leftarrow 2\nabla x_2 = (0, 2)^*$$

$$\nabla x_4 \leftarrow \nabla x_1 + \nabla x_3 = (1, 0)^* + (0, 2)^* = (1, 2)^*$$

$$\nabla x_5 \leftarrow x_2\nabla x_1 + x_1\nabla x_2 = (x_2, x_2)^*$$

$$\nabla x_6 \leftarrow \cos x_5 \nabla x_5 = \cos x_5 (x_2, x_2)^*$$

$$\nabla x_7 \leftarrow x_6\nabla x_4 + x_4\nabla x_6 = (x_6 + x_2x_4\cos x_5, 2x_6 + x_1x_4\cos x_5)^*$$

Let us now outline the reverse mode of automatic differentiation. For this purpose, we view the output y with varying dependencies, namely

$$y = h_i(x_1, \ldots, x_i) \quad \text{for } i = m, m-1, \ldots, n,$$

where h_i denotes the result of eliminating the variables in Algorithm 16.1 in reverse order. In our example:

$$h_7 = x_7.$$

$$h_6 = x_6 x_4.$$

$$h_5 = x_4 \sin x_5.$$

$$h_4 = x_4 \sin(x_1 x_2).$$

$$h_3 = (x_1 + x_3) \sin(x_1 x_2).$$

$$h_2 = (x_1 + 2x_2) \sin(x_1 x_2).$$

Note that

$$h_{i-1}(x_1, \ldots, x_{i-1}) = h_i\big(x_1, \ldots, x_{i-1}, f_i(\{x_j\}_{j \in I_i})\big),$$

and therefore the chain rule gives

$$\partial_j h_{i-1} = \partial_j h_i + (\partial_i h_i)(\partial_j f_i)$$

for $j = 1, \ldots, i-1$. Of course, the second term vanishes for $j \notin I_i$. This leads to the following algorithm:

ALGORITHM 16.3 (*Function and gradient, reverse mode*).

input $x_1, \ldots, x_n$
output $y = f(x_1, \ldots, x_n)$, $g = \nabla f(x_1, \ldots, x_n)$
for $i = n+1, \ldots, m$
    $x_i \leftarrow f_i(\{x_j\}_{j \in I_i})$
end for
$y \leftarrow x_m$ % *first output*
for $j = 1, \ldots, m$
    $\partial_j h_m = \delta_{mj}$ % *initialization*
end for
for $i = m, m-1, \ldots, n+1$
    for $j = 1, \ldots, i-1$

$$\partial_j h_{i-1} \leftarrow \begin{cases} \partial_j h_i + \partial_j f_i\left(\{x_j\}_{j \in I_i}\right) \partial_i h_i, & j \in I_i \\ \partial_j h_i & \text{otherwise} \end{cases}$$

    end for
end for
$g \leftarrow (\partial_1 h_n, \ldots, \partial_n h_n)^*$ % *second output*

The starting quantities $\{\partial_j h_m\}$ can be overwritten by the $\{\partial_j h_{i-1}\}$ such that only the elements with $j \in I_i$ have to be updated in the i-loop. So the computational cost of the reverse mode is reduced by the factor n compared to the forward mode. Indeed, it can be shown that the computational effort of evaluating the gradient compared to evaluating the original function is multiplied only by a number $C \leqslant 5$ (which is independent of n). On the other hand, the values $\{x_i\}$ from the forward sweep and the information about the arguments of f_i have to be saved for the reverse sweep which, in general, causes high additional bookkeeping costs. In fact, the decision of what mode should be used depends on the class of functions, especially on the relation between n and m, and on the specific implementation.

In our example, the calculation of

$$(\partial_1 h_i, \dots, \partial_i h_i)$$

for $i = 7, 6, \dots, 2$ in reverse mode takes the form:

i	$\partial_1 h_i$	$\partial_2 h_i$	$\partial_3 h_i$	$\partial_4 h_i$	$\partial_5 h_i$	$\partial_6 h_i$	$\partial_7 h_i$
7	0	0	0	0	0	0	1
6	0	0	0	x_6	0	x_4	
5	0	0	0	x_6	$x_4 \cos x_5$		
4	$x_2 x_4 \cos x_5$	$x_1 x_4 \cos x_5$	0	x_6			
3	$x_6 + x_2 x_4 \cos x_5$	$x_1 x_4 \cos x_5$	x_6				
2	$x_6 + x_2 x_4 \cos x_5$	$2x_6 + x_1 x_4 \cos x_5$					

Of course, the end result is the same as for the forward mode.

Forward evaluations of higher order derivatives can be performed similarly to Algorithm 16.1. The reverse mode is not so simple. However, often in applications only an action of a higher derivative on vectors $v_1, v_2, \dots, v_p$ is needed, and these can be obtained via successive applications of the gradient, i.e.,

$$\begin{aligned} \nabla^{1+p} f(x) v_1 v_2 \cdots v_p &= \nabla\big(\nabla^p f(x) v_1 v_2 \cdots v_p\big) \\ &= v_p^*\big(\nabla\big(\nabla^p f(x) v_1 v_2 \cdots v_{p-1}\big)\big). \end{aligned}$$

The above ideas could be implemented by a user whenever implementing a function subroutine. However, though the implementation is straightforward, typical users are likely to find this onerous. Therefore, in recent years, extensive efforts have been made to let a precompiler handle the extensions, i.e., a user only needs to furnish a function in some standard way (say, in a C^{++} or FORTRAN function subroutine), and a precompiler automatically extends the subroutine to include either a forward or a reverse mode calculation of the derivatives. For details and an extensive bibliography on the subject including available software such as, e.g., ADOL-C or ADIFOR, we refer to GRIEWANK and CORLISS [1991].

16.2. Finite difference approximations

A straightforward finite difference approximation of the Jacobian H' would be quite expensive ($N + 1$ evaluations of H even for the coarse forward difference formula).

However, in some cases, it is only necessary to calculate an action $H'(u)p$ of H' on a vector p. To evaluate this action, we may use a central difference approach

$$H'(u)p = \frac{H(u+\varepsilon p) - H(u-\varepsilon p)}{2\varepsilon} + O(\varepsilon^2). \tag{16.1}$$

This formula involves only two evaluations of H. Decreasing ε reduces the truncation error, but unfortunately increases the cancellation error.

For $\|H(u)\| = O(1)$ a balance is reached at $\varepsilon \approx \sqrt[3]{\delta}$ where δ indicates the unit round-off of the computer. This is seen from the following argument. The cancellation error can be estimated by $O(\delta/\varepsilon)$. Setting the cancellation error equal to the truncation error leads to $O(\delta/\varepsilon) = O(\varepsilon^2)$ from which $\varepsilon = O(\sqrt[3]{\delta})$ is readily obtained.

The resulting precision is approximately $O(\delta^{2/3})$ which means that about $\frac{2}{3}$ of the significant digits are valid in the approximation of the directional derivative (16.1). If this precision is still unsatisfactory, a higher order formula can be employed which is easily obtained from (16.1) via Richardson extrapolation, since the truncation error in (16.1) has an expansion in powers of ε^2. If, for example, a formula of order $2k$ is employed, then $2k$ evaluations of H are necessary, a balanced stepsize for the finite difference formula would lead to $\varepsilon \approx \delta^{1/(2k+1)}$, and the resulting precision would mean that about $2k/(2k+1)$ of the significant digits are valid.

A similar analysis can be done for the second derivative using the central difference approximations:

$$H''(u)[p,p] = \frac{H(u+\varepsilon p) + H(u-\varepsilon p) - 2H(u)}{\varepsilon^2} + O(\varepsilon^2), \tag{16.2}$$

$$\begin{aligned} H''(u)[p,q] &= \frac{1}{4}\big(H''(u)[p+q,p+q] - H''[p-q,p-q]\big) \\ &= \frac{1}{4}\frac{H(u+\varepsilon p+\varepsilon q) + H(u-\varepsilon p-\varepsilon q)}{\varepsilon^2} \\ &\quad - \frac{1}{4}\frac{H(u+\varepsilon p-\varepsilon q) + H(u-\varepsilon p+\varepsilon q)}{\varepsilon^2} + O(\varepsilon^2). \end{aligned}$$

The truncation errors are again expandable in terms of powers of ε^2, and the corresponding statements about choosing the adequate stepsize ε and about Richardson's extrapolation are the same as those above for the first derivative. There is only one change: the cancellation error can now be estimated by $O(\delta/\varepsilon^2)$, from which $\varepsilon = O(\sqrt[4]{\delta})$ is obtained, see, e.g., MACKENS ([1989], p. 248).

17. Large scale problems

One of the primary applications of continuation methods involves the numerical solution of nonlinear eigenvalue problems. Such problems are likely to have arisen from a discretization of an operator equation in a Banach space context, which also involves an additional bifurcation parameter. For some specific examples see Section 25. As a result of the discretization and the wish to maintain a reasonably low truncation error, the corresponding finite dimensional problem $H(u) = 0$ where $H: \mathbb{R}^{N+1} \to \mathbb{R}^N$, may require that N be quite large. This then leads to the task of solving large scale continuation problems.

An area in which a considerable amount of experience concerning large scale continuation methods exists is structural mechanics, see, e.g., RHEINBOLDT [1981, 1986] and the further references cited therein. There has also been work done on combining continuation methods with multigrid methods for solving large scale continuation problems arising from discretization of elliptic problems via finite differences, see, e.g., CHAN and KELLER [1982], BANK and CHAN [1986], MITTELMANN and ROOSE [1990], and further literature cited therein. Another area where large scale continuation problems have been treated concerns finite element discretizations of elliptic problems, which are then combined with a conjugate gradient solver in the continuation algorithm, see GLOWINSKI, KELLER and REINHART [1985].

It is clear that an endless variety of combinations can be made of continuation algorithms and sparse solvers. In view of this, following Eq. (12.7) in Section 12 we have discussed how any sparse solver process can be incorporated into the general scheme of continuation methods which we have been describing. Here we will indicate more specifically how to incorporate a conjugate gradient method or a Krylov subspace method. The following discussion is still at an early stage of development.

Let us recall the nonlinear conjugate gradient method of POLAK and RIBIÈRE [1969]. This choice of nonlinear minimizer is based upon reports that in numerical practice it has generally yielded the best results, see, e.g., POWELL [1977] or BERTSEKAS [1984]. To outline the method, let us assume that the problem to be solved is

$$\min_u \{\varphi(u):\ u \in \mathbb{R}^{N+1}\} \tag{17.1}$$

where $\varphi : \mathbb{R}^{N+1} \to \mathbb{R}$ is a smooth nonlinear functional, usually having an isolated local minimal point $\bar{u}$ which we desire to approximate.

The following is an outline of the conjugate gradient method due to POLAK and RIBIÈRE [1969].

ALGORITHM 17.1 (*Nonlinear CG*).

input: $u_0 \in \mathbb{R}^{N+1}$ % *initial point*
output: $u_0, u_1, \ldots$ % *converging to* $\bar{u}$ *(hopefully)*

$g_0 \leftarrow \nabla\varphi(u_0)$; $d_0 \leftarrow g_0$; % *initial gradients*
for $n = 0, 1, \ldots$
  $\rho_n \leftarrow \arg\min_{\rho>0} \varphi(u_n - \rho d_n)$; % *line search*
  $u_{n+1} \leftarrow u_n - \rho_n d_n$
  $g_{n+1} \leftarrow \nabla\varphi(u_{n+1})$
  $\gamma_n \leftarrow \dfrac{[(g_{n+1} - g_n)^* g_{n+1}]}{\|g_n\|^2}$
  $d_{n+1} \leftarrow g_n + \gamma_n d_n$ % *new conjugate gradient*
end for

Since our aim here is merely to make an application of a conjugate gradient method, we will not give a detailed account concerning conjugate gradient methods for nonlinear problems. However, we shall recall some of their properties. For more details we suggest, e.g., the book of FLETCHER [1987].

The main theoretical justification of the (nonlinear) conjugate gradient algorithm lies in its properties when $\varphi(u)$ is a uniformly convex quadratic functional. In this special case the algorithm becomes theoretically equivalent to the familiar (linear) conjugate gradient method due to HESTENES and STIEFEL [1952]. Furthermore, in this case the steplength ρ_n in Algorithm 17.1 is readily calculated to be

$$\rho_n = \frac{\varphi'(u_n)d_n}{d_n^* \nabla\varphi'(u_n)d_n} \tag{17.2}$$

where $\nabla\varphi'$ denotes the Hessian of φ.

The following two statements summarize the main convergence results on conjugate gradient methods, see GOLUB and VAN LOAN ([1989], Section 10.2.8) or STOER ([1983], Section 1).

Let φ be a uniformly convex quadratic form. If the Hessian $\nabla\varphi'$ has exactly k distinct eigenvalues, then the conjugate gradient method stops at the solution $\bar{u}$ after k steps.

Let φ be a uniformly convex quadratic form. If κ denotes the condition of the Hessian $\nabla\varphi'$, then

$$\|u_n - \bar{u}\| \leqslant \frac{\|u_0 - \bar{u}\|}{T_n\left(\frac{\kappa+1}{\kappa-1}\right)} \leqslant 2\left(\frac{\sqrt{\kappa}-1}{\sqrt{\kappa}+1}\right)^n \|u_0 - \bar{u}\|, \tag{17.3}$$

where $T_n(x) = \cos(n \arccos(x))$ is the nth Chebyshev polynomial.

The conclusion to be drawn from the above results is that initially, the convergence of the conjugate gradient method may be slow because of (17.3), but by the kth step a very substantial improvement in the approximation of the solution has been obtained. This appears to hold even in the general case where the functional φ is no longer quadratic. To be more specific, let us assume that $\bar{u}$ is a local minimal solution point of the problem (17.1) at which the Hessian $\nabla\varphi'(\bar{u})$ is positive definite. There are several results concerning the convergence of the conjugate gradient method which essentially state that local superlinear convergence towards $\bar{u}$ holds, see, e.g., COHEN [1972] or MCCORMICK and RITTER [1974]. However it appears that as of this date, the convergence results are somewhat unsatisfactory. One of the difficulties is that there are various possibilities for obtaining the factors γ_n in Algorithm 17.1. Another difficulty is that in practice, we do not want to perform a very precise one-dimensional minimization in the line search of Algorithm 17.1 in order to obtain an acceptable ρ_n since this is costly. Most of the convergence rate proofs require cyclic reloading, i.e., setting $\gamma_n = 0$ after every $N + 1$ steps. The general idea of such proofs involves the approximation of $\varphi(u)$ via Taylor's formula by

$$\varphi(u) \approx \varphi(\bar{u}) + \varphi'(\bar{u})(u - \bar{u}) + (u - \bar{u})^* \nabla\varphi'(\bar{u})(u - \bar{u}),$$

and then to use the convergence results for the quadratic case. Actually, even in the quadratic case, because of the presence of rounding errors, we cannot expect that stopping will occur after k steps. Instead, we should regard the conjugate gradient method even in this case as an iterative method which makes a substantial improvement after k steps.

The ideal convergence (one step!) would occur when the condition number $\kappa = 1$, i.e., when all the eigenvalues of $\nabla\varphi'$ are equal. Intuitively, the next best situation would occur when the eigenvalues have as few cluster points as possible. This observation may be used to motivate the idea of preconditioning for the conjugate gradient method. For more details (in case of a quadratic functional) the reader may refer to GOLUB and VAN LOAN ([1989], Section 10.3).

Let us now turn to the case which actually concerns us, namely minimizing the functional

$$\varphi(u) := \tfrac{1}{2}\|H(u)\|^2. \tag{17.4}$$

The minimal points of (17.4) obviously form the solution curve $c(s)$ which we are considering throughout the paper. We have

$$\begin{aligned}\nabla\varphi(u) &= H'(u)^* H(u);\\ \nabla\varphi'(u) &= H'(u)^* H'(u) + \mathrm{O}(\|H(u)\|).\end{aligned}$$

The gradient $\nabla\varphi(u) = H'(u)^* H(u)$ is orthogonal to the tangent vector $t(H'(u))$. This motivates the idea for implementing the conjugate gradient method (Algorithm 17.1) as a corrector into a continuation method. Analogously to the case when the minimization problem has isolated solutions at which the Hessian is positive definite, we may also expect local superlinear convergence of the conjugate gradient method for the functional (17.4).

The solution will be a point $\bar{u} \in H^{-1}(0)$ which is essentially nearest to a predictor point v which is taken as the starting point for the (nonlinear) conjugate gradient method. Numerical experience suggests that local superlinear convergence holds, as in the case of functionals φ with an isolated minimal point. However, to our knowledge even in the latter case a general proof of this has not been given. We propose the above conjugate gradient method as a reasonable corrector procedure nevertheless, provided once again, that an effective preconditioning is incorporated. In the present context, we propose a preconditioning of the dependent variables, which we now describe in more detail.

Let us consider the following transformation of φ:

$$\tilde{\varphi}(u) = \tfrac{1}{2}\|L^{-1}H(u)\|^2$$

where L is an as yet unspecified nonsingular (N, N)-matrix. Then we have

$$\begin{aligned}\nabla\tilde{\varphi}(u) &= H'(u)^*(LL^*)^{-1}H(u);\\ \nabla\tilde{\varphi}'(u) &= H'(u)^*(LL^*)^{-1}H'(u) + \mathrm{O}(\|H(u)\|).\end{aligned} \tag{17.5}$$

If we assume that our continuation method furnishes predictor points which are already near $H^{-1}(0)$, we may neglect the $O(\|H(u)\|)$ term in (17.5).

Furthermore, if $H(u) \approx 0$, then an ideal choice would be to take L such that

$$LL^* = H'(u)H'(u)^*$$

is the Cholesky decomposition. We then have

$$\begin{aligned}\nabla\tilde{\varphi}(u) &= H'(u)^*(LL^*)^{-1}H(u)\\ &= H'(u)^*\left(H'(u)H'(u)^*\right)^{-1}H(u)\\ &= H'(u)^+H(u).\end{aligned}$$

Hence in this case, the gradient $\nabla\tilde{\varphi}(u) = H'(u)^+H(u)$ coincides with the usual Newton direction which we have discussed as the standard corrector.

Of course, if we really want to use the Cholesky decomposition, then we would relinquish whatever advantage sparseness may have offered. Thus, we want to determine L also with a small computational expense and in such a way that linear equations $LL^*x = y$ are cheaply solved for x. This leads naturally to a recommendation of using an incomplete Cholesky factorization, see, e.g., GOLUB and VAN LOAN ([1989], Section 10.3.2).

The paper of GLOWINSKI, KELLER and REINHART [1985] treats nonlinear elliptic eigenvalue problems via a combination of finite element discretization, formulating a pseudo-arclength continuation method via an augmented system of equations and applying the Polak–Ribière nonlinear conjugate gradient method to a least squares formulation of the augmented systems of equations. In this situation the minimization problem generally has isolated solutions.

In the paper by ALLGOWER, CHIEN and GEORG [1989] a version of a continuation algorithm incorporating a nonlinear conjugate gradient method as a corrector is given. Applications are made to a nonlinear elliptic boundary value problem on a variety of domains having symmetries.

Several of the recently developed large scale continuation methods deal with discretizations of nonlinear elliptic eigenvalue problems. In these cases, usually the $N \times N$-submatrix of $H'(u)$ obtained by deleting the column corresponding to the eigenvalue parameter is positive definite or even diagonally dominant. Let us stress here that Algorithm 17.2 does not make use of any such property and is meant to apply to more general situations. This greater versatility may be obtained at the cost of a greater computational effort.

We now outline an algorithm which incorporates a nonlinear conjugate gradient method as corrector.

ALGORITHM 17.2 (*CG predictor corrector method*).

input $u_0 \in \mathbb{R}^{N+1}$ % *such that* $H(u_0) \approx 0$ (*initial point*)
    $t \in \mathbb{R}^{N+1}$ % *initial approximation to* $t(H'(u_0))$
    $h > 0$ % *initial steplength*

output u_i, $i = 0, 1, 2, \ldots$ % *approximating the solution curve*

for $i = 1, 2, \ldots$
  $v \leftarrow u_{i-1} + ht$ % *predictor step*
  calculate $LL^* \approx H'(v)H'(v)^*$ % *preconditioner*
  $g_v \leftarrow H'(v)^*(LL^*)^{-1}H(v), \quad d \leftarrow g_v$ % *gradients*
  repeat % *corrector loop*
    $\bar{\rho} \leftarrow \arg\min_{\rho \geqslant 0} \|L^{-1}H(v - \rho d)\|^2$
    $w \leftarrow v - \bar{\rho} d$ % *corrector step, nonlinear CG*
    $g_w \leftarrow H'(w)^*(LL^*)^{-1}H(w)$ % *new gradient*
    $\gamma \leftarrow \dfrac{(g_w - g_v)^* g_w}{\|g_v\|^2}$ % *Polak–Ribière*
    $d \leftarrow g_w + \gamma d$ % *new conjugate gradient*
    $v \leftarrow w, \quad g_v \leftarrow g_w$
  until convergence
  adapt steplength $h > 0$
  $t \leftarrow \dfrac{u_{i-1} - w}{\|u_{i-1} - w\|}$ % *approximation to* $t(H'(w))$
  $u_i \leftarrow w$ % *new point approximately on* $H^{-1}(0)$
end for

Let us make some remarks concerning details and modifications of the above algorithm. First of all, if the evaluation of $H'(w)$ is very costly, one may prefer to hold it fixed in the corrector loop. Furthermore, let us mention several possibilities for solving the line search problem

$$\min_{\rho \geqslant 0} \left\|L^{-1}H(v - \rho d)\right\|^2. \tag{17.6}$$

Recalling (17.2), let us approximate the functional $\tilde{\varphi}(v - \rho d)$, which is to be minimized, by its truncated Taylor expansion:

$$\tilde{\varphi}(v) - \rho \tilde{\varphi}'(v) d + \tfrac{1}{2}\rho^2 d^* \nabla\tilde{\varphi}(v)' d.$$

This is minimized exactly when

$$\bar{\rho} = \frac{\tilde{\varphi}'(v) d}{d^* \nabla\tilde{\varphi}(v)' d} \tag{17.7}$$

provided $d^* \nabla\tilde{\varphi}(v)' d > 0$. For the case at hand, this is a reasonable assumption, because of (17.9) and the fact that the conjugate gradient d is chosen essentially orthogonal to $\ker H'(v)$. Furthermore, for

$$\tilde{\varphi}(v - \rho d) = \tfrac{1}{2}\left\|L^{-1}H(v - \rho d)\right\|^2$$

we have

$$\tilde{\varphi}'(v)d = H(v)^*(LL^*)^{-1}H'(v)d = g_v^* d \tag{17.8}$$

and

$$\begin{aligned} d^*\nabla\tilde{\varphi}(v)'d &= d^*H'(v)^*(LL^*)^{-1}H'(v)d + \mathrm{O}(\|H(v)\|\ \|d\|) \\ &\approx \|L^{-1}H'(v)d\|^2. \end{aligned} \tag{17.9}$$

Since the evaluation of $H'(v)d$ may be costly for large scale problems, an inexpensive approximation of $H'(v)d$ may be made by using the central difference formula

$$H'(v)d = \frac{H\left(v+\varepsilon\frac{d}{\|d\|}\right) - H\left(v-\varepsilon\frac{d}{\|d\|}\right)}{2\varepsilon}\|d\| + \mathrm{O}(\varepsilon^2) \tag{17.10}$$

for an appropriate discretization step $\varepsilon > 0$, see Section 16.2. Usually, the predictor corrector steps of a continuation method are performed in such a way that all generated points are close to the solution curve in $H^{-1}(0)$. Hence, the quadratic approximation considered in (17.7) will give good results in the situation which we are presently considering. Thus we recommend substituting (17.8)–(17.10) into (17.7).

A second possibility for solving (17.6) is to merely use a standard line search algorithm which does not require the evaluation of $\nabla\tilde{\varphi}$ such as a quadratic fit or golden section algorithm, see, e.g., FLETCHER [1987]. The disadvantage of this approach is that it may require many evaluations of $L^{-1}H$.

Recently, some classes of generalized conjugate direction methods or Krylov subspace methods have been developed to solve systems of linear equations $Mx = b$ where the matrix M is not necessarily assumed to be positive definite or even symmetric, see, e.g., the survey FREUND, GOLUB and NACHTIGAL [1992]. The generalized minimal residual algorithm of SAAD and SCHULTZ [1986] seems to be of particular interest in our context, since it only uses multiplications by M. If we take

$$M = \begin{pmatrix} H'(u) \\ t^* \end{pmatrix},$$

where t is some suitable approximation of $t(H'(u))$, e.g., given by a secant, then it is easy to program a multiplication Mx. In fact, the multiplication $H'(u)x$ may be approximated by a forward or central difference formula for the directional derivative as in (17.10), see also Section 16.2, so that one multiplication by M essentially involves one scalar product and one or two evaluations of the map H. Note that for this choice of M the solution s of the equation

$$Ms = \begin{pmatrix} H(u) \\ 0 \end{pmatrix}$$

is the Gauss–Newton increment $s = H'(u)^{+}H(u)$ which is used in our path following algorithms.

An early survey of the incorporation of fast solvers into continuation codes has been given by CHAN [1984c]. Some implementations and applications are discussed in ALLGOWER, CHIEN, GEORG and WANG [1991d]. DESA, IRANI, RIBBENS, WATSON and WALKER [1992] have incorporated Craig's variant of the conjugate gradient method and the SYMMLQ algorithm for symmetric indefinite problems into a homotopy program. However, one may prefer an incorporation of a transpose-free method, see FREUND, GOLUB and NACHTIGAL ([1992], Section 3.4), since we are not aware of any cheap finite difference approximation of the transpose action $H'(u)^{*}x$.

CHAPTER V

Applications

In this chapter we present a selection of examples where path following methods have been used. Many more specific examples exist in the literature. Our discussion of applications concentrates to a large extent on cases in which the predictor corrector methods apply. Applications in which the dimension is relatively low but smoothness does not hold can be handled by the piecewise linear methods discussed in Chapter VI.

18. Sard's theorem

One frequent application of path following methods involves homotopy methods. These methods are often resorted to in zero point or fixed point problems when no suitable starting point is available for an iterative method such as Newton's method, or when contraction methods do not apply. In these situations, one constructs a homotopy map which deforms from a map which is trivial (or at least its solution points are known) to the map of interest. By the implicit function theorem, a curve emanates from each of the trivial starting points. These curves are then numerically traced in the expectation that a solution point of the map of interest will be reached. In many applications of the numerical homotopy methods, it is possible to avoid degeneracies in the solution curve by introducing suitable parameters (perturbations). The theoretical basis of this approach lies in Sard's theorem for maps with additional (perturbation) parameters, see, e.g., ABRAHAM and ROBBIN [1967] or HIRSCH [1976]. We consider the following general form:

THEOREM 18.1 (Sard). *Let A, B, C be smooth manifolds of finite dimensions with* $\dim A \geqslant \dim C$, *and let $F: A \times B \to C$ be a smooth map. Assume that $c \in C$ is a regular value of F, i.e., for $F(a,b) = c$ we have that the total derivative $F'(a,b): T_a A \times T_b B \to T_c C$ has maximal rank. Here $T_a A$ denotes the tangent space of A at a, etc. Then for almost all $b \in B$ (in the sense of some Lebesgue measure on B) the restricted map $F(\bullet, b): A \to C$ has c as a regular*

The global version of the implicit function theorem, see Theorem 1.1, now implies that for such b and c the set $\{a \in A\colon\ F(a,b) = c\}$ is a smooth manifold of dimension $\dim A - \dim C$. In particular, for the case $\dim A = \dim C + 1$, this perturbation technique is a basic tool in homotopy continuation methods to guarantee the existence of a smooth solution curve (without singularities) which is then traced numerically.

The parameter set B is typically considered to consist of perturbation parameters in this case, and a choice of b as in the above theorem is considered to be a *generic* choice.

19. Fixed point problems

To illustrate the use of Sard's theorem, let us consider a homotopy arising from a fixed point problem. Let $f:\mathbb{R}^N \to \mathbb{R}^N$ be a smooth map which is bounded. According to the theorem of BROUWER [1912], the map f has at least one fixed point. To simplify the discussion, let us make the assumption that the map $x \mapsto x - f(x)$ has zero as a regular value. This implies that the fixed points of f are isolated, and that Newton's method converges locally. However, the global convergence of Newton's method is by no means guaranteed.

We therefore consider the homotopy

$$H(x,\lambda,p) = x - p - \lambda(f(x) - p). \tag{19.1}$$

For the *trivial level* $\lambda = 0$, we obtain the *trivial map* $H(x,0,p) = x - p$ which has the unique zero point p, our *starting point*. On the *target level* $\lambda = 1$, we obtain the *target map* $H(x,1,p) = x - f(x)$ whose zero points are our points of interest, i.e., the fixed points of f.

Let us illustrate via this example how Sard's theorem is typically employed: The Jacobian of H is given by

$$H'(x,\lambda,p) = \big(\operatorname{Id} - \lambda f'(x),\ p - f(x),\ (\lambda - 1)\operatorname{Id}\big).$$

Due to our assumptions, the first N columns of the Jacobian are linearly independent for $H(x,\lambda,p) = 0$ and $\lambda = 1$, and clearly the last N columns are linearly independent for $\lambda \neq 1$. Consequently, by Sard's theorem we can conclude that for almost all $p \in \mathbb{R}^N$ zero is a regular value of the restricted map $H(\bullet,\bullet,p)$.

For such a generic choice of p, the solution manifold $H(\bullet,\bullet,p)^{-1}(0)$ consists of smooth curves which are either diffeomorphic to the circle or to the real line, see Theorem 1.3. Consider the solution curve $c(s) = (x(s),\lambda(s))$ (parametrized for convenience with respect to arclength) such that $c(0) = (p,0)$. It is easy to see that the initial tangent vector in the direction of increasing λ has the form

$$\dot c(0) = \big(1 + \|f(p) - p\|^2\big)^{-1/2} \begin{pmatrix} f(p) - p \\ 1 \end{pmatrix},$$

and hence the curve is not tangent to the plane $\lambda = 0$.

Since the solution point $(p,0)$ is unique for $\lambda = 0$, it follows that c cannot be closed and hence is diffeomorphic to the real line. Furthermore, the boundedness of f implies that $x(s)$ is bounded for $0 \leqslant \lambda(s) \leqslant 1$.

Let us assume (for contradiction) that $\lambda(s) < 1$ for all $s > 0$. By the above boundedness, the theorem of Bolzano–Weierstrass implies that there is a sequence

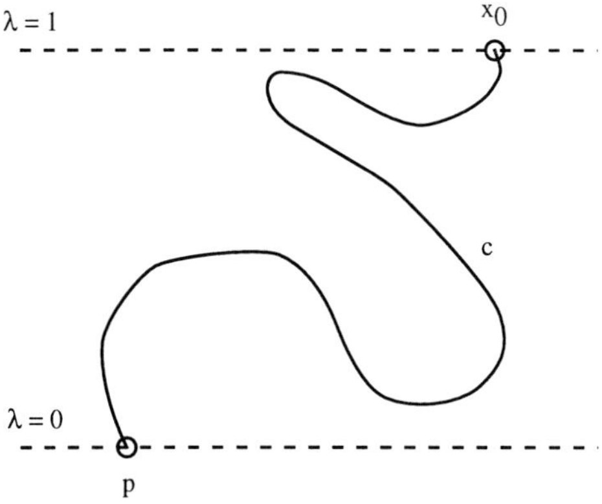

FIG. 19.1. Homotopy fixed point method.

$s_i \to \infty$ and a $\bar{u} \in \mathbb{R}^{N+1}$ such that $c(s_i) \to \bar{u}$. It follows that $H(\bar{u}, p) = 0$, and by the implicit function theorem the connected component $\mathcal{C} \subset \{u \in \mathbb{R}^{N+1}\colon\ H(u, p) = 0\}$ containing c also has to contain $\bar{u}$. It follows that for each i there is a parameter value $\bar{s}_i > s_i$ such that $c(\bar{s}_i) = \bar{u}$. This implies that the curve c is closed, a contradiction.

It follows that the curve c reaches the level $\lambda = 1$ after a finite arclength s_0, i.e., $c(s_0) = (x_0, 1)$, and hence x_0 is a fixed point of f which can be approximated by tracing the curve c, see Fig. 19.1.

Let us note that

$$\big(\mathrm{Id} - f'(x_0)\big)\dot{x}(s_0) = \dot{\lambda}(s_0)\big(f(x_0) - p\big),$$

and our above assumption on f implies that $(\mathrm{Id} - f'(x_0))$ cannot have a nontrivial kernel, and hence $\dot{\lambda}(s_0) \neq 0$, i.e., the curve c is tranversal to the level $\lambda = 1$ at any solution.

The above discussion is in the spirit of CHOW, MALLET-PARET and YORKE [1978]. An earlier approach based on the nonretraction principle of HIRSCH [1963] was given by KELLOGG, LI and YORKE [1976]. Since the appearance of the constructive proofs of the Brouwer fixed point theorem many other constructive existence proofs have been described. Further references may be found in ALLGOWER and GEORG ([1990], Section 11.1).

20. Global Newton methods

Newton's method is a popular method for numerically calculating a zero point of a smooth map $G\colon \mathbb{R}^N \to \mathbb{R}^N$. As is well known, this method may diverge if the starting point p is not sufficiently near to a zero point $\bar{x}$ of G. Often one would like to determine whether a certain open bounded region $\Omega \subset \mathbb{R}^N$ contains a zero point $\bar{x}$ of G and furthermore, for which starting values p this solution $\bar{x}$ can be obtained by Newton's method. The so-called global Newton methods offer a possibility of answering such questions.

One may interpret Newton's method as the numerical integration of the differential equation

$$\dot{x} = -G'(x)^{-1}G(x) \tag{20.1}$$

using Euler's method with unit step size. The idea of using the above flow to find zero points of G was exploited by BRANIN [1972]. SMALE [1976] gave conditions on $\partial\Omega$ under which the flow leads to a zero point of G in Ω. Such numerical methods have been referred to as *global Newton methods*. However, we note that this method becomes numerically unstable near singularities, i.e., at points x where the condition number of $G'(x)$ is large. KELLER [1978] observed that the above flow can also be obtained in a *numerically stable* way from a homotopy equation which he consequently named the *global homotopy* method. Independently, GARCIA and GOULD [1978, 1980] discussed this flow.

We briefly sketch Keller's approach. The global homotopy method involves tracing the curve defined by the equation $G(x) - (1-\lambda)G(p) = 0$ starting from $(x, \lambda) = (p, 0) \in \partial\Omega \times \{0\}$ inward into $\Omega \times \mathbb{R}$. If the level $\Omega \times \{1\}$ is encountered, then a zero point of G has been found. Hence, the *global homotopy* $H : \mathbb{R}^N \times \mathbb{R} \times \partial\Omega \longrightarrow \mathbb{R}^N$ is defined by

$$H(x, \lambda, p) := G(x) - (1-\lambda)G(p). \tag{20.2}$$

The following theorem is essentially due to SMALE [1976].

THEOREM 20.1. *Let the following conditions be satisfied for the smooth map $G : \mathbb{R}^N \to \mathbb{R}^N$:*

(1) *$\Omega \subset \mathbb{R}^N$ is open and bounded and $\partial\Omega$ is a connected smooth submanifold of $\mathbb{R}^N$,*

(2) *zero is a regular value of G,*

(3) *$G(q) \neq 0$ for $q \in \partial\Omega$,*

(4) *the Jacobian $G'(q)$ is nonsingular for $q \in \partial\Omega$,*

(5) *the Newton direction $-G'(q)^{-1}G(q)$ is not tangent to $\partial\Omega$ at q for $q \in \partial\Omega$.*

Then for almost all initial points $p \in \partial\Omega$, there is a smooth curve $c(s) = (x(s), \lambda(s))$ emanating from $c(0) := (p, 0)$. Here we assume for simplicity that s is arclength. The derivative $\dot{x}(0)$ is not tangent to $\partial\Omega$, and hence we orient c in such a way that $\dot{x}(0)$ points into Ω. Furthermore, c hits the target level $\Omega \times \{1\}$ at an odd number of points $x(s_1), \dots, x(s_k)$ with $s_1 < \dots < s_k$.

This possibility of obtaining more than one solution was first observed by BRANIN and HOO [1972].

PROOF. We give a modified version of KELLER's [1978] proof. Since p varies over the $(N-1)$-dimensional surface $\partial\Omega$, it is somewhat difficult to apply Sard's theorem.

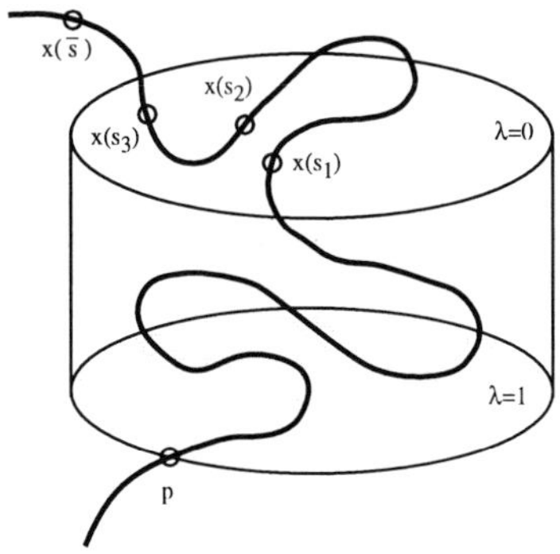

FIG. 20.1. Illustration of global homotopy.

This task was achieved by PERCELL [1980] who showed that for almost all $p \in \partial\Omega$ the global homotopy H has 0 as a regular value. Let p be such a generic choice.

We consider the solution curve $c(s) = (x(s), \lambda(s))$ in $H(\bullet, \bullet, p)^{-1}(0)$ such that $c(0) = (p, 0)$ and such that s represents arclength, see Fig. 20.1.

Since $\partial\Omega$ is connected, we can infer from assumption (5) without loss of generality that the Newton direction

$$G'(q)^{-1}G(q) \quad \text{always points out of } \Omega \text{ for } q \in \partial\Omega. \tag{20.3}$$

The other case, i.e., that the Newton direction $G'(q)^{-1}G(q)$ always points into Ω for $q \in \partial\Omega$, is treated in a similar way by switching some signs.

We differentiate the homotopy equation

$$G\big(x(s)\big) - \big(1 - \lambda(s)\big)G(p) = 0 \tag{20.4}$$

and obtain

$$G'\big(x(s)\big)\dot{x}(s) + \dot{\lambda}(s)G(p) = 0, \tag{20.5}$$

and by substituting $(1 - \lambda(s))^{-1}G(x(s))$ for $G(p)$ we obtain

$$\lambda(s) \neq 1 \Rightarrow \; G'\big(x(s)\big)\,\dot{x}(s) + \frac{\dot{\lambda}(s)}{1 - \lambda(s)}\,G\big(x(s)\big) = 0. \tag{20.6}$$

If $\dot{\lambda}(s) = 0$, then $\|\dot{x}(s)\| = 1$ since s is arclength, and the above equation would imply that $G'(x(s))$ is singular. Hence, from assumption (4) it follows that $\dot{\lambda}(s) \neq 0$ for $x(s) \in \partial\Omega$. Furthermore, assumption (3) implies that $\lambda(s) \neq 1$ for $x(s) \in \partial\Omega$. And finally, from (20.3) and the above equation, it now follows that $\dot{x}(s)$ points into Ω or out of Ω for $x(s) \in \partial\Omega$, according as the factor $\dot{\lambda}(s)/(1 - \lambda(s))$ is positive or negative.

In particular, initially, we choose the orientation of c in such a way that

$$\dot{\lambda}(0) > 0, \tag{20.7}$$

and hence the above considerations imply that $\dot{x}(0)$ points into Ω.

Since $G(p) \neq 0$ and Ω is bounded, we see that the set

$$\{\lambda\colon\ G(x) = (1-\lambda)G(p),\ x \in \overline{\Omega}\}$$

is bounded. Hence the curve c must exit from $\Omega \times \mathbb{R}$ at some parameter value $\bar{s} > 0$, i.e., $x(\bar{s}) \in \partial\Omega$. Since $x(\bar{s})$ points out of Ω, it follows that

$$\frac{\dot{\lambda}(\bar{s})}{1-\lambda(\bar{s})} < 0. \tag{20.8}$$

Now consider the augmented Jacobian

$$A(s) := \begin{pmatrix} G'(x(s)) & G(p) \\ \dot{x}(s)^* & \dot{\lambda}(s) \end{pmatrix}$$

of the homotopy (20.2). We obtain

$$A(s)\begin{pmatrix} \mathrm{Id} & \dot{x}(s) \\ 0^* & \dot{\lambda}(s) \end{pmatrix} = \begin{pmatrix} G'(x(s)) & 0 \\ \dot{x}(s)^* & 1 \end{pmatrix}$$

and consequently

$$\det A(s)\, \dot{\lambda}(s) = \det G'\big(x(s)\big). \tag{20.9}$$

By assumption (4), and since $\partial\Omega$ is connected, the function $\det G'(x)$ does not change sign on $\partial\Omega$. On the other hand, the function $\det A(s)$ does not change sign along the path c. Consequently, $\operatorname{sign} \dot{\lambda}(0) = \operatorname{sign} \dot{\lambda}(\bar{s})$, and hence (20.7) implies $\dot{\lambda}(\bar{s}) > 0$. From (20.8) we obtain

$$\lambda(\bar{s}) > 1, \tag{20.10}$$

and hence there exists at least one s_1 with $0 < s_1 < \bar{s}$ such that $\lambda(s_1) = 1$.

Let $\lambda(s_i) = 1$ for $0 < s_i < \bar{s}$. If $\dot{\lambda}(s_i) = 0$, then $G(x(s_i)) = 0$ by (20.4), and (20.5) would imply $G'\big(x(s)\big)\dot{x}(s) = 0$ with $\|\dot{x}(s)\| = 1$, contradictory to assumption (2). Hence, $\dot{\lambda}(s_i) \neq 0$.

Since $\lambda(0) = 0$ and $\lambda(\bar{s}) > 1$, it now follows that there are an odd number of values $0 < s_1 < \cdots < s_k < \bar{s}$ such that $\lambda(s_i) = 1$ for $i = 1, \ldots, k$. □

If $\tilde{x}$ is a zero point of G such that $G'(\tilde{x})$ is nonsingular, then the assumptions of Theorem 20.1 hold for a sufficiently small ball Ω around $\tilde{x}$. Thus, in a certain sense the global homotopy extends the well-known Newton–Kantorovich type theorems concerning the local convergence of Newton's method, see, e.g., ORTEGA and RHEINBOLDT [1970].

20.1. Some applications to chemical engineering and circuit analysis

The principal application considered in the papers by BRANIN [1972] and BRANIN and HOO [1972] was in the area of electrical circuit design and analysis. Since then homotopy continuation methods have frequently been employed as a tool in the computer aided design of electronic circuits and computer chips. The paper of CHAO and SAEKS [1977] surveys continuation methods in circuit analysis and cites many papers on the topic up to 1977. Recently, MELVILLE, TRAJKOVIC, FANG and WATSON [1993b] have applied a homotopy continuation algorithm to compute multiple solutions of systems of equations arising in the simulation of integrated circuits, see also MCQUAIN, RIBBENS, WATSON and MELVILLE [1993] and MELVILLE, MOINIAN, FELDMANN and WATSON [1993a]. The paper MELVILLE, TRAJKOVIC, FANG and WATSON [1993b] also contains an updated survey of the literature in which continuation methods are used in circuit analysis. Watson, the principal author of HOMPACK, see WATSON, BILLUPS and MORGAN [1987], reports that homotopy methods are on the verge of replacing damped Newton methods in the circuit simulation programs of at least one of the major industrial producers of integrated circuits. In addition, homotopy methods have become a standard tool in the numerical simulation of robotics in several major industrial firms. The latter application deals primarily with polynomial systems and so we will discuss this application further in Section 22.

The fundamental nonlinear term arising in the network problems, is of the form $\alpha(\mathrm{e}^{v/\beta}-1)$ where α and β are constants and v denotes voltage. In CHUA and USHIDA [1976] an example is given of a system of four nonlinear equations modelling a four-transistor multi-state circuit, which after some simplifications assumes the following general form:

$$A\begin{pmatrix} v_1 \\ \vdots \\ v_N \end{pmatrix} + B\begin{pmatrix} \mathrm{e}^{v_1/\beta} \\ \vdots \\ \mathrm{e}^{v_N/\beta} \end{pmatrix} + \begin{pmatrix} c_1 \\ \vdots \\ c_N \end{pmatrix} = 0$$

where N is the number of nodes of the circuit, A, B are (N, N)-matrices, the c_i are constants and the unknown v_i are voltages.

In general, A and B are large sparse nonsymmetric matrices. The example of this form which was numerically treated in CHUA and USHIDA [1976] (for $N = 4$) essentially by means of the global Newton method was found to have nine solutions.

In addition to the application of the predictor corrector homotopy methods in the area of circuit analysis, there is also a body of work in which piecewise linear methods have been applied. Such works can be traced back at least to the paper of KATZENELSON [1965]. We will discuss this topic further in Chapter VI. In any event, the application of continuation methods in the area of electronic circuit analysis seems currently to be a busy scene of activity.

Another active area in which path following methods are being currently used concerns the numerical simulation of chemical processes. The paper of SEIDER, BRENGEL and WIDAGDO [1991] reviews the use of nonlinear analysis in chemical process design and includes a comprehensive bibliography concerning the use of continuation methods in the field of chemical engineering. The paper of BYRNE and BAIRD [1985] appears to be one of the earliest in which a continuation method was applied to solve difficult distillation problems where the determination of a suitable starting point for a Newton type method presents considerable trouble. The paper by WAYBURN and SEADER [1987] gives a general discussion outlining how and when homotopy continuation might be applied to chemical engineering problems. WAYBURN [1982] reviews the literature on the application of continuation methods to separation problems up to 1988. The paper of KOVACH and SEIDER [1987] presents an example of a distillation system in which five steady state solutions were found via a homotopy continuation where previously only three solutions had been found by Newton type methods. The paper of CHAVEZ, SEADER and WAYBURN [1986] demonstrated via homotopy continuation the presence of multiple solutions in interlinked separation systems.

It should perhaps be noted that in the application of homotopy methods there is in general no distinction made as to the stability of the solutions which are found. Thus a user may still need to do some analysis to sort out the unstable solutions among the solutions which are found by the homotopy method. In the next section we will see that successive solutions which lie on a homotopy curve are necessarily of opposite topological index, see Theorem 21.3. Hence in the case of critical points, a homotopy method cannot yield successive minima or successive maxima. This suggests that in some cases successive solutions found on a continuation curve may not both be stable.

21. Multiple solutions

In the previous section it was observed that the global homotopy method might in fact yield more than one zero point of the map G in a bounded region Ω. This raises the question whether one might be able to compute more zero points of G in Ω in addition to those which lie on the global homotopy path. To be more precise, let us suppose that $\Omega \subset \mathbb{R}^N$ is an open bounded region, and that $G : \mathbb{R}^N \to \mathbb{R}^N$ is a smooth map having a zero point $z_0 \in \Omega$. The task is now to find additional zero points of G in Ω, provided they exist. One method which has often been used for handling this problem is deflation, see, e.g., BROWN and GEARHART [1971]. In this method a *deflated map* $G_1 : \mathbb{R}^N \setminus \{z_0\} \to \mathbb{R}^N$ is defined by

$$G_1(x) = G(x)/\|x - z_0\|. \tag{21.1}$$

One then applies an iterative method to try to find a zero point of G_1. Numerical experience with deflation has shown that it is often a matter of seeming chance whether one obtains an additional solution and if one is obtained, it is very often not the one which is nearest to z_0.

By utilizing homotopy-type methods we can give some conditions which will guarantee the existence of an additional solution and yield insights into the behavior of deflation. This additional solution will lie on a homotopy path. We illustrate this approach with a discussion of the *d-homotopy*. Let us consider the homotopy map $H_d : \mathbb{R}^N \times \mathbb{R} \to \mathbb{R}$ defined by

$$H_d(x, \lambda) := G(x) - \lambda d$$

where $d \in \mathbb{R}^N$ is some fixed vector with $d \neq 0$. Since we assume that a zero point z_0 is already given, we have $H_d(z_0, 0) = 0$. Let us further assume zero is a regular value of G. Then it follows from Sard's theorem that zero is also a regular value of H_d for almost all $d \in \mathbb{R}^N$. In order to assure that the solution curve c in $H_d^{-1}(0)$ which contains $(z_0, 0)$ again reaches the level $\lambda = 0$, we need to impose a boundary condition. The following theorem uses a boundary condition which is motivated by a simple degree consideration.

THEOREM 21.1. *Let the following hypotheses hold:*

(1) $G : \mathbb{R}^N \to \mathbb{R}^N$ *is a smooth map with zero as a regular value;*

(2) $d \in \mathbb{R}^N \setminus \{0\}$ *is a point such that the homotopy* H_d *also has zero as a regular value;*

(3) $\Omega \subset \mathbb{R}^N$ *is a bounded open set which contains a (known) initial zero point* z_0 *of* G;

(4) *the* boundary condition $H_d(x, \lambda) = G(x) - \lambda d \neq 0$ *holds for all* $x \in \partial\Omega$, $\lambda \in \mathbb{R}$.

Then the curve c *in* $H_d^{-1}(0)$ *which contains* $(z_0, 0)$ *intersects the level* $\Omega \times \{0\}$ *an even number of times at points* $(z_i, 0)$, $i = 0, \dots, n$, *at which* $G(z_i) = 0$.

PROOF. The boundary condition (item (4)) implies that the curve c lies strictly inside the cylinder $\partial\Omega \times \mathbb{R}$. A point (x, λ) on the curve c satisfies $G(x) = \lambda d$ and $x \in \Omega$, and hence $|\lambda| = \|G(x)\|/\|d\|$ remains bounded. Recalling that c is homeomorphic either to the line $\mathbb{R}$ or to the circle S^1, the boundedness of c implies that $c \simeq S^1$. Since 0 is a regular value of G, it is easily seen that c intersects the level $\Omega \times \{0\}$ transversely, and the assertion follows immediately. □

The boundary condition (item (4)) can be relaxed:

THEOREM 21.2. *The conclusion of the above theorem remains true if the boundary condition (item (4)) is replaced by either of*

(4-1) $H_d(x, \lambda) = G(x) - \lambda d \neq 0$ *for all* $x \in \partial\Omega$, $\lambda \geqslant 0$;

(4-2) $H_d(x, \lambda) = G(x) - \lambda d \neq 0$ *for all* $x \in \partial\Omega$, $\lambda \leqslant 0$.

PROOF. We consider only the case (4-1). If (x, λ) on c is a solution with $\lambda \geqslant 0$, the same argument as in the above proof shows that $\lambda = \|G(x)\|/\|d\|$ remains bounded. Hence, starting at a solution point and traversing the curve initially in the positive λ-direction gives the desired assertion. □

To continue our discussion, we now introduce the following definition: Let z_0 be a zero point of the smooth map $G : \mathbb{R}^N \to \mathbb{R}^N$ such that $\det G'(z_0) \neq 0$. Then the *index* of z_0 is defined to be the sign of $\det G'(z_0)$.

THEOREM 21.3. *Under the hypotheses of Theorem* 21.1 *or* 21.2, *any two zero points of* G *which are consecutively obtained by traversing the curve* c *have opposite index.*

PROOF. Let $(x(s), \lambda(s)) = c(s)$ be parametrized according to arclength. Differentiating

$$H_d\big(x(s), \lambda(s)\big) = G\big(x(s)\big) - \lambda(s)d = 0$$

we obtain

$$G'\big(x(s)\big)\dot{x}(s) - \dot{\lambda}(s)d = 0$$

and hence

$$\begin{pmatrix} G'(x(s)) & -d \\ \dot{x}(s)^* & \dot{\lambda}(s) \end{pmatrix} \begin{pmatrix} \mathrm{Id} & \dot{x}(s) \\ 0^* & \dot{\lambda}(s) \end{pmatrix} = \begin{pmatrix} G'(x(s)) & 0 \\ \dot{x}(s)^* & 1 \end{pmatrix}.$$

We note that the left matrix in the latter equation is the augmented Jacobian of H_d. Its determinant is of constant sign $\varepsilon \in \{-1, 1\}$ on c. Hence by the product rule of determinants we obtain that

$$\operatorname{sign} \dot{\lambda}(s) \bullet \operatorname{sign} \det G'\big(x(s)\big) = \varepsilon.$$

Let $(x(s_1), 0)$ and $(x(s_2), 0)$ be two consecutive solution points on c. It is clear that they are traversed in opposite λ-directions, i.e., $\dot{\lambda}(s_1)\dot{\lambda}(s_2) < 0$ and the assertion follows. □

The above theorems enable us to make conclusions concerning the relationship between deflation and homotopy methods, see ALLGOWER and GEORG [1983b] for a more detailed discussion. In the context of the above discussion of deflation, see Eq. (21.1), let us consider the global homotopy $H : (\mathbb{R}^N \setminus \{z_0\}) \times \mathbb{R} \to \mathbb{R}^N$ defined by

$$H(x, \lambda) := G_1(x) - \lambda G_1(x_0). \tag{21.2}$$

In view of our discussions following Eq. (20.1), performing Newton's method on G_1 starting at x_0 amounts to a particular integration of the global homotopy (21.2) and also starting at $(x_0, 1)$. Thus we see that in general successive deflation will at best produce the zeros of G which lie on $H^{-1}(0)$. However, because the Newton steps represent an Euler method for integrating Eq. (20.1) with possibly large steps, some iterate may get far enough away from $H^{-1}(0)$ that the Newton method might diverge or possibly accidently converge to a zero point not on $H^{-1}(0)$. Numerical experience, see BROWN and GEARHART [1971], with deflation confirms the above analysis in the sense that, in general, zero points which are successively obtained via deflation have opposite index, and zero points even in close proximity are not successively obtained by deflation if they have the same index.

In a recent paper LAGARIAS, MELVILLE and GEOGHEGAN [1993] make the observation based on the proof of Theorem 21.3 that $\deg G \in \{0, 1, -1\}$ is a necessary condition for the solution curve of a homotopy H to pass through all of the zero points of G, provided that zero is a regular value of G. Conversely, as a corollary of the Whitney lemma (see, e.g., the paper of JEZIERSKI [1993]), if $\deg G \in \{0, 1, -1\}$, and if $N \neq 2$, then there exists a homotopy H of the same smoothness class as G for which the solution curve passes through all of the zero points of G. Except for the trivial case $N = 1$, there does not seem to be any explicit construction of the desired homotopy available. The authors discuss some applications to the DC operating point problem for nonlinear circuits (see also Section 20.1). For certain types of circuits a coercivity condition can be verified from which the condition $\deg G \in \{0, 1, -1\}$ can in turn be concluded.

21.1. A nonlinear boundary value problem

To give an illustration of how Theorem 21.1 can be applied, we consider a system of equations arising from a discretization of a nonlinear elliptic boundary value problem

$$\mathcal{L}u(\xi) = \mu f\big(u(\xi)\big), \quad \xi \in \mathcal{D};$$

$$u(\xi) = 0, \quad \xi \in \partial\mathcal{D}.$$

Here $\mathcal{D} \subset \mathbb{R}^m$ is a bounded domain, $\mathcal{L}$ is a linear elliptic differential operator, and f is a smooth nonlinear function which is bounded from below and satisfies

$$\lim_{u\to\infty} \frac{f(u)}{u} = \infty.$$

Problems of this general form are discussed in the survey paper of AMANN [1976], the problem from which our particular example derives has been discussed by AMBROSETTI and HESS [1980].

Discretizations of the above problem generally take the form

$$G(x) := Ax - \mu F(x) = 0 \tag{21.3}$$

where A is a positive definite (N, N)-matrix such that A^{-1} has only positive entries, and $F : \mathbb{R}^N \to \mathbb{R}^N$ is a smooth map whose co-ordinates are bounded from below by a constant $C < 0$ and satisfy

$$\lim_{x[i]\to\infty} \frac{F(x)[i]}{x[i]} = \infty$$

for each co-ordinate $x[i]$ of the vector x.

From the contraction principle it follows that for small $\mu > 0$ the fixed point iteration

$$x_{k+1} = \mu A^{-1} F(x_k), \qquad x_0 = 0$$

converges to a zero point z_0 of G. We choose a directional vector $d \in \mathbb{R}^N$ whose co-ordinates are all positive, and a set

$$\Omega := \{x \in \mathbb{R}^N \colon \|x\|_\infty < \beta\}$$

where $\beta > 0$ is chosen so large that the following conditions are satisfied:

$$\beta > \mu C \|A^{-1}\|_\infty;$$

$$F(x)[i] > \frac{\beta}{\mu a} \quad \text{for } x[i] = \beta \text{ and all } i.$$

In the latter inequality $a > 0$ denotes the smallest entry of A^{-1}. Both inequalities can be satisfied, because of the assumptions on F.

Let us show that for the above choices the assumptions of Theorem 21.2 and in particular the boundary condition (4-1) are satisfied. If $x \in \partial\Omega$, then either $x[i] = -\beta$ or $x[i] = \beta$ for some co-ordinate i. In the first case we estimate

$$\begin{aligned} A^{-1}\big(G(x) - \lambda d\big)[i] &= -\beta - \mu A^{-1} F(x)[i] - \lambda A^{-1} d[i] \\ &\leqslant -\beta + \mu C \|A^{-1}\|_\infty < 0. \end{aligned}$$

In the second case we obtain

$$\begin{aligned} A^{-1}\big(G(x) - \lambda d\big)[i] &= \beta - \mu A^{-1} F(x)[i] - \lambda A^{-1} d[i] \\ &\leqslant \beta - \mu a F(x)[i] < 0. \end{aligned}$$

Hence, in both cases we have $G(x) - \lambda d \neq 0$ for $\lambda \geqslant 0$.

This application of the d-homotopy makes it possible to reach an additional solution z_1 via the homotopy equation

$$Ax - \mu F(x) - \lambda d = 0$$

for a fixed μ and varying λ. We emphasize that z_0, z_1 do not necessarily lie on the same solution branch of the equation

$$Ax - \mu F(x) = 0$$

for varying μ. Hence, the d-homotopy can permit moving between disjoint solution branches of the nonlinear eigenvalue problem (21.3).

21.2. Periodic solutions of a Duffing equation

Recently a constructive proof of the Poincaré–Birkhoff theorem has been given by LI and LIN [1994]. Their proof is in the spirit of the methods discussed in this chapter. The statement of the theorem is as follows: Let A denote an annular region in $\mathbb{R}^2-\{0\}$, whose inner boundary Γ_1 and outer boundary Γ_2 are two disjoint simple closed curves. Let D_i denote the open region bounded by Γ_i, $i=1,2$. Then $A=\bar{D}_2-D_1$ and $0\in D_1\subset D_2$.

THEOREM 21.4 (Poincaré–Birkhoff theorem). *Let $T: A\to T(A)\subset\mathbb{R}^2-\{0\}$ be an area preserving homeomorphism. Suppose:*

(i) *$T:(r,\theta)\mapsto(r^*,\theta^*)$ has the polar co-ordinate representation $r^*=f(r,\theta)$, $\theta^*=\theta+g(r,\theta)$ such that $g(r,\theta)>0$ on Γ_1 and $g(r,\theta)<0$ on Γ_2, where f and g are C^2 and 2π periodic in θ.*

(ii) *There exists a continuous area preserving map $T_1:\bar{D}_2\to\mathbb{R}^2$ such that $T_1|_A=T$ and $0\in T_1(D_1)$.*

Then T has at least two fixed points in A.

The Poincaré–Birkhoff theorem has been applied on many ocassions in the study of dynamical systems and periodic solutions of Duffing's equation. Let us briefly show how the Poincaré–Birkhoff theorem together with the homotopy method is applied to calculate periodic solutions of a Duffing equation. We consider an equation of the form

$$x''+F(x,t)=0 \tag{21.4}$$

where $F:\mathbb{R}^2\to\mathbb{R}$ is a continuous function which is twice continuously differentiable and 2π periodic in the second variable. Equation (21.4) has the equivalent form

$$x'=y,\qquad y'=-F(x,t). \tag{21.5}$$

Let $(x(t,x_0,y_0),y(t,x_0,y_0))$ denote the unique solution of (21.5) for the initial value $(x(0),y(0))=(x_0,y_0)$. Then the Poincaré map $T:\mathbb{R}^2\to\mathbb{R}^2$ defined by

$$T(x_0,y_0)=\bigl(x(2\pi,x_0,y_0),y(2\pi,x_0,y_0)\bigr)$$

is an area preserving homeomorphism (see JACOBOWITZ [1976]). More precisely, T satisfies the assumptions of Theorem 21.4.

Upon setting $x=r\cos\theta$, $y=r\sin\theta$, (21.5) assumes the form

$$\begin{aligned} r'&=r\sin\theta\cos\theta-\sin\theta\;F(r\cos\theta,t),\\ \theta'&=-\sin^2\theta-\frac{1}{r}\cos\theta\;F(r\cos\theta,t). \end{aligned} \tag{21.6}$$

Denote by $(r(t,r_0,\theta_0),\theta(t,r_0,\theta_0))$ the unique solution of (21.6) with the initial value $(r(0),\theta(0)) = (r_0,\theta_0)$. Using the notation of Theorem 21.4, we have

$$r(2\pi,r_0,\theta_0) = f(r_0,\theta_0),$$
$$\theta(2\pi,r_0,\theta_0) = g(r_0,\theta_0) + \theta_0 \pmod{2\pi}.$$

To compute the periodic solutions of (21.6), we may fix a value $\theta_0 \in [0,2\pi)$ and find a point r_0 near the set $\{r\colon\ g(r,\theta_0) = 0\}$ by, e.g., a Newton-type method. By Theorem 18.1 (Sard), the probability is unity that (r_0,θ_0) is such a point that zero is a regular value of the global homotopy

$$H(r,\theta,\lambda) := \begin{pmatrix} f(r,\theta) - r \\ g(r,\theta) \end{pmatrix} - (1-\lambda)\begin{pmatrix} f(r_0,\theta_0) - r_0 \\ 0 \end{pmatrix} = 0. \tag{21.7}$$

When $\lambda = 0$, the solution of $H(r,\theta,\lambda) = 0$ is $(r_0,\theta_0,0)$ and when $\lambda \to 1$, any limit point (r^*,θ^*) of the solutions of (21.7) satisfies $f(r^*,\theta^*) = r^*$, $g(r^*,\theta^*) = \theta^* \pmod{2\pi}$.

LI and LIN [1994] show (in the general context of Theorem 21.4) that the solution curve of (21.7), when started from the initial point $(r_0,\theta_0,0)$ in the two possible directions leads to two different solutions at $\lambda = 1$.

For actual numerical implementations the function values $f(r,\theta), g(r,\theta)$ can be calculated via an initial value solving routine, such as a Runge–Kutta method.

22. Polynomial systems

In the preceding section we considered the task of computing multiple zero points of general smooth maps. In the case of complex polynomial systems it is possible to compute (at least in principle) all of the zero points by means of homotopy methods. This subject has received considerable attention in recent years. The book of MORGAN [1987] deals extensively with this topic, in connection with the path following approach. It also contains a number of interesting applications to robotics and other fields.

We consider a system of complex polynomials $P\colon \mathbb{C}^N \to \mathbb{C}^N$. The task is to find *all* solutions of the equation $P(z) = 0$. Let us review some of the standard terminology in this context. If a term of the kth component P_k of P has the form

$$a z_1^{r_1} z_2^{r_2} \cdots z_N^{r_N},$$

then its degree is $r_1 + r_2 + \cdots + r_N$. The degree d_k of P_k is the maximum of the degrees of its terms. The *homogeneous part* $\hat{P}$ of P is obtained by deleting in each component P_k all terms having degree less than d_k. The *homogenization* $\tilde{P}$ of P is obtained by multiplying each term of each component P_k with an appropriate power z_0^r such that its degree is d_k. Note that the homogenization $\tilde{P}\colon \mathbb{C}^{N+1} \to \mathbb{C}^N$ involves one more variable z_0. If

$$(w_0,\ldots,w_N) \neq 0$$

is a zero point of $\tilde{P}$, then the entire ray

$$[w_0 : \cdots : w_N] := \{(\xi w_0, \ldots, \xi w_N) \mid \xi \in \mathbb{C}\}$$

consists of zero points of $\tilde{P}$. Usually, $[w_0 : \cdots : w_N]$ is regarded as a point in the complex projective space $\mathbf{CP}^N$. There are two cases to consider:

(1) The solution $[w_0 : \cdots : w_N]$ intersects the hyperplane $z_0 = 0$ transversely, i.e., $w_0 \neq 0$, and hence, without loss of generality, we may take $w_0 = 1$. This corresponds to a zero point $(w_1, \ldots, w_N)$ of P. Conversely, each zero point $(w_1, \ldots, w_N)$ of P corresponds to a solution $[1 : w_1 : \cdots : w_N]$ of $\tilde{P}$.

(2) The solution $[w_0 : \cdots : w_N]$ lies in the hyperplane $z_0 = 0$, i.e., $w_0 = 0$. This corresponds to a *nontrivial* solution $[w_1 : \cdots : w_N]$ of the homogeneous part $\hat{P}$, and such solutions are called *zero points of P at infinity*.

As in the case of one variable, it is possible to define the multiplicity of a solution. However, this is a more complicated matter than it is in one dimension and requires some deeper ideas of algebra and analysis. We will give a brief sketch. If $[w_0 : \cdots : w_N]$ is an isolated solution of the homogenization $\tilde{P}(z_0, \ldots, z_N) = 0$ with respect to the topology of $\mathbf{CP}^N$, we can define a multiplicity of $[w_0 : \cdots : w_N]$ in two different ways. However, it is not a trivial exercise to show that these definitions are equivalent.

(1) Consider a co-ordinate w_k which is different from zero. Without loss of generality we can assume $w_k = 1$. If we fix the variable $z_k = 1$ in the homogenization $\tilde{P}(z_0, \ldots, z_N) = 0$, then we have N complex equations in N complex variables or $2n$ real equations in $2n$ real variables with the complex solution $z_j = w_j$, $j = 0, \ldots, N$, $j \neq k$. The multiplicity is now defined by the local topological degree of this solution, see, e.g., MILNOR [1968]. It can be shown that this definition is independent of the special choice of the nonvanishing co-ordinate w_k.

(2) As above, consider a co-ordinate $w_k = 1$. Again, we fix the variable $z_k = 1$ in the homogenization $\tilde{P}(z_0, \ldots, z_N) = 0$, and after a translation obtain new equations

$$F(z_0, \ldots, \widehat{z_k}, \ldots, z_N) := \tilde{P}(z_0 + w_0, \ldots, 1, \ldots, z_N + w_N) = 0$$

in the variables $z_0, \ldots, \widehat{z_k}, \ldots, z_N$ where $\widehat{\ }$ denotes omission of the term beneath it. These new equations have a zero point at the origin. Now, the multiplicity of the solution is defined as the dimension of the quotient

$$\frac{\mathbb{C}[[z_0, \ldots, \widehat{z_k}, \ldots, z_N]]}{(F_1, \ldots, \widehat{F_k}, \ldots, F_N)}$$

where $\mathbb{C}[[z_0, \ldots, \widehat{z_k}, \ldots, z_N]]$ is the usual power series ring and the symbol $(F_1, \ldots, \widehat{F_k}, \ldots, F_N)$ denotes the ideal generated by the corresponding polynomials $F_1, \ldots, \widehat{F_k}, \ldots, F_N$, see, e.g., FULTON [1984] or VAN DER WAERDEN [1953]. It can be shown that also this definition is independent of the special choice of the nonvanishing co-ordinate w_k.

The higher dimensional analogue of the fundamental theorem of algebra is Bezout's theorem, which states that the number of zero points of P (counting their multiplicities and zeros at infinity) equals the product $d = d_1 d_2 \cdots d_N$, provided all solutions are isolated. This number d is often called the *Bezout number* of the system.

As an illustration, let us examine the polynomial system

$$\begin{aligned} z_1^3 - z_1 &= 0, \\ z_1^2 z_2 + 1 &= 0, \end{aligned}$$

which has the homogeneous part

$$\begin{aligned} z_1^3 &= 0, \\ z_1^2 z_2 &= 0, \end{aligned}$$

and the homogenization

$$\begin{aligned} z_1^3 - z_0^2 z_1 &= 0, \\ z_1^2 z_2 + z_0^3 &= 0. \end{aligned}$$

It has the three isolated solutions $[1:\pm 1:-1]$ and $[0:0:1]$. According to Bezout's theorem, this system has nine roots. It is routine to see that $\det P'(z) \neq 0$ at the zero points $(\pm 1, -1)$. Hence they are simple, and by Bezout's theorem the zero point at infinity has multiplicity seven. Let us show that this is true by using the second definition of multiplicity. Setting $z_2 = 1$ in the homogenization, we obtain

$$\begin{aligned} z_1^3 - z_0^2 z_1 &= 0, \\ z_1^2 + z_0^3 &= 0. \end{aligned}$$

Since $z_1^3 - z_0^2 z_1 = z_1(z_1 + z_0)(z_1 - z_0)$, a factorization theorem yields

$$\begin{aligned} \dim \frac{\mathbb{C}[[z_0, z_1]]}{(z_1^3 - z_0^2 z_1, z_1^2 + z_0^3)} &= \dim \frac{\mathbb{C}[[z_0, z_1]]}{(z_1, z_1^2 + z_0^3)} + \dim \frac{\mathbb{C}[[z_0, z_1]]}{(z_1 + z_0, z_1^2 + z_0^3)} \\ &\quad + \dim \frac{\mathbb{C}[[z_0, z_1]]}{(z_1 - z_0, z_1^2 + z_0^3)} \\ &= 3 + 2 + 2. \end{aligned}$$

GARCIA and ZANGWILL [1979] and CHOW, MALLET-PARET and YORKE [1979] introduced homotopy methods in $\mathbf{C}^N \times \mathbb{R}$ for finding all solutions of the equation $P = 0$. A difficulty they had to overcome was the handling of solution paths which connect to zero points at infinity. WRIGHT [1985] realized that their approaches could be simplified by going into the complex projective space $\mathbf{CP}^N$. We use his approach to illustrate the homotopy idea for polynomial systems.

Define a homotopy $H = (H_1, \ldots, H_N)$ by involving the homogenization $\tilde{P}$ of P via

$$H_k(z_0, \ldots, z_N, \lambda) = (1 - \lambda)\big(a_k z_k^{d_k} - b_k z_0^{d_k}\big) + \lambda \tilde{P}_k(z_0, \ldots, z_N).$$

Wright shows by Sard-type arguments that for almost all coefficients a_k, b_k in $\mathbb{C}$ the restricted homotopies $H^{(j)}$ which are obtained from H by fixing $z_j = 1$ for

$j = 0, \ldots, N$ have zero as a regular value for $\lambda < 1$. Given such a generic choice of coefficients, he concludes that for $\lambda < 1$, the homogeneous system of polynomials H has exactly d simple zero point curves $c_i(\lambda) \in \mathbf{CP}^N$, $i = 1, \ldots, d$, in complex projective N-space. On the trivial level $\lambda = 0$, the d solutions are obvious, and it is possible to trace the d curves emanating from these solutions into the direction of increasing λ. The solution curves are monotone in λ, and hence all have to reach the target level $\lambda = 1$ on the compact manifold $\mathbf{CP}^N$. Thus, in this approach solutions at infinity are treated no differently than finite solutions. The solution curves are traced in the projective space $\mathbf{CP}^N$, and from the numerical point of view we have the slight drawback that occasionally a chart in $\mathbf{CP}^N$ has to be switched.

Recently, attention has been given to the task of trying to formulate homotopies which eliminate the sometimes wasteful effort involved in tracing paths which go to solutions of $P(z_1, \ldots, z_N) = 0$ at infinity. Work in this direction has been done in MORGAN [1986], LI, SAUER and YORKE [1987, 1989] and LI and WANG [1993a,b]. MORGAN and SOMMESE [1987] describe the easily implemented "projective transformation" which allows the user to avoid the drawback of changing co-ordinate charts on $\mathbf{CP}^N$. MORGAN and SOMMESE [1989] show how to exploit relations among the system coefficients, via "coefficient parameter continuation". Such relations occur commonly in engineering problems, as described in WAMPLER and MORGAN [1991], WAMPLER, MORGAN and SOMMESE [1990, 1992]. The papers MORGAN, SOMMESE and WAMPLER [1991, 1992, 1993] combine a homotopy method with contour integrals to calculate singular solutions to polynomial and nonlinear analytic systems. MORGAN, SOMMESE and WATSON [1989] documented that HOMPACK, see WATSON, BILLUPS and MORGAN [1987], in the case of polynomial systems has some stability issues that CONSOL8, see MORGAN [1987], does not have. The path following approach to systems of polynomial equations is particularly suited for parallel processing, see ALLISON, HARIMOTO and WATSON [1989].

23. Sparse polynomial systems

In many applications, the Bezout number of a system of polynomial equations is prohibitively high, so that the path following codes mentioned in the previous section run into serious problems because of the many paths that need to be followed. Recently, a new approach has been initiated to solve sparse polynomial systems via path following, see HUBER and STURMFELS [1993]. This approach looks very promising and is based on Bernstein's theorem, see Theorem 23.1 below. The method is too technical even to sketch, but let us try to at least give a few hints.

Let us denote by $x^a := x_1^{a_1} \cdots x_N^{a_N}$ a monomial in N variables having integer exponents $a = (a_1, \ldots, a_N) \in \mathbb{Z}^N$. A sparse system $f = (f_1, \ldots, f_N)$ is a collection of Laurent polynomials

$$f_i(x) = \sum_{a \in A_i} c_{i,a} x^a, \quad i = 1, \ldots, N,$$

where the A_i are (usually small) finite subsets of $\mathbb{Z}^N$ and are called the support of f_i.

We denote by $Q_i := \operatorname{conv} A_i$ the convex hull of A_i in $\mathbb{R}^N$, and by "vol" the usual Euclidean volume in $\mathbb{R}^N$. It can be seen that

$$R(\lambda_1, \dots, \lambda_N) := \operatorname{vol}(\lambda_1 Q_1 + \cdots + \lambda_N Q_N)$$

is a polynomial in $\{\lambda_1, \dots, \lambda_N\}$, and the coefficient of $\lambda_1 \cdots \lambda_N$ is called the *mixed volume* $\mathcal{M}(Q_1, \dots, Q_N)$. If $\mathbb{C}^*$ denotes the nonzero complex numbers, then the theorem of BERNSTEIN [1975] can be stated in the following way, see also CANNY and ROJAS [1991].

THEOREM 23.1. *For almost all coefficients* $c_{i,a} \in \mathbb{C}^*$, *the number of zeros of* $f(x) = 0$ *in* $(\mathbb{C}^*)^N$ *equals the mixed volume* $\mathcal{M}(Q_1, \dots, Q_N)$.

If each polynomial f_i has only a few terms, which is typical for applications, then the mixed volume is a much smaller number than the Bezout number. For example, if $N = 2$ and

$$f_1(x) = a_0 + a_1 x_1 + a_2 x_1^n x_2^n,$$
$$f_2(x) = b_0 + b_1 x_1 + b_2 x_1^n x_2^n,$$

then the mixed volume is $2n$ and the Bezout number is $(2n)^2$.

The qualifier *almost all* in the theorem is very technical, it cannot be explained by a simple perturbation as in Sard's theorem. However, it is easily overcome numerically.

The point of departure for constructing a homotopy is a *lifting* function $\omega = (\omega_1, \dots, \omega_N)$ such that $\omega_i : A_i \to \mathbb{R}$. This defines a homotopy $\hat{f}(x,t)$ via

$$\hat{f}_i(x,t) := \sum_{q \in A_i} c_{i,q} x^q t^{\omega_i(q)}, \quad i = 1, \dots, N.$$

The lifting has to be chosen in a *sufficiently generic* way so that a certain finite subdivision $\{S_\omega = (A_{1,j}, \dots, A_{N,j})\}_{j \in J}$ of the set $A := (A_1, \dots, A_N)$ which is generated via ω has adequate properties (called a *fine mixed subdivision*).

In principle, the path following technique consists in following a solution branch of $f(x,t) = 0$ from $t = 0$ to $t = 1$. However, the branches are singular at $t = 0$, and transformations depending on the member $j \in J$ of the subdivision are used. The main point of the approach is that the number of branches to be followed equals the mixed volume. Furthermore, it is possible to exploit additional special structure in the polynomial system which arises from some of the exponent sets A_i being equal.

Other recent methods related to this approach have been discussed in VERLINDEN and HAEGEMANS [1994], VERSCHELDE and COOLS [1992, 1994a,b], VERSCHELDE and GATERMANN [1994], VERSCHELDE and HAEGEMANS [1993, 1994], VERSCHELDE, VERLINDEN and COOLS [1994].

24. Nonlinear eigenvalue problems, bifurcation

Path following methods are frequently applied in numerical studies of bifurcation problems. For purposes of curve tracing we have so far assumed that zero is a regular value of the smooth mapping $H : \mathbb{R}^{N+1} \to \mathbb{R}^N$. However, in the context of nonlinear eigenvalue problems, bifurcation points play an important role, and they are singular points on $H^{-1}(0)$. Hence, if path following algorithms are applied, some special adaptations are required. In typical applications, bifurcation points are defined in a Banach space context, see for example the book of CHOW and HALE [1982]. In the case that H represents a mapping arising from a discretization of an operator of the form $\mathcal{H} : E \times \mathbb{R} \to F$ where E and F represent appropriate Banach spaces, it is usually of interest to approximate bifurcation points of the operator equation $\mathcal{H} = 0$. Often one can make the discretization H in such a way that the resulting discretized equation $H = 0$ also has a corresponding bifurcation point. Under reasonable nondegeneracy assumptions it is possible to obtain error estimates for the bifurcation point of the original problem $\mathcal{H} = 0$. Such studies are made in the papers BREZZI, RAPPAZ and RAVIART [1980a,b, 1981], CROUZEIX and RAPPAZ [1990], FINK and RHEINBOLDT [1983, 1984, 1985], LIU and RHEINBOLDT [1991].

Since we are primarily concerned with bifurcation in the numerical curve following context, we confine our discussion to the case of the finite dimensional (discretized) equation $H = 0$. However, we note that the theoretical discussion below will essentially extend to the Banach space context if we assume that H is a Fredholm operator of index one. We will discuss how certain types of bifurcation points along a solution curve c can be detected, and having detected a bifurcation point, how one can numerically switch from c onto a bifurcating branch.

Some of the fundamental results on the numerical solution of bifurcation problems are due to KELLER [1970], see also KEENER and KELLER [1974] and KELLER [1977]. The recent literature on the numerical treatment of bifurcation is very extensive. For an introduction into the field we suggest the lecture notes of KELLER [1987], see also the two articles DOEDEL, KELLER and KERNÉVEZ [1991a,b] which discuss the use of the software package AUTO. For surveys and bibliography we suggest the recent book SEYDEL [1988] and the recent proceedings MITTELMANN and ROOSE [1990], ROOSE, DE DIER and SPENCE [1990], SEYDEL, SCHNEIDER, KÜPPER and TROGER [1991].

Most authors study bifurcation problems in the context of a nonlinear eigenvalue problem

$$H(x, \lambda) = 0$$

where λ is the eigenvalue parameter which usually has some physical significance. Conventionally, the solution branches are parametrized according to λ. We have taken the viewpoint that the solution branches c_i are parametrized with respect to arclength. There is only one essential difference, namely that the former approach considers folds with respect to λ also as singularities, i.e., points such that $H(x, \lambda) = 0$ and $H'(x, \lambda)$ has full rank, but $H_x(x, \lambda)$ is rank deficient.

Such folds frequently are of intrinsic interest, and there are special algorithms for detecting and calculating them. We refer the interested reader to, e.g., BOLSTAD and KELLER [1986], CHAN [1984b], FINK and RHEINBOLDT [1986, 1987], MELHEM and RHEINBOLDT [1982], PÖNISCH and SCHWETLICK [1981], SCHWETLICK [1984a,b], USHIDA and CHUA [1984]. See also the remarks on this topic in Section 8 and in Section 25.

A standard approach to the determination of bifurcation or other singular points is to directly characterize such points by adjoining additional equations to $H = 0$ and handling the resulting new set of equations by some special iterative method. In this context, continuation methods often are used to obtain starting points for these direct methods, see, e.g., GRIEWANK [1985], MOORE and SPENCE [1980], YANG and KELLER [1986], see also Section 25. Hybrid and recursive projection methods for handling unstable branches have been developed by SHROFF and KELLER [1991, 1993]. See also JARAUSCH and MACKENS [1984, 1987].

In view of the extensive literature we can only touch upon the problem here, and we will confine our discussion primarily to the task of detecting a simple bifurcation point along a solution curve c and effecting a branch switching numerically. We will see that the detection of simple bifurcation points requires only minor modifications of predictor corrector algorithms.

Suppose that $c: J \to \mathbb{R}^{N+1}$ is a smooth curve, defined on an open interval J containing zero, and parametrized (for reasons of simplicity) with respect to arclength such that $H(c(s)) = 0$ for $s \in J$. The point $c(0)$ is called a *bifurcation point* of the equation $H = 0$ if there exists an $\varepsilon > 0$ such that every neighborhood of $c(0)$ contains zero-points z of H which are not on $c(-\varepsilon, \varepsilon)$.

An immediate consequence of this definition is that a bifurcation point of $H = 0$ must be a singular point of H. Hence the Jacobian $H'(c(0))$ must have a kernel of dimension at least two. We consider the simplest case:

A point $\bar{u} \in \mathbb{R}^{N+1}$ is called a *simple bifurcation point* of the equation $H = 0$ if the following conditions hold:

$$H(\bar{u}) = 0, \quad \dim\ker H'(\bar{u}) = 2, \quad \text{and} \quad \bar{e}^* H''(\bar{u})\big|_{(\ker H'(\bar{u}))^2} \tag{24.1}$$

has one positive and one negative eigenvalue, where $\bar{e}$ spans $\ker H'(\bar{u})^*$. For convenience we normalize $\|\bar{e}\| = 1$.

Using the well-known Liapunov–Schmidt reduction, the following theorem can be shown, which is essentially a restatement of a famous result of CRANDALL and RABINOWITZ [1971].

THEOREM 24.1. *Let $\bar{u} \in \mathbb{R}^{N+1}$ be a simple bifurcation point of the equation $H = 0$. Then there exist two smooth curves $c_1(s), c_2(s) \in \mathbb{R}^{N+1}$, parametrized with respect to arclength s, defined for $s \in (-\varepsilon, \varepsilon)$ and ε sufficiently small, such that the following holds:*

(1) $H(c_i(s)) = 0$, $i \in \{1, 2\}$, $s \in (-\varepsilon, \varepsilon)$,

(2) $c_i(0) = \bar{u}$, $i \in \{1, 2\}$,

(3) $\dot{c}_1(0)$, $\dot{c}_2(0)$ *are linearly independent,*

(4) $H^{-1}(0)$ *coincides locally with* $\text{range}(c_1) \cup \text{range}(c_2)$, *more precisely:* $\bar{u}$ *is not in the closure of* $H^{-1}(0) \setminus (\text{range}(c_1) \cup \text{range}(c_2))$.

PROOF. Let us introduce the decompositions

$$\mathbb{R}^{N+1} = E_1 \oplus E_2 \quad \text{and} \quad \mathbb{R}^N = F_1 \oplus F_2,$$

where

$$E_1 := \ker H'(\bar{u}), \qquad E_2 := E_1^{\perp}, \qquad F_2 := \text{range}\, H'(\bar{u}), \qquad F_1 := F_2^{\perp}.$$

Consequently,

$$\dim E_1 = 2, \qquad \dim E_2 = N-1, \qquad \dim F_1 = 1, \qquad \dim F_2 = N-1.$$

By introducing bases in $\mathbb{R}^{N+1}$ and $\mathbb{R}^N$ which respect these direct sums, we may view H and its derivatives in the following block form:

$$H(u) = H(u_1, u_2) = \begin{pmatrix} H_1(u_1, u_2) \\ H_2(u_1, u_2) \end{pmatrix}$$

where $u_i \in E_i$ and $H_i : E_i \to F_i$, $i = 1, 2$. From the above choice of decompositions, we have

$$H'(u) = \begin{pmatrix} \partial_1 H_1(u_1, u_2) & \partial_2 H_1(u_1, u_2) \\ \partial_1 H_2(u_1, u_2) & \partial_2 H_2(u_1, u_2) \end{pmatrix}$$

and in particular,

$$H'(\bar{u}) = \begin{pmatrix} 0 & 0 \\ 0 & \partial_2 H_2(\bar{u}_1, \bar{u}_2) \end{pmatrix}. \tag{24.2}$$

Here ∂_1, ∂_2 denote the partial derivative operators with respect to the parameters of E_1 and E_2, respectively.

Note that $\partial_2 H_2(\bar{u})$ is a nonsingular $(N-1, N-1)$-matrix. Since the equation $H_2(u_1, u_2) = 0$ has the solution point $(\bar{u}_1, \bar{u}_2)$, by the implicit function theorem, there exist neighborhoods U_1 of $\bar{u}_1$ in E_1 and U_2 of $\bar{u}_2$ in E_2 and a smooth map $\varphi : U_1 \to U_2$ such that

$$H_2(u_1, u_2) = 0 \quad \text{if and only if} \quad u_2 = \varphi(u_1) \tag{24.3}$$

holds for all $u_1 \in U_1$, $u_2 \in U_2$. Thus, we have a local parametrization of the equation $H_2(u_1, u_2) = 0$ in terms of the variable u_1 in the 2-dimensional space E_1. Consequently, for all $u_1 \in U_1$, $u_2 \in U_2$, the equation $H(u) = 0$ is equivalent to $u_2 = \varphi(u_1)$ and $H_1\big(u_1, u_2\big) = 0$, or

$$b(u_1) := H_1\big(u_1, \varphi(u_1)\big) = 0. \tag{24.4}$$

This is called the *bifurcation equation* for $H(u) = 0$ at the singular point $\bar{u}$.

By differentiating the equation $H_2(u_1, \varphi(u_1)) = 0$ arising from (24.3) we have by the chain rule

$$\partial_1 H_2(\bar{u}) + \partial_2 H_2(\bar{u})\varphi'(\bar{u}_1) = 0.$$

Since $\partial_1 H_2(\bar{u}) = 0$ and $\partial_2 H_2(\bar{u})$ is nonsingular, it follows that

$$\varphi'(\bar{u}_1) = 0.$$

Differentiating $b(u_1) = H_1(u_1, \varphi(u_1))$ twice we obtain for $u = (u_1, \varphi(u_1))$,

$$\begin{aligned} b'(u_1) &= \partial_1 H_1(u) + \partial_2 H_1(u)\varphi'(u_1), \\ b''(u_1) &= \partial_1^2 H_1(u) + 2\partial_1\partial_2 H_1(u)\varphi'(u_1) \\ &\qquad + \partial_2^2 H_1(u)\big[\varphi'(u_1), \varphi'(u_1)\big] + \partial_2 H_1(u)\varphi''(u_1). \end{aligned}$$

Setting $u_1 := \bar{u}_1$ and taking into account that $\varphi'(\bar{u}_1) = 0$, $\partial_1 H_1(\bar{u}) = 0$, $\partial_2 H_1(\bar{u}) = 0$, we obtain

$$b(\bar{u}_1) = 0, \qquad b'(\bar{u}_1) = 0, \qquad b''(\bar{u}_1) = \partial_1^2 H_1(\bar{u}).$$

The simplest (generic) case is that the $(2,2)$ Hessian matrix $b''(\bar{u}_1)$ is nonsingular, i.e., both eigenvalues are different from zero. We use the following 2-dimensional version of a celebrated theorem of Morse, see, e.g., the book HIRSCH ([1976], p. 145]), in order to characterize the local structure of the solution set $b^{-1}(0)$.

Let $\bar{u}_1 \in \mathbb{R}^2$, and let $b : \mathbb{R}^2 \to \mathbb{R}$ be a smooth function such that $b(\bar{u}_1) = 0$, $b'(\bar{u}_1) = 0$ and the Hessian $b''(\bar{u}_1)$ has nonzero eigenvalues λ_1, λ_2. Then there are open neighborhoods U of $0 \in \mathbb{R}^2$ and V of $\bar{u}_1 \in \mathbb{R}^2$ and a diffeomorphism $\psi : U \to V$ such that $\psi(0) = \bar{u}_1$ and $b(\psi(\xi_1, \xi_2)) = \lambda_1\xi_1^2 + \lambda_2\xi_2^2$ where $(\xi_1, \xi_2) \in U$.

If both eigenvalues have the same sign, then $\bar{u}_1$ is an isolated zero point of b and consequently $\bar{u}$ is an isolated zero point of H. Such points are of no interest to us, since they cannot be obtained by traversing a solution curve of the equation $H = 0$.

If the eigenvalues are of opposite sign, then the local structure of $b^{-1}(0)$ near $\bar{u}_1$ and consequently the local structure of $H^{-1}(0)$ near $\bar{u}$ are described by two curves, intersecting transversely at $\bar{u}_1$ and $\bar{u}$, respectively. Note that definition (24.1) generates exactly this case since the component map H_1 corresponds to $\bar{e}^*H$. □

By differentiating the equation $\bar{e}^* H\big(c_i(s)\big) = 0$ twice and evaluating the result at $s = 0$, we obtain the following equations:

$$\ker H'(\bar{u}) = \operatorname{span}\big\{\dot{c}_1(0), \dot{c}_2(0)\big\}, \tag{24.5}$$

$$\bar{e}^* H''(\bar{u})\big[\dot{c}_i(0), \dot{c}_i(0)\big] = 0 \quad \text{for } i \in \{1,2\}. \tag{24.6}$$

The following theorem reflects the well-known fact, see KRASNOSEL'SKIĬ [1964] or RABINOWITZ [1971], that simple bifurcation points cause a switch of orientation along the solution branches. This furnishes a numerically implementable criterion for detecting a simple bifurcation point when traversing one of the curves c_i.

THEOREM 24.2. *Let $\bar{u} \in \mathbb{R}^{N+1}$ be a simple bifurcation point of the equation $H = 0$. Then the determinant of the augmented Jacobian*

$$\det \begin{pmatrix} H'(c_i(s)) \\ \dot{c}_i(s)^* \end{pmatrix}$$

changes sign at $s = 0$ for $i \in \{1,2\}$.

PROOF. We treat the case $i = 1$. It is more convenient for the proof to use the permuted matrix:

$$A(s) := \begin{pmatrix} \dot{c}_1(s)^* \\ H'(c_1(s)) \end{pmatrix}.$$

Consider an orthogonal $(N+1, N+1)$-matrix $V = (v_1, \dots, v_{N+1})$ where $v_1 := \dot{c}_1(0)$, $\operatorname{span}\{v_1, v_2\} = \ker H'(\bar{u})$, and an orthogonal (N, N)-matrix $W = (w_1, \dots, w_N)$ where $w_1 := e$ spans $\ker H'(\bar{u})^*$ as in definition (24.1). Since

$$\begin{aligned} \dot{c}_1(s)^* v_j &= \dot{c}_1(0)^* v_j + O(s), \\ w_k^* H'\big(c_1(s)\big) v_j &= w_k^* H'(\bar{u}) v_j + w_k^* H''(\bar{u})\big[\dot{c}_1(0), v_j\big] s + O(s^2), \end{aligned}$$

we obtain:

$$\begin{pmatrix} 1 & 0^* \\ 0 & W^* \end{pmatrix} A(s) V = \begin{pmatrix} 1 + O(s) & O(s) & O(s) \\ O(s^2) & \rho s + O(s^2) & O(s) \\ O(s) & O(s) & B + O(s) \end{pmatrix}. \tag{24.7}$$

The $(N-1, N-1)$ block matrix B in (24.7) is nonsingular, see (24.2) and the remarks thereafter. The scalar ρ in (24.7) is given as the off-diagonal entry of the following symmetric $(2,2)$-matrix

$$\begin{pmatrix} \bar{e}^* H''(\bar{u})[v_1, v_1] & \bar{e}^* H''(\bar{u})[v_1, v_2] \\ \bar{e}^* H''(\bar{u})[v_2, v_1] & \bar{e}^* H''(\bar{u})[v_2, v_2] \end{pmatrix}.$$

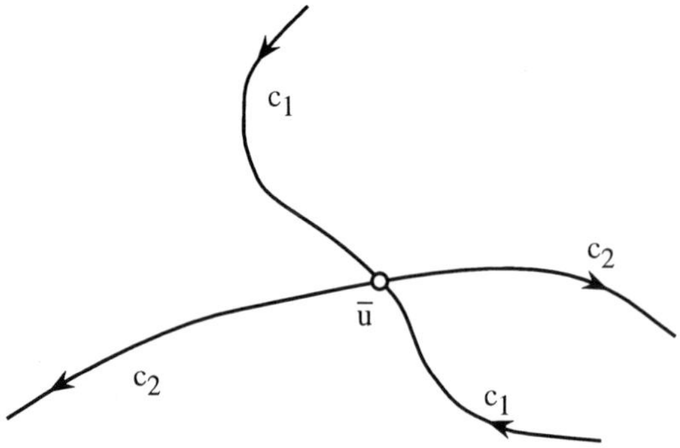

FIG. 24.1. Change of orientation.

Since this matrix is nonsingular, see (24.1), and since the diagonal entry

$$\bar{e}^* H''(\bar{u})[v_1, v_1]$$

vanishes, see (24.6), it follows that $\rho \neq 0$. Now by performing Gaussian elimination upon the first two columns of (24.7), we obtain a reduced form

$$\begin{pmatrix} 1 + \mathrm{O}(s) & \mathrm{O}(1) & \mathrm{O}(s) \\ 0 & \rho s + \mathrm{O}(s^2) & \mathrm{O}(s) \\ 0 & 0 & B + \mathrm{O}(s) \end{pmatrix}$$

which clearly has a determinant of the form

$$\rho \det(B) s + \mathrm{O}(s^2).$$

It follows that the determinant of $A(s)$ changes sign at $s = 0$. □

The above theorem implies that when traversing a solution curve c, a simple bifurcation point is detected by a change in orientation, see Fig. 24.1. Depending upon the method used to perform the decomposition of the Jacobian during path following, the above orientation can often be calculated at very small additional cost. A predictor corrector algorithm generally has no difficulty in *jumping over*, i.e., proceeding beyond the bifurcation point $\bar{u}$. That is, KELLER [1977] has shown that for sufficiently small steplength h, the predictor point will fall into the *cone of attraction* of the Newton corrector. See JEPSON and DECKER [1986] for further studies. See also ALLGOWER and GEORG ([1990], Lemma (8.1.16)).

Conversely, suppose that a smooth c in $H^{-1}(0)$ is traversed and that $c(0)$ is an isolated singular point of H such that the above determinant changes sign at $s = 0$. Then, using a standard argument in degree theory, see KRASNOSEL'SKIĬ [1964] or RABINOWITZ [1971], it can be shown that $c(0)$ is a bifurcation point of $H = 0$. However, $c(0)$ is not necessarily a simple bifurcation point.

When using a predictor corrector method with Newton-type correctors along the solution curve, then generally the Jacobian H' may be replaced by an approximation, e.g., a finite difference approximation, or an approximation generated by an update method, see Section 14. Thus a chord method may be implemented. We emphasize, however, that it is necessary to obtain a good approximation of the Jacobian at least once at the predictor point since otherwise the local convergence of the Newton corrector iterations cannot be guaranteed when jumping over a simple bifurcation point. This is explained in ALLGOWER and GEORG ([1990], p. 84).

The determinant in Theorem 24.2 is only the simplest example of a so-called *test function*. Such test functions are real functions defined on a neighborhood of the curve c and are monitored during path following to reveal certain types of singular points by a change of sign. In the case of Hopf bifurcation, the determinant is not an adequate test function. Recently, several authors have proposed and studied classes of test functions for various types of singular points, see, e.g., CHU, GOVAERTS and SPENCE [1994], DAI and RHEINBOLDT [1990], GARRATT, MOORE and SPENCE [1991], GRIEWANK and REDDIEN [1984], SEYDEL [1991b], WERNER [1992]. A different approach for the prediction of singular points along the path c has been given by HUITFELDT and RUHE [1990].

24.1. Switching branches via perturbation

In the previous section we have seen that it is possible to detect and jump over simple bifurcation points while numerically tracing a solution curve c via a predictor corrector method. The more difficult task is to numerically branch off onto the second solution curve at the detected bifurcation point $\bar{u}$. The simplest device for branching off numerically rests upon Sard's theorem 18.1. If a small perturbation vector $d \in \mathbb{R}^N$ is chosen at random, then the probability that d is a regular value of H is unity. Of course, in this case $H^{-1}(d)$ has no bifurcation point. Since $d \in \mathbb{R}^N$ is chosen so that $\|d\|$ is small, the solution sets $H^{-1}(0)$ and $H^{-1}(d)$ are close together. On $H^{-1}(d)$, no change of orientation can occur. Therefore, corresponding solution curves in $H^{-1}(d)$ must branch off near the bifurcation point $\bar{u}$, see Fig. 24.2. It is easy to implement this idea, see, e.g., ALLGOWER and CHIEN [1986], ALLGOWER, CHIEN, GEORG and WANG [1991d], CHIEN [1989], GEORG [1981b], GLOWINSKI, KELLER and REINHART [1985].

Recently, an interesting variation on this idea has been proposed by HUITFELDT [1991]. He introduces an additional parameter on the perturbation and an additional constraint equation to obtain the *branch connecting equation*

$$\mathcal{B}(u,\tau) := \begin{pmatrix} H(u) + \tau d \\ \|u - \hat{u}\|^2 + \tau^2 - \varepsilon^2 \end{pmatrix} = 0, \tag{24.8}$$

where $\hat{u}$ is an approximation to the bifurcation point $\bar{u}$. Such approximations are easily obtained via path following together with test function monitoring as described above. Note the relationship between the above homotopy and the d-homotopy discussed in Section 21 in connection with finding multiple solutions.

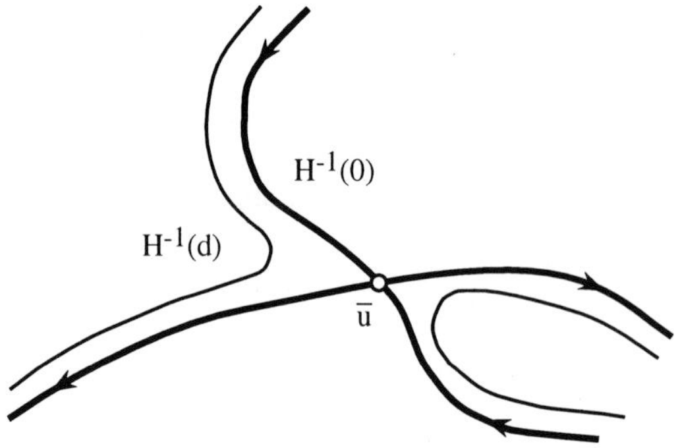

FIG. 24.2. Perturbation of a simple bifurcation point.

It is not difficult to see that for almost all d and $\varepsilon > 0$, zero is a regular value of $\mathcal{B}$, provided that $\bar{u}$ is an isolated singular point of H in $H^{-1}(0)$. Let us assume that such a generic choice of d and ε has been made.

Then the solution manifold $\mathcal{B}^{-1}(0)$ splits into one or more simple closed curves of the form $(b(s), \tau(s))$. For $\tau(s) = 0$ we obtain $H(b(s)) = 0$. Hence the curves connect points in the intersection of $H^{-1}(0)$ with the sphere $\|u - \hat{u}\|^2 = \varepsilon^2$. Starting points for a path following of $(b(s), \tau(s))$ are available from the tracing of the current solution curve c of $H = 0$. Let $b_i = b(s_i)$, $i = 0, 1, \ldots$, be successively obtained points such that $\tau(s_i) = 0$. It remains to be demonstrated that b_i and b_{i+1} are on different solution branches of the equation $H = 0$.

Since this seems to have been omitted in the paper of HUITFELDT [1991], we sketch a proof. It is easily seen that the determinant of the matrix

$$\begin{pmatrix} H'(b(s)) & d \\ (b(s) - \hat{u})^* & \tau(s) \\ \dot{b}(s)^* & \dot{\tau}(s) \end{pmatrix}$$

never changes sign since it never becomes singular. By multiplying the above matrix on the right with

$$\begin{pmatrix} \mathrm{Id} & \dot{b}(s) \\ 0^* & \dot{\tau}(s) \end{pmatrix}$$

we obtain

$$\begin{pmatrix} H'(b(s)) & 0 \\ (b(s) - \hat{u})^* & 0 \\ \dot{b}(s)^* & 1 \end{pmatrix}.$$

Since $\dot{\tau}(s_i)$ changes sign for successive i, we obtain that the determinant of

$$\begin{pmatrix} H'(b_i) \\ (b_i - \hat{u})^* \end{pmatrix}$$

changes sign for successive i. Under reasonable assumptions this implies that $t(H'(b_{i+1}))$ points out of the sphere $\|u - \hat{u}\|^2 = \varepsilon^2$ if $t(H'(b_i))$ points into it. For a simple bifurcation point (or more generally for a bifurcation point which is detected by a change of determinant in the sense of Theorem 24.2), this means that b_i and b_{i+1} cannot lie on the same solution branch.

Huitfeldt reports very successful numerical tests on some interesting problems of applied mathematics: the Taylor problem, and the Von Karman plate equations. In his experiments he succeeded in obtaining all of the bifurcating branches at several multiple bifurcation points, i.e., the 1-manifold $\mathcal{B}^{-1}(0)$ was connected in all cases he considered. However, it does not seem that this should always be the case. Advantages of this approach are that no a priori information concerning the multiplicity of the bifurcation is needed, and that it enjoys better numerical stability properties than ordinary perturbation. It should, however, be emphasized that any existing symmetries leading to higher multiplicities ought to be taken into account initially, i.e., by using group actions in the formulation of the problem, see GOLUBITSKY, STEWART and SCHAEFFER [1988] and the discussion in Section 24.3.

24.2. Branching off via minimally extended systems

Although the branching off via perturbation techniques works effectively, this approach has shortcomings. In general, it cannot be decided in advance which of the two possible directions along the bifurcating branch will be taken. Furthermore, if the perturbation vector d is not chosen correctly (and it is not always clear how this is to be done), one may still have some difficulty in tracing the resulting path (because of instabilities). Furthermore, the solution set $H^{-1}(0)$ can be approximated near the bifurcation point $\bar{u}$ only after an additional bifurcating branch has been approximated.

To obtain an approximation of $H^{-1}(0)$ near a simple bifurcation point $\bar{u}$, the alternative is a direct approach. After a coarse (initial) approximation of the bifurcation point has been obtained via a determinant (or some other) test, the following methods could be employed:

(1) The nonlinear system $H(u) = 0$ is extended in order to directly characterize bifurcation points. A rapidly convergent Newton-like iteration can then be applied to the extended system of equations. We refer to SEYDEL [1979] and WEBER [1981] as early papers on this method.

(2) A numerical model for the so-called bifurcation equation is constructed in order to approximate all tangents of the bifurcating branches in $\bar{u}$. Equation (24.6) describes such an equation for the case of a simple bifurcation point. The approaches, e.g., in KELLER [1977, 1987] and RHEINBOLDT [1978, 1986] deal with this idea.

We will discuss in some detail the first approach, in particular for the case that only a minimal number of auxiliary equations are added. The first minimally extended system for bifurcation points appears to go back to GRIEWANK and REDDIEN [1984] where auxiliary quantities are defined by nonsingular $(n+2)$-dimensional linear systems. It seems, however, that this system (which occurs there as a special case of a more general approach) has not fully been recognized as a tool for computing simple bifurcation points.

A different system has been proposed by PÖNISCH [1985] where the auxiliary quantities are obtained from $(n+1)$-dimensional linear systems. However, there the restricting assumption

$$\operatorname{rank} \partial_x H(\underbrace{\bar{x}, \bar{\lambda}}_{\bar{u}}) = N - 1 \tag{24.9}$$

is posed. Later JANOVSKÝ [1989] gave a simplification of Pönisch's system which reduces the cost of the linear algebra involved. Since in both systems the freedom in the choice of certain parameters which specify the method was unnecessarily restricted and, in particular, excluded bifurcation points with $\operatorname{rank} \partial_x F(\bar{x}, \bar{\lambda}) = N - 2$, ALLGOWER and SCHWETLICK [1993] proposed a generalization of the systems considered by Pönisch and Janovský which does not require assumption (24.9). This allows one to compute arbitrary simple bifurcation points and to choose the parameters in an optimal way.

Though there seems to be no difference in the computational cost between the method of Griewank and Reddien and the one of Allgower and Schwetlick, we follow the latter approach, because it seems to be somewhat better suited to be integrated into the path following situation.

Let us assume in the following that $\bar{u}$ is a simple bifurcation point of the equation $H(u) = 0$. In particular, see (24.1), there exist vectors $\bar{v}_1, \bar{v}_2$ spanning $\ker H'(\bar{u})$, and a vector $\bar{\psi}$ spanning $\ker H'(\bar{u})^*$.

Given a fixed $d \in \mathbb{R}^N$ and sufficiently smooth $f_i : \mathbb{R}^{N+1} \to \mathbb{R}$, $i = 1, 2$, we call the system

$$G(u, \mu) := \begin{pmatrix} H(u) + \mu d \\ f_1(u) \\ f_2(u) \end{pmatrix} = 0 \tag{24.10}$$

a *minimally extended system* for $\bar{u}$ of $H(u) = 0$ if $u = \bar{u}$ and $\mu = 0$ is a regular zero point of this system. The idea is that rapidly convergent Newton-like iterations can be employed on (24.10) to approximate $\bar{u}$.

Given fixed vectors $w, r \in \mathbb{R}^{N+1}$, it is not difficult to see that the matrices

$$C_1(u) := \begin{pmatrix} H'(u) + dr^* \\ w^* \end{pmatrix}, \qquad C_2(u) := \begin{pmatrix} H'(u) + dw^* \\ r^* \end{pmatrix} \tag{24.11}$$

are nonsingular at $u = \bar{u}$ if the conditions

$$d^*\bar{\psi} \neq 0, \qquad \det\left[\begin{pmatrix} r^* \\ w^* \end{pmatrix}\begin{pmatrix}\bar{v}_1 & \bar{v}_2\end{pmatrix}\right] \neq 0 \tag{24.12}$$

hold.

Let us assume that these (generic) conditions hold for d, w, r. Then the auxiliary variables $v_i(u) \in \mathbb{R}^{N+1}$ are implicitly defined and smooth on some neighborhood of $\bar{u}$ as solutions of the linear systems

$$C_i(u)v_i(u) = e_{N+1}. \tag{24.13}$$

It is not difficult to see that the quantities d and

$$f_1(u) := r^*v_1(u), \qquad f_2(u) := w^*v_2(u) \tag{24.14}$$

minimally extend $H(u) = 0$ in the sense of (24.10). Furthermore, the $v_i(u)$ are linearly independent at $u = \bar{u}$ and therefore on some neighborhood of $\bar{u}$.

Let us consider Newton's method for solving (24.10). If we denote by (u, μ) the current approximation of the solution, and by $(u+\Delta u, \mu+\Delta\mu)$ its Newton correction, then we obtain the following equations for Δu and $\Delta\mu$:

$$H'(u)\Delta u + d\Delta\mu = -H(u) - \mu d, \tag{24.15}$$

$$f_i'(u)\Delta u = -f_i(u). \tag{24.16}$$

If we denote by $v_0(u)$ the solution of the equation

$$C_2(u)v_0(u) = \begin{pmatrix} -H(u) - d\mu \\ 0 \end{pmatrix} \tag{24.17}$$

(which is defined on some neighborhood of $\bar{u}$), then we see that $(\Delta u, \Delta\mu) = (v_0(u), w^*v_0(u))$ satisfies (24.15). Furthermore, the $v_i(u)$ are solutions of the homogeneous part of (24.15), and therefore a solution of (24.15)–(24.16) has the form

$$\begin{pmatrix} \Delta u \\ \Delta\mu \end{pmatrix} = \begin{pmatrix} v_0(u) + \xi_1 v_1(u) + \xi_2 v_2(u) \\ w^*v_0(u) + \xi_1 f_1(u) + \xi_2 f_2(u) \end{pmatrix}. \tag{24.18}$$

It remains to find the values ξ_1, ξ_2. They are determined by (24.16).

To do this, it is helpful to see that (24.13)–(24.14) implies

$$\begin{aligned} f_1'(u)\Delta u &= -\underbrace{r^*C_1(u)^{-1}}_{=:\,(\psi_1(u)^*,g_1(u))}\; C_1'(u)\Delta u\; \underbrace{C_1(u)^{-1}e_{N+1}}_{=v_1(u)} \\ f_2'(u)\Delta u &= -\underbrace{w^*C_2(u)^{-1}}_{=:\,(\psi_2(u)^*,g_2(u))}\; C_2'(u)\Delta u\; \underbrace{C_2(u)^{-1}e_{N+1}}_{=v_2(u)} \end{aligned} \tag{24.19}$$

and consequently by (24.11):

$$f_i'(u)\Delta u = -\psi_i^* H''(u)\big[v_i(u), \Delta u\big]. \tag{24.20}$$

By using the definitions in (24.13)–(24.14) and (24.19), it can be shown $f_i(u) = g_i(u)$, $i = 1, 2$, and

$$\begin{aligned} \psi_2(u)^* H'(u) &= \frac{f_1(u)(f_2(u) - 1)}{1 - f_1(u)} w^* - f_2(u) r^*, \\ \psi_2(u)^* d &= \frac{1 - f_1(u) f_2(u)}{1 - f_1(u)}. \end{aligned} \tag{24.21}$$

Finally, we note that

$$C_1(u) - C_2(u) = \begin{pmatrix} d \\ -1 \end{pmatrix} (r^* - w^*)$$

has rank one, and hence by the Sherman–Morrison formula

$$C_1(u)^{-1} = C_2(u)^{-1} - \frac{\overbrace{C_2(u)^{-1}\begin{pmatrix} d \\ -1 \end{pmatrix}}^{=:\ \hat v(u)} (r^* - w^*) C_2(u)^{-1}}{\underbrace{1 + (r^* - w^*) C_2(u)^{-1}\begin{pmatrix} d \\ -1 \end{pmatrix}}_{=-w^* \hat v(u)}}. \tag{24.22}$$

It can be seen from the definition of the $v_i(u)$ in (24.13) and $\hat v(u)$ in (24.22) that $\hat v(\bar u) = v_1(\bar u) - v_2(\bar u)$ holds, and consequently $w^* \hat v(\bar u) = 1$. Hence the update formula (24.22) for $C_1(u)^{-1}$ is numerically stable on some neighborhood of $\bar u$.

Multiplying (24.22) from the right with e_{N+1} gives

$$v_1(u) = v_2(u) + \frac{1 - f_2(u)}{w^* \hat v(u)} \hat v(u).$$

Multiplying (24.22) from the left with r^* and taking the first n components gives

$$\psi_1(u)^* = \frac{\psi_2(u)^*}{w^* \hat v(u)}.$$

Hence the calculation of $v_1(u)$ and $\psi_1(u)$ can be cheaply obtained from $v_2(u)$ and $\psi_2(u)$, respectively.

Since it is known that the solution of (24.10) is $u = \bar u$ and $\mu = 0$, it makes sense to reset $\mu = 0$ after each Newton step. This clearly can only improve the rate of convergence. The above ideas are summarized in the following algorithm for computing a simple bifurcation point with Newton's method, see also ALLGOWER and SCHWETLICK [1993].

ALGORITHM 24.1 (*Simple bifurcation point*).

input initial point $u \in \mathbb{R}^{N+1}$
parameters $d \in \mathbb{R}^N$, $w, r \in \mathbb{R}^{N+1}$

repeat until convergence

$$J \leftarrow H'(u),\ C \leftarrow \begin{pmatrix} J + dw^* \\ r^* \end{pmatrix}$$

$$\text{solve } C(v_2 \quad \hat{v} \quad v_0) = \begin{pmatrix} 0 & d & -H(u) \\ 1 & -1 & 0 \end{pmatrix} \text{ for } v_2,\ \hat{v},\ v_0$$

$$v_1 \leftarrow v_2 + \frac{1 - w^* v_2}{w^* \hat{v}} \hat{v}$$

$$\text{solve } C^* \begin{pmatrix} \psi_2 \\ g_2 \end{pmatrix} = w \text{ for } \psi_2$$

$$\eta_i \leftarrow \psi_2^* H''(u)[v_i, v_0],\ i = 1, 2$$

$$M_{ij} \leftarrow \psi_2^* H''(u)[v_i, v_j],\ M \leftarrow (M_{ij}),\ i, j = 1, 2$$

$$\text{solve } M \begin{pmatrix} \xi_1 \\ \xi_2 \end{pmatrix} = \begin{pmatrix} (w^* \hat{v})(r^* v_2) - \eta_1 \\ w^* v_2 - \eta_2 \end{pmatrix}$$

$$u \leftarrow u + v_0 + \xi_1 v_1 + \xi_2 v_2$$

end repeat

Note that the second derivatives involved in the calculation of the η and M values can be obtained via finite difference formulas, i.e., requiring only a few evaluations of H, see (16.2).

The original approach of GRIEWANK and REDDIEN [1984] can be implemented in a similar way. The computational cost is essentially the same as in Algorithm 24.1. However, when a bifurcation point is approached by path following along one of the paths using predictor corrector methods, then usually linear systems with bordered matrices, e.g.,

$$A := \begin{pmatrix} J \\ r^* \end{pmatrix} \quad \text{with } J \approx H'(u) \text{ and } r \approx t\big(H'(u)\big) \tag{24.23}$$

have to be solved. So switching to Algorithm 24.1 requires only modifying the upper block J of C by the rank one term dw^*, but not to change the dimension of the linear systems. Therefore, the approach of ALLGOWER and SCHWETLICK [1993] is somewhat more suitable in the context of path following than the method of GRIEWANK and REDDIEN [1984].

Switching branches
Let us now turn to the problem of switching branches after a bifurcation point has been located with the above method. Let us recall the definition (24.1) of a simple bifurcation point, and note that $\bar{e}$ is now called $\bar{\psi}$. Furthermore, (24.21) implies that $\psi_2(\bar{u})$ is a multiple of $\bar{\psi}$. Hence, the above algorithm furnishes a good approximation of $\bar{\psi}$.

Let us also recall that the tangents t_i of the two bifurcating branches at $\bar{u}$ were characterized in (24.6). This leads to

$$\bar{\psi}^* H''(\bar{u})\big[t_i, t_i\big] = 0.$$

Since the matrix M in Algorithm 24.1 approximates $\bar{\psi}^* H''(\bar{u})$ on $\ker H'(\bar{u})$ (which is spanned by $\bar{v}_1, \bar{v}_2$), we have

$$\tilde{t}_i := \beta_1\bar{v}_1 + \beta_2\bar{v}_2, \qquad t_i \approx \frac{\tilde{t}_i}{\|\tilde{t}_i\|}$$

for a nontrivial solution of the quadratic equations $\beta^* M\beta = 0$. Hence, the above algorithm furnishes without further computational expense the starting direction for following the bifurcating branch.

Choosing the parameters
Because of (24.12), the parameters d and $\{r, w\}$ should be chosen in such a way that $\ker(H'(\bar{u})^*) \approx \operatorname{span}\{d\}$ and $\ker H'(\bar{u}) \approx \operatorname{span}\{r, w\}$ with r, w approximately orthogonal, respectively. When starting Algorithm 24.1 from a predictor corrector method, we have the following situation: One of the two bifurcating curves is being traced, say c_1. An approximation u of $\bar{u}$ is known, e.g., by interpolating between the last two points on the curve c_1 for which a change of sign in the sense of Theorem 24.2 was detected. More generally, some test function might have been used for this detection, see the remarks before Section 24.1. Also, the choices $r = t(H'(u))$ and $J = H'(u)$ are already available. Then, setting A as in (24.23), the missing parameters d and w can be obtained by $A^*d \approx 0$, $Aw \approx 0$. Since u is only an approximation of $\bar{u}$, these conditions can be implemented in a numerically stable way, see ALLGOWER and GEORG ([1990], (8.3.6)), by approximating $\min_d\{\|A^*d\|\colon \|d\| = 1\}$ and $\min_w\{\|Aw\|\colon \|w\| = 1\}$ either via inverse iteration or via a Lanczos method for approximating the smallest eigenvalues and corresponding eigenvectors of AA^* and A^*A, see GOLUB and VAN LOAN [1989]. Both problems can be solved simultaneously at the same cost.

24.3. Multiple bifurcation and symmetry

Multiple bifurcations often arise from symmetries with respect to certain group actions, i.e., H satisfies an *equivariance* condition

$$\mathcal{H}(\gamma x, \lambda) = \gamma\mathcal{H}(x, \lambda) \tag{24.24}$$

for γ in a group Γ. See the books of GOLUBITSKY and SCHAEFFER [1985], GOLUBITSKY, STEWART and SCHAEFFER [1988], and VANDERBAUWHEDE [1982], and the recent monograph of IZE [1993]. These symmetries can also be exploited numerically, see, e.g., ALLGOWER, BÖHMER and MEI [1991a,c], ALLGOWER, BÖHMER, GEORG and MIRANDA [1992a], CLIFFE and WINTERS [1986], DELLNITZ and WERNER [1989], GEORG and MIRANDA [1992, 1993], JEPSON, SPENCE and CLIFFE [1991], HEALEY [1988a,b, 1989], HEALEY and TREACY [1991], HONG [1993], see also the proceedings ALLGOWER, BÖHMER and GOLUBITSKY [1992b]. As the above list suggests, there is currently very much interest in this topic.

Let us sketch briefly the main ideas involved in equivariant branching. We consider a nonlinear eigenvalue problem

$$\mathcal{H}(u, \lambda) = 0 \tag{24.25}$$

where $\mathcal{H}: E \times \mathbb{R} \to F$ is sufficiently smooth. Here E, F denote Banach spaces of functions which are defined on a domain Ω. We further assume that Γ is a finite group of transformations leaving Ω invariant such that $\mathcal{H}$ has the equivariance property (24.24) for all $\lambda \in \mathbb{R}$ and all $\gamma \in \Gamma$. Here the group Γ acts on both spaces E and F according to $\gamma u(z) := u(\gamma^{-1} z)$ for $z \in \Omega$.

Under a suitable discretization of (24.25) which respects these symmetry properties a corresponding finite-dimensional nonlinear eigenvalue problem is obtained:

$$H(x, \lambda) = 0 \tag{24.26}$$

where $H: \mathbb{R}^{n+1} \to \mathbb{R}^n$ has an equivariance property

$$H(gx, \lambda) = gH(x, \lambda) \tag{24.27}$$

for all $g \in G$, where now G is a group of permutations.

An immediate conclusion which can be drawn from the equivariance property is that if $\mathcal{H}(u, \lambda) = 0$, then $\mathcal{H}(\gamma u, \lambda) = 0$. That is, if we define the *orbit* of u under Γ by $\mathrm{or}(u) := \{\gamma u\colon\ \gamma \in \Gamma\}$, then $\mathcal{H}$ vanishes on the entire orbit. On the other hand, we only gain a new solution in $\mathrm{or}(u)$ if $\gamma u \neq u$ for some $\gamma \in \Gamma$. The elements of Γ which leave u fixed form the *isotropy subgroup*

$$\Sigma_u := \{\gamma \in \Gamma\colon\ \gamma u = u\}.$$

The elements of E which are fixed under a subgroup Σ of Γ are the *fixed-point subspace*

$$E^{\Sigma} := \{u \in E\colon\ \gamma u = u \text{ for } \gamma \in \Sigma\}.$$

A subspace $X_1 \subset E$ is called *Σ-invariant* for a subgroup Σ of Γ if $\sigma u \in X_1$ for every $\sigma \in \Sigma$ and $u \in X_1$. An example is the above subspace E^{Σ}.

Now $\mathcal{H}$ maps $E^{\Sigma} \times \mathbb{R}$ into F^{Σ}, since

$$\gamma\mathcal{H}(u,\lambda) = \mathcal{H}(\gamma u,\lambda) = \mathcal{H}(u,\lambda)$$

for $\gamma \in \Sigma$, $u \in E^{\Sigma}$, and $\lambda \in \mathbb{R}$. Hence it is often convenient to replace the original problem (24.25) by corresponding reduced problems

$$\mathcal{H}^{\Sigma}\left(u^{\Sigma},\lambda\right) = 0 \tag{24.28}$$

where $\mathcal{H}^{\Sigma} : E^{\Sigma} \times \mathbb{R} \to F^{\Sigma}$ denotes the restriction of $\mathcal{H}$ on $E^{\Sigma} \times \mathbb{R}$. A projection $P^{\Sigma} : E \to E^{\Sigma}$ is given by

$$P^{\Sigma}u := \frac{1}{|\Sigma|}\sum_{\sigma\in\Sigma}\sigma u$$

where $|\Sigma|$ denotes the order of the subgroup Σ.

Now let (u_0,λ_0) be a bifurcation point of (24.25) and let $\mathcal{H}_0$, $D_u\mathcal{H}_0$, $D_\lambda\mathcal{H}_0$, etc. represent the evaluations of $\mathcal{H}$, $D_u\mathcal{H}$, $D_\lambda\mathcal{H}$ at (u_0,λ_0), respectively. We assume further that E, F are Hilbert spaces and $D_u\mathcal{H}_0$ is a Fredholm operator which hence admits the following orthogonal decompositions:

$$E = N(D_u\mathcal{H}_0) \oplus R\big((D_u\mathcal{H}_0)^*\big), \tag{24.29}$$

$$F = N\big((D_u\mathcal{H}_0)^*\big) \oplus R(D_u\mathcal{H}_0), \tag{24.30}$$

where * denotes forming the adjoint operator, and N, R denote the null space and range, respectively.

From the equivariance of $\mathcal{H}$ and the chain rule we see that $N(D_u\mathcal{H}_0)$ and $R((D_u\mathcal{H}_0)^*)$ are Σ_{u_0}-invariant spaces and hence also

$$E^{\Sigma} = N\big(D_u\mathcal{H}_0^{\Sigma}\big) \oplus R\big((D_u\mathcal{H}_0^{\Sigma})^*\big)$$

for any subgroup Σ of Σ_{u_0}.

If $D_\lambda\mathcal{H}_0 \in R(D_u\mathcal{H}_0)$, then there is a $w_0 \in E$ such that $D_u\mathcal{H}_0 w_0 + D_\lambda\mathcal{H}_0 = 0$. If $v_0 \in R((D_u\mathcal{H}_0)^*)$ is the second component of w_0 with respect to (24.29), then

$$D_u\mathcal{H}_0 v_0 + D_\lambda\mathcal{H}_0 = 0$$

with $v_0 \in R((D_u\mathcal{H}_0)^*)$ being unique.

On the other hand, by the Γ-equivariance of $\mathcal{H}$, we have $D_u\mathcal{H}_0\sigma v_0 + D_\lambda\mathcal{H}_0 = 0$ for every $\sigma \in \Sigma_{u_0}$. Since $R((D_u\mathcal{H}_0^{\Sigma})^*)$ is Σ_{u_0}-invariant, the uniqueness of v_0 yields $\sigma v_0 = v_0$ for all $\sigma \in \Sigma_{u_0}$, i.e., $v_0 \in E^{\Sigma_{u_0}}$.

If $D_\lambda\mathcal{H}_0 \notin R(D_u\mathcal{H}_0)$, then the null space of $D\mathcal{H}(u_0,\lambda_0)$ is given by

$$N\big(D\mathcal{H}(u_0,\lambda_0)\big) = N(D_u\mathcal{H}_0) \times \{0\},$$

and hence (u_0, λ_0) is only a fold point.

A subgroup $\Sigma \subset \Sigma_{u_0}$ is called a *bifurcation subgroup* of Σ_{u_0} with respect to a subspace $Y \subset E \times \mathbb{R}$ if:

(i) Y is Σ_{u_0}-invariant;
(ii) there is a pair $(y, c) \in Y$ such that $\Sigma_y = \Sigma$;
(iii) dim $[Y \cap (E^\Sigma \times \mathbb{R})] = 2$.

The following results are shown in ALLGOWER, BÖHMER and MEI [1991c] and may be shown in a fashion similar to the proofs in CRANDALL and RABINOWITZ [1971] or in Section 8.3 of ALLGOWER and GEORG [1990].

THEOREM 24.3. *Let Σ be a bifurcation subgroup of Σ_{u_0} with respect to $N(D_u\mathcal{H}_0)$ and let $D_\lambda\mathcal{H}_0 \in R(D_u\mathcal{H}_0)$. Then there are $(\phi, 0), (v_0, 1) \in N(D\mathcal{H}_0)$ such that*

$$N(D\mathcal{H}_0) \cap (E^\Sigma \times \mathbb{R}) = \text{span}\,\{(\phi, 0), (v_0, 1)\} \tag{24.31}$$

and

(a) $N(D_u\mathcal{H}_0^\Sigma) = \text{span}\{\phi\}$,
(b) $E^\Sigma = \text{span}\{\phi\} \oplus R((D_u\mathcal{H}_0^\Sigma)^*)$.

For ϕ, v_0 as in (24.31) and $\tilde{\phi} \neq 0$ in $N((D_u\mathcal{H}_0^\Sigma)^*)$, define:

$$a := (\tilde{\phi}, D_{uu}\mathcal{H}_0^\Sigma \phi\phi),$$
$$b := (\tilde{\phi}, DD_u\mathcal{H}_0^\Sigma (v_0, 1)\phi),$$
$$c := (\tilde{\phi}, D^2\mathcal{H}_0^\Sigma (v_0, 1)^2).$$

THEOREM 24.4. *Let (u_0, λ_0) be a bifurcation point of $\mathcal{H}$ and let $D_\lambda\mathcal{H}_0 \in R(D_u\mathcal{H}_0)$. Let Σ be a bifurcation subgroup of Σ_{u_0} with respect to $N(D_u\mathcal{H}_0)$ for the Γ-equivariant operator $\mathcal{H}$ such that* (24.29) *holds and $b^2 - ac \geqslant 0$. Then (u_0, λ_0) is a simple bifurcation point of the reduced problem* (24.28).

Theorem 24.4 is a version of the equivariant branching lemma of CICOGNA [1981] and VANDERBAUWHEDE [1982], see also GOLUBITSKY, STEWART, and SCHAEFFER [1988] or DELLNITZ and WERNER [1989]. The result shows that for every bifurcation subgroup $\Sigma \subset$ of the isotropy subgroup Σ_{u_0} the point (u_0, λ_0) is a simple bifurcation point in the invariant subspace E^Σ. Hence the equivariance property (24.24) often splits multiple bifurcations into simple bifurcations, which can be more easily detected by path following, see the preceding sections.

The above discussion can also be applied to finite-dimensional maps $H : \mathbb{R}^n \times \mathbb{R} \to \mathbb{R}^n$. As mentioned above, the corresponding discretization (24.26) then usually has an equivariance (24.27) provided the discretization is chosen suitably, i.e., conforming to the symmetry structure. Now the numerical linear algebra calculations in the corrector steps of the path following can be decomposed into linear subproblems as outlined in Section 15. These subproblems can possibly be handled in parallel. Such an implementation has been given by GATERMANN and HOHMANN [1991].

It has been observed in the case of a number of semilinear elliptic problems (see, e.g., ALLGOWER, BÖHMER and MEI [1991b,c]) that the subproblems often correspond to the discrete analogues arising from the symmetries corresponding to the bifurcation subgroups of Γ. In this way the multiple bifurcations are split up so that the subproblems have simple bifurcations which are again easily detected via, e.g., the determinant criterion (24.2).

25. Applications to fluid dynamics

Researchers in computational fluid dynamics, computational physics and combustion have in recent years made extensive use of numerical continuation methods. In these applications one generally deals with systems of nonlinear partial differential equations. Typically, the systems are of the Navier–Stokes type. A research group at Harwell Laboratory involving Cliffe, Winters et al. has given numerous reports of their experiences, see, e.g., WINTERS [1987, 1991], WINTERS and CLIFFE [1985], WINTERS, CLIFFE and JACKSON [1987] WINTERS, MYERSCOUGH and MAINI [1990]. We will sketch here a sample of such examples.

After the partial differential equations have been discretized, for example by finite elements, one obtains a system of equations of the form:

$$H(x, \lambda, \alpha) = 0, \tag{25.1}$$

where $H : \mathbb{R}^N \times \mathbb{R} \times \mathbb{R}^p \to \mathbb{R}^N$ is a smooth map. In general, x is a state variable, N is the number of degrees of freedom in the discretization, λ is a bifurcation parameter, and α is a vector of control parameters of dimension p. The *bifurcation parameter* λ is singled out, because one wants to study the change in behavior of the solution as this parameter is varied. Thus this parameter is regarded as free to be varied, e.g., as in an experiment, whereas the x and α parameters are more geometric in nature, or at least more or less fixed at the outset. The bifurcation parameter may for example, be the Reynolds number or the Rayleigh number in problems arising from fluid mechanics.

In many of the applications the points of greatest physical interest on the solution curves are thoses at which a change of stability occurs. Typically, these points are fold points and bifurcation points. Very frequently the physical problem also exhibits geometrical symmetries arising from cell structure or geometrical symmetry of vessels, pipes, etc. In such cases of geometric symmetry, if the differential operator also enjoys equivariance properties, see (24.24), with respect to the elements of the symmetry group, multiple bifurcation may arise. Situations of this sort may be theoretically described via the theory of equivariant branching which stems from VANDERBAUWHEDE [1982] and CICOGNA [1981], see also GOLUBITSKY, STEWART and SCHAEFFER [1988] for an extensive discussion of the equivariant branching lemma. See also the discussion in Section 24.3.

The numerical treatment of the discretized problem (25.1) involves in addition to the handling of relatively large degrees of freedom the need to detect the points where changes of stability, i.e., fold or bifurcation points, occur. At the end of Section 24 we touched upon some functionals for detecting and calculating special points on the solution curve.

The paper WINTERS, CLIFFE and JACKSON [1987] summarizes a number of applications in fluid dynamics. In their numerical studies the authors report experience using an Euler–Newton method involving LU factorizations of $H_x = \partial H/\partial x$ and a simple steplength control of halving or doubling governed by the behavior of the corrector. Since the bifurcation parameter λ has the greatest physical meaning in these applications, singularity and stability is regarded with respect to it. As an alternative to switching to Newton steplength control as discussed in Algorithm 8.1, singular points are calculated directly by introducing augmented maps in the spirit of MOORE and SPENCE [1980], GRIEWANK and REDDIEN [1984], RABIER and REDDIEN [1986]. Section 24.2 contains details concerning such a method.

Let us briefly illustrate the main points in this application. A point $(x_0, \lambda_0, \alpha_0)$ is a simple singular point of Eq. (25.1) if the null space of $H_x(x_0, \lambda_0, \alpha_0)$ is of dimension 1, i.e., is spanned by a vector $\varphi_0 \neq 0$. The range is then given by

$$\text{range}\, H_x(x_0, \lambda_0, \alpha_0) := \left\{ y \in \mathbb{R}^N\colon\ \psi_0^* y = 0 \right\}$$

for some $\psi_0 \neq 0$.

A fold point of H is a simple singular point which satisfies

$$\psi_0^* H_\lambda(x_0, \lambda_0, \alpha_0) \neq 0.$$

If in addition

$$\psi_0^* H_{xx}(x_0, \lambda_0, \alpha_0)[\varphi_0, \varphi_0] \neq 0,$$

then it is called a *quadratic fold point*. The augmented system of equations used for calculating a fold point is that given by MOORE and SPENCE [1980]:

$$F(x, \varphi, \lambda, \alpha) := \begin{pmatrix} H(x, \lambda, \alpha) \\ H_x(x, \lambda, \alpha)\varphi \\ l^*\varphi - 1 \end{pmatrix} = 0 \tag{25.2}$$

where $l \in \mathbb{R}^N$ is not in the null space of $H_x(x_0, \lambda_0, \alpha_0)$. In general, from the curve following technique, good starting values for a Newton-type method for solving (25.2) are already available.

Equation (25.2) has a regular zero point at $(x_0, \varphi_0, \lambda_0, \alpha_0)$ if and only if it is a quadratic fold point of H in the above sense. A path of fold points can now be numerically traced by applying a continuation method with respect to one of the α parameters, see, e.g., RHEINBOLDT [1986, 1993].

The above described approach was applied by WINTERS and CLIFFE [1985] to the equations modelling thermal ignition in a finite cylinder with aspect ratio $\gamma = R/L$ where R and L denote the radius and length of the cylinder. Assume the cylinder has the origin as center and the z-axis as its rotational axis. Let us sketch their discussion as an example.

The nondimensional form of the thermal equation takes the form

$$-\Delta u = \delta \exp\left(\frac{u}{1+\varepsilon u}\right)$$

where Δ is the Laplacian operator in cylindrical co-ordinates, δ is the exothermicity, and ε is the inverse activation energy. The conditions on the boundary of the cylinder are $u = 0$. In addition, the following symmetry conditions are imposed: $\partial u/\partial z = 0$ on the plane $z = 0$, and $\partial u/\partial r = 0$ on the axis $r = 0$. The replacement of z by γz leads to the equation

$$-\frac{1}{r}\frac{\partial}{\partial r}r\frac{\partial u}{\partial r} - \frac{1}{\gamma^2}\frac{\partial^2 u}{\partial z^2} = \delta \exp\left(\frac{u}{1+\varepsilon u}\right).$$

For a fixed value of γ, a starting point $\delta = 0$ and $u = 0$ is available on the solution curve for the discretized problem.

The peak temperature $\hat{u}$ of the discretized solution when plotted against δ has a characteristic S-shaped curve whose two fold points correspond to the points of ignition and extinction of combustion. The Euler–Newton continuation method was used to calculate the $(\hat{u}, \delta)$-solution curve for fixed values of ε, γ. The Moore–Spence system was used to calculate the fold points. Then in turn the Euler–Newton method was used to calculate curves of fold points.

An analogous approach may be used to handle many bifurcation problems. Let us illustrate this in the case of symmetry breaking bifurcation points. As a simple example we take the symmetry group $\mathbb{Z}_2 = \{1, S\}$ where S represents, e.g., a reflection. We assume that equivariance holds:

$$H(Sx, \lambda, \alpha) = SH(x, \lambda, \alpha).$$

The reflection mapping induces a natural decomposition

$$\mathbb{R}^N = X_s \oplus X_a$$

where $X_s := \{x \in \mathbb{R}^N\colon\ Sx = x\}$, $X_a := \{x \in \mathbb{R}^N\colon\ Sx = -x\}$ represent symmetric and anti-symmetric vectors of $\mathbb{R}^N$.

Now X_s is invariant under H and H_x, and so the analysis of the symmetric solutions of (25.1) can be carried out in this restricted context. However, the stability can be affected by the unstable eigenvalues corresponding to an anti-symmetric eigenvector. At a symmetric solution, the eigenvectors are either symmetric or anti-symmetric. A simple singular point having an anti-symmetric eigenvector is called a *symmetry-breaking bifurcation point* provided

$$\psi_0^* H_{x\lambda}(x_0, \lambda_0, \alpha_0)\varphi_0 + H_{xx}(x_0, \lambda_0, \alpha_0)[\varphi_0, v_\lambda] \neq 0$$

where v_λ satisfies $H_x(x_0, \lambda_0, \alpha_0)v_\lambda + H_\lambda(x_0, \lambda_0, \alpha_0) = 0$. At a symmetry-breaking bifurcation point a curve of anti-symmetric solutions bifurcates off from a curve of symmetric solutions.

The extended system of equations for calculating a symmetry-breaking bifurcation point proposed by WERNER and SPENCE [1984] takes the form

$$G(x, \varphi, \lambda, \alpha) := \begin{pmatrix} H(x, \lambda, \alpha) \\ H_x(x, \lambda, \alpha)\varphi \\ l^*\varphi - 1 \end{pmatrix} = 0 \tag{25.3}$$

where

$$G: X_s \times X_a \times \mathbb{R} \times \mathbb{R}^p \to X_s \times X_a \times \mathbb{R}$$

and $l \in \mathbb{R}^N$ is not in the null space of $H_x(x_0, \lambda_0, \alpha_0)$. The main difference between the maps in (25.3) and (25.2) is that now in (25.3), $x \in X_s$ and $\phi \in X_a$.

In WINTERS, CLIFFE and JACKSON [1987] the successful application of the above described techniques to problems modelling Bénard convection, flow in curved tubes, Taylor flow between concentric rotating cylinders, periodic flow in semiconductor crystal melt, and flow past a cylinder are reported. In general, the extended system (25.3) is used to accurately locate a symmetry-breaking bifurcation point (in a few Newton iterations), and then the continuation method is again employed to track a bifurcating solution curve.

Other recent numerical experience involving the application of path following techniques to the Navier–Stokes equations and nonlinear biharmonic equations have been given by GLOWINSKI, KELLER and REINHART [1985] and HUITFELDT and RUHE [1990], see also Sections 17, 12.

26. Critical points

Among the applications of numerical continuation methods is the calculation of critical points of a smooth mapping $f: \mathbb{R}^N \to \mathbb{R}$. In general, one chooses a smooth mapping $g: \mathbb{R}^N \to \mathbb{R}$ with known regular *critical points* $a \in \mathbb{R}^N$, i.e., $\nabla g(a) = 0$ for the gradient and the Hessian $\nabla g'(a)$ has full rank. One then formulates a smooth homotopy map $H: \mathbb{R}^{N+1} \to \mathbb{R}^N$ such that

$$H(0, x) = \nabla g(x) \quad \text{and} \quad H(1, x) = \nabla f(x).$$

Typically, one uses the convex homotopy

$$H(\lambda, x) := (1 - \lambda)\nabla g(x) + \lambda \nabla f(x). \tag{26.1}$$

The numerical aspect then consists of tracing a smooth curve

$$c(s) = (\lambda(s), x(s)) \in H^{-1}(0)$$

with starting point $c(0) = (0, a)$ for some given critical point a of g, and starting tangent $\dot{c}(0) = (\dot{\lambda}(0), \dot{x}(0))$ with $\dot{\lambda}(0) > 0$. The aim of course is to trace the curve c until the homotopy level $\lambda = 1$ is reached, at which a critical point of f is obtained. If all critical points of f are regular, then by Sard's theorem 18.1 it is generally possible to make a choice of g such that zero is a regular value of H. The following result of ALLGOWER and GEORG ([1980], Section 2.17) indicates that the continuation method has an appealing property which can permit targeting critical points having a specific Morse index.

THEOREM 26.1. *Let $f, g : \mathbb{R}^N \to \mathbb{R}$ be smooth functions and let H be the convex homotopy of* (26.1) *which has zero as a regular value. Let $c(s) = (\lambda(s), x(s)) \in H^{-1}(0)$ be the smooth solution curve such that $c(0) = (0, a)$ where a is a regular critical point of g. Suppose that $\lambda(s)$ is increasing for $s \in [0, \bar{s}]$, $\lambda(\bar{s}) = 1$, and that the critical point $b := x(\bar{s})$ of ∇f is regular. Then the critical points a, b of g and f, respectively, have the same Morse index, i.e., the Hessians $\nabla g'(a)$ and $\nabla f'(b)$ have the same number of negative eigenvalues.*

PROOF. By differentiating $H(c(s)) = 0$ we have that the augmented Jacobian, see (1.7), satisfies

$$\begin{pmatrix} \dot{\lambda}(s) & \dot{x}(s)^* \\ H_\lambda(c(s)) & H_x(c(s)) \end{pmatrix} \begin{pmatrix} \dot{\lambda}(s) \\ \dot{x}(s) \end{pmatrix} = \begin{pmatrix} 1 \\ 0 \end{pmatrix}.$$

This implies that

$$\begin{pmatrix} \dot{\lambda}(s) & \dot{x}(s)^* \\ H_\lambda(c(s)) & H_x(c(s)) \end{pmatrix} \begin{pmatrix} \dot{\lambda}(s) & 0^* \\ \dot{x}(s) & \mathrm{Id} \end{pmatrix} = \begin{pmatrix} 1 & \dot{x}(s)^* \\ 0 & H_x(c(s)) \end{pmatrix}.$$

Since the determinant of the augmented Jacobian never changes sign, it follows from the above equation that $\det H_x(c(s))$ changes sign exactly when $\dot{\lambda}(s)$ changes sign. The latter does not occur, and $H_x(c(s))$ can have at most one eigenvalue equal to zero, since all points $c(s)$ are regular points of H. Using a result of perturbation theory, see, e.g., DUNFORD and SCHWARTZ ([1963], p. 922), namely that an isolated eigenvalue of the symmetric matrix $H_x(c(s))$ depends smoothly on s, we conclude that no eigenvalue of $H_x(c(s))$ can change sign for increasing s. Hence $H_x(0, a) = \nabla g'(a)$ and $H_x(1, b) = \nabla f'(b)$ have the same number of negative eigenvalues. □

From the above proof it is clear that the convex homotopy (26.1) can be replaced by any other homotopy linking f and g.

More general studies of the behavior of critical points in parametric optimization are to be found in the books of JONGEN, JONKER and TWILT [1983, 1986]. Also semi-infinite problems can in principle be regarded as parametric optimization problems, see, e.g., JONGEN and ZWIER [1985].

27. Complex bifurcation

It has been observed by ALLGOWER [1984] and ALLGOWER and GEORG [1983a] that folds in the λ co-ordinate of solution curves of $H(x, \lambda) = 0$ lead to bifurcation points in a setting of complex extension. This observation can be used to connect separated real components of $H^{-1}(0)$, and hence may serve as a tool to find additional solutions of the equation $H = 0$. HENDERSON [1985] and HENDERSON and KELLER [1990] study complex bifurcation in a general Banach space setting.

For simplicity, let us discuss some results for the finite-dimensional setting, see also ALLGOWER and GEORG ([1990], Section 11.8).

Consider a smooth nonlinear problem of the form

$$H(\lambda, z) = 0, \qquad H : \mathbb{R} \times \mathbb{C}^N \to \mathbb{C}^N,$$

where H is analytic in the complex variable $z = x + \mathrm{i}y \in \mathbb{C}^N$. We denote the real and imaginary parts of H by H^{r}, H^{i}, respectively. Hence,

$$H^{\mathrm{r}} = \tfrac{1}{2}(H + \bar{H}), \qquad H^{\mathrm{i}} = \frac{1}{2\mathrm{i}}(H - \bar{H}).$$

THEOREM 27.1. *Let $c(s) = (\lambda(s), z(s)) = (\lambda(s), x(s) + \mathrm{i}y(s))$ ($s =$ arclength) be a smooth solution curve of the equation $H = 0$, and let $c(0)$ be a* fold, *i.e., $\dot\lambda(s)$ changes signs at $s = 0$ such that $c(s)$ is a regular point of H for $s \neq 0$ and sufficiently small. Then $c(0)$ is a bifurcation point of the equation $H = 0$.*

PROOF. We use the Cauchy–Riemann equations

$$H^{\mathrm{r}}_y = -H^{\mathrm{i}}_x \quad \text{and} \quad H^{\mathrm{i}}_y = H^{\mathrm{r}}_x .$$

In the following matrix representations we consider the block structure implied by the λ-axis, the real part of z, and the imaginary part of z, respectively.

The augmented Jacobian satisfies the equations

$$\begin{pmatrix} \dot\lambda & \dot x^* & \dot y^* \\ H_\lambda & H_x & H_y \end{pmatrix} \begin{pmatrix} \dot\lambda & 0^* & 0^* \\ \dot x & \mathrm{Id} & 0 \\ \dot y & 0 & \mathrm{Id} \end{pmatrix} = \begin{pmatrix} 1 & \dot x^* & \dot y^* \\ 0 & H^{\mathrm{r}}_x & H^{\mathrm{i}}_x \\ 0 & -H^{\mathrm{i}}_x & H^{\mathrm{r}}_x \end{pmatrix}$$

and

$$\dot\lambda \det \begin{pmatrix} \dot\lambda & \dot x & \dot y \\ H_\lambda & H_x & H_y \end{pmatrix} = \det \begin{pmatrix} H^{\mathrm{r}}_x & H^{\mathrm{i}}_x \\ -H^{\mathrm{i}}_x & H^{\mathrm{r}}_x \end{pmatrix} = \det H_{(x,y)}.$$

A standard argument shows that the last determinant is always nonnegative. Hence, the last equation implies that $\dot\lambda(c(s))$ changes sign at $s = 0$, and consequently the determinant of

$$\begin{pmatrix} \dot\lambda & \dot x^* & \dot y^* \\ H_\lambda & H_x & H_y \end{pmatrix}$$

evaluated at $c(s)$ does the same.

Now, using a standard argument in degree theory, see KRASNOSEL'SKIĬ [1964] or RABINOWITZ [1971], it can be shown that $c(0)$ is a bifurcation point of $H = 0$. However, $c(0)$ is not necessarily a simple bifurcation point. □

Let us now also assume that H is real for real arguments, i.e.,

$$\overline{H(\lambda, z)} = H(\lambda, \bar z),$$

and denote the restriction of H to real arguments by $\tilde H$.

THEOREM 27.2. *Let the curve* $s \longmapsto c(s) = (\lambda(s), x(s), 0)$ *be a "real" solution curve of* $H^{-1}(0)$ *such that the point* $(\lambda(s), x(s))$ *is a regular point of the real homotopy* $\tilde H$ *for all* s (including $s = 0$). *Suppose that* $(\lambda(0), x(0))$ *is a* simple fold *of the equation* $H = 0$, *i.e.,* $\dot\lambda(0) = 0$ *and* $\ddot\lambda(0) \neq 0$. *Then* $c(0)$ *is a simple bifurcation point of the equation* $H = 0$.

PROOF. Since H is real, it is easy to see that

$$H^{\mathfrak i}_x(\lambda, x, 0) = 0, \qquad H^{\mathfrak i}_{xx}(\lambda, x, 0) = 0, \qquad H^{\mathfrak i}_\lambda(\lambda, x, 0) = 0 \tag{27.1}$$

holds for $x \in \mathbb{R}^N$ and $\lambda \in \mathbb{R}$. Hence, using the Cauchy–Riemann equations, the augmented Jacobian satisfies the equation

$$\begin{pmatrix} \dot\lambda & \dot x^* & \dot y^* \\ H_\lambda & H_x & H_y \end{pmatrix} = \begin{pmatrix} 0 & \dot x^* & 0 \\ H^{\mathfrak r}_\lambda & H^{\mathfrak r}_x & 0 \\ 0 & 0 & H^{\mathfrak r}_x \end{pmatrix} \tag{27.2}$$

at $c(0)$.

Since the rank of

$$\begin{pmatrix} 0 & \dot x^* \\ \tilde H_\lambda & \tilde H_x \end{pmatrix}$$

at $(\lambda(0), x(0))$ is $N + 1$, we conclude that the matrix

$$H^{\mathfrak r}_x(c(0)) = \tilde H_x(\lambda(0), x(0))$$

has rank at least $N-1$. On the other hand, differentiating $H(c(s))=0$ at $s=0$ implies that $H_x^{\mathrm{r}}(c(0))\dot{x}(0)=0$, and hence $\operatorname{rank} H_x^{\mathrm{r}}(c(0))=N-1$. Therefore the matrix in (27.2) has rank $2N$, and the Jacobian $H'(c(0))$ has a two-dimensional kernel spanned by the vectors

$$\begin{pmatrix} 0 \\ \dot{x}(0) \\ 0 \end{pmatrix}, \quad \begin{pmatrix} 0 \\ 0 \\ \dot{x}(0) \end{pmatrix}.$$

It remains to show that the nondegeneracy conditions for the second derivatives hold, see (24.1). Let e span the kernel of $H_x^{\mathrm{r}}(c(0))^*$. Then $\binom{0}{e}$ spans the kernel of $H'(c(0))^*$. Furthermore,

$$e^* H_\lambda^{\mathrm{r}}\big(c(0)\big) \neq 0 \tag{27.3}$$

since otherwise the above kernel would not have dimension one.

We now investigate whether the bilinear form

$$(\xi,\eta) \longmapsto (0,e^*)H''\big(c(0)\big)\left[\xi\begin{pmatrix} 0 \\ \dot{x}(0) \\ 0 \end{pmatrix}, \ \eta\begin{pmatrix} 0 \\ 0 \\ \dot{x}(0) \end{pmatrix}\right]$$

has one positive and one negative eigenvalue. Using (27.1) and the Cauchy–Riemann equations, a straightforward calculation shows that the above bilinear form reduces to

$$(\xi,\eta) \longmapsto 2\xi\eta\ e^* H_{xy}^{\mathrm{i}}\big(c(0)\big)\big[\dot{x}(0),\dot{x}(0)\big]. \tag{27.4}$$

It is clear that the simple bilinear form (27.4) has one positive and one negative eigenvalue if and only if

$$e^* H_{xy}^{\mathrm{i}}\big(c(0)\big)\big[\dot{x}(0),\dot{x}(0)\big] \neq 0. \tag{27.5}$$

To show this, let us differentiate the equation $e^*\tilde{H}(\lambda(s),x(s))=0$ twice. Using the facts $e^*\tilde{H}_x(\lambda(0),x(0))=0$, $\tilde{H}_x(\lambda(0),x(0))\dot{x}(0)=0$ and $\dot{\lambda}(0)=0$, we obtain

$$e^*\tilde{H}_\lambda\big(\lambda(0),x(0)\big)\ddot{\lambda}(0) + e^*\tilde{H}_{xx}\big(\lambda(0),x(0)\big)\big[\dot{x}(0),\dot{x}(0)\big]=0.$$

Since $\ddot{\lambda}(0)\neq 0$, we can conclude from (27.3) that $e^*\tilde{H}_{xx}(\lambda(0),x(0))\neq 0$. Now (27.5) follows from the Cauchy–Riemann equations. □

Let us finally show that at bifurcation points at which one of the two solution branches is real, the corresponding curvatures are of opposite sign, and hence at such points a choice of following branches is available so that the λ co-ordinate is increasing.

THEOREM 27.3. *Under the assumptions of the previous theorem, let us now denote the "real" solution curve in* $H^{-1}(0)$ *by* $c_1(s) =: (\lambda_1(s), x_1(s), 0)$ *and the bifurcating solution curve by* $c_2(s) =: (\lambda_2(s), x_2(s), y_2(s))$. *The curves are defined for* s *near* 0, *and* $\bar{u} := c_1(0) = c_2(0)$ *is the bifurcation point. Then* $\ddot{\lambda}_1(0) = -\ddot{\lambda}_2(0)$.

PROOF. Let us denote by $c(s) =: (\lambda(s), x(s), y(s))$ either of the two solution curves c_1 or c_2. Differentiating $H(c(s)) = 0$ twice with respect to s and taking $\dot{\lambda}(0) = 0$ into account yields

$$\begin{aligned} &H_\lambda(\bar{u})\ddot{\lambda}(0) + H_{xx}(\bar{u})\big[\dot{x}(0), \dot{x}(0)\big] + 2H_{xy}(\bar{u})\big[\dot{x}(0), \dot{y}(0)\big] \\ &\quad + H_{yy}(\bar{u})\big[\dot{y}(0), \dot{y}(0)\big] + H_x(\bar{u})\ddot{x}(0) + H_y(\bar{u})\ddot{y}(0) = 0. \end{aligned} \tag{27.6}$$

Let e span the kernel of $H_x^{\mathrm{r}}(c(0))^*$ as in the previous proof. Multiplying (27.6) from the left with $(e^*, 0)$ and taking the Cauchy–Riemann equations and (27.1) into account, we obtain

$$e^* H_\lambda^{\mathrm{r}}(\bar{u})\ddot{\lambda}(0) + e^* H_{xx}^{\mathrm{r}}(\bar{u})\big[\dot{x}(0), \dot{x}(0)\big] + e^* H_{yy}^{\mathrm{r}}(\bar{u})\big[\dot{y}(0), \dot{y}(0)\big] = 0. \tag{27.7}$$

Since

$$\dot{c}_1(0) = \big(0, \dot{x}_1(0), 0\big) \tag{27.8}$$

holds, it can be seen from (27.4) and (24.6) that

$$\dot{c}_2(0) = \big(0, 0, \pm\dot{x}_1(0)\big). \tag{27.9}$$

Substituting (27.8), (27.9) into (27.7) we obtain

$$\begin{aligned} &e^* H_\lambda^{\mathrm{r}}(\bar{u})\ddot{\lambda}_1(0) + e^* H_{xx}^{\mathrm{r}}(\bar{u})\big[\dot{x}_1(0), \dot{x}_1(0)\big] = 0, \\ &e^* H_\lambda^{\mathrm{r}}(\bar{u})\ddot{\lambda}_2(0) + e^* H_{yy}^{\mathrm{r}}(\bar{u})\big[\dot{x}_1(0), \dot{x}_1(0)\big] = 0, \end{aligned}$$

respectively. Since $e^* H_\lambda^{\mathrm{r}}(\bar{u}) \neq 0$, see (27.3), and since by the Cauchy–Riemann equations $H_{xx}^{\mathrm{r}}(\bar{u}) = -H_{yy}^{\mathrm{r}}(\bar{u})$ holds, the assertion follows. □

The last two theorems have been generalized by LI and WANG ([1993b], Proposition 2.1) for complex folds.

28. Linear eigenvalue problems

In recent years many of the classical problems of numerical linear algebra have been re-examined in the context of homotopies and path following. One of the earliest contributors has been CHU [1984a,b, 1986, 1988, 1990, 1991]. In these papers iterative processes and matrix factorizations have been studied in the context of flows satisfying various differential equations. A typical example is the Toda flow which has been studied as a continuous analogue of the QR algorithm. A survey of these ideas has been given by WATKINS [1984]. According to Watkins, although it seems that the Toda flow and related flows yield insight into the workings of algorithms, they do not necessarily directly offer algorithms which are competitive with standard library algorithms that have been developed and refined over numerous years.

Surprisingly, LI and LI [1993], LI and RHEE [1989], LI, ZENG and CONG [1992], LI, ZHANG and SUN [1991] have been able to construct special implementations of homotopy methods which are now at least competitive with the library routines of EISPACK and IMSL for linear eigenvalue problems.

The versatility of homotopy methods permits their application also to generalized eigenvalue problems, see CHU, LI and SAUER [1988] and nonsymmetric matrices, see LI and ZENG [1992], LI, ZENG and CONG [1992]. In this case complex eigenvalues are likely to arise, and it is necessary to invoke the idea of complex bifurcation, see Section 27.

As an example, let us briefly discuss the homotopy approach given by LI, ZHANG and SUN [1991]. Consider a real symmetric tridiagonal matrix A. We assume that A is irreducible, since otherwise one off-diagonal element $A[i+1,i] = A[i,i+1]$ would vanish and the matrix A would split into two blocks which can be treated independently. We consider a homotopy $H : \mathbb{R}^N \times \mathbb{R} \times [0,1] \to \mathbb{R}^N \times \mathbb{R}$ defined by

$$H(x,\lambda,s) = \begin{pmatrix} \lambda x - [(1-s)D + sA]x \\ x^*x - 1 \end{pmatrix}.$$

Here D is a real symmetric reducible tridiagonal matrix which is generated from A by setting some of the off-diagonal entries of A to zero. The simplest example for D would be to set all off-diagonal entries to zero. However, it is advantageous to only reduce D to tridiagonal block structure with relatively small blocks, e.g., of size < 50. This technique is referred to as *divide and conquer*.

Since $A(s) := (1-s)D + sA$ is irreducible for all $s > 0$, the solution set of $H = 0$ consists of $2n$ disjoint smooth curves c (*eigenpaths*) which can be parametrized with respect to s. Note that s is not the arclength, but the homotopy parameter. Hence $c(s) = (\pm x(s), \lambda(s))$ for $0 \leqslant s \leqslant 1$. The curves obviously occur in pairs, and of course only one of each pair needs to be traced. At the level $s = 0$, initial values on the curves can be obtained by approximating all eigenvectors and eigenvalues of all small blocks in D. If D is diagonal, this is trivial, and otherwise a QR routine has to be employed.

Let us sketch a typical step of the predictor corrector method. We note first that it follows from differentiation of $H(c(s)) = 0$ with respect to s that

$$\dot\lambda(s) = x(s)^*(A - D)x(s).$$

Assume that $(x(s), \lambda(s))$ is (approximately) known. After having decided on a stepsize h (we are not going to discuss this feature), a predicted eigenvalue $\tilde\lambda(s+h)$ is obtained from the above differential equation by a two-step ODE method. Now a predicted eigenvector $\tilde{x}(s+h)$ is obtained by one step of the inverse power method with shift, i.e., solve $(A(s+h) - \tilde{\lambda}(s+h)\,\mathrm{Id})y = x(s)$ for y and set $\tilde{x}(s+h) = y/\|y\|$. Then a Rayleigh quotient iteration is performed as a corrector to approximate $(x(s+h), \lambda(s+h))$.

There are some stability problems for the case that different eigenvalues become close. Sturm sequences are computed to stabilize the procedure.

Let us finally note that the above homotopy method has an order preserving property, i.e., different λ-paths can never cross. Hence the jth eigenvalue of A can be calculated without calculating any other eigenvalues. This is very often an advantageous feature for applications. On the other hand, the homotopy method lends itself conveniently to parallelization, since each solution path can be traced independently of the others and hence also simultaneously.

29. Mathematical programming

29.1. Generalized equations

The paper of REINOZA [1985] deals with homotopy methods for generalized equations. The concept of generalized equations has been introduced to describe inclusion relations involving multivalued functions, in particular normal cone operators. The type of *generalized equations* which are considered are of the form:

$$0 \in f(x) + N_C(x), \tag{29.1}$$

where $f : \Omega \to \mathbb{R}^N$ is a C^2 function on an open set $\Omega \subset \mathbb{R}^N$, C is a convex polyhedral set in $\mathbb{R}^N$ and $N_C(x)$ is the *normal cone operator*

$$N_C(x) := \begin{cases} \{v \in \mathbb{R}^N \colon \forall c \in C \ \ (v, c - x) \leqslant 0\}, \\ \varnothing \quad \text{otherwise.} \end{cases}$$

Many problems arising in nonlinear programming can be formulated as generalized equations. To illustrate this, note that if (29.1) holds, then the sum is nonempty since it contains 0, and so $x \in C$. Also then $-f(x)$ must belong to $N_C(x)$ and so for each $c \in C$,

$$\big(-f(x), c - x\big) \leqslant 0.$$

Thus we see that (29.1) is equivalent to the *variational inequality*

$$x \in C, \quad \text{and} \quad \forall c \in C \ \ \big(f(x), c - x\big) \geqslant 0.$$

It is sometimes useful to formulate problems in mathematical programming in terms of generalized equations to facilitate their analysis. The survey paper by ROBINSON [1983] gives an extensive review of the theory and applications of generalized equations.

A generalized homotopy equation is a relation of the form

$$0 \in \Phi(x,\lambda) := H(x,\lambda) + N_C(x) \tag{29.2}$$

where $H : \Omega \to \mathbb{R}^N$ is a C^2 function on an open set $\Omega \subset \mathbb{R}^{N+1}$, C is a convex polyhedral set in $\mathbb{R}^N$ and $N_C(x)$ is the normal cone operator.

It turns out that under appropriate modifications and extensions of the ideas in Section 1 an analogue of Theorem 1.3 can be obtained. Let us give a brief synopsis of Reinoza's results. Let $K \subset \mathbb{R}^{N+1}$ be a cylinder defined by $K := (C \cap P) \times [-l, l]$ where $P \subset \mathbb{R}^N$ is a compact convex set with smooth boundary ∂P and $l \in \mathbb{R}$ is a positive number. A curve $\Gamma \subset K$ is called a *path* if it starts and ends on the boundary of K. The curve Γ is called a *closed loop* if it does not intersect the boundary of K. Under a somewhat technical definition for $0 \in \mathbb{R}^N$ to be a regular value of Φ, the following theorem is shown:

THEOREM 29.1. *Let* 0 *be a regular value of* Φ. *Then* $\Phi^{-1}(0) \cap K$ *is a finite union of disjoint paths and closed loops in* K.

In this more general context an analogue of Theorem 20.1 is also shown. An analogue of Newton's method for generalized equations can be formulated by considering the analogue of the linearization of (29.1):

$$0 \in f(x_0) + f'(x_0)(x - x_0) + N_C(x). \tag{29.3}$$

For such a Newton's method JOSEPHY [1979] gave a version of the Newton–Kantorovich theory, see also ROBINSON [1983]. REINOZA [1985] shows for the generalized global homotopy

$$0 \in \Phi(x,\lambda) := f(x) + \lambda f(x_0) + N_C(x) \tag{29.4}$$

that a sequence of points generated by the generalized homotopy method coincides with the sequence of points generated by Josephy's method. SELLAMI and ROBINSON [1993] recently presented an implementation of a continuation method for normal maps.

The *differentiable* techniques for generalized equations discussed above represent a bridge linking the predictor corrector methods to the piecewise linear methods which we will discuss in Chapter VI. It will be seen that piecewise linear methods also apply to certain multi-valued maps.

29.2. Parametric programming problems

Parametric programming problems and sensitivity analysis also can be studied in the context of continuation methods. Consider the problem

$$\min\{f(x,\alpha):\ c_i(x,\alpha) = 0,\ i \in E,\ c_i(x,\alpha) \leqslant 0,\ i \in I\}, \tag{29.5}$$

where $f, c_i : \mathbb{R}^{N+1} \to \mathbb{R}$ are smooth functions. Here

$$E = \{1, \dots, q\} \quad \text{and} \quad I = \{q+1, \dots, q+p\}$$

denote the index sets for the equality and inequality constraints, respectively. The local sensitivity of such systems has been analyzed, e.g., in FIACCO [1983, 1984] and ROBINSON [1987]. Many authors have used bifurcation and singularity theory to investigate the local behavior and persistence of minima at the singular points of the above system, see, e.g., BANK, GUDDAT, KLATTE, KUMMER and TAMMER [1983], GFRERER, GUDDAT and WACKER [1983], GFRERER, GUDDAT, WACKER and ZULEHNER [1985], GUDDAT, GUERRA VASQUEZ and JONGEN [1990], GUDDAT, JONGEN, KUMMER and NOŽIČKA [1987], JONGEN, JONKER and TWILT [1983, 1986], JONGEN and WEBER [1990], KOJIMA and HIRABAYASHI [1984], POORE and TIAHRT [1987, 1990]. RAKOWSKA, HAFTKA and WATSON [1991] discuss algorithms for tracking paths of optimal solutions. LUNDBERG and POORE [1993] report on a numerical implementation of a path following method for the above problem. Our discussion is motivated by their exposition.

The Fritz John first order necessary conditions for (29.5) imply the existence of $(\lambda, \nu) \in \mathbb{R}^{p+q} \times \mathbb{R}$ such that

$$\mathcal{L}_x(x, \lambda, \nu, \alpha) = 0, \tag{29.6}$$

$$c_i(x, \alpha) = 0, \quad i \in E, \tag{29.7}$$

$$\lambda_i c_i(x, \alpha) = 0, \quad i \in I, \tag{29.8}$$

$$\nu \geqslant 0, \qquad c_i(x, \alpha) \leqslant 0, \quad \lambda_i \geqslant 0, \qquad i \in I, \tag{29.9}$$

where $\mathcal{L}(x, \lambda, \nu, \alpha) = \nu f(x, \alpha) + \sum \lambda_i c_i(x, \alpha)$ is the Lagrangian.

Now an active set strategy is implemented by using the following homotopy equation for a path following algorithm:

$$H\left(x, \{\lambda_i\}_{i \in \mathcal{A}}, \nu, \alpha\right) = \begin{pmatrix} \mathcal{L}_x(x, \{\lambda_i\}_{i \in \mathcal{A}}, \nu, \alpha) \\ c_i(x, \alpha), \ i \in \mathcal{A} \\ \nu^2 + \sum \lambda_i^2 - 1 \end{pmatrix} = 0, \tag{29.10}$$

where $\mathcal{A}$ is the set of active constraints. Hence $\mathcal{A}$ includes all of the indices E and some of the indices I. During the path following procedure, this active set is adapted in such a way that the inequalities (29.9) are respected.

There are various technical difficulties (such as handling singularities or efficiently adapting the active set) which have to be overcome in a successful implementation.

Continuous deformations of semi-infinite optimization problems have been studied by HETTICH and JONGEN [1994].

29.3. Interior point methods

KHACHIYAN [1979] started a new class of polynomial time algorithms for solving the linear programming problem. KARMARKAR [1984] subsequently gave a much noted polynomial time algorithm based upon projective rescaling. GILL, MURRAY, SAUNDERS, TOMLIN and WRIGHT [1986] noted that Karmarkar's algorithm is equivalent to a projected Newton barrier method which in turn is closely related to a recent class of polynomial time methods involving a continuation method, namely the tracing of the "central path". This last technique can be extended to quadratic programming problems, and both linear and nonlinear complementarity problems. Typically, algorithms of this nature are now referred to as *interior point* methods.

The presentation of a continuous trajectory (central path) of the iterative Karmarkar method was extensively studied by BAYER and LAGARIAS [1989], see also SONNEVEND [1985]. MEGIDDO [1988] related this path to the classical barrier path of nonlinear optimization, FIACCO and MCCORMICK [1968]. Several authors have proposed algorithms that generally follow the central path to a solution, see, e.g., GONZAGA [1988], KOJIMA, MIZUNO and YOSHISE [1988, 1989], MONTEIRO and ADLER [1989], NAZARETH [1986, 1991, 1994a,b], RENEGAR [1988a], VAIDYA [1990].

To make the algorithms more efficient, variable steplength and/or higher order predictor algorithms have been proposed in ADLER, RESENDE, VEIGA and KARMARKAR [1989], MIZUNO, TODD and YE [1992], SONNEVEND, STOER and ZHAO [1989, 1991]. The algorithm of MIZUNO, TODD and YE [1992] has subsequently been shown by YE, GÜLER, TAPIA and ZHANG [1991] to have both polynomial time complexity and quadratic convergence. KOJIMA, MEGIDDO and MIZUNO [1991a] believe that there still remain differences between the theoretical primal-dual algorithms which enjoy global and/or polynomial-time convergence and the efficient implementations of primal-dual algorithms, e.g., MARSTEN, SUBRAMANIAN, SALTZMAN, LUSTIG and SHANNO [1990] and MCSHANE, MONMA and SHANNO [1989].

ADLER, RESENDE, VEIGA and KARMARKAR [1989] report extensive computational experiments for an interior point implementation with solution times being in most cases less than those required by a state-of-the-art simplex method MINOS, see MURTAGH and SAUNDERS [1987]. KARMARKAR and RAMAKRISHNAN [1991] report computational experience on large scale problems which are representative of large classes of applications of current interest. Their interior point implementation incorporates a preconditioned conjugate gradient method as a corrector step and is consistently faster than MINOS by orders of magnitude. Further computational experience comparing an interior point method `OB1` and a simplex method CPLEX is reported in the technical reports by BIXBY, GREGORY, LUSTIG, MARSTEN and SHANNO [1991], CARPENTER and SHANNO [1991] and LUSTIG, MARSTEN and SHANNO [1991]. POLAK, HIGGINS and MAYNE [1992] have given an algorithm for solving semi-infinite minimax problems which bears a resemblance to the interior penalty function methods. They report numerical results which show that the algorithm is extremely robust and its performance is at least comparable to that of current first-order minimax algorithms.

There is currently immense activity in studying and developing implementations of interior point algorithms. It is to be expected that our brief account will be outdated in a few years. For further details and literature, we refer to the recent surveys of GONZAGA [1992], KOJIMA, MEGIDDO, NOMA and YOSHISE [1991c], TODD [1989], WRIGHT [1992], and the proceedings edited by ROOS and VIAL [1991]. As an example, we outline the central path approach for a primal-dual linear programming problem, following the introductory parts of MONTEIRO and ADLER [1989] and MIZUNO, TODD and YE [1992].

Consider the following linear programming problem and its corresponding dual form:

PROBLEM 29.1.

$$\min_x\{c^*x\colon\ Ax = b,\ x \geqslant 0\},\tag{29.11}$$

$$\max_y\{b^*y\colon\ A^*y + z = c,\ z \geqslant 0\}.\tag{29.12}$$

We make the following standard.

ASSUMPTION 29.1. The rank of A equals the number of its rows, and the interior feasible set of the primal–dual problem

$$\mathcal{F}^o := \bigl\{(x,z)\colon\ x,z > 0,\ Ax = b,\ A^*y + z = c \text{ for some } y\bigr\}$$

is not empty.

It is well known that the linear programming problem has a unique solution under the above assumptions. The logarithmic barrier function method associated with Problem 29.1 is

$$\min_x\Bigl\{c^*x - \mu\sum_j \ln x_j\colon\ Ax = b,\ x > 0\Bigr\},\tag{29.13}$$

where $\mu > 0$ is the barrier penalty parameter. Under Assumption 29.1, the logarithmic barrier function is strictly convex and has a unique minimal point $x(\mu)$ for all $\mu > 0$. Moreover, $x(\mu)$ tends to the unique solution of Problem 29.1 as μ tends to zero.

The Karush–Kuhn–Tucker optimality condition which characterizes the solution $x(\mu)$ can be expressed in the following way: $(x(\mu), z)$ must belong to the set

$$\mathcal{C} := \bigl\{(x,z) \in \mathcal{F}^o\colon\ \operatorname{diag}(x)z = \mu e\bigr\},\tag{29.14}$$

where e denotes the column of ones. In fact, $\mathcal{C}$ is parametrized by μ and is commonly called the *central path* of the problem. It turns out that μ is related to the so-called *duality gap*: $c^*x - b^*y = x^*z$ via

$$\mu = \frac{x^*z}{N}\tag{29.15}$$

for $(x,z) \in \mathcal{C}$, where N is the number of columns of A.

By the above remarks, it is clear that the objective now is to follow the central path $\mathcal{C}$ as μ tends to zero. In fact, most interior point methods can be viewed, one way or another, as a special path following method along these lines. The methods differ in the choice of predictor step, corrector procedure (usually one or several Newton type iterations) and predictor steplength control. Many papers discussing such methods or introducing new methods also contain a sophisticated complexity analysis, see, e.g., Chapter VII.

The above interior point algorithms typically require a phase I in which a feasible starting point is generated. A somewhat different approach is taken by FREUND [1991] who introduces a shifted barrier function approach so that the need for phase I is obviated.

Finally, the above technique is quite general and can be extended to quadratic programming problems and linear and nonlinear complementarity problems, see, e.g., KOJIMA, MEGIDDO and YE [1992]. The literature on interior methods is rapidly increasing, and the subject has become one of the major topics of mathematical programming. In our opinion, it is only a question of time until the venerable simplex methods will be superceded by interior point implementations.

30. Numerical integration over curves

Although the predictor corrector curve tracing method has found a wide variety of significant applications, the calculation of line integals over implicitly defined curves seems to have been overlooked. In view of the interesting possiblities of applications involving Stoke's theorem, complex integration and boundary integrals, this omission seems somewhat surprising. The recent dissertation of SIYYAM [1994] investigates the efficient blending of predictor corrector methods and numerical quadrature methods. In this section we outline some of the ideas for this task and discuss some of the results of the Siyyam dissertation.

One of the interesting cases for the line integral is the integration over a closed curve. In order to handle this case for an implicitly defined curve, it is necessary to develop a reliable numerical method for determining when the curve has been completely traversed. This can be achieved by means of an adaptation of the methods for computing special points on an implicitly defined curve which were discussed in Section 8.

Another aspect which has to be addressed is how to deal with the lack of an explicit parametrization of the curve in the numerical quadrature for the line integral. We will discuss two basic approaches for handling this aspect. One involves essentially constructing a local high order approximation via, e.g., the ideas given by MACKENS [1989], see also Section 11. The other approach involves adapting a modified trapezoidal rule to the numerical integration over curves. Such a modified trapezoidal rule was developed by GEORG [1991] to approximate surface integrals for implicitly defined surfaces, see also Section 41, but it can obviously be adapted to the easier task of handling curve integrals.

Let us first deal briefly with the task of giving a numerical method for determining when an implicitly defined closed curve has been completely traversed. We suppose that $H:\mathbb{R}^{N+1}\to\mathbb{R}^N$ is a smooth map and $\mathcal{C}\subset H^{-1}(0)$ is a smooth closed curve with a given starting point $u\in\mathcal{C}$. We suppose further that t is the tangent to $\mathcal{C}$ at the starting point u in the direction in which the traversing of $\mathcal{C}$ begins. The points z on the hyperplane which is normal to C at u satisfy the equation

$$f(z) := t^*(z - u) = 0. \tag{30.1}$$

Thus the functional f defined in Eq. (30.1) changes sign each time the curve C passes through the hyperplane $f(z) = 0$ during the traversing process. Each time f changes sign for two successive points u_n, u_{n+1}, the curve tracing algorithm is switched over to a Newton steplength control, see (8.2), using the functional $f(z)$ in the context of the discussion in Section 8.1. In general, the Newton steplength control algorithm will converge within very few iterations to a point w which satisfies both $H(w) = 0$ and $f(w) = 0$. When in addition, $\|w - u\|$ is within a chosen tolerance, then u_{n+1} is replaced by the starting point u and the traversing is stopped. Otherwise the curve tracing is resumed at u_{n+1} in the same direction as before.

30.1. The modified trapezoidal rule for curves

We assume that F is a continuous vector field in $\mathbb{R}^{N+1}$ which is defined on a neighborhood of C. The line integral which we want to numerically approximate is:

$$\int_C F \bullet \mathrm{d}C, \tag{30.2}$$

where a particular orientation of C will hereafter be assumed. If u_i and u_{i+1} are two consecutive points on C which have been generated by a numerical traversing, and C_i is the section of C between u_i and u_{i+1}, the *modified trapezoidal rule* for curves is given by:

$$\int_{C_i} F \bullet \mathrm{d}C \approx \tfrac{1}{2}\big[F(u_i) + F(u_{i+1})\big] \bullet [u_{i+1} - u_i]. \tag{30.3}$$

For a sequence of points $\{u_i\}_{i=1}^n$ obtained via a traversing of C, the composite version of the modified trapezoidal rule takes the form:

$$\int_C F \bullet \mathrm{d}C \approx \tfrac{1}{2}\sum_{i=1}^{n-1} \big[F(u_i) + F(u_{i+1})\big] \bullet [u_{i+1} - u_i], \tag{30.4}$$

where u_1 and u_n are the end points of C, and both are identified with the starting point if C is a closed curve.

Let us point out the distinction between the modified trapezoidal rule and the standard trapezoidal rule for integrating over a curve C which has an explicit parametrization, say $C = \{c(s)\colon\ 0 \leqslant s \leqslant 1\}$. Then the line integral (30.2) can be written as an ordinary integral, i.e.,

$$\int_C F \bullet \mathrm{d}C = \int_0^1 F\big(c(s)\big) \bullet c'(s)\,\mathrm{d}s. \tag{30.5}$$

For the integral on the right hand side of (30.5) the standard composite trapezoidal rule can be applied over any partitioning of the interval $[0,1]$. For this case, if $c'(s)$ is smooth on $[0,1]$ and if the interval is uniformly partitioned with a mesh length, say $1/n$, it is well known that the truncation error has an expansion in even powers of $1/n$, and hence extrapolation methods (in the sense of Romberg integration) can be applied.

The advantage of the modified trapezoidal rule (30.3) is that it obviates the need for derivatives, which are generally only available at an additional computational expense. However, it is now no longer clear that the use of extrapolation methods are justified for the modified trapezoidal rule.

Let us begin by formulating what needs to be shown. We suppose again that u and w are two consecutive points on $\mathcal{C}$ which have been obtained in the numerical curve tracing process. More precisely, we suppose that $v = u + ht$ is the predictor point for which the corrector process yields the limit point w. Similarly, the corrector process applied to the points $v_j := u + t(jh/n)$ for any positive integer n and $j = 0, \dots, n$ yields corresponding limit points w_j, $j = 1, \dots, n$, on $\mathcal{C}$ to which the composite version of the modified trapezoidal rule (30.4) can then be applied for integrating over the section $\mathcal{C}_i$ of the curve $\mathcal{C}$ which lies between the points u and w. Now the truncation formula which needs to be shown is essentially

$$\int_{\mathcal{C}_i} F \bullet \mathrm{d}\mathcal{C} - \tfrac{1}{2}\sum_{j=1}^{n-1}\left[F(w_j) + F(w_{j+1})\right] \bullet [w_{j+1} - w_j] \sim \sum_{k=1}^{\infty} c_k \left(\frac{1}{n}\right)^{2k} \tag{30.6}$$

for some constants c_k which are independent of n. In fact, it turns out that the c_k are polynomials in h. This was proven by SIYYAM [1994], whose proof is a modification for the one-dimensional case of a result of VERLINDEN and COOLS [1994].

Siyyam has made a Matlab implementation of a Romberg-like method for the modified trapezoidal rule and performed successful tests on a number of examples. As one would expect, it is not crucial to obtain an intial approximation with many points on the implicitly defined curve over which the integration is to be carried out, since successive refinements are made anyway in performing the Romberg procedure.

30.2. Taylor approximation methods

Other approaches to the task of numerically integrating over an implicitly defined curve generally involve obtaining a high order local approximation to the curve and then performing a highly accurate integration method such as Gaussian quadrature. We mention two such approximation methods. One is to consider the Taylor expansion of $c(h)$ about $h = 0$, where $c(h)$ is defined as the point on the curve such that $c(h) - (u + ht)$ is orthogonal to t. In Section 11 we describe how the Taylor coefficients in

$$c(h) \sim \sum_{i=0}^{k} \frac{1}{i!} c^{(i)}(0) h^i \tag{30.7}$$

can be successively approximated using finite difference formulae for higher derivatives. Finally, the Taylor approximation (30.7) is inserted into the line integral (30.5) and this is in turn approximated via Gaussian quadrature.

This method has been implemented and studied by Siyyam. He also implemented a steplength and order control in the spirit of Section 9. The general conclusions reached in his study were as follows:

(1) Both of the above mentioned implementations succeeded in yielding highly accurate approximations to line integrals over smooth implicitly defined curves.

(2) Romberg extrapolation in connection with the modified trapezoidal rule generally required the least amount of computational effort to obtain a highly accurate numerical approximation of the line integral.

(3) The method employing the variable order predictor and variable steplength control in connection with the Taylor polynomial approximation generally performed well for a moderately accurate numerical approximation of the line integral.

A second possible approach is to use interpolation predictors, see Section 10, instead of Taylor polynomials. We do not have any numerical experience with this approach.

CHAPTER VI

Piecewise Linear Methods

Up to now we have assumed that the map $H:\mathbb{R}^{N+1} \to \mathbb{R}^N$ was smooth. Next we will discuss piecewise linear methods which can again be viewed as curve tracing methods, but which can be applied to nonsmooth situations. The piecewise linear methods trace a polygonal path which is obtained by successively stepping through certain "transversal" cells of a piecewise linear manifold. The first and most prominent example of a piecewise linear algorithm was designed by LEMKE and HOWSON [1964] and LEMKE [1965] to calculate a solution of the linear complementarity problem, see Section 38. This algorithm played a crucial role in the development of subsequent piecewise linear algorithms. SCARF [1967] gave a numerically implementable proof of the Brouwer fixed point theorem, based upon Lemke's algorithm. EAVES [1972] observed that a related class of algorithms can be obtained by considering piecewise linear approximations of homotopy maps. Thus the piecewise linear continuation methods began to emerge as a parallel to the classical embedding or predictor corrector methods.

The piecewise linear methods require no smoothness of the underlying equations and hence have, at least in theory, a more general range of applicability than classical embedding methods. In fact, they can be used to calculate fixed points of set-valued maps. They are more combinatorial in nature and are closely related to the topological degree, see PEITGEN and SIEGBERG [1981]. Piecewise linear continuation methods are usually considered to be less efficient than the predictor corrector methods when the latter are applicable, especially in higher dimensions. The reasons for this lie in the fact that steplength adaptation and exploitation of special structure are more difficult to implement for piecewise linear methods.

EAVES [1976] has given a very elegant geometric approach to general piecewise-linear methods, see also EAVES and SCARF [1976]. We adopt this point of view and cast the notion of piecewise linear algorithms into the general setting of subdivided manifolds which we will call *piecewise linear manifolds*. Our exposition follows the introduction of GEORG [1990] to some extent.

31. Basic facts

Let $\mathbf{E}$ denote some ambient finite dimensional Euclidean space which contains all points arising in the sequel. A *half-space* η and the corresponding *hyperplane* $\partial\eta$ are defined by $\eta = \{y \in \mathbf{E}\colon\ x^*y \leqslant \alpha\}$ and $\partial\eta = \{y \in \mathbf{E}\colon\ x^*y = \alpha\}$, respectively, for

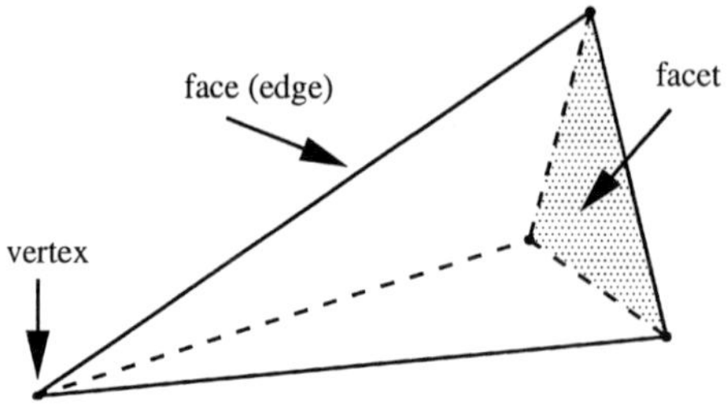

FIG. 31.1. A 3-simplex, a special 3-cell.

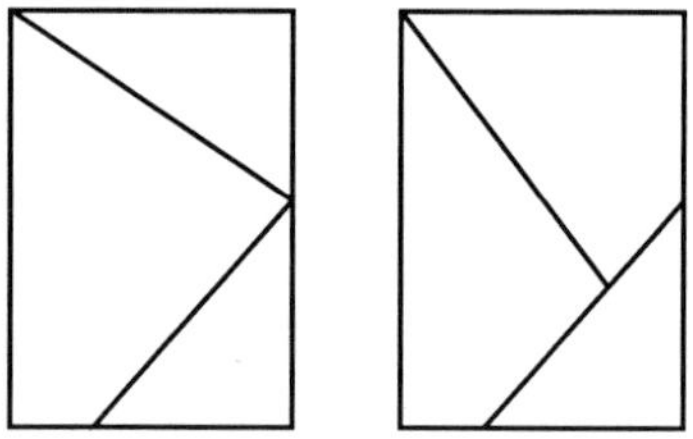

FIG. 31.2. Condition 1 is not satisfied.

some $x \in \mathbf{E}$ with $x \neq 0$ and some $\alpha \in \mathbb{R}$. A finite intersection of half-spaces is called a *cell*. If σ is a cell and ξ a half-space such that $\sigma \subset \xi$ and $\tau := \sigma \cap \partial\xi \neq \varnothing$, then the cell τ is called a *face* of σ. For reasons of notation we consider σ also to be a face of itself, and all other faces are *proper* faces of σ. The *dimension* of a cell is the dimension of its affine hull. In particular, the dimension of a singleton is 0 and the dimension of the empty set is -1. If the singleton $\{v\}$ is a face of σ, then v is called a *vertex* of σ. If τ is a face of σ such that $\dim \tau = \dim \sigma - 1$, then τ is called a *facet* of σ, see Fig. 31.1. The *interior* of a cell σ consists of all points of σ which do not belong to a proper face of σ.

A *piecewise linear manifold* of dimension N is a system $\mathcal{M} \neq \varnothing$ of cells of dimension N such that the following conditions hold, see Fig. 31.2:

(1) If $\sigma_1, \sigma_2 \in \mathcal{M}$, then $\sigma_1 \cap \sigma_2$ is a common face of σ_1 and σ_2.

(2) A cell τ of dimension $N-1$ can be a facet of at most two cells in $\mathcal{M}$.

(3) The family $\mathcal{M}$ is *locally finite*, i.e., any relatively compact subset of

$$|\mathcal{M}| := \bigcup_{\sigma \in \mathcal{M}} \sigma \tag{31.1}$$

meets only finitely many cells $\sigma \in \mathcal{M}$.

The simplest example of a piecewise linear manifold is $\mathbb{R}^N$ subdivided into unit cubes with integer vertices.

We introduce the *boundary* $\partial\mathcal{M}$ of $\mathcal{M}$ as the system of facets which are common to exactly one cell of $\mathcal{M}$. Generally, we cannot expect $\partial\mathcal{M}$ to again be a piecewise linear manifold. However, this is true for the case that $|\mathcal{M}|$ is convex. Two cells which

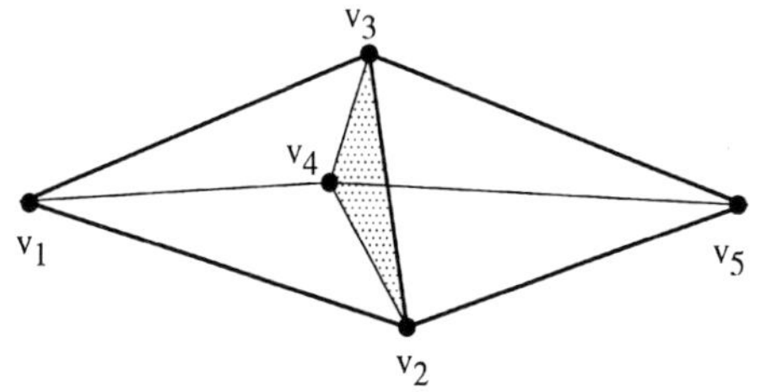

FIG. 31.3. The simplex $[v_1, v_2, v_3, v_4]$ is pivoted into the simplex $[v_5, v_2, v_3, v_4]$ across the facet $[v_2, v_3, v_4]$, and the vertex v_1 is pivoted into the vertex v_5.

have a common facet τ are called *adjacent*. We say that one cell is *pivoted* into the other cell across the facet τ. We will see that piecewise linear algorithms perform pivoting steps.

Typical for piecewise linear path following is that only one current cell is stored in the computer, along with some additional data, and the pivoting step is performed by calling a subroutine which makes use of the data to determine an adjacent cell which then becomes the new current cell.

A cell of particular interest is a *simplex* $\sigma = [v_1, v_2, \dots, v_{N+1}]$ of dimension N which is defined as the convex hull of $N + 1$ affinely independent points $v_1, v_2, \dots, v_{N+1} \in \mathbf{E}$. These points are the vertices of σ. If a piecewise linear manifold $\mathcal{M}$ of dimension N consists only of simplices, then $\mathcal{M}$ is called a *pseudo manifold* of dimension N. Such manifolds are of special importance, see, e.g., GOULD and TOLLE [1983], TODD [1976a]. If a pseudo manifold $\mathcal{T}$ subdivides a set $|\mathcal{T}|$, then we also say that $\mathcal{T}$ *triangulates* $|\mathcal{T}|$. We will use the notions pseudo manifold and triangulation somewhat synonymously. Some triangulations of $\mathbb{R}^N$ of practical importance were already considered by COXETER [1934] and FREUDENTHAL [1942], see also TODD [1976a]. EAVES [1984] gave an overview of standard triangulations.

If σ is a simplex in a pseudo manifold $\mathcal{T}$ and τ a facet of σ which is not in the boundary of $\mathcal{T}$, then there is exactly one simplex $\tilde{\sigma}$ in $\mathcal{T}$ which is different from σ but has the same facet τ, and there is exactly one vertex v of σ which is not a vertex of $\tilde{\sigma}$. We call v the vertex of σ opposite τ. There is also exactly one vertex $\tilde{v}$ of $\tilde{\sigma}$ opposite τ. We say that σ is pivoted across τ into $\tilde{\sigma}$, and that the vertex v of σ is pivoted into $\tilde{v}$. See Fig. 31.3.

A simple triangulation can be generated by the following pivoting rule (pivoting by reflection), see ALLGOWER and GEORG [1979] or COXETER [1963]: if

$$\sigma = [v_1, v_2, \dots, v_i, \dots, v_{N+1}]$$

is a simplex in $\mathbb{R}^N$, and τ is the facet opposite a vertex v_i, then σ is pivoted across τ into $\tilde{\sigma} = [v_1, v_2, \dots, \tilde{v}_i, \dots, v_{N+1}]$ by setting

$$\tilde{v}_i = \begin{cases} v_{i+1} + v_{i-1} - v_i & \text{for } 1 < i < N+1, \\ v_2 + v_{N+1} - v_1 & \text{for } i = 1, \\ v_N + v_1 - v_{N+1} & \text{for } i = N+1. \end{cases} \tag{31.2}$$

In fact, a minimal (nonempty) system of N-simplices in $\mathbb{R}^N$ which is closed under the above pivoting rule is a triangulation of $\mathbb{R}^N$.

32. Special triangulations

In this section we present the details for pivoting in two important examples of triangulations:

(1) The Coxeter–Freudenthal–Kuhn (CFK) triangulation K_1, see COXETER [1934], FREUDENTHAL [1942], KUHN [1968],

(2) The Union Jack triangulation J_1 of TODD [1976a].

We will describe these triangulations and formulate their pivoting rules via pseudo codes in the general context of the space $\mathbb{R}^q$. For an extensive discussion of the triangulations which are used in the context of piecewise linear homotopy methods, we refer the reader to the monograph of EAVES [1984]. For the purpose of our discussion, we will ignore the mesh size and present the triangulations via integer nodes $v \in \mathbb{Z}^q$.

A simplex $\sigma \in K_1$ (the CFK triangulation) of $\mathbb{R}^q$ can be expressed as the convex hull of affinely independent vertices v_i, $i = 1, \dots, q+1$, which satisfy a recursion formula

$$v_{i+1} = v_i + e_{\pi(i)}, \quad i = 1, \dots, q, \tag{32.1}$$

for a permutation $\pi : \{1, \dots, q\} \to \{1, \dots, q\}$. The simplex σ can be compactly stored via the pair of integer tuples: $(v_1; \pi)$. We denote the q-simplex σ with the vertices v_i, $i = 1, \dots, q+1$, by $\sigma = [v_1, \dots, v_{q+1}]$. It is easy to see that the set of all q-simplices σ of this form is a pseudo manifold which triangulates $\mathbb{R}^q$, and that the simplices are all geometrically isometric. See Fig. 32.1.

The pivoting steps in K_1 can be equivalently described in terms of either reflections, or in terms of the permutation representation (32.1).

The pseudo codes below are formulated under the assumption that at each step a decision has been made for determining which vertex is to be pivoted next, e.g., the linear programming steps of Section 34 may furnish such a decision. Our first

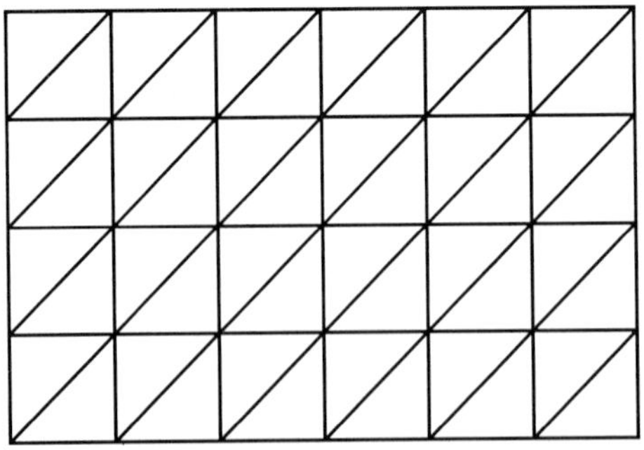

FIG. 32.1. The CFK triangulation in $\mathbb{R}^2$.

pseudo code is based on pivoting by reflection. We denote the cyclic permutation $(1, 2, \ldots, q+1)$ by ρ.

ALGORITHM 32.1 (*Pivoting in K_1 by reflection*).

> input $[v_1, v_2, \ldots, v_{q+1}] \subset \mathbb{R}^q$ % *starting simplex*
> repeat
>   enter $i \in \{1, 2, \ldots, q+1\}$ % *index of vertex to be pivoted next*
> $v_i \leftarrow v_{\rho^{-1}(i)} - v_i + v_{\rho(i)}$ % *reflection rule*
> until pivoting is stopped

The pivoting rules in K_1 can also be equivalently performed by interchange permutations.

ALGORITHM 32.2 (*Pivoting in K_1 by interchange permutations*).

> input $[v_1, v_2, \ldots, v_{q+1}] \subset \mathbb{R}^q$ % *starting simplex*
> $$u_j \leftarrow \begin{cases} v_{j+1} - v_j & \text{for } j = 1, \ldots, q \\ v_1 - v_{q+1} & \text{for } j = q+1 \end{cases} \quad \% \textit{ standard axes}$$
> for $j = 1, \ldots, q+1$ do $\pi(j) \leftarrow j$ % *initial permutation*
> repeat
>   enter $i \in \{1, 2, \ldots, q+1\}$ % *index of vertex to be pivoted next*
> $v_i \leftarrow v_{\rho^{-1}(i)} + u_{\pi(i)}$ % *pivoting rule*
>   interchange $\pi(\rho^{-1}(i))$ and $\pi(i)$
> until pivoting is stopped

The above pseudo codes illustrate that pivoting in K_1 can be performed in integer arithmetic, and in the case of a piecewise linear homotopy algorithm, at any given stage, only one simplex has to be stored. On the other hand, in the case of a piecewise linear algorithm for approximating a surface or manifold of dimension greater than 1, the generated simplices can be cheaply stored in terms of integer vectors (see, e.g., Section 40.2).

In our next example, we give similar descriptions of the triangulation J_1. One of the advantageous features of J_1 over K_1 is that it carries less directional bias and offers more symmetry. This is useful in the piecewise linear approximation of surfaces which have symmetry. The nodes of J_1 are again given by the points $v \in \mathbb{Z}^q$, i.e., having integer co-ordinates. A q-simplex $\sigma \subset \mathbb{R}^q$ belongs to the triangulation J_1 if the following rules are obeyed:

(1) the vertices of σ are nodes of J_1 in the above sense;

(2) the vertices of σ can be ordered in such a way that they are given by the following recursion formula

$$v_{j+1} = v_j + s(j)\, e_{\pi(j)}, \quad j = 1, \ldots, q, \tag{32.2}$$

where $e_1, \ldots, e_q$ is the standard unit basis of $\mathbb{R}^q$, $\pi : \{1, 2, \ldots, q\} \to \{1, 2, \ldots, q\}$ is a permutation and $s : \{1, 2, \ldots, q\} \to \{+1, -1\}$ is a sign function;

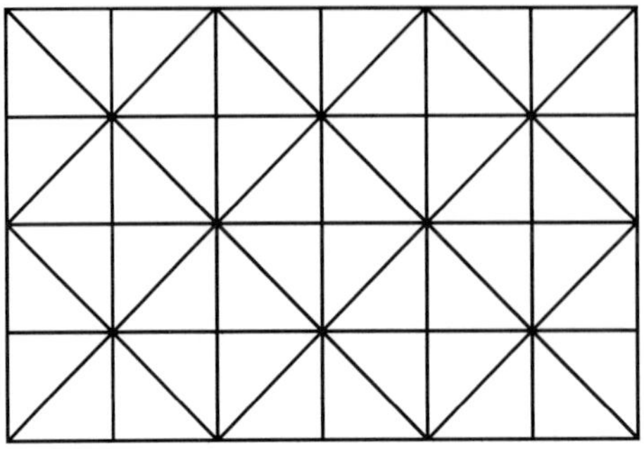

FIG. 32.2. The Union Jack triangulation in $\mathbb{R}^2$.

(3) the *central vertex* v_1 has odd integer co-ordinates.

From Fig. 32.2 it is apparent why TODD [1976a] names this triangulation the *Union Jack triangulation*.

From the description of the pivoting rules below it is evident that these conditions define a pseudo manifold $\mathcal{T}$ which triangulates $\mathbb{R}^q$. The formal proof is however somewhat technical, and we refer the reader to TODD [1976a]. Analogously to Algorithms 32.1 and 32.2, let us now describe the pivoting rules in J_1:

ALGORITHM 32.3 (*Pivoting in J_1 by reflections*).

input $[v_1, v_2, \ldots, v_{q+1}] \subset \mathbb{R}^q$ % *starting simplex*

repeat

enter $i \in \{1, 2, \ldots, q+1\}$ % *index of vertex to be pivoted next*

$$v_i \leftarrow \begin{cases} 2v_2 - v_1 & \text{for } i = 1 \\ 2v_q - v_{q+1} & \text{for } i = q+1 \\ v_{i-1} - v_i + v_{i+1} & \text{else} \end{cases}$$

until pivoting is stopped

Similarly to the discussion for pivoting in Freudenthal's triangulation, the pivoting rules for J_1 can also be obtained by interchange permutations:

ALGORITHM 32.4 (*Pivoting in J_1 by interchange permutations*).

input $[v_1, v_2, \ldots, v_{q+1}] \subset \mathbb{R}^q$ % *starting simplex*

for $j = 1$ to q do

$u_j \leftarrow v_{j+1} - v_j$ % *standard axes*

$\pi(j) \leftarrow j$ % *initial permutation*

$s(j) \leftarrow 1$ % *initial sign function*

end for

repeat

enter $i \in \{1, 2, \ldots, q+1\}$ % *index of vertex to be pivoted next*

if $i = 1$ % *consider different cases*

$v_1 \leftarrow v_2 + s(1)\, u_{\pi(1)}$
$s(1) \leftarrow -s(1)$
else if $i = q+1$
$v_{q+1} \leftarrow v_q - s(q)\, u_{\pi(q)}$
$s(q) \leftarrow -s(q)$
else $v_i \leftarrow v_{i-1} + s(i)\, u_{\pi(i)}$
interchange $s(i-1)$ and $s(i)$
interchange $\pi(i-1)$ and $\pi(i)$
end if
until pivoting is stopped

It is possible to compare different triangulations via various ways of measuring their efficiency. Such results can be found in TODD [1976a], SAIGAL [1977, 1979], VAN DER LAAN and TALMAN [1980], ALEXANDER and SLUD [1983], ALEXANDER [1987], EAVES and YORKE [1984], EAVES [1984]. One of these measures is the *thickness of a triangulation* (see also (40.6)).

33. Piecewise linear algorithms

Let $\mathcal{M}$ be a piecewise linear manifold of dimension $N+1$. We call $H : |\mathcal{M}| \to \mathbb{R}^N$ a *piecewise linear map* if the restriction $H_\sigma : \sigma \to \mathbb{R}^N$ of H to σ is an affine map for all $\sigma \in \mathcal{M}$. In this case, H_σ can be uniquely extended to an affine map on the affine space spanned by σ. The Jacobian H'_σ has the property $H'_\sigma(x-y) = H_\sigma(x) - H_\sigma(y)$ for x, y in this affine space. Note that under an appropriate choice of basis H'_σ corresponds to an $(N, N+1)$-matrix which has a one-dimensional kernel in case of nondegeneracy, i.e., if its rank is maximal.

If $\mathcal{M}$ is a pseudo manifold triangulating a set $X = |\mathcal{M}|$, and if $\tilde{H} : X \to \mathbb{R}^k$ is a map, then the *piecewise linear approximation* of $\tilde{H}$ (with respect to $\mathcal{M}$) is defined as the unique piecewise linear map $H : X \to \mathbb{R}^k$ which coincides with $\tilde{H}$ on all vertices of $\mathcal{M}$, i.e., $\tilde{H}$ is affinely interpolated on the simplices of $\mathcal{M}$.

A piecewise linear algorithm is a method for following a polygonal path in $H^{-1}(0)$. To avoid degeneracies, we introduce a concept of regularity, see EAVES [1976]. A point $x \in |\mathcal{M}|$ is called a *regular point* of H if x is not contained in any face of dimension $< N$, and if H'_τ has maximal rank N for all facets τ. A value $y \in \mathbb{R}^N$ is a *regular value* of H if all points in $H^{-1}(y)$ are regular. By definition, y is vacuously a regular value if it is not contained in the range of H. If a point or value is not regular it is called *singular*.

The following analogue of Sard's theorem 18.1 holds for piecewise linear maps, see, e.g., EAVES [1976] or PEITGEN and SIEGBERG [1981] for details. This enables us to confine ourselves to regular values. We note that degeneracies could be handled via the concept of lexicographical ordering, see DANTZIG [1963], TODD [1976a].

THEOREM 33.1 (Perturbation theorem). *Let $H : \mathcal{M} \to \mathbb{R}^N$ be a piecewise linear map where $\mathcal{M}$ is a piecewise linear manifold of dimension $N+1$. Then for any relatively compact subset $C \subset |\mathcal{M}|$ there are at most finitely many $\varepsilon > 0$ such that $C \cap H^{-1}(\vec{\varepsilon})$*

contains a singular point of H. Consequently, $\vec{\varepsilon}$ is a regular value of H for almost all $\varepsilon > 0$. Here we use the notation

$$\vec{\varepsilon} := \begin{pmatrix} \varepsilon \\ \varepsilon^2 \\ \vdots \\ \varepsilon^N \end{pmatrix}.$$

PROOF. The proof will be given by contradiction. Let us assume that there is a strictly decreasing sequence $\{\varepsilon_i\}_{i\in\mathcal{N}}$ of positive numbers, converging to zero, for which a bounded sequence $\{x_i\}_{i\in\mathcal{N}} \subset |\mathcal{M}|$ of singular points can be found such that the equations

$$H(x_i) = \vec{\varepsilon_i} \tag{33.1}$$

for $i \in \mathcal{N}$ are satisfied. For any subset $I \subset \mathcal{N}$ of cardinality $N+1$ we see that the $\{\vec{\varepsilon_i}\}_{i\in I}$ are affinely independent, and by (33.1) and the piecewise linearity of H the $\{x_i\}_{i\in I}$ cannot all be contained in the same lower dimensional face $\tau \in \mathcal{M}^k$ for $k < N$. Since this holds for all index sets I, we use this argument repeatedly, and the local finiteness of $\mathcal{M}$ permits us to find a strictly increasing function $\nu : \mathcal{N} \to \mathcal{N}$ (to generate a subsequence), and to find a face $\sigma \in \mathcal{M}^{N+1} \cup \mathcal{M}^N$ such that the subsequence $\{x_{\nu(i)}\}_{i\in\mathcal{N}}$ is contained in the interior of σ. But now we can again use the above argument: for an index set $I \subset \nu(\mathcal{N})$ of cardinality $N+1$ the $\{\vec{\varepsilon_i}\}_{i\in I}$ are affinely independent, and we conclude that H'_σ has maximal rank N. However, this means that all points $\{x_{\nu(i)}\}_{i\in\mathcal{N}}$ are regular, a contradiction to the choice of $\{x_i\}_{i\in\mathcal{N}}$. The last assertion of the perturbation theorem follows since $|\mathcal{M}|$ can be written as a countable union of relatively compact subsets. □

Let 0 be a regular value of H. This implies that $H^{-1}(0)$ consists of polygonal paths whose vertices are always in the interior of some facet. If σ is a cell, then $\sigma \cap H^{-1}(0)$ is a segment (two end points), a ray (one end point), a line (no end point) or empty. The latter two cases are not of interest for piecewise linear path following. A step of the method consists of following the ray or segment from one cell into a uniquely determined adjacent cell. The method is typically started at a point of the boundary or on a ray (coming from infinity), and it is typically terminated at a point of the boundary or in a ray (going to infinity), see Fig. 33.1. The numerical linear algebra required to perform one step of the method is typical for linear programming and usually involves $O(N^2)$ operations for dense matrices.

On the other hand, even if 0 is not a regular value of H, the above theorem helps us to do something similar. Namely, $\sigma \cap H^{-1}(\vec{\varepsilon})$ is a segment (two end points) for all sufficiently small $\varepsilon > 0$, a ray (one end point) for all sufficiently small $\varepsilon > 0$, a line (no end point) for all sufficiently small $\varepsilon > 0$ or empty for all sufficiently small $\varepsilon > 0$. This leads us to the following definition: we call a facet τ *completely labeled* with respect to H, if $\tau \cap H^{-1}(\vec{\varepsilon}) \neq \varnothing$ for all sufficiently small $\varepsilon > 0$. We call a cell

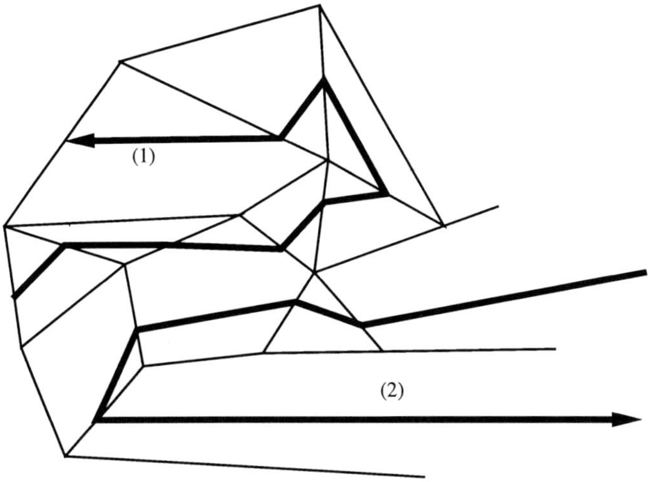

FIG. 33.1. (1) Boundary start with boundary termination. (2) Ray start with ray termination.

σ *transverse* with respect to H, if $\sigma \cap H^{-1}(\vec{\varepsilon}) \neq \varnothing$ for all sufficiently small $\varepsilon > 0$. Instead of following the paths $H^{-1}(0)$ for a regular value 0, we now follow more specifically the *regularized paths*

$$\bigcup \{H^{-1}(0) \cap \sigma\colon \ \sigma \text{ transverse}\}.$$

Of course, this set coincides with $H^{-1}(0)$ for the case that 0 is a regular value of H.

For $\varepsilon > 0$ sufficiently small and $\vec{\varepsilon}$ a regular value of H, a node of the polygonal paths $H^{-1}(\vec{\varepsilon})$ corresponds to a completely labeled facet (which is intersected), and hence the piecewise linear algorithm traces such completely labeled facets belonging to the same cell. The method is usually started either on the boundary, i.e., in a completely labeled facet $\tau \in \partial\mathcal{M}$, or on a ray, i.e., in a transverse cell $\sigma \in \mathcal{M}$ which has only one completely labeled facet. We are thus led to the following two generic versions of a piecewise linear algorithm.

ALGORITHM 33.1 (*PL algorithm with boundary start*).

input $\tau_1 \in \partial\mathcal{M}$ completely labeled % *starting facet*
find the unique $\sigma_1 \in \mathcal{M}$ such that $\tau_1 \subset \sigma_1$
for $i = 1, 2, 3, \ldots$
    if τ_i is the only completely labeled facet of σ_i
        stop % *ray termination*
    else % *piecewise linear step*
        find the other completely labeled facet τ_{i+1} of σ_i
    end if
    if $\tau_{i+1} \in \partial\mathcal{M}$
        stop % *boundary termination*

else % *pivoting step*
    pivot σ_i across τ_{i+1} into σ_{i+1}
end if
end for

ALGORITHM 33.2 (*PL algorithm with ray start*).

input $\sigma_1 \in \mathcal{M}$ transverse % *starting cell*
    having exactly one completely labeled facet

find the completely labeled facet τ_2 of σ_1% *initial PL step*
for $i = 2, 3, \ldots$
    if $\tau_i \in \partial\mathcal{M}$
        stop % *boundary termination*
    else % *pivoting step*
        pivot σ_{i-1} across τ_i into σ_i
    end if
    if τ_i is the only completely labeled facet of σ_i
        stop % *ray termination*
    else % *piecewise linear step*
        find the other completely labeled facet τ_{i+1} of σ_i
    end if
end for

34. Numerical considerations

From a numerical point of view, two steps of a piecewise linear algorithm have to be efficiently implemented. Usually, a current cell σ and a completely labeled facet τ of σ is stored via some characteristic data.

A pivoting step consists of finding the adjacent cell $\tilde{\sigma}$ with the same facet τ. The implementation of this step is dependent on the special piecewise linear manifold under consideration. But typically this step is performed by only a few operations. The pivoting rule (31.2) is a simple example, see also Section 32.

A piecewise linear step consists of finding a second completely labeled facet $\tilde{\tau}$ of σ (if it exists, otherwise we have ray termination). This is usually computationally more expensive than the pivoting rule and typically involves some numerical linear algebra.

Let us consider an example. We assume that a cell of dimension $N+1$ is given by

$$\sigma := \{x \in \mathbb{R}^{N+1}\colon\ Lx \geqslant c\},$$

where $L\colon \mathbb{R}^{N+1} \to \mathbb{R}^m$ is a linear map and $c \in \mathbb{R}^m$ is a given value. Furthermore, let us assume that

$$\tau_i := \{x \in \mathbb{R}^{N+1}\colon\ Lx \geqslant c,\ e_i^* Lx = e_i^* c\},$$

for $i = 1, 2, \ldots, m$ is a numbering of all the facets of σ.

On the cell σ, the piecewise linear map $H : \mathcal{M} \to \mathbb{R}^N$ reduces to an affine map, and hence there is a linear map $A : \mathbb{R}^{N+1} \to \mathbb{R}^N$ and a vector $b \in \mathbb{R}^N$ such that the segment of the path in σ can be written as

$$\sigma \cap H^{-1}(0) = \left\{x \in \mathbb{R}^{N+1}\colon\ Ax = b,\ Lx \geqslant c\right\}. \tag{34.1}$$

Let τ_i be completely labeled. This implies that the rank of A is N. If we exclude degeneracies, then $\tau_i \cap H^{-1}(0) = \{x_0\}$ is a singleton, and there is a unique vector t in the one-dimensional kernel $A^{-1}(0)$ such that $e_i^* Lt = -1$. Since x_0 is in the interior of τ_i (by excluding degeneracies), we have $e_j^* L x_0 > e_j^* c$ for $j = 1, \ldots, m$, $j \neq i$, and hence $x_0 - \lambda t$ is in the interior of σ for small $\lambda > 0$.

If (34.1) is a ray, then $e_j^* L(x_0 - \lambda t) > e_j^* c$ for all $\lambda > 0$. Otherwise we have $e_j^* Lt > 0$ for at least one index j, and since we are excluding degeneracies, the minimization

$$k := \arg\min \left\{ \frac{e_j^*(Lx_0 - c)}{e_j^* Lt}\colon\ j = 1, \ldots, m,\ e_j^* Lt > 0 \right\} \tag{34.2}$$

yields the unique completely labeled facet τ_k of σ with $k \neq i$. For the minimum

$$\lambda_0 := \frac{e_k^*(Lx_0 - c)}{e_k^* Lt} > 0$$

we obtain: $\sigma \cap H^{-1}(0) = \{x_0 - \lambda t\colon\ 0 \leqslant \lambda \leqslant \lambda_0\}$.

Minimizations such as (34.2) are typical for linear programming, and the numerical linear algebra can be efficiently handled by standard routines. Successive linear programming steps can often make use of previous matrix factorizations via update methods, see, e.g., GILL, GOLUB, MURRAY and SAUNDERS [1974]. In the case of a pseudo manifold $\mathcal{M}$ where the cell σ is a simplex, it is convenient to handle the numerical linear algebra with respect to the barycentric co-ordinates based on the vertices of σ. Then the equations become particularly simple, see, e.g., ALLGOWER and GEORG ([1990], Sections 12.2–12.4) or TODD [1976a] for details.

We describe the simplest case (without taking degeneracies into account) which is sufficient for most implementations. Let $\sigma = [v_1, \ldots, v_{N+2}]$ in $\mathbb{R}^{N+1}$ be a simplex of dimension $N + 1$, and let us denote by τ_i the facet obtained via deleting vertex v_i in σ. We further introduce the labeling matrix

$$L := \begin{pmatrix} 1 & \ldots & 1 & 1 \\ H(v_1) & \ldots & H(v_{N+1}) & H(v_{N+2}) \end{pmatrix}$$

and the submatrix L_i obtained via deleting column i in L. Then τ_i is completely labeled with respect to H if and only if L_i^{-1} exists and is *lexicographically positive*, i.e., the first nonzero entry in each row is positive. For numerical purposes, it can usually be assumed that the situation is nondegenerate, i.e., that the first column of

L_i^{-1} consists of positive entries. In this case, we have a solution $L\alpha = e_1$ such that $\alpha(j) > 0$ for $j \neq i$ and $\alpha(i) = 0$. Let γ be a solution of $L\gamma = 0$ such that $\gamma(i) < 0$. The idea now is to find a largest possible $t > 0$ such that $\alpha - t\gamma$ has nonnegative entries. This leads to the minimization

$$k := \arg\min_j \left\{ \frac{\alpha(j)}{\gamma(j)} \colon \gamma(j) > 0 \right\}.$$

If this minimization yields a unique k, then τ_k is the other completely labeled facet of σ.

We now give some examples of how the piecewise linear path following methods are used.

35. Piecewise linear homotopy algorithms

Let us see how the above ideas can be used to approximate zero points of a map $G\colon \mathbb{R}^N \to \mathbb{R}^N$ by applying piecewise linear methods to an appropriate homotopy map. In order to also allow for applications to optimization problems or other nonlinear programming problems, we consider the case where G is not necessarily continuous, e.g., G might be a selection of a multi-valued map. For the case that $\bar{x}$ is a point of discontinuity of G, we have to generalize the notion of a zero point in an appropriate way, i.e., $0 \in G_\Sigma(\bar{x})$ in the terminology introduced below.

EAVES [1972] presented the first piecewise linear homotopy method. A restart method based on somewhat similar ideas was developed by MERRILL [1972]. A number of authors have studied the efficiency and complexity of piecewise linear homotopy algorithms, see, e.g., ALEXANDER [1987], EAVES and YORKE [1984], SAIGAL [1977, 1984], SAIGAL and TODD [1978], SAUPE [1982], TODD [1982, 1986].

As an example of a piecewise linear homotopy algorithm, let us sketch the algorithm of EAVES and SAIGAL [1972]. We consider a triangulation $\mathcal{T}$ of $\mathbb{R}^N \times (0,1]$ into $(N+1)$-simplices σ such that every simplex is contained in some slab $\mathbb{R}^N \times [2^{-k}, 2^{-k-1}]$ for $k = 0, 1, \dots$. Let us call the maximum of the last co-ordinates of all vertices of σ the *level* of σ. We call $\mathcal{T}$ a *refining* triangulation if for $\sigma \in \mathcal{T}$, the diameter of σ tends to zero as the level of σ tends to zero. Of course, the main point here is to obtain a triangulation which is easily implemented. The first such triangulation was proposed by EAVES [1972]. TODD [1976a] gave a triangulation with refining factor $\frac{1}{2}$, see Fig. 35.1. Subsequently, many triangulations with arbitrary refining factors were developed, see EAVES [1984]. To insure success (i.e., convergence) of the algorithms, it is necessary to assume a boundary condition. For this we will follow a presentation of GEORG [1982] which uses a quite general boundary condition extending somewhat that of MERRILL [1972].

Let us first introduce some notation. For $x \in \mathbb{R}^N$ we denote by $\mathcal{U}(x)$ the system of neighborhoods of x. By $\overline{\text{co}}(X)$ we denote the closed convex hull of a set $X \subset \mathbb{R}^N$. By $\mathbb{R}^N_\Sigma$ we denote the system of compact convex nonempty subsets of $\mathbb{R}^N$. We call the map $G\colon \mathbb{R}^N \to \mathbb{R}^N$ *asymptotically linear* if the following three conditions hold:

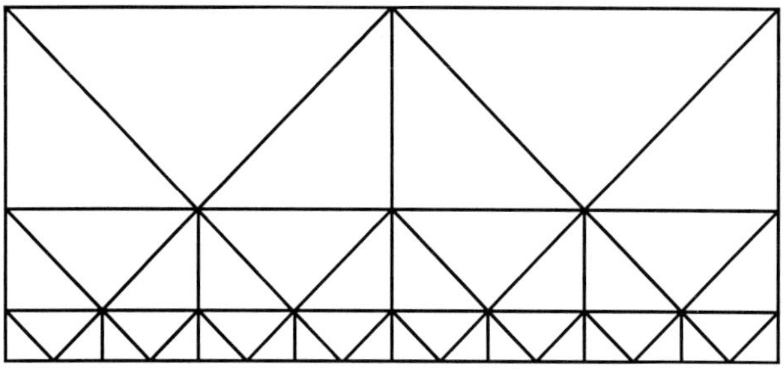

FIG. 35.1. Refining triangulation.

(1) G is *locally bounded*, i.e., each point $x \in \mathbb{R}^N$ has a neighborhood $U \in \mathcal{U}(x)$ such that $G(U)$ is a bounded set.

(2) G is *differentiable at* ∞, i.e., there exists a linear map $G'_\infty : \mathbb{R}^N \to \mathbb{R}^N$ such that $\|x\|^{-1}\|G(x) - G'_\infty x\| \to 0$ for $\|x\| \to \infty$.

(3) G'_∞ is nonsingular.

If a map $G : \mathbb{R}^N \to \mathbb{R}^N$ is locally bounded, then we can define its *set-valued hull* $G_\Sigma : \mathbb{R}^N \to \mathbb{R}^N_\Sigma$ by setting

$$G_\Sigma(x) := \bigcap_{U \in \mathcal{U}(x)} \overline{\text{co}}\,(G(U)).$$

It is not difficult to see that G_Σ is upper semicontinuous, and that G is continuous at x if and only if $G_\Sigma(x)$ is a singleton. By using a degree argument, see, e.g., GÓRNIEWICZ [1976], on the set-valued homotopy

$$H_\Sigma(x, \lambda) := (1 - \lambda)G'_\infty x + \lambda G_\Sigma(x),$$

it can be seen that G_Σ has at least one zero point, i.e., a point $\bar{x}$ such that $0 \in G_\Sigma(\bar{x})$. It is known, see, e.g., PEITGEN and PRÜFER [1979], PEITGEN and SIEGBERG [1981], that degree arguments in nonlinear analysis are essentially constructive. Our aim here is to approximate this solution numerically.

We now construct a piecewise linear homotopy for an asymptotically linear map $G : \mathbb{R}^N \to \mathbb{R}^N$. First we define $\tilde{H} : \mathbb{R}^N \times [0, \infty) \to \mathbb{R}^N$ by setting

$$\tilde{H}(x, \lambda) := \begin{cases} G'_\infty(x - x_1) & \text{for } \lambda = 1, \\ G(x) & \text{for } \lambda < 1. \end{cases}$$

Here x_1 is a chosen starting point of the method. Then we consider a refining triangulation $\mathcal{T}$ of $\mathbb{R}^N \times (0, 1]$ as above, and we use the piecewise linear approximation H of $\tilde{H}$ (with respect to $\mathcal{T}$) to trace the polygonal path in $H^{-1}(0)$ which contains the starting point $(x_1, 1)$.

The boundary $\partial\mathcal{T}$ is a pseudo manifold which triangulates the sheet $\mathbb{R}^N \times \{1\}$. If we assume that the starting point $u_1 := (x_1, 1)$ is in the interior of a facet $\tau_1 \in \partial\mathcal{T}$, then it is immediately clear that τ_1 is the only completely labeled facet of $\partial\mathcal{T}$. Hence, the piecewise linear algorithm started in τ_1 cannot terminate in the boundary, and since all cells of $\mathcal{T}$ are compact, it cannot terminate in a ray. Hence, it has no termination. Thus the piecewise linear algorithm generates a sequence $\tau_1, \tau_2, \ldots$ of completely labeled facets of $\mathcal{T}$. Let us also consider the polygonal path generated by the piecewise linear algorithm. This path is characterized by the nodes $(x_1, \lambda_1), (x_2, \lambda_2), \ldots$ such that (x_i, λ_i) is the unique zero point of the piecewise linear homotopy H in τ_i for $i = 1, 2, \ldots$. The resulting algorithm, i.e., applying Algorithm 33.1 to the above homotopy H, is due to EAVES [1972] and EAVES and SAIGAL [1972].

We call $\bar{x} \in \mathbb{R}^N$ an *accumulation point* of the algorithm if

$$\liminf_{i\to\infty} \|x_i - \bar{x}\| = 0.$$

The following convergence theorem holds:

THEOREM 35.1. *The set A of accumulation points of the Eaves–Saigal algorithm is compact, connected and nonempty. Each point $\bar{x} \in A$ is a zero point of G_Σ, i.e., we have $0 \in G_\Sigma(\bar{x})$*

PROOF. From the construction of the piecewise linear map H it follows that

$$\lim_{\|x\|\to\infty} \|x\|^{-1} \big\| H(x, \lambda) - G'_\infty x \big\| = 0$$

uniformly for $\lambda \in (0, 1]$. Since G'_∞ is nonsingular, $H(x_i, \lambda_i) = 0$ implies that the sequence x_i is bounded. Hence the set A is nonempty and compact.

Let us assume that A can be written as a disjoint union of two nonempty compact sets A_1 and A_2. Then $\operatorname{dist}(A_1, A_2) > 0$, and

$$\liminf_{i\to\infty} \operatorname{dist}(x_i, A_j) = 0 \quad \text{for } j = 1, 2.$$

On the other hand, $\lim_{i\to\infty} \operatorname{dist}(x_i, A) = 0$, and the refining property of the triangulation $\mathcal{T}$ implies that $\lim_{i\to\infty} \|x_i - x_{i+1}\| = 0$. This leads to a contradiction, and hence A is connected.

Since a piecewise linear manifold is locally bounded and the τ_i stay in a bounded set, it follows that the level of the τ_i tends to 0 for $i \to \infty$. Hence, for i sufficiently large, the definition of H and the fact that the facets τ_i are completely labeled implies that $0 \in \overline{\text{co}}\, G(\pi_N(\tau_i))$. Here $\pi_N : \mathbb{R}^N \times (0, 1] \to \mathbb{R}^N$ denotes the canonical projection.

Since a point $\bar{x} \in A$ is an accumulation point of the sequence x_i, and since $\lim_{i\to\infty} \operatorname{diam}(\tau_i) = 0$, we have that for each neighborhood $U \in \mathcal{U}(\bar{x})$ there is an i (arbitrarily large) such that $\pi_N(\tau_i) \subset U$ and hence

$$0 \in \overline{\text{co}}\, G\big(\pi_N(\tau_i)\big) \subset \overline{\text{co}}\, G(U).$$

Intersecting over all $U \in \mathcal{U}(\bar{x})$ gives $0 \in G_\Sigma(\bar{x})$. □

As a consequence, if the set-valued hull G_Σ has only isolated zero points, then the sequence x_i generated by the Eaves–Saigal algorithm converges to a zero point of G_Σ.

As a simple example, we consider the situation of the celebrated Brouwer fixed point theorem, see BROUWER [1912]. Let $F: C \to C$ be a continuous map on a convex, compact, nonempty subset $C \subset \mathbb{R}^N$ with nonempty interior. We define an asymptotically linear map $G: \mathbb{R}^N \to \mathbb{R}^N$ by setting

$$G(x) := \begin{cases} x - F(x) & \text{for } x \in C, \\ x - x_1 & \text{for } x \notin C. \end{cases}$$

Here, a point x_1 in the interior of C is used as a starting point. The above piecewise linear algorithm generates a point $\bar{x} \in \mathbb{R}^N$ such that $0 \in G_\Sigma(\bar{x})$. If $\bar{x} \notin C$, then $G_\Sigma(\bar{x}) = \{\bar{x} - x_1\}$, but $\bar{x} \neq x_1$ implies that this case is impossible. If $\bar{x}$ is an interior point of C, then $G_\Sigma(\bar{x}) = \{\bar{x} - F(\bar{x})\}$, and hence $\bar{x}$ is a fixed point of F. If $\bar{x}$ is in the boundary ∂C, then $G_\Sigma(\bar{x})$ is the convex hull of $\bar{x} - x_1$ and $\bar{x} - F(\bar{x})$, and hence $\bar{x} = (1 - \lambda)x_1 + \lambda F(\bar{x})$ for some $0 \leqslant \lambda \leqslant 1$. But $\lambda < 1$ would imply that $\bar{x}$ is an interior point of C, and hence we have $\lambda = 1$, and again $\bar{x}$ is a fixed point of F. Hence, the above piecewise linear homotopy algorithm generates a fixed point of F in either case. Many similar asymptotically linear maps can be constructed which correspond to important nonlinear problems, see, e.g., ALLGOWER and GEORG ([1990], Chapter 13).

36. Mixing PL and Newton steps

As we have seen above, the refining triangulation used in the Eaves–Saigal algorithm is very convenient for discussing the question of convergence. If no stopping is allowed, it generates a sequence of nodes (x_n, λ_n) for $n = 0, 1, 2, \ldots$. We have seen that x_n converges to a zero point $\bar{x}$ of G_Σ under reasonable and very weak assumptions. Without additional assumptions on G however, nothing can be said about the rate of convergence of the x_n. BROOKS [1980] has shown that infinite retrogression (yoyoing) can occur. To ensure linear convergence, assumptions in the spirit of the Newton–Kantorovitch theorems, see ORTEGA and RHEINBOLDT [1970], are necessary. Such convergence discussions have been given by SAIGAL [1977]. It can be seen that there is a close interrelationship between piecewise linear steps and Newton's method. Several other papers discuss techniques of mixing piecewise linear and Newton steps in order to accelerate a piecewise linear homotopy algorithm, see, e.g., SAIGAL and TODD [1978] and TODD [1978b, 1980]. In the context of piecewise linear continuation methods, i.e., when a whole curve $c(s)$ is to be approximated by a polygonal path. SAUPE [1982] has considered a mixing of piecewise linear and predictor corrector steps.

An elementary way of mixing piecewise linear and Newton steps was given by GEORG [1982], see also ALLGOWER and GEORG ([1990], Section 13.5). It is based on the simple observation that a modified Newton's method expressed in barycentric co-ordinates leads to a system of linear equations which is closely related to the linear equations obtained in a piecewise linear step described in Section 34.

More precisely, the following can be seen. Let $G:\mathbb{R}^N \to \mathbb{R}^N$ be a map and $\tau = [z_1, z_2, \ldots, z_{N+1}] \subset \mathbb{R}^N$ an N-simplex. Let $B := G'_\tau$ denote the Jacobian of the piecewise linear approximation G_τ of G with respect to τ, i.e., B is the finite difference approximation of G' using the values of G on $z_1, z_2, \ldots, z_{N+1}$. We assume that B is nonsingular and define a modified Newton step $\mathcal{N}:\mathbb{R}^N \to \mathbb{R}^N$ by

$$\mathcal{N}(x) := x - B^{-1}G(x)\,. \tag{36.1}$$

Then

$$G_\tau \mathcal{N}(z_i) = 0 \quad \text{for } i = 1, 2, \ldots, N+1\,.$$

Furthermore, for any $z_{N+2} \in \mathbb{R}^N$, let

$$L(\tau, z_{N+2}) := \begin{pmatrix} 1 & \ldots & 1 & 1 \\ G(z_1) & \ldots & G(z_{N+1}) & G(z_{N+2}) \end{pmatrix}$$

and consider the barycentric co-ordinates β such that

$$L\beta = e_1, \qquad \beta(N+2) = -1.$$

Then

$$\mathcal{N}(z_{N+2}) = \sum_{j=1}^{N+2} \beta(j) z_j\,.$$

The above equations show that the piecewise linear steps of the Eaves–Saigal algorithm, see the end of Section 34, can be combined with Newton-type steps (36.1), since they are based on the same labeling matrix. In particular, no initialization is necessary when switching between the two methods. Thus it is possible to design an algorithm which enjoys both the global convergence features mentioned in Theorem 35.1 and the fast local convergence of Newton-like methods (if the map G is sufficiently smooth and regular at the solution point). See ALLGOWER and GEORG ([1990], Algorithm 13.5.2) for more details.

37. Index and orientation

Nearly all piecewise linear manifolds $\mathcal{M}$ which are of importance for practical implementations, are orientable. If $\mathcal{M}$ is orientable and of dimension $N+1$, and if $H:\mathcal{M} \to \mathbb{R}^N$ is a piecewise linear map, then it is possible to introduce an index for the piecewise linear solution manifold $H^{-1}(0)$ which has important invariance properties and occasionally yields some useful information, see EAVES [1976], EAVES and SCARF [1976], LEMKE and GROTZINGER [1976], SHAPLEY [1974], TODD [1976c]. It should be noted that this index is closely related, see, e.g., PEITGEN [1982], to the topological index which is a standard tool in topology and nonlinear analysis. Occasionally, index arguments are used to guarantee a certain qualitative behavior of the solution path. There are many ways to introduce the index. Our discussion is similar to that in ALLGOWER and GEORG ([1990], Section 14.2).

We begin with some basic definitions. Let $\mathbf{F}$ be a linear space of dimension k. An *orientation* of $\mathbf{F}$ is a function $\text{or}: \mathbf{F}^k \to \{-1, 0, 1\}$ such that the following conditions hold:

(1) $\text{or}(b_1, \ldots, b_k) \neq 0$ if and only if $b_1, \ldots, b_k$ are linearly independent.

(2) $\text{or}(b_1, \ldots, b_k) = \text{or}(c_1, \ldots, c_k) \neq 0$ if and only if the transformation matrix between $b_1, \ldots, b_k$ and $c_1, \ldots, c_k$ has positive determinant.

It is clear from the basic facts of linear algebra that any finite dimensional linear space permits exactly two orientations.

Let σ be a cell of dimension k and aff σ its affine hull. We introduce the k-dimensional linear space $\text{tng}\, \sigma := \{x - y\colon\ x, y \in \text{aff}\, \sigma\}$ as the *tangent space* of σ. The cell σ is oriented by orienting this tangent space. Such an orientation or_σ of σ *induces an orientation* $\text{or}_{\tau,\sigma}$ on a facet τ of σ by the following convention:

$$\text{or}_{\tau,\sigma}(b_1, \ldots, b_{k-1}) := \text{or}_\sigma(b_1, \ldots, b_k)$$

whenever b_k points from τ into the interior of the cell σ. It is routine to check that the above definition of $\text{or}_{\tau,\sigma}$ indeed satisfies the definition of an orientation.

If $\mathcal{M}$ is a piecewise linear manifold of dimension $N+1$, then an *orientation of* $\mathcal{M}$ is a choice of orientations $\{\text{or}_\sigma\}_{\sigma \in \mathcal{M}}$ such that

$$\text{or}_{\tau,\sigma_1} = -\,\text{or}_{\tau,\sigma_2} \tag{37.1}$$

for each τ which is a facet of two different cells $\sigma_1, \sigma_2 \in \mathcal{M}$. By making use of the standard orientation

$$\text{or}(b_1, \ldots, b_N) := \text{sign}\det(b_1, \ldots, b_N)$$

of $\mathbb{R}^N$, it is clear that any piecewise linear manifold of dimension N which subdivides a subset of $\mathbb{R}^N$ is oriented in a natural way.

If $H: \mathcal{M} \to \mathbb{R}^N$ is a piecewise linear map on a piecewise linear manifold of dimension $N+1$ such that zero is a regular value of H, then it is clear that the system

$$\ker H := \{\sigma \cap H^{-1}(0)\}_{\sigma \in \mathcal{M}}$$

is a 1-dimensional piecewise linear manifold which subdivides the solution set $H^{-1}(0)$. For the case that $\mathcal{M}$ is oriented, the orientation of $\mathcal{M}$ and the natural orientation of $\mathbb{R}^N$ induce an orientation of $\ker H$. Namely, for $\xi \in \ker H$, $v \in \operatorname{tng}(\xi)$ and $\sigma \in \mathcal{M}$ such that $\xi \subset \sigma$, the definition

$$\operatorname{or}_\xi(v) := \operatorname{or}_\sigma(b_1, \ldots, b_N, v) \operatorname{sign} \det(H'_\sigma b_1, \ldots, H'_\sigma b_N) \tag{37.2}$$

is independent of the special choice of $b_1, \ldots, b_N \in \operatorname{tng}(\sigma)$, provided the $b_1, \ldots, b_N$ are linearly independent. Clearly, an orientation of the 1-dimensional manifold $\ker H$ is just a rule which indicates a direction for traversing each connected component of $\ker H$. Keeping this in mind, we now briefly indicate why the above definition indeed yields an orientation for $\ker H$.

Let τ be a facet of $\mathcal{M}$ which meets $H^{-1}(0)$ and does not belong to the boundary $\partial\mathcal{M}$, let $\sigma_1, \sigma_2 \in \mathcal{M}$ be the two cells containing τ, and let $\xi_j := H^{-1}(0) \cap \sigma_j \in \ker H$ for $j = 1, 2$. If $b_1, \ldots, b_N$ is a basis of $\operatorname{tng}(\tau)$, and if $a_j \in \operatorname{tng}(\xi_j)$ points from τ into σ_j, then from condition (37.1) it follows that

$$\operatorname{or}_{\sigma_1}(b_1, \ldots, b_N, a_1) = -\operatorname{or}_{\sigma_2}(b_1, \ldots, b_N, a_2),$$

and hence (37.2) implies that

$$\operatorname{or}_{\xi_1}(a_1) = -\operatorname{or}_{\xi_2}(a_2),$$

which is exactly the right condition in the sense of (37.1) to ensure that the manifold $\ker H$ is oriented.

Let $H : \mathcal{M} \to \mathbb{R}^N$ be a piecewise linear map on an oriented piecewise linear manifold $\mathcal{M}$ of dimension $N+1$. Given a facet τ of a cell $\sigma \in \mathcal{M}$, we can define the *index* of H at τ with respect to σ by setting

$$\operatorname{index}_{\tau,\sigma}(H) := \operatorname{or}_{\tau,\sigma}(b_1, \ldots, b_N) \operatorname{sign} \det(H'_\sigma b_1, \ldots, H'_\sigma b_N)$$

if τ is completely labeled with respect to H, and $\operatorname{index}_{\tau,\sigma}(H) := 0$ otherwise. It is clear that this definition is independent of the special choice of the basis $b_1, \ldots, b_N$ of $\operatorname{tng}(\tau)$. Furthermore, if zero is a regular value of H, τ is completely labeled and $\xi := H^{-1}(0) \cap \sigma \in \ker H$, then (37.2) implies that $\operatorname{index}_{\tau,\sigma}(H) = 1$ if and only if a positively oriented vector in $\operatorname{tng}(\xi)$ points from τ into σ, i.e., the path runs from τ into σ. By using the perturbation technique (if necessary), see Theorem 33.1 and the definitions thereafter, we obtain

$$\operatorname{index}_{\tau_1,\sigma}(H) = -\operatorname{index}_{\tau_2,\sigma}(H)$$

for the case that τ_1 and τ_2 are two different completely labeled facets of a cell σ (see the piecewise linear step in Algorithms 33.1, 33.2). Similarly we obtain

$$\operatorname{index}_{\tau,\sigma_1}(H) = -\operatorname{index}_{\tau,\sigma_2}(H)$$

for the case that τ is a completely labeled facet of two different cells σ_1 and σ_2 (see the pivoting step in Algorithm 33.1). A case of special importance is a facet τ in the boundary $\partial\mathcal{M}$, since then we do not have to specify the cell σ which contains τ because σ is unique. If the piecewise linear Algorithm 33.1 starts on the boundary in a completely labeled facet τ_1 and stops again in the boundary in a completely labeled facet τ_k, then the above formulae imply that

$$\mathrm{index}_{\tau_1}(H) = -\,\mathrm{index}_{\tau_k}(H)$$

holds. Hence, for a compact piecewise linear manifold (i.e., if $|\mathcal{M}|$ is compact) where only a boundary start or termination is possible, we obtain the following celebrated *index formula*

$$\sum_{\tau\in\partial\mathcal{M}} \mathrm{index}_{\tau}(H) = 0\,.$$

38. Lemke's algorithm

The first and most prominent example of a piecewise linear algorithm was designed by LEMKE [1965] and LEMKE and HOWSON [1964] to calculate a solution of the linear complementarity problem. Subsequently, several authors have studied complementarity problems from the standpoint of piecewise linear homotopy methods, see, e.g., KOJIMA [1974, 1979], KOJIMA, NISHINO and SEKINE [1976], SAIGAL [1971, 1976], TODD [1976b]. Complementarity problems can also be considered from an interior point algorithm viewpoint, see Section 29.3, hence by following a smooth path, see, e.g., KOJIMA, MIZUNO and NOMA [1990b], KOJIMA, MIZUNO and YOSHISE [1991], KOJIMA, MEGIDDO and NOMA [1991b], KOJIMA, MEGIDDO and MIZUNO [1990a], MIZUNO [1992].

We present the Lemke algorithm as an example of a piecewise linear algorithm since it played a crucial role in the development of subsequent piecewise linear algorithms. Let us consider the following *linear complementarity problem*: Given an affine map $g:\mathbb{R}^N \to \mathbb{R}^N$, find an $x \in \mathbb{R}^N$ such that

$$x \in \mathbb{R}^N_+; \qquad g(x) \in \mathbb{R}^N_+; \qquad x^* g(x) = 0. \tag{38.1}$$

Here $\mathbb{R}_+$ denotes the set of nonnegative real numbers, and in the sequel we also denote the set of positive real numbers by $\mathbb{R}_{++}$. If $g(0) \in \mathbb{R}^N_+$, then $x = 0$ is a trivial solution to the problem. Hence this trivial case is always excluded and the additional assumption

$$g(0) \notin \mathbb{R}^N_+$$

is made.

Linear complementarity problems arise in quadratic programming, bimatrix games, variational inequalities and economic equilibria problems. Hence numerical methods for their solution have been of considerable interest, see, e.g., COTTLE [1974], COTTLE and DANTZIG [1968], COTTLE, GOLUB and SACHER [1978], LEMKE [1980]. See also the recent book COTTLE, PANG and STONE [1992] for further references.

For $x \in \mathbb{R}^N$ we introduce the positive part $x_+ \in \mathbb{R}^N_+$ by setting

$$e_i^* x_+ := \max\{e_i^* x, 0\}, \quad i = 1, \ldots, N,$$

and the negative part $x_- \in \mathbb{R}^N_+$ by $x_- := (-x)_+$. The following formulae are then obvious: $x = x_+ - x_-$, $(x_+)^*(x_-) = 0$.

It is not difficult to show the following: Define $f : \mathbb{R}^N \to \mathbb{R}^N$ by $f(z) := g(z_+) - z_-$. If x is a solution of the linear complementarity problem, then $z := x - g(x)$ is a zero point of f. Conversely, if z is a zero point of f, then $x := z_+$ solves the linear complementarity problem.

The advantage which f provides is that it is obviously a piecewise linear map if we subdivide $\mathbb{R}^N$ into orthants. This is the basis for our description of Lemke's algorithm. For a fixed $d \in \mathbb{R}^N_{++}$ we define the homotopy $H : \mathbb{R}^N \times [0, \infty) \to \mathbb{R}^N$ by

$$H(x, \lambda) := f(x) + \lambda d. \tag{38.2}$$

For a given subset $I \subset \{1, 2, \ldots, N\}$ an orthant can be written in the form

$$\sigma_I := \{ (x, \lambda) \colon \lambda \geqslant 0,\ e_i^* x \geqslant 0 \text{ for } i \in I,\ e_i^* x \leqslant 0 \text{ for } i \in I' \}, \tag{38.3}$$

where I' denotes the complement of I. The collection of all such orthants forms a piecewise linear manifold $\mathcal{M}$ (of dimension $N + 1$) which subdivides $\mathbb{R}^N \times [0, \infty)$. Furthermore it is clear that $H : \mathcal{M} \to \mathbb{R}^N$ is a piecewise linear map since $x \mapsto x_+$ switches its linearity character only at the co-ordinate hyperplanes.

Let us assume for simplicity (as usual) that zero is a regular value of H. We note however, that the case of a singular value is treated in the same way by using the perturbation techniques. Lemke's algorithm is started on a ray: if $\lambda > 0$ is sufficiently large, then

$$\big(- g(0) - \lambda d \big)_+ = 0 \quad \text{and} \quad \big(- g(0) - \lambda d \big)_- = g(0) + \lambda d \in \mathbb{R}^N_{++},$$

and consequently

$$H\big(- g(0) - \lambda d\, ,\, \lambda \big) = 0.$$

Hence, the ray defined by

$$\lambda \in [\lambda_0, \infty) \longmapsto -g(0) - \lambda d \in \sigma_\varnothing \tag{38.4}$$

for

$$\lambda_0 := \max_{i=1,\ldots,N} \frac{-g(0)[i]}{d[i]} \tag{38.5}$$

is used (for decreasing λ-values) to start the path following. Since the piecewise linear manifold $\mathcal{M}$ consists of the orthants of $\mathbb{R}^N \times [0,\infty)$, it is finite, and there are only two possibilities:

(1) The algorithm terminates on the boundary $|\partial\mathcal{M}| = \mathbb{R}^N \times \{0\}$ at a point $(z,0)$. Then z is a zero point of f, and hence z_+ solves the linear complementarity problem.

(2) The algorithm terminates on a secondary ray. Then it can be shown, see COTTLE [1974], that the linear complementarity problem has no solution, at least if the Jacobian g' belongs to a certain class of matrices.

Let us illustrate the use of index and orientation by showing that the algorithm generates a solution in the sense that it terminates on the boundary under the assumption that all principle minors of the Jacobian g' are positive. Note that the Jacobian g' is a constant matrix since g is affine.

For $\sigma_I \in \mathcal{M}$, see (38.3), we immediately calculate the Jacobian

$$H'_{\sigma_I} = (f'_{\sigma_I}, d),$$

where

$$f'_{\sigma_I} e_i = \begin{cases} g' e_i & \text{for } i \in I, \\ e_i & \text{for } i \in I'. \end{cases} \tag{38.6}$$

If $\xi \in \ker H$ is a solution path in σ_I, then formula (37.2) yields

$$\mathrm{or}_\xi(v) = \operatorname{sign}\det f'_{\sigma_I}\, \mathrm{or}_{\sigma_I}(e_1,\ldots,e_N,v),$$

and since $\mathrm{or}_{\sigma_I}(e_1,\ldots,e_N,v) = \operatorname{sign}(v^* e_{N+1})$ by the standard orientation in $\mathbb{R}^{N+1}$, we have that $\det f'_{\sigma_I}$ is positive or negative if and only if the λ-direction is increasing or decreasing, respectively, while ξ is traversed according to its orientation. It is immediately seen from (38.6) that $\det f'_{\sigma_I}$ is obtained as a *principle minor of* g', i.e., by deleting all columns and rows of g' with index $i \in I'$ and taking the determinant of the resulting matrix (where the determinant of the "empty matrix" is assumed to be 1). Since we start in the negative orthant $\sigma_\varnothing$ where the principle minor is 1, we see that the algorithm traverses the primary ray against its orientation, because the λ-values are initially decreased. Hence, the algorithm continues to traverse $\ker H$ against its orientation. For the important case that all principle minors of g' are positive, the algorithm must continue to decrease the λ-values and thus stops at the boundary $|\partial\mathcal{M}| = \mathbb{R}^N \times \{0\}$. Hence, in this case the algorithm finds a solution. Furthermore, it is clear that this solution is unique, since $\ker H$ can contain no other ray than the primary ray.

39. Variable dimension algorithms

In recent years, a new class of piecewise linear algorithms has attracted considerable attention. They are called *variable dimension algorithms* since they all start from a single point, a zero dimensional simplex, and successively generate simplices of varying dimension, until a completely labeled simplex is found. Numerical results of KOJIMA and YAMAMOTO [1984] indicate that these algorithms improve the computational efficiency of piecewise linear homotopy methods. The first variable dimension algorithm is due to KUHN [1969]. However, this algorithm had the disadvantage that it could only be started from a vertex of a large triangulated standard simplex S, and therefore piecewise linear homotopy algorithms were preferred. By increasing the sophistication of Kuhn's algorithm considerably, VAN DER LAAN and TALMAN [1979] developed an algorithm which could start from any point inside S. It soon became clear, see TODD [1978a], that this algorithm could be interpreted as a homotopy algorithm. Numerous other variable dimension algorithms were developed. Some of the latest are due to DAI, SEKITANI and YAMAMOTO [1992], DAI and YAMAMOTO [1989], KAMIYA and TALMAN [1990], TALMAN and YAMAMOTO [1989]. Two unifying approaches have been given, one due to KOJIMA and YAMAMOTO [1982], the other due to FREUND [1984a,b]. A variable dimension algorithm which is easy to comprehend and may serve the reader as a gateway is the octahedral algorithm of WRIGHT [1981], see also Section 39.2.

We present here a modified version of the approach described by KOJIMA and YAMAMOTO [1982]. The modification consists of introducing a cone construction for dealing with the homotopy parameter. In a special case, this construction was also used by Kojima and Yamamoto, see their Lemma 5.13.

Before we can give a description of these algorithms, we introduce the notion of a primal–dual pair of piecewise linear manifolds due to Kojima and Yamamoto. In fact, we only need a special case. Let $\mathcal{P}$ and $\mathcal{D}$ be two piecewise linear manifolds of dimension N. Let us denote the set of k-dimensional faces of cells in $\mathcal{P}$ by $\mathcal{P}^k$. In particular, $\mathcal{P}^0$ is the set of vertices of $\mathcal{P}$, and $\mathcal{P}^N = \mathcal{P}$.

We call $(\mathcal{P}, \mathcal{D})$ a *primal–dual pair* if there is a bijective map

$$\tau \in \mathcal{P}^k \longmapsto \tau^d \in \mathcal{D}^{N-k}, \quad k = 0, 1, \ldots, N,$$

such that

$$\tau_1 \subset \tau_2 \iff \tau_2^d \subset \tau_1^d \text{ holds for all } \tau_1, \tau_2 \in \bigcup_{k=0}^{N} \mathcal{P}^k.$$

We will deal with a homotopy parameter via the following cone construction. Throughout the rest of this chapter, ω denotes a point which is affinely independent from all cells under consideration. The introduction of ω is only formal and may be obtained by, e.g., increasing the dimension of the ambient finite dimensional Euclidean space $\mathbf{E}$ introduced at the beginning of Section 31. If σ is a cell, then

$$\sigma^\omega := \{(1-\lambda)\omega + \lambda x \colon \ x \in \sigma, \ \lambda \geqslant 0\}$$

denotes the cone with vertex ω generated by σ. Clearly, σ^ω is again a cell and $\dim \sigma^\omega = \dim \sigma + 1$. If $H : \sigma \to \mathbb{R}^k$ is an affine map, then the affine extension $H^\omega : \sigma^\omega \to \mathbb{R}^k$ is defined by

$$H^\omega\big((1-\lambda)\omega + \lambda x\big) := \lambda H(x)$$

for $x \in \sigma$ and $\lambda \geqslant 0$, i.e., H is extended in such a way that $H^\omega(\omega) = 0$.

If $\mathcal{M}$ is a piecewise linear manifold of dimension N, then

$$\mathcal{M}^\omega := \{\sigma^\omega \colon \sigma \in \mathcal{M}\}$$

is a piecewise linear manifold of dimension $N + 1$, and a piecewise linear map $H : \mathcal{M} \to \mathbb{R}^k$ is extended to a piecewise linear map $H^\omega : \mathcal{M}^\omega \to \mathbb{R}^k$.

We will be interested below in rays traversing a cone σ^ω, and we therefore collect some formulae. A ray in σ^ω is given as

$$\big\{(1-\varepsilon)z_1 + \varepsilon z_2 \colon \varepsilon \geqslant 0\big\} \subset \sigma^\omega,$$

where

$$z_j = (1-\lambda_j)\omega + \lambda_j x_j, \quad j = 1, 2,$$

for some suitable $\lambda_1, \lambda_2 \geqslant 0$ and $x_1, x_2 \in \sigma$. A simple calculation using the affine independence of ω from σ yields

$$(1-\varepsilon)z_1 + \varepsilon z_2 = (1-\lambda_\varepsilon)\omega + \lambda_\varepsilon x_\varepsilon,$$

where

$$\lambda_\varepsilon = (1-\varepsilon)\lambda_1 + \varepsilon\lambda_2$$

and

$$x_\varepsilon = \frac{(1-\varepsilon)\lambda_1 x_1 + \varepsilon\lambda_2 x_2}{\lambda_\varepsilon}.$$

Since $\lambda_\varepsilon \geqslant 0$ for all $\varepsilon \geqslant 0$, it follows that $\lambda_2 \geqslant \lambda_1$. This leaves two cases to consider:

$$\lambda_2 > \lambda_1 \geqslant 0 \Rightarrow \lim_{\varepsilon \to \infty} x_\varepsilon = \frac{\lambda_2 x_2 - \lambda_1 x_1}{\lambda_2 - \lambda_1} \in \sigma, \tag{39.1}$$

$$\lambda_2 = \lambda_1 > 0 \Rightarrow x_1 \neq x_2,$$

$$x_\varepsilon = (1-\varepsilon)x_1 + \varepsilon x_2 \in \sigma \quad \text{for } \varepsilon \geqslant 0.$$

The second case is only possible if the cell σ is unbounded.

Let $\mathcal{T}$ and $\mathcal{M}$ be piecewise linear manifolds of dimension N. We call $\mathcal{T}$ a *refinement* of $\mathcal{M}$ if for all $\sigma \in \mathcal{M}$ the restricted piecewise linear manifold $\mathcal{T}_\sigma := \{\xi\colon \xi \in \mathcal{T},\ \xi \subset \sigma\}$ subdivides σ.

We are now in a position to introduce primal-dual manifolds. Let $(\mathcal{P}, \mathcal{D})$ be a primal-dual pair of N-dimensional piecewise linear manifolds, and let $\mathcal{T}$ be a refinement of $\mathcal{P}$. Then

$$\mathcal{T} \otimes \mathcal{D} := \{\xi \times \tau^d\colon\ \xi \in \mathcal{T}^k,\ \tau \in \mathcal{M}^k,\ \xi \subset \tau,\ 0 \leqslant k \leqslant N\},$$

is an N-dimensional piecewise linear manifold with empty boundary. A proof of this and related results was given by Kojima and Yamamoto. We call $\mathcal{T} \otimes \mathcal{D}$ the *primal–dual manifold* generated by $\mathcal{T}$ and $\mathcal{D}$. In particular, $\mathcal{T} = \mathcal{P}$ is one of the standard choices. An essential part of the proof consists of discussing the possible pivoting steps:

Let $\xi \times \tau^d \in \mathcal{T} \otimes \mathcal{D}$ with $k = \dim \xi = \dim \tau$ as above, and let κ be a facet of $\xi \times \tau^d$. We now describe the pivoting of $\xi \times \tau^d$ across the facet κ, i.e., we have to find a cell $\eta \in \mathcal{T} \otimes \mathcal{D}$ such that $\eta \neq \xi \times \tau^d$ and $\kappa \subset \eta$. There are three possible cases:

Increasing the dimension. Let $\kappa = \xi \times \sigma^d$ such that $\sigma \in \mathcal{M}^{k+1}$ contains τ. Then there is exactly one $\rho \in \mathcal{T}^{k+1}$ such that $\xi \subset \rho$ and $\rho \subset \sigma$. This is a consequence of the fact that $\mathcal{T}$ refines $\mathcal{P}$ and is not difficult to prove. Then $\eta := \rho \times \sigma$ is the desired second cell. In this case the dimension k of the primal cell ξ is increased when performing the pivoting step.

Decreasing the dimension. Let $\kappa = \delta \times \tau^d$ such that $\delta \in \mathcal{T}^{k-1}$ is a facet of ξ. If $\delta \subset \partial\tau$, then there exists exactly one facet $\nu \in \mathcal{M}^{k-1}$ of τ such that $\delta \subset \nu$, and $\eta := \delta \times \nu^d$ is the desired second cell. In this case the dimension k of the primal cell ξ is decreased when performing the pivoting step.

Keeping the dimension. Let $\kappa = \delta \times \tau^d$ such that $\delta \in \mathcal{T}^{k-1}$ is a facet of ξ. If $\delta \not\subset \partial\tau$, then there exists exactly one cell $\xi' \in \mathcal{T}^k$ such that $\xi' \neq \xi$, $\xi' \subset \tau$ and $\delta \subset \xi'$. This is again a consequence of the fact that $\mathcal{T}$ refines $\mathcal{P}$ and is not difficult to prove. Now $\eta := \xi' \times \tau$ is the desired second cell. In this case the dimension k of the primal cell ξ is left invariant when performing the pivoting step.

The main point for practical purposes is that the above three different kinds of pivoting steps must be easy to implement. This is of course mainly a question of choosing a simple primal–dual pair $(\mathcal{P}, \mathcal{D})$ and either $\mathcal{T} = \mathcal{P}$ or some standard refinement $\mathcal{T}$ of $\mathcal{P}$ which can be handled conveniently.

We now slightly modify the construction of primal–dual manifolds to include cones for the refinement $\mathcal{T}$ of the primal manifold:

$$\mathcal{T}^\omega \otimes \mathcal{D} := \{\xi^\omega \times \tau^d\colon\ \xi \in \mathcal{T}^k,\ \tau \in \mathcal{M}^k,\ \xi \subset \tau,\ 0 \leqslant k \leqslant N\}.$$

If $\dim \xi = k > 0$, then the facets of ξ^ω are simply the ρ^ω where $\rho \in \mathcal{T}^{k-1}$ is a facet of ξ, and it is readily seen that the above pivoting steps apply correspondingly. The only exception is the case $\dim \xi = k = 0$. In this case it follows that $\xi = \tau$, and ξ

is a vertex of the piecewise linear manifold $\mathcal{D}$, but ξ^ω is a ray which has one vertex, namely $\{\omega\}$. Hence, we now have a boundary

$$\partial(\mathcal{T}^\omega \otimes \mathcal{D}) = \big\{ \{\omega\} \times \{v\}^d \colon \{v\} \in \mathcal{P}^0 \big\}.$$

Clearly, such a boundary facet $\{\omega\} \times \{v\}^d$ belongs to the $(N+1)$-dimensional cell $\{v\}^\omega \times \{v\}^d \in \mathcal{T}^\omega \otimes \mathcal{D}$. We will later see that such boundary facets are used for starting a piecewise linear algorithm. This corresponds to starting a homotopy method on the trivial level $\lambda = 0$ at the point v. We will now apply the above concept of primal–dual manifolds in order to describe some piecewise linear algorithms.

In particular, we want to illustrate that the concept of primal–dual manifolds allows for a unifying description of many variable dimension algorithms. The homotopy methods where the homotopy parameter induces a cone construction serve as one example, see Section 39.2. An important feature of primal–dual manifolds is that a complementarity property of the variables (x, y) may be incorporated into the construction of the primal–dual manifolds so that this property needs not to be assumed by extra conditions or constructions. This is a very convenient trick for dealing with complementarity problems or related questions, and will be illustrated here for the case of the linear complementarity problem in the next section. Many more applications have been considered, see the literature cited in ALLGOWER and GEORG [1990].

39.1. Lemke's algorithm revisited

We consider again the linear complementarity problem (38.1) and choose a primal–dual pair $(\mathcal{P}, \mathcal{D})$ by defining for $I \subset \{1, 2, \ldots, N\}$ and $I' := \{1, 2, \ldots, N\} \setminus I$ the primal and dual faces

$$\alpha_I := \big\{x \in \mathbb{R}^N\colon\ e_i^* x \geqslant 0 \text{ for } i \in I,\ e_i^* x = 0 \text{ for } i \in I'\big\},$$

$$\alpha_I^d := \alpha_{I'}\,.$$

The primal and dual manifolds consist of just one cell: $\mathcal{P} = \mathcal{D} = \{\mathbb{R}^n_+\}$. We now define a piecewise linear map $H\colon \mathcal{P} \otimes \mathcal{D} \times [0, \infty) \longrightarrow \mathbb{R}^N$ by $H(x, y, \lambda) := y - g(x) - \lambda d$ where $d \in \mathbb{R}^N_{++}$ is fixed. Note that the variables x and y are placed into complementarity with each other by the construction of $\mathcal{P} \otimes \mathcal{D}$, and hence a more complex definition of H as in (38.2) is not necessary. For sufficiently large $\lambda > 0$ the solutions of $H(x, y, \lambda) = 0$ are given by the primary ray $(x, y, \lambda) = (0\,, g(0) + \lambda d\,, \lambda)$. Here the piecewise linear algorithm following $H^{-1}(0)$ is started in the negative λ-direction. If the level $\lambda = 0$ is reached, a solution $H(x, y, 0) = 0$ solves the linear complementarity problem since the complementarity $x \in \mathbb{R}^N_+$, $y = g(x) \in \mathbb{R}^N_+$, $x^* y = 0$ holds by the construction of $\mathcal{P} \otimes \mathcal{D}$.

39.2. The octahedral algorithm

As a typical representative of the class of variable dimension algorithms for approximating a zero point of a map we present the *octahedral algorithm* of WRIGHT [1981], since numerical experiments indicate that it performs favorably, see KOJIMA and YAMAMOTO [1984], and since it can be described in a reasonably simple way. Let us point out that similar discussions also hold for many other algorithms where the refinement $\mathcal{T}$ of the primal manifold $\mathcal{P}$ is a pseudo manifold which triangulates $\mathbb{R}^N$, and where the dual manifold $\mathcal{D}$ subdivides a compact subset of $\mathbb{R}^N$, see ALLGOWER and GEORG [1990], FREUND [1984a,b], KOJIMA and YAMAMOTO [1982, 1984], VAN DER LAAN and TALMAN [1979].

We denote by $\Sigma := \{+1, 0, -1\}^N \setminus \{0\}$ the set of all nonzero sign vectors. For two vectors $s, p \in \Sigma$ we introduce the relation

$$s \prec p :\Longleftrightarrow \forall i = 1, \ldots, N \ \left(e_i^* s \neq 0 \Rightarrow e_i^* s = e_i^* p\right),$$

i.e., s and p coincide except that s may have additional zeros. Then we define a primal–dual pair $(\mathcal{P}, \mathcal{D})$ of N-dimensional manifolds by introducing the following duality:

$$\alpha_0 := \{0\}, \qquad \alpha_0^d := \left\{ y \in \mathbb{R}^N \colon \ \|y\|_1 \leqslant 1 \right\},$$

and for $s \in \Sigma$ we consider

$$\alpha_s := \left\{ \sum_{\substack{p \in \Sigma \\ s \prec p}} \lambda_p p \colon \ \lambda_p \geqslant 0 \right\},$$

$$\alpha_s^d := \left\{ y \in \mathbb{R}^N \colon \ \|y\|_1 \leqslant 1, \ s^* y = 1 \right\},$$

i.e., α_s is a cone spanned by the unit basis vectors $\{s(i)\, e_i\}_{s(i) \neq 0}$, and $y \in \alpha_s^d$ if and only if

$$\begin{aligned} y(i) = 0 &\iff s(i) = 0, \\ s(i) y(i) > 0 &\iff s(i) \neq 0, \\ \sum_i |y(i)| = 1. \end{aligned}$$

Hence, the primal manifold $\mathcal{P}$ subdivides $\mathbb{R}^N$ into 2^N cones spanned by the unit basis vectors $\pm e_i$ for $i = 1, 2, \ldots, N$, and the dual manifold $\mathcal{D}$ just consists of the unit ball with respect to the $\|.\|_1$-norm. We easily check that

$$y \in \alpha_s^d, \ s \prec p \Rightarrow y^* p \geqslant 0$$

and hence

$$(x, y) \in \mathcal{P} \otimes \mathcal{D} \Rightarrow x^* y \geqslant 0. \tag{39.2}$$

We now consider a pseudo manifold $\mathcal{T}$ which is a refinement of $\mathcal{P}$, for example it is easy to see that the Union Jack triangulation of $\mathbb{R}^N$, see TODD [1976a], has this property, see Section 32.

The aim now is to find an approximate zero point of an asymptotically linear map $G:\mathbb{R}^N \to \mathbb{R}^N$. To do this, let $G_\mathcal{T}$ be the piecewise linear approximation of G with respect to the pseudo manifold $\mathcal{T}$ triangulating $\mathbb{R}^N$, see the beginning of Section 33. If $\sigma = [v_1, \ldots, v_{N+1}] \in \mathcal{T}$, and if a point $u \in \sigma$ is expanded into its barycentric coordinates $u = \sum_{i=1}^{N+1} c_i v_i$, then $G_\mathcal{T}(u) = \sum_{i=1}^{N+1} c_i G(v_i)$. It is clear that $G_\mathcal{T}$ is also asymptotically linear and $G'_\mathcal{T}(\infty) = G'(\infty)$. A homotopy $\tilde{H}:\mathcal{T}\otimes\mathcal{D}\times[0,\infty)\to\mathbb{R}^N$ is introduced by setting

$$\tilde{H}(x, y, \lambda) := G'(\infty)y + \lambda G_\mathcal{T}(x). \tag{39.3}$$

Here, for simplicity, $y = 0$ plays the role of a starting point. Unfortunately, $\tilde{H}$ is not piecewise linear. Hence, we use the cone construction to identify $\tilde{H}$ with a piecewise linear map $H:\mathcal{P}^\omega \otimes \mathcal{D} \longrightarrow \mathbb{R}^N$ by collecting the variables in a different way:

$$H(z, y) := G'(\infty)y + G_\mathcal{T}^\omega(z). \tag{39.4}$$

For $z = \omega$, which corresponds to $\lambda = 0$, there is exactly one solution of $H(z, y) = 0$, namely $(z, y) = (\omega, 0)$. Hence $H^{-1}(0)$ intersects the boundary $\partial(\mathcal{P}^\omega \otimes \mathcal{D})$ at just a single point. This is the starting point for our piecewise linear algorithm which traces $H^{-1}(0)$.

Let us first show that there is a constant $C > 0$ such that $\tilde{H}(x, y, \lambda) = 0$ implies $\|x\| < C$. Indeed, otherwise we could find a sequence

$$\{(x_k, y_k, \lambda_k)\}_{k=1,2,\ldots} \subset H^{-1}(0)$$

such that $\lim_{k\to\infty} \|x_k\| = \infty$. It follows from $\tilde{H}(x_k, y_k, \lambda_k) = 0$ and (39.3) that

$$\lambda_k^{-1} y_k + G'(\infty)^{-1} G_\mathcal{T}(x_k) = 0. \tag{39.5}$$

If we multiply this equation from the left with x_k^* and divide by $\|x_k\|^2$, the asymptotic linearity of $G_\mathcal{T}$ yields

$$\lim_{k\to\infty} \|x_k\|^{-2} x_k^* G'(\infty)^{-1} G_\mathcal{T}(x_k) = 1,$$

and the boundedness $\|y_k\| \leqslant 1$ implies that

$$x_k^* G'(\infty)^{-1} G_\mathcal{T}(x_k) > 0 \tag{39.6}$$

for all sufficiently large k. Now by (39.5), (39.6) we have that $x_k^* y_k > 0$ for all sufficiently large k, which is a contradiction to (39.2).

Hence the above boundedness implies that the algorithm can only traverse finitely many cells, and since the solution on the boundary $\partial(\mathcal{T}^\omega \otimes \mathcal{D})$ is unique, it can only terminate in a ray

$$\{((1-\varepsilon)z_1 + \varepsilon z_2,\ (1-\varepsilon)y_1 + \varepsilon y_2):\ \varepsilon \geqslant 0\} \subset \tau^\omega \times \alpha_I^d \in \mathcal{T}^\omega \otimes \mathcal{D},$$

where $\tau \in \mathcal{T}^k$ such that $\tau \subset \alpha_I$ and k is the number of elements of I. We refer to (39.2) and the notation and remarks preceding it. It follows from

$$H\big((1-\varepsilon)z_1 + \varepsilon z_2,\ (1-\varepsilon)y_1 + \varepsilon y_2\big) = 0 \tag{39.7}$$

and (39.3)–(39.4) that

$$(1-\varepsilon)y_1 + \varepsilon y_2 + \lambda_\varepsilon G'(\infty)^{-1} G_{\mathcal{T}}(x_\varepsilon) = 0 \quad \text{for } \varepsilon \geqslant 0\,.$$

Since the k-cell τ is bounded, we only have to consider the case $\lambda_2 > \lambda_1 \geqslant 0$, see (39.2). Dividing Eq. (39.7) by $\varepsilon > 0$ and letting $\varepsilon \to \infty$ yields

$$G_{\mathcal{T}}(x) = 0\,, \quad \text{where } x := \frac{\lambda_2 x_2 - \lambda_1 x_1}{\lambda_2 - \lambda_1} \in \tau$$

is the desired approximate zero point of G.

40. Approximating manifolds

The ideas of predictor corrector and piecewise linear curve tracing can be extended to the approximation of implicitly defined manifolds $\tilde{H}^{-1}(0)$ where $\tilde{H}: \mathbb{R}^{N+K} \to \mathbb{R}^N$.

For simplicity, we assume in this section that zero is a regular value of the smooth map $\tilde{H}: \mathbb{R}^{N+K} \to \mathbb{R}^N$. Hence $\tilde{\mathcal{M}} := \tilde{H}^{-1}(0)$ is a smooth K-dimensional manifold. Before we discuss the methods for obtaining piecewise linear approximations of $\tilde{\mathcal{M}}$, let us briefly indicate that the Gauss–Newton method can be used as a corrector in the sense of Section 4 also in this more general setting.

If B is an $N \times (N+K)$-matrix with maximal rank, then in analogy to Section 3, the Moore–Penrose inverse B^+ of B is given by, e.g., $B^+ = B^*(BB^*)^{-1}$. The product BB^+ is the identity on $\mathbb{R}^N$, and $\mathrm{Id} - B^+B$ is the orthogonal projection onto $\ker(B)$.

In analogy to Theorem 4.1, there exists an open neighborhood U of $\tilde{\mathcal{M}}$ such that the Gauss–Newton method

$$v_{i+1} = v_i - \tilde{H}'(v_i)^+ \tilde{H}(v_i), \quad i = 0, 1, \ldots\ , \tag{40.1}$$

converges quadratically to a point $v_\infty \in \tilde{\mathcal{M}}$ whenever the starting point v_0 is in U. Since the evaluation and decomposition of the Jacobian matrix $\tilde{H}'(v_i)$ may be costly, one often modifies (40.1) to the so-called *chord method*

$$v_{i+1} = v_i - B^+ \tilde{H}(v_i), \quad i = 0, 1, \ldots\ , \tag{40.2}$$

where B is some fixed approximation of $\tilde{H}'(v_0)$. It is well known that the above mentioned quadratic convergence reduces to linear convergence in the latter case.

Orthogonal decompositions are particularly useful in this context. If Q is an orthogonal $(N+K, N+K)$-matrix such that $BQ = (L, 0)$ for some lower triangular (N, N)-matrix L, and if we split the orthogonal matrix $Q = (Q_N, Q_K)$ into the first

N and the last K columns, then it is straightforward to see that $B^+ = Q_N L^{-1}$, and the columns of Q_K provide an orthonormal basis for $\ker(B)$.

Unfortunately, for all known decomposition methods, this basis matrix Q_K does not depend continuously on the choice of the matrix B, and this is a fact which complicates matters in constructing the moving frame algorithm, see Section 40.1. The remedy is to introduce a reference $(N+K, K)$-matrix T_K whose columns form an orthonormal system (i.e., $T_K^* T_K = \mathrm{Id}$) and to use the singular value decomposition

$$V_1^* T_K^* Q_K V_2 = \Sigma \tag{40.3}$$

see, e.g., GOLUB and VAN LOAN [1989]. RHEINBOLDT [1987] shows that the map

$$B \mapsto W_K := Q_K V_1 V_2^* \tag{40.4}$$

is smooth if B varies over the open set of $(N+K, K)$-matrices which have full rank and a kernel such that $T_K^* Q_K$ is nonsingular. We simplify our discussion by slightly abusing the notation of Rheinboldt and calling the new matrix W_K the *moving frame* of the kernel of B with respect to the *reference matrix* T_K.

There are two basic types of algorithms for approximating manifolds with $K > 1$: one is the moving frame algorithm of RHEINBOLDT [1987, 1988b], see also BRODZIK and RHEINBOLDT [1994], which is a higher dimensional analogue of the predictor corrector method, the other is a piecewise linear algorithm which has been developed in ALLGOWER and GNUTZMANN [1987], ALLGOWER and SCHMIDT [1985], GNUTZMANN [1989], WIDMANN [1990a,b], see also Chapter 15 of ALLGOWER and GEORG [1990].

40.1. The moving frame algorithm

The moving frame algorithm involves predictors arising from a local triangulation of the tangent space at a current point. The corrector consists of using the Gauss–Newton method for projecting the generated mesh back to the manifold. This method is well-suited for smooth manifolds in which the dimension N is large, such as in multiple parameter nonlinear eigenvalue problems, see, e.g., RHEINBOLDT [1988b, 1993]. It has been applied to the calculation of fold curves and to differential–algebraic equations, see DAI and RHEINBOLDT [1990], RHEINBOLDT [1986, 1992, 1993].

To motivate the idea of the moving frame algorithm, we first give a very heuristic description. At some starting point $p \in \tilde{\mathcal{M}}$, we triangulate the tangent space of $\tilde{\mathcal{M}}$ at p using some standard triangulation $\mathcal{T}$. The tangent space is of course isometric to $\mathbb{R}^K$. We now imagine that the manifold is "rolled" over the tangent space, thereby "imprinting" a triangulation on $\tilde{\mathcal{M}}$. This is used to provide an approximation of some part of $\tilde{\mathcal{M}}$ by a pseudo manifold. Actually, the "imprinting" is carried out via a chord method (40.2) which must eventually be restarted from time to time with a new tangent space. The moving frame idea keeps the fitting together of the imprinted triangulation in the "rolling" process consistent when restarts are made. The following pseudo algorithm sketches the essential ideas of Rheinboldt's method. Given a triangulation

$\mathcal{T}$ of $\mathbb{R}^K$, the algorithm constructs an "imprint" $\varphi: \mathcal{X} \to \mathcal{M}$ where $\mathcal{X} \subset \mathcal{T}^0$ is a subset of "marked" nodes of $\mathcal{T}$, which is successively enlarged.

ALGORITHM 40.1 (*Moving frame algorithm*).

```
input  s ∈ T^0 % initial vertex
       φ(s) ∈ M̃ % starting point on M̃, imprint of s
       T_K % reference matrix
       h > 0 % steplength,
       % should be much bigger than the meshsize of the triangulation
X ← {s} % initial set of marked nodes
repeat
       % begin building a new frame:
       get x ∈ X such that dist(x, T^0 \ X) < h
       % new Jacobian which will generally be decomposed at this point:
       B ← H̃'(φ(x))
       % moving frame of ker(B) with respect to T_K:
       calculate W_K according to (40.4)
       while  dist(x, T^0 \ X) < h  do
           % new marked node:
             get y ∈ T^0 \ X such that ||x − y|| < h
             v ← W_K(y − x) + φ(x) % predictor for imprint of y
             repat  v ← v − B^+ H̃(v) % chord corrector method
             until convergence
             φ(y) ← v % imprint of y
             X ← X ∪ {y} % set of marked nodes is augmented
       end while
until  a stopping criterion is satisfied.
```

By examining the construction of the moving frame at the end of the previous section it becomes evident that we have to make the following technical restriction for nodes $x \in \mathcal{X}$ where we begin a new frame: let $\tilde{\mathcal{M}}_0$ be the set of points where the reference matrix T_K induces a local co-ordinate system on $\tilde{\mathcal{M}}$, i.e.,

$$\tilde{\mathcal{M}}_0 := \left\{ z \in \tilde{\mathcal{M}} \colon \det \begin{pmatrix} \tilde{H}'(z) \\ T_K^* \end{pmatrix} \neq 0 \right\}.$$

Then a point x is only permitted if its imprint $\varphi(x)$ is in the connected component of $\tilde{\mathcal{M}}_0$ which contains the starting point $\varphi(s)$. It is possible to relax this restriction, but this is usually done at the cost of having an overlapping approximation of $\tilde{\mathcal{M}}$ by a pseudo manifold.

In typical applications of the above method, the dimension N will be significantly larger than K, and hence the computational cost of the singular value decomposition (40.3) is comparatively small.

The above algorithm can be regarded as a higher dimensional analogue of the predictor corrector continuation methods. The predictor step is more complicated than for $K = 1$, since a triangulation of $\mathbb{R}^K$ is mapped onto the tangent space of $\tilde{\mathcal{M}}$ at x via the moving frame device. For the case $K = 1$, the moving frame idea coincides with the concept of orientation as described, e.g., in (1.5). For $K > 1$ however, the moving frame device induces more structure than just orientation. The corrector process is quite analogous to the case $K = 1$. Some topics which remain to be investigated further are:

Globalization: If $\tilde{\mathcal{M}}$ is a compact manifold without boundary, it would be desirable to adapt the construction of the marked nodes $\mathcal{X}$ and the imprint $\varphi(\mathcal{X})$ in such a way that $\varphi(\mathcal{X})$ can be regarded as a compact pseudo manifold without boundary (by declaring appropriate K-simplices). HOHMANN [1991] has made a start on this for $K = 2$.

Steplength adaptation: As in the case $K = 1$, it is possible to vary the steplength h in the above algorithm according to the performance of the Gauss–Newton corrector (and possibly other factors.)

Handling singular points: It would be desirable to incorporate techniques for detecting, classifying and handling singularities on the manifold (e.g., bifurcation points). This is a much more complex problem than for the case $K = 1$.

40.2. Approximating manifolds via PL methods

The difficulty of obtaining a global approximation of an implicitly defined compact manifold without boundary does not exist for piecewise linear algorithms. That is, it is possible to obtain an approximating piecewise linear manifold having no holes or overlappings. However, these algorithms become extremely costly for large N. The piecewise linear algorithms have been applied to the visualization of body surfaces, see ALLGOWER and GNUTZMANN [1991], and to the approximation of surface and body integrals, see ALLGOWER, GEORG and WIDMANN [1991], GEORG and WIDMANN [1993]. They can also be used as automatic mesh generators for boundary element methods, see GEORG [1991]. For software for surface and volume approximation via piecewise linear methods, see the software "pla_s_k" in Chapter VIII.

We begin with a description of the underlying ideas. Let us suppose that the space $\mathbb{R}^{N+K}$ is triangulated by a triangulation $\mathcal{T}$. In our earlier piecewise linear algorithms there was not much reason to store any simplices. In the present situation, however, we will need for certain reasons to store simplices. An important advantage of the usual standard triangulations is that any simplex can be very compactly stored and cheaply recovered by means of an $(N + K)$-tuple of integers corresponding to its barycenter. Let us illustrate this for the example of the Coxeter–Freudenthal–Kuhn triangulation, see Section 32 and the notation therein.

Let $\sigma = [v_1, \ldots, v_{N+K+1}]$ be an $(N+K)$-simplex whose vertices have integer coordinates such that $v_{i+1} = v_i + e_{\pi(i)}$, $i = 1, \ldots, N+K$, where π is some permutation of the numbers $\{1, \ldots, N+K\}$. Adding up all vertices defines the vector

$$m = \sum_{i=1}^{N+K+1} v_i,$$

which can be uniquely decomposed in the following way: $m = (N+K+1)z + \lambda$ where z has integer co-ordinates, and the co-ordinates of λ are a permutation of $\{1, \ldots, N+K\}$. It is easy to see that the vertex v_1 and the permutation π can be easily recovered from m via $v_1 = z$ and $\lambda(\pi(1)) > \lambda(\pi(2)) > \cdots > \lambda(\pi(N+K))$. Note also that $(N+K+1)^{-1}m$ is the barycenter of σ. Hence, it is possible to compactly store the information characterizing the simplex σ in the integer vector m.

It is also possible to perform the pivoting steps directly on the integer vector m and thereby to save some arithmetic operations. The following rules are immediately recovered by translating the pivoting rules of the Coxeter–Freudenthal–Kuhn triangulation for m:

(1) Pivoting the leading vertex v_1 of σ generates a simplex $\tilde{\sigma}$ whose integer vector $\tilde{m}$ is obtained by adding 1 to all components of m and an additional 1 to the component $m_{\pi(1)}$, which otherwise would have a remainder 0 modulo $N+K+1$.

(2) Conversely, if the last vertex v_{N+K+1} of σ is pivoted, a simplex $\tilde{\sigma}$ is generated whose integer vector $\tilde{m}$ is obtained by subtracting 1 from all components of m and an additional 1 from the component $m_{\pi(N+K)}$, which otherwise would have a remainder 0 modulo $N+K+1$.

(3) Pivoting one of the other vertices v_q, $1 < q < N+K+1$, of σ generates a simplex $\tilde{\sigma}$ whose integer vector $\tilde{m}$ is obtained by adding 1 to the component $m_{\pi(q)}$ and subtracting 1 from the component $m_{\pi(q-1)}$.

The Union Jack triangulation of TODD [1976a], see also COXETER [1963], similarly offers compact storing and pivoting in integer arithmetic.

As in Section 33, let H denote the piecewise linear approximation of $\tilde{H}$ with respect to $\mathcal{T}$. The definitions of regular points and regular values extend analogously to this context. We again obtain a perturbation theorem, i.e., the proof of Theorem 33.1 involving ε-perturbations, generalizes verbatim if 1 is replaced by K.

If zero is a regular value of H, the zero set $H^{-1}(0)$ carries the structure of a K-dimensional piecewise linear manifold. We formulate this last remark more precisely:

THEOREM 40.1. *Let zero be a regular value of H. If $\sigma \in \mathcal{T}$ has a nonempty intersection with $H^{-1}(0)$, then $\mathcal{M}_\sigma := \sigma \cap H^{-1}(0)$ is a K-dimensional polytope, and the family*

$$\mathcal{M} := \{\mathcal{M}_\sigma \colon \sigma \in \mathcal{T},\ \sigma \cap H^{-1}(0) \neq \varnothing\}$$

is a K-dimensional piecewise linear manifold.

The following algorithm describes the fundamental steps for obtaining the piecewise linear manifold $\mathcal{M}$ approximating $\tilde{\mathcal{M}}$. We again make the assumptions that $\tilde{H} : \mathbb{R}^{N+K} \to \mathbb{R}^N$ is a smooth map, $\mathcal{T}$ is a triangulation of $\mathbb{R}^{N+K}$, and zero is a regular value of both $\tilde{H}$ and its piecewise linear approximation H. Analogously to the definitions preceding Algorithm 33.1, we call a simplex $\sigma \in \mathcal{T}$ *transverse* if it contains an N-face which is completely labeled with respect to H. In the algorithm below, (see the update step), the dynamically varying set $V(\sigma)$ keeps track of all vertices of the transverse simplex σ which remain to be checked in order to find all possible new transverse simplices by pivoting.

ALGORITHM 40.2 (*PL approximation of a manifold*).

```
input  σ ∈ 𝒯 transverse % starting simplex
       D ⊂ ℝ^{N+K} compact % bounds the search
Σ := {σ} % list of transverse simplices
V(σ) := set of vertices of σ
while V(σ) ≠ ∅ for some σ ∈ Σ do
      get σ ∈ Σ such that V(σ) ≠ ∅
      get v ∈ V(σ)
      pivot σ → σ' via v → v'
      if σ' is not transverse  or  σ' ∩ D = ∅
         delete v from V(σ) % update step
      else if σ' ∈ Σ % σ' is not new
         delete v from V(σ) and v' from V(σ') % update step
      else % σ' is added to the list Σ in this case
         Σ := Σ ∪ {σ'}
         V(σ') := set of vertices of σ'
         delete v from V(σ) and v' from V(σ') % update step
      end if
end while
```

For purposes of exposition we have formulated the above algorithm in a very general way. One may regard the algorithm as a draft for the "outer loop" of the method. A number of items remain to be discussed. We will show below how a starting simplex can be obtained in the neighborhood of a point $x \in \tilde{\mathcal{M}}$. The list Σ can be used to generate a K-dimensional connected piecewise linear manifold

$$\mathcal{M} := \{\mathcal{M}_\sigma\}_{\sigma \in \Sigma},$$

see Theorem 40.1. This piecewise linear manifold approximates $\tilde{\mathcal{M}}$ quadratically in the mesh size of $\mathcal{T}$, as will be seen from the error estimates which will be given in the next section. If $\tilde{\mathcal{M}}$ is compact, the generated piecewise linear manifold will be compact without boundary, provided the mesh of the triangulation is sufficiently small. It is not really necessary to perform the pivot $\sigma \to \sigma'$ if σ' is not transverse, since it will already be known from the current data whether the facet $\sigma \cap \sigma'$ is transverse. In the above comparing process called "σ' is not new", it is crucial that compact exact storing is possible for standard triangulations. The list searching can be performed via efficient binary tree searching. Implementations using such ideas has been given by GNUTZMANN [1989] and WIDMANNN [1990a,b].

The piecewise linear manifold $\mathcal{M}$ furnishes an initial coarse piecewise linear approximation of $\tilde{\mathcal{M}}$. Several improvements are possible. The first is that a Gauss–Newton type method (see (40.1) or (40.2)) can be used to project the nodes of $\mathcal{M}$ onto $\tilde{\mathcal{M}}$. Thus a new piecewise linear manifold $\mathcal{M}_1$ is generated which inherits the adjacency structure of the nodes from $\mathcal{M}$ and has nodes on $\tilde{\mathcal{M}}$.

In many applications (e.g., boundary element methods) it is desirable to uniformize the mesh $\mathcal{M}_1$. A very simple and successful means of doing this is "mesh smoothing". One such possible method consists of replacing each node of the mesh by the average of the nodes with which it shares an edge and by using the resulting point as a starting value for a Gauss–Newton type process to iterate back to $\tilde{\mathcal{M}}$. The edges or nodal adjacencies are maintained as before. Three or four sweeps of this smoothing process over all of the nodes of $\mathcal{M}_1$ generally yields a very uniform piecewise linear approximation of $\tilde{\mathcal{M}}$. Mesh smoothing has been implemented in the programs developed by Gnutzmann and by Widmann.

Another step which is useful for applications such as boundary elements is to locally subdivide the cells of the piecewise linear manifolds $\mathcal{M}$ or $\mathcal{M}_1$ into simplices in such a way that the resulting manifold can be given the structure of a pseudo manifold $\mathcal{M}_2$. This is a technical problem which for $K = 2$ is easy to implement, and this has been done in the above mentioned programs.

Once an approximating pseudo manifold $\mathcal{M}_2$ has been generated, it is easy to refine it by, e.g., the well-known construction of halving all edges of each simplex $\tau \in \mathcal{M}_2$, triangulating it into 2^K subsimplices and projecting the new nodes back onto $\tilde{\mathcal{M}}$.

We have assumed that zero is a regular value of H. In fact, as in the Perturbation theorem 33.1 and following remarks, $\vec{\varepsilon}$-perturbations and the corresponding general definition "completely labeled" automatically resolves singularities even if zero is not a regular value of H. The situation is similar to the case $K = 1$ which has been discussed by PEITGEN [1982], PEITGEN and SCHMITT [1983], see also GNUTZMANN [1989] where the general case is treated.

Let us next address the question of obtaining a transverse starting simplex. If we assume that a point x on $\tilde{\mathcal{M}}$ is given, then it can be shown that any $(N+K)$-simplex with barycenter x and sufficiently small diameter is transverse, see Theorem 40.5.

There is a better way, namely to construct a completely labeled N-face and a transverse simplex containing it. Let us assume that x is a point in $\tilde{\mathcal{M}}$. The normal space of $\tilde{\mathcal{M}}$ at x is given by the orthogonal complement $\ker \tilde{H}'(x)^{\perp}$ of $\ker \tilde{H}'(x)$. From the inverse function theorem, it is clear that the restriction of $\tilde{H}$ to the affine subspace $x + \ker \tilde{H}'(x)^{\perp}$ has x as a regular isolated zero point. Hence, if $\tau \subset x + \ker \tilde{H}'(x)^{\perp}$ is an N-simplex with barycenter x and sufficiently small diameter, then it is completely labeled. Error estimates implying this will be given in the next section. Hence, we only have to construct an affine map Φ sending an N-face of some standard triangulation $\tilde{\mathcal{T}}$ of $\mathbb{R}^{N+K}$ onto τ, and then using the triangulation $\mathcal{T} = \Phi(\tilde{\mathcal{T}})$, see ALLGOWER and GEORG ([1990], p. 242) for more details.

Algorithm 40.2 merely generates a list Σ of transverse simplices. For particular purposes such as boundary element methods, computer graphics, etc., a user will wish to have more information concerning the structure of the piecewise linear manifold $\mathcal{M}$, e.g., all nodes of the piecewise linear manifold $\mathcal{M}$ together with their adjacency structure. Hence, to meet such requirements, it is necessary to "customize" the above algorithm by, e.g., incorporating inner loops which serve to yield such information. This is illustrated in ALLGOWER and GEORG ([1990], (15.4.7–8)).

40.3. Approximation estimates

We conclude this section with some error estimates concerning the quality of the above piecewise linear approximations. Although some of the piecewise linear algorithms are useful under much weaker assumptions on the map $\tilde{H}$, in order to obtain error estimates, it is necessary to make some smoothness assumptions regarding the first and second derivatives of $\tilde{H}$. The results in this section are analogous to results given in GNUTZMANN [1989] and ALLGOWER and GEORG [1989, 1990]. For reasons of simplicity, in this section we make the following assumptions:

$$\begin{aligned} &\tilde{H} : \mathbb{R}^{N+K} \to \mathbb{R}^N \text{ is a smooth map with zero a regular value,} \\ &\|\tilde{H}'(x)^+\| \leqslant \kappa \quad \text{for all } x \in \tilde{\mathcal{M}} := \tilde{H}^{-1}(0), \\ &\|\tilde{H}''(x)\| \leqslant \alpha \quad \text{for all } x \in \mathbb{R}^{N+K}. \end{aligned} \tag{40.5}$$

Actually, the bounds in (40.5) need only to hold in a convex region containing all of the points considered in the following discussion. We remark also that it would be sufficient to assume that the Jacobian $\tilde{H}'(x)$ is Lipschitz continuous with constant α. The above assumptions only serve to make our proofs less technical, however the results are essentially the same.

Let $\mathcal{T}$ be a triangulation of $\mathbb{R}^{N+K}$ having mesh size $\delta > 0$. As in the preceding section we let H denote the piecewise linear approximation of $\tilde{H}$ with respect to $\mathcal{T}$. Our first result concerns the accuracy with which H approximates $\tilde{H}$.

THEOREM 40.2. *Under the assumptions* (40.5), *there holds:*

$$\|\tilde{H}(x) - H(x)\| \leqslant \tfrac{1}{2}\alpha\delta^2 \quad \textit{for } x \in \mathbb{R}^{N+K}.$$

PROOF. Let $\sigma = [v_1, v_2, \ldots, v_{N+K+1}] \in \mathcal{T}$ be an $(N+K)$-simplex such that

$$x = \sum_{i=1}^{N+K+1} \gamma_i v_i \in \sigma.$$

From Taylor's formula we have

$$\tilde{H}(v_i) = \tilde{H}(x) + \tilde{H}'(x)(v_i - x) + \tfrac{1}{2}A_i[v_i - x, v_i - x]$$

for $i = 1, 2, \ldots, N+K+1$, where we use the mean values

$$A_i := \int_0^1 \tilde{H}''\big(x + t(v_i - x)\big)2(1-t)\,\mathrm{d}t$$

of $\tilde{H}''$. Multiplying these equations with the corresponding barycentric co-ordinates γ_i, summing and taking norms yields

$$\left\| \tilde{H}(x) - \sum_{i=1}^{N+K+1} \gamma_i \tilde{H}(v_i) \right\| \leqslant \tfrac{1}{2}\alpha\delta^2$$

as a consequence of (40.5). The result now follows since

$$H(x) = \sum_{i=1}^{N+K+1} \gamma_i \tilde{H}(v_i). \qquad \square$$

In the next estimate the thickness of a simplex has a meaningful role. One possible measure of thickness is the following definition:

Let σ be a simplex with diameter δ and barycenter x. Let ρ be the radius of the largest ball having center x and being contained in σ. Then the *measure of thickness* of σ is defined by

$$\theta(\sigma) := \frac{\rho}{\delta}. \tag{40.6}$$

The *measure of thickness of a triangulation* $\mathcal{T}$ is defined by

$$\theta(\mathcal{T}) := \inf\{\theta(\sigma) \colon \sigma \in \mathcal{T}\}.$$

For standard triangulations, such measures are well known and > 0, see, e.g., KOJIMA [1978] or SAIGAL [1979]. For example, the Coxeter–Freudenthal–Kuhn triangulation of $\mathbb{R}^q$ has thickness $\theta = 1/((q+1)\sqrt{2})$.

The next estimate gives a measure of how well H'_σ approximates the derivative $\tilde{H}'$.

THEOREM 40.3. *Let $\sigma \subset \mathbb{R}^{N+K}$ be an $(N+K)$-simplex having diameter δ and thickness θ. If $x \in \sigma$, then $\|\tilde{H}'(x) - H'_\sigma(x)\| \leqslant \delta\alpha/\theta$.*

PROOF. Let $\sigma = [v_1, v_2, \ldots, v_{N+K+1}]$. From Taylor's formula we have

$$\begin{aligned} \tilde{H}'(x)(v_i - v_j) &= \tilde{H}'(x)(v_i - x) - \tilde{H}'(x)(v_j - x) \\ &= \tilde{H}(v_i) - \tilde{H}(v_j) \\ &\quad - \tfrac{1}{2}A_i[v_i - x, v_i - x] + \tfrac{1}{2}A_j[v_j - x, v_j - x] \end{aligned}$$

for $i, j = 1, 2, \ldots, N+K+1$, where the mean values A_i of $\tilde{H}''$ are defined as in the previous proof. From the definition of the piecewise linear approximation we immediately obtain

$$H'_\sigma(x)(v_i - v_j) = \tilde{H}(v_i) - \tilde{H}(v_j).$$

Subtracting corresponding sides of the above equations and taking norms and using Theorem 40.2 yields

$$\left\| \left(\tilde{H}'(x) - H'_\sigma(x)\right)(v_i - v_j)\right\| \leqslant \alpha\delta^2.$$

By making convex combinations with this last estimate, we obtain

$$\left\| \left(\tilde{H}'(x) - H'_\sigma(x)\right)(u - v)\right\| \leqslant \alpha\delta^2$$

for all $u, v \in \sigma$. From the definition of thickness (40.6) it follows that the set $\{u - v\colon\ u, v \in \sigma\}$ contains the ball with radius $\theta\delta$ and center zero. Thus the above estimate extends to the corresponding matrix norms

$$\theta\delta\left\| \left(\tilde{H}'(x) - H'_\sigma(x)\right)\right\| \leqslant \alpha\delta^2,$$

and the assertion follows. □

The next theorem is a useful characterization of transverse simplices. We employ the perturbation notation of Theorem 33.1.

THEOREM 40.4. *A simplex $\sigma \in \mathcal{T}$ is transverse if and only if it contains solutions v_ε of $H(v) = \vec{\varepsilon}$ for sufficiently small $\varepsilon > 0$.*

PROOF. The proof is obtained by modifying the arguments in the Perturbation theorem 33.1 and the piecewise linear step (see, e.g., Algorithm 33.1). If σ does not contain the asserted solutions v_ε for sufficiently small $\varepsilon > 0$, then by the definition of a completely labeled facet and a transverse cell, it cannot be transverse. On the other hand, if σ contains solutions v_ε for sufficiently small $\varepsilon > 0$, then by an obvious generalization of Theorem 33.1, the solution set consists of regular points of H for sufficiently small $\varepsilon > 0$. Hence, if ε varies, no faces of σ of dimension $< N$ can be intersected, and hence always the same N-faces of σ have to be intersected by this solution set. Clearly, those are the completely labeled N-faces of σ. □

The following theorem guarantees that all regular zero points of $\tilde{H}$ can be approximated by transverse simplices. In particular, such estimates as these may be used for obtaining the starting simplices for the piecewise linear algorithms discussed in this chapter.

THEOREM 40.5. *Let $\sigma \subset \mathbb{R}^{N+K}$ be an $(N + K)$-simplex having vertices v_i, $i = 1, \dots, N + K + 1$, diameter δ, thickness θ and barycenter x such that $\tilde{H}(x) = 0$. If*

$$\frac{\kappa\alpha\delta}{\theta} < \tfrac{1}{2},$$

then σ is transverse.

PROOF. In view of the previous theorem, it suffices to show that the affine approximation H_σ has a solution point $x_\varepsilon \in \sigma$ such that

$$H_\sigma(x_\varepsilon) = \vec{\varepsilon} \tag{40.7}$$

for sufficiently small $\varepsilon > 0$. Since H_σ is affine, any point given by a generalized Newton step

$$x_\varepsilon := x - B\big(H_\sigma(x) - \vec{\varepsilon}\big)$$

satisfies Eq. (40.7), provided that B is a right inverse of H'_σ. If we show that the essential part of the Newton term satisfies the estimate

$$\big\|BH_\sigma(x)\big\| < \theta\delta \tag{40.8}$$

for a particular B, then we conclude from the definition of thickness of a simplex that $x_\varepsilon \in \sigma$ for sufficiently small $\varepsilon > 0$, and the assertion follows. From Theorem 40.3 we have

$$\big\|\tilde{H}'(x) - H'_\sigma(x)\big\| \leqslant \frac{\delta\alpha}{\theta}$$

and hence by the bounds (40.5) and the hypothesis,

$$\big\|\tilde{H}'(x)^+\big(\tilde{H}'(x) - H'_\sigma(x)\big)\big\| \leqslant \frac{\kappa\delta\alpha}{\theta} < \tfrac{1}{2}.$$

We can now define B via the Neumann series

$$B := \sum_{i=0}^{\infty} \big(\tilde{H}'(x)^+\big(\tilde{H}'(x) - H'_\sigma(x)\big)\big)^i \tilde{H}'(x)^+.$$

Multiplying the identity

$$H'_\sigma(x) = \tilde{H}'(x)\big(\mathrm{Id} - \tilde{H}'(x)^+\big(\tilde{H}'(x) - H'_\sigma(x)\big)\big)$$

from the right by B verifies that B is indeed a right inverse of H'_σ. From the Neumann series we can also see that the estimate

$$\|B\| \leqslant \frac{\kappa}{1 - \kappa\alpha\delta/\theta} < 2\kappa$$

holds. On the other hand, Theorem 40.2 implies

$$\big\|H_\sigma(x)\big\| = \big\|H_\sigma(x) - \tilde{H}(x)\big\| \leqslant \tfrac{1}{2}\alpha\delta^2.$$

Combining the last two estimates yields the estimate (40.8) and hence the assertion follows. □

The next theorem shows that the piecewise linear manifold $\mathcal{M} = H^{-1}(0)$ approximates the given manifold $\tilde{\mathcal{M}} = \tilde{H}^{-1}(0)$ quadratically in the meshsize.

THEOREM 40.6. *Let $x \in \mathbb{R}^{N+K}$ be such that* $\operatorname{dist}(x, \tilde{\mathcal{M}}) < (\kappa\alpha)^{-1}$. *Let $w \in \tilde{\mathcal{M}}$ be a nearest point to x, i.e.,* $\|x - w\| = \operatorname{dist}(x, \tilde{\mathcal{M}})$. *If $H(x) = 0$, then $\|x - w\| \leqslant \kappa\alpha\delta^2$.*

PROOF. Since w satisfies the optimization problem

$$\min_w \left\{\|x - w\|\colon\ \tilde{H}(w) = 0\right\},$$

the Lagrange equations yield

$$x - w \in \operatorname{range}\left(\tilde{H}'(w)^*\right) \quad \text{or equivalently,} \quad (x - w) \perp \ker\left(\tilde{H}'(w)\right).$$

From Taylor's formula we have

$$\tilde{H}(x) - \tilde{H}(w) = \tilde{H}'(w)(x - w) + \tfrac{1}{2}A[x - w, x - w],$$

where

$$A = \int_0^1 \tilde{H}''\big(w + t(x - w)\big)2(1 - t)\,\mathrm{d}t$$

again denotes a mean value of $\tilde{H}''$. Since $(x-w) \perp \ker(\tilde{H}'(w))$, and since the Moore–Penrose inverse performs the inversion orthogonally to $\ker(\tilde{H}'(w))$, we have

$$\tilde{H}'(w)^+\tilde{H}(x) = x - w + \tfrac{1}{2}\tilde{H}'(w)^+A[x - w, x - w].$$

From Theorem 40.2 we have

$$\|\tilde{H}(x)\| = \|\tilde{H}(x) - H(x)\| \leqslant \tfrac{1}{2}\alpha\delta^2.$$

From these last two statements and the assumptions (40.5) we obtain

$$\|x - w\| \leqslant \tfrac{1}{2}\kappa\alpha\delta^2 + \tfrac{1}{2}\kappa\alpha\|x - w\|^2 \leqslant \tfrac{1}{2}\kappa\alpha\delta^2 + \tfrac{1}{2}\|x - w\|,$$

and the assertion follows. □

Up to now our approximation estimates have been of a local nature. In order to obtain global approximation results we need to apply more sophisticated tools and technical arguments. One such tool is the Brouwer degree, which for $K = 1$ may be used in a manner similar to that of RABINOWITZ [1971] to obtain the existence of global continua. PEITGEN and PRÜFER [1979] and also PEITGEN [1982] have given extensive discussions of the constructive role the piecewise linear methods play in connection with such arguments.

For our purpose the continuous Newton method seems to be a suitable tool. We consider the autonomous differential equation

$$\dot{x} = -\tilde{H}'(x)^{+}\tilde{H}(x). \tag{40.9}$$

If an initial point x_0 for (40.9) is sufficiently close to $\tilde{\mathcal{M}} = \tilde{H}^{-1}(0)$, then the flow initiating at x_0 has an exponentially asymptotic limit $x_\infty \in \tilde{\mathcal{M}}$, and the map $x_0 \mapsto x_\infty$ is smooth, see, e.g., TANABE [1979]. Analogously, if zero is a regular value of H and the meshsize of $\mathcal{T}$ is sufficiently small, then we may consider the flow defined by

$$\dot{x} = -H'(x)^{+}H(x). \tag{40.10}$$

Note that the right hand of (40.10) is piecewise affine, but not continuous, and that a solution path consists of a polygonal path having nodes on lower dimensional faces $\tau \in \mathcal{T}^{N+K-1}$.

To see this, we concentrate on one simplex σ where the map is affine and consider the auxiliary flow

$$\dot{x} = -A^{+}(Ax - b)$$

for some $(N, N+K)$-matrix A with maximal rank. Integrating this equation leads to

$$x = \mathrm{e}^{-A^{+}At}\big(x_0 - A^{+}b\big) + A^{+}b$$

where x_0 is an initial point on one facet of σ. Of course, the solution is only valid until the simplex σ is exited. We see that the above solution describes a straight line, though the parametrization is not linear.

It is possible by use of some technical arguments to show that the piecewise linear case (40.10) has results analogous to (40.9), i.e., if an initial point x_0 for (40.10) is sufficiently close to $\mathcal{M} = H^{-1}(0)$, then the flow initiating at x_0 has an exponentially asymptotic limit $x_\infty \in \tilde{\mathcal{M}}$, and the map $x_0 \mapsto x_\infty$ is absolutely continuous. The detailed arguments are omitted. We shall only sketch how this technique may be used to obtain the following two theorems.

THEOREM 40.7. *If $x_0 \in \tilde{\mathcal{M}}$ and the meshsize divided by the measure of thickness δ/θ of $\mathcal{T}$ is sufficiently small, then there exists a transverse $\sigma \in \mathcal{T}$ such that* $\operatorname{dist}(x_0, \sigma) \leqslant \kappa\alpha\delta^2$.

PROOF. Consider the initial value problem (40.10) with initial value x_0 and asymptotic limit $x_\infty \in \mathcal{M}$. A full initial Newton step is given by

$$x_1 = x_0 - H'(x_0)^{+}H(x_0).$$

From Theorem 40.2 we obtain the estimate $\|H(x_0)\| \leqslant \frac{1}{2}\alpha\delta^2$. From Theorem 40.3 and (40.5) we obtain $\|H'(x_0)^+\| \approx \|\tilde{H}'(x_0)^+\| \leqslant \kappa$. Thus a rough bound for the stepsize of the full initial Newton step is given by $\frac{1}{2}\kappa\alpha\delta^2$. Hence to obtain the assertion we estimate $\|x_0 - x_\infty\|$ by twice this steplength. □

The algorithms in this section generate connected components of the piecewise linear manifold $\mathcal{M}$. The following theorem assures that such a connected component approximates the entire manifold $\tilde{\mathcal{M}}$ if it is compact and connected.

THEOREM 40.8. *Let zero also be a regular value of H. Let $C \subset \tilde{\mathcal{M}}$ be a compact connected subset (which could be all of $\tilde{\mathcal{M}}$). Then for any triangulation $\mathcal{T}$ for which the meshsize divided by the measure of thickness δ/θ is sufficiently small, there is a connected compact piecewise linear submanifold $C_{\mathcal{T}} \subset \mathcal{M}$ such that for every $x_0 \in C$ there is an $x_\infty \in C_{\mathcal{T}}$ for which $\|x_0 - x_\infty\| < \kappa\alpha\delta^2$ holds.*

PROOF. Consider the Newton map $x_0 \in C \mapsto x_\infty \in \mathcal{M}$ introduced above. Since this map is continuous, and since the continuous image of a compact and connected set is compact and connected, the piecewise linear submanifold

$$C_{\mathcal{T}} := \{\mathcal{M}_\sigma\colon\ \sigma \in \mathcal{T} \text{ and } x_\infty \in \sigma \text{ for some } x_0 \in C\}$$

is compact and connected. Now the assertion follows from estimates in Theorem 40.7. □

It is now clear from the preceding discussion that if $\tilde{\mathcal{M}}$ is compact and connected, then a connected component of $\mathcal{M}$ approximates $\tilde{\mathcal{M}}$ globally and quadratically for sufficiently small meshsize, provided the measure of thickness of $\mathcal{T}$ stays bounded away from zero.

It is also possible to formulate measures of efficiency for piecewise linear approximations of K-manifolds. Analogously to corresponding results for $K = 1$ as cited in the section on piecewise linear homotopy methods, ALEXANDER [1987] has studied the average intersection density for several triangulations in the context of piecewise linear approximations of K-manifolds.

If zero is a regular value of $\tilde{H}$ and H, then the smooth manifold $\tilde{\mathcal{M}}$ and the approximating manifold $\mathcal{M}$ inherit a natural orientation which in the former case is a basic concept of differential geometry and in the latter case is analogous to the orientation described in Section 37. It can be shown that these orientations are consistent with each other for sufficiently fine mesh size, see GNUTZMANN [1989].

41. Numerical integration over surfaces

Recently, GEORG [1991] introduced a new approach to the numerical quadrature of surface integrals for compact surfaces $\mathcal{B}$ which are only implicitly defined, for example, by $\mathcal{B} = \{x \in \mathbb{R}^3\colon\ H(x) = 0\}$ where $H : \mathbb{R}^3 \to \mathbb{R}$. It is assumed that the surface $\mathcal{B}$ is globally approximated via a pseudo manifold $\mathcal{T}$ (piecewise linear approximation) such as was described in the preceding section. Such piecewise linear approximations are typically used in the panel method for solving boundary integral equations related to three-dimensional partial differential equations, see, e.g., BALLMAN, EPPLER and HACKBUSCH [1988] or HACKBUSCH [1989]. Since (in the context of the last section) the approximation is given automatically, there is typically no explicit parametrization of $\mathcal{B}$ available.

Hence, Georg assumed that a parametrization of the surface $\mathcal{B}$ is only indirectly given via a piecewise smooth isomorphism

$$m : \mathcal{T} \to \mathcal{B},$$

e.g., via an iterative method such as Newton's method.

For example, the boundary element package of ATKINSON [1993] asks the user to write a subroutine for defining a parametrization m which seems quite different from the above assumptions. However, it can be shown that both concepts are equivalent. Atkinson handles the difficulty that the parametrization is not given explicitly by using cubic interpolation. The purpose of Georg's approach was to avoid the handling of the partial derivatives of m at all, even via finite differences or interpolation. This leads to a quadrature method which is potentially as accurate as wanted or needed.

Let σ denote a triangle of $\mathcal{T}$. Hence $m(\sigma)$ denotes a smooth piece of the surface $\mathcal{B}$ and $m : \sigma \to m(\sigma)$ denotes a smooth parametrization of this piece. For the purposes of our discussion we can assume without loss of generality that σ is the standard triangle $\sigma = \{(s,t)\colon\ 0 \leqslant s, t, s+t \leqslant 1\}$. We consider the task of numerically approximating

$$\int_{m(\sigma)} f(x)\,\mu(\mathrm{d}x). \tag{41.1}$$

Here μ indicates the usual measure on $\mathcal{B}$ (i.e., the so-called surface element) and f is a given integrand.

The standard approach to the numerical quadrature of (41.1) is to consider the equivalent integral in planar coordinates

$$\int_{m(\sigma)} f(s)\,\mu(\mathrm{d}x) = \int_{\sigma} f\big(m(s,t)\big)\,\|m_s \times m_t\|\,\mathrm{d}s\,\mathrm{d}t. \tag{41.2}$$

Here m_s, m_t denote the partial derivatives of m with respect to the parameters s, t. Many quadrature rules are known for the right-hand side of (41.2), see, e.g., DAVIS and RABINOWITZ [1984] or STROUD [1971]. A simple approach to numerically approximate (41.2) is to subdivide σ into small triangles σ_i (this corresponds to a subdivision of the surface $m(\sigma)$ into the pieces $m(\sigma_i)$) and then to approximate the integral via one of the composite rules:

$$\int_{m(\sigma)} f(x)\,\mu(\mathrm{d}x) = \sum_i \int_{\sigma_i} f\big(m(s,t)\big)\,\gamma(s,t)\,\mathrm{d}s\,\mathrm{d}t$$

$$\approx \tfrac{1}{3} \sum_{i} \sum_{j=0}^{2} f\big(m(v_{i,j})\big)\, \gamma(v_{i,j})\, \mathcal{A}(\mathcal{V}_i), \tag{41.3}$$

$$\int_{m(\sigma)} f(x)\, \mu(\mathrm{d}x) \approx \sum_{i} f\big(m(b_i)\big)\, \gamma(b_i)\, \mathcal{A}(\mathcal{V}_i). \tag{41.4}$$

Here, the vertices and the barycenter of σ_i are denoted by $\mathcal{V}_i := \{v_{i,j}\colon\ j = 0, 1, 2\}$ and b_i, respectively, and $\gamma := \|m_s \times m_t\|$. $\mathcal{A}(\mathcal{V}_i)$ denotes the area of a triangle with vertices $\mathcal{V}_i$. Methods (41.3) and (41.4) are the respective extensions of the composite trapezoidal or midpoint rule for integrating over a triangle.

Under adequate smoothness assumptions on the integrand f, the local error, i.e., the error of each summand, is $\mathrm{O}(h^4)$ where $h = \max_i \operatorname{diam} \sigma_i$. Under the summation of the composite rule, these local errors lead to a global error $\mathrm{O}(h^2)$. It is also known (in the case of an equidistant subdivision) that the global error can be expanded in terms of h^2, see LYNESS [1978]. This is important for applying extrapolation methods to increase the accuracy (e.g., Romberg's scheme).

An alternative approach to approximate (41.1), which does not use the partial derivatives of m, was proposed by GEORG [1991]:

$$\int_{m(\sigma)} f(x)\, \mu(\mathrm{d}x) \approx \tfrac{1}{3} \sum_{i} \sum_{j=0}^{2} f\big(m(v_{i,j})\big)\, \mathcal{A}\big(m(\mathcal{V}_i)\big), \tag{41.5}$$

$$\int_{m(\sigma)} f(x)\, \mu(\mathrm{d}x) \approx \sum_{i} f\big(m(b_i)\big)\, \mathcal{A}\big(m(\mathcal{V}_i)\big). \tag{41.6}$$

We refer to the quadrature formulae (41.5), (41.6) as the *modified trapezoidal rule* and the *modified midpoint rule* for surface integrals, respectively.

The analogues of these quadrature rules for line integrals were discussed in Section 30. As in the case of the line integrals, the modified quadrature versions offer the significant advantage of avoidance of the need to evaluate derivatives.

Until very recently, it was not known whether the modified rules would also permit an asymptotic expansion of the global error in terms of h^2 (in the case of an equidistant subdivision). GEORG [1991] conjectured this expansion, GEORG and TAUSCH [1993] established that at least the leading error term is $\mathrm{O}(h^2)$. Very recently, VERLINDEN and COOLS [1994], and LYNESS [1993] have independently proven the full validity of the asymptotic expansion.

This justifies the incorporation of extrapolation steps in an adaptive numerical integration method for surface integrals which was suggested by GEORG and WIDMANN [1993]. The authors assumed an integrand with a weak singularity in an unknown position and developed a strategy which mixes cautious extrapolation steps (when indicators signal smoothness) with an adaptive strategy (when indicators signal the vicinity of a singularity).

We present the third example taken from their paper. The numerical integration method was programmed in C and was run on a PC with the 80386/387 processors.

We integrate over the surface of a ring cyclide, i.e., a torus with varying radius. The surface $\mathcal{B}$ is defined by the equation

$$\left(x_1^2 + x_2^2 + x_3^2 + R^2 - b^2 - k^2\right)^2 - 4(Rx_1 + kb)^2 - 4\left(R^2 - b^2\right)x_2^2 = 0.$$

Here, R defines the "big" radius, and the small radius r satisfies $k - b \leqslant r \leqslant k + b$. We choose the parameters $R = 1,\ k = 0.3$ and $b = 0.15$. The approximating pseudo manifold $\mathcal{T}$ consists of 4486 triangles and was generated with our software pla_s_k written by WIDMANN [1990a], see Chapter VIII, using a mesh size of $\delta = 0.15$.

We use the integrand

$$f(x) = \frac{\nu(x) \bullet (x - e)}{\|x - e\|^3}$$

with the weak singularity $e := (1 + k + b)e_1$, where $\nu(x)$ denotes the outer normal of length 1 at $x \in \mathcal{B}$. The resulting integral is typical for boundary element methods. In fact, this particular integral has an interpretation as a solid angle and hence is known explicitly, i.e.,

$$\int_{\mathcal{B}} f(x)\, \mu(\mathrm{d}x) = 2\pi.$$

The table below lists the performance of the numerical integration with increasing precision. "Tol" is a tolerance parameter used to monitor the adaptive and extrapolation steps. "Rel. error" gives the relative error of the numerical integral, "Time" gives the run time of the integration in seconds.

Tol	Rel. error	Time
1e-2	2.7e-03	117.10
1e-3	8.5e-05	117.70
1e-4	2.4e-05	120.89
1e-5	6.6e-06	129.30
1e-6	4.2e-06	151.82
1e-7	2.5e-06	208.83
1e-8	4.3e-08	357.46
1e-9	3.2e-08	722.33
1e-10	1.4e-08	1363.09
1e-11	1.4e-08	3801.83
1e-12	3.0e-09	10872.55

CHAPTER VII

Complexity

42. Smale's approach

In modern complexity investigations of continuation-type methods the so-called α-theory of SMALE [1986] is a convenient tool. This theory is closely related to the classical Kantorovich estimates for Newton iterations, see, e.g., ORTEGA and RHEINBOLDT [1970] and DEUFLHARD and HEINDL [1979]. In contrast to the Kantorovich estimates, Smale's estimates are based on information at only one point, involving however all derivatives. The maps under consideration have to be analytic.

On the other hand, an analytic map is characterized by all its derivatives at one point. In fact, RHEINBOLDT [1988a] showed that Smale's estimates can be derived from the Kantorovich estimates. However, for complexity considerations, it is more convenient to have all the relevant information situated at only one point. Let us briefly present Smale's estimates and show how they are used for complexity discussions. Our presentation is based on the introductory parts of the papers of SHUB and SMALE [1991] and RENEGAR and SHUB [1992].

Let E, F be complex Banach spaces and $f : E \to F$ an analytic map. It would be possible to assume that f is given only on some open domain, but for reasons of simplicity of exposition we assume f to be defined on all of E. Then for each point $x \in E$ such that $Df(x) : E \to F$ is an isomorphism the following quantities are defined:

$$\beta(f, x) = \|Df(x)^{-1} f(x)\|, \tag{42.1}$$

$$\gamma(f, x) = \sup_{k>1} \left\| \frac{1}{k!} Df(x)^{-1} D^k f(x) \right\|^{1/(k-1)}, \tag{42.2}$$

$$\alpha(f, x) = \beta(f, x)\gamma(f, x), \tag{42.3}$$

$$\mathcal{N}_f(x) = x - Df(x)^{-1} f(x). \tag{42.4}$$

Note that $\mathcal{N}_f(x)$ is the Newton iterate of x. It is also convenient to introduce the notation

$$\mathcal{N}_f^{\infty}(x) = \lim_{i\to\infty} \mathcal{N}_f^i(x) \tag{42.5}$$

provided Newton's method (started at x) is convergent.

A related one-dimensional "control" Newton method is occasionally generated from the following family of functions

$$h_{\beta,\gamma}(t) = \beta - t + \frac{\gamma t^2}{1 - \gamma t}. \tag{42.6}$$

For $0 < \alpha < 3 - 2\sqrt{2} \approx 0.1716$, the function $h_{\beta,\gamma}$ has two real positive roots, the smaller one being

$$\frac{\tau(\alpha)}{\gamma} = \frac{(\alpha+1) - \sqrt{(\alpha+1)^2 - 8\alpha}}{4\gamma}. \tag{42.7}$$

Moreover, $h''_{\beta,\gamma} > 0$ on the interval $(0, 1/\gamma)$. Thus, Newton's method starting at zero generates a strictly increasing sequence $t_i(\beta,\gamma) = \mathcal{N}^i_{h_{\beta,\gamma}}(0)$ converging to this root.

Occasionally, a slightly smaller upper bound for α is used, namely $\alpha_0 = \frac{1}{4}(13 - 3\sqrt{17}) \approx 0.1577$.

The following is a modification of Smale's α-theorem.

THEOREM 42.1. *Let $x_0 \in E$, $\alpha = \alpha(f, x_0)$, $\gamma = \gamma(f, x_0)$. If $\alpha \leqslant \alpha_0 \approx 0.1577$, then the iterates $x_{i+1} = \mathcal{N}_f(x_i)$ are defined and converge to a zero point $x_\infty = \mathcal{N}^\infty_f(x_0) \in E$ with the rate*

$$\|x_{i+1} - x_i\| \leqslant \left(\tfrac{1}{2}\right)^{2^i - 1} \|x_1 - x_0\|.$$

Moreover, the following estimates hold:

$$\|x_\infty - x_0\| \leqslant \frac{\tau(\alpha)}{\gamma}, \qquad \|x_\infty - x_1\| \leqslant \frac{\tau(\alpha) - \alpha}{\gamma}.$$

An easy consequence is

COROLLARY 42.1.

$$\|x_\infty - x_i\| \leqslant \varepsilon \quad \text{for } i \geqslant 1 + \log\left|\log \frac{\tau(\alpha)}{\varepsilon\gamma}\right|.$$

Furthermore, by using the control Newton iterates $t_i = t_i(\beta,\gamma)$, a stricter estimate can be obtained under the same hypotheses:

THEOREM 42.2. $\|x_i - x_{i-1}\| \leqslant t_i - t_{i-1}$.

Another property which is important for complexity discussions is the fact that α is upper semicontinuous, more precisely:

THEOREM 42.3. *Let* $\psi(u) := 2u^2 - 4u + 1$ *and* $u := \gamma(f, x_0)\|x_0 - x\|$. *Then*

$$\alpha(f, x) \leqslant \frac{\alpha(f, x_0)(1-u) + u}{\psi(u)^2}.$$

From the previous theorem it is possible to obtain a uniform estimate for Newton steps:

THEOREM 42.4. *There are universal constants* $\bar{\alpha} \approx 0.0802$ *and* $\bar{u} \approx 0.0221$ *with the following property: Let* $\bar{\gamma} > 0$ *and* $x, \zeta \in E$. *If* $\beta(f, \zeta) \leqslant \bar{\alpha}/\bar{\gamma}$ *and* $\|x - \zeta\| \leqslant \bar{u}/\bar{\gamma}$, *then* $\|\mathcal{N}_f(x) - \mathcal{N}_f^\infty(\zeta)\| \leqslant \bar{u}/\bar{\gamma}$.

This theorem is used to investigate the complexity of path following in the following way: Let $H : [0, 1] \times E \to F$ be a continuous (homotopy) map which is analytic in the second argument. We further assume that a continuous solution path $\zeta : [0, 1] \to E$ exists, i.e., $H(t, \zeta(t)) = 0$ for $t \in [0, 1]$, such that the derivative $H_\zeta(t, \zeta(t))$ is an isomorphism. The following crude path-following method can be designed: choose a subdivision $0 = t_0 < t_1 < \cdots < t_k = 1$ and define

$$x_i := \mathcal{N}_{H(t_i, .)}(x_{i-1}) \quad \text{for } i = 1, \ldots, k. \tag{42.8}$$

It is clear that this method follows the solution curve if $\|x_0 - \zeta(0)\|$ and $|t_i - t_{i-1}|$ are small enough. Of course, the crucial number for complexity considerations is the number k of Newton steps involved in the above embedding method. If it is wished to obtain some points of the solution curve with high accuracy, then the complexity described in Theorem 42.1 has to be added.

The preceding analysis immediately furnishes a tool to determine the estimates necessary for a successful tracing of the solution curve:

THEOREM 42.5. *Let* $\|x_0 - \zeta(0)\| \leqslant \bar{u}/\bar{\gamma}$, *and let the mesh* t_i *be so fine that*

$$\beta\big(H(t_i, .), \zeta(t_{i-1})\big) \leqslant \bar{\alpha}/\bar{\gamma} \quad \textit{and} \quad \gamma\big(H(t_i, .), \zeta(t_{i-1})\big) \leqslant \bar{\gamma}.$$

Then the embedding method (42.8) *follows the solution path* ζ. *In fact,* $\|x_i - \zeta(t_i)\| \leqslant \bar{u}/\bar{\gamma}$.

43. Notes and remarks

To summarize, we have outlined a program for approaching complexity investigations when Newton steps are the primary tool of path following methods. As can be seen from the last theorem, the success of the approach depends heavily on the availability of estimates $\beta(H(t, .), \zeta(s)) \leqslant C_1|t-s|$ and $\gamma(H(t, .), \zeta(s)) \leqslant C_2|t-s|$ with explicit constants C_1 and C_2.

This program was carried out by SHUB and SMALE [1991] for the case of a homotopy method for calculating all solutions of a system of polynomial equations (Bezout's theorem). A previous effort along similar lines was given by RENEGAR [1987].

Recently, this approach has also been used by RENEGAR and SHUB [1992] for a unified complexity analysis of various interior methods designed for solving linear and convex quadratic programming problems. They obtain and rederive various "polynomial time" estimates. The linear programming barrier method was first analysed by GONZAGA [1988]. The quadratic programming barrier method was analysed by GOLDFARB and LIU [1991]. A primal–dual linear programming algorithm was investigated by KOJIMA, MIZUNO and YOSHISE [1988] and MONTEIRO and ADLER [1989]. The algorithm has roots in MEGIDDO [1988]. Primal–dual linear complementarity and quadratic programming algorithms were discussed by KOJIMA, MIZUNO and YOSHISE [1989] and MONTEIRO and ADLER [1989]. All of the above algorithms follow the *central trajectory* studied by BAYER and LAGARIAS [1989] and MEGIDDO and SHUB [1989]. For the case of the linear complementarity problem, MIZUNO, YOSHISE and KIKUCHI [1989] present several implementations and report computational experience which confirms the polynomial complexity.

The above discussion involved path following methods of Newton type. RENEGAR [1985, 1988b] has made complexity investigations for piecewise linear path following methods.

CHAPTER VIII

Available Software

We conclude the paper by listing some available software related to path following and indicate how the reader might access these codes. No attempt to compare or evaluate the various codes is offered. In any case, our opinion is that path following codes always need to be considerably adapted to the special purposes for which they are designed. The path following literature offers various tools for accomplishing such tasks. Although there are some general purpose codes, probably none will slay every dragon.

RHEINBOLDT, ROOSE and SEYDEL [1990] present a list of features and options that appear to be necessary or desirable for continuation codes. This should be viewed as a guideline for people who want to create a new code.

Several of the codes can be accessed via *netlib*: The best way to obtain them is to ftp into `netlib@research.att.com`, login as netlib, password = your e-mail address. It is also possible to e-mail to netlib by writing *send index*. Information on how to proceed will then be e-mailed back to you.

ABCON

This is a predictor corrector continuation algorithm using variable order Adams–Bashforth predictors written by LUNDBERG and POORE [1991], see also Section 9. It can be obtained via anonymous ftp from

```
netlib@research.att.com
(ftp 192.20.225.2, ..., bin, get contin/abcon.f.Z).
```

ALCON

This package has been written by DEUFLHARD, FIEDLER and KUNKEL [1987]. It is a continuation method for algebraic equations $f(x,\tau) = 0$, based on QR factorization as a solver for the equations arising in the Gauss–Newton iteration of the corrector step. Turning points and simple bifurcations can be computed on demand. It can be found in the electronic library of the Konrad Zuse Zentrum für Informationstechnik in Berlin. The reader may telnet or ftp to `sc.ZIB-Berlin.de (130.73.108.11)` and login under the user identification elib, no password is required. The sources can be found in the directory `/pub/ELIB/codelib` either in unpacked form or as a `tar.Z` file.

AUTO

This is a software package written by E. Doedel. It is mainly intended to investigate bifurcation phenomena. There is a charge of $175 for the software, a manual by DOEDEL and KERNÉVEZ [1986] is also available, contact:

S.K. Shull
Applied Mathematics, 217-50
California Institute of Technology
Pasadena, CA 91125
phone: (818) 356-4560

BIFPACK

This package has been written by SEYDEL [1991a]. It is meant primarily for bifurcation analysis of ODEs. This is not public domain software. However, as a research tool, it is freely distributed for *noncommercial* use, except for a $20 contribution for handling. Indicate whether you prefer `BIFPACK` on 5.25" or on 3.5" diskette (1.4 MB, DOS double-density). Contact:

Prof. Rüdiger Seydel
Abt. Mathematik VI, Universität Ulm
Helmholtzstr. 18
D-89081 Ulm/Donau, Germany
e-mail: `seydel@mathematik.uni-ulm.de`

CANDYS/QA

This is a software system for the qualitative analysis of nonlinear dynamical systems by FEUDEL and JANSEN [1992] of the Arbeitsgruppe "Nichtlineare Dynamik", Max-Planck-Gesellschaft, Universität Potsdam, Germany.

CONKUB

This is an interactive program for continuation and bifurcation of large systems of nonlinear equations written by MEJIA [1986], see also MEJIA [1990]. It is currently available from him via e-mail: `ray@helix.nih.gov`.

DERPAR

This package was written by KUBÍČEK [1976] and HOLODNIOK and KUBÍČEK [1984]. This is a Fortran subprogram for the evaluation of the dependence of the solution of a nonlinear system on a parameter. The modified method of Davidenko, which applies the implicit function theorem, is used in combination with Newton's method and Adam's integration formulas. The program can be accessed via netlib, see number 502 in the directory *toms*.

DSTOOL

This is a computer assisted exploration package for dynamical systems by BACK, GUCKENHEIMER, MYERS, WICLIN and WORFOLK [1992] of the Center for Applied Mathematics, Cornell University, Ithaca.

DYNAMICS

This is software for the numerical exploration of chaotic systems developed by NUSSE and YORKE [1992] of the University of Maryland.

HOMPACK

This is a suite of FORTRAN 77 subroutines for solving nonlinear systems of equations by homotopy methods, written by L.T. Watson, see WATSON, BILLUPS and MORGAN [1987]. There are subroutines for fixed point, zero finding, and general homotopy curve tracking problems, utilizing both dense and sparse Jacobian matrices, and implementing three different algorithms: ODE-based, normal flow, and augmented Jacobian. The program can be accessed via netlib under the directory *hompack*. See also number 652 in the directory *toms*.

INSITE

This is a set of practical numerical algorithms for chaotic systems written by PARKER and CHUA [1989] of INSITE Software, Berkeley.

KAOS

This is a computational environment for exploring dynamical systems developed by GUCKENHEIMER and KIM [1991] of the Center for Applied Mathematics, Cornell University, Ithaca.

LOCBIF

A. Khibnik and collaborators in Moscow have developed several codes for path following and bifurcation analysis. `CYCLE` is a one-parameter continuation program for limit cycles. `LINLBF` has been designed for multi-parameter bifircation analysis of equilibrium points, limit cycles, fixed points of maps, respectively. LOCBIF, developed by KHIBNIK, KUZNETSOV, LEVITIN and NIKOLAEV [1993], is a package involving contination techniques and interactive software for bifurcation analysis of ODEs and iterated maps, built originally on top of LINLBF. People interested in trying this software should contact A. Khibnik via e-mail: na.khibnik@na-net.ornl.gov or khibnik@impb.serpukhov.su.

MacMath

This is a dynamical systems software package for the Macintosh, developed by HUBBARD and WEST [1992] of Cornell University, Ithaca.

OB1

This interior point method has been written by I.J. Lustig, R.E. Marsten, and D.F. Shanno. The version of OB1 that implements a primal–dual algorithm for linear programming is available in source code form to academics from Roy Marsten at Georgia Tech. This is the December 1989 version, also known as the WRIP (Workshop on Research in Programming) version. The current version of OB1 is commercial. It implements a primal–dual predictor corrector algorithm for linear programming and is available from XMP Software at prices ranging from $15 000 to $100 000.

XMP Software
Suite 279, Bldg 802
930 Tahoe Blvd
Incline Village, NV 89451
phone: (702) 831- 4XMP
e-mail: tlowe@mcimail.com

PATH

This software package for dynamical systems was originally coded in FORTRAN 77 by KAAS-PETERSEN [1989], and is currently modified to include a graphical interface. According to the workers at the Technical University of Denmark, it seems to be able to handle much larger systems of ODEs than AUTO. For more details and availability, readers may contact Michael Rose via e-mail: lamfmr@lamf.dth.dk.

Phase Plane—XPP

This is a dynamical systems tool developed by ERMENTROUT [1990] of the University of Pittsburgh.

PITCON

This is a Fortran subprogram for continuation and limit points, written by RHEINBOLDT and BURKARDT [1983a,b]. It is used for computing solutions of a nonlinear system of equations containing a parameter. The location of target points where a given variable has a specified value can be located. Limit points are also identified. It uses a local parameterization based on curvature estimates to control the choice of parameter value. The program can be accessed via netlib under the directory *contin*. See also number 596 in the directory *toms*.

PLALGO

This is a software for piecewise linear homotopy methods developed by TODD [1981]. It can be obtained from him via e-mail: `miketodd@orie.cornell.edu`. No support is available, and he says that on-line documentation is weak, although he can send a hard copy.

pla_s_k

This is a C program, written by WIDMANN [1990a], for triangulating surfaces in $\mathbb{R}^3$ which are implicitly defined, see Section 40. It incorporates mesh smoothing and some other features. It is particularly suited for mesh generation (e.g., for boundary element methods) and for visualization purposes. The program can be obtained via e-mail (`Georg@Math.ColoState.Edu`).

PLTMG

This package has been written by R.E. Bank, see also the paper of BANK and CHAN [1986]. It solves elliptic partial differential equations in general regions of the plane. It features adaptive local mesh refinement, multigrid iteration, and a pseudo arclength continuation option for parameter dependencies. The package includes an initial mesh generator and several graphics packages. Full documentation can be obtained in the `PLTMG` User's Guide by R.E. Bank, available from SIAM publications (e-mail: `SIAMPUBS@wharton.upenn.edu`). The program can be accessed via netlib under the directory *pltmg*.

SYMCON

This is a special path following program featuring symbolic exploitation of symmetry developed by GATERMANN and HOHMANN [1991] of the Konrad-Zuse-Zentrum für Informationstechnik, Berlin, Germany. SYMCON is available by anonymous ftp from `elib.zib-berlin.de (130.73.108.11)` in the subdirectory `pub/symcon`.

TraX

This is a program for the simulation and analysis of dynamical systems developed by Levitin and Khibnik of the Institute of Mathematical Problems in Biology, Russian

Academy of Sciences, Pushchino, see KHIBNIK [1990]. It is available from Exeter Software, Setauket, NY.

Last and least

The book ALLGOWER and GEORG [1990] contains several Fortran codes for path following which are to be regarded primarily as illustrations. The intention was to encourage the readers to experiment and be led to make improvements and adaptations suited to their particular applications. We emphasize that these programs should not be regarded as programs of library quality. They can be obtained via e-mail (`Georg@Math.ColoState.Edu`).

References

ABRAHAM, R. and J. ROBBIN (1967), *Transversal Mappings and Flows* (W.A. Benjamin, New York).

ADLER, I., G.C. RESENDE, G. VEIGA and N. KARMARKAR (1989), An implementation of Karmarkar's algorithm for linear programming, *Math. Programming* **44**, 297–335.

ALEXANDER, J.C. (1987), Average intersection and pivoting densities, *SIAM J. Numer. Anal.* **24**, 129–146.

ALEXANDER, J.C. and E.V. SLUD (1983), Global convergence rates of piecewise-linear continuation methods: A probabilistic approach, in: B.C. Eaves, F.J. Gould, H.-O. Peitgen and M.J. Todd, eds., *Homotopy Methods and Global Convergence* (Plenum Press, New York) 15–30.

ALLGOWER, E.L. (1984), Bifurcations arising in the calculation of critical points via homotopy methods, in: T. Küpper, H.D. Mittelmann and H. Weber, eds., *Numerical Methods for Bifurcation Problems*, ISNM **70** (Birkhäuser, Basel) 15–28.

ALLGOWER, E.L., K. BÖHMER, K. GEORG and R. MIRANDA (1992a), Exploiting symmetry in boundary element methods, *SIAM J. Numer. Anal.* **29**, 534–552.

ALLGOWER, E.L., K. BÖHMER and M. GOLUBITSKY, eds. (1992), *Bifurcation and Symmetry*, ISNM **104** (Birkhäuser, Basel).

ALLGOWER, E.L., K. BÖHMER and Z. MEI (1991a), A complete bifurcation scenario for the 2-D nonlinear Laplacian with Neumann boundary conditions, in: R. Seydel, F.W. Schneider, T. Küpper and H. Troger, eds., *Bifurcation and Chaos: Analysis, Algorithms, Applications*, ISNM **97** (Birkhäuser, Basel) 1–18.

ALLGOWER, E.L., K. BÖHMER and Z. MEI (1991b), On a problem decomposition for semi-linear nearly symmetric elliptic problems, in: W. Hackbusch, ed., *Parallel Algorithms for PDE's*, Notes on Numerical Fluid Mechanics **31** (Vieweg Verlag, Braunschweig), 1–17.

ALLGOWER, E.L., K. BÖHMER and Z. MEI (1991c), On new bifurcation results for semi-linear elliptic equations with symmetries, in: J. Whiteman, ed., *The Mathematics of Finite Elements and Applications* **VII** (Academic Press, Brunel) 487–494.

ALLGOWER, E.L. and C.-S. CHIEN (1986), Continuation and local perturbation for multiple bifurcations, *SIAM J. Sci. Statist. Comput.* **7**, 1265–1281.

ALLGOWER, E.L., C.-S. CHIEN and K. GEORG (1989), Large sparse continuation problems, *J. Comput. Appl. Math.* **26**, 3–21.

ALLGOWER, E.L., C.-S. CHIEN, K. GEORG and C.-F. WANG (1991d), Conjugate gradient methods for continuation problems, *J. Comput. Appl. Math.* **38**, 1–16.

ALLGOWER, E.L. and A.F. FÄSSLER (1993), Blockstructure and equivariance of matrices, Submitted.

ALLGOWER, E.L. and K. GEORG (1979), Generation of triangulations by reflections, *Utilitas Math.* **16**, 123–129.

ALLGOWER, E.L. and K. GEORG (1980), Homotopy methods for approximating several solutions to nonlinear systems of equations, in: W. Forster, ed., *Numerical Solution of Highly Nonlinear Problems* (North-Holland, Amsterdam) 253–270.

ALLGOWER, E.L. and K. GEORG (1983a), Predictor-corrector and simplicial methods for approximating fixed points and zero points of nonlinear mappings, in: A. Bachem, M. Grötschel and B. Korte, eds., *Mathematical Programming: The State of the Art* (Springer, Berlin) 15–56.

ALLGOWER, E.L. and K. GEORG (1983b), Relationships between deflation and global methods in the problem of approximating additional zeros of a system of nonlinear equations, in: B.C. Eaves, F.J. Gould, H.-O. Peitgen and M.J. Todd, eds., *Homotopy Methods and Global Convergence* (Plenum Press, New York) 31–42.

ALLGOWER, E.L. and K. GEORG (1989), Estimates for piecewise linear approximations of implicitly defined manifolds, *Appl. Math. Lett.* **1**(5), 1–7.

ALLGOWER, E.L. and K. GEORG (1990), *Numerical Continuation Methods: An Introduction*, Series in Computational Mathematics **13** (Springer, Berlin), 388.

ALLGOWER, E.L. and K. GEORG (1993), Continuation and path following, *Acta Numer.* **2**, 1–64.

ALLGOWER, E.L., K. GEORG and R. WIDMANN (1991), Volume integrals for boundary element methods, *J. Comput. Appl. Math.* **38**, 17–29.

ALLGOWER, E.L., K. GEORG and R. MIRANDA (1993), Exploiting permutation symmetry with fixed points in linear equations, in: E.L. Allgower, K. Georg and R. Miranda, eds., *Exploiting Symmetry in Applied and Numerical Analysis*, Lectures in Applied Mathematics **29** (American Mathematical Society, Providence, RI) 23–36.

ALLGOWER, E.L., K. GEORG and J. WALKER (1993), Exploiting symmetry in 3D boundary element methods, in: R.P. Agarwal, ed., *Contributions in Numerical Mathematics*, World Scientific Series in Applicable Analysis **2** (World Scientific, Singapore) 15–25.

ALLGOWER, E.L. and S. GNUTZMANN (1987), An algorithm for piecewise linear approximation of implicitly defined two-dimensional surfaces, *SIAM J. Numer. Anal.* **24**, 452–469.

ALLGOWER, E.L. and S. GNUTZMANN (1991), Simplicial pivoting for mesh generation of implicitly defined surfaces, *Comput. Aided Geom. Design* **8**, 305–325.

ALLGOWER, E.L. and P.H. SCHMIDT (1985), An algorithm for piecewise-linear approximation of an implicitly defined manifold, *SIAM J. Numer. Anal.* **22**, 322–346.

ALLGOWER, E.L. and H. SCHWETLICK (1993), On minimally extended systems for simple bifurcation points, Preprint.

ALLISON, D.C.S., S. HARIMOTO and L.T. WATSON (1989), The granularity of parallel homotopy algorithms for polynomial systems of equations, *Intern. J. Comput. Math.* **29**, 21–37.

AMANN, H. (1976), Fixed point equations and nonlinear eigenvalue problems in ordered Banach spaces, *SIAM Rev.* **18**, 620–709.

AMBROSETTI, A. and P. HESS (1980), Positive solutions of asymptotically linear elliptic eigenvalue problems, *J. Math. Anal. Appl.* **73**, 411–422.

ATKINSON, K.E. (1993), User's Guide to a boundary element package for solving integral equations on piecewise smooth surfaces, Report 44, Univ. of Iowa.

BACK, A., J. GUCKENHEIMER, M. MYERS, F. WICLIN and P. WORFOLK (1992), Dstool; computer assisted exploration of dynamical systems, *Notices ACM* **39**, 303–309.

BALLISTI, R., C. HAFNER and P. LEUCHTMAN (1982), Application of the representation theory of finite groups to field computation problems with symmetrical boundaries, *IEEE Trans. Magnetics* **18**, 584–587.

BALLMAN, J., R. EPPLER and W. HACKBUSCH (1988), Panel method in fluid mechanics with emphasis on aerodynamics, in: W. Hackbusch, ed., *Notes on Numerical Fluid Dynamics* **21** (Vieweg, Braunschweig).

BANK, R.E. and T.F. CHAN (1986), PLTMGC: A multi-grid continuation program for parameterized nonlinear elliptic systems, *SIAM J. Sci. Statist. Comput.* **7**, 540–559.

BANK, B., J. GUDDAT, D. KLATTE, B. KUMMER and K. TAMMER (1983), *Non-Linear Parametric Optimization* (Birkhäuser, Basel).

BAYER, D. and J.C. LAGARIAS (1989), The nonlinear geometry of linear programming, I: Affine and projective scaling trajectories, II: Legendre transform coordinates and central trajectories, *Trans. Amer. Math. Soc.* **314**, 499–581.

BEN-ISRAEL, A. and T.N.E. GREVILLE (1974), *Generalized Inverses: Theory and Applications* (Wiley, New York).

BERNSTEIN, D.N. (1975), The number of roots of a system of equations, *Functional Anal. Appl.* **9**, 1–4.

BERNSTEIN, S. (1910), Sur la généralisation du problème de Dirichlet, *Math. Ann.* **69**, 82–136.

BERTSEKAS, D.P. (1984), *Constrained Optimization and Lagrange Multiplier Methods* (Academic Press, New York).

BEYN, W.-J. (1992), On smoothness and invariance properties of the Gauss–Newton method, Preprint, Univ. of Bielefeld, Germany.

BIXBY, R.E., J.W. GREGORY, I.J. LUSTIG, R.E. MARSTEN and D.F. SHANNO (1991), Very large-scale linear programming: A case study in combining interior point and simplex methods, Technical Report RRR 34-91, Rutgers Univ., New Brunswick, NJ.

BOLSTAD, J.H. and H.B. KELLER (1986), A multigrid continuation method for elliptic problems with folds, *SIAM J. Sci. Statist. Comput.* **7**, 1081–1104.

BOSSAVIT, A. (1986), Symmetry, groups, and boundary value problems. A progressive introduction to non-commutative harmonic analysis of partial differential equations in domains with geometrical symmetry, *Comput. Methods Appl. Mech. Engrg.* **56**, 167–215.

BOSSAVIT, A. (1993), Boundary value problems with symmetry and their approximation by finite elements, *SIAM J. Appl. Math.* **53**, 1352–1380.

BOURJI, S.K. and H.F. WALKER (1990), Least-change secant updates of nonsquare matrices, *SIAM J. Numer. Math.* **27**, 1263–1294.

BRANIN, F.H. (1972), Widely convergent method for finding multiple solutions of simultaneous nonlinear equations, *IBM J. Res. Develop.* **16**, 504–522.

BRANIN, F.H. and S.K. HOO (1972), A method for finding multiple extrema of a function of n variables, in: F.A. Lootsma, ed., *Numerical Methods for Non-Linear Optimization* (Academic Press, New York) 231–237.

BREZZI, F., J. RAPPAZ and P.A. RAVIART (1980a), Finite dimensional approximation of nonlinear problems. Part 1: Branches of nonsingular solutions, *Numer. Math.* **36**, 1–25.

BREZZI, F., J. RAPPAZ and P.A. RAVIART (1980b), Finite dimensional approximation of nonlinear problems. Part 2: Limit points, *Numer. Math.* **37**, 1–28.

BREZZI, F., J. RAPPAZ and P.A. RAVIART (1981), Finite dimensional approximation of nonlinear problems. Part 3: Simple bifurcation points, *Numer. Math.* **38**, 1–30.

BRODZIK, M.L. and W.C. RHEINBOLDT (1994), On the computation of simplicial approximations of implicitly defined two-dimensional manifolds, Preprint, University of Pittsburgh.

BROOKS, P.S. (1980), Infinite retrogression in the Eaves–Saigal algorithm, *Math. Programming* **19**, 313–327.

BROUWER, L.E.J. (1912), Über Abbildung von Mannigfaltigkeiten, *Math. Ann.* **71**, 97–115.

BROWN, K.M. and W.B. GEARHART (1971), Deflation techniques for the calculation of further solutions of a nonlinear system, *Numer. Math.* **16**, 334–342.

BROYDEN, C.G. (1965), A class of methods for solving nonlinear simultaneous equations, *Math. Comp.* **19**, 577–593.

BYRNE, G.D. and L.A. BAIRD (1985), Distillation calculations using a locally parametrized continuation method, *Comput. Chem. Engrg.* **9**, 593–599.

CANNY, J. and J.M. ROJAS (1991), An optimal condition for determining the exact number of roots of a polynomial system, in: *Proceedings of the ISSAC 91*, Bonn, Germany 96–102.

CARPENTER, T.J. and D.F. SHANNO (1991), An interior point method for quadratic programs based on conjugate projected gradients, Technical Report RRR 55-91, Rutgers Univ., New Brunswick, NJ.

CHAN, T.F. (1984a), Deflation techniques and block-elimination algorithms for solving bordered singular systems, *SIAM J. Sci. Stat. Comp.* **5**, 121–134.

CHAN, T.F. (1984b), Newton-like pseudo-arclength methods for computing simple turning points, *SIAM J. Sci. Statist. Comput.* **5**, 135–148.

CHAN, T.F. (1984c), Techniques for large sparse systems arising from continuation methods, in: T. Küpper, H.D. Mittelmann and H. Weber, eds., *Numerical Methods for Bifurcation Problems*, ISNM **70** (Birkhäuser, Basel) 116–128.

CHAN, T.F. and H.B. KELLER (1982), Arc-length continuation and multi-grid techniques for nonlinear eigenvalue problems, *SIAM J. Sci. Statist. Comput.* **3**, 173–194.

CHAN, T.F. and Y. SAAD (1985), Iterative methods for solving bordered systems with applications to continuation methods, *SIAM J. Sci. Statist. Comput.* **6**, 438–451.

CHAO, K.S. and R. SAEKS (1977), Continuation methods in circuit analysis, *Proc. IEEE* **65**, 1187–1194.

CHAVEZ, R., J.D. SEADER and T.L. WAYBURN (1986), Multiple steady state solutions for interlinked separation systems, *Ind. Eng. Chem. Fundam.* **25**, 566–576.

CHIEN, C.-S. (1989), Secondary bifurcations in the buckling problem, *J. Comput. Appl. Math.* **25**, 277–287.

CHOW, S.N. and J.K. HALE (1982), *Methods of Bifurcation Theory* (Springer, New York).

CHOW, S.N., J. MALLET-PARET and J.A. YORKE (1978), Finding zeros of maps: Homotopy methods that are constructive with probability one, *Math. Comp.* **32**, 887–899.

CHOW, S.N., J. MALLET-PARET and J.A. YORKE (1979), A homotopy method for locating all zeros of a system of polynomials, in: H.-O. Peitgen and H.-O. Walther, eds, *Functional Differential Equations and Approximation of Fixed Points*, Lecture Notes in Math. **730** (Springer, Berlin) 77–88.

CHU, K.-W.E., W. GOVAERTS and A. SPENCE (1994), Matrices with rank deficiency two in eigenvalue problems and dynamical systems, *SIAM J. Num. Anal.* **31**, 524–539.

CHU, M.T. (1984a), The generalized Toda flow, the QR algorithm and the center manifold theory, *SIAM J. Alg. Disc. Meth.* **5**, 187–201.

CHU, M.T. (1984b), A simple application of the homotopy method to symmetric eigenvalue problems, *Linear Algebra Appl.* **59**, 85–90.

CHU, M.T. (1986), A continuous approximation to the generalized Schur's decomposition, *Linear Algebra Appl.* **78**, 119–132.

CHU, M.T. (1988), On the continuous realization of iterative processes, *SIAM Rev.* **30**, 375–387.

CHU, M.T. (1990), Solving additive inverse eigenvalue problems for symmetric matrices by the homotopy method, *IMA J. Numer. Anal.* **9**, 331–342.

CHU, M.T. (1991), A continuous Jacobi-like approach to the simultaneous reduction of real matrices, *Linear Algebra Appl.* **147**, 75–96.

CHU, M.T., T.-Y. LI and T. SAUER (1988), Homotopy methods for general λ-matrix problems, *SIAM J. Matrix Anal. Appl.* **9**, 528–536.

CHUA, L.O. and A. USHIDA (1976), A switching-parameter algorithm for finding multiple solutions of nonlinear resistive circuits, *Circuit Theory and Appl.* **4**, 215–239.

CICOGNA, G. (1981), Symmetry breakdown from bifurcation, *Lett. Nuovo Cimento* **31**, 600–602.

CLIFFE, K.A. and K.H. WINTERS (1986), The use of symmetry in bifurcation calculations and its application to the Bénard problem, *J. Comput. Phys.* **67**, 310–326.

COHEN, A.I. (1972), Rate of convergence of several conjugate gradient algorithms, *SIAM J. Numer. Anal.* **9**, 248–259.

COTTLE, R.W. (1974), Solution rays for a class of complementarity problems, *Math. Programming Study* **1**, 58–70.

COTTLE, R.W. and G.B. DANTZIG (1968), Complementary pivot theory of mathematical programming, *Linear Algebra Appl.* **1**, 103–125.

COTTLE, R.W., G.H. GOLUB and R.S. SACHER (1978), On the solution of large structured linear complementarity problems: The block partitioned case, *Appl. Math. Optim.* **4**, 347–363.

COTTLE, R.W., J.-S. PANG and R.E. STONE (1992), *The Linear Complementarity Problem* (Academic Press, New York).

COXETER, H.S.M. (1934), Discrete groups generated by reflections, *Ann. of Math.* **6**, 13–29.

COXETER, H.S.M. (1963), *Regular Polytopes* (MacMillan, New York, 3rd ed.).

CRANDALL, M.G. and P.H. RABINOWITZ (1971), Bifurcation from simple eigenvalues, *J. Funct. Anal.* **8**, 321–340.

CROUZEIX, M. and J. RAPPAZ (1990), *On Numerical Approximation in Bifurcation Theory* (RMA, Masson, Paris).

DAI, R.-X. and W.C. RHEINBOLDT (1990), On the computation of manifolds of fold points for parameter-dependent problems, *SIAM J. Numer. Anal.* **27**, 437–446.

DAI, Y., K. SEKITANI and Y. YAMAMOTO (1992), A variable dimension algorithm with the Dantzig–Wolfe decomposition for structured stationary point problems, *Z. Oper. Res.* **36**, 23–53.

DAI, Y. and Y. YAMAMOTO (1989), The path following algorithm for stationary point problems on polyhedral cones, *J. Oper. Res. Soc. Japan* **32**, 286–309.

DANTZIG, G.B. (1963), *Linear Programming and Extensions* (Princeton Univ. Press, Princeton, NJ).

DAVIDENKO, D. (1953), On a new method of numerical solution of systems of nonlinear equations, *Dokl. Akad. Nauk USSR* **88**, 601–602, in Russian.

DAVIS, P.J. (1979), *Circulant Matrices* (Wiley, New York).

DAVIS, P.J. and P. RABINOWITZ (1984), *Methods of Numerical Integration* (Academic Press, Orlando, 2nd ed.).

DELLNITZ, M. and B. WERNER (1989), Computational methods for bifurcation problems with symmetries – with special attention to steady state and Hopf bifurcation points, *J. Comput. Appl. Math.* **26**, 97–123.

DEN HEIJER, C. and W.C. RHEINBOLDT (1981), On steplength algorithms for a class of continuation methods, *SIAM J. Numer. Anal.* **18**, 925–948.

DENNIS, J.E. and R.B. SCHNABEL (1983), *Numerical Methods for Unconstrained Optimization and Nonlinear Equations* (Prentice-Hall, Englewood Cliffs, NJ).

DESA, C., K.M. IRANI, C.G. RIBBENS, L.T. WATSON and H.F. WALKER (1992), Preconditioned iterative methods for homotopy curve tracking, *SIAM J. Sci. Stat. Comput.* **13**, 30–46.

DEUFLHARD, P., B. FIEDLER and P. KUNKEL (1987), Efficient numerical pathfollowing beyond critical points, *SIAM J. Numer. Anal.* **24**, 912–927.

DEUFLHARD, P. and G. HEINDL (1979), Affine invariant convergence theorems for Newton's method and extensions to related methods, *SIAM Numer. Anal.* **16**, 1–10.

DOEDEL, E., H.B. KELLER and J.P. KERNÉVEZ (1991a), Numerical analysis and control of bifurcation problems. Part I: Bifurcation in finite dimensions, *Internat. J. Bifurcation Chaos* **1**, 493–520.

DOEDEL, E., H.B. KELLER and J.P. KERNÉVEZ (1991b), Numerical analysis and control of bifurcation problems. Part II: Bifurcation in infinite dimensions, *Internat. J. Bifurcation Chaos* **1**, 745–772.

DOEDEL, E. and J.P. KERNÉVEZ (1986), AUTO: Software for continuation and bifurcation problems in ordinary differential equations, California Institute of Technology.

DUNFORD, N. and J.T. SCHWARTZ (1963), *Linear Operators. Part II: Spectral Theory* (Interscience, New York).

EAVES, B.C. (1972), Homotopies for the computation of fixed points, *Math. Programming* **3**, 1–22.

EAVES, B.C. (1976), A short course in solving equations with PL homotopies, in: R.W. Cottle and C.E. Lemke, eds., *Nonlinear Programming*, SIAM-AMS Proc. **9** (Providence, RI) 73–143.

EAVES, B.C. (1984), *A Course in Triangulations for Solving Equations with Deformations*, Lecture Notes in Economics and Mathematical Systems **234** (Springer, Berlin).

EAVES, B.C. and R. SAIGAL (1972), Homotopies for computation of fixed points on unbounded regions, *Math. Programming* **3**, 225–237.

EAVES, B.C. and H. SCARF (1976), The solution of systems of piecewise linear equations, *Math. Oper. Res.* **1**, 1–27.

EAVES, B.C. and J.A. YORKE (1984), Equivalence of surface density and average directional density, *Math. Oper. Res.* **9**, 363–375.

ERMENTROUT, B. (1990), *PhasePlane: The Dynamical Systems Tool* (Brooks/Cole Publishing).

FÄSSLER, A. (1976), Application of group theory to the method of finite elements, PhD thesis, ETH Zürich, Switzerland.

FEUDEL, U. and W. JANSEN (1992), CANDYS/QA – A software system for the qualitative analysis of nonlinear dynamical systems, *Bifurcation & Chaos*, to appear.

FIACCO, A.V. (1983), *Introduction to Sensitivity and Stability Analysis in Nonlinear Programming* (Academic Press, New York).

FIACCO, A.V., ed., (1984), *Sensitivity, Stability and Parametric Analysis*, Mathematical Programming Study **21** (North-Holland, Amsterdam).

FIACCO, A.V. and G. MCCORMICK (1968), *Nonlinear Programming: Sequential Unconstrained Minimization Techniques* (Wiley, New York, NY).

FINK, J.P. and W.C. RHEINBOLDT (1983), On the discretization error of parametrized nonlinear equations, *SIAM J. Numer. Anal.* **20**, 732–746.

FINK, J.P. and W.C. RHEINBOLDT (1984), Solution manifolds and submanifolds of parametrized equations and their discretization errors, *Numer. Math.* **45**, 323–343.

FINK, J.P. and W.C. RHEINBOLDT (1985), Local error estimates for parametrized nonlinear equations, *SIAM J. Numer. Anal.* **22**, 729–735.

FINK, J.P. and W.C. RHEINBOLDT (1986), Folds on the solution manifold of a parametrized equation, *SIAM J. Numer. Anal.* **23**, 693–706.

FINK, J.P. and W.C. RHEINBOLDT (1987), A geometric framework for the numerical study of singular points, *SIAM J. Numer. Anal.* **24**, 618–633.

FLETCHER, R. (1987), *Practical Methods of Optimization* (Wiley, New York, 2nd ed.).

FREUDENTHAL, H. (1942), Simplizialzerlegungen von beschränkter Flachheit, *Ann. of Math.* **43**, 580–582.

FREUND, R.M. (1984a), Variable dimension complexes. Part I: Basic theory, *Math. Oper. Res.* **9**, 479–497.

FREUND, R.M. (1984b), Variable dimension complexes. Part II: A unified approach to some combinatorial lemmas in topology, *Math. Oper. Res.* **9**, 498–509.

FREUND, R.M. (1991), A potential-function reduction algorithm for solving a linear program directly from an infeasible "warm start", in: C. Roos and J.-P. Vial, eds., *Interior Point Methods for Linear Programming: Theory and Practice*, Math. Programming, Ser. B **52** (Mathematical Programming Society, North-Holland, Amsterdam) 441–446.

FREUND, R.W., G.H. GOLUB and N.M. NACHTIGAL (1992), Iterative solution of linear systems, *Acta Numerica* **1**, 57–100.

FULTON, W. (1984), *Intersection Theory*, (Springer, Berlin).

GARCIA, C.B. and F.J. GOULD (1978), A theorem on homotopy paths, *Math. Oper. Res.* **3**, 282–289.

GARCIA, C.B. and F.J. GOULD (1980), Relations between several path following algorithms and local and global Newton methods, *SIAM Rev.* **22**, 263–274.

GARCIA, C.B. and W.I. ZANGWILL (1979), Finding all solutions to polynomial systems and other systems of equations, *Math. Programming* **16**, 159–176.

GARCIA, C.B. and W.I. ZANGWILL (1981), *Pathways to Solutions, Fixed Points, and Equilibria* (Prentice-Hall, Englewood Cliffs, NJ).

GARRATT, T.J., G. MOORE and A. SPENCE (1991), Two methods for the numerical detection of Hopf bifurcations, in: R. Seydel, F.W. Schneider, T. Küpper and H. Troger, eds., *Bifurcation and Chaos: Analysis, Algorithms, Applications*, ISNM **97** (Birkhäuser, Basel) 129–134.

GATERMANN, K. and A. HOHMANN (1991), Symbolic exploitation of symmetry in numerical pathfollowing, *Impact of Computing in Science and Engineering* **3**, 330–365.

GEORG, K. (1981a), A numerically stable update for simplicial algorithms, in: E.L. Allgower, K. Glashoff and H.-O. Peitgen, eds., *Numerical Solution of Nonlinear Equations*, Lecture Notes in Math. **878** (Springer, Berlin) 117–127.

GEORG, K. (1981b), On tracing an implicitly defined curve by quasi-Newton steps and calculating bifurcation by local perturbation, *SIAM J. Sci. Stat. Comput.* **2**, 35–50.

GEORG, K. (1982), Zur numerischen Realisierung von Kontinuitätsmethoden mit Prädiktor-Korrektor- oder simplizialen Verfahren, Habilitationsschrift, Univ. Bonn, Germany.

GEORG, K. (1983), A note on stepsize control for numerical curve following, in: B.C. Eaves, F.J. Gould, H.-O. Peitgen and M.J. Todd, eds., *Homotopy Methods and Global Convergence* (Plenum Press, New York) 145–154.

GEORG, K. (1990), An introduction to PL algorithms, in: E.L. Allgower and K. Georg, eds., *Computational Solution of Nonlinear Systems of Equations*, Lectures in Applied Mathematics **26** (American Mathematical Society, Providence, RI) 207–236.

GEORG, K. (1991), Approximation of integrals for boundary element methods, *SIAM J. Sci. Stat. Comput.* **12**, 443–453.

GEORG, K. and R. MIRANDA (1992), Exploiting symmetry in solving linear equations, in: E.L. Allgower, K. Böhmer and M. Golubitsky, eds., *Bifurcation and Symmetry*, ISNM **104** (Birkhäuser, Basel) 157–168.

GEORG, K. and R. MIRANDA (1993), Symmetry aspects in numerical linear algebra with applications to boundary element methods, in: E.L. Allgower, K. Georg and R. Miranda, eds., *Exploiting Symmetry in Applied and Numerical Analysis*, Lectures in Applied Mathematics **29** (American Mathematical Society, Providence, RI) 213–228.

GEORG, K. and J. TAUSCH (1993), Some error estimates for the numerical approximation of surface integrals, *Math. Comp.* **62**, 755–763.

GEORG, K. and R. WIDMANN (1993), Adaptive quadratures over surfaces, Preprint, Colorado State University.

GFRERER, H., J. GUDDAT and H.-J. WACKER (1983), A globally convergent algorithm based on imbedding and parametric optimization, *Computing* **30**, 225–252.

GFRERER, H., J. GUDDAT, H.-J. WACKER and W. ZULEHNER (1985), Path-following for Kuhn–Tucker curves by an active set strategy, in: A. Baghi and H.T. Jongen, eds., *System and Optimization*, Lect. Notes Contr. Inf. Sc. **66** (Springer, Berlin) 111–132.

GILL, P.E., G.H. GOLUB, W. MURRAY and M.A. SAUNDERS (1974), Methods for modifying matrix factorizations, *Math. Comp.* **28**, 505–535.

GILL, P.E., W. MURRAY, M.A. SAUNDERS, J.A. TOMLIN and M.H. WRIGHT (1986), On projected Newton barrier methods for linear programming and an equivalence to Karmarkar's projective method, *Math. Programming* **36**, 183–209.

GLOWINSKI, R., H.B. KELLER and L. REINHART (1985), Continuation-conjugate gradient methods for the least squares solution of nonlinear boundary value problems, *SIAM J. Sci. Statist. Comput.* **6**, 793–832.

GNUTZMANN, S. (1989), Stückweise lineare Approximation implizit definierter Mannigfaltigkeiten, PhD thesis, University of Hamburg, Fed. Rep. Germany.

GOLDFARB, D. and S. LIU (1991), An algorithm for solving linear programming problems in $O(n^3L)$ operations, *Math. Programming* **49**, 325–340.

GOLUB, G.H. and C.F. VAN LOAN (1989), *Matrix Computations* (J. Hopkins University Press, Baltimore, 2nd ed.).

GOLUBITSKY, M. and D.G. SCHAEFFER (1985), *Singularities and Groups in Bifurcation Theory* **1** (Springer, Berlin).

GOLUBITSKY, M., I. STEWART and D.G. SCHAEFFER (1988), *Singularities and Groups in Bifurcation Theory* **2** (Springer, Berlin).

GONZAGA, C.C. (1988), An algorithm for solving linear programming problems in $O(n^3L)$ operations, in: N. Megiddo, ed., *Progress in Mathematical Programming, Interior Point and Related Methods* (Springer, New York) 1–28.

GONZAGA, C.C. (1992), Path-following methods for linear programming, *SIAM Rev.* **34**, 167–224.

GÓRNIEWICZ, L. (1976), Homological methods in fixed point theory of multivalued maps, *Diss. Math.* **129**.

GOULD, F.J. and J.W. TOLLE (1983), *Complementary Pivoting on a Pseudomanifold Structure with Applications on the Decision Sciences*, Sigma Series in Applied Mathematics **2** (Heldermann, Berlin).

GRIEWANK, A. (1985), On solving nonlinear equations with simple singularities or nearly singular solutions, *SIAM Rev.* **27**, 537–563.

GRIEWANK, A. (1989), On automatic differentiation, in: M. Iri and K. Tanabe, eds., *Mathematical Programming: Recent Developments and Applications* (Kluwer, Dordrecht) 83–108.

GRIEWANK, A. and G. CORLISS, eds. (1991), *Automatic Differentiation of Algorithms: Theory, Implementation, and Application* (SIAM, Philadelphia).

GRIEWANK, A. and G.W. REDDIEN (1984), Characterization and computation of generalized turning points, *SIAM J. Numer. Anal.* **21**, 176–185.

GUCKENHEIMER, J. and S. KIM (1991), Computational environments for exploring dynamical systems, *Bifurcation & Chaos* **1**, 269–276.

GUDDAT, J., F. GUERRA VASQUEZ and H.T. JONGEN (1990), *Parametric Optimization: Singularities, Path Following, and Jumps* (Wiley, Chichester, England).

GUDDAT, J., H.T. JONGEN, B. KUMMER and F. NOŽIČKA, eds. (1987), *Parametric Optimization and Related Topics* (Akademie, Berlin).

GUY, J. and B. MANGEOT (1981), Use of group theory in various integral equations, *SIAM J. Appl. Math.* **40**, 390–399.

HACKBUSCH, W. (1989), *Integralgleichungen. Theorie und Numerik* (B.G. Teubner, Stuttgart).

HACKBUSCH, W. (1994), *Iterative Solution of Large Sparse Systems of Equations*, Applied Mathematical Sciences **95** (Springer, Berlin).

HASELGROVE, C.B. (1961), The solution of nonlinear equations and of differential equations with two-point boundary conditions, *Comput. J.* **4**, 255–259.

HEALEY, T.J. (1988a), Global bifurcation and continuation in the presence of symmetry with an application to solid mechanics, *SIAM J. Math. Anal.* **19**, 824–840.

HEALEY, T.J. (1988b), A group theoretic approach to computational bifurcation problems with symmetry, *Comput. Meth. Appl. Mech. Engrg.* **67**, 257–295.

HEALEY, T.J. (1989), Symmetry and equivariance in nonlinear elastostatics. Part I, *Arch. Rational Mech. Anal.* **105**, 205–228.

HEALEY, T.J. and J.A. TREACY (1991), Exact block diagonalization of large eigenvalue problems for structures with symmetry, *Internat. J. Numer. Methods Engrg.* **31**, 265–285.

HENDERSON, M.E. (1985), Complex bifurcation, PhD thesis, CALTECH, Pasadena.

HENDERSON, M.E. and H.B. KELLER (1990), Complex bifurcation from real paths, *SIAM J. Appl. Math.* **50**, 460–482.

HESTENES, M.R. and E. STIEFEL (1952), Methods of conjugate gradients for solving linear systems, *J. Res. Nat. Bur. Standards* **49**, 409–436.

HETTICH, R. and H.T. JONGEN (1994), On continuous deformations of semi-infinite optimization problems, Preprint 94-02, Mathematics, University of Trier, Trier, Germany.

HIRSCH, M.W. (1963), A proof of the nonretractibility of a cell onto its boundary, *Proc. Amer. Math. Soc.* **14**, 364–365.

HIRSCH, M.W. (1976), *Differential Topology* (Springer, Berlin).

HOHMANN, A. (1991), An adaptive continuation method for implicitly defined surfaces, Preprint, Konrad-Zuse-Zentrum Berlin, SC-91-20.

HOLODNIOK, M. and M. KUBÍČEK (1984), DERPER – An algorithm for the continuation of periodic solutions in ordinary differential equations, *J. Comput. Phys.* **55**, 254–267.

HONG, B. (1993), Computational methods for bifurcation problems with symmetries on the manifold, *SIAM J. Numer. Anal.* **30**, 1134–1154.

HUBBARD, J. and B. WEST (1992), *A Dynamical Systems Software Package for the Macintosh* (Springer, Berlin).

HUBER, B. and B. STURMFELS (1993), A polyhedral method for solving sparse polynomial systems, Preprint, Cornell Univ.

HUITFELDT, J. (1991), Nonlinear eigenvalue problems – Prediction of bifurcation points and branch switching, Technical Report 17, Univ. of Göteborg, Sweden.

HUITFELDT, J. and A. RUHE (1990), A new algorithm for numerical path following applied to an example from hydrodynamical flow, *SIAM J. Sci. Statist. Comput.* **11**, 1181–1192.

IRI, M. and K. KUBOTA (1987), Methods of fast automatic differentiation and applications, Research Memorandum RMI 87–02, Department of Mathematical Engineering and Information Physics, Faculty of Engineering, University of Tokyo.

IZE, J. (1993), Topological bifurcation, in: M. Matzeu and A. Vignoli, eds., *Nonlinear Analysis* (Birkhäuser, Boston, MA).

JACOBOWITZ, H. (1976), Periodic solutions of $x'' + f(x,t) = 0$ via the Poincaré–Birkhoff theorem, *J. Differential Equations* **20**, 37–52.

JANOVSKÝ, V. (1989), A note on computing simple bifurcation points, *Computing* **43**, 27–36.

JARAUSCH, H. and W. MACKENS (1984), Numerical treatment of bifurcation branches by adaptive condensation, in: T. Küpper, H.D. Mittelmann and H. Weber, eds., *Numerical Methods for Bifurcation Problems*, ISNM **70** (Birkhäuser, Basel).

JARAUSCH, H. and W. MACKENS (1987), Solving large nonlinear systems of equations by an adaptive condensation process, *Numer. Math.* **50**, 633–653.

JEPSON, A.D. and D.W. DECKER (1986), Convergence cones near bifurcation, *SIAM J. Numer. Anal.* **23**, 959–975.

JEPSON, A.D., A. SPENCE and K.A. CLIFFE (1991), The numerical solution of nonlinear equations having several parameters. Part III: Equations with Z_2-symmetry, *SIAM J. Numer. Anal.* **28**, 809–832.

JEZIERSKI, J. (1993), One codimensional Wecken type theorems, *Forum Math.*, 421–439.

JONGEN, H.T., P. JONKER and F. TWILT (1983), *Nonlinear Optimization in R^N*, **I**. *Morse Theory, Chebyshev Approximation* (Peter Lang, New York).

JONGEN, H.T., P. JONKER and F. TWILT (1986), *Nonlinear Optimization in R^N*, **II**. *Transversality, Flows, Parametric Aspects* (Peter Lang, New York).

JONGEN, H.T. and G.-W. WEBER (1990), On parametric nonlinear programming, *Ann. Oper. Res.* **27**, 253–284.

JONGEN, H.T. and G. ZWIER (1985), On the local structure of the feasible set in semi-infinite optimization, ISNM **72** (Birkhäuser, Basel) 185–202.

JOSEPHY, N.H. (1979), Newton's method for generalized equations, Technical Report 1965, University of Wisconsin, Madison, WI, available from National Technical Information Service under accession number AD A077 096.

KAAS-PETERSEN, C. (1989), PATH – User's Guide, University of Leeds, England.

KAMIYA, K. and A.J.J. TALMAN (1990), Variable dimension simplicial algorithm for balanced games, Technical Report 9025, Tilburg Univ., The Netherlands.

KARMARKAR, N.K. (1984), A new polynomial-time algorithm for linear programming, *Combinatorica* **4**, 373–395.

KARMARKAR, N.K. and K.G. RAMAKRISHNAN (1991), Computational results of an interior point algorithm for large scale linear programming, in: C. Roos and J.-P. Vial, eds., *Interior Point Methods for Linear Programming: Theory and Practice*, Math. Programming, Ser. B **52** (Mathematical Programming Society, North-Holland, Amsterdam) 555–586.

KATZENELSON, J. (1965), An algorithm for solving nonlinear resistive networks, *Bell System Tech. J.* **44**, 1605–1620.

KEARFOTT, R.B. (1989), An interval step control for continuation methods, *Math. Comp.*, to appear.

KEARFOTT, R.B. (1990), Interval arithmetic techniques in the computational solution of nonlinear systems of equations: Introduction, examples and comparisons, in: E.L. Allgower and K. Georg, eds., *Computational Solution of Nonlinear Systems of Equations*, Lectures in Applied Mathematics **26** (American Mathematical Society, Providence, RI) 337–357.

KEENER, J.P. and H.B. KELLER (1974), Perturbed bifurcation theory, *Arch. Rational Mech. Anal.* **50**, 159–175.

KELLER, H.B. (1970), Nonlinear bifurcation, *J. Differential Equations* **7**, 417–434.

KELLER, H.B. (1977), Numerical solution of bifurcation and nonlinear eigenvalue problems, in: P.H. Rabinowitz, ed., *Applications of Bifurcation Theory* (Academic Press, New York) 359–384.

KELLER, H.B. (1978), Global homotopies and Newton methods, in: C. de Boor and G.H. Golub, eds., *Recent Advances in Numerical Analysis* (Academic Press, New York) 73–94.

KELLER, H.B. (1982), Practical procedures in path following near limit points, in: R. Glowinski and J.-L. Lions, eds., *Computing Methods in Applied Sciences and Engineering* (North-Holland, Amsterdam).

KELLER, H.B. (1983), The bordering algorithm and path following near singular points of higher nullity, *SIAM J. Sci. Statist. Comput* **4**, 573–582.

KELLER, H.B. (1987), *Lectures on Numerical Methods in Bifurcation Problems* (Springer, Berlin).

KELLOGG, R.B., T.-Y. LI and J.A. YORKE (1976), A constructive proof of the Brouwer fixed point theorem and computational results, *SIAM J. Numer. Anal.* **13**, 473–483.

KHACHIYAN, L.G. (1979), A polynomial algorithm in linear programming, *Soviet Math. Dokl.* **20**, 191–194.

KHIBNIK, A. (1990), *Using TraX, a Tutorial to Accompany TraX, a Program for the Simulation and Analysis of Dynamical Systems* (Exeter Software, Setauket, NY).

KHIBNIK, A.I., Y.A. KUZNETSOV, V.V. LEVITIN and E.V. NIKOLAEV (1993), Continuation techniques and interactive software for bifurcation analysis of odes and iterated maps, *Phys. D*, 360–371.

KLEIN, F. (1882–1883), Neue Beiträge zur Riemannschen Funktionentheorie, *Math. Ann.* **21**.

KOJIMA, M. (1974), Computational methods for solving the nonlinear complementarity problem, *Keio Engrg. Rep.* **27**, 1–41.

KOJIMA, M. (1978), Studies on piecewise-linear approximations of piecewise-C^1 mappings in fixed points and complementarity theory, *Math. Oper. Res.* **3**, 17–36.

KOJIMA, M. (1979), A complementarity pivoting approach to parametric programming, *Math. Oper. Res.* **4**, 464–477.

KOJIMA, M. and R. HIRABAYASHI (1984), Sensitivity, stability and parametric analysis, in: *Mathematical Programming Study* **21** (North-Holland, Amsterdam) 150–198.

KOJIMA, M., N. MEGIDDO and S. MIZUNO (1990a), A general framework of continuation methods for complementarity problems, Technical report, IBM Almaden Research Center, San Jose, CA.

KOJIMA, M., N. MEGIDDO and S. MIZUNO (1991a), Theoretical convergence of large-step primal-dual interior point algorithms for linear programming, Preprint, Tokyo Inst. of Techn.

KOJIMA, M., N. MEGIDDO and T. NOMA (1991b), Homotopy continuation methods for nonlinear complementarity problems, *Math. Oper. Res.* **16**, 754–774.

KOJIMA, M., N. MEGIDDO, T. NOMA and A. YOSHISE (1991c), *A Unified Approach to Interior Point Algorithms for Linear Complementarity Problems*, Lecture Notes in Computer Science **538** (Springer, Berlin).

KOJIMA, M., N. MEGIDDO and Y. YE (1992), An interior point potential reduction algorithm for the linear complementarity problem, *Math. Programming* **54**, 267–279.

KOJIMA, M., S. MIZUNO and T. NOMA (1990b), Limiting behavior of trajectories generated by a continuation method for monotone complementarity problems, *Math. Oper. Res.* **15**, 662–675.

KOJIMA, M., S. MIZUNO and A. YOSHISE (1988), A primal-dual interior point algorithm for linear programming, in: N. Megiddo, ed., *Progress in Mathematical Programming, Interior Point and Related Methods* (Springer, New York) 29–47.

KOJIMA, M., S. MIZUNO and A. YOSHISE (1989), A polynomial-time algorithm for a class of linear complementarity problems, *Math. Programming* **44**, 1–26.

KOJIMA, M., S. MIZUNO and A. YOSHISE (1991), An $O(\sqrt{n}L)$ iteration potential reduction algorithm for linear complementarity problems, *Math. Programming* **50**, 331–342.

KOJIMA, M., H. NISHINO and T. SEKINE (1976), An extension of Lemke's method to the piecewise linear complementarity problem, *SIAM J. Appl. Math.* **31**, 600–613.

KOJIMA, M. and Y. YAMAMOTO (1982), Variable dimension algorithms: Basic theory, interpretation, and extensions of some existing methods, *Math. Programming* **24**, 177–215.

KOJIMA, M. and Y. YAMAMOTO (1984), A unified approach to the implementation of several restart fixed point algorithms and a new variable dimension algorithm, *Math. Programming* **28**, 288–328.

KOVACH, J.W. and W.D. SEIDER (1987), Heterogeneous azeotropic distillation – Homotopy-continuation methods, *Comput. Chem. Engrg.* **11**, 593–605.

KRASNOSEL'SKIĬ, M.A. (1964), *Topological Methods in the Theory of Nonlinear Integral Equations* (Pergamon Press, New York).

KUBÍČEK, M. (1976), Algorithm 502. Dependence of solutions of nonlinear systems on a parameter, *ACM Trans. Math. Software* **2**, 98–107.

KUHN, H.W. (1968), Simplicial approximation of fixed points, *Proc. Nat. Acad. Sci. USA* **61**, 1238–1242.

KUHN, H.W. (1969), Approximate search for fixed points, in: L.A. Zadek, L.W. Neustat and A.V. Balakrishnan, eds., *Computing Methods in Optimization Problems* **2** (Academic Press, New York) 199–211.

LAGARIAS, J.C., R.C. MELVILLE and R. GEOGHEGAN (1993), λ-threading homotopies and DC operating points of nonlinear circuits, Preprint.

LEMKE, C.E. (1965), Bimatrix equilibrium points and mathematical programming, *Management Sci.* **11**, 681–689.

LEMKE, C.E. (1980), A survey of complementarity theory, in: R.W. Cottle, F. Gianessi and J.L. Lions, eds., *Variational Inequalities and Complentarity Problems* (Wiley, London).

LEMKE, C.E. and S.J. GROTZINGER (1976), On generalizing Shapley's index theory to labelled pseudo manifolds, *Math. Programming* **10**, 245–262.

LEMKE, C.E. and J.T. HOWSON (1964), Equilibrium points of bimatrix games, *SIAM J. Appl. Math.* **12**, 413–423.

LI, K. and T.-Y. LI (1993), An algorithm for symmetric tridiagonal eigen-problems – Divide and conquer with homotopy continuation, *SIAM J. Sci. Statist. Comput.* **14**, 735–751.

LI, T.-Y. and N. H. RHEE (1989), Homotopy algorithm for symmetric eigenvalue problems, *Numer. Math.* **55**, 265–280.

LI, T.-Y., T. SAUER and J.A. YORKE (1987), Numerical solution of a class of deficient polynomial systems, *SIAM J. Numer. Anal.* **24**, 435–451.

LI, T.-Y., T. SAUER and J.A. YORKE (1989), The cheater's homotopy: An efficient procedure for solving systems of polynomial equations, *SIAM J. Numer. Anal.* **26**, 1241–1251.

LI, T.-Y. and X. WANG (1993a), Nonlinear homotopies for solving deficient polynomial systems with parameters, *SIAM J. Numer. Anal.* **29**, 1104–1118.

LI, T.-Y. and X. WANG (1993b), Solving real polynomial systems with real homotopies, *Math. Comp.* **60**, 669–680.

LI, T.-Y. and Z. ZENG (1992), Homotopy-determinant algorithm for solving nonsymmetric eigenvalue problems, *Math. Comp.*, to appear.

LI, T.-Y., Z. ZENG and L. CONG (1992), Solving eigenvalue problems of real nonsymmetric matrices with real homotopies, *SIAM J. Numer. Anal.* **29**, 229–248.

LI, T.-Y., H. ZHANG and X.-H. SUN (1991), Parallel homotopy algorithm for the symmetric tridiagonal eigenvalue problem, *SIAM J. Sci. Statist. Comput.* **12**, 469–487.

LI, Y. and Z. LIN (1994), Constructive proof of the Poincaré–Birkhoff theorem, Preprint, Jilin University, China.

LIU, J.L. and W.C. RHEINBOLDT (1991), A posteriori error estimates for parametrized nonlinear equations, in: P. Wriggers and W. Wagner, eds., *Nonlinear Computational Mechanics* (Springer, Heidelberg) 31–46.

LUNDBERG, B.N. and A.B. POORE (1991), Variable order Adams–Bashforth predictors with an error-stepsize control for continuation methods, *SIAM J. Sci. Statist. Comput.* **12**, 695–723.

LUNDBERG, B.N. and A.B. POORE (1993), Numerical continuation and singularity detection methods for parametric nonlinear programming, *SIAM J. Optim.* **3**, 134–154.

LUSTIG, I.J., R.E. MARSTEN and D.F. SHANNO (1991), The interaction of algorithms and architectures for interior point methods, Technical Report RRR 36-91, Rutgers Univ., New Brunswick, NJ.

LYNESS, J.N. (1978), Quadrature over a simplex: Part 2. A representation for the error functional, *SIAM J. Numer. Anal.* **15**, 870–887.

LYNESS, J.N. (1993), Quadrature over curved surfaces by extrapolation, *Math. Comp.*, to appear.

MACKENS, W. (1989), Numerical differentiation of implicitly defined space curves, *Computing* **41**, 237–260.

MARSTEN, R., R. SUBRAMANIAN, M. SALTZMAN, I.J. LUSTIG and D.F. SHANNO (1990), Interior point methods for linear programming: Just call Newton, Lagrange and Fiacco and Mccormick!, *Interfaces* **20**, 105–116.

MCCORMICK, G.P. and K. RITTER (1974), Alternate proofs of the convergence properties of the conjugate-gradient method, *J. Optim. Theory Appl.* **13**, 497–518.

MCQUAIN, W.D., C.J. RIBBENS, L.T. WATSON and R.C. MELVILLE (1993), Preconditioned iterative methods for sparse linear algebra problems arising in circuit simulation, *Comput. Math. Appl.*, to appear.

MCSHANE, K.A., C.L. MONMA and D.F. SHANNO (1989), An implementation of a primal-dual interior point method for linear programming, *ORSA J. Comput.* **1**, 70–83.

MEGIDDO, N. (1988), Pathways to the optimal set in linear programming, in: N. Megiddo, ed., *Progress in Mathematical Programming, Interior Point and Related Methods* (Springer, New York) 131–158.

MEGIDDO, N. and M. SHUB (1989), Boundary behavior of interior point algorithms in linear programming, *Math. Oper. Res.* **14**, 97–146.

MEJIA, R. (1986), CONKUB: A conversational path-follower for systems of nonlinear equations, *J. Comput. Phys.* **63**, 67–84.

MEJIA, R. (1990), Interactive program for continuation of solutions of large systems of nonlinear equations, in: E.L. Allgower and K. Georg, eds., *Computational Solution of Nonlinear Systems of Equations*, Lectures in Applied Mathematics **26** (American Mathematical Society, Providence, RI) 429–449.

MELHEM, R.G. and W.C. RHEINBOLDT (1982), A comparison of methods for determining turning points of nonlinear equations, *Computing* **29**, 201–226.

MELVILLE, R., S. MOINIAN, P. FELDMANN and L.T. WATSON (1993a), Sframe: An efficient system for detailed DC simulation of bipolar analog integrated circuits using continuation methods, *Analog Integr. Circuits Signal Process.* **3**, 163–180.

MELVILLE, R.C., L. TRAJKOVIC, S.-C. FANG and L.T. WATSON (1993b), Globally convergent homotopy methods for the DC operating point problem, *IEEE Trans. Comput. Aided Design*, to appear.

MENZEL, R. and H. SCHWETLICK (1978), Zur Lösung parameterabhängiger nichtlinearer Gleichungen mit singulären Jacobi-Matrizen, *Numer. Math.* **30**, 65–79.

MENZEL, R. and H. SCHWETLICK (1985), Parametrization via secant length and application to path following, *Numer. Math.* **47**, 401–412.

MERRILL, O. (1972), Applications and extensions of an algorithm that computes fixed points of a certain upper semi-continuous point to set mapping, PhD thesis, Univ. of Michigan, Ann Arbor, MI.

MILNOR, J.W. (1968), *Singular Points of Complex Hypersurfaces* (Princeton Univ. Press, Princeton, NJ).

MILNOR, J.W. (1969), *Topology from the Differentiable Viewpoint* (Univ. Press of Virginia, Charlottesville, VA).

MITTELMANN, H.D. and D. ROOSE, eds. (1990), *Continuation Techniques and Bifurcation Problems*, ISNM **92** (Birkhäuser, Basel).

MIZUNO, S. (1992), A new polynomial time method for a linear complementarity problem, *Math. Programming* **56**, 31–43.

MIZUNO, S., M.J. TODD and Y. YE (1992), On adaptive-step primal-dual interior-point algorithms for linear programming, *Math. Oper. Res.*, to appear.

MIZUNO, S., A. YOSHISE and T. KIKUCHI (1989), Practical polynomial time algorithms for linear complementarity problems, *J. Oper. Res. Soc. Japan* **32**, 75–92.

MONTEIRO, R.C. and I. ADLER (1989), Interior path following primal-dual algorithms, I: Linear programming, II: Convex quadratic programming, *Math. Programming* **44**, 27–66.

MOORE, G. and A. SPENCE (1980), The calculation of turning points of nonlinear equations, *SIAM J. Numer. Anal.* **17**, 567–576.

MORGAN, A.P. (1986), A transformation to avoid solutions at infinity for polynomial systems, *Appl. Math. Comput.* **18**, 77–86.

MORGAN, A.P. (1987), *Solving Polynomial Systems Using Continuation for Engineering and Scientific Problems* (Prentice-Hall, Englewood Cliffs, NJ).

MORGAN, A.P. and A.J. SOMMESE (1987), A homotopy for solving general polynomial systems that respects m-homogeneous structures, *Appl. Math. Comput.* **24**, 101–113.

MORGAN, A.P. and A.J. SOMMESE (1989), Coefficient parameter polynomial continuation, *Appl. Math. Comput.* **29**, 123–160.

MORGAN, A.P., A.J. SOMMESE and C.W. WAMPLER (1991), Computing singular solutions to nonlinear analytic systems, *Numer. Math.* **58**, 669–684.

MORGAN, A.P., A.J. SOMMESE and C.W. WAMPLER (1992), Computing singular solutions to polynomial systems, *Adv. in Appl. Math.* **13**, 305–327.

MORGAN, A.P., A.J. SOMMESE and C.W. WAMPLER (1993), A power series method for computing singular solutions to nonlinear analytic systems, *Numer. Math.* **63**, 391–409.

MORGAN, A.P., A.J. SOMMESE and L.T. WATSON (1989), Finding all isolated solutions to polynomial systems using HOMPACK, *ACM Trans. Math. Software* **15**, 93–122.

MURTAGH, B.A. and M.A. SAUNDERS (1987), *MINOS 5.1 Users Guide*, Stanford Univ., CA.

NAZARETH, J.L. (1986), Homotopy techniques in linear programming, *Algorithmica* **1**, 529–535.

NAZARETH, J.L. (1991), The homotopy principle and algorithms for linear programming, *SIAM J. Optim.* **1**, 316–332.

NAZARETH, J.L. (1994a), A concise introduction to interior point methods for linear programming, Preprint.

NAZARETH, J.L. (1994b), The implementation of linear programming algorithms based on homotopies, *Algorithmica* **15**, 332–350.

NUSSE, H.E. and J.A. YORKE (1994), *DYNAMICS: Numerical Explorations* (Springer, New York).

ORTEGA, J.M. and W.C. RHEINBOLDT (1970), *Iterative Solution of Nonlinear Equations in Several Variables* (Academic Press, New York).

PARKER, T. and L. CHUA (1989), *Practical Numerical Algorithms for Chaotic Systems* (Springer, Berlin).

PEITGEN, H.-O. (1982), Topologische Perturbationen beim globalen numerischen Studium nichtlinearer Eigenwert- und Verzweigungsprobleme, *Jahresber. Deutsch. Math.-Verein.* **84**, 107–162.

PEITGEN, H.-O. and M. PRÜFER (1979), The Leray–Schauder continuation method is a constructive element in the numerical study of nonlinear eigenvalue and bifurcation problems, in: H.-O. Peitgen and H.-O. Walther, eds., *Functional Differential Equations and Approximation of Fixed Points*, Lecture Notes in Math. **730** (Springer, Berlin) 326–409.

PEITGEN, H.-O. and K. SCHMITT (1983), Global topological perturbations in the study of nonlinear eigenvalue problems, *Math. Methods Appl. Sci.* **5**, 376–388.

PEITGEN, H.-O. and H.W. SIEGBERG (1981), An $\bar{\varepsilon}$-perturbation of Brouwer's definition of degree, in: E. Fadell and G. Fournier, eds., *Fixed Point Theory*, Lecture Notes in Math. **886** (Springer, Berlin) 331–366.

PERCELL, P. (1980), Note on a global homotopy, *Numer. Funct. Anal. Optim.* **2**, 99–106.

POINCARÉ, H. (1881–1886), *Sur les Courbes Definé par une Équation Differentielle.* I–IV, *Oeuvres* **I** (Gauthier-Villars, Paris).

POLAK, E., J.E. HIGGINS and D.Q. MAYNE (1992), A barrier function method for minimax problems, *Math. Programming* **54**, 155–176.

POLAK, E. and G. RIBIÈRE (1969), Note sur la convergence de méthodes de directions conjugées, *Rev. Française Informat. Recherche Opérationelle* **3**, 35–43.

PÖNISCH, G. (1985), Computing simple bifurcation points using a minimally extended system of nonlinear equations, *Computing* **35**, 277–294.

PÖNISCH, G. and H. SCHWETLICK (1981), Computing turning points of curves implicitly defined by nonlinear equations depending on a parameter, *Computing* **26**, 107–121.

PÖNISCH, G. and H. SCHWETLICK (1982), Ein überlinear konvergentes Verfahren zur Bestimmung von Rückkehrpunkten implizit definierter Raumkurven, *Numer. Math.* **38**, 455–466.

POORE, A.B. and C.A. TIAHRT (1987), Bifurcation problems in nonlinear parametric programming, *Math. Programming* **39**, 189–205.

POORE, A.B. and C.A. TIAHRT (1990), A bifurcation analysis of the nonlinear parametric programming problem, *Math. Programming* **47**, 117–141.

POWELL, M.J.D. (1970a), A Fortran subroutine for solving nonlinear algebraic equations, in: P. Rabinowitz, ed., *Numerical Methods for Nonlinear Algebraic Equations* (Gordon and Breach, New York) 115–161.

POWELL, M.J.D. (1970b), A hybrid method for nonlinear equations, in: P. Rabinowitz, ed., *Numerical Methods for Nonlinear Algebraic Equations* (Gordon and Breach, New York) 87–114.

POWELL, M.J.D. (1977), Restart procedures for the conjugate gradient method, *Math. Programming* **12**, 241–254.

RABIER, P.J. and G.W. REDDIEN (1986), Characterization and computation of singular points with maximum rank deficiency, *SIAM J. Numer. Anal.* **23**, 1040–1051.

RABINOWITZ, P.H. (1971), Some global results for nonlinear eigenvalue problems, *J. Funct. Anal.* **7**, 487–513.

RAKOWSKA, J., R.T. HAFTKA and L.T. WATSON (1991), An active set algorithm for tracing parametrized optima, *Struct. Optim.* **3**, 29–44.

REINOZA, A. (1985), Solving generalized equations via homotopies, *Math. Programming* **31**, 307–320.

RENEGAR, J. (1985), On the complexity of a piecewise linear algorithm for approximating roots of complex polynomials, *Math. Programming* **32**, 301–318.

RENEGAR, J. (1987), On the efficiency of Newton's method in approximating all zeros of systems of complex polynomials, *Math. Oper. Res.* **12**, 121–148.

RENEGAR, J. (1988a), A polynomial-time algorithm, based on Newton's method, for linear programming, *Math. Programming* **40**, 59–93.

RENEGAR, J. (1988b), Rudiments of an average case complexity theory for piecewise-linear path following algorithms, *Math. Programming* **40**, 113–163.

RENEGAR, J. and M. SHUB (1992), Unified complexity analysis for Newton LP methods, *Math. Programming* **53**, 1–16.

RHEINBOLDT, W.C. (1978), Numerical methods for a class of finite dimensional bifurcation problems, *SIAM J. Numer. Anal.* **15**, 1–11.

RHEINBOLDT, W.C. (1980), Solution fields of nonlinear equations and continuation methods, *SIAM J. Numer. Anal.* **17**, 221–237.

RHEINBOLDT, W.C. (1981), Numerical analysis of continuation methods for nonlinear structural problems, *Comput. & Structures* **13**, 103–113.

RHEINBOLDT, W.C. (1986), *Numerical Analysis of Parametrized Nonlinear Equations* (Wiley, New York).

RHEINBOLDT, W.C. (1987), On a moving-frame algorithm and the triangulation of equilibrium manifolds, in: T. Küpper, R. Seydel and H. Troger, eds., *Bifurcation: Analysis, Algorithms, Applications*, ISNM **79** (Birkhäuser, Basel) 256–267.

RHEINBOLDT, W.C. (1988a), On a theorem of S. Smale about Newton's method for analytic mappings, *Appl. Math. Lett.* **1**, 69–72.

RHEINBOLDT, W.C. (1988b), On the computation of multi-dimensional solution manifolds of parametrized equations, *Numer. Math.* **53**, 165–182.

RHEINBOLDT, W.C. (1992), On the theory and numerics of differential-algebraic equations, in: W. Light, ed., *Advances in Numerical Analysis* (Oxford Univ. Press, Oxford, UK) 237–275.

RHEINBOLDT, W.C. (1993), On the sensitivity of solutions of parametrized equations, *SIAM J. Numer. Anal.* **30**, 305 320.

RHEINBOLDT, W.C. and J.V. BURKARDT (1983a), Algorithm 596: A program for a locally-parametrized continuation process, *ACM Trans. Math. Software* **9**, 236–241.

RHEINBOLDT, W.C. and J.V. BURKARDT (1983b), A locally-parametrized continuation process, *ACM Trans. Math. Software* **9**, 215–235.

RHEINBOLDT, W.C., D. ROOSE and R. SEYDEL (1990), Aspects of continuation software, in: D. Roose, B. De Dier and A. Spence, eds., *Continuation and Bifurcations: Numerical Techniques and Applications*, NATO ASI Series C **313** (Kluwer, Dordrecht) 261–268.

ROBINSON, S.M. (1983), Generalized equations, in: A. Bachem, M. Grötschel and B. Korte, eds., *Mathematical Programming – The State of the Art* (Springer, Berlin) 346–367.

ROBINSON, S.M. (1987), Local structure of feasible sets in nonlinear programming, Part III: Stability and sensitivity, *Math. Programming Study* **30**, 45–66.

ROOS, C. and J.-P. VIAL, eds. (1991), *Interior Point Methods for Linear Programming: Theory and Practice*, Math. Programming, Ser. B **52** (Mathematical Programming Society, North-Holland, Amsterdam).

ROOSE, D., B. DE DIER and A. SPENCE, eds. (1990), *Continuation and Bifurcations: Numerical Techniques and Applications*, NATO ASI Series C **313** (Kluwer, Dordrecht).

SAAD, Y. and M. SCHULTZ (1986), GMRES: A generalized minimal residual method for solving nonsymmetric linear systems, *SIAM J. Sci. Statist. Comp.* **7**, 856–869.

SAIGAL, R. (1971), Lemke's algorithm and a special linear complementarity problem, *Oper. Res.* **8**, 201–208.

SAIGAL, R. (1976), Extension of the generalized complementarity problem, *Math. Oper. Res.* **1**, 260–266.

SAIGAL, R. (1977), On the convergence rate of algorithms for solving equations that are based on methods of complementary pivoting, *Math. Oper. Res.* **2**, 108–124.

SAIGAL, R. (1979), On piecewise linear approximations to smooth mappings, *Math. Oper. Res.* **4**, 153–161.

SAIGAL, R. (1984), Computational complexity of a piecewise linear homotopy algorithm, *Math. Programming* **28**, 164–173.

SAIGAL, R. and M.J. TODD (1978), Efficient acceleration techniques for fixed point algorithms, *SIAM J. Numer. Anal.* **15**, 997–1007.

SAUPE, D. (1982), On accelerating PL continuation algorithms by predictor-corrector methods, *Math. Programming* **23**, 87–110.

SCARF, H.E. (1967), The approximation of fixed points of a continuous mapping, *SIAM J. Appl. Math.* **15**, 1328–1343.

SCHWETLICK, H. (1984a), Algorithms for finite-dimensional turning point problems from viewpoint to relationships with constrained optimization methods, in: T. Küpper, H. Mittelmann and H. Weber, eds., *Numerical Methods for Bifurcation Problems* (Birkhäuser, Basel) 459–479.

SCHWETLICK, H. (1984b), Effective methods for computing turning points of curves implicitly defined by nonlinear equations, in: A. Wakulicz, ed., *Computational Mathematics*, Banach Center Publications **13** (PWN, Warsaw) 623–645.

SCHWETLICK, H. and J. CLEVE (1987), Higher order predictors and adaptive stepsize control in path following algorithms, *SIAM J. Numer. Anal.* **24**, 1382–1393.

SEIDER, W.D., D.D. BRENGEL and S. WIDAGDO (1991), Nonlinear analysis in process design, *AIChE J.* **37**, 1–38.

SELLAMI, H. and S.M. ROBINSON (1993), A continuation method for normal maps, Technical Report 93-8, Department of Industrial Engineering, University of Wisconsin–Madison, Madison, WI.

SERRE, J.-P. (1977), *Linear Representations of Finite Groups*, Graduate Texts in Mathematics **42** (Springer, Berlin).

SEYDEL, R. (1979), Numerical computation of branch points of nonlinear equations, *Numer. Math.* **33**, 339–352.

SEYDEL, R. (1988), *From Equilibrium to Chaos. Practical Bifurcation and Stability Analysis* (Elsevier, New York).

SEYDEL, R. (1991a), BIFPACK: A program package for continuation, bifurcation and stability analysis, Version 2.3+, University of Ulm, Germany.

SEYDEL, R. (1991b), On detecting stationary bifurcations, *Internat. J. Bifurcation Chaos* **1**, 335–337.

SEYDEL, R., F.W. SCHNEIDER, T. KÜPPER and H. TROGER, eds. (1991), *Bifurcation and Chaos: Analysis, Algorithms, Applications*, ISNM **97** (Birkhäuser, Basel).

SHAMPINE, L.F. and M.K. GORDON (1975), *Computer Solutions of Ordinary Differential Equations. The Initial Value Problem* (Freeman and Co., San Francisco).

SHAPLEY, L.S. (1974), A note on the Lemke–Howson algorithm, in: M.L. Balinski, ed., *Pivoting and Extensions: In Honor of A.W. Tucker*, Math. Programming Study **1** (North-Holland, New York) 175–189.

SHROFF, G.M. and H.B. KELLER (1991), Stabilization of unstable procedures: A hybrid algorithm for continuation, Preprint.

SHROFF, G.M. and H.B. KELLER (1993), Stabilization of unstable procedures: The recursive projection method, *SIAM J. Numer. Anal.* **30**, 1099–1120.

SHUB, M. and S. SMALE (1991), Complexity of Bezout's theorem, I: Geometric aspects, IBM Research Report.

SIYYAM, H.I. (1994), Numerical integration over implicitly defined curves, PhD thesis, Colorado State University, Fort Collins, CO.

SMALE, S. (1976), A convergent process of price adjustement and global Newton methods, *J. Math. Econom* **3**, 1–14.

SMALE, S. (1986), Newton's method estimates from data at one point, in: *The Merging of Disciplines in Pure, Applied and Computational Mathematics* (Springer, New York) 185–196.

SONNEVEND, G. (1985), An analytical center for polyhedrons and new classes of global algorithms for linear (smooth, convex) programming, in: *Lecture Notes in Control and Information Sciences* **84** (Springer, New York) 866–876.

SONNEVEND, G., J. STOER and G. ZHAO (1989), On the complexity of following the central path of linear programs by linear extrapolation, *Methods Oper. Res.* **63**, 19–31.

SONNEVEND, G., J. STOER and G. ZHAO (1991), On the complexity of following the central path of linear programs by linear extrapolation II, in: C. Roos and J.-P. Vial, eds., *Interior Point Methods for Linear Programming: Theory and Practice*, Math. Programming, Ser. B **52** (Mathematical Programming Society, North-Holland, Amsterdam) 527–553.

STIEFEL, E. (1952), Two applications of group characters to the solution of boundary-value problems, *J. Res. Nat. Bur. Standards* **6**, 424–427.

STIEFEL, E. and A. FÄSSLER (1979), *Gruppentheorethische Methoden und ihre Anwendung* (Teubner, Stuttgart).

STOER, J. (1983), Solution of large linear systems of equations by conjugate gradient type methods, in: A. Bachem, M. Grötschel and B. Korte, eds., *Mathematical Programming: The State of the Art* (Springer, Berlin) 540–565.

STROUD, A.H. (1971), *Approximate Calculation of Multiple Integrals* (Prentice-Hall, Englewood Cliffs, NJ).

TALMAN, A.J.J. and Y. YAMAMOTO (1989), A simplicial algorithm for stationary point problems on polytopes, *Math. Oper. Res.* **14**, 383–399.

TANABE, K. (1979), Continuous Newton–Raphson method for solving an underdetermined system of nonlinear equations, *Nonlinear Anal.* **3**, 495–503.

TAUSCH, J. (1993), A generalization of the discrete Fourier transformation, in: E.L. Allgower, K. Georg and R. Miranda, eds., *Exploiting Symmetry in Applied and Numerical Analysis*, Lectures in Applied Mathematics **29** (American Mathematical Society, Providence, RI) 405–412.

TODD, M.J. (1976a), *The Computation of Fixed Points and Applications*, Lecture Notes in Economics and Mathematical Systems **124** (Springer, Berlin).

TODD, M.J. (1976b), Extensions of Lemke's algorithm for the linear complementarity problem, *J. Optim. Theory Appl.* **20**, 397–416.

TODD, M.J. (1976c), Orientation in complementary pivot algorithms, *Math. Oper. Res.* **1**, 54–66.

TODD, M.J. (1978a), Fixed-point algorithms that allow restarting without extra dimension, Technical Report, Cornell University, Ithaca, NY.

TODD, M.J. (1978b), On the Jacobian of a function at a zero computed by a fixed point algorithm, *Math. Oper. Res.* **3**, 126–132.

TODD, M.J. (1980), A quadratically convergent fixed point algorithm for economic equilibria and linearly constrained optimization, *Math. Programming* **18**, 111–126.

TODD, M.J. (1981), PLALGO: A FORTRAN implementation of a piecewise-linear homotopy algorithm for solving systems of nonlinear equations, School of Operations Research and Industrial Engineering, Cornell University, Ithaca, NY.

TODD, M.J. (1982), On the computational complexity of piecewise-linear homotopy algorithms, *Math. Programming* **24**, 216–224.

TODD, M.J. (1986), Polynomial expected behavior of a pivoting algorithm for linear complementarity and linear programming problems, *Math. Programming* **35**, 173–192.

TODD, M.J. (1989), Recent developments and new directions in linear programming, in: N. Iri and K. Tanabe, eds., *Mathematical Programming, Recent Developments and Applications* (Kluwer, London) 109–157.

USHIDA, A. and L.O. CHUA (1984), Tracing solution curves of nonlinear equations with sharp turning points, *Internat. J. Circuit Theory Appl.* **12**, 1–21.

VAIDYA, P.M. (1990), An algorithm for linear programming which requires $O(((m+n)n^2+(m+n)^{1.5}n)L)$ arithmetic operations, *Math. Programming* **47**, 175–202.

VANDERBAUWHEDE, A. (1982), *Local Bifurcation Theory and Symmetry* (Pitman, London).

VAN DER LAAN, G. and A.J.J. TALMAN (1979), A restart algorithm for computing fixed points without an extra dimension, *Math. Programming* **17**, 74–84.

VAN DER LAAN, G. and A.J.J. TALMAN (1980), An improvement of fixed point algorithms by using a good triangulation, *Math. Programming* **18**, 274–285.

VAN DER WAERDEN, B.L. (1953), *Modern Algebra*, Vols. **I** and **II** (Ungar, New York).

VAN LOAN, C. (1992), *Computational Frameworks for the Fast Fourier Transform*, Frontiers in Applied Mathematics **10** (SIAM, Philadelphia, PA).

VERLINDEN, P. and R. COOLS (1994), Proof of a conjectured asymptotic expansion for the approximation of surface integrals, *Math. Comp.* **63**, 717–725.

VERLINDEN, P. and A. HAEGEMANS (1994), Solving very deficient systems of polynomial equations by implicit elimination, Report TW 207, Katholieke Universiteit Leuven, Department of Computing Science, Leuven, Belgium.

VERSCHELDE, J. and R. COOLS (1992), Nonlinear reduction for solving deficient polynomial systems by continuation methods, *Numer. Math.* **63**, 263–282.

VERSCHELDE, J. and R. COOLS (1994a), Symbolic homotopy construction, *Appl. Algebra Engrg., Commun. Comput.* **4**, 169–183.

VERSCHELDE, J. and R. COOLS (1994b), Symmetric homotopy construction, *J. Comput. Appl. Math.* **50**, 575–592.

VERSCHELDE, J. and K. GATERMANN (1994), Symmetric Newton polytopes for solving sparse polynomial systems, Preprint, Konrad-Zuse-Zentrum Berlin, SC 94-3.

VERSCHELDE, J. and A. HAEGEMANS (1993), The GBQ-algorithm for constructing start systems of homotopies for polynomial systems, *SIAM J. Numer. Anal.* **30**, 583–594.

VERSCHELDE, J. and A. HAEGEMANS (1994), Homotopies for solving polynomial systems within a bounded domain, *Theoret. Comput. Sci.* **133**, 165–185.

VERSCHELDE, J., P. VERLINDEN and R. COOLS (1994), Homotopies exploiting Newton polytopes for solving sparse polynomial systems, *SIAM J. Numer. Anal.* **31**, 915–930.

WALKER, H.F. (1990), Newton-like methods for underdetermined systems, in: E.L. Allgower and K. Georg, eds., *Computational Solution of Nonlinear Systems of Equations*, Lectures in Applied Mathematics **26** (American Mathematical Society, Providence, RI) 679–700.

WAMPLER, C.W. and A.P. MORGAN (1991), Solving the $6R$ inverse position problem using a generic-case solution methodology, *Mech. Mach. Theory* **26**, 91–106.

WAMPLER, C.W., A.P. MORGAN and A.J. SOMMESE (1990), Numerical continuation methods for solving polynomial systems arising in kinematics, *ASME J. Design* **112**, 59–68.

WAMPLER, C.W., A.P. MORGAN and A.J. SOMMESE (1992), Complete solution of the nine-point path synthesis problem for four-bar linkages, *ASME J. Mech. Design* **114**, 153–159.

WATKINS, D.S. (1984), Isospectral flows, *SIAM Rev.* **26**, 379–391.

WATSON, L.T., S.C. BILLUPS and A.P. MORGAN (1987), HOMPACK: A suite of codes for globally convergent homotopy algorithms, *ACM Trans. Math. Software* **13**, 281–310.

WAYBURN, T.L. (1982), A review of continuation methods and their application to separations problems, *AIChE Comput. Sys. Tech.* **11**, 8–22.

WAYBURN, T.L. and J.D. SEADER (1987), Homotopy continuation methods for computer-aided process design, *Comput. Chem. Engrg.* **11**, 7–25.

WEBER, H. (1981), On the numerical approximation of secondary bifurcation problems, in: E.L. Allgower, K. Glashoff and H.-O. Peitgen, eds., *Numerical Solution of Nonlinear Equations*, Lecture Notes in Math. **878** (Springer, New York) 407–425.

WERNER, B. (1992), Test functions for bifurcation points and Hopf points in problems with symmetries, in: E.L. Allgower, K. Böhmer and M. Golubitsky, eds., *Bifurcation and Symmetry*, ISNM **104** (Birkhäuser, Basel) 317–327.

WERNER, B. and A. SPENCE (1984), The computation of symmetry-breaking bifurcation points, *SIAM J. Numer. Anal.* **21**, 388–399.

WIDMANN, R. (1990a), An efficient algorithm for the triangulation of surfaces in $\mathbf{R}^3$, Preprint, Colorado State University.

WIDMANN, R. (1990b), Efficient triangulation of 3-dimensional domains, Preprint, Colorado State University.

WINTERS, K.H. (1987), A bifurcation study of laminar flow in a curved rectangular cross-section, *J. Fluid Mech.* **180**, 343–369.

WINTERS, K.H. (1991), Bifurcation and stability: A computational approach, *Comput. Phys. Commun.* **65**, 299–309.

WINTERS, K.H. and K.A. (1985), The prediction of critical points for thermal explosions in a finite volume, *Combust. Flame* **62**, 13–20.

WINTERS, K.H., K.A. CLIFFE and C.P. JACKSON (1987), The prediction of instabilities using bifurcation theory, in: R.W. Lewis, E.P. Hinton, P. Bettess and B.A. Schrefler, eds., *Numerical Methods for Transient and Coupled Problems* (Wiley, New York) 179–198.

WINTERS, K.H., M.R. MYERSCOUGH and P.K. MAINI (1990), Tracking bifurcating solutions of a model biological pattern generator, *Impact Comput. Sci. Eng.* **2**, 355–371.

WRIGHT, A.H. (1981), The octahedral algorithm, a new simplicial fixed point algorithm, *Math. Programming* **21**, 47–69.

WRIGHT, A.H. (1985), Finding all solutions to a system of polynomial equations, *Math. Comp* **44**, 125–133.

WRIGHT, M.H. (1992), Interior methods for constrained optimization, *Acta Numer.* **1**, 341–407.

YANG, Z.-H. and H.B. KELLER (1986), A direct method for computing higher order folds, *SIAM J. Sci. Statist. Comput.* **7**, 351–361.

YE, Y., O. GÜLER, R.A. TAPIA and Y. ZHANG (1991), A quadratically convergent $O(\sqrt{n}L)$-iteration algorithm for linear programming, *Math. Programming* **59**, 151–162.

Subject Index

action of an inverse matrix, 51
almost all, 79
asymptotically linear, 144
augmented matrix, 51
automatic differentiation, 67
  forward mode, 68
  reverse mode, 68

barycenter, 164
Bezout
  number, 94
  theorem, 93
bifurcation
  complex, 119
  equation, 100
  parameter, 114
  point, 98
  subgroup, 113
bordering algorithm, 53
boundary condition, 87
boundary value problem
  elliptic, 89
branch connecting equation, 103
Broyden's good update, 56

cell, 134
  dimension of, 134
  facet of, 134
  interior of, 134
  transverse, 141
cells
  adjacent, 135
central path, 128
central vertex
  of J_1, 138
CFK triangulation K_1, 136
change in orientation, 102
chord method, 160
circuit analysis, 85
circulant matrix, 62
closed loop, 125
complementarity problem
  linear, 151
complex projective space, 93
cone construction, 155
cone of attraction, 102
conjugate gradient method
  nonlinear, 72
contraction rate, 35
corrector, 16
critical points, 117

Davidenko, 14
deflated map, 86
deflation, 86
differentiable at ∞, 145
divide and conquer, 123
duality gap, 128

Eaves–Saigal algorithm, 146
embedding methods, 16
equivariance, 61, 110
equivariant
  matrix, 61
  operator, 61
Euler–Newton method, 26
  using updates, 56
extremal points, 39

face, 134
  proper, 134
facet
  completely labeled, 140
finite Fourier transform, 63
  generalized, 64
fixed-point subspace, 111
fold, 17, 119
  simple, 120
fold point, 39
  quadratic, 115
Frobenius norm, 56

Gauss–Newton map, 22
Gauss–Newton method, 21
generalized equations, 124
generic, 80

half-space, 133
homogeneous part, 92
homogenization, 92
homotopy, 80
  convex, 117
  d-homotopy, 87
  global, 82
hyperplane, 133

implicit function theorem, 13
index, 87
  of a map, 150
index formula, 151
interior point methods, 127
interpolation polynomial, 45
irreducible representation, 63
isotropy subgroup, 111

Jacobian, 13
  augmented, 14
  of a piecewise linear map, 139

Krylov method, 72

least change principle, 55
Lemke's algorithm, 151
lexicographically positive, 143
local uniqueness, 15
locally bounded, 145
locally finite, 134

midpoint rule, 175
  modified, 175
minimally extended system, 106
mixed volume, 96
modified trapezoidal rule
  for curves, 130
Moore–Penrose inverse, 22
moving frame, 161
multiplicity of a zero point, 93

Newton method
  chord method, 54
  global, 81
normal cone operator, 124

orbit, 111
orientation, 14, 149
  induced, 149
  of a PL manifold, 149
oriented tangent, 14

parametric programming, 125
path, 125
piecewise linear
  approximation, 139
  manifold, 134
    boundary of, 134
  map, 139
pivoting
  of cells, 135
pivoting by permutations, 137
pivoting by reflection, 135
positive part
  of a vector, 152
predictor, 16
predictor corrector, 16
primal–dual manifold, 156
primal–dual pair, 154
pseudo arclength, 18
pseudo manifold, 135

quasi-Newton methods, 54

ray start, 152
reference matrix, 161
refinement
  of a PL manifold, 156
regular
  point, 13
  value, 13
    of a PL map, 139
regularized paths, 141

Σ-invariant, 111
Schur complement, 51
secant equation, 55
set-valued hull, 145
simple bifurcation point, 98
simplex, 135
  thickness of, 168
  transverse, 165
singular
  point, 13
  value, 13
    of a PL map, 139
solution curve, 13
starting point, 80
steplength adaptation, 33
  Newton, 40
symmetry
  breaking, 116
  of equations, 60

tangent space
  of a cell, 149
tangent vector, 14
target level, 80
target map, 80
test function, 103
theorem of Bernstein, 96
trapezoidal rule, 175
  modified, 175
triangulation, 135
  refining, 144
  thickness of, 139, 168
trivial level, 80
trivial map, 80
turning point, 17

Union Jack triangulation, 138
Union Jack triangulation J_1, 136

variable dimension algorithms, 153
variational inequality, 124
vertex
  of a simplex, 135
  of a cell, 134

zero point at infinity, 93

Spectral Methods

Christine Bernardi and Yvon Maday

Laboratoire d'Analyse Numérique
Université Pierre et Marie Curie
Tour 55–65, 5ème étage
4 place Jussieu
75252 Paris Cédex 05
France

HANDBOOK OF NUMERICAL ANALYSIS, VOL. V
Techniques of Scientific Computing (Part 2)
Edited by P.G. Ciarlet and J.L. Lions

Contents

FOREWORD 213

CHAPTER I. Preliminaries 219

1. Sobolev spaces 219
2. Lax–Milgram lemma and applications 228
3. Legendre orthogonal polynomials 233
4. Quadrature formulas 240
5. Polynomial inverse inequalities 252

CHAPTER II. Polynomial Approximation Error, the Galerkin Method 257

6. Polynomial approximation in an interval 257
7. Polynomial approximation on tensorized domains 268
8. Galerkin method for the second-order Dirichlet problem 273
9. Galerkin method for the fourth-order Dirichlet problem 276
10. Galerkin method for the second-order Neumann problem 279
11. The eigenvalue Dirichlet problem 281
12. Polynomial approximation in the half-line 286

CHAPTER III. Polynomial Interpolation Error, the Collocation Method 293

13. Polynomial interpolation in an interval 293
14. Polynomial interpolation on tensorized domains 304
15. Collocation method for the second-order Dirichlet problem 308
16. Galerkin methods with numerical integration for the fourth-order Dirichlet problem 318
17. Galerkin method with numerical integration for the second-order Neumann problem 328

CHAPTER IV. Spectral Methods in Weighted Sobolev Spaces 333

18. Weighted Sobolev spaces 333
19. Jacobi orthogonal polynomials 348
20. Polynomial approximation in weighted norms 356
21. Polynomial interpolation in weighted norms 360
22. Weighted collocation methods 365

CHAPTER V. Spectral Discretizations of the Navier–Stokes Equations 373

23. The general saddle-point problem 375
24. The single grid method for the Stokes problem 388
25. The $\mathbb{P}_N - \mathbb{P}_{N-2}$ method for the Stokes problem 415
26. The staggered grid method for the Stokes problem 425
27. Extension to the Navier–Stokes equations 432

28. The stream-function and vorticity method 441

CHAPTER VI. Spectral Discretizations of Hyperbolic Conservation Laws 449

29. Filtering procedures 449
30. The cell averaging method 458
31. The spectral viscosity method 463

REFERENCES 471

LIST OF SYMBOLS 477

SUBJECT INDEX 483

Foreword

"Ah! monsieur, c'est un spectre, je le reconnais au marcher."
This article is dedicated to Professor P.-A. Raviart
who once upon a time told us
"Spectre, fantôme, ou diable, je veux voir ce que c'est."
Molière, *Dom Juan, ou le festin de pierre.*

Spectral methods are recent techniques for the discretization of partial differential equations. They rely on the approximation of the exact solution by polynomials. As a consequence and in contrast to other methods such as finite differences or finite elements, they are able to achieve an infinite degree of accuracy. From the theoretical point of view, it is related to the fact that any analytic function can be approximated by polynomials in an exponential way. From the numerical point of view, it means that the order of the convergence of the discrete solution to the exact one is limited only by the regularity of the exact solution. From the practical point of view, fewer degrees of freedom are necessary in many numerical simulations to obtain a given accuracy. This high order justifies the ever increasing use of spectral methods, at least for elliptic or parabolic problems, in a large number of university research laboratories and also in industrial research departments, in parallel to the development of fast computers.

In most cases, the basic idea to compute the approximate solution simply relies on the variational formulation of the continuous problem. Then, in the discrete problem, the test functions vary in a finite-dimensional space of polynomials and the integrals are replaced by appropriate quadrature formulas. Consequently, the spectral methods can easily be described in a few lines. Let us consider for instance a linear problem

$$Lu = f,$$

provided with a boundary condition (which is taken homogeneous for the sake of simplicity)

$$Bu = 0.$$

A variational formulation of this problem is assumed to be known: *find a function u in X such that*

$$\forall v \in X, \quad l(u, v) = \langle f, v \rangle$$

where the bilinear form $l(\bullet, \bullet)$ is associated with the operator L in a natural way. The discrete solution is sought in a finite-dimensional space of polynomials Y_N, where the discretization parameter N is a positive integer related to the maximal degree of the polynomials in the space. Then, X_N stands for the space $Y_N \cap X$ (i.e., the polynomials of X_N satisfy either all the boundary conditions or at least their essential part). Three methods can be used to define the discrete problem:

(i) The *tau-method*, now seldom used, consists in the minimization of the residual in the following way: *find u_N in X_N such that*

$$\forall v_N \in Z_N, \quad \langle Lu_N - f, v_N \rangle = 0,$$

where Z_N does not necessarily coincide with X_N but has the same dimension (typically, if X_N is the space of polynomials with degree $\leqslant N$ on an interval which vanish at the end points, Z_N can be chosen as the space of polynomials with degree $\leqslant N-2$ on this interval). However, it can be observed that, for the same number of degrees of freedom, the accuracy of this technique is less than the accuracy of the two following ones. Hence, it will not be considered in this article.

(ii) The *Galerkin method* consists in replacing X by X_N in the variational formulation, so that the discrete problem is written: *find u_N in X_N such that*

$$\forall v_N \in X_N, \quad l(u_N, v_N) = \langle f, v_N \rangle.$$

However, this technique can only be implemented in very few special cases, since generally the integrals which are involved in the right-hand side cannot be computed in an analytical way. In this article, its interest is only theoretical: it simplifies the analysis by focusing on the main difficulties.

(iii) If *numerical integration*, combined with the variational formulation, is used, not only X is replaced by X_N but also the integrals that are involved in the quantities $l(u, v)$ and $\langle f, v \rangle$ are computed by Gauss type quadrature formulas. Then the discrete problem is written: *find u_N in X_N such that*

$$\forall v_N \in X_N, \quad l_N(u_N, v_N) = \langle f, v_N \rangle_N.$$

A very important consequence is that, in most cases, this problem can be written equivalently as a system of *collocation equations*: *find u_N in Y_N such that*

$$\begin{cases} (L_N u_N)(\boldsymbol{x}) = f(\boldsymbol{x}), & \boldsymbol{x} \in \Xi_N, \\ (B_N u_N)(\boldsymbol{x}) = 0, & \boldsymbol{x} \in \Xi'_N, \end{cases}$$

where L_N and B_N are approximations of the operators L and B respectively, while Ξ_N and Ξ'_N are finite sets of points which are built from the nodes of the quadrature formulas. The sum of the cardinalities of Ξ_N and Ξ'_N is equal to the dimension of Y_N. This last method works for a number of important physical problems, and it will be thoroughly analyzed in this article.

From a historical point of view, it must be recalled that spectral methods were born twenty years ago, when the use of truncated Fourier series to discretize problems provided with periodic boundary conditions (asymptotic methods) was combined with the application of the recent *Fast Fourier Transform* of COOLEY and TURKEY [1965]. The first ideas are due to KREISS and OLIGER [1972] and, independently, to ORSZAG [1972] (see the book of GOTTLIEB and ORSZAG [1977] and the paper of KREISS and OLIGER [1979] for the basic results of numerical analysis). Since the periodic conditions are not so important in the applications, except perhaps in the direct simulation of turbulence, Fourier series will no longer be mentioned in this article.

Going back to polynomials and, of course, to standard boundary conditions of Dirichlet, Neumann or mixed type, it must be said that, in this case, two types of spectral techniques coexist: they rely respectively on the quadrature formulas of Legendre type and of Chebyshev type. Indeed, the drawback of the polynomial approximation is that it is expensive. So, the first idea to reduce the computation cost was to use Chebyshev techniques as a direct extension of Fourier spectral methods, since the fast Fourier transform can also be employed in this framework. Most codes still rely on Chebyshev quadrature formulas. The second idea, suggested by ORSZAG [1980], consists in using tensorized bases of polynomials, to take advantage of the properties of one-dimensional operators when possible. This is the main difference with the p-version of the finite element method (see BABUŠKA, SZABÓ and KATZ [1981]), and also the most important factor in the cost reduction, because of the huge growth of computing power.

Moreover, as suggested by GOTTLIEB [1981] and MERCIER [1982], the nodes of the collocation spectral method are chosen as cosines of nearly equidistant points, and it is well known that the Lagrange interpolation operator at these points has good approximation properties. Then, the most natural choice was the set of nodes of a Gauss type formula: either the Gauss formula or the Gauss–Lobatto formula to take the boundary conditions into account. These formulas exist for a number of weighted measures but, there also, it was natural to consider only two of them: the Gauss–Chebyshev formula, since the nodes are the cosines of exactly equidistant points already used in the Fourier spectral method, and the Gauss–Legendre formula since the associated measure is the Lebesgue measure. As a consequence, the Legendre type techniques are simpler to analyze and they most often lead to linear systems with a symmetric stiffness matrix, which of course is another important factor for reducing the computation cost.

Due to their growing importance, the Legendre spectral methods will be extensively analyzed in this article. The use of Chebyshev methods is still justified by the fast Fourier transform, so they are worth being presented and analyzed. Since their numerical analysis involves weighted Sobolev spaces with Chebyshev weight, we shall work in the more general case of Sobolev spaces with a Jacobi type weight, for the sake of generality and to include both Legendre and Chebyshev approximations in a unified framework. It will be proven that most properties of these methods, such as approximation or interpolation errors, are similar. However a number of new and important results will only be stated or proven for the Legendre techniques.

In this article, in order to prove the theoretical and numerical advantage of the tensorization properties, we analyze the basic spectral methods only for tensorized domains: intervals, squares and cubes. The methods can trivially be extended to rectangles or parallelepipeds. However, to handle more complex geometries, it becomes necessary to use deformations (see ORSZAG [1980]) or decompositions of the initial domain. This last idea results into the *spectral element method*, introduced by PATERA [1984], which will be presented by BERNARDI, MADAY and PATERA, in the future, in this Handbook.

It seems impossible to quote the hundreds of papers devoted to spectral methods that were published in the last twenty years, concerning either the numerical analysis or the computational experiments. However, it must be recalled that the results of CANUTO and QUARTERONI [1982] are the basis of the analysis of polynomial methods in Sobolev spaces. Note that several books about spectral methods are available: the book of GOTTLIEB and ORSZAG [1977] has already been quoted. Both Fourier and polynomial discretizations are described in the book of MERCIER [1989]. The books of CANUTO, HUSSAINI, QUARTERONI and ZANG [1987], of BOYD [1989], and more recently of FUNARO [1992], present new results with particular applications. BERNARDI and MADAY [1992b] already published, in French, the text of a doctoral course that was given to the students of the Laboratoire d'Analyse Numérique de l'Université Pierre et Marie Curie in Paris from 1988 to 1992.

Most of the problems that are considered in this article are elliptic partial differential equations in a square or a cube: they are associated with the Laplace or bilaplacian operators and provided with either Dirichlet or Neumann boundary conditions. The numerical analysis of their spectral discretization is performed in a detailed way, and optimal or nearly optimal error estimates are proven in each case. A first application is given and thoroughly analyzed: it deals with the spectral discretization of the Navier–Stokes equations that govern the flow of an incompressible viscous fluid. A very recent application is the discretization of nonlinear hyperbolic equations: the first ideas and basic results will be presented. Of course, it is a very restricted framework for spectral methods, and it will be completed in the article of BERNARDI, MADAY and PATERA [1995] in this Handbook. Let us however quote some already known extensions:

(i) handling complex geometries in two or three dimensions: transformation of domains, decomposition of domains (spectral element method, mortar method);

(ii) discretizing elliptic equations with nonconstant coefficients, nonlinear elliptic equations;

(iii) using the Laguerre polynomials in exterior complex geometries;

(iv) coupling spectral discretizations with finite element methods.

Here is an outline of what follows. In the first chapter, we recall some basic and standard results about the Sobolev spaces, the Legendre orthogonal polynomials, the Gauss type quadrature formulas. Chapter II is devoted to the study of polynomial approximation: it means that the distance of a function to a space of polynomials with respect to a Sobolev norm is estimated, depending on the regularity of the function. As a straightforward consequence, the numerical analysis of the Galerkin method is performed for three model equations: the Dirichlet problem for the Laplace or bilaplacian operators when the boundary conditions are homogeneous, the Neumann problem for the Laplace operator. The behavior of the discrete eigenvalues is described for a model problem, in comparison with the exact ones. Basic results on polynomial approximation in unbounded domains are also presented.

Chapter III is very similar. First, the distance between a function and its interpolate at the nodes of the quadrature formula is estimated; it is proven to be of the same order as the best approximation. Second, as an application, the spectral method with numerical integration is analyzed for the same problems as in Chapter II and with similar results.

Chapter IV is devoted to weighted spectral methods, following the same lines as previously. The Sobolev spaces with the Jacobi weight are introduced, together with the associated family of Jacobi polynomials. Next, in these spaces, estimates are given for the distance of a function to its best approximation and also to an appropriate Lagrange interpolate. As a consequence, weighted spectral discretizations of Dirichlet problems are studied.

Chapter V presents the spectral approximation of the Navier–Stokes equations. Three methods in the primitive variables of velocity and pressure are proposed: for each of them, the problem of spurious modes on the pressure is fully investigated, and error estimates are given both for the velocity and the pressure. The framework of the analysis is essentially the Legendre type methods, but an analogous weighted spectral method is also presented. The numerical analysis of the three methods is performed for the Stokes problem, and the extension of two of them to the full Navier–Stokes equations is studied in a separate section. Finally, a discretization relying on the stream function and vorticity formulation is analyzed.

Chapter VI deals with the spectral discretization of hyperbolic systems of conservation laws. Three ideas are presented, concerning these very recent techniques: the approximation of discontinuous functions by polynomials, a technique of cell averaging for the discrete solution and finally the spectral viscosity method.

The authors are very grateful to Professors P.G. Ciarlet and J.L. Lions for accepting this article in the Handbook of Numerical Analysis. They thank all their colleagues and students of the Laboratoire d'Analyse Numérique, who were of very great help for the scientific results and the publication of this book. They thank in advance their readers for their remarks and comments.

CHAPTER I

Preliminaries

The chapter begins with some basic notation and properties concerning the standard Sobolev spaces. Indeed, these spaces are involved in the variational formulation of the elliptic problems. Combined with the Lax–Milgram lemma which is presented in the second section, the variational formulation leads to existence and uniqueness results and is the basis of the numerical analysis of spectral discretizations. As a second tool for spectral methods, the family of Legendre polynomials is introduced in the third section: their properties are standard, but the basic ones will be proven since they are the cornerstone of the approximation results. The Legendre polynomials are used firstly for the construction of Gauss type quadrature formulas in Section 4 and secondly for the proof of some important inverse inequalities in spaces of polynomials in Section 5.

1. Sobolev spaces

In order to make precise the functional framework of spectral methods and to provide the tools which are required for their numerical analysis, we recall the basic notation and the main properties of Sobolev spaces. These results are standard, and their proofs can be found in the following reference books for instance: ADAMS [1975], BREZIS [1983], DAUTRAY and LIONS [1987], GRISVARD [1985], LIONS and MAGENES [1968] and NEČAS [1967].

In all that follows, the positive integer d stands for the dimension of the space. The notation ∂ followed by the name of an open set, denotes its boundary. For the sake of completeness, we need two definitions about the geometry of the domains.

DEFINITION. A bounded open set $\mathcal{O}$ in $\mathbb{R}^d$ is said to be *Lipschitz-continuous* if, for any point $\boldsymbol{x}$ in $\partial\mathcal{O}$, there exists a system of orthogonal coordinates $(y_1, \dots, y_d)$, an hypercube $U^{\boldsymbol{x}} = \prod_{i=1}^{d}]-a_i, a_i[$ and a Lipschitz-continuous mapping $\Phi^{\boldsymbol{x}}$ from $\prod_{i=1}^{d-1}]-a_i, a_i[$ into $]-a_d/2, a_d/2[$ such that

$$\mathcal{O} \cap U^{\boldsymbol{x}} = \{(y_1, \dots, y_d) \in U^{\boldsymbol{x}}\colon\ y_d > \Phi^{\boldsymbol{x}}(y_1, \dots, y_{d-1})\},$$
$$\partial\mathcal{O} \cap U^{\boldsymbol{x}} = \{(y_1, \dots, y_d) \in U^{\boldsymbol{x}}\colon\ y_d = \Phi^{\boldsymbol{x}}(y_1, \dots, y_{d-1})\}.$$

This property means that the boundary locally coincides with the graph of a Lipschitz-continuous function. It is satisfied by all the domains in this work, such as polygons. It is recalled that any bounded convex open set is Lipschitz-continuous (see GRISVARD ([1985], Corollary 1.2.2.3)). In this work, the basic domains are squares, for which the notation Ω is reserved, or hypercubes in $\mathbb{R}^d$, which are denoted by Ω_d.

DEFINITION. Let m be a nonnegative integer. A relatively open part Γ of the boundary of a Lipschitz-continuous bounded open set is said to be *of class* $\mathscr{C}^{m,1}$ if, for any point $\boldsymbol{x}$ in Γ, the mapping $\Phi^{\boldsymbol{x}}$ introduced in the previous definition can be chosen differentiable up to the order m with a Lipschitz-continuous derivative of order m.

In this section, $\mathcal{O}$ always denotes a Lipschitz-continuous connected bounded open set in $\mathbb{R}^d$. We recall that $\mathscr{D}(\mathcal{O})$ stands for the space of infinitely differentiable functions with a compact support in $\mathcal{O}$ and that $\mathscr{C}^\infty(\overline{\mathcal{O}})$ stands for the space of infinitely differentiable functions on $\overline{\mathcal{O}}$. The dual space $\mathscr{D}'(\mathcal{O})$ of $\mathscr{D}(\mathcal{O})$ is the space of distributions on $\mathcal{O}$. Then, let $L^2(\mathcal{O})$ denote the space of functions on $\mathcal{O}$ with real values, which are measurable with respect to the Lebesgue measure $\mathrm{d}\boldsymbol{x}$ and such that their square is integrable. It is provided with the norm

$$\|v\|_{L^2(\mathcal{O})} = \left(\int_{\mathcal{O}} v^2(\boldsymbol{x})\,\mathrm{d}\boldsymbol{x}\right)^{1/2}. \tag{1.1}$$

The space $L^2(\mathcal{O})$ is a Hilbert space, it contains both $\mathscr{D}(\mathcal{O})$ and $\mathscr{C}^\infty(\overline{\mathcal{O}})$ as dense subspaces and it is contained in $\mathscr{D}'(\mathcal{O})$. Hence the duality pairing between the spaces $\mathscr{D}(\mathcal{O})$ and $\mathscr{D}'(\mathcal{O})$ is an extension of the scalar product in $L^2(\mathcal{O})$. The theory of distributions (see SCHWARTZ [1966]) allows for associating partial derivatives of any order in $\mathscr{D}'(\mathcal{O})$ with each function in $L^2(\mathcal{O})$: for any d-tuple $\boldsymbol{k} = (k_1, \ldots, k_d)$ in $\mathbb{N}^d$, if $|\boldsymbol{k}|$ stands for the length $k_1 + \cdots + k_d$, the symbol $\partial^{\boldsymbol{k}}$ denotes the derivative of total order $|\boldsymbol{k}|$, and of order k_j with respect to the jth variable. The notation $\partial/\partial x_1, \ldots, \partial/\partial x_d$ will also be used for the partial derivatives of order 1 with respect to the d variables $x_1, \ldots, x_d$, and the term **grad** denotes the vector whose components are these d derivatives. When d is equal to 1, for any positive integer k, the derivative of order k is simply written $\mathrm{d}^k/\mathrm{d}\zeta^k$ and the first three derivatives are also denoted by $'$, $''$ and $'''$.

DEFINITION. Let m be a positive integer. The Sobolev space $H^m(\mathcal{O})$ is defined by

$$H^m(\mathcal{O}) = \{v \in L^2(\mathcal{O}):\ \forall \boldsymbol{k} \in \mathbb{N}^d,\ |\boldsymbol{k}| \leqslant m,\ \partial^{\boldsymbol{k}} v \in L^2(\mathcal{O})\}. \tag{1.2}$$

It is provided with the norm

$$\|v\|_{H^m(\mathcal{O})} = \left(\int_{\mathcal{O}} \sum_{|\boldsymbol{k}|\leqslant m} (\partial^{\boldsymbol{k}} v)^2(\boldsymbol{x})\,\mathrm{d}\boldsymbol{x}\right)^{1/2}, \tag{1.3}$$

and it is a Hilbert space for the associated scalar product.

The first basic property is recalled in the following theorem (see for instance ADAMS ([1975], Theorem 3.16) for the proof).

THEOREM 1.1. *For any positive integer* m, *the space* $\mathscr{C}^\infty(\overline{\mathcal{O}})$ *is dense in the space* $H^m(\mathcal{O})$.

The following definition is standard.

DEFINITION. For any positive integer m, the space $H_0^m(\mathcal{O})$ is the closure of $\mathscr{D}(\mathcal{O})$ in $H^m(\mathcal{O})$.

The second basic result concerning these spaces is known as the *Poincaré–Friedrichs inequality*: there exists a positive constant $\mathscr{P}$ depending only on the geometry of $\mathcal{O}$ such that any function v in $H_0^1(\mathcal{O})$ satisfies

$$\|v\|_{L^2(\mathcal{O})} \leqslant \mathscr{P}\left(\int_{\mathcal{O}} \sum_{j=1}^{d} \left(\frac{\partial v}{\partial x_j}\right)^2 (\boldsymbol{x})\,\mathrm{d}\boldsymbol{x}\right)^{1/2}. \tag{1.4}$$

A straightforward consequence is stated in the following theorem.

THEOREM 1.2. *For any positive integer* m, *the seminorm*

$$|v|_{H^m(\mathcal{O})} = \left(\int_{\mathcal{O}} \sum_{|\boldsymbol{k}|=m} (\partial^{\boldsymbol{k}} v)^2(\boldsymbol{x})\,\mathrm{d}\boldsymbol{x}\right)^{1/2} \tag{1.5}$$

is a norm on the space $H_0^m(\mathcal{O})$, *equivalent to the norm* $\|\cdot\|_{H^m(\mathcal{O})}$.

DEFINITION. For any positive integer m, the dual space of $H_0^m(\mathcal{O})$ is denoted by $H^{-m}(\mathcal{O})$. It is provided with the norm

$$\|f\|_{H^{-m}(\mathcal{O})} = \sup_{v \in H_0^m(\mathcal{O})} \frac{\langle f, v\rangle}{|v|_{H^m(\mathcal{O})}}, \tag{1.6}$$

where $\langle\cdot,\cdot\rangle$ stands the duality pairing between $H^{-m}(\mathcal{O})$ and $H_0^m(\mathcal{O})$.

We also recall a simple version of the Bramble–Hilbert lemma (see NEČAS ([1967], Chapter 1, §1.7), for the proof and also CIARLET ([1978], Theorem 4.1.3), for a general statement).

THEOREM 1.3. *The seminorm* $|\cdot|_{H^1(\mathcal{O})}$ *is a norm, equivalent to the norm* $\|\cdot\|_{H^1(\mathcal{O})}$, *on the space* $H^1(\mathcal{O})/\mathbb{R}$ *quotient of* $H^1(\mathcal{O})$ *by the constant functions.*

Next, we wish to define the Sobolev spaces $H^s(\mathcal{O})$ for any real value of s. In order to do that, we first recall some basic results of the theory of interpolation between Hilbert spaces, which can be found in BERGH and LÖFSTRÖM [1976] and LIONS and

MAGENES [1968] for instance. Let X and Y be two separable Hilbert spaces such that X is continuously imbedded and dense in Y. Denoting respectively by $(\bullet,\bullet)_X$ and $(\bullet,\bullet)_Y$ the scalar products on X and Y, we introduce the subspace $D(S)$ of all functions u in X such that the linear form: $v \mapsto (u,v)_X$ can be extended into a continuous form on Y, and the operator S defined by

$$\forall u \in D(S),\ \forall v \in Y, \quad (Su,v)_Y = (u,v)_X. \tag{1.7}$$

Clearly, the operator S is self-adjoint and positive. Thus, setting $T = S^{1/2}$, we note that X coincides with the domain $D(T)$ and that the norm on X is equivalent to the graph norm of T. The operator T is not unique, however the following definition does not depend on the choice of T.

DEFINITION. For any real number θ, $0 < \theta < 1$, the *interpolation space* $[X,Y]_\theta$ *of index* θ *between* X *and* Y is the domain $D(T^{1-\theta})$ of the operator $T^{1-\theta}$, provided with the graph norm.

REMARK 1.1. Since X and Y are Hilbert spaces, there exist other equivalent definitions for the space $[X,Y]_\theta$: for instance, $[X,Y]_\theta$ coincides with the space of the values $v(0)$ of measurable functions v from $]0,1[$ into Y such that

$$\int_0^1 (\|v(t)\|_X^2 + \|v'(t)\|_Y^2)\, t^{2\theta-1}\, \mathrm{d}t < +\infty, \tag{1.8}$$

with equivalent norms.

Finally, it should be noted that for any θ and η, $0 < \theta < \eta < 1$, the following imbeddings hold:

$$X = [X,Y]_0 \subset [X,Y]_\theta \subset [X,Y]_\eta \subset [X,Y]_1 = Y, \tag{1.9}$$

where each space is dense in the following one. Also, we have the inequality, for any θ, $0 < \theta < 1$:

$$\forall v \in X, \quad \|v\|_{[X,Y]_\theta} \leqslant \|v\|_X^{1-\theta} \|v\|_Y^{\theta}. \tag{1.10}$$

Two basic properties are recalled in the following theorems. The first one states the reiteration result.

THEOREM 1.4. *Let X and Y be two Hilbert spaces such that X is continuously imbedded and dense in Y. Then the following identity holds for any real numbers θ, λ and μ, $0 < \theta < 1$ and $0 \leqslant \lambda < \mu \leqslant 1$:*

$$\big[[X,Y]_\lambda, [X,Y]_\mu\big]_\theta = [X,Y]_{(1-\theta)\lambda+\theta\mu}, \tag{1.11}$$

with equivalent norms.

The second theorem is called the principal theorem of interpolation.

THEOREM 1.5. *Let X and Y (resp. X^* and Y^*) be separable Hilbert spaces such that X (resp. X^*) is continuously imbedded and dense in Y (resp. Y^*). If a linear operator is continuous from X into X^* with norm a and from Y into Y^* with norm b, it is continuous from $[X, Y]_\theta$ into $[X^*, Y^*]_\theta$ with norm $\leqslant a^{1-\theta} b^\theta$, for any θ, $0 < \theta < 1$.*

Next, we apply this theory to the Sobolev spaces.

DEFINITION. Let s be a positive real number. The space $H^s(\mathcal{O})$ is defined as the interpolation space of index $1 - s/m$ between $H^m(\mathcal{O})$ and $L^2(\mathcal{O})$, for an integer $m \geqslant s$.

As a consequence of Theorem 1.4, this property is independent of m. Indeed, for any nonnegative integers k and m, $k \leqslant m$, it is well known that the interpolation space $[H^m(\mathcal{O}), L^2(\mathcal{O})]_{1-k/m}$ coincides with $H^k(\mathcal{O})$. Two more general results are stated as follows.

THEOREM 1.6. *For any nonnegative real numbers r and s, $s \leqslant r$, and for any θ, $0 < \theta < 1$, the following identity holds*

$$[H^r(\mathcal{O}), H^s(\mathcal{O})]_\theta = H^{(1-\theta)r+\theta s}(\mathcal{O}). \tag{1.12}$$

THEOREM 1.7. *For any nonnegative real number s with integral part $[s]$ and fractional part σ, $0 < \sigma < 1$, the following identity holds*

$$H^s(\mathcal{O}) = \{v \in H^{[s]}(\mathcal{O}) \colon \forall \boldsymbol{k},\ |\boldsymbol{k}| = [s],\ \partial^{\boldsymbol{k}} v \in H^\sigma(\mathcal{O})\}, \tag{1.13}$$

with equivalent norms.

This allows for writing an intrinsic norm on the space $H^s(\mathcal{O})$. Indeed, for $0 < \sigma < 1$, it can be proven (see ADAMS ([1975], Theorem 7.48) for instance) that the norm

$$\|v\|_{H^\sigma(\mathcal{O})} = \left(\|v\|^2_{L^2(\mathcal{O})} + \int_{\mathcal{O}} \int_{\mathcal{O}} \frac{|v(\boldsymbol{x}) - v(\boldsymbol{y})|^2}{|\boldsymbol{x} - \boldsymbol{y}|^{2\sigma+d}} \, \mathrm{d}\boldsymbol{x} \, \mathrm{d}\boldsymbol{y} \right)^{1/2}, \tag{1.14}$$

is equivalent to the interpolation norm on $H^\sigma(\mathcal{O})$. So, in the general case, we have the following result: for any nonnegative real number s with integral part $[s]$ and fractional part σ, $0 < \sigma < 1$, the norm

$$\|v\|_{H^s(\mathcal{O})} = \left(\|v\|^2_{H^{[s]}(\mathcal{O})} + \sum_{|\boldsymbol{k}|=m} \int_{\mathcal{O}} \int_{\mathcal{O}} \frac{|\partial^{\boldsymbol{k}} v(\boldsymbol{x}) - \partial^{\boldsymbol{k}} v(\boldsymbol{y})|^2}{|\boldsymbol{x} - \boldsymbol{y}|^{2\sigma+d}} \, \mathrm{d}\boldsymbol{x} \, \mathrm{d}\boldsymbol{y} \right)^{1/2}, \tag{1.15}$$

is a norm on $H^s(\mathcal{O})$, equivalent to the interpolation norm.

The previous definition implies that $\mathscr{C}^\infty(\overline{\mathcal{O}})$ is dense in $H^s(\mathcal{O})$. In analogy to the case of integral order spaces, this leads to the introduction of the following spaces.

DEFINITION. For any positive real number s, the space $H_0^s(\mathcal{O})$ is the closure of $\mathscr{D}(\mathcal{O})$ in $H^s(\mathcal{O})$, while the space $H^{-s}(\mathcal{O})$ is the dual space of $H_0^s(\mathcal{O})$.

A general interpolation theorem also exists for these spaces. Note however that the following result does not hold when some parameters are half-integers.

THEOREM 1.8. *For any nonnegative real numbers q, r and s, $s \leqslant r$, such that neither $r - \frac{1}{2}$ nor $s - \frac{1}{2}$ is an integer and for any θ, $0 < \theta < 1$, such that $(1-\theta)r + \theta s - \frac{1}{2}$ is not an integer, the following identity holds*

$$\left[H_0^r(\mathcal{O}) \cap H^q(\mathcal{O}), H_0^s(\mathcal{O}) \cap H^q(\mathcal{O})\right]_\theta = H_0^{(1-\theta)r+\theta s}(\mathcal{O}) \cap H^q(\mathcal{O}). \tag{1.16}$$

Next, we give a simple version of the Sobolev imbedding theorem which allows for linking the Sobolev spaces with the space of continuous functions.

THEOREM 1.9. *Let s be a positive real number. The space $H^s(\mathcal{O})$ is contained, with a continuous imbedding, in the space of continuous functions on $\overline{\mathcal{O}}$ if and only if*

$$2s > d. \tag{1.17}$$

REMARK 1.2. Let E be a separable Hilbert space for the norm $\|\cdot\|_E$. All the previous statements can be extended to the functions defined on $\mathcal{O}$ with values in E. More precisely, we denote by $L^2(\mathcal{O}; E)$ the space of measurable functions v from $\mathcal{O}$ into E such that the function: $v \mapsto \|v\|_E$ belongs to $L^2(\mathcal{O})$. For any positive integer m, $H^m(\mathcal{O}; E)$ stands for the space of functions in $L^2(\mathcal{O}; E)$ such that all their partial derivatives with order $\leqslant m$ belong to $L^2(\mathcal{O}; E)$. Then, for any positive real number s, $H^s(\mathcal{O}; E)$ is the space $[H^m(\mathcal{O}; E), L^2(\mathcal{O}; E)]_{1-s/m}$, where m is an integer $\geqslant s$. Finally, we define $H_0^s(\mathcal{O}; E)$ as the closure in $H^s(\mathcal{O}; E)$ of the space of infinitely differentiable functions from $\mathcal{O}$ into E with a compact support in $\mathcal{O}$, and $H^{-s}(\mathcal{O}; E)$ as its dual space. For any nonnegative integer m, $H^m(\mathcal{O}; E)$ is provided with the norm

$$\|v\|_{H^m(\mathcal{O};E)} = \left(\int_\mathcal{O} \sum_{|\boldsymbol{k}| \leqslant m} \|(\partial^{\boldsymbol{k}} v)(\boldsymbol{x})\|_E^2 \, \mathrm{d}\boldsymbol{x}\right)^{1/2} \tag{1.18}$$

and with the seminorm

$$|v|_{H^m(\mathcal{O};E)} = \left(\int_\mathcal{O} \sum_{|\boldsymbol{k}| = m} \|(\partial^{\boldsymbol{k}} v)(\boldsymbol{x})\|_E^2 \, \mathrm{d}\boldsymbol{x}\right)^{1/2}. \tag{1.19}$$

The trace properties of the Sobolev spaces are necessary to handle the Dirichlet and Neumann boundary conditions. Their proofs, for nonsmooth domains, can be found in GRISVARD [1985] for instance. Let us recall that, since the domain $\mathcal{O}$ is Lipschitz-continuous, there exists a unit outward normal vector $\boldsymbol{n}$ to $\partial\mathcal{O}$ almost everywhere along the boundary $\partial\mathcal{O}$. The normal derivative operator $n_1(\partial/\partial x_1)+\cdots+n_d(\partial/\partial x_d)$, where $n_1, \ldots$ and n_d are the components of $\boldsymbol{n}$, is denoted by $\partial/\partial n$.

THEOREM 1.10. *Let m be a positive integer, and let Γ be a relatively open part of $\partial\mathcal{O}$, which is of class $\mathscr{C}^{m-1,1}$. The trace mapping T_m^Γ:*

$$v \mapsto T_m^\Gamma v = \left(v_{|\Gamma}, \left(\frac{\partial v}{\partial n}\right)\Big|_\Gamma, \ldots, \left(\frac{\partial^{m-1} v}{\partial n^{m-1}}\right)\Big|_\Gamma \right), \tag{1.20}$$

defined on $\mathscr{C}^\infty(\overline{\mathcal{O}})$ with values in $L^2(\Gamma)^m$, can be extended by density to any space $H^s(\mathcal{O})$, where s is a real number $> m - \frac{1}{2}$.

The previous theorem allows for the characterization of the spaces $H_0^s(\mathcal{O})$.

THEOREM 1.11. *Let m be a positive integer. Let us assume that the boundary $\partial\mathcal{O}$ is the union of a finite number of $\overline{\Gamma}_j$, where the Γ_j are relatively open parts of class $\mathscr{C}^{m-1,1}$ in $\partial\mathcal{O}$. For any positive real number $s < m + \frac{1}{2}$ such that $s - \frac{1}{2}$ is not an integer, the following characterization holds:*

$$H_0^s(\mathcal{O}) = \left\{ v \in H^s(\mathcal{O}) \colon \forall j,\ T_k^{\Gamma_j} v = 0 \right\}, \tag{1.21}$$

where k is the integral part of $s + \frac{1}{2}$.

When a part Γ of $\partial\mathcal{O}$ is contained in an hyperplane of $\mathbb{R}^d$, the scale of Sobolev spaces $H^s(\Gamma)$ is defined on Γ by restriction of functions in $H^s(\mathbb{R}^{d-1})$. Then, as already hinted in (1.8), it can be proven that, for any real number $s > \frac{1}{2}$, the image of the space $H^s(\mathcal{O})$ by the trace mapping T_1^Γ is the space $H^{s-1/2}(\Gamma)$. This leads to the following definition (we refer to LIONS and MAGENES [1968] for the complete justification).

DEFINITION. Let Γ be a relatively open part of $\partial\mathcal{O}$. For any real number $s > \frac{1}{2}$, the image of the space $H^s(\mathcal{O})$ by the trace mapping T_1^Γ is denoted by $H^{s-1/2}(\Gamma)$. It is provided with the norm

$$\|\varphi\|_{H^{s-1/2}(\Gamma)} = \inf \left\{ \|v\|_{H^s(\mathcal{O})},\ v \in H^s(\mathcal{O}) \text{ and } v_{|\Gamma} = \varphi \right\}. \tag{1.22}$$

When the whole boundary $\partial\mathcal{O}$ is of class $\mathscr{C}^{m-1,1}$, it is possible (see GRISVARD ([1985], Theorem 1.5.1.2)) to build a *lifting operator*, i.e., an operator which maps traces on the boundary $\partial\mathcal{O}$ onto a function defined on $\mathcal{O}$, as stated in the following theorem. Note that the lifting operator of $T_m^{\partial\mathcal{O}}$ is continuous on Sobolev spaces of different orders, up to a limit related to the regularity of the boundary only.

THEOREM 1.12. *Let m and m_0 be two positive integers, $m \leqslant m_0$. Let us assume that the boundary $\partial\mathcal{O}$ is of class $\mathscr{C}^{m_0-1,1}$. There exists an operator R_m which is continuous* *from* $\prod_{l=0}^{m-1} H^{s-l-1/2}(\partial\mathcal{O})$ *into* $H^s(\mathcal{O})$ *for any real number* s, $m-\frac{1}{2} < s \leqslant m_0$, *and such that* $T_m^{\partial\mathcal{O}} \circ R_m$ *is the identity operator.*

This result is stated differently when the boundary $\partial\mathcal{O}$ is not regular but is the union of a finite number of regular parts Γ_j. In particular, in the two-dimensional case, when traces are given on these regular parts, compatibility conditions should be satisfied at their end points in order that a lifting function exists. We refer to GRISVARD [1985] and BERNARDI, DAUGE and MADAY [1992b] for complete results. Here, we consider the case of a square and of integral order spaces on this square, since only these results are needed for what follows.

Let Ω be the square $]-1,1[^2$. The corners $(-1,-1)$, $(1,-1)$, $(1,1)$ and $(-1,1)$ are denoted respectively by $\boldsymbol{a}_1$, $\boldsymbol{a}_2$, $\boldsymbol{a}_3$ and $\boldsymbol{a}_4 = \boldsymbol{a}_0$. Then, for $J = 1,2,3,4$, Γ_J stands for the edge with end points $\boldsymbol{a}_{J-1}$ and $\boldsymbol{a}_J$; the unit outward normal to Γ_J is denoted by $\boldsymbol{n}_J$ and the unit vector orthogonal to $\boldsymbol{n}_J$ and directed from $\boldsymbol{a}_{J-1}$ to $\boldsymbol{a}_J$ is denoted by $\boldsymbol{\tau}_J$. As previously, we define the normal and tangential derivative operators, more precisely, we set

$$\frac{\partial}{\partial n_1} = -\frac{\partial}{\partial x}, \quad \frac{\partial}{\partial n_2} = -\frac{\partial}{\partial y}, \quad \frac{\partial}{\partial n_3} = \frac{\partial}{\partial x}, \quad \frac{\partial}{\partial n_4} = \frac{\partial}{\partial y},$$
$$\frac{\partial}{\partial \tau_1} = -\frac{\partial}{\partial y}, \quad \frac{\partial}{\partial \tau_2} = \frac{\partial}{\partial x}, \quad \frac{\partial}{\partial \tau_3} = \frac{\partial}{\partial y}, \quad \frac{\partial}{\partial \tau_4} = -\frac{\partial}{\partial x}.$$

When a function v belongs to $H^s(\Omega)$ for $s > 1$, it is continuous on $\overline{\Omega}$, so that, if φ_J denotes its trace on Γ_J, $J = 1,2,3,4$, (with the convention $\varphi_5 = \varphi_1$), the following compatibility conditions hold:

$$\varphi_J(\boldsymbol{a}_J) = \varphi_{J+1}(\boldsymbol{a}_J), \quad J = 1,2,3,4. \tag{1.23}$$

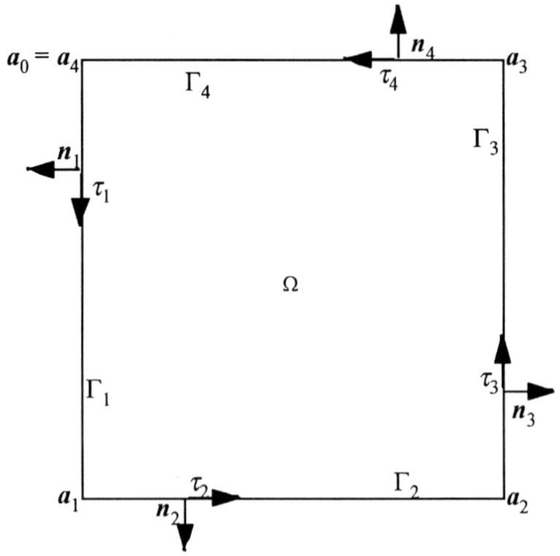

FIG. 1.1. The geometry of the square Ω.

However, if the function v only belongs to $H^1(\Omega)$, these equations have no meaning in the general case and must be replaced by weaker ones, namely

$$\int_0^2 \left|\varphi_J(\boldsymbol{a}_J - t\boldsymbol{\tau}_J) - \varphi_{J+1}(\boldsymbol{a}_J + t\boldsymbol{\tau}_{J+1})\right|^2 \frac{\mathrm{d}t}{t} < +\infty, \quad J = 1,2,3,4, \tag{1.24}$$

as stated in the following theorem.

THEOREM 1.13. *Let m be a positive integer. For any $4m$-tuple $(\varphi_J^0, \dots, \varphi_J^{m-1})_{J=1,2,3,4}$ in $\prod_{J=1}^4 \prod_{k=0}^{m-1} H^{m-k-1/2}(\Gamma_J)$ such that the following compatibility conditions hold:*

$$\left(\frac{\mathrm{d}^l \varphi_J^k}{\mathrm{d}\tau_J^l}\right)(\boldsymbol{a}_J) = (-1)^k \left(\frac{\mathrm{d}^k \varphi_{J+1}^l}{\mathrm{d}\tau_{J+1}^k}\right)(\boldsymbol{a}_J), \quad 0 \leqslant k + l \leqslant m-2, \ J = 1,2,3,4,$$

$$\begin{aligned} \mathcal{A}_J^{k,l} = \int_0^2 \Bigg| \left(\frac{\mathrm{d}^l \varphi_J^k}{\mathrm{d}\tau_J^l}\right)(\boldsymbol{a}_J - t\boldsymbol{\tau}_J) \\ -(-1)^k \left(\frac{\mathrm{d}^k \varphi_{J+1}^l}{\mathrm{d}\tau_{J+1}^k}\right)(\boldsymbol{a}_J + t\boldsymbol{\tau}_{J+1}) \Bigg|^2 \frac{\mathrm{d}t}{t} < +\infty, \\ k + l = m-1, \ J = 1,2,3,4, \end{aligned} \tag{1.25}$$

there exists a function v in $H^m(\Omega)$ such that

$$\frac{\partial^k v}{\partial n_J^k} = \varphi_J^k \quad \text{on } \Gamma_J, \quad 0 \leqslant k \leqslant m-1, \ J = 1,2,3,4, \tag{1.26}$$

and such that the following stability condition holds:

$$\|v\|_{H^m(\Omega)} \leqslant c \sum_{J=1}^{4} \sum_{k=0}^{m-1} \left(\|\varphi_J^k\|^2_{H^{m-k-1/2}(\Gamma_J)} + \mathcal{A}_J^{k,m-1-k}\right)^{1/2}. \tag{1.27}$$

REMARK 1.3. In the special case $m = 1$ and in view of Theorem 1.12, Theorem 1.13 can be written more precisely as follows

$$H^{1/2}(\partial\Omega) = \left\{\varphi\colon \varphi_J = \varphi|_{\Gamma_J} \in H^{1/2}(\Gamma_J) \text{ and (1.24) holds}\right\}, \tag{1.28}$$

and the norm:

$$\begin{aligned} \varphi \mapsto \Bigg(\sum_{J=1}^{4} \Bigg(\|\varphi|_{\Gamma_J}\|^2_{H^{1/2}(\Gamma_J)} \\ + \int_0^2 \left|\varphi|_{\Gamma_J}(\boldsymbol{a}_J - t\boldsymbol{\tau}_J) - \varphi|_{\Gamma_{J+1}}(\boldsymbol{a}_J + t\boldsymbol{\tau}_{J+1})\right|^2 \frac{\mathrm{d}t}{t} \Bigg) \Bigg)^{1/2}, \end{aligned} \tag{1.29}$$

is equivalent to the norm of $H^{1/2}(\partial\Omega)$.

Note that, when all the functions $\mathrm{d}^l\varphi_J^k/\mathrm{d}\tau_J^l$, $k+l=m-1$, belong to $H^{1/2}(\Gamma_J)$ and satisfy a Hölder property, conditions (1.25) are replaced by the simpler (and natural) ones:

$$\left(\frac{\mathrm{d}^l\varphi_J^k}{\mathrm{d}\tau_J^l}\right)(\boldsymbol{a}_J)=(-1)^k\left(\frac{\mathrm{d}^k\varphi_{J+1}^l}{\mathrm{d}\tau_{J+1}^k}\right)(\boldsymbol{a}_J),$$
$$0\leqslant k+l\leqslant m-1,\quad J=1,2,3,4. \tag{1.30}$$

Of course, Theorem 1.13 can be extended to the hypercube $\Omega_d=]-1,1[^d$, however we prefer not to write the compatibility conditions in the general case on each $(d-k)$-face, for $0\leqslant k\leqslant d-1$. We refer to BEN BELGACEM [1994] for further results in the three-dimensional geometry and we only write the theorem in the case $m=1$. Let Γ_j, $1\leqslant j\leqslant 2d$, be the $(d-1)$-faces of Ω_d, and $\boldsymbol{n}_j$ be the unit outward normal vector on Γ_j.

THEOREM 1.14. *For any 2d-tuple $(\varphi_j)_{1\leqslant j\leqslant 2d}$ in $\prod_{j=1}^{2d}H^{1/2}(\Gamma_j)$ such that the following compatibility conditions hold:*

$$\mathcal{B}_{j,k}=\int_{\overline{\Gamma}_j\cap\overline{\Gamma}_k}\int_0^2\left|\varphi_j(\boldsymbol{a}-t\boldsymbol{n}_k)-\varphi_k(\boldsymbol{a}-t\boldsymbol{n}_j)\right|^2\frac{\mathrm{d}t}{t}d\boldsymbol{a}<+\infty,$$
$$1\leqslant j,k\leqslant 2d, \tag{1.31}$$

there exists a function v in $H^1(\Omega_d)$ such that

$$v=\varphi_j\quad \text{on } \Gamma_j,\ 1\leqslant j\leqslant 2d,$$

and such that the following stability condition holds:

$$\|v\|_{H^1(\Omega)}\leqslant c\sum_{j=1}^{2d}\left(\|\varphi_j\|^2_{H^{1/2}(\Gamma_j)}+\sum_{k=j+1}^{2d}\mathcal{B}_{j,k}\right)^{1/2}. \tag{1.32}$$

2. Lax–Milgram lemma and applications

Firstly, we recall the lemma, due to LAX and MILGRAM [1954] and basic to the analysis of elliptic partial differential equations.

THEOREM 2.1 (Lax–Milgram lemma). *Let X be a reflexive Banach space. Let $\|\bullet\|_X$ denote its norm and let $\langle\bullet,\bullet\rangle$ be the duality pairing between the dual space X' of X and X. Let $a(\bullet,\bullet)$ be a continuous bilinear form on $X\times X$ which is elliptic on X, i.e., there exists a constant $\alpha>0$ such that*

$$\forall v\in X,\quad a(v,v)\geqslant\alpha\|v\|_X^2. \tag{2.1}$$

Then, for any f in X', the problem: find u in X such that:

$$\forall v \in X, \quad a(u,v) = \langle f, v\rangle, \tag{2.2}$$

has a unique solution u. Moreover, this solution satisfies

$$\|u\|_X \leqslant \frac{1}{\alpha} \sup_{v\in X,\ v\neq 0} \frac{\langle f, v\rangle}{\|v\|_X}. \tag{2.3}$$

Three applications of this lemma to model problems are presented successively: the Dirichlet problem for the Laplace operator, the Dirichlet problem for the bilaplacian operator, the Neumann problem for the Laplace operator.

EXAMPLE 2.1. On a Lipschitz-continuous bounded open set $\mathcal{O}$ in $\mathbb{R}^d$, we consider the equation:

$$\begin{cases} -\Delta u = f & \text{in } \mathcal{O}, \\ u = g & \text{on } \partial\mathcal{O}. \end{cases} \tag{2.4}$$

The distribution f is supposed to be in $H^{-1}(\mathcal{O})$ and the boundary data g are in $H^{1/2}(\partial\mathcal{O})$. In order to derive the variational formulation of this problem, the first equation is multiplied by a function v in $\mathscr{D}(\mathcal{O})$ (indeed, this equation is satisfied in the distribution sense), so that integrating by parts and using the density of $\mathscr{D}(\mathcal{O})$ in $H_0^1(\mathcal{O})$ lead to, for all v in $H_0^1(\mathcal{O})$

$$\int_{\mathcal{O}} (\mathbf{grad}\, u)(\boldsymbol{x}) \cdot (\mathbf{grad}\, v)(\boldsymbol{x})\, \mathrm{d}\boldsymbol{x} = \langle f, v\rangle.$$

The boundary condition is equivalent to the fact that $u - u_\flat$ vanishes on $\partial\mathcal{O}$, where the first trace of $u_\flat$ on the boundary $\partial\mathcal{O}$ coincides with g; such a function $u_\flat$ exists in $H^1(\mathcal{O})$ by Theorem 1.12, and the solution u is also sought in $H^1(\mathcal{O})$. Consequently, problem (2.4) admits the following equivalent variational formulation: *find u in $H^1(\mathcal{O})$, with $u - u_\flat$ in $H_0^1(\mathcal{O})$, such that*

$$\forall v \in H_0^1(\mathcal{O}), \quad \int_{\mathcal{O}} (\mathbf{grad}\, u)(\boldsymbol{x}) \cdot (\mathbf{grad}\, v)(\boldsymbol{x})\, \mathrm{d}\boldsymbol{x} = \langle f, v\rangle. \tag{2.5}$$

In order to apply the Lax–Milgram lemma, we note that the function u is a solution of (2.5) if and only if the function $u^* = u - u_\flat$ is a solution in $H_0^1(\mathcal{O})$ of the problem:

$$\begin{aligned} &\forall v \in H_0^1(\mathcal{O}), \\ &\quad \int_{\mathcal{O}} (\mathbf{grad}\, u^*)(\boldsymbol{x}) \cdot (\mathbf{grad}\, v)(\boldsymbol{x})\, \mathrm{d}\boldsymbol{x} \\ &\qquad = \langle f, v\rangle - \int_{\mathcal{O}} (\mathbf{grad}\, u_\flat)(\boldsymbol{x}) \cdot (\mathbf{grad}\, v)(\boldsymbol{x})\, \mathrm{d}\boldsymbol{x}. \end{aligned} \tag{2.6}$$

Clearly, the right-hand side is continuous on $H_0^1(\mathcal{O})$ and the ellipticity property is given in Theorem 1.2, so that the following theorem is straightforward.

THEOREM 2.2. *For any distribution f in $H^{-1}(\mathcal{O})$ and for any function g in $H^{1/2}(\partial\mathcal{O})$, problem* (2.4) *has a unique solution u in $H^1(\mathcal{O})$. Moreover, this solution satisfies*

$$\|u\|_{H^1(\mathcal{O})} \leqslant c\big(\|f\|_{H^{-1}(\mathcal{O})} + \|g\|_{H^{1/2}(\partial\mathcal{O})}\big). \tag{2.7}$$

EXAMPLE 2.2. Now, we consider the equation:

$$\begin{cases} \Delta^2 u = f & \text{in } \mathcal{O}, \\ u = 0 & \text{on } \partial\mathcal{O}, \\ \dfrac{\partial u}{\partial n} = 0 & \text{on } \partial\mathcal{O}, \end{cases} \tag{2.8}$$

where the distribution f belongs to $H^{-2}(\mathcal{O})$. Clearly, problem (2.8) admits the variational formulation: *find u in $H_0^2(\mathcal{O})$ such that*

$$\forall v \in H_0^2(\mathcal{O}), \qquad \int_{\mathcal{O}} (\Delta u)(\boldsymbol{x})(\Delta v)(\boldsymbol{x})\,\mathrm{d}\boldsymbol{x} = \langle f, v\rangle. \tag{2.9}$$

The ellipticity of the bilinear form follows from Theorem 1.2 together with the formula, valid for any u in $H_0^2(\mathcal{O})$,

$$\int_{\mathcal{O}} (\Delta u)^2(\boldsymbol{x})\,\mathrm{d}\boldsymbol{x} = \sum_{i=1}^{d}\left(\int_{\mathcal{O}}\left(\frac{\partial^2 u}{\partial x_i^2}\right)^2(\boldsymbol{x})\,\mathrm{d}\boldsymbol{x} + 2\sum_{j=i+1}^{d}\int_{\mathcal{O}}\left(\frac{\partial^2 u}{\partial x_i \partial x_j}\right)^2(\boldsymbol{x})\,\mathrm{d}\boldsymbol{x}\right),$$

so that applying the Lax–Milgram lemma leads to the wellposedness result.

THEOREM 2.3. *For any distribution f in $H^{-2}(\mathcal{O})$, problem* (2.8) *has a unique solution u in $H_0^2(\mathcal{O})$. Moreover, this solution satisfies*

$$\|u\|_{H^2(\mathcal{O})} \leqslant c\|f\|_{H^{-2}(\mathcal{O})}. \tag{2.10}$$

EXAMPLE 2.2$'$. Assuming moreover that the Lipschitz-continuous bounded open set $\mathcal{O}$ in $\mathbb{R}^d$ has its whole boundary of class $\mathscr{C}^{1,1}$, we consider the equation with inhomogeneous boundary conditions:

$$\begin{cases} \Delta^2 u = f & \text{in } \mathcal{O}, \\ u = g & \text{on } \partial\mathcal{O}, \\ \dfrac{\partial u}{\partial n} = h & \text{on } \partial\mathcal{O}. \end{cases} \tag{2.11}$$

The distribution f belongs to $H^{-2}(\mathcal{O})$ and the pair (g,h) is taken in $H^{3/2}(\partial\mathcal{O}) \times H^{1/2}(\partial\mathcal{O})$. It follows from Theorem 1.12 that there exists a function $u_\flat$ in $H^2(\mathcal{O})$ such that

$$u_\flat = g \quad \text{and} \quad \frac{\partial u_\flat}{\partial n} = h \quad \text{on } \partial\mathcal{O}. \tag{2.12}$$

Thus, problem (2.11) admits the equivalent, variational and straightforward formulation: *find u in $H^2(\mathcal{O})$, with $u - u_\flat$ in $H_0^2(\mathcal{O})$, such that*

$$\forall v \in H_0^2(\mathcal{O}), \quad \int_{\mathcal{O}} (\Delta u)(\boldsymbol{x})(\Delta v)(\boldsymbol{x})\,\mathrm{d}\boldsymbol{x} = \langle f, v\rangle. \tag{2.13}$$

Here again, introducing the equivalent variational problem satisfied by the function $u^\circ = u - u_\flat$ and applying the Lax–Milgram lemma leads to the following result.

THEOREM 2.4. *For any distribution f in $H^{-2}(\mathcal{O})$ and for any pair (g,h) in $H^{3/2}(\partial\mathcal{O}) \times H^{1/2}(\partial\mathcal{O})$, problem* (2.11) *has a unique solution u in $H^2(\mathcal{O})$. Moreover, this solution satisfies*

$$\|u\|_{H^2(\mathcal{O})} \leqslant c\big(\|f\|_{H^{-2}(\mathcal{O})} + \|g\|_{H^{3/2}(\partial\mathcal{O})} + \|h\|_{H^{1/2}(\partial\mathcal{O})}\big). \tag{2.14}$$

EXAMPLE 2.2″. We are also interested in the case where the domain $\mathcal{O}$ is the square $\Omega =]-1,1[^2$. Here, the problem is written as

$$\begin{cases} \Delta^2 u = f & \text{in } \Omega, \\ u = g_J & \text{on } \Gamma_J,\ J = 1,2,3,4, \\ \dfrac{\partial u}{\partial n_J} = h_J & \text{on } \Gamma_J,\ J = 1,2,3,4. \end{cases} \tag{2.15}$$

The distribution f is still supposed to be in $H^{-2}(\Omega)$, while the 8-tuple (g_J, h_J) belongs to $\prod_{J=1}^4 H^{3/2}(\Gamma_J) \times H^{1/2}(\Gamma_J)$ and satisfies

$$\begin{aligned} & g_J(\boldsymbol{a}_J) = g_{J+1}(\boldsymbol{a}_J), \quad J = 1,2,3,4, \\ & \mathcal{A}_J^{1,0} = \int_0^2 \left| h_J(\boldsymbol{a}_J - t\boldsymbol{\tau}_J) + \left(\frac{\mathrm{d}g_{J+1}}{\mathrm{d}\tau_{J+1}}\right)(\boldsymbol{a}_J + t\boldsymbol{\tau}_{J+1})\right|^2 \frac{\mathrm{d}t}{t} < +\infty, \\ & \quad J = 1,2,3,4, \\ & \mathcal{A}_J^{0,1} = \int_0^2 \left| \left(\frac{\mathrm{d}g_J}{\mathrm{d}\tau_J}\right)(\boldsymbol{a}_J - t\boldsymbol{\tau}_J) - h_{J+1}(\boldsymbol{a}_J + t\boldsymbol{\tau}_{J+1})\right|^2 \frac{\mathrm{d}t}{t} < +\infty, \\ & \quad J = 1,2,3,4. \end{aligned} \tag{2.16}$$

In this case, it follows from Theorem 1.13 that there exists a function $u_\flat$ in $H^2(\Omega)$ such that

$$u_\flat = g_J \quad \text{and} \quad \frac{\partial u_\flat}{\partial n_J} = h_J \quad \text{on } \Gamma_J, \quad J = 1,2,3,4, \tag{2.17}$$

so that the problem admits the same variational formulation (2.13), with $\mathcal{O}$ replaced by Ω. This leads to the following theorem.

THEOREM 2.5. *For any distribution f in $H^{-2}(\Omega)$ and for any 8-tuple $(g_J, h_J)_{J=1,2,3,4}$ in $\prod_{J=1}^4 H^{3/2}(\Gamma_J) \times H^{1/2}(\Gamma_J)$ which satisfies* (2.16), *problem* (2.15) *has a unique solution u in $H^2(\Omega)$. Moreover, this solution satisfies*

$$\|u\|_{H^2(\Omega)} \leqslant c\Bigg(\|f\|_{H^{-2}(\Omega)} + \sum_{J=1}^{4} \big(\|g_J\|^2_{H^{3/2}(\Gamma_J)} + \|h_J\|^2_{H^{1/2}(\Gamma_J)} + \mathcal{A}_J^{1,0} + \mathcal{A}_J^{0,1}\big)^{1/2}\Bigg). \tag{2.18}$$

EXAMPLE 2.3. Let us go back to the case of a Lipschitz-continuous bounded open set $\mathcal{O}$ in $\mathbb{R}^d$, for which we consider the equation:

$$\begin{cases} -\Delta u = f & \text{in } \mathcal{O}, \\ \dfrac{\partial u}{\partial n} = h & \text{on } \partial\mathcal{O}. \end{cases} \tag{2.19}$$

In this case, the function f is assumed to be in $L^2(\mathcal{O})$ and the boundary data h belong to the dual space $(H^{1/2}(\partial\mathcal{O}))'$ of $H^{1/2}(\partial\mathcal{O})$. Since the solution u is defined only up to an additive constant, it can be sought in the quotient-space $H^1(\mathcal{O})/\mathbb{R}$ of $H^1(\mathcal{O})$ by constant functions. So, it is straightforward to check that Eq. (2.19) admits the equivalent variational formulation: *find u in $H^1(\mathcal{O})/\mathbb{R}$ such that*

$$\forall v \in H^1(\mathcal{O})/\mathbb{R}, \quad \int_{\mathcal{O}} (\mathbf{grad}\, u)(\boldsymbol{x}) \cdot (\mathbf{grad}\, v)(\boldsymbol{x})\, \mathrm{d}\boldsymbol{x} = \int_{\mathcal{O}} f(\boldsymbol{x}) v(\boldsymbol{x})\, \mathrm{d}\boldsymbol{x} + \langle h, v\rangle_{\partial\mathcal{O}}, \tag{2.20}$$

where $\langle \cdot, \cdot \rangle_{\partial\mathcal{O}}$ stands for the duality pairing between $H^{1/2}(\partial\mathcal{O})$ and its dual space. In order that the right-hand side is defined on $H^1(\mathcal{O})/\mathbb{R}$, we are led to assume that the following condition holds

$$\int_{\mathcal{O}} f(\boldsymbol{x})\, \mathrm{d}\boldsymbol{x} + \langle h, 1\rangle_{\partial\mathcal{O}} = 0. \tag{2.21}$$

Then, since the ellipticity property is a consequence of Theorem 1.3, applying the Lax–Milgram lemma immediately leads to the result.

THEOREM 2.6. *For any f in $L^2(\mathcal{O})$ and for any distribution h in $(H^{1/2}(\partial\mathcal{O}))'$ such that condition* (2.21) *holds, problem* (2.19) *has a unique solution u in $H^1(\mathcal{O})/\mathbb{R}$. Moreover, this solution satisfies*

$$\|u\|_{H^1(\mathcal{O})/\mathbb{R}} \leqslant c\big(\|f\|_{L^2(\mathcal{O})} + \|h\|_{(H^{1/2}(\partial\mathcal{O}))'}\big). \tag{2.22}$$

REMARK 2.1. A very important property is the regularity of the solution of problems (2.4), (2.8) and (2.19). For the sake of clarity, we prefer to state the result only in the case of homogeneous boundary conditions; indeed, the complete result can immediately be derived from this case combined with the trace lifting theorems. In other words the name of the game is: let us introduce the mappings T_{D} (resp. T_{D4} and T_{N}) $: f \mapsto u$, where u is the solution of problem (2.4) (resp. (2.8) and (2.19)) with homogeneous boundary conditions $g = h = 0$. For which values of a real number $s \geqslant 2$ do these applications satisfy the continuity properties:

$$\begin{aligned} &T_{\mathrm{D}}: H^{s-2}(\mathcal{O}) \to H^s(\mathcal{O}), \qquad T_{\mathrm{D4}}: H^{s-3}(\mathcal{O}) \to H^{s+1}(\mathcal{O}),\\ &T_{\mathrm{N}}: H^{s-2}(\mathcal{O}) \to H^s(\mathcal{O})? \end{aligned} \tag{2.23}$$

In general, this is not true for high values of s, since the singularities of the boundary $\partial\mathcal{O}$ lead to singularities of the solution, even for smooth data. We only give a partial answer to the question, which can be found in GRISVARD [1985], in three special cases:

(i) when the boundary $\partial\mathcal{O}$ is of class $\mathscr{C}^{m-1,1}$, the property holds for any real number $s \leqslant m$;

(ii) when the domain $\mathcal{O}$ is convex, the property holds for any real number $s \leqslant 2$ in the case of problems (2.4) and (2.19);

(iii) when the domain $\mathcal{O}$ is a polygon with largest angle equal to ω, the property holds for any real number $s < 1 + \pi/\omega$ in the case of problems (2.4) and (2.19) and for any real number $s < 1 + \eta(\omega)$ in the case of problem (2.8), where $\eta(\omega)$ is the smallest positive imaginary part of the solutions z, $z \neq +\mathrm{i}$, of the equation

$$\operatorname{sh}^2(\omega z) = z^2 \sin^2 \omega. \tag{2.24}$$

Two values of the function η will be needed later on:

$$\eta\left(\frac{\pi}{2}\right) \simeq 2.7395933563 \quad \text{and} \quad \eta\left(\frac{3\pi}{2}\right) \simeq 0.5444837368. \tag{2.25}$$

3. Legendre orthogonal polynomials

This section is devoted to the properties of the Legendre polynomials, which are the basic tool for spectral methods. These properties are well known, so that only the more important ones will be explicitly proven. We refer to SZEGÖ [1978] for more complete results. It should also be noted that most of these properties have their analogue for other families of orthogonal polynomials and they will be stated in the general case of Jacobi polynomials in Section 19. However, we first limit ourselves to the case of Legendre polynomials for the sake of clarity.

In all that follows, Λ stands for the open interval $]-1,1[$. By using the Gram–Schmidt procedure, we can construct a family $(\tilde{L}_n)_{n\geqslant 0}$ of orthogonal polynomials in $L^2(\Lambda)$ such that $\tilde{L}_n$ has degree n and the coefficient of ζ^n in $\tilde{L}_n$ is equal to 1. This family is unique and, as standard, none of the $\tilde{L}_n$ vanishes in 1. This allows for defining the Legendre polynomials in a straightforward way.

DEFINITION. The family of *Legendre polynomials* is the family $(L_n)_{n\geqslant 0}$ of polynomials with one variable, which are orthogonal to each other in $L^2(\Lambda)$ and such that, for any integer $n \geqslant 0$, the polynomial L_n has degree n and satisfies: $L_n(1) = 1$. The coefficient of ζ^n in L_n is denoted by k_n.

Let us recall two basic properties that characterize these polynomials.

(i) For any positive integer n, the zeros of L_n are distinct real numbers in Λ.

(ii) For any nonnegative integer n, the polynomial L_n is even if n is even and odd if n is odd.

Indeed, property (i) holds for any family of orthogonal polynomials and property (ii) follows from the symmetry of Λ.

The next formula is essential for spectral methods.

THEOREM 3.1 (Differential equation). *For any integer $n \geqslant 0$, the polynomial L_n satisfies the differential equation:*

$$\frac{\mathrm{d}}{\mathrm{d}\zeta}\big((1-\zeta^2)L_n'\big) + n(n+1)L_n = 0. \tag{3.1}$$

PROOF. The polynomial $\mathrm{d}((1-\zeta^2)L_n')/\mathrm{d}\zeta$ has degree $\leqslant n$ and, for any polynomial φ with degree $\leqslant n-1$, it satisfies

$$\begin{aligned}\int_{-1}^{1}\frac{\mathrm{d}}{\mathrm{d}\zeta}\big((1-\zeta^2)L_n'\big)(\zeta)\varphi(\zeta)\,\mathrm{d}\zeta &= -\int_{-1}^{1}(1-\zeta^2)L_n'(\zeta)\varphi'(\zeta)\,\mathrm{d}\zeta\\ &= \int_{-1}^{1}L_n(\zeta)\frac{\mathrm{d}}{\mathrm{d}\zeta}\big((1-\zeta^2)\varphi'\big)(\zeta)\,\mathrm{d}\zeta = 0.\end{aligned}$$

Consequently, there exists a real number λ_n such that

$$\frac{\mathrm{d}}{\mathrm{d}\zeta}\big((1-\zeta^2)L_n'\big) + \lambda_n L_n = 0.$$

Comparing the coefficients of ζ^n in the two members of this identity leads to

$$-k_n n(n+1) + k_n\lambda_n = 0,$$

whence the result.

REMARK 3.1. The operator A defined by

$$A\varphi = -\frac{\mathrm{d}}{\mathrm{d}\zeta}\big((1-\zeta^2)\varphi'\big), \tag{3.2}$$

is of Sturm–Liouville type (see DAUTRAY and LIONS ([1987], Chapter 8, §2)). It is self-adjoint and positive. Equation (3.1) means that the Legendre polynomials are eigenfunctions of this operator. This is the origin of the word *spectral* which designates the methods described in this article.

REMARK 3.2. As an immediate consequence of (3.1), we derive by integration by parts, for any integers $n \geqslant 0$ and $k \geqslant 0$:

$$\int_{-1}^{1} L_n'(\zeta)L_k'(\zeta)(1-\zeta^2)\,\mathrm{d}\zeta = n(n+1)\int_{-1}^{1} L_n(\zeta)L_k(\zeta)\,\mathrm{d}\zeta. \tag{3.3}$$

It means that the polynomials L_n', $n \geqslant 1$, are orthogonal for the measure $(1-\zeta^2)\,\mathrm{d}\zeta$ in Λ. This will be widely used later on. Another consequence of (3.1) is written

$$L_n'(1) = \frac{n(n+1)}{2}. \tag{3.4}$$

In order to be more general, we note that differentiating Eq. (3.1) leads to

$$(1-\zeta^2)L_n''' - 4\zeta L_n'' + (n-1)(n+2)L_n' = 0,$$

or equivalently to

$$\frac{\mathrm{d}}{\mathrm{d}\zeta}\big((1-\zeta^2)^2 L_n''\big) + (n-1)(n+2)(1-\zeta^2)L_n' = 0, \tag{3.5}$$

$$\frac{\mathrm{d}^2}{\mathrm{d}\zeta^2}\big((1-\zeta^2)^2 L_n''\big) - (n-1)n(n+1)(n+2)L_n = 0. \tag{3.6}$$

Following the same lines together with an induction argument, we derive the following identity, which holds for any positive integer m:

$$\begin{aligned}&\frac{\mathrm{d}^m}{\mathrm{d}\zeta^m}\left((1-\zeta^2)^m\left(\frac{\mathrm{d}^m L_n}{\mathrm{d}\zeta^m}\right)\right)\\ &\quad + (-1)^{m+1}(n-m+1)(n-m+2)\cdots n(n+1)\times\cdots\\ &\quad \times\cdots(n+m-1)(n+m)L_n = 0.\end{aligned} \tag{3.7}$$

Consequently, the $\mathrm{d}^m L_n/\mathrm{d}\zeta^m$, $n \geqslant m$, form an orthogonal family of polynomials for the measure $(1-\zeta^2)^m\,\mathrm{d}\zeta$ in Λ. They also satisfy

$$\left(\frac{\mathrm{d}^m L_n}{\mathrm{d}\zeta^m}\right)(1) = \frac{(n-m+1)(n-m+2)\cdots n(n+1)\cdots(n+m-1)(n+m)}{2^m m!}. \tag{3.8}$$

Noting that $\mathrm{d}^n(1-\zeta^2)^n/\mathrm{d}\zeta^n$ is a polynomial of degree $\leqslant n$ which is orthogonal to any polynomial with degree $\leqslant n-1$ in $L^2(\Lambda)$ and computing its value in 1, we derive the following identity, called *Rodrigues' formula*, which holds for any nonnegative integer n:

$$L_n = \frac{(-1)^n}{2^n n!}\left(\frac{\mathrm{d}^n}{\mathrm{d}\zeta^n}\right)\left((1-\zeta^2)^n\right). \tag{3.9}$$

As an immediate consequence, we obtain the value of k_n:

$$k_n = \frac{(2n)!}{2^n (n!)^2}. \tag{3.10}$$

We can also compute the norm of L_n in $L^2(\Lambda)$.

THEOREM 3.2. *For any integer $n \geqslant 0$, the polynomial L_n satisfies*

$$\int_{-1}^{1} L_n^2(\zeta)\,\mathrm{d}\zeta = \frac{1}{n+1/2}. \tag{3.11}$$

PROOF. Using (3.9) and integrating n times by parts, we obtain

$$\begin{aligned}\int_{-1}^{1} L_n^2(\zeta)\,\mathrm{d}\zeta &= \frac{(-1)^n}{2^n n!}\int_{-1}^{1}\left(\frac{\mathrm{d}^n}{\mathrm{d}\zeta^n}\right)\left((1-\zeta^2)^n\right)(\zeta) L_n(\zeta)\,\mathrm{d}\zeta \\ &= \frac{1}{2^n n!}\int_{-1}^{1}(1-\zeta^2)^n k_n n!\,\mathrm{d}\zeta \\ &= \frac{(2n)!}{2^{2n}(n!)^2}\int_{-1}^{1}(1-\zeta^2)^n\,\mathrm{d}\zeta.\end{aligned}$$

The last integral is a Wallis integral, which is computed by induction on n:

$$\int_{-1}^{1}(1-\zeta^2)^n\,\mathrm{d}\zeta = 2\int_0^1(1-\zeta^2)^n\,\mathrm{d}\zeta = 2\int_0^{\pi/2}(\sin\theta)^{2n+1}\,\mathrm{d}\theta = \frac{2^{2n+1}(n!)^2}{(2n+1)!}.$$

THEOREM 3.3. *The following formula holds for any positive integer n:*

$$(2n+1)L_n = L'_{n+1} - L'_{n-1}. \tag{3.12}$$

PROOF. Let us set

$$K_{n+1}(\zeta) = \int_{-1}^{\zeta} L_n(\xi)\,\mathrm{d}\xi.$$

Clearly, K_{n+1} is a polynomial with degree $\leqslant n+1$ which vanishes in -1 and in $+1$ (due to the definition of L_n, $n \geqslant 1$). Hence, the following identity is valid for any $k \geqslant 0$:

$$\int_{-1}^{1} K_{n+1}(\zeta)L_k(\zeta)\,\mathrm{d}\zeta = \int_{-1}^{1} L_n(\zeta)K_{k+1}(\zeta)\,\mathrm{d}\zeta.$$

This quantity is equal to 0 for $k > n+1$ and for $n > k+1$, so that K_{n+1} admits the expansion

$$K_{n+1} = \alpha_{n+1}L_{n+1} + \alpha_n L_n + \alpha_{n-1}L_{n-1}. \tag{3.13}$$

Since K_{n+1}, L_{n-1} and L_{n+1} are odd when L_n is even and vice versa, the coefficient α_n is equal to 0. Comparing the coefficient of ζ^{n+1} in the two sides of (3.13) implies that

$$\frac{k_n}{n+1} = \alpha_{n+1}k_{n+1},$$

hence, from (3.10), α_{n+1} is equal to $1/(2n+1)$. Finally, K_{n+1} vanishes in 1, so that

$$0 = K_{n+1}(1) = \frac{1}{2n+1}L_{n+1}(1) + \alpha_{n-1}L_{n-1}(1),$$

so that α_{n-1} is equal to $-1/(2n+1)$. That concludes the proof.

The next formula allows for computing the Legendre polynomials by induction on their degree.

THEOREM 3.4 (Induction formula). *The family of Legendre polynomials is given by*

$$L_0(\zeta) = 1 \quad \textit{and} \quad L_1(\zeta) = \zeta,$$
$$(n+1)L_{n+1}(\zeta) = (2n+1)\zeta L_n(\zeta) - nL_{n-1}(\zeta), \quad n \geqslant 1. \tag{3.14}$$

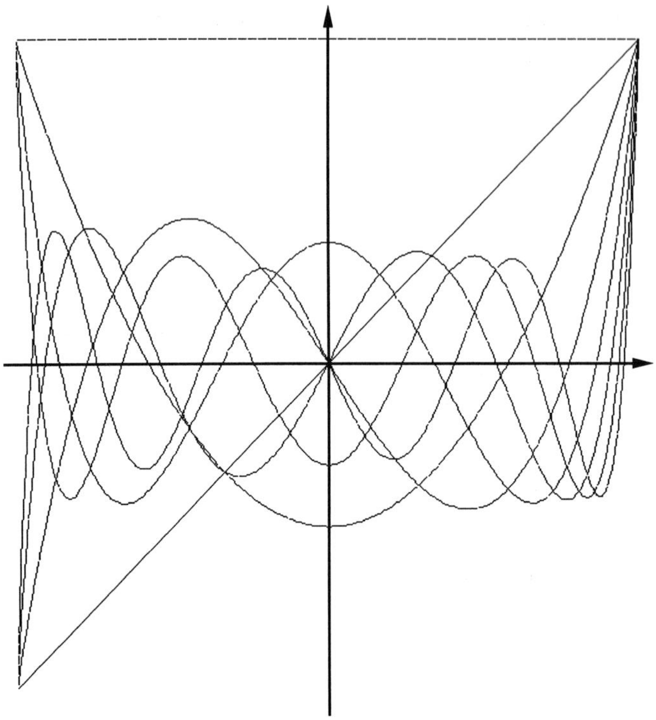

FIG. 3.1. The first eight Legendre polynomials on Λ.

PROOF. It is readily checked that the polynomial $L_{n+1} - (k_{n+1}/k_n)\zeta L_n$ is orthogonal to any polynomial with degree $\leqslant n-2$ in $L^2(\Lambda)$, so that

$$L_{n+1} - \frac{k_{n+1}}{k_n}\zeta L_n = \mu_n L_n - \nu_n L_{n-1}.$$

Next, μ_n is equal to 0 since L_n has not the same parity as L_{n+1}, ζL_n and L_{n-1}. Finally, computing the value in 1, we have

$$1 - \frac{k_{n+1}}{k_n} = -\nu_n.$$

Making use of (3.10) gives the desired result.

The first eight Legendre polynomials are drawn in Fig. 3.1.

REMARK 3.3. Differentiating (3.14) and combining it with (3.12) leads to the *induction formula* for the polynomials L'_n, $n \geqslant 2$:

$$nL'_{n+1} = (2n+1)\zeta L'_n - (n+1)L'_{n-1}. \tag{3.15}$$

More generally, the formula for the polynomials $\mathrm{d}^m L_n/\mathrm{d}\zeta^m$, $n \geqslant m+1$, follows by an induction argument:

$$(n-m+1)\left(\frac{\mathrm{d}^m L_{n+1}}{\mathrm{d}\zeta^m}\right) = (2n+1)\zeta\left(\frac{\mathrm{d}^m L_n}{\mathrm{d}\zeta^m}\right) - (n+m)\left(\frac{\mathrm{d}^m L_{n-1}}{\mathrm{d}\zeta^m}\right). \tag{3.16}$$

REMARK 3.4. It is sometimes useful to consider the orthonormal family $(L_n^*)_n$ in $L^2(\Lambda)$:

$$L_n^* = \frac{L_n}{\|L_n\|_{L^2(\Lambda)}} = \sqrt{n+\frac{1}{2}}\,L_n. \tag{3.17}$$

The induction formula is obtained by combining (3.14) with (3.17):

$$\frac{n+1}{\sqrt{(2n+1)(2n+3)}}L_{n+1}^* = \zeta L_n^* - \frac{n}{\sqrt{(2n-1)(2n+1)}}L_{n-1}^*. \tag{3.18}$$

REMARK 3.5. Using formula (3.14) with an induction argument leads to the so-called *Christoffel–Darboux formula*, valid for any positive integer n:

$$\forall\zeta \in \Lambda,\ \forall\eta \in \Lambda,\quad L_0(\zeta)L_0(\eta) + 3L_1(\zeta)L_1(\eta) + \cdots + (2n-1)L_{n-1}(\zeta)L_{n-1}(\eta) = n\,\frac{L_n(\zeta)L_{n-1}(\eta) - L_n(\eta)L_{n-1}(\zeta)}{\zeta-\eta}. \tag{3.19}$$

A similar formula for the $\mathrm{d}^m L_n/\mathrm{d}\zeta^m$, $n \geqslant m+1$, can be derived from formula (3.16): setting

$$\alpha_n^0 = 1 \quad\text{and}\quad \alpha_n^m = \left(\prod_{k=-(m-1)}^{m}(n+k)\right)^{-1} = \frac{(n-m)!}{(n+m)!},\qquad m \geqslant 1, \tag{3.20}$$

we have

$$\forall\zeta \in \Lambda,\ \forall\eta \in \Lambda,\quad \sum_{k=m}^{n-1}\alpha_k^m(2k+1)\left(\frac{\mathrm{d}^m L_k}{\mathrm{d}\zeta^m}\right)(\zeta)\left(\frac{\mathrm{d}^m L_k}{\mathrm{d}\zeta^m}\right)(\eta) = \alpha_n^m(n+m)\frac{\left(\frac{\mathrm{d}^m L_n}{\mathrm{d}\zeta^m}\right)(\zeta)\left(\frac{\mathrm{d}^m L_{n-1}}{\mathrm{d}\zeta^m}\right)(\eta) - \left(\frac{\mathrm{d}^m L_n}{\mathrm{d}\zeta^m}\right)(\eta)\left(\frac{\mathrm{d}^m L_{n-1}}{\mathrm{d}\zeta^m}\right)(\zeta)}{\zeta-\eta}. \tag{3.21}$$

4. Quadrature formulas

It is well known that the zeros and the extrema of Legendre polynomials (or of elements of any family of orthogonal polynomials) are used for the construction of very accurate quadrature formulas (i.e., which are exact on a space of high degree polynomials). The basic formulas are those of Gauss and Gauss–Lobatto. We refer for instance to CROUZEIX and MIGNOT [1984] and DAVIS and RABINOWITZ [1985] for their numerical analysis. A generalized formula is studied by BERNARDI and MADAY [1991]. The estimates concerning the nodes and the weights are standard, see for instance SZEGÖ [1978].

In all that follows, N is a fixed positive integer. First, denoting by ζ_j, $1 \leqslant j \leqslant N$, the zeros of L_N (we recall that they are distinct and all in Λ), we introduce the standard Gauss formula:

$$\int_{-1}^{1} \Phi(\zeta)\,\mathrm{d}\zeta \simeq \sum_{j=1}^{N} \Phi(\zeta_j)\omega_j.$$

Secondly, denoting by ξ_j, $1 \leqslant j \leqslant N-1$, the zeros of L'_N, we write the Gauss–Lobatto formula:

$$\int_{-1}^{1} \Phi(\zeta)\,\mathrm{d}\zeta \simeq \sum_{j=1}^{N-1} \Phi(\xi_j)\rho_j + \Phi(-1)\rho_0 + \Phi(1)\rho_N.$$

In the particular case where Φ is equal to $(1-\zeta^2)\Psi$, this is written

$$\int_{-1}^{1} \Psi(\zeta)\left(1-\zeta^2\right)\mathrm{d}\zeta \simeq \sum_{j=1}^{N-1} \Psi(\xi_j)\left(1-\xi_j^2\right)\rho_j.$$

Comparing these three formulas suggests the generalized ones, which are presented in the next two theorems. The first one deals with the Gauss formula with respect to the measure $(1-\zeta^2)^m\,\mathrm{d}\zeta$.

DEFINITION. For any nonnegative integer n, $\mathbb{P}_n(\Lambda)$ stands for the space of restrictions to Λ of polynomials of one variable with degree $\leqslant n$.

THEOREM 4.1. *Let m be a nonnegative integer, and let N be a fixed integer $\geqslant m$. There exist:*

(i) *a unique set of $N-m$ points ζ_j^m, $1 \leqslant j \leqslant N-m$, in Λ,*

(ii) *a unique set of $N-m$ real numbers ω_j^m, $1 \leqslant j \leqslant N-m$,*

such that the following equality holds for any polynomial Ψ in $\mathbb{P}_{2N-2m-1}(\Lambda)$:

$$\int_{-1}^{1} \Psi(\zeta)\left(1-\zeta^2\right)^m \mathrm{d}\zeta = \sum_{j=1}^{N-m} \Psi(\zeta_j^m)\omega_j^m. \tag{4.1}$$

The ζ_j^m, $1 \leqslant j \leqslant N-m$, *are the zeros of the polynomial* $\mathrm{d}^m L_N/\mathrm{d}\zeta^m$.

PROOF. With the ζ_j^m, $1 \leqslant j \leqslant N-m$, we associate the Lagrange polynomials h_j^m: each h_j^m belongs to $\mathbb{P}_{N-m-1}(\Lambda)$, is equal to 1 in ζ_j^m and vanishes in ζ_i^m, $1 \leqslant i \leqslant N-m$, $i \neq j$. Then, setting:

$$\omega_j^m = \int_{-1}^{1} h_j^m(\zeta)\big(1-\zeta^2\big)^m \,\mathrm{d}\zeta, \tag{4.2}$$

we observe that the quadrature formula (4.1) is exact for Ψ equal to h_j^m. Hence, since the polynomials h_j^m, $1 \leqslant j \leqslant N-m$, form a basis of the space $\mathbb{P}_{N-m-1}(\Lambda)$, the formula is exact on this space. Next, making an Euclidean division by the product $(\zeta-\zeta_1^m)\cdots(\zeta-\zeta_{N-m}^m)$, we can write each polynomial Ψ in $\mathbb{P}_{2N-2m-1}(\Lambda)$ as

$$\Psi(\zeta) = X(\zeta)(\zeta-\zeta_1^m)\cdots(\zeta-\zeta_{N-m}^m) + \psi(\zeta),$$

where both X and ψ belong to $\mathbb{P}_{N-m-1}(\Lambda)$, and we have

$$\sum_{j=1}^{N-m} \Psi(\zeta_j^m)\omega_j^m = \sum_{j=1}^{N-m} \psi(\zeta_j^m)\omega_j^m = \int_{-1}^{1} \psi(\zeta)\,\mathrm{d}\zeta.$$

Thus, the theorem reduces to the existence of nodes ζ_j^m such that

$$\forall X \in \mathbb{P}_{N-m-1}(\Lambda), \qquad \int_{-1}^{1} X(\zeta)(\zeta-\zeta_1^m)\cdots(\zeta-\zeta_{N-m}^m)\big(1-\zeta^2\big)^m \,\mathrm{d}\zeta = 0.$$

This is equivalent to saying that the polynomial $(\zeta-\zeta_1^m)\cdots(\zeta-\zeta_{N-m}^m)$, whose degree is $N-m$, belongs to a family of orthogonal polynomials with respect to the measure $(1-\zeta^2)^m$ on Λ. As noted in Remark 3.2, the derivatives $\mathrm{d}^m L_n/\mathrm{d}\zeta^m$ of the Legendre polynomials, $n \geqslant m$, form such a family (and have $N-m$ distinct zeros in Λ), so that $(\zeta-\zeta_1^m)\cdots(\zeta-\zeta_{N-m}^m)$ is equal to a constant times $\mathrm{d}^m L_N/\mathrm{d}\zeta^m$. This completes the proof of the theorem.

The Gauss formulas for the measures $(1-\zeta^2)^m\,\mathrm{d}\zeta$ are used to construct generalized formulas for the Lebesgue measure.

THEOREM 4.2. *Let* m *be a nonnegative integer, and let* N *be a fixed integer* $\geqslant m$. *There exist:*

(i) *a unique set of* $N-m$ *points* ζ_j^m, $1 \leqslant j \leqslant N-m$, *in* Λ,

(ii) *a unique set of* $N-m$ *real numbers* σ_j^m, $1 \leqslant j \leqslant N-m$,

(iii) *a unique set of* $2m$ *real numbers* $\sigma_-^{m,k}$ *and* $\sigma_+^{m,k}$, $0 \leqslant k \leqslant m-1$, *such that the following equality holds for any polynomial* Φ *in* $\mathbb{P}_{2N-1}(\Lambda)$:

$$\int_{-1}^{1} \Phi(\zeta)\,\mathrm{d}\zeta = \sum_{j=1}^{N-m} \Phi(\zeta_j^m)\sigma_j^m + \sum_{k=0}^{m-1}\left(\left(\frac{\mathrm{d}^k\Phi}{\mathrm{d}\zeta^k}\right)(-1)\sigma_-^{m,k} + \left(\frac{\mathrm{d}^k\Phi}{\mathrm{d}\zeta^k}\right)(+1)\sigma_+^{m,k}\right). \tag{4.3}$$

The ζ_j^m, $1 \leqslant j \leqslant N-m$, *are the zeros of the polynomial* $\mathrm{d}^m L_N/\mathrm{d}\zeta^m$.

PROOF. We observe that each polynomial Φ in $\mathbb{P}_{2N-1}(\Lambda)$ can be written as

$$\Phi(\zeta) = \Psi(\zeta)\big(1-\zeta^2\big)^m + Y(\zeta)\big(\zeta-\zeta_1^m\big)\cdots\big(\zeta-\zeta_{N-m}^m\big)$$

where Ψ belongs to $\mathbb{P}_{2N-2m-1}(\Lambda)$ and Y belongs to $\mathbb{P}_{2m-1}(\Lambda)$. Firstly, by setting

$$\sigma_j^m = \big(1-\big(\zeta_j^m\big)^2\big)^{-m}\omega_j^m, \quad 1 \leqslant j \leqslant N-m, \tag{4.4}$$

we deduce from the previous Theorem 4.1 that the formula is exact when applied to the first term $\Psi(\zeta)(1-\zeta^2)^m$. Secondly, applying the quadrature formula to the polynomials

$$\big(\zeta-\zeta_1^m\big)\cdots\big(\zeta-\zeta_{N-m}^m\big)(1-\zeta)^m(1+\zeta)^l$$

and

$$\big(\zeta-\zeta_1^m\big)\cdots\big(\zeta-\zeta_{N-m}^m\big)(1-\zeta)^l(1+\zeta)^m,$$

for l decreasing from $m-1$ to 0, we can compute the σ_-^k and σ_+^k in order that the formula is also exact when applied to $Y(\zeta)(\zeta-\zeta_1^m)\cdots(\zeta-\zeta_{N-m}^m)$. This gives the desired result.

REMARK 4.1. It is readily checked by applying formula (4.1) to the squares $(h_j^m)^2$ of the Lagrange polynomials, that the ω_j^m, and consequently the σ_j^m, $1 \leqslant j \leqslant N-m$, are positive.

Now that the general result is proven, we go back to the special cases $m=0$, $m=1$, $m=2$ and $m=3$, the only ones which will be used in this article.

We begin with the *case* $m=0$. In all that follows, we denote by ζ_j, $1 \leqslant j \leqslant N$, the zeros of L_N (the index N is omitted for the sake of clarity) and by ω_j, $1 \leqslant j \leqslant N$, the weights $\omega_j^0 = \sigma_j^0$ which are associated with these nodes in Theorems 4.1 and 4.2. Thus, we obtain the *Gauss–Legendre formula*:

$$\int_{-1}^{1} \Phi(\zeta)\,\mathrm{d}\zeta \simeq \sum_{j=1}^{N} \Phi(\zeta_j)\omega_j. \tag{4.5}$$

To conclude the study of this formula, we need to compute the weights ω_j for $1 \leqslant j \leqslant N$. Using the Christoffel–Darboux formula (3.19) with η equal to ζ_j, leads to:

$$L_0(\zeta)L_0(\zeta_j) + \cdots + (2N-1)L_{N-1}(\zeta)L_{N-1}(\zeta_j) = N\frac{L_N(\zeta)L_{N-1}(\zeta_j)}{\zeta - \zeta_j}.$$

Hence, we obtain by integrating this equation and using the Gauss formula:

$$2 = NL_{N-1}(\zeta_j)\int_{-1}^{1} \frac{L_N(\zeta)}{\zeta - \zeta_j}\,\mathrm{d}\zeta = NL_N'(\zeta_j)L_{N-1}(\zeta_j)\omega_j,$$

so that

$$\omega_j = \frac{2}{NL_N'(\zeta_j)L_{N-1}(\zeta_j)}, \qquad 1 \leqslant j \leqslant N. \tag{4.6}$$

It follows from formula (3.14) that

$$(N+1)L_{N+1}(\zeta_j) = -NL_{N-1}(\zeta_j),$$

and from formulas (3.1) and (3.12) that

$$\begin{aligned}(1-\zeta^2)L_N'(\zeta) &= -N(N+1)\int_{-1}^{\zeta} L_N(\xi)\,\mathrm{d}\xi \\ &= -\frac{N(N+1)}{2N+1}(L_{N+1} - L_{N-1})(\zeta),\end{aligned} \tag{4.7}$$

so that $(1-\zeta_j^2)L_N'(\zeta_j)$ is equal to $NL_{N-1}(\zeta_j)$. Hence, formula (4.6) can be written equivalently

$$\omega_j = \frac{2}{(1-\zeta_j^2)L_N'^2(\zeta_j)}, \qquad 1 \leqslant j \leqslant N. \tag{4.8}$$

REMARK 4.2. Applying formula (3.21) with $\eta = \zeta_j^m$ and integrating it on Λ with respect to the measure $(1-\zeta^2)^m\,\mathrm{d}\zeta$, we obtain exactly in the same way the formula for the ω_j^m of Theorem 4.1, $1 \leqslant j \leqslant N-m$:

$$\omega_j^m = \frac{2}{\alpha_N^m (N+m)\left(\frac{\mathrm{d}^{m+1}L_N}{\mathrm{d}\zeta^{m+1}}\right)(\zeta_j^m)\left(\frac{\mathrm{d}^m L_{N-1}}{\mathrm{d}\zeta^m}\right)(\zeta_j^m)}. \tag{4.9}$$

REMARK 4.3. An efficient computation of the nodes of the quadrature formula is the first step for using spectral techniques. For high values of the parameter N, it seems that the faster algorithm to compute the ζ_j, $1 \leqslant j \leqslant N$, is derived by writing formula (3.18) under the form

$$\zeta \begin{pmatrix} L_0^* \\ L_1^* \\ \cdots \\ L_{N-2}^* \\ L_{N-1}^* \end{pmatrix} = \begin{pmatrix} 0 & \beta_1 & \cdots & 0 & 0 \\ \beta_1 & 0 & \cdots & 0 & 0 \\ \cdots\cdots & & & \cdots & \cdots \\ 0 & 0 & \cdots & 0 & \beta_{N-1} \\ 0 & 0 & \cdots & \beta_{N-1} & 0 \end{pmatrix} \begin{pmatrix} L_0^* \\ L_1^* \\ \cdots \\ L_{N-2}^* \\ L_{N-1}^* \end{pmatrix} + \beta_N \begin{pmatrix} 0 \\ 0 \\ \cdots \\ 0 \\ L_N^* \end{pmatrix},$$

where the coefficients β_n are defined by

$$\beta_n = \frac{n}{\sqrt{4n^2-1}}, \qquad n \geqslant 1; \tag{4.10}$$

hence, the ζ_j, $1 \leqslant j \leqslant N$, are the eigenvalues of the matrix

$$M = \begin{pmatrix} 0 & \beta_1 & \cdots & 0 & 0 \\ \beta_1 & 0 & \cdots & 0 & 0 \\ \cdots\cdots & & & \cdots & \cdots \\ 0 & 0 & \cdots & 0 & \beta_{N-1} \\ 0 & 0 & \cdots & \beta_{N-1} & 0 \end{pmatrix}. \tag{4.11}$$

Since this matrix is tridiagonal and symmetric with a null diagonal, it is easy to compute the eigenvalues (for instance by the Givens–Householder algorithm, see CIARLET [1982]). Letting η tend to ζ in the Christoffel–Darboux formula (3.19) and combining the resulting equation with (4.6), we also have:

$$\omega_j = \left(L_0^{*2}(\zeta_j) + \cdots + L_{N-1}^{*2}(\zeta_j)\right)^{-1},$$

which allows for computing the ω_j, $1 \leqslant j \leqslant N$, from the eigenvectors of the matrix M with first component $L_0^* = 1/\sqrt{2}$.

The study of the Lagrange interpolation operator at the nodes ζ_j, $1 \leqslant j \leqslant N$, which will be performed in Section 13, requires a precise statement about the location of these nodes. This is stated in the following theorem. Assuming that the ζ_j, $1 \leqslant j \leqslant N$, are in increasing order, we set

$$\theta_j = \arccos \zeta_j, \quad 1 \leqslant j \leqslant N. \tag{4.12}$$

The symbol $[\bullet]$ denotes the integral part.

THEOREM 4.3. *The angles θ_j, $1 \leqslant j \leqslant N$, satisfy the following inequalities:*

$$\begin{aligned} &\frac{(2j-1)\pi}{2N} < \theta_{N-j+1} < \frac{(2j+1)\pi}{2N}, \quad 1 \leqslant j \leqslant \left[\frac{N}{2}\right] - 1, \\ &\frac{(2j-3)\pi}{2N} < \theta_{N-j+1} < \frac{(2j-1)\pi}{2N}, \quad \left[\frac{N+1}{2}\right] + 2 \leqslant j \leqslant N, \end{aligned} \tag{4.13}$$

and, moreover,

(i) *when N is even,*

$$\frac{(N-1)\pi}{2N} < \theta_{N/2+1} < \frac{\pi}{2} < \theta_{N/2} < \frac{(N+1)\pi}{2N}, \tag{4.14}$$

(ii) *when N is odd,*

$$\frac{(N-1)\pi}{2N} < \theta_{(N+3)/2} < \theta_{(N+1)/2} = \frac{\pi}{2} < \theta_{(N-1)/2} < \frac{(N+1)\pi}{2N}. \tag{4.15}$$

PROOF. Let us introduce the functions

$$\varphi(\zeta) = (1-\zeta^2)^{1/2} L_N(\zeta) \quad \text{and} \quad \psi(\zeta) = (1-\zeta^2)^{1/4} \cos(N \arccos \zeta).$$

Thanks to (3.1), it is readily checked that they satisfy the differential equations

$$\varphi'' + \frac{N(N+1)(1-\zeta^2)+1}{(1-\zeta^2)^2}\varphi = 0 \quad \text{and} \quad \psi'' + \frac{N^2(1-\zeta^2)+\frac{1}{2}+\frac{1}{4}\zeta^2}{(1-\zeta^2)^2}\psi = 0,$$

so that

$$(\varphi\psi' - \varphi'\psi)' = \lambda\varphi\psi,$$

for a positive function λ. Let α and β be two consecutive zeros of ψ in Λ. Integrating the previous equation gives

$$\varphi(\beta)\psi'(\beta) - \varphi(\alpha)\psi'(\alpha) = \int_\alpha^\beta \lambda(\zeta)\varphi(\zeta)\psi(\zeta)\,d\zeta.$$

TABLE 4.1

j	ζ_j	θ_j	j	ζ_j	θ_j
1	−0.97390653	0.927124777π	6	0.14887434	0.452435017π
2	−0.86506337	0.832721907π	7	0.43339539	0.357315377π
3	−0.67940957	0.737764017π	8	0.67940957	0.262235997π
4	−0.43339539	0.642684637π	9	0.86506337	0.167278107π
5	−0.14887434	0.547564997π	10	0.97390653	0.072875237π

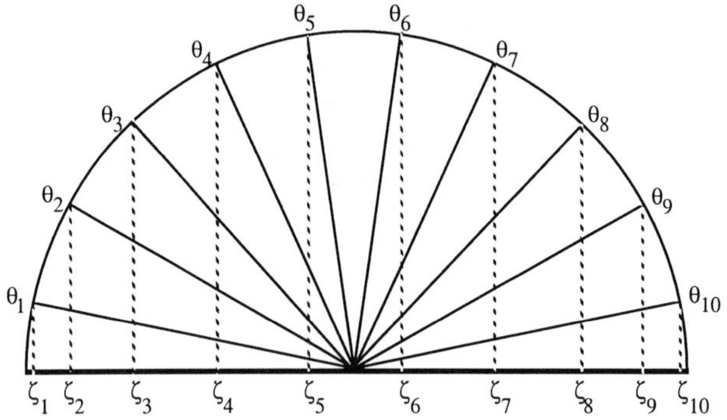

FIG. 4.1. Location of the zeros of L_{10}.

It yields that the quantities $\varphi(\beta)|\psi'(\beta)|+\varphi(\alpha)|\psi'(\alpha)|$ and $\int_\alpha^\beta \lambda(\zeta)\varphi(\zeta)|\psi(\zeta)|\,\mathrm{d}\zeta$ have different signs (recall that the zeros are simple), so that φ necessarily vanishes on the open interval $]\alpha,\beta[$. Hence, going back to the polynomials, we have proven that, for any j, $1 \leqslant j \leqslant N-1$, there exists a ζ_i, $1 \leqslant i \leqslant N$, such that

$$\cos\frac{(2j+1)\pi}{2N} < \zeta_i \leqslant \cos\frac{(2j-1)\pi}{2N}.$$

Thus, to obtain the desired inequalities, it suffices to note that

(i) when N is even, there exists two distinct zeros $\zeta_{N/2}$ and $\zeta_{N/2+1} = -\zeta_{N/2}$ between $\cos{(N+1)\pi/2N}$ and $\cos{(N-1)\pi/2N}$,

(ii) when N is odd, both L_N and $\cos(N \arccos\zeta)$ vanish in 0.

There exist more precise results about the location of the θ_j, see for instance SZEGÖ ([1978], Theorem 6.21.3). However, inequalities (4.13) to (4.15) are sufficient to state a very important property: for large values of the parameter N, the nodes ζ_j are the cosines of nearly equidistant numbers in $(0,\pi)$. This implies that these points are clustered in a neighborhood of the end points of the interval Λ: the distance between ζ_1 and ζ_2 for instance behaves like c/N^2, while (assuming that N is even) the distance between $\zeta_{N/2}$ and $\zeta_{N/2+1}$ behaves like c/N. This property is illustrated in Table 4.1 and Fig. 4.1, which present the ζ_j and the θ_j, $1 \leqslant j \leqslant N$, for $N=10$. Note that the differences $\theta_j - \theta_{j+1}$ are of order $0.095\,\pi$, which is close to π/N.

The next theorem gives a useful estimate for the ω_j.

THEOREM 4.4. *The weights* ω_j *satisfy the following inequalities*

$$\omega_j \leqslant cN^{-1}\big(1-\zeta_j^2\big)^{1/2}, \quad 1 \leqslant j \leqslant N. \tag{4.16}$$

PROOF. The fact that L_N is either odd or even combined with (4.8) implies that

$$\omega_j = \omega_{N+1-j}, \quad 1 \leqslant j \leqslant N. \tag{4.17}$$

Hence, it suffices to prove the result for $\zeta_j \geqslant 0$. Here, we consider the function

$$\chi(\theta) = (\sin\theta)^{1/2} L_N(\cos\theta),$$

and we check from (3.1) that it satisfies the differential equation

$$\chi'' + \mu\chi = 0, \quad \text{with } \mu(\theta) = N(N+1) + \tfrac{1}{4} + \tfrac{1}{4}(\sin\theta)^{-2}. \tag{4.18}$$

Next, we consider two zeros ζ_i and ζ_j, $0 \leqslant \zeta_i < \zeta_j$. Multiplying (4.18) by χ' and integrating from θ_j to θ_i, we obtain that

$$\chi'^2(\theta_i) - \chi'^2(\theta_j) = -\int_{\theta_j}^{\theta_i} \mu(\theta)\big(\chi^2\big)'(\theta)\,\mathrm{d}\theta = \int_{\theta_j}^{\theta_i} \mu'(\theta)\chi^2(\theta)\,\mathrm{d}\theta;$$

since $\mu' \leqslant 0$, $\chi'^2(\theta_i)$ is smaller than $\chi'^2(\theta_j)$. Note that, at each node ζ_j, we have by formula (4.8),

$$\chi'^2(\theta_j) = (\sin\theta_j)^3 L_N'^2(\cos\theta_j) = \frac{2(1-\zeta_j^2)^{1/2}}{\omega_j}.$$

Consequently, the quantities $\omega_j(1-\zeta_j^2)^{-1/2}$ for $\zeta_j \geqslant 0$, decrease with j and it suffices to check that $\omega_{j_0}(1-\zeta_{j_0}^2)^{-1/2}$ is bounded by cN^{-1}, with $j_0 = [N/2]+1$.

(i) When N is odd, ζ_{j_0} is equal to 0 and $\omega_{j_0}(1-\zeta_{j_0}^2)^{-1/2}$ is equal to $2/L_N'^2(0)$. It follows from formula (3.15) that

$$\begin{aligned} L_N'(0) &= -\frac{N}{N-1}L_{N-2}'(0) = \cdots = (-1)^{(N-1)/2}\frac{N(N-2)\cdots 3}{(N-1)(N-3)\cdots 2}L_1'(0) \\ &= (-1)^{(N-1)/2}\frac{N!}{2^{N-1}\big(\big(\frac{N-1}{2}\big)!\big)^2}. \end{aligned}$$

Stirling's formula:

$$k! = \sqrt{2\pi}\,\mathrm{e}^{-k}k^{k+1/2}\big(1+\mathrm{O}(k^{-1})\big),$$

shows that $|L_N'(0)|$ is larger than a constant times $N^{1/2}$, and the result follows.

TABLE 4.2

j	ω_j	$\sin\theta_j$	j
1	0.06667134	0.22694950	10
2	0.14945135	0.50166261	9
3	0.21908636	0.73375925	8
4	0.26926672	0.90120388	7
5	0.29552422	0.98885612	6

(ii) When N is even, using once more (4.18), multiplying it by χ' and integrating it now from θ_{j_0} to $\pi/2$, we derive that

$$\chi'^2\left(\frac{\pi}{2}\right) - \chi'^2(\theta_{j_0}^2) = -\int_{\theta_{j_0}}^{\pi/2} \mu(\theta)(\chi^2)'(\theta)\,\mathrm{d}\theta$$

$$= -\mu\left(\frac{\pi}{2}\right)\chi^2\left(\frac{\pi}{2}\right) + \mu(\theta_{j_0})\chi^2(\theta_{j_0}) + \int_{\theta_{j_0}}^{\pi/2} \mu'(\theta)\chi^2(\theta)\,\mathrm{d}\theta.$$

Hence, since μ' is negative, we deduce that

$$\frac{2(1-\zeta_{j_0})^{1/2}}{\omega_{j_0}} = \chi'^2(\theta_{j_0}^2) \geqslant \mu\left(\frac{\pi}{2}\right)\chi^2\left(\frac{\pi}{2}\right) = \left(N^2 + N + \frac{1}{2}\right)L_N^2(0).$$

In this case, we have to compute

$$L_N(0) = -\frac{N-1}{N}L_{N-2}(0) = \cdots = (-1)^{N/2}\frac{N!}{2^N\left(\left(\frac{N}{2}\right)!\right)^2}.$$

Hence, by Stirling's formula, $|L_N(0)|$ is larger than $cN^{-1/2}$, which proves the result.

REMARK 4.4. The result of Theorem 4.4 is optimal: indeed, the converse estimate is stated by SZEGÖ ([1978], (15.3.14)):

$$\omega_j \geqslant c'N^{-1}\left(1-\zeta_j^2\right)^{1/2}. \tag{4.19}$$

It can be proven from the previous arguments by checking that ω_N is larger than cN^{-2}.

Table 4.2 compares the ω_j with $\sin\theta_j = (1-\zeta_j^2)^{1/2}$, for $N = 10$. Clearly, estimate (4.16) is satisfied in the case $N = 10$ with the constant c equal to 3.

Next, we consider the second quadrature formula, i.e., the special *case* $m = 1$ of formula (4.3): the main difference with the Gauss formula is that now the end points of the interval are nodes of the formula. In all that follows, we denote by ξ_j, $0 \leqslant j \leqslant N$, the zeros of the polynomial $(1-\zeta^2)L_N'$ arranged in increasing order (so that $\xi_0 = -1$ and $\xi_N = 1$) and respectively by ρ_j, $0 \leqslant j \leqslant N$, the weights $\sigma_-^{1,0}$, ω_j^1 and $\sigma_+^{1,0}$ which are associated with these nodes in Theorem 4.2. Thus, we obtain the *Gauss–Lobatto–Legendre formula*:

$$\int_{-1}^{1} \Phi(\zeta)\,\mathrm{d}\zeta \simeq \sum_{j=0}^{N} \Phi(\xi_j)\rho_j. \tag{4.20}$$

REMARK 4.5. Exactly as for the nodes of the Gauss formula, a simple and efficient way to compute the ξ_j, $1 \leqslant j \leqslant N-1$, is to prove that they are the eigenvalues of a symmetric matrix. This is performed by making use of (3.15) and writing the corresponding formula for the polynomials

$$J_n^* = \sqrt{\frac{n+\frac{1}{2}}{n(n+1)}}\,L_n'$$

(which form an orthonormal basis for the measure $(1-\zeta^2)\,\mathrm{d}\zeta$). Thus, we prove that the ξ_j, $1 \leqslant j \leqslant N-1$, are the eigenvalues of the matrix

$$\begin{pmatrix} 0 & \gamma_1 & \cdots & 0 & 0 \\ \gamma_1 & 0 & \cdots & 0 & 0 \\ \cdots & \cdots & \cdots & \cdots & \cdots \\ 0 & 0 & \cdots & 0 & \gamma_{N-2} \\ 0 & 0 & \cdots & \gamma_{N-2} & 0 \end{pmatrix}, \tag{4.21}$$

with

$$\gamma_n = \sqrt{\frac{n(n+2)}{\left(n+\frac{1}{2}\right)\left(n+\frac{3}{2}\right)}}, \qquad 1 \leqslant n \leqslant N-2. \tag{4.22}$$

From (4.4) and (4.9), the weights ρ_j, $1 \leqslant j \leqslant N-1$, are given by

$$\rho_j = \left(1-\xi_j^2\right)^{-1}\omega_j^1 = \frac{2N}{(1-\xi_j^2)L_N''(\xi_j)L_{N-1}'(\xi_j)}.$$

From the differential equation (3.1), we have

$$\left(1-\xi_j^2\right)L_N''(\xi_j) = -N(N+1)L_N(\xi_j),$$

while combining (3.15) and (3.12) leads to

$$NL_{N+1}'(\xi_j) = -(N+1)L_{N-1}'(\xi_j),$$
$$(2N+1)L_N(\xi_j) = L_{N+1}'(\xi_j) - L_{N-1}'(\xi_j) = -\frac{2N+1}{N}L_{N-1}'(\xi_j).$$

This allows for writing ρ_j as a function of $L_N(\xi_j)$:

$$\rho_j = \frac{2}{N(N+1)L_N^2(\xi_j)}.$$

Recalling that

$$\rho_0 = \frac{1}{2L_N'(-1)} \int_{-1}^{1} (1-\zeta)L_N'(\zeta)\,\mathrm{d}\zeta,$$

$$\rho_N = \frac{1}{2L_N'(1)} \int_{-1}^{1} (1+\zeta)L_N'(\zeta)\,\mathrm{d}\zeta,$$

integrating by parts and using (3.4) gives

$$\rho_0 = -\frac{2L_N(-1)}{2L_N'(-1)} = \frac{2}{N(N+1)} \quad \text{and} \quad \rho_N = \frac{2L_N(1)}{2L_N'(1)} = \frac{2}{N(N+1)}.$$

Consequently, we derive the general formula

$$\rho_j = \frac{2}{N(N+1)L_N^2(\xi_j)}, \quad 0 \leqslant j \leqslant N. \tag{4.23}$$

It must be observed that the location of the points ξ_j, $1 \leqslant j \leqslant N-1$, is given by Theorem 4.3 together with the inequality

$$\zeta_j \leqslant \xi_j \leqslant \zeta_{j+1}, \quad 1 \leqslant j \leqslant N-1. \tag{4.24}$$

The bound for the weights is stated in the following theorem.

THEOREM 4.5. *The weights ρ_j satisfy the following inequalities*

$$\rho_j \leqslant cN^{-1}\big(1-\xi_j^2\big)^{1/2}, \quad 1 \leqslant j \leqslant N-1. \tag{4.25}$$

PROOF. Here also, we only have to prove this inequality for the nonnegative ξ_j. We note that, from the previous formulas, the weight ρ_j can also be written as

$$\rho_j = \frac{2N(N+1)}{(1-\xi_j^2)^2 L_N''^2(\xi_j)}.$$

We introduce the function

$$\chi_1(\theta) = (\sin\theta)^{3/2} L_N'(\cos\theta),$$

and we check that

$$\chi_1'' + \mu_1\chi_1 = 0, \quad \text{with } \mu_1(\theta) = N(N+1) + \tfrac{1}{4} - \tfrac{3}{4}(\sin\theta)^{-2}. \tag{4.26}$$

In contrast to the case $m = 0$, μ_1' is positive on $]0, \pi/2[$, so that multiplying the previous line by χ_1' and integrating from $\arccos \xi_j$ to $\arccos \xi_i$, with $0 \leqslant \xi_i < \xi_j$, yield

$$\begin{aligned} 2N(N+1)\left(\frac{(1-\xi_i^2)^{1/2}}{\rho_i} - \frac{(1-\xi_j^2)^{1/2}}{\rho_j}\right) &= \chi_1'^2(\arccos \xi_i) - \chi_1'^2(\arccos \xi_j) \\ &= \int_{\arccos \xi_j}^{\arccos \xi_i} \mu_1'(\theta)\chi_1^2(\theta)\,\mathrm{d}\theta > 0. \end{aligned}$$

Hence, the quantities $\rho_j(1-\xi_j^2)^{-1/2}$ are increasing with the ξ_j, so that it remains to check that ρ_{N-1} is bounded by cN^{-2} or equivalently that $L_N(\xi_{N-1})$ is larger than a constant independent of N. In order to do that, we use Rodrigues' formula (3.9) together with the Leibnitz identity:

$$\begin{aligned} &L_N(\xi_{N-1}) \\ &\quad = \frac{(-1)^N}{2^N N!}\sum_{k=0}^{N}(-1)^k \frac{N!}{k!(N-k)!}\frac{N!}{k!}(1-\xi_{N-1})^k \frac{N!}{(N-k)!}(1+\xi_{N-1})^{N-k} \\ &\quad = (-1)^N (N!)^2\left(\frac{1+\xi_{N-1}}{2}\right)^N \sum_{k=0}^{N}(-1)^k\left(\frac{1}{k!(N-k)!}\right)^2\left(\frac{1-\xi_{N-1}}{1+\xi_{N-1}}\right)^k. \end{aligned}$$

Using Theorem 4.3 with (4.24), we firstly see that $((1+\xi_{N-1})/2)^N$ is larger than a constant independent of N and, secondly, we can check that the term

$$\left(\frac{1}{k!(N-k)!}\right)^2\left(\frac{1-\xi_{N-1}}{1+\xi_{N-1}}\right)^k$$

is a decreasing function of k for $k \geqslant 2$ and N large enough. Computing the two first terms of the sum leads to the desired result.

We also need formula (4.3) in the *cases* $m = 2$ *and* $m = 3$. Here, we do not introduce special notation. Then, the nodes are the zeros ζ_j^m of $\mathrm{d}^m L_N/\mathrm{d}\zeta^m$. Starting from (4.4) and (4.9) and using exactly the same arguments as for the Gauss–Lobatto formula leads to the following formula for the internal weights:

$$\sigma_j^m = \frac{2}{\alpha_N^m (N-m+1)^2(N+m)^2\big(1-(\zeta_j^m)^2\big)^{m-1}\big(\frac{\mathrm{d}^{m-1}L_N}{\mathrm{d}\zeta^{m-1}}\big)^2(\zeta_j^m)}, \quad 1 \leqslant j \leqslant N-m. \tag{4.27}$$

REMARK 4.6. Exactly as previously, studying the function

$$\chi_m(\theta) = (\sin\theta)^{m+1/2}\left(\frac{\mathrm{d}^m L_N}{\mathrm{d}\zeta^m}\right)(\cos\theta),$$

leads to the following estimate

$$\sigma_j^m \leqslant cN^{-1}\left(1 - (\zeta_j^m)^2\right)^{1/2}, \quad 1 \leqslant j \leqslant N - m. \tag{4.28}$$

It remains to compute the boundary weights $\sigma_-^{m,k}$ and $\sigma_+^{m,k}$. As already hinted, this is performed by applying the quadrature formula (4.3) to the functions

$$\left(\frac{\mathrm{d}^m L_N}{\mathrm{d}\zeta^m}\right)(\zeta)(1-\zeta)^m(1+\zeta)^l \quad \text{and} \quad \left(\frac{\mathrm{d}^m L_N}{\mathrm{d}\zeta^m}\right)(\zeta)(1-\zeta)^l(1+\zeta)^m,$$

for l decreasing from $m-1$ to 0. Indeed, the integral of these functions is easily computed by integration by parts. So, let us only give the final formulas:

$$\begin{aligned}
\sigma_-^{2,1} &= -\sigma_+^{2,1} = \frac{8}{(N-1)N(N+1)(N+2)},\\
\sigma_-^{2,0} &= \sigma_+^{2,0} = \frac{8(2N^2+2N-3)}{3(N-1)N(N+1)(N+2)},
\end{aligned} \tag{4.29}$$

and

$$\begin{aligned}
\sigma_-^{3,2} &= \sigma_+^{3,2} = \frac{48}{(N-2)(N-1)N(N+1)(N+2)(N+3)},\\
\sigma_-^{3,1} &= -\sigma_+^{3,1} = \frac{36(N^2+N-4)}{(N-2)(N-1)N(N+1)(N+2)(N+3)},\\
\sigma_-^{3,0} &= \sigma_+^{3,0} = \frac{9(11N^4+22N^3-61N^2-72N+80)}{10(N-2)(N-1)N(N+1)(N+2)(N+3)}.
\end{aligned} \tag{4.30}$$

5. Polynomial inverse inequalities

Let us recall two important observations:

(i) for any domain $\mathcal{O}$, the space $H^1(\mathcal{O})$ is strictly contained in $L^2(\mathcal{O})$ with a continuous imbedding;

(ii) on a finite-dimensional space, all norms are equivalent.

This means for instance that there exists a constant $c(N)$ depending on N such that any polynomial φ_N with degree $\leqslant N$ on the interval Λ satisfies the inequality

$$\|\varphi_N\|_{H^1(\Lambda)} \leqslant c(N)\|\varphi_N\|_{L^2(\Lambda)}.$$

More generally, for any real numbers r and s, $r \geqslant s$, there exists a constant $c_{r,s}(N)$ such that any polynomial φ_N in $\mathbb{P}_N(\Lambda)$ satisfies the inequality

$$\|\varphi_N\|_{H^r(\Lambda)} \leqslant c_{r,s}(N)\|\varphi_N\|_{H^s(\Lambda)}. \tag{5.1}$$

The previous inequality is called *inverse*, and we are interested in estimating the smallest constant $c_{r,s}(N)$ such that (5.1) holds. This result is not so easy for non-integral values of s and we refer to BERNARDI, DAUGE and MADAY [1992b] for the complete proofs. However, we are going to state the first estimates and also some useful extensions which involve weighted measures.

The first and basic estimate is due to CANUTO and QUARTERONI [1982].

THEOREM 5.1. *The following inequality holds for any positive integer N and for any polynomial φ_N in $\mathbb{P}_N(\Lambda)$:*

$$|\varphi_N|_{H^1(\Lambda)} \leqslant \sqrt{3}N^2\|\varphi_N\|_{L^2(\Lambda)}. \tag{5.2}$$

Before proving the theorem, let us recall the useful formula, valid for any nonnegative integer n:

$$\int_{-1}^{1} L_n'^2(\zeta)\,\mathrm{d}\zeta = n(n+1). \tag{5.3}$$

It is established by integrating by parts, the value of $L'_n(1) = (-1)^{n-1}L'_n(-1)$ being given in (3.4).

PROOF. Let φ_N be any polynomial in $\mathbb{P}_N(\Lambda)$. We write its expansion in the basis of Legendre polynomials:

$$\varphi_N = \sum_{n=0}^{N} \varphi^n L_n,$$

so that

$$\|\varphi_N\|^2_{L^2(\Lambda)} = \sum_{n=0}^{N} \frac{(\varphi^n)^2}{n+\frac{1}{2}}.$$

From the previous inequality (5.3), we have

$$|\varphi_N|_{H^1(\Lambda)} \leqslant \sum_{n=0}^{N} |\varphi^n||L_n|_{H^1(\Lambda)} \leqslant \sum_{n=0}^{N} |\varphi^n|\sqrt{n(n+1)}.$$

Then, using the Cauchy–Schwarz inequality leads to

$$|\varphi_N|_{H^1(\Lambda)} \leqslant \left(\sum_{n=0}^{N} \frac{(\varphi^n)^2}{n+\frac{1}{2}}\right)^{1/2} \left(\sum_{n=0}^{N} n(n+1)\left(n+\frac{1}{2}\right)\right)^{1/2}$$

$$\leqslant N^2 \|\varphi_N\|_{L^2(\Lambda)} \left(\sup_{1\leqslant n\leqslant N} \frac{n}{N}\left(\frac{n}{N}+1\right)\left(\frac{n}{N}+\frac{1}{2}\right)\right)^{1/2},$$

which is the desired result.

Comparing the norm $\|\bullet\|_{L^2(\Lambda)}$ and the seminorm $|\bullet|_{H^1(\Lambda)}$ of simple polynomials, for instance

$$\|\zeta^N\|_{L^2(\Lambda)} = \frac{1}{\sqrt{N+\frac{1}{2}}} \quad \text{and} \quad |\zeta^N|_{H^1(\Lambda)} = \frac{N}{\sqrt{N-\frac{1}{2}}},$$

one could expect that the power of the parameter N in (5.2) is equal to 1, or at least smaller than 2. We intend to prove that the power 2 is the smallest possible, by exhibiting for any value of N a polynomial φ_N in $\mathbb{P}_N(\Lambda)$ such that

$$|\varphi_N|_{H^1(\Lambda)} \geqslant cN^2 \|\varphi_N\|_{L^2(\Lambda)}. \tag{5.4}$$

Indeed, let us take $\varphi_N = L'_N$, and recall that $\|\varphi_N\|_{L^2(\Lambda)}$ is equal to $\sqrt{N(N+1)}$, hence is smaller than $\sqrt{2}N$. To compute $|\varphi_N|_{H^1(\Lambda)} = \|L''_N\|_{L^2(\Lambda)}$, we use the quadrature formula (4.20), the fact that the weights ρ_j, $1 \leqslant j \leqslant N-1$, are positive and formulas (4.23) and (3.8):

$$|\varphi_N|^2_{H^1(\Lambda)} = \sum_{j=0}^{N} L''^2_N(\xi_j)\rho_j$$

$$\geqslant L''^2_N(-1)\rho_0 + L''^2_N(1)\rho_N \geqslant \frac{2}{N(N+1)}\left(L''^2_N(-1) + L''^2_N(1)\right)$$

$$\geqslant \frac{(N-1)^2 N(N+1)(N+2)^2}{16}.$$

Thus, $|\varphi_N|_{H^1(\Lambda)} \geqslant cN^3$ for $N \geqslant 2$. Comparing the two norms shows that the polynomial φ_N satisfies (5.4).

Next, we state the general result which can be derived from Theorem 5.1.

THEOREM 5.2. *Let m be an integer and r be a real number, $0 \leqslant m \leqslant r$. The following inequality holds for any positive integer N and for any polynomial φ_N in $\mathbb{P}_N(\Lambda)$:*

$$\|\varphi_N\|_{H^r(\Lambda)} \leqslant cN^{2(r-m)} \|\varphi_N\|_{H^m(\Lambda)}. \tag{5.5}$$

PROOF. The proof has two steps.

(1) When r is an integer, applying Theorem 5.1 to the derivatives of φ_N gives the result at once.

(2) If r is not an integer, we deduce from the result for integral values that the identity mapping is continuous from $\mathbb{P}_N(\Lambda)$ provided with the norm $\|\bullet\|_{H^m(\Lambda)}$ both into $H^{[r]}(\Lambda)$ with norm smaller than $c_1 N^{2([r]-m)}$ and into $H^{[r]+1}(\Lambda)$ with norm smaller than $c_2 N^{2([r]+1-m)}$. Next, recall (see Theorem 1.6) that $H^r(\Lambda)$ is the interpolation space $[H^{[r]+1}(\Lambda), H^{[r]}(\Lambda)]_\theta$ with θ equal to $[r]+1-r$. Thus, applying Theorem 1.5 yields that the identity mapping is continuous from $\mathbb{P}_N(\Lambda)$ provided with the norm $\|\bullet\|_{H^m(\Lambda)}$ into $H^r(\Lambda)$ and that its norm is smaller than

$$\left(c_2 N^{2([r]+1-m)}\right)^{1-\theta}\left(c_1 N^{2([r]-m)}\right)^{\theta}.$$

That is the desired result.

As previously, inequality (5.5) cannot be improved since the polynomial $\varphi_N = L'_N$ satisfies

$$\|\varphi_N\|_{H^r(\Lambda)} \geqslant c N^{2(r-m)} \|\varphi_N\|_{H^m(\Lambda)}, \tag{5.6}$$

for any real number r and any integer $m \leqslant r$ (indeed, $|L_N|_{H^l(\Lambda)}$ is of order N^{2l-1} for any positive integer l).

REMARK 5.1. Let r and s be two real numbers, $0 \leqslant s \leqslant r$. Applying Theorem 1.5 with the same arguments as in the previous proof leads to the following result: any polynomial φ_N in $\mathbb{P}_N(\Lambda)$ satisfies

$$\|\varphi_N\|_{H^r(\Lambda)} \leqslant c N^{2(r-s)} \|\varphi_N\|_{X_N^s}, \tag{5.7}$$

where X_N^s is the interpolation space of index $[s]+1-s$ between $\mathbb{P}_N(\Lambda)$ provided with the norm $\|\bullet\|_{H^{[s]+1}(\Lambda)}$ and this same space provided with the norm $\|\bullet\|_{H^{[s]}(\Lambda)}$. Clearly, the space X_N^s is the space $\mathbb{P}_N(\Lambda)$. Clearly also, the norm of X_N^s is equivalent to the norm of $H^s(\Lambda)$, i.e., for each integer N, there exists two constants $c_1(N)$ and $c_2(N)$ such that

$$\forall \varphi_N \subset \mathbb{P}_N(\Lambda), \quad c_1(N)\|\varphi_N\|_{H^s(\Lambda)} \leqslant \|\varphi_N\|_{X_N^s} \leqslant c_2(N)\|\varphi_N\|_{H^s(\Lambda)}. \tag{5.8}$$

But it is not clear at all to prove that the constants $c_1(N)$ and $c_2(N)$ are bounded independently of the maximal degree N of the polynomials. This requires a much more sophisticated theory about traces of polynomials, which is developed in BERNARDI, DAUGE and MADAY [1992a,b] and MADAY [1989].

We conclude with two "weighted" inverse inequalities where the power of N is smaller than in Theorem 5.1.

THEOREM 5.3. *The following inequality holds for any positive integer N and for any polynomial φ_N in $\mathbb{P}_N(\Lambda)$:*

$$\left(\int_{-1}^{1} \varphi_N'^2(\zeta)\left(1-\zeta^2\right) \mathrm{d}\zeta\right)^{1/2} \leqslant \sqrt{2} N \|\varphi_N\|_{L^2(\Lambda)}. \tag{5.9}$$

PROOF. Writing the expansion: $\varphi_N = \sum_{n=0}^{N} \varphi^n L_n$, we check that

$$\int_{-1}^{1} \varphi_N'^2(\zeta)\left(1-\zeta^2\right) \mathrm{d}\zeta = \sum_{m=0}^{N} \sum_{n=0}^{N} \varphi^m \varphi^n \int_{-1}^{1} L_m'(\zeta) L_n'(\zeta)\left(1-\zeta^2\right) \mathrm{d}\zeta,$$

whence, by (3.3),

$$\int_{-1}^{1} \varphi_N'^2(\zeta)\left(1-\zeta^2\right) \mathrm{d}\zeta = \sum_{n=0}^{N} \left(\varphi^n\right)^2 n(n+1) \int_{-1}^{1} L_n^2(\zeta) \, \mathrm{d}\zeta.$$

The conclusion follows.

THEOREM 5.4. *The following inequality holds for any positive integer N and for any polynomial φ_N in $\mathbb{P}_N(\Lambda)$ that vanishes at -1 and $+1$:*

$$|\varphi_N|_{H^1(\Lambda)} \leqslant \sqrt{2} N \left(\int_{-1}^{1} \varphi_N^2(\zeta)\left(1-\zeta^2\right)^{-1} \mathrm{d}\zeta\right)^{1/2}. \tag{5.10}$$

PROOF. Since φ_N vanishes in ± 1, we can write

$$\varphi_N(\zeta) = \left(1-\zeta^2\right) \sum_{n=1}^{N-1} \varphi^{*n} L_n'(\zeta).$$

Using (3.1) and (3.3), we have to compare

$$|\varphi_N|_{H^1(\Lambda)}^2 = \sum_{n=1}^{N-1} \left(\varphi^{*n}\right)^2 n^2(n+1)^2 \|L_n\|_{L^2(\Lambda)}^2$$

and

$$\int_{-1}^{1} \varphi_N^2(\zeta)\left(1-\zeta^2\right)^{-1} \mathrm{d}\zeta = \sum_{n=1}^{N-1} \left(\varphi^{*n}\right)^2 n(n+1) \|L_n\|_{L^2(\Lambda)}^2.$$

The result is straightforward.

CHAPTER II

Polynomial Approximation Error, the Galerkin Method

This chapter begins with the analysis of the best approximation by polynomials of a function with a given regularity in Sobolev spaces. Since the spaces we are concerned with, are Hilbert spaces, the best approximation is obtained via orthogonal projection operators. We study these operators in a fairly general way, firstly on an interval, secondly on a hypercube by tensorization arguments.

The numerical analysis of the Galerkin discretization relies on the previous approximation results. Indeed, for elliptic problems, the Galerkin method, which consists in replacing the Sobolev space involved in the variational formulation by a space of polynomials, provides a discrete solution such that its distance to the exact one is of the same size as the distance to the best fit in the discrete space. The Galerkin problem is presented for second- and fourth-order Dirichlet model problems, and also for the Neumann second-order problem.

NOTATION. In the following sections, the discretization parameter is a positive integer N. The symbol c (or c', c'') stands for a positive constant which can change from one line to the other but is always independent of N.

6. Polynomial approximation in an interval

This section is devoted to the analysis of orthogonal projection operators in the Sobolev spaces $H^r(\Lambda)$, where Λ stands for the open interval $]-1,1[$. The main results are stated in four theorems with increasing generality: they deal respectively with the case where r is equal to 0, where r is an integer successively for the spaces $H_0^r(\Lambda)$ and $H^r(\Lambda)$, and where r is any nonnegative real number. We recall that $\mathbb{P}_N(\Lambda)$ denotes the space of polynomials with degree $\leqslant N$ on Λ.

So, let us begin with the case $r=0$ of the projection in $L^2(\Lambda)$.

NOTATION. Let π_N be the orthogonal projection operator from $L^2(\Lambda)$ onto $\mathbb{P}_N(\Lambda)$.

Equivalently, it means that, for any function φ in $L^2(\Lambda)$, $\pi_N\varphi$ belongs to $\mathbb{P}_N(\Lambda)$ and satisfies

$$\forall\psi_N \in \mathbb{P}_N(\Lambda), \quad \int_{-1}^{1} (\varphi - \pi_N\varphi)(\zeta)\psi_N(\zeta)\,\mathrm{d}\zeta = 0. \tag{6.1}$$

It can also be noted that, since the space of all polynomials on Λ is dense in the space of continuous functions on $\overline{\Lambda}$, hence in $L^2(\Lambda)$, the family $(L_n)_n$ of Legendre polynomials is an orthogonal basis of $L^2(\Lambda)$. Consequently, any function φ in the space $L^2(\Lambda)$ admits the expansion

$$\varphi = \sum_{n=0}^{+\infty} \varphi^n L_n, \quad \text{with } \varphi^n = \frac{1}{\|L_n\|^2_{L^2(\Lambda)}} \int_{-1}^{1} \varphi(\zeta)L_n(\zeta)\,\mathrm{d}\zeta, \tag{6.2}$$

and it is readily checked that

$$\pi_N\varphi = \sum_{n=0}^{N} \varphi^n L_n.$$

Let us point out the importance of the next theorem, which is due to CANUTO and QUARTERONI [1982]: anything in the numerical analysis of spectral methods for partial differential equations relies on the following basic result.

THEOREM 6.1. *For any nonnegative real number s, there exists a positive constant c depending only on s such that, for any function φ in $H^s(\Lambda)$, the following estimate holds*

$$\|\varphi - \pi_N\varphi\|_{L^2(\Lambda)} \leqslant cN^{-s}\|\varphi\|_{H^s(\Lambda)}. \tag{6.3}$$

The first step in the proof is a continuity result for the Sturm–Liouville operator introduced in (3.2):

$$A\varphi = -\frac{\mathrm{d}}{\mathrm{d}\zeta}\big((1-\zeta^2)\varphi'\big).$$

LEMMA. *For any nonnegative integer l, the operator A is continuous from $H^{l+2}(\Lambda)$ into $H^l(\Lambda)$.*

PROOF. It is readily checked by induction on k that, for any nonnegative integer k,

$$\frac{\mathrm{d}^k(A\varphi)}{\mathrm{d}\zeta^k} = -(1-\zeta^2)\,\frac{\mathrm{d}^{k+2}\varphi}{\mathrm{d}\zeta^{k+2}} + 2(k+1)\zeta\,\frac{\mathrm{d}^{k+1}\varphi}{\mathrm{d}\zeta^{k+1}} + k(k+1)\,\frac{\mathrm{d}^k\varphi}{\mathrm{d}\zeta^k}.$$

This formula implies that, for any k, $0 \leqslant k \leqslant l$,

$$\left\| \frac{\mathrm{d}^k (A\varphi)}{\mathrm{d}\zeta^k} \right\|_{L^2(\Lambda)} \leqslant c \left(\left\| \frac{\mathrm{d}^{k+2}\varphi}{\mathrm{d}\zeta^{k+2}} \right\|_{L^2(\Lambda)} + \left\| \frac{\mathrm{d}^{k+1}\varphi}{\mathrm{d}\zeta^{k+1}} \right\|_{L^2(\Lambda)} + \left\| \frac{\mathrm{d}^{k}\varphi}{\mathrm{d}\zeta^{k}} \right\|_{L^2(\Lambda)} \right),$$

which gives the lemma.

PROOF OF THE THEOREM. We first prove the result when s is an even integer $2m$. For any function in $H^{2m}(\Lambda)$, we have

$$\|\varphi - \pi_N \varphi\|^2_{L^2(\Lambda)} = \sum_{n=N+1}^{+\infty} (\varphi^n)^2 \|L_n\|^2_{L^2(\Lambda)}.$$

Since the Legendre polynomials satisfy the differential equation (3.1), the Legendre coefficient φ^n, $n \geqslant 0$, can be written as

$$\begin{aligned} \varphi^n &= \frac{1}{\|L_n\|^2_{L^2(\Lambda)}} \int_{-1}^{1} \varphi(\zeta) L_n(\zeta)\,\mathrm{d}\zeta \\ &= \frac{1}{\|L_n\|^2_{L^2(\Lambda)}} \frac{1}{n(n+1)} \int_{-1}^{1} \varphi(\zeta)(AL_n)(\zeta)\,\mathrm{d}\zeta. \end{aligned}$$

Recalling that the operator A is self-adjoint in $L^2(\Lambda)$ leads to

$$\varphi^n = \frac{1}{\|L_n\|^2_{L^2(\Lambda)}} \frac{1}{n(n+1)} \int_{-1}^{1} (A\varphi)(\zeta) L_n(\zeta)\,\mathrm{d}\zeta.$$

Iterating m times this argument, we derive

$$\varphi^n = \frac{1}{\|L_n\|^2_{L^2(\Lambda)}} \frac{1}{(n(n+1))^m} \int_{-1}^{1} (A^m\varphi)(\zeta) L_n(\zeta)\,\mathrm{d}\zeta.$$

Hence, we obtain

$$\begin{aligned} &\|\varphi - \pi_N \varphi\|^2_{L^2(\Lambda)} \\ &\quad = \sum_{n=N+1}^{+\infty} \frac{1}{(n(n+1))^{2m}} \left(\frac{\int_{-1}^{1} (A^m\varphi)(\zeta) L_n(\zeta)\,\mathrm{d}\zeta}{\|L_n\|^2_{L^2(\Lambda)}} \right)^2 \|L_n\|^2_{L^2(\Lambda)}. \end{aligned}$$

Since the $n(n+1)$ are larger than N^2, it yields

$$\|\varphi - \pi_N \varphi\|^2_{L^2(\Lambda)} \leqslant N^{-4m} \sum_{n=0}^{+\infty} \left(\frac{\int_{-1}^{1} (A^m\varphi)(\zeta) L_n(\zeta)\,\mathrm{d}\zeta}{\|L_n\|^2_{L^2(\Lambda)}} \right)^2 \|L_n\|^2_{L^2(\Lambda)},$$

or equivalently

$$\|\varphi - \pi_N\varphi\|^2_{L^2(\Lambda)} \leqslant N^{-4m}\|A^m\varphi\|^2_{L^2(\Lambda)}. \tag{6.4}$$

Due to the previous lemma, we obtain the desired inequality (6.3) when s is an even integer. Now, let s be a nonnegative real number. If m is an integer such that $2m$ is larger than s, we know that the operator $\mathrm{id} - \pi_N$ is linear continuous from $L^2(\Lambda)$ into itself with norm 1 and from $H^{2m}(\Lambda)$ into $L^2(\Lambda)$ with norm $\leqslant cN^{-2m}$. By combining Theorems 1.5 and 1.6, we obtain that it is linear continuous from $H^s(\Lambda)$ into $L^2(\Lambda)$ with norm $\leqslant cN^{-s}$, which is the desired result.

REMARK 6.1. By a very simple duality argument, we also derive estimates in the Sobolev spaces with negative order from Theorem 6.1: for any nonnegative real numbers r and s, and for any function φ in $H^s(\Lambda)$, the following estimate holds

$$\|\varphi - \pi_N\varphi\|_{H^{-r}(\Lambda)} \leqslant cN^{-r-s}\|\varphi\|_{H^s(\Lambda)}. \tag{6.5}$$

REMARK 6.2. It should be noted that Theorem 6.1 provides an optimal estimate of the distance of a function in $L^2(\Lambda)$ to $\mathbb{P}_N(\Lambda)$, in the sense that the power of $1/N$ in formula (6.3) is equal to the difference of the orders of the Sobolev spaces in the norms of the left- and right-hand sides of the inequality. Indeed, it is easy to check that this power is the largest possible one: if a function φ is written $\sum_{n=0}^{+\infty}\alpha_n(L_{n+1} - L_{n-1})$, we have from formula (3.12):

$$\|\varphi - \pi_N\varphi\|^2_{L^2(\Lambda)} = 2\sum_{n=N+1}^{+\infty}\frac{(\alpha_{n+1} - \alpha_{n-1})^2}{2n+1}$$

and

$$|\varphi|^2_{H^1(\Lambda)} = 2\sum_{n=0}^{+\infty}\alpha_n^2(2n+1);$$

taking

$$\alpha_n = \begin{cases} (2n+1)^{-\gamma} & \text{if 4 divides } n,\\ 0 & \text{elsewhere,}\end{cases}$$

we check that φ belongs to $H^1(\Lambda)$ whenever $\gamma > 1$ and that $\|\varphi - \pi_N\varphi\|_{L^2(\Lambda)}$ behaves like a constant times $N^{-\gamma}$, i.e., essentially like a constant times $N^{-1}|\varphi|_{H^1(\Lambda)}$.

REMARK 6.3. However, the result of Theorem 6.1 can be improved by the following argument, which was first used by DORR [1984, 1986] for the p-version of the finite element method. Let $D(A)$ be the domain of the operator A in $L^2(\Lambda)$, i.e., the subspace of functions φ in $L^2(\Lambda)$ such that $A\varphi$ belongs to $L^2(\Lambda)$. Since A is positive self-adjoint, we can define the subspace $D(A^{s/2})$ for any nonnegative real number s. We denote by $\|\cdot\|_{D(A^{s/2})}$ the associated graph norm. It can be observed that the spaces

$D(A^{s/2})$, $s \geqslant 0$, form a scale of weighted Sobolev spaces. Thus, inequality (6.4) reads

$$\|\varphi - \pi_N \varphi\|_{L^2(\Lambda)} \leqslant N^{-2m} \|\varphi\|_{D(A^m)},$$

so that an interpolation argument leads to the following inequality, which holds for any nonnegative integer s and for any function φ in $D(A^{s/2})$:

$$\|\varphi - \pi_N \varphi\|_{L^2(\Lambda)} \leqslant N^{-s} \|\varphi\|_{D(A^{s/2})}. \tag{6.6}$$

Since the space $H^s(\Lambda)$ is strictly imbedded in $D(A^{s/2})$, this allows to improve the estimate (6.3) for some special functions: for instance, the function $\varphi = (1 - \zeta^2)^\gamma$ belongs to $H^s(\Lambda)$ for $s < \gamma + \frac{1}{2}$ and to $D(A^{s/2})$ for $s < 2\gamma + 1$ (see BERNARDI and MADAY [1992a]), so that $\|\varphi - \pi_N \varphi\|_{L^2(\Lambda)}$ is less than $cN^{\varepsilon - \gamma - 1/2}$ from formula (6.3) and less than $cN^{\varepsilon - 2\gamma - 1}$ from formula (6.6), for any $\varepsilon > 0$. However, the space $H^s(\Lambda)$ is not imbedded in any space $D(A^{r/2})$ with $r > s$, so that Theorem 6.1 is optimal with respect to the number of derivatives of the function, according to Remark 6.2, but it is not optimal with respect to a singular behavior of the function near the boundary.

It must be noted that using the $L^2(\Lambda)$-projection operator does not lead to optimal estimates in Sobolev spaces with higher order. Indeed, for any positive integer N, there exists a function φ_N in $H_0^1(\Lambda)$ such that

$$|\varphi_N - \pi_N \varphi_N|_{H^1(\Lambda)} \geqslant cN^{1/2} |\varphi_N|_{H^1(\Lambda)} \tag{6.7}$$

(take for instance the function $\varphi_N = L_{N+1} - L_{N-1}$, so that $\pi_N \varphi_N = -L_{N-1}$). The best possible estimate was proven by CANUTO and QUARTERONI [1982], it is written in the following way: for any nonnegative real numbers r and s, $r \leqslant s$, and for any function φ in $H^s(\Lambda)$,

$$\|\varphi - \pi_N \varphi\|_{H^r(\Lambda)} \leqslant \begin{cases} cN^{(3r/2)-s} \|\varphi\|_{H^s(\Lambda)} & \text{if } r \leqslant 1, \\ cN^{2r-(1/2)-s} \|\varphi\|_{H^s(\Lambda)} & \text{if } r \geqslant 1. \end{cases} \tag{6.8}$$

Hence, we are led to introduce other projection operators to derive optimal estimates. The next step is to study the polynomial approximation of functions in $H_0^k(\Lambda)$ for a fixed positive integer k.

NOTATION. Let k be a positive integer. For any integer $N \geqslant 2k - 1$, $\mathbb{P}_N^{k,0}(\Lambda)$ stands for the space $\mathbb{P}_N(\Lambda) \cap H_0^k(\Lambda)$, i.e., for the space of polynomials in $\mathbb{P}_N(\Lambda)$ which vanish in ± 1 together with their derivatives up to order $k - 1$. The space $\mathbb{P}_N^{1,0}(\Lambda)$ is also denoted by $\mathbb{P}_N^0(\Lambda)$ for the sake of simplicity.

NOTATION. For any positive integer k, let $\pi_N^{k,0}$ be the orthogonal projection operator from $H_0^k(\Lambda)$ onto $\mathbb{P}_N^{k,0}(\Lambda)$ for the scalar product associated with the norm $|\cdot|_{H^k(\Lambda)}$.

This is equivalent to write that, for any function φ in $H_0^k(\Lambda)$, $\pi_N^{k,0}\varphi$ belongs to $\mathbb{P}_N^{k,0}(\Lambda)$ and satisfies:

$$\forall \psi_N \in \mathbb{P}_N^{k,0}(\Lambda), \quad \int_{-1}^{1} \left(\frac{\mathrm{d}^k\varphi}{\mathrm{d}\zeta^k} - \frac{\mathrm{d}^k(\pi_N^{k,0}\varphi)}{\mathrm{d}\zeta^k}\right)(\zeta)\left(\frac{\mathrm{d}^k\psi_N}{\mathrm{d}\zeta^k}\right)(\zeta)\,\mathrm{d}\zeta = 0. \tag{6.9}$$

The operator $\pi_N^{1,0}$ was firstly studied by MADAY and QUARTERONI [1981].

THEOREM 6.2. *Let k be a positive integer. For any nonnegative real numbers r and s, $0 \leqslant r \leqslant k \leqslant s$, there exists a positive constant c depending only on s such that, for any function φ in $H^s(\Lambda) \cap H_0^k(\Lambda)$, the following estimate holds*

$$\|\varphi - \pi_N^{k,0}\varphi\|_{H^r(\Lambda)} \leqslant cN^{r-s}\|\varphi\|_{H^s(\Lambda)}. \tag{6.10}$$

PROOF. It is divided into three steps, according as r is equal to k, equal to 0 or between 0 and k.

(1) Since the constant c which appears in (6.10) depends on k, there is no restriction to assume that N is larger than $2k-1$ (the estimate is straightforward for $N < 2k-1$). Let us for a while define $\pi_N^{0,0}$ as the orthogonal projection operator π_N in $L^2(\Lambda)$. Then, we are going to check the identity for any $k \geqslant 1$:

$$\forall \varphi \in H_0^k(\Lambda), \quad (\pi_N^{k,0}\varphi)(\zeta) = \int_{-1}^{\zeta} (\pi_{N-1}^{k-1,0}\varphi')(\xi)\,\mathrm{d}\xi. \tag{6.11}$$

Indeed, the function defined on the right-hand side satisfies equation (6.9). To check that it belongs to $\mathbb{P}_N^{k,0}(\Lambda)$, we note that it belongs to $\mathbb{P}_N(\Lambda)$, that it vanishes in -1 and that its derivatives up to order $k-1$ vanish in ± 1, so that it suffices to prove that it also vanishes in 1. Applying the definition of the operator $\pi_{N-1}^{k,0}$ to the function $(1-\zeta^2)^{k-1}$ gives

$$\begin{aligned}\int_{-1}^{1} &\left(\frac{\mathrm{d}^{k-1}}{\mathrm{d}\zeta^{k-1}}\right)(\pi_{N-1}^{k-1,0}\varphi')(\zeta)\left(\frac{\mathrm{d}^{k-1}}{\mathrm{d}\zeta^{k-1}}\right)\left((1-\zeta^2)^{k-1}\right)\mathrm{d}\zeta \\ &= \int_{-1}^{1}\left(\frac{\mathrm{d}^{k-1}\varphi'}{\mathrm{d}\zeta^{k-1}}\right)(\zeta)\left(\frac{\mathrm{d}^{k-1}}{\mathrm{d}\zeta^{k-1}}\right)\left((1-\zeta^2)^{k-1}\right)\mathrm{d}\zeta.\end{aligned}$$

Integrating $k-1$ times by parts, we derive

$$\begin{aligned}(2(k-1))!\int_{-1}^{1}(\pi_{N-1}^{k-1,0}\varphi')(\zeta)\,\mathrm{d}\zeta &= (2(k-1))!\int_{-1}^{1}\varphi'(\zeta)\,\mathrm{d}\zeta \\ &= (2(k-1))!\big(\varphi(1)-\varphi(-1)\big) = 0,\end{aligned}$$

which ends the proof of (6.11). Finally, we have

$$|\varphi - \pi_N^{k,0}\varphi|_{H^k(\Lambda)} = |\varphi' - \pi_{N-1}^{k-1,0}\varphi'|_{H^{k-1}(\Lambda)} = \cdots$$
$$= \left\| \frac{\mathrm{d}^k\varphi}{\mathrm{d}\zeta^k} - \pi_{N-k}\left(\frac{\mathrm{d}^k\varphi}{\mathrm{d}\zeta^k}\right)\right\|_{L^2(\Lambda)},$$

so that Theorem 6.1 implies

$$|\varphi - \pi_N^{k,0}\varphi|_{H^k(\Lambda)} \leqslant c(N-k)^{k-s}\left\|\frac{\mathrm{d}^k\varphi}{\mathrm{d}\zeta^k}\right\|_{H^{s-k}(\Lambda)} \leqslant c' N^{k-s}\|\varphi\|_{H^s(\Lambda)}.$$

Combining this estimate with the Poincaré–Friedrichs inequality yields (6.10) for r equal to k.

(2) The estimate for $r = 0$ relies on a duality argument. Indeed, we have

$$\|\varphi - \pi_N^{k,0}\varphi\|_{L^2(\Lambda)} = \sup_{g\in L^2(\Lambda)} \frac{\int_{-1}^1 (\varphi - \pi_N^{k,0}\varphi)(\zeta)g(\zeta)\,\mathrm{d}\zeta}{\|g\|_{L^2(\Lambda)}}. \tag{6.12}$$

For any g in $L^2(\Lambda)$, we consider the solution χ in $H_0^k(\Lambda)$ of the problem

$$\forall\psi \in H_0^k(\Lambda), \quad \int_{-1}^1 \left(\frac{\mathrm{d}^k\chi}{\mathrm{d}\zeta^k}\right)(\zeta)\left(\frac{\mathrm{d}^k\psi}{\mathrm{d}\zeta^k}\right)(\zeta)\,\mathrm{d}\zeta = \int_{-1}^1 g(\zeta)\psi(\zeta)\,\mathrm{d}\zeta. \tag{6.13}$$

The existence and uniqueness of this solution follow from the Lax–Milgram Theorem 2.1. Moreover, since $\mathrm{d}^{2k}\chi/\mathrm{d}\zeta^{2k} = (-1)^k g$ belongs to $L^2(\Lambda)$, it is readily checked that the solution χ belongs to $H^{2k}(\Lambda)$ and satisfies

$$\|\chi\|_{H^{2k}(\Lambda)} \leqslant c\|g\|_{L^2(\Lambda)}. \tag{6.14}$$

Then we compute

$$\int_{-1}^1 (\varphi - \pi_N^{k,0}\varphi)(\zeta)g(\zeta)\,\mathrm{d}\zeta = \int_{-1}^1 \left(\frac{\mathrm{d}^k}{\mathrm{d}\zeta^k}\right)(\varphi - \pi_N^{k,0}\varphi)(\zeta)\left(\frac{\mathrm{d}^k\chi}{\mathrm{d}\zeta^k}\right)(\zeta)\,\mathrm{d}\zeta$$
$$= \int_{-1}^1 \left(\frac{\mathrm{d}^k}{\mathrm{d}\zeta^k}\right)(\varphi - \pi_N^{k,0}\varphi)(\zeta)\left(\frac{\mathrm{d}^k}{\mathrm{d}\zeta^k}\right)(\chi - \pi_N^{k,0}\chi)(\zeta)\,\mathrm{d}\zeta$$
$$\leqslant |\varphi - \pi_N^{k,0}\varphi|_{H^k(\Lambda)}|\chi - \pi_N^{k,0}\chi|_{H^k(\Lambda)}.$$

Applying twice the estimate (6.10) for $r = k$, we obtain

$$\int_{-1}^1 (\varphi - \pi_N^{k,0}\varphi)(\zeta)g(\zeta)\,\mathrm{d}\zeta \leqslant cN^{k-s}\|\varphi\|_{H^s(\Lambda)}N^{-k}\,\|\chi\|_{H^{2k}(\Lambda)},$$

and combining this result with (6.12) and (6.14) gives (6.10) for $r = 0$.

(3) Finally, we know from (1.10) and (1.12) that, for $0 \leqslant r \leqslant k$,

$$\|\varphi - \pi_N^{k,0}\varphi\|_{H^r(\Lambda)} \leqslant \|\varphi - \pi_N^{k,0}\varphi\|_{L^2(\Lambda)}^{1-r/k} \|\varphi - \pi_N^{k,0}\varphi\|_{H^k(\Lambda)}^{r/k},$$

so that the estimates for $r = k$ and $r = 0$ lead to the general result.

REMARK 6.4. Let k, r and s satisfy the hypotheses of the theorem. Exactly as in Remark 6.3, we can deduce from (6.11) that the modified estimate holds for any function φ in $H_0^k(\Lambda)$ such that $\mathrm{d}^k\varphi/\mathrm{d}\zeta^k$ belongs to $D(A^{(s-k)/2})$:

$$\|\varphi - \pi_N^{k,0}\varphi\|_{H^r(\Lambda)} \leqslant cN^{r-s}\left\|\frac{\mathrm{d}^k\varphi}{\mathrm{d}\zeta^k}\right\|_{D(A^{(s-k)/2})}. \tag{6.15}$$

Of course, we are also interested with the approximation in $H^k(\Lambda)$ of functions which do not vanish at the end points. To this aim, we introduce a set of polynomials $\chi_{k,l}$, $0 \leqslant l \leqslant k-1$: $\chi_{k,l}$ stands for the unique polynomial in $\mathbb{P}_{2k-1}(\Lambda)$ which satisfies

$$\left(\frac{\mathrm{d}^l\chi_{k,l}}{\mathrm{d}\zeta^l}\right)(-1) = 1 \quad \text{and} \quad \left(\frac{\mathrm{d}^m\chi_{k,l}}{\mathrm{d}\zeta^m}\right)(-1) = 0, \quad 0 \leqslant m \leqslant k-1, m \neq l,$$
$$\chi_{k,l}(1) = \chi'_{k,l}(1) = \cdots = \left(\frac{\mathrm{d}^{k-1}\chi_{k,l}}{\mathrm{d}\zeta^{k-1}}\right)(1) = 0. \tag{6.16}$$

Next, with each function φ in $H^k(\Lambda)$, we associate a function $\widetilde{\varphi}_k$ in $H_0^k(\Lambda)$ by

$$\widetilde{\varphi}_k(\zeta) = \varphi(\zeta) - \sum_{l=0}^{k-1}\left(\frac{\mathrm{d}^l\varphi}{\mathrm{d}\zeta^l}\right)(-1)\chi_{k,l}(\zeta)$$
$$- \sum_{l=0}^{k-1}(-1)^l\left(\frac{\mathrm{d}^l\varphi}{\mathrm{d}\zeta^l}\right)(1)\chi_{k,l}(-\zeta). \tag{6.17}$$

Note that, due the Sobolev Theorem 1.9, the quantities $(\mathrm{d}^l\varphi/\mathrm{d}\zeta^l)(\pm 1)$ satisfy

$$\left|\left(\frac{\mathrm{d}^l\varphi}{\mathrm{d}\zeta^l}\right)(-1)\right| + \left|\left(\frac{\mathrm{d}^l\varphi}{\mathrm{d}\zeta^l}\right)(1)\right| \leqslant c\|\varphi\|_{H^k(\Lambda)}, \quad 0 \leqslant l \leqslant k-1, \tag{6.18}$$

so that the function $\widetilde{\varphi}_k$ is well defined and satisfies for any real number $s \geqslant k$:

$$\|\widetilde{\varphi}_k\|_{H^s(\Lambda)} \leqslant c\|\varphi\|_{H^s(\Lambda)}. \tag{6.19}$$

This allows for defining an operator on $H^k(\Lambda)$.

NOTATION. For any positive integer k, assuming that N is $\geqslant 2k-1$, the operator $\tilde{\pi}_N^k$ is defined by associating with each function φ in $H^k(\Lambda)$ the function $\widetilde{\varphi}_k$ of (6.17) and setting

$$(\tilde{\pi}_N^k\varphi)(\zeta) = (\pi_N^{k,0}\widetilde{\varphi}_k)(\zeta) + \sum_{l=0}^{k-1}\left(\frac{\mathrm{d}^l\varphi}{\mathrm{d}\zeta^l}\right)(-1)\chi_{k,l}(\zeta)$$
$$+\sum_{l=0}^{k-1}(-1)^l\left(\frac{\mathrm{d}^l\varphi}{\mathrm{d}\zeta^l}\right)(1)\chi_{k,l}(-\zeta). \tag{6.20}$$

As an immediate consequence of this definition, we have the identity

$$\varphi - \tilde{\pi}_N^k\varphi = \widetilde{\varphi}_k - \pi_N^{k,0}\widetilde{\varphi}_k. \tag{6.21}$$

Hence, the next result follows from Theorem 6.2 combined with (6.19).

THEOREM 6.3. *Let k be a positive integer. For any real numbers r and s satisfying $0 \leqslant r \leqslant k \leqslant s$, there exists a positive constant c depending only on s such that, for any function φ in $H^s(\Lambda)$, the following estimate holds:*

$$\|\varphi - \tilde{\pi}_N^k\varphi\|_{H^r(\Lambda)} \leqslant cN^{r-s}\|\varphi\|_{H^s(\Lambda)}. \tag{6.22}$$

REMARK 6.5. The operator $\tilde{\pi}_N^k$ is not the orthogonal projection operator from $H^k(\Lambda)$ onto $\mathbb{P}_N(\Lambda)$ (that is why it is overlined with a tilde). Estimate (6.22) also holds with $\tilde{\pi}_N^k$ replaced by the orthogonal projection operator. However, by construction, the operator $\tilde{\pi}_N^k$ has the additional property of preserving the values of the function and of its derivatives up to order $k-1$ at the end points of the interval, which is important for a number of applications.

Finally, we present the completely general statement which is due to MADAY [1990]. It includes the results of the three previous theorems, however the construction of the operator is less explicit than for the $\tilde{\pi}_N^k$.

THEOREM 6.4. *Let t and p be two nonnegative real numbers such that neither $t-\frac{1}{2}$ nor $p-\frac{1}{2}$ is an integer. There exists an operator $\tilde{\pi}_N^{t,p,0}$ from $H^t(\Lambda)\cap H_0^p(\Lambda)$ into $\mathbb{P}_N(\Lambda)\cap H_0^p(\Lambda)$ such that, for any real numbers r and s, $0\leqslant r\leqslant t\leqslant s$, and for any function φ in $H^s(\Lambda)\cap H_0^p(\Lambda)$, the following estimate holds:*

$$\|\varphi - \tilde{\pi}_N^{t,p,0}\varphi\|_{H^r(\Lambda)} \leqslant cN^{r-s}\|\varphi\|_{H^s(\Lambda)}. \tag{6.23}$$

PROOF. When t is an integer, the desired property is satisfied by the restriction to $H_0^p(\Lambda)$ of the operator $\tilde{\pi}_N^t$, since it preserves the values of the function and of its derivatives at ± 1 up to order $t-1$. So, it remains to study the case where neither t nor $t-\frac{1}{2}$ is an integer. We begin by the case where p is equal to t. Then, it follows from LIONS and MAGENES ([1968], Chapter 1, Theorem 11.6) that $H_0^t(\Lambda)$ is the interpolation space of index $\frac{1}{2}$ between $H_0^{2t}(\Lambda)$ and $L^2(\Lambda)$ if $2t-\frac{1}{2}$ is not an integer and the interpolation space of index $\frac{2}{3}$ between $H_0^{3t}(\Lambda)$ and $L^2(\Lambda)$ if $2t-\frac{1}{2}$ is an integer. Equivalently, there exists a positive self-adjoint operator T_0 in $L^2(\Lambda)$ such that the domain of T_0 coincides with $H_0^t(\Lambda)$, with the graph norm equivalent to the norm on $H_0^t(\Lambda)$, and that the domain of T_0^2 (resp. the domain of T_0^3) coincides with $H_0^{2t}(\Lambda)$ if $2t - \frac{1}{2}$ is not an integer (resp. with $H_0^{3t}(\Lambda)$ if $2t - \frac{1}{2}$ is an integer). Now, let $\pi_N^{t,0}$ be the orthogonal projection operator for the scalar product associated with the graph norm of T_0 onto $\mathbb{P}_N(\Lambda) \cap H_0^t(\Lambda)$. More explicitly, for any function φ in $H_0^t(\Lambda)$, $\pi_N^{t,0}$ belongs to $\mathbb{P}_N(\Lambda) \cap H_0^t(\Lambda)$ and satisfies

$$\forall \psi_N \in \mathbb{P}_N(\Lambda) \cap H_0^t(\Lambda),$$
$$\int_{-1}^{1} (\varphi - \pi_N^{t,0}\varphi)(\zeta)\psi_N(\zeta)\,\mathrm{d}\zeta + \int_{-1}^{1} \big(T_0(\varphi - \pi_N^{t,0}\varphi)\big)(\zeta)(T_0\psi_N)(\zeta)\,\mathrm{d}\zeta = 0. \tag{6.24}$$

Let us prove the estimate (6.23) for the operator $\pi_N^{t,0}$ and for the different values of r.

(1) From definition (6.24), we have

$$\|\varphi - \pi_N^{t,0}\varphi\|_{H^t(\Lambda)} \leqslant c \inf_{\varphi_N \in \mathbb{P}_N(\Lambda) \cap H_0^t(\Lambda)} \|\varphi - \varphi_N\|_{H^t(\Lambda)}.$$

Assuming that the function φ belongs to $H^{[t]+1}(\Lambda)$ and choosing $\varphi_N = \tilde{\pi}_N^{[t]+1}\varphi$, we check that it belongs to $\mathbb{P}_N(\Lambda) \cap H_0^t(\Lambda)$ and we obtain from (6.22)

$$\|\varphi - \pi_N^{t,0}\varphi\|_{H^t(\Lambda)} \leqslant cN^{t-s}\|\varphi\|_{H^s(\Lambda)},$$

which is (6.23) in the case $r = t$ and for $s \geqslant [t] + 1$. Next, a simple interpolation argument between this inequality and the property

$$\|\varphi - \pi_N^{t,0}\varphi\|_{H^t(\Lambda)} \leqslant c\|\varphi\|_{H^t(\Lambda)},$$

gives the result for $r = t$ and any $s \geqslant t$.

(2) Let us now turn to the case $r = 0$. The duality argument is very similar to the argument for integral values of t. Indeed we have

$$\varphi - \pi_N^{t,0}\varphi = \sup_{g \in L^2(\Lambda)} \frac{\int_{-1}^{1} (\varphi - \pi_N^{t,0}\varphi)(\zeta) g(\zeta)\,\mathrm{d}\zeta}{\|g\|_{L^2(\Lambda)}}. \tag{6.25}$$

Due to the Riesz theorem, since the operator T_0 is positive, with any g in $L^2(\Lambda)$, we can associate a unique function χ in $D(T_0) = H_0^t(\Lambda)$ such that

$$\forall \psi \in D(T_0), \quad \int_{-1}^{1} \chi(\zeta)\psi(\zeta)\,\mathrm{d}\zeta + \int_{-1}^{1} (T_0\chi)(\zeta)(T_0\psi)(\zeta)\,\mathrm{d}\zeta = \int_{-1}^{1} g(\zeta)\psi(\zeta)\,\mathrm{d}\zeta. \tag{6.26}$$

It is readily checked that the linear form:

$$\psi \mapsto \int_{-1}^{1} (T_0\chi)(\zeta)(T_0\psi)(\zeta)\,\mathrm{d}\zeta = \int_{-1}^{1} g(\zeta)\psi(\zeta)\,\mathrm{d}\zeta - \int_{-1}^{1} \chi(\zeta)\psi(\zeta)\,\mathrm{d}\zeta,$$

is continuous on $L^2(\Lambda)$, so that the function χ belongs to $D(T_0^2)$. Since this last space is equal to $H_0^{2t}(\Lambda)$ when $2t - \frac{1}{2}$ is not an integer (resp. imbedded in $H_0^{2t}(\Lambda)$ when $2t - \frac{1}{2}$ is an integer), the function χ satisfies

$$\|\chi\|_{H^{2t}(\Lambda)} \leqslant c\|g\|_{L^2(\Lambda)}. \tag{6.27}$$

Then, a standard computation leads to

$$\begin{aligned}
&\int_{-1}^{1} (\varphi - \pi_N^{t,0}\varphi)(\zeta)g(\zeta)\,\mathrm{d}\zeta \\
&\quad = \int_{-1}^{1} (\varphi - \pi_N^{t,0}\varphi)(\zeta)(\chi - \pi_N^{t,0}\chi)(\zeta)\,\mathrm{d}\zeta \\
&\qquad + \int_{-1}^{1} (T_0(\varphi - \pi_N^{t,0}\varphi))(\zeta)(T_0(\chi - \pi_N^{t,0}\chi))(\zeta)\,\mathrm{d}\zeta \\
&\quad \leqslant \|\varphi - \pi_N^{t,0}\varphi\|_{H^t(\Lambda)}\|\chi - \pi_N^{t,0}\chi\|_{H^t(\Lambda)},
\end{aligned}$$

so that combining the estimate (6.23) for $r = t$ with (6.25) and (6.27) leads to the estimate for $r = 0$.

(3) The same interpolation argument as for Theorem 6.2 allows for deriving the estimate for any value of r, $0 \leqslant r \leqslant t$, from the estimates for $r = 0$ and $r = t$, when p is equal to t.

Finally, denoting respectively by k and m the integral parts of $t + \frac{1}{2}$ and $p + \frac{1}{2}$, with each function φ in $H^t(\Lambda) \cap H_0^p(\Lambda)$, we associate the function $\widetilde{\varphi}_k$ of (6.17) and we set, (nearly) as in (6.20),

$$\begin{aligned}
(\tilde{\pi}_N^{t,p,0}\varphi)(\zeta) = (\pi_N^{t,0}\widetilde{\varphi}_k)(\zeta) &+ \sum_{l=m}^{k-1} \left(\frac{\mathrm{d}^l\varphi}{\mathrm{d}\zeta^l}\right)(-1)\chi_{k,l}(\zeta) \\
&+ \sum_{l=m}^{k-1} (-1)^l \left(\frac{\mathrm{d}^l\varphi}{\mathrm{d}\zeta^l}\right)(1)\chi_{k,l}(-\zeta).
\end{aligned} \tag{6.28}$$

The operator $\tilde{\pi}_N^{t,p,0}$ satisfies the desired properties!

REMARK 6.6. Note that the operator $\tilde{\pi}_N^{t,0,0}$ preserves the values of the lth derivative of the function at the end points, for any integer $l < t - \frac{1}{2}$. This will be used in Section 13.

7. Polynomial approximation on tensorized domains

In all that follows, Ω_d stands for the open hypercube $]-1,1[^d$. The generic point in Ω_d is denoted by $\boldsymbol{x}=(x_1,\ldots,x_d)$. The aim of this section is to extend the results of the previous Section 6 to multidimensional domains, so the way is similar. We study orthogonal projection operators successively in $L^2(\Omega_d)$, Sobolev spaces of integral order and general Sobolev spaces.

The main argument of this section is that, since the domain is the product of d times the same interval Λ, the Sobolev spaces are tensor products of Sobolev spaces on Λ. This explains why we need some special notation.

NOTATION. With each function v in $L^2(\Omega_d)$, we associate the d-functions v_j defined by

$$v_j(x_j)(x_1,\ldots,x_{j-1},x_{j+1},\ldots,x_d)=v(x_1,\ldots,x_d),\quad 1\leqslant j\leqslant d.$$

Thus, for any j, $1\leqslant j\leqslant d$, and for any nonnegative real numbers r and s, we define the Sobolev space

$$H^r\big(\Lambda_j;H^s(\Omega_{d-1})\big)=\big\{v\in L^2(\Omega_d)\colon\ v_j\in H^r\big(\Lambda;H^s\big(\Lambda^{d-1}\big)\big)\big\} \tag{7.1}$$

(recall that the Sobolev spaces of vector-valued functions have been introduced in Remark 1.2). By straightforward extensions and an induction argument, it is also possible to define the spaces $H^t(\Lambda_i;H^r(\Lambda_j;H^s(\Omega_{d-2})))$, and so on. When an operator concerning the functions defined on Λ, is applied with respect to the x_j variable, it will appear with the exponent ${}^{(j)}$, in order to avoid confusion.

The basic property of the tensorized space is that, since the measure $\mathrm{d}\boldsymbol{x}$ is a product $\mathrm{d}x_1\cdots\mathrm{d}x_d$, the following identity holds for any j, $1\leqslant j\leqslant d$:

$$L^2(\Omega_d)=L^2\big(\Lambda_j;L^2(\Omega_{d-1})\big). \tag{7.2}$$

Two further properties of the tensorized spaces will be used in the section:

(i) for any integers k and m, $k\leqslant m$, the following imbedding is obvious from the definitions of the norms:

$$H^m(\Omega_d)\subset H^k\big(\Lambda_j;H^{m-k}(\Omega_{d-1})\big).$$

Using the fact that the Sobolev space $H^s(\Lambda_j)$ is the domain of an operator with respect to the x_j variable and that two operators with respect to different variables commute, leads to the general result: for any nonnegative real numbers r and s, $r\leqslant s$, the following imbedding holds:

$$H^s(\Omega_d)\subset H^r\big(\Lambda_j;H^{s-r}(\Omega_{d-1})\big). \tag{7.3}$$

(ii) The next result is proven for instance by LIONS and MAGENES ([1968], Chapter 4): for any nonnegative integer m, we have the identity

$$H^m(\Omega_d) = \bigcap_{j=1}^{d} H^m\big(\Lambda_j; L^2(\Omega_{d-1})\big).$$

There also, a simple interpolation argument allows for deriving the general result: for any nonnegative real number s, the following identity holds

$$H^s(\Omega_d) = \bigcap_{j=1}^{d} H^s\big(\Lambda_j; L^2(\Omega_{d-1})\big). \tag{7.4}$$

NOTATION. For any nonnegative integer n, $\mathbb{P}_n(\Omega_d)$ stands for the space of restrictions to Ω_d of polynomials with d variables and with degree $\leqslant n$ with respect to each variable x_j, $1 \leqslant j \leqslant d$.

This space is also very appropriate for the use of tensorization properties, since it is the tensor product of d times the space $\mathbb{P}_N(\Lambda)$ of polynomials with one variable. Moreover, it is also natural to use the tensorized basis

$$\big\{L_{n_1}(x_1)\cdots L_{n_d}(x_d),\ 0 \leqslant n_1, \ldots, n_d \leqslant N\big\},$$

as a basis of $\mathbb{P}_N(\Omega_d)$. Now that all the tools are presented, we begin the analysis of the $L^2(\Omega_d)$ projection operator.

NOTATION. Let Π_N be the orthogonal projection operator from $L^2(\Omega_d)$ onto $\mathbb{P}_N(\Omega_d)$.

Now, let us note that each function v in $L^2(\Omega_d)$ satisfies, for $1 \leqslant j \leqslant d$,

$$\int_{-1}^{1} v(\boldsymbol{x}) L_{n_j}(x_j)\,\mathrm{d}x_j = \int_{-1}^{1} \big(\pi_N^{(j)} v\big)(\boldsymbol{x}) L_{n_j}(x_j)\,\mathrm{d}x_j, \quad 0 \leqslant n_j \leqslant N.$$

More generally, it is easy to check that

$$\begin{aligned}
&\int_{\Omega_d} v(\boldsymbol{x}) L_{n_1}(x_1)\cdots L_{n_d}(x_d)\,\mathrm{d}\boldsymbol{x} \\
&\quad = \int_{\Omega_d} \big(\pi_N^{(1)} \circ \cdots \circ \pi_N^{(d)} v\big)(\boldsymbol{x}) L_{n_1}(x_1)\cdots L_{n_d}(x_d)\,\mathrm{d}\boldsymbol{x}, \\
&\quad 0 \leqslant n_1, \ldots, n_d \leqslant N.
\end{aligned}$$

Since the previous equation is satisfied for all the polynomials running through the tensorized basis of $\mathbb{P}_N(\Omega_d)$, we derive at once the basic identity

$$\Pi_N = \pi_N^{(1)} \circ \cdots \circ \pi_N^{(d)}, \tag{7.5}$$

and we note that all the operators $\pi_N^{(j)}$ commute.

THEOREM 7.1. *For any nonnegative real number s, there exists a positive constant c depending only on s such that, for any function v in $H^s(\Omega_d)$, the following estimate holds*

$$\|v - \Pi_N v\|_{L^2(\Omega_d)} \leqslant cN^{-s}\|v\|_{H^s(\Omega_d)}. \tag{7.6}$$

PROOF. From (7.5), it comes

$$\|v - \Pi_N v\|_{L^2(\Omega_d)} \leqslant \|v - \pi_N^{(1)} v\|_{L^2(\Lambda_1;L^2(\Omega_{d-1}))} + \|\pi_N^{(1)}(v - \pi_N^{(2)} \circ \cdots \circ \pi_N^{(d)} v)\|_{L^2(\Lambda_1;L^2(\Omega_{d-1}))}.$$

Thus, applying Theorem 6.1 and using the fact that $\pi_N^{(1)}$ is an orthogonal projection operator in $L^2(\Lambda_1)$ yield

$$\|v - \Pi_N v\|_{L^2(\Omega_d)} \leqslant cN^{-s}\|v\|_{H^s(\Lambda_1;L^2(\Omega_{d-1}))} + \|v - \pi_N^{(2)} \circ \cdots \circ \pi_N^{(d)} v\|_{L^2(\Lambda_1;L^2(\Omega_{d-1}))}.$$

Iterating d times this argument gives

$$\|v - \Pi_N v\|_{L^2(\Omega_d)} \leqslant cN^{-s} \sum_{j=1}^{d} \|v\|_{H^s(\Lambda_j;L^2(\Omega_{d-1}))},$$

so that the desired estimate follows from the identity (7.4).

Here also, the power of N in the right-hand side of (7.6) is optimal with respect to the regularity of the function. And it is still necessary to introduce other operators in order to obtain optimal estimates in higher order norms.

NOTATION. Let k be a positive integer. For any integer $N \geqslant 2k-1$, $\mathbb{P}_N^{k,0}(\Omega_d)$ stands for the space $\mathbb{P}_N(\Omega_d) \cap H_0^k(\Omega_d)$. The space $\mathbb{P}_N^{1,0}(\Omega_d)$ is also denoted by $\mathbb{P}_N^0(\Omega_d)$ for the sake of simplicity.

NOTATION. For any positive integer k, let $\Pi_N^{k,0}$ be the orthogonal projection operator from $H_0^k(\Omega_d)$ onto $\mathbb{P}_N^{k,0}(\Omega_d)$ for the scalar product associated with the norm $|\cdot|_{H^k(\Omega_d)}$.

THEOREM 7.2. *Let k be a positive integer. For any nonnegative real number $s \geqslant k$, there exists a positive constant c depending only on s such that, for any function v in $H^s(\Omega_d) \cap H_0^k(\Omega_d)$, the following estimate holds*

$$\|v - \Pi_N^{k,0} v\|_{H^k(\Omega_d)} \leqslant cN^{k-s}\|v\|_{H^s(\Omega_d)}. \tag{7.7}$$

PROOF. The first step consists in establishing the estimate when the function v is smooth enough, namely in $H^s(\Omega_d) \cap H_0^k(\Omega_d)$, for $s \geqslant dk$. Indeed, we know from the definition of $\Pi_N^{k,0}$ that

$$\left|v - \Pi_N^{k,0} v\right|_{H^k(\Omega_d)} = \inf_{v_N \in \mathbb{P}_N^{k,0}(\Omega_d)} |v - v_N|_{H^k(\Omega_d)}. \tag{7.8}$$

Using the Poincaré–Friedrichs inequality and the imbedding

$$H_0^k(\Omega_d) \cap H^{dk}(\Omega) \subset H_0^k\big(\Lambda_j; H^{(d-1)k}(\Omega_{d-1}) \cap H_0^k(\Omega_{d-1})\big),$$

we derive

$$\left\|v - \Pi_N^{k,0} v\right\|_{H^k(\Omega_d)} \leqslant c\left|v - \pi_N^{k,0(1)} \circ \cdots \circ \pi_N^{k,0(d)} v\right|_{H^k(\Omega_d)},$$

and, from (7.4),

$$\begin{aligned}
&\left\|v - \Pi_N^{k,0} v\right\|_{H^k(\Omega_d)} \\
&\quad \leqslant c \sum_{j=1}^{d} \left\| \left(\frac{\partial^k}{\partial x_j^k}\right)\left(v - \pi_N^{k,0(1)} \circ \cdots \circ \pi_N^{k,0(d)} v\right)\right\|_{L^2(\Lambda_j; L^2(\Omega_{d-1}))} \\
&\quad \leqslant c \sum_{j=1}^{d} \left(\left\| \left(\frac{\partial^k}{\partial x_j^k}\right)\left(v - \pi_N^{k,0(j)} v\right)\right\|_{L^2(\Lambda_j; L^2(\Omega_{d-1}))} \right. \\
&\qquad + \left\| \left(\frac{\partial^k}{\partial x_j^k}\right) \pi_N^{k,0(j)} \left(v - \pi_N^{k,0(1)} \circ \cdots \circ \pi_N^{k,0(j-1)} \right.\right. \\
&\qquad \left.\left.\left. \circ\, \pi_N^{k,0(j+1)} \circ \cdots \circ \pi_N^{k,0(d)} v\right)\right\|_{L^2(\Lambda_j; L^2(\Omega_{d-1}))} \right).
\end{aligned}$$

Next, we apply inequality (6.10) with $r = k$:

$$\begin{aligned}
\left\|v - \Pi_N^{k,0} v\right\|_{H^k(\Omega_d)} &\leqslant c N^{k-s} \sum_{j=1}^{d} \|v\|_{H^s(\Lambda_j; L^2(\Omega_{d-1}))} \\
&\quad + c' \sum_{j=1}^{d} \left\|v - \pi_N^{k,0(1)} \circ \cdots \circ \pi_N^{k,0(j-1)} \circ \pi_N^{k,0(j+1)} \circ \cdots \right. \\
&\quad \left. \circ\, \pi_N^{k,0(d)} v\right\|_{H^k(\Lambda_j; L^2(\Omega_{d-1}))},
\end{aligned}$$

so that, taking for instance $j = d$, it suffices to prove the estimate

$$\left\|v - \pi_N^{k,0(1)} \circ \cdots \circ \pi_N^{k,0(d-1)} v\right\|_{L^2(\Omega_{d-1})} \leqslant c N^{k-s} \|v\|_{H^{s-k}(\Omega_{d-1})}. \tag{7.9}$$

To this aim, we write the inequality

$$\begin{aligned}
&\left\|v - \pi_N^{k,0(1)} \circ \cdots \circ \pi_N^{k,0(d-1)} v\right\|_{L^2(\Omega_{d-1})} \\
&\quad \leqslant \left\|v - \pi_N^{k,0(d-1)} v\right\|_{L^2(\Omega_{d-1})} + \left\|v - \pi_N^{k,0(1)} \circ \cdots \circ \pi_N^{k,0(d-2)} v\right\|_{L^2(\Omega_{d-1})} \\
&\qquad + \left\|\left(\mathrm{id} - \pi_N^{k,0(d-1)}\right)\left(v - \pi_N^{k,0(1)} \circ \cdots \circ \pi_N^{k,0(d-2)} v\right)\right\|_{L^2(\Omega_{d-1})}.
\end{aligned}$$

Using now (6.10) with $r = 0$ gives

$$\begin{aligned}
&\left\|v - \pi_N^{k,0(1)} \circ \cdots \circ \pi_N^{k,0(d-1)} v\right\|_{L^2(\Omega_{d-1})} \\
&\quad \leqslant cN^{k-s}\|v\|_{H^{s-k}(\Lambda_{d-1};L^2(\Omega_{d-2}))} \\
&\qquad + \left\|v - \pi_N^{k,0(1)} \circ \cdots \circ \pi_N^{k,0(d-2)} v\right\|_{L^2(\Lambda_{d-1};L^2(\Omega_{d-2}))} \\
&\qquad + cN^{-k}\left\|v - \pi_N^{k,0(1)} \circ \cdots \circ \pi_N^{k,0(d-2)} v\right\|_{H^k(\Lambda_{d-1};L^2(\Omega_{d-2}))},
\end{aligned}$$

and applying iteratively this argument leads to (7.9). Hence, we obtain the estimate when $s \geqslant dk$. Using it when s is equal to dk together with the stability inequality

$$\left\|v - \Pi_N^{k,0} v\right\|_{H^k(\Omega_d)} \leqslant c\|v\|_{H^k(\Omega_d)},$$

and an interpolation argument relying on Theorem 1.5 leads to the general estimate (7.7).

In order to study the approximation of functions which do not vanish on the boundary, we introduce the orthogonal projection operator in the Sobolev spaces of integral order.

NOTATION. For any positive integer k, let Π_N^k be the orthogonal projection operator from $H^k(\Omega_d)$ onto $\mathbb{P}_N(\Omega_d)$ for the scalar product associated with the norm $\|\bullet\|_{H^k(\Omega_d)}$.

The proof of the following theorem is very similar to the proof of Theorem 7.2: only the operator $\pi_N^{k,0}$ is replaced by the operator $\tilde{\pi}_N^k$ and (6.10) is replaced by (6.22).

THEOREM 7.3. *Let k be a positive integer. For any nonnegative real number $s \geqslant k$, there exists a positive constant c depending only on s such that, for any function v in $H^s(\Omega_d)$, the following estimate holds*

$$\left\|v - \Pi_N^k v\right\|_{H^k(\Omega_d)} \leqslant cN^{k-s}\|v\|_{H^s(\Omega_d)}. \tag{7.10}$$

REMARK 7.1. In d-dimensional domains, $d \geqslant 2$, we prefer to consider the orthogonal projection operator Π_N^k, instead of the operator

$$\tilde{\Pi}_N^k = \tilde{\pi}_N^{k(1)} \circ \cdots \circ \tilde{\pi}_N^{k(d)}, \tag{7.11}$$

which is only defined on $H^{dk}(\Omega_d)$. However, as it could be seen in the missing proof, this last operator also satisfies the estimate

$$\|v - \widetilde{\Pi}_N^k v\|_{H^k(\Omega_d)} \leqslant cN^{k-s}\|v\|_{H^s(\Omega_d)}, \tag{7.12}$$

for any function v in $H^s(\Omega_d)$, $s \geqslant dk$, and it has the further property of preserving the values of the functions and of its derivatives up to order $k-1$ at the corners of the domain. Also, the normal derivatives of $\widetilde{\Pi}_N^k v$ on a $(d-1)$-face up to order $k-1$ only depend on the values of the normal derivatives of the function v on this face (for instance, the trace of $\widetilde{\Pi}_N^k v$ on each face $x_j = \pm 1$ is the image of the trace of v on this face by the product of the $\tilde{\pi}_N^{k(i)}$, $i \neq j$).

In Theorems 7.2 and 7.3, we did not state any estimate in lower order norms. Indeed, such estimates would follow from a duality argument relying on the regularity of the operator Δ^k (or a similar one) from $L^2(\Omega_d)$ into $H^{2k}(\Omega_d)$. Such a result is very standard for $k=1$ and $d=2$ (see GRISVARD [1985] and Remark 2.1), less standard but true for $k=1$ and $d \geqslant 3$, and already known for $k=2$ and $d=2$. However, the general result is unknown. That is why we prefer to build a general operator by abstract arguments similar to that of Theorem 6.4, to obtain the general results. We do not write down the proof which is the same as for Theorem 6.4 (see MADAY ([1987], Chapter 2)).

THEOREM 7.4. *Let t and p be nonnegative real numbers such that neither $t-\frac{1}{2}$ nor $p-\frac{1}{2}$ is an integer. There exists an operator $\widetilde{\Pi}_N^{t,p,0}$ from $H^t(\Omega_d) \cap H_0^p(\Omega_d)$ into $\mathbb{P}_N(\Omega_d) \cap H_0^p(\Omega_d)$ such that, for any real numbers r and s, $0 \leqslant r \leqslant t \leqslant s$, and for any function φ in $H^s(\Omega_d) \cap H_0^p(\Omega_d)$, the following estimate holds*

$$\|v - \widetilde{\Pi}_N^{t,p,0} v\|_{H^r(\Omega_d)} \leqslant cN^{r-s}\|v\|_{H^s(\Omega_d)}. \tag{7.13}$$

REMARK 7.2. As hinted before the proof, estimate (7.13) holds with the operator $\widetilde{\Pi}_N^{1,1,0}$ replaced by $\Pi_N^{1,0}$, and also with the operator $\widetilde{\Pi}_N^{2,2,0}$ replaced by $\Pi_N^{2,0}$ for d equal to 2.

8. Galerkin method for the second-order Dirichlet problem

In this section, we consider the second-order Dirichlet problem with homogeneous boundary conditions in the hypercube $\Omega_d =]-1,1[^d$, which is written

$$\begin{cases} -\Delta u = f & \text{in } \Omega_d, \\ u = 0 & \text{on } \partial\Omega_d. \end{cases} \tag{8.1}$$

We recall (see Section 2) that, if the distribution f is given in $H^{-1}(\Omega_d)$, it admits the following equivalent variational formulation: *find u in $H_0^1(\Omega_d)$ such that*

$$\forall v \in H_0^1(\Omega_d), \quad a(u,v) = \langle f, v \rangle, \tag{8.2}$$

where the bilinear form $a(\bullet,\bullet)$ is defined by

$$\forall u \in H^1(\Omega_d),\ \forall v \in H^1(\Omega_d),$$
$$a(u,v) = \int_{\Omega_d} (\mathbf{grad}\, u)(\boldsymbol{x}) \bullet (\mathbf{grad}\, v)(\boldsymbol{x})\, \mathrm{d}\boldsymbol{x}. \tag{8.3}$$

Then, the existence and uniqueness of the solution of this problem are a special case of Theorem 2.2.

Let N be a positive integer. The Galerkin method consists in replacing the space $H_0^1(\Omega_d)$ by its finite-dimensional subspace $\mathbb{P}_N^0(\Omega_d) = \mathbb{P}_N(\Omega_d) \cap H_0^1(\Omega_d)$. This leads to the following discrete problem: *find u_N in $\mathbb{P}_N^0(\Omega_d)$ such that*

$$\forall v_N \in \mathbb{P}_N^0(\Omega_d), \quad a(u_N, v_N) = \langle f, v_N \rangle. \tag{8.4}$$

Equivalently, combining equations (8.2) and (8.4), we note that

$$\forall v_N \in \mathbb{P}_N^0(\Omega_d), \quad \int_{\Omega_d} \big(\mathbf{grad}(u - u_N)\big)(\boldsymbol{x}) \bullet (\mathbf{grad}\, v_N)(\boldsymbol{x})\, \mathrm{d}\boldsymbol{x} = 0. \tag{8.5}$$

This leads to the conclusion that u_N coincides with the orthogonal projection $\Pi_N^{1,0} u$, where the operator $\Pi_N^{1,0}$ was introduced in Section 7. Consequently, the numerical analysis of problem (8.4) is essentially contained in Theorem 7.2. Let us state the results in two theorems.

THEOREM 8.1. *For any distribution f in $H^{-1}(\Omega_d)$, problem* (8.4) *has a unique solution u_N in $\mathbb{P}_N^0(\Omega_d)$. Moreover, this solution satisfies*

$$\|u_N\|_{H^1(\Omega_d)} \leqslant c \|f\|_{H^{-1}(\Omega_d)}. \tag{8.6}$$

THEOREM 8.2. *Let us assume that the solution u of problem* (8.1) *belongs to $H^s(\Omega_d)$ for a real number $s \geqslant 1$. Then, the following error estimate holds for problem* (8.4)*:*

$$|u - u_N|_{H^1(\Omega_d)} \leqslant c N^{1-s} \|u\|_{H^s(\Omega_d)}. \tag{8.7}$$

So, only one result is missing: the estimate in $L^2(\Omega_d)$. Its proof relies on the duality technique of Aubin and Nitsche.

THEOREM 8.3. *If the assumptions of Theorem* 8.2 *are satisfied, the following error estimate holds for problem* (8.4)*:*

$$\|u - u_N\|_{L^2(\Omega_d)} \leqslant c N^{-s} \|u\|_{H^s(\Omega_d)}. \tag{8.8}$$

PROOF. The argument is the following

$$\|u - u_N\|_{L^2(\Omega_d)} = \sup_{g \in L^2(\Omega_d)} \frac{\int_{\Omega_d} (u - u_N)(\boldsymbol{x}) g(\boldsymbol{x})\, \mathrm{d}\boldsymbol{x}}{\|g\|_{L^2(\Omega_d)}}. \tag{8.9}$$

Then, for any function g in $L^2(\Omega_d)$, we consider the problem:

$$\begin{cases} -\Delta w = g & \text{in } \Omega_d, \\ w = 0 & \text{on } \partial\Omega_d. \end{cases} \tag{8.10}$$

By Theorem 2.2, it has a unique solution in $H_0^1(\Omega_d)$ and, as already stated in Remark 2.1, since the hypercube Ω_d is convex, it can be checked (see GRISVARD ([1985], Theorem 3.2.1.2)) that the solution w belongs to $H^2(\Omega_d)$ and satisfies

$$\|w\|_{H^2(\Omega_d)} \leqslant c\|g\|_{L^2(\Omega_d)}. \tag{8.11}$$

We have to compute

$$\int_{\Omega_d} (u - u_N)(\boldsymbol{x}) g(\boldsymbol{x})\, \mathrm{d}\boldsymbol{x} = \int_{\Omega_d} \big(\mathbf{grad}(u - u_N)\big)(\boldsymbol{x}) \bullet (\mathbf{grad}\, w)(\boldsymbol{x})\, \mathrm{d}\boldsymbol{x},$$

so that, from (8.5) and for any w_N in $\mathbb{P}_N^0(\Omega_d)$,

$$\begin{aligned} \int_{\Omega_d} (u - u_N)(\boldsymbol{x}) g(\boldsymbol{x})\, \mathrm{d}\boldsymbol{x} &= \int_{\Omega_d} \big(\mathbf{grad}(u - u_N)\big)(\boldsymbol{x}) \bullet \big(\mathbf{grad}(w - w_N)\big)(\boldsymbol{x})\, \mathrm{d}\boldsymbol{x} \\ &\leqslant |u - u_N|_{H^1(\Omega_d)} |w - w_N|_{H^1(\Omega_d)}. \end{aligned}$$

Choosing w_N equal to $\Pi_N^{1,0} w$ and combining the results of Theorems 7.2 and 8.2 with (8.9) and (8.11), we obtain the desired estimate.

Combining the estimates (8.7) and (8.8) via an interpolation argument, we derive the general estimate in $H^r(\Omega_d)$, $0 \leqslant r \leqslant 1$:

$$\|u - u_N\|_{H^r(\Omega_d)} \leqslant c N^{r-s} \|u\|_{H^s(\Omega_d)}. \tag{8.12}$$

REMARK 8.1. Clearly, the method can be extended to the case of inhomogeneous boundary conditions. However, we prefer to handle inhomogeneous boundary data with the collocation method, as will be presented in Section 15.

REMARK 8.2. Let us for a while consider the case $d = 2$ of the square domain Ω, and let us assume that the right-hand member f belongs to $L^2(\Omega)$. Since a possible basis of the space $\mathbb{P}_N^0(\Lambda)$ is given by the $(1-\zeta^2)L_n'$, $1 \leqslant n \leqslant N-1$, we can write the expansion of the discrete solution u_N in this basis:

$$u_N(x,y) = (1-x^2)(1-y^2) \sum_{m=1}^{N-1} \sum_{n=1}^{N-1} u^{mn} L_m'(x) L_n'(y).$$

Thus, due to the differential equation (3.1), it can be observed that u_N is a solution of problem (8.4) if and only if the u^{mn}, $1 \leqslant m, n \leqslant N-1$, are the solution of the linear system of $(N-1)^2$ equations with $(N-1)^2$ unknowns:

$$\begin{aligned}
&\frac{k^2(k+1)^2}{k+\frac{1}{2}} \sum_{n=1}^{N-1} u^{kn} \int_{-1}^{1} L_n'(y) L_l'(y) (1-y^2)^2 \, \mathrm{d}y \\
&\quad + \frac{l^2(l+1)^2}{l+\frac{1}{2}} \sum_{m=1}^{N-1} u^{ml} \int_{-1}^{1} L_m'(x) L_k'(x) (1-x^2)^2 \, \mathrm{d}x \\
&\quad = \int_{-1}^{1} \int_{-1}^{1} f(x,y) L_k'(x) L_l'(y) (1-x^2)(1-y^2) \, \mathrm{d}x \, \mathrm{d}y, \quad 1 \leqslant k, l \leqslant N-1.
\end{aligned} \tag{8.13}$$

To solve this system, we need to know the $(N-1)^2$ quantities

$$\int_{-1}^{1} \int_{-1}^{1} f(x,y) L_k'(x) L_l'(y) (1-x^2)(1-y^2) \, \mathrm{d}x \, \mathrm{d}y$$

which form the right-hand member. In fact, whatever the choice of the basis of $\mathbb{P}_N^0(\Omega_d)$ which is used for the test functions, the exact computation of the integrals in the right-hand member of problem (8.4) is often impossible and, in any case, very expensive. For these reasons, the Galerkin method cannot be used in practical situations. But it has a real theoretical interest, since its very simple numerical analysis focuses on the importance of polynomial approximation results.

9. Galerkin method for the fourth-order Dirichlet problem

This section is devoted to the fourth-order Dirichlet problem in the hypercube Ω_d, when it is provided with homogeneous Dirichlet conditions:

$$\begin{cases}
\Delta^2 u = f & \text{in } \Omega_d, \\
u = 0 & \text{on } \partial\Omega_d, \\
\dfrac{\partial u}{\partial n} = 0 & \text{on } \partial\Omega_d.
\end{cases} \tag{9.1}$$

When the distribution f is given in $H^{-2}(\Omega_d)$, it admits the following equivalent variational formulation: *find u in $H_0^2(\Omega_d)$ such that*

$$\forall v \in H_0^2(\Omega_d), \quad d(u,v) = \langle f, v\rangle, \tag{9.2}$$

where the bilinear form $d(\bullet,\bullet)$ is defined by

$$\forall u \in H^2(\Omega_d),\ \forall v \in H^2(\Omega_d), \quad d(u,v) = \int_{\Omega_d} (\Delta u)(\boldsymbol{x})(\Delta v)(\boldsymbol{x})\,\mathrm{d}\boldsymbol{x}. \tag{9.3}$$

As proven in Theorem 2.3, it has a unique solution.

Let N be a positive integer. Here, the Galerkin method consists in replacing the space $H_0^2(\Omega_d)$ by its subspace $\mathbb{P}_N^{2,0}(\Omega_d)$, which leads to the following discrete problem: *find u_N in $\mathbb{P}_N^{2,0}(\Omega_d)$ such that*

$$\forall v_N \in \mathbb{P}_N^{2,0}(\Omega_d), \quad d(u_N, v_N) = \langle f, v_N\rangle. \tag{9.4}$$

In this case, the form $d(\bullet,\bullet)$ is not exactly the same as the scalar product associated with the norm $|\bullet|_{H^2(\Omega_d)}$, hence the numerical analysis will be slightly different. However, the ellipticity of the form $d(\bullet,\bullet)$ on $H_0^2(\Omega_d)$ and its continuity on $H^2(\Omega_d) \times H^2(\Omega_d)$, combined with the Lax–Milgram Theorem 2.1, leads to the following statement.

THEOREM 9.1. *For any distribution f in $H^{-2}(\Omega_d)$, problem* (9.4) *has a unique solution u_N in $\mathbb{P}_N^{2,0}(\Omega_d)$. Moreover, this solution satisfies*

$$\|u_N\|_{H^2(\Omega_d)} \leqslant c\|f\|_{H^{-2}(\Omega_d)}. \tag{9.5}$$

We state the error estimates in two steps, firstly in $H^2(\Omega_d)$, secondly in $L^2(\Omega_d)$.

THEOREM 9.2. *Let us assume that the solution u of problem* (9.1) *belongs to $H^s(\Omega_d)$ for a real number $s \geqslant 2$. Then, the following error estimate holds for problem* (9.4)*:*

$$|u - u_N|_{H^2(\Omega_d)} \leqslant cN^{2-s}\|u\|_{H^s(\Omega_d)}. \tag{9.6}$$

PROOF. From (9.2) and (9.4), we derive the identity

$$\forall v_N \in \mathbb{P}_N^{2,0}(\Omega_d), \quad d(u - u_N, v_N) = 0,$$

so that, taking v_N equal to $u_N - z_N$ for any z_N in $\mathbb{P}_N^{2,0}(\Omega_d)$, we have

$$\begin{aligned} c|u - u_N|^2_{H^2(\Omega_d)} &\leqslant d(u - u_N, u - u_N) = d(u - u_N, u - z_N) \\ &\leqslant c'|u - u_N|_{H^2(\Omega_d)}|u - z_N|_{H^2(\Omega_d)}. \end{aligned}$$

Taking now z_N equal to $\Pi_N^{2,0}u$ and using Theorem 7.2 leads to the desired result.

THEOREM 9.3. *In the case* $d = 2$ *of the square domain* Ω, *if the assumptions of Theorem* 9.2 *are satisfied, the following error estimate holds for problem* (9.4)*:*

$$\|u - u_N\|_{L^2(\Omega)} \leqslant cN^{-s}\|u\|_{H^s(\Omega)}. \tag{9.7}$$

PROOF. The Aubin–Nitsche duality argument relies on the following identity

$$\|u - u_N\|_{L^2(\Omega)} = \sup_{g\in L^2(\Omega)} \frac{\int_\Omega (u - u_N)(\boldsymbol{x})g(\boldsymbol{x})\,\mathrm{d}\boldsymbol{x}}{\|g\|_{L^2(\Omega)}}. \tag{9.8}$$

Next, for any function g in $L^2(\Omega)$, we consider the following fourth-order problem:

$$\begin{cases} \Delta^2 w = g & \text{in } \Omega,\\ w = 0 & \text{on } \partial\Omega,\\ \dfrac{\partial w}{\partial n} = 0 & \text{on } \partial\Omega. \end{cases} \tag{9.9}$$

By Theorem 2.3, it has a unique solution in $H_0^2(\Omega)$ and, in the square Ω, it can be checked (see Remark 2.1 and GRISVARD ([1985], Theorem 7.2.2.3) for the proof) that the solution w belongs to $H^4(\Omega)$ and satisfies

$$\|w\|_{H^4(\Omega)} \leqslant c\|g\|_{L^2(\Omega)}. \tag{9.10}$$

Then, we obviously have for any w_N in $\mathbb{P}_N^{2,0}(\Omega)$

$$\begin{aligned} \int_\Omega (u - u_N)(\boldsymbol{x})g(\boldsymbol{x})\,\mathrm{d}\boldsymbol{x} &= \int_\Omega \big(\Delta(u - u_N)\big)(\boldsymbol{x})\big(\Delta(w - w_N)\big)(\boldsymbol{x})\,\mathrm{d}\boldsymbol{x}\\ &\leqslant |u - u_N|_{H^2(\Omega)}|w - w_N|_{H^2(\Omega)}, \end{aligned}$$

and choosing w_N equal to $\Pi_N^{2,0}w$ gives the estimate.

There also, in the square Ω, the general result is derived from (9.6) and (9.7) via an interpolation argument: for any real number r, $0 \leqslant r \leqslant 2$, we have the following estimate

$$\|u - u_N\|_{H^r(\Omega)} \leqslant cN^{r-s}\|u\|_{H^s(\Omega)}. \tag{9.11}$$

REMARK 9.1. Clearly, the estimates (9.7) and (9.11) can be extended to any hypercube Ω_d such that the mapping which, with g, associates the solution w of problem (9.9) is continuous from $L^2(\Omega_d)$ into $H^4(\Omega_d)$. However, this regularity property is not presently known, even in the case $d = 3$ of the cube. By using the arguments of BABUŠKA and SURI [1990] or those of Remark 6.3 (see BERNARDI and MADAY [1992a]), a slightly weaker regularity property would be sufficient, however it is still not known.

10. Galerkin method for the second-order Neumann problem

As a third example, we now consider the Neumann problem in the hypercube Ω_d:

$$\begin{cases} -\Delta u = f & \text{in } \Omega_d, \\ \dfrac{\partial u}{\partial n} = h & \text{on } \partial\Omega_d. \end{cases} \tag{10.1}$$

We assume that the function f belongs to $L^2(\Omega_d)$, that the distribution h belongs to $(H^{1/2}(\partial\Omega_d))'$ and that they satisfy the compatibility condition

$$\int_{\Omega_d} f(\boldsymbol{x})\,\mathrm{d}\boldsymbol{x} + \langle h, 1\rangle_{\partial\Omega_d} = 0. \tag{10.2}$$

Then, problem (10.1) admits the equivalent variational formulation: *find* u *in* $H^1(\Omega_d)/\mathbb{R}$ *such that*

$$\forall v \in H^1(\Omega_d)/\mathbb{R}, \quad a(u, v) = \int_{\Omega_d} f(\boldsymbol{x}) v(\boldsymbol{x})\,\mathrm{d}\boldsymbol{x} + \langle h, v\rangle_{\partial\Omega_d}, \tag{10.3}$$

where the bilinear form $a(\cdot,\cdot)$ is defined in (8.3). As stated in Theorem 2.6, it has a unique solution.

Here also, since $\mathbb{P}_N(\Omega_d)/\mathbb{R}$ is a finite-dimensional subspace of $H^1(\Omega_d)/\mathbb{R}$ (its dimension is $(N+1)^d - 1$), the Galerkin discretized problem is easy to write: *find* u_N *in* $\mathbb{P}_N(\Omega_d)/\mathbb{R}$ *such that*

$$\forall v_N \in \mathbb{P}_N(\Omega_d)/\mathbb{R}, \quad a(u_N, v_N) = \int_{\Omega_d} f(\boldsymbol{x}) v_N(\boldsymbol{x})\,\mathrm{d}\boldsymbol{x} + \langle h, v_N\rangle_{\partial\Omega_d}. \tag{10.4}$$

From the ellipticity of the form $a(\cdot,\cdot)$ on $H^1(\Omega_d)/\mathbb{R}$ (see Theorem 1.3), we deduce the first result.

THEOREM 10.1. *For any function* f *in* $L^2(\Omega_d)$ *and for any distribution* h *in* $(H^{1/2}(\partial\Omega_d))'$ *satisfying the compatibility condition* (10.2), *problem* (10.4) *has a unique solution* u_N *in* $\mathbb{P}_N(\Omega_d)/\mathbb{R}$. *Moreover, this solution satisfies*

$$\|u_N\|_{H^1(\Omega_d)/\mathbb{R}} \leqslant c\big(\|f\|_{L^2(\Omega_d)} + \|h\|_{(H^{1/2}(\partial\Omega_d))'}\big). \tag{10.5}$$

As in Section 8, we immediately observe that

$$\forall v_N \in \mathbb{P}_N(\Omega_d), \quad \int_{\Omega_d} \big(\mathbf{grad}(u - u_N)\big)(\boldsymbol{x}) \cdot (\mathbf{grad}\, v_N)(\boldsymbol{x})\,\mathrm{d}\boldsymbol{x} = 0, \tag{10.6}$$

whence

$$|u - u_N|_{H^1(\Omega_d)} = \inf_{v_N \in \mathbb{P}_N(\Omega_d)} |u - v_N|_{H^1(\Omega_d)}. \tag{10.7}$$

Using the operator Π_N^1 and estimate (7.10), we derive the following result.

THEOREM 10.2. *Let us assume that the solution u of problem* (10.1) *belongs to $H^s(\Omega_d)$ for a real number $s \geqslant 1$. Then, the following error estimate holds for problem* (10.4)*:*

$$|u - u_N|_{H^1(\Omega_d)} \leqslant cN^{1-s} \|u\|_{H^s(\Omega_d)}. \tag{10.8}$$

The final result is the estimate in $L^2(\Omega_d)$. Its proof is very similar to that of Theorem 8.3. However, till now, the solutions of problems (10.1) and (10.4) are defined only up to an additive constant. In order to fix the constants in a compatible way, we enforce the additional condition

$$\int_{\Omega_d} u(\boldsymbol{x})\,\mathrm{d}\boldsymbol{x} = 0 \quad \text{and} \quad \int_{\Omega_d} u_N(\boldsymbol{x})\,\mathrm{d}\boldsymbol{x} = 0. \tag{10.9}$$

THEOREM 10.3. *Assume that condition* (10.9) *holds. If the assumptions of Theorem* 10.2 *are satisfied, the following error estimate holds for problem* (10.4)*:*

$$\|u - u_N\|_{L^2(\Omega_d)} \leqslant cN^{-s} \|u\|_{H^s(\Omega_d)}. \tag{10.10}$$

PROOF. We start from the identity

$$\|u - u_N\|_{L^2(\Omega_d)} = \sup_{g \in L^2(\Omega_d)} \frac{\int_{\Omega_d} (u - u_N)(\boldsymbol{x}) g(\boldsymbol{x})\,\mathrm{d}\boldsymbol{x}}{\|g\|_{L^2(\Omega_d)}}. \tag{10.11}$$

Next, for any function g in $L^2(\Omega_d)$, we set

$$g_0 = g - 2^{-d} \int_{\Omega_d} g(\boldsymbol{x})\,\mathrm{d}\boldsymbol{x}.$$

Then, the following problem:

$$\begin{cases} -\Delta w = g_0 & \text{in } \Omega_d, \\ \dfrac{\partial w}{\partial n} = 0 & \text{on } \partial\Omega_d, \end{cases} \tag{10.12}$$

has a unique solution in $H^1(\Omega_d)/\mathbb{R}$ (see Theorem 2.6) and the convexity of Ω_d yields (see Remark 2.1 and GRISVARD ([1985], Theorem 3.2.1.3)) that the solution w belongs to $H^2(\Omega_d)$ and satisfies

$$\|w\|_{H^2(\Omega_d)} \leqslant c\|g_0\|_{L^2(\Omega_d)} \leqslant c'\|g\|_{L^2(\Omega_d)}. \tag{10.13}$$

From (10.9) and (10.6), we compute for any w_N in $\mathbb{P}_N(\Omega_d)$,

$$\int_{\Omega_d} (u - u_N)(\boldsymbol{x}) g(\boldsymbol{x})\, \mathrm{d}\boldsymbol{x} = \int_{\Omega_d} \big(\mathbf{grad}(u - u_N)\big)(\boldsymbol{x}) \cdot \big(\mathbf{grad}(w - w_N)\big)(\boldsymbol{x})\, \mathrm{d}\boldsymbol{x}$$
$$\leqslant |u - u_N|_{H^1(\Omega_d)} |w - w_N|_{H^1(\Omega_d)},$$

so that choosing w_N equal to $\Pi^1_N w$ leads to the desired estimate.

The general estimate in $H^r(\Omega_d)$, $0 \leqslant r \leqslant 1$, is derived from (10.8) and (10.10) via an interpolation argument:

$$\|u - u_N\|_{H^r(\Omega_d)} \leqslant c N^{r-s} \|u\|_{H^s(\Omega_d)}. \tag{10.14}$$

11. The eigenvalue Dirichlet problem

Let us consider the eigenvalue problem with Dirichlet boundary conditions on the interval Λ: *find a complex eigenvalue* λ *and a function* φ *in* $H^1_0(\Lambda)$, $\varphi \neq 0$, *such that*

$$\forall \psi \in H^1_0(\Lambda), \quad \int_{-1}^{1} \varphi'(\zeta)\psi'(\zeta)\, \mathrm{d}\zeta = \lambda \int_{-1}^{1} \varphi(\zeta)\psi(\zeta)\, \mathrm{d}\zeta. \tag{11.1}$$

Since the bilinear form on the left-hand side is symmetric and elliptic, it can be observed that the eigenvalues λ are necessarily positive real numbers. More precisely, since equation (11.1) is equivalent to the following differential equation

$$-\varphi'' = \lambda \varphi \quad \text{in } \Lambda,$$

provided with the boundary conditions $\varphi(-1) = \varphi(1) = 0$, it is readily checked that the eigenvalues are $\lambda^k = k^2\pi^2/4$, where k is any positive integer, and that the corresponding eigenfunctions are, e.g.,

$$\varphi^k(\zeta) = \begin{cases} \sin\left(\frac{k\pi\zeta}{2}\right) & \text{if } k \text{ is even,} \\ \cos\left(\frac{k\pi\zeta}{2}\right) & \text{if } k \text{ is odd.} \end{cases} \tag{11.2}$$

Next, we introduce the Galerkin approximation of problem (11.1): *find a complex eigenvalue* λ_N *and a function* φ_N *in* $\mathbb{P}^0_N(\Lambda)$, $\varphi_N \neq 0$, *such that*

$$\forall \psi_N \in \mathbb{P}^0_N(\Lambda), \quad \int_{-1}^{1} \varphi_N'(\zeta)\psi_N'(\zeta)\, \mathrm{d}\zeta = \lambda_N \int_{-1}^{1} \varphi_N(\zeta)\psi_N(\zeta)\, \mathrm{d}\zeta. \tag{11.3}$$

Still using the properties of the bilinear form in the left-hand side, we see that this problem has $N-1$ real positive eigenvalues, which we denote by λ^k_N, $1 \leqslant k \leqslant N-1$, in increasing order, and that there exists an orthonormal basis $\{\varphi^1_N, \ldots, \varphi^{N-1}_N\}$ of

$\mathbb{P}_N^0(\Lambda)$ made of eigenfunctions φ_N^k associated with each λ_N^k. The essential question is now to compare the λ_N^k with the λ^k.

REMARK 11.1. In the case of the hypercube Ω_d, the eigenvalues of the problem: *find a complex number λ and a function u in $H_0^1(\Omega_d)$ such that*

$$\forall v \in H_0^1(\Omega_d), \quad a(u,v) = \lambda \int_{\Omega_d} u(\boldsymbol{x}) v(\boldsymbol{x}) \, \mathrm{d}\boldsymbol{x}, \tag{11.4}$$

are $(k_1^2 + \cdots + k_d^2)\pi^2/4$, where $k_1, \ldots$ and k_d are positive integers, and the corresponding eigenfunctions are, e.g., $\varphi^{k_1}(x_1) \cdots \varphi^{k_d}(x_d)$. Similarly, due to the tensorization properties of the space of polynomials, the eigenvalues of the discrete problem: *find a complex number λ and a polynomial u_N in $\mathbb{P}_N^0(\Omega_d)$ such that*

$$\forall v_N \in \mathbb{P}_N^0(\Omega_d), \quad a(u_N, v_N) = \lambda \int_{\Omega_d} u_N(\boldsymbol{x}) v_N(\boldsymbol{x}) \, \mathrm{d}\boldsymbol{x}, \tag{11.5}$$

are all the sums of d eigenvalues λ_N^k, say $\lambda_N^{k_1} + \cdots + \lambda_N^{k_d}$, $1 \leqslant k_1, \ldots, k_d \leqslant N-1$, and the corresponding eigenfunctions are, e.g., $\varphi_N^{k_1}(x_1) \cdots \varphi_N^{k_d}(x_d)$. Consequently, we only present the analysis of the one-dimensional problem (11.3), without any loss of generality.

Note that each polynomial φ_N in $\mathbb{P}_N^0(\Lambda)$ can be written in the form

$$\varphi_N = \sum_{k=1}^{N-1} \alpha_k \varphi_N^k, \tag{11.6}$$

with

$$\|\varphi_N\|_{L^2(\Lambda)} = \left(\sum_{k=1}^{N-1} \alpha_k^2 \right)^{1/2} \quad \text{and} \quad |\varphi_N|_{H^1(\Lambda)} = \left(\sum_{k=1}^{N-1} \lambda_N^k \alpha_k^2 \right)^{1/2}. \tag{11.7}$$

Due to the inverse inequality (5.2) which is proven to be optimal in (5.4) for polynomials in $\mathbb{P}_N(\Lambda)$ (and is still optimal even if restricted to polynomials of $\mathbb{P}_N^0(\Lambda)$), at least the largest eigenvalue λ_N^{N-1} must be larger than cN^4, hence it cannot be a good approximation of $\lambda^{N-1} = (N-1)^2\pi^2/4$.

The behavior of the eigenvalues is completely described in the following theorem which is due to VANDEVEN [1990, 1991], we only present the basic idea of the proof.

THEOREM 11.1. *The eigenvalues λ_N^k, $1 \leqslant k \leqslant N-1$, of problem (11.3) satisfy:*

(i) *There exists a function L on $[0,1[$ such that, for any sequence $(k_N)_N$ of integers, $1 \leqslant k_N \leqslant N-1$, the following result holds:*

$$\lim_{N \to +\infty} \frac{k_N}{N} = \gamma < 1 \Rightarrow \lim_{N \to +\infty} \frac{\lambda_N^{k_N}}{\lambda^{k_N}} = L(\gamma). \tag{11.8}$$

The function L is equal to 1 *on* $[0, 2/\pi]$ *and is increasing on* $[2/\pi, 1[$. *Moreover the following error estimate holds for the eigenvalues* λ_N^k, $1 \leqslant k \leqslant \gamma N$ *with* $0 \leqslant \gamma < 2/\pi$:

$$|\lambda^k - \lambda_N^k| \leqslant c\rho^{N^{2/3}}, \tag{11.9}$$

where the positive constants c and $\rho < 1$ *only depend on* γ.

(ii) *For any sequence* $(k_N)_N$ *of integers,* $1 \leqslant k_N \leqslant N-1$, *the following result holds:*

$$\lim_{N\to+\infty} \frac{k_N}{N} = 1 \Rightarrow \lim_{N\to+\infty} \frac{\lambda_N^{k_N}}{\lambda^{k_N}} = +\infty. \tag{11.10}$$

The general idea of the proof is the following. Problem (11.3) can equivalently be written

$$\forall \psi_N \in \mathbb{P}_N^0(\Lambda), \quad \int_{-1}^{1} (\varphi_N'' + \lambda_N \varphi_N)(\zeta)\psi_N(\zeta)\,\mathrm{d}\zeta = 0. \tag{11.11}$$

Since it is readily checked from the differential equation (3.1) that the orthogonal of $\mathbb{P}_N^0(\Lambda)$ (which is of dimension $N-1$) in $\mathbb{P}_N(\Lambda)$ (which is of dimension $N+1$) is spanned by the polynomials L_N' and L_{N+1}', λ_N is an eigenvalue of problem (11.3) if and only if there exists φ_N in $\mathbb{P}_N^0(\Lambda)$ (not equal to 0) such that $\varphi_N'' + \lambda_N\varphi_N$ is a linear combination of L_N' and L_{N+1}'. Now, it is readily checked that, among the eigenfunctions φ_N, $N/2$ are even and $N/2-1$ are odd if N is even ($(N-1)/2$ are even and $(N-1)/2$ are odd if N is odd). For the sake of brevity, we only consider the case when N is even, since the study of the other case is completely similar. Then, λ_N is an eigenvalue of problem (11.3), corresponding to odd eigenfunctions φ_N, if one of these satisfies

$$\varphi_N'' + \lambda_N \varphi_N = L_N'. \tag{11.12}$$

Inverting the operator $\mathbb{I} + (1/\lambda_N)(\mathrm{d}^2/\mathrm{d}\zeta^2)$ in $\mathbb{P}_N(\Lambda)$, we note that this equation is equivalent to

$$\varphi_N = \sum_{k=0}^{N/2-1} \frac{(-1)^k}{\lambda_N^{k+1}} \frac{\mathrm{d}^{2k+1} L_N}{\mathrm{d}\zeta^{2k+1}}.$$

Equivalently, λ_N is an eigenvalue of problem (11.3), corresponding to odd eigenfunctions, if and only if it is a root of the polynomial with respect to $1/\lambda$:

$$A_N(\lambda) = \sum_{k=0}^{N/2-1} \frac{(-1)^k}{\lambda^{k+1}} \left(\frac{\mathrm{d}^{2k+1} L_N}{\mathrm{d}\zeta^{2k+1}}\right)(1). \tag{11.13}$$

Next, setting

$$B_N(\lambda) = \sum_{k=0}^{N/2} \frac{(-1)^k}{\lambda^{k+1/2}} \left(\frac{\mathrm{d}^{2k} L_N}{\mathrm{d}\zeta^{2k}} \right)(1), \tag{11.14}$$

and integrating by parts, we derive the identity

$$\int_{-1}^{1} L_N(\zeta) \cos\left(\sqrt{\lambda}\zeta\right) \mathrm{d}\zeta = 2A_N(\lambda) \cos\sqrt{\lambda} + 2B_N(\lambda) \sin\sqrt{\lambda}$$

(the left-hand side is equal to the real part of $2(-1)^{N/2} j_N(\sqrt{\lambda})$, where $j_N(z)$ stands for the spherical Bessel function of the first kind, see for instance ABRAMOWITZ and STEGUN ([1970], formula (10.1.14))). This formula also reads

$$\frac{A_N(\lambda)}{B_N(\lambda)} = -\tan\sqrt{\lambda} + \frac{1}{2B_N(\lambda)\cos\sqrt{\lambda}} \int_{-1}^{1} L_N(\zeta) \cos\left(\sqrt{\lambda}\zeta\right) \mathrm{d}\zeta.$$

As an easy consequence of the Rodrigues formula (3.9), we deduce the following property, when λ tends to 0:

$$\left| \frac{A_N(\lambda)}{B_N(\lambda)} + \tan\sqrt{\lambda} \right| \leqslant c\left(\sqrt{\lambda}\right)^{N+1/2}. \tag{11.15}$$

So, the quantity $A_N(\lambda)/B_N(\lambda)$ is nothing but the $[N-1, N]$ Padé approximant of the function $-\tan\sqrt{\lambda}$ (see BREZINSKI [1980] for instance). Since half of the eigenvalues of problem (11.1), those which correspond to odd eigenfunctions, are the roots of the function $\tan\sqrt{\lambda}$, all the results of Theorem 11.1 are derived from the properties of the fraction A_N/B_N.

For example, setting $\lambda = \omega^2$, we define the rational fraction

$$F_N(\omega) = \frac{A_N(\omega^2)}{B_N(\omega^2)}. \tag{11.16}$$

A rather technical computation relying on the properties of the spherical Bessel functions, or equivalently of the Legendre polynomials, leads to the following formula:

$$\frac{F_N'(\omega)}{1 + F_N^2(\omega)} = -1 + \frac{1}{\varepsilon_N(\omega)},$$

$$\text{with } \varepsilon_N(\omega) = 1 + \sum_{m=1}^{N} \frac{(1 \times 3 \times (2m-1))^2}{(2m)!} \frac{\alpha_N^m}{\omega^{2m}}, \tag{11.17}$$

where the constants α_N^m have been introduced in (3.20). It is readily checked from (11.17) that the function F_N is decreasing, so that its zeros are alternate with its poles. By setting $\omega_N^k = (\lambda_N^k)^{1/2}$, we also derive that

$$\omega_N^k - \frac{k\pi}{2} = \int_0^{\omega_N^k} \frac{d\omega}{\varepsilon_N(\omega)}. \tag{11.18}$$

Now, let γ be a real number, $0 \leqslant \gamma < 1$, and let $(k_N)_N$ be a sequence of integers between 1 and $N-1$ such that the sequence $(k_N/N)_N$ tends to γ. There exists a subsequence, which we still denote by $(k_N)_N$, such that the sequence $(\omega_N^{k_N}/N)_N$ tends to a limit Z. Next, since $\gamma < 1$, it can be checked that this limit is $< +\infty$. Passing to the limit in the formula

$$\frac{1}{N}\left(\omega_N^{k_N} - k_N\frac{\pi}{2}\right) = \int_0^{\omega_N^{k_N}/N} \frac{dt}{\varepsilon_N(Nt)},$$

we obtain

$$Z - \gamma\frac{\pi}{2} = \int_0^Z \frac{dt}{\varepsilon(t)},$$
$$\text{with } \varepsilon(t) = 1 + \sum_{m=1}^{+\infty} \frac{(1\times 3\times(2m-1))^2}{(2m)!}\frac{1}{t^{2m}}. \tag{11.19}$$

Hence, Z is a fixed point of the mapping:

$$z \mapsto \gamma\frac{\pi}{2} + \int_0^z \frac{dt}{\varepsilon(t)}.$$

Clearly, this mapping is a contraction of $\mathbb{R}$, so that it has a unique fixed point. As a consequence, the whole sequence $(\omega_N^{k_N}/N)_N$ converges to Z. Moreover, since the radius of the convergence of the series in (11.19) is equal to 1, Z is equal to $\gamma\pi/2$ if and only if $Z \leqslant 1$, i.e., if and only if $\gamma \leqslant 2/\pi$. Also, an estimate of the function $\varepsilon_N(Nt) - \varepsilon(t)$ leads to (11.9). This gives the first part of the theorem.

As a conclusion, only the first $2N/\pi$ discrete eigenvalues converge to the corresponding eigenvalues of the continuous problem. However, their convergence is exponential, and the same type of convergence can easily be derived for appropriate corresponding eigenfunctions. The exact behavior of the largest eigenvalues is given by VANDEVEN [1990, 1991], for instance the last ones satisfy

$$\lim_{N\to+\infty} \frac{\lambda_N^{N-1}}{N^4} = \lim_{N\to+\infty} \frac{\lambda_N^{N-2}}{N^4} = \frac{1}{4\pi^2},$$
$$\lim_{N\to+\infty} \frac{\lambda_N^{N-3}}{N^4} = \lim_{N\to+\infty} \frac{\lambda_N^{N-4}}{N^4} = \frac{1}{16\pi^2}, \quad \ldots \tag{11.20}$$

REMARK 11.2. The eigenvalues of the Neumann problem: *find a complex eigenvalue* λ *and a function* φ *in* $H^1(\Lambda)$ *such that*

$$\forall \psi \in H^1(\Lambda), \quad \int_{-1}^{1} \varphi'(\zeta)\psi'(\zeta)\,\mathrm{d}\zeta = \lambda \int_{-1}^{1} \varphi(\zeta)\psi(\zeta)\,\mathrm{d}\zeta, \tag{11.21}$$

are the same as those of the Dirichlet problem. A similar discretization can be used, and the results of Theorem 11.1 still hold for the discrete eigenvalues (indeed, if N is even, the eigenvalues associated with even eigenfunctions are exactly the roots of the function $B_N(\lambda)$ introduced in (11.14)).

Knowing the asymptotic behavior of the largest eigenvalue allows for an improvement of the inverse inequality constant. Recalling inequality (5.2), we derive an asymptotic limit for the best constant. Indeed, it follows from (11.20) that

$$\lim_{N \to +\infty} \sup_{\varphi_N \in \mathbb{P}_N^0(\Lambda)} \frac{|\varphi_N|_{H^1(\Lambda)}}{N^2 \|\varphi_N\|_{L^2(\Lambda)}} = \frac{1}{2\pi}, \tag{11.22}$$

however the similar result on the whole space $\mathbb{P}_N(\Lambda)$ involves the eigenvalues of the discrete Neumann problem, it reads:

$$\lim_{N \to +\infty} \sup_{\varphi_N \in \mathbb{P}_N(\Lambda)} \frac{|\varphi_N|_{H^1(\Lambda)}}{N^2 \|\varphi_N\|_{L^2(\Lambda)}} = \frac{1}{\pi}. \tag{11.23}$$

These asymptotic estimates have a number of applications when implementing spectral methods. Among them, let us quote:

(i) computing the exact condition number of the stiffness matrix derived from a spectral discretization,

(ii) computing the optimal constant in the Courant–Friedrichs–Lévy condition when an unsteady equation is discretized by a spectral technique for the space variables and an explicit or semi-explicit scheme with respect to the time variable.

12. Polynomial approximation in the half-line

A number of physical problems are set in unbounded domains, for instance in exterior domains. The first remark with spectral methods on these problems, is that polynomials are not integrable on unbounded domains. So, the idea is to work with weighted Sobolev spaces, where the weight must be of exponential type in order to avoid restrictions on the degree of the polynomials.

In one dimension, there are two kinds of unbounded connected domains: the whole line $\mathbb{R}$ and half-lines, for example the set $\mathbb{R}_+$ of positive real numbers. The families of orthogonal polynomials on these domains and for appropriate weighted measures are well known: the family of *Hermite polynomials* on $\mathbb{R}$ for the measure $\mathrm{e}^{-\zeta^2/4}\,\mathrm{d}\zeta$, the family of *Laguerre polynomials* on $\mathbb{R}_+$ for the measure $\mathrm{e}^{-\zeta}\,\mathrm{d}\zeta$ (families of orthogonal polynomials for the measure $e^{-\zeta^2/4\alpha}\,d\zeta$ or $\zeta^\alpha\,e^{-\zeta}\,d\zeta$ also exist, however we limit ourselves to the basic case).

Both Hermite and Laguerre type spectral discretizations are used, we refer for instance to AHJAOU [1994], COULAUD, FUNARO and KAVIAN [1990], FUNARO and KAVIAN [1990], MADAY, PERNAUD-THOMAS and VANDEVEN [1985] and MAVRIPLIS [1989]. Here, we only present the Laguerre spectral methods, for the sake of brevity and also because we have more applications in this case. Indeed, the results which follow can easily be extended to tensorized domains of the type $\mathbb{R}_+^d$ or $\Lambda^k \times \mathbb{R}_+^{d-k}$, $1 \leqslant k \leqslant d-1$, which allows for discretizations in a large number of exterior domains by the spectral element method (see BERNARDI, MADAY and PATERA, to appear).

We begin with the weighted Sobolev spaces which are needed for spectral approximations. We first introduce the space

$$L^2_\diamond(\mathbb{R}_+) = \left\{\varphi\colon \mathbb{R}_+ \to \mathbb{R} \text{ measurable: } \int_0^{+\infty} v^2(\zeta)\,e^{-\zeta}\,d\zeta < +\infty\right\}, \tag{12.1}$$

provided with the norm

$$\|\varphi\|_{L^2_\diamond(\mathbb{R}_+)} = \left(\int_0^{+\infty} v^2(\zeta)\,e^{-\zeta}\,d\zeta\right)^{1/2}. \tag{12.2}$$

Next, for any integer $m \geqslant 0$, we define $H^m_\diamond(\mathbb{R}_+)$ as the space of functions in $L^2_\diamond(\mathbb{R}_+)$ such that all their derivatives up to order m belong to $L^2_\diamond(\mathbb{R}_+)$. It is provided with the norm

$$\|\varphi\|_{H^m_\diamond(\mathbb{R}_+)} = \left(\sum_{k=0}^{m}\int_0^{+\infty} \left(\frac{d^k\varphi}{d\zeta^k}\right)^2(\zeta)\,e^{-\zeta}\,d\zeta\right)^{1/2}. \tag{12.3}$$

For any positive real number s, we define $H^s_\diamond(\mathbb{R}_+)$ as the interpolate space between $H^m_\diamond(\mathbb{R}_+)$ and $L^2_\diamond(\mathbb{R}_+)$ with index $1 - s/m$, where m is any integer $\geqslant s$.

It can be noted that, for any bounded interval I contained in $\mathbb{R}_+$, the restrictions to I of all functions in $H^s_\diamond(\mathbb{R}_+)$ belong to $H^s(I)$. So, we are not interested in the behavior of these functions far from infinity, for instance with their traces in 0. Other properties of the spaces $H^s_\diamond(\mathbb{R}_+)$ are easy to derive, we quote the essential ones:

(i) for any $s \geqslant 0$, the space $\mathscr{C}^\infty(\overline{\mathbb{R}}_+)$ of infinitely differentiable functions in $\overline{\mathbb{R}}_+$ is dense in $H^s_\diamond(\mathbb{R}_+)$;

(ii) for any $s \geqslant 0$, the mapping: $\varphi \mapsto \varphi\,e^{-\zeta/2}$ is an isomorphism from $H^s_\diamond(\mathbb{R}_+)$ onto $H^s(\mathbb{R}_+)$;

(iii) for any real numbers r and s, $0 \leqslant s < r$, and for any θ, $0 < \theta < 1$, the following interpolation identity holds

$$\left[H^r_\diamond(\mathbb{R}_+), H^s_\diamond(\mathbb{R}_+)\right]_\theta = H^{(1-\theta)r+\theta s}_\diamond(\mathbb{R}_+). \tag{12.4}$$

(iv) for any $s > 1/2$, the space $H^s_\diamond(\mathbb{R}_+)$ is contained in the space of continuous functions in $\overline{\mathbb{R}}_+$ and the following inequality holds for any function φ in $H^s(\mathbb{R}_+)$:

$$\sup_{\zeta \geqslant 0} \left|\varphi(\zeta)\,\mathrm{e}^{-\zeta/2}\right| \leqslant c\|\varphi\|_{H^s_\diamond(\mathbb{R}_+)}. \tag{12.5}$$

For technical reasons, we also need another scale of Sobolev spaces, associated with a positive integer α:

$$H^s_{\diamond\alpha}(\mathbb{R}_+) = \{\varphi \in H^s_\diamond(\mathbb{R}_+)\colon\ \varphi\zeta^{\alpha/2} \in H^s_\diamond(\mathbb{R}_+)\}. \tag{12.6}$$

These spaces are provided with the associated natural norm $\|(1+\zeta)^{\alpha/2}\varphi\|_{H^s_\diamond(\mathbb{R}_+)}$.

Let us now introduce the family of Laguerre polynomials: for each $n \geqslant 0$, the polynomial $\mathcal{L}_n$ has degree n, is orthogonal to any other polynomial $\mathcal{L}_m$, $m \neq n$, in $L^2_\diamond(\mathbb{R}_+)$ and satisfies $\mathcal{L}_n(0) = 1$. The properties of Laguerre polynomials are very similar to those of Legendre polynomials, and they are derived by similar arguments. Indeed, they satisfy the differential equation

$$\frac{\mathrm{d}}{\mathrm{d}\zeta}\left(\zeta \mathrm{e}^{-\zeta}\mathcal{L}_n'\right) + n\,\mathrm{e}^{-\zeta}\mathcal{L}_n = 0, \quad n \geqslant 0. \tag{12.7}$$

They also satisfy

$$\int_0^{+\infty} \mathcal{L}_n^2(\zeta)\,\mathrm{e}^{-\zeta}\,\mathrm{d}\zeta = 1, \quad n \geqslant 0, \tag{12.8}$$

and

$$\mathcal{L}_n = \mathcal{L}_n' - \mathcal{L}_{n+1}', \quad n \geqslant 0. \tag{12.9}$$

The induction formula reads

$$\begin{cases} \mathcal{L}_0(\zeta) = 1 \quad \text{and} \quad \mathcal{L}_1(\zeta) = 1 - \zeta, \\ (n+1)\mathcal{L}_{n+1}(\zeta) = (2n+1-\zeta)\mathcal{L}_n(\zeta) - (n-1)\mathcal{L}_{n-1}(\zeta), \quad n \geqslant 1. \end{cases} \tag{12.10}$$

The Sturm–Liouville type operator which appears in formula (12.7) is defined by

$$A_\diamond\varphi = -\mathrm{e}^{\zeta}\frac{\mathrm{d}}{\mathrm{d}\zeta}\left(\zeta\,\mathrm{e}^{-\zeta}\varphi'\right). \tag{12.11}$$

This can also be written as

$$A_\diamond\varphi = -\zeta\varphi'' - (1-\zeta)\varphi', \tag{12.12}$$

whence the following statement.

LEMMA. *For any nonnegative integer l and for any nonnegative real number α, the operator $A_\diamond$ is continuous from $H^{l+2}_{\diamond\alpha+2}(\mathbb{R}_+)$ into $H^l_{\diamond\alpha}(\mathbb{R}_+)$.*

From this lemma, it is readily checked that, for any positive integer m, the operator $A_\diamond^m$ is continuous, for instance, from $H^{2m}_{\diamond 2m}(\mathbb{R}_+)$ into $L^2_\diamond(\mathbb{R}_+)$. This is not completely analogous to the property of the operator A defined in (3.2) since the operator $A_\diamond$ does not act between Sobolev spaces with the same weight. However, it is sufficient to prove the basic approximation result.

NOTATION. For each integer $n \geqslant 0$, let $\mathbb{P}_n(\mathbb{R}_+)$ be the space of restrictions to $\mathbb{R}_+$ of polynomials with one variable and degree $\leqslant n$. For a fixed positive integer N, let $\pi_N^\diamond$ be the orthogonal projection operator from $L^2_\diamond(\mathbb{R}_+)$ onto $\mathbb{P}_N(\mathbb{R}_+)$.

The following result is due to MADAY, PERNAUD-THOMAS and VANDEVEN [1985], see also FUNARO [1991].

THEOREM 12.1. *For any nonnegative real number s, there exists a positive constant c depending only on s such that, for any function φ in $H^s_{\diamond k}(\mathbb{R}_+)$, the following estimate holds*

$$\|\varphi - \pi_N^\diamond \varphi\|_{L^2_\diamond(\mathbb{R}_+)} \leqslant c N^{-s/2} \|\varphi\|_{H^s_{\diamond k}(\mathbb{R}_+)}, \tag{12.13}$$

where k is the largest integer $< s + 1$.

PROOF. Any function φ in $L^2_\diamond(\mathbb{R}_+)$ can be written as

$$\varphi = \sum_{n=0}^{+\infty} \varphi^{\diamond n} \mathcal{L}_n,$$

and we have

$$\|\varphi - \pi_N^\diamond \varphi\|^2_{L^2_\diamond(\mathbb{R}_+)} = \sum_{n=N+1}^{+\infty} (\varphi^{\diamond n})^2. \tag{12.14}$$

(1) When $s = k$ is an even integer, we compute from (12.7)

$$\begin{aligned}
\varphi^{\diamond n} &= \int_0^{+\infty} \varphi(\zeta) \mathcal{L}_n(\zeta)\, \mathrm{e}^{-\zeta}\, \mathrm{d}\zeta \\
&= (-1)^{k/2} n^{-k/2} \int_0^{+\infty} (A_\diamond^{k/2} \varphi)(\zeta) \mathcal{L}_n(\zeta)\, \mathrm{e}^{-\zeta}\, \mathrm{d}\zeta,
\end{aligned}$$

whence

$$\begin{aligned}
\|\varphi - \pi_N^\diamond \varphi\|^2_{L^2_\diamond(\mathbb{R}_+)} &\leqslant N^{-k} \sum_{n=N+1}^{+\infty} \left(\int_0^{+\infty} (A_\diamond^{k/2} \varphi)(\zeta) \mathcal{L}_n(\zeta)\, \mathrm{e}^{-\zeta}\, \mathrm{d}\zeta \right)^2 \\
&\leqslant N^{-k} \|A_\diamond^{k/2} \varphi\|^2_{L^2_\diamond(\mathbb{R}_+)}.
\end{aligned}$$

Thus, the desired result follows from the lemma.

(2) When $s = k$ is an odd integer, we write

$$\begin{aligned} \varphi^{\diamond n} &= (-1)^{(k-1)/2} n^{-(k-1)/2} \int_0^{+\infty} \big(A_\diamond^{(k-1)/2}\varphi\big)(\zeta)\mathcal{L}_n(\zeta)\,\mathrm{e}^{-\zeta}\,\mathrm{d}\zeta \\ &= (-1)^{(k-1)/2} n^{-(k+1)/2} \int_0^{+\infty} \big(A_\diamond^{(k-1)/2}\varphi\big)'(\zeta)\mathcal{L}_n'(\zeta)\zeta\,\mathrm{e}^{-\zeta}\,\mathrm{d}\zeta. \end{aligned}$$

Next, from equation (12.7), we observe that the polynomials $\mathcal{L}_n'$ are orthogonal to each other for the measure $\zeta\,\mathrm{e}^{-\zeta}\,\mathrm{d}\zeta$ and satisfy

$$\int_0^{+\infty} \mathcal{L}_n'^2(\zeta)\zeta\,\mathrm{e}^{-\zeta}\,\mathrm{d}\zeta = n.$$

Hence, we deduce

$$\begin{aligned} \|\varphi - \pi_N^\diamond \varphi\|^2_{L^2_\diamond(\mathbb{R}_+)} &\leqslant N^{-k} \sum_{n=N+1}^{\infty} \frac{\big(\int_0^{+\infty} (A_\diamond^{(k-1)/2}\varphi)'(\zeta)\mathcal{L}_n'(\zeta)\zeta\,\mathrm{e}^{-\zeta}\,\mathrm{d}\zeta\big)^2}{\int_0^{+\infty} \mathcal{L}_n'^2(\zeta)\zeta\,\mathrm{e}^{-\zeta}\,\mathrm{d}\zeta} \\ &\leqslant N^{-k} \big\| \big(A_\diamond^{(k-1)/2}\varphi\big)'\zeta^{1/2}\big\|_{L^2_\diamond(\mathbb{R}_+)}. \end{aligned}$$

Here also, applying the lemma gives the desired result.

(3) The general result follows by an interpolation argument between the spaces $H^{k+1}_{\diamond k+1}(\mathbb{R}_+)$ and $H^k_{\diamond k+1}(\mathbb{R}_+)$ (which contains $H^k_{\diamond k}(\mathbb{R}_+)$). Indeed, the interpolation properties of these spaces are readily checked from (12.4).

REMARK 12.1. The complete result would be that, for any nonnegative integer s and for any function φ in $H^s_{\diamond s}(\mathbb{R}_+)$,

$$\|\varphi - \pi_N^\diamond \varphi\|_{L^2_\diamond(\mathbb{R}_+)} \leqslant cN^{-s/2}\|\varphi\|_{H^s_{\diamond s}(\mathbb{R}_+)}.$$

However, we do not presently know the appropriate interpolation result on the scale of spaces $H^s_{\diamond\alpha}(\mathbb{R}_+)$.

Let us now prove a basic inverse inequality, which is better than the corresponding inequality (5.2) on Λ.

THEOREM 12.2. *The following inequality holds for any positive integer N and for any polynomial φ_N in $\mathbb{P}_N(\mathbb{R}_+)$:*

$$\|\varphi_N'\|_{L^2_\diamond(\mathbb{R}_+)} \leqslant N\|\varphi_N\|_{L^2_\diamond(\mathbb{R}_+)}. \tag{12.15}$$

PROOF. Writing $\varphi_N = \sum_{n=0}^{N} \varphi^{\diamond n} \mathcal{L}_n$, we derive from formula (12.9) that

$$\varphi_N' = \sum_{n=0}^{N} \varphi^{\diamond n} \mathcal{L}_n' = -\sum_{n=1}^{N} \varphi^{\diamond n} \left(\sum_{k=0}^{n-1} \mathcal{L}_k \right) = -\sum_{k=0}^{N-1} \left(\sum_{n=k+1}^{N} \varphi^{\diamond n} \right) \mathcal{L}_k,$$

so that

$$\|\varphi_N'\|^2_{L^2_\diamond(\mathbb{R}_+)} = \sum_{k=0}^{N-1} \left(\sum_{n=k+1}^{N} \varphi^{\diamond n} \right)^2 \leqslant \sum_{n=0}^{N} (\varphi^{\diamond n})^2 \sum_{k=0}^{N-1} (N-k).$$

This gives the result.

Using the inverse inequality (12.15), we can prove a more general approximation result.

THEOREM 12.3. *For any nonnegative real numbers r and s, $r \leqslant s$, there exists a positive constant c depending only on s such that, for any function φ in $H^s_{\diamond k}(\mathbb{R}_+)$, the following estimate holds*

$$\|\varphi - \pi_N^\diamond \varphi\|_{H^r_\diamond(\mathbb{R}_+)} \leqslant c N^{r-s/2} \|\varphi\|_{H^s_{\diamond k}(\mathbb{R}_+)}, \tag{12.16}$$

where k is the largest integer $< s+1$.

PROOF. We first study the case where r is an integer and we are going to prove (12.16) by induction. Since it holds for $r = 0$ by Theorem 12.1, we assume that it holds for r and we write

$$\begin{aligned} \|\varphi - \pi_N^\diamond \varphi\|_{H^{r+1}_\diamond(\mathbb{R}_+)} &\leqslant \|\varphi - \pi_N^\diamond \varphi\|_{L^2_\diamond(\mathbb{R}_+)} + \|\varphi' - \pi_N^\diamond(\varphi')\|_{H^r_\diamond(\mathbb{R}_+)} \\ &\quad + \|\pi_N^\diamond(\varphi') - (\pi_N^\diamond \varphi)'\|_{H^r_\diamond(\mathbb{R}_+)}. \end{aligned} \tag{12.17}$$

Let us consider the last term. Writing

$$\varphi = \sum_{n=0}^{+\infty} \varphi^{\diamond n} \mathcal{L}_n,$$

we observe from (12.9) that

$$\pi_N^\diamond(\varphi') = -\sum_{k=0}^{N} \left(\sum_{n=k+1}^{+\infty} \varphi^{\diamond n} \right) \mathcal{L}_k \quad \text{and} \quad (\pi_N^\diamond \varphi)' = -\sum_{k=0}^{N-1} \left(\sum_{n=k+1}^{N} \varphi^{\diamond n} \right) \mathcal{L}_k,$$

so that

$$\pi_N^\diamond(\varphi') - (\pi_N^\diamond \varphi)' = -\left(\sum_{k=0}^{N} \mathcal{L}_k \right) \left(\sum_{n=N+1}^{+\infty} \varphi^{\diamond n} \right) = \mathcal{L}_{N+1}' \left(\sum_{n=N+1}^{+\infty} \varphi^{\diamond n} \right).$$

As in the proof of Theorem 12.1, we observe that, for $n > N$,

$$\varphi^{\diamond n} = \begin{cases} (-1)^{k/2} n^{-k/2} \int_0^{+\infty} (A_\diamond^{k/2} \varphi)(\zeta) \mathcal{L}_n(\zeta)\, \mathrm{e}^{-\zeta}\, \mathrm{d}\zeta \\ \quad \text{if } s = k \text{ is even,} \\ (-1)^{(k-1)/2} n^{-(k+1)/2} \int_0^{+\infty} (A_\diamond^{(k-1)/2} \varphi)'(\zeta) \mathcal{L}_n'(\zeta) \zeta\, \mathrm{e}^{-\zeta}\, \mathrm{d}\zeta \\ \quad \text{if } s = k \text{ is odd.} \end{cases} \tag{12.18}$$

Moreover, using r times inequality (12.15), we obtain

$$\|\mathcal{L}_{N+1}'\|_{H_\diamond^r(\mathbb{R}_+)} \leqslant N^r \|\mathcal{L}_{N+1}'\|_{L_\diamond^2(\mathbb{R}_+)} \leqslant N^{r+1/2}.$$

Finally, we obtain, from (12.18) if s is an integer $\geqslant r$ and by an interpolation argument if not,

$$\|\pi_N^\diamond(\varphi') - (\pi_N^\diamond \varphi)'\|_{H_\diamond^r(\mathbb{R}_+)} \leqslant cN^{r+1/2-s/2} \|\varphi\|_{H_{\diamond k+1}^{s+1}(\mathbb{R}_+)}.$$

Combining this line with (12.17) and Theorem 12.1 applied to φ and φ', proves (12.16) with r replaced by $r+1$ and s replaced by $s+1$, hence for all integral values of r.

(2) When r is not an integer, the result is proven by an interpolation argument between the two inequalities

$$\|\varphi - \pi_N^\diamond \varphi\|_{H_\diamond^{[r]+1}(\mathbb{R}_+)} \leqslant cN^{[r]+1-([s]+1)/2} \|\varphi\|_{H_{\diamond k+1}^{[s]+1}(\mathbb{R}_+)},$$

$$\|\varphi - \pi_N^\diamond \varphi\|_{H_\diamond^{[r]}(\mathbb{R}_+)} \leqslant cN^{[r]-s/2} \|\varphi\|_{H_{\diamond k+1}^{s}(\mathbb{R}_+)}.$$

Of course, estimate (12.13) is the basis of the numerical analysis of spectral methods in unbounded domains, and more general projection operators can be defined to obtain optimal estimates in higher order norms. Since most often spectral elements are used to handle unbounded geometries, we refer to BERNARDI, MADAY and PATERA, to appear, for further results. Only preliminary numerical experiments are available concerning these techniques.

REMARK 12.2. Another type of spectral methods exists, relying on the approximation by functions $\varphi_N\, \mathrm{e}^{-\zeta}$, where φ_N varies in the space of polynomials. Then, the numerical analysis involves the scale of weighted spaces associated with

$$L_\clubsuit^2(\mathbb{R}_+) = \left\{ \varphi\colon \mathbb{R}_+ \to \mathbb{R} \text{ measurable: } \int_0^{+\infty} v^2(\zeta)\, \mathrm{e}^{\zeta}\, \mathrm{d}\zeta < +\infty \right\}. \tag{12.19}$$

The analogue for Hermite polynomials also exists, and the corresponding approximation functions, known as Hermite functions, are often involved in physical problems (see AHJAOU [1994] and FUNARO and KAVIAN [1990]). The choice between the two types of methods depends on the behavior at infinity which is enforced on the solution of the exact problem.

CHAPTER III

Polynomial Interpolation Error, the Collocation Method

From the previous chapter, we derive the conclusion that the Galerkin method for model elliptic differential equations leads to optimal error estimates. However, it cannot be implemented in most cases, since the computation of the right-hand side terms is not easy. The idea is now to replace all the integrals which are involved in the Galerkin discrete problem by appropriate quadrature formulas. To preserve the ellipticity of the problem, very accurate formulas should be used, namely the Gauss type formulas which were described in Section 4. As will appear later on, the precise choice of the formula, which can be a Gauss formula, a Gauss–Lobatto formula or a generalized Gauss type formula, essentially depends on the boundary conditions which are enforced on the exact solution. In most cases, the corresponding discrete problem is equivalent to a system of collocation equations, which allows for an easy extension of the technique to problems with nonconstant coefficients or even to nonlinear problems.

In these methods, the right-hand side is only involved by its values at the nodes of the formula. So, the first step in their numerical analysis is to derive estimates of the distance of a function with a given regularity to its Lagrange interpolate at the nodes. This will be performed in Section 13 for the interpolation on the interval and in Section 14 for the interpolation on the hypercube. Next, the methods with numerical integration are studied in Sections 15 to 17. The first two ones lead to a system of collocation equations, they are concerned with the discretization of the second-order and fourth-order Dirichlet problems respectively. The third one also deals with the fourth-order Dirichlet problem, but it is no longer a collocation method. And the last one leads to a discretization of the second-order Neumann problem in which the boundary conditions are not enforced by collocation equations.

NOTATION. In the following sections, the discretization parameter is a positive integer N. The symbol c (or c', c'') stands for a positive constant which can change from one line to the other but is always independent of N.

13. Polynomial interpolation in an interval

The aim of this section is to study the approximation properties of the interpolation operators at the nodes of the quadrature formulas which are studied in Section 4, namely the Gauss formula, the Gauss–Lobatto formula and the generalized Gauss formula. For all of them, thanks to the approximation results of Section 6, estimating the interpolation error is a simple consequence of a stability inequality which is stated first.

We begin by studying the interpolation operator at the nodes of the Gauss formula. The operator j_{N-1} is defined in the following way: for any continuous function φ on Λ, $j_{N-1}\varphi$ is the only polynomial in $\mathbb{P}_{N-1}(\Lambda)$ which satisfies

$$(j_{N-1}\varphi)(\zeta_j) = \varphi(\zeta_j), \quad 1 \leqslant j \leqslant N, \tag{13.1}$$

where the ζ_j, $1 \leqslant j \leqslant N$, are the zeros of L_N.

Of course, the properties of this operator rely on the Gauss quadrature formula, i.e., on the fact (see Theorem 4.1) that there exist N positive weights ω_j, $1 \leqslant j \leqslant N$, such that the following equality holds:

$$\forall \Phi \in \mathbb{P}_{2N-1}(\Lambda), \qquad \int_{-1}^{1} \Phi(\zeta)\,\mathrm{d}\zeta = \sum_{j=1}^{N} \Phi(\zeta_j)\,\omega_j. \tag{13.2}$$

The first step is to prove the stability of the operator j_{N-1} in the following sense: the norm of $j_{N-1}\varphi$ in $L^2(\Lambda)$ is bounded independently of N for any function φ in $H^r(\Lambda)$, $r > \frac{1}{2}$ (note that this regularity assumption is minimum, due to the Sobolev Theorem 1.9). The first proof of this result is due to MADAY [1991], it involves some arguments of PASCIAK [1980] which have also been used by QUARTERONI [1987] for Chebyshev polynomials.

THEOREM 13.1. *For any real number r, $\frac{1}{2} < r \leqslant 1$, there exists a positive constant c such that, for any function φ in $H_0^r(\Lambda)$, the following estimate holds*

$$\|j_{N-1}\varphi\|_{L^2(\Lambda)} \leqslant c\big(\|\varphi\|_{L^2(\Lambda)} + N^{-r}\|\varphi\|_{H^r(\Lambda)}\big). \tag{13.3}$$

Before giving the proof, we recall a Hardy's type inequality, which will be a useful tool in what follows:

(i) for any real number $\beta < 1$, the inequality

$$\int_{-1}^{1} \varphi^2(\zeta)\big(1-\zeta^2\big)^{\beta-2}\,\mathrm{d}\zeta$$
$$\leqslant \sup\left\{\frac{1}{1-\beta}, \frac{1}{(1-\beta)^2}\right\} \int_{-1}^{1} \varphi'^2(\zeta)\big(1-\zeta^2\big)^{\beta}\,\mathrm{d}\zeta, \tag{13.4}$$

holds for any function φ in $H^1(\Lambda)$ which vanishes in ± 1 and such that $\varphi'(1-\zeta^2)^{\beta/2}$ belongs to $L^2(\Lambda)$,

(ii) for any real number $\beta > 1$, the inequality

$$\int_{-1}^{1} \varphi^2(\zeta)\big(1-\zeta^2\big)^{\beta-2}\,\mathrm{d}\zeta \leqslant \frac{\sup\{\beta^2, 2\}}{(\beta-1)^2}\int_{-1}^{1}\big(\varphi'^2(\zeta)+\varphi^2(\zeta)\big)\big(1-\zeta^2\big)^{\beta}\,\mathrm{d}\zeta, \tag{13.5}$$

holds for any function φ such that both $\varphi(1-\zeta^2)^{\beta/2}$ and $\varphi'(1-\zeta^2)^{\beta/2}$ belong to $L^2(\Lambda)$. Both inequalities (13.4) and (13.5) are derived from the following lines

$$\begin{aligned}
0 &\leqslant \int_{-1}^{1}\big(\varphi'(\zeta)+(1-\beta)\zeta\varphi(\zeta)\big(1-\zeta^2\big)^{-1}\big)^2\big(1-\zeta^2\big)^{\beta}\,\mathrm{d}\zeta\\
&\leqslant \int_{-1}^{1}\varphi'^2(\zeta)\big(1-\zeta^2\big)^{\beta}+(1-\beta)^2\zeta^2\varphi^2(\zeta)\big(1-\zeta^2\big)^{\beta-2}\\
&\quad + (\varphi^2)'(\zeta)(1-\beta)\zeta\big(1-\zeta^2\big)^{\beta-1}\big)\,\mathrm{d}\zeta\\
&\leqslant \int_{-1}^{1}\varphi'^2(\zeta)\big(1-\zeta^2\big)^{\beta}\,\mathrm{d}\zeta - \int_{-1}^{1}\varphi^2(\zeta)(1-\beta)\big(1-\beta\zeta^2\big)\big(1-\zeta^2\big)^{\beta-2}\,\mathrm{d}\zeta,
\end{aligned}$$

with adding and subtracting the term

$$\frac{\beta^2}{2}\int_{-1}^{1}\varphi^2(\zeta)\big(1-\zeta^2\big)^{\beta}(\zeta)\,\mathrm{d}\zeta$$

when $\beta > 1$. It will be used for integral values of β in this section, and for general values of β in Section 18.

PROOF. Since the quadrature formula is exact on $\mathbb{P}_{2N-1}(\Lambda)$, the definition of j_{N-1} implies that

$$\|j_{N-1}\varphi\|^2_{L^2(\Lambda)} = \sum_{j=1}^{N}(j_{N-1}\varphi)^2(\zeta_j)\omega_j = \sum_{j=1}^{N}\varphi^2(\zeta_j)\omega_j.$$

Next, we use Theorem 4.4:

$$\|j_{N-1}\varphi\|^2_{L^2(\Lambda)} \leqslant cN^{-1}\sum_{j=1}^{N}\varphi^2(\zeta_j)\big(1-\zeta_j^2\big)^{1/2}.$$

By the change of variable $\zeta = \cos\theta$, setting: $\widehat{\varphi}(\theta) = \varphi(\zeta)$ and using notation (4.12), we derive

$$\|j_{N-1}\varphi\|^2_{L^2(\Lambda)} \leqslant cN^{-1}\sum_{j=1}^{N}\widehat{\varphi}^2(\theta_j)\sin\theta_j.$$

It follows from Theorem 4.3 that there exist intervals K_j, $1 \leqslant j \leqslant N$, of length π/N, such that θ_j belongs to K_j and the intersection of two intervals K_i and K_j is empty when j differs from $i-1$, i and $i+1$. This yields

$$\|j_{N-1}\varphi\|^2_{L^2(\Lambda)} \leqslant cN^{-1} \sum_{j=1}^{N} \sup_{\theta \in \overline{K}_j} |\widehat{\varphi}(\theta)(\sin\theta)^{1/2}|^2. \tag{13.6}$$

Next, we have to consider separately the two cases $r=1$ and $\frac{1}{2} < r < 1$.

(1) We begin with the case $r=1$. Let c_1 denote the norm of the Sobolev imbedding of $H^1(0,1)$ into $L^\infty(0,1)$ (see Theorem 1.9). Using the appropriate scaling implies that, on a finite interval $]a,b[$, $a<b$, any function ψ in $H^1(a,b)$ satisfies

$$\sup_{a \leqslant \theta \leqslant b} |\psi(\theta)|^2 \leqslant c_1^2 \left(\frac{1}{b-a} \|\psi\|^2_{L^2(a,b)} + (b-a)|\psi|^2_{H^1(a,b)} \right). \tag{13.7}$$

Applying this inequality to the function $\widehat{\varphi}(\sin\theta)^{1/2}$ on each interval K_j leads to

$$\|j_{N-1}\varphi\|^2_{L^2(\Lambda)} \leqslant c \sum_{j=1}^{N} \left(\|\widehat{\varphi}(\sin\theta)^{1/2}\|^2_{L^2(K_j)} + N^{-2}|\widehat{\varphi}(\sin\theta)^{1/2}|^2_{H^1(K_j)} \right),$$

and, since each point of Λ belongs to at most two intervals K_j, to

$$\|j_{N-1}\varphi\|^2_{L^2(\Lambda)} \leqslant c \left(\|\widehat{\varphi}(\sin\theta)^{1/2}\|^2_{L^2(0,\pi)} + N^{-2}|\widehat{\varphi}(\sin\theta)^{1/2}|^2_{H^1(0,\pi)} \right). \tag{13.8}$$

This can be written equivalently as

$$\|j_{N-1}\varphi\|^2_{L^2(\Lambda)} \leqslant c \int_0^\pi \left(\widehat{\varphi}^2(\theta)\sin\theta + N^{-2}\widehat{\varphi}^2(\theta)\frac{\cos^2\theta}{\sin\theta} + N^{-2}\widehat{\varphi}'^2(\theta)\sin\theta \right) \mathrm{d}\theta,$$

so that the inverse change of variable yields

$$\begin{aligned} &\|j_{N-1}\varphi\|^2_{L^2(\Lambda)} \\ &\quad \leqslant c \int_{-1}^{1} \left(\varphi^2(\zeta) + N^{-2}\varphi^2(\zeta)\zeta^2(1-\zeta^2)^{-1} + N^{-2}\varphi'^2(\zeta)(1-\zeta^2)\right) \mathrm{d}\zeta \\ &\quad \leqslant c \int_{-1}^{1} \left(\varphi^2(\zeta) + N^{-2}\varphi^2(\zeta)(1-\zeta^2)^{-2} + N^{-2}\varphi'^2(\zeta)\right) \mathrm{d}\zeta. \end{aligned}$$

Using Hardy's inequality (13.4) leads to

$$\|j_{N-1}\varphi\|^2_{L^2(\Lambda)} \leqslant c\left(\|\varphi\|^2_{L^2(\Lambda)} + N^{-2}|\varphi|^2_{H^1(\Lambda)}\right),$$

which is the desired estimate.

(2) In the case $\frac{1}{2} < r < 1$, we must use the intrinsic norm of $H^r(\Lambda)$ (see formula (1.14)). Indeed, from Theorem 1.9, there exists a positive constant c_r such that any function χ in $H^r(0,1)$ satisfies

$$\sup_{0\leqslant\zeta\leqslant 1} \left|\chi(\zeta)\right|^2 \leqslant c_r^2\left(\|\chi\|^2_{L^2(0,1)} + \int_0^1\int_0^1 \frac{|\chi(\zeta)-\chi(\xi)|^2}{|\zeta-\xi|^{2r+1}}\,\mathrm{d}\zeta\,\mathrm{d}\xi\right),$$

so that the appropriate scaling implies that, for any function ψ in $H^r(a,b)$, $a<b$:

$$\sup_{a\leqslant\theta\leqslant b}\left|\psi(\theta)\right|^2 \leqslant c_r^2\left(\frac{1}{b-a}\,\|\psi\|^2_{L^2(a,b)} + (b-a)^{2r-1}\int_a^b\int_a^b \frac{|\psi(\theta)-\psi(\eta)|^2}{|\theta-\eta|^{2r+1}}\,\mathrm{d}\theta\,\mathrm{d}\eta\right). \quad (13.9)$$

Applying this inequality in (13.6) gives

$$\|j_{N-1}\varphi\|^2_{L^2(\Lambda)} \leqslant c\left\|\widehat{\varphi}(\sin\theta)^{1/2}\right\|^2_{L^2(0,\pi)} + c'N^{-2r}\sum_{j=1}^{N}\int_{K_j}\int_{K_j}\frac{|\widehat{\varphi}(\theta)(\sin\theta)^{1/2}-\widehat{\varphi}(\eta)(\sin\eta)^{1/2}|^2}{|\theta-\eta|^{2r+1}}\,\mathrm{d}\theta\,\mathrm{d}\eta. \quad (13.10)$$

In order to bound the last term, we define the pair of functions

$$\big(\alpha(\theta,\eta),\beta(\theta,\eta)\big) = \begin{cases} (\theta,\eta) & \text{if } \sin\theta \leqslant \sin\eta,\\ (\eta,\theta) & \text{if } \sin\theta > \sin\eta,\end{cases}$$

and we use the identity

$$\left|\widehat{\varphi}(\theta)(\sin\theta)^{1/2} - \widehat{\varphi}(\eta)(\sin\eta)^{1/2}\right| = \left|\big(\widehat{\varphi}(\alpha)-\widehat{\varphi}(\beta)\big)(\sin\alpha)^{1/2} + \widehat{\varphi}(\beta)\big((\sin\alpha)^{1/2}-(\sin\beta)^{1/2}\big)\right|.$$

We obtain

$$\begin{aligned}\int_{K_j}\int_{K_j}&\frac{|\widehat{\varphi}(\theta)(\sin\theta)^{1/2}-\widehat{\varphi}(\eta)(\sin\eta)^{1/2}|^2}{|\theta-\eta|^{2r+1}}\,\mathrm{d}\theta\,\mathrm{d}\eta\\ &\leqslant 2\int_{K_j}\int_{K_j}\frac{|\widehat{\varphi}(\alpha)-\widehat{\varphi}(\beta)|^2}{|\theta-\eta|^{2r+1}}(\sin\alpha)\,\mathrm{d}\theta\,\mathrm{d}\eta\\ &\quad + 2\int_{K_j}\int_{K_j}\widehat{\varphi}^2(\beta)\frac{|(\sin\alpha)^{1/2}-(\sin\beta)^{1/2}|^2}{|\theta-\eta|^{2r+1}}\,\mathrm{d}\theta\,\mathrm{d}\eta.\end{aligned}$$

In the first term, we use the formula

$$\cos\theta - \cos\eta = -(\theta - \eta)\sin\gamma,$$

where γ is between α and β. We derive

$$\int_{K_j}\int_{K_j} \frac{|\widehat{\varphi}(\alpha) - \widehat{\varphi}(\beta)|^2}{|\theta - \eta|^{2r+1}} (\sin\alpha)\,\mathrm{d}\theta\,\mathrm{d}\eta$$
$$\leqslant \int_{K_j}\int_{K_j} \frac{|\widehat{\varphi}(\theta) - \widehat{\varphi}(\eta)|^2}{|\cos\theta - \cos\eta|^{2r+1}} \frac{|\sin\gamma|^{2r+1}}{\sin\beta} (\sin\alpha)(\sin\beta)\,\mathrm{d}\theta\,\mathrm{d}\eta,$$

and we observe that the quantity $|\sin\gamma|^{2r+1}/\sin\beta$ is bounded since $\sin\gamma$ is positive and $\leqslant \sin\beta$ if K_j does not contain $\pi/2$, and $\sin\beta$ cannot tend to 0 if K_j contains $\pi/2$. For the second term, we can check the inequality

$$\int_{K_j} \frac{|(\sin\alpha)^{1/2} - (\sin\beta)^{1/2}|^2}{|\alpha - \beta|^{2r+1}}\,\mathrm{d}\alpha \leqslant \frac{c}{\sin\beta}.$$

Hence, we have

$$\sum_{j=1}^{N} \int_{K_j}\int_{K_j} \frac{|\widehat{\varphi}(\theta)(\sin\theta)^{1/2} - \widehat{\varphi}(\eta)(\sin\eta)^{1/2}|^2}{|\theta - \eta|^{2r+1}}\,\mathrm{d}\theta\,\mathrm{d}\eta$$
$$\leqslant c\int_0^{\pi}\int_0^{\pi} \frac{|\widehat{\varphi}(\theta) - \widehat{\varphi}(\eta)|^2}{|\cos\theta - \cos\eta|^{2r+1}} (\sin\theta)(\sin\eta)\,\mathrm{d}\theta\,\mathrm{d}\eta$$
$$+ c'\int_0^{\pi} \widehat{\varphi}^2(\beta)(\sin\beta)^{-1}\,\mathrm{d}\beta.$$

Finally, we make the inverse change of variables

$$\sum_{j=1}^{N} \int_{K_j}\int_{K_j} \frac{|\widehat{\varphi}(\theta)(\sin\theta)^{1/2} - \widehat{\varphi}(\eta)(\sin\eta)^{1/2}|^2}{|\theta - \eta|^{2r+1}}\,\mathrm{d}\theta\,\mathrm{d}\eta$$
$$\leqslant c\int_{-1}^{1}\int_{-1}^{1} \frac{|\varphi(\zeta) - \varphi(\xi)|^2}{|\zeta - \xi|^{2r+1}}\,\mathrm{d}\zeta\,\mathrm{d}\xi + c'\int_{-1}^{1} \varphi^2(\zeta)(1 - \zeta^2)^{-1}\,\mathrm{d}\zeta.$$

Recalling (LIONS and MAGENES ([1968], Theorem 11.3)) that the multiplication by $(1 - \zeta^2)^{-1/2}$ is continuous from $H_0^r(\Lambda)$ into $L^2(\Lambda)$, we obtain the desired result.

REMARK 13.1. Applying the same arguments as in the previous proof, we also derive that, for any polynomial φ_M in $\mathbb{P}_M(\Lambda)$,

$$\|j_{N-1}\varphi_M\|_{L^2(\Lambda)} \leqslant c\left(1 + \frac{M}{N}\right)\|\varphi_M\|_{L^2(\Lambda)}. \tag{13.11}$$

Note that the power of M in this inequality is optimal. An important consequence is the stability of the operator j_{N-1} applied to polynomials of the form $(\varphi_N)^k$ for any fixed positive integer k, with φ_N in $\mathbb{P}_N(\Lambda)$. This is an essential argument for the discretization of nonlinear equations.

As a consequence, we derive an estimate of the distance between a function and its interpolate polynomial in $L^2(\Lambda)$.

THEOREM 13.2. *For any real number $s > \frac{1}{2}$, there exists a positive constant c depending only on s such that, for any function φ in $H^s(\Lambda)$, the following estimate holds*

$$\|\varphi - j_{N-1}\varphi\|_{L^2(\Lambda)} \leqslant cN^{-s}\|\varphi\|_{H^s(\Lambda)}. \tag{13.12}$$

PROOF. Let φ_{N-1} stands for the polynomial $\tilde{\pi}^1_{N-1}\varphi$ when $s \geqslant 1$ and for the polynomial $\tilde{\pi}^{s,0,0}_{N-1}\varphi$ when $s < 1$. Setting $r = \inf\{s, 1\}$ and applying Theorem 13.1 to the function $\varphi - \varphi_{N-1}$ together with the triangular inequality, we have

$$\|\varphi - j_{N-1}\varphi\|_{L^2(\Lambda)} \leqslant c\|\varphi - \varphi_{N-1}\|_{L^2(\Lambda)} + c'N^{-r}\|\varphi - \varphi_{N-1}\|_{H^r(\Lambda)},$$

so that the result follows from (6.22) and (6.23) (see Remark 6.6).

The essential point in this theorem is that the distance between a function φ and its interpolate polynomial $j_{N-1}\varphi$ is exactly of the same order as the distance of this function to its best approximation $\pi_{N-1}\varphi$. Of course, this cannot be improved and the estimate is optimal.

REMARK 13.2. As in Remark 6.3 and with the same notation, the estimate (13.12) can be improved for some special functions. Indeed, it follows from the proof of Theorem 13.1 that, for any function φ in $D(A^{s/2})$, $s > 1$, the following estimate holds

$$\|\varphi - j_{N-1}\varphi\|_{L^2(\Lambda)} \leqslant cN^{-s}(\log N)\|\varphi\|_{D(A^{s/2})} \tag{13.13}$$

we refer to BERNARDI and MADAY ([1992c], §6) for the detailed proof.

However, using the operator j_{N-1} does not lead to optimal estimates in the $H^1(\Lambda)$ norm. For instance, taking the function φ_N equal to $L_{N+1} - L_{N-1}$ as in (6.7), we deduce from equation (3.14) that $j_{N-1}\varphi_N$ is equal to $-(2N+1)L_{N-1}/(N+1)$, so that

$$|\varphi_N - j_{N-1}\varphi_N|_{H^1(\Lambda)} \geqslant cN^{1/2}|\varphi_N|_{H^1(\Lambda)}. \tag{13.14}$$

It is only possible to prove that, for any real numbers r and s, $r \leqslant s$ and $s \geqslant 1$, and for any function φ in $H^s(\Lambda)$,

$$\|\varphi - j_{N-1}\varphi\|_{H^r(\Lambda)} \leqslant \begin{cases} cN^{3r/2-s}\|\varphi\|_{H^s(\Lambda)} & \text{if } r \leqslant 1, \\ cN^{2r-1/2-s}\|\varphi\|_{H^s(\Lambda)} & \text{if } r \geqslant 1. \end{cases} \tag{13.15}$$

So, in order to obtain optimal estimates in $H^1(\Lambda)$, we introduce new interpolation operators.

Let us recall that, for any nonnegative integer m, the nodes of the generalized Gauss type formula (4.3) are the zeros ζ_j^m of the polynomial $\mathrm{d}^m L_N/\mathrm{d}\zeta^m$, $1 \leqslant j \leqslant N-m$. The associated "internal" weights are denoted by σ_j^m, $1 \leqslant j \leqslant N-m$, and the "boundary" weights by $\sigma_\pm^{m,k}$, $0 \leqslant k \leqslant m-1$. For the sake of simplicity, we introduce the bilinear form

$$[\varphi, \psi]_{N,m} = \sum_{j=1}^{N-m} \varphi(\zeta_j^m)\psi(\zeta_j^m)\sigma_j^m + \sum_{k=0}^{m-1} \left(\left(\frac{\mathrm{d}^k(\varphi\psi)}{\mathrm{d}\zeta^k}\right)(-1)\sigma_-^{m,k} + \left(\frac{\mathrm{d}^k(\varphi\psi)}{\mathrm{d}\zeta^k}\right)(+1)\sigma_+^{m,k}\right). \quad (13.16)$$

The operator k_{N+m-1}^m is defined on the space of continuous functions on $\overline{\Lambda}$ which have derivatives up to order $m-1$ in ± 1, with values in $\mathbb{P}_{N+m-1}(\Lambda)$, by the equations

$$\begin{aligned} &\left(k_{N+m-1}^m\varphi\right)\left(\zeta_j^m\right) = \varphi(\zeta_j^m), \quad 1 \leqslant j \leqslant N-m, \\ &\left(\frac{\mathrm{d}^k k_{N+m-1}^m \varphi}{\mathrm{d}\zeta^k}\right)(-1) = \left(\frac{\mathrm{d}^k\varphi}{\mathrm{d}\zeta^k}\right)(-1), \quad 0 \leqslant k \leqslant m-1, \\ &\left(\frac{\mathrm{d}^k k_{N+m-1}^m \varphi}{\mathrm{d}\zeta^k}\right)(+1) = \left(\frac{\mathrm{d}^k\varphi}{\mathrm{d}\zeta^k}\right)(+1), \quad 0 \leqslant k \leqslant m-1. \end{aligned} \quad (13.17)$$

The first step is to derive a stability estimate similar to (13.3) for the operator k_{N+m-1}^m. However, since the quadrature formula is exact on $\mathbb{P}_{2N-1}(\Lambda)$ while the interpolate polynomials k_{N+m-1}^m are of degree $N+m-1$, such an estimate can a priori only be obtained for the function $(k_{N+m-1}^m\varphi)(1-\zeta^2)^{-m/2}$. The following lemma allows to derive the same result for the function $(k_{N+m-1}^m\varphi)(1-\zeta^2)^{-(m-1)/2}$.

LEMMA. *For any positive integer m, any polynomial φ_N in $\mathbb{P}_{N+m-1}^{m-1,0}(\Lambda)$ satisfies*

$$\begin{aligned} \left\|\varphi_N\left(1-\zeta^2\right)^{-(m-1)/2}\right\|_{L^2(\Lambda)}^2 &\leqslant \left[\varphi_N\left(1-\zeta^2\right)^{-(m-1)}, \varphi_N\right]_{N,m} \\ &\leqslant c\left\|\varphi_N\left(1-\zeta^2\right)^{-(m-1)/2}\right\|_{L^2(\Lambda)}^2. \end{aligned} \quad (13.18)$$

PROOF. Any polynomial in $\mathbb{P}_{N+m-1}^{m-1,0}(\Lambda)$ can be written in the basis

$$\left\{ \left(1-\zeta^2\right)^{m-1}\frac{\mathrm{d}^{m-1}L_n}{\mathrm{d}\zeta^{m-1}}, \ m-1 \leqslant n \leqslant N\right\}.$$

Since this basis is orthogonal for the measure $(1-\zeta^2)^{-(m-1)}\,\mathrm{d}\zeta$ and due to the exactness of the quadrature formula on $\mathbb{P}_{2N-1}(\Lambda)$, it suffices to check that (13.18) holds when φ_N is equal to

$$(1-\zeta^2)^{m-1}\frac{\mathrm{d}^{m-1}L_N}{\mathrm{d}\zeta^{m-1}},$$

or equivalently that

$$\int_{-1}^{1}\left(\frac{\mathrm{d}^{m-1}L_N}{\mathrm{d}\zeta^{m-1}}\right)^2(\zeta)(1-\zeta^2)^{m-1}\,\mathrm{d}\zeta \leqslant \left[\frac{\mathrm{d}^{m-1}L_N}{\mathrm{d}\zeta^{m-1}}(1-\zeta^2)^{m-1}, \frac{\mathrm{d}^{m-1}L_N}{\mathrm{d}\zeta^{m-1}}\right]_{N,m} \leqslant c\int_{-1}^{1}\left(\frac{\mathrm{d}^{m-1}L_N}{\mathrm{d}\zeta^{m-1}}\right)^2(\zeta)(1-\zeta^2)^{m-1}\,\mathrm{d}\zeta.$$

Since the polynomial

$$\left(\frac{\mathrm{d}^{m-1}L_N}{\mathrm{d}\zeta^{m-1}}\right)^2 + \frac{1}{(N-m+1)^2}(1-\zeta^2)\left(\frac{\mathrm{d}^{m}L_N}{\mathrm{d}\zeta^{m}}\right)^2$$

has degree $2N-2m+1$, it follows from the properties of the quadrature formula that

$$\left[\frac{\mathrm{d}^{m-1}L_N}{\mathrm{d}\zeta^{m-1}}(1-\zeta^2)^{m-1}, \frac{\mathrm{d}^{m-1}L_N}{\mathrm{d}\zeta^{m-1}}\right]_{N,m} = \int_{-1}^{1}\left(\frac{\mathrm{d}^{m-1}L_N}{\mathrm{d}\zeta^{m-1}}\right)^2(\zeta)(1-\zeta^2)^{m-1}\,\mathrm{d}\zeta + \frac{1}{(N-m+1)^2}\int_{-1}^{1}\left(\frac{\mathrm{d}^{m}L_N}{\mathrm{d}\zeta^{m}}\right)^2(\zeta)(1-\zeta^2)^{m}\,\mathrm{d}\zeta.$$

Making use of (3.7) implies that

$$\left[\frac{\mathrm{d}^{m-1}L_N}{\mathrm{d}\zeta^{m-1}}(1-\zeta^2)^{m-1}, \frac{\mathrm{d}^{m-1}L_N}{\mathrm{d}\zeta^{m-1}}\right]_{N,m} = \int_{-1}^{1}\left(\frac{\mathrm{d}^{m-1}L_N}{\mathrm{d}\zeta^{m-1}}\right)^2(\zeta)(1-\zeta^2)^{m-1}\,\mathrm{d}\zeta \times\left(1+\frac{1}{(N-m+1)^2}(N-m+1)(N+m)\right).$$

Finally, we obtain

$$\left[\frac{\mathrm{d}^{m-1}L_N}{\mathrm{d}\zeta^{m-1}}(1-\zeta^2)^{m-1}, \frac{\mathrm{d}^{m-1}L_N}{\mathrm{d}\zeta^{m-1}}\right]_{N,m} = \frac{2N+1}{N-m+1}\int_{-1}^{1}\left(\frac{\mathrm{d}^{m-1}L_N}{\mathrm{d}\zeta^{m-1}}\right)^2(\zeta)(1-\zeta^2)^{m-1}\,\mathrm{d}\zeta, \qquad (13.19)$$

which gives the desired inequalities.

REMARK 13.3. As a particular case of (13.18) and more precisely of (13.19), we derive

$$\forall \varphi_N \in \mathbb{P}_N(\Lambda), \quad \|\varphi_N\|^2_{L^2(\Lambda)} \leqslant [\varphi_N, \varphi_N]_{N,1} \leqslant 3\|\varphi_N\|^2_{L^2(\Lambda)}. \tag{13.20}$$

This inequality, which was firstly proven by CANUTO and QUARTERONI [1982], will be of great use in what follows.

We are now in a position to state the general stability theorem.

THEOREM 13.3. *Let m be a positive integer, and let k be equal to m or $m-1$. For any real number r, $\frac{1}{2} < r \leqslant 1$, such that $r-k$ is not equal to 1, there exists a positive constant c such that, for any function φ in $H_0^{m-1+r}(\Lambda)$, the following estimate holds*

$$\begin{aligned} &\big\|\big(k^m_{N+m-1}\varphi\big)\big(1-\zeta^2\big)^{-k/2}\big\|_{L^2(\Lambda)} \\ &\quad\leqslant c\big(\big\|\varphi\big(1-\zeta^2\big)^{-k/2}\big\|_{L^2(\Lambda)} + N^{-r}\big\|\varphi\big(1-\zeta^2\big)^{(r-k)/2}\big\|_{H^r(\Lambda)}\big). \end{aligned} \tag{13.21}$$

PROOF. The proof is very similar to that of Theorem 13.1 with the function φ replaced by the function $\psi = \varphi(1-\zeta^2)^{-k/2}$. Indeed, we deduce from the previous lemma that

$$\big\|\big(k^m_{N+m-1}\varphi\big)\big(1-\zeta^2\big)^{-k/2}\big\|^2_{L^2(\Lambda)} \leqslant \sum_{j=1}^{N-m} \psi^2\big(\zeta_j^m\big)\sigma_j^m.$$

Next, we use the bound of the σ_j^m which is given in Remark 4.6 and we observe that, due to the location of the zeros of L_N, the arccosine of each ζ_j^m belongs to an interval K_j^m of length $(m+1)\pi/N$ and that each K_j^m intersects at most $m+2$ other intervals K_i^m. Hence, using the change of variable $\zeta = \cos\theta$ and applying (13.7) or (13.9) leads to the conclusion by the same arguments as in the proof of Theorem 13.1.

Note that the regularity assumption on φ can be relaxed when the function φ and its derivatives up to the order $m-1$ vanish in ± 1. Note also that the case $r-k=1$ only occurs for $r=1$ and $k=0$, hence for m is equal to 1. Then, we are in the limit case of the Hardy's inequality, so that (13.21) is replaced by

$$\begin{aligned} \|k^1_N\varphi\|_{L^2(\Lambda)} &\leqslant c\big(\|\varphi\|_{L^2(\Lambda)} + N^{-1}\big\|\varphi\big(1-\zeta^2\big)^{-1/2}\big\|_{L^2(\Lambda)} \\ &\quad + N^{-1}\big\|\varphi'\big(1-\zeta^2\big)^{1/2}\big\|_{L^2(\Lambda)}\big). \end{aligned} \tag{13.22}$$

REMARK 13.4. By a slightly different proof, we can check that inequalities (13.18) still hold in the case $m=0$ for any φ_N in $\mathbb{P}_{N-1}(\Lambda)$ (this will be used in Section 26), which leads to an additional estimate for the Gauss operator j_{N-1}. Indeed, inequality (13.21) with $k=-1$ holds in this case for any function φ in $H^s(\Lambda)$, $s > \frac{1}{2}$, and applying it to the function $\tilde{\pi}^1_N\varphi$ or $\tilde{\pi}^{s,0,0}_N\varphi$ according to the value of s leads to the following estimate which holds for any function φ such that $\varphi(1-\zeta^2)^{1/2}$ belongs to $H^s(\Lambda)$, $s > \frac{1}{2}$:

$$\left\|(\varphi - j_{N-1}\varphi)\left(1-\zeta^2\right)^{1/2}\right\|_{L^2(\Lambda)} \leqslant cN^{-s}\left\|\varphi\left(1-\zeta^2\right)^{1/2}\right\|_{H^s(\Lambda)}. \tag{13.23}$$

Theorem 13.3 leads to estimates in higher order norms, when combined with the inverse inequality (5.10). We are going to state these results in the cases which are useful for the numerical analysis of the collocation methods. We begin with the Gauss–Lobatto interpolation operator.

The operator i_N is defined as follows: for any continuous function φ on $\overline{\Lambda}$, $i_N\varphi$ is the only polynomial in $\mathbb{P}_N(\Lambda)$ which satisfies

$$(i_N\varphi)(\xi_j) = \varphi(\xi_j), \quad 0 \leqslant j \leqslant N, \tag{13.24}$$

where the ξ_j, $0 \leqslant j \leqslant N$, are the zeros of $(1-\zeta^2)L_N'$. Let us state the result for this operator.

THEOREM 13.4. *For any real numbers r and s satisfying $s > (1+r)/2$ and $0 \leqslant r \leqslant 1$, there exists a positive constant c depending only on s such that, for any function φ in $H^s(\Lambda)$, the following estimates hold*

$$\|\varphi - i_N\varphi\|_{H^r(\Lambda)} \leqslant cN^{r-s}\|\varphi\|_{H^s(\Lambda)}. \tag{13.25}$$

For any real number $s \geqslant 1$, there exists a positive constant c depending only on s such that, for any function φ in $H^s(\Lambda)$, the following estimates hold

$$\left\|(\varphi - i_N\varphi)\left(1-\zeta^2\right)^{-1/2}\right\|_{L^2(\Lambda)} \leqslant cN^{-s}\|\varphi\|_{H^s(\Lambda)}. \tag{13.26}$$

PROOF. Firstly, we apply (13.22) to the function $\varphi - \tilde{\pi}_N^1\varphi$ (bounding $(1-\zeta^2)^{1/2}$ by 1) or Theorem 13.3 for $k = 0$ to $\varphi - \tilde{\pi}_N^{s,0,0}\varphi$ according to the value of s, to obtain the estimate of $\|\varphi - i_N\varphi\|_{L^2(\Lambda)}$. Secondly, we apply Theorem 13.3 for $k = 1$ to the function $(\varphi - \tilde{\pi}_N^1\varphi)(1-\zeta^2)^{-1/2}$, which gives (13.26), and using the inverse inequality of Theorem 5.4 for the interpolate polynomial, we derive the estimate in the $H^1(\Lambda)$-norm. The estimate in the $H^r(\Lambda)$-norm follows by an interpolation argument.

Noting that estimate (13.25) also holds with $r = s = 1$, we obtain the stability estimate, which is valid for any function φ in $H^1(\Lambda)$:

$$\|i_N\varphi\|_{H^1(\Lambda)} \leqslant c\|\varphi\|_{H^1(\Lambda)}. \tag{13.27}$$

Hence, the derivative of a Lagrange interpolation polynomial can be stable!

REMARK 13.5. As in Remark 13.1, we can also prove the following estimate, for any polynomial φ_M in $\mathbb{P}_M(\Lambda)$,

$$\|i_N\varphi_M\|_{L^2(\Lambda)} \leqslant c\left(1 + \frac{M}{N}\right)\|\varphi_M\|_{L^2(\Lambda)}. \tag{13.28}$$

Similar results hold in the general case. We shall only use the following ones:
(i) for any function φ such that $\varphi(1-\zeta^2)^{-1}$ belongs to $H^s(\Lambda)$, $s > \frac{1}{2}$,

$$\left\| \left(\varphi - k_{N+1}^2 \varphi\right)\left(1-\zeta^2\right)^{-1} \right\|_{L^2(\Lambda)} \leqslant c N^{-s} \left\| \varphi\left(1-\zeta^2\right)^{-1} \right\|_{H^s(\Lambda)}; \tag{13.29}$$

(ii) for any real number r, $0 \leqslant r \leqslant 2$, and for any function φ in $H^s(\Lambda)$, $s > (6+r)/4$,

$$\left\| \varphi - k_{N+1}^2 \varphi \right\|_{H^r(\Lambda)} \leqslant c N^{r-s} \|\varphi\|_{H^s(\Lambda)} \tag{13.30}$$

(this estimate also holds with $r = s = 2$);
(iii) for any function φ such that $\varphi(1-\zeta^2)^{-1}$ belongs to $H^s(\Lambda)$, $s > \frac{3}{2}$,

$$\left\| \left(\varphi - k_{N+2}^3 \varphi\right)\left(1-\zeta^2\right)^{-1} \right\|_{L^2(\Lambda)} \leqslant c N^{-s} \left\| \varphi\left(1-\zeta^2\right)^{-1} \right\|_{H^s(\Lambda)}. \tag{13.31}$$

However, more precise results can be derived for the polynomial interpolation of functions with singularities in ± 1.

14. Polynomial interpolation on tensorized domains

Both notation and proofs will be very similar to those of Section 7. Indeed, the interpolation points in the hypercube $\overline{\Omega}_d$ are chosen such that the coordinates of each point are nodes of a fixed one-dimensional quadrature formula. As a consequence, the Lagrange interpolation operator at these points is the product of one-dimensional interpolation operators with respect to each variable.

We begin by proving estimates for the operator related to the Gauss formula.

NOTATION. The *Gauss grid* of Legendre type Σ_N is defined by

$$\Sigma_N = \left\{ \boldsymbol{x} = (\zeta_{j_1}, \ldots, \zeta_{j_d}) \colon \ 1 \leqslant j_1, \ldots, j_d \leqslant N \right\}. \tag{14.1}$$

NOTATION. Let $\mathcal{J}_{N-1}$ be the Lagrange interpolation operator at the points of the grid Σ_N: for any continuous function v on Ω_d, $\mathcal{J}_{N-1} v$ belongs to $\mathbb{P}_{N-1}(\Omega_d)$ and satisfies

$$(\mathcal{J}_{N-1} v)(\boldsymbol{x}) = v(\boldsymbol{x}), \quad \boldsymbol{x} \in \Sigma_N. \tag{14.2}$$

Of course and with the notation of Section 7, this definition can be written as

$$\mathcal{J}_{N-1} = j_{N-1}^{(1)} \circ \cdots \circ j_{N-1}^{(d)}. \tag{14.3}$$

Then, the proof of the following theorem is straightforward.

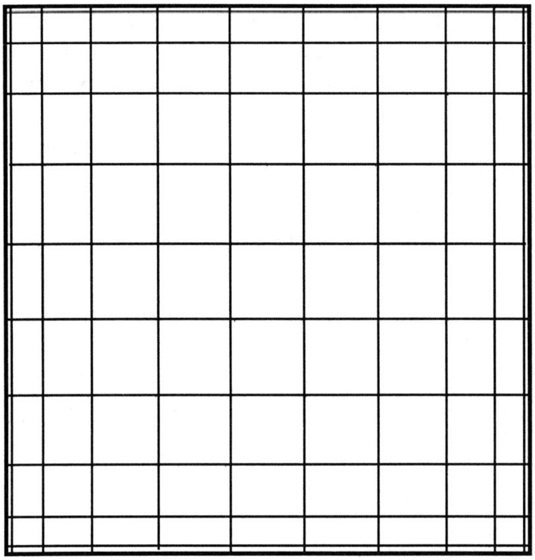

FIG. 14.1. The Gauss grid of Legendre type for $N = 10$.

THEOREM 14.1. *For any real number $s > d/2$, there exists a positive constant c depending only on s such that, for any function v in $H^s(\Omega_d)$, the following estimate holds*

$$\|v - \mathcal{J}_{N-1}v\|_{L^2(\Omega_d)} \leqslant cN^{-s}\|v\|_{H^s(\Omega_d)}. \tag{14.4}$$

PROOF. From (14.3), using the basic identity (7.2), we can write

$$\begin{aligned}
&\|v - \mathcal{J}_{N-1}v\|_{L^2(\Omega_d)} \\
&\quad \leqslant \left\|v - j_{N-1}^{(1)}v\right\|_{L^2(\Lambda_1;L^2(\Omega_{d-1}))} + \left\|v - j_{N-1}^{(2)} \circ \cdots \circ j_{N-1}^{d}v\right\|_{L^2(\Lambda_1;L^2(\Omega_{d-1}))} \\
&\qquad + \left\|\left(\mathrm{id} - j_{N-1}^{1}\right)\left(v - j_{N-1}^{(2)} \circ \cdots \circ j_{N-1}^{d}v\right)\right\|_{L^2(\Lambda_1;L^2(\Omega_{d-1}))},
\end{aligned}$$

and using Theorem 13.2 yields

$$\begin{aligned}
&\|v - \mathcal{J}_{N-1}v\|_{L^2(\Omega_d)} \\
&\quad \leqslant cN^{-s}\|v\|_{H^s(\Lambda_1;L^2(\Omega_{d-1}))} + \left\|v - j_{N-1}^{(2)} \circ \cdots \circ j_{N-1}^{d}v\right\|_{L^2(\Lambda_1;L^2(\Omega_{d-1}))} \\
&\qquad + c'N^{-s/d}\left\|v - j_{N-1}^{(2)} \circ \cdots \circ j_{N-1}^{d}v\right\|_{H^{s/d}(\Lambda_1;L^2(\Omega_{d-1}))}.
\end{aligned} \tag{14.5}$$

Iterating d times this argument and making use of (7.3) gives the desired result.

REMARK 14.1. Estimate (14.4) is also of optimal order, in comparison with the distance to the best fit given in Theorem 7.1. Moreover, the hypothesis $s > d/2$ is necessary since the operator $\mathcal{J}_{N-1}$ is only defined on continuous functions.

Now, we consider the Gauss–Lobatto interpolation operator, which satisfies very similar properties.

NOTATION. The *Gauss–Lobatto grid* of Legendre type Ξ_N is defined by

$$\Xi_N = \{\boldsymbol{x} = (\xi_{j_1}, \dots, \xi_{j_d}) \colon\ 0 \leqslant j_1, \dots, j_d \leqslant N\}. \tag{14.6}$$

We denote by $\mathcal{I}_N$ the Lagrange interpolation operator at the points of the grid Ξ_N: for any continuous function v on $\overline{\Omega}_d$, $\mathcal{I}_N v$ belongs to $\mathbb{P}_N(\Omega_d)$ and satisfies

$$(\mathcal{I}_N v)(\boldsymbol{x}) = v(\boldsymbol{x}), \quad \boldsymbol{x} \in \Xi_N. \tag{14.7}$$

Here also, we have the identity

$$\mathcal{I}_N = i_N^{(1)} \circ \cdots \circ i_N^{(d)}, \tag{14.8}$$

where all the operators $i_N^{(j)}$ commute. This allows for an easy proof of the following theorem.

THEOREM 14.2. *For any real numbers r and s satisfying $s > (d+r)/2$ and $0 \leqslant r \leqslant 1$, there exists a positive constant c depending only on s such that, for any function v in $H^s(\Omega_d)$, the following estimate holds*

$$\|v - \mathcal{I}_N v\|_{H^r(\Omega_d)} \leqslant c N^{r-s} \|v\|_{H^s(\Omega_d)}. \tag{14.9}$$

PROOF. The proof of inequality (14.9) for $r = 0$ is exactly the same as for Theorem 14.1. To derive the inequality for $r = 1$, we note that, for $1 \leqslant j \leqslant d$,

$$\begin{aligned}\|v - \mathcal{I}_N v\|_{H^1(\Lambda_j; L^2(\Omega_{d-1}))} &\leqslant \|v - i_N^{(j)} v\|_{H^1(\Lambda_j; L^2(\Omega_{d-1}))} \\ &+ \|i_N^{(j)}(v - i_N^{(1)} \circ \cdots \circ i_N^{(j-1)} \circ i_N^{(j+1)} \circ i_N^d v)\|_{H^1(\Lambda_j; L^2(\Omega_{d-1}))}.\end{aligned}$$

Combining the stability result (13.27) with Theorem 13.4, we obtain

$$\begin{aligned}\|v - \mathcal{I}_N v\|_{H^1(\Lambda_j; L^2(\Omega_{d_1}))} &\leqslant c N^{1-s} \|v\|_{H^s(\Lambda_j; L^2(\Omega_{d-1}))} \\ &+ \|v - i_N^{(1)} \circ \cdots \circ i_N^{(j-1)} \circ i_N^{(j+1)} \circ i_N^d v\|_{H^1(\Lambda_j; L^2(\Omega_{d-1}))},\end{aligned}$$

and the result for $r = 0$ and d replaced by $d - 1$, combined with (7.3), yields the inequality (14.9) for $r = 1$. The general result follows by an interpolation argument.

It can be observed from the previous proof that estimate (14.9) also holds in the case $s = (d+1)/2$ and $r = 1$.

REMARK 14.2. Of course, more complicated operators can be constructed by combining the Gauss and Gauss–Lobatto interpolation operators: for instance, in Section 26, we shall need the following operator

$$\mathcal{L}_{N,j} = j_{N-1}^{(1)} \circ \cdots \circ j_{N-1}^{(j-1)} \circ i_N^{(j)} \circ j_{N-1}^{(j+1)} \circ \cdots \circ j_{N-1}^{(d)}. \tag{14.10}$$

Exactly as previously, we derive from Theorems 13.2 and 13.4 the following estimate, which is valid for any function v in $H^s(\Omega_d)$, $s > d/2$:

$$\|v - \mathcal{L}_{N,j} v\|_{L^2(\Omega_d)} \leqslant c N^{-s} \|v\|_{H^s(\Omega_d)}. \tag{14.11}$$

Let us finally consider the interpolation operators which are associated with the generalized quadrature formula of order $m \geqslant 2$. They are more complicated, since they are no longer of Lagrange type. This justifies the following definition.

NOTATION. Let m be a positive integer. Let Σ_N^m be the grid

$$\Sigma_N^m = \left\{ \boldsymbol{x} = \left(\zeta_{i_1}^m, \ldots, \zeta_{i_d}^m\right) \colon \ 0 \leqslant i_1, \ldots, i_d \leqslant N - m + 1 \right\} \tag{14.12}$$

with $\zeta_0^m = -1$ and $\zeta_{N-m+1}^m = +1$. We define the interpolation operator $\mathcal{K}_{N+m-1}^m$ by

$$\mathcal{K}_{N+m-1}^m = k_{N+m-1}^{m(1)} \circ \cdots \circ k_{N+m-1}^{m(d)}. \tag{14.13}$$

Let us for a while consider the simplest case of the square Ω in the case $m = 2$. The definition of $\mathcal{K}_{N+1}^2$ in this case can equivalently be written: for any function v which is continuous on Ω, has continuous derivatives on each edge of Ω and second order continuous derivatives at the corners of Ω, $\mathcal{K}_{N+1}^2 v$ is the only polynomial in $\mathbb{P}_{N+1}(\Omega)$ which satisfies

(i) $(N-2)^2$ conditions in the interior of Ω:

$$\left(\mathcal{K}_{N+1}^2 v\right)(\boldsymbol{x}) = v(\boldsymbol{x}), \quad \boldsymbol{x} \in \Sigma_N^m, \tag{14.14}$$

(ii) $2(N-2)$ conditions on each edge Γ_j of Ω, $j = 1, 2, 3, 4$:

$$\left.\begin{aligned}
&\left(\mathcal{K}_{N+1}^2 v\right)\left(\pm 1, \zeta_i^2\right) = v\left(\pm 1, \zeta_i^2\right) \quad \text{and} \\
&\left(\frac{\partial(\mathcal{K}_{N+1}^2 v)}{\partial x}\right)\left(\pm 1, \zeta_i^2\right) = \left(\frac{\partial v}{\partial x}\right)\left(\pm 1, \zeta_i^2\right), \\
&\left(\mathcal{K}_{N+1}^2 v\right)\left(\zeta_i^2, \pm 1\right) = v\left(\zeta_i^2, \pm 1\right) \quad \text{and} \\
&\left(\frac{\partial(\mathcal{K}_{N+1}^2 v)}{\partial y}\right)\left(\zeta_i^2, \pm 1\right) = \left(\frac{\partial v}{\partial y}\right)\left(\zeta_i^2, \pm 1\right),
\end{aligned}\right\} \quad 1 \leqslant i \leqslant N-2. \tag{14.15}$$

(iii) four conditions at each corner $\boldsymbol{a}_J$ of Ω, $J = 1,2,3,4$:

$$\begin{aligned}
&\left(\mathcal{K}_{N+1}^2 v\right)(\boldsymbol{a}_J) = v(\boldsymbol{a}_J),\\
&\left(\frac{\partial(\mathcal{K}_{N+1}^2 v)}{\partial x}\right)(\boldsymbol{a}_J) = \left(\frac{\partial v}{\partial x}\right)(\boldsymbol{a}_J), \quad \left(\frac{\partial(\mathcal{K}_{N+1}^2 v)}{\partial y}\right)(\boldsymbol{a}_J) = \left(\frac{\partial v}{\partial y}\right)(\boldsymbol{a}_J),\\
&\left(\frac{\partial^2(\mathcal{K}_{N+1}^2 v)}{\partial x \partial y}\right)(\boldsymbol{a}_J) = \left(\frac{\partial^2 v}{\partial x \partial y}\right)(\boldsymbol{a}_J).
\end{aligned} \tag{14.16}$$

So, the complete definition of the operator $\mathcal{K}_{N+1}^2$ is not so easy to write when the function v does not belong to $H_0^2(\Omega)$. Note that, in the general case, the operator $\mathcal{K}_{N+m-1}^m$ on the hypercube Ω_d, involves crossed derivatives of order $d(m-1)$ of the function at the corners of Ω_d.

However, we shall only need a few properties of these operators, which can be derived from estimates (13.29) to (13.31) by exactly the same arguments as previously.

(i) for any function v such that $v(1-x_1^2)^{-1}\cdots(1-x_d^2)^{-1}$ belongs to $H^s(\Omega_d)$, $s > d/2$,

$$\begin{aligned}
&\left\|\left(v - \mathcal{K}_{N+1}^2 v\right)\left(1-x_1^2\right)^{-1}\cdots\left(1-x_d^2\right)^{-1}\right\|_{L^2(\Omega_d)}\\
&\quad \leqslant c N^{-s}\left\|v\left(1-x_1^2\right)^{-1}\cdots\left(1-x_d^2\right)^{-1}\right\|_{H^s(\Omega_d)};
\end{aligned} \tag{14.17}$$

(ii) for any real number r, $0 \leqslant r \leqslant 2$, and for any function v in $H^s(\Omega_d)$, $s > (6d+r)/4$,

$$\left\|v - \mathcal{K}_{N+1}^2 v\right\|_{H^r(\Omega_d)} \leqslant c N^{r-s}\,\|v\|_{H^s(\Omega_d)}; \tag{14.18}$$

(iii) for any function v such that $v(1-x_1^2)^{-1}\cdots(1-x_d^2)^{-1}$ belongs to $H^s(\Omega_d)$, $s > 3d/2$,

$$\begin{aligned}
&\left\|\left(v - \mathcal{K}_{N+2}^3 v\right)\left(1-x_1^2\right)^{-1}\cdots\left(1-x_d^2\right)^{-1}\right\|_{L^2(\Omega_d)}\\
&\quad \leqslant c N^{-s}\left\|v\left(1-x_1^2\right)^{-1}\cdots\left(1-x_d^2\right)^{-1}\right\|_{H^s(\Omega_d)}.
\end{aligned} \tag{14.19}$$

15. Collocation method for the second-order Dirichlet problem

The aim of this section is to present a basic and easy to implement, discretization of the Dirichlet problem. The first analysis of this discretization is due to CANUTO and QUARTERONI [1984]. We begin with the case of homogeneous boundary conditions, next we extend the method to general Dirichlet conditions.

The homogeneous Dirichlet problem is written

$$\begin{cases} -\Delta u = f & \text{in } \Omega_d,\\ u = 0 & \text{on } \partial\Omega_d. \end{cases} \tag{15.1}$$

In this section, we assume that the function f is continuous on Ω_d. We recall (see Section 2) that problem (15.1) admits the following equivalent variational formulation: *find u in $H_0^1(\Omega_d)$ such that*

$$\forall v \in H_0^1(\Omega_d), \quad a(u, v) = \int_{\Omega_d} f(\boldsymbol{x}) v(\boldsymbol{x}) \, \mathrm{d}\boldsymbol{x}, \tag{15.2}$$

where the form $a(\cdot, \cdot)$ is defined in (8.3).

The Galerkin discretization is constructed by replacing $H_0^1(\Omega_d)$ by $\mathbb{P}_N^0(\Omega_d)$. In the collocation method, also integrals are replaced by quadrature formulas and, in order to take the boundary conditions into account, a Gauss–Lobatto formula is chosen. So, we define for any continuous functions u and v on $\overline{\Omega}_d$,

$$(u, v)_N = \sum_{i_1=0}^{N} \cdots \sum_{i_d=0}^{N} u(\xi_{i_1}, \ldots, \xi_{i_d}) v(\xi_{i_1}, \ldots, \xi_{i_d}) \, \rho_{i_1} \cdots \rho_{i_d}. \tag{15.3}$$

We also set, for any polynomials u_N and v_N in $\mathbb{P}_N(\Omega_d)$,

$$a_N(u_N, v_N) = (\mathbf{grad}\, u_N, \mathbf{grad}\, v_N)_N. \tag{15.4}$$

Thus, the discrete problem is written: *find u_N in $\mathbb{P}_N^0(\Omega_d)$ such that*

$$\forall v_N \in \mathbb{P}_N^0(\Omega_d), \quad a_N(u_N, v_N) = (f, v_N)_N. \tag{15.5}$$

Of course, due to the Lax–Milgram lemma, the wellposedness of this problem relies on the ellipticity of the discrete bilinear form $a_N(\cdot, \cdot)$, which is stated in the following theorem.

THEOREM 15.1. *The bilinear form $a_N(\cdot, \cdot)$ satisfies the following properties of continuity*

$$\forall u_N \in \mathbb{P}_N(\Omega_d),\ \forall v_N \in \mathbb{P}_N(\Omega_d), \quad \left|a_N(u_N, v_N)\right| \leqslant 3^{d-1} |u_N|_{H^1(\Omega_d)} |v_N|_{H^1(\Omega_d)}, \tag{15.6}$$

and of ellipticity

$$\forall v_N \in \mathbb{P}_N(\Omega_d), \quad a_N(v_N, v_N) \geqslant |v_N|^2_{H^1(\Omega_d)}. \tag{15.7}$$

PROOF. The bilinear form $a_N(\cdot, \cdot)$ is written

$$a_N(u_N, v_N) = \sum_{j=1}^{d} \left(\frac{\partial u_N}{\partial x_j}, \frac{\partial v_N}{\partial x_j} \right)_N. \tag{15.8}$$

Since the quadrature formula is exact on $\mathbb{P}_{2N-1}(\Lambda)$, in the jth term of this sum, the summation on i_j can be replaced by the integral with respect to x_j and the $d-1$ other summations are of type $[\bullet,\bullet]_{N,1}$ defined in (13.16). Then, it follows at once from Remark 13.3 that

$$a_N(v_N, v_N) \geqslant \sum_{j=1}^{d} \left\| \frac{\partial v_N}{\partial x_j} \right\|^2_{L^2(\Omega_d)} = |v_N|^2_{H^1(\Omega)},$$

and from Remark 13.3 together with the Cauchy–Schwarz inequality that

$$|a_N(u_N, v_N)| \leqslant 3^{d-1} \sum_{j=1}^{d} \left\| \frac{\partial u_N}{\partial x_j} \right\|_{L^2(\Omega_d)} \left\| \frac{\partial v_N}{\partial x_j} \right\|_{L^2(\Omega_d)}.$$

This ends the proof.

As far as the right-hand side is concerned, we note that

$$(f, v_N)_N = (\mathcal{I}_N f, v_N)_N,$$

so that Remark 13.3 yields for any v_N in $\mathbb{P}_N(\Lambda)$,

$$(f, v_N)_N \leqslant 3^d \|\mathcal{I}_N f\|_{L^2(\Omega_d)} \|v_N\|_{L^2(\Omega_d)}. \tag{15.9}$$

Then, Theorem 15.1, together with the Poincaré–Friedrichs inequality (1.4) and the Lax–Milgram Theorem 2.1, leads to the following result.

THEOREM 15.2. *For any continuous function f on Ω_d, problem* (15.5) *has a unique solution u_N in $\mathbb{P}^0_N(\Omega_d)$. Moreover, this solution satisfies*

$$\|u_N\|_{H^1(\Omega_d)} \leqslant c\|\mathcal{I}_N f\|_{L^2(\Omega_d)}. \tag{15.10}$$

The next step is to establish a bound for the error $|u - u_N|_{H^1(\Omega_d)}$. The general framework is that of an elliptic problem in a reflexive Banach space X with norm $\|\bullet\|_X$: *find u in X such that*

$$\forall v \in X, \quad a(u, v) = \langle f, v \rangle, \tag{15.11}$$

which is discretized in the following way: *find u_δ in X_δ such that*

$$\forall v_\delta \in X_\delta, \quad a_\delta(u_\delta, v_\delta) = (f, v_\delta)_\delta. \tag{15.12}$$

The following result is known as Strang's lemma: if the space X_δ is contained in X and if the form $a_\delta(\bullet,\bullet)$ satisfies the ellipticity property

$$\forall v_\delta \in X_\delta, \quad a_\delta(v_\delta, v_\delta) \geqslant \alpha_\delta \|v_\delta\|^2_X, \tag{15.13}$$

for a positive constant α_δ, then the following error estimate holds:

$$\begin{aligned}
&\|u - u_\delta\|_X \\
&\leqslant \left(1 + \frac{\gamma}{\alpha_\delta}\right)\left(\inf_{v_\delta \in X_\delta} \left(\|u - v_\delta\|_X + \sup_{w_\delta \in X_\delta} \frac{a(v_\delta, w_\delta) - a_\delta(v_\delta, w_\delta)}{\|w_\delta\|_X} \right) \right. \\
&\qquad \left. + \sup_{w_\delta \in X_\delta} \frac{\langle f, w_\delta \rangle - (f, w_\delta)_\delta}{\|w_\delta\|_X} \right),
\end{aligned} \tag{15.14}$$

where γ stands for the norm of $a(\bullet,\bullet)$ on $X \times X$.

THEOREM 15.3. *The function f is supposed to be in a space $H^\sigma(\Omega_d)$, $\sigma > d/2$. Let us assume that the solution u of problem* (15.1) *belongs to $H^s(\Omega_d)$ for a real number $s \geqslant 1$. Then, the following error estimate holds for problem* (15.5)*:*

$$|u - u_N|_{H^1(\Omega_d)} \leqslant c\big(N^{1-s}\,\|u\|_{H^s(\Omega_d)} + N^{-\sigma}\|f\|_{H^\sigma(\Omega_d)}\big). \tag{15.15}$$

PROOF. Since the ellipticity constant in (15.7) is independent of N, applying (15.14) leads to

$$\begin{aligned}
|u - u_N|_{H^1(\Omega_d)} \leqslant c\Bigg(&|u - v_N|_{H^1(\Omega)} + \sup_{w_N \in \mathbb{P}_N^0(\Omega_d)} \frac{a(v_N, w_N) - a_N(v_N, w_N)}{|w_N|_{H^1(\Omega_d)}} \\
&+ \sup_{w_N \in \mathbb{P}_N^0(\Omega_d)} \frac{\int_{\Omega_d} f(\boldsymbol{x}) w_N(\boldsymbol{x})\,\mathrm{d}\boldsymbol{x} - (f, w_N)_N}{|w_N|_{H^1(\Omega_d)}} \Bigg),
\end{aligned} \tag{15.16}$$

for any v_N in $\mathbb{P}_N^0(\Omega_d)$. Here, the key idea is to choose v_N in $\mathbb{P}_{N-1}^0(\Omega_d)$, since the exactness of the quadrature formula implies that

$$\forall v_N \in \mathbb{P}_{N-1}(\Omega_d),\ \forall w_N \in \mathbb{P}_N(\Omega_d), \quad a_N(v_N, w_N) = a(v_N, w_N). \tag{15.17}$$

So, v_N can be taken equal to $\Pi_{N-1}^{1,0} u$ for instance. Similarly, we observe that, for any w_N in $\mathbb{P}_N(\Omega_d)$,

$$\begin{aligned}
&\int_{\Omega_d} f(\boldsymbol{x}) w_N(\boldsymbol{x})\,\mathrm{d}\boldsymbol{x} - (f, w_N)_N \\
&\quad = \int_{\Omega_d} (f - \Pi_{N-1} f)(\boldsymbol{x}) w_N(\boldsymbol{x})\,\mathrm{d}\boldsymbol{x} - (\mathcal{I}_N f - \Pi_{N-1} f, w_N)_N \\
&\quad \leqslant \big(\|f - \Pi_{N-1} f\|_{L^2(\Omega_d)} + 3^d \|\mathcal{I}_N f - \Pi_{N-1} f\|_{L^2(\Omega_d)}\big) \|w_N\|_{L^2(\Omega)}.
\end{aligned}$$

This yields

$$\begin{aligned}
&\sup_{w_N \in \mathbb{P}_N(\Omega_d)} \frac{\int_{\Omega_d} f(\boldsymbol{x}) w_N(\boldsymbol{x})\,\mathrm{d}\boldsymbol{x} - (f, w_N)_N}{\|w_N\|_{L^2(\Omega_d)}} \\
&\quad \leqslant c\big(\|f - \Pi_{N-1} f\|_{L^2(\Omega_d)} + \|f - \mathcal{I}_N f\|_{L^2(\Omega_d)}\big).
\end{aligned} \tag{15.18}$$

Using this estimate and the Poincaré–Friedrichs inequality (1.4), we obtain

$$|u-u_N|_{H^1(\Omega_d)} \leqslant c\big(|u-\Pi_{N-1}^{1,0}u|_{H^1(\Omega_d)} + \|f-\Pi_{N-1}f\|_{L^2(\Omega_d)} + \|f-\mathcal{I}_N f\|_{L^2(\Omega_d)}\big), \tag{15.19}$$

and Theorems 7.1, 7.2 and 14.2 give the conclusion.

For this simple problem and for smooth data f, the estimate is exactly of the same order as for the Galerkin discretization. Of course, different discrete problems could be defined by using more accurate quadrature formulas; their numerical analysis leads to exactly the same error estimates and the convergence curves for several test cases are exactly the same as for problem (15.5) (see MADAY and RØNQUIST [1990]). So, it is completely needless to use more accurate quadrature formulas, which are more expensive for complex equations.

REMARK 15.1. The previous comment also holds for problems with nonconstant coefficients. Indeed, let us for a while consider the problem:

$$\begin{cases} -\displaystyle\sum_{i=1}^{d}\sum_{j=1}^{d}\left(\frac{\partial}{\partial x_i}\right)\left(a_{ij}\frac{\partial u}{\partial x_j}\right) = f & \text{in } \Omega_d, \\ u = 0 & \text{on } \partial\Omega_d, \end{cases} \tag{15.20}$$

for given smooth functions a_{ij} on $\overline{\Omega}_d$ which satisfy the ellipticity property for a positive constant α:

$$\forall \boldsymbol{x} \in \overline{\Omega}_d,\ \forall(\xi_1,\ldots,\xi_d) \in \mathbb{R}^d, \quad \sum_{i=1}^{d}\sum_{j=1}^{d} a_{ij}(\boldsymbol{x})\xi_i\xi_j \geqslant \alpha\sum_{j=1}^{d}\xi_j^2. \tag{15.21}$$

It follows from the Lax–Milgram Theorem 2.1 that this problem has a unique solution. Then, its spectral discretization reads: *find* u_N *in* $\mathbb{P}_N^0(\Omega_d)$ *such that*

$$\forall v_N \in \mathbb{P}_N^0(\Omega_d), \quad \sum_{i=1}^{d}\sum_{j=1}^{d}\left(a_{ij}\frac{\partial u_N}{\partial x_j}, \frac{\partial v_N}{\partial x_i}\right)_N = (f, v_N)_N. \tag{15.22}$$

It is proven in MADAY and RØNQUIST [1990] that this discrete problem also has a unique solution and that the error estimate is very similar to (15.15):

$$|u-u_N|_{H^1(\Omega_d)} \leqslant c\big(N^{1-s}\|u\|_{H^s(\Omega_d)} + N^{-\sigma}\|f\|_{H^\sigma(\Omega_d)} + c_a N^{-\tau_a}\big), \tag{15.23}$$

where the power τ_a and the constant c_a only depend on the coefficients a_{ij}. There also, in the general case, there is no advantage in using more accurate formulas for the numerical integration.

The next step is to derive an estimate in $L^2(\Omega_d)$, by the Aubin–Nitsche duality method. The arguments are very similar to those of Theorem 8.3.

THEOREM 15.4. *If the assumptions of Theorem* 15.3 *are satisfied, the following error estimate holds for problem* (15.5)*:*

$$\|u - u_N\|_{L^2(\Omega_d)} \leqslant c\big(N^{-s}\|u\|_{H^s(\Omega_d)} + N^{-\sigma}\|f\|_{H^\sigma(\Omega_d)}\big). \tag{15.24}$$

PROOF. Exactly as in (8.9), we have

$$\|u - u_N\|_{L^2(\Omega_d)} = \sup_{g\in L^2(\Omega_d)} \frac{\int_{\Omega_d}(u - u_N)(\boldsymbol{x})g(\boldsymbol{x})\,\mathrm{d}\boldsymbol{x}}{\|g\|_{L^2(\Omega_d)}}. \tag{15.25}$$

Hence, for any g in $L^2(\Omega_d)$, we introduce the solution w of problem (8.10) and we recall that it satisfies (8.11). In order to compute

$$\int_{\Omega_d} (u - u_N)(\boldsymbol{x})g(\boldsymbol{x})\,\mathrm{d}\boldsymbol{x} = a(u - u_N, w),$$

we observe that, for all w_{N-1} in $\mathbb{P}_{N-1}(\Omega_d)$,

$$\begin{aligned} a(u_N, w_{N-1}) &= a_N(u_N, w_{N-1}) = (f, w_{N-1})_N \\ &= (\mathcal{I}_N f, w_{N-1})_N = \int_{\Omega_d} (\mathcal{I}_N f)(\boldsymbol{x})w_{N-1}(\boldsymbol{x})\,\mathrm{d}\boldsymbol{x}. \end{aligned}$$

Inserting this equation in the previous one and choosing w_{N-1} equal to $\Pi^{1,0}_{N-1}w$, we obtain

$$\begin{aligned} \int_{\Omega_d} (u - u_N)(\boldsymbol{x})g(\boldsymbol{x})\,\mathrm{d}\boldsymbol{x} &= a\big(u - u_N, w - \Pi^{1,0}_{N-1}w\big) \\ &\quad + \int_{\Omega_d} (f - \mathcal{I}_N f)(\boldsymbol{x})\big(\Pi^{1,0}_{N-1}w\big)(\boldsymbol{x}). \end{aligned}$$

This leads to the inequality

$$\begin{aligned} \int_{\Omega_d} (u - u_N)(\boldsymbol{x})g(\boldsymbol{x})\,\mathrm{d}\boldsymbol{x} &\leqslant |u - u_N|_{H^1(\Omega_d)}\big|w - \Pi^{1,0}_{N-1}w\big|_{H^1(\Omega_d)} \\ &\quad + \|f - \mathcal{I}_N f\|_{L^2(\Omega_d)}\big\|\Pi^{1,0}_{N-1}w\big\|_{L^2(\Omega_d)}. \end{aligned}$$

Theorems 7.2, 14.2 and 15.3 lead to the conclusion.

As standard, an interpolation argument allows for deriving from (15.15) and (15.24) the general estimate, for $0 \leqslant r \leqslant 1$, which holds when the assumptions of Theorem 15.3 are satisfied:

$$\|u - u_N\|_{H^r(\Omega_d)} \leqslant c\big(N^{r-s}\|u\|_{H^s(\Omega_d)} + N^{-\sigma}\|f\|_{H^\sigma(\Omega_d)}\big). \tag{15.26}$$

No doubt that the reader is now wondering about the title of this section, and especially about the word "collocation", since nothing but a Galerkin method with numerical integration has been presented. And this is an important property of spectral methods for most simple problems, that Galerkin methods with the integrals replaced by appropriate quadrature formulas coincide with collocation type discretizations. In the case of problem (15.5), we recall that, in the term $(\partial u_N/\partial x_j, \partial u_N/\partial x_j)_N$, the quadrature formula with respect to the jth variable is exact, so that an integration by parts can be made. If v_N vanishes for $x_j = \pm 1$, this leads to

$$\begin{aligned}\left[\frac{\partial u_N}{\partial x_j}, \frac{\partial v_N}{\partial x_j}\right]_{N,1} &= \int_{-1}^{1} \left(\frac{\partial u_N}{\partial x_j}\right)(x_j)\left(\frac{\partial v_N}{\partial x_j}\right)(x_j)\,\mathrm{d}x_j \\ &= -\int_{-1}^{1} \left(\frac{\partial^2 u_N}{\partial x_j^2}\right)(x_j) v_N(x_j)\,\mathrm{d}x_j = -\left[\frac{\partial^2 u_N}{\partial x_j^2}, v_N\right]_{N,1},\end{aligned}$$

where $[\bullet,\bullet]_{N,1}$ denotes the bilinear form introduced in (13.16) applied with respect to x_j. Hence, we derive

$$\forall u_N \in \mathbb{P}_N(\Omega_d),\ \forall v_N \in \mathbb{P}_N^0(\Omega_d), \quad a_N(u_N, v_N) = -(\Delta u_N, v_N)_N. \tag{15.27}$$

Let us introduce the Lagrange polynomials l_j that are associated with the set of points ξ_j, $0 \leqslant j \leqslant N$: for any j, $0 \leqslant j \leqslant N$, l_j belongs to $\mathbb{P}_N(\Lambda)$ and satisfies

$$l_j(\xi_j) = 1 \quad \text{and} \quad l_j(\xi_i) = 0, \quad 0 \leqslant i \leqslant j,\ i \neq j. \tag{15.28}$$

Then, applying (15.27) to the solution u_N of problem (15.5), with v_N equal to the polynomial $l_{i_1}(x_1)\cdots l_{i_d}(x_d)$, $1 \leqslant i_1, \ldots, i_d \leqslant N-1$, implies that

$$-(\Delta u_N)(\xi_{i_1}, \ldots, \xi_{i_d}) = f(\xi_{i_1}, \ldots, \xi_{i_d}), \quad 1 \leqslant i_1, \ldots, i_d \leqslant N-1. \tag{15.29}$$

This means that $-\Delta u_N$ coincides with f at the $(N-1)^d$ points of the grid Ξ_N which are located in the interior of Ω_d. Conversely, since the polynomials $l_{i_1}(x_1)\cdots l_{i_d}(x_d)$, $1 \leqslant i_1, \ldots, i_d \leqslant N-1$, form a basis of $\mathbb{P}_N^0(\Omega_d)$, (15.29) implies (15.5). Moreover, the fact that u_N vanishes on $\partial\Omega_d$ is equivalent to the fact that it vanishes in $(N+1)^{d-1}$ distinct points on each $\overline{\Gamma}_j$ where the Γ_j, $1 \leqslant j \leqslant 2d$, are the open faces of Ω_d, and these points can be chosen as the nodes of $\Xi_N \cap \overline{\Gamma}_j$. As a conclusion, problem (15.5), which is built by a Galerkin method with numerical integration, is equivalent to the following collocation system: find u_N in $\mathbb{P}_N(\Omega_d)$ such that

$$\begin{cases} -\Delta u_N(\boldsymbol{x}) = f(\boldsymbol{x}), & \boldsymbol{x} \in \Xi_N \cap \Omega_d, \\ u_N(\boldsymbol{x}) = 0, & \boldsymbol{x} \in \Xi_N \cap \overline{\Gamma}_j,\ 1 \leqslant j \leqslant 2d. \end{cases} \tag{15.30}$$

May be, this is a more natural discretization of the initial problem (15.1). This formulation also allows for handling easily more complex problems, and in particular, nonlinear equations.

REMARK 15.2. Note that the collocation form of problem (15.22) reads: find u_N in $\mathbb{P}_N(\Omega_d)$ such that

$$\begin{cases} -\sum_{i=1}^{d}\sum_{j=1}^{d}\left(\frac{\partial}{\partial x_i}\right)\mathcal{I}_N\left(a_{ij}\frac{\partial u_N}{\partial x_j}\right)(\boldsymbol{x}) = f(\boldsymbol{x}), & \boldsymbol{x} \in \Xi_N \cap \Omega_d, \\ u_N(\boldsymbol{x}) = 0, \quad \boldsymbol{x} \in \Xi_N \cap \overline{\Gamma}_j,\ 1 \leqslant j \leqslant 2d. \end{cases} \tag{15.31}$$

This is the right collocation formulation of problems with nonconstant coefficients, since it is easy to implement: only the derivation of polynomials of degree N are required.

The collocation formulation (15.30) of problem (15.5) suggests to extend the method to inhomogeneous Dirichlet conditions. So let us consider the problem:

$$\begin{cases} -\Delta u = f & \text{in } \Omega_d, \\ u = g_j & \text{on } \Gamma_j,\ 1 \leqslant j \leqslant 2d. \end{cases} \tag{15.32}$$

Here also, the data are continuous: the function f is continuous on Ω, and the functions g_j, $1 \leqslant j \leqslant 2d$, are continuous on $\overline{\Gamma}_j$ and satisfy the compatibility conditions

$$\forall \boldsymbol{x} \in \overline{\Gamma}_j \cap \overline{\Gamma}_k,\ g_j(\boldsymbol{x}) = g_k(\boldsymbol{x}), \quad 1 \leqslant j < k \leqslant 2d \tag{15.33}$$

(as a special case of this condition, the g_j define a unique value on each $(d-2)$-face of the hypercube).

Then, the idea is to take the following discrete problem: *find u_N in $\mathbb{P}_N(\Omega_d)$ such that*

$$\begin{cases} -\Delta u_N(\boldsymbol{x}) = f(\boldsymbol{x}), & \boldsymbol{x} \in \Xi_N \cap \Omega_d, \\ u_N(\boldsymbol{x}) = g_j(\boldsymbol{x}), & \boldsymbol{x} \in \Xi_N \cap \overline{\Gamma}_j,\ 1 \leqslant j \leqslant 2d. \end{cases} \tag{15.34}$$

The discrete solution u_N is equal to $i_N^j g_j$ on each Γ_j, where i_N^j denotes the $(d-1)$-dimensional Lagrange interpolation operator at the Gauss–Lobatto nodes on the face $\overline{\Gamma}_j$, $1 \leqslant j \leqslant 2d$. If we assume moreover that the functions g_j belong to $H^{1/2}(\Gamma_j)$

and are Hölder–continuous in a neighbourhood of the $(d-2)$-faces, it follows from Theorem 1.14 and (15.33), that there exists a function u_b in $H^1(\Omega_d)$ such that

$$u_b = g_j \quad \text{on } \Gamma_j,\ 1 \leqslant j \leqslant 2d. \tag{15.35}$$

Thus, if the function u_b belongs to $H^s(\Omega_d)$ for a real number $s > d/2$, the boundary conditions in (15.34) are equivalent to the fact that $u_N - \mathcal{I}_N u_b$ belongs to $H_0^1(\Omega_d)$. As previously, using the fact that the polynomials $l_{i_1}(x_1)\cdots l_{i_d}(x_d)$, $1 \leqslant i_1,\dots,i_d \leqslant N-1$, form a basis of $\mathbb{P}_N^0(\Omega_d)$, we derive that the collocation equations in (15.34) are equivalent to

$$\forall v_N \in \mathbb{P}_N^0(\Omega_d), \quad -(\Delta u_N, v_N)_N = (f, v_N)_N. \tag{15.36}$$

Hence, problem (15.34) has the following equivalent variational formulation: *find u_N in $\mathbb{P}_N(\Omega_d)$, with $u_N - \mathcal{I}_N u_b$ in $\mathbb{P}_N^0(\Omega_d)$, such that*

$$\forall v_N \in \mathbb{P}_N^0(\Omega_d), \quad a_N(u_N, v_N) = (f, v_N)_N. \tag{15.37}$$

This variational formulation allows for a very simple numerical analysis of the problem, from Theorem 15.1.

THEOREM 15.5. *For any continuous function f on Ω_d and for any continuous functions g_j, $1 \leqslant j \leqslant 2d$, in $H^{1/2}(\Gamma_j)$ satisfying* (15.33) *such that there exists a continuous function u_b satisfying* (15.35), *problem* (15.34) *has a unique solution u_N in $\mathbb{P}_N(\Omega_d)$.*

PROOF. The idea is to note that the polynomial $u_N^* = u_N - \mathcal{I}_N u_b$ belongs to $\mathbb{P}_N^0(\Omega_d)$ and satisfies

$$\forall v_N \in \mathbb{P}_N^0(\Omega_d), \quad a_N(u_N^*, v_N)_N = (f, v_N)_N - a_N(\mathcal{I}_N u_b, v_N). \tag{15.38}$$

From Theorem 15.1 and the Lax–Milgram Theorem 2.1, we deduce that this problem has a unique solution u_N^* in $\mathbb{P}_N^0(\Omega_d)$, whence the existence of a solution u_N of problem (15.34). The uniqueness of the solution also follows from the ellipticity property (15.7).

REMARK 15.3. The natural stability property here would be

$$\|u_N\|_{H^1(\Omega_d)} \leqslant c\left(\|\mathcal{I}_N f\|_{L^2(\Omega_d)} + \sum_{j=1}^{2d} \|i_N^j g_j\|_{H^{1/2}(\Gamma_j)}\right). \tag{15.39}$$

However, its proof is not trivial, since it requires the existence of an operator which lifts given polynomials on the Γ_j satisfying (15.33) into a polynomial on Ω_d, and which is stable from $\prod_{j=1}^{2d} H^{1/2}(\Gamma_j)$ into $H^1(\Omega_d)$. Such an operator is constructed in BERNARDI, DAUGE and MADAY [1992b] in the two-dimensional case only, we refer to BEN BELGACEM [1994] for an extension to the case $d = 3$.

THEOREM 15.6. *The function f is supposed to be in a space $H^\sigma(\Omega_d)$, $\sigma > d/2$. Let us assume that the solution u of problem* (15.32) *belongs to $H^s(\Omega_d)$ for a real number $s \geqslant (d+1)/2$. Then, the following error estimate holds for problem* (15.34):

$$|u - u_N|_{H^1(\Omega_d)} \leqslant c\big(N^{1-s}\|u\|_{H^s(\Omega_d)} + N^{-\sigma}\|f\|_{H^\sigma(\Omega_d)}\big). \tag{15.40}$$

PROOF. The idea is to apply the ellipticity property to the function $u_N - \mathcal{I}_N u$ which of course belongs to $\mathbb{P}_N^0(\Omega_d)$. This yields

$$\begin{aligned}|u_N - \mathcal{I}_N u|^2_{H^1(\Omega_d)} &\leqslant a_N(u_N - \mathcal{I}_N u, u_N - \mathcal{I}_N u)\\ &\leqslant (f, u_N - \mathcal{I}_N u)_N - a_N(\mathcal{I}_N u, u_N - \mathcal{I}_N u)\\ &\leqslant (f, u_N - \mathcal{I}_N u)_N - \langle f, u_N - \mathcal{I}_N u\rangle\\ &\quad + a(u, u_N - \mathcal{I}_N u) - a_N(\mathcal{I}_N u, u_N - \mathcal{I}_N u).\end{aligned}$$

By a triangular inequality, we derive

$$\begin{aligned}|u - u_N|_{H^1(\Omega_d)} &\leqslant |u - \mathcal{I}_N u|_{H^1(\Omega_d)}\\ &\quad + \sup_{w_N \in \mathbb{P}_N^0(\Omega_d)} \frac{\int_{\Omega_d} f(\boldsymbol{x}) w_N(\boldsymbol{x})\,\mathrm{d}\boldsymbol{x} - (f, w_N)_N}{|w_N|_{H_0^1(\Omega_d)}}\\ &\quad + \sup_{w_N \in \mathbb{P}_N^0(\Omega_d)} \frac{a(u, w_N) - a_N(\mathcal{I}_N u, w_N)}{|w_N|_{H^1(\Omega_d)}}.\end{aligned} \tag{15.41}$$

We estimate the first term thanks to Theorem 14.2, the second term thanks to (15.18) combined with Theorems 7.1 and 14.2, and the third term by noting that

$$\begin{aligned}a(u, w_N) - a_N(\mathcal{I}_N u, w_N) &= a\big(u - \Pi_{N-1}^{1,0} u, w_N\big) - a_N\big(\mathcal{I}_N u - \Pi_{N-1}^{1,0} u, w_N\big)\\ &\leqslant c\big(\big|u - \Pi_{N-1}^{1,0} u\big|_{H^1(\Omega_d)} + |u - \mathcal{I}_N u|_{H^1(\Omega_d)}\big)|w_N|_{H^1(\Omega_d)},\end{aligned}$$

and by making use of Theorems 7.2 and 14.2.

The estimate in $L^2(\Omega_d)$ is proven thanks to the same duality argument as Theorem 15.4.

THEOREM 15.7. *The function f is supposed to be in a space $H^\sigma(\Omega_d)$, $\sigma > d/2$, and the boundary functions g_j, $1 \leqslant j \leqslant 2d$, are supposed to be in spaces $H^{\tau_j}(\Gamma_j)$, $\tau_j > (d-1)/2$. Let us assume that the solution u of problem* (15.32) *belongs to $H^s(\Omega_d)$ for a real number $s \geqslant (d+1)/2$. Then, the following error estimate holds for problem* (15.34):

$$\begin{aligned}&\|u - u_N\|_{L^2(\Omega_d)}\\ &\quad \leqslant c\left(N^{-s}\|u\|_{H^s(\Omega_d)} + N^{-\sigma}\|f\|_{H^\sigma(\Omega_d)} + \sum_{j=1}^{2d} N^{-\tau_j}\|g_j\|_{H^{\tau_j}(\Gamma_j)}\right).\end{aligned} \tag{15.42}$$

PROOF. As previously, we have

$$\|u-u_N\|_{L^2(\Omega)} = \sup_{g\in L^2(\Omega)} \frac{\int_\Omega (u-u_N)(\boldsymbol{x})g(\boldsymbol{x})\,\mathrm{d}\boldsymbol{x}}{\|g\|_{L^2(\Omega)}}. \tag{15.43}$$

Again, we consider the solution w of problem (8.10) and we compute

$$\int_{\Omega_d} (u-u_N)(\boldsymbol{x})g(\boldsymbol{x})\,\mathrm{d}\boldsymbol{x} = a(u-u_N,w) - \sum_{j=1}^{2d}\int_{\Gamma_j} (g_j - i_N^j g_j)(\tau)\frac{\partial w}{\partial n_j}(\tau)\,\mathrm{d}\tau.$$

By subtracting the polynomial $\Pi_{N-1}^{1,0}w$ in the first term, we obtain

$$\begin{aligned}\int_{\Omega_d} (u-u_N)(\boldsymbol{x})g(\boldsymbol{x})\,\mathrm{d}\boldsymbol{x} &\leqslant |u-u_N|_{H^1(\Omega)}\big|w-\Pi_{N-1}^{1,0}w\big|_{H^1(\Omega)} \\ &+\big(\|f-\Pi_{N-1}f\|_{L^2(\Omega)} + \|f-\mathcal{I}_N f\|_{L^2(\Omega)}\big)\big\|\Pi_{N-1}^{1,0}w\big\|_{L^2(\Omega)} \\ &+\sum_{j=1}^{2d}\|g_j - i_N^j g_j\|_{L^2(\Gamma_j)}\left\|\frac{\partial w}{\partial n_j}\right\|_{H^{1/2}(\Gamma_j)},\end{aligned}$$

and the end of the proof is the same as for Theorem 15.4.

This ends the analysis of the spectral discretization for the second-order Dirichlet problem, which is the basic example of spectral methods. Due to the collocation formulation of the discrete problem, the extension to more complex equations is easy, to nonlinear equations in particular. Due to its variational form, it results into a linear system where the vector of unknowns is made by the values of the discrete solution on the points of the grid. It is readily checked that the mass matrix is diagonal and that the stiffness matrix is symmetric and positive. Other details on the implementation are given by CANUTO, HUSSAINI, QUARTERONI and ZANG [1987], MADAY and PATERA [1989], RØNQUIST [1988], see also BERNARDI, MADAY and PATERA to appear in this Handbook.

16. Galerkin methods with numerical integration for the fourth-order Dirichlet problem

Here, we are interested in the discretization of the fourth-order Dirichlet problem, which we provide with homogeneous boundary conditions in a first step:

$$\begin{cases} \Delta^2 u = f & \text{in } \Omega_d, \\ u = 0 & \text{on } \partial\Omega_d, \\ \dfrac{\partial u}{\partial n} = 0 & \text{on } \partial\Omega_d. \end{cases} \tag{16.1}$$

The function f is assumed to be continuous on Ω_d. Thus, problem (16.1) admits the equivalent variational formulation: *find u in $H_0^2(\Omega_d)$ such that*

$$\forall v \in H_0^2(\Omega_d), \quad d(u,v) = \int_{\Omega_d} f(\boldsymbol{x}) v(\boldsymbol{x})\, \mathrm{d}\boldsymbol{x}, \tag{16.2}$$

where the form $d(\bullet,\bullet)$ is defined in (9.3), and it has a unique solution in $H_0^2(\Omega)$.

The discrete solution is sought in the space $\mathbb{P}_N^{2,0}(\Omega_d) = \mathbb{P}_N(\Omega_d) \cap H_0^2(\Omega_d)$. So, the construction of the discrete problem only depends on the choice of an appropriate quadrature formula. Since there are two boundary conditions on each face, the idea is to use the generalized formula (4.3) with $m = 2$. Moreover, the dimension of $\mathbb{P}_N^{2,0}(\Omega_d)$ is equal to $(N-3)^d$, hence only $(N-3)^d$ collocation equations are necessary in the interior of Ω_d, so that we shall use formula (4.3) with N replaced by $N-1$. Thus, we denote by $\zeta_i^{2(N-1)}$, $1 \leqslant i \leqslant N-3$, the zeros of the polynomial L''_{N-1}, and by $\sigma_j^{2(N-1)}$, the corresponding weights in formula (4.3). Next, we introduce the bilinear form $(\bullet,\bullet)_{N-1,2}$, which is defined for any sufficiently smooth functions u and v in the following way: in $\int_{\Omega_d} u(\boldsymbol{x})v(\boldsymbol{x})\,\mathrm{d}\boldsymbol{x}$, each integral with respect to x_j, $1 \leqslant j \leqslant d$, is replaced by the quadrature formula $[u,v]_{N-1,2}$ with notation (13.16). When at least one of two functions u or v belongs to $H_0^2(\Omega_d)$, this is written as

$$\begin{aligned}(u,v)_{N-1,2} = \sum_{i_1=1}^{N-3} \cdots \sum_{i_d=1}^{N-3} u\big(\zeta_{i_1}^{2(N-1)}, \ldots, \zeta_{i_d}^{2(N-1)}\big) \\ v\big(\zeta_{i_1}^{2(N-1)}, \ldots, \zeta_{i_d}^{2(N-1)}\big)\sigma_{i_1}^{2(N-1)} \cdots \sigma_{i_d}^{2(N-1)},\end{aligned} \tag{16.3}$$

but try to write it in the general case! In particular, crossed derivatives $\partial^d/\partial x_1 \cdots \partial x_d$ of order d of the functions are involved at each corner of the domain.

We also set, for any polynomials u_N in $\mathbb{P}_N(\Omega_d)$ and v_N in $\mathbb{P}_N(\Omega_d)$,

$$d_{N-1,2}(u_N, v_N) = \big(\Delta^2 u_N, v_N\big)_{N-1,2}. \tag{16.4}$$

Then, the discrete problem is the following one: *find u_N in $\mathbb{P}_N^{2,0}(\Omega_d)$ such that*

$$\forall v_N \in \mathbb{P}_N^{2,0}(\Omega_d), \quad d_{N-1,2}(u_N, v_N) = (f, v_N)_{N-1,2}. \tag{16.5}$$

For $1 \leqslant j \leqslant N-3$, let us define r_j as a polynomial in $\mathbb{P}_{N-4}(\Lambda)$ which vanishes in all the points $\zeta_i^{2(N-1)}$, $1 \leqslant j \leqslant N-3$, but not in $\zeta_j^{2(N-1)}$. Letting the function v_N run through the basis

$$\big\{(1-x_1)^2 \cdots (1-x_d)^2\, r_{i_1}(x_1) \cdots r_{i_d}(x_d)\colon\ 1 \leqslant i_1, \ldots, i_d \leqslant N-3\big\}$$

of $\mathbb{P}_N^{2,0}(\Omega_d)$, we observe that equation (16.5) is completely equivalent to (see notation (14.12))

$$(\Delta^2 u_N)(\boldsymbol{x}) = f(\boldsymbol{x}), \quad \boldsymbol{x} \in \Sigma_{N-1}^2. \tag{16.6}$$

The boundary conditions can also be written in the following way: the functions u_N and $\partial u_N/\partial n$, together with their first tangential derivatives, satisfy $(N+1)^2$ cancellation conditions on each face $\overline{\Gamma}_j$, $1 \leqslant j \leqslant 2d$. Hence, problem (16.5) is equivalent to a system of $(N+1)^d$ collocation equations. Let us make it explicit in the case $d = 2$ of the square Ω:

$$\begin{cases} -(\Delta^2 u_N)(\boldsymbol{x}) = f(\boldsymbol{x}), \quad \boldsymbol{x} \in \Sigma_{N-1}^2 \cap \Omega, \\ u_N(\boldsymbol{x}) = 0 \text{ and } \left(\dfrac{\partial u_N}{\partial n_J}\right)(\boldsymbol{x}) = 0, \quad \boldsymbol{x} \in \Sigma_{N-1}^2 \cap \Gamma_J,\ J = 1,2,3,4, \\ u_N(\boldsymbol{a}_J) = \left(\dfrac{\partial u_N}{\partial x}\right)(\boldsymbol{a}_J) = \left(\dfrac{\partial u_N}{\partial y}\right)(\boldsymbol{a}_J) = \left(\dfrac{\partial^2 u_N}{\partial x \partial y}\right)(\boldsymbol{a}_J) = 0, \\ \quad J = 1,2,3,4. \end{cases} \tag{16.7}$$

This problem involves second-order crossed derivatives at the corners of the square.

Since the quadrature formula which is used for this problem, is exact only on the space $\mathbb{P}_{2N-3}(\Lambda)$, it can be checked that the bilinear form $d_{N-1,2}(\bullet,\bullet)$ does not coincide with the form:

$$(u_N, v_N) \mapsto (\Delta u_N, \Delta v_N)_{N-1,2}.$$

However, using an appropriate basis allows for proving that the form $d_{N-1,2}(\bullet,\bullet)$ is symmetric on $\mathbb{P}_N^{2,0}(\Omega_d) \times \mathbb{P}_N^{2,0}(\Omega_d)$. The properties of this form are stated in the following lemma. Since they rely on rather technical arguments, we skip the proof here, we refer to BERNARDI, COPPOLETTA and MADAY ([1990], §3), [1992]) for the detailed proof in the two-dimensional case, its extension to higher dimensions being straightforward.

THEOREM 16.1. *The bilinear form $d_{N-1,2}(\bullet,\bullet)$ satisfies the following properties of continuity*

$$\begin{aligned} &\forall u_N \in \mathbb{P}_N(\Omega_d),\ \forall v_N \in \mathbb{P}_N(\Omega_d), \\ &\quad \left|d_{N-1,2}(u_N, v_N)\right| \leqslant c|u_N|_{H^2(\Omega_d)}|v_N|_{H^2(\Omega_d)}, \end{aligned} \tag{16.8}$$

and of ellipticity

$$\forall v_N \in \mathbb{P}_N^{2,0}(\Omega_d), \quad d_{N-1,2}(v_N, v_N) \geqslant cN^{1-d}|v_N|^2_{H^2(\Omega)}. \tag{16.9}$$

This theorem is very disappointing, since the constant of ellipticity tends to 0 when N tends to $+\infty$, for $d \geqslant 2$. This lack of ellipticity relies on the one-dimensional result:

$$\forall \varphi_N \in \mathbb{P}_N^{2,0}(\Lambda),$$
$$cN^{-1}\,\|\varphi_N\|_{L^2(\Lambda)}^2 \leqslant [\varphi_N, \varphi_N]_{N-1,2} \leqslant c'\|\varphi_N\|_{L^2(\Lambda)}^2. \qquad (16.10)$$

It can be checked that the result cannot be improved: taking

$$v_N = \varphi_N(x_1)\cdots\varphi_N(x_{d-1})\psi_N(x_d),$$

where φ_N is the polynomial $(1-\zeta^2)^2 L_{N-1}''''$ and ψ_N is any polynomial in $\mathbb{P}_{N-3}^{2,0}(\Lambda)$ satisfying

$$\|\psi_N\|_{H^2(\Lambda)} \geqslant cN^4\|\psi_N\|_{L^2(\Lambda)}, \qquad (16.11)$$

we derive that

$$d_{N-1,2}(v_N, v_N) \leqslant cN^{1-d}\,|v_N|_{H^2(\Omega)}^2. \qquad (16.12)$$

An example of polynomial ψ_N satisfying (16.11) and the detailed study of the counter-example is given in BERNARDI and MADAY ([1992b], Chapter V, Remark 2.8).

However, combining Theorem 16.1 with the Lax–Milgram Theorem 2.1 leads to the existence and uniqueness result. Note that the continuity of the right-hand side is given by (16.10), since it yields that

$$\forall u_N \in \mathbb{P}_N(\Omega_d), \forall v_N \in \mathbb{P}_N(\Omega_d),$$
$$(u_N, v_N)_{N-1,2} \leqslant c\|u_N\|_{L^2(\Omega_d)}\|v_N\|_{L^2(\Omega_d)}. \qquad (16.13)$$

THEOREM 16.2. *For any continuous function f on Ω_d, problem* (16.5) *has a unique solution u_N in $\mathbb{P}_N^{2,0}(\Omega_d)$. Moreover, if the function f is sufficiently smooth in a neighborhood of the boundary $\partial\Omega_d$, this solution satisfies*

$$\|u_N\|_{H^2(\Omega_d)} \leqslant cN^{d-1}\|\mathcal{K}_N^2 f\|_{L^2(\Omega_d)}. \qquad (16.14)$$

The assumption on the function f is due to the fact that $\mathcal{K}_N^2 f$ is defined only if the function f has a continuous normal derivative on each face, a continuous second-order normal derivative on each $(d-2)$-face, and so on. Now, we prove two error estimates concerning problem (16.5), one in $H^2(\Omega_d)$ and one in $L^2(\Omega_d)$ by the Aubin–Nitsche method.

THEOREM 16.3. *The function f is supposed to be in a space $H^\sigma(\Omega_d)$, $\sigma > d/2$. Let us assume that the solution u of problem* (16.1) *belongs to $H^s(\Omega_d)$ for a real number $s \geqslant 2$. Then, the following error estimate holds for problem* (16.5):

$$|u - u_N|_{H^2(\Omega_d)} \leqslant c\big(N^{d+1-s}\|u\|_{H^s(\Omega_d)} + N^{d-1-\sigma}\|f\|_{H^\sigma(\Omega_d)}\big). \qquad (16.15)$$

PROOF. Since the ellipticity constant in (16.9) is larger than cN^{1-d}, applying (15.14) and taking v_N equal to $\Pi^{2,0}_{N-3}u$ gives

$$|u-u_N|_{H^2(\Omega_d)} \leqslant cN^{d-1}\Bigg(\big|u-\Pi^{2,0}_{N-3}u\big|_{H^2(\Omega)} + \sup_{w_N\in\mathbb{P}^{2,0}_N(\Omega_d)} \frac{\int_{\Omega_d} f(\boldsymbol{x})w_N(\boldsymbol{x})\,\mathrm{d}\boldsymbol{x} - (f,w_N)_{N-1,2}}{|w_N|_{H^2(\Omega_d)}}\Bigg). \tag{16.16}$$

The first term is estimated thanks to Theorem 7.2. As in (15.18), since the quadrature formula is exact on $\mathbb{P}_{2N-3}(\Lambda)$, setting $\tilde{f} = f(1-x_1^2)\cdots(1-x_d^2)$ and using (16.13), we observe that

$$\begin{aligned}\sup_{w_N\in\mathbb{P}_N(\Omega_d)} &\frac{\int_{\Omega_d} f(\boldsymbol{x})w_N(\boldsymbol{x})\,\mathrm{d}\boldsymbol{x} - (f,w_N)_{N-1,2}}{\|w_N\|_{L^2(\Omega_d)}}\\ &\leqslant c\Big(\big\|\big(\tilde{f}-\mathcal{K}^2_{N-1}\tilde{f}\big)\big(1-x_1^2\big)^{-1}\cdots\big(1-x_d^2\big)^{-1}\big\|_{L^2(\Omega_d)}\\ &\quad+\big\|\big(\tilde{f}-\mathcal{K}^2_{N}\tilde{f}\big)\big(1-x_1^2\big)^{-1}\cdots\big(1-x_d^2\big)^{-1}\big\|_{L^2(\Omega_d)}\Big),\end{aligned} \tag{16.17}$$

so that the estimate of the second term follows from (14.17).

We prove the estimate in $L^2(\Omega_d)$ in the case $d=2$ of the square Ω, since appropriate regularity results are known only in this case.

THEOREM 16.4. *In the case* $d=2$ *of the square* Ω, *if the assumptions of Theorem* 16.3 *are satisfied, the following error estimate holds for problem* (16.5):

$$\|u-u_N\|_{L^2(\Omega)} \leqslant c\big(N^{1-s}\|u\|_{H^s(\Omega)} + N^{1-\sigma}\|f\|_{H^\sigma(\Omega)}\big). \tag{16.18}$$

PROOF. As in (9.8), due to the formula

$$\|u-u_N\|_{L^2(\Omega)} = \sup_{g\in L^2(\Omega)} \frac{\int_\Omega (u-u_N)(\boldsymbol{x})g(\boldsymbol{x})\,\mathrm{d}\boldsymbol{x}}{\|g\|_{L^2(\Omega)}}, \tag{16.19}$$

for any g in $L^2(\Omega)$, we define the solution w of problem (9.9), which satisfies (9.10). Thus, we have

$$\begin{aligned}\int_\Omega (u-u_N)(\boldsymbol{x})g(\boldsymbol{x})\,\mathrm{d}\boldsymbol{x} &= \int_\Omega (u-u_N)(\boldsymbol{x})\big(\Delta^2 w\big)(\boldsymbol{x})\,\mathrm{d}\boldsymbol{x} = d(u-u_N,w)\\ &= d\big(u-u_N, w-\Pi^{2,0}_{N-3}w\big) + \int_\Omega f(\boldsymbol{x})\big(\Pi^{2,0}_{N-3}w\big)(\boldsymbol{x})\,\mathrm{d}\boldsymbol{x}\\ &\quad-\big(f,\Pi^{2,0}_{N-3}w\big)_{N-1,2},\end{aligned}$$

so that, with the same notation as previously,

$$\int_\Omega (u-u_N)(\boldsymbol{x})g(\boldsymbol{x})\,\mathrm{d}\boldsymbol{x} \leqslant c\Big(|u-u_N|_{H^2(\Omega)}|w-\Pi^{2,0}_{N-3}w|_{H^2(\Omega)} + \|(\tilde f-\mathcal{K}^2_N\tilde f)(1-x^2)^{-1}(1-y^2)^{-1}\|_{L^2(\Omega)}\|\Pi^{2,0}_{N-3}w\|_{L^2(\Omega)}\Big).$$

The result follows from Theorems 7.2 and 16.3, and (14.17).

Of course, an interpolation argument between estimates (16.15) and (16.18) leads to the following inequality for any r, $0\leqslant r\leqslant 2$:

$$\|u-u_N\|_{H^r(\Omega)} \leqslant c\big(N^{1+r-s}\|u\|_{H^s(\Omega)} + N^{1-\sigma}\|f\|_{H^\sigma(\Omega)}\big). \tag{16.20}$$

Now, we present the extension of this collocation method to the case of inhomogeneous Dirichlet boundary conditions. Only for the sake of clarity, we limit ourselves to the case $d=2$ of the square Ω. Then, the problem we intend to approximate reads:

$$\begin{cases} \Delta^2 u = f & \text{in } \Omega_d,\\ u = g_J & \text{on } \Gamma_J,\ J=1,2,3,4,\\ \dfrac{\partial u}{\partial n} = h_J & \text{on } \Gamma_J,\ J=1,2,3,4. \end{cases} \tag{16.21}$$

Here, the function f is supposed to be continuous on Ω_d. The functions g_J and h_J, $J=1,2,3,4$, are supposed to be in $H^{3/2}(\Gamma_J)$ and $H^{1/2}(\Gamma_J)$ respectively, to be continuous on $\overline{\Gamma}_J$ with Hölder-continuous first derivatives at the end points and to satisfy the following compatibility conditions:

$$\begin{aligned} &g_J(\boldsymbol{a}_J) = g_{J+1}(\boldsymbol{a}_J),\\ &h_J(\boldsymbol{a}_J) = -\Big(\frac{\mathrm{d}g_{J+1}}{\mathrm{d}\tau_{J+1}}\Big)(\boldsymbol{a}_J),\quad \Big(\frac{\mathrm{d}g_J}{\mathrm{d}\tau_J}\Big)(\boldsymbol{a}_J) = h_{J+1}(\boldsymbol{a}_J),\\ &\Big(\frac{\mathrm{d}h_J}{\mathrm{d}\tau_J}\Big)(\boldsymbol{a}_J) = -\Big(\frac{\mathrm{d}h_{J+1}}{\mathrm{d}\tau_{J+1}}\Big)(\boldsymbol{a}_J),\quad J=1,2,3,4. \end{aligned} \tag{16.22}$$

From Theorem 1.13 and (1.30), there exists a function u_b in $H^2(\Omega)$ such that

$$u_b = g_J \quad\text{and}\quad \frac{\partial u_b}{\partial n_J} = h_J \quad \text{on } \Gamma_J,\ J=1,2,3,4, \tag{16.23}$$

so that problem (16.21) admits the equivalent variational formulation: *find u in $H^2(\Omega)$, with $u-u_b$ in $H^2_0(\Omega)$, such that*

$$\forall v\in H^2_0(\Omega),\quad d(u,v) = \int_\Omega f(\boldsymbol{x})v(\boldsymbol{x})\,\mathrm{d}\boldsymbol{x}. \tag{16.24}$$

Consequently, it has a unique solution in $H^2(\Omega)$.

Next, assuming that the function u_b is sufficiently smooth on Ω, we define the discrete problem: *find* u_N *in* $\mathbb{P}_N(\Omega)$, *with* $u_N - \mathcal{K}_N^2 u_b$ *in* $\mathbb{P}_N(\Omega)$, *such that*

$$\forall v_N \in \mathbb{P}_N^{2,0}(\Omega), \quad d_{N-1,2}(u_N, v_N) = (f, v_N)_{N-1,2}. \tag{16.25}$$

Exactly as in (16.7), it is easily checked that this problem consists equivalently in finding a polynomial u_N in $\mathbb{P}_N(\Omega)$ such that

$$\begin{cases} (\Delta^2 u_N)(\boldsymbol{x}) = f(\boldsymbol{x}), \quad \boldsymbol{x} \in \Sigma_{N-1}^2 \cap \Omega, \\ u_N(\boldsymbol{x}) = g_J(\boldsymbol{x}) \quad \text{and} \quad \left(\dfrac{\partial u_N}{\partial n_J}\right)(\boldsymbol{x}) = h_J(\boldsymbol{x}), \\ \quad \boldsymbol{x} \in \Sigma_{N-1}^2 \cap \Gamma_J, \ J = 1,2,3,4, \\ u_N(\boldsymbol{a}_J) = g_J(\boldsymbol{a}_J), \quad \left(\dfrac{\partial u_N}{\partial n_J}\right)(\boldsymbol{a}_J) = h_J(\boldsymbol{a}_J), \\ \left(\dfrac{\partial u_N}{\partial n_{J+1}}\right)(\boldsymbol{a}_J) = h_{J+1}(\boldsymbol{a}_J), \quad \left(\dfrac{\partial^2 u_N}{\partial n_J \partial n_{J+1}}\right)(\boldsymbol{a}_J) = \left(\dfrac{\partial h_J}{\partial \tau_J}\right)(\boldsymbol{a}_J), \\ \quad J = 1,2,3,4, \end{cases} \tag{16.26}$$

which is a system of collocation equations.

The first result concerning problem (16.25) follows from Theorem 16.1 by considering the problem which is solved by the polynomial $u_N - \mathcal{K}_N^2 u_b$.

THEOREM 16.5. *For any continuous function* f *on* Ω *and for any 8-tuple functions* (g_J, h_J)*in* $\prod_{J=1}^4 H^{3/2}(\Gamma_J) \times H^{1/2}(\Gamma_J)$ *satisfying* (16.22) *such that there exists a sufficiently smooth function* u_b *satisfying* (16.23), *problem* (16.25) *has a unique solution* u_N *in* $\mathbb{P}_N(\Omega)$.

Of course, a sufficient condition on the regularity of u_b is that it is of class $\mathscr{C}^2$ on $\overline{\Omega}$. It is proven in BERNARDI, DAUGE and MADAY [1992b] for instance that this property holds when the functions g_J and h_J, $J = 1,2,3,4$, belong to $H^{\tau_J}(\Gamma_J)$ and $H^{\rho_J}(\Gamma_J)$ respectively, $\tau_J > \frac{5}{2}$ and $\rho_J > \frac{3}{2}$, but these conditions are not necessary. For the following error estimates, we assume that the solution u of problem (16.21) belongs to $H^s(\Omega)$, $s > 3$, so that the assumptions of Theorem 16.5 are satisfied with $u_b = u$.

THEOREM 16.6. *The function* f *is supposed to be in a space* $H^\sigma(\Omega)$, $\sigma > 1$. *Let us assume that the solution* u *of problem* (16.21) *belongs to* $H^s(\Omega_d)$ *for a real number* $s > \frac{7}{2}$. *Then, the following error estimate holds for problem* (16.25):

$$|u - u_N|_{H^2(\Omega)} \leqslant c\big(N^{3-s}\|u\|_{H^s(\Omega)} + N^{1-\sigma}\|f\|_{H^\sigma(\Omega)}\big). \tag{16.27}$$

PROOF. The function $u_N^\circ = u_N - \mathcal{K}_N^2 u$ satisfies

$$\forall v_N \in \mathbb{P}_N^{2,0}(\Omega), \quad d_N\big(u_N^\circ, v_N\big) = (f, v_N)_{N-1,2} - d_N\big(\mathcal{K}_N^2 u, v_N\big),$$

so that applying the ellipticity property (16.9) yields

$$
\begin{aligned}
|u_N^\circ|^2_{H^2(\Omega)} &\leqslant cN d_N\big(u_N - \mathcal{K}_N^2 u, u_N^\circ\big) \\
&\leqslant cN\bigg(d\big(u - \Pi_{N-3}^2 u, u_N^\circ\big) + d_N\big(\Pi_{N-3}^2 u - \mathcal{K}_N^2 u, u_N^\circ\big) \\
&\quad - \int_\Omega f(\boldsymbol{x}) u_N^\circ(\boldsymbol{x})\,\mathrm{d}\boldsymbol{x} + (f, u_N^\circ)_{N-1,2}\bigg).
\end{aligned}
$$

Then, applying the continuity property (16.8) and a triangular inequality gives

$$
\begin{aligned}
|u - u_N|_{H^2(\Omega_d)} &\leqslant cN\bigg(\big\|u - \Pi_{N-3}^2 u\big\|_{H^2(\Omega)} + \big\|u - \mathcal{K}_N^2 u\big\|_{H^2(\Omega)} \\
&\quad + \sup_{w_N \in \mathbb{P}_N^{2,0}(\Omega_d)} \frac{\int_{\Omega_d} f(\boldsymbol{x}) w_N(\boldsymbol{x})\,\mathrm{d}\boldsymbol{x} - (f, w_N)_{N-1,2}}{|w_N|_{H^2(\Omega_d)}}\bigg).
\end{aligned}
\tag{16.28}
$$

Applying Theorems 7.1, 7.3, (14.17) and (16.17) gives the result.

THEOREM 16.7. *The function f is supposed to be in a space $H^\sigma(\Omega)$, $\sigma > 1$, and the boundary functions g_J and h_J, $J = 1,2,3,4$, are supposed to be in spaces $H^{\tau_J}(\Gamma_J)$ and $H^{\rho_J}(\Gamma_J)$, $\tau_J \geqslant \frac{3}{2}$ and $\rho_J > \frac{1}{2}$. Let us assume that the solution u of problem* (16.21) *belongs to $H^s(\Omega)$ for a real number $s > \frac{7}{2}$. Then, the following error estimate holds for problem* (16.25):

$$
\begin{aligned}
\|u - u_N\|_{L^2(\Omega)} &\leqslant c\bigg(N^{1-s}\|u\|_{H^s(\Omega)} + N^{1-\sigma}\|f\|_{H^\sigma(\Omega)} \\
&\quad + \sum_{J=1}^{4}\big(N^{-\tau_J}\|g_J\|_{H^{\tau_J}(\Gamma_J)} + N^{-\rho_J}\|h_J\|_{H^{\rho_J}(\Gamma_J)}\big)\bigg).
\end{aligned}
\tag{16.29}
$$

PROOF. Exactly as in the proof of Theorem 16.4, for any g in $L^2(\Omega)$, we consider the solution w of problem (9.9) and we compute

$$
\begin{aligned}
&\int_\Omega (u - u_N)(\boldsymbol{x}) g(\boldsymbol{x})\,\mathrm{d}\boldsymbol{x} \\
&\quad = d\big(u - u_N, w - \Pi_{N-3}^{2,0} w\big) \\
&\quad\quad - \int_\Omega f(\boldsymbol{x})\big(\Pi_{N-3}^{2,0} w\big)(\boldsymbol{x})\,\mathrm{d}\boldsymbol{x} + \big(f, \Pi_{N-3}^{2,0} w\big)_{N-1,2} \\
&\quad\quad + \sum_{J=1}^{4}\bigg(\int_{\Gamma_J} \big(g_J - k_N^{2J} g_J\big)(\tau)\Big(\frac{\partial w}{\partial n_J}\Big)(\tau)\,\mathrm{d}\tau \\
&\quad\quad - \int_{\Gamma_J} \big(h_J - k_N^{2J} h_J\big)(\tau) w(\tau)\,\mathrm{d}\tau\bigg),
\end{aligned}
$$

where k_N^{2J} stands for the interpolation operator k_N^2 translated onto the edge Γ_J. From Theorems 7.2 and 16.6, (16.17) (and its one-dimensional analogue), (14.17) and (13.29), the desired result is derived in a standard way.

The collocation method is not optimal, however it allows for a natural treatment of inhomogeneous boundary conditions and also for easy extensions to less simple equations. But, since it is sometimes better to have an optimal method, we are going to present two of them, in the d-dimensional case and for homogeneous boundary conditions. So, we again consider problem (16.1) and we use other quadrature formulas to approximate the integrals in the form $d(\cdot,\cdot)$ and in the right-hand side. In view of (13.20), these quadrature formulas are formulas (4.3) with $m = 1$ and $m = 3$.

So, for sufficiently smooth functions u and v, we define the quantities $(u, v)_{N,m}$, $m = 1$ and 3, by replacing in $\int_{\Omega_d} u(\boldsymbol{x})v(\boldsymbol{x})\,\mathrm{d}\boldsymbol{x}$, each integral with respect to x_j, $1 \leqslant j \leqslant d$, by the quadrature formula (4.3) with $m = 1$ or 3. Of course, the bilinear form $(\cdot,\cdot)_{N,1}$ coincides with the form $(\cdot,\cdot)_N$ defined in (15.3). When one of the functions u and v belongs to $H_0^1(\Omega_d)$ in the case $m = 1$ and when both functions u and v belongs to $H_0^2(\Omega_d)$ in the case $m = 3$, the quantity $(u, v)_{N,m}$ is written easily:

$$(u,v)_{N,m} = \sum_{i_1=1}^{N-m} \cdots \sum_{i_d=1}^{N-m} u(\zeta_{i_1}^m, \ldots, \zeta_{i_d}^m)\, v(\zeta_{i_1}^m, \ldots, \zeta_{i_d}^m)\, \sigma_{i_1}^m \cdots \sigma_{i_d}^m. \tag{16.30}$$

Next, we set, for any polynomials u_N and v_N on Ω_d,

$$d_{N,m}(u_N, v_N) = (\Delta u_N, \Delta v_N)_{N,m}. \tag{16.31}$$

The discrete problem is the following one: *find* u_N *in* $\mathbb{P}_N^{2,0}(\Omega_d)$ *such that*

$$\forall v_N \in \mathbb{P}_N^{2,0}(\Omega_d), \quad d_{N,m}(u_N, v_N) = (f, v_N)_{N,m}. \tag{16.32}$$

It can be observed that, since the quadrature formulas are exact on $\mathbb{P}_{2N-1}(\Lambda)$, in each term $(\partial^2 u_N/\partial x_i^2, \partial^2 v_N/\partial x_j^2)_{N,m}$, the integrals with respect to x_i and x_j are exact, so that integrations by parts can be made in these directions. It yields that

$$\begin{aligned}&\forall u_N \in \mathbb{P}_N(\Omega_d),\ \forall v_N \in \mathbb{P}_N^{2,0}(\Omega_d),\\ &\quad d_{N,m}(u_N, v_N) = \left(\Delta^2 u_N, v_N\right)_{N,m}.\end{aligned} \tag{16.33}$$

However, the Lagrange polynomials associated with the nodes ζ_i^m, $1 \leqslant i \leqslant N - m$, do not form a basis of $\mathbb{P}_N^{2,0}(\Lambda)$, so that problem (16.32) cannot be written as a system of collocation equations.

The properties of the form $d_{N,m}(\cdot,\cdot)$ are a straightforward consequence of the one-dimensional inequalities (see notation (13.16)):

$$\forall \varphi_N \in \mathbb{P}_N^{(m-1)/2,0}(\Lambda), \quad \|\varphi_N\|_{L^2(\Lambda)}^2 \leqslant [\varphi_N, \varphi_N]_{N,m} \leqslant c\|\varphi_N\|_{L^2(\Lambda)}^2. \tag{16.34}$$

In the case $m = 1$, these inequalities are written in (13.20) as a special case of (13.18). In the case $m = 3$, they can be derived from (13.18) by observing that the mapping: $\varphi_N \mapsto \varphi_N(1-\zeta^2)^{-1}$ is one-to-one from $\mathbb{P}_{N+2}^{2,0}(\Lambda)$ onto $\mathbb{P}_N^0(\Lambda)$. When combined with the Lax–Milgram Theorem 2.1, this leads to the following theorems.

THEOREM 16.8. *The bilinear form $d_{N,m}(\bullet,\bullet)$ satisfies the following properties of continuity*

$$\forall u_N \in \mathbb{P}_N^0(\Omega_d),\ \forall v_N \in \mathbb{P}_N^0(\Omega_d),$$
$$|d_{N,m}(u_N,v_N)| \leqslant c|u_N|_{H^2(\Omega_d)}|v_N|_{H^2(\Omega_d)}, \tag{16.35}$$

and of ellipticity

$$\forall v_N \in \mathbb{P}_N^0(\Omega_d), \quad d_{N,m}(v_N,v_N) \geqslant c|v_N|_{H^2(\Omega_d)}^2. \tag{16.36}$$

THEOREM 16.9. *For any continuous function f on Ω_d, problem* (16.32) *has a unique solution u_N in $\mathbb{P}_N^{2,0}(\Omega_d)$.*

REMARK 16.1. Of course, another discrete problem can be constructed by using the quadrature formula (4.3) with $m = 2$. However, a counter-example given in BERNARDI and MADAY [1992b] proves that the constant of ellipticity of the associated bilinear form tends to 0 when N tends to $+\infty$. Indeed, (16.36) is not valid in the case $m = 2$. Since this method is not of collocation type, there is no advantage to use it.

From Theorem 16.8 combined with the previous results, we derive the error estimates in a very similar way to (16.15) and (16.18). We refer to BERNARDI, COPPOLETTA and MADAY [1990, 1992] for the detailed proofs which we skip here.

THEOREM 16.10. *The function f is supposed to be in a space $H^\sigma(\Omega_d)$, $\sigma > md/2$. Let us assume that the solution u of problem* (16.1) *belongs to $H^s(\Omega_d)$ for a real number $s \geqslant 2$. Then, the following error estimate holds for problem* (16.32)*:*

$$|u - u_N|_{H^2(\Omega_d)} \leqslant c\big(N^{2-s}\|u\|_{H^s(\Omega_d)} + N^{-\sigma}\|f\|_{H^\sigma(\Omega_d)}\big). \tag{16.37}$$

THEOREM 16.11. *In the case $d = 2$ of the square Ω, if the assumptions of Theorem* 16.10 *are satisfied, the following error estimate holds for problem* (16.32)*:*

$$\|u - u_N\|_{L^2(\Omega)} \leqslant c\big(N^{-s}\|u\|_{H^s(\Omega)} + N^{-\sigma}\|f\|_{H^\sigma(\Omega)}\big). \tag{16.38}$$

We do not present the case of general Dirichlet boundary conditions here, we refer to BERNARDI and MADAY ([1992b], §4.3) for its numerical analysis.

As a conclusion, we have proposed three spectral methods for the Dirichlet fourth-order problem. The first one is of collocation type, but does not lead to optimal order error estimates. Error estimates are of optimal order for the two last ones. So, only numerical experiments can make the difference between them: the convergence is slightly faster in the case $m = 3$ than for $m = 1$ (note that the mass matrix is diagonal for $m = 3$).

17. Galerkin method with numerical integration for the second-order Neumann problem

The idea to discretize the Neumann problem is strictly the same as for the second-order Dirichlet problem. So, we consider the problem:

$$\begin{cases} -\Delta u = f & \text{in } \Omega_d, \\ \dfrac{\partial u}{\partial n} = h_j & \text{on } \Gamma_j,\ 1 \leqslant j \leqslant 2d. \end{cases} \tag{17.1}$$

Here, the functions f and h_j, $1 \leqslant j \leqslant 2d$, are supposed to be continuous on $\overline{\Omega}_d$ and $\overline{\Gamma}_j$ respectively, and to satisfy

$$\int_{\Omega_d} f(\boldsymbol{x})\,\mathrm{d}\boldsymbol{x} + \sum_{j=1}^{2d} \int_{\Gamma_j} h_j(\tau)\,\mathrm{d}\tau = 0, \tag{17.2}$$

so that problem (17.1) admits the following equivalent formulation: *find* u *in* $H^1(\Omega_d)/\mathbb{R}$ *such that*

$$\forall v \in H^1(\Omega_d)/\mathbb{R},$$
$$a(u,v) = \int_{\Omega_d} f(\boldsymbol{x})v(\boldsymbol{x})\,\mathrm{d}\boldsymbol{x} + \sum_{j=1}^{2d} \int_{\Gamma_j} h_j(\tau)v(\tau)\,\mathrm{d}\tau, \tag{17.3}$$

where the form $a(\cdot,\cdot)$ is defined in (8.3).

In order to approximate the integrals on the Γ_j, we consider the $(d-1)$-dimensional quadrature formula on $]-1,1[^{d-1}$, and we define the discrete bilinear form $(u,v)_N^j$ for continuous functions u and v on $\overline{\Gamma}_j$, from the bilinear form

$$\sum_{i_1=0}^{N} \cdots \sum_{i_{d-1}=0}^{N} u(\xi_{i_1},\ldots,\xi_{i_{d-1}})v(\xi_{i_1},\ldots,\xi_{i_{d-1}})\,\rho_{i_1}\cdots\rho_{i_{d-1}}, \tag{17.4}$$

by appropriate translation and rotation. As previously, we also define the operators i_N^j, $1 \leqslant j \leqslant 2d$, as the Lagrange interpolation operators at the points of $\Xi_N \cap \overline{\Gamma}_j$.

Next, the idea is to make use of the variational formulation (17.3), to replace the quotient space $H^1(\Omega_d)/\mathbb{R}$ by $\mathbb{P}_N(\Omega_d)/\mathbb{R}$, the form $a(\cdot,\cdot)$ by the form $a_N(\cdot,\cdot)$ defined in (15.4) and the integrals on Ω_d and Γ_j respectively by $(\cdot,\cdot)_N$ and $(\cdot,\cdot)_N^j$. However, a difficulty appears: hypothesis (17.2) does not imply that the constant

$$\lambda_N = (f,1)_N + \sum_{j=1}^{2d} (h_j,1)_N^j, \tag{17.5}$$

or equivalently

$$\lambda_N = -\int_{\Omega_d} (f - \mathcal{I}_N f)(\boldsymbol{x})\,\mathrm{d}\boldsymbol{x} - \sum_{j=1}^{2d} \int_{\Gamma_j} (h_j - i_N^j h_j)(\tau)\,\mathrm{d}\tau, \tag{17.6}$$

is equal to 0. So, the discrete problem has to be defined as follows: *find* u_N *in* $\mathbb{P}_N(\Omega_d)/\mathbb{R}$ *such that*

$$\forall v_N \in \mathbb{P}_N(\Omega_d)/\mathbb{R},$$
$$a_N(u_N, v_N) = \left(f - 2^{-d}\lambda_N, v_N\right)_N + \sum_{j=1}^{2d} (h_j, v_N)_N^j. \tag{17.7}$$

The continuity of the form $a_N(\bullet,\bullet)$ on $\mathbb{P}_N(\Omega_d)/\mathbb{R} \times \mathbb{P}_N(\Omega_d)/\mathbb{R}$ is proven in Theorem 15.1, its ellipticity on $\mathbb{P}_N(\Omega_d)/\mathbb{R}$ is established by combining Theorems 15.1 and 1.3. The continuity of the right-hand side on $\mathbb{P}_N(\Omega_d)/\mathbb{R}$ follows from the choice (17.5) of λ_N. Hence, the following existence result is derived from the Lax–Milgram Theorem 2.1.

THEOREM 17.1. *For any continuous function f on $\overline{\Omega}_d$ and for any continuous functions h_j, $1 \leqslant j \leqslant 2d$, on $\overline{\Gamma}_j$, problem* (17.7) *has a unique solution u_N in $\mathbb{P}_N(\Omega_d)/\mathbb{R}$. Moreover, this solution satisfies*

$$|u_N|_{H^1(\Omega_d)} \leqslant c\left(\|\mathcal{I}_N f\|_{L^2(\Omega_d)} + \sum_{j=1}^{2d} \|i_N^j h_j\|_{L^2(\Gamma_j)} \right). \tag{17.8}$$

Note that, in opposite to the case of Dirichlet boundary conditions (see Remark 15.3), the stability estimate is straightforward. The first error estimate is derived in a very standard way.

THEOREM 17.2. *The function f is supposed to be in a space $H^\sigma(\Omega_d)$, $\sigma > d/2$, and the boundary functions h_j, $1 \leqslant j \leqslant 2d$, are supposed to be in spaces $H^{\tau_j}(\Gamma_j)$, $\tau_j > (d-1)/2$. Let us assume that the solution u of problem* (17.1) *belongs to $H^s(\Omega_d)$ for a real number $s \geqslant 1$. Then, the following error estimate holds for problem* (17.7)*:*

$$|u - u_N|_{H^1(\Omega_d)}$$
$$\leqslant c\left(N^{1-s}\|u\|_{H^s(\Omega_d)} + N^{-\sigma}\|f\|_{H^\sigma(\Omega_d)} + \sum_{j=1}^{2d} N^{-\tau_j}\|h_j\|_{H^{\tau_j}(\Gamma_j)} \right). \tag{17.9}$$

PROOF. The idea is to apply (15.14) with v_δ equal to $\Pi^1_{N-1}u$, which gives

$$|u-u_N|_{H^1(\Omega_d)} \leqslant c\Bigg(|u-\Pi^1_{N-1}u|_{H^1(\Omega)} + |\lambda_N| \\ + \sup_{w_N\in\mathbb{P}_N(\Omega_d)} \frac{\int_{\Omega_d} f(\boldsymbol{x})w_N(\boldsymbol{x})\,\mathrm{d}\boldsymbol{x} - (f,w_N)_N}{|w_N|_{H^1(\Omega_d)}} \\ + \sup_{w_N\in\mathbb{P}_N(\Omega_d)} \sum_{j=1}^{2d} \frac{\int_{\Gamma_j} h_j(\tau)w_N(\tau)\,\mathrm{d}\tau - (h_j,w_N)_N^j}{|w_N|_{H^1(\Omega_d)}}\Bigg). \tag{17.10}$$

The first line is estimated from Theorem 7.3 and (17.6), the second line thanks to (15.18) and Theorems 7.1 and 14.2. To bound the third line, we observe that

$$\sup_{w_N\in\mathbb{P}_N(\Omega_d)} \frac{\int_{\Gamma_j} h_j(\tau)w_N(\tau)\,\mathrm{d}\tau - (h_j,w_N)_N^j}{|w_N|_{H^1(\Omega_d)}} \\ \leqslant c\big(\|h_j-\pi^j_{N-1}h_j\|_{L^2(\Gamma_j)} + \|h_j - i^j_N h_j\|_{L^2(\Gamma_j)}\big), \tag{17.11}$$

where π^j_{N-1} stands for the orthogonal projection operator from $L^2(\Gamma_j)$ onto $\mathbb{P}_{N-1}(\Gamma_j)$. Then, according to the dimension d which can be equal to 2 or larger, it suffices to apply Theorems 6.1 and 13.4, or Theorems 7.1 and 14.2 to conclude.

The solutions u and u_N are defined up to an additive constant. In order to derive estimates in $L^2(\Omega_d)$, we have to fix these constants. For instance, we take

$$\int_{\Omega_d} u(\boldsymbol{x})\,\mathrm{d}\boldsymbol{x} = 0 \quad\text{and}\quad \int_{\Omega_d} u_N(\boldsymbol{x})\,\mathrm{d}\boldsymbol{x} = 0. \tag{17.12}$$

THEOREM 17.3. *If the assumptions of Theorem* 17.2 *are satisfied, and if the solutions* u *and* u_N *satisfy* (17.12), *the following error estimate holds for problem* (17.7):

$$\|u-u_N\|_{L^2(\Omega_d)} \\ \leqslant c\Bigg(N^{-s}\|u\|_{H^s(\Omega_d)} + N^{-\sigma}\|f\|_{H^\sigma(\Omega_d)} + \sum_{j=1}^{2d} N^{-\tau_j}\|h_j\|_{H^{\tau_j}(\Gamma_j)}\Bigg). \tag{17.13}$$

PROOF. Following the Aubin–Nitsche method, we write

$$\|u-u_N\|_{L^2(\Omega_d)} = \sup_{g\in L^2(\Omega_d)} \frac{\int_{\Omega_d}(u-u_N)(\boldsymbol{x})g(\boldsymbol{x})\,\mathrm{d}\boldsymbol{x}}{\|g\|_{L^2(\Omega_d)}}. \tag{17.14}$$

As in the proof of Theorem 10.3, for any function g in $L^2(\Omega_d)$, we set

$$g_0 = g - 2^{-d} \int_{\Omega_d} g(\boldsymbol{x})\,\mathrm{d}\boldsymbol{x}.$$

Next, we consider the solution w of problem (10.12), which satisfies (10.13), and, using (17.12), we compute

$$\begin{aligned}
&\int_{\Omega_d} (u - u_N)(\boldsymbol{x}) g(\boldsymbol{x})\,\mathrm{d}\boldsymbol{x} \\
&\quad = \int_{\Omega_d} (u - u_N)(\boldsymbol{x}) g_0(\boldsymbol{x})\,\mathrm{d}\boldsymbol{x} = a(u - u_N, w - \Pi^1_{N-1} w) \\
&\quad + \int_{\Omega_d} \big(f - \mathcal{I}_N f - 2^{-d}\lambda_N\big)(\boldsymbol{x}) \big(\Pi^1_{N-1} w\big)(\boldsymbol{x})\,\mathrm{d}\boldsymbol{x} \\
&\quad + \sum_{j=1}^{2d} \int_{\Gamma_j} \big(h_j - i^j_N h_j\big)(\tau) \big(\Pi^1_{N-1} w\big)(\tau)\,\mathrm{d}\tau,
\end{aligned}$$

and we conclude in a standard way.

As a conclusion, in opposite to the Dirichlet discrete problems (15.5) or (15.37), problem (17.7) cannot be written as a system of collocation equations. Indeed, we first note that equation (17.7) is always satisfied when v_N is a constant (this is due to (17.5)), hence for any v_N in $\mathbb{P}_N(\Omega_d)$. Letting v_N running through the Lagrange polynomials $l_{i_1}(x_1)\cdots l_{i_d}(x_d)$, $1 \leqslant i_1, \dots, i_d \leqslant N-1$, which form a basis of $\mathbb{P}^0_N(\Omega_d)$, we derive exactly as in (15.29)

$$-(\Delta u_N)(\xi_{i_1}, \dots, \xi_{i_d}) = f(\xi_{i_1}, \dots, \xi_{i_d}), \quad 1 \leqslant i_1, \dots, i_d \leqslant N-1. \tag{17.15}$$

These equations are of collocation type, at the interior of Ω_d. But, taking now v_N equal to $l_{i_1}(x_1)\cdots l_{i_{d-1}}(x_{d-1}) l_0(x_d)$, $1 \leqslant i_1, \dots, i_{d-1} \leqslant N-1$, and assuming for instance that the point $(\xi_{i_1}, \dots, \xi_{i_{d-1}}, -1)$ belongs to Γ_1, we obtain the equation

$$\begin{aligned}
&-(\Delta u_N)(\xi_{i_1}, \dots, \xi_{i_{d-1}}, -1)\rho_0 + \left(\frac{\partial u_N}{\partial n}\right)(\xi_{i_1}, \dots, \xi_{i_{d-1}}, -1) \\
&\quad = f(\xi_{i_1}, \dots, \xi_{i_{d-1}}, -1)\,\rho_0 + h_1(\xi_{i_1}, \dots, \xi_{i_{d-1}}, -1),
\end{aligned}$$

or equivalently

$$\begin{aligned}
\left(\frac{\partial u_N}{\partial n}\right)(\xi_{i_1}, \dots, \xi_{i_{d-1}}, -1) &= h_1(\xi_{i_1}, \dots, \xi_{i_{d-1}}, -1) \\
&\quad + (\Delta u_N + f)(\xi_{i_1}, \dots, \xi_{i_{d-1}}, -1)\rho_0.
\end{aligned} \tag{17.16}$$

This equation is no longer of collocation type: the Neumann boundary condition is satisfied in $(\xi_{i_1}, \ldots, \xi_{i_{d-1}}, -1)$ up to the residual $\Delta u_N + f$ of the equation multiplied by $\rho_0 = 2/N(N+1)$. The complete system of equations is very complex to write in the general case, but in the two-dimensional case of the square Ω and with the notation of Section 14, it reads:

(i) in the internal points of the domain,

$$-(\Delta u_N)(\boldsymbol{x}) = f(\boldsymbol{x}), \quad \boldsymbol{x} \in \Xi_N \cap \Omega; \tag{17.17}$$

(ii) in the points of the boundary which are not corners,

$$\left(\frac{\partial u_N}{\partial n_J}\right)(\boldsymbol{x}) = h_J(\boldsymbol{x}) + \frac{2}{N(N+1)}(f + \Delta u_N)(\boldsymbol{x}), \quad \boldsymbol{x} \in \Xi_N \cap \Gamma_J,$$
$$J = 1, 2, 3, 4; \tag{17.18}$$

(iii) in the corners of the domain

$$\left(\frac{\partial u_N}{\partial n_J} + \frac{\partial u_N}{\partial n_{J+1}}\right)(\boldsymbol{a}_J) = (h_J + h_{J+1})(\boldsymbol{a}_J) + \frac{2}{N(N+1)}(f + \Delta u_N)(\boldsymbol{a}_J),$$
$$J = 1, 2, 3, 4. \tag{17.19}$$

No optimal collocation method is known for the Neumann problem.

CHAPTER IV

Spectral Methods in Weighted Sobolev Spaces

Spectral methods were born with the use of truncated Fourier series combined with the Fast Fourier Transform of COOLEY and TUKEY [1965], to approximate periodic solutions of partial differential equations. So, the idea to discretize problems with nonperiodic boundary conditions such as homogeneous Dirichlet conditions, was a change of variables: setting $\zeta = \cos\theta$ transforms a truncated Fourier series into a polynomial which is written in the basis of Chebyshev polynomials. The Chebyshev polynomials are orthogonal in the interval $\Lambda =]-1, 1[$ with respect to the measure $(1-\zeta^2)^{-1/2}\,d\zeta$. So, the numerical analysis of this discretization involves weighted Sobolev spaces relying on the weight $(1-\zeta^2)^{-1/2}$ in one dimension and on tensorized weights in higher dimension; for duality arguments, it also requires spaces with the "dual" weight, i.e., $(1-\zeta^2)^{1/2}$ in one dimension. For all these reasons, this chapter is devoted to problems set in the framework of weighted Sobolev spaces for which the weight is $(1-\zeta^2)^{\alpha}$ in one dimension and its tensorized analogue in higher dimension, where α is a real parameter > -1: the corresponding orthogonal polynomials are the *Jacobi polynomials*. Thus, it simultaneously handles the discretizations of Legendre type and of Chebyshev type, as well as the duality results. Due to the general trace theorem which is stated in Section 18, spectral methods for positive values of α also have their own applications.

The weighted Sobolev spaces are described in Section 18 and the properties of Jacobi polynomials are stated in Section 19. This is used for the analysis of polynomial approximation in Section 20 and of Lagrange polynomial interpolation in Section 21. The application to new collocation methods for the basic Dirichlet problems is presented in Section 22.

18. Weighted Sobolev spaces

The weighted Sobolev spaces are only introduced on the interval $\Lambda =]-1, 1[$ and on the hypercube $\Omega_d =]-1, 1[^d$. The basic properties of these spaces were established by CANUTO and QUARTERONI [1984], MADAY and QUARTERONI [1981] and QUARTERONI [1987] in the Chebyshev case, and extended to the case of Jacobi weight by

BERNARDI and MADAY [1989]. Some proofs are skipped in this work, we refer to BERNARDI, DAUGE and MADAY [1992b] for the complete analysis.

Let us begin with the one-dimensional case. The weight on the interval Λ is defined by

$$\rho_\alpha(\zeta) = \left(1 - \zeta^2\right)^\alpha, \tag{18.1}$$

where α is a real parameter > -1. This last condition is necessary for the weight to be integrable, hence for the polynomials to belong to the corresponding weighted spaces. No other restriction on α is really justified, however we are more specifically interested in the case $-1 < \alpha < 1$.

We define the basic space

$$L^2_\alpha(\Lambda) = \left\{\varphi \in \mathscr{D}'(\Lambda)\colon \int_{-1}^{1} \varphi^2(\zeta)\rho_\alpha(\zeta)\,\mathrm{d}\zeta < +\infty\right\}, \tag{18.2}$$

which is provided with the norm

$$\|\varphi\|_{L^2_\alpha(\Lambda)} = \left(\int_{-1}^{1} \varphi^2(\zeta)\,\rho_\alpha(\zeta)\,\mathrm{d}\zeta\right)^{1/2}. \tag{18.3}$$

DEFINITION. Let m be a positive integer. The Sobolev space $H^m_\alpha(\Lambda)$ is defined by

$$H^m_\alpha(\Lambda) = \left\{\varphi \in L^2_\alpha(\Lambda)\colon \frac{\mathrm{d}^k\varphi}{\mathrm{d}\zeta^k} \in L^2_\alpha(\Lambda),\ 1 \leqslant k \leqslant m\right\}. \tag{18.4}$$

It is provided with the norm

$$\|\varphi\|_{H^m_\alpha(\Lambda)} = \left(\int_{-1}^{1} \sum_{k=0}^{m} \left(\frac{\mathrm{d}^k\varphi}{\mathrm{d}\zeta^k}\right)^2(\zeta)\rho_\alpha(\zeta)\,\mathrm{d}\zeta\right)^{1/2}. \tag{18.5}$$

The space $H^m_{\alpha,0}(\Lambda)$ is the closure of $\mathscr{D}(\Lambda)$ in $H^m_\alpha(\Lambda)$, and $H^{-m}_\alpha(\Lambda)$ stands for its dual space. The duality pairing between $H^{-m}_\alpha(\Lambda)$ and $H^m_{\alpha,0}(\Lambda)$ is denoted by $\langle\bullet,\bullet\rangle_\alpha$, it is an extension of the scalar product in $L^2_\alpha(\Lambda)$.

DEFINITION. Let s be a positive real number which is not an integer. The space $H^s_\alpha(\Lambda)$ is defined as the interpolate space of index $1 - s/m$ between $H^m_\alpha(\Lambda)$ and $L^2_\alpha(\Lambda)$, for an integer $m \geqslant s$. The space $H^s_{\alpha,0}(\Lambda)$ is the closure of $\mathscr{D}(\Lambda)$ in $H^s_\alpha(\Lambda)$, and $H^{-s}_\alpha(\Lambda)$ denotes its dual space.

Of course, the definition of $H^s_\alpha(\Lambda)$ is independent of the choice of m, this is a consequence of the following theorem which can be found in BERNARDI, DAUGE and MADAY ([1992b], Theorem 1.e.7).

THEOREM 18.1. *Let α be a real number > -1. For any nonnegative real numbers r and s, $s \leqslant r$, and for any θ, $0 < \theta < 1$, the following identity holds*

$$\big[H^r_\alpha(\Lambda), H^s_\alpha(\Lambda)\big]_\theta = H^{(1-\theta)r+\theta s}_\alpha(\Lambda). \tag{18.6}$$

It is readily checked that, for any value of α and for any nonnegative real number s, the restrictions of functions of $H^s_\alpha(\Lambda)$ to any interval $\Lambda_\varepsilon = (-1+\varepsilon, 1-\varepsilon)$, $0 < \varepsilon < 1$, belong to $H^s(\Lambda_\varepsilon)$. Hence, the properties of functions of the weighted spaces in the interior of the interval, are the same as for standard Sobolev spaces. Only the trace properties are different, they will be stated in what follows.

For any nonnegative real number s and for two real numbers α and β, $-1 < \alpha < \beta$, the space $H^s_\alpha(\Lambda)$ is contained in the space $H^s_\beta(\Lambda)$. In particular, we have the following imbeddings:

$$H^s_\alpha(\Lambda) \subset H^s(\Lambda) \text{ if } \alpha \leqslant 0 \quad \text{and} \quad H^s(\Lambda) \subset H^s_\alpha(\Lambda) \text{ if } \alpha \geqslant 0. \tag{18.7}$$

We begin by stating inverse imbeddings, which were firstly proven by CANUTO and QUARTERONI [1984].

THEOREM 18.2. *The following imbeddings hold for any nonnegative real number s:*

(i) *when α satisfies $-1 < \alpha \leqslant 0$, the space $H^{s-\alpha/2}(\Lambda)$ is contained in the space $H^s_\alpha(\Lambda)$;*

(ii) *when α satisfies $0 \leqslant \alpha < 1$, the space $H^s_\alpha(\Lambda)$ is contained in the space $H^{s-\alpha/2}(\Lambda)$.*

PROOF. The theorem relies on a standard result (see for instance LIONS and MAGENES ([1968], Chapter 1, Theorem 11.2)): for any real number α, $-1 < \alpha < 0$, the mapping: $\varphi \mapsto \varphi t^{\alpha/2}$ is continuous from $H^{-\alpha/2}(0,1)$ into $L^2(0,1)$. Applying this result together with two changes of variables and noting the inequalities (α is negative!)

$$\forall \zeta \in \Lambda,\ \zeta < 0, \quad 2^\alpha(1+\zeta)^\alpha < \rho_\alpha(\zeta) < (1+\zeta)^\alpha,$$
$$\forall \zeta \in \Lambda,\ \zeta > 0, \quad 2^\alpha(1-\zeta)^\alpha < \rho_\alpha(\zeta) < (1-\zeta)^\alpha,$$

we derive that, for $-1 < \alpha < 0$, the space $H^{-\alpha/2}(\Lambda)$ is contained in $L^2_\alpha(\Lambda)$, which is the case (i) of the theorem when s is equal to 0. A duality argument also yields that, for any real number α, $0 < \alpha < 1$, the space $L^2_\alpha(\Lambda)$ is contained in $H^{-\alpha/2}(\Lambda)$, which is the case (ii) of the theorem for $s = 0$. Next, applying this result to the derivatives of the functions gives the theorem when s is an integer. Finally, an interpolation argument relying on Theorem 1.5 leads to the general result.

The next theorem makes use of Hardy's inequality, which is a basic tool for the analysis of the weighted spaces and was proven in (13.4) and (13.5).

THEOREM 18.3. *Let α be a real number, $-1 < \alpha < 1$. For any nonnegative integer m, the mapping $M_\alpha : \varphi \mapsto \varphi(1-\zeta^2)^{\alpha/2}$ is an isomorphism from $H^m_{\alpha,0}(\Lambda)$ onto $H^m_0(\Lambda)$.*

PROOF. It can be checked by induction over k that any function φ in $\mathscr{D}(\Lambda)$ satisfies

$$\left(\frac{\mathrm{d}^k}{\mathrm{d}\zeta^k}\right)(M_\alpha\varphi) = \sum_{r=0}^{k} p_r(\zeta)\left(\frac{\mathrm{d}^{k-r}\varphi}{\mathrm{d}\zeta^{k-r}}\right)(\zeta)\left(1-\zeta^2\right)^{\alpha/2-r},$$

where the p_r are polynomials with degree $\leqslant r$, hence bounded on Λ. So, applying the inequality (13.4) or (13.5) r times to each integral

$$\int_{-1}^{1}\left(\frac{\mathrm{d}^{k-r}\varphi}{\mathrm{d}\zeta^{k-r}}\right)^2(\zeta)\rho_{\alpha-2r}\,\mathrm{d}\zeta$$

and using a density argument imply that the mapping M_α is linear continuous from $H^m_{\alpha,0}(\Lambda)$ into $H^m_0(\Lambda)$. Similarly, the mapping $M_{-\alpha}$ is linear continuous from $H^m_0(\Lambda)$ into $H^m_{\alpha,0}(\Lambda)$, which completes the proof.

As another consequence of Hardy's inequality (13.4), we derive that, when α is such that $-1 < \alpha < 1$, the seminorm defined by

$$|\varphi|_{H^m_\alpha(\Lambda)} = \left(\int_{-1}^{1}\left(\frac{\mathrm{d}^m\varphi}{\mathrm{d}\zeta^m}\right)^2(\zeta)\rho_\alpha(\zeta)\,\mathrm{d}\zeta\right)^{1/2}, \tag{18.8}$$

is a norm on the space $H^m_{\alpha,0}(\Lambda)$, equivalent to the norm $\|\bullet\|_{H^m_\alpha(\Lambda)}$. The second consequence is the following identity, valid for any nonnegative real number s with integral part $[s]$ and fractional part σ, $0 < \sigma < 1$:

$$H^s_\alpha(\Lambda) = \left\{\varphi \in H^{[s]}_\alpha(\Lambda) \colon \frac{\mathrm{d}^{[s]}\varphi}{\mathrm{d}\zeta^{[s]}} \in H^\sigma_\alpha(\Lambda)\right\}. \tag{18.9}$$

The existence of a trace mapping $T^\pm_m$:

$$\varphi \mapsto T^\pm_m\varphi = \left(\varphi(\pm 1), \varphi'(\pm 1), \ldots, \left(\frac{\mathrm{d}^{m-1}\varphi}{\mathrm{d}\zeta^{m-1}}\right)(\pm 1)\right), \tag{18.10}$$

on the space $H^s_\alpha(\Lambda)$ is proven by GRISVARD ([1963], Theorem 7.1), for any real number $s > m - (1-\alpha)/2$; in the case $0 < \alpha < 1$, it is a simple consequence of

Theorem 18.2 combined with Theorem 1.10. It allows for the characterization of the subspaces $H^s_{\alpha,0}(\Lambda)$.

THEOREM 18.4. *Let α be a real number > -1. For any positive real number s such that $s + (1-\alpha)/2$ is not an integer, the following characterization holds*

$$H^s_{\alpha,0}(\Lambda) = \{\varphi \in H^s_\alpha(\Lambda)\colon\ T_k^- \varphi = T_k^+ \varphi = 0\}, \tag{18.11}$$

where k is the integral part of $s + (1-\alpha)/2$.

We conclude the study of the one-dimensional case with a second interpolation result which can be found in BERNARDI, DAUGE and MADAY ([1992b], Proposition 1.e.6).

THEOREM 18.5. *Let α be a real number > -1. For any nonnegative real numbers r and s, $s \leqslant r$, such that neither $r - (1+\alpha)/2$ nor $s - (1+\alpha)/2$ is an integer, and for any θ, $0 < \theta < 1$, such that $(1-\theta)r + \theta s - (1+\alpha)/2$ is not an integer, the following identity holds*

$$\left[H^r_{\alpha,0}(\Lambda), H^s_{\alpha,0}(\Lambda)\right]_\theta = H^{(1-\theta)r+\theta s}_{\alpha,0}(\Lambda). \tag{18.12}$$

The weighted Sobolev spaces on the hypercube $\Omega_d =]-1,1[^d$ are defined in a similar way. Assuming that α is a real number > -1, we introduce the weight

$$\varpi_\alpha(\boldsymbol{x}) = \rho_\alpha(x_1)\cdots\rho_\alpha(x_d). \tag{18.13}$$

Next, we define the space

$$L^2_\alpha(\Omega_d) = \left\{v \in \mathscr{D}'(\Omega_d)\colon\ \int_{\Omega_d} v^2(\boldsymbol{x})\,\varpi_\alpha(\boldsymbol{x})\,\mathrm{d}\boldsymbol{x} < +\infty\right\}, \tag{18.14}$$

provided with the norm

$$\|\varphi\|_{L^2_\alpha(\Omega_d)} = \left(\int_{\Omega_d} v^2(\boldsymbol{x})\varpi_\alpha(\boldsymbol{x})\,\mathrm{d}\boldsymbol{x}\right)^{1/2}. \tag{18.15}$$

DEFINITION. Let m be a positive integer. The Sobolev space $H^m_\alpha(\Omega_d)$ is defined by

$$H^m_\alpha(\Omega_d) = \{v \in L^2_\alpha(\Omega_d);\ \forall \boldsymbol{k} \in \mathbb{N}^d,\ |\boldsymbol{k}| \leqslant m,\ \partial^{\boldsymbol{k}} v \in L^2_\alpha(\Omega_d)\}. \tag{18.16}$$

It is provided with the norm

$$\|\varphi\|_{H^m_\alpha(\Omega_d)} = \left(\int_{-1}^1 \sum_{|\boldsymbol{k}|\leqslant m} (\partial^{\boldsymbol{k}} v)^2(\boldsymbol{x})\,\varpi_\alpha(\boldsymbol{x})\,\mathrm{d}\boldsymbol{x}\right)^{1/2}. \tag{18.17}$$

The space $H^m_{\alpha,0}(\Omega_d)$ is the closure of $\mathscr{D}(\Omega_d)$ in $H^m_\alpha(\Omega_d)$, and $H^{-m}_\alpha(\Omega_d)$ stands for its dual space. The duality pairing between $H^{-m}_\alpha(\Omega_d)$ and $H^m_{\alpha,0}(\Omega_d)$ is still denoted by $\langle\bullet,\bullet\rangle_\alpha$.

DEFINITION. Let s be a positive real number which is not an integer. The space $H^s_\alpha(\Omega_d)$ is defined as the interpolate space of index $1-s/m$ between $H^m_\alpha(\Omega_d)$ and $L^2_\alpha(\Omega_d)$, for an integer $m \geqslant s$ (it can be checked that this definition is independent of m). The space $H^s_{\alpha,0}(\Omega_d)$ is the closure of $\mathscr{D}(\Omega_d)$ in $H^s_\alpha(\Omega_d)$, and $H^{-s}_\alpha(\Omega_d)$ denotes its dual space.

For $1 \leqslant k \leqslant d$, observing that each $(d-k)$-face of Ω_d coincides with Ω_{d-k} up to appropriate translation and rotation, we can extend the previous definitions of weighted Sobolev spaces to each $(d-k)$-face in a trivial way.

Here also, the restrictions of functions of $H^s_\alpha(\Omega_d)$ to any open set Ω_* such that $\overline{\Omega}_*$ is contained in Ω_d, belong to $H^s(\Omega_*)$. Hence, we are mainly interested in the trace properties of the space $H^s_\alpha(\Omega_d)$. First, let us state the d-dimensional analogue of Theorem 18.3, which is derived by combining the result of this theorem with (7.4).

THEOREM 18.6. *Let α be a real number, $-1<\alpha<1$. For any nonnegative integer m, the mapping $P_\alpha : v \mapsto v\,\varpi_{\alpha/2}$ is an isomorphism from $H^m_{\alpha,0}(\Omega_d)$ onto $H^m_0(\Omega_d)$.*

Also, we derive from (7.4) that, for $-1<\alpha<1$, the seminorm

$$|v|_{H^m_\alpha(\Omega_d)} = \left(\int_{\Omega_d} \sum_{|\boldsymbol{k}|=m} \left(\partial^{\boldsymbol{k}} v\right)^2(\boldsymbol{x})\varpi_\alpha(\boldsymbol{x})\,\mathrm{d}\boldsymbol{x}\right)^{1/2}, \tag{18.18}$$

is equivalent to the norm $\|\bullet\|_{H^m_\alpha(\Omega_d)}$ on $H^m_{\alpha,0}(\Omega_d)$.

Now, let us consider the trace problem. With an obvious extension of the notation of Section 7, we observe that, for any nonnegative integer m, the space $H^m_\alpha(\Omega_d)$ is contained in each space $H^k_\alpha(\Lambda_j; H^{m-k}_\alpha(\Omega_{d-1}))$, $0 \leqslant k \leqslant m$. Hence, the space $H^s_\alpha(\Omega_d)$ is contained in any space $H^p_\alpha(\Lambda_j; H^q_\alpha(\Omega_{d-1}))$ for any nonnegative real numbers p and q such that $p+q \leqslant s$. As a special case, when $s \geqslant 1$, we obtain the imbedding

$$H^s_\alpha(\Omega_d) \subset L^2_\alpha\left(\Lambda_j; H^s_\alpha(\Omega_{d-1})\right) \cap H^1_\alpha\left(\Lambda_j; H^{s-1}_\alpha(\Omega_{d-1})\right).$$

Equivalently, any function v in $H^s_\alpha(\Omega_d)$ satisfies

$$\left(1-x_j^2\right)^{\alpha/2} v \in L^2\left(\Lambda_j; H^s_\alpha(\Omega_{d-1})\right)$$

and

$$\left(1-x_j^2\right)^{\alpha/2} \frac{\partial v}{\partial x_j} \in L^2\left(\Lambda_j; H^{s-1}_\alpha(\Omega_{d-1})\right).$$

So, if Γ denotes for instance the $(d-1)$-face contained either in the hyperplane $x_j = -1$ or in the hyperplane $x_j = 1$, it follows from Remark 1.1 that the trace of the function v on Γ belongs to $[H^s_\alpha(\Gamma), H^{s-1}_\alpha(\Gamma)]_\theta$, with $2\theta - 1 = \alpha$. Applying Theorem 18.1 together with the fact that operators with respect to different variables commute, we deduce that this space coincides with $H^{s-(1+\alpha)/2}_\alpha(\Gamma)$ when $s - (1+\alpha)/2$ is not an integer. This result can be extended to the case $(1+\alpha)/2 < s < 1$ by Theorem 18.2 and it can be applied to the derivatives of the function to obtain normal derivatives on the edges. Finally, we derive the general statement related to each $(d-1)$-face Γ_j of Ω_d, $1 \leqslant j \leqslant 2d$. We denote by $\boldsymbol{n}_j$ the unit outward normal vector to the $(d-1)$-face Γ_j.

THEOREM 18.7. *Let α be a real number > -1, and let m be a positive integer. For any j, $1 \leqslant j \leqslant 2d$, the trace mapping $T^{\Gamma_j}_m$:*

$$v \mapsto T^{\Gamma_j}_m v = \left(v_{|\Gamma_j}, \left(\frac{\partial v}{\partial n_j}\right), \ldots, \left(\frac{\partial^{m-1} v}{\partial n_j^{m-1}}\right)\right), \tag{18.19}$$

defined on $\mathscr{D}(\overline{\Omega}_d)$ with values in $L^2(\Gamma_j)^m$, can be extended by density to any space $H^s_\alpha(\Omega_d)$, where s is a real number $> m - (1-\alpha)/2$. The image of $H^s_\alpha(\Omega_d)$ by $T^{\Gamma_j}_m$ is contained in $\prod_{k=0}^{m-1} H^{s-k-(1+\alpha)/2}_\alpha(\Gamma_j)$.

Now, it is possible to derive the following characterization.

THEOREM 18.8. *Let α be a real number > -1. For any positive real number s such that $s + (1-\alpha)/2$ is not an integer, the following characterization holds:*

$$H^s_{\alpha,0}(\Omega_d) = \left\{ v \in H^s_\alpha(\Omega_d) \colon\ T^{\Gamma_j}_K v = 0,\ 1 \leqslant j \leqslant 2d \right\}, \tag{18.20}$$

where K is the integral part of $s + (1-\alpha)/2$.

Of course, an induction argument allows for extending the trace theorem to faces of any dimension.

THEOREM 18.9. *Let α be a real number > -1, and let m be a positive integer. For any $(d-l)$-face Γ, $1 \leqslant l \leqslant d-1$, and for any unit outward normal vector $\boldsymbol{n}$ to Γ, the trace mapping T^Γ_m:*

$$v \mapsto T^\Gamma_m v = \left(v_{|\Gamma}, \left(\frac{\partial v}{\partial n}\right), \ldots, \left(\frac{\partial^{m-1} v}{\partial n^{m-1}}\right)\right), \tag{18.21}$$

defined on $\mathscr{D}(\overline{\Omega}_d)$ with values in $L^2(\Gamma)^m$, can be extended by density to any space $H^s_\alpha(\Omega_d)$, where s is a real number $> m - 1 + l(1+\alpha)/2$. The image of $H^s_\alpha(\Omega_d)$ by T^Γ_m is contained in $\prod_{k=0}^{m-1} H^{s-k-l(1+\alpha)/2}_\alpha(\Gamma)$.

The final theorems state the existence of a lifting operator, they are the weighted analogues of Theorems 1.13 and 1.14. Also, like in Theorem 1.13, the statement of Theorem 18.11 is limited to the square $\Omega =]-1,1[^2$ for the sake of brevity.

THEOREM 18.10. *Let α a real number such that $-1 < \alpha < 1$, $\alpha \neq 0$. For any 2d-tuple $(\varphi_j)_{1\leqslant j\leqslant 2d}$ in $\prod_{j=1}^{2d} H_\alpha^{(1-\alpha)/2}(\Gamma_j)$ such that the following compatibility conditions hold when $\alpha < 0$:*

$$\varphi_j = \varphi_k \quad \text{a.e. on } \overline{\Gamma}_j \cap \overline{\Gamma}_k,\ 1 \leqslant j < k \leqslant 2d, \tag{18.22}$$

there exists a function v in $H_\alpha^1(\Omega_d)$ such that

$$v = \varphi_j \quad \text{on } \Gamma_j,\ 1 \leqslant j \leqslant 2d, \tag{18.23}$$

and that the following stability condition holds:

$$\|v\|_{H_\alpha^1(\Omega_d)} \leqslant c\left(\sum_{j=1}^{2d} \|\varphi_j\|^2_{H_\alpha^{(1-\alpha)/2}(\Gamma_j)}\right)^{1/2}. \tag{18.24}$$

THEOREM 18.11. *Let α be a real number such that $-1 < \alpha < 1$, $\alpha \neq 0$. Let m be a positive integer. For any 4m-tuple $(\varphi_J^0,\dots,\varphi_J^{m-1})_{J=1,2,3,4}$ in*

$$\prod_{J=1}^{4}\prod_{k=0}^{m-1} H_\alpha^{m-k-(1+\alpha)/2}(\Gamma_J)$$

such that the following compatibility conditions hold:

$$\left(\frac{\mathrm{d}^l\varphi_J^k}{\mathrm{d}\tau_J^l}\right)(\boldsymbol{a}_J) = (-1)^k\left(\frac{\mathrm{d}^k\varphi_{J+1}^l}{\mathrm{d}\tau_{J+1}^k}\right)(\boldsymbol{a}_J),$$
$$0 \leqslant k+l \leqslant K,\ J = 1,2,3,4, \tag{18.25}$$

where K is the integral part of $m-1-\alpha$, there exists a function v in $H_\alpha^m(\Omega_d)$ such that

$$\frac{\partial^k v}{\partial n_J^k} = \varphi_J^k \quad \text{on } \Gamma_J,\ 0 \leqslant k \leqslant m-1,\ J = 1,2,3,4, \tag{18.26}$$

and that the following stability condition holds:

$$\|v\|_{H_\alpha^m(\Omega_d)} \leqslant c\left(\sum_{J=1}^{4}\sum_{k=0}^{m-1} \|\varphi_J^k\|^2_{H_\alpha^{m-k-(1+\alpha)/2}(\Gamma_J)}\right)^{1/2}. \tag{18.27}$$

From the previous theorem, it can be checked that the case $\alpha = 0$ of standard Sobolev spaces is a limit case: indeed, the number of compatibility conditions (18.25) at each corner $\boldsymbol{a}_J$ which are necessary for the existence of a lifting functions is $m(m-1)/2$ when α is positive (and < 1), and $m(m+1)/2$ when α is negative. A more general result is proven by BERNARDI, DAUGE and MADAY ([1992b], §3), however we shall not need it.

Using weighted Sobolev spaces leads to new (and equivalent for regular data) variational formulations of the basic Dirichlet problems in the hypercube Ω_d. We are going to state and analyze these formulations.

EXAMPLE 18.1. We consider the equation:

$$\begin{cases} -\Delta u = f & \text{in } \Omega_d, \\ u = g_j & \text{on } \Gamma_j, \ 1 \leqslant j \leqslant 2d. \end{cases} \tag{18.28}$$

Now, for a fixed real number α, $-1 < \alpha < 1$, $\alpha \neq 0$, the distribution f is supposed to be in $H_\alpha^{-1}(\Omega_d)$ while the boundary data g_j, $1 \leqslant j \leqslant 2d$, are in $H^{(1-\alpha)/2}(\Gamma_j)$ and satisfy when α is negative:

$$g_j = g_k \quad \text{a.e. on } \overline{\Gamma}_j \cap \overline{\Gamma}_k, \ 1 \leqslant j < k \leqslant 2d. \tag{18.29}$$

To derive the variational formulation of problem (18.28), we multiply the first line by $v\,\varpi_\alpha$, where v is any function of $\mathscr{D}(\Omega_d)$, we integrate by parts and we use the density of $\mathscr{D}(\Omega_d)$ in $H^1_{\alpha,0}(\Omega_d)$. We also note from Theorem 18.10 and the previous hypotheses that there exists a function $u_\mathfrak{b}$ in $H^1_\alpha(\Omega_d)$ such that the trace of $u_\mathfrak{b}$ on Γ_j coincides with g_j, $1 \leqslant j \leqslant 2d$. Consequently, problem (18.28) admits the following equivalent variational formulation: *find u in $H^1_\alpha(\Omega_d)$, with $u - u_\mathfrak{b}$ in $H^1_{\alpha,0}(\Omega_d)$, such that*

$$\forall v \in H^1_{\alpha,0}(\Omega_d), \quad a_\alpha(u,v) = \langle f, v\rangle_\alpha, \tag{18.30}$$

where the bilinear form $a_\alpha(\cdot,\cdot)$ is defined by

$$a_\alpha(u,v) = \int_{\Omega_d} (\mathbf{grad}\, u)(\boldsymbol{x}) \cdot \big(\mathbf{grad}(v\varpi_\alpha)\big)(\boldsymbol{x})\,\mathrm{d}\boldsymbol{x}. \tag{18.31}$$

Of course, the analysis of formulation (18.30) relies on the properties of the form $a_\alpha(\cdot,\cdot)$. These properties were firstly proven by CANUTO and QUARTERONI [1984], with slightly different arguments, in the case $\alpha = -\frac{1}{2}$.

THEOREM 18.12. *Let α be a real number which satisfies $-1 < \alpha < 1$. The form $a_\alpha(\cdot,\cdot)$ is continuous on $H^1_\alpha(\Omega_d) \times H^1_{\alpha,0}(\Omega_d)$ and elliptic on $H^1_{\alpha,0}(\Omega_d)$.*

PROOF. We have

$$a_\alpha(u,v) = \int_{\Omega_d} (\mathbf{grad}\, u)(\boldsymbol{x}) \cdot (\mathbf{grad}\, v)(\boldsymbol{x}) \varpi_\alpha(\boldsymbol{x})\, \mathrm{d}\boldsymbol{x}$$
$$- 2\alpha \sum_{j=1}^{2d} \int_{\Omega_d} \left(\frac{\partial u}{\partial x_j}\right)(\boldsymbol{x}) v(\boldsymbol{x})\, x_j \left(1 - x_j^2\right)^{-1} \varpi_\alpha(\boldsymbol{x})\, \mathrm{d}\boldsymbol{x},$$

so that the Cauchy–Schwarz inequality yields

$$a_\alpha(u,v) \leqslant |u|_{H^1_\alpha(\Omega_d)} |v|_{H^1_\alpha(\Omega_d)}$$
$$+ 2|\alpha|\, |u|_{H^1_\alpha(\Omega_d)} \sum_{j=1}^{2d} \int_{\Omega_d} v^2(\boldsymbol{x}) \left(1 - x_j^2\right)^{-2} \varpi_\alpha(\boldsymbol{x})\, \mathrm{d}\boldsymbol{x}.$$

Using Hardy's inequality (13.4) with respect to each x_j, proves the continuity when v belongs to $H^1_{\alpha,0}(\Omega_d)$. On the other hand, for any u in $H^1_{\alpha,0}(\Omega_d)$, we write

$$a_\alpha(u,u) = |u|^2_{H^1_\alpha(\Omega_d)} - \alpha \sum_{j=1}^{2d} \int_{\Omega_d} \left(\frac{\partial(u^2)}{\partial x_j}\right)(\boldsymbol{x})\, x_j \left(1 - x_j^2\right)^{-1} \varpi_\alpha(\boldsymbol{x})\, \mathrm{d}\boldsymbol{x}.$$

Next, we integrate by parts with respect to each variable, which gives

$$a_\alpha(u,u) = |u|^2_{H^1_\alpha(\Omega_d)}$$
$$+ \alpha \sum_{j=1}^{2d} \int_{\Omega_d} u^2(\boldsymbol{x}) \left(1 + (1-2\alpha)x_j^2\right) \left(1 - x_j^2\right)^{-2} \varpi_\alpha(\boldsymbol{x})\, \mathrm{d}\boldsymbol{x}.$$

For $0 < \alpha < 1$, we note that $(1 + (1-2\alpha)\zeta^2)$ is larger than $1 - \zeta^2$, so that the last integral is positive. Since the seminorm $|\bullet|_{H^1_\alpha(\Omega_d)}$ is a norm equivalent to $\|\bullet\|_{H^1_\alpha(\Omega_d)}$ on $H^1_{\alpha,0}(\Omega_d)$, this inequality proves the ellipticity for $\alpha > 0$. When α is negative, setting $w = u\varpi_\alpha$, we know by two applications of Theorem 18.6 that w belongs to $H^1_{-\alpha,0}(\Omega_d)$. Moreover, we check that

$$a_\alpha(u,u) = a_{-\alpha}(w,w).$$

Since the ellipticity result is proven for positive values of the parameter, we derive

$$a_\alpha(u,u) \geqslant c\|w\|^2_{H^1_{-\alpha}(\Omega_d)},$$

and Theorem 18.6 leads to

$$a_\alpha(u,u) \geqslant c\|u\|^2_{H^1_\alpha(\Omega_d)},$$

which is the desired result.

REMARK 18.1. It is readily checked that the form $a_\alpha(\bullet,\bullet)$ is not continuous on $H^1_\alpha(\Omega_d)\times H^1_\alpha(\Omega_d)$, so that the analogue of (18.30) does not exist for the Laplace equation with Neumann boundary conditions.

As already recalled, it follows from Theorem 18.10 that, if conditions (18.30) hold when $\alpha<0$, there exists a function u_b in $H^1_\alpha(\Omega_d)$ such that the trace of u_b on Γ_j, $1\leqslant j\leqslant 2d$, coincides with g_j and that

$$\|u_b\|_{H^1_\alpha(\Omega_d)}\leqslant c\sum_{j=1}^{2d}\|g_j\|_{H^{(1-\alpha)/2}_\alpha(\Gamma_j)}. \tag{18.32}$$

Then, combining the Lax–Milgram Theorem 2.1 with Theorem 18.12 implies that there exists a solution u^* in $H^1_{\alpha,0}(\Omega_d)$ of the problem:

$$\forall v\in H^1_{\alpha,0}(\Omega_d),\quad a_\alpha(u^*,v)=\langle f,v\rangle_\alpha-a_\alpha(u_b,v). \tag{18.33}$$

Setting $u=u^*+u_b$ leads to the following theorem.

THEOREM 18.13. *Let α be a real number which satisfies $-1<\alpha<1$, $\alpha\neq 0$. For any distribution f in $H^{-1}_\alpha(\Omega_d)$ and for any 2d-tuple $(g_j)_{1\leqslant j\leqslant 2d}$ in $\prod_{j=1}^{2d}H^{(1-\alpha)/2}_\alpha(\Gamma_j)$ which satisfies* (18.29) *when α is negative, problem* (18.30) *has a unique solution u in $H^1_\alpha(\Omega_d)$. Moreover, this solution satisfies*

$$\|u\|_{H^1_\alpha(\Omega_d)}\leqslant c\left(\|f\|_{H^{-1}_\alpha(\Omega_d)}+\sum_{j=1}^{2d}\|g_j\|_{H^{(1-\alpha)/2}_\alpha(\Gamma_j)}\right). \tag{18.34}$$

REMARK 18.2. When the data f and g_j, $1\leqslant j\leqslant 2d$, are smooth enough, we have obtained an equivalent variational formulation of the same problem (18.28) for each value of α, $-1<\alpha<1$. Of course, the solutions of all these variational problems coincide.

REMARK 18.3. We give at once a first application of the previous theorem. Let us consider the problem

$$\begin{cases}-\Delta u=0 & \text{in } \Omega_d,\\ u=0 & \text{on } \Gamma_j,\ 1\leqslant j\leqslant 2d-1,\\ u=1 & \text{on } \Gamma_{2d}.\end{cases} \tag{18.35}$$

This can be considered as a basic step for the well-known driven cavity problem. Theorem 18.13 states that this problem has a unique solution in $H^1_\alpha(\Omega_d)$, for any value of the parameter α such that $0<\alpha<1$. As a consequence, this solution belongs to $H^1(\Omega_*)$, for any open domain Ω_* such that $\overline{\Omega}_*$ is contained in Ω_d.

The properties of regularity of the solution of problem (18.28) are not fully established. In the case $d = 2$, the continuity of the operator which associates the solution u of problem (18.28) with the data f and $(g_J)_{J=1,2,3,4}$, from

$$H_\alpha^{s-2}(\Omega) \times \prod_{J=1}^{4} H_\alpha^{s-(1+\alpha)/2}(\Gamma_J)$$

into $H_\alpha^s(\Omega)$ is conjectured (but not proven) in BERNARDI and MADAY [1989] for any α, $-1 < \alpha < 1$, and for any real number $s < 3 + \alpha$. We only give a basic result, in the case of homogeneous boundary conditions, which will be useful in the following. Let T_{D}^α denote the operator which associates the solution u of problem (18.28) with the data f when all the functions g_j, $1 \leqslant j \leqslant 2d$, are equal to 0.

THEOREM 18.14. *Let α be a real number which satisfies $-\frac{1}{2} \leqslant \alpha \leqslant \frac{1}{2}$. In the cases of dimension $d = 2$ and $d = 3$, the operator T_{D}^α is continuous from $L_\alpha^2(\Omega_d)$ into $H_\alpha^2(\Omega_d)$.*

PROOF. The first step is to prove that any function in $H_\alpha^2(\Omega_d) \cap H_{\alpha,0}^1(\Omega_d)$ (and for any $d \geqslant 2$) satisfies

$$|u|_{H_\alpha^2(\Omega_d)} \leqslant c\|\Delta u\|_{L_\alpha^2(\Omega_d)}. \tag{18.36}$$

For any u in $H_\alpha^2(\Omega_d) \cap H_{\alpha,0}^1(\Omega_d)$, we start from the equation

$$\int_{\Omega_d} (\Delta u)^2(\boldsymbol{x})\varpi_\alpha(\boldsymbol{x})\,\mathrm{d}\boldsymbol{x} = \sum_{j=1}^{d} \int_{\Omega_d} \left(\frac{\partial^2 u}{\partial x_j^2}\right)^2 (\boldsymbol{x})\varpi_\alpha(\boldsymbol{x})\,\mathrm{d}\boldsymbol{x}$$
$$+ 2\sum_{i=1}^{d} \sum_{j=i+1}^{d} \int_{\Omega_d} \left(\frac{\partial^2 u}{\partial x_i^2}\right)(\boldsymbol{x})\left(\frac{\partial^2 u}{\partial x_j^2}\right)(\boldsymbol{x})\varpi_\alpha(\boldsymbol{x})\,\mathrm{d}\boldsymbol{x}. \tag{18.37}$$

We observe that on each $(d-1)$-face $x_j = \pm 1$, all the derivatives $\partial^k u/\partial x_i^k$, $i \neq j$, are equal to 0. Thus, integrating twice by parts gives for $i \neq j$

$$\int_{\Omega_d} \left(\frac{\partial^2 u}{\partial x_i^2}\right)(\boldsymbol{x})\left(\frac{\partial^2 u}{\partial x_j^2}\right)(\boldsymbol{x})\varpi_\alpha(\boldsymbol{x})\,\mathrm{d}\boldsymbol{x}$$
$$= \int_{\Omega_d} \left(\frac{\partial^2(u\rho_\alpha(x_i))}{\partial x_i \partial x_j}\right)(\boldsymbol{x})\left(\frac{\partial^2(u\rho_\alpha(x_j))}{\partial x_i \partial x_j}\right)(\boldsymbol{x})$$
$$\varpi_\alpha(\boldsymbol{x})\rho_{-\alpha}(x_i)\rho_{-\alpha}(x_j)\,\mathrm{d}\boldsymbol{x}. \tag{18.38}$$

Now, we observe that

$$\int_{-1}^{1}\int_{-1}^{1} \left(\frac{\partial^2(u\rho_\alpha(x_i))}{\partial x_i \partial x_j}\right)(x_i, x_j)\left(\frac{\partial^2(u\rho_\alpha(x_j))}{\partial x_i \partial x_j}\right)(x_i, x_j)\,\mathrm{d}x_i\,\mathrm{d}x_j$$

$$= \int_{-1}^{1}\int_{-1}^{1}\left(\frac{\partial^2 u}{\partial x_i \partial x_j}\right)^2 (x_i, x_j)\rho_\alpha(x_i)\rho_\alpha(x_j)\,\mathrm{d}x_i\,\mathrm{d}x_j$$
$$+ \int_{-1}^{1}\int_{-1}^{1}\left(\frac{\partial u}{\partial x_j}\right)(x_i, x_j)\left(\frac{\partial^2 u}{\partial x_i \partial x_j}\right)(x_i, x_j)\rho'_\alpha(x_i)\rho_\alpha(x_j)\,\mathrm{d}x_i\,\mathrm{d}x_j$$
$$+ \int_{-1}^{1}\int_{-1}^{1}\left(\frac{\partial^2 u}{\partial x_i \partial x_j}\right)(x_i, x_j)\left(\frac{\partial u}{\partial x_i}\right)(x_i, x_j)\rho_\alpha(x_i)\rho'_\alpha(x_j)\,\mathrm{d}x_i\,\mathrm{d}x_j$$
$$+ \int_{-1}^{1}\int_{-1}^{1}\left(\frac{\partial u}{\partial x_i}\right)(x_i, x_j)\left(\frac{\partial u}{\partial x_j}\right)(x_i, x_j)\rho'_\alpha(x_i)\rho'_\alpha(x_j)\,\mathrm{d}x_i\,\mathrm{d}x_j.$$

Two additional integrations by parts lead to the formula

$$\int_{-1}^{1}\int_{-1}^{1}\left(\frac{\partial^2 (u\rho_\alpha(x_i))}{\partial x_i \partial x_j}\right)(x_i, x_j)\left(\frac{\partial^2 (u\rho_\alpha(x_j))}{\partial x_i \partial x_j}\right)(x_i, x_j)\,\mathrm{d}x_i\,\mathrm{d}x_j$$
$$= \int_{-1}^{1}\int_{-1}^{1}\left(\frac{\partial^2 u}{\partial x_i \partial x_j}\right)^2 (x_i, x_j)\rho_\alpha(x_i)\,\rho_\alpha(x_j)\,\mathrm{d}x_i\,\mathrm{d}x_j$$
$$- \frac{1}{2}\int_{-1}^{1}\int_{-1}^{1}\left(\frac{\partial u}{\partial x_j}\right)^2 (x_i, x_j)\rho''_\alpha(x_i)\rho_\alpha(x_j)\,\mathrm{d}x_i\,\mathrm{d}x_j$$
$$- \frac{1}{2}\int_{-1}^{1}\int_{-1}^{1}\left(\frac{\partial u}{\partial x_i}\right)^2 (x_i, x_j)\rho_\alpha(x_i)\rho''_\alpha(x_j)\,\mathrm{d}x_i\,\mathrm{d}x_j$$
$$+ \int_{-1}^{1}\int_{-1}^{1}\left(\frac{\partial u}{\partial x_i}\right)(x_i, x_j)\left(\frac{\partial u}{\partial x_j}\right)(x_i, x_j)\rho'_\alpha(x_i)\rho'_\alpha(x_j)\,\mathrm{d}x_i\,\mathrm{d}x_j.$$

Next, we use an appropriate inequality in the last line to derive

$$\int_{-1}^{1}\int_{-1}^{1}\left(\frac{\partial^2 (u\rho_\alpha(x_i))}{\partial x_i \partial x_j}\right)(x_i, x_j)\left(\frac{\partial^2 (u\rho_\alpha(x_j))}{\partial x_i \partial x_j}\right)(x_i, x_j)\,\mathrm{d}x_i\,\mathrm{d}x_j$$
$$\geqslant \int_{-1}^{1}\int_{-1}^{1}\left(\frac{\partial^2 u}{\partial x_i \partial x_j}\right)^2 (x_i, x_j)\rho_\alpha(x_i)\rho_\alpha(x_j)\,\mathrm{d}x_i\,\mathrm{d}x_j$$
$$+ \frac{1}{2}\int_{-1}^{1}\int_{-1}^{1}\left(\frac{\partial u}{\partial x_j}\right)^2 (x_i, x_j)\big(-\rho''_\alpha(x_i)\rho_\alpha^{-1}(x_i)$$
$$- \rho'^2_\alpha(x_i)\rho_\alpha^{-2}(x_i)\big)\rho_\alpha(x_i)\rho_\alpha(x_j)\,\mathrm{d}x_i\,\mathrm{d}x_j$$
$$+ \frac{1}{2}\int_{-1}^{1}\int_{-1}^{1}\left(\frac{\partial u}{\partial x_i}\right)^2 (x_i, x_j)\big(-\rho''_\alpha(x_j)\rho_\alpha^{-1}(x_j)$$
$$- \rho'^2_\alpha(x_j)\rho_\alpha^{-2}(x_j)\big)\rho_\alpha(x_i)\rho_\alpha(x_j)\,\mathrm{d}x_i\,\mathrm{d}x_j.$$

(1) In the case $0 \leqslant \alpha \leqslant \frac{1}{2}$, we have

$$-\rho_\alpha''(\zeta)\rho_\alpha^{-1}(\zeta) - \rho_\alpha'^2(\zeta)\rho_\alpha^{-2}(\zeta) = 2\alpha\big(1 + (1-4\alpha)\zeta^2\big)\big(1-\zeta^2\big)^{-2} \geqslant 0,$$

so that (18.36) follows from (18.37) and (18.38).

(2) If α is negative, setting $w = u\varpi_\alpha$, we observe that

$$\begin{aligned}\int_{-1}^{1}\int_{-1}^{1}\left(\frac{\partial^2(u\rho_\alpha(x_j))}{\partial x_i \partial x_j}\right)(x_i, x_j)\left(\frac{\partial^2(u\rho_\alpha(x_i))}{\partial x_i \partial x_j}\right)(x_i, x_j)\,\mathrm{d}x_i\,\mathrm{d}x_j \\ = \int_{-1}^{1}\int_{-1}^{1}\left(\frac{\partial^2(w\rho_{-\alpha}(x_i))}{\partial x_i \partial x_j}\right)(x_i, x_j)\left(\frac{\partial^2(w\rho_{-\alpha}(x_j))}{\partial x_i \partial x_j}\right)(x_i, x_j)\,\mathrm{d}x_i\,\mathrm{d}x_j.\end{aligned}$$

From the last lines, this quantity is positive. Next, it follows from (18.37) and (18.38) that the norm of u in $H_\alpha^2(\Lambda_j; L_\alpha^2(\Omega_{d-1}))$ for any j, $1 \leqslant j \leqslant 2d$, is bounded by a constant times $\|\Delta u\|_{L_\alpha^2(\Omega_d)}$. An interpolation argument yields that, for $1 \leqslant i < j \leqslant d$,

$$H_\alpha^2\big(\Lambda_i; L_\alpha^2(\Omega_{d-1})\big) \cap H_\alpha^2\big(\Lambda_j; L_\alpha^2(\Omega_{d-1})\big) \subset H_\alpha^1\big(\Lambda_i; H_\alpha^1\big(\Lambda_j; L_\alpha^2(\Omega_{d-2})\big)\big),$$

which completes the proof of (18.36).

Finally, for any function f in $L_\alpha^2(\Omega_d)$, there exists a sequence $(f_n)_n$ of functions in $\mathscr{C}^\infty(\overline{\Omega}_d)$ converging to f in $L_\alpha^2(\Omega_d)$. Let u_n be the solution of problem (18.28) with f replaced by f_n and all the g_j, $1 \leqslant j \leqslant 2d$, equal to 0. In the cases $d = 2$ and $d = 3$, it follows from DAUGE ([1988], §18 and 19) that u_n belongs to $H^s(\Omega_d)$ for any real number $s < 3$ (see also Remark 2.1), hence to $H_\alpha^2(\Omega_d)$ by (18.7) and Theorem 18.2. Passing to the limit in (18.36) applied to each function u_n gives the desired regularity result.

EXAMPLE 18.2. We are also interested in the basic fourth-order problem. We begin with the case of homogeneous boundary conditions:

$$\begin{cases} \Delta^2 u = f & \text{in } \Omega_d, \\ u = 0 & \text{on } \partial\Omega_d, \\ \dfrac{\partial u}{\partial n} = 0 & \text{on } \partial\Omega_d. \end{cases} \tag{18.39}$$

Now, the distribution f is supposed to be in $H_\alpha^{-2}(\Omega_d)$, so that we can write the variational formulation of problem (18.39): *find u in $H_{\alpha,0}^2(\Omega_d)$, such that*

$$\forall v \in H_{\alpha,0}^2(\Omega_d), \quad d_\alpha(u, v) = \langle f, v\rangle_\alpha, \tag{18.40}$$

where the bilinear form $d_\alpha(\bullet, \bullet)$ is defined by

$$d_\alpha(u, v) = \int_{\Omega_d} (\Delta u)(\boldsymbol{x})\big(\Delta(v\varpi_\alpha)\big)(\boldsymbol{x})\,\mathrm{d}\boldsymbol{x}. \tag{18.41}$$

The analysis of the variational problem (18.40) relies on the properties of the form $d_\alpha(\bullet,\bullet)$. The proof of these properties is rather technical and is written by MADAY and MÉTIVET [1986] for $\alpha = -\frac{1}{2}$ and by BERNARDI, COPPOLETTA and MADAY ([1990], Appendix) in the case $d = 2$, that is why we prefer to skip it over.

THEOREM 18.15. *Let α be a real number which satisfies $-\frac{1}{2} \leqslant \alpha \leqslant \frac{1}{2}$. The form $d_\alpha(\bullet,\bullet)$ is continuous on $H^2_\alpha(\Omega_d) \times H^2_{\alpha,0}(\Omega_d)$ and elliptic on $H^2_{\alpha,0}(\Omega_d)$.*

As an immediate conclusion, we derive the following result.

THEOREM 18.16. *Let α be a real number which satisfies $-\frac{1}{2} \leqslant \alpha \leqslant \frac{1}{2}$. For any distribution f in $H^{-2}_\alpha(\Omega_d)$, problem (18.40) has a unique solution u in $H^2_{\alpha,0}(\Omega_d)$. Moreover, this solution satisfies*

$$\|u\|_{H^2_\alpha(\Omega_d)} \leqslant c\|f\|_{H^{-2}_\alpha(\Omega_d)}. \tag{18.42}$$

EXAMPLE 18.2′. Next, we consider the problem with inhomogeneous boundary conditions, but only in the square Ω for the sake of brevity:

$$\begin{cases} \Delta^2 u = f & \text{in } \Omega, \\ u = g_J & \text{on } \Gamma_J, \quad J = 1,2,3,4. \\ \dfrac{\partial u}{\partial n_J} = h_J & \text{on } \Gamma_J, \end{cases} \tag{18.43}$$

Now, the distribution f is supposed to be in $H^{-2}_\alpha(\Omega)$, while the 8-tuple (g_J, h_J) belongs to $\prod_{J=1}^4 H_\alpha^{(3-\alpha)/2}(\Gamma_J) \times H_\alpha^{(1-\alpha)/2}(\Gamma_J)$ and satisfies

$$g_J(\boldsymbol{a}_J) = g_{J+1}(\boldsymbol{a}_J), \quad J = 1,2,3,4, \tag{18.44}$$

when α is positive, and

$$\begin{aligned} & g_J(\boldsymbol{a}_J) = g_{J+1}(\boldsymbol{a}_J), \quad J = 1,2,3,4, \\ & \frac{\mathrm{d}g_J}{\mathrm{d}\tau_J}(\boldsymbol{a}_J) = h_{J+1}(\boldsymbol{a}_J) \text{ and } h_J(\boldsymbol{a}_J) = -\frac{\mathrm{d}g_{J+1}}{\mathrm{d}\tau_{J+1}}(\boldsymbol{a}_J), \quad J = 1,2,3,4, \end{aligned} \tag{18.45}$$

when α is negative. In this case, it follows from Theorem 18.11 that there exists a function u_b in $H^2_\alpha(\Omega)$ such that

$$u_\mathrm{b} = g_J \text{ and } \frac{\partial u_\mathrm{b}}{\partial n_J} = h_J \quad \text{on } \Gamma_J, \ J - 1,2,3,4. \tag{18.46}$$

This allows for writing the variational formulation of problem (18.43): *find u in $H^2_\alpha(\Omega)$, with $u - u_\mathrm{b}$ in $H^2_{\alpha,0}(\Omega)$, such that*

$$\forall v \in H^2_{\alpha,0}(\Omega), \quad d_\alpha(u,v) = \langle f, v\rangle_\alpha. \tag{18.47}$$

The wellposedness of this problem follows from Theorem 18.15.

THEOREM 18.17. *Let α be a real number which satisfies $-\frac{1}{2} \leqslant \alpha \leqslant \frac{1}{2}$, $\alpha \neq 0$. For any distribution f in $H_\alpha^{-2}(\Omega)$ and for any 8-tuple $(g_J, h_J)_{J=1,2,3,4}$ in*

$$\prod_{J=1}^{4} H_\alpha^{(3-\alpha)/2}(\Gamma_J) \times H_\alpha^{(1-\alpha)/2}(\Gamma_J)$$

which satisfies condition (18.44) when α is positive or condition (18.45) when α is negative, problem (18.47) has a unique solution u in $H_\alpha^2(\Omega)$. Moreover, this solution satisfies

$$\|u\|_{H_\alpha^2(\Omega)} \leqslant c\left(\|f\|_{H_\alpha^{-2}(\Omega)} + \sum_{J=1}^{4} \left(\|g_J\|_{H_\alpha^{(3-\alpha)/2}(\Gamma_J)} + \|h_J\|_{H_\alpha^{(1-\alpha)/2}(\Gamma_J)}\right)\right). \tag{18.48}$$

An application of this theorem is the existence of a solution for the two-dimensional driven cavity problem when formulated with the stream-function variable.

19. Jacobi orthogonal polynomials

The idea of working with weighted Sobolev spaces is related to the use of a family of orthogonal polynomials for the corresponding weighted measure. It is thus natural to introduce the orthogonal polynomials for the measure $\rho_\alpha(\zeta)\,d\zeta$: they are a special case of Jacobi polynomials. In this section, α stands for a real parameter > -1 (which is necessary for polynomials to belong to $L_\alpha^2(\Lambda)$).

DEFINITION. The family of *Jacobi polynomials* is the family $(J_n^\alpha)_{n\geqslant 0}$ of polynomials with one variable, which are orthogonal to each other in $L_\alpha^2(\Lambda)$ and such that, for any nonnegative integer n, the polynomial J_n^α has degree n and satisfies:

$$J_n^\alpha(1) = \frac{\Gamma(n+\alpha+1)}{n!\,\Gamma(\alpha+1)}, \tag{19.1}$$

where Γ stands for the Euler's Gamma function. The coefficient of ζ^n in J_n^α is denoted by k_n^α.

The Jacobi polynomials J_n^α are a special case of the Jacobi polynomials $J_n^{\alpha,\beta}$ which are orthogonal to each other for the measure $(1-\zeta)^\alpha(1+\zeta)^\beta\,d\zeta$. They coincide, up to multiplicative constants, with the ultraspherical polynomials $C_n^{\alpha+1/2}$. More important are the identities:

$$L_n = J_n^0 \quad \text{and} \quad T_n = \frac{4^n (n!)^2}{(2n)!} J_n^{-1/2}, \quad n \geqslant 0, \tag{19.2}$$

where T_n is the Chebyshev polynomial of degree n, defined by

$$T_n(\zeta) = \cos(n \arccos \zeta).$$

Most properties of the Jacobi polynomials are a simple extension of the corresponding properties of the Legendre polynomials and their proofs are completely similar, that is why we do not write them down.

THEOREM 19.1 (Differential equation). *For any nonnegative integer n, the polynomial J_n^α satisfies the differential equation:*

$$\frac{\mathrm{d}}{\mathrm{d}\zeta}\big((1-\zeta^2)^{\alpha+1} J_n^{\alpha\prime}\big) + n(n+2\alpha+1)\big(1-\zeta^2\big)^{\alpha} J_n^\alpha = 0. \tag{19.3}$$

An equivalent statement of Theorem 19.1 is that the Jacobi polynomials J_n^α are eigenfunctions of the Sturm–Liouville operator A_α defined by

$$A_\alpha \varphi = -\big(1-\zeta^2\big)^{-\alpha} \frac{\mathrm{d}}{\mathrm{d}\zeta}\big((1-\zeta^2)^{\alpha+1}\varphi'\big). \tag{19.4}$$

REMARK 19.1. As an immediate consequence of (19.3), we also derive that the $J_n^{\alpha\prime}$, $n \geqslant 1$, are a family of orthogonal polynomials for the measure $(1-\zeta^2)^{\alpha+1}\,\mathrm{d}\zeta$ in Λ. Computing the value in 1 of the polynomial $J_n^{\alpha\prime}$, which can also be derived from formula (19.3), we obtain the general formula:

$$J_n^{\alpha\prime} = \frac{n+2\alpha+1}{2} J_{n-1}^{\alpha+1}, \quad n \geqslant 1. \tag{19.5}$$

The Jacobi polynomials also satisfy a *Rodrigues formula*, which holds for any nonnegative integer n:

$$J_n^\alpha = \frac{(-1)^n}{2^n n!}\big(1-\zeta^2\big)^{-\alpha}\left(\frac{\mathrm{d}^n}{\mathrm{d}\zeta^n}\right)\big((1-\zeta^2)^{n+\alpha}\big). \tag{19.6}$$

Thus, we derive the value of k_n^α:

$$k_n^\alpha = \frac{\Gamma(2n+2\alpha+1)}{2^n n!\,\Gamma(n+2\alpha+1)}, \tag{19.7}$$

and we can also compute the norm of J_n^α in $L^2_\alpha(\Lambda)$.

THEOREM 19.2. *For any nonnegative integer n, the polynomial J_n^α satisfies*

$$\int_{-1}^{1} \big(J_n^\alpha\big)^2(\zeta)\rho_\alpha(\zeta)\,\mathrm{d}\zeta = \frac{2^{2\alpha+1}\,\Gamma(n+\alpha+1)^2}{(2n+2\alpha+1)\,n!\,\Gamma(n+2\alpha+1)}. \tag{19.8}$$

This formula is not so simple as the corresponding formula (3.11) for Legendre polynomials, however the estimate

$$\frac{c}{n+\frac{1}{2}} \leqslant \int_{-1}^{1} \left(J_n^\alpha\right)^2(\zeta)\rho_\alpha(\zeta)\,\mathrm{d}\zeta \leqslant \frac{c'}{n+\frac{1}{2}}, \tag{19.9}$$

can easily be derived from (19.8) by using Stirling's formula, which we recall for the sake of completeness:

$$\Gamma(t) = \sqrt{2\pi}\,\mathrm{e}^{-t}\,t^{t-1/2}\left(1+\mathrm{O}\left(\frac{1}{t}\right)\right), \quad t \to +\infty. \tag{19.10}$$

Formula (3.12) has two different analogues for $\alpha \neq 0$, but the proof relies on the same arguments.

THEOREM 19.3. *The following formula holds for any positive integer n:*

$$(2n+2\alpha+1)J_n^\alpha = \frac{n+2\alpha+1}{n+\alpha+1}J_{n+1}^{\alpha\prime} - \frac{n+\alpha}{n+2\alpha}J_{n-1}^{\alpha\prime}, \tag{19.11}$$

$$(2n+2\alpha+1)J_n^\alpha\rho_\alpha = \frac{n+1}{n+\alpha+1}\left(J_{n+1}^\alpha\rho_\alpha\right)' - \frac{n+\alpha}{n}\left(J_{n-1}^\alpha\rho_\alpha\right)'. \tag{19.12}$$

We are now in a position to state the induction formula for the J_n^α.

THEOREM 19.4 (Induction formula). *The family of Jacobi polynomials is given by*

$$\begin{cases} J_0^\alpha(\zeta) = 1 \quad \text{and} \quad J_1^\alpha(\zeta) = (\alpha+1)\zeta, \\ (n+1)(n+2\alpha+1)\,J_{n+1}^\alpha(\zeta) \\ \quad = (2n+2\alpha+1)(n+\alpha+1)\zeta J_n^\alpha(\zeta) \\ \qquad -(n+\alpha)(n+\alpha+1)J_{n-1}^\alpha(\zeta), \quad n \geqslant 1. \end{cases} \tag{19.13}$$

Taking α equal to m in this formula and using iteratively formula (19.5) gives another proof of the induction formula (3.16).

Using formula (19.13) with an induction argument, we can also derive the *Christoffel–Darboux formula* for the Jacobi polynomials, however we shall not give it since the coefficients are not simple.

REMARK 19.2. Let us summarize these results in the Chebyshev case $\alpha = -\frac{1}{2}$, which is the more useful and the simplest one. The Chebyshev polynomials T_n, $n \geqslant 0$, are given by

$$T_n(\zeta) = \cos(n \arccos \zeta). \tag{19.14}$$

They are orthogonal to each other for the measure $(1-\zeta^2)^{-1/2}\,\mathrm{d}\zeta$. They satisfy the differential equation

$$\frac{\mathrm{d}}{\mathrm{d}\zeta}\left((1-\zeta^2)^{1/2}T_n'\right)+n^2(1-\zeta^2)^{-1/2}T_n=0. \tag{19.15}$$

Their norm is given by

$$\int_{-1}^{1}T_0^2(\zeta)(1-\zeta^2)^{-1/2}\,\mathrm{d}\zeta=\pi$$

and

$$\int_{-1}^{1}T_n^2(\zeta)(1-\zeta^2)^{-1/2}\,\mathrm{d}\zeta=\pi/2,\quad n\geqslant 1. \tag{19.16}$$

They satisfy the induction formula:

$$\begin{cases}T_0(\zeta)=1 \quad\text{and}\quad T_1(\zeta)=\zeta,\\ T_{n+1}(\zeta)=2\zeta T_n(\zeta)-T_{n-1}(\zeta),\quad n\geqslant 1.\end{cases} \tag{19.17}$$

The next step is to consider quadrature formulas for approximating the integrals with respect to the measure $\rho_\alpha(\zeta)\,\mathrm{d}\zeta$. The following theorems are proven exactly as in Section 4 (see Theorems 4.1 and 4.2). We begin with the Gauss type formula.

THEOREM 19.5. *Let N be a fixed positive integer. There exist:*
(i) *a unique set of N points ζ_j^α, $1\leqslant j\leqslant N$, in Λ,*
(ii) *a unique set of N real numbers ω_j^α, $1\leqslant j\leqslant N$,*
such that the following equality holds for any polynomial Ψ in $\mathbb{P}_{2N-1}(\Lambda)$:

$$\int_{-1}^{1}\Psi(\zeta)\rho_\alpha(\zeta)\,\mathrm{d}\zeta=\sum_{j=1}^{N}\Psi(\zeta_j^\alpha)\omega_j^\alpha. \tag{19.18}$$

The ζ_j^α, $1\leqslant j\leqslant N$, are the zeros of the polynomial J_N^α.

In what follows, the zeros of J_N^α in increasing order are denoted by ζ_j^α, $1\leqslant j\leqslant N$, and the weights which are associated with them in the previous theorem are denoted by ω_j^α. Formula (19.18) is the Gauss–Jacobi formula. It should be noted that the statement of Theorem 4.1 coincides exactly with the previous one if α is chosen equal to m and N is replaced by $N-m$.

The weights ω_j^α, $1\leqslant j\leqslant N$, are computed from the identity

$$\omega_j^\alpha=\frac{1}{J_N^{\alpha\prime}(\zeta_j^\alpha)}\int_{-1}^{1}\frac{J_N^\alpha(\zeta)}{\zeta-\zeta_j^\alpha}(1-\zeta^2)^\alpha\,\mathrm{d}\zeta. \tag{19.19}$$

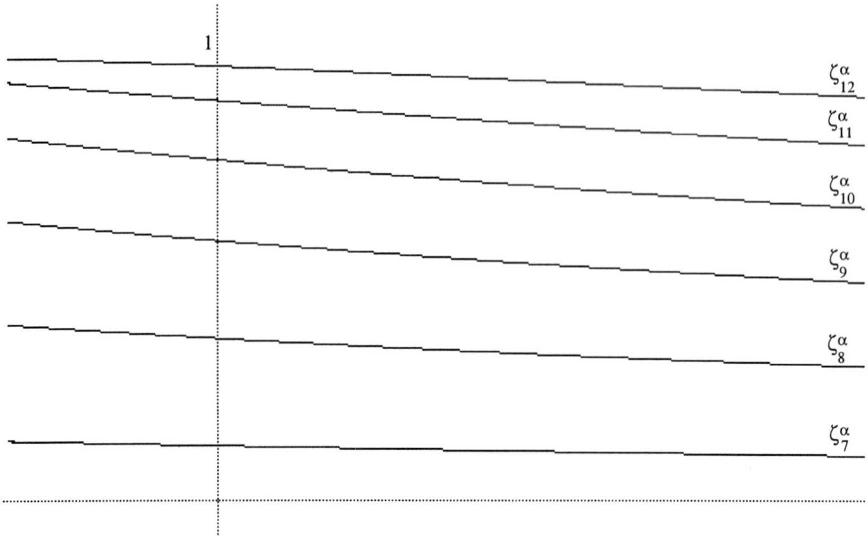

FIG. 19.1. The location of the positive zeros of J^α_{12}, $-1 < \alpha < 3$.

They are equal to

$$\omega_j^\alpha = \frac{2^{2\alpha+1}\Gamma(N+\alpha)\Gamma(N+\alpha+1)}{N!\Gamma(N+2\alpha+1)} \frac{1}{J_N^{\alpha\prime}(\zeta_j^\alpha)J_{N-1}^\alpha(\zeta_j^\alpha)} \tag{19.20}$$

(this equality is derived from the Christoffel–Darboux formula). Equivalently, combining (19.3), (19.12) and (19.13) yields that $(1-(\zeta_j^\alpha)^2)J_N^{\alpha\prime}(\zeta_j^\alpha)$ is equal to $(N+\alpha)J_{N-1}^\alpha(\zeta_j^\alpha)$, whence we also obtain

$$\omega_j^\alpha = \frac{2^{2\alpha+1}\Gamma(N+\alpha+1)^2}{N!\,\Gamma(N+2\alpha+1)} \frac{1}{(1-(\zeta_j^\alpha)^2)(J_N^{\alpha\prime})^2(\zeta_j^\alpha)}. \tag{19.21}$$

Let us note that the location of the zeros ζ_j^α is explicitly known from formula (19.14) in both cases $\alpha = -\frac{1}{2}$ and $\alpha = \frac{1}{2}$:

$$\zeta_j^{-1/2} = \cos\frac{(2N-2j+1)\pi}{2N} \text{ and } \zeta_j^{1/2} = \cos\frac{(N+1-j)\pi}{N+1}, \quad 1 \leqslant j \leqslant N. \tag{19.22}$$

The location of these zeros for any value of $\alpha > -1$ will be derived from the following theorem.

THEOREM 19.6. *Let α and β be two real numbers such that $-1 < \beta < \alpha \leqslant 0$. Then, the following inequalities hold*

$$\begin{cases} \zeta_{j+1}^{\beta} > \zeta_j^{\alpha} > \zeta_j^{\beta} & \textit{if } 1 \leqslant j \leqslant \left[\frac{N}{2}\right], \\ \zeta_{j-1}^{\beta} < \zeta_j^{\alpha} < \zeta_j^{\beta} & \textit{if } \left[\frac{N+2}{2}\right] \leqslant j \leqslant N. \end{cases} \tag{19.23}$$

PROOF. Introducing the function

$$\varphi_\alpha(\zeta) = \left(1-\zeta^2\right)^{(\alpha+1)/2} J_N^{\alpha}(\zeta),$$

we obtain the equation

$$\varphi_\alpha'' + \lambda_\alpha \varphi_\alpha = 0,$$

with

$$\lambda_\alpha(\zeta) = \frac{N(N+2\alpha+1)(1-\zeta^2) + (\alpha+1)(1-\alpha\zeta^2)}{(1-\zeta^2)^2}.$$

From this, we deduce that

$$(\varphi_\alpha\varphi_\beta' - \varphi_\alpha'\varphi_\beta)' = (\lambda_\alpha - \lambda_\beta)\varphi^\alpha\varphi^\beta,$$

where the function $\lambda_\alpha - \lambda_\beta$ is positive since it is equal to

$$(\lambda_\alpha - \lambda_\beta)(\zeta) = \frac{(2N+1)(\alpha-\beta)(1-\zeta^2) + (\beta^2-\alpha^2)\zeta^2}{(1-\zeta^2)^2}.$$

Integrating this formula between ζ_{j-1}^{β} and ζ_j^{β}, $[(N+3)/2] \leqslant j \leqslant N$, proves that J_N^{α} vanishes in this interval. Recalling the symmetry property $\zeta_{N+1-j}^{\alpha} = -\zeta_j^{\alpha}$ and looking at the location of the ζ_j^{α} which are the nearest to 0, allow for concluding.

What is the consequence for the location of the zeros? Due to the symmetry relation:

$$\zeta_j^{\alpha} = -\zeta_{N+1-j}^{\alpha},$$

we only consider the positive ζ_j^{α}. Combining Theorem 19.6 with (19.22) gives the following inequalities when α is negative:

$$\begin{cases} \zeta_{j-1}^{\alpha} \leqslant \cos\frac{(2N-2j+1)\pi}{2N} \leqslant \zeta_j^{\alpha} \\ \quad \text{if } \alpha < -\frac{1}{2}, \\ \cos\frac{(2N-2j+3)\pi}{2N} \leqslant \zeta_j^{\alpha} \leqslant \cos\frac{(2N-2j+1)\pi}{2N}, \\ \quad \text{if } \alpha > -\frac{1}{2}, \end{cases} \qquad \left[\frac{N+2}{2}\right] \leqslant j \leqslant N. \tag{19.24}$$

When α is positive, there exists an integer m such that $\alpha - m$ is between $-1 < \alpha \leqslant 0$, so that the location of the zeros of $J_{N+m}^{\alpha-m}$ is known. This allows for locating the zeros of $J_{N+m-1}^{\alpha-m+1}$, and finally of J_N^α.

The estimate for the weights is stated in the next theorem. Thanks to (19.21), its proof is similar to that of Theorem 4.4, it relies on the differential equation which is satisfied by the function

$$\chi^\alpha(\theta) = (\sin\theta)^{\alpha+1/2} J_N^\alpha(\cos\theta).$$

THEOREM 19.7. *The weights ω_j^α, $1 \leqslant j \leqslant N$, satisfy the following inequality*

$$\omega_j^\alpha \leqslant cN^{-1}\bigl(1 - (\zeta_j^\alpha)^2\bigr)^{\alpha+1/2}. \tag{19.25}$$

Next, we study the Gauss–Lobatto formula.

THEOREM 19.8. *Let N be a fixed positive integer. There exist:*

(i) *a unique set of $N-1$ points ξ_j^α, $1 \leqslant j \leqslant N-1$, in Λ,*

(ii) *a unique set of $N+1$ real numbers ρ_j^α, $0 \leqslant j \leqslant N$,*

such that the following equality holds for any polynomial Ψ in $\mathbb{P}_{2N-1}(\Lambda)$:

$$\int_{-1}^{1} \Psi(\zeta)\rho_\alpha(\zeta)\,\mathrm{d}\zeta = \sum_{j=1}^{N-1} \Psi(\xi_j^\alpha)\rho_j^\alpha + \Psi(-1)\rho_0^\alpha + \Psi(1)\rho_N^\alpha. \tag{19.26}$$

The ξ_j^α, $1 \leqslant j \leqslant N-1$, are the zeros of the polynomial $J_N^{\alpha\prime}$.

In what follows, the zeros of $(1-\zeta^2)\,J_N^{\alpha\prime}$ in increasing order are denoted by ξ_j^α, $0 \leqslant j \leqslant N$, and the weights which are associated with them are denoted by ρ_j^α. Formula (19.26) is the Gauss–Lobatto–Jacobi formula.

Note that the Gauss–Lobatto–Jacobi formula for the measure $(1-\zeta^2)^\alpha\,\mathrm{d}\zeta$ is strongly linked to the Gauss–Jacobi formula for the measure $(1-\zeta^2)^{\alpha+1}\,\mathrm{d}\zeta$ but with $N-1$ points instead of N. More precisely, let us for a while denote by $\zeta_j^{\alpha(N)}$ and $\omega_j^{\alpha(N)}$, $1 \leqslant j \leqslant N$, the nodes and weights of the Gauss–Jacobi formula (19.18). Then, it is readily checked that

$$\xi_j^\alpha = \zeta_j^{\alpha+1(N-1)} \quad \text{and} \quad \rho_j^\alpha = \bigl(1 - \bigl(\zeta_j^{\alpha+1(N-1)}\bigr)^2\bigr)^{-1}\omega_j^{\alpha+1(N-1)},$$
$$1 \leqslant j \leqslant N-1. \tag{19.27}$$

So the location of the nodes ξ_j^α, $1 \leqslant j \leqslant N-1$, is also given by Theorem 19.6, while the weights ρ_j^α, $1 \leqslant j \leqslant N-1$, also satisfy the inequality

$$\rho_j^\alpha \leqslant cN^{-1}\bigl(1 - \bigl(\xi_j^\alpha\bigr)^2\bigr)^{\alpha+1/2}. \tag{19.28}$$

As another consequence of (19.27), we derive the explicit value for the weights ρ_j^α from formula (19.20):

$$\rho_j^\alpha = \frac{2^{2\alpha+3}\Gamma(N+\alpha)\Gamma(N+\alpha+1)}{(N-1)!\Gamma(N+2\alpha+2)} \frac{1}{(1-(\xi_j^\alpha)^2)(J_{N-1}^{\alpha+1\prime})(\xi_j^\alpha)(J_{N-2}^{\alpha+1})(\xi_j^\alpha)}.$$

Using the fact that the ξ_j^α are the zeros of $J_{N-1}^{\alpha+1}$ together with formulas (19.3), (19.11) and (19.13), we derive, for $1 \leqslant j \leqslant N-1$,

$$\rho_j^\alpha = \frac{2^{2\alpha+1}\Gamma(N+\alpha)\Gamma(N+\alpha+1)}{(N-1)!\Gamma(N+2\alpha)} \frac{1}{(1-(\xi_j^\alpha)^2)J_N^{\alpha\prime\prime}(\xi_j^\alpha)J_{N-1}^{\alpha\prime}(\xi_j^\alpha)}, \tag{19.29}$$

or, equivalently,

$$\rho_j^\alpha = \frac{2^{2\alpha+1}\Gamma(N+\alpha+1)^2}{NN!\Gamma(N+2\alpha+2)} \frac{1}{(J_N^\alpha)^2(\xi_j^\alpha)}. \tag{19.30}$$

It remains to compute the boundary weights

$$\rho_0^\alpha = \frac{1}{2J_N^{\alpha\prime}(-1)} \int_{-1}^{1} (1-\zeta)J_N^{\alpha\prime}(\zeta)(1-\zeta^2)^\alpha \,\mathrm{d}\zeta,$$

$$\rho_N^\alpha = \frac{1}{2J_N^{\alpha\prime}(1)} \int_{-1}^{1} (1+\zeta)J_N^{\alpha\prime}(\zeta)(1-\zeta^2)^\alpha \,\mathrm{d}\zeta.$$

Using iteratively (19.11) and noting that the integrals $\int_{-1}^{1} J_n^\alpha(\zeta)\,\rho_\alpha(\zeta)\,\mathrm{d}\zeta$ are equal to 0 for $n \geqslant 1$, we derive the formulas

$$\rho_0^\alpha = \rho_N^\alpha = 2^{2\alpha+1}\Gamma(\alpha+1)\Gamma(\alpha+2)\frac{(N-1)!}{\Gamma(N+2\alpha+2)}. \tag{19.31}$$

The generalized Gauss type formula exists for the weighted measure $(1-\zeta^2)^\alpha\,\mathrm{d}\zeta$, it is thoroughly analyzed by BERNARDI and MADAY ([1991], §2). Even if we only use the formula for $m = 2$ in this work, we state the general result for the sake of completeness.

THEOREM 19.9. *Let m be a nonnegative integer, and let N be a fixed integer $\geqslant m$. There exist:*

(i) *a unique set of $N-m$ points $\zeta_j^{\alpha,m}$, $1 \leqslant j \leqslant N-m$, in Λ,*

(ii) *a unique set of $N-m$ real numbers $\sigma_j^{\alpha,m}$, $1 \leqslant j \leqslant N-m$,*

(iii) *a unique set of* $2m$ *real numbers* $\sigma_-^{\alpha,m,k}$ *and* $\sigma_+^{\alpha,m,k}$, $0 \leqslant k \leqslant m-1$, *such that the following equality holds for any polynomial* Φ *in* $\mathbb{P}_{2N-1}(\Lambda)$*:*

$$\int_{-1}^{1} \Phi(\zeta)\rho_\alpha(\zeta)\,\mathrm{d}\zeta = \sum_{j=1}^{N-m} \Phi(\zeta_j^{\alpha,m})\sigma_j^{\alpha,m}$$

$$+\sum_{k=0}^{m-1}\left(\left(\frac{\mathrm{d}^k\Phi}{\mathrm{d}\zeta^k}\right)(-1)\sigma_-^{\alpha,m,k} + \left(\frac{\mathrm{d}^k\Phi}{\mathrm{d}\zeta^k}\right)(+1)\sigma_+^{\alpha,m,k}\right). \tag{19.32}$$

The $\zeta_j^{\alpha,m}$, $1 \leqslant j \leqslant N-m$, *are the zeros of the polynomial* $\mathrm{d}^m J_N^\alpha/\mathrm{d}\zeta^m$.

We refer to BERNARDI and MADAY ([1991], §2) for the exact formulas which allow for computing the weights $\sigma_j^{\alpha,m}$ and $\sigma_\pm^{\alpha,m,k}$, but we observe that a relation analogous to (19.27) leads to the inequality

$$\sigma_j^{\alpha,m} \leqslant cN^{-1}\big(1-(\zeta_j^{\alpha,m})^2\big)^{\alpha+1/2}, \quad 1 \leqslant j \leqslant N-m. \tag{19.33}$$

20. Polynomial approximation in weighted norms

The results of this section are the weighted analogues of the results of Sections 6 for the interval and 7 for multi-dimensional domains. The proofs are very similar and will be skipped for this reason, except for duality results. Indeed, these ones are not so obvious as in the standard case and the way of turning around the difficulty is explained. From now on, we assume that $-1 < \alpha < 1$.

NOTATION. Let π_N^α be the orthogonal projection operator from $L^2_\alpha(\Lambda)$ onto $\mathbb{P}_N(\Lambda)$.

Any function φ in $L^2_\alpha(\Lambda)$ admits the expansion

$$\varphi = \sum_{n=0}^{+\infty} \varphi^n J_n^\alpha, \quad \text{with } \varphi^n = \frac{1}{\|J_n^\alpha\|^2_{L^2_\alpha(\Lambda)}} \int_{-1}^{1} \varphi(\zeta) J_n^\alpha(\zeta)\rho_\alpha(\zeta)\,\mathrm{d}\zeta, \tag{20.1}$$

and the polynomial $\pi_N^\alpha \varphi$ is given by

$$\pi_N^\alpha \varphi = \sum_{n=0}^{N} \varphi^n J_n^\alpha. \tag{20.2}$$

Next, since the J_n^α, $n \geqslant 0$, are the eigenfunctions of the operator A_α defined in (19.4) associated with the eigenvalues $n(n+2\alpha+1)$, using exactly the same arguments as for Theorem 6.1 leads to the following result.

THEOREM 20.1. *For any nonnegative real number* s, *there exists a positive constant* c *depending only on* s *such that, for any function* φ *in* $H^s_\alpha(\Lambda)$, *the following estimate holds*

$$\|\varphi - \pi_N^\alpha \varphi\|_{L^2_\alpha(\Lambda)} \leqslant cN^{-s}\|\varphi\|_{H^s_\alpha(\Lambda)}. \tag{20.3}$$

REMARK 20.1. As in Remark 6.3, using the regularity of the function to be approximated in the domain of the operator A_α up to a fixed power, leads to an improved estimate for functions which are singular in a neighborhood of ± 1.

Next, we wish to prove estimates in high order norms. To this aim, we introduce other projection operators.

NOTATION. For any positive integer k, let $\pi_N^{\alpha,k,0}$ be the orthogonal projection operator from $H^k_{\alpha,0}(\Lambda)$ onto $\mathbb{P}_N^{k,0}(\Lambda)$ for the scalar product associated with the norm $|\bullet|_{H^k_\alpha(\Lambda)}$.

From this definition, it is readily checked that the following formula holds for any integer $k \geqslant 1$ and for any $N \geqslant 2k-1$:

$$\begin{aligned}(\pi_N^{\alpha,k,0}\varphi)(\zeta) = \int_{-1}^{\zeta} \Bigg(&(\pi_{N-1}^{\alpha,k-1,0}\varphi')(\xi) \\ &- \frac{\int_{-1}^{1} (\pi_{N-1}^{\alpha,k-1,0}\varphi')(\eta)\,\mathrm{d}\eta}{\int_{-1}^{1}(1-\eta^2)^{k-1}\,\mathrm{d}\eta}\left(1-\xi^2\right)^{k-1}\Bigg)\,\mathrm{d}\xi. \end{aligned} \tag{20.4}$$

So, an induction argument leads to the theorem that is stated below.

THEOREM 20.2. *Let k be a positive integer. For any nonnegative real number $s \geqslant k$, there exists a positive constant c depending only on s such that, for any function φ in $H^s_\alpha(\Lambda) \cap H^k_{\alpha,0}(\Lambda)$, the following estimate holds*

$$\left\|\varphi - \pi_N^{\alpha,k,0}\varphi\right\|_{H^k_\alpha(\Lambda)} \leqslant cN^{k-s}\|\varphi\|_{H^s_\alpha(\Lambda)}. \tag{20.5}$$

However, estimates in $H^r_\alpha(\Lambda)$, $0 \leqslant r < k$, are not so simple to establish for the operator $\pi_N^{\alpha,k,0}$, since the weight does not allow for simple duality arguments. The following result is nevertheless proven in BERNARDI and MADAY ([1989], Theorem 4.2): for any function φ in $H^s_\alpha(\Lambda) \cap H^1_{\alpha,0}(\Lambda)$, $s \geqslant 1$,

$$\left\|\varphi - \pi_N^{\alpha,1,0}\varphi\right\|_{L^2_\alpha(\Lambda)} \leqslant cN^{\sup\{\alpha,0\}-s}\|\varphi\|_{H^s_\alpha(\Lambda)}. \tag{20.6}$$

Estimates in $H^r_\alpha(\Lambda)$, $0 \leqslant r \leqslant 1$, can of course be derived from (20.6) by an interpolation argument. However, it must be observed that (20.6) is not optimal when α is positive.

Since $\alpha < 1$ and exactly as in the case of standard Sobolev spaces, using the polynomials $\chi_{k,l}$, $0 \leqslant l \leqslant k-1$, defined in (6.16), we can associate with each function φ in $H^s_\alpha(\Lambda)$, $s \geqslant k$, the function $\widetilde{\varphi}_k$ defined in (6.17), which belongs to $H^s_\alpha(\Lambda) \cap H^k_{\alpha,0}(\Lambda)$ and satisfies

$$\|\widetilde{\varphi}_k\|_{H^s_\alpha(\Lambda)} \leqslant c\|\varphi\|_{H^s_\alpha(\Lambda)}. \tag{20.7}$$

Thus, setting

$$\tilde{\pi}_N^{\alpha,k}\varphi = \pi_N^{\alpha,k,0}\widetilde{\varphi}_k + \varphi - \widetilde{\varphi}_k, \tag{20.8}$$

we observe that the operator $\tilde{\pi}_N^{\alpha,k}$ is defined on $H_\alpha^k(\Lambda)$ and that the following estimate holds for any function φ in $H_\alpha^s(\Lambda)$, $s \geqslant k$:

$$\|\varphi - \tilde{\pi}_N^{\alpha,k}\varphi\|_{H_\alpha^k(\Lambda)} \leqslant cN^{k-s}\|\varphi\|_{H_\alpha^s(\Lambda)}. \tag{20.9}$$

REMARK 20.2. To obtain duality results in a simpler and optimal way, the idea is to introduce modified operators by using the ellipticity properties stated in Theorems 18.12 and 18.15: indeed, for k equal to 1 or 2, it is possible to define an operator $\pi_N^{\alpha,k,0*}$ from $H_{\alpha,0}^k(\Lambda)$ into $\mathbb{P}_N^{k,0}(\Lambda)$ by the equation:

(i) for k equal to 1, when α satisfies $-1 < \alpha < 1$,

$$\forall \psi_N \in \mathbb{P}_N^0(\Lambda), \quad \int_{-1}^1 \psi_N'(\zeta)\big((\varphi - \pi_N^{\alpha,1,0*}\varphi)\rho_\alpha\big)'(\zeta)\,\mathrm{d}\zeta = 0, \tag{20.10}$$

(ii) for k equal to 2, when α satisfies $-\frac{1}{2} \leqslant \alpha \leqslant \frac{1}{2}$,

$$\forall \psi_N \in \mathbb{P}_N^{2,0}(\Lambda), \quad \int_{-1}^1 \psi_N''(\zeta)\big((\varphi - \pi_N^{\alpha,2,0*}\varphi)\rho_\alpha\big)''(\zeta)\,\mathrm{d}\zeta = 0. \tag{20.11}$$

Indeed, the associated dual problems are standard Dirichlet problems, so that the usual duality arguments yield the following estimates which hold for any real numbers r and s, $0 \leqslant r \leqslant k \leqslant s$, and for any functions φ in $H_\alpha^s(\Lambda) \cap H_{\alpha,0}^k(\Lambda)$:

$$\|\varphi - \pi_N^{\alpha,k,0*}\varphi\|_{H_\alpha^r(\Lambda)} \leqslant cN^{r-s}\|\varphi\|_{H_\alpha^s(\Lambda)}. \tag{20.12}$$

The corresponding operators $\tilde{\pi}_N^{\alpha,k*}$, defined by

$$\tilde{\pi}_N^{\alpha,k*}\varphi = \pi_N^{\alpha,k,0*}\widetilde{\varphi}_k + \varphi - \widetilde{\varphi}_k, \tag{20.13}$$

satisfy for any real numbers r and s, $0 \leqslant r \leqslant k \leqslant s$, and for any function φ in $H_\alpha^s(\Lambda)$:

$$\|\varphi - \tilde{\pi}_N^{\alpha,k*}\varphi\|_{H_\alpha^r(\Lambda)} \leqslant cN^{r-s}\|\varphi\|_{H_\alpha^s(\Lambda)}. \tag{20.14}$$

To conclude with projection operators on the interval Λ, we state the general result which is proven in MADAY [1990] by arguments very similar to those of Theorem 6.4 relying on the interpolation properties of the weighted spaces.

THEOREM 20.3. *Let t and p be nonnegative real numbers such that neither $t - (1+\alpha)/2$ nor $p - (1+\alpha)/2$ is an integer. There exists an operator $\tilde{\pi}_N^{\alpha,t,p,0}$ from*

$H^t_\alpha(\Lambda) \cap H^p_{\alpha,0}(\Lambda)$ *into* $\mathbb{P}_N(\Lambda) \cap H^p_{\alpha,0}(\Lambda)$ *such that, for any real numbers* r *and* s, $0 \leqslant r \leqslant t \leqslant s$, *and for any function* φ *in* $H^s_\alpha(\Lambda) \cap H^p_{\alpha,0}(\Lambda)$, *the following estimate holds*

$$\left\|\varphi - \tilde{\pi}_N^{\alpha,t,p,0}\varphi\right\|_{H^r_\alpha(\Lambda)} \leqslant cN^{r-s}\|\varphi\|_{H^s_\alpha(\Lambda)}. \tag{20.15}$$

The results concerning polynomial approximation can easily be extended to multidimensional domains since the weighted Sobolev spaces on tensorized domains satisfy the same tensorization properties as the standard ones. Recall, with obvious notation,

(i) for any real numbers r and s, $0 \leqslant r \leqslant s$, the imbedding

$$H^s_\alpha(\Omega) \subset H^r_\alpha\big(\Lambda_j; H^{s-r}_\alpha(\Omega_{d-1})\big), \quad 1 \leqslant j \leqslant d, \tag{20.16}$$

(ii) for any nonnegative real number s, the identity

$$H^s_\alpha(\Omega) = \bigcap_{j=1}^{d} L^2_\alpha\big(\Lambda_j; H^s_\alpha(\Omega_{d-1})\big). \tag{20.17}$$

Hence, combining the first theorems of this section with the arguments of Section 7, we obtain the approximation results in the hypercube. We refer to Chapter 2 of MADAY [1987] and BERNARDI and MADAY [1989] for the complete proofs.

NOTATION. Let Π^α_N be the orthogonal projection operator from $L^2_\alpha(\Omega_d)$ onto $\mathbb{P}_N(\Omega_d)$.

THEOREM 20.4. *For any nonnegative real number* s, *there exists a positive constant* c *depending only on* s *such that, for any function* v *in* $H^s_\alpha(\Omega_d)$, *the following estimate holds*

$$\left\|v - \Pi_N v\right\|_{L^2_\alpha(\Omega_d)} \leqslant cN^{-s}\|v\|_{H^s_\alpha(\Omega_d)}. \tag{20.18}$$

NOTATION. For any positive integer k, let $\Pi_N^{\alpha,k,0}$ be the orthogonal projection operator from $H^k_{\alpha,0}(\Omega_d)$ onto $\mathbb{P}_N^{k,0}(\Omega_d)$ for the scalar product associated with the norm $|\cdot|_{H^k_\alpha(\Omega_d)}$.

THEOREM 20.5. *Let* k *be a positive integer. For any nonnegative real number* $s \geqslant k$, *there exists a positive constant* c *depending only on* s *such that, for any function* v *in* $H^s_\alpha(\Omega_d) \cap H^k_{\alpha,0}(\Omega_d)$, *the following estimate holds*

$$\left\|v - \Pi_N^{\alpha,k,0}v\right\|_{H^k_\alpha(\Omega_d)} \leqslant c\,N^{k-s}\,\|v\|_{H^s_\alpha(\Omega_d)}. \tag{20.19}$$

NOTATION. For any positive integer k, let $\Pi_N^{\alpha,k}$ be the orthogonal projection operator from $H^k_\alpha(\Omega_d)$ onto $\mathbb{P}_N(\Omega_d)$ for the scalar product associated with the norm $\|\cdot\|_{H^k_\alpha(\Omega_d)}$.

THEOREM 20.6. *Let k be a positive integer. For any nonnegative real number $s \geqslant k$, there exists a positive constant c depending only on s such that, for any function v in $H^s_\alpha(\Omega_d)$, the following estimate holds*

$$\|v - \Pi_N^{\alpha,k} v\|_{H^k_\alpha(\Omega_d)} \leqslant c N^{k-s} \|v\|_{H^s_\alpha(\Omega_d)}. \tag{20.20}$$

We conclude with the general result, nothing in its proof is really new in comparison with those of Theorems 7.4 and 20.3.

THEOREM 20.7. *Let t and p be nonnegative real numbers such that neither $t - (1+\alpha)/2$ nor $p - (1+\alpha)/2$ is an integer. There exists an operator $\widetilde{\Pi}_N^{\alpha,t,p,0}$ from $H^t_\alpha(\Omega_d) \cap H^p_{\alpha,0}(\Omega_d)$ into $\mathbb{P}_N(\Omega_d) \cap H^p_{\alpha,0}(\Omega_d)$ such that, for any real numbers r and s, $0 \leqslant r \leqslant t \leqslant s$, and for any function v in $H^s_\alpha(\Omega_d) \cap H^p_{\alpha,0}(\Omega_d)$, the following estimate holds*

$$\|v - \widetilde{\Pi}_N^{\alpha,t,p,0} v\|_{H^r_\alpha(\Omega_d)} \leqslant c N^{r-s} \|v\|_{H^s_\alpha(\Omega_d)}. \tag{20.21}$$

Of course, from the weighted variational formulations (18.30) and (18.40) of problems (18.28) and (18.39), discrete problems can be built by the Galerkin method. The previous approximation results allow for the numerical analysis of the method and lead to optimal error estimates. For the sake of brevity, we limit ourselves to the analysis of the more general method with numerical integration, so that we also need some results about polynomial interpolation in weighted spaces.

21. Polynomial interpolation in weighted norms

The analysis of the interpolation at the nodes of the quadrature formulas introduced in Section 19, is very similar to Sections 13 and 14, due to the results of Theorems 19.6 and 19.7. So, the properties of the interpolation operators are stated firstly in the interval Λ, secondly in the hypercube Ω_d, and a number of proofs are skipped. We still assume that $-1 < \alpha < 1$.

We begin with the operator related to the Gauss–Jacobi formula. The operator j^α_{N-1} is defined in the following way: for any continuous function φ on Λ, $j^\alpha_{N-1}\varphi$ is the only polynomial in $\mathbb{P}_{N-1}(\Lambda)$ which satisfies

$$(j^\alpha_{N-1}\varphi)(\zeta^\alpha_j) = \varphi(\zeta^\alpha_j), \quad 1 \leqslant j \leqslant N, \tag{21.1}$$

where the ζ^α_j, $1 \leqslant j \leqslant N$, are the zeros of J^α_N. The first theorem proves a stability result for this operator.

THEOREM 21.1. *For any real number r, $\frac{1}{2} < r \leqslant 1$, there exists a positive constant c such that, for any function φ in $H^r_{\alpha,0}(\Lambda)$, the following estimate holds*

$$\|j^\alpha_{N-1}\varphi\|_{L^2_\alpha(\Lambda)} \leqslant c\big(\|\varphi\|_{L^2_\alpha(\Lambda)} + N^{-r}\|\varphi\|_{H^r_\alpha(\Lambda)}\big). \tag{21.2}$$

We only prove the result for $r = 1$, since the proof for $r < 1$ is very technical, not so different from the proof of Theorem 13.1 and presented in a detailed way in BERNARDI and MADAY ([1992c], Theorem 3.3). It relies on Theorems 19.6 and 19.7 for the location of the interpolation nodes and the estimate of the weights of the Gauss–Jacobi quadrature formula.

PROOF. In the case $r = 1$, we compute

$$\|j_{N-1}^{\alpha}\varphi\|_{L_{\alpha}^{2}(\Lambda)}^{2} = \sum_{j=1}^{N} \left(j_{N-1}^{\alpha}\varphi\right)^{2}(\zeta_j^{\alpha})\omega_j^{\alpha} = \sum_{j=1}^{N} \varphi^{2}(\zeta_j^{\alpha})\omega_j^{\alpha}.$$

From Theorem 19.7, we derive

$$\|j_{N-1}^{\alpha}\varphi\|_{L_{\alpha}^{2}(\Lambda)}^{2} \leqslant cN^{-1}\sum_{j=1}^{N} \varphi^{2}(\zeta_j^{\alpha})\left(1-(\zeta_j^{\alpha})^{2}\right)^{\alpha+1/2}.$$

Next, we set $\zeta_j^{\alpha} = \cos\theta_j^{\alpha}$, $1 \leqslant j \leqslant N$. Using the change of variable $\zeta = \cos\theta$, together with the notation: $\widehat{\varphi}(\theta) = \varphi(\zeta)$, we obtain

$$\|j_{N-1}^{\alpha}\varphi\|_{L_{\alpha}^{2}(\Lambda)}^{2} \leqslant cN^{-1}\sum_{j=1}^{N} \widehat{\varphi}^{2}(\theta_j^{\alpha})\left(\sin\theta_j^{\alpha}\right)^{2\alpha+1}.$$

Due to Theorem 19.6, there exist intervals K_j^{α} of length c/N containing ζ_j^{α} such that each K_j^{α} meets a finite number (depending only on α) of other intervals K_i^{α}. This yields

$$\|j_{N-1}^{\alpha}\varphi\|_{L_{\alpha}^{2}(\Lambda)}^{2} \leqslant cN^{-1}\sum_{j=1}^{N} \sup_{\theta\in\overline{K}_j^{\alpha}} \left|\widehat{\varphi}(\theta)(\sin\theta)^{\alpha+1/2}\right|^{2}. \tag{21.3}$$

Making use of (13.7) and adding up on all intervals K_j^{α}, we derive

$$\begin{aligned}\|j_{N-1}^{\alpha}\varphi\|_{L_{\alpha}^{2}(\Lambda)}^{2} \leqslant{}& c\Big(\|\widehat{\varphi}(\sin\theta)^{\alpha+1/2}\|_{L^{2}(0,\pi)}^{2} \\ &+ N^{-2}|\widehat{\varphi}(\sin\theta)^{\alpha+1/2}|_{H^{1}(0,\pi)}^{2}\Big).\end{aligned} \tag{21.4}$$

The inverse change of variable yields

$$\begin{aligned}\|j_{N-1}^{\alpha}\varphi\|_{L_{\alpha}^{2}(\Lambda)}^{2} \leqslant{}& c\int_{-1}^{1} \Big(\varphi^{2}(\zeta)\left(1-\zeta^{2}\right)^{\alpha} + N^{-2}\varphi^{2}(\zeta)\zeta^{2}\left(1-\zeta^{2}\right)^{\alpha-1} \\ &+ N^{-2}\varphi'^{2}(\zeta)\left(1-\zeta^{2}\right)^{\alpha+1}\Big)\,\mathrm{d}\zeta.\end{aligned}$$

We conclude by using Hardy's inequality (13.4) or (13.5).

The following result is a consequence of the previous theorem combined with Theorem 20.3 in a standard way.

THEOREM 21.2. *For any real number* $s > \frac{1}{2}$, *there exists a positive constant c depending only on s such that, for any function* φ *in* $H^s_\alpha(\Lambda)$, *the following estimate holds*

$$\|\varphi - j^\alpha_{N-1}\varphi\|_{L^2_\alpha(\Lambda)} \leqslant c N^{-s} \|\varphi\|_{H^s_\alpha(\Lambda)}. \tag{21.5}$$

REMARK 21.1. As in Remarks 13.2 and 20.1, the previous estimate (21.5) can be improved for functions which are singular in a neighborhood of ± 1.

However, as in the Legendre case, this operator does not allow for optimal estimates in high order norms. Hence, we now consider the operator associated with the Gauss–Lobatto–Jacobi formula. The operator i^α_N is defined in the following way: for any continuous function φ on $\overline{\Lambda}$, $i^\alpha_N\varphi$ is the only polynomial in $\mathbb{P}_N(\Lambda)$ which satisfies

$$(i^\alpha_N\varphi)(\xi^\alpha_j) = \varphi(\xi^\alpha_j), \quad 0 \leqslant j \leqslant N, \tag{21.6}$$

where the ξ^α_j, $0 \leqslant j \leqslant N$, are the zeros of $(1-\zeta^2) J^{\alpha\,\prime}_N$.

The first stability estimate for the operator i^α_N relies on exactly the same arguments as Theorem 21.1, together with the weighted analogue of (13.20):

$$\forall \varphi_N \in \mathbb{P}_N(\Lambda),$$
$$\|\varphi_N\|^2_{L^2_\alpha(\Lambda)} \leqslant \sum_{j=0}^{N} \varphi^2_N(\xi^\alpha_j)\rho^\alpha_j \leqslant (3+2\alpha)\, \|\varphi_N\|^2_{L^2_\alpha(\Lambda)}. \tag{21.7}$$

These estimates are established by writing φ_N in the basis of Jacobi polynomials and using the fact that the quadrature formula is exact on $\mathbb{P}_{2N-1}(\Lambda)$, together with the identity

$$\sum_{j=0}^{N} (J^\alpha_N)^2(\xi^\alpha_j)\rho^\alpha_j = \left(2 + \frac{2\alpha+1}{N}\right) \|J^\alpha_N\|^2_{L^2_\alpha(\Lambda)}, \tag{21.8}$$

which follows from (19.30), (19.31), (19.3) and (19.8). The proof of the second stability estimate is very similar (see Theorem 13.3 for the analogue in the Legendre case).

THEOREM 21.3. *For any real number r*, $\sup\{\frac{1}{2}, (1+\alpha)/2\} < r \leqslant 1$, *there exists a positive constant c such that, for any function* φ *in* $H^r_{\alpha,0}(\Lambda)$, *the following estimates hold*

$$\|i^\alpha_N\varphi\|_{L^2_\alpha(\Lambda)} \leqslant c\big(\|\varphi\|_{L^2_\alpha(\Lambda)} + N^{-r}|\varphi|_{H^r_\alpha(\Lambda)}\big), \tag{21.9}$$

and

$$\|(i_N^\alpha \varphi)(1-\zeta^2)^{-1/2}\|_{L^2_\alpha(\Lambda)} \leqslant c\Big(\|\varphi(1-\zeta^2)^{-1/2}\|_{L^2_\alpha(\Lambda)} + N^{-r}|\varphi(1-\zeta^2)^{(r-1)/2}|_{H^r_\alpha(\Lambda)}\Big). \tag{21.10}$$

The following inverse inequality is proven by writing the polynomial φ_N in the basis $\{(1-\zeta^2)J_n^{\alpha\prime}\colon\ 1 \leqslant n \leqslant N-1\}$:

$$\forall \varphi_N \in \mathbb{P}_N^0(\Lambda), \quad \|\varphi_N'\|_{L^2_\alpha(\Lambda)} \leqslant cN\|\varphi_N(1-\zeta^2)^{-1/2}\|_{L^2_\alpha(\Lambda)}. \tag{21.11}$$

Combined with Theorems 21.3 and 20.3, it allows for deriving the following estimates.

THEOREM 21.4. *For any real numbers r and s satisfying $0 \leqslant r \leqslant 1$ and*

$$s > \sup\{(1+r)/2, (1+\alpha+r)/2\},$$

there exists a positive constant c depending only on s such that, for any function φ in $H^s_\alpha(\Lambda)$, the following estimate holds

$$\|\varphi - i_N^\alpha \varphi\|_{H^r_\alpha(\Lambda)} \leqslant cN^{r-s}\|\varphi\|_{H^s_\alpha(\Lambda)}. \tag{21.12}$$

Of course, Hermite type interpolation operators can also be associated with the generalized quadrature formula (19.32). We only look at the case $m = 2$. The operator $k_{N+1}^{\alpha,2}$ is defined in the following way: for any continuous function φ on Λ with continuous derivatives in ± 1, $k_{N+1}^{\alpha,2}\varphi$ is the only polynomial in $\mathbb{P}_{N+1}(\Lambda)$ which satisfies

$$\begin{aligned} &(k_{N+1}^{\alpha,2}\varphi)(\zeta_j^{\alpha,2}) = \varphi(\zeta_j^{\alpha,2}), \quad 1 \leqslant j \leqslant N-2,\\ &(k_{N+1}^{\alpha,2}\varphi)(-1) = \varphi(-1), \quad (k_{N+1}^{\alpha,2}\varphi)(+1) = \varphi(+1),\\ &(k_{N+1}^{\alpha,2}\varphi)'(-1) = \varphi'(-1) \quad \text{and} \quad (k_{N+1}^{\alpha,2}\varphi)'(+1) = \varphi'(+1), \end{aligned} \tag{21.13}$$

where the $\zeta_j^{\alpha,2}$, $1 \leqslant j \leqslant N-2$, are the zeros of $J_N^{\alpha\prime\prime}$. The following results are proven with the same arguments as previously (we refer to BERNARDI and MADAY [1992c] for the details):

(i) for any function φ such that $\varphi(1-\zeta^2)^{-1}$ belongs to $H^s_\alpha(\Lambda)$ with $s > \sup\{\frac{1}{2}, (1+\alpha)/2\}$,

$$\|(\varphi - k_{N+2}^{\alpha,2}\varphi)(1-\zeta^2)^{-1}\|_{L^2_\alpha(\Lambda)} \leqslant cN^{-s}\|\varphi(1-\zeta^2)^{-1}\|_{H^s_\alpha(\Lambda)}; \tag{21.14}$$

(ii) for any real number r, $0 \leqslant r \leqslant 2$, and for any function φ in $H^s_\alpha(\Lambda)$ with $s > \sup\{(6+r)/4, (3+\alpha)/2\}$,

$$\|\varphi - k^{\alpha,2}_{N+2}\varphi\|_{H^r_\alpha(\Lambda)} \leqslant cN^{r-s}\|\varphi\|_{H^s_\alpha(\Lambda)}. \tag{21.15}$$

Next, we need to extend these results to the hypercube Ω_d. Since the proofs here are strictly the same as in Section 14, we only recall the results.

NOTATION. The *Gauss grid* of Jacobi type Σ^α_N is defined by

$$\Sigma^\alpha_N = \{\boldsymbol{x} = (\zeta^\alpha_{i_1}, \dots, \zeta^\alpha_{i_d}) \colon 1 \leqslant i_1, \dots, i_d \leqslant N\}. \tag{21.16}$$

We denote by $\mathcal{J}^\alpha_{N-1}$ the Lagrange interpolation operator at the points of the grid Σ^α_N: for any continuous function v on Ω_d, $\mathcal{J}^\alpha_{N-1}v$ belongs to $\mathbb{P}_{N-1}(\Omega_d)$ and satisfies

$$(\mathcal{J}^\alpha_{N-1}v)(\boldsymbol{x}) = v(\boldsymbol{x}), \quad \boldsymbol{x} \in \Sigma^\alpha_N. \tag{21.17}$$

Since the formula

$$\mathcal{J}^\alpha_{N-1} = j^{\alpha(1)}_{N-1} \circ \cdots \circ j^{\alpha(d)}_{N-1}, \tag{21.18}$$

also holds (the exponent $^{(j)}$ represents the variable x_j for which the operator is applied), the proof of the following theorem is straightforward.

THEOREM 21.5. *For any real number $s > d/2$, there exists a positive constant c depending only on s such that, for any function v in $H^s_\alpha(\Omega_d)$, the following estimate holds*

$$\|v - \mathcal{J}^\alpha_{N-1}v\|_{L^2_\alpha(\Omega_d)} \leqslant cN^{-s}\|v\|_{H^s_\alpha(\Omega_d)}. \tag{21.19}$$

We now consider the interpolation operator related to the Gauss–Lobatto formula.

NOTATION. The *Gauss–Lobatto grid* of Jacobi type Ξ^α_N is defined by

$$\Xi^\alpha_N = \{\boldsymbol{x} = (\xi^\alpha_{i_1}, \dots, \xi^\alpha_{i_d}) \colon 0 \leqslant i_1, \dots, i_d \leqslant N\}. \tag{21.20}$$

We denote by $\mathcal{I}^\alpha_N$ the Lagrange interpolation operator at the points of the grid Ξ^α_N: for any continuous function v on $\overline{\Omega}_d$, $\mathcal{I}^\alpha_N v$ belongs to $\mathbb{P}_N(\Omega_d)$ and satisfies

$$(\mathcal{I}^\alpha_N v)(\boldsymbol{x}) = v(\boldsymbol{x}), \quad \boldsymbol{x} \in \Xi^\alpha_N. \tag{21.21}$$

Here also, the identity

$$\mathcal{I}^\alpha_N = i^{\alpha(1)}_N \circ \cdots \circ i^{\alpha(d)}_N, \tag{21.22}$$

leads to the following theorem.

THEOREM 21.6. *For any real numbers r and s satisfying $0 \leqslant r \leqslant 1$ and*

$$s > \sup\big\{(d+r)/2, \big(d(1+\alpha)+r\big)/2\big\},$$

there exists a positive constant c depending only on s such that, for any function v in $H^s_\alpha(\Omega_d)$, the following estimate holds

$$\big\|v - \mathcal{I}^\alpha_N v\big\|_{H^r_\alpha(\Omega_d)} \leqslant cN^{r-s}\|v\|_{H^s_\alpha(\Omega_d)}. \tag{21.23}$$

Finally, defining the grid

$$\Sigma^{\alpha,2}_N = \big\{\boldsymbol{x} = \big(\zeta^{\alpha,2}_{i_1}, \ldots, \zeta^{\alpha,2}_{i_d}\big)\colon\ 0 \leqslant i_1, \ldots, i_d \leqslant N-1\big\}, \tag{21.24}$$

with $\zeta^{\alpha,2}_0 = -1$ and $\zeta^{\alpha,2}_{N-1} = +1$ and the operator $\mathcal{K}^{\alpha,2}_{N+1}$ as the Hermite interpolation operator on this grid, i.e., by the formula

$$\mathcal{K}^{\alpha,2}_{N+1} = k^{\alpha,2(1)}_{N+1} \circ \cdots \circ k^{\alpha,2(d)}_{N+1}, \tag{21.25}$$

we obtain the following estimates:

(i) for any function v such that $v(1-x_1^2)^{-1}\cdots(1-x_d^2)^{-1}$ belongs to $H^s_\alpha(\Omega_d)$, where $s > \sup\{d/2, d(1+\alpha)/2\}$,

$$\begin{aligned}&\big\|\big(v - \mathcal{K}^{\alpha,2}_{N+1}v\big)\big(1-x_1^2\big)^{-1}\cdots\big(1-x_d^2\big)^{-1}\big\|_{L^2_\alpha(\Omega_d)}\\ &\quad \leqslant cN^{-s}\big\|v\big(1-x_1^2\big)^{-1}\cdots\big(1-x_d^2\big)^{-1}\big\|_{H^s_\alpha(\Omega_d)};\end{aligned} \tag{21.26}$$

(ii) for any real number r, $0 \leqslant r \leqslant 2$, and for any function v in $H^s(\Omega_d)$, where $s > \sup\{(6d+r)/4, (2d(3+\alpha)+r)/4\}$,

$$\big\|v - \mathcal{K}^{\alpha,2}_{N+1}v\big\|_{H^r_\alpha(\Omega_d)} \leqslant cN^{r-s}\|v\|_{H^s_\alpha(\Omega_d)}. \tag{21.27}$$

The interpolation error estimates in the hypercube find their applications in the collocation methods which are presented in the next section.

22. Weighted collocation methods

The discrete problems which are described here are equivalent to a system of collocation equations. These systems are very similar to (15.34) or (16.26), only the nodes are changing. However, the numerical analysis now involves the weighted Sobolev spaces introduced in Section 18 and the approximation and interpolation results of Sections 20 and 21.

We begin with the second-order Dirichlet problem:

$$\begin{cases} -\Delta u = f & \text{in } \Omega_d, \\ u = g_j & \text{on } \Gamma_j,\ 1 \leqslant j \leqslant 2d. \end{cases} \tag{22.1}$$

Even if this is not necessary for the existence result of Theorem 18.13, the function f is supposed to be continuous on Ω_d, and the functions g_j, $1 \leqslant j \leqslant 2d$, are supposed to be continuous on $\overline{\Gamma}_j$ and to satisfy the compatibility conditions (18.29) for any value of α, $-1 < \alpha < 1$.

Now, the collocation problem reads: *find u_N in $\mathbb{P}_N(\Omega_d)$ such that*

$$\begin{cases} -\Delta u_N(\boldsymbol{x}) = f(\boldsymbol{x}), & \boldsymbol{x} \in \Xi_N^\alpha \cap \Omega_d, \\ u_N(\boldsymbol{x}) = g_j(\boldsymbol{x}), & \boldsymbol{x} \in \Xi_N^\alpha \cap \overline{\Gamma}_j,\ 1 \leqslant j \leqslant 2d. \end{cases} \tag{22.2}$$

This problem has the same formulation as the first discrete problem (15.34), except that the collocation points are more general.

REMARK 22.1. This method is largely used in the Chebyshev case, i.e., with $\alpha = -\frac{1}{2}$. Indeed, in this case, the nodes $\xi_j^{-1/2}$, $1 \leqslant j \leqslant N-1$, are explicitly known as cosines of equidistant points:

$$\xi_j^{-1/2} = \cos \frac{(N-j)\pi}{N}, \tag{22.3}$$

which allows for the use of the Fast Fourier Transform. Indeed, the unknowns of problem (22.2) are the values $u_N(\xi_{i_1}^\alpha, \ldots, \xi_{i_d}^\alpha)$, $1 \leqslant i_1, \ldots, i_d \leqslant N-1$, so that an iterative resolution of the linear system equivalent to (22.2) requires to compute at each step k, the residuals $(\Delta u_N^k + f)(\xi_{i_1}^\alpha, \ldots, \xi_{i_d}^\alpha)$ as a function of the $u_N^k(\xi_{i_1}^\alpha, \ldots, \xi_{i_d}^\alpha)$. The basic process which is involved in the computation of the residuals, associates with the values $\varphi_N(\xi_j^\alpha)$, $0 \leqslant j \leqslant N$, of a polynomial in $\mathbb{P}_N(\Lambda)$, the values $\varphi_N'(\xi_j^\alpha)$ of its derivative at the same nodes. From the numerical point of view, it has three steps:

(i) computing the coefficients φ_n, $0 \leqslant n \leqslant N$, of φ_N in the basis made by the J_n^α,
(ii) deriving from these, the coefficients of φ_N' in the same basis by using (19.11),
(iii) computing from these last coefficients the values $\varphi_N'(\xi_j^\alpha)$, $0 \leqslant j \leqslant N$.

The first step:

$$\big(\varphi_N(\xi_j^\alpha)\big)_{0 \leqslant j \leqslant N} \mapsto (\varphi_n)_{0 \leqslant n \leqslant N}, \tag{22.4}$$

and the third step, which is the inverse one, requires cN^2 elementary operations in the general case and only $cN \log N$ in the case $\alpha = -\frac{1}{2}$ when the Fast Fourier Transform is used. This is presently the main advantage of using Chebyshev type spectral methods. However, the advantage diminishes when the dimension d grows up: indeed, due to the tensorization properties of the basis of polynomials, the analogue of (22.4) in the hypercube Ω_d requires only cN^{d+1} elementary operations versus $cN^d \log N$ with the Fast Fourier Transform.

First, we note that the boundary conditions in (22.2) are equivalent to the fact that u_N is equal to $\mathcal{I}_N^\alpha u_\flat$ on Γ_j, $1 \leqslant j \leqslant 2d$, where $u_\flat$ is a continuous function which lifts the traces g_j. Next, we introduce the bilinear form, for continuous functions u and v on $\overline{\Omega}_d$,

$$(u, v)_{\alpha,N} = \sum_{i_1=0}^{N} \cdots \sum_{i_d=0}^{N} u(\xi_{i_1}^\alpha, \ldots, \xi_{i_d}^\alpha) v(\xi_{i_1}^\alpha, \ldots, \xi_{i_d}^\alpha) \rho_{i_1}^\alpha \cdots \rho_{i_d}^\alpha. \tag{22.5}$$

Multiplying the equation at the point $(\xi_{i_1}^\alpha, \ldots, \xi_{i_d}^\alpha)$ in the first line of (22.2) by the quantity $v_N(\xi_{i_1}^\alpha, \ldots, \xi_{i_d}^\alpha)\rho_{i_1}^\alpha \cdots \rho_{i_d}^\alpha$, $1 \leqslant i_1, \ldots, i_d \leqslant N-1$, we observe that the $(N-1)^d$ collocation equations at the interior nodes of the domain read

$$\forall v_N \in \mathbb{P}_N^0(\Omega_d), \quad -(\Delta u_N, v_N)_{\alpha,N} = (f, v)_{\alpha,N}. \tag{22.6}$$

Since the quadrature formula is exact on $\mathbb{P}_{2N-1}(\Lambda)$, an integration by parts can be made in each term $(\partial^2 u_N/\partial x_j^2, v_N)_{\alpha,N}$ with respect to x_j: it follows from the one-dimensional argument, valid for any polynomials φ_N in $\mathbb{P}_N(\Lambda)$ and ψ_N in $\mathbb{P}_N^0(\Lambda)$,

$$\begin{aligned} -\sum_{j=0}^{N} \varphi_N''(\xi_j^\alpha)\psi_N(\xi_j^\alpha)\rho_j^\alpha &= -\int_{-1}^{1} \varphi_N''(\zeta)\psi_N(\zeta)\rho_\alpha(\zeta)\,\mathrm{d}\zeta \\ &= \int_{-1}^{1} \varphi_N'(\zeta)(\psi_N\,\rho_\alpha)'(\zeta)\,\mathrm{d}\zeta, \end{aligned}$$

and, since $\rho_{-\alpha}(\psi_N\rho_\alpha)'$ belongs to $\mathbb{P}_{N-1}(\Lambda)$,

$$-\sum_{j=0}^{N} \varphi_N''(\xi_j^\alpha)\psi_N(\xi_j^\alpha)\rho_j^\alpha = \sum_{j=0}^{N} \varphi_N'(\xi_j^\alpha)(\psi_N\rho_\alpha)'(\xi_j^\alpha)\rho_{-\alpha}(\xi_j^\alpha)\rho_j^\alpha. \tag{22.7}$$

This leads to the equivalent formulation of problem (22.2): *find u_N in $\mathbb{P}_N(\Omega_d)$, with $u_N - \mathcal{I}_N u_\flat$ in $\mathbb{P}_N^0(\Omega_d)$, such that*

$$\forall v_N \in \mathbb{P}_N(\Omega_d), \quad a_{\alpha,N}(u_N, v_N) = (f, v_N)_{\alpha,N}, \tag{22.8}$$

where the bilinear form $a_{\alpha,N}(\cdot,\cdot)$ is defined for any polynomials u_N in $\mathbb{P}_N(\Omega_d)$ and v_N in $\mathbb{P}_N^0(\Omega_d)$, by

$$a_{\alpha,N}(u_N, v_N) = \big(\mathbf{grad}\, u_N, \varpi_\alpha^{-1}\mathbf{grad}(v_N\varpi_\alpha)\big)_{\alpha,N}. \tag{22.9}$$

Of course, the numerical analysis of problem (22.2) relies on the variational formulation (22.8), hence on the properties of the form $a_{\alpha,N}(\cdot,\cdot)$. These are derived exactly as in Theorem 15.1, by combining the exactness of the quadrature formula on $\mathbb{P}_{2N-1}(\Lambda)$ and the additional inequalities (21.7) with Theorem 18.12. Note however that the bilinear form $a_{\alpha,N}(\cdot,\cdot)$ is not symmetric for $\alpha \neq 0$.

THEOREM 22.1. *Let α be such that $-1 < \alpha < 1$. The bilinear form $a_{\alpha,N}(\cdot,\cdot)$ satisfies the following properties of continuity*

$$\forall u_N \in \mathbb{P}_N(\Omega_d),\ \forall v_N \in \mathbb{P}_N^0(\Omega_d),$$
$$\big|a_{\alpha,N}(u_N, v_N)\big| \leqslant c\|u_N\|_{H^1_\alpha(\Omega_d)}\|v_N\|_{H^1_\alpha(\Omega_d)}, \tag{22.10}$$

and of ellipticity

$$\forall v_N \in \mathbb{P}_N^0(\Omega_d),\quad a_{\alpha,N}(v_N, v_N) \geqslant c\|v_N\|^2_{H^1_\alpha(\Omega_d)}. \tag{22.11}$$

Next, it suffices to use the Lax–Milgram Theorem 2.1 to prove the wellposedness of the discrete problem.

THEOREM 22.2. *Let α be such that $-1 < \alpha < 1$. For any continuous function f on Ω_d and for any continuous functions g_j, $1 \leqslant j \leqslant 2d$, in $H_\alpha^{(1-\alpha)/2}(\Gamma_j)$ satisfying* (18.29), *problem* (22.8) *has a unique solution u_N in $\mathbb{P}_N(\Omega_d)$.*

The error estimate in $H^1_\alpha(\Omega_d)$ follows from Theorem 22.1 combined with Theorems 20.6 and 21.6.

THEOREM 22.3. *Let α be such that $-1 < \alpha < 1$. The function f is supposed to be in a space $H^\sigma(\Omega_d)$, $\sigma > \sup\{d/2, d(1+\alpha)/2\}$. Let us assume that the solution u of problem* (22.1) *belongs to $H^s_\alpha(\Omega_d)$ for a real number*

$$s > \sup\big\{(d+1)/2, \big(d(1+\alpha)+1\big)/2\big\}.$$

Then, the following error estimate holds for problem (22.8)*:*

$$|u - u_N|_{H^1_\alpha(\Omega_d)} \leqslant c\big(N^{1-s}\|u\|_{H^s_\alpha(\Omega_d)} + N^{-\sigma}\|f\|_{H^\sigma_\alpha(\Omega_d)}\big). \tag{22.12}$$

Let us have a look at the duality argument which is needed for the estimate in $L^2_\alpha(\Omega_d)$. Of course, we have

$$\|u - u_N\|_{L^2_\alpha(\Omega_d)} = \sup_{g \in L^2_\alpha(\Omega_d)} \frac{\int_{\Omega_d}(u - u_N)(\boldsymbol{x})g(\boldsymbol{x})\varpi_\alpha(\boldsymbol{x})\,\mathrm{d}\boldsymbol{x}}{\|g\|_{L^2_\alpha(\Omega_d)}}. \tag{22.13}$$

Here, we have to consider the solution w of the following problem, for a fixed g in $L^2_\alpha(\Omega_d)$: *find* w *in* $H^1_{\alpha,0}(\Omega_d)$ *such that*

$$\forall v \in H^1_{\alpha,0}(\Omega_d), \quad a_\alpha(v,w) = \int_\Omega g(\boldsymbol{x})v(\boldsymbol{x})\varpi_\alpha(\boldsymbol{x})\,\mathrm{d}\boldsymbol{x}. \tag{22.14}$$

The existence and uniqueness of the solution w follow from Theorem 18.12. Note that problem (22.14) is equivalent to

$$\begin{cases} -\Delta(w\varpi_\alpha) = g\varpi_\alpha & \text{in } \Omega_d, \\ w = 0 & \text{on } \partial\Omega_d, \end{cases} \tag{22.15}$$

and the following regularity property is proven in Theorem 18.14 for $-\frac{1}{2} \leqslant \alpha \leqslant \frac{1}{2}$: the function $w\,\varpi_\alpha$ belongs to $H^2_{-\alpha}(\Omega_d)$ and satisfies

$$\|w\varpi_\alpha\|_{H^2_{-\alpha}(\Omega_d)} \leqslant c\|g\varpi_\alpha\|_{L^2_{-\alpha}(\Omega_d)} = c\|g\|_{L^2_\alpha(\Omega_d)}.$$

We have, for any polynomial w_{N-1} in $\mathbb{P}^0_{N-1}(\Omega_d)$,

$$\begin{aligned} \int_{\Omega_d} (u-u_N)(\boldsymbol{x})g(\boldsymbol{x})\,\varpi_\alpha(\boldsymbol{x})\,\mathrm{d}\boldsymbol{x} &= a_\alpha(u-u_N, w-w_{N-1}) \\ &\quad - \sum_{j=1}^{2d} \int_{\Gamma_j} (g_j - i^j_N g_j)(\tau)\Big(\frac{\partial(w\varpi_\alpha)}{\partial n_j}\Big)(\tau)\,\mathrm{d}\tau \\ &\quad + \int_{\Omega_d} (f - \mathcal{I}_N f)(\boldsymbol{x})w_{N-1}(\boldsymbol{x})\,\mathrm{d}\boldsymbol{x}. \end{aligned} \tag{22.16}$$

So, the idea is to estimate the quantity

$$\inf_{w_{N-1}\in\mathbb{P}_{N-1}(\Omega_d)} \|w - w_{N-1}\|_{H^1_\alpha(\Omega_d)}$$

when only the function $w\,\varpi_\alpha$ belongs to $H^2_{-\alpha}(\Omega_d)$. The proof of this estimate is rather technical, it is given in BERNARDI and MADAY ([1989], §4) in the case $d = 2$, and leads to a nonoptimal result for positive values of α, as it appears in the following theorem.

THEOREM 22.4. *Let* α *be a real number such that* $-\frac{1}{2} \leqslant \alpha \leqslant \frac{1}{2}$. *The function* f *is supposed to be in a space* $H^\sigma_\alpha(\Omega_d)$, $\sigma > \sup\{d/2, d(1+\alpha)/2\}$ *and the boundary functions* g_j, $1 \leqslant j \leqslant 2d$, *are supposed to be in spaces* $H^{\tau_j}_\alpha(\Gamma_j)$, $\tau_j > \sup\{(d-1)/2, (d-1)(1+\alpha)/2\}$. *Let us assume that the solution* u *of problem*

(22.1) *belongs to* $H^s_\alpha(\Omega_d)$ *for a real number* $s \geqslant \sup\{(d+1)/2, (d(1+\alpha)+1)/2\}$. *Then, the following error estimate holds for problem* (22.8):

$$\|u-u_N\|_{L^2_\alpha(\Omega_d)} \leqslant c\Bigg(N^{\sup\{\alpha,0\}-s}\|u\|_{H^s_\alpha(\Omega_d)} + N^{-\sigma}\|f\|_{H^\sigma_\alpha(\Omega_d)} + \sum_{j=1}^{2d} N^{-\tau_j}\|g_j\|_{H^{\tau_j}_\alpha(\Gamma_j)}\Bigg). \tag{22.17}$$

Of course, the fourth-order problem

$$\begin{cases} \Delta^2 u = f & \text{in } \Omega_d, \\ u = 0 & \text{on } \partial\Omega_d, \\ \dfrac{\partial u}{\partial n} = 0 & \text{on } \partial\Omega_d, \end{cases} \tag{22.18}$$

also has a collocation discretization when the function f is continuous on Ω_d.

Here, the discrete problem consists in finding a polynomial u_N in $\mathbb{P}^{2,0}_N(\Omega_d)$ such that

$$(\Delta^2 u_N)(\boldsymbol{x}) = f(\boldsymbol{x}), \quad \boldsymbol{x} \in \Sigma^{\alpha,2}_{N-1} \cap \Omega_d. \tag{22.19}$$

For any functions u in $H^2(\Omega_d)$ and v in $H^2_0(\Omega_d)$, we define the quantity $(u,v)_{\alpha,N-1,2}$ by applying the quadrature formula (19.32) with m equal to 2 and N replaced by $N-1$, to compute the integral of uv in each direction x_j. And, setting for any polynomials u_N in $\mathbb{P}_N(\Lambda)$ and v_N in $\mathbb{P}^{2,0}_N(\Lambda)$,

$$d_{\alpha,N-1,2}(u_N, v_N) = \left(\Delta^2 u_N, v_N\right)_{\alpha,N-1,2}, \tag{22.20}$$

we see that problem (22.19) admits the following equivalent variational formulation: *find* u_N *in* $\mathbb{P}^{2,0}_N(\Omega_d)$ *such that*

$$\forall v_N \in \mathbb{P}^{2,0}_N(\Omega_d), \quad d_{\alpha,N-1,2}(u_N, v_N) = (f, v_N)_{N-1,2}. \tag{22.21}$$

However, the nodes of the quadrature formula, i.e., the zeros of $J^{\alpha\prime\prime}_{N-1}$ or equivalently of $J^{\alpha+2}_{N-3}$ (see (19.5)), are no longer the cosines of equidistant points, so that the Fast Fourier Transform cannot be used. Moreover, the properties of the form $d_{\alpha,N-1,2}(\bullet,\bullet)$ are not better than for $d_{N-1,2}(\bullet,\bullet)$: this means that its constant of ellipticity behaves like cN^{1-d}. Hence, we have no real application for this method. Only for the sake of completeness, we state the main result.

THEOREM 22.5. *Let* α *be a real number which satisfies* $-\frac{1}{2} \leqslant \alpha \leqslant \frac{1}{2}$. *The function* f *is supposed to be in a space* $H^\sigma_\alpha(\Omega_d)$, $\sigma > \sup\{d/2, d(1+\alpha)/2\}$. *Let us assume that* *the solution u of problem (22.18) belongs to $H^s_\alpha(\Omega_d)$ for a real number $s \geqslant 2$. Then, problem (22.21) has a unique solution u_N in $\mathbb{P}_N^{2,0}(\Omega_d)$. Moreover, the following error estimate holds for problem (22.21):*

$$|u - u_N|_{H^2_\alpha(\Omega_d)} \leqslant c\big(N^{d+1-s}\|u\|_{H^s_\alpha(\Omega_d)} + N^{d-1-\sigma}\|f\|_{H^\sigma_\alpha(\Omega_d)}\big). \tag{22.22}$$

In the case of the square Ω, the fourth-order problem with inhomogeneous boundary conditions:

$$\begin{cases} \Delta^2 u = f & \text{in } \Omega, \\ u = g_J & \text{on } \Gamma_J,\ J = 1,2,3,4, \\ \dfrac{\partial u}{\partial n} = h_J & \text{on } \Gamma_J,\ J = 1,2,3,4, \end{cases} \tag{22.23}$$

also has a collocation discretization when the function f is continuous on Ω and the functions g_J and h_J, $J = 1,2,3,4$, are continuous on $\overline{\Gamma}_J$, with their first derivatives continuous at the corners of Ω, and satisfy (16.22).

The discrete problem consists in finding a polynomial u_N of $\mathbb{P}_N(\Omega)$ such that

$$\begin{cases} (\Delta^2 u_N)(\boldsymbol{x}) = f(\boldsymbol{x}),\ \boldsymbol{x} \in \Sigma^{\alpha,2}_{N-1} \cap \Omega, \\ u_N(\boldsymbol{x}) = g_J(\boldsymbol{x}) \text{ and } \left(\dfrac{\partial u_N}{\partial n_J}\right)(\boldsymbol{x}) = h_J(\boldsymbol{x}),\ \boldsymbol{x} \in \Sigma^{\alpha,2}_{N-1} \cap \Gamma_J, \\ \quad J = 1,2,3,4, \\ u_N(\boldsymbol{a}_J) = g_J(\boldsymbol{a}_J),\ \left(\dfrac{\partial u_N}{\partial n_J}\right)(\boldsymbol{a}_J) = h_J(\boldsymbol{a}_J), \\ \left(\dfrac{\partial u_N}{\partial n_{J+1}}\right)(\boldsymbol{a}_J) = h_{J+1}(\boldsymbol{a}_J),\ \left(\dfrac{\partial^2 u_N}{\partial n_J \partial n_{J+1}}\right)(\boldsymbol{a}_J) = \left(\dfrac{\partial h_J}{\partial \tau_J}\right)(\boldsymbol{a}_J), \\ \quad J = 1,2,3,4. \end{cases} \tag{22.24}$$

Introducing a function $u_\flat$ which satisfies (16.23), we obtain the variational formulation of the problem: *find u_N in $\mathbb{P}_N(\Omega)$, with $u - \mathcal{K}_N^{\alpha,2} u_\flat$ in $\mathbb{P}_N^{2,0}(\Omega_d)$, such that*

$$\forall v_N \in \mathbb{P}_N^{2,0}(\Omega_d),\quad d_{\alpha,N-1,2}(u_N, v_N) = (f, v_N)_{N-1,2}. \tag{22.25}$$

This allows for deriving the following results.

THEOREM 22.6. *Let α be such that $-\frac{1}{2} \leqslant \alpha \leqslant \frac{1}{2}$. The function f is supposed to be in a space $H^\sigma_\alpha(\Omega)$, $\sigma > 3$. Let us assume that the solution u of problem (22.23) belongs to $H^s_\alpha(\Omega)$ for a real number $s > \sup\{\frac{7}{2}, (2(3+\alpha)+1)/2\}$. Then, problem (22.25) has a unique solution u_N in $\mathbb{P}_N(\Omega)$. Moreover, the following error estimate holds for problem (22.25):*

$$|u - u_N|_{H^2_\alpha(\Omega)} \leqslant c\big(N^{3-s}\|u\|_{H^s_\alpha(\Omega)} + N^{1-\sigma}\|f\|_{H^\sigma_\alpha(\Omega)}\big). \tag{22.26}$$

In the case of fourth-order problems, the estimate in $L^2_\alpha(\Omega)$ cannot be established by the usual duality argument, since the regularity result which is involved in the proof, is presently unknown.

CHAPTER V

Spectral Discretizations of the Navier–Stokes Equations

The first application of spectral methods to the real world dealt with fluid mechanics. A huge amount of work on this subject has been done during the ten last years, we refer to the book of CANUTO, HUSSAINI, QUARTERONI and ZANG [1987] and to the references therein for a (nearly) complete review. Here, we intend to present the basic results of numerical analysis for incompressible viscous flows, which are governed by the Navier–Stokes equations.

In the square or cubic domain Ω_d, $d = 2$ or 3, we consider the Navier–Stokes equations

$$\begin{cases} -\nu\Delta \boldsymbol{u} + (\boldsymbol{u} \bullet \nabla)\boldsymbol{u} + \mathbf{grad}\, p = \boldsymbol{f} & \text{in } \Omega_d, \\ \operatorname{div} \boldsymbol{u} = 0 & \text{in } \Omega_d. \end{cases}$$

In this system, the unknowns are the velocity $\boldsymbol{u}$ and the pressure p. The quantity ν is a positive parameter which represents the kinematic viscosity. The data are a density of body forces $\boldsymbol{f}$ on the whole domain. When the velocity is small enough, the nonlinear term $(\boldsymbol{u} \bullet \nabla)\boldsymbol{u}$ can be neglected with respect to the diffusion term, so that these equations reduce to the Stokes problem

$$\begin{cases} -\nu\Delta \boldsymbol{u} + \mathbf{grad}\, p = \boldsymbol{f} & \text{in } \Omega_d, \\ \operatorname{div} \boldsymbol{u} = 0 & \text{in } \Omega_d. \end{cases}$$

The spectral discretizations are firstly analyzed for this problem, indeed the major difficulties already appear in the linear case.

The Navier–Stokes equations can be provided with several types of boundary conditions, as can be seen in PIRONNEAU [1988] and CANUTO, HUSSAINI, QUARTERONI and ZANG [1987] (see also BERNARDI, CANUTO and MADAY [1991] for the numerical analysis of spectral discretizations with boundary conditions on the pressure). Here, we limit ourselves to the basic cases of Dirichlet boundary conditions on the velocity:

$$\boldsymbol{u} = \boldsymbol{g} \quad \text{on } \partial\Omega_d,$$

and even, most of the time, to the case of homogeneous Dirichlet boundary conditions

$$\boldsymbol{u} = \mathbf{0} \quad \text{on } \partial\Omega_d.$$

Indeed, the treatment of inhomogeneous boundary conditions has been thoroughly explained in Section 15 in the case of the Dirichlet problem for the Laplace operator, the extension to the Stokes problem is not very difficult.

In standard Sobolev spaces, the variational formulation of the Stokes problem is known to be of saddle-point type, an abstract form of such problems is presented for instance in GIRAULT and RAVIART ([1986], Chapter I, §4), and conditions for their wellposedness are given. However, spectral discretizations of the Stokes problem are both of Chebyshev and Legendre type, and Chebyshev formulation in weighted Sobolev spaces requires a generalized form of saddle-point problems which can be found in NICOLAIDES [1982], with additional results by BERNARDI, CANUTO and MADAY [1988]. As in the standard case, the wellposedness of the problem relies on the existence of inf–sup conditions. Conditions of this type are now well known, they were introduced independently by BABUŠKA [1973] and BREZZI [1974]. This generalized formulation is presented in the next Section 23 in an abstract framework, and then applied to the Stokes problem. The abstract discretization of saddle-point problems is also described, and the extension of Strang's lemma to this situation is given.

Three kinds of spectral discretizations exist for the Stokes problem when it is formulated in the primitive variables of velocity and pressure, indeed the choice of compatible discrete spaces for the two unknowns is not obvious:

(i) the first method, denoted as "single grid", is a collocation one, it uses the same degree of polynomials for the velocity and the pressure, however spurious modes for the pressure have to be filtered in an explicit way;

(ii) the second method is no longer of collocation type, however it solves the problem of spurious modes by using different degrees of polynomials for the velocity and the pressure, namely N for the velocity and $N-2$ for the pressure;

(iii) the third method is inherited from the finite difference techniques, it is still of collocation type and relies on the use of staggered grids for the velocity and the pressure in order to avoid the problem of spurious modes: three grids in dimension $d = 2$ and four grids in dimension $d = 3$.

The advantages and drawbacks of these three methods will appear from their numerical analysis, which is performed in Sections 24, 25 and 26 respectively. The main difficulty here is to derive error estimates for the pressure. This is achieved via discrete inf–sup conditions, however the constant which is involved in these conditions in the case of spectral methods is, as far as we know, never independent of the discretization parameter N. The dependency of the constant with respect to N is thoroughly investigated.

In Section 27, the methods are extended to the full Navier–Stokes equations with similar results, the analysis relies on the discrete implicit function theorem of BREZZI, RAPPAZ and RAVIART [1980] (see also CROUZEIX and RAPPAZ [1989] for an improved version).

A last spectral discretization of the Navier–Stokes equations is presented in Section 28, which is valid in dimension $d = 2$. It is based on a different formulation of the problem, where the unknowns are the stream-function and the vorticity, and is often preferred for its low computation cost.

23. The general saddle-point problem

We present the saddle-point problems in the following abstract framework, which is required for our analysis. Let X_1 and X_2, M_1 and M_2 be four reflexive Banach spaces. We denote respectively by X_1' and X_2', M_1' and M_2' their dual spaces and, for the sake of simplicity, we denote by $\langle\bullet,\bullet\rangle$ all the duality pairings. We also introduce three bilinear forms:

(i) the form $a(\bullet,\bullet)$ which is continuous on $X_2 \times X_1$,

(ii) the forms $b_i(\bullet,\bullet)$, $i = 1, 2$, which are continuous on $X_i \times M_i$.

For any pair (f, g) of $X_1' \times M_2'$, we consider the problem: *find a pair (u, p) in $X_2 \times M_1$ such that*

$$\begin{aligned} &\forall v \in X_1, \quad a(u,v) + b_1(v,p) = \langle f, v\rangle, \\ &\forall q \in M_2, \quad b_2(u,q) = \langle g, q\rangle. \end{aligned} \tag{23.1}$$

We are going to state the appropriate conditions for the wellposedness of this problem.

The inf–sup condition has been introduced by BABUŠKA [1973] and BREZZI [1974] to handle problems with Lagrange multipliers (in the Stokes problem, the pressure can be considered as the Lagrange multiplier of the incompressibility condition). More generally, it allows for treating variational problems where the bilinear form is defined on the product of two different Banach spaces, which means that the space where the solution is sought, does not coincide with the space of test functions. We firstly state the general result for the inf–sup condition, we refer for instance to GIRAULT and RAVIART ([1986], Chapter I, Lemma 4.1) for the complete proof which relies on Banach's closed range theorem.

THEOREM 23.1. *Let X and M be two reflexive Banach spaces. Let $b(\bullet,\bullet)$ be a continuous bilinear form on $X \times M$, we define the kernel*

$$K = \{v \in X\colon \forall q \in M,\ b(v,q) = 0\}. \tag{23.2}$$

The following propositions are equivalent:

(i) *there exists a constant $\beta > 0$ such that*

$$\inf_{q\in M} \sup_{v\in X} \frac{b(v,w)}{\|v\|_X \|q\|_M} \geqslant \beta, \tag{23.3}$$

(ii) *for any f in X' which satisfies*

$$\forall v \in K, \quad \langle f, v\rangle = 0,$$

the problem: find p *in* M *such that*

$$\forall v \in X, \quad b(v,p) = \langle f, v\rangle, \tag{23.4}$$

has a unique solution p *in* M. *Moreover, this solution satisfies*

$$\|p\|_M \leqslant \frac{1}{\beta}\|f\|_{X'}, \tag{23.5}$$

(iii) *for any* g *in* M', *there exists a unique* v *in the orthogonal of* K *in* X *such that*

$$\forall q \in M, \quad b(v,q) = \langle g, q\rangle, \tag{23.6}$$

Moreover, this element v *satisfies*

$$\|v\|_X \leqslant \frac{1}{\beta}\|g\|_{M'}. \tag{23.7}$$

The following theorem is now proven by repetitive application of Theorem 23.1 (see BERNARDI, CANUTO and MADAY ([1988], Theorem 2.1)), indeed three inf–sup conditions and a half are involved in its statement. Let us denote by K_i, $i = 1, 2$, the kernels

$$K_i = \{v \in X_i \colon \forall q \in M_i,\ b_i(v,q) = 0\}. \tag{23.8}$$

THEOREM 23.2. *Problem* (23.1) *has a unique solution* (u,p) *in* $X_2 \times M_1$ *for any pair* (f,g) *in* $X_1' \times M_2'$ *if and only if the following conditions hold:*

(i) *there exists a positive constant* α_1 *such that*

$$\inf_{u\in K_2}\sup_{v\in K_1}\frac{a(u,v)}{\|u\|_{X_2}\|v\|_{X_1}} \geqslant \alpha_1 \quad \text{and} \quad \inf_{v\in K_1}\sup_{u\in K_2}\frac{a(u,v)}{\|u\|_{X_2}\|v\|_{X_1}} > 0, \tag{23.9}$$

(ii) *for* $i = 1, 2$, *there exists a positive constant* β_i *such that*

$$\inf_{q\in M_i}\sup_{v\in X_i}\frac{b_i(v,q)}{\|v\|_{X_i}\|q\|_{M_i}} \geqslant \beta_i. \tag{23.10}$$

Moreover this solution satisfies

$$\|u\|_{X_2} + \beta_1\|p\|_{M_1} \leqslant \left(1 + \frac{1}{\alpha_1}\right)\|f\|_{X_1'} + \frac{\gamma}{\beta_2}\|g\|_{M_2'}, \tag{23.11}$$

where γ *denotes the norm of* $a(\bullet,\bullet)$.

Note that condition (i) in the theorem can equivalently be replaced by the existence of a positive constant α_2 such that

$$\inf_{u \in K_2} \sup_{v \in K_1} \frac{a(u,v)}{\|u\|_{X_2}\|v\|_{X_1}} > 0 \quad \text{and} \quad \inf_{v \in K_1} \sup_{u \in K_2} \frac{a(u,v)}{\|u\|_{X_2}\|v\|_{X_1}} \geqslant \alpha_2, \tag{23.12}$$

however it is the constant α_1 which appears in the right-hand side of (23.11).

REMARK 23.1. The standard formulation of saddle-point problems in general and of the Stokes problem in particular, is set in the case where

$$X_1 = X_2 = X, \qquad M_1 = M_2 = M, \qquad b_1(\bullet,\bullet) = b_2(\bullet,\bullet) = b(\bullet,\bullet).$$

So, for any (f,g) in $X' \times M'$, the problem is written: *find a pair* (u,p) *in* $X \times M$ *such that*

$$\begin{aligned} &\forall v \in X, \quad a(u,v) + b(v,p) = \langle f, v\rangle, \\ &\forall q \in M, \quad b(u,q) = \langle g, q\rangle. \end{aligned} \tag{23.13}$$

Then, the two inf–sup conditions in (23.10) coincide, they are reduced to condition (23.3). Moreover, in most cases, the two conditions in (23.9) are replaced by an ellipticity property on the kernel K defined in (23.2): there exists a positive constant α such that

$$\forall v \in K, \quad a(v,v) \geqslant \alpha\|v\|_X^2. \tag{23.14}$$

When the form $a(\bullet,\bullet)$ is symmetric and elliptic, the following property holds

$$\alpha \leqslant \inf_{u \in K} \sup_{v \in K} \frac{a(u,v)}{\|u\|_X\|v\|_X} = \inf_{v \in K} \sup_{u \in K} \frac{a(u,v)}{\|u\|_X\|v\|_X} \leqslant \sqrt{\alpha\gamma}, \tag{23.15}$$

where α is the largest possible constant of ellipticity and γ denotes the norm of $a(\bullet,\bullet)$.

As a conclusion, let us consider the simpler (and basic) problem where the function g is equal to 0. In this case, problem (23.1) can easily be decoupled into two equations: *find* u *in* X_2 *such that*

$$\forall v \in K_2, \quad a(u,v) = \langle f, v\rangle, \tag{23.16}$$

then *find* p *in* M_1 *such that*

$$\forall v \in X_1, \quad b_1(v,p) = \langle f, v\rangle - a(u,v). \tag{23.17}$$

Problem (23.16) has a unique solution from (23.9), next problem (23.17) has a unique solution from (23.10).

Now, we apply the previous theory to the Stokes problem with Dirichlet boundary conditions on the velocity. Let $\mathcal{O}$ be a Lipschitz-continuous bounded open set in $\mathbb{R}^d$, $d = 2$ or 3. The Stokes problem reads

$$\begin{cases} -\nu\Delta \boldsymbol{u} + \mathbf{grad}\, p = \boldsymbol{f} & \text{in } \mathcal{O}, \\ \operatorname{div} \boldsymbol{u} = 0 & \text{in } \mathcal{O}, \\ \boldsymbol{u} = \boldsymbol{g} & \text{on } \partial\mathcal{O}. \end{cases} \tag{23.18}$$

We consider firstly the formulation in standard Sobolev spaces which will be used for Legendre type discretization, secondly the formulation in weighted Sobolev spaces when the domain $\mathcal{O}$ is a square or a cube. The main application of this second formulation is the Chebyshev type discretization.

Let us consider the case of standard Sobolev spaces. We assume that the boundary condition $\boldsymbol{g}$ belongs to $H^{1/2}(\partial\mathcal{O})^d$, so that there exists a function $\boldsymbol{u}_\flat$ in $H^1(\mathcal{O})^d$ such that its trace on $\partial\mathcal{O}$ coincides with $\boldsymbol{g}$ and which satisfies

$$\|\boldsymbol{u}_\flat\|_{H^1(\mathcal{O})^d} \leqslant c\|\boldsymbol{g}\|_{H^{1/2}(\partial\mathcal{O})^d}. \tag{23.19}$$

Moreover, we assume that the function $\boldsymbol{g}$ satisfies

$$\int_{\partial\mathcal{O}} \boldsymbol{g} \cdot \boldsymbol{n}\, \mathrm{d}\tau = 0. \tag{23.20}$$

Indeed, it follows from the Stokes formula

$$\int_{\mathcal{O}} (\operatorname{div} \boldsymbol{u})(\boldsymbol{x}) q(\boldsymbol{x})\, \mathrm{d}\boldsymbol{x} = -\int_{\mathcal{O}} \boldsymbol{u}(\boldsymbol{x}) \cdot (\mathbf{grad}\, q)(\boldsymbol{x})\, \mathrm{d}\boldsymbol{x} + \int_{\partial\mathcal{O}} (\boldsymbol{u} \cdot \boldsymbol{n})(\tau) q(\tau)\, \mathrm{d}\tau, \tag{23.21}$$

that condition (23.20) is necessary for problem (23.18) to have a solution. When this condition holds, the function $\boldsymbol{u}_\flat$ can be chosen divergence-free (see GIRAULT and RAVIART ([1986], Chapter I, Lemma 2.2) for details), however we do not need this property in this section.

Due to the boundary condition, we multiply the first equation by $\boldsymbol{v}$ in $\mathscr{D}(\mathcal{O})^d$, we integrate by parts and we use the density of $\mathscr{D}(\mathcal{O})$ in $H_0^1(\mathcal{O})$. The second equation can be multiplied by any function q in $L^2(\mathcal{O})$, however the pressure p is defined only up to an additive constant and it follows from the Stokes formula (23.21) combined with hypothesis (23.20) that the divergence of the function $\boldsymbol{u}$ is orthogonal to the constants. Hence, we define the space

$$L_0^2(\mathcal{O}) = \left\{ q \in L^2(\mathcal{O}) \colon \int_{\mathcal{O}} q(\boldsymbol{x})\, \mathrm{d}\boldsymbol{x} = 0 \right\}. \tag{23.22}$$

Thus, we observe that, when the data $\boldsymbol{f}$ are in $H^{-1}(\mathcal{O})^d$, the Stokes problem (23.18) admits the equivalent variational formulation: *find a pair* $(\boldsymbol{u}, p)$ *in* $H^1(\mathcal{O})^d \times L^2_0(\mathcal{O})$, *with* $\boldsymbol{u} - \boldsymbol{u}_b$ *in* $H^1_0(\mathcal{O})^d$, *such that*

$$\begin{aligned} &\forall \boldsymbol{v} \in H^1_0(\mathcal{O})^d, \quad a(\boldsymbol{u}, \boldsymbol{v}) + b(\boldsymbol{v}, p) = \langle \boldsymbol{f}, \boldsymbol{v} \rangle, \\ &\forall q \in L^2_0(\mathcal{O}), \qquad b(\boldsymbol{u}, q) = 0, \end{aligned} \tag{23.23}$$

where the forms $a(\bullet, \bullet)$ and $b(\bullet, \bullet)$ are defined by

$$\forall \boldsymbol{u} = (u_1, \ldots, u_d) \in H^1(\mathcal{O})^d, \ \forall \boldsymbol{v} = (v_1, \ldots, v_d) \in H^1(\mathcal{O})^d,$$

$$a(\boldsymbol{u}, \boldsymbol{v}) = \nu \sum_{i=1}^{d} \int_{\mathcal{O}} (\mathbf{grad}\, u_i)(\boldsymbol{x}) \cdot (\mathbf{grad}\, v_i)(\boldsymbol{x}) \, \mathrm{d}\boldsymbol{x}, \tag{23.24}$$

$$\forall \boldsymbol{v} \in H^1(\mathcal{O})^d, \ \forall q \in L^2(\mathcal{O}), \quad b(\boldsymbol{v}, \boldsymbol{q}) = -\int_{\mathcal{O}} (\operatorname{div} \boldsymbol{v})(\boldsymbol{x}) q(\boldsymbol{x}) \, \mathrm{d}\boldsymbol{x}. \tag{23.25}$$

Setting $\boldsymbol{u}^* = \boldsymbol{u} - \boldsymbol{u}_b$, we observe that the problem which is satisfied by the pair $(\boldsymbol{u}^*, p)$ is exactly of type (23.13), with $X = H^1_0(\mathcal{O})^d$ and $M = L^2_0(\mathcal{O})$. So, we need to check the hypotheses of Theorem 23.2 in this case. The continuity of the forms $a(\bullet, \bullet)$ and $b(\bullet, \bullet)$ on $H^1(\mathcal{O})^d \times H^1(\mathcal{O})^d$ and $H^1(\mathcal{O})^d \times L^2(\mathcal{O})$ is obvious, the ellipticity of the form $a(\bullet, \bullet)$ on $H^1_0(\mathcal{O})^d$ is a straightforward consequence of the Poincaré–Friedrichs inequality (1.4). So, the crucial point is the inf–sup condition (23.3). It is a direct consequence of the following statement.

THEOREM 23.3. *The divergence operator from* $H^1_0(\mathcal{O})^d$ *in* $L^2_0(\mathcal{O})$ *is onto.*

This result is well known for general Lipschitz-continuous domains, we refer to the book of GIRAULT and RAVIART ([1986], Chapter I, Corollary 2.4) for a detailed proof. However, we prefer here to give an alternative proof which is valid only for two-dimensional domains with a boundary of class $\mathscr{C}^{1,1}$ or convex polygons, but which will have an extension to the case of weighted spaces and a discrete analogue in Section 24.

PROOF FOR REGULAR PLANE DOMAINS OR CONVEX POLYGONS. Let q be any function in $L^2_0(\mathcal{O})$. The first step is to consider the problem:

$$\begin{cases} -\Delta\varphi = -q & \text{in } \mathcal{O}, \\ \dfrac{\partial\varphi}{\partial n} = 0 & \text{on } \partial\mathcal{O}. \end{cases} \tag{23.26}$$

Since q has a null integral on $\mathcal{O}$, it follows from Theorem 2.6 that this problem has a unique solution in $H^1(\mathcal{O}) \cap L^2_0(\mathcal{O})$. As stated in Remark 2.1, if the domain has a boundary of class $\mathscr{C}^{1,1}$ or is convex, φ belongs in fact to $H^2(\mathcal{O})$. The idea is now to take

$$\boldsymbol{v} = \mathbf{grad}\, \varphi + \mathbf{curl}\, \psi,$$

where ψ belongs to $H^2(\mathcal{O})$ and satisfies

$$\psi = 0 \text{ on } \partial\mathcal{O} \quad \text{and} \quad \frac{\partial\psi}{\partial n} = -\frac{\partial\varphi}{\partial\tau} \text{ on } \partial\mathcal{O}.$$

Such a function exists either from Theorem 1.12 in the case of a $\mathscr{C}^{1,1}$ boundary or from an extension of Theorem 1.13 in the case of a polygon, since the compatibility conditions hold (see BERNARDI, DAUGE and MADAY ([1992b], Chapter 2)). The function $\boldsymbol{v}$ belongs to $H_0^1(\mathcal{O})^d$ and satisfies $\operatorname{div} \boldsymbol{v} = q$.

Now, the wellposedness of problem (23.23) is a straightforward application of Theorem 23.2.

THEOREM 23.4. *For any $\boldsymbol{f}$ in $H^{-1}(\mathcal{O})^d$ and for any $\boldsymbol{g}$ in $H^{1/2}(\partial\mathcal{O})^d$ which satisfies condition* (23.20)*, problem* (23.23) *has a unique solution $(\boldsymbol{u}, p)$ in $H^1(\mathcal{O})^d \times L_0^2(\mathcal{O})$. Moreover, this solution satisfies*

$$\|\boldsymbol{u}\|_{H^1(\mathcal{O})^d} + \|p\|_{L^2(\mathcal{O})} \leqslant c\left(1 + \frac{1}{\nu}\right)\left(\|\boldsymbol{f}\|_{H^{-1}(\mathcal{O})^d} + \|\boldsymbol{g}\|_{H^{1/2}(\partial\mathcal{O})^d}\right). \tag{23.27}$$

REMARK 23.2. In our case, the kernel K is defined by

$$K = \left\{\boldsymbol{v} \in H_0^1(\mathcal{O})^d \colon \operatorname{div} \boldsymbol{v} = 0 \text{ in } \mathcal{O}\right\}. \tag{23.28}$$

We refer to GIRAULT and RAVIART ([1986], Chapter I, Theorems 3.1 and 3.4) for the following characterization of the kernel, the statement of which depends on the dimension. For the sake of simplicity, we assume here that the domain $\mathcal{O}$ has a connected boundary.

(i) When the dimension d is equal to 2, a function $\boldsymbol{u}$ belongs to K if and only if there exists a stream-function ψ in $H_0^2(\mathcal{O})$ such that

$$\boldsymbol{u} = \mathbf{curl}\,\psi \quad \text{in } \mathcal{O}. \tag{23.29}$$

Moreover, this function satisfies

$$\|\psi\|_{H^2(\mathcal{O})} \leqslant c\|\boldsymbol{u}\|_{H^1(\mathcal{O})^2}. \tag{23.30}$$

This function will be used in Section 28 for writing a new formulation of the Stokes problem, leading to a different type of spectral discretization.

(ii) When the dimension d is equal to 3, if a function $\boldsymbol{u}$ belongs to K, there exists a vector potential $\boldsymbol{\psi}$ in $L^2(\mathcal{O})^3$ such that

$$\boldsymbol{u} = \mathbf{curl}\,\boldsymbol{\psi} \text{ in } \mathcal{O} \quad \text{and} \quad \operatorname{div} \boldsymbol{\psi} = 0 \text{ in } \mathcal{O}. \tag{23.31}$$

When the domain is simply-connected, this function can satisfy either $\psi \cdot n = 0$ on $\partial\mathcal{O}$ or $\psi \times n = 0$ on $\partial\mathcal{O}$, the simplest conditions on ψ which imply the boundary conditions on u are written

$$\psi \times n = \mathbf{curl}\,\psi \times n = \mathbf{0} \quad \text{on } \partial\mathcal{O}. \tag{23.32}$$

REMARK 23.3. As in Remark 2.1, we have to state the regularity properties of the Stokes problem, or at least the properties that are presently known. We limit ourselves to the case of homogeneous boundary conditions. More precisely, let T_S be the operator which associates with f the solution (u, p) of problem (23.18) when g is equal to $\mathbf{0}$. From the previous Remark 23.2 and in the case of dimension $d = 2$, it can be observed that the regularity of u is derived from the regularity of the associated stream-function ψ which is the solution of a Dirichlet problem for the bilaplacian, hence follows from the results of Remark 2.1. The answer is not so clear in three-dimensional domains. We only state two simple results: the operator T_S is continuous from $H^{s-2}(\mathcal{O})^d$ into $H^s(\mathcal{O})^d \times H^{s-1}(\mathcal{O})$

(i) for any real number $s \leqslant m$ when the boundary of $\partial\mathcal{O}$ is of class $\mathscr{C}^{m-1,1}$ for a positive integer m;

(ii) for any real number $s \leqslant 2$ in the cube;

(iii) for any real number $s < 1+\eta(\omega)$ in the case of a polygon with largest angle ω, where the function η is defined in Remark 2.1.

Let us now consider the weighted formulations. The results in this case are due to BERNARDI, CANUTO and MADAY [1988]. In the hypercube Ω_d, $d = 2$ or 3, we consider the problem

$$\begin{cases} -\nu\Delta u + \mathbf{grad}\, p = f & \text{in } \Omega_d, \\ \operatorname{div} u = 0 & \text{in } \Omega_d, \\ u = g_j & \text{on } \Gamma_j,\ 1 \leqslant j \leqslant 2d. \end{cases} \tag{23.33}$$

Let α be a real number, $-1 < \alpha < 1$, $\alpha \neq 0$. Here, we assume that the functions g_j belong to $H^{(1-\alpha)/2}(\Gamma_j)^d$, $1 \leqslant j \leqslant 2d$, and moreover, if $\alpha < 0$, that they satisfy the compatibility conditions

$$g_j = g_k \quad \text{on } \overline{\Gamma}_j \cap \overline{\Gamma}_k,\ 1 \leqslant j < k \leqslant 2d. \tag{23.34}$$

Then, it follows from Theorem 18.10 that there exists a function u_b in $H^1_\alpha(\Omega_d)^d$ such that its trace on Γ_d coincides with g_j, $1 \leqslant j \leqslant 2d$, and which satisfies

$$\|u_b\|_{H^1_\alpha(\Omega_d)^d} \leqslant c \sum_{j=1}^{2d} \|g_j\|_{H^{(1-\alpha)/2}_\alpha(\Gamma_j)^d}. \tag{23.35}$$

As previously, we assume that these functions satisfy

$$\sum_{j=1}^{2d} \int_{\Gamma_j} \boldsymbol{g}_j \cdot \boldsymbol{n}_j \, \mathrm{d}\tau = 0. \tag{23.36}$$

To derive the variational formulation, we multiply now the first equation of (23.33) by $\boldsymbol{v}\varpi_\alpha$ and the second one by $q\varpi_\alpha$ for some functions $\boldsymbol{v}$ in $\mathscr{D}(\Omega_d)^d$ and q in $\mathscr{D}(\Omega_d)$. Here, we have to introduce two subspaces of $L^2_\alpha(\Omega_d)$, precisely

$$\begin{aligned} L^2_{\alpha 1}(\Omega_d) &= \left\{ q \in L^2_\alpha(\Omega_d) \colon \int_{\Omega_d} q(\boldsymbol{x}) \varpi_\alpha(\boldsymbol{x}) \, \mathrm{d}\boldsymbol{x} = 0 \right\}, \\ L^2_{\alpha 2}(\Omega_d) &= \left\{ q \in L^2_\alpha(\Omega_d) \colon \int_{\Omega_d} q(\boldsymbol{x}) \, \mathrm{d}\boldsymbol{x} = 0 \right\}. \end{aligned} \tag{23.37}$$

Thus, when the data f are in $H^{-1}_\alpha(\Omega_d)^d$, problem (23.33) admits the following equivalent formulation: *find a pair* $(\boldsymbol{u}, p)$ *in* $H^1_\alpha(\Omega_d)^d \times L^2_{\alpha 1}(\Omega_d)$, *with* $\boldsymbol{u} - \boldsymbol{u}_\mathrm{b}$ *in* $H^1_{\alpha,0}(\Omega_d)^d$, *such that*

$$\begin{aligned} &\forall \boldsymbol{v} \in H^1_{\alpha,0}(\Omega_d)^d, \quad a_\alpha(\boldsymbol{u}, \boldsymbol{v}) + b_{\alpha 1}(\boldsymbol{v}, p) = \langle \boldsymbol{f}, \boldsymbol{v} \rangle_\alpha, \\ &\forall q \in L^2_{\alpha 2}(\Omega_d), \qquad b_{\alpha 2}(\boldsymbol{u}, q) = 0, \end{aligned} \tag{23.38}$$

where the forms $a_\alpha(\bullet, \bullet)$ and $b_{\alpha i}(\bullet, \bullet)$, $i = 1, 2$, are defined by

$$\forall \boldsymbol{u} = (u_1, \dots, u_d) \in H^1_\alpha(\Omega_d)^d, \ \forall \boldsymbol{v} = (v_1, \dots, v_d) \in H^1_{\alpha,0}(\Omega_d)^d,$$

$$a_\alpha(\boldsymbol{u}, \boldsymbol{v}) = \nu \sum_{i=1}^{d} \int_{\Omega_d} (\mathbf{grad}\, u_i)(\boldsymbol{x}) \cdot \big(\mathbf{grad}(v_i \varpi_\alpha)\big)(\boldsymbol{x}) \, \mathrm{d}\boldsymbol{x}, \tag{23.39}$$

$$\forall \boldsymbol{v} \in H^1_{\alpha,0}(\Omega_d)^d, \ \forall q \in L^2_\alpha(\Omega_d),$$

$$b_{\alpha 1}(\boldsymbol{v}, q) = -\int_{\Omega_d} \big(\mathrm{div}(\boldsymbol{v}\varpi_\alpha)\big)(\boldsymbol{x}) q(\boldsymbol{x}) \, \mathrm{d}\boldsymbol{x},$$

$$\forall \boldsymbol{u} \in H^1_\alpha(\Omega_d)^d, \ \forall q \in L^2_\alpha(\Omega_d),$$

$$b_{\alpha 2}(\boldsymbol{u}, q) = -\int_{\Omega_d} (\mathrm{div}\, \boldsymbol{u})(\boldsymbol{x}) q(\boldsymbol{x}) \varpi_\alpha(\boldsymbol{x}) \, \mathrm{d}\boldsymbol{x}. \tag{23.40}$$

We set $\boldsymbol{u}^* = \boldsymbol{u} - \boldsymbol{u}_\mathrm{b}$. There, the pair $(\boldsymbol{u}^*, p)$ is the solution of a problem which is now of the generalized form (23.1), with $X_1 = X_2 = H^1_{\alpha,0}(\Omega_d)^d$ and $M_i = L^2_{\alpha i}(\Omega_d)$, $i = 1, 2$. We are going to check the hypotheses of Theorem 23.2. The continuity of the form $a_\alpha(\bullet, \bullet)$ on $H^1_\alpha(\Omega_d)^d \times H^1_{\alpha,0}(\Omega_d)^d$ can be proven exactly as in Theorem 18.12, the continuity of the form $b_{\alpha 1}(\bullet, \bullet)$ on $H^1_{\alpha,0}(\Omega_d)^d \times L^2_\alpha(\Omega_d)$ is established in the same way, thanks to Hardy's inequality (13.4), and the continuity of the form $b_{\alpha 2}(\bullet, \bullet)$ on

$H^1_\alpha(\Omega_d)^d \times L^2_\alpha(\Omega_d)$ is obvious. So, it suffices to check the inf–sup conditions. We begin by the conditions on the forms $b_{\alpha i}(\bullet,\bullet)$, since the proof is somewhat similar to that of Theorem 23.3. We limit ourselves to the case $d=2$, and we refer to BERNARDI, DAUGE and MADAY [1992b] for results that allow for the extension of the proof to the cube.

THEOREM 23.5. *Let α be a real number which satisfies $-\frac{1}{2} \leqslant \alpha \leqslant \frac{1}{2}$, $\alpha \neq 0$. For $i=1,2$, there exists a positive constant β_i such that*

$$\inf_{q\in L^2_{\alpha i}(\Omega_d)} \sup_{\boldsymbol{v}\in H^1_{\alpha,0}(\Omega_d)^d} \frac{b_{\alpha i}(\boldsymbol{v},q)}{\|\boldsymbol{v}\|_{H^1_{\alpha,0}(\Omega_d)^d}\|q\|_{L^2_\alpha(\Omega_d)}} \geqslant \beta_i. \tag{23.41}$$

PROOF FOR THE SQUARE. We prove the inequality successively for $i=2$ and $i=1$.

(1) For any function q in $L^2_{\alpha 2}(\Omega)$, we consider the problem with Dirichlet boundary conditions:

$$\begin{cases} -\Delta\varphi = -q & \text{in } \Omega, \\ \varphi = 0 & \text{on } \Gamma_J,\ J=1,2,3,4. \end{cases} \tag{23.42}$$

From Theorem 18.13, it has a unique solution φ which, as stated in Theorem 18.14, belongs to $H^2_\alpha(\Omega)$ and satisfies

$$\|\varphi\|_{H^2_\alpha(\Omega)} \leqslant c\|q\|_{L^2_\alpha(\Omega)}. \tag{23.43}$$

Next, setting $\psi_0(\boldsymbol{a}_0)=0$, we define successively for $J=1,2,3,4$:

$$\forall \boldsymbol{x}\in\Gamma_J,\quad \psi_J(\boldsymbol{x}) = \psi_{J-1}(\boldsymbol{a}_{J-1}) - \int_{\Gamma_J}\left(\frac{\partial\varphi}{\partial n_J}\right)(\tau)\chi^{\boldsymbol{x}}_J(\tau)\,\mathrm{d}\tau,$$

where $\chi^{\boldsymbol{x}}_J$ is the characteristic function of the segment $[\boldsymbol{a}_{J-1},\boldsymbol{x}]$. Since the appropriate compatibility conditions are satisfied (indeed, the fact that q belongs to $L^2_{\alpha 2}(\Omega)$ together with (23.42), implies that $\psi_4(\boldsymbol{a}_4)=\psi_0(\boldsymbol{a}_0)$), Theorem 18.11 yields that there exists a function ψ in $H^2_\alpha(\Omega)$ such that

$$\psi = \psi_J \text{ on } \Gamma_J \quad\text{and}\quad \frac{\partial\psi}{\partial n_J} = 0 \text{ on } \Gamma_J,$$

together with the stability property

$$\|\psi\|_{H^2_\alpha(\Omega)} \leqslant c\sum_{J=1}^{4}\left\|\frac{\partial\varphi}{\partial n_J}\right\|_{H^{(1-\alpha)/2}_\alpha(\Gamma_J)}. \tag{23.44}$$

Setting $\boldsymbol{v} = \mathbf{grad}\,\varphi + \mathbf{curl}\,\psi$, we have

$$b_{\alpha 2}(\boldsymbol{v}, q) = \int_\Omega q^2(\boldsymbol{x})\varpi_\alpha(\boldsymbol{x})\,\mathrm{d}\boldsymbol{x},$$

and also

$$\|\boldsymbol{v}\|_{H^1_\alpha(\Omega)^2} \leqslant c\|q\|_{L^2_\alpha(\Omega)}.$$

Hence, we obtain the desired inf–sup condition.

(2) To prove the condition for $i = 1$, we recall from Theorem 18.6 that the mapping $\boldsymbol{v} \mapsto \boldsymbol{w} = \boldsymbol{v}\varpi_\alpha$ is an isomorphism from $H^1_{\alpha,0}(\Omega)^2$ onto $H^1_{-\alpha,0}(\Omega)^2$. Since the mapping $q \mapsto r = q\varpi_\alpha$ is also an isomorphism from $L^2_{\alpha 1}(\Omega)$ onto $L^2_{-\alpha 2}(\Omega)$, we have

$$\inf_{q\in L^2_{\alpha 1}(\Omega)} \sup_{\boldsymbol{v}\in H^1_{\alpha,0}(\Omega)^2} \frac{b_{\alpha 1}(\boldsymbol{v}, q)}{\|\boldsymbol{v}\|_{H^1_{\alpha,0}(\Omega)^2}\|q\|_{L^2_\alpha(\Omega)}}$$
$$= \inf_{r\in L^2_{-\alpha 2}(\Omega)} \sup_{\boldsymbol{w}\in H^1_{-\alpha,0}(\Omega)^2} \frac{b_{-\alpha 2}(\boldsymbol{w}, r)}{\|\boldsymbol{w}\|_{H^1_{-\alpha,0}(\Omega)^2}\|r\|_{L^2_{-\alpha}(\Omega)}},$$

and this last inf–sup condition has been proven for $-\frac{1}{2} \leqslant -\alpha \leqslant \frac{1}{2}$ in the first part of the proof.

The ellipticity of the form $a_\alpha(\bullet,\bullet)$ follows from Theorem 18.12, however this property is not sufficient to apply Theorem 23.2. Clearly, in the present case, the kernels K_i are given by

$$K_1 = \{\boldsymbol{v} \in H^1_{\alpha,0}(\Omega_d)^d \colon \operatorname{div}(\boldsymbol{v}\varpi_\alpha) = 0 \text{ in } \Omega_d\},$$
$$K_2 = \{\boldsymbol{u} \in H^1_{\alpha,0}(\Omega_d)^d \colon \operatorname{div}\boldsymbol{u} = 0 \text{ in } \Omega_d\}. \tag{23.45}$$

This allows for proving the inf–sup conditions on the form $a_\alpha(\bullet,\bullet)$.

THEOREM 23.6. *Let α be a real number which satisfies $-\frac{1}{2} \leqslant \alpha \leqslant \frac{1}{2}$. There exists a positive constant c such that, for $i = 1, 2$,*

$$\inf_{\boldsymbol{u}\in K_2} \sup_{\boldsymbol{v}\in K_1} \frac{a_\alpha(\boldsymbol{u}, \boldsymbol{v})}{\|\boldsymbol{u}\|_{H^1_\alpha(\Omega_d)^d}\|\boldsymbol{v}\|_{H^1_\alpha(\Omega_d)^d}} \geqslant c\nu,$$
$$\inf_{\boldsymbol{v}\in K_1} \sup_{\boldsymbol{u}\in K_2} \frac{a_\alpha(\boldsymbol{u}, \boldsymbol{v})}{\|\boldsymbol{u}\|_{H^1_\alpha(\Omega_d)^d}\|\boldsymbol{v}\|_{H^1_\alpha(\Omega_d)^d}} \geqslant c\nu. \tag{23.46}$$

PROOF. In the case $d = 2$ and in analogy to Remark 23.2, it is readily checked that each function $\boldsymbol{u}$ in K_2 satisfies $\boldsymbol{u} = \mathbf{curl}\,\psi$, where ψ belongs to $H^2_{\alpha,0}(\Omega)$; similarly, for any function $\boldsymbol{v}$ in K_1, the function $\boldsymbol{v}\varpi_\alpha$ belongs to $H^1_{-\alpha,0}(\Omega)^2$ by Theorem 18.6 and is divergence-free; hence another application of Theorem 18.6 yields that it can be written $\boldsymbol{v} = \varpi_{-\alpha}\mathbf{curl}(\chi\varpi_\alpha)$, where χ also belongs to $H^2_{\alpha,0}(\Omega)$. Next, we associate with each $\boldsymbol{u} = \mathbf{curl}\,\psi$ in K_2 the function $\boldsymbol{v} = \varpi_{-\alpha}\mathbf{curl}(\psi\varpi_\alpha)$ and conversely, so that both inf–sup conditions reduce to the ellipticity property

$$\nu \int_\Omega (\Delta\psi)(\boldsymbol{x})\Delta(\psi\varpi_\alpha)(\boldsymbol{x})\,\mathrm{d}\boldsymbol{x} \geqslant c\|\psi\|^2_{H^2_\alpha(\Omega)}. \tag{23.47}$$

This has been stated in Theorem 18.15. The result can be easily extended to the case $d = 3$ by introducing a divergence-free vector potential $\boldsymbol{\psi}$ whose tangential components vanishes on each face Γ_j, $1 \leqslant j \leqslant 6$.

Thus, the following theorem is a direct application of Theorem 23.2.

THEOREM 23.7. *Let α be a real number which satisfies $-\frac{1}{2} \leqslant \alpha \leqslant \frac{1}{2}$, $\alpha \neq 0$. For any $\boldsymbol{f}$ in $H^{-1}_\alpha(\Omega_d)^d$ and for any $2d$-tuple $(\boldsymbol{g}_j)_{1\leqslant j\leqslant 2d}$ in $\prod_{j=1}^{2d} H^{(1-\alpha)/2}_\alpha(\Gamma_j)^d$ which satisfies conditions (23.34) if $\alpha < 0$ and (23.36), problem (23.38) has a unique solution $(\boldsymbol{u}, p)$ in $H^1_\alpha(\Omega_d)^d \times L^2_{\alpha 1}(\Omega_d)$. Moreover, this solution satisfies*

$$\|\boldsymbol{u}\|_{H^1_\alpha(\Omega_d)^d} + \|p\|_{L^2_\alpha(\Omega_d)} \leqslant c\left(1 + \frac{1}{\nu}\right)\left(\|\boldsymbol{f}\|_{H^{-1}_\alpha(\Omega_d)^d} + \sum_{j=1}^{2d} \|\boldsymbol{g}_j\|_{H^{(1-\alpha)/2}_\alpha(\Gamma_j)^d}\right). \tag{23.48}$$

Note that the previous estimate does not hold when α is equal to 0, since this is a limit case for compatibility conditions (23.34), and additional terms should be added in the estimate, like in (1.32). It can also be observed that no compatibility condition is required between the $\boldsymbol{g}_j$ when α is positive. So, Theorem 23.7 gives the existence of a solution which belongs to $H^1(\Omega_*)$ for any open set Ω_* such that $\overline{\Omega}_*$ is contained in Ω_d in the case of driven cavity problem for instance (i.e., for $\boldsymbol{g}_j = \boldsymbol{0}$, $1 \leqslant j \leqslant 2d-1$, and $\boldsymbol{g}_{2d} = (1,0)$ or $(1,0,0)$).

REMARK 23.4. Let us for a while denote by $(\boldsymbol{u}_\alpha, p_\alpha)$ the solution of problem (23.38). Let α and β be real numbers such that $-\frac{1}{2} \leqslant \beta < \alpha \leqslant \frac{1}{2}$. If the functions $\boldsymbol{f}$ and $\boldsymbol{g}_J$ satisfy the hypotheses of Theorem 23.7, the solutions $(\boldsymbol{u}_\alpha, p_\alpha)$ and $(\boldsymbol{u}_\beta, p_\beta)$ are equal up to an additive constant on the pressure, i.e.,

$$\boldsymbol{u}_\alpha = \boldsymbol{u}_\beta \quad \text{and} \quad p_\alpha = p_\beta - \frac{\int_{\Omega_d} p_\beta(\boldsymbol{x})\,\varpi_\alpha(\boldsymbol{x})\,\mathrm{d}\boldsymbol{x}}{\int_{\Omega_d} \varpi_\alpha(\boldsymbol{x})\,\mathrm{d}\boldsymbol{x}}. \tag{23.49}$$

We conclude this section by studying an abstract approximation of problem (23.1), in a general form which will be needed later on. Let δ be the discretization parameter, we are given closed subspaces $X_{i\delta}$ and $M_{i\delta}$ of X_i and M_i respectively, $i = 1, 2$. We also introduce three "discrete" bilinear forms:

(i) the form $a_\delta(\cdot,\cdot)$ which is defined on $X_{2\delta} \times X_{1\delta}$,

(ii) the forms $b_{i\delta}(\bullet,\bullet)$ which are defined on $X_{i\delta} \times M_{i\delta}$.
For a pair (f_δ, g_δ) of $X_1' \times M_2'$, we consider the following discrete problem: *find a pair* (u_δ, p_δ) *in* $X_{2\delta} \times M_{1\delta}$ *such that*

$$\begin{aligned} &\forall v_\delta \in X_{1\delta}, \quad a_\delta(u_\delta, v_\delta) + b_{1\delta}(v_\delta, p_\delta) = \langle f_\delta, v_\delta \rangle, \\ &\forall q_\delta \in M_{2\delta}, \quad b_{2\delta}(u_\delta, q_\delta) = \langle g_\delta, q_\delta \rangle. \end{aligned} \tag{23.50}$$

The conditions for its wellposedness are strictly the same as for the continuous problem (23.1), so that the following theorem is in fact a special case of Theorem 23.2. Let us denote by $K_{i\delta}$, $i = 1, 2$, the kernels

$$K_{i\delta} = \{ v_\delta \in X_{i\delta} \colon \forall q_\delta \in M_{i\delta},\ b_{i\delta}(v_\delta, q_\delta) = 0 \}. \tag{23.51}$$

THEOREM 23.8. *Assume that the following conditions hold:*
(i) *there exists a positive constant* $\alpha_{1\delta}$ *such that*

$$\begin{aligned} &\inf_{u_\delta \in K_{2\delta}} \sup_{v_\delta \in K_{1\delta}} \frac{a_\delta(u_\delta, v_\delta)}{\|u_\delta\|_{X_2} \|v_\delta\|_{X_1}} \geqslant \alpha_{1\delta} \quad \textit{and} \\ &\inf_{v_\delta \in K_{1\delta}} \sup_{u_\delta \in K_{2\delta}} \frac{a_\delta(u_\delta, v_\delta)}{\|u_\delta\|_{X_2} \|v_\delta\|_{X_1}} > 0, \end{aligned} \tag{23.52}$$

(ii) *for* $i = 1,\ 2$, *there exists a positive constant* $\beta_{i\delta}$ *such that*

$$\inf_{q_\delta \in M_{i\delta}} \sup_{v_\delta \in X_{i\delta}} \frac{b_{i\delta}(v_\delta, q_\delta)}{\|v_\delta\|_{X_i} \|q_\delta\|_{M_i}} \geqslant \beta_{i\delta}. \tag{23.53}$$

Then, for any pair (f_δ, g_δ) *in* $X_1' \times M_2'$, *problem* (23.50) *has a unique solution* (u_δ, p_δ) *in* $X_{2\delta} \times M_{1\delta}$. *Moreover this solution satisfies*

$$\|u_\delta\|_{X_2} + \beta_{1\delta} \|p_\delta\|_{M_1} \leqslant \left(1 + \frac{1}{\alpha_{1\delta}}\right) \|f_\delta\|_{X_1'} + \frac{\gamma_\delta}{\beta_{2\delta}} \|g_\delta\|_{M_2'}, \tag{23.54}$$

where γ_δ *denotes the norm of* $a_\delta(\bullet,\bullet)$.

Note that, when the kernels $K_{1\delta}$ and $K_{2\delta}$ are finite-dimensional and have the same dimension, each inf–sup condition in (23.52) is equivalent to the other one, so that only one has to be checked.

We are now in a position to derive a general estimate for the error between the solutions of problems (23.1) and (23.50) respectively (we refer to BERNARDI, CANUTO and MADAY ([1988], Theorems 2.2 and 2.3) for the proof and more detailed results). We introduce the affine subspace

$$K_{2\delta}(g_\delta) = \{ w_\delta \in X_{2\delta} \colon \forall q_\delta \in M_{2\delta},\ b_{2\delta}(w_\delta, q_\delta) = \langle g_\delta, q_\delta \rangle \}. \tag{23.55}$$

THEOREM 23.9. *Assume that the hypotheses of Theorem* 23.8 *hold. Then, the following error estimate holds for problem* (23.50):

$$\begin{aligned}
&\|u-u_\delta\|_{X_2}+\frac{\beta_{1\delta}}{1+\gamma_\delta}\|p-p_\delta\|_{M_1}\\
&\quad\leqslant c\Big(1+\frac{1}{\alpha_{1\delta}}\Big)\Big\{(1+\gamma_\delta)\inf_{w_\delta\in K_{2\delta}(g_\delta)}\|u-w_\delta\|_{X_2}\\
&\qquad+\inf_{v_\delta\in X_{2\delta}}\Big((1+\gamma_\delta)\|u-v_\delta\|_{X_2}+\sup_{z_\delta\in X_{1\delta}}\frac{(a-a_\delta)(v_\delta,z_\delta)}{\|z_\delta\|_{X_1}}\Big)\\
&\qquad+\inf_{q_\delta\in M_{1\delta}}\Big(\|p-q_\delta\|_{M_1}+\sup_{z_\delta\in X_{1\delta}}\frac{(b_1-b_{1\delta})(z_\delta,q_\delta)}{\|z_\delta\|_{X_1}}\Big)\\
&\qquad+\sup_{z_\delta\in X_{1\delta}}\frac{\langle f-f_\delta,z_\delta\rangle}{\|z_\delta\|_{X_1}}\Big\}. \qquad (23.56)
\end{aligned}$$

Since the size of this estimate is very impressive, we present two standard situations where it can be simplified:

(i) when the discrete kernel $K_{1\delta}$ is contained in K_1 (it is not always the case!), the following shorter estimate on $\|u-u_\delta\|_{X_2}$ can be derived:

$$\begin{aligned}
&\|u-u_\delta\|_{X_2}\\
&\quad\leqslant c\Big(1+\frac{1}{\alpha_{1\delta}}\Big)\Big\{(1+\gamma_\delta)\inf_{w_\delta\in K_{2\delta}(g_\delta)}\|u-w_\delta\|_{X_2}\\
&\qquad+\inf_{v_\delta\in X_{2\delta}}\Big((1+\gamma_\delta)\,\|u-v_\delta\|_{X_2}+\sup_{z_\delta\in X_{1\delta}}\frac{(a-a_\delta)(v_\delta,z_\delta)}{\|z_\delta\|_{X_1}}\Big)\\
&\qquad+\sup_{z_\delta\in X_{1\delta}}\frac{\langle f-f_\delta,z_\delta\rangle}{\|z_\delta\|_{X_1}}\Big\}; \qquad (23.57)
\end{aligned}$$

(ii) when the forms $a(\bullet,\bullet)$ and $a_\delta(\bullet,\bullet)$ coincide, as well as the forms $b_i(\bullet,\bullet)$ and $b_{i\delta}(\bullet,\bullet)$, $i=1,2$, f and f_δ, and g and g_δ, the estimate is written

$$\begin{aligned}
&\|u-u_\delta\|_{X_2}+\frac{\beta_{1\delta}}{1+\gamma_\delta}\|p-p_\delta\|_{M_1} \qquad (23.58)\\
&\quad\leqslant c\Big(1+\frac{1}{\alpha_{1\delta}}\Big)\Big\{(1+\gamma_\delta)\inf_{w_\delta\in K_{2\delta}(g_\delta)}\|u-w_\delta\|_{X_2}+\inf_{q_\delta\in M_{1\delta}}\|p-q_\delta\|_{M_1}\Big\}.
\end{aligned}$$

For the sake of completeness, we end this section by a standard result about the approximation of functions in X_2 by functions in $K_{2\delta}(g_\delta)$. However, it does not seem worth to use it when the constant $\beta_{2\delta}$ is not bounded independently of δ, since other types of arguments can lead to optimal estimates even with a wrong behavior of $\beta_{2\delta}$.

THEOREM 23.10. *Assume that hypothesis* (23.53) *holds for* $i = 2$. *Then, for any* g *in* M_2' *and for any function* u *in* X_2 *satisfying*

$$\forall q \in M_2, \quad b_2(u, q) = \langle g, q \rangle, \tag{23.59}$$

the following estimate holds:

$$\begin{aligned} &\inf_{w_\delta \in K_{2\delta}(g_\delta)} \|u - w_\delta\|_{X_2} \\ &\quad \leqslant c\left(1 + \frac{1}{\beta_{2\delta}}\right)\left\{\inf_{v_\delta \in X_{2\delta}} \left(\|u - v_\delta\|_{X_2} + \sup_{q_\delta \in M_{2\delta}} \frac{(b_2 - b_{2\delta})(v_\delta, q_\delta)}{\|q_\delta\|_{M_2}}\right)\right. \\ &\qquad \left. + \sup_{q_\delta \in M_{2\delta}} \frac{\langle g - g_\delta, q_\delta \rangle}{\|q_\delta\|_{M_2}}\right\}. \end{aligned} \tag{23.60}$$

REMARK 23.5. The following converse result is proven by SCOTT and VOGELIUS [1985], when the forms $b_2(\bullet, \bullet)$ and $b_{2\delta}(\bullet, \bullet)$ coincide: the estimate

$$\inf_{w_\delta \in K_{2\delta}} \|u - w_\delta\|_{X_2} \leqslant c \inf_{v_\delta \in X_{2\delta}} \|u - v_\delta\|_{X_2},$$

for all $u \in K_2$ with a constant c independent of δ, implies that $\beta_{2\delta}$ is larger than a positive constant which is also independent of δ.

In the next sections, these abstract results will be applied to several spectral discretizations of the Stokes problem. It appears that the main problem is to choose compatible discrete spaces for the velocity and the pressure in order that the discrete inf–sup conditions hold.

24. The single grid method for the Stokes problem

To focus on the main difficulties, we firstly present the discretization of the Stokes problem in the case of homogeneous boundary conditions and by a Legendre type method. So, we start from the variational formulation: *find a pair* $(\boldsymbol{u}, p)$ *in* $H_0^1(\Omega_d)^d \times L_0^2(\Omega_d)$ *such that*

$$\begin{aligned} &\forall \boldsymbol{v} \in H_0^1(\Omega_d)^d, \quad a(\boldsymbol{u}, \boldsymbol{v}) + b(\boldsymbol{v}, p) = \langle \boldsymbol{f}, \boldsymbol{v} \rangle, \\ &\forall q \in L_0^2(\Omega_d), \qquad b(\boldsymbol{u}, q) = 0. \end{aligned} \tag{24.1}$$

From now on, the data $\boldsymbol{f}$ are assumed to be continuous on Ω_d. An integer $N \geqslant 3$ being fixed, it seems natural to look for a discrete velocity in $\mathbb{P}_N^0(\Omega_d)^2$ and for a discrete pressure of the same degree, i.e., in a subspace M_N of $\mathbb{P}_N(\Omega_d)$. This first method was proposed by MORCHOISNE [1983] and its first implementation was performed by MÉTIVET [1987].

Like in Section 15, we define, for continuous functions u and v on $\overline{\Omega}_d$,

$$(u,v)_N = \sum_{i_1=0}^{N} \cdots \sum_{i_d=0}^{N} u(\xi_{i_1},\ldots,\xi_{i_d})v(\xi_{i_1},\ldots,\xi_{i_d})\rho_{i_1}\cdots\rho_{i_d}, \tag{24.2}$$

and we set, for any polynomials $\boldsymbol{u}_N$ and $\boldsymbol{v}_N$ in $\mathbb{P}_N(\Omega_d)^d$ and q_N in $\mathbb{P}_N(\Omega)$,

$$a_N(\boldsymbol{u}_N,\boldsymbol{v}_N) = \nu(\mathbf{grad}\,\boldsymbol{u}_N, \mathbf{grad}\,\boldsymbol{v}_N)_N, \tag{24.3}$$

$$b_N(\boldsymbol{v}_N,q_N) = -(\operatorname{div}\boldsymbol{v}_N,q_N)_N. \tag{24.4}$$

Then, the discrete problem reads: *find a pair* $(\boldsymbol{u}_N,p_N)$ *in* $\mathbb{P}_N^0(\Omega_d)^d \times M_N$ *such that*

$$\begin{aligned} &\forall \boldsymbol{v}_N \in \mathbb{P}_N^0(\Omega_d)^d, \quad a_N(\boldsymbol{u}_N,\boldsymbol{v}_N) + b_N(\boldsymbol{v}_N,p_N) = (\boldsymbol{f},\boldsymbol{v}_N)_N,\\ &\forall q_N \in M_N, \qquad b_N(\boldsymbol{u}_N,q_N) = 0. \end{aligned} \tag{24.5}$$

Let us define the kernel

$$K_N = \{\boldsymbol{v}_N \in \mathbb{P}_N^0(\Omega_d)^d\colon\ \forall q_N \in M_N,\ b_N(\boldsymbol{v}_N,q_N) = 0\}. \tag{24.6}$$

It is readily checked that, if $(\boldsymbol{u}_N,p_N)$ is a solution of problem (24.5), then the polynomial $\boldsymbol{u}_N$ belongs to K_N and satisfies

$$\forall \boldsymbol{v}_N \in K_N, \quad a_N(\boldsymbol{u}_N,\boldsymbol{v}_N) = (f,\boldsymbol{v}_N)_N. \tag{24.7}$$

Since the ellipticity of the form $a_N(\cdot,\cdot)$ is stated in Theorem 15.1 (up to the viscosity!), problem (24.7) has a unique solution $\boldsymbol{u}_N$. Hence, we only have to consider the uniqueness and stability of the pressure. This is the crucial problem, and it requires some notation.

DEFINITION. Is called a *spurious mode* for the pressure, any polynomial q_N in $\mathbb{P}_N(\Omega_d)$ such that

$$\forall \boldsymbol{v}_N \in \mathbb{P}_N^0(\Omega_d)^d, \quad b_N(\boldsymbol{v}_N,q_N) = 0. \tag{24.8}$$

The subspace of $\mathbb{P}_N(\Omega_d)$ made of these spurious modes is denoted by Z_N. The image of the space $\mathbb{P}_N^0(\Omega_d)^d$ by the divergence operator is denoted by D_N.

Clearly, the subspace Z_N is the orthogonal of D_N in $\mathbb{P}_N(\Omega_d)$ for the scalar product $(\cdot,\cdot)_N$. We now have to characterize Z_N. For any sets A_i, $1 \leqslant i \leqslant d$, of polynomials with one variable, we agree to denote by $A_1 \otimes \cdots \otimes A_d$ the set of polynomials

$$a_1(x_1)\cdots a_d(x_d), \quad a_1 \in A_1,\ldots,a_d \in A_d, \tag{24.9}$$

while $A^{\otimes d}$ stands for $A_1 \otimes \cdots \otimes A_d$ when all the A_i coincide with A.

THEOREM 24.1. *In the case of dimension $d = 2$, the space Z_N has dimension 8, and it is spanned by the polynomials in*

$$\{L_0, L_N\}^{\otimes 2} \quad \textit{and} \quad \{(1-\zeta)L'_N, (1+\zeta)L'_N\}^{\otimes 2}. \tag{24.10}$$

In the case of dimension $d = 3$, the space Z_N has dimension $12N + 4$, and it is spanned by the polynomials in

$$\begin{aligned}
&\{L_0, L_N\}^{\otimes 3} \quad \textit{and} \quad \{(1-\zeta)L'_N, (1+\zeta)L'_N\}^{\otimes 2} \otimes \mathbb{P}_N(\Lambda), \\
&\{(1-\zeta)L'_N, (1+\zeta)L'_N\} \otimes \mathbb{P}_N(\Lambda) \otimes \{(1-\zeta)L'_N, (1+\zeta)L'_N\}, \\
&\mathbb{P}_N(\Lambda) \otimes \{(1-\zeta)L'_N, (1+\zeta)L'_N\}^{\otimes 2}.
\end{aligned} \tag{24.11}$$

PROOF. The proof proceeds in three steps.

(1) It is readily checked that the polynomials given in (24.10) and (24.11) are spurious modes. Indeed, for any polynomial v_N in $\mathbb{P}_N^0(\Omega_d)$, $\partial v_N/\partial x_j$ has degree $\leqslant N-1$ with respect to x_j, hence is orthogonal to $L_N(x_j)$, and is also orthogonal to $L_0(x_j)$, as can be seen by integrating by parts. Moreover, for any $\boldsymbol{v}_N$ in $\mathbb{P}_N^0(\Omega_d)^d$, $\operatorname{div} \boldsymbol{v}_N$ vanishes on the $(d-2)$-faces of Ω_d (vertices for $d = 2$, edges for $d = 3$), and the polynomials of the second type in (24.10) and (24.11) span exactly the same subspace as the Lagrange polynomials associated with the nodes of Ξ_N which belong to these $(d-2)$-faces.

(2) Conversely, in the case $d = 2$, we fix a polynomial q_N in Z_N. Starting from the identity

$$b(\boldsymbol{v}_N, q_N) = (\boldsymbol{v}_N, \mathbf{grad}\, q_N)_N,$$

we begin by choosing

$$\begin{aligned}
&\boldsymbol{v}_N = (v_N, 0), \quad \text{with} \\
&\qquad v_N(x,y) = (1-x^2)L'_m(x)(1-y^2)L'_n(y), \quad 1 \leqslant m, n \leqslant N-1.
\end{aligned}$$

We observe from (3.1) that $\partial v_N/\partial x$ is equal to $-m(m+1)L_m(x)(1-y^2)L'_n(y)$, secondly that the orthogonal of the subspace spanned by $L_1, \ldots$ and L_{N-1} for the discrete scalar product is spanned by L_0 and L_N and, similarly, that the orthogonal of the subspace $\mathbb{P}_N^0(\Lambda)$ is spanned by $(1-\zeta)L'_N$ and $(1+\zeta)L'_N$. Thus, we deduce that q_N admits the expansion

$$\begin{aligned}
q_N(x,y) = {}& \alpha_N(y)L_0(x) + \beta_N(y)L_N(x) \\
&+ \gamma_N(x)(1-y)L'_N(y) + \delta_N(x)(1+y)L'_N(y),
\end{aligned} \tag{24.12}$$

where α_N, β_N, γ_N and δ_N belong to $\mathbb{P}_N(\Lambda)$. In the same way, exchanging the roles of the variables x and y, we obtain that q_N admits the expansion

$$q_N(x,y) = \tilde{\alpha}_N(x)L_0(y) + \tilde{\beta}_N(x)L_N(y) + \tilde{\gamma}_N(y)(1-x)L'_N(x) + \tilde{\delta}_N(y)(1+x)L'_N(x). \tag{24.13}$$

Consequently, it belongs to the subspace spanned by the polynomials given in (24.10). Similar arguments allow for proving the analogous result in the case of dimension $d=3$.

(3) In the case $d=2$, it remains to check that the polynomials given in (24.10) are linearly independent. Clearly, this property follows from the fact that the polynomials L_0, L_N, L'_N and $\zeta\, L'_N$ are independent in $\mathbb{P}_N(\Lambda)$, which we prove now. Let us assume that

$$\alpha L_0(\zeta) + \beta L_N(\zeta) + \gamma L'_N(\zeta) + \delta\zeta L'_N(\zeta) = 0.$$

Looking at the coefficient of ζ^{N-1} and using a parity argument, we observe that γ is equal to 0. Thus, thanks to (3.12) and (3.15), the previous line can also be written as

$$\alpha L_0(\zeta) + \delta L'_{N-1}(\zeta) + (\beta + N\delta)L_N(\zeta) = 0,$$

which implies that α, β and δ are equal to 0. Hence, the dimension of Z_N is equal to 8. In the case $d=3$, the same arguments imply that the polynomials in

$$\begin{aligned}
&\{L_0, L_N\}^{\otimes 3} \quad \text{and} \quad \{(1-\zeta)L'_N, (1+\zeta)L'_N\}^{\otimes 3},\\
&\{(1-\zeta)L'_N, (1+\zeta)L'_N\}^{\otimes 2} \otimes \{(1-\zeta^2)L'_k,\ 1 \leqslant k \leqslant N-1\},\\
&\{(1-\zeta)L'_N, (1+\zeta)L'_N\} \otimes \{(1-\zeta^2)L'_k,\ 1 \leqslant k \leqslant N-1\}\\
&\qquad\qquad \otimes \{(1-\zeta)L'_N, (1+\zeta)L'_N\},\\
&\{(1-\zeta^2)L'_k,\ 1 \leqslant k \leqslant N-1\} \otimes \{(1-\zeta)L'_N, (1+\zeta)L'_N\}^{\otimes 2},
\end{aligned}$$

are linearly independent, whence the result.

Note that the mode $L_0(x)L_0(y) = 1$ or $L_0(x)L_0(y)L_0(z) = 1$ is not due to the numerical method, since it already appears in the continuous problem (23.23), leading to take $L^2_0(\Omega_d)$ for the pressure space instead of $L^2(\Omega_d)$. However, we leave it in Z_N for symmetry reasons. Indeed, in the case $d=2$ for instance, when the polynomials of the second type in (24.10) are the Lagrange polynomials associated with the corners of the square, the polynomials of the first type can be seen as the corner polynomials in the spectral plane.

REMARK 24.1. Since the discrete form $(\cdot,\cdot)_N$ is a scalar product on $\mathbb{P}_N(\Omega_d)$, Theorem 24.1 states that the codimension of the subspace D_N in $\mathbb{P}_N(\Omega_d)$, is equal to 8 for

$d=2$ and to $12N+4$ in the case $d=3$. From this observation, it is easy to prove that the space of polynomials q_N in $\mathbb{P}_N(\Omega_d)$ such that

$$\forall \boldsymbol{v}_N \in \mathbb{P}_N^0(\Omega_d)^d, \quad b(\boldsymbol{v}_N, q_N) = 0, \tag{24.14}$$

has the same dimension as Z_N and that it is spanned:

(i) in the case $d=2$ by the polynomials given in

$$\{L_0, L_N\}^{\otimes 2} \quad \text{and} \quad \{L'_N, L'_{N+1}\}^{\otimes 2}. \tag{24.15}$$

(ii) in the case $d=3$ by the polynomials given in

$$\begin{aligned}&\{L_0, L_N\}^{\otimes 3} \quad \text{and} \quad \{L'_N, L'_{N+1}\}^{\otimes 2} \otimes \mathbb{P}_N(\Lambda),\\ &\{L'_N, L'_{N+1}\} \otimes \mathbb{P}_N(\Lambda) \otimes \{L'_N, L'_{N+1}\}, \quad \mathbb{P}_N(\Lambda) \otimes \{L'_N, L'_{N+1}\}^{\otimes 2}.\end{aligned} \tag{24.16}$$

Now, the idea is to choose for M_N a supplementary subspace of Z_N in $\mathbb{P}_N(\Omega_d)$, i.e., to assume that

$$\mathbb{P}_N(\Omega_d) = Z_N \oplus M_N. \tag{24.17}$$

An example of space M_N satisfying this condition is of course the space D_N, however we will see later on that it is not the right one for a good approximation of the pressure. It has to be observed that hypothesis (24.17) implies the existence of a constant $\beta_N > 0$ such that

$$\inf_{q_N \in M_N} \sup_{\boldsymbol{v}_N \in \mathbb{P}_N^0(\Omega_d)^d} \frac{b_N(\boldsymbol{v}_N, q_N)}{\|\boldsymbol{v}_N\|_{H^1(\Omega_d)^d} \|q_N\|_{L^2(\Omega_d)}} \geqslant \beta_N, \tag{24.18}$$

even if its dependency with respect to N is still unknown. Moreover, since the ellipticity of $a_N(\cdot,\cdot)$ follows from Theorem 15.1, applying Theorem 23.8, we can write the following theorem.

THEOREM 24.2. *If hypothesis* (24.17) *holds, for any continuous function* $\boldsymbol{f}$ *on* Ω_d*, problem* (24.5) *has a unique solution* $(\boldsymbol{u}_N, p_N)$ *in* $\mathbb{P}_N^0(\Omega_d)^d \times M_N$.

The idea is now to derive error estimates from the general inequality (23.56), so we have to characterize the kernel K_N defined in (24.6). Using once more hypothesis (24.17), we observe that its definition can equivalently be written as

$$K_N = \{\boldsymbol{v}_N \in \mathbb{P}_N^0(\Omega_d)^2;\ \forall q_N \in \mathbb{P}_N(\Omega_d),\ b_N(\boldsymbol{v}_N, q_N) = 0\},$$

or also

$$K_N = \{\boldsymbol{v}_N \in \mathbb{P}_N^0(\Omega_d)^2;\ \operatorname{div} \boldsymbol{v}_N = 0 \text{ in } \Omega_d\}. \tag{24.19}$$

As a first consequence, for any choice of M_N such that hypothesis (24.17) is satisfied, the element $\boldsymbol{u}_N$ in the solution $(\boldsymbol{u}_N, p_N)$ of problem (24.5) is the solution in K_N of the problem

$$\forall \boldsymbol{v}_N \in \mathbb{P}_N^0(\Omega_d)^d, \quad a_N(\boldsymbol{u}_N, \boldsymbol{v}_N) = \langle \boldsymbol{f}, \boldsymbol{v}_N \rangle, \tag{24.20}$$

and, since K_N is independent of the choice of M_N, so is $\boldsymbol{u}_N$. Consequently, the error between $\boldsymbol{u}$ and $\boldsymbol{u}_N$ does not depend of the choice of the space M_N satisfying (24.17).

The next theorem states a result of approximation of divergence-free functions by divergence-free polynomials. A complete proof is given only in the two-dimensional case, since the case $d = 3$ is much more difficult.

THEOREM 24.3. *If hypothesis* (24.17) *holds, for any real number* $s \geqslant 1$, *there exists a positive constant* c *such that, for any divergence-free function* $\boldsymbol{u}$ *in* $H^s(\Omega_d)^d \cap H_0^1(\Omega_d)^d$, *the following inequality holds:*

$$\inf_{\boldsymbol{v}_N \in K_N} \|\boldsymbol{u} - \boldsymbol{v}_N\|_{H^1(\Omega_d)^d} \leqslant cN^{1-s} \|\boldsymbol{u}\|_{H^s(\Omega_d)^d}. \tag{24.21}$$

PROOF. We distinguish the two cases $d = 2$ and $d = 3$.

(1) In the case $d = 2$, the proof relies on the use of the stream-function associated with $\boldsymbol{u}$, as explained in Remark 23.2: there exists a function ψ in $H_0^2(\Omega)$ such that $\boldsymbol{u}$ is equal to $\mathbf{curl}\,\psi$. Moreover, if $\boldsymbol{u}$ belongs to $H^s(\Omega)^2$, it is readily checked that ψ belongs to $H^{s+1}(\Omega)$ and satisfies

$$\|\psi\|_{H^{s+1}(\Omega)} \leqslant c\|\boldsymbol{u}\|_{H^s(\Omega)^2}. \tag{24.22}$$

Next, we approximate $\boldsymbol{u}$ by $\mathbf{curl}(\Pi_N^{2,0}\psi)$, so that

$$\inf_{\boldsymbol{v}_N \in K_N} \|\boldsymbol{u} - \boldsymbol{v}_N\|_{H^1(\Omega)^2} \leqslant \|\psi - \Pi_N^{2,0}\psi\|_{H^2(\Omega)}.$$

Thus, applying Theorem 7.2 leads to

$$\inf_{\boldsymbol{v}_N \in K_N} \|\boldsymbol{u} - \boldsymbol{v}_N\|_{H^1(\Omega)^2} \leqslant cN^{1-s}\, \|\psi\|_{H^{s+1}(\Omega)},$$

which, together with (24.22), gives the desired result.

(2) In the case $d = 3$, SACCHI-LANDRIANI and VANDEVEN [1989] have proven (24.21) if $s \geqslant 3$. So, if Π_{K_N} denotes the orthogonal projection from K onto K_N, the estimate

$$\forall \boldsymbol{u} \in K \cap H^s(\Omega_3)^3, \quad \|\boldsymbol{u} - \Pi_{K_N}\boldsymbol{u}\|_{H^1(\Omega_3)^3} \leqslant cN^{1-s}\|\boldsymbol{u}\|_{H^s(\Omega_3)^3},$$

holds for $s = 1$ and $s \geqslant 3$. Thus, the general result follows from the interpolation result

$$\left[K \cap H^3(\Omega_3)^3, K \cap H^1(\Omega_3)^3\right]_{(3-s)/2} = K \cap H^s(\Omega_3)^3,$$

which is established in BERNARDI, DAUGE and MADAY [1994].

However, it can be observed from the previous proof that the following result also holds with the same assumptions:

$$\inf_{\boldsymbol{v}_N \in K_N \cap \mathbb{P}_{N-1}(\Omega_d)^d} \|\boldsymbol{u} - \boldsymbol{v}_N\|_{H^1(\Omega_d)^d} \leqslant cN^{1-s} \|\boldsymbol{u}\|_{H^s(\Omega_d)^d}. \tag{24.23}$$

This is necessary for what follows.

THEOREM 24.4. *The function $\boldsymbol{f}$ is supposed to be in a space $H^\sigma(\Omega_d)^d$, $\sigma > d/2$. Let us assume that the solution $(\boldsymbol{u}, p)$ of problem* (24.1) *belongs to $H^s(\Omega_d)^d \times H^{s-1}(\Omega_d)$ for a real number $s \geqslant 1$. Then, if hypothesis* (24.17) *holds, the following error estimate holds for problem* (24.5):

$$|\boldsymbol{u} - \boldsymbol{u}_N|_{H^1(\Omega_d)^d} \leqslant c\big(N^{1-s} \|\boldsymbol{u}\|_{H^s(\Omega_d)^d} + N^{-\sigma} \|\boldsymbol{f}\|_{H^\sigma(\Omega_d)^d}\big). \tag{24.24}$$

PROOF. Applying either (15.14) or (23.57) and using the exactness of the quadrature formula on $\mathbb{P}_{2N-1}(\Lambda)$, we obtain for any $\boldsymbol{v}_N$ in $K_N \cap \mathbb{P}_{N-1}(\Omega_d)^d$,

$$|\boldsymbol{u} - \boldsymbol{u}_N|_{H^1(\Omega_d)^d} \leqslant c\Bigg(\|\boldsymbol{u} - \boldsymbol{v}_N\|_{H^1(\Omega_d)^d} + \sup_{\boldsymbol{w}_N \in \mathbb{P}_N^0(\Omega_d)^d} \frac{\int_{\Omega_d} \boldsymbol{f}(\boldsymbol{x}) \cdot \boldsymbol{w}_N(\boldsymbol{x})\,\mathrm{d}\boldsymbol{x} - (\boldsymbol{f}, \boldsymbol{w}_N)_N}{|\boldsymbol{w}_N|_{H^1(\Omega_d)^d}} \Bigg), \tag{24.25}$$

so that we use (24.23) to bound the first term. The estimate of the last term follows from (15.18), combined with Theorems 7.1 and 14.2.

The standard duality argument again leads to an improved estimate for the velocity in $L^2(\Omega_d)^d$.

THEOREM 24.5. *If the assumptions of Theorem* 24.4 *are satisfied, the following error estimate holds for problem* (24.5):

$$\|\boldsymbol{u} - \boldsymbol{u}_N\|_{L^2(\Omega_d)^d} \leqslant c\big(N^{-s} \|\boldsymbol{u}\|_{H^s(\Omega_d)^d} + N^{-\sigma} \|\boldsymbol{f}\|_{H^\sigma(\Omega_d)^d}\big). \tag{24.26}$$

PROOF. Here, the proof relies on the identity

$$\|\boldsymbol{u} - \boldsymbol{u}_N\|_{L^2(\Omega_d)^d} = \sup_{\boldsymbol{g} \in L^2(\Omega_d)^d} \frac{\int_{\Omega_d} (\boldsymbol{u} - \boldsymbol{u}_N)(\boldsymbol{x}) \cdot \boldsymbol{g}(\boldsymbol{x})\,\mathrm{d}\boldsymbol{x}}{\|\boldsymbol{g}\|_{L^2(\Omega_d)^d}}. \tag{24.27}$$

For any $\boldsymbol{g}$ in $L^2(\Omega_d)^d$, the Stokes problem:

$$\begin{cases} -\nu\Delta\boldsymbol{w} + \mathbf{grad}\, r = \boldsymbol{g} & \text{in } \Omega_d, \\ \operatorname{div} \boldsymbol{w} = 0 & \text{in } \Omega_d, \\ \boldsymbol{w} = \mathbf{0} & \text{on } \partial\Omega_d, \end{cases} \tag{24.28}$$

has a unique solution $(\boldsymbol{w}, r)$ in $H_0^1(\Omega_d)^d \times L_0^2(\Omega_d)$. Moreover, as already stated in Remark 23.3, the pair $(\boldsymbol{w}, r)$ belongs to $H^2(\Omega_d)^d \times H^1(\Omega_d)$ and satisfies

$$\|\boldsymbol{w}\|_{H^2(\Omega_d)^d} + \|r\|_{H^1(\Omega_d)} \leqslant c\|\boldsymbol{g}\|_{L^2(\Omega_d)^d}. \tag{24.29}$$

Next, since both $\boldsymbol{u}$ and $\boldsymbol{u}_N$ are divergence-free, a simple calculation proves that, for any $\boldsymbol{w}_{N-1}$ in $K_N \cap \mathbb{P}_{N-1}(\Omega_d)^d$,

$$\begin{aligned}
&\int_{\Omega_d} (\boldsymbol{u} - \boldsymbol{u}_N)(\boldsymbol{x}) \cdot \boldsymbol{g}(\boldsymbol{x})\,\mathrm{d}\boldsymbol{x} \\
&\quad = a(\boldsymbol{u} - \boldsymbol{u}_N, \boldsymbol{w}) = a(\boldsymbol{u} - \boldsymbol{u}_N, \boldsymbol{w} - \boldsymbol{w}_{N-1}) + \langle \boldsymbol{f}, \boldsymbol{w}_{N-1}\rangle - (\mathcal{I}_N \boldsymbol{f}, \boldsymbol{w}_{N-1})_N \\
&\quad \leqslant \nu\,|\boldsymbol{u} - \boldsymbol{u}_N|_{H^1(\Omega_d)^d}|\boldsymbol{w} - \boldsymbol{w}_{N-1}|_{H^1(\Omega_d)^d} \\
&\qquad + \|\boldsymbol{f} - \mathcal{I}_N \boldsymbol{f}\|_{L^2(\Omega_d)^d}\|\boldsymbol{w}_{N-1}\|_{L^2(\Omega_d)^d},
\end{aligned}$$

and the result follows from Theorems 14.2 and 24.4 and (24.23), combined with (24.27) and (24.29).

The final step for the analysis of problem (24.5) is to derive an error estimate on the pressure. It relies on the estimate of the best possible constant β_N in (24.18), which will be deduced of the next two theorems. First estimates were given by BERNARDI, MADAY and MÉTIVET [1987] and BERNARDI and MADAY [1990]. We also refer to JENSEN and VOGELIUS [1990] for the idea of the final proof (and to JENSEN and ZHANG [1994] for the three-dimensional extension).

THEOREM 24.6. *For any function q_N in $\mathbb{P}_N(\Omega_d)$ which is orthogonal to the polynomials in $\{L_0, L_N\}^{\otimes d}$, there exists a polynomial $\boldsymbol{w}_N$ in $\mathbb{P}_N(\Omega_d)^d$ which satisfies*

$$\operatorname{div} \boldsymbol{w}_N = q_N \text{ in } \Omega_d \quad \text{and} \quad \boldsymbol{w}_N \cdot \boldsymbol{n}_j = 0 \text{ on } \Gamma_j,\ 1 \leqslant j \leqslant 2d, \tag{24.30}$$

and

$$\begin{aligned}
&\|\boldsymbol{w}_N\|_{H^1(\Omega_d)^d} \leqslant cN\|q_N\|_{L^2(\Omega_d)} \quad \text{and} \\
&\|\boldsymbol{w}_N\|_{L^2(\Gamma_j)^d} \leqslant c\|q_N\|_{L^2(\Omega_d)}, \quad 1 \leqslant j \leqslant 2d.
\end{aligned} \tag{24.31}$$

PROOF. The proof is given only in the case $d = 2$, since the corresponding proof in the case $d = 3$ is completely similar. Any function q_N in $L^2(\Omega)$ orthogonal to the polynomials in $\{L_0, L_N\}^{\otimes 2}$ has the expansion

$$\begin{aligned}
q_N(x, y) = &\sum_{m=0}^{N-1}\sum_{n=0}^{N-1} \alpha_{mn}\, L_m(x)L_n(y) + \sum_{m=1}^{N-1} \beta_m L_m(x)\big(L_N(y) - L_{N-2}(y)\big) \\
&+ \sum_{n=1}^{N-1} \gamma_n \big(L_N(x) - L_{N-2}(x)\big)L_n(y),
\end{aligned} \tag{24.32}$$

with α_{00} equal to 0. Next, we observe that any polynomial $\varphi_N = \lambda L_{N-2} + \mu(L_N - L_{N-2})$ satisfies

$$\|\varphi_N\|^2_{L^2(\Lambda)} = \frac{(\lambda-\mu)^2}{N-\frac{3}{2}} + \frac{\mu^2}{N+\frac{1}{2}} \geqslant \frac{1}{N+\frac{1}{2}}\left(\frac{1}{3}\lambda^2 + \frac{1}{2}\mu^2\right),$$

from which we deduce that

$$\|q_N\|^2_{L^2(\Omega_d)} \geqslant c\left(\sum_{m=0}^{N-1}\sum_{n=0}^{N-1} \alpha^2_{mn} \frac{1}{(m+\frac{1}{2})(n+\frac{1}{2})} + \frac{1}{N+\frac{1}{2}}\sum_{m=1}^{N-1}\beta^2_m \frac{1}{m+\frac{1}{2}} + \frac{1}{N+\frac{1}{2}}\sum_{n=1}^{N-1}\gamma^2_n \frac{1}{n+\frac{1}{2}}\right). \tag{24.33}$$

The idea is now to choose $\boldsymbol{w}_N = (w_N, z_N)$, with

$$w_N(x,y) = \sum_{m=0}^{N-1}\sum_{n=0}^{m} \alpha_{mn} \frac{L_{m+1}(x) - L_{m-1}(x)}{2m+1} L_n(y) + \sum_{m=1}^{N-1} \beta_m \frac{L_{m+1}(x) - L_{m-1}(x)}{2m+1}\big(L_N(y) - L_{N-2}(y)\big) \tag{24.34}$$

and

$$z_N(x,y) = \sum_{m=0}^{N-1}\sum_{n=m+1}^{N-1} \alpha_{mn} L_m(x) \frac{L_{n+1}(y) - L_{n-1}(y)}{2n+1} + \sum_{n=1}^{N-1} \gamma_n \big(L_N(x) - L_{N-2}(x)\big) \frac{L_{n+1}(y) - L_{n-1}(y)}{2n+1}. \tag{24.35}$$

Indeed, by (3.12), the two equations in (24.30) are obviously satisfied with this choice. So, it remains to check (24.31).

(1) First, we note that

$$\left(\frac{\partial w_N}{\partial x}\right)(x,y) = \sum_{m=0}^{N-1}\sum_{n=0}^{m} \alpha_{mn} L_m(x) L_n(y) + \sum_{m=1}^{N-1} \beta_m L_m(x)\big(L_N(y) - L_{N-2}(y)\big), \tag{24.36}$$

which allows for bounding $\|\partial w_N/\partial x\|_{L^2(\Omega)}$ by comparing with (24.33). Using a similar argument for $\partial z_N/\partial y$, we derive that

$$\left\|\frac{\partial w_N}{\partial x}\right\|_{L^2(\Omega)} + \left\|\frac{\partial z_N}{\partial y}\right\|_{L^2(\Omega)} \leqslant c\|q_N\|_{L^2(\Omega)}, \tag{24.37}$$

and also, thanks to the Poincaré–Friedrichs inequality (1.4) applied with respect either to x or y,

$$\|w_N\|_{L^2(\Omega)} + \|z_N\|_{L^2(\Omega)} \leqslant c\|q_N\|_{L^2(\Omega)}. \tag{24.38}$$

(2) The next estimate is more technical, due to the lack of orthogonality properties in the formula

$$\begin{aligned}\left(\frac{\partial w_N}{\partial y}\right)(x,y) = &\sum_{m=0}^{N-1}\sum_{n=0}^{m} \alpha_{mn}\,\frac{L_{m+1}(x)-L_{m-1}(x)}{2m+1}L_n'(y)\\ &+\sum_{m=1}^{N-1}\beta_m\frac{L_{m+1}(x)-L_{m-1}(x)}{2m+1}(2N-1)L_{N-1}(y).\end{aligned}$$

However, since $\|L_n'\|_{L^2(\Lambda)}$ is equal to $\sqrt{n(n+1)}$ from (5.3), we compute

$$\begin{aligned}\left\|\frac{\partial w_N}{\partial y}\right\|^2_{L^2(\Omega)} \leqslant\; &c\sum_{m=0}^{N-1}\frac{1}{(m+\frac12)^3}\left(\sum_{n=0}^{m}|\alpha_{mn}|\sqrt{n(n+1)}\right)^2\\ &+c'\left(N-\frac12\right)\sum_{m=1}^{N-1}\beta_m^2\frac{1}{(m+\frac12)^3}.\end{aligned}$$

Next, we apply the Cauchy–Schwarz inequality:

$$\begin{aligned}\left\|\frac{\partial w_N}{\partial y}\right\|^2_{L^2(\Omega)} \leqslant\; &c\sum_{m=0}^{N-1}\frac{1}{(m+\frac12)^3}\left(\sum_{n=0}^{m}\alpha_{mn}^2\frac{1}{n+\frac12}\right)\left(\sum_{n=0}^{m}n(n+1)\left(n+\frac12\right)\right)\\ &+c'\left(N-\frac12\right)\sum_{m=1}^{N-1}\beta_m^2\frac{1}{(m+\frac12)^3},\end{aligned}$$

Since $n \leqslant m$, this gives

$$\left\|\frac{\partial w_N}{\partial y}\right\|^2_{L^2(\Omega)} \leqslant c\sum_{m=0}^{N-1}\sum_{n=0}^{m}\alpha_{mn}^2\frac{m+\frac12}{n+\frac12} + c'\left(N-\frac12\right)\sum_{m=1}^{N-1}\beta_m^2\frac{1}{(m+\frac12)^3}.$$

Comparing this inequality with (24.33), we derive that

$$\left\| \frac{\partial w_N}{\partial y} \right\|^2_{L^2(\Omega)} \leqslant cN^2 \|q_N\|^2_{L^2(\Omega)}.$$

Finally, applying the same arguments to $\partial z_N/\partial x$, we obtain

$$\left\| \frac{\partial w_N}{\partial y} \right\|_{L^2(\Omega_d)} + \left\| \frac{\partial z_N}{\partial x} \right\|_{L^2(\Omega_d)} \leqslant cN \|q_N\|_{L^2(\Omega_d)}. \tag{24.39}$$

The inequalities (24.37), (24.38) and (24.39) give the first inequality in (24.31).

(4) In order to estimate $\|\boldsymbol{w}_N \bullet \boldsymbol{\tau}_J\|_{L^2(\Gamma_J)}$, we observe that, on the even edges for instance,

$$\boldsymbol{w}_N \bullet \boldsymbol{\tau}_J = \pm w_N(x, \pm 1) = \pm \sum_{m=0}^{N-1} \sum_{n=0}^{m} \alpha_{mn} \frac{L_{m+1}(x) - L_{m-1}(x)}{2m+1} (\pm 1)^n,$$

so that

$$\begin{aligned} \|\boldsymbol{w}_N \bullet \boldsymbol{\tau}_J\|^2_{L^2(\Gamma_J)} &\leqslant c \sum_{m=0}^{N-1} \frac{1}{(m+\frac{1}{2})^3} \left(\sum_{n=0}^{m} |\alpha_{mn}| \right)^2 \\ &\leqslant c \sum_{m=0}^{N-1} \frac{1}{(m+\frac{1}{2})^3} \left(\sum_{n=0}^{m} \alpha_{mn}^2 \frac{1}{n+\frac{1}{2}} \right) \left(\sum_{n=0}^{m} \left(n + \frac{1}{2} \right) \right). \end{aligned}$$

Since $n \leqslant m$, this leads to the desired result on the even edges and a similar argument applied to $z_N(\pm 1, y)$ also gives the result on the odd edges.

The only drawback of the polynomial $\boldsymbol{w}_N$ which has been constructed in the previous theorem is that it does not vanish on the boundary of Ω_d. So, a new polynomial has to be constructed. Here, the polynomial q_N must clearly be taken in the subspace D_N.

THEOREM 24.7. *For any function q_N in D_N, there exists a polynomial $\boldsymbol{v}_N$ in $\mathbb{P}^0_N(\Omega_d)^d$ which satisfies*

$$\operatorname{div} \boldsymbol{v}_N = q_N, \tag{24.40}$$

and

$$\|\boldsymbol{v}_N\|_{H^1(\Omega_d)^d} \leqslant cN \|q_N\|_{L^2(\Omega_d)}. \tag{24.41}$$

PROOF. We only give the proof in the case $d = 2$, since the proof for $d = 3$ is analogous and rather technical (we refer to MADAY, PATERA and RØNQUIST ([1992],

Lemma 4.3) for the ideas which allow for this extension). In the case $d = 2$, we choose

$$\boldsymbol{v}_N = \boldsymbol{w}_N + \mathbf{curl}\, \psi_N, \tag{24.42}$$

where the polynomial $\boldsymbol{w}_N$ is exhibited in Theorem 24.6 and ψ_N belongs to $\mathbb{P}_N(\Omega)$. With this choice, condition (24.40) is obviously satisfied and $\boldsymbol{v}_N$ belongs to $\mathbb{P}_N(\Omega)^2$. Sufficient conditions for this polynomial to vanish on Γ_J, $J = 1, 2, 3, 4$, is that

$$\psi_N = 0 \text{ on } \Gamma_J \quad \text{and} \quad \frac{\partial \psi_N}{\partial n_J} = \boldsymbol{w}_N \cdot \boldsymbol{\tau}_J \text{ on } \Gamma_J, \; J = 1, 2, 3, 4. \tag{24.43}$$

Then, we introduce the polynomial

$$\omega_N(\zeta) = \frac{(1-\zeta)^2}{4}(1+\zeta)\frac{L_N'''(\zeta)}{L_N'''(-1)}, \tag{24.44}$$

which vanishes in -1 and 1, while its first derivative is equal to 1 in -1 and vanishes in 1. We take $\psi_N = \psi_1 + \psi_2 + \psi_3 + \psi_4$, with

$$\begin{cases} \psi_1(x,y) = \big(z_N(-1,y) + (\frac{\partial w_N}{\partial x})(-1,-1)\omega_N(y) \\ \qquad\qquad - (\frac{\partial w_N}{\partial x})(-1,1)\omega_N(-y)\big)\omega_N(x), \\ \psi_2(x,y) = -w_N(x,-1)\omega_N(y), \\ \psi_3(x,y) = -\big(z_N(1,y) + (\frac{\partial w_N}{\partial x})(1,-1)\omega_N(y) \\ \qquad\qquad - (\frac{\partial w_N}{\partial x})(1,1)\omega_N(-y)\big)\omega_N(-x), \\ \psi_4(x,y) = w_N(x,1)\omega_N(-y). \end{cases} \tag{24.45}$$

From the compatibility conditions

$$\boldsymbol{w}_N(\boldsymbol{a}_J) = \boldsymbol{0} \quad \text{and} \quad (\operatorname{div} \boldsymbol{w}_N)(\boldsymbol{a}_J) = q_N(\boldsymbol{a}_J) = 0, \; J = 1, 2, 3, 4,$$

it is readily checked that ψ_N satisfies the conditions (24.43). Moreover, since the polynomial $(1-\zeta^2)\, L_N'''$ is equal to $4\zeta L_N'' - (N-1)(N+2)L_N'$ by (3.5), we derive from (3.8) together with the inverse inequality (5.2) and (5.3) that

$$\|\omega_N\|_{L^2(\Lambda)} \leqslant cN^{-3}. \tag{24.46}$$

We have for instance

$$\|\psi_1\|_{H^2(\Omega)} \leqslant \big\|z_N(-1,\cdot)\big\|_{H^2(\Lambda)}\|\omega_N\|_{L^2(\Lambda)} + \big\|z_N(-1,\cdot)\big\|_{H^1(\Lambda)}\|\omega_N\|_{H^1(\Lambda)}$$
$$+ \big\|z_N(-1,\cdot)\big\|_{L^2(\Lambda)}\|\omega_N\|_{H^2(\Lambda)},$$

so that applying six times the inverse inequality (5.2) leads to

$$\|\psi_1\|_{H^2(\Omega)} \leqslant cN^4 \big\|z_N(-1,\bullet)\big\|_{L^2(\Lambda)} \|\omega_N\|_{L^2(\Lambda)}.$$

As a conclusion, thanks to (24.46), we obtain

$$\|\psi_N\|_{H^2(\Omega)} \leqslant cN \sum_{J=1}^{4} \|\boldsymbol{w}_N \bullet \boldsymbol{\tau}_J\|_{L^2(\Gamma_J)}. \tag{24.47}$$

Combining this last estimate with (24.31) and (24.42) gives the desired result.

The following inf–sup condition is a straightforward consequence of Theorem 24.7: there exists a positive constant c independent of N such that

$$\inf_{q_N \in D_N} \sup_{\boldsymbol{v}_N \in \mathbb{P}_N^0(\Omega_d)^d} \frac{b_N(\boldsymbol{v}_N, q_N)}{\|\boldsymbol{v}_N\|_{H^1(\Omega_d)^d} \|q_N\|_{L^2(\Omega_d)}} \geqslant cN^{-1}. \tag{24.48}$$

REMARK 24.2. Condition (24.48) is a little disappointing in comparison with finite elements where the constants of the inf–sup condition are usually independent of the discretization parameter. However, the power of N in the right-hand side of (24.48) cannot be improved, as it can be seen from the following counter-example: taking, in the two-dimensional case for instance,

$$q_N(x,y) = x(1-x^2)L_N(y), \tag{24.49}$$

we see that q_N belongs to D_N whenever $N \geqslant 4$. Moreover, defining $\boldsymbol{w}_N = (w_N, z_N)$ as the only solution in $\mathbb{P}_N^0(\Omega)^2$ of the problem

$$\forall \boldsymbol{v}_N \in \mathbb{P}_N^0(\Omega)^2, \quad \int_\Omega (\mathbf{grad}\, \boldsymbol{w}_N)(\boldsymbol{x}) \bullet (\mathbf{grad}\, \boldsymbol{v}_N)(\boldsymbol{x})\, \mathrm{d}\boldsymbol{x} = b(\boldsymbol{v}_N, q_N),$$

we observe that

$$\sup_{\boldsymbol{v}_N \in \mathbb{P}_N^0(\Omega)^2} \frac{b(\boldsymbol{v}_N, q_N)}{|\boldsymbol{v}_N|_{H^1(\Omega)^2}} = |\boldsymbol{w}_N|_{H^1(\Omega)^2} \tag{24.50}$$

(this is the usual trick for checking the optimality of an inf–sup condition, see VANDEVEN [1989b]). It is easy to see that z_N is equal to 0 and that w_N satisfies

$$\begin{aligned} |w_N|_{H^1(\Omega)^2} &= -\int_\Omega \left(\frac{\partial w_N}{\partial x}\right)(x,y)x(1-x^2)L_N(y)\,\mathrm{d}x\,\mathrm{d}y \\ &= \int_\Omega w_N(x,y)(1-3x^2)L_N(y)\,\mathrm{d}x\,\mathrm{d}y \end{aligned}$$

$$= -\int_{\Omega} \left(\frac{\partial w_N}{\partial y}\right)(x,y)\big(1-3x^2\big)\frac{L_{N+1}(y)-L_{N-1}(y)}{2N+1}\,\mathrm{d}x\,\mathrm{d}y$$
$$\leqslant c|w_N|_{H^1(\Omega)}\big(N+\tfrac{1}{2}\big)^{-3/2}.$$

Hence, $|w_N|_{H^1(\Omega)}$ is smaller than cN^{-1} times the norm $\|q_N\|_{L^2(\Omega)}$, so that

$$\sup_{\boldsymbol{v}_N \in \mathbb{P}_N^0(\Omega)^2} \frac{b(\boldsymbol{v}_N, q_N)}{|\boldsymbol{v}_N|_{H^1(\Omega)^2}} \leqslant cN^{-1}. \tag{24.51}$$

Since the same inequality holds with $b(\cdot,\cdot)$ replaced by $b_N(\cdot,\cdot)$, this proves that inequality (24.48) cannot be improved.

In view of (24.48), it seems that taking M_N equal to D_N would be an appropriate choice for the pressure space. But, to estimate the error on the pressure, we intend to apply the inequality (23.56) which involves the term $\inf_{q_N \in M_N} \|p - q_N\|_{L^2(\Omega_d)}$ and the space D_N has no good approximation properties for all pressures p: for instance, as already noted, all the polynomials in D_N vanish in the corners of the square or on the edges of the cube, and there is no reason for the pressure to be small in the corners or on the edges. Also, we note that the polynomials L_N' and $\zeta\, L_N'$ have the expansion (see (3.12) and (3.15))

$$\begin{aligned} L_N' &= (2N-1)L_{N-1} + (2N-5)L_{N-3} + \cdots, \\ \zeta L_N' &= NL_N + (2N-3)L_{N-2} + (2N-7)L_{N-4} + \cdots, \end{aligned} \tag{24.52}$$

so that all polynomials of low degree do not belong to D_N. This leads to choose M_N not too far from D_N in order that the same inf–sup condition holds (this implies (24.17)), but such that all polynomials with degree $\leqslant [\lambda N]$ belong to M_N, where λ is between 0 and 1.

We denote by Π_{D_N} the orthogonal projection operator from $\mathbb{P}_N(\Omega_d)$ onto D_N for the discrete scalar product $(\cdot,\cdot)_N$.

THEOREM 24.8. *Let λ be a fixed real number, $0 < \lambda < 1$, and let M_N be a subspace of $\mathbb{P}_N(\Omega_d)$ which contains $\mathbb{P}_{[\lambda N]}(\Omega_d) \cap L_0^2(\Omega)$ and such that the following property holds:*

$$\forall q_N \in M_N, \quad \|q_N\|_{L^2(\Omega_d)} \leqslant c\,\|\Pi_{D_N} q_N\|_{L^2(\Omega_d)}. \tag{24.53}$$

If the assumptions of Theorem 24.4 are satisfied, the following error estimate holds for problem (24.5):

$$\begin{aligned} \|p - p_N\|_{L^2(\Omega_d)} \leqslant c\big(&N^{2-s}\big(\|\boldsymbol{u}\|_{H^s(\Omega_d)^d} + \|p\|_{H^{s-1}(\Omega_d)}\big) \\ &+ N^{1-\sigma}\|\boldsymbol{f}\|_{H^\sigma(\Omega_d)^d}\big). \end{aligned} \tag{24.54}$$

PROOF. We first observe from (24.48) that, for any q_N in M_N,

$$\sup_{v_N\in\mathbb{P}_N^0(\Omega_d)^2}\frac{b_N(v_N,q_N)}{|v_N|_{H^1(\Omega_d)^2}\|q_N\|_{L^2(\Omega_d)}}=\sup_{v_N\in\mathbb{P}_N^0(\Omega_d)^2}\frac{b_N(v_N,\Pi_{D_N}q_N)}{|v_N|_{H^1(\Omega_d)^2}\|q_N\|_{L^2(\Omega_d)}}$$
$$\geqslant cN^{-1}\frac{\|\Pi_{D_N}q_N\|_{L^2(\Omega_d)}}{\|q_N\|_{L^2(\Omega_d)}},$$

so that, by hypothesis (24.53),

$$\sup_{v_N\in\mathbb{P}_N^0(\Omega_d)^2}\frac{b_N(v_N,q_N)}{|v_N|_{H^1(\Omega_d)^2}}\geqslant cN^{-1}. \tag{24.55}$$

Of course, we now apply the abstract estimate (23.56) for w_N in $K_N\cap\mathbb{P}_{N-1}(\Omega_d)^d$ and q_N equal to $\Pi_{[\lambda N]}p$:

$$\|p-p_N\|_{L^2(\Omega_d)}\leqslant cN\Bigg(\|u-w_N\|_{H^1(\Omega_d)^d}+\|p-\Pi_{[\lambda N]}p\|_{L^2(\Omega_d)}$$
$$+\sup_{z_N\in\mathbb{P}_N(\Omega_d)^d}\frac{\langle f,z_N\rangle-(f,z_N)_N}{\|z_N\|_{L^2(\Omega_d)^d}}\Bigg).$$

The desired estimate follows from Theorems 7.1 and 24.3, together with (15.18) combined with Theorems 7.1 and 14.2.

It remains to exhibit examples of spaces M_N such that the hypotheses of Theorem 24.8 are satisfied. As proposed firstly by BERNARDI, MADAY and MÉTIVET [1987], we choose to replace, in the orthogonality condition which is satisfied by the subspace D_N, the orthogonality to the polynomials L_N' and $\zeta L_N'$ by the orthogonality to the leading modes. More precisely, we set

$$A_{N-1}=L_N'-\pi_{[\lambda N]}L_N' \quad\text{and}\quad A_N=\pi_{N-1}(\zeta L_N')-\pi_{[\lambda N]}(\zeta L_N'). \tag{24.56}$$

THEOREM 24.9. *In the case of dimension $d=2$, the space M_N which is orthogonal in $\mathbb{P}_N(\Omega)$ to the polynomials in*

$$\{L_0,L_N\}^{\otimes 2}\quad\textit{and}\quad\{A_{N-1},A_N\}^{\otimes 2},$$

contains $\mathbb{P}_{[\lambda N]}(\Omega)\cap L_0^2(\Omega)$ and satisfies (24.53). *In the case of dimension $d=3$, the space M_N which is orthogonal in $\mathbb{P}_N(\Omega_3)$ to the polynomials in*

$$\{L_0,L_N\}^{\otimes 3}\quad\textit{and}\quad\{A_{N-1},A_N\}^{\otimes 2}\otimes\mathbb{P}_N(\Lambda),$$
$$\{A_{N-1},A_N\}\otimes\mathbb{P}_N(\Lambda)\otimes\{A_{N-1},A_N\},\quad\mathbb{P}_N(\Lambda)\otimes\{A_{N-1},A_N\}^{\otimes 2}, \tag{24.57}$$

contains $\mathbb{P}_{[\lambda N]}(\Omega_3)\cap L_0^2(\Omega_3)$ and satisfies (24.53).

PROOF. From the definition, it is obvious that the space M_N contains $\mathbb{P}_{[\lambda N]}(\Omega_d)$. We only give the proof of inequality (24.53) in the case $d = 2$, since the extension to dimension $d = 3$ is very similar but technical. For a while, we denote by $\mathcal{Q}$ the set of polynomials

$$L_N'(x)L_N'(y),\ xL_N'(x)L_N'(y),\ L_N'(x)yL_N'(y),\ xL_N'(x)yL_N'(y). \tag{24.58}$$

With each polynomial q in $\mathcal{Q}$, we associate its projection $\overline{q}$ onto the orthogonal of the polynomials in $\{L_0, L_N\}^{\otimes 2}$, and its corresponding polynomial $\tilde{q}$ in $\{A_{N-1}, A_N\}^{\otimes 2}$. This notation allows for writing the formula

$$\forall q_N \in M_N, \quad q_N = \pi_{D_N} q_N - \sum_{q \in \mathcal{Q}} \frac{\int_\Omega (\Pi_{D_N} q_N)(\boldsymbol{x})\tilde{q}(\boldsymbol{x})\,\mathrm{d}\boldsymbol{x}}{\int_\Omega \overline{q}(\boldsymbol{x})\tilde{q}(\boldsymbol{x})\,\mathrm{d}\boldsymbol{x}}\,\overline{q},$$

from which we deduce, for all q_N in M_N,

$$\|q_N\|_{L^2(\Omega_d)} \leqslant \|\Pi_{D_N} q_N\|_{L^2(\Omega)} \left(1 + \sum_{q \in \mathcal{Q}} \frac{\|q\|_{L^2(\Omega)}}{\|\tilde{q}\|_{L^2(\Omega)}}\right).$$

Thus, the estimate (24.53) follows from the fact that

$$\|\zeta L_N'\|_{L^2(\Lambda)} \leqslant \|L_N'\|_{L^2(\Lambda)} = \sqrt{N(N+1)},$$

and from expansion (24.52), from which it is readily checked that

$$\|A_{N-1}\|_{L^2(\Lambda)} \geqslant c(1-\lambda)N \quad \text{and} \quad \|A_N\|_{L^2(\Lambda)} \geqslant c(1-\lambda)N.$$

As a final step for this method, we observe that, like in (15.27), the exactness of the quadrature formula on $\mathbb{P}_{2N-1}(\Lambda)$ yields that, for any $\boldsymbol{u}_N$ in $\mathbb{P}_N(\Omega_d)^d$, $\boldsymbol{v}_N$ in $\mathbb{P}_N^0(\Omega_d)^d$ and q_N in $\mathbb{P}_N(\Omega_d)$,

$$a_N(\boldsymbol{u}_N, \boldsymbol{v}_N) = -\nu(\Delta \boldsymbol{u}_N, \boldsymbol{v}_N)_N \quad \text{and} \quad b_N(\boldsymbol{v}_N, q_N) = (\boldsymbol{v}_N, \mathbf{grad}\, q_N)_N.$$

So, we let $\boldsymbol{v}_N$ run through a basis of $\mathbb{P}_N^0(\Omega_d)^d$ made of Lagrange polynomials and q_N run through a basis of $\mathbb{P}_N(\Omega_d)$ also made of Lagrange polynomials (recall that, by (24.17), the second equation in (24.5) is satisfied for all q_N in M_N if and only if it is satisfied by all q_N in $\mathbb{P}_N(\Omega_d)$). Thus, we obtain the following system of collocation equations which is equivalent to problem (24.5): find a pair $(\boldsymbol{u}_N, p_N)$ in $\mathbb{P}_N(\Omega_d)^d \times M_N$ such that

$$\begin{cases} -\nu(\Delta \boldsymbol{u}_N)(\boldsymbol{x}) + (\mathbf{grad}\, p_N)(\boldsymbol{x}) = \boldsymbol{f}(\boldsymbol{x}), & \boldsymbol{x} \in \Xi_N \cap \Omega_d, \\ (\operatorname{div} \boldsymbol{u}_N)(\boldsymbol{x}) = 0, & \boldsymbol{x} \in \Xi_N, \\ \boldsymbol{u}_N(\boldsymbol{x}) = \boldsymbol{0}, & \boldsymbol{x} \in \Xi_N \cap \partial\Omega_d. \end{cases} \tag{24.59}$$

REMARK 24.3. It can be noted that the system (24.59) is a rectangular system since there are eight more equations than unknowns in dimension $d = 2$ and $12N + 4$ more equations in dimension $d = 3$: these "spurious" equations are of course linked to the spurious modes on the pressure. Thus, it is readily checked that the system (24.59) can equivalently be written

$$\begin{cases} -\nu(\Delta \boldsymbol{u}_N)(\boldsymbol{x}) + (\mathbf{grad}\, p_N)(\boldsymbol{x}) = \boldsymbol{f}(\boldsymbol{x}), & \boldsymbol{x} \in \Xi_N \cap \Omega_d, \\ (\operatorname{div} \boldsymbol{u}_N)(\boldsymbol{x}) = 0, & \boldsymbol{x} \in \widetilde{\Xi}_N, \\ \boldsymbol{u}_N(\boldsymbol{x}) = \mathbf{0}, & \boldsymbol{x} \in \Xi_N \cap \partial\Omega_d, \end{cases} \tag{24.60}$$

where the grid $\widetilde{\Xi}_N$ is

(i) in the case $d = 2$, the grid Ξ_N minus the four corners $\boldsymbol{a}_J$, $J = 1, 2, 3, 4$, and four other points $\boldsymbol{b}_J$, $J = 1, 2, 3, 4$, such that the following condition holds: the determinant made by the $q(\boldsymbol{b}_J)$, where q runs through the set $\{L_0, L_N\}^{\otimes 2}$, is not equal to 0;

(ii) in the case $d = 3$, the grid Ξ_N minus the eight corners, the $12(N-1)$ points on the interior of the 12 edges and eight other points $\boldsymbol{b}_j$, $1 \leqslant j \leqslant 8$, such that the following condition holds: the determinant made by the $q(\boldsymbol{b}_j)$, where q runs through the set $\{L_0, L_N\}^{\otimes 3}$, is not equal to 0.

The collocation form (24.59) suggests to apply the same method to the problem with inhomogeneous boundary conditions on the velocity:

$$\begin{cases} -\nu\Delta \boldsymbol{u} + \mathbf{grad}\, p = \boldsymbol{f} & \text{in } \Omega_d, \\ \operatorname{div} \boldsymbol{u} = 0 & \text{in } \Omega_d, \\ \boldsymbol{u} = \boldsymbol{g}_j & \text{on } \Gamma_j,\ 1 \leqslant j \leqslant 2d. \end{cases} \tag{24.61}$$

Here, the function $\boldsymbol{f}$ is supposed to be continuous on Ω_d, while the functions $\boldsymbol{g}_j$, $1 \leqslant j \leqslant 2d$, are supposed to be in $H^{1/2}(\Gamma_j)^d$, Hölder-continuous on $\overline{\Gamma}_j$, and to satisfy the compatibility conditions

$$\boldsymbol{g}_j = \boldsymbol{g}_k \quad \text{on } \overline{\Gamma}_j \cap \overline{\Gamma}_k,\ 1 \leqslant j < k \leqslant 2d, \tag{24.62}$$

and also

$$\sum_{j=1}^{2d} \int_{\Gamma_j} \boldsymbol{g}_j(\tau) \cdot \boldsymbol{n}_j \,\mathrm{d}\tau = 0. \tag{24.63}$$

Thanks to (24.62), applying Theorem 1.14, we know that there exists a function $\boldsymbol{u}_\mathrm{b}$ in $H^1(\Omega_d)^d$ such that the trace of $\boldsymbol{u}_\mathrm{b}$ on each Γ_j, $1 \leqslant j \leqslant 2d$, coincides with $\boldsymbol{g}_j$.

The first idea for defining the discrete problem consists in using the collocation form (24.59), replacing the last line by

$$\boldsymbol{u}_N(\boldsymbol{x}) = \boldsymbol{g}_j(\boldsymbol{x}), \quad \boldsymbol{x} \in \Xi_N \cap \overline{\Gamma}_j,\ 1 \leqslant j \leqslant 2d. \tag{24.64}$$

However, an important property of the discretization is that it preserves some conservation property: more precisely, the discrete velocity is supposed to satisfy

$$\int_{\Omega_d} (\operatorname{div} \boldsymbol{u}_N)(\boldsymbol{x})\, \mathrm{d}\boldsymbol{x} = 0,$$

or equivalently, by the Stokes formula,

$$\sum_{j=1}^{2d} \int_{\Gamma_j} \boldsymbol{u}_N(\tau) \bullet \boldsymbol{n}_j \, \mathrm{d}\tau = 0.$$

But, if condition (24.64) holds, $\boldsymbol{u}_N$ is equal to $i_N^j \boldsymbol{g}_j$ on Γ_j, and the polynomials $i_N^j \boldsymbol{g}_j$ do not satisfy this compatibility condition, even if the functions $\boldsymbol{g}_j$ do. So, there is a little problem. The solution is similar to (17.5): we introduce the constant

$$\mu_N = \frac{1}{2^d d} \sum_{j=1}^{2d} \int_{\Gamma_j} (i_N^j \boldsymbol{g}_j)(\tau) \bullet \boldsymbol{n}_j \, \mathrm{d}\tau, \tag{24.65}$$

and we replace condition (24.64) by

$$\boldsymbol{u}_N(\boldsymbol{x}) = \boldsymbol{g}_j(\boldsymbol{x}) - \mu_N \boldsymbol{\chi}, \quad \boldsymbol{x} \in \Xi_N \cap \overline{\Gamma}_j, \ 1 \leqslant j \leqslant 2d, \tag{24.66}$$

where $\boldsymbol{\chi}$ denotes the function with values in $\mathbb{R}^d$ such that all its components are constant equal to 1. Also, we observe that

$$|\mu_N| \leqslant c \sum_{j=1}^{2d} \|\boldsymbol{g}_j - i_N^j \boldsymbol{g}_j\|_{L^2(\Gamma_j)^d} \leqslant c \|\boldsymbol{u} - \mathcal{I}_N \boldsymbol{u}\|_{H^1(\Omega_d)^d}. \tag{24.67}$$

Let M_N be a subspace of $\mathbb{P}_N(\Omega_d)$ which satisfies (24.17). From the previous considerations, the discrete problem is written: *find a pair* $(\boldsymbol{u}_N, p_N)$ *in* $\mathbb{P}_N(\Omega_d)^d \times M_N$, *with* $\boldsymbol{u}_N - \mathcal{I}_N \boldsymbol{u}_\mathrm{b} + \mu_N \boldsymbol{\chi}$ *in* $\mathbb{P}_N^0(\Omega_d)^d$, *such that*

$$\begin{aligned} &\forall \boldsymbol{v}_N \in \mathbb{P}_N^0(\Omega_d)^d, \quad a_N(\boldsymbol{u}_N, \boldsymbol{v}_N) + b_N(\boldsymbol{v}_N, p_N) = (\boldsymbol{f}, \boldsymbol{v}_N)_N, \\ &\forall q_N \in M_N, \qquad b_N(\boldsymbol{u}_N, q_N) = 0. \end{aligned} \tag{24.68}$$

Of course, the first line in this problem is equivalent to $d(N-1)^d$ collocation equations:

$$-\nu(\Delta \boldsymbol{u}_N)(\boldsymbol{x}) + (\mathbf{grad}\, p_N)(\boldsymbol{x}) = \boldsymbol{f}(\boldsymbol{x}), \quad \boldsymbol{x} \in \Xi_N \cap \Omega_d, \tag{24.69}$$

but the discrete solution $\boldsymbol{u}_N$ is enforced to satisfy the boundary conditions (24.66) (and not (24.64)) and the second equation in (24.68) is not equivalent to a set of collocation equations for $\operatorname{div} \boldsymbol{u}_N$ on Ξ_N.

A large part of the work for the numerical analysis has already been done in the beginning of this section. Indeed, let us set: $\boldsymbol{u}_N^* = \boldsymbol{u}_N - \mathcal{I}_N \boldsymbol{u}_\mathrm{b} + \mu_N \boldsymbol{\chi}$. The pair $(\boldsymbol{u}_N^*, p_N)$ belongs to $\mathbb{P}_N^0(\Omega_d)^d \times M_N$ and satisfies

$$\begin{aligned}
&\forall \boldsymbol{v}_N \in \mathbb{P}_N^0(\Omega_d)^d,\\
&\qquad a_N(\boldsymbol{u}_N^*, \boldsymbol{v}_N) + b_N(\boldsymbol{v}_N, p_N) = (\boldsymbol{f}, \boldsymbol{v}_N)_N - a_N(\mathcal{I}_N \boldsymbol{u}_\mathrm{b}, \boldsymbol{v}_N),\\
&\forall q_N \in M_N, \quad b_N(\boldsymbol{u}_N^*, q_N) = -b_N(\mathcal{I}_N \boldsymbol{u}_\mathrm{b}, q_N).
\end{aligned} \tag{24.70}$$

Since this problem is exactly of type (23.50), Theorem 23.8 leads to the following result.

THEOREM 24.10. *If hypothesis* (24.17) *holds, for any continuous function* $\boldsymbol{f}$ *on* Ω_d *and for any Hölder-continuous functions* $\boldsymbol{g}_j$ *in* $H^{1/2}(\Gamma_j)^d$, $1 \leqslant j \leqslant 2d$, *such that* (24.62) *holds, problem* (24.68) *has a unique solution* $(\boldsymbol{u}_N, p_N)$ *in* $\mathbb{P}_N(\Omega_d)^d \times M_N$.

We are now in a position to derive the error estimates on the velocity and the pressure.

THEOREM 24.11. *Let* λ *be a fixed real number,* $0 < \lambda < 1$, *and let* M_N *be a subspace of* $\mathbb{P}_N(\Omega_d)$ *which contains* $\mathbb{P}_{[\lambda N]}(\Omega_d) \cap L_0^2(\Omega_d)$ *and such that property* (24.53) *holds. The function* $\boldsymbol{f}$ *is supposed to be in a space* $H^\sigma(\Omega_d)^d$, $\sigma > d/2$. *Let us assume that the solution* $(\boldsymbol{u}, p)$ *of problem* (24.61) *belongs to* $H^s(\Omega_d)^d \times H^{s-1}(\Omega_d)$ *for a real number* $s \geqslant (d+1)/2$. *Then, the following error estimate holds for problem* (24.68):

$$\begin{aligned}
&\|\boldsymbol{u} - \boldsymbol{u}_N\|_{H^1(\Omega_d)^d} + N^{-1}\|p - p_N\|_{L^2(\Omega_d)}\\
&\qquad \leqslant c\big(N^{2-s}\big(\|\boldsymbol{u}\|_{H^s(\Omega_d)^d} + \|p\|_{H^{s-1}(\Omega_d)}\big) + N^{-\sigma}\|\boldsymbol{f}\|_{H^\sigma(\Omega_d)^d}\big).
\end{aligned} \tag{24.71}$$

PROOF. We observe that the pair $(\boldsymbol{0}, p)$ is a solution in $H_0^1(\Omega_d)^d \times L_0^2(\Omega_d)$ of the Stokes problem (23.23) with $\langle \boldsymbol{f}, \boldsymbol{v}\rangle$ replaced by $\langle \boldsymbol{f}, \boldsymbol{v}\rangle - a(\boldsymbol{u}, \boldsymbol{v})$. On the other hand, the pair $(\boldsymbol{u}_N - \mathcal{I}_N \boldsymbol{u} + \mu_N \boldsymbol{\chi}, p_N)$ is a solution in $\mathbb{P}_N^0(\Omega_d)^d \times M_N$ of the discrete problem (24.5) with $(\boldsymbol{f}, \boldsymbol{v})_N$ replaced by $(\boldsymbol{f}, \boldsymbol{v})_N - a_N(\mathcal{I}_N \boldsymbol{u}, \boldsymbol{v})$ and a right-hand member in the second equation equal to $-b_N(\mathcal{I}_N \boldsymbol{u}, q_N)$. Then, applying the abstract estimate (23.56) (with $v_\delta = 0$) leads to the following inequality

$$\begin{aligned}
&\|\boldsymbol{u}_N - \mathcal{I}_N \boldsymbol{u} + \mu_N \boldsymbol{\chi}\|_{H^1(\Omega_d)^d} + N^{-1}\|p - p_N\|_{L^2(\Omega_d)}\\
&\quad \leqslant c\bigg(\inf_{\boldsymbol{w}_N \in K_N(\operatorname{div}(\mathcal{I}_N \boldsymbol{u}))} \|\boldsymbol{w}_N\|_{H^1(\Omega_d)^d} + \|p - \Pi_{N-1} p\|_{L^2(\Omega_d)}\\
&\qquad + \sup_{\boldsymbol{z}_N \in \mathbb{P}_N(\Omega_d)^d} \frac{\langle \boldsymbol{f}, \boldsymbol{z}_N\rangle - (\boldsymbol{f}, \boldsymbol{z}_N)_N}{\|\boldsymbol{z}_N\|_{H^1(\Omega_d)^d}}\\
&\qquad + \sup_{\boldsymbol{z}_N \in \mathbb{P}_N(\Omega_d)^d} \frac{a(\boldsymbol{u}, \boldsymbol{z}_N) - a_N(\mathcal{I}_N \boldsymbol{u}, \boldsymbol{z}_N)}{\|\boldsymbol{z}_N\|_{H^1(\Omega_d)^d}}\bigg),
\end{aligned}$$

where $K_N(\mathrm{div}(\mathcal{I}_N \boldsymbol{u}))$ stands for the subspace of all polynomials in $\mathbb{P}_N^0(\Omega_d)^d$ such that

$$\forall q_N \in M_N, \quad b_N(\boldsymbol{v}_N, q_N) = -b_N(\mathcal{I}_N \boldsymbol{u}, q_N).$$

Using now (23.60) (also with $v_\delta = 0$) gives the inequality

$$\begin{aligned}
&\inf_{\boldsymbol{w}_N \in K_N(\mathrm{div}(\mathcal{I}_N \boldsymbol{u}))} \|\boldsymbol{w}_N\|_{H^1(\Omega_d)^d} \\
&\quad \leqslant cN \sup_{q_N \in M_N} \frac{b_N(\mathcal{I}_N \boldsymbol{u}, q_N)}{\|q_N\|_{L^2(\Omega_d)^d}} \\
&\quad \leqslant cN\big(\|\boldsymbol{u} - \Pi_{N-1}^{1,0}\boldsymbol{u}\|_{H^1(\Omega_d)^d} + \|\boldsymbol{u} - \mathcal{I}_N \boldsymbol{u}\|_{H^1(\Omega_d)^d}\big).
\end{aligned}$$

Combining the two previous formulas and using a triangular inequality lead to

$$\begin{aligned}
&\|\boldsymbol{u} - \boldsymbol{u}_N\|_{H^1(\Omega_d)^d} + N^{-1}\|p - p_N\|_{L^2(\Omega_d)} \\
&\quad \leqslant c\big(N\|\boldsymbol{u} - \Pi_{N-1}^{1,0}\boldsymbol{u}\|_{H^1(\Omega^d)^d} + N\|\boldsymbol{u} - \mathcal{I}_N \boldsymbol{u}\|_{H^1(\Omega_d)^d} + |\mu_N| \\
&\qquad + \|p - \Pi_{N-1} p\|_{L^2(\Omega_d)} + \|\boldsymbol{f} - \Pi_{N-1}\boldsymbol{f}\|_{L^2(\Omega^d)^d} + \|\boldsymbol{f} - \mathcal{I}_N \boldsymbol{f}\|_{L^2(\Omega_d)^d}\big).
\end{aligned}$$

Hence, the desired result follows from Theorems 7.1, 7.2 and 14.2 together with (24.67).

REMARK 24.4. In the case $d = 2$ of the square, a slightly different estimate is proven by BERNARDI, DAUGE and MADAY [1992b], relying on the construction of a lifting operator with values in divergence-free polynomials. If the assumptions of Theorem 24.11 are satisfied, if moreover the solution $(\boldsymbol{u}, p)$ of problem (24.61) belongs to $H^s(\Omega)^2 \times H^{s-1}(\Omega)$ for a real number $s \geqslant 3$ and the boundary data $\boldsymbol{g}_J$, $J = 1, 2, 3, 4$, to $H^{\tau_J}(\Gamma_J)^2$ for real numbers $\tau_J > \frac{3}{2}$, the following estimate holds

$$\begin{aligned}
&\|\boldsymbol{u} - \boldsymbol{u}_N\|_{H^1(\Omega)^2} + N^{-1}\|p - p_N\|_{L^2(\Omega)} \\
&\quad \leqslant c\Bigg(N^{1-s}\big(\|\boldsymbol{u}\|_{H^s(\Omega)^2} + \|p\|_{H^{s-1}(\Omega)}\big) \\
&\qquad + N^{-\sigma}\|\boldsymbol{f}\|_{H^\sigma(\Omega)^2} + \sum_{J=1}^{4} N^{3/2-\tau_J}\|\boldsymbol{g}_J\|_{H^{\tau_J}(\Gamma_J)^2}\Bigg).
\end{aligned} \tag{24.72}$$

For smooth boundary data, this estimate is better than (24.71), but still nonoptimal. We refer to BERNARDI, DAUGE and MADAY [1992b] for a modified treatment of the boundary conditions which leads to optimal estimates (at least for the velocity) in the case of the square.

There also, a duality argument leads to the error estimate on the velocity in $L^2(\Omega_d)^d$.

THEOREM 24.12. *If the assumptions of Theorem* 24.11 *are satisfied, the following error estimate holds for problem* (24.68):

$$\|\boldsymbol{u}-\boldsymbol{u}_N\|_{L^2(\Omega_d)^d} \leqslant c\Bigg(N^{1-s}\big(\|\boldsymbol{u}\|_{H^s(\Omega_d)^d}+\|p\|_{H^{s-1}(\Omega_d)}\big) + N^{-\sigma}\|\boldsymbol{f}\|_{H^\sigma(\Omega_d)^d}+\sum_{j=1}^{2d}N^{-\tau_j}\|\boldsymbol{g}_j\|_{H^{\tau_j}(\Gamma_j)^d}\Bigg). \tag{24.73}$$

PROOF. It is very similar to that of Theorem 24.5. Indeed, we again have the identity

$$\|\boldsymbol{u}-\boldsymbol{u}_N\|_{L^2(\Omega_d)^d} = \sup_{\boldsymbol{g}\in L^2(\Omega_d)^d}\frac{\int_{\Omega_d}(\boldsymbol{u}-\boldsymbol{u}_N)(\boldsymbol{x})\cdot\boldsymbol{g}(\boldsymbol{x})\,\mathrm{d}\boldsymbol{x}}{\|\boldsymbol{g}\|_{L^2(\Omega_d)^d}}, \tag{24.74}$$

and, for any $\boldsymbol{g}$ in $L^2(\Omega_d)^d$, we consider the Stokes problem (24.28) and we recall that its solution $(\boldsymbol{w},r)$ belongs to $H^2(\Omega_d)^d\times H^1(\Omega_d)$ and satisfies (24.29). Next, we have to compute, for any pair $(\boldsymbol{w}_{N-1},r_{N-1})$ in $(K_N\cap\mathbb{P}_{N-1}(\Omega_d)^d)\times\mathbb{P}_{N-1}(\Omega_d)$,

$$\begin{aligned}
&\int_{\Omega_d}(\boldsymbol{u}-\boldsymbol{u}_N)(\boldsymbol{x})\cdot\boldsymbol{g}(\boldsymbol{x})\,\mathrm{d}\boldsymbol{x}\\
&\quad= a(\boldsymbol{u}-\boldsymbol{u}_N,\boldsymbol{w})+b(\boldsymbol{u}-\boldsymbol{u}_N,r)\\
&\qquad-\sum_{j=1}^{2d}\int_{\Gamma_j}(\boldsymbol{g}_j-i_N^j\boldsymbol{g}_j+\mu_N\boldsymbol{\chi})(\boldsymbol{\tau})\cdot\left(\frac{\partial\boldsymbol{w}}{\partial n_j}-r\boldsymbol{n}_j\right)(\boldsymbol{\tau})\,\mathrm{d}\boldsymbol{\tau}\\
&\quad= a(\boldsymbol{u}-\boldsymbol{u}_N,\boldsymbol{w}-\boldsymbol{w}_{N-1})+b(\boldsymbol{u}-\boldsymbol{u}_N,r-r_{N-1})\\
&\qquad+(\boldsymbol{f}-\mathcal{I}_N\boldsymbol{f},w_{N-1})_N\\
&\qquad-\sum_{j=1}^{2d}\int_{\Gamma_j}(\boldsymbol{g}_j-i_N^j\boldsymbol{g}_j+\mu_N\boldsymbol{\chi})(\boldsymbol{\tau})\cdot\left(\frac{\partial\boldsymbol{w}}{\partial n_j}-r\boldsymbol{n}_j\right)(\boldsymbol{\tau})\,\mathrm{d}\boldsymbol{\tau},
\end{aligned}$$

and the result follows from Theorems 7.1, 13.4, 14.2 and 24.11, (24.23) and (24.67), combined with (24.74) and (24.29).

Let us now consider the weighted discretization of problem (24.1). As far as we know, only the Chebyshev type discretization of this problem is used for practical computation, we present the discretization for all values of α since there is no additional complexity. As previously, to focus on the main difficulties, we study the case of homogeneous boundary conditions. We recall that problem (24.1) admits the variational formulation, for any data $\boldsymbol{f}$ in $H_\alpha^{-1}(\Omega_d)^d$: *find a pair* $(\boldsymbol{u},p)$ *in* $H^1_{\alpha,0}(\Omega_d)^d\times L^2_{\alpha 1}(\Omega_d)$ *such that*

$$\begin{aligned}
&\forall\boldsymbol{v}\in H^1_{\alpha,0}(\Omega_d)^d,\quad a_\alpha(\boldsymbol{u},\boldsymbol{v})+b_{\alpha 1}(\boldsymbol{v},p)=\langle\boldsymbol{f},\boldsymbol{v}\rangle_\alpha,\\
&\forall q\in L^2_{\alpha 2}(\Omega_d),\qquad b_{\alpha 2}(\boldsymbol{u},q)=0,
\end{aligned}\tag{24.75}$$

where the forms $a_\alpha(\cdot,\cdot)$ and $b_{i\alpha}(\cdot,\cdot)$ are defined in (23.39) and (23.40) respectively.
Like in Section 22, we set

$$(u,v)_{\alpha,N} = \sum_{i_1=0}^{N} \cdots \sum_{i_d=0}^{N} u(\xi_{i_1}^\alpha,\ldots,\xi_{i_d}^\alpha)\,v(\xi_{i_1}^\alpha,\ldots,\xi_{i_d}^\alpha)\,\rho_{i_1}^\alpha \cdots \rho_{i_d}^\alpha, \tag{24.76}$$

where the ξ_i^α, $0 \leqslant i \leqslant N$, are the zeros of $(1-\zeta^2)J_N^{\alpha\prime}$ and the ρ_i^α, $0 \leqslant i \leqslant N$, are the weights associated with these nodes in formula (19.26). Next, we set, for any polynomials $\boldsymbol{u}_N$ in $\mathbb{P}_N(\Omega_d)^d$, $\boldsymbol{v}_N$ in $\mathbb{P}_N^0(\Omega_d)^d$ and q_N in $\mathbb{P}_N(\Omega)$,

$$a_{\alpha,N}(\boldsymbol{u}_N,\boldsymbol{v}_N) = \nu\big(\mathbf{grad}\,\boldsymbol{u}_N, \varpi_{-\alpha}\,\mathbf{grad}(\boldsymbol{v}_N\,\varpi_\alpha)\big)_{\alpha,N}, \tag{24.77}$$

$$b_{\alpha 1,N}(\boldsymbol{v}_N,q_N) = -\big(\varpi_{-\alpha}\,\mathrm{div}(\boldsymbol{v}_N\,\varpi_\alpha), q_N\big)_{\alpha,N},$$
$$b_{\alpha 2,N}(\boldsymbol{v}_N,q_N) = -(\mathrm{div}\,\boldsymbol{v}_N, q_N)_{\alpha,N}. \tag{24.78}$$

As previously, we fix a subspace M_N^α of $\mathbb{P}_N(\Omega_d) \cap L^2_{\alpha 1}(\Omega_d)$. Then, the discrete problem reads: *find a pair* $(\boldsymbol{u}_N,p_N)$ *in* $\mathbb{P}_N^0(\Omega_d)^d \times M_N^\alpha$ *such that*

$$\begin{aligned} &\forall \boldsymbol{v}_N \in \mathbb{P}_N^0(\Omega_d)^d, \quad a_{\alpha,N}(\boldsymbol{u}_N,\boldsymbol{v}_N) + b_{\alpha 1,N}(\boldsymbol{v}_N,p_N) = (\boldsymbol{f},\boldsymbol{v}_N)_{\alpha,N},\\ &\forall q_N \in \mathbb{P}_N(\Omega_d), \quad b_{\alpha 2,N}(\boldsymbol{u}_N,q_N) = 0. \end{aligned} \tag{24.79}$$

This problem is exactly of type (23.50), which suggests the successive steps for its numerical analysis with an appropriate choice of M_N^α.
Let us first define the kernels, for $i = 1,2$,

$$K_{\alpha i,N} = \big\{\boldsymbol{v}_N \in \mathbb{P}_N^0(\Omega_d)^d\colon\ \forall q_N \in \mathbb{P}_N(\Omega_d), b_{\alpha i,N}(\boldsymbol{v}_N,q_N) = 0\big\}. \tag{24.80}$$

It is readily checked that

$$\begin{aligned} K_{\alpha 1,N} &= \big\{\boldsymbol{v}_N \in \mathbb{P}_N^0(\Omega_d)^d\colon\ \mathrm{div}(\boldsymbol{v}_N\varpi_\alpha) = 0 \text{ in } \Omega_d\big\},\\ K_{\alpha 2,N} &= \big\{\boldsymbol{v}_N \in \mathbb{P}_N^0(\Omega_d)^d\colon\ \mathrm{div}\,\boldsymbol{v}_N = 0 \text{ in } \Omega_d\big\}. \end{aligned} \tag{24.81}$$

Consequently, the velocity $\boldsymbol{u}_N$ of any solution $(\boldsymbol{u}_N,p_N)$ of problem (24.79) is exactly divergence-free.

DEFINITION. For $i = 1,2$, is called a spurious mode, any polynomial q_N in $\mathbb{P}_N(\Omega_d)$ such that

$$\forall \boldsymbol{v}_N \in \mathbb{P}_N^0(\Omega_d)^d, \quad b_{\alpha i,N}(\boldsymbol{v}_N,q_N) = 0. \tag{24.82}$$

The subspace of $\mathbb{P}_N(\Omega_d)$ made of these spurious modes is denoted by $Z_{\alpha i,N}$. In fact, $Z_{\alpha 1,N}$ is the space of spurious modes for the pressure, while $Z_{\alpha 2,N}$ is the space of spurious modes for the test functions.

We have to characterize successively $Z_{\alpha 1,N}$ and $Z_{\alpha 2,N}$. The proof of the next theorem is completely similar to that of Theorem 24.1.

THEOREM 24.13. *In the case of dimension* $d = 2$, *the space* $Z_{\alpha 1,N}$ *has dimension* 8, *and it is spanned by the polynomials in*

$$\{J_0^\alpha, J_N^\alpha\}^{\otimes 2} \quad \textit{and} \quad \{(1-\zeta)J_N^{\alpha\prime}, (1+\zeta)J_N^{\alpha\prime}\}^{\otimes 2}. \tag{24.83}$$

In the case of dimension $d = 3$, *the space* $Z_{\alpha 1,N}$ *has dimension* $12N + 4$, *and it is spanned by the polynomials in*

$$\begin{aligned} &\{J_0^\alpha, J_N^\alpha\}^{\otimes 3} \quad \textit{and} \quad \{(1-\zeta)J_N^{\alpha\prime}, (1+\zeta)J_N^{\alpha\prime}\}^{\otimes 2} \otimes \mathbb{P}_N(\Lambda), \\ &\{(1-\zeta)J_N^{\alpha\prime}, (1+\zeta)J_N^{\alpha\prime}\} \otimes \mathbb{P}_N(\Lambda) \otimes \{(1-\zeta)J_N^{\alpha\prime}, (1+\zeta)J_N^{\alpha\prime}\}, \\ &\mathbb{P}_N(\Lambda) \otimes \{(1-\zeta)J_N^{\alpha\prime}, (1+\zeta)J_N^{\alpha\prime}\}^{\otimes 2}. \end{aligned} \tag{24.84}$$

Clearly, the subspace $Z_{\alpha 2,N}$ is the orthogonal of D_N in $\mathbb{P}_N(\Omega_d)$ for the scalar product $(\cdot,\cdot)_{\alpha,N}$. Since the codimension of D_N in $\mathbb{P}_N(\Omega_d)$ is known from Theorem 24.1, the proof of the following theorem is straightforward. Let us introduce the polynomial q_N^α in $\mathbb{P}_N(\Lambda)$ which satisfies

$$\forall \varphi_N \in \mathbb{P}_N(\Lambda), \quad (q_N^\alpha, \varphi_N)_{\alpha,N} = \int_{-1}^{1} \varphi_N(\zeta)\,\mathrm{d}\zeta. \tag{24.85}$$

THEOREM 24.14. *In the case of dimension* $d = 2$, *the space* $Z_{\alpha 2,N}$ *has dimension* 8, *and it is spanned by the polynomials in*

$$\{q_N^\alpha, J_N^\alpha\}^{\otimes 2} \quad \textit{and} \quad \{(1-\zeta)J_N^{\alpha\prime}, (1+\zeta)J_N^{\alpha\prime}\}^{\otimes 2}. \tag{24.86}$$

In the case of dimension $d = 3$, *the space* $Z_{\alpha 2,N}$ *has dimension* $12N + 4$, *and it is spanned by the polynomials in*

$$\begin{aligned} &\{q_N^\alpha, J_N^\alpha\}^{\otimes 3} \quad \textit{and} \quad \{(1-\zeta)J_N^{\alpha\prime}, (1+\zeta)J_N^{\alpha\prime}\}^{\otimes 2} \otimes \mathbb{P}_N(\Lambda), \\ &\{(1-\zeta)J_N^{\alpha\prime}, (1+\zeta)J_N^{\alpha\prime}\} \otimes \mathbb{P}_N(\Lambda) \otimes \{(1-\zeta)J_N^{\alpha\prime}, (1+\zeta)J_N^{\alpha\prime}\}, \\ &\mathbb{P}_N(\Lambda) \otimes \{(1-\zeta)J_N^{\alpha\prime}, (1+\zeta)J_N^{\alpha\prime}\}^{\otimes 2}. \end{aligned} \tag{24.87}$$

The characterization of $Z_{\alpha 2,N}$ is not so important: the only consequence is that the second equation in (24.79) can equivalently be enforced for all polynomials in $\mathbb{P}_N(\Omega_d)$ or for all polynomials in a supplementary space of $Z_{\alpha 2,N}$ in $\mathbb{P}_N(\Omega_d)$: this leads to the fact that the inf–sup condition on the form $b_{\alpha 2,N}(\cdot,\cdot)$ is satisfied with a positive constant β_{2N}. But the characterization of $Z_{\alpha 1,N}$ allows for an appropriate choice of M_N^α: in what follows, we always assume that

$$\mathbb{P}_N(\Omega_d) = Z_{\alpha 1,N} \oplus M_N^\alpha. \tag{24.88}$$

This implies that there exists a positive constant β_{1N} such that

$$\inf_{q_N \in M_N^\alpha} \sup_{\boldsymbol{v}_N \in \mathbb{P}_N^0(\Omega_d)^d} \frac{b_{\alpha 1,N}(\boldsymbol{v}_N, q_N)}{\|\boldsymbol{v}_N\|_{H^1_\alpha(\Omega_d)^d} \|q_N\|_{L^2_\alpha(\Omega_d)}} \geqslant \beta_{1N}. \tag{24.89}$$

The continuity and the ellipticity of the form $a_{\alpha,N}(\cdot,\cdot)$ have been proven in Theorem 22.1, up to the viscosity. However, we need here an inf–sup condition on this form and between the kernels $K_{\alpha 1,N}$ and $K_{\alpha 2,N}$.

THEOREM 24.15. *Let α be a real number which satisfies $-\frac{1}{2} \leqslant \alpha \leqslant \frac{1}{2}$. There exists a positive constant c such that, for $i = 1, 2$,*

$$\inf_{\boldsymbol{u}_N \in K_{\alpha 2,N}} \sup_{\boldsymbol{v}_N \in K_{\alpha 1,N}} \frac{a_{\alpha,N}(\boldsymbol{u}_N, \boldsymbol{v}_N)}{\|\boldsymbol{u}_N\|_{H^1_\alpha(\Omega_d)^d} \|\boldsymbol{v}_N\|_{H^1_\alpha(\Omega_d)^d}} \geqslant c\nu,$$

$$\inf_{\boldsymbol{v}_N \in K_{\alpha 1,N}} \sup_{\boldsymbol{u}_N \in K_{\alpha 2,N}} \frac{a_{\alpha,N}(\boldsymbol{u}_N, \boldsymbol{v}_N)}{\|\boldsymbol{u}_N\|_{H^1_\alpha(\Omega_d)^d} \|\boldsymbol{v}_N\|_{H^1_\alpha(\Omega_d)^d}} \geqslant c\nu. \tag{24.90}$$

PROOF. The argument is very similar to the proof of Theorem 23.6. In the case $d = 2$, each function $\boldsymbol{u}_N$ in $K_{\alpha 2,N}$ can be written $\mathbf{curl}\,\psi_N$, where ψ_N belongs to $\mathbb{P}_N^{2,0}(\Omega)$ and each function $\boldsymbol{v}_N$ in $K_{\alpha 1,N}$ can be written $\varpi_{-\alpha}\mathbf{curl}(\chi_N\,\varpi_\alpha)$, where χ_N also belongs to $\mathbb{P}_N^{2,0}(\Omega)$. Associating with each function $\boldsymbol{u}_N = \mathbf{curl}\,\psi_N$ in $K_{\alpha 2,N}$ the function $\boldsymbol{v}_N = \varpi_{-\alpha}\mathbf{curl}(\psi_N \varpi_\alpha)$ in $K_{\alpha 1,N}$, we observe that both inf–sup conditions are a consequence of the ellipticity property

$$a_{\alpha,N}(\boldsymbol{u}_N, \boldsymbol{v}_N) = \big(\Delta\psi_N, \varpi_{-\alpha}\Delta(\psi_N\,\varpi_\alpha)\big)_{\alpha,N} \geqslant c|\psi_N|^2_{H^2_\alpha(\Omega)}.$$

The proof of this property is due to MADAY and MÉTIVET [1986], it relies on the same arguments as Theorem 23.6.

As an immediate consequence of these properties, we derive from Theorem 23.8 the first result concerning problem (24.79).

THEOREM 24.16. *If hypothesis* (24.88) *holds, for any continuous function $\boldsymbol{f}$ on Ω_d, problem* (24.79) *has a unique solution $(\boldsymbol{u}_N, p_N)$ in $\mathbb{P}_N^0(\Omega_d)^d \times M_N^\alpha$.*

The proof of the following approximation result is strictly the same as for Theorem 24.3, with the operator $\Pi_N^{2,0}$ replaced by the operator $\Pi_N^{\alpha,2,0}$ in the case $d = 2$, and the extension of the estimate of SACCHI-LANDRIANI and VANDEVEN [1989] to any value of α in the case $d = 3$.

THEOREM 24.17. *Let α be a real number which satisfies $-1 < \alpha < 1$. For any real number $s \geqslant 1$, there exists a positive constant c such that, for any divergence-free function $\boldsymbol{u}$ in $H^1_{\alpha,0}(\Omega_d)^d \cap H^s(\Omega_d)^d$, the following inequality holds:*

$$\inf_{\boldsymbol{v}_N \in K_{\alpha 2,N}} \|\boldsymbol{u} - \boldsymbol{v}_N\|_{H^1_\alpha(\Omega_d)^d} \leqslant cN^{1-s}\|\boldsymbol{u}\|_{H^s_\alpha(\Omega_d)^d}. \tag{24.91}$$

Combining this result with the abstract inequality (23.56), we derive the first error estimate.

THEOREM 24.18. *Let α be a real number which satisfies $-\frac{1}{2} \leqslant \alpha \leqslant \frac{1}{2}$. The function $\boldsymbol{f}$ is supposed to be in a space $H_\alpha^\sigma(\Omega_d)^d$, $\sigma > \sup\{d/2, d(1+\alpha)/2\}$. Let us assume that the solution $(\boldsymbol{u}, p)$ of problem* (24.1) *belongs to $H_\alpha^s(\Omega_d)^d \times H_\alpha^{s-1}(\Omega_d)$ for a real number $s \geqslant 1$. Then, the following error estimate holds for problem* (24.79):

$$|\boldsymbol{u} - \boldsymbol{u}_N|_{H_\alpha^1(\Omega_d)^d} \leqslant c\big(N^{1-s}\|\boldsymbol{u}\|_{H_\alpha^s(\Omega_d)^d} + N^{-\sigma}\|\boldsymbol{f}\|_{H_\alpha^\sigma(\Omega_d)^d}\big). \tag{24.92}$$

No duality result exists since the regularity properties of the Stokes problem in weighted spaces are presently unknown. So, it remains to establish error estimates for the pressure. In order to do that, we need to derive a lower bound for the constant β_{1N} of the inf–sup condition (24.89). The proof of this result is very similar to the proof of Theorems 24.6 and 24.7.

THEOREM 24.19. *For any function q_N in $\mathbb{P}_N(\Omega_d)$ which is orthogonal to the polynomials in $\{J_0^\alpha, J_N^\alpha\}^{\otimes 2}$ in the case $d = 2$ and in $\{J_0^\alpha, J_N^\alpha\}^{\otimes 3}$ in the case $d = 3$, there exists a polynomial $\boldsymbol{w}_N$ in $\mathbb{P}_N(\Omega_d)^d$ which satisfies*

$$\operatorname{div}(\boldsymbol{w}_N \varpi_\alpha) = q_N \varpi_\alpha \ \text{ in } \Omega_d \quad \text{and}$$
$$\boldsymbol{w}_N \cdot \boldsymbol{n}_j = 0 \quad \text{on } \Gamma_j, \quad 1 \leqslant j \leqslant 2d, \tag{24.93}$$

and

$$\|\boldsymbol{w}_N\|_{H_\alpha^1(\Omega_d)^d} \leqslant cN\|q_N\|_{L_\alpha^2(\Omega_d)} \quad \text{and}$$
$$\|\boldsymbol{w}_N\|_{L_\alpha^2(\Gamma_j)^d} \leqslant c\|q_N\|_{L_\alpha^2(\Omega_d)}, \qquad 1 \leqslant j \leqslant 2d. \tag{24.94}$$

PROOF. We only give an abridged proof in the case $d = 2$. Like in (24.31), the idea is to write

$$\begin{aligned} q_N(x,y) &= \sum_{m=0}^{N-1}\sum_{n=0}^{N-1} \alpha_{mn} J_m^\alpha(x) J_n^\alpha(y) \\ &\quad + \sum_{m=1}^{N-1} \beta_m J_m^\alpha(x)\left(\frac{N}{N+\alpha} J_N^\alpha(y) - \frac{N+\alpha-1}{N-1} J_{N-2}^\alpha(y)\right) \\ &\quad + \sum_{n=1}^{N-1} \gamma_n \left(\frac{N}{N+\alpha} J_N^\alpha(x) - \frac{N+\alpha-1}{N-1} J_{N-2}^\alpha(x)\right) J_n^\alpha(y), \end{aligned} \tag{24.95}$$

with α_{00} equal to 0. Next, we choose $\boldsymbol{w}_N = (w_N, z_N)$, with

$$w_N(x,y) = \sum_{m=0}^{N-1}\sum_{n=0}^{m} \alpha_{mn} \frac{\frac{m+1}{m+\alpha+1} J_{m+1}^\alpha(x) - \frac{m+\alpha}{m} J_{m-1}^\alpha(x)}{2m+2\alpha+1} J_n^\alpha(y)$$

$$+\sum_{m=1}^{N-1}\beta_m \frac{\frac{m+1}{m+\alpha+1}J_{m+1}^\alpha(x)-\frac{m+\alpha}{m}J_{m-1}^\alpha(x)}{2m+2\alpha+1}$$
$$\times\left(\frac{N}{N+\alpha}J_N^\alpha(y)-\frac{N+\alpha-1}{N-1}J_{N-2}^\alpha(y)\right) \tag{24.96}$$

and

$$z_N(x,y)=\sum_{m=0}^{N-1}\sum_{n=m+1}^{N-1}\alpha_{mn}\,J_m^\alpha(x)\frac{\frac{n+1}{n+\alpha+1}J_{n+1}^\alpha(y)-\frac{n+\alpha}{n}J_{n-1}^\alpha(y)}{2n+2\alpha+1}$$
$$+\sum_{n=1}^{N-1}\gamma_n\left(\frac{N}{N+\alpha}J_N^\alpha(x)-\frac{N+\alpha-1}{N-1}J_{N-2}^\alpha(x)\right)$$
$$\times\frac{\frac{n+1}{n+\alpha+1}J_{n+1}^\alpha(y)-\frac{n+\alpha}{n}J_{n-1}^\alpha(y)}{2n+2\alpha+1}. \tag{24.97}$$

By formula (19.12), the polynomial $\boldsymbol{w}_N$ obviously satisfies (24.93), and the stability properties (24.94) are proven exactly like (24.31).

THEOREM 24.20. *For any function q_N in $\mathbb{P}_N(\Omega_d)$ which is orthogonal to the subspace $Z_{\alpha 1,N}$, there exists a polynomial $\boldsymbol{v}_N$ in $\mathbb{P}_N^0(\Omega_d)^d$ which satisfies*

$$\operatorname{div}(\boldsymbol{v}_N\,\varpi_\alpha)=q_N\,\varpi_\alpha, \tag{24.98}$$

and

$$\|\boldsymbol{v}_N\|_{H^1_\alpha(\Omega_d)^d}\leqslant c\,N\,\|q_N\|_{L^2_\alpha(\Omega_d)}. \tag{24.99}$$

PROOF. The proof is completely similar to that of Theorem 24.7. Only the polynomial ω_N has to be replaced by

$$\omega_N^\alpha(\zeta)=\frac{(1-\zeta)^2}{4}(1+\zeta)\frac{J_N^{\alpha\prime\prime\prime}(\zeta)}{J_N^{\alpha\prime\prime\prime}(1)}. \tag{24.100}$$

This last result immediately leads to the inf–sup condition

$$\inf_{q_N\in D_N^\alpha}\ \sup_{\boldsymbol{v}_N\in\mathbb{P}_N^0(\Omega_d)^d}\frac{b_{\alpha 1,N}(\boldsymbol{v}_N,q_N)}{\|\boldsymbol{v}_N\|_{H^1_\alpha(\Omega_d)^d}\|q_N\|_{L^2_\alpha(\Omega_d)}}\geqslant cN^{-1}, \tag{24.101}$$

where D_N^α stands for the orthogonal in $\mathbb{P}_N(\Omega_d)$ to the subspace $Z_{\alpha 1,N}$. However, for the same reasons as previously, this space has no good approximation properties, and we have to introduce a more suitable one.

REMARK 24.5. By using the same arguments as previously (replacing formula (19.12) by formula (19.11)), we derive the following inf–sup condition on the form $b_{\alpha 2,N}(\cdot,\cdot)$:

$$\inf_{q_N\in D_N}\ \sup_{\boldsymbol{v}_N\in\mathbb{P}_N^0(\Omega_d)^d}\frac{b_{\alpha 2,N}(\boldsymbol{v}_N,q_N)}{\|\boldsymbol{v}_N\|_{H^1_\alpha(\Omega_d)^d}\|q_N\|_{L^2_\alpha(\Omega_d)}}\geqslant c\,N^{-1},\tag{24.102}$$

however we do not need it.

We denote by $\Pi^{\alpha}_{D_N}$ the orthogonal projection operator from $\mathbb{P}_N(\Omega_d)$ onto D_N^{α} for the discrete scalar product $(\cdot,\cdot)_{\alpha,N}$.

THEOREM 24.21. *Let λ be a fixed real number, $0<\lambda<1$, and let M_N^{α} be a subspace of $\mathbb{P}_N(\Omega_d)$ which contains $\mathbb{P}_{[\lambda N]}(\Omega_d)\cap L^2_{\alpha 1}(\Omega_d)$ and such that the following property holds:*

$$\forall q_N\in M_N^{\alpha},\quad \|q_N\|_{L^2_\alpha(\Omega_d)}\leqslant c\,\|\Pi^{\alpha}_{D_N}q_N\|_{L^2_\alpha(\Omega_d)}.\tag{24.103}$$

If the assumptions of Theorem 24.18 are satisfied, the following error estimate holds for problem (24.79):

$$\begin{aligned}\|p-p_N\|_{L^2_\alpha(\Omega_d)}\leqslant c\big(&N^{2-s}\big(\|\boldsymbol{u}\|_{H^s_\alpha(\Omega_d)^d}+\|p\|_{H^{s-1}_\alpha(\Omega_d)}\big)\\&+N^{1-\sigma}\|\boldsymbol{f}\|_{H^\sigma_\alpha(\Omega_d)^d}\big).\end{aligned}\tag{24.104}$$

An example of space M_N^{α} satisfying the hypotheses of Theorem 24.21 is constructed exactly as in Theorem 24.9. We set

$$A^{\alpha}_{N-1}=J_N^{\alpha\prime}-\pi^{\alpha}_{[\lambda N]}J_N^{\alpha\prime}\ \text{and}\ A_N^{\alpha}=\pi^{\alpha}_{N-1}\big(\zeta J_N^{\alpha\prime}\big)-\pi^{\alpha}_{[\lambda N]}\big(\zeta J_N^{\alpha\prime}\big).\tag{24.105}$$

THEOREM 24.22. *In the case of dimension $d=2$, the space M_N^{α} which is orthogonal in $\mathbb{P}_N(\Omega)$ to the polynomials in*

$$\{J_0^{\alpha},J_N^{\alpha}\}^{\otimes 2}\quad\text{and}\quad\{A^{\alpha}_{N-1},A^{\alpha}_N\}^{\otimes 2},$$

for the scalar product $(\cdot,\cdot)_{\alpha,N}$, contains $\mathbb{P}_{[\lambda N]}(\Omega)\cap L^2_{\alpha 1}(\Omega)$ and satisfies (24.103). In the case of dimension $d=3$, the space M_N^{α} which is orthogonal in $\mathbb{P}_N(\Omega_3)$ to the polynomials in

$$\begin{aligned}&\{J_0^{\alpha},J_N^{\alpha}\}^{\otimes 3}\quad\text{and}\quad\{A^{\alpha}_{N-1},A^{\alpha}_N\}^{\otimes 2}\otimes\mathbb{P}_N(\Lambda),\\&\{A^{\alpha}_{N-1},A^{\alpha}_N\}\otimes\mathbb{P}_N(\Lambda)\otimes\{A^{\alpha}_{N-1},A^{\alpha}_N\},\quad\mathbb{P}_N(\Lambda)\otimes\{A^{\alpha}_{N-1},A^{\alpha}_N\}^{\otimes 2},\end{aligned}\tag{24.106}$$

contains $\mathbb{P}_{[\lambda N]}(\Omega_3)\cap L^2_{\alpha 1}(\Omega_3)$ and satisfies (24.103).

REMARK 24.6. Exactly the same type of weighted approximation of the Stokes problem can be applied in the case of inhomogeneous boundary conditions on the velocity and leads to similar results.

This concludes the numerical analysis of the first spectral method for the discretization of the Stokes problem. Let us point out the main features of this method:

(i) the velocity is exactly divergence-free;

(ii) the method is of collocation type, at least for homogeneous boundary conditions;

(iii) the spurious modes for the pressure are fully identified, they do not pollute the discrete velocity but a special choice of pressure space is required to eliminate them;

(iv) the best constant of the inf–sup condition on the pressure is of order N^{-1} in both cases $d=2$ and $d=3$.

25. The $\mathbb{P}_N - \mathbb{P}_{N-2}$ method for the Stokes problem

Here, we present a second spectral method to discretize the Stokes problem (24.1), which is due to MADAY, PATERA and RØNQUIST [1987] and was aimed to eliminate the spurious modes on the pressure. The discrete problem is simply written: *find a pair* $(\boldsymbol{u}_N, p_N)$ *in* $\mathbb{P}_N^0(\Omega_d)^d \times (\mathbb{P}_{N-2}(\Omega_d) \cap L_0^2(\Omega_d))$ *such that*

$$\begin{aligned} &\forall \boldsymbol{v}_N \in \mathbb{P}_N^0(\Omega_d)^d, \quad a_N(\boldsymbol{u}_N, \boldsymbol{v}_N) + b_N(\boldsymbol{v}_N, p_N) = (\boldsymbol{f}, \boldsymbol{v}_N)_N, \\ &\forall q_N \in \mathbb{P}_{N-2}(\Omega_d), \quad b_N(\boldsymbol{u}_N, q_N) = 0. \end{aligned} \tag{25.1}$$

Note that, due to the maximal degree of polynomials in the pressure space, the form $b_N(\bullet,\bullet)$ can equivalently be replaced by the form $b(\bullet,\bullet)$ in this problem, and also that the second equation is equivalently satisfied for any polynomial q_N in $\mathbb{P}_{N-2}(\Omega_d)$ or in $\mathbb{P}_{N-2}(\Omega_d) \cap L_0^2(\Omega_d)$. We are going to perform the numerical analysis of this problem.

Most of the work is already done. The form $b(\bullet,\bullet)$ is clearly continuous on $H^1(\Omega_d) \times L^2(\Omega_d)$ and, from Theorem 15.1, the form $a_N(\bullet,\bullet)$ is continuous on $\mathbb{P}_N(\Omega_d)^d \times \mathbb{P}_N(\Omega_d)^d$ and elliptic on $\mathbb{P}_N^0(\Omega_d)^d$. So, only the inf–sup condition on the form $b(\bullet,\bullet)$ is a problem. However, the first result in this direction is a straightforward consequence of the characterization of the space Z_N which is given in Theorem 24.1.

THEOREM 25.1. *The set of polynomials* q_N *in* $\mathbb{P}_{N-2}(\Omega_d) \cap L_0^2(\Omega_d)$ *such that*

$$\forall \boldsymbol{v}_N \in \mathbb{P}_N^0(\Omega_d)^d, \quad b(\boldsymbol{v}_N, q_N) = 0, \tag{25.2}$$

is reduced to $\{0\}$.

This theorem states that there is no spurious modes for the pressure in the new method. Hence, an inf–sup condition

$$\inf_{q_N \in \mathbb{P}_{N-2}(\Omega_d) \cap L_0^2(\Omega_d)} \sup_{\boldsymbol{v}_N \in \mathbb{P}_N^0(\Omega_d)^d} \frac{b(\boldsymbol{v}_N, q_N)}{\|\boldsymbol{v}_N\|_{H^1(\Omega_d)^d} \|q_N\|_{L^2(\Omega_d)}} \geqslant \beta_N, \tag{25.3}$$

holds for a positive constant β_N. It leads to the first result on the $\mathbb{P}_N - \mathbb{P}_{N-2}$ method.

THEOREM 25.2. *For any continuous function* $\boldsymbol{f}$ *on* Ω_d, *problem* (25.1) *has a unique solution* $(\boldsymbol{u}_N, p_N)$ *in* $\mathbb{P}_N^0(\Omega_d)^d \times (\mathbb{P}_{N-2}(\Omega_d) \cap L_0^2(\Omega_d))$.

Note that the functions $\boldsymbol{v}_N$ in $\mathbb{P}_N^0(\Omega_d)^d$ which satisfy

$$\forall q_N \in \mathbb{P}_{N-2}(\Omega_d), \quad b(\boldsymbol{v}_N, q_N) = 0, \tag{25.4}$$

are not exactly divergence-free, but that the divergence-free polynomials in $\mathbb{P}_N^0(\Omega_d)^d$ satisfy (25.4). This allows for stating the first error estimate.

THEOREM 25.3. *The function* $\boldsymbol{f}$ *is supposed to be in a space* $H^\sigma(\Omega_d)^d$, $\sigma > d/2$. *Let us assume that the solution* $(\boldsymbol{u}, p)$ *of problem* (24.1) *belongs to* $H^s(\Omega_d)^d \times H^{s-1}(\Omega_d)$ *for a real number* $s \geqslant 1$. *Then, the following error estimate holds for problem* (25.1):

$$\begin{aligned} |\boldsymbol{u} - \boldsymbol{u}_N|_{H^1(\Omega_d)^d} \leqslant c\big(& N^{1-s}\big(\|\boldsymbol{u}\|_{H^s(\Omega_d)^d} + \|p\|_{H^{s-1}(\Omega_d)}\big) \\ &+ N^{-\sigma}\|\boldsymbol{f}\|_{H^\sigma(\Omega_d)^d}\big). \end{aligned} \tag{25.5}$$

PROOF. From estimate (23.56), we derive that, for any divergence-free polynomial $\boldsymbol{v}_N$ in $\mathbb{P}_{N-1}^0(\Omega_d)^d$,

$$\begin{aligned} \|\boldsymbol{u} - \boldsymbol{u}_N\|_{H^1(\Omega_d)^d} \leqslant c\Bigg(& \|\boldsymbol{u} - \boldsymbol{v}_N\|_{H^1(\Omega_d)^d} + \|p - \Pi_{N-2} p\|_{L^2(\Omega_d)} \\ & + \sup_{\boldsymbol{w}_N \in \mathbb{P}_N^0(\Omega_d)^d} \frac{\int_{\Omega_d} \boldsymbol{f}(\boldsymbol{x}) \cdot \boldsymbol{w}_N(\boldsymbol{x})\, \mathrm{d}\boldsymbol{x} - (\boldsymbol{f}, \boldsymbol{w}_N)_N}{|\boldsymbol{w}_N|_{H^1(\Omega_d)^d}}\Bigg). \end{aligned} \tag{25.6}$$

Then, the result follows from Theorems 24.3 and 7.1, and (15.18) combined with Theorems 7.1 and 14.2.

We can now derive an estimate for the velocity in $L^2(\Omega_d)^d$. The duality argument is the same as for Theorem 24.5, and the only difference in the proof comes from the fact that $\boldsymbol{u}_N$ is no longer divergence-free.

THEOREM 25.4. *If the hypotheses of Theorem* 25.3 *are satisfied, the following error estimate holds for problem* (25.1):

$$\begin{aligned} \|\boldsymbol{u} - \boldsymbol{u}_N\|_{L^2(\Omega_d)^d} \leqslant c\big(& N^{-s}\big(\|\boldsymbol{u}\|_{H^s(\Omega_d)^d} + \|p\|_{H^{s-1}(\Omega_d)}\big) \\ &+ N^{-\sigma}\|\boldsymbol{f}\|_{H^\sigma(\Omega_d)^d}\big). \end{aligned} \tag{25.7}$$

PROOF. Once more, we have

$$\|\boldsymbol{u} - \boldsymbol{u}_N\|_{L^2(\Omega_d)^d} = \sup_{\boldsymbol{g} \in L^2(\Omega_d)^d} \frac{\int_{\Omega_d} (\boldsymbol{u} - \boldsymbol{u}_N)(\boldsymbol{x}) \cdot \boldsymbol{g}(\boldsymbol{x})\, \mathrm{d}\boldsymbol{x}}{\|\boldsymbol{g}\|_{L^2(\Omega_d)^d}}. \tag{25.8}$$

So, for any g in $L^2(\Omega_d)^d$, we solve problem (24.28) and we compute, for any divergence-free polynomial w_N in $K_N \cap \mathbb{P}^0_{N-1}(\Omega_d)^d$,

$$\begin{aligned}\int_{\Omega_d} (u - u_N)(x) \cdot g(x)\,\mathrm{d}x &= a(w, u - u_N) + b(u - u_N, r) \\ &= a(u - u_N, w - w_N) + b(u - u_N, r - \Pi_{N-2}r) \\ &\quad + \langle f, w_N\rangle - (f, w_N)_N,\end{aligned}$$

so that we derive the following inequality

$$\begin{aligned}&\int_{\Omega_d} (u - u_N)(x) \cdot g(x)\,\mathrm{d}x \\ &\quad \leqslant |u - u_N|_{H^1(\Omega_d)^d}\big(|w - w_N|_{H^1(\Omega_d)^d} + \|r - \Pi_{N-2}r\|_{L^2(\Omega_d)}\big) \\ &\qquad + \|f - \mathcal{I}_N f\|_{L^2(\Omega_d)^d}\|w_N\|_{L^2(\Omega_d)^d}.\end{aligned}$$

The estimate now follows from Theorems 25.3, 24.3, 7.1 and 14.2, combined with the regularity result (24.29) and (25.8).

So, it remains to obtain an error estimate on the pressure, and the first step for this result is to establish a bound for the constant β_N of the inf–sup condition (25.3) (of course, since $\mathbb{P}_{N-2}(\Omega_d)$ is contained in $\mathbb{P}_N(\Omega_d)$, we deduce from Section 24 that β_N is larger than $c\,N^{-1}$). Estimating the constant β_N is the object of the three following theorems.

THEOREM 25.5. *For any polynomial q_N in $\mathbb{P}_{N-2}(\Omega_d)$, there exists a polynomial w_N in $\mathbb{P}_N(\Omega_d)^d$ which satisfies*

$$\forall r_N \in \mathbb{P}_{N-2}(\Omega_d), \quad \int_{\Omega_d} (\operatorname{div} w_N + q_N)(x) r_N(x)\,\mathrm{d}x = 0, \tag{25.9}$$

$$w_N \times n_j = \mathbf{0}, \quad 1 \leqslant j \leqslant 2d, \tag{25.10}$$

and

$$\|w_N\|_{H^1(\Omega_d)^d} \leqslant cN^{(d-1)/2}\|q_N\|_{L^2(\Omega_d)}. \tag{25.11}$$

The proof of this theorem requires a technical result.

LEMMA. *Any polynomial φ_N in $\mathbb{P}^0_N(\Lambda)$ satisfies*

$$\|\varphi_N\|_{L^2(\Lambda)} \leqslant cN^{1/2}\|\pi_{N-2}\varphi_N\|_{L^2(\Lambda)}. \tag{25.12}$$

PROOF. Let us write the polynomial φ_N as

$$\varphi_N = \sum_{n=0}^{N} \alpha_n L_n,$$

so that

$$\|\varphi_N\|_{L^2(\Lambda)} \leqslant \|\pi_{N-2}\varphi_N\|_{L^2(\Lambda)} + |\alpha_{N-1}|\big(N - \tfrac{1}{2}\big)^{-1/2} + |\alpha_N|\big(N + \tfrac{1}{2}\big)^{-1/2}. \tag{25.13}$$

If the polynomial φ_N vanishes in ± 1, we have

$$\varphi_N(1) = \alpha_N + \alpha_{N-1} + \alpha_{N-2} + \cdots = 0,$$
$$\varphi_N(-1) = (-1)^N \alpha_N + (-1)^{N-1}\alpha_{N-1} + (-1)^{N-2}\alpha_{N-2} + \cdots = 0,$$

so that, for $i = 0$ or 1,

$$\begin{aligned}|\alpha_{N-i}| &= |\alpha_{N-i-2} + \alpha_{N-i-4} + \cdots|\\ &\leqslant \left(\sum_{N=0}^{N-2} \alpha_n^2 \big(n + \tfrac{1}{2}\big)^{-1}\right)^{1/2} \left(\sum_{n=0}^{N} \big(n + \tfrac{1}{2}\big)\right)^{1/2}\\ &\leqslant cN\|\pi_{N-2}\varphi_N\|_{L^2(\Lambda)}.\end{aligned}$$

Inserting this last inequality in (25.13) gives the desired result (which can be proven to be optimal).

PROOF OF THEOREM 25.5. We choose $\boldsymbol{w}_N = \mathbf{grad}\,\psi_N$, where ψ_N is the polynomial in $\mathbb{P}_N^0(\Omega_d)$ which satisfies

$$\forall r_N \in \mathbb{P}_{N-2}(\Omega_d), \quad \int_{\Omega_d} (\Delta\psi_N + q_N)(\boldsymbol{x}) r_N(\boldsymbol{x})\,\mathrm{d}\boldsymbol{x} = 0. \tag{25.14}$$

Since this is a linear system of $(N-1)^d$ equations with $(N-1)^d$ unknowns, the existence, uniqueness and the properties that are stated in the theorem follow from the inequality

$$\|\psi_N\|_{H^2(\Omega_d)} \leqslant cN^{(d-1)/2}\|q_N\|_{L^2(\Omega_d)}, \tag{25.15}$$

which we prove now. In the case $d = 2$ for instance, equation (25.14) infers that,

$$q_N = \Pi_{N-2}(\Delta\psi_N) = \pi_{N-2}^{(y)}\left(\frac{\partial^2\psi_N}{\partial x^2}\right) + \pi_{N-2}^{(x)}\left(\frac{\partial^2\psi_N}{\partial y^2}\right).$$

Integrating by parts, we check that

$$\begin{aligned}\|q_N\|^2_{L^2(\Omega)} &= \int_\Omega \Big(\pi^{(y)}_{N-2}\Big(\frac{\partial^2\psi_N}{\partial x^2}\Big)\Big)^2(\boldsymbol{x})\,\mathrm{d}\boldsymbol{x} + \int_\Omega \Big(\pi^{(x)}_{N-2}\Big(\frac{\partial^2\psi_N}{\partial y^2}\Big)\Big)^2(\boldsymbol{x})\,\mathrm{d}\boldsymbol{x}\\ &\quad + 2\int_\Omega \Big(\frac{\partial^2\psi_N}{\partial x^2}\Big)(\boldsymbol{x})\Big(\frac{\partial^2\psi_N}{\partial y^2}\Big)(\boldsymbol{x})\,\mathrm{d}\boldsymbol{x}\\ &= \int_\Omega \Big(\pi^{(y)}_{N-2}\Big(\frac{\partial^2\psi_N}{\partial x^2}\Big)\Big)^2(\boldsymbol{x})\,\mathrm{d}\boldsymbol{x} + \int_\Omega \Big(\pi^{(x)}_{N-2}\Big(\frac{\partial^2\psi_N}{\partial y^2}\Big)\Big)^2(\boldsymbol{x})\,\mathrm{d}\boldsymbol{x}\\ &\quad + 2\int_\Omega \Big(\frac{\partial^2\psi_N}{\partial x\partial y}\Big)^2(\boldsymbol{x})\,\mathrm{d}\boldsymbol{x}.\end{aligned}$$

Applying the previous lemma once with respect to x and once with respect to y leads to inequality (25.15). The argument is strictly the same in the case $d = 3$.

In order to state the next theorem, we introduce the spaces $\mathbb{P}_{n_1,\dots,n_d}(\Omega_d)$ of polynomials with degree $\leqslant n_j$ with respect to each variable x_j, $1 \leqslant j \leqslant d$.

THEOREM 25.6. *For any polynomial q_N in $\mathbb{P}_{N-2}(\Omega_d) \cap L^2_0(\Omega_d)$, there exists a polynomial $\boldsymbol{z}_N$ in*

$$\begin{aligned}&\big(\mathbb{P}_{N+1,N}(\Omega) \cap H^1_0(\Omega)\big) \times \big(\mathbb{P}_{N,N+1}(\Omega) \cap H^1_0(\Omega)\big) \quad \textit{in the case } d = 2,\\ &\big(\mathbb{P}_{N+1,N,N}(\Omega_3) \cap H^1_0(\Omega_3)\big) \times \big(\mathbb{P}_{N,N+1,N}(\Omega_3) \cap H^1_0(\Omega_3)\big)\\ &\quad \times \big(\mathbb{P}_{N,N,N+1}(\Omega_3) \cap H^1_0(\Omega_3)\big) \quad \textit{in the case } d = 3,\end{aligned}$$

which satisfies

$$\forall r_N \in \mathbb{P}_{N-2}(\Omega_d), \quad \int_{\Omega_d} (\operatorname{div} \boldsymbol{z}_N + q_N)(\boldsymbol{x}) r_N(\boldsymbol{x})\,\mathrm{d}\boldsymbol{x} = 0, \tag{25.16}$$

and

$$\|\boldsymbol{z}_N\|_{H^1(\Omega_d)^d} \leqslant c N^{(d-1)/2} \|q_N\|_{L^2(\Omega_d)}. \tag{25.17}$$

PROOF. We study separately the cases $d = 2$ and $d = 3$.

(1) In the case $d = 2$, the idea is to take $\boldsymbol{z}_N = \boldsymbol{w}_N + \mathbf{curl}\,\chi_N$, where $\boldsymbol{w}_N$ has been introduced in Theorem 25.5 while the polynomial χ_N satisfies

$$\frac{\partial \chi_N}{\partial \tau_J} = -\boldsymbol{w}_N \cdot \boldsymbol{n}_J \quad \text{and} \quad \frac{\partial \chi_N}{\partial n_J} = 0 \quad \text{on } \Gamma_J, \quad J = 1, 2, 3, 4. \tag{25.18}$$

Indeed, setting $\varphi^0_N(\boldsymbol{a}_0) = 0$ and, successively for $J = 1, 2, 3, 4$,

$$\varphi^J_N(\boldsymbol{x}) = \varphi^{J-1}_N(\boldsymbol{a}_{J-1}) - \int_{\boldsymbol{a}_{J-1}}^{\boldsymbol{x}} (\boldsymbol{w}_N \cdot \boldsymbol{n}_J)(\tau)\,\mathrm{d}\tau,$$

we observe that the φ_N^J, $J = 1, 2, 3, 4$, belong to $\mathbb{P}_{N+1}(\Gamma_J)$, that

$$\varphi_N^4(\boldsymbol{a}_4) = \varphi_N^0(\boldsymbol{a}_0) - \sum_{J=1}^{4} \int_{\Gamma_J} (\boldsymbol{w}_N \cdot \boldsymbol{n}_J)(\tau)\,\mathrm{d}\tau = -\int_{\Omega} (\operatorname{div} \boldsymbol{w}_N)(\boldsymbol{x})\,\mathrm{d}\boldsymbol{x} = 0,$$

and also that

$$\left(\frac{\partial \varphi_N^J}{\partial \tau_J}\right)(\boldsymbol{a}_J) = \left(\frac{\partial \varphi_N^{J+1}}{\partial \tau_{J+1}}\right)(\boldsymbol{a}_J) = 0,$$
$$\left(\frac{\partial^2 \varphi_N^J}{\partial x \partial y}\right)(\boldsymbol{a}_J) = \left(\frac{\partial^2 \varphi_N^{J+1}}{\partial x \partial y}\right)(\boldsymbol{a}_J) = 0, \quad J = 1, 2, 3, 4.$$

Due to these sixteen compatibility conditions, it follows from BERNARDI, DAUGE and MADAY ([1992b], Chapter 3) that there exists a polynomial χ_N in $\mathbb{P}_{N+1}(\Omega)$ which satisfies (25.18) and such that

$$\|\chi_N\|_{H^2(\Omega)} \leqslant c \sum_{J=1}^{4} \|\varphi_N^J\|_{H^{3/2}(\Gamma_J)} \leqslant c' \|\boldsymbol{w}_N\|_{H^1(\Omega)^2},$$

which gives the desired result.

(2) In the case $d = 3$, we have to choose $\boldsymbol{z}_N = \boldsymbol{w}_N + \mathbf{curl}\,\boldsymbol{\chi}_N$, where the polynomial $\boldsymbol{\chi}_N$ belongs to $\mathbb{P}_{N+1}(\Omega_3)^3$ and satisfies

$$(\mathbf{curl}\,\boldsymbol{\chi}_N) \cdot \boldsymbol{n}_j = -\boldsymbol{w}_N \cdot \boldsymbol{n}_j \quad \text{and}$$
$$(\mathbf{curl}\,\boldsymbol{\chi}_N) \times \boldsymbol{n}_j = \mathbf{0} \text{ on } \Gamma_j,\ 1 \leqslant j \leqslant 6. \tag{25.19}$$

Such a polynomial can be built by similar arguments and chosen in order to satisfy the same stability estimate, we refer to MADAY, PATERA and RØNQUIST ([1992], Lemma 4.3) for the complete proof.

THEOREM 25.7. *For any polynomial q_N in $\mathbb{P}_{N-2}(\Omega_d) \cap L_0^2(\Omega_d)$, there exists a polynomial $\boldsymbol{v}_N$ in $\mathbb{P}_N^0(\Omega_d)^d$ which satisfies*

$$\forall r_N \in \mathbb{P}_{N-2}(\Omega_d), \quad \int_{\Omega_d} (\operatorname{div} \boldsymbol{v}_N + q_N)(\boldsymbol{x}) r_N(\boldsymbol{x})\,\mathrm{d}\boldsymbol{x} = 0, \tag{25.20}$$

and

$$\|\boldsymbol{v}_N\|_{H^1(\Omega_d)^d} \leqslant c N^{(d-1)/2} \|q_N\|_{L^2(\Omega_d)}. \tag{25.21}$$

PROOF. We only give the proof in the case $d = 2$, since it can be extended to the case $d = 3$ in a straightforward way. Let q_N be in $\mathbb{P}_{N-2}(\Omega_d) \cap L_0^2(\Omega_d)$, and let

$\boldsymbol{z}_N = (z_N, t_N)$ stand for the polynomial which is associated with it in Theorem 25.6. We write

$$z_N(x,y) = \sum_{n=1}^{N} \alpha_n(y)(L_{n+1} - L_{n-1})(x) \quad \text{and}$$
$$t_N(x,y) = \sum_{n=1}^{N} \beta_n(x)(L_{n+1} - L_{n-1})(y). \tag{25.22}$$

Next, we set

$$e_N(x,y) = \alpha_N(y)(L_{N+1} - L_{N-1})(x) \quad \text{and}$$
$$f_N(x,y) = \beta_N(x)(L_{N+1} - L_{N-1})(y), \tag{25.23}$$

and we take $\boldsymbol{v}_N$ equal to $(z_N - e_N, t_N - f_N)$. Firstly, we observe that, since q_N has degree $\leqslant N-2$,

$$\begin{aligned}\int_\Omega (\operatorname{div} \boldsymbol{v}_N + q_N)(\boldsymbol{x}) r_N(\boldsymbol{x})\, \mathrm{d}\boldsymbol{x} &= \int_\Omega (\operatorname{div} \boldsymbol{z}_N + q_N)(\boldsymbol{x}) r_N(\boldsymbol{x})\, \mathrm{d}\boldsymbol{x} \\ &\quad - (2N+1) \int_\Omega \alpha_N(y) L_N(x) q_N(x,y)\, \mathrm{d}\boldsymbol{x} \\ &\quad - (2N+1) \int_\Omega \beta_N(x) L_N(y) q_N(x,y)\, \mathrm{d}\boldsymbol{x} \\ &= \int_\Omega (\operatorname{div} \boldsymbol{z}_N + q_N)(\boldsymbol{x}) r_N(\boldsymbol{x})\, \mathrm{d}\boldsymbol{x},\end{aligned}$$

so that equation (25.20) is satisfied. Secondly, we note that

$$z_N(x,y) = \alpha_N(y) L_{N+1}(x) + \alpha_{N-1}(y) L_N(x) + \cdots,$$

from which we derive

$$\|z_N\|_{L^2(\Omega)} \geqslant \|\alpha_N\|_{L^2(\Lambda)} \left(N + \tfrac{3}{2}\right)^{-1/2} \geqslant c \|e_N\|_{L^2(\Omega)},$$

and

$$\left\| \frac{\partial z_N}{\partial y} \right\|_{L^2(\Omega)} \geqslant \|\alpha_N'\|_{L^2(\Lambda)} \left(N + \tfrac{3}{2}\right)^{-1/2} \geqslant c \left\| \frac{\partial e_N}{\partial y} \right\|_{L^2(\Omega)}.$$

We also have

$$\left(\frac{\partial z_N}{\partial x}\right)(x,y) = (2N+1)\alpha_N(y) L_N(x) + (2N-1)\alpha_{N-1}(y) L_{N-1}(x) + \cdots,$$

so that

$$\left\| \frac{\partial z_N}{\partial x} \right\|_{L^2(\Omega)} \geqslant 2\|\alpha_N\|_{L^2(\Lambda)} \left(N+\tfrac{1}{2}\right)^{1/2} \geqslant \left\| \frac{\partial e_N}{\partial x} \right\|_{L^2(\Omega)}.$$

Putting together all these estimates and applying the same arguments to f_N, we obtain that

$$\|e_N\|_{H^1(\Omega)} + \|f_N\|_{H^1(\Omega)} \leqslant \|\boldsymbol{z}_N\|_{H^1(\Omega)^2},$$

which of course leads to (25.21).

REMARK 25.1. Using exactly the same arguments, we can derive a slightly improved version of Theorem 25.7, where the polynomial $\boldsymbol{v}_N$ still satisfies the same properties but belongs to

$$\begin{aligned}
&\big(\mathbb{P}_{N-1,N}(\Omega) \cap H_0^1(\Omega)\big) \times \big(\mathbb{P}_{N,N-1}(\Omega) \cap H_0^1(\Omega)\big) \quad \text{in the case } d=2,\\
&\big(\mathbb{P}_{N-1,N,N}(\Omega_3) \cap H_0^1(\Omega_3)\big) \times \big(\mathbb{P}_{N,N-1,N}(\Omega_3) \cap H_0^1(\Omega_3)\big)\\
&\quad \times \big(\mathbb{P}_{N,N,N-1}(\Omega_3) \cap H_0^1(\Omega_3)\big) \quad \text{in the case } d=3.
\end{aligned}$$

From Theorem 25.7, we immediately derive the inf–sup condition

$$\begin{aligned}
&\inf_{q_N \in \mathbb{P}_{N-2}(\Omega_d) \cap L_0^2(\Omega_d)} \ \sup_{\boldsymbol{v}_N \in \mathbb{P}_N^0(\Omega_d)^d} \frac{b(\boldsymbol{v}_N, q_N)}{\|\boldsymbol{v}_N\|_{H^1(\Omega_d)^d} \|q_N\|_{L^2(\Omega_d)}}\\
&\quad \geqslant cN^{-(d-1)/2}.
\end{aligned} \tag{25.24}$$

The constant is of better order than in (24.48) in the case $d=2$.

REMARK 25.2. The constant in the inf–sup condition (25.24) is not independent of N, and it cannot be! Indeed, in the case $d=2$, taking

$$q_N(x,y) = L'_{N-1}(x) L'_{N-2}(y),$$

we observe from Remark 24.1 that, for any $\boldsymbol{v}_N$ in $\mathbb{P}_N^0(\Omega)^2$,

$$b(\boldsymbol{v}_N, q_N) = b\big(\boldsymbol{v}_N, L'_{N-1}(x) L'_{N-2}(y) - L'_{N+1}(x) L'_N(y)\big),$$

so that

$$\begin{aligned}
&\sup_{\boldsymbol{v}_N \in \mathbb{P}_N^0(\Omega)^2} \frac{b(\boldsymbol{v}_N, q_N)}{\|\boldsymbol{v}_N\|_{H^1(\Omega)^2} \|q_N\|_{L^2(\Omega)}}\\
&\quad \leqslant \frac{\|L'_{N-1}(x) L'_{N-2}(y) - L'_{N+1}(x) L'_N(y)\|_{L^2(\Omega)}}{\|L'_{N-1}(x) L'_{N-2}(y)\|_{L^2(\Omega)}}.
\end{aligned}$$

Using the expansion

$$L'_{N-1}(x)L'_{N-2}(y) - L'_{N+1}(x)L'_N(y)$$
$$= -(2N+1)L_N(x)L'_{N-2}(y) - (2N-1)L'_{N+1}(x)L_{N-1}(y),$$

we easily check from (3.11) that this last ratio is less than a constant times $N^{-1/2}$. Applying the same argument in the case $d = 3$ to the polynomial

$$q_N(x,y,z) = L'_{N-1}(x)L'_{N-2}(y)L'_{N-2}(z), \tag{25.25}$$

we prove that there exists a polynomial q_N in $\mathbb{P}_{N-2}(\Omega_d)$ such that

$$\sup_{v_N \in \mathbb{P}_N^0(\Omega_d)^d} \frac{b(v_N, q_N)}{\|v_N\|_{H^1(\Omega_d)^d}\|q_N\|_{L^2(\Omega_d)}} \leqslant cN^{-(d-1)/2}, \tag{25.26}$$

which implies the optimality of (25.24).

At last, we know the error estimate on the pressure, which follows from estimate (23.56).

THEOREM 25.8. *If the hypotheses of Theorem 25.3 are satisfied, the following error estimate holds for problem* (25.1):

$$\|p - p_N\|_{L^2(\Omega_d)} \leqslant c\big(N^{(d+1)/2-s}\big(\|u\|_{H^s(\Omega_d)^d} + \|p\|_{H^{s-1}(\Omega_d)}\big)$$
$$+ N^{(d-1)/2-\sigma}\|f\|_{H^\sigma(\Omega_d)^d}\big). \tag{25.27}$$

Finally, we observe that the method can easily be extended to the problem with inhomogeneous boundary conditions. Like in Section 24, we assume that the boundary data g_j, $1 \leqslant j \leqslant 2d$, of problem (24.61) belong to $H^{1/2}(\Gamma_j)^d$, are Hölder-continuous on $\overline{\Gamma}_j$, and satisfy the conditions (24.62) and (24.63). Then the discrete problem reads: *find a pair* (u_N, p_N) *in* $\mathbb{P}_N(\Omega_d)^d \times (\mathbb{P}_{N-2}(\Omega_d) \cap L_0^2(\Omega_d))$, *with* $u_N - \mathcal{I}_N u_b + \mu_N \chi$ *in* $\mathbb{P}_N^0(\Omega_d)^d$, *such that*

$$\forall v_N \in \mathbb{P}_N^0(\Omega_d)^d, \quad a_N(u_N, v_N) + b_N(v_N, p_N) = (f, v_N)_N,$$
$$\forall q_N \in \mathbb{P}_{N-2}(\Omega_d), \quad b_N(u_N, q_N) = 0, \tag{25.28}$$

where u_b is any continuous function in $H^1(\Omega_d)^d$ equal to g_j on Γ_j, $1 \leqslant j \leqslant 2d$.

Writing the discrete problem which is satisfied by the pair $(u_N - \mathcal{I}_N u_b + \mu_N \chi, p_N)$ gives the wellposedness result.

THEOREM 25.9. *For any continuous function* f *on* Ω_d *and for any Hölder-continuous functions* g_j *in* $H^{1/2}(\Gamma_j)^d$, $1 \leqslant j \leqslant 2d$, *such that* (24.62) *holds, problem* (25.28) *has a unique solution* (u_N, p_N) *in* $\mathbb{P}_N(\Omega_d)^d \times (\mathbb{P}_{N-2}(\Omega_d) \cap L_0^2(\Omega_d))$.

When the regularity assumptions of Theorem 24.11 are satisfied, exactly the same arguments lead to the following (and nonoptimal) estimate

$$\begin{aligned}&\|\boldsymbol{u}-\boldsymbol{u}_N\|_{H^1(\Omega_d)^d}+N^{(1-d)/2}\|p-p_N\|_{L^2(\Omega_d)}\\&\quad\leqslant c\big(N^{(d+1)/2-s}\big(\|\boldsymbol{u}\|_{H^s(\Omega_d)^d}+\|p\|_{H^{s-1}(\Omega_d)}\big)+N^{-\sigma}\|\boldsymbol{f}\|_{H^\sigma(\Omega_d)^d}\big).\end{aligned}\tag{25.29}$$

However, we prefer to state an optimal estimate in the case $d=2$, which is proven in BERNARDI, DAUGE and MADAY [1992b] by sophisticated arguments (the main idea is to construct a stable lifting of the traces $\boldsymbol{g}_J-\mu_N\,\boldsymbol{\chi}$ by a polynomial such that its divergence is orthogonal to M_N).

THEOREM 25.10. *In the case $d=2$ of the square Ω, the function $\boldsymbol{f}$ is supposed to be in a space $H^\sigma(\Omega)^2$, $\sigma>1$, and the boundary data $\boldsymbol{g}_J$, $J=1,2,3,4$, are supposed to be in a space $H^{\tau_J}(\Gamma_J)^2$ for real numbers $\tau_J>\frac{1}{2}$. Let us assume that the solution $(\boldsymbol{u},p)$ of problem (24.61) belongs to $H^s(\Omega)^2\times H^{s-1}(\Omega)$ for a real number $s\geqslant 1$. Then, the following error estimate holds for problem (25.28):*

$$\begin{aligned}&\|\boldsymbol{u}-\boldsymbol{u}_N\|_{H^1(\Omega)^2}+N^{-1/2}\|p-p_N\|_{L^2(\Omega)}\\&\quad\leqslant c\Bigg(N^{1-s}\big(\|\boldsymbol{u}\|_{H^s(\Omega)^2}+\|p\|_{H^{s-1}(\Omega)}\big)\\&\qquad+N^{-\sigma}\|\boldsymbol{f}\|_{H^\sigma(\Omega)^2}+\sum_{J=1}^{4}N^{1/2-\tau_J}\|\boldsymbol{g}_J\|_{H^{\tau_J}(\Gamma_J)^2}\Bigg).\end{aligned}\tag{25.30}$$

We state (without proof) the estimate in $L^2(\Omega)^2$ which is derived by the same duality argument as for Theorem 24.12.

THEOREM 25.11. *If the assumptions of Theorem 25.10 are satisfied, the following error estimate holds for problem (25.28):*

$$\begin{aligned}\|\boldsymbol{u}-\boldsymbol{u}_N\|_{L^2(\Omega)^2}\leqslant c\Bigg(&N^{-s}\big(\|\boldsymbol{u}\|_{H^s(\Omega)^2}+\|p\|_{H^{s-1}(\Omega)}\big)\\&+N^{-\sigma}\|\boldsymbol{f}\|_{H^\sigma(\Omega)^2}+\sum_{J=1}^{4}N^{-\tau_J}\|\boldsymbol{g}_J\|_{H^{\tau_J}(\Gamma_J)^2}\Bigg).\end{aligned}\tag{25.31}$$

Weighted versions of the $\mathbb{P}_N-\mathbb{P}_{N-2}$ method can be defined in a simple manner, in order to avoid the problem of spurious modes, but no numerical analysis is currently available.

As a conclusion, we state the main features of the second spectral discretization of the Stokes problem:

(i) the velocity is not exactly divergence-free;
(ii) the method is not of collocation type;

(iii) but there is no spurious mode for the pressure;

(iv) the best constant of the inf–sup condition on the pressure is of order $N^{-(d-1)/2}$.

For implementation reasons, the advantages of points (iii) and (iv) are relevant as is explained by MADAY, MEIRON, PATERA and RØNQUIST [1993]. Many applications of the method can be found in, e.g., AZAÏEZ [1990], FISHER and RØNQUIST [1994], RØNQUIST [1988] (where different choices of quadrature formulas are proposed for the approximation of the form $b(\bullet, \bullet)$).

26. The staggered grid method for the Stokes problem

The advantage of the third method is to be of collocation type without spurious modes on the pressure. The idea to obtain these properties is inherited from the finite difference techniques, it relies on the use of different grids for the velocity and the pressure, which are called *staggered grids*.

Indeed, the grid for the pressure is the Gauss grid defined in (14.1):

$$\Sigma_N = \{\boldsymbol{x} = (\zeta_{j_1}, \ldots, \zeta_{j_d}):\ 1 \leqslant j_1, \ldots, j_d \leqslant N\}, \tag{26.1}$$

but other grids are introduced, one for each component of the velocity: in the case $d = 2$,

$$\begin{aligned} \Xi_N^{(x)} &= \{\boldsymbol{x} = (\xi_i, \zeta_j):\ 1 \leqslant i \leqslant N-1,\ 1 \leqslant j \leqslant N\}, \\ \Xi_N^{(y)} &= \{\boldsymbol{x} = (\zeta_i, \xi_j):\ 1 \leqslant i \leqslant N,\ 1 \leqslant j \leqslant N-1\}, \end{aligned} \tag{26.2}$$

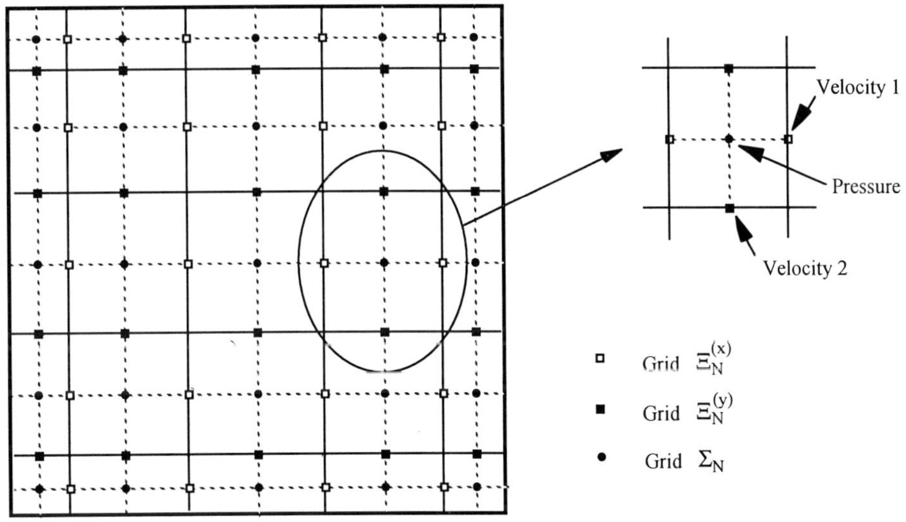

FIG. 26.1. The three grids in the square Ω.

in the case $d = 3$,

$$\begin{aligned}
\Xi_N^{(x)} &= \{\boldsymbol{x} = (\xi_i, \zeta_j, \zeta_k):\ 1 \leqslant i \leqslant N-1,\ 1 \leqslant j,k \leqslant N\},\\
\Xi_N^{(y)} &= \{\boldsymbol{x} = (\zeta_i, \xi_j, \zeta_k):\ 1 \leqslant i,k \leqslant N,\ 1 \leqslant j \leqslant N-1\},\\
\Xi_N^{(z)} &= \{\boldsymbol{x} = (\zeta_i, \zeta_j, \xi_k):\ 1 \leqslant i,j \leqslant N,\ 1 \leqslant k \leqslant N-1\}.
\end{aligned} \tag{26.3}$$

Note that the term "staggered" is completely justified, since the zeros ζ_j of L_N are alternate with the zeros ξ_j of L'_N, as illustrated in Fig. 26.1.

As in the previous section, we introduce the spaces $\mathbb{P}_{n_1,\ldots,n_d}(\Omega_d)$ of polynomials with degree $\leqslant n_j$ with respect to x_j. We firstly define the discrete space of pressures

$$M_N = \mathbb{P}_{N-1}(\Omega_d) \cap L_0^2(\Omega_d), \tag{26.4}$$

next the discrete space of velocity: in the case $d = 2$,

$$X_N = \big(\mathbb{P}_{N,N+1}(\Omega) \cap H_0^1(\Omega)\big) \times \big(\mathbb{P}_{N+1,N}(\Omega) \cap H_0^1(\Omega)\big), \tag{26.5}$$

in the case $d = 3$,

$$\begin{aligned}
X_N = {}& \big(\mathbb{P}_{N,N+1,N+1}(\Omega_3) \cap H_0^1(\Omega_3)\big) \times \big(\mathbb{P}_{N+1,N,N+1}(\Omega_3) \cap H_0^1(\Omega_3)\big)\\
& \times \big(\mathbb{P}_{N+1,N+1,N}(\Omega_3) \cap H_0^1(\Omega_3)\big).
\end{aligned} \tag{26.6}$$

Then, the discrete problem reads, in collocation formulation:

(i) in the case $d = 2$: *find a pair* $(\boldsymbol{u}_N, p_N)$ *in* $X_N \times M_N$, $\boldsymbol{u}_N = (u_N, v_N)$, *such that*

$$\begin{cases}
-\nu(\Delta u_N)(\boldsymbol{x}) + \big(\frac{\partial p_N}{\partial x}\big)(\boldsymbol{x}) = f(\boldsymbol{x}), & \boldsymbol{x} \in \Xi_N^{(x)},\\
-\nu(\Delta v_N)(\boldsymbol{x}) + \big(\frac{\partial p_N}{\partial y}\big)(\boldsymbol{x}) = g(\boldsymbol{x}), & \boldsymbol{x} \in \Xi_N^{(y)},\\
(\operatorname{div} \boldsymbol{u}_N)(\boldsymbol{x}) = 0, & \boldsymbol{x} \in \Sigma_N,
\end{cases} \tag{26.7}$$

where f and g are the components of the data $\boldsymbol{f}$,

(ii) in the case $d = 3$: *find a pair* $(\boldsymbol{u}_N, p_N)$ *in* $X_N \times M_N$, $\boldsymbol{u}_N = (u_N, v_N, w_N)$, *such that*

$$\begin{cases}
-\nu(\Delta u_N)(\boldsymbol{x}) + \big(\frac{\partial p_N}{\partial x}\big)(\boldsymbol{x}) = f(\boldsymbol{x}), & \boldsymbol{x} \in \Xi_N^{(x)},\\
-\nu(\Delta v_N)(\boldsymbol{x}) + \big(\frac{\partial p_N}{\partial y}\big)(\boldsymbol{x}) = g(\boldsymbol{x}), & \boldsymbol{x} \in \Xi_N^{(y)},\\
-\nu(\Delta w_N)(\boldsymbol{x}) + \big(\frac{\partial p_N}{\partial z}\big)(\boldsymbol{x}) = h(\boldsymbol{x}), & \boldsymbol{x} \in \Xi_N^{(z)},\\
(\operatorname{div} \boldsymbol{u}_N)(\boldsymbol{x}) = 0, & \boldsymbol{x} \in \Sigma_N,
\end{cases} \tag{26.8}$$

where f, g and h are the components of the data $\boldsymbol{f}$.

We must now give a variational formulation of problem (26.7) or (26.8). So, we introduce the following bilinear form, related to the previous grids: in the case $d = 2$, for any functions $\boldsymbol{u} = (u, v)$ and $\boldsymbol{z} = (z, t)$,

$$(\boldsymbol{u}, \boldsymbol{z})_{N\sim} = \sum_{i=0}^{N}\sum_{j=1}^{N} u(\xi_i, \zeta_j) z(\xi_i, \zeta_j)\, \rho_i \omega_j + \sum_{i=1}^{N}\sum_{j=0}^{N} v(\zeta_i, \xi_j) t(\zeta_i, \xi_j) \omega_i \rho_j, \tag{26.9}$$

in the case $d = 3$, for any functions $\boldsymbol{u} = (u, v, w)$ and $\boldsymbol{z} = (z, t, s)$,

$$\begin{aligned}(\boldsymbol{u}, \boldsymbol{z})_{N\sim} = &\sum_{i=0}^{N}\sum_{j=1}^{N}\sum_{k=1}^{N} u(\xi_i, \zeta_j, \zeta_k) z(\xi_i, \zeta_j, \zeta_k) \rho_i \omega_j \omega_k \\ &+ \sum_{i=1}^{N}\sum_{j=0}^{N}\sum_{k=1}^{N} v(\zeta_i, \xi_j, \zeta_k) t(\zeta_i, \xi_j, \zeta_k) \omega_i \rho_j \omega_k \\ &+ \sum_{i=1}^{N}\sum_{j=1}^{N}\sum_{k=0}^{N} w(\zeta_i, \zeta_j, \xi_k) s(\zeta_i, \zeta_j, \xi_k) \omega_i \omega_j \rho_k.\end{aligned} \tag{26.10}$$

And, of course, we set for any $\boldsymbol{u}_N$ and $\boldsymbol{v}_N$ in $\mathbb{P}_{N+1}(\Omega_d)^d$

$$\tilde{a}_N(\boldsymbol{u}_N, \boldsymbol{v}_N) = -\nu(\Delta \boldsymbol{u}_N, \boldsymbol{v}_N)_{N\sim}. \tag{26.11}$$

Now, it is readily checked from the properties of the Gauss and Gauss–Lobatto formulas that, for any polynomials $\boldsymbol{v}_N$ in X_N and p_N in M_N,

$$\begin{aligned}(\boldsymbol{v}_N, \mathbf{grad}\, p_N)_{N\sim} &= -(\operatorname{div} \boldsymbol{v}_N, p_N)_{N\sim} \\ &= -\sum_{i_1=1}^{N} \cdots \sum_{i_d=1}^{N} (\operatorname{div} \boldsymbol{v}_N)(\zeta_{i_1}, \ldots, \zeta_{i_1}) \omega_{i_1} \cdots \omega_{i_d}.\end{aligned}$$

So, we set

$$\tilde{b}_N(\boldsymbol{v}_N, q_N) = -\sum_{i_1=1}^{N} \cdots \sum_{i_d=1}^{N} (\operatorname{div} \boldsymbol{v}_N)(\zeta_{i_1}, \ldots, \zeta_{i_1}) \omega_{i_1} \cdots \omega_{i_d}.$$

Then problem (26.7) or (26.8) clearly admits the following equivalent variational formulation: *find a pair* $(\boldsymbol{u}_N, p_N)$ *in* $X_N \times M_N$ *such that*

$$\begin{aligned}&\forall \boldsymbol{v}_N \in X_N, \quad \tilde{a}_N(\boldsymbol{u}_N, \boldsymbol{v}_N) + \tilde{b}_N(\boldsymbol{v}_N, p_N) = (\boldsymbol{f}, \boldsymbol{v}_N)_{N\sim}, \\ &\forall q_N \in M_N, \quad \tilde{b}_N(\boldsymbol{u}_N, q_N) = 0.\end{aligned} \tag{26.12}$$

As always, the analysis of this problem relies on the properties of the forms $\tilde{a}_N(\bullet,\bullet)$ and $\tilde{b}_N(\bullet,\bullet)$.

Here also, it is readily checked that there is no spurious modes on the pressure, as stated in the following theorem.

THEOREM 26.1. *The set of polynomials q_N in M_N such that*

$$\forall \boldsymbol{v}_N \in X_N, \quad \tilde{b}_N(\boldsymbol{v}_N, q_N) = 0, \tag{26.13}$$

is reduced to $\{0\}$.

PROOF. We only give the proof in the case $d = 2$. In (26.13), we let $\boldsymbol{v}_N$ vary in $\mathbb{P}_N^0(\Omega)^2$. Since the forms $b(\bullet,\bullet)$ and $\tilde{b}_N(\bullet,\bullet)$ coincide on $\mathbb{P}_N^0(\Omega)^2 \times \mathbb{P}_{N-1}(\Omega)$, it follows from Remark 24.1 that the only possible spurious mode is $L_N'(x)L_N'(y)$. However, taking for instance $\boldsymbol{v}_N$ equal to $(0,(1-x^2)L_N'(x)(L_N - L_{N-2})(y))$ in (26.13), proves that $L_N'(x)L_N'(y)$ is not a spurious mode, whence the result.

REMARK 26.1. A first staggered grid method was proposed by MONTIGNY-RANNOU and MORCHOISNE [1987]: in the case $d = 2$ for instance, the problem consists in finding a pair $(\boldsymbol{u}_N, p_N)$ in $\mathbb{P}_N^0(\Omega)^2 \times M_N$ such that

$$\begin{cases} -\nu(\Delta \boldsymbol{u}_N)(\boldsymbol{x}) + (\mathbf{grad}\, p_N)(\boldsymbol{x}) = \boldsymbol{f}(\boldsymbol{x}), & \boldsymbol{x} \in \Xi_N, \\ (\operatorname{div} \boldsymbol{u}_N)(\boldsymbol{x}) = 0, & \boldsymbol{x} \in \Sigma_N. \end{cases} \tag{26.14}$$

Only two grids are involved in this formulation, independently of the dimension, however spurious modes appear in this problem, the mode $L_N'(x)L_N'(y)$ in the case $d = 2$ for instance. That is why it does not seem worth to use it.

We now study the continuity and the ellipticity of the form $\tilde{a}_N(\bullet,\bullet)$ which are stated in the following theorem.

THEOREM 26.2. *The bilinear form $\tilde{a}_N(\bullet,\bullet)$ satisfies the following properties of continuity*

$$\begin{aligned} &\forall \boldsymbol{u}_N \in \mathbb{P}_N^0(\Omega_d)^d,\ \forall \boldsymbol{z}_N \in \mathbb{P}_N^0(\Omega_d)^d, \\ &\qquad |\tilde{a}_N(\boldsymbol{u}_N, \boldsymbol{z}_N)| \leqslant c\,\|\boldsymbol{u}_N\|_{H^1(\Omega_d)^d}\|\boldsymbol{v}_N\|_{H^1(\Omega_d)^d}, \end{aligned} \tag{26.15}$$

and of ellipticity

$$\forall \boldsymbol{u}_N \in \mathbb{P}_N^0(\Omega_d)^d, \quad \tilde{a}_N(\boldsymbol{u}_N, \boldsymbol{u}_N) \geqslant cN^{1-d}\|\boldsymbol{u}_N\|^2_{H^1(\Omega_d)^d}. \tag{26.16}$$

This analysis relies on several one-dimensional inequalities which hold for any polynomial φ_N in $\mathbb{P}_{N-1}(\Lambda)$:

$$\int_{-1}^{1} \varphi_N^2(\zeta)(1-\zeta^2)\,\mathrm{d}\zeta \leqslant \sum_{j=1}^{N} \varphi_N^2(\zeta_j)(1-\zeta_j^2)\omega_j \leqslant 2\int_{-1}^{1} \varphi_N^2(\zeta)(1-\zeta^2)\,\mathrm{d}\zeta, \tag{26.17}$$

$$cN^{-1}\int_{-1}^{1} \varphi_N^2(\zeta)(1-\zeta^2)^2\,\mathrm{d}\zeta \leqslant \sum_{j=1}^{N} \varphi_N^2(\zeta_j)(1-\zeta_j^2)^2\omega_j \leqslant c'\int_{-1}^{1} \varphi_N^2(\zeta)(1-\zeta^2)^2\,\mathrm{d}\zeta. \tag{26.18}$$

These inequalities are proven by writing φ_N in the basis $\{L'_n,\ 1 \leqslant n \leqslant N\}$ for (26.17) and in the basis $\{L''_n,\ 2 \leqslant n \leqslant N+1\}$ for (26.18). Since the Gauss quadrature formula is exact on $\mathbb{P}_{2N-1}(\Lambda)$, it suffices to compute (see (4.8))

$$\sum_{j=1}^{N} L_N'^2(\zeta_j)(1-\zeta_j^2)\omega_j = 2N = \frac{2N+1}{N+1}\int_{-1}^{1} L_N'^2(\zeta)(1-\zeta^2)\,\mathrm{d}\zeta,$$

to derive (26.17), but we refer to BERNARDI and MADAY ([1988], Lemma 2) for the complete proof of (26.18) which is much more technical.

PROOF OF THEOREM 26.2. We give the proof only in the case $d = 2$. We have to check for instance that, for any u_N and z_N in $\mathbb{P}_{N,N+1}(\Omega)$,

$$\left|\sum_{i=0}^{N}\sum_{j=1}^{N}\left(\frac{\partial^2 u_N}{\partial x^2}\right)(\xi_i,\zeta_j)z_N(\xi_i,\zeta_j)\rho_i\omega_j\right| \leqslant c\left\|\frac{\partial u_N}{\partial x}\right\|_{L^2(\Omega)}\left\|\frac{\partial z_N}{\partial x}\right\|_{L^2(\Omega)}, \tag{26.19}$$

$$\left|\sum_{i=0}^{N}\sum_{j=1}^{N}\left(\frac{\partial^2 u_N}{\partial y^2}\right)(\xi_i,\zeta_j)z_N(\xi_i,\zeta_j)\rho_i\omega_j\right| \leqslant c\left\|\frac{\partial u_N}{\partial y}\right\|_{L^2(\Omega)}\left\|\frac{\partial z_N}{\partial y}\right\|_{L^2(\Omega)}, \tag{26.20}$$

and

$$\sum_{i=0}^{N}\sum_{j=1}^{N}\left(\frac{\partial^2 u_N}{\partial x^2}\right)(\xi_i,\zeta_j)u_N(\xi_i,\zeta_j)\rho_i\omega_j \geqslant cN^{-1}\left\|\frac{\partial u_N}{\partial x}\right\|^2_{L^2(\Omega)}, \tag{26.21}$$

$$\sum_{i=0}^{N}\sum_{j=1}^{N}\left(\frac{\partial^2 u_N}{\partial y^2}\right)(\xi_i,\zeta_j)u_N(\xi_i,\zeta_j)\,\rho_i\omega_j \geqslant c\left\|\frac{\partial u_N}{\partial y}\right\|^2_{L^2(\Omega)}. \tag{26.22}$$

(1) Due to the accuracy of the quadrature formula, we derive that

$$\sum_{i=0}^{N}\sum_{j=1}^{N}\left(\frac{\partial^2 u_N}{\partial x^2}\right)(\xi_i,\zeta_j)z_N(\xi_i,\zeta_j)\rho_i\omega_j$$
$$= \sum_{i=0}^{N}\sum_{j=1}^{N}\left(\frac{\partial u_N}{\partial x}\right)(\xi_i,\zeta_j)\left(\frac{\partial z_N}{\partial x}\right)(\xi_i,\zeta_j)\,\rho_i\omega_j,$$

so that both estimates (26.19) and (26.21) follow from (26.18), by observing that both

$$\frac{\partial u_N}{\partial x}(1-y^2)^{-1} \quad \text{and} \quad \frac{\partial z_N}{\partial x}(1-y^2)^{-1}$$

belong to $\mathbb{P}_{N-1}(\Omega)$.

(2) Here, the product $(\partial^2 u_N/\partial y^2)z_N$ belongs to $\mathbb{P}_{2N}(\Omega)$. The idea is to write

$$u_N(x,y) = (1-y^2)\sum_{n=1}^{N}\alpha_n(x)L'_n(y) \quad \text{and}$$
$$z_N(x,y) = (1-y^2)\sum_{n=1}^{N}\beta_n(x)\,L'_n(y), \tag{26.23}$$

so that

$$\left\|\frac{\partial u_N}{\partial y}\right\|^2_{L^2(\Omega)} = \sum_{n=1}^{N}\|\alpha_n\|^2_{L^2(\Lambda)}\frac{n^2(n+1)^2}{n+\frac{1}{2}} \quad \text{and}$$
$$\left\|\frac{\partial z_N}{\partial y}\right\|^2_{L^2(\Omega)} = \sum_{n=1}^{N}\|\beta_n\|^2_{L^2(\Lambda)}\frac{n^2(n+1)^2}{n+\frac{1}{2}}.$$

Using notation (13.16), we also have

$$\sum_{i=0}^{N}\sum_{j=1}^{N}\left(\frac{\partial^2 u_N}{\partial y^2}\right)(\xi_i,\zeta_j)z_N(\xi_i,\zeta_j)\rho_i\omega_j$$
$$= \sum_{n=1}^{N-1}[\alpha_n,\beta_n]_{N,1}\frac{n^2(n+1)^2}{n+\frac{1}{2}}$$
$$+ [\alpha_N,\beta_N]_{N,1}N(N+1)\sum_{j=0}^{N}L'^2_N(\zeta_j)(1-\zeta_j^2)\omega_j.$$

Then, inequalities (26.20) and (26.22) are derived by applying a Cauchy–Schwarz inequality and (13.18) in each term, and also (26.17) with respect to y in the last term.

REMARK 26.2. The constant of ellipticity in (26.16) tends to 0 when N tends to $+\infty$, which is clearly an important drawback as far as the stability of problem (26.12) is concerned. However, the order of N in this constant cannot be improved as can be seen by taking $\boldsymbol{u}_N = (u_N, 0)$ or $\boldsymbol{u}_N = (u_N, 0, 0)$ with

$$u_N(x_1, \ldots, x_d) = t_N(x_1) \cdots t_N(x_{d-1})\big(1 - x_d^2\big) L''_{N+1}(x_d), \tag{26.24}$$

where t_N is any polynomial in $\mathbb{P}_N^0(\Lambda)$ satisfying

$$c_1 \leqslant \|t_N\|_{L^2(\Lambda)} + N^{-2}\|t'_N\|_{L^2(\Lambda)} \leqslant c_2. \tag{26.25}$$

Such a polynomial exists, take for instance

$$t_N(\zeta) = N^{-1}\left(\sum_{n=1}^{[N/2]} n^2 (L_{n+1} - L_{n-1})(\zeta) \right.$$
$$\left. + \sum_{n=[N/2+1]}^{N} (N-n)^2 (L_{n+1} - L_{n-1})(\zeta)\right). \tag{26.26}$$

Combining Theorems 26.1 and 26.2 leads to the first results for problem (26.12).

THEOREM 26.3. *For any continuous function $\boldsymbol{f}$ on Ω_d, problem* (26.12) *has a unique solution $(\boldsymbol{u}_N, p_N)$ in $X_N \times (\mathbb{P}_{N-1}(\Omega_d) \cap L_0^2(\Omega_d))$.*

THEOREM 26.4. *The function $\boldsymbol{f}$ is supposed to be in a space $H^\sigma(\Omega_d)^d$, $\sigma > d/2$. Let us assume that the solution $(\boldsymbol{u}, p)$ of problem* (24.1) *belongs to $H^s(\Omega_d)^d \times H^{s-1}(\Omega_d)$ for a real number $s \geqslant 1$. Then, the following error estimate holds for problem* (26.12):

$$|\boldsymbol{u} - \boldsymbol{u}_N|_{H^1(\Omega_d)^d} \leqslant c\big(N^{d-s}\big(\|\boldsymbol{u}\|_{H^s(\Omega_d)^d} + \|p\|_{H^{s-1}(\Omega_d)}\big)$$
$$+ N^{d-1-\sigma}\|\boldsymbol{f}\|_{H^\sigma(\Omega_d)^d}\big). \tag{26.27}$$

Also, the same duality argument as in Theorem 25.4 allows for deriving the estimate in $L^2(\Omega_d)$.

THEOREM 26.5. *If the hypotheses of Theorem* 26.4 *are satisfied, the following error estimate holds for problem* (26.12):

$$\|\boldsymbol{u} - \boldsymbol{u}_N\|_{L^2(\Omega_d)^d} \leqslant c\big(N^{d-1-s}\big(\|\boldsymbol{u}\|_{H^s(\Omega_d)^d} + \|p\|_{H^{s-1}(\Omega_d)}\big)$$
$$+ N^{d-1-\sigma}\|\boldsymbol{f}\|_{H^\sigma(\Omega_d)^d}\big). \tag{26.28}$$

It remains to estimate the inf–sup condition on the form $\tilde{b}_N(\cdot,\cdot)$. Note that, from Remark 25.1 with N replaced by $N+1$, we know that

$$\inf_{q_N\in M_N}\sup_{\boldsymbol{v}_N\in X_N}\frac{b(\boldsymbol{v}_N,q_N)}{\|\boldsymbol{v}_N\|_{H^1(\Omega_d)^d}\|q_N\|_{L^2(\Omega_d)}}\geqslant cN^{-(d-1)/2}. \tag{26.29}$$

Hence, exactly the same arguments as in the proofs of Theorems 25.5 to 25.7 with the integral on Ω_d replaced by the Gauss quadrature formula, lead to the following result.

THEOREM 26.6. *The bilinear from $\tilde{b}_N(\cdot,\cdot)$ satisfies the following inf–sup condition*

$$\inf_{q_N\in M_N}\sup_{\boldsymbol{v}_N\in X_N}\frac{\tilde{b}_N(\boldsymbol{v}_N,q_N)}{\|\boldsymbol{v}_N\|_{H^1(\Omega_d)^d}\|q_N\|_{L^2(\Omega_d)}}\geqslant cN^{-(d-1)/2}. \tag{26.30}$$

This allows for writing the estimate on the pressure.

THEOREM 26.7. *If the hypotheses of Theorem* 26.4 *are satisfied, the following error estimate holds for problem* (26.12):

$$\begin{aligned}\|p-p_N\|_{L^2(\Omega_d)}\leqslant c\big(&N^{(3d-1)/2-s}\big(\|\boldsymbol{u}\|_{H^s(\Omega_d)^d}+\|p\|_{H^{s-1}(\Omega_d)}\big)\\&+N^{3(d-1)/2-\sigma}\|\boldsymbol{f}\|_{H^\sigma(\Omega_d)^d}\big).\end{aligned} \tag{26.31}$$

As a conclusion, this method is of collocation type and it has no spurious modes on the pressure. But, due to its loss of optimality (N^3 on the pressure in the case $d=3$!), it seems not worth to use it, in comparison with the two methods presented in Sections 24 and 25. We refer to LEQUÉRÉ and PÉCHEUX [1990] for details on the implementation.

27. Extension to the Navier–Stokes equations

Our aim now is to extend the results of Sections 24 and 25 to the full nonlinear Navier–Stokes equations. For the sake of simplicity, we limit ourselves to the case of homogeneous boundary conditions on the velocity, i.e., we consider the problem:

$$\begin{cases}-\nu\Delta\boldsymbol{u}+(\boldsymbol{u}\cdot\nabla)\boldsymbol{u}+\mathbf{grad}\,p=\boldsymbol{f} & \text{in }\Omega_d,\\ \operatorname{div}\boldsymbol{u}=0 & \text{in }\Omega_d,\\ \boldsymbol{u}=\mathbf{0} & \text{on }\partial\Omega_d.\end{cases} \tag{27.1}$$

Let us introduce the trilinear form $c(\cdot,\cdot,\cdot)$ on $H^1(\Omega_d)^d\times H^1(\Omega_d)^d\times H^1(\Omega_d)^d$:

$$\begin{aligned}c(\boldsymbol{w},\boldsymbol{u},\boldsymbol{v})&=\int_{\Omega_d}(\boldsymbol{w}\cdot\nabla)\boldsymbol{u}\cdot\boldsymbol{v}\,\mathrm{d}\boldsymbol{x}\\&=\sum_{i=1}^d\sum_{j=1}^d\int_{\Omega_d}w_j(\boldsymbol{x})\Big(\frac{\partial u_i}{\partial x_j}\Big)(\boldsymbol{x})v_i(\boldsymbol{x})\,\mathrm{d}\boldsymbol{x},\end{aligned} \tag{27.2}$$

with obvious notation for the components of $\boldsymbol{u}$, $\boldsymbol{v}$ and $\boldsymbol{w}$. Since $H^1(\Omega_d)$ is imbedded into $L^4(\Omega_d)$ with a continuous and compact imbedding in the cases of dimension $d = 2$ and $d = 3$ (this comes from the Sobolev imbedding theorem, see ADAMS ([1975], Theorem 5.4) and NEČAS ([1967], Chapter 2, Theorem 3.4) for instance), this form is continuous. Note the important symmetry property (recall that K is the space of divergence-free functions in $H_0^1(\Omega_d)^d$)

$$\forall \boldsymbol{v} \in H^1(\Omega_d)^d,\ \forall \boldsymbol{w} \in K, \quad c(\boldsymbol{w}, \boldsymbol{v}, \boldsymbol{v}) = 0. \tag{27.3}$$

When the data $\boldsymbol{f}$ belong to $H^{-1}(\Omega_d)^d$, it is readily checked that problem (27.1) admits the following equivalent formulation: *find a pair* $(\boldsymbol{u}, p)$ *in* $H_0^1(\Omega_d)^d \times L_0^2(\Omega_d)$ *such that*

$$\begin{aligned} &\forall \boldsymbol{v} \in H_0^1(\Omega_d)^d, \quad a(\boldsymbol{u}, \boldsymbol{v}) + c(\boldsymbol{u}, \boldsymbol{u}, \boldsymbol{v}) + b(\boldsymbol{v}, p) = \langle \boldsymbol{f}, \boldsymbol{v} \rangle, \\ &\forall q \in L^2(\Omega_d), \qquad b(\boldsymbol{u}, q) = 0. \end{aligned} \tag{27.4}$$

Since studying the Navier–Stokes equations is not the subject of this article, we refer to GIRAULT and RAVIART [1986] and TEMAM [1977] for the following basic results on problem (27.1).

THEOREM 27.1. *For any* $\boldsymbol{f}$ *in* $H^{-1}(\Omega_d)^d$, *problem* (27.1) *has at least one solution* $(\boldsymbol{u}, p)$ *in* $H_0^1(\Omega_d)^d \times L_0^2(\Omega_d)$. *Moreover, this solution satisfies*

$$\|\boldsymbol{u}\|_{H^1(\Omega_d)^d} + \|p\|_{L^2(\Omega_d)} \leqslant c\|\boldsymbol{f}\|_{H^{-1}(\Omega_d)^d}. \tag{27.5}$$

Next, let δ be the norm of the trilinear form $c(\bullet, \bullet, \bullet)$ on $H_0^1(\Omega_d)^d \times H_0^1(\Omega_d)^d \times H_0^1(\Omega_d)^d$.

THEOREM 27.2. *For any* $\boldsymbol{f}$ *in* $H^{-1}(\Omega_d)^d$ *such that*

$$\frac{\delta}{\nu^2}\|\boldsymbol{f}\|_{H^{-1}(\Omega_d)^d} < 1, \tag{27.6}$$

problem (27.1) *has a unique solution* $(\boldsymbol{u}, p)$ *in* $H_0^1(\Omega_d)^d \times L_0^2(\Omega_d)$.

However, the hypothesis which ensures the uniqueness of the solution is too restrictive. In order to make a weaker assumption, we introduce the following formulation. Let us now denote by T_S the operator from $H^{-1}(\Omega_d)^d$ into $H_0^1(\Omega_d)^d$ which associates with $\boldsymbol{f}$ the velocity $\boldsymbol{u}$ in the solution $(\boldsymbol{u}, p)$ of problem (24.1). Setting

$$\forall \boldsymbol{v} \in H_0^1(\Omega_d)^d, \quad \langle \boldsymbol{G}(\boldsymbol{u}), \boldsymbol{v} \rangle = c(\boldsymbol{u}, \boldsymbol{u}, \boldsymbol{v}) - \langle \boldsymbol{f}, \boldsymbol{v} \rangle, \tag{27.7}$$

we observe that problem (27.4) can equivalently be written

$$\boldsymbol{u} + T_S \boldsymbol{G}(\boldsymbol{u}) = \boldsymbol{0}. \tag{27.8}$$

The mapping $\boldsymbol{G}$ is continuously differentiable from $H_0^1(\Omega_d)^d$ into $H^{-1}(\Omega_d)^d$, with a compact derivative. In all that follows, we assume that

$$\text{the operator } \mathbb{I} + T_{\mathrm{S}}\mathrm{D}\boldsymbol{G}(\boldsymbol{u}) \text{ is an isomorphism of } H_0^1(\Omega_d)^d. \tag{27.9}$$

Due to the local inversion theorem, this hypothesis ensures that the solution $(\boldsymbol{u}, p)$ of problem (27.1) is locally unique. This is sufficient for the numerical analysis of its discretization.

REMARK 27.1. By a boot-strap argument relying on equation (27.8), it is readily checked that the regularity results of Remark 23.3 can be extended to problem (27.1).

The idea to construct the discretization of problem (27.1) is of course to combine one of the methods which have been studied for the Stokes problem with an appropriate treatment of the nonlinear term. In order to do that, we denote by T_{SN} any of the two following operators:

(i) the operator T_{SN}^1 associates with any function $\boldsymbol{f}$ in $H^{-1}(\Omega_d)^d$ the polynomial $\boldsymbol{u}_N$, where $(\boldsymbol{u}_N, p_N)$ is the solution in $\mathbb{P}_N^0(\Omega_d)^d \times M_N$ of the problem

$$\begin{aligned} &\forall \boldsymbol{v}_N \in \mathbb{P}_N^0(\Omega_d)^d, \quad a_N(\boldsymbol{u}_N, \boldsymbol{v}_N) + b_N(\boldsymbol{v}_N, p_N) = \langle \boldsymbol{f}, \boldsymbol{v}_N \rangle, \\ &\forall q_N \in M_N, \qquad b_N(\boldsymbol{u}_N, q_N) = 0, \end{aligned} \tag{27.10}$$

where M_N satisfies the hypothesis (24.17) (this is a modified version of problem (24.5));

(ii) the operator T_{SN}^2 associates with any function $\boldsymbol{f}$ in $H^{-1}(\Omega_d)^d$ the polynomial $\boldsymbol{u}_N$, where $(\boldsymbol{u}_N, p_N)$ is the solution in $\mathbb{P}_N^0(\Omega_d)^d \times (\mathbb{P}_{N-2}(\Omega_d) \cap L_0^2(\Omega_d))$ of the problem

$$\begin{aligned} &\forall \boldsymbol{v}_N \in \mathbb{P}_N^0(\Omega_d)^d, \quad a_N(\boldsymbol{u}_N, \boldsymbol{v}_N) + b(\boldsymbol{v}_N, p_N) = \langle \boldsymbol{f}, \boldsymbol{v}_N \rangle, \\ &\forall q_N \in \mathbb{P}_{N-2}(\Omega_d), \quad b(\boldsymbol{u}_N, q_N) = 0 \end{aligned} \tag{27.11}$$

(this is a modified version of problem (25.1)).
We state the two basic properties of these operators which will be needed in this section and can easily be derived from Sections 24 and 25: the first one is a stability estimate

$$\|T_{\mathrm{SN}}\boldsymbol{f}\|_{H^1(\Omega_d)^d} \leqslant c \sup_{\boldsymbol{w}_N \in \mathbb{P}_N^0(\Omega_d)^d} \frac{\langle \boldsymbol{f}, \boldsymbol{w}_N \rangle}{\|\boldsymbol{w}_N\|_{H^1(\Omega_d)^d}}, \tag{27.12}$$

the second one is an error estimate which holds when the function $T_{\mathrm{S}}\boldsymbol{f}$ belongs to $H^s(\Omega_d)^d$ for a real number $s \geqslant 1$,

$$\|(T_{\mathrm{S}} - T_{\mathrm{SN}})\boldsymbol{f}\|_{H^1(\Omega_d)^d} \leqslant cN^{1-s}\|T_{\mathrm{S}}\boldsymbol{f}\|_{H^s(\Omega_d)^d}. \tag{27.13}$$

In formulation (27.4), from the symmetry property (27.3), the form $c(\boldsymbol{u}, \boldsymbol{u}, \boldsymbol{v})$ can equivalently be written

$$c(\boldsymbol{u}, \boldsymbol{u}, \boldsymbol{v}) = -\sum_{i=1}^{d}\sum_{j=1}^{d}\int_{\Omega_d} u_i(\boldsymbol{x})u_j(\boldsymbol{x})\left(\frac{\partial v_i}{\partial x_j}\right)(\boldsymbol{x})\,\mathrm{d}\boldsymbol{x}, \tag{27.14}$$

or also

$$c(\boldsymbol{u}, \boldsymbol{u}, \boldsymbol{v}) = \int_{\Omega_d} (\mathbf{curl}\,\boldsymbol{u})(\boldsymbol{x}) \cdot (\boldsymbol{u} \times \boldsymbol{v})(\boldsymbol{x})\,\mathrm{d}\boldsymbol{x}.$$

But, if these forms are equivalent for the continuous problem, their approximations lead to slightly different discrete problems. Several choices are possible and give equivalent results. One of them is to define the trilinear form $c_N(\cdot,\cdot,\cdot)$ on polynomials in $\mathbb{P}_N^0(\Omega_d)^d$ by the formula

$$c_N(\boldsymbol{w}_N, \boldsymbol{u}_N, \boldsymbol{v}_N) = -\sum_{i=1}^{d}\sum_{j=1}^{d}\left(w_{Ni}u_{Nj}, \frac{\partial v_{Ni}}{\partial x_j}\right)_N. \tag{27.15}$$

Next, we set

$$\langle \boldsymbol{G}_N(\boldsymbol{u}), \boldsymbol{v}_N\rangle = c_N(\boldsymbol{u}_N, \boldsymbol{u}_N, \boldsymbol{v}_N) - (\boldsymbol{f}, \boldsymbol{v}_N)_N. \tag{27.16}$$

With the previous notation, we simply define the discrete problem for the velocity by

$$\boldsymbol{u}_N + T_{\mathrm{SN}}\boldsymbol{G}(\boldsymbol{u}_N) = 0. \tag{27.17}$$

Let us make explicit its variational form:

(i) the first discretization reads: *find a pair* $(\boldsymbol{u}_N, p_N)$ *in* $\mathbb{P}_N^0(\Omega_d)^d \times M_N$ *such that*

$$\begin{aligned} &\forall \boldsymbol{v}_N \in \mathbb{P}_N^0(\Omega_d)^d,\\ &\qquad a_N(\boldsymbol{u}_N, \boldsymbol{v}_N) + c_N(\boldsymbol{u}_N, \boldsymbol{u}_N, \boldsymbol{v}_N) + b_N(\boldsymbol{v}_N, p_N) = (\boldsymbol{f}, \boldsymbol{v}_N)_N,\\ &\forall q_N \in M_N, \quad b_N(\boldsymbol{u}_N, q_N) = 0; \end{aligned} \tag{27.18}$$

(ii) the second discretization reads: *find a pair* $(\boldsymbol{u}_N, p_N)$ *in*

$$\mathbb{P}_N^0(\Omega_d)^d \times (\mathbb{P}_{N-2}(\Omega_d) \cap L_0^2(\Omega_d))$$

such that

$$\begin{aligned}&\forall \boldsymbol{v}_N \in \mathbb{P}_N^0(\Omega_d)^d,\\&\qquad a_N(\boldsymbol{u}_N,\boldsymbol{v}_N)+c_N(\boldsymbol{u}_N,\boldsymbol{u}_N,\boldsymbol{v}_N)+b(\boldsymbol{v}_N,p_N)=(\boldsymbol{f},\boldsymbol{v}_N)_N,\\&\forall q_N\in\mathbb{P}_{N-2}(\Omega_d),\quad b(\boldsymbol{u}_N,q_N)=0.\end{aligned} \tag{27.19}$$

REMARK 27.2. In the case of the first discretization and when hypothesis (24.17) holds, problem (27.18) admits the equivalent collocation formulation: *find a pair* $(\boldsymbol{u}_N,p_N)$ *in* $\mathbb{P}_N(\Omega_d)^d\times M_N$ *such that*

$$\begin{cases}-\nu(\Delta\boldsymbol{u}_N)(\boldsymbol{x})+\displaystyle\sum_{j=1}^d\left(\frac{\partial}{\partial x_j}\mathcal{I}_N(u_{Nj}\boldsymbol{u}_N)\right)(\boldsymbol{x})+(\mathbf{grad}\,p_N)(\boldsymbol{x})\\\quad=\boldsymbol{f}(\boldsymbol{x}),\ \boldsymbol{x}\in\Xi_N\cap\Omega_d,\\(\operatorname{div}\boldsymbol{u}_N)(\boldsymbol{x})=0,\quad \boldsymbol{x}\in\Xi_N,\\\boldsymbol{u}_N(\boldsymbol{x})=\boldsymbol{0},\quad \boldsymbol{x}\in\Xi_N\cap\partial\Omega_d.\end{cases} \tag{27.20}$$

The computation of the nonlinear term only requires the derivation of a polynomial in $\mathbb{P}_N(\Omega_d)^{d^2}$ which is known by its values on the collocation grid.

The numerical analysis of problem (27.17) relies on the discrete implicit function theorem of BREZZI, RAPPAZ and RAVIART [1980] (see also CROUZEIX and RAPPAZ ([1989], Theorem 3.1)). Let us just state this result in the special case that we need: we introduce a polynomial $\tilde{\boldsymbol{u}}_N$ in $\mathbb{P}_N^0(\Omega_d)^d$ such that $\mathbb{I}+T_{\mathrm{SN}}\mathrm{D}\boldsymbol{G}_N(\tilde{u}_N)$ is an isomorphism of $\mathbb{P}_N^0(\Omega_d)^d$, and we set

$$\begin{aligned}&\varepsilon_N=\left\|\tilde{u}_N+T_{\mathrm{SN}}\boldsymbol{G}_N(\tilde{\boldsymbol{u}}_N)\right\|_{H^1(\Omega_d)^d},\\&\gamma_N=\left\|\left(\mathbb{I}+T_{\mathrm{SN}}\mathrm{D}\boldsymbol{G}_N(\tilde{\boldsymbol{u}}_N)\right)^{-1}\right\|_{\mathscr{L}(\mathbb{P}_N^0(\Omega_d)^d,\mathbb{P}_N^0(\Omega_d)^d)},\\&C_N(\alpha)=\sup\left\{\left\|T_{\mathrm{SN}}\mathrm{D}\boldsymbol{G}_N(\tilde{\boldsymbol{u}}_N)-T_{\mathrm{SN}}\mathrm{D}\boldsymbol{G}_N(\boldsymbol{v}_N)\right\|_{\mathscr{L}(\mathbb{P}_N^0(\Omega_d)^d,\mathbb{P}_N^0(\Omega_d)^d)}:\right.\\&\qquad\qquad\left.\boldsymbol{v}_N\in\mathbb{P}_N^0(\Omega_d)^d,\ \|\tilde{\boldsymbol{u}}_N-\boldsymbol{v}_N\|_{H^1(\Omega_d)^d}\leqslant\alpha\right\}.\end{aligned} \tag{27.21}$$

If the quantity $2\gamma_N C_N(2\gamma_N\varepsilon_N)<1$, for all $\alpha>2\gamma_N\varepsilon_N$ such that $\gamma_N C_N(\alpha)<1$, there exists a unique solution of problem (27.17) in the ball

$$\left\{\boldsymbol{v}_N\in\mathbb{P}_N^0(\Omega_d)^d:\ \|\tilde{\boldsymbol{u}}_N-\boldsymbol{v}_N\|_{H^1(\Omega_d)^d}\leqslant\alpha\right\}. \tag{27.22}$$

Moreover, this solution satisfies

$$\|\boldsymbol{u}_N-\tilde{\boldsymbol{u}}_N\|_{H^1(\Omega_d)^d}\leqslant\frac{\gamma_N}{1-\gamma_N C_N(\alpha)}\varepsilon_N. \tag{27.23}$$

So, we choose $\tilde{\boldsymbol{u}}_N$ equal to $\Pi_M^{1,0}\boldsymbol{u}$, where M stands for the integral part of $(N-1)/2$ and we are going to apply this result to study problem (27.17). Firstly, we state the main result.

THEOREM 27.3. *The function $\boldsymbol{f}$ is supposed to be in a space $H^\sigma(\Omega_d)^d$, $\sigma > d/2$. If hypothesis (27.9) holds, there exists a positive constant α such that, for N large enough, problem (27.17) has a unique solution in the ball (27.22). Moreover, let us assume that the solution $(\boldsymbol{u}, p)$ of problem (27.1) belongs to $H^s(\Omega_d)^d \times H^{s-1}(\Omega_d)$ for a real number $s \geqslant 1$. Then, the following error estimate holds for problem (27.17):*

$$\begin{aligned}\|\boldsymbol{u}-\boldsymbol{u}_N\|_{H^1(\Omega_d)^d} \leqslant c\big(&N^{1-s}\left(\|\boldsymbol{u}\|_{H^s(\Omega_d)^d}+\|p\|_{H^{s-1}(\Omega_d)}\right)\\ &+N^{-\sigma}\|\boldsymbol{f}\|_{H^\sigma(\Omega_d)^d}\big).\end{aligned} \tag{27.24}$$

We begin with a technical estimate, which shows that the treatment of the nonlinear term is correct.

LEMMA. *For any polynomials $\boldsymbol{u}_N$, $\boldsymbol{v}_N$ and $\boldsymbol{w}_N$ in $\mathbb{P}_N^0(\Omega_d)^d$, the following estimate holds*

$$\begin{aligned}&c(\boldsymbol{w}_N, \boldsymbol{u}_N, \boldsymbol{v}_N) - c_N(\boldsymbol{w}_N, \boldsymbol{u}_N, \boldsymbol{v}_N)\\ &\quad \leqslant N^{-1/4}\|\boldsymbol{u}_N\|_{H^1(\Omega_d)^d}\|\boldsymbol{v}_N\|_{H^1(\Omega_d)^d}\|\boldsymbol{w}_N\|_{H^1(\Omega_d)^d}.\end{aligned} \tag{27.25}$$

PROOF. From the identity

$$\begin{aligned}&c(\boldsymbol{w}_N, \boldsymbol{u}_N, \boldsymbol{v}_N) - c_N(\boldsymbol{w}_N, \boldsymbol{u}_N, \boldsymbol{v}_N)\\ &\quad = -\sum_{i=1}^{d}\sum_{j=1}^{d}\bigg(\int_{\Omega_d}\big(w_{Ni}u_{Nj} - \Pi_{N-1}(w_{Ni}u_{Nj})\big)(\boldsymbol{x})\left(\frac{\partial v_{Ni}}{\partial x_j}\right)(\boldsymbol{x})\,\mathrm{d}\boldsymbol{x}\\ &\qquad - \left(\mathcal{I}_N(w_{Ni}u_{Nj}) - \Pi_{N-1}(w_{Ni}u_{Nj}), \frac{\partial v_{Ni}}{\partial x_j}\right)_N,\end{aligned}$$

we derive that

$$\begin{aligned}&c(\boldsymbol{w}_N, \boldsymbol{u}_N, \boldsymbol{v}_N) - c_N(\boldsymbol{w}_N, \boldsymbol{u}_N, \boldsymbol{v}_N)\\ &\quad \leqslant c\sum_{i=1}^{d}\sum_{j=1}^{d}\big(\|w_{Ni}u_{Nj} - \Pi_{N-1}(w_{Ni}u_{Nj})\|_{L^2(\Omega_d)}\\ &\qquad + \|\mathcal{I}_N\big(w_{Ni}u_{Nj} - \Pi_{N-1}(w_{Ni}u_{Nj})\big)\|_{L^2(\Omega_d)}\big)\left\|\frac{\partial \boldsymbol{v}_{Ni}}{\partial x_j}\right\|_{L^2(\Omega_d)}.\end{aligned}$$

Next, we know from (13.28) that for any polynomial φ_{2N} in $\mathbb{P}_{2N}^0(\Lambda)$,

$$\|i_N\varphi_{2N}\|_{L^2(\Lambda)} \leqslant c\|\varphi_{2N}\|_{L^2(\Lambda)}. \tag{27.26}$$

Applying this stability estimate to each polynomial $w_{Ni}u_{Nj}$ in each direction, we simply obtain

$$\begin{aligned}
&c(\boldsymbol{w}_N, \boldsymbol{u}_N, \boldsymbol{v}_N) - c_N(\boldsymbol{w}_N, \boldsymbol{u}_N, \boldsymbol{v}_N) \\
&\quad \leqslant c \sum_{i=1}^{d} \sum_{j=1}^{d} \big\| w_{Ni}u_{Nj} - \Pi_{N-1}(w_{Ni}u_{Nj}) \big\|_{L^2(\Omega_d)} \left\| \frac{\partial v_{Ni}}{\partial x_j} \right\|_{L^2(\Omega_d)}.
\end{aligned}$$

Since the multiplication is continuous from $H^1(\Omega_d) \times H^1(\Omega_d)$ into $H^{1/4}(\Omega_d)$ for d equal to 2 or 3, the result follows from Theorem 7.1.

PROOF OF THEOREM 27.3. We have to compute the quantities ε_N, γ_N and $C_N(\alpha)$.

(1) Let us begin with γ_N. Due to hypothesis (27.9), we write

$$\begin{aligned}
\mathbb{I} + T_{\mathrm{SN}}\mathrm{D}\boldsymbol{G}_N(\tilde{\boldsymbol{u}}_N) &= \mathbb{I} + T_{\mathrm{S}}\mathrm{D}\boldsymbol{G}(\boldsymbol{u}) + (T_{\mathrm{S}} - T_{\mathrm{SN}})\mathrm{D}\boldsymbol{G}(\boldsymbol{u}) \\
&\quad + T_{\mathrm{SN}}\big(\mathrm{D}\boldsymbol{G}(\boldsymbol{u}) - \mathrm{D}\boldsymbol{G}(\tilde{\boldsymbol{u}}_N)\big) + T_{\mathrm{SN}}\big(\mathrm{D}\boldsymbol{G}(\tilde{\boldsymbol{u}}_N) - D\boldsymbol{G}_N(\tilde{\boldsymbol{u}}_N)\big). \qquad (27.27)
\end{aligned}$$

By hypothesis (27.9), the operator $\mathbb{I} + T_{\mathrm{S}}\mathrm{D}\boldsymbol{G}(\boldsymbol{u})$ is an isomorphism of $H_0^1(\Omega_d)^d$. Furthermore, the term $(T_{\mathrm{S}} - T_{\mathrm{SN}})\mathrm{D}\boldsymbol{G}(\boldsymbol{u})$ tends to $\mathbf{0}$ in $H^1(\Omega_d)^d$ by (27.13), while the term $T_{\mathrm{SN}}(\mathrm{D}\boldsymbol{G}(\boldsymbol{u}) - \mathrm{D}\boldsymbol{G}(\tilde{\boldsymbol{u}}_N))$ also tends to $\mathbf{0}$ by (27.12) and the continuity of the linear mapping $\mathrm{D}\boldsymbol{G}$. It can finally be checked that the term $T_{\mathrm{SN}}(\mathrm{D}\boldsymbol{G}(\tilde{\boldsymbol{u}}_N) - \mathrm{D}\boldsymbol{G}_N(\tilde{\boldsymbol{u}}_N))$ tends to $\mathbf{0}$, this comes from the formula

$$\begin{aligned}
&\langle \mathrm{D}\boldsymbol{G}(\tilde{\boldsymbol{u}}_N) - \mathrm{D}\boldsymbol{G}_N(\tilde{\boldsymbol{u}}_N) \cdot \boldsymbol{w}_N, \boldsymbol{v}_N \rangle \\
&\quad = c(\boldsymbol{w}_N, \tilde{\boldsymbol{u}}_N, \boldsymbol{v}_N) - c_N(\boldsymbol{w}_N, \tilde{\boldsymbol{u}}_N, \boldsymbol{v}_N) \\
&\qquad + c(\tilde{\boldsymbol{u}}_N, \boldsymbol{w}_N, \boldsymbol{v}_N) - c_N(\tilde{\boldsymbol{u}}_N, \boldsymbol{w}_N, \boldsymbol{v}_N), \qquad (27.28)
\end{aligned}$$

combined with estimates (27.25) and (27.12). Putting together all these considerations, we derive that, for N large enough, the operator $\mathbb{I} + T_{\mathrm{SN}}\mathrm{D}\boldsymbol{G}_N(\tilde{\boldsymbol{u}}_N)$ is an isomorphism of $\mathbb{P}_N^0(\Omega_d)^d$ and that the norm γ_N of its inverse is bounded by a constant γ independent of N.

(2) To bound $C_N(\alpha)$, we make use of the decomposition

$$\begin{aligned}
&\mathrm{D}\boldsymbol{G}_N(\tilde{\boldsymbol{u}}_N) - \mathrm{D}\boldsymbol{G}_N(\boldsymbol{v}_N) \\
&\quad = -\big(\mathrm{D}\boldsymbol{G}(\tilde{\boldsymbol{u}}_N) - \mathrm{D}\boldsymbol{G}_N(\tilde{\boldsymbol{u}}_N)\big) + \big(\mathrm{D}\boldsymbol{G}(\tilde{\boldsymbol{u}}_N) - \mathrm{D}\boldsymbol{G}(\boldsymbol{v}_N)\big) \\
&\qquad + \big(\mathrm{D}\boldsymbol{G}(\boldsymbol{v}_N) - \mathrm{D}\boldsymbol{G}_N(\boldsymbol{v}_N)\big),
\end{aligned}$$

we derive from the continuity of $\mathrm{D}\boldsymbol{G}$ that $\mathrm{D}\boldsymbol{G}(\tilde{\boldsymbol{u}}_N) - \mathrm{D}\boldsymbol{G}(\boldsymbol{v}_N)$ is bounded by a constant times the norm $\|\tilde{\boldsymbol{u}}_N - \boldsymbol{v}_N\|_{H^1(\Omega_d)^d}$ and we bound the other differences by the same arguments as previously. We obtain that $C_N(\alpha)$ is bounded by $c\alpha$, where c is a positive constant independent of N.

(3) It remains to compute ε_N. We observe from (27.8) that

$$\tilde{\boldsymbol{u}}_N + T_{\mathrm{SN}}\boldsymbol{G}_N(\tilde{\boldsymbol{u}}_N) = \tilde{\boldsymbol{u}}_N - \boldsymbol{u} + (T_{\mathrm{SN}} - T_{\mathrm{S}})\boldsymbol{G}(\boldsymbol{u}) + T_{\mathrm{SN}}\big(\boldsymbol{G}(\tilde{\boldsymbol{u}}_N) - \boldsymbol{G}(\boldsymbol{u})\big) \\ + T_{\mathrm{SN}}\big(\boldsymbol{G}_N(\tilde{\boldsymbol{u}}_N) - \boldsymbol{G}(\tilde{\boldsymbol{u}}_N)\big).$$

It follows from the choice of $\tilde{\boldsymbol{u}}_N$ in $\mathbb{P}_M(\Omega)^d$ that, for any $\boldsymbol{v}_N$ in $\mathbb{P}_N(\Omega_d)^d$,

$$c_N(\tilde{\boldsymbol{u}}_N, \tilde{\boldsymbol{u}}_N, \boldsymbol{v}_N) = c(\tilde{\boldsymbol{u}}_N, \tilde{\boldsymbol{u}}_N, \boldsymbol{v}_N),$$

so that we obtain from (27.12)

$$\varepsilon_N \leqslant c\bigg(\|\boldsymbol{u} - \tilde{\boldsymbol{u}}_N\|_{H^1(\Omega_d)^d} + \|(T_{\mathrm{S}} - T_{\mathrm{SN}})\boldsymbol{G}(\boldsymbol{u})\|_{H^1(\Omega_d)^d} \\ + \sup_{\boldsymbol{v}_N \in \mathbb{P}_N^0(\Omega_d)^d} \frac{\int_{\Omega_d} \boldsymbol{f}(\boldsymbol{x}) \cdot \boldsymbol{v}_N(\boldsymbol{x})\,\mathrm{d}\boldsymbol{x} - (\boldsymbol{f}, \boldsymbol{v}_N)_N}{\|\boldsymbol{v}_N\|_{H^1(\Omega_d)^d}}\bigg), \tag{27.29}$$

where the constant c depends on $\boldsymbol{u}$. All these terms can be estimated from Theorem 7.2, (27.13) and (15.18) together with Theorems 7.1 and 14.2. They tend to 0 when N tends to $+\infty$ whenever the data $\boldsymbol{f}$ are smooth enough, and the order of convergence only depends on the regularity of the solution $\boldsymbol{u} = -T_{\mathrm{S}}\boldsymbol{G}(\boldsymbol{u})$ and $\boldsymbol{f}$.

Combining all these estimates leads to the desired existence and local uniqueness result, together with the estimate

$$\|\boldsymbol{u}_N - \Pi_M^{1,0}\boldsymbol{u}\|_{H^1(\Omega_d)^d} \leqslant c\varepsilon_N.$$

Estimate (27.24) follows by a triangular inequality.

It must be observed that this error estimate is strictly the same as for the Stokes problem, and that it does not require any additional regularity on the solution $(\boldsymbol{u}, p)$ of the Navier–Stokes equations.

An estimate in $L^2(\Omega_d)^d$ is derived by a standard duality argument which involves the linearized Stokes problem in the velocity $\boldsymbol{u}$. We refer to BERNARDI, CANUTO, MADAY and MÉTIVET [1990] for the complete proof.

THEOREM 27.4. *If the hypotheses of Theorem 27.3 are satisfied, the following error estimate holds for problem* (27.17):

$$\|\boldsymbol{u} - \boldsymbol{u}_N\|_{L^2(\Omega_d)^d} \leqslant c\big(N^{-s}\big(\|\boldsymbol{u}\|_{H^s(\Omega_d)^d} + \|p\|_{H^{s-1}(\Omega_d)}\big) \\ + N^{-\sigma}\|\boldsymbol{f}\|_{H^\sigma(\Omega_d)^d}\big). \tag{27.30}$$

In view of the formulations (27.18) and (27.19) of the discrete problem, the existence and the uniqueness of the discrete pressure, and the corresponding error estimate, can easily be derived from the inf–sup condition.

THEOREM 27.5. *Let us assume that the hypotheses of Theorem 27.3 are satisfied.*

(i) *If the space* M_N *contains* $\mathbb{P}_{[\lambda N]}(\Omega_d) \cap L_0^2(\Omega_d)$ *for a real number* λ, $0 < \lambda < 1$, *and satisfies* (24.53), *for each discrete velocity* $\boldsymbol{u}_N$ *that is a solution of problem* (27.17) *with* $T_{\mathrm{SN}} = T_{\mathrm{SN}}^1$, *there exists a unique discrete pressure* p_N *in* M_N *such that the pair* $(\boldsymbol{u}_N, p_N)$ *is a solution of problem* (27.18). *Moreover, the following error estimate holds for problem* (27.18):

$$\begin{aligned}\|p - p_N\|_{L^2(\Omega_d)} \leqslant c\big(&N^{2-s}\big(\|\boldsymbol{u}\|_{H^s(\Omega_d)^d} + \|p\|_{H^{s-1}(\Omega_d)}\big)\\ &+ N^{1-\sigma}\|\boldsymbol{f}\|_{H^\sigma(\Omega_d)^d}\big).\end{aligned} \tag{27.31}$$

(ii) *For each discrete velocity* $\boldsymbol{u}_N$ *that is a solution of problem* (27.17) *with* $T_{\mathrm{SN}} = T_{\mathrm{SN}}^2$, *there exists a unique discrete pressure* p_N *in* $\mathbb{P}_{N-2}(\Omega_d) \cap L_0^2(\Omega_d)$ *such that the pair* $(\boldsymbol{u}_N, p_N)$ *is a solution of problem* (27.19). *Moreover, the following error estimate holds for problem* (27.19):

$$\begin{aligned}\|p - p_N\|_{L^2(\Omega_d)} \leqslant c\big(&N^{(d+1)/2-s}\big(\|\boldsymbol{u}\|_{H^s(\Omega_d)^d} + \|p\|_{H^{s-1}(\Omega_d)}\big)\\ &+ N^{(d-1)/2-\sigma}\|\boldsymbol{f}\|_{H^\sigma(\Omega_d)^d}\big).\end{aligned} \tag{27.32}$$

REMARK 27.3. A discrete problem relying on the formulation of the Navier–Stokes equations in weighted Sobolev spaces, can also be written. Its numerical analysis is performed in BERNARDI, CANUTO, MADAY and MÉTIVET [1990] in the case $d = 2$ of the square, for α equal to $-\frac{1}{2}$. Indeed, let α be any real number, $-\frac{1}{2} \leqslant \alpha \leqslant \frac{1}{2}$. Setting for any polynomials $\boldsymbol{u}_N$, $\boldsymbol{v}_N$ and $\boldsymbol{w}_N$ in $\mathbb{P}_N^0(\Omega_d)$,

$$\begin{aligned}&c_{\alpha,N}(\boldsymbol{w}_N, \boldsymbol{u}_N, \boldsymbol{v}_N)\\ &\quad= -\sum_{i=1}^{d}\sum_{j=1}^{d}\left(w_{Ni}u_{Nj}, \rho_{-\alpha}(x_j)\frac{\partial(v_{Ni}\rho_\alpha(x_j))}{\partial x_j}\right)_{\alpha,N},\end{aligned} \tag{27.33}$$

we can consider the problem: *find a pair* $(\boldsymbol{u}_N, p_N)$ *in* $\mathbb{P}_N^0(\Omega_d)^d \times M_N^\alpha$ *such that*

$$\begin{aligned}&\forall \boldsymbol{v}_N \in \mathbb{P}_N^0(\Omega_d)^d,\\ &\qquad a_{\alpha,N}(\boldsymbol{u}_N, \boldsymbol{v}_N) + c_{\alpha,N}(\boldsymbol{u}_N, \boldsymbol{u}_N, \boldsymbol{v}_N) + b_{\alpha 1,N}(\boldsymbol{v}_N, p_N) = (\boldsymbol{f}, \boldsymbol{v}_N)_{\alpha,N},\\ &\forall q_N \in M_N^\alpha, \quad b_{\alpha 2,N}(\boldsymbol{u}_N, q_N) = 0.\end{aligned} \tag{27.34}$$

Then, the continuity of the form $c_{\alpha,N}(\cdot,\cdot,\cdot)$ on $\mathbb{P}_N^0(\Omega_d)^d \times \mathbb{P}_N^0(\Omega_d)^d \times \mathbb{P}_N^0(\Omega_d)^d$ can be checked, together with the weighted analogue of estimate (27.25), so that the existence and uniqueness results are similar to those of Theorem 27.3 while the error estimates are strictly the same as in Theorems 24.18 and 24.21, with slightly more technical proofs.

28. The stream-function and vorticity method

In this section, we only consider the two-dimensional case of the square Ω. Indeed, it is well known that, in two-dimensional domains, a new variational formulation of the Stokes and Navier–Stokes equations can be given. Firstly, as stated in Remark 23.2, the incompressibility condition $\operatorname{div} \boldsymbol{u} = 0$, is equivalent to the existence of a stream-function ψ such that $\boldsymbol{u}$ coincides with $\mathbf{curl}\,\psi$, and the boundary conditions on the velocity $\boldsymbol{u}$ can be written as boundary conditions on the function ψ (for instance, if $\boldsymbol{u}$ vanishes on the boundary, both ψ and $\partial\psi/\partial n$ vanish on the boundary for simply-connected domains). Secondly, the vorticity

$$\omega = \operatorname{curl} \boldsymbol{u} = -\Delta\psi,$$

is introduced as a new variable, and taking the curl of the momentum equation, for instance in the Stokes problem, leads to

$$-\Delta\omega = \operatorname{curl} \boldsymbol{f}.$$

Hence, the Stokes problem is equivalent to a system of two Laplace equations on scalar unknowns, which are coupled by double boundary conditions on one unknown and no boundary condition on the other one. The following analysis is due to BERNARDI, GIRAULT and MADAY [1992].

More precisely, we consider the Stokes problem (with homogeneous boundary conditions, only for the sake of simplicity):

$$\begin{cases} -\Delta\boldsymbol{u} + \mathbf{grad}\, p = \boldsymbol{f} & \text{in } \Omega, \\ \operatorname{div} \boldsymbol{u} = 0 & \text{in } \Omega, \\ \boldsymbol{u} = \mathbf{0} & \text{on } \partial\Omega, \end{cases} \tag{28.1}$$

and we assume that the data $\boldsymbol{f}$ belongs to $L^2(\Omega)^2$. We introduce the space

$$\mathscr{H} = \{\mu \in L^2(\Omega) \colon \Delta\mu \in H^{-1}(\Omega)\}, \tag{28.2}$$

provided with the norm

$$\|\mu\|_{\mathscr{H}} = \left(\|\mu\|^2_{L^2(\Omega)} + \|\Delta\mu\|^2_{H^{-1}(\Omega)}\right)^{1/2}. \tag{28.3}$$

We also define two bilinear forms, the first one on $\mathscr{H} \times \mathscr{H}$:

$$\forall\theta \in \mathscr{H},\ \forall\mu \in \mathscr{H},\quad a_*(\theta, \mu) = \int_\Omega \theta(\boldsymbol{x})\mu(\boldsymbol{x})\,\mathrm{d}\boldsymbol{x}, \tag{28.4}$$

the second one on $\mathscr{H} \times H^1_0(\Omega)$:

$$\forall\mu \in \mathscr{H}, \forall\varphi \in H^1_0(\Omega),\quad b_*(\mu, \varphi) = \langle\Delta\mu, \varphi\rangle, \tag{28.5}$$

where $\langle \cdot, \cdot \rangle$ stands for the duality pairing between $H^{-1}(\Omega)$ and $H_0^1(\Omega)$.

Now, we consider the problem: *find a pair* (ω, ψ) *in* $\mathscr{H} \times H_0^1(\Omega)$ *such that*

$$\begin{aligned} &\forall \mu \in \mathscr{H}, \qquad a_*(\omega, \mu) + b_*(\mu, \psi) = 0, \\ &\forall \varphi \in H_0^1(\Omega), \quad b_*(\omega, \varphi) = -\frac{1}{\nu} \int_\Omega \boldsymbol{f}(\boldsymbol{x}) \cdot (\mathbf{curl}\, \varphi)(\boldsymbol{x})\, \mathrm{d}\boldsymbol{x}. \end{aligned} \tag{28.6}$$

It is readily checked that, if $(\boldsymbol{u}, p)$ in $H_0^1(\Omega)^2 \times L_0^2(\Omega)$ is a solution of the Stokes problem (28.1), the pair (ω, ψ) in $\mathscr{H} \times H_0^1(\Omega)$, with $\boldsymbol{u} = \mathbf{curl}\, \psi$ and $\omega = \operatorname{curl} \boldsymbol{u}$, is a solution of problem (28.6). Conversely, if (ψ, ω) is a solution of (28.6), note that ψ belongs to $H_0^2(\Omega)$, indeed the regularity of the function ψ follows from the convexity of Ω and the fact that $\partial\psi/\partial n$ vanishes on $\partial\Omega$ is enforced in the variational formulation (28.6) in a weak way. Then, for any solution (ω, ψ) of problem (28.6), there exists a unique pressure p in $L_0^2(\Omega)$ such that the pair $(\boldsymbol{u} = \mathbf{curl}\, \psi, p)$ is a solution of problem (28.1).

Clearly, problem (28.6) is of type (23.13), so that the next result follows from the general theory of Section 23.

THEOREM 28.1. *For any* $\boldsymbol{f}$ *in* $L^2(\Omega)^2$, *problem* (28.6) *has a unique solution* (ω, ψ) *in* $\mathscr{H} \times H_0^1(\Omega)$. *Moreover, this solution satisfies*

$$\|\omega\|_{\mathscr{H}} + \|\psi\|_{H^1(\Omega)} \leqslant c\|\boldsymbol{f}\|_{L^2(\Omega)^2}, \tag{28.7}$$

and the function ψ *belongs to* $H_0^2(\Omega)$.

PROOF. The continuity of the forms $a_*(\cdot, \cdot)$ and $b_*(\cdot, \cdot)$ being obvious, it remains to check an ellipticity property for $a_*(\cdot, \cdot)$ and an inf–sup condition for $b_*(\cdot, \cdot)$.

(1) First, we observe that the kernel

$$K_* = \{\mu \in \mathscr{H}\colon \forall \varphi \in H_0^1(\Omega),\ b_*(\mu, \varphi) = 0\},$$

is the space of harmonic functions in $L^2(\Omega)$. Thus, for any μ in K_*, we have

$$a_*(\mu, \mu) = \|\mu\|_{L^2(\Omega)}^2 \geqslant \|\mu\|_{\mathscr{H}}^2,$$

which is the ellipticity property.

(2) Second, for any φ in $H_0^1(\Omega)$, taking $\mu = -\varphi$ gives (note that $\|\mu\|_{\mathscr{H}}$ is smaller than $\|\varphi\|_{H^1(\Omega)}$)

$$b_*(\mu, \varphi) = -\langle \Delta\varphi, \varphi \rangle = |\varphi|_{H^1(\Omega)}^2 \geqslant c|\varphi|_{H^1(\Omega)} \|\mu\|_{\mathscr{H}},$$

which is the inf–sup condition.

Hence, we derive the wellposedness of problem (28.6) by applying Theorem 23.2. The properties of ψ have already been observed.

We now assume that the data $\boldsymbol{f}$ are continuous on $\overline{\Omega}$. Writing the discrete problem is very natural. We define the bilinear forms: for any polynomials θ_N, μ_N and φ_N in $\mathbb{P}_N(\Omega)$,

$$a_{*N}(\theta_N, \mu_N) = (\theta_N, \mu_N)_N, \tag{28.8}$$

$$b_{*N}(\mu_N, \varphi_N) = -(\mathbf{curl}\, \mu_N, \mathbf{curl}\, \varphi_N)_N. \tag{28.9}$$

Then, the discrete problems reads: *find a pair* (ω_N, ψ_N) *in* $\mathbb{P}_N(\Omega) \times \mathbb{P}_N^0(\Omega)$ *such that*

$$\begin{aligned} &\forall \mu_N \in \mathbb{P}_N(\Omega), \quad a_{*N}(\omega_N, \mu_N) + b_{*N}(\mu_N, \psi_N) = 0, \\ &\forall \varphi_N \in \mathbb{P}_N^0(\Omega), \quad b_{*N}(\omega_N, \varphi_N) = -\frac{1}{\nu}(\boldsymbol{f}, \mathbf{curl}\, \varphi_N)_N. \end{aligned} \tag{28.10}$$

The first step is to check the wellposedness of the problem.

THEOREM 28.2. *For any continuous function* $\boldsymbol{f}$ *on* $\overline{\Omega}$, *problem* (28.10) *has a unique solution* (ω_N, ψ_N) *in* $\mathbb{P}_N(\Omega) \times \mathbb{P}_N^0(\Omega)$. *Moreover, this solution satisfies*

$$\|\omega_N\|_{L^2(\Omega)} + \|\psi_N\|_{H^1(\Omega)} \leqslant c\|\mathcal{I}_N \boldsymbol{f}\|_{L^2(\Omega)^2}. \tag{28.11}$$

PROOF. As previously, we have to check the ellipticity of the form $a_{*N}(\bullet,\bullet)$ and an inf–sup condition for the form $b_{*N}(\bullet,\bullet)$.

(1) As a simple application of (13.18), we have for any μ_N in $\mathbb{P}_N(\Omega)$,

$$a_{*N}(\mu_N, \mu_N) = (\mu_N, \mu_N)_N \geqslant \|\mu_N\|^2_{L^2(\Omega)}.$$

Next, from the inverse inequality (5.2) applied in each direction, we deduce that

$$\begin{aligned} \|\Delta\mu_N\|_{H^{-1}(\Omega)} &= \sup_{\varphi \in H_0^1(\Omega)} \frac{\int_\Omega \mathbf{curl}\, \mu_N \cdot \mathbf{curl}\, \varphi \, \mathrm{d}\boldsymbol{x}}{|\varphi|_{H^1(\Omega)}} \\ &\leqslant |\mu_N|_{H^1(\Omega)} \leqslant cN^2 \|\mu_N\|_{L^2(\Omega)}, \end{aligned}$$

whence the ellipticity inequality

$$a_{*N}(\mu_N, \mu_N) \geqslant c\big(\|\mu_N\|_{L^2(\Omega)} + N^{-2}\,\|\Delta\mu_N\|_{H^{-1}(\Omega)}\big). \tag{28.12}$$

(2) We also observe from (13.18) that, for φ_N in $\mathbb{P}_N^0(\Omega)$, taking μ_N equal to $-\varphi_N$ gives

$$b_{*N}(\mu_N, \varphi_N) = (\varphi_N, \varphi_N)_N \geqslant |\varphi_N|^2_{H^1(\Omega)},$$

and, since $\|\mu_N\|_{\mathscr{H}}$ is smaller than $\|\mu_N\|_{H^1(\Omega)}$, we obtain

$$\sup_{\mu_N \in \mathbb{P}_N(\Omega)} \frac{b_{*N}(\mu_N, \varphi_N)}{|\varphi_N|_{H^1(\Omega)}} \geqslant c\|\mu_N\|_{\mathscr{H}}. \tag{28.13}$$

Combining these properties proves the theorem.

REMARK 28.1. In contrast to finite element type methods, a discrete velocity $\boldsymbol{u}_N$ in $\mathbb{P}_N^0(\Omega)^2$ can easily be computed from the stream-function by a standard polynomial derivation process.

The first idea to derive error bounds is to make use of estimate (23.56), however it appears in (28.12) that the constant of ellipticity is only of order N^{-2}. So, we have to work a little more to derive an optimal estimate. The proof requires the following operator Π_N^{1*}, which is slightly different from the orthogonal projection operator Π_N^1 introduced in Section 7: for any function θ in $H^1(\Omega)$, $\Pi_N^{1*}\theta$ belongs to $\mathbb{P}_N(\Omega)$ and satisfies

$$\begin{aligned} &\forall \mu_N \in \mathbb{P}_N(\Omega),\\ &\quad \big(\mathbf{curl}(\Pi_N^{1*}\theta), \mathbf{curl}\,\mu_N\big)_N = \int_\Omega (\mathbf{curl}\,\theta)(\boldsymbol{x}) \cdot (\mathbf{curl}\,\mu_N)(\boldsymbol{x})\,\mathrm{d}\boldsymbol{x},\\ &\quad \int_\Omega (\Pi_N^{1*}\theta)(\boldsymbol{x})\,\mathrm{d}\boldsymbol{x} = \int_\Omega \theta(\boldsymbol{x})\,\mathrm{d}\boldsymbol{x}. \end{aligned} \tag{28.14}$$

Clearly, these equations define $\Pi_N^{1*}\theta$ in a unique way. The first theorem deals with the approximation properties of this operator.

THEOREM 28.3. *For any real number $s \geqslant 1$, there exists a positive constant c such that the following inequality holds for any function θ in $H^s(\Omega)$:*

$$|\theta - \Pi_N^{1*}\theta|_{H^1(\Omega)} + N\|\theta - \Pi_N^{1*}\theta\|_{L^2(\Omega)} \leqslant cN^{1-s}\|\theta\|_{H^s(\Omega)}. \tag{28.15}$$

PROOF. We prove the estimates for $|\theta - \Pi_N^{1*}\theta|_{H^1(\Omega)}$ and for $\|\theta - \Pi_N^{1*}\theta\|_{L^2(\Omega)}$ successively.

(1) From equation (28.14), we have

$$\begin{aligned} |\Pi_N^{1*}\theta - \Pi_{N-1}^1\theta|_{H^1(\Omega)}^2 &\leqslant \big(\mathbf{curl}(\Pi_N^{1*}\theta - \Pi_{N-1}^1\theta), \mathbf{curl}(\Pi_N^{1*}\theta - \Pi_{N-1}^1\theta)\big)_N\\ &\leqslant \int_\Omega \big(\mathbf{curl}(\theta - \Pi_{N-1}^1\theta)\big)(\boldsymbol{x}) \cdot \big(\mathbf{curl}(\Pi_N^{1*}\theta - \Pi_{N-1}^1\theta)\big)(\boldsymbol{x})\,\mathrm{d}\boldsymbol{x}, \end{aligned}$$

so that

$$|\theta - \Pi_N^{1*}\theta|_{H^1(\Omega)} \leqslant 2|\theta - \Pi_{N-1}^1\theta|_{H^1(\Omega)}.$$

Hence, the first estimate follows from Theorem 7.3.

(2) To obtain the second estimate, we use a duality argument. We solve the problem

$$\begin{cases} -\Delta w = \theta - \Pi_N^{1*}\theta & \text{in } \Omega, \\ \dfrac{\partial w}{\partial n} = 0 & \text{on } \partial\Omega, \end{cases} \tag{28.16}$$

which is well posed since $\theta - \Pi_N^{1*}\theta$ has a null integral, and we derive from the convexity of Ω that

$$\|w\|_{H^2(\Omega)} \leqslant c\big\|\theta - \Pi_N^{1*}\theta\big\|_{L^2(\Omega)}. \tag{28.17}$$

Next, we compute

$$\begin{aligned} \big\|\theta - \Pi_N^{1*}\theta\big\|^2_{L^2(\Omega)} &= \int_\Omega \big(\mathbf{curl}(\theta - \Pi_N^{1*}\theta)\big)(\boldsymbol{x}) \cdot (\mathbf{curl}\, w)(\boldsymbol{x}) \\ &= \int_\Omega \big(\mathbf{curl}(\theta - \Pi_N^{1*}\theta)\big)(\boldsymbol{x}) \cdot \big(\mathbf{curl}(w - \Pi_{N-1}^1 w)\big)(\boldsymbol{x}) \\ &\leqslant \big|\theta - \Pi_N^{1*}\theta\big|_{H^1(\Omega)} \big|w - \Pi_{N-1}^1 w\big|_{H^1(\Omega)}, \end{aligned}$$

so that, by Theorem 7.3 and (28.17),

$$\big\|\theta - \Pi_N^{1*}\theta\big\|_{L^2(\Omega)} \leqslant cN^{-1}\big|\theta - \Pi_N^{1*}\theta\big|_{H^1(\Omega)}.$$

This concludes the proof.

THEOREM 28.4. *The function $\boldsymbol{f}$ is supposed to be in a space $H^\sigma(\Omega)^2$, $\sigma > d/2$. Let us assume that the solution (ω, ψ) of problem* (28.6) *belongs to $H^{s-1}(\Omega) \times H^{s+1}(\Omega)$ for a real number $s \geqslant 2$. Then, the following error estimate holds for problem* (28.10):

$$\begin{aligned} &\|\omega - \omega_N\|_{L^2(\Omega)} + \|\psi - \psi_N\|_{H^1(\Omega)} \\ &\quad\leqslant c\big(N^{1-s}\big(\|\omega\|_{H^{s-1}(\Omega)} + \|\psi\|_{H^{s+1}(\Omega)}\big) + N^{-\sigma}\|\boldsymbol{f}\|_{H^\sigma(\Omega)^2}\big). \end{aligned} \tag{28.18}$$

PROOF. Two steps are necessary.

(1) Firstly, from problems (28.6) and (28.10), we compute

$$\begin{aligned} &\big|\psi_N - \Pi_{N-1}^{2,0}\psi\big|^2_{H^1(\Omega)} \\ &\quad\leqslant b_{*N}\big(\psi_N - \Pi_{N-1}^{2,0}\psi, \psi_N - \Pi_{N-1}^{2,0}\psi\big) \\ &\quad\leqslant -a_{*N}\big(\omega_N, \psi_N - \Pi_{N-1}^{2,0}\psi\big) - b_*\big(\Pi_{N-1}^{2,0}\psi, \psi_N - \Pi_{N-1}^{2,0}\psi\big) \\ &\quad\leqslant a_*\big(\omega - \Pi_{N-1}\omega, \psi_N - \Pi_{N-1}^{2,0}\psi\big) - a_{*N}\big(\omega_N - \Pi_{N-1}\omega, \psi_N - \Pi_{N-1}^{2,0}\psi\big) \\ &\qquad + b_*\big(\psi - \Pi_{N-1}^{2,0}\psi, \psi_N - \Pi_{N-1}^{2,0}\psi\big). \end{aligned}$$

This leads to

$$\begin{aligned}\left|\psi_N - \Pi_{N-1}^{2,0}\psi\right|_{H^1(\Omega)} \leqslant c\big(&\|\omega - \omega_N\|_{L^2(\Omega)} + \|\omega - \Pi_{N-1}\omega\|_{L^2(\Omega)} \\ &+ \left|\psi - \Pi_{N-1}^{2,0}\psi\right|_{H^1(\Omega)}\big)\end{aligned} \tag{28.19}$$

and the last two terms in the right-hand side are easy to estimate.

(2) The second step is to take μ_N in problem (28.10) equal to $\omega_N - \Pi_N^{1*}\omega$ and to observe that, for any φ_N in $\mathbb{P}_N^0(\Omega)$,

$$\begin{aligned}b_{*N}\big(\omega_N - \Pi_N^{1*}\omega, \varphi_N\big) &= b_{*N}(\omega_N, \varphi_N) - b_*(\omega, \varphi_N) \\ &= \frac{1}{\nu}\int_\Omega \boldsymbol{f}(\boldsymbol{x}) \cdot (\mathbf{curl}\,\varphi_N)(\boldsymbol{x})\,\mathrm{d}\boldsymbol{x} - \frac{1}{\nu}(\boldsymbol{f}, \mathbf{curl}\,\varphi_N)_N.\end{aligned}$$

Thus, standard arguments yield

$$\begin{aligned}b_{*N}\big(\omega_N - \Pi_N^{1*}\omega, \varphi_N\big) \leqslant\; &\frac{c}{\varepsilon^2}\big(\|\boldsymbol{f} - \Pi_{N-1}\boldsymbol{f}\|_{L^2(\Omega)^2} + \|\boldsymbol{f} - \mathcal{I}_N\boldsymbol{f}\|_{L^2(\Omega)^2}\big)^2 \\ &+ \varepsilon^2|\varphi_N|^2_{H^1(\Omega)},\end{aligned} \tag{28.20}$$

where ε is any positive real number. We also derive from problems (28.6) and (28.10) that

$$\begin{aligned}&\left\|\omega_N - \Pi_N^{1*}\omega\right\|^2_{L^2(\Omega)} \\ &\quad\leqslant a_{*N}\big(\omega_N - \Pi_N^{1*}\omega, \omega_N - \Pi_N^{1*}\omega\big) \\ &\quad\leqslant -b_{*N}\big(\omega_N - \Pi_N^{1*}\omega, \psi_N\big) - a_{*N}\big(\Pi_N^{1*}\omega, \omega_N - \Pi_N^{1*}\omega\big) \\ &\quad\leqslant a_*\big(\omega, \omega_N - \Pi_N^{1*}\omega\big) - a_{*N}\big(\Pi_N^{1*}\omega, \omega_N - \Pi_N^{1*}\omega\big) \\ &\qquad + b_*\big(\omega_N - \Pi_N^{1*}\omega, \psi\big) - b_{*N}\big(\omega_N - \Pi_N^{1*}\omega, \psi_N\big) \\ &\quad\leqslant a_*\big(\omega - \Pi_{N-1}\omega, \omega_N - \Pi_N^{1*}\omega\big) - a_{*N}\big(\Pi_N^{1*}\omega - \Pi_{N-1}\omega, \omega_N - \Pi_N^{1*}\omega\big) \\ &\qquad + b_*\big(\omega_N - \Pi_N^{1*}\omega, \psi - \Pi_{N-1}^{2,0}\psi\big) - b_{*N}\big(\omega_N - \Pi_N^{1*}\omega, \psi_N - \Pi_{N-1}^{2,0}\psi\big).\end{aligned}$$

Combining this last inequality with (28.20) and using the fact that ψ belongs to $H_0^2(\Omega)$ to integrate by parts, we obtain

$$\begin{aligned}&\|\omega - \omega_N\|_{L^2(\Omega)} \\ &\quad\leqslant c\big(\|\omega - \Pi_{N-1}\omega\|_{L^2(\Omega)} + \left\|\omega - \Pi_N^{1*}\omega\right\|_{L^2(\Omega)} + \left|\psi - \Pi_{N-1}^{2,0}\psi\right|_{H^2(\Omega)}\big) \\ &\qquad + \frac{c}{\varepsilon}\big(\|\boldsymbol{f} - \Pi_{N-1}\boldsymbol{f}\|_{L^2(\Omega)^2} + \|\boldsymbol{f} - \mathcal{I}_N\boldsymbol{f}\|_{L^2(\Omega)^2}\big) \\ &\qquad + \varepsilon\left|\psi_N - \Pi_{N-1}^{2,0}\psi\right|_{H^1(\Omega)}.\end{aligned} \tag{28.21}$$

The end of the proof consists in choosing ε equal to $\frac{1}{2}$, putting together the inequalities (28.19) and (28.21) and applying Theorems 7.1, 7.2, 14.2 and 28.3.

Clearly, estimate (28.18) is optimal for the vorticity but not for the stream-function. A duality argument is necessary to recover the optimality.

THEOREM 28.5. *Under the hypotheses of Theorem* 28.4, *the following error estimate holds for problem* (28.10):

$$\|\psi - \psi_N\|_{H^1(\Omega)} \leqslant c\big(N^{-s}\big(\|\omega\|_{H^{s-1}(\Omega)} + \|\psi\|_{H^{s+1}(\Omega)}\big) + N^{-\sigma}\|\boldsymbol{f}\|_{H^\sigma(\Omega)^2}\big). \tag{28.22}$$

PROOF. We start from the equality

$$|\psi - \psi_N|_{H^1(\Omega)} = \sup_{\boldsymbol{g}\in L^2(\Omega)^2} \frac{\int_\Omega \boldsymbol{g}(\boldsymbol{x}) \cdot (\mathbf{curl}(\psi - \psi_N))(\boldsymbol{x})\,\mathrm{d}\boldsymbol{x}}{\|\boldsymbol{g}\|_{L^2(\Omega)^2}}. \tag{28.23}$$

Then, for any $\boldsymbol{g}$ in $L^2(\Omega)^2$, we solve the Stokes problem: *find* (θ, χ) *in* $\mathscr{H} \times H_0^1(\Omega)$ *such that*

$$\begin{aligned} &\forall \mu \in \mathscr{H}, && a_*(\theta, \mu) + b_*(\mu, \chi) = 0, \\ &\forall \varphi \in H_0^1(\Omega), && b_*(\theta, \varphi) = \int_\Omega \boldsymbol{g}(\boldsymbol{x}) \cdot (\mathbf{curl}\varphi)(\boldsymbol{x})\,\mathrm{d}\boldsymbol{x}. \end{aligned} \tag{28.24}$$

From the regularity result of Remark 23.3, it can be proven that the pair (θ, χ) belongs to $H^1(\Omega) \times (H^3(\Omega) \cap H_0^2(\Omega))$ and satisfies

$$\|\theta\|_{H^1(\Omega)} + \|\chi\|_{H^3(\Omega)} \leqslant c\,\|\boldsymbol{g}\|_{L^2(\Omega)^2}. \tag{28.25}$$

Now, we compute

$$\int_\Omega \boldsymbol{g}(\boldsymbol{x}) \cdot (\mathbf{curl}(\psi - \psi_N))(\boldsymbol{x})\,\mathrm{d}\boldsymbol{x} = b_*(\theta, \psi - \psi_N) = b_*(\theta, \psi) - b_{*N}(\Pi_N^{1*}\theta, \psi_N),$$

so that

$$\int_\Omega \boldsymbol{g}(\boldsymbol{x}) \cdot (\mathbf{curl}(\psi - \psi_N))(\boldsymbol{x})\,\mathrm{d}\boldsymbol{x} = b_*(\theta - \Pi_N^{1*}\theta, \psi - \Pi_{N-1}^{2,0}\psi) + b_*(\Pi_N^{1*}\theta, \psi) - b_{*N}(\Pi_N^{1*}\theta, \psi_N).$$

Next, we check that

$$\begin{aligned}
&b_*\big(\Pi_N^{1*}\theta, \psi\big) - b_{*N}\big(\Pi_N^{1*}\theta, \psi_N\big) \\
&\quad = -a_*\big(\omega, \Pi_N^{1*}\theta\big) + a_{*N}\big(\omega_N, \Pi_N^{1*}\theta\big) \\
&\quad = a_*\big(\omega - \omega_N, \theta - \Pi_N^{1*}\theta\big) - a_*(\omega - \omega_N, \theta) \\
&\qquad -a_*\big(\omega_N, \Pi_N^{1*}\theta\big) + a_{*N}\big(\omega_N, \Pi_N^{1*}\theta\big) \\
&\quad = a_*\big(\omega - \omega_N, \theta - \Pi_N^{1*}\theta\big) + b_*\big(\omega - \omega_N, \chi - \Pi_N^{2,0}\chi\big) \\
&\qquad -\frac{1}{\nu}\int_\Omega (\boldsymbol{f} - \mathcal{I}_N\boldsymbol{f})(\boldsymbol{x}) \cdot \big(\mathbf{curl}\big(\Pi_N^{2,0}\chi\big)\big)(\boldsymbol{x})\,\mathrm{d}\boldsymbol{x} \\
&\qquad - a_*\big(\omega_N - \Pi_{N-1}\omega, \Pi_N^{1*}\theta - \Pi_{N-1}\theta\big) \\
&\qquad + a_{*N}\big(\omega_N - \Pi_{N-1}\omega, \Pi_N^{1*}\theta - \Pi_{N-1}\theta\big).
\end{aligned}$$

Using (28.25), we deduce from Theorems 7.1, 7.2, 14.2 and 28.3 that each term in the right-hand side of this equation is bounded by a constant times N^{-s} or $N^{-\sigma}$, which ends this rather technical proof.

REMARK 28.2. Using formulation (28.6) does not allow for computing directly the pressure. However, a discrete pressure can be recovered as a solution of the problem: *find* p_N *in* $\mathbb{P}_N(\Omega) \cap L_0^2(\Omega)$ *such that*

$$\begin{aligned}
&\forall q_N \in \mathbb{P}_N(\Omega), \\
&\quad (\mathbf{grad}\, p_N, \mathbf{grad}\, q_N)_N = (\boldsymbol{f}, \mathbf{grad}\, q_N)_N - \nu(\mathbf{curl}\, \omega_N, \mathbf{grad}\, q_N)_N.
\end{aligned} \tag{28.26}$$

It is readily checked that this problem has a unique solution. Moreover, under the assumptions of Theorem 28.4, the pressure p in the Stokes problem (28.1) also belongs to $H^{s-1}(\Omega)$ and the following error estimate is proven in BERNARDI, GIRAULT and MADAY ([1992], Corollary 4.7):

$$\begin{aligned}
\|p - p_N\|_{L^2(\Omega)} \leqslant c\big(&N^{1-s}\|p\|_{H^{s-1}(\Omega)} + N^{2-s}\big(\|\omega\|_{H^{s-1}(\Omega)} + \|\psi\|_{H^{s+1}(\Omega)}\big) \\
&+ N^{1-\sigma}\|\boldsymbol{f}\|_{H^\sigma(\Omega)^2}\big).
\end{aligned} \tag{28.27}$$

Here also, like in every spectral discretization, the estimate for the pressure is not optimal.

The method has been extended to the full Navier–Stokes equations in BERNARDI, GIRAULT and MADAY [1992], with completely similar results.

CHAPTER VI

Spectral Discretizations of Hyperbolic Conservation Laws

Nonlinear hyperbolic equations generally have discontinuous solutions. It may seem to be a crazy idea to use high-order methods to globally approximate these solutions as everyone knows that second- or third-order methods are better than first-order methods *only* when some regularity is assumed to hold on the function which has to be approximated. Like other types of high-order methods, spectral discretizations of problems with discontinuous solutions are only very slowly convergent and apparently, not better than the finite-order methods. Indeed the best polynomial approximation of a function which is discontinuous at only one point presents global oscillations everywhere in the domain of computation. Nevertheless, following MOCK and LAX [1978], GOTTLIEB and TADMOR [1985] have put forward the idea that from these oscillations, some (spectrally) accurate information can be recovered thanks to appropriate smart tools if the irregularities of the function are only the exception (see for instance ABARBANEL, GOTTLIEB and TADMOR [1986] and GOTTLIEB and TADMOR [1985]). These tools are presented in this chapter.

In Section 29, recent filtering procedures for recovering accurate polynomial approximations of discontinuous functions are described. Two techniques for the discretization of hyperbolic equations are studied in Sections 30 and 31 respectively: the cell averaging method and the spectral viscosity method.

29. Filtering procedures

The best polynomial approximation of discontinuous functions presents global oscillations everywhere on the domain of computation even if the solution is discontinuous at only one point and regular everywhere else: this is known as the *Gibbs phenomenon.* Lagrange interpolation polynomials of these functions have the same behavior, as it can be observed in the next two figures.

In all that follows, the functions we deal with, have discontinuities at a finite number of points in dimension $d = 1$ or along some lines in dimension $d = 2$, and are more regular everywhere else. The idea is to derive from a spectral polynomial approximation some pointwise information that, at least far away from the discontinuities, is as good as the spectral approximation for a globally regular function. The tools that are

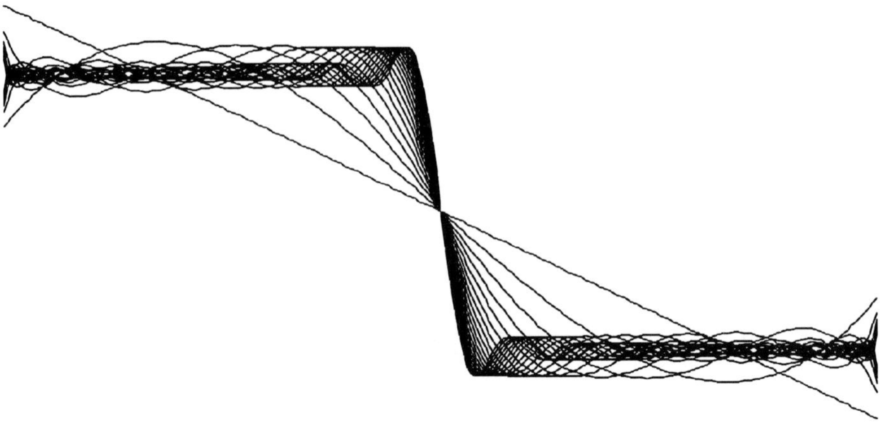

FIG. 29.1. Approximation of the Heaviside function via the projection operator π_{2N-1}, $1 \leqslant N \leqslant 20$.

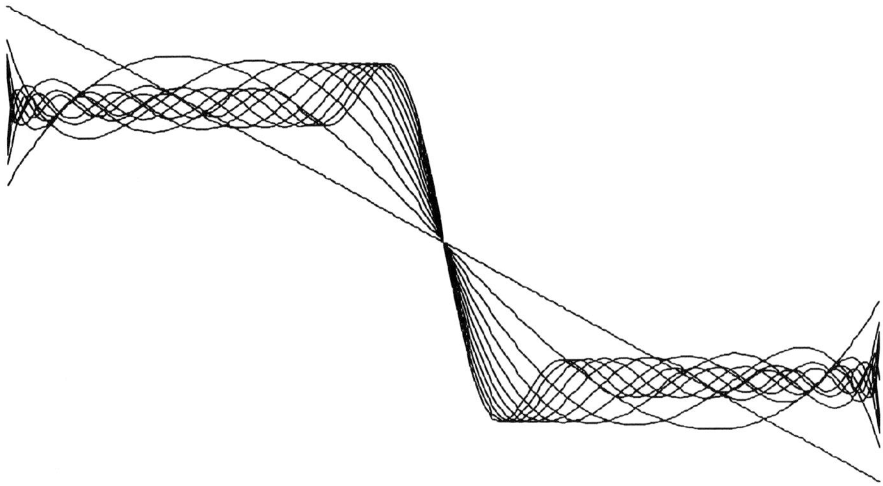

FIG. 29.2. Approximation of the Heaviside function via the interpolation operator j_{2N-1}, $1 \leqslant N \leqslant 20$.

used to derive this spectral accuracy, are extensions of Cesàro averaging, algorithms to compute the location of the discontinuities or convolution products by an appropriate kernel.

Let us recall the concept of Cesàro averaging. It relies on the decomposition of the orthogonal projections of the discontinuous function φ in the orthogonal basis of Legendre polynomials:

$$\pi_N\varphi = \sum_{n=0}^{N} \widehat{\varphi}^n \left(n + \tfrac{1}{2}\right) L_n.$$

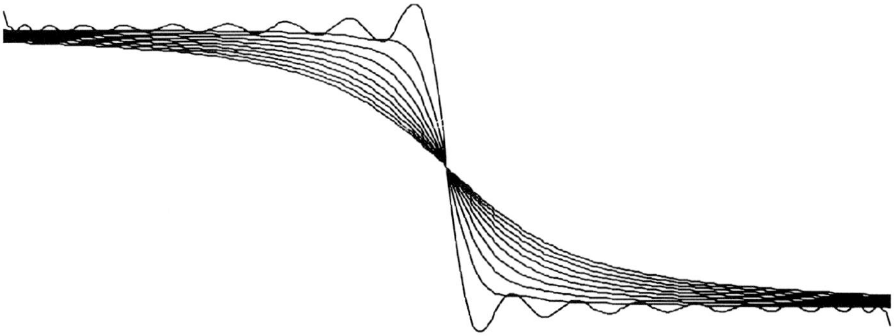

FIG. 29.3. Approximation of the Heaviside function via the Cesàro means $C_{39,j}$, $0 \leqslant j \leqslant 9$.

The idea is to take the mean of the first $N+1$ projections: so, the first Cesàro mean $C_N\varphi$ is defined by

$$C_N\varphi = \frac{1}{N+1}\sum_{n=0}^{N}\pi_n\varphi = \sum_{n=0}^{N}\left(1-\frac{n}{N+1}\right)\widehat{\varphi}^{\,n}\left(n+\tfrac{1}{2}\right)L_n.$$

More generally, applying iteratively this procedure, we define for any positive integer j the Cesàro mean of order j:

$$C_{N,j}\varphi = \sum_{n=0}^{N}\left(1-\frac{n}{N+1}\right)\left(1-\frac{n}{N+2}\right)\cdots\left(1-\frac{n}{N+j}\right)\widehat{\varphi}^{\,n}\left(n+\tfrac{1}{2}\right)L_n.$$

We only recall the basic result about the first Cesàro mean: there exists a positive constant c such that, for any bounded function φ on Λ,

$$\sup_{-1\leqslant\zeta\leqslant1}\left|(C_N\varphi)(\zeta)\right| \leqslant c \sup_{-1\leqslant\zeta\leqslant1}\left|\varphi(\zeta)\right|. \tag{29.1}$$

Figure 29.3 represents the action of the Cesàro averaging on the best polynomial approximation in $L^2(\Lambda)$ of the Heaviside function. It can be noticed that the oscillations disappear and that the convergence is better away from the discontinuity. This is still not high-order but it is an improvement in comparison with the original best approximation. This better than best behavior is due to the fact that the error is then localized around the discontinuity, resulting in a worse global behavior in $L^2(\Lambda)$ but in a better local behavior in the maximum norm.

From the previous considerations, it can be noticed that the oscillations in $\pi_N\varphi$ are due to the high modes represented by the $\widehat{\varphi}^n$ with n close to N, so that damping these coefficients in a suitable way allows for recovering the spectral accuracy, at least away from the discontinuity. Actually, a recent work of VANDEVEN [1993] shows that this accuracy can also be recovered at the level of the discontinuity if its position is known with sufficient precision. In this direction, as it was firstly achieved by CAI, GOTTLIEB and SHU [1989] in the Fourier framework, another way to retrieve a good accuracy is to recognize in a function on a one-dimensional domain that is regular except in one point s where it is discontinuous, the sum of a Heaviside (or step) function and of a continuous function. As a consequence, the Legendre coefficients are the sum of those of the Heaviside function and of the continuous function. The following proposition, proven by OULD KABER and VANDEVEN [1993], gives the behavior of the Legendre coefficients of a function regular in $\overline{\Lambda}$ except at some point s in the interior of Λ. We denote by $[\varphi(s)]$ the jump of some given function φ at a point s of discontinuity.

THEOREM 29.1. *Let us assume that the function φ has the expansion*

$$\varphi = \sum_{n=0}^{+\infty} \widehat{\varphi}^n \left(n+\tfrac{1}{2}\right) L_n,$$

and that, for a fixed point s in Λ, its restrictions to each interval $]-1,s[$ and $]s,1[$ are the restrictions of continuously differentiable functions up to order m over $\overline{\Lambda}$. Then, the following formula holds for any integer $n \geqslant m$

$$\begin{aligned}\widehat{\varphi}^n = \sum_{k=1}^{m} \frac{(n-k)!}{(n+k)!} \left[\left(\frac{\mathrm{d}^{k-1}\varphi}{\mathrm{d}\zeta^{k-1}}\right)(s)\right] \left(1-s^2\right)^k \left(\frac{\mathrm{d}^k L_n}{\mathrm{d}\zeta^k}\right)(s) \\ + (-1)^m \left(\int_{-1}^{s} \left(\frac{\mathrm{d}^m\varphi}{\mathrm{d}\zeta^m}\right)(\zeta) L_{n,m}(\zeta)\,\mathrm{d}\zeta \right. \\ \left. + \int_{s}^{1} \left(\frac{\mathrm{d}^m\varphi}{\mathrm{d}\zeta^m}\right)(\zeta) L_{n,m}(\zeta)\,\mathrm{d}\zeta \right),\end{aligned} \tag{29.2}$$

where $L_{n,m}$ stands for the mth primitive of L_n which vanishes in ± 1 together with its derivatives up to order $m-1$.

By using the asymptotic behavior of the derivatives and primitives of the Legendre polynomials, we deduce that

$$\widehat{\varphi}^n = [\varphi(s)] \frac{(1-s^2) L_n'(s)}{n(n+1)} + \mathrm{O}\left(n^{-5/2}\right).$$

Hence, the coefficients $\widehat{\varphi}^n$ are bounded by $cn^{-3/2}$ whenever the function φ is discontinuous. If s is a zero of L_n', the previous formula shows that no information on the size of the jump $[\varphi(s)]$ can be recovered from $\widehat{\varphi}^n$. The interesting feature of the derivatives of two successive Legendre polynomials is that they do not vanish at the same point, since it is proven also in OULD KABER and VANDEVEN [1993] that there exist two positive constants c and c' such that, for any $n \geqslant 0$,

$$\forall x \in \Lambda, \quad cn \leqslant L_n'^2(x) + L_{n+1}'^2(x) \leqslant c'\, n\left(1-x^2\right)^{-3/2}.$$

From these estimates, it is not too difficult to derive the following theorem.

THEOREM 29.2. *Assume that the hypotheses of Theorem* 29.1 *hold. For any positive integer n, the quantity s_n defined by*

$$\begin{aligned} s_n = &\frac{(n+3)\widehat{\varphi}^{\,n+2} + n\widehat{\varphi}^{\,n}}{(2n+3)((\widehat{\varphi}^{\,n})^2 + (\widehat{\varphi}^{\,n+1})^2)}\widehat{\varphi}^{\,n+1} \\ &+ \frac{(n+2)\widehat{\varphi}^{\,n+1} + (n-1)\widehat{\varphi}^{\,n-1}}{(2n+1)((\widehat{\varphi}^{\,n})^2 + (\widehat{\varphi}^{\,n+1})^2)}\widehat{\varphi}^{\,n}, \end{aligned} \tag{29.3}$$

satisfies

$$s = s_n + \frac{[\varphi'(s)]}{[\varphi(s)]}\frac{1-s^2}{n^2} + \mathrm{O}(n^{-3}). \tag{29.4}$$

This theorem states that s_n is a second-order approximation of the discontinuity location s. We also deduce from (29.4) that a Richardson acceleration process can improve the accuracy so that, with eight successive Legendre coefficients of the function φ, we can derive a third-order approximation s_n^* of s. Then, OULD KABER and VANDEVEN [1993] have deduced from (29.4) that, from such a third-order approximation of s, a second-order approximation of $[\varphi(s)]$ can be computed. The step function that depends on these two parameters, s and $[\varphi(s)]$, is then well approximated and *all* its Legendre coefficients can be computed accurately, in fact with a second-order accuracy.

Recall now that the discontinuous function φ is the sum of a step function H and of a continuous (or more regular) function ψ. From the Legendre coefficients of φ and the approximation of the Legendre coefficients of H (with a second-order accuracy), we obtain by subtraction (with a second-order accuracy as well) the coefficients of ψ in the basis made of the $(n+\frac{1}{2})L_n$, $0 \leqslant n \leqslant N$, that we denote $\widehat{\psi}^{\,n}$. The partial sum $\sum_{n=0}^{N}\widehat{\psi}^{\,n}(n+\frac{1}{2})L_n$ converges to ψ faster than the partial sum $\sum_{n=0}^{N}\widehat{\varphi}^{\,n}(n+\frac{1}{2})L_n$ converges to φ: indeed, by setting

$$\widetilde{\varphi}_N = \widetilde{H} + \sum_{n=0}^{N}\widehat{\psi}^{\,n}(n+\tfrac{1}{2})L_n,$$

where $\widetilde{H}$ is the second-order approximation of H, we obtain a better approximation of φ that satisfies, for any $\varepsilon > 0$,

$$\|\varphi - \widetilde{\varphi}_N\|_{L^2(\Lambda)} \leqslant c(\varepsilon)N^{-3/2+\varepsilon} \tag{29.5}$$

(this can be derived from (6.3) since ψ belongs at least to $H^{3/2-\varepsilon}(\Lambda)$).

This result is not yet of spectral accuracy but it certainly proves that a better convergence rate than expected can be obtained by an appropriate combination of the

Legendre coefficients of φ. Thus, the idea to use high-order methods even for discontinuous data may not be so crazy. The first results in this direction are due to MAJDA, MCDONOUGH and OSHER [1978] and have been extended to spectral computations of discontinuous solutions of conservation laws by ABARBANEL, GOTTLIEB and TADMOR [1986] and GOTTLIEB and TADMOR [1985]. All this work was related to the periodic form of spectral methods based on Fourier approximation and the proofs are then slightly less technical than in the Legendre or Chebyshev case, nevertheless the main ideas already appear in these references.

To get a better convergence, one has to make a better use of the information that is contained within the Legendre coefficients of the discontinuous function. These two procedures of filtering are based on averaging the Legendre coefficients. For chronological reasons much of the work has been done on the truncated Fourier series and the work of GOTTLIEB and TADMOR [1985] has to be quoted here, it can be easily extended to Chebyshev projections. The work of VANDEVEN [1993] is of prior importance in this direction since it provides a precise analysis of the convergence rate of the filtering procedure with respect to some simple criteria to verify. Here, we present a filter in the Legendre framework that has been developed and analyzed by OULD KABER [1991] in the same spirit as the filters introduced by GOTTLIEB and TADMOR [1985].

Let us first define, for any nonnegative integer n, the convolution kernel $\mathcal{K}_n$ defined by

$$\mathcal{K}_n(\zeta,\eta) = \sum_{k=0}^{n} \frac{L_k(\zeta)L_k(\eta)}{\|L_k\|^2_{L^2(\Lambda)}}. \tag{29.6}$$

This kernel allows for writing the projection of any function φ in $L^2(\Lambda)$ as follows

$$(\pi_N\varphi)(\zeta) = \int_{-1}^{1} \varphi(\eta)\mathcal{K}_N(\zeta,\eta)\,\mathrm{d}\eta, \tag{29.7}$$

but this is not a good approximation of φ when the function φ is not regular. The idea is to introduce a different kernel that allows for more accurate results.

Let ρ be a cut-off function, even and positive with a compact support in Λ; we also assume that $\rho(0)$ is equal to 1. Let δ be a positive real parameter, and let M be a positive integer. The filtering procedure consists in associating with any function φ in $L^2(\Lambda)$ the function $\mathcal{F}^{\delta,M}(\pi_N\varphi)$ defined by

$$\mathcal{F}^{\delta,M}(\pi_N\varphi)(\zeta) = \int_{-1}^{1} (\pi_N\varphi)(\eta)\rho\left(\frac{\zeta-\eta}{\delta}\right)\mathcal{K}_M(\zeta,\eta)\,\mathrm{d}\eta. \tag{29.8}$$

Introducing the notation

$$\psi^{\delta,M}(\zeta,\eta) = \rho\left(\frac{\zeta-\eta}{\delta}\right)\mathcal{K}_M(\zeta,\eta), \tag{29.9}$$

will be useful in the sequel, it allows for writing (29.8) in an abridged manner

$$\mathcal{F}^{\delta,M}(\pi_N\varphi)(\zeta) = \int_{-1}^{1} (\pi_N\varphi)(\eta)\psi^{\delta,M}(\zeta,\eta)\,\mathrm{d}\eta. \tag{29.10}$$

We are going to prove that this function is a good approximation of φ for suitable choices of the parameters δ and M. Let us set

$$e(\zeta) = \mathcal{F}^{\delta,M}(\pi_N\varphi)(\zeta) - \varphi(\zeta).$$

In order to analyze this error, we assume as previously that, for a fixed point s in Λ, the restrictions of the function φ to each interval $]-1,s[$ and $]s,1[$ are the restrictions of continuously differentiable functions up to the order m over $\overline{\Lambda}$. We first note that (as previously, the exponent $^{(\eta)}$ after an operator means that it is applied with respect to η)

$$e(\zeta) = e_1(\zeta) + e_2(\zeta),$$

with

$$e_1(\zeta) = \int_{-1}^{1} \varphi(\eta)\psi^{\delta,M}(\zeta,\eta)\,\mathrm{d}\eta - \varphi(\zeta), \tag{29.11}$$

$$e_2(\zeta) = \int_{-1}^{1} (\varphi - \pi_N\varphi)(\eta)\big(\pi_N^{(\eta)}\psi^{\delta,M} - \psi^{\delta,M}\big)(\zeta,\eta)\,\mathrm{d}\eta, \tag{29.12}$$

indeed it follows from the definition of the operator π_N that

$$\int_{-1}^{1} (\pi_N\varphi - \varphi)(\eta)\big(\pi_N^{(\eta)}\psi^{\delta,M}\big)(\zeta,\eta)\,\mathrm{d}\eta = 0.$$

To bound the first term, we remark from the definition (29.9) that

$$e_1(\zeta) = \int_{-1}^{1} \Phi^{\delta,\zeta}(\eta)\mathcal{K}_M(\zeta,\eta)\,\mathrm{d}\eta,$$

$$\text{with } \Phi^{\delta,\zeta}(\eta) = \varphi(\eta)\rho\left(\frac{\zeta-\eta}{\delta}\right) - \varphi(\zeta). \tag{29.13}$$

The function $\Phi^{\delta,\zeta}$ is regular for any point ζ such that $|\zeta - s| > \delta$. So, for $|\zeta - s| > \delta$, we deduce from (29.7) and (6.8), together with the standard Gagliardo–Nirenberg inequality which holds for any function χ in $H^1(\Lambda)$:

$$\sup_{-1\leqslant\zeta\leqslant 1} |\chi(\zeta)| \leqslant \|\chi\|_{L^2(\Lambda)}^{1/2}\|\chi\|_{H^1(\Lambda)}^{1/2},$$

that

$$\forall \zeta' \in \Lambda, \quad \left| \Phi^{\delta,\varsigma}(\zeta') - \int_{-1}^{1} \Phi^{\delta,\varsigma}(\eta) \mathcal{K}_M(\zeta', \eta)\, \mathrm{d}\eta \right| \leqslant c M^{3/4-m} \left\| \Phi^{\delta,\zeta} \right\|_{H^m(\Lambda)}. \tag{29.14}$$

Recalling now that $\rho(0) = 1$, we deduce that $\Phi^{\delta,\varsigma}(\zeta)$ is equal to 0, so that

$$\forall \zeta \in \Lambda, \quad |\zeta - s| > \delta, \quad \left| e_1(\zeta) \right| \leqslant c M^{3/4-m} \left\| \Phi^{\delta,\zeta} \right\|_{H^m(\Lambda)}. \tag{29.15}$$

In order to bound the second term, we first make use of the Cauchy–Schwarz inequality to derive

$$\left| e_2(\zeta) \right| \leqslant \left\| \psi^{\delta,M}(\zeta, \bullet) - \pi_N^{(\eta)} \psi^{\delta,M}(\zeta, \bullet) \right\|_{L^2(\Lambda)} \| \varphi - \pi_N \varphi \|_{L^2(\Lambda)},$$

from which we easily deduce that

$$\left| e_2(\zeta) \right| \leqslant c N^{-(m+\sigma)} \left\| \psi^{\delta,M}(\zeta, \bullet) \right\|_{H^m(\Lambda)} \| \varphi \|_{H^\sigma(\Lambda)} \tag{29.16}$$

whenever φ belongs to $H^\sigma(\Lambda)$. We are left to analyze the behavior of $\|\Phi^{\delta,\zeta}\|_{H^m(\Lambda)}$ and $\|\psi^{\delta,M}(\zeta, \bullet)\|_{H^m(\Lambda)}$. We only state the result since the proof is quite tedious and we refer to OULD KABER ([1991], Chapter 2, Lemmas 3.2 and 3.3) for the complete analysis:

$$\|\Phi^{\delta,\zeta}\|_{H^m(\Lambda)} \leqslant c(m) \left(1 + \frac{1}{\delta}\right)^m \|\rho\|_{H^m(\Lambda)} \sup_{0 \leqslant j \leqslant m} \Big(\sup_{|\zeta - \eta| < \delta} \left| D^j \varphi(\eta) \right| \Big), \tag{29.17}$$

while

$$\left\| \psi^{\delta,M}(\zeta, \bullet) \right\|_{H^m(\Lambda)} \leqslant c(m)\, \|\rho\|_{H^m(\Lambda)} M^2 \left(M^2 + \frac{1}{\delta} \right)^m. \tag{29.18}$$

The previous estimates allow for deriving a proper approximation of φ by $\mathcal{F}^{\delta,M}(\pi_N \varphi)$. For instance, if we choose $M = N^{1/3}$, we derive

$$\begin{aligned} \left| \mathcal{F}^{\delta,M}(\pi_N \varphi)(\zeta) - \varphi(\zeta) \right| &\leqslant c(m) \|\rho\|_{H^m(\Lambda)} \left(1 + \frac{1}{\delta}\right)^m N^{-m/3} \\ &\quad \times \Big(N^{1/4} \sup_{0 \leqslant j \leqslant m} \sup_{|\zeta - \eta| < \delta} \left| D^j \varphi(\eta) \right| + N^{2/3-\sigma} \|\varphi\|_{H^\sigma(\Lambda)} \Big), \end{aligned} \tag{29.19}$$

which gives the spectral convergence away from the discontinuity.

As a conclusion, note that the previous filter is tuned to the Legendre decomposition and has an equivalent in the Chebyshev framework. In opposite to Cesàro type filters, it is directly defined in the physical space, since it is a convolution product of the function $\pi_N\varphi$ by an appropriate kernel.

Another extension of Cesàro averaging was introduced and analyzed in VANDEVEN [1989a], in the framework of approximation by Fourier series. However, these results can be translated directly to the Chebyshev framework thanks to the change of variable $\zeta = \cos\theta$. The new family of filters allows for improving the convergence of the Chebyshev orthogonal expansion

$$\pi_N^{-1/2}\varphi = \sum_{n=0}^{N} \varphi^n T_n,$$

by setting

$$C_{N,\sigma}^{-1/2}\varphi = \sum_{n=0}^{N} \sigma\left(\frac{n}{N+1}\right)\varphi^n\, T_n, \tag{29.20}$$

where σ is an even cut-off function defined on $\overline{\Lambda}$. The choice of the function σ which provides optimal approximation is given in VANDEVEN [1989a], by the polynomials

$$\sigma_k(\zeta) = 1 - \frac{(2k-1)!}{(k-1)!^2}\int_0^{\zeta} \big(t(1-t)\big)^{k-1}\,\mathrm{d}t \tag{29.21}$$

for a fixed positive integer k (note that the choice $\sigma = \sigma_1$ leads to an approximation operator $C_{N,\sigma_1}^{-1/2}$ which is nothing but the first Cesàro mean). The following theorem states the corresponding approximation results.

THEOREM 29.3. *Let us assume that, for a fixed point s in Λ, the restrictions of a function φ on $[-1,s[$ and $]s,1]$ are the restrictions of analytic functions on neighborhoods in $\mathbb{C}$ of $[-1,s]$ and $[s,1]$ respectively. For any positive real number ε and for any integer $k \leqslant c\,N^{\frac{3}{4}}$, the following estimate holds*

$$\sup_{|\zeta - s| > N^{-1-\varepsilon}} \left|\varphi(\zeta) - C_{N,\sigma_k}^{-1/2}\varphi(\zeta)\right| \leqslant N^{\beta}\left(c(\varphi)N^{-\varepsilon/2}\right)^{N^{\varepsilon/4}}, \tag{29.22}$$

for positive constants $c(\varphi)$ independent of N and β independent of φ and N.

In order to be complete, we quote again the work of GOTTLIEB, SHU, SOLOMONOFF and VANDEVEN [1992] and more recently VANDEVEN [1993] who manage to recover a spectral accuracy up to the discontinuity point if this one is known. Of course, the results of Theorem 29.2 allow for approximating this position in an appropriate way. Since the construction of these filters is highly technical, we refer to the original papers for their definition.

The main question now is to be able to compute accurately the first Legendre coefficients of the discontinuous solution of some hyperbolic conservation law. We are going to describe two different techniques which have proven to be quite successful, both in dimensions $d = 1$ and $d = 2$.

30. The cell averaging method

This first method is, as almost every modern finite difference scheme, based on the cell averaged form of the equations. Let f be a given continuous function on $\mathbb{R}$. We consider the nonlinear conservation law

$$\frac{\partial}{\partial t}u(x,t) + \frac{\partial}{\partial x}f\big(u(x,t)\big) = 0, \quad (x,t) \in \Lambda \times [0,\infty), \tag{30.1}$$

provided with appropriate boundary conditions:

$$\begin{cases} u(-1,t) = g_-(t) & \text{if } f'\big(u(-1,t)\big) > 0, \\ u(+1,t) = g_+(t) & \text{if } f'\big(u(+1,t)\big) < 0, \end{cases} \quad t \in [0,\infty), \tag{30.2}$$

and initial value:

$$u(x,0) = u_0(x), \quad x \in \Lambda. \tag{30.3}$$

We recall (see GODLEWSKI and RAVIART [1991] and SMOLLER [1983]) the basic results concerning this problem. Let us assume that the function f is locally Lipschitz-continuous, that the initial condition u_0 is bounded on Λ together with the necessary boundary conditions for all t in $[0,+\infty)$. Then, there exists a unique solution to problem (30.1) provided with conditions (30.2) and (30.3) which satisfies the entropy condition

$$\frac{\partial U(u)}{\partial t} + \frac{\partial F(u)}{\partial x} \leqslant 0,$$

for any choice (U,F) of entropy pair, i.e., for any convex function U and for any function F such that $F' = U'\,f'$.

Now, let x_j, $0 \leqslant j \leqslant N$, be an increasing sequence of discretization points in Λ. For $0 \leqslant j \leqslant N-1$, we define the average $\mu_{j+1/2}(t)$ of the solution $u(\bullet,t)$ over $]x_j, x_{j+1}[$:

$$\mu_{j+1/2}(t) = \frac{1}{x_{j+1} - x_j}\int_{x_j}^{x_{j+1}} u(x,t)\,\mathrm{d}x. \tag{30.4}$$

The cell averaged form of the equation is obtained by integrating equation (30.1) between any two successive points x_j and x_{j+1}, in order to derive an equation on the average as follows

$$\frac{\mathrm{d}\mu_{j+1/2}}{\mathrm{d}t} = -\frac{1}{x_{j+1} - x_j}\big(f\big(u(x_{j+1})\big) - f\big(u(x_j)\big)\big). \tag{30.5}$$

Instead of the initial problem, this system, which is a weak statement of equation (30.1), is solved. This approach, introduced by LAX and WENDROFF [1960] has been extensively used in the finite difference framework (see GODLEWSKI and RAVIART [1991], HARTEN [1986], OSHER [1985] and SHU and OSHER [1989] for instance).

In order to have an accurate approximation for the computation of the averages $\mu_{j+1/2}$, one has to be able to compute $f(u(x_j))$ with a good precision. It is thus necessary to be able to recover, from the averages $\mu_{j+1/2}$, good approximations of $u(x_j)$. This reconstruction procedure is the cornerstone of many cell average schemes in the literature. The first contribution in this direction for spectral approximations is the work of CAI, GOTTLIEB and SHU [1989] in the periodic Fourier spectral framework. CAI, GOTTLIEB and HARTEN [1990] have then extended the scheme in the spectral Chebyshev framework and these ideas have been generalized to the multidomain case by SIDILKOVER and KARNIADAKIS [1992] and to the Legendre case by OULD KABER [1991].

Here, we present the method both in the Legendre and Chebyshev frameworks. The discretization points x_j are then the Gauss–Lobatto nodes ξ_j^α, $0 \leqslant j \leqslant N$, with $\alpha = 0$ (in the Legendre case) or $\alpha = -\frac{1}{2}$ (in the Chebyshev case). Let $\mathcal{A}_N$ denote the averaging operator, i.e., the operator which with any function u in $L^2(\Lambda)$, associates its averages

$$\mu_{j+1/2}, \quad 0 \leqslant j \leqslant N-1,$$

defined in (30.4). The first point is to construct, from these averages, a polynomial of degree N that coincides with u when u is a polynomial of $\mathbb{P}_N(\Lambda)$. However, the first observation is that we do not have enough information for that since only N averages are given and $N+1$ coefficients have to be computed. The information that is lacking traduces the fact that:

$$\begin{cases} \mathcal{A}_N(L_N) = \mathcal{A}_N(((1-\zeta^2)L_N')') = 0 & \text{if } \alpha = 0, \\ \mathcal{A}_N(((1-\zeta^2)T_N')') = 0 & \text{if } \alpha = -\frac{1}{2} \end{cases}$$

and actually this will be useful to enforce one inflow boundary condition (30.2). In what follows, we assume that this boundary condition is required at -1. The following theorem states that the mapping

$$u \mapsto \big(\mathcal{A}_N(u), u(-1)\big)$$

is linear and one-to-one from $\mathbb{P}_N(\Lambda)$ onto $\mathbb{R}^{N+1}$.

THEOREM 30.1. *There exists an operator $\mathcal{R}_N$ from $\mathbb{R}^{N+1}$ onto $\mathbb{P}_N(\Lambda)$ such that, for any $(N+1)$-tuple $a_N = ((\mu_{j+1/2})_{0\leqslant j\leqslant N-1}, \lambda)$ of real values, the following identity holds*

$$\mathcal{A}_N(\mathcal{R}_N a_N) = (\mu_{j+1/2})_{0\leqslant j\leqslant N-1} \quad \textit{and} \quad (\mathcal{R}_N a_N)(-1) = \lambda. \tag{30.6}$$

The discretization can now proceed by using some time marching technique on equation (30.5) which is satisfied by the averages. Let δt stand for the time step. Assume that $(\mu^k_{j+1/2})_{0\leqslant j\leqslant N-1}$ is a computed accurate approximation of the averages of $u(\bullet, k\delta t)$. In order to compute the approximation $(\mu^{k+1}_{j+1/2})_{0\leqslant j\leqslant N-1}$ at the time $(k+1)\delta t$, we use equation (30.5) and need to evaluate the values of f at $u(\xi_j, k\delta t)$. These values are derived via the reconstruction operator $\mathcal{R}_N$ described in the previous theorem that provides the first $N+1$ coefficients of the approximation of $u(\bullet, k\delta t)$ either in the Legendre or the Chebyshev basis.

Here, it is important to note that, if the averages are accurately computed, the accurate information which is given by the reconstruction, does not consist of the pointwise values but the coefficients. Indeed, we have explained in Section 29 that the best polynomial approximation in the $L^2(\Lambda)$ norm is not directly the right approximation in the physical space, though the coefficients are the good ones. In order to compute now the increment on the averages, one has to derive, from these accurate coefficients, an accurate value (in the physical space) of the solution at time $k\delta t$. It is common to need a nonoscillatory reconstruction of the solution in order, among other reasons, to have shocks that are well located. The pure filtering procedure explained in Section 29 is not sufficient for this purpose since the accuracy of the filters that are currently implemented is not high at the level of the discontinuity.

In the present implementation of this type of scheme, before filtering, the possible discontinuity of the solution must be located, which can be done thanks to Theorem 29.2 or to similar arguments. The step function responsible of the discontinuities of the solution is thus determined from the last part of the coefficients, the coefficients of the continuous contribution v^k of the approximate solution can then be computed, and this last function is filtered out to obtain a good accuracy of both v^k (which is not of real interest), and of the approximate solution at time $k\delta t$ at the points ξ^α_j, $0\leqslant j\leqslant N$.

Such a reconstruction in the physical space as the sum of a Heaviside function and the approximation of a continuous part is proven to conserve the boundedness of the total variation in the case $\alpha = -\frac{1}{2}$ by CAI, GOTTLIEB and HARTEN [1990]. These authors also establish that the corresponding scheme is stable, in some weighted $L^2(\Lambda)$ type norm, when applied on the linear equation (30.1) with $f(v) = v$. In the case of nonlinear fluxes, adding some limiters, which is current in such cell averaging contexts, is sometimes helpful in order to stabilize the process, and leads to accurate results.

Figures 30.1 and 30.2 are provided by SIDILKOVER and KARNIADAKIS [1992], where the extension of these ideas to the multidomain case is undertaken. The discrete function is no more a polynomial on the domain but a continuous function that is piecewise polynomial of high degree on intervals that form a disjoint partition of

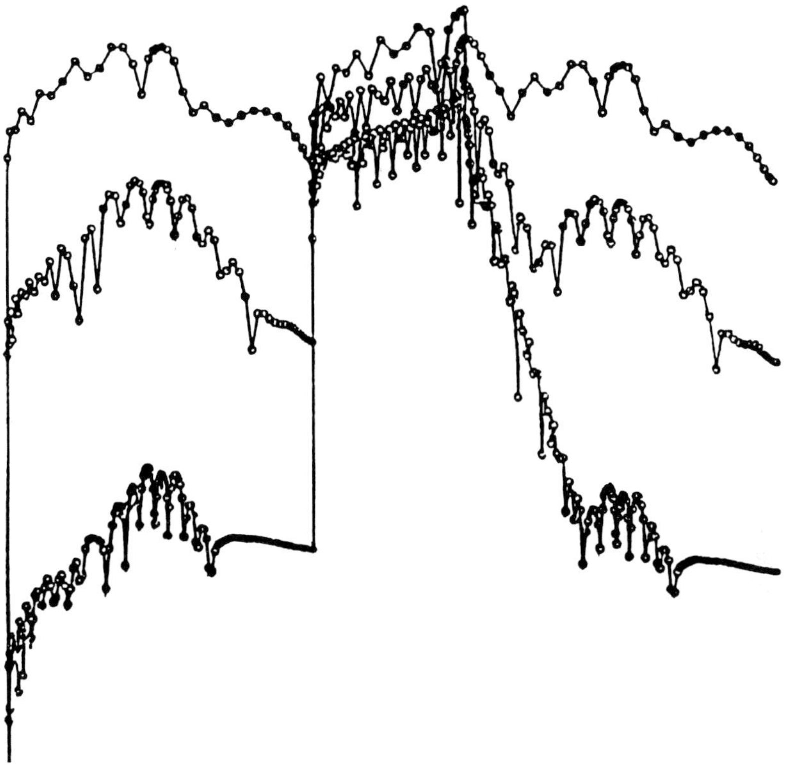

FIG. 30.1. Pointwise error in the discrete solution of the Bürgers' equation via the cell averaging method.

the computational domain. The additional ingredient is the treatment of the interface conditions which, like the boundary condition in the case of one domain, are the additional pieces of information that allow for reconstructing, element by element, the coefficients of the discrete function from its given averages.

The first example deals with the Bürgers' equation:

$$\frac{\partial}{\partial t}u(x,t) + \frac{\partial}{\partial x}\left(\frac{u^2}{2}\right)(x,t) = 0, \quad (x,t) \in \Lambda \times [0,\infty), \tag{30.7}$$

provided with initial condition

$$u_0(\zeta) = 1 + \tfrac{1}{2}\sin\zeta$$

and periodic boundary conditions. The initial smooth problem develops a shock discontinuity. The exact solution is easily computed and the accuracy of the proposed algorithm can be evaluated. In Fig. 30.1, we plot the pointwise error on a logarithmic scale for different values of the degree N of the polynomial approximation. The

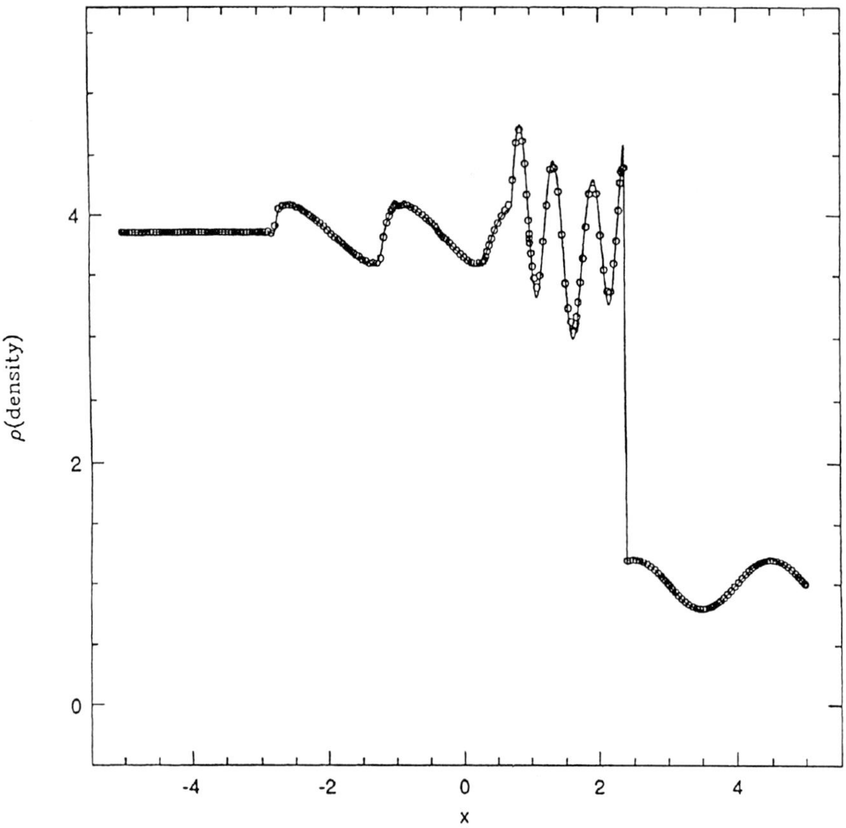

FIG. 30.2. Discrete density of Euler's equations via the cell averaging method.

method uses a domain decomposition with eight subdomains and the interface conditions are those explained for the reconstruction procedure. The numerical solution is filtered out with the Vandeven filter (courtesy of Sidilkover and Karniadakis).

The second example deals with the Euler's equations in a conservative form:

$$\frac{\partial}{\partial t}U(x,t) + \frac{\partial}{\partial x}F\big(U(x,t)\big) = 0, \quad (x,t) \in \Lambda \times [0,\infty), \tag{30.8}$$

with

$$U = (\rho, \rho u, e) \quad \text{and} \quad F(U) = \big(\rho u, \rho u^2 + p, (e+p)u\big).$$

Here, ρ is the density of the fluid, u is its velocity, p is its pressure and e is the specific total energy, given by:

$$e = \frac{p}{\gamma - 1} + \frac{1}{2}\rho u^2, \tag{30.9}$$

where the coefficient γ is taken equal to 1.4 (indeed, the gas is supposed to be perfect and polytropic).

The problem that is solved is a test problem representing a shock–turbulence interaction. It has been proposed in SHU and OSHER [1989] since a high accuracy is necessary to capture the small scales behind the shock. The initial condition on Λ is

$$(u,\rho,p)_0 = \begin{cases} (2.629369, 3.857143, 10.333333) & \text{if } -1 < x < -0.8, \\ (0, 1, 1 + 0.2\sin(5\pi x)) & \text{if } -0.8 < x < 1. \end{cases} \tag{30.10}$$

As is shown in Fig. 30.2, a lot of fine structures are present and high-order methods are necessary to represent the solution well, indeed a second-order scheme smears out all the structure (see CAI, GOTTLIEB and HARTEN [1990]).

31. The spectral viscosity method

We explain this discretization on the same model problem (30.1) as previously. Let us first recall that the Galerkin approximation of this equation cannot converge to the right solution since the entropy of the discrete solution remains constant.

A simple way to derive a solution is to add some dissipation to the equation, i.e., instead of solving (30.1), we discretize the parabolic equation

$$\frac{\partial}{\partial t}\tilde{u}(x,t) + \frac{\partial}{\partial x} f\big(\tilde{u}(x,t)\big) = \varepsilon \frac{\partial^2}{\partial x^2}\tilde{u}(x,t), \quad (x,t) \in \Lambda \times [0,\infty). \tag{31.1}$$

If the discretization parameter N is not too small, the associated Galerkin discrete problem has a solution that will be not too far from $\tilde{u}$ and, if the parameter ε is not too large, $\tilde{u}$ itself is not too far from the actual solution u of problem (30.1). This procedure would certainly provide an approximation, but not a high-order approximation of u, since an error of size ε remains.

We are concerned here with the *spectral viscosity* method, which is due to TADMOR [1989] and MADAY and TADMOR [1989]: it relies on the same idea as equation (31.1) but the difference is that now the parameter ε is strongly linked to the discretization parameter N. As in Sections 15 to 17, we concentrate on the Legendre spectral method with numerical integration and we show that, by adding a small amount of spectral viscosity to the high modes, we can construct a discretization of problem (30.1) which has the right stability (hence convergence) properties, without sacrificing the spectral accuracy of the underlying approximation. This method has been analyzed and numerically tested in MADAY, OULD KABER and TADMOR [1993].

From now on, we denote by $(\bullet,\bullet)_N$ the one-dimensional product, defined for continuous functions u and v on $\overline{\Lambda}$:

$$(u,v)_N = \sum_{j=0}^{N} u(\xi_j)v(\xi_j)\rho_j.$$

In the spectral viscosity approximation of equation (30.1), we seek a polynomial u_N in $\mathbb{P}_N(\Lambda)$ which satisfies

$$\begin{aligned}
&\forall v_N \in \mathbb{P}_N(\Lambda),\\
&\quad\left(\frac{\partial u_N}{\partial t} + \frac{\partial}{\partial x}(\mathcal{I}_N f)(u_N), v_N\right)_N\\
&\qquad = -\varepsilon_N\left(Q_N\left(\frac{\partial u_N}{\partial x}\right), \frac{\partial v_N}{\partial x}\right)_N + \big(B_N(u_N), v_N\big)_N.
\end{aligned} \tag{31.2}$$

Here, $B_N(u_N)$ is a forcing polynomial in $\mathbb{P}_N(\Lambda)$ of the form

$$B_N(u_N) = \big(\lambda(t)(1-\zeta) + \mu(t)(1+\zeta)\big)L_N'(\zeta), \tag{31.3}$$

involving (at most) two nonzero free parameters, $\lambda(t)$ and $\mu(t)$, which should enable $u_N(x,t)$ to match the *inflow* boundary data (30.2) which is prescribed at $x = -1$ whenever $f'(u_N(-1,t))$ is positive and at $x = +1$ whenever $f'(u_N(+1,t))$ is negative. Finally, Q_N denotes the spectral viscosity operator, defined for any function $\varphi = \sum_{n=0}^{+\infty} \varphi^n L_n$ by the formula

$$Q_N\varphi = \sum_{n=0}^{N} Q^n \varphi^n L_n,$$

with bounded viscosity coefficients

$$\begin{cases} Q^n = 0 & \text{if } n \leqslant m_N,\\ 1 - (m_N/n)^4 \leqslant Q^n \leqslant 1 & \text{if } n > m_N. \end{cases} \tag{31.4}$$

The free pair (ε_N, m_N) of spectral viscosity parameters will be chosen later on, with ε_N tending to 0 and m_N tending to $+\infty$ when N tends to $+\infty$, in order to retain the formal spectral accuracy of problem (31.2) as a discretization of problem (30.1).

REMARK 31.1. Let us describe how the spectral viscosity method can be implemented as a usual collocation method. We let the test function v_N run through the basis made of the Lagrange polynomials l_j associated with the nodes ξ_j, $0 \leqslant j \leqslant N$, which have been introduced in (15.28). At the interior nodes, we obtain

$$\begin{aligned}
&\frac{\partial}{\partial t}u_N(\xi_j,t) + \frac{\partial}{\partial x}(\mathcal{I}_N f)(u_N)(\xi_j,t)\\
&\quad = \varepsilon_N \frac{\partial}{\partial x} Q_N\left(\frac{\partial}{\partial x}u_N\right)(\xi_j,t), \quad 1 \leqslant j \leqslant N-1.
\end{aligned} \tag{31.5}$$

At the outflow boundaries, say at $x = +1$ for instance, the boundary equation reads

$$\frac{\partial}{\partial t}u_N(+1,t) + \frac{\partial}{\partial x}(\mathcal{I}_N f)(u_N)(+1,t)$$
$$= \varepsilon_N \frac{\partial}{\partial x} Q_N\left(\frac{\partial}{\partial x}u_N\right)(+1,t) - \varepsilon_N \rho_N^{-1} Q_N\left(\frac{\partial}{\partial x}u_N\right)(+1,t). \tag{31.6}$$

Note that the last term on the right of (31.6) prevents the creation of a boundary layer. Equations (31.5) and (31.6), together with the prescribed inflow data, furnish a complete equivalent statement of the viscosity approximation (31.2).

Now, we present the numerical analysis of problem (31.2). For the sake of simplicity, we deal with the prototype case where one boundary, say $x = -1$, is an inflow boundary, while $x = +1$ is an outflow one. In this case, the boundary operator $B_N(\bullet)$ takes the form

$$B_N(u_N) = \lambda(t)(1-x)L'_N(x), \tag{31.7}$$

where $\lambda(t)$ is a free Lagrange multiplier parameter which enables $u_N(x,t)$ to match the prescribed inflow boundary data

$$u_N(-1,t) = g_-(t) \tag{31.8}$$

(see MADAY, OULD KABER and TADMOR [1993] for details).

We begin by noting that, in view of

$$(1-x_j)L'_N(x_j) = \begin{cases} 0 & \text{when } 1 \leqslant j \leqslant N, \\ -2(-1)^N/\rho_0 & \text{when } j = 0, \end{cases}$$

the contribution of the boundary operator $B_N(u_N)$ to problem (31.2), amounts to

$$\big(B_N(u_N), v_N\big)_N = -2(-1)^N \lambda(t) v_N(-1).$$

To have a better insight on the influence of $\lambda(t)$, we take v_N equal to 1 in (31.2), and we obtain

$$\frac{\partial}{\partial t}(u_N(+1,t) + f\big(u_N(+1,t)\big) - f\big(u_N(-1,t)\big) = 2(-1)^N\lambda(t). \tag{31.9}$$

Thus, $\lambda(t)$ measures the rate of change of total mass over the whole interval $\overline{\Lambda}$.

Let T be a fixed positive real number. To go further, we are led to make the following assumption of boundedness on the discrete solution u_N of problem (31.2): there exists a constant M such that

$$\sup_{0 \leqslant t \leqslant T} \sup_{x \in \overline{\Lambda}} \big|u_N(x,t)\big| \leqslant M \tag{31.10}$$

(this boundedness is proven by MADAY and TADMOR [1989] in the case of periodic boundary conditions).

Then, it follows from (31.9) that

$$\left|\lambda(t)\right| \leqslant \frac{1}{\sqrt{2}}\left\|\left(\frac{\partial u_N}{\partial t}\right)(\cdot,t)\right\|_{L^2(\Lambda)} + \sup_{|v|\leqslant M}\left|f(v)\right|. \tag{31.11}$$

Also, a pessimistic bound on the total mass is

$$\left|\int_0^t \lambda(s)\,\mathrm{d}s\right| \leqslant \frac{1}{\sqrt{2}}\left(\left\|u_N(\cdot,t)\right\|_{L^2(\Lambda)} + \|u_0\|_{L^2(\Lambda)}\right) + t\sup_{|v|\leqslant M}\left|f(v)\right|. \tag{31.12}$$

This in turn, leads to the following result.

THEOREM 31.1. *Assume that hypothesis* (31.10) *holds and that there exist real numbers p and q, $0 < q < p \leqslant 1$, such that the pair (ε_N, m_N) of spectral viscosity parameters satisfies*

$$cN^{-p} \leqslant \varepsilon_N \leqslant c'N^{-p} \quad \textit{and} \quad m_N < c''\,N^{q/4}. \tag{31.13}$$

Then, for any inflow boundary data $g_\pm$ in $H^1(0,T)$, the following estimate holds:

$$\left\|\frac{\partial u_N}{\partial x}\right\|^2_{L^2(0,T;\Lambda)} + \left\|\frac{\partial u_N}{\partial t}\right\|^2_{L^2(0,T;\Lambda)} \leqslant \frac{c}{\varepsilon_N}\left(1 + \|g_\pm\|^2_{H^1(0,T)}\right). \tag{31.14}$$

The next tool for understanding why the spectral viscosity scheme provides a good approximation of problem (30.1), is a weak representation of the truncation error of problem (31.2). Indeed, a rather technical computation (see MADAY, OULD KABER and TADMOR [1993]) allows for deriving that the solution u_N of problem (31.2) satisfies, for any function φ in $\mathscr{D}(\Lambda\times\,]0,T[\,)$,

$$\left\langle \frac{\partial u_N}{\partial t} + \frac{\partial}{\partial x}f(u_N), \varphi\right\rangle = \sum_{j=1}^{4} I_j(\varphi), \tag{31.15}$$

where the following estimates hold for any polynomial φ_N in $\mathbb{P}_N(\Lambda)$,

$$\begin{aligned}&\left|I_1(\varphi)(t)\right| \\ &\quad\leqslant \left(\frac{c}{\sqrt{\varepsilon_N}}\right)\left(\left\|(\varphi-\varphi_N)(\cdot,t)\right\|_{L^2(\Lambda)} + N^{-1}\left\|\left(\frac{\partial\varphi_N}{\partial x}\right)(\cdot,t)\right\|_{L^2(\Lambda)}\right),\end{aligned} \tag{31.16}$$

$$\left|I_2(\varphi)(t)\right| \leqslant c\,\varepsilon_N\, m_N^2(\log N)\left\|\left(\frac{\partial\varphi_N}{\partial x}\right)(\cdot,t)\right\|_{L^2(\Lambda)}, \tag{31.17}$$

$$I_3(\varphi)(t) = -\varepsilon_N \left(\frac{\partial u_N}{\partial x}, \frac{\partial \varphi_N}{\partial x} \right)_N, \tag{31.18}$$

$$I_4(\varphi)(t) = -2(-1)^N \int_0^T \lambda(t)\varphi_N(-1,t)\,\mathrm{d}t. \tag{31.19}$$

Then, the choice $\varphi_N = i_N\varphi$ seems appropriate to our purpose. Indeed, it gives the estimates

$$|I_1(\varphi)(t)| \leqslant \left(\frac{c}{N\sqrt{\varepsilon_N}} \right) \left\| \left(\frac{\partial \varphi}{\partial x} \right)(\cdot, t) \right\|_{L^2(\Lambda)},$$

$$|I_2(\varphi)(t)| \leqslant c\varepsilon_N m_N^2 (\log N) \left\| \left(\frac{\partial \varphi}{\partial x} \right)(\cdot, t) \right\|_{L^2(\Lambda)},$$

$$\|I_3(\varphi)\|_{L^2(0,T)} \leqslant c\sqrt{\varepsilon_N} \sup_{0 \leqslant t \leqslant T} \left\| \left(\frac{\partial \varphi}{\partial x} \right)(\cdot, t) \right\|_{L^2(\Lambda)},$$

while $I_4(\varphi)(t)$ is equal to 0.

We conclude that the spectral viscosity discretization (31.2) satisfies

$$\left\| \frac{\partial u_N}{\partial t} + \frac{\partial}{\partial x} f(u_N) \right\|_{L^2(0,T;H^{-1}(\Lambda))} \leqslant c \left(\frac{1}{N\sqrt{\varepsilon_N}} + \varepsilon_N m_N^2 (\log N) + \sqrt{\varepsilon_N} \right).$$

Since the right-hand side tends to 0 when N tends to $+\infty$ and when assumption (31.13) is satisfied, for all values of N, the distribution $\partial u_N/\partial t + (\partial/\partial x) f(u_N)$ belong to a compact subset of $L^2(0,T;H^{-1}(\Lambda))$. Using compensated compactness arguments, we derive the following theorem.

THEOREM 31.2. *If the assumptions of Theorem* 31.1 *are satisfied, there exists a subsequence of the solutions* $(u_N)_N$ *of problem* (31.2) *which converges to a weak solution of problem* (30.1), *strongly in* $L^r(\Lambda \times]0,T[)$, *where* r *is any real number* $< +\infty$. *Moreover, if the parameter* p *in* (31.13) *is* < 1, *the whole sequence* $(u_N)_N$ *converges strongly to the unique entropy solution of problem* (30.1).

We end this section by presenting some numerical results which are obtained by applying the spectral viscosity method. The test cases are exactly the same as in Section 30: Bürgers' equation (30.7) with initial condition $u_0(\zeta) = 1 + \frac{1}{2}\sin\zeta$ and periodic boundary conditions in Fig. 31.1, Euler's equations (30.8) with initial condition (30.10) in Fig. 31.2. The discretization with respect to the time variable is performed via the (second-order) Adams–Bashforth scheme. The spectral viscosity method is applied with N equal to 300 (courtesy of Ould Kaber).

Of course, in this case as in any other, the pure solution only converges in $L^2(\Lambda)$ at each time step to the solution as it has been proven in the previous theorem.

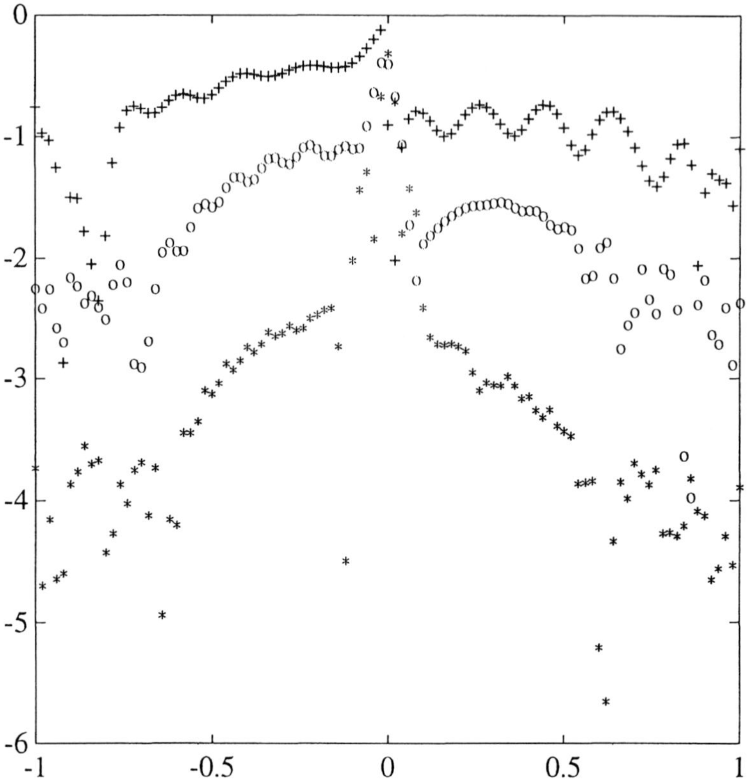

FIG. 31.1. Pointwise error in the discrete solution of Bürgers' equation via the spectral viscosity method.

The discontinuity induces oscillations all over the domain of computation, which is a good sign since it means that the viscosity has not polluted the solution too much. The spectral viscosity method is not supposed to produce an oscillation free approximation. It is supposed to provide stable and accurate results, in the sense that the discrete solution u_N is close to the $L^2(\Lambda)$ orthogonal projection of the solution. This means that it has to be filtered out, by using one of the procedures that were presented in Section 29. The filter that has been used is not optimal at the level of the discontinuity and the pick close to the discontinuity, though present in the unfiltered solution, is smeared out in the filtered representation. Another type of filter, better suited to the capture of the discontinuity would certainly improve the representation at this level.

Theorem 31.2 does not provide either a convergence rate of the discrete solution towards the exact one. However, a convergence rate better than $N^{-1/4}$ is established in SCHOCHET [1989].

The strategy is easily extendable to multidimensional problems, and the spectral viscosity is then a Poisson equation on the high modes obtained by the tensorization of the one-dimensional spectral viscosity operator. The example that has been treated

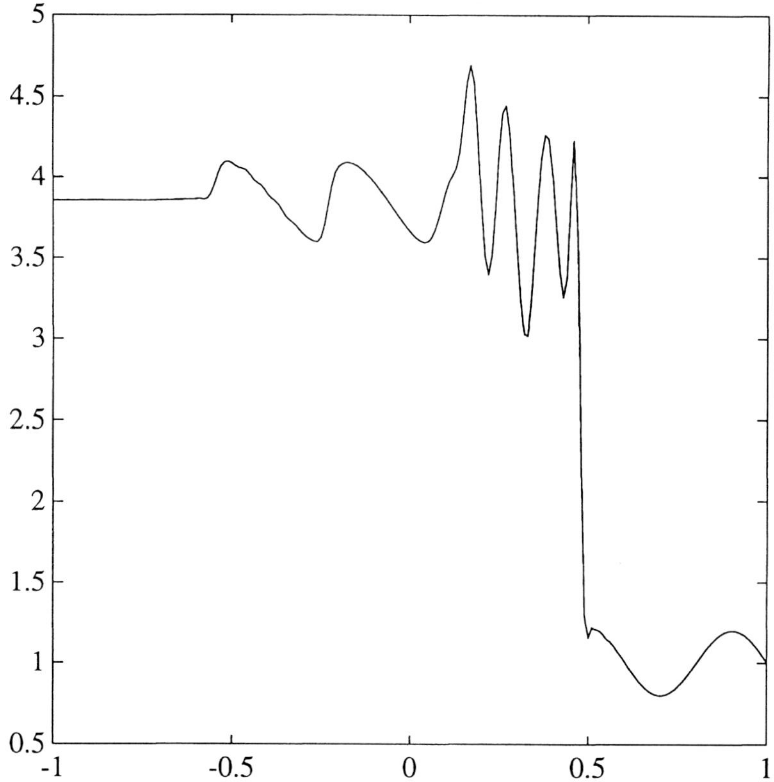

FIG. 31.2. Discrete density of Euler's equations via the spectral viscosity method.

by ORSZAG [1980] is the reflexion of a shock, for which the method is not competitive. What we only want to show here, is that the method can handle problems that are quite difficult. In fact, the method is designed to tackle problems that cannot be handled by other techniques like the interaction shock-turbulence that is now under investigation.

References

ABARBANEL, S., D. GOTTLIEB and E. TADMOR (1986), Spectral methods for discontinuous problems, in: K.W. Morton and M.J. Baines, eds., *Numerical Methods for Fluid Dynamics* **II** (Oxford Univ. Press, Oxford).

ABRAMOWITZ, M. and I.A. STEGUN (1970), *Handbook of Mathematical Functions* (Dover, New York).

ADAMS, R.A. (1975), *Sobolev Spaces* (Academic Press, New York).

AHJAOU, A. (1994), Approximation de l'équation de Bürgers dans $\mathbb{R}$ par les fonctions d'Hermite, in: Approximation numérique de certaines EDP non-linéaires en domaines non bornés par des méthodes spectrales de type Hermite, Thèse de l'Université de Nancy I, France.

AZAÏEZ, M. (1990), Calcul de la pression dans le problème de Stokes pour des fluides visqueux incompressibles par une méthode spectrale de collocation, Thesis, Université de Paris-Sud.

BABUŠKA, I. (1973), The finite element method with Lagrangian multipliers, *Numer. Math.* **20**, 179–192.

BABUŠKA, I. and M. SURI (1990), The p and the $h-p$-versions of the finite element method, an overview, *Comput. Methods Appl. Mech. Engrg.* **80**, 5–26.

BABUŠKA, I., B.A. SZABÓ and I.N. KATZ (1981), The p-version of the finite element method, *SIAM J. Numer. Anal.* **18**, 512–545.

BEN BELGACEM, F. (1994), Polynomial extensions of polynomial traces in 3D, in: C. Bernardi and Y. Maday, eds., *International Conference On Spectral And High Order Methods 1992* (North-Holland, Amsterdam) 235–241.

BERGH, J. and J. LÖFSTRÖM (1976), *Interpolation Spaces: An Introduction* (Springer, Berlin).

BERNARDI, C., C. CANUTO and Y. MADAY (1988), Generalized inf–sup conditions for Chebyshev spectral approximation of the Stokes problem, *SIAM J. Numer. Anal.* **25**, 1237–1271.

BERNARDI, C., C. CANUTO and Y. MADAY (1991), Spectral approximations of the Stokes equations with boundary conditions on the pressure, *SIAM J. Numer. Anal.* **28**, 333–362.

BERNARDI, C., C. CANUTO, Y. MADAY and B. MÉTIVET (1990), Single-grid spectral collocation for the Navier–Stokes equations, *IMA J. Numer. Anal.* **10**, 253–297.

BERNARDI, C., G. COPPOLETTA and Y. MADAY (1990), Some spectral approximations of multi-dimensional fourth-order problems, Internal Report of the Laboratoire d'Analyse Numérique, Université Pierre et Marie Curie, Paris.

BERNARDI, C., G. COPPOLETTA and Y. MADAY (1992), Some spectral approximations of bidimensional fourth-order problems, *Math. Comput.* **59**, 63–76.

BERNARDI, C., M. DAUGE and Y. MADAY (1992a), Relèvements de traces préservant les polynômes, *C.R. Acad. Sci. Paris Sér. I. Math.* **315**, 333–338.

BERNARDI, C., M. DAUGE and Y. MADAY (1992b), Polynomials in weighted Sobolev spaces, Internal Report of the Laboratoire d'Analyse Numérique, Université Pierre et Marie Curie, Paris.

BERNARDI, C., M. DAUGE and Y. MADAY (1994), Interpolation of nullspaces for polynomial approximation of divergence-free functions in a cube, in: M. Costabel, M. Dauge and S. Nicaise, eds., *Proc. Conf. Boundary Values Problems and Integral Equations in Non-smooth Domains*, Lecture Notes in Pure and Applied Mathematics **167** (Dekker) 27–46.

BERNARDI, C., V. GIRAULT and Y. MADAY (1992), Mixed spectral element approximation of the Navier–Stokes equations in the stream-function and vorticity formulation, *IMA J. Numer. Anal.* **12**, 565–608.

BERNARDI, C. and Y. MADAY (1988), A collocation method over staggered grids for the Stokes problem, *Internat. J. Numer. Methods Fluids* **8**, 537–557.

BERNARDI, C. and Y. MADAY (1989), Properties of some weighted Sobolev spaces, and application to spectral approximations, *SIAM J. Numer. Anal.* **26**, 769–829.

BERNARDI, C. and Y. MADAY (1990), Relèvement polynômial de traces et applications, *Modél. Math. Anal. Numér.* **24**, 557–611.

BERNARDI, C. and Y. MADAY (1991), Some spectral approximations of one-dimensional fourth-order problems, in: P. Nevai and A. Pinkus, eds., *Progress in Approximation Theory* (Academic Press, San Diego) 43–116.

BERNARDI, C. and Y. MADAY (1992a), Polynomial approximation of some singular functions, *Appl. Anal.* **42**, 1–32.

BERNARDI, C. and Y. MADAY (1992b), *Approximations Spectrales de Problèmes aux Limites Elliptiques*, Collection Mathématiques et Applications **10** (Springer, Paris).

BERNARDI, C. and Y. MADAY (1992c), Polynomial interpolation results in Sobolev spaces, *J. Comput. Appl. Math.* **43**, 53–80.

BERNARDI, C., Y. MADAY and B. MÉTIVET (1987), Calcul de la pression dans la résolution spectrale du problème de Stokes, *Rech. Aérospat.* **1**, 1–21.

BERNARDI, C., Y. MADAY and A.T. PATERA, Spectral element methods, in: P.G. Ciarlet and J.-L. Lions, eds., *Handbook of Numerical Analysis* (North-Holland, Amsterdam) to appear.

BOYD, J.P. (1989), *Chebyshev and Fourier Spectral Methods*, Lectures Notes in Engineering **49** (Springer, Berlin).

BREZINSKI, C. (1980), *Padé Type Approximation and General Orthogonal Polynomials* (Birkhäuser, Basel).

BREZIS, H. (1983), *Analyse Fonctionnelle: Théorie et Applications* (Masson, Paris).

BREZZI, F. (1974), On the existence, uniqueness and approximation of saddle-point problems arising from Lagrange multipliers, *R.A.I.R.O. Anal. Numér.* **8** R2, 129–151.

BREZZI, F., J. RAPPAZ and P.-A. RAVIART (1980), Finite dimensional approximation of nonlinear problems, Part I: Branches of nonlinear solutions, *Numer. Math.* **36**, 1–25.

CAI, W., D. GOTTLIEB and A. HARTEN (1990), Cell averaging Chebyshev methods for hyperbolic problems, ICASE Report 90–27.

CAI, W., D. GOTTLIEB and C.-W. SHU (1989), Essentially nonoscillatory spectral Fourier methods for shock wave calculations, *Math. Comput.* **52**, 389–410.

CANUTO, C., M.Y. HUSSAINI, A. QUARTERONI and T.A. ZANG (1987), *Spectral Methods in Fluid Dynamics* (Springer, Berlin).

CANUTO, C. and A. QUARTERONI (1982), Approximation results for orthogonal polynomials in Sobolev spaces, *Math. Comput.* **38**, 67–86.

CANUTO, C. and A. QUARTERONI (1984), Variational methods in the theoretical analysis of spectral approximations, in: R.G. Voigt, D. Gottlieb and M.Y. Hussaini, eds., *Spectral Methods for Partial Differential Equations* (SIAM, Philadelphia, PA) 55–78.

CIARLET, P.G. (1978), *The Finite Element Method for Elliptic Problems* (North-Holland, Amsterdam).

CIARLET, P.G. (1982), *Introduction à l'Analyse Numérique Matricielle et à l'Optimisation* (Masson, Paris), English translation: *Introduction to Numerical Linear Algebra and Optimisation* (Cambridge Univ. Press, Cambridge, 1989).

COOLEY, J.W. and J.W. TUKEY (1965), An algorithm for the machine calculation of complex Fourier series, *Math. Comput.* **19**, 297–301.

COULAUD, O., D. FUNARO and O. KAVIAN (1990), Laguerre spectral approximation of elliptic problems in exterior domains, *Comput. Methods Appl. Mech. Engrg.* **80**, 451–458.

CROUZEIX, M. and A. MIGNOT (1984), *Analyse Numérique des Équations Différentielles* (Masson, Paris).

CROUZEIX, M. and J. RAPPAZ (1989), *On Numerical Approximation in Bifurcation Theory*, Collection Recherches en Mathématiques Appliquées **13** (Masson, Paris).

DAUGE, M. (1988), *Elliptic Boundary Value Problems on Corner Domains*, Lecture Notes in Mathematics **1341** (Springer, Berlin).

DAUTRAY, R. and J.L. LIONS (1987), *Analyse Mathématique et Calcul Numérique pour les Sciences et les Techniques* (Masson, Paris).

DAVIS, P.J. and P. RABINOWITZ (1985), *Methods of Numerical Integration* (Academic Press, Orlando).

DORR, M.R. (1984), The approximation theory for the p-version of the finite element method, *SIAM J. Numer. Anal.* **21**, 1180–1207.

DORR, M.R. (1986), The approximation of solutions of elliptic boundary-value problems via the p-version of the finite element method, *SIAM J. Numer. Anal.* **23**, 58–77.

FISHER, P. and E.M. RØNQUIST (1994), Spectral element methods for large scale parallel Navier–Stokes calculations, in: C. Bernardi and Y. Maday, eds., *International Conference on Spectral and High Order Methods 1992* (North-Holland, Amsterdam) 69–76.

FUNARO, D. (1991), Estimates of Laguerre spectral projectors in Sobolev spaces, in: C. Brezinski, L. Gori and A. Ronveaux, eds., *Orthogonal Polynomials and their Applications* (IMACS, New Brunswick, NJ) 263–266.

FUNARO, D. (1992), *Polynomial Approximation of Differential Equations*, Lecture Notes in Physics, New Series m: Monographs **8** (Springer, Berlin).

FUNARO, D. and O. KAVIAN (1990), Approximation of some diffusion evolution equations in unbounded domains by Hermite functions, *Math. Comput.* **57**, 597–619.

GIRAULT, V. and P.-A. RAVIART (1986), *Finite Element Methods for Navier–Stokes Equations, Theory and Algorithms* (Springer, Berlin).

GODLEWSKI, E. and P.-A. RAVIART (1991), *Hyperbolic Systems of Conservation Laws*, Collection Mathématiques et Applications **3–4** (Ellipses, Paris).

GOTTLIEB, D. (1981), The stability of pseudospectral Chebyshev methods, *Math. Comput.* **36**, 107–118.

GOTTLIEB, D. and S.A. ORSZAG (1977), *Numerical Analysis of Spectral Methods, Theory and Applications* (SIAM, Philadelphia, PA).

GOTTLIEB, D., C.W. SHU, A. SOLOMONOFF and H. VANDEVEN (1992), On the Gibbs phenomenon, I: Recovering exponential accuracy from the Fourier partial sum of a non-periodic analytic function, ICASE Report 92–4.

GOTTLIEB, D. and E. TADMOR (1985), Recovering pointwise values of discontinuous data within spectral accuracy, in: E.M. Murman and S. Abarbanel, eds., *Progress in Supercomputing in Computational Fluid Dynamics, Progress in Scientific Computing* **6** (Birkhäuser, Boston) 357–375.

GRISVARD, P. (1963), Espaces intermédiaires entre espaces de Sobolev avec poids, *Ann. Scuola Norm. Sup. Pisa* **17**, 255–296.

GRISVARD, P. (1985), *Elliptic Problems in Nonsmooth Domains* (Pitman, Boston).

HARTEN, A. (1986), On the nonlinearity of modern shock-capturing schemes, ICASE Report 86–69.

JENSEN, S. and S. ZHANG (1994), The p and $h-p$ versions of some finite element methods for Stokes problem, in: C. Bernardi and Y. Maday, eds., *International Conference on Spectral and High Order Methods 1992* (North-Holland, Amsterdam) 147–155.

JENSEN, S. and M. VOGELIUS (1990), Divergence stability in connection with the p-version of the finite element method, *Modél. Math. Anal. Numér.* **24**, 737–764.

KREISS, H.O. and J. OLIGER (1972), Comparison of accurate methods for the integration of hyperbolic equations, *Tellus* **24**, 199–215.

KREISS, H.O. and J. OLIGER (1979), Stability of the Fourier method, *SIAM J. Numer. Anal.* **16**, 421–433.

LAX, P. and N. MILGRAM (1954), *Parabolic Equations. Contribution to the Theory of Partial Differential Equations* (Princeton Univ. Press, Princeton, NJ).

LAX, P. and B. WENDROFF (1960), System of conservation laws, *Comm. Pure Appl. Math.* **13**, 217–237.

LEQUÉRÉ, P. and J. PÉCHEUX (1990), A three-dimensional pseudo-spectral algorithm for the computation of convection in a rotating annulus, *Comput. Methods Appl. Mech. Engrg.* **80**, 261–271.

LIONS, J.L. and E. MAGENES (1968), *Problèmes aux Limites non Homogènes et Applications* **I** (Dunod, Paris).

MADAY, Y. (1987), Contributions à l'analyse numérique des méthodes spectrales, Thesis, Université Pierre et Marie Curie, Paris.

MADAY, Y. (1989), Relèvement de traces polynômiales et interpolations hilbertiennes entre espaces de polynômes, *C.R. Acad. Sci. Paris Sér. I* **309**, 463–468.

MADAY, Y. (1990), Analysis of spectral projectors in one-dimensional domains, *Math. Comput.* **55**, 537–562.

MADAY, Y. (1991), Résultats d'approximation optimaux pour les opérateurs d'interpolation polynômiale, *C.R. Acad. Sci. Paris Sér. I* **312**, 705–710.

MADAY, Y., D. MEIRON, A.T. PATERA and E.M. RØNQUIST (1993), Analysis of iterative methods for the steady and unsteady Stokes problem: Application to spectral element discretizations, *SIAM J. Numer. Anal.* **14**, 310–337.

MADAY, Y. and B. MÉTIVET (1986), Chebyshev spectral approximation of Navier–Stokes equations in a two-dimensional domain, *Modél. Math. Anal. Numér.* **21**, 93–123.

MADAY, Y., S.M. OULD KABER and E. TADMOR (1993), Analysis of the Legendre pseudospectral approximation of a nonlinear conservation law, *SIAM J. Numer. Anal.* **30**, 321–343.

MADAY, Y. and A.T. PATERA (1989), Spectral element methods for the incompressible Navier–Stokes equations, in: A.K. Noor, ed., *State-of-the-Art Surveys in Computational Mechanics* (ASME, New York).

MADAY, Y., A.T. PATERA and E.M. RØNQUIST (1987), A well-posed optimal Legendre spectral element approximation for the Stokes semi-periodic problem, ICASE Report 87–48.

MADAY, Y., A.T. PATERA and E.M. RØNQUIST (1992), The $\mathbb{P}_N-\mathbb{P}_{N-2}$ method for the approximation of the Stokes problem, Internal Report of the Laboratoire d'Analyse Numérique, Université Pierre et Marie Curie, Paris.

MADAY, Y., B. PERNAUD-THOMAS and H. VANDEVEN (1985), Une réhabilitation des méthodes spectrales de type Laguerre, *Rech. Aérospat.* **6**, 353–375.

MADAY, Y. and A. QUARTERONI (1981), Legendre and Chebyshev spectral approximations of Bürgers' equation, *Numer. Math.* **37**, 321–332.

MADAY, Y. and E.M. RØNQUIST (1990), Optimal error analysis of spectral methods with emphasis on non-constant coefficients and deformed geometries, *Comput. Methods Appl. Mech. Engrg.* **80**, 91–115.

MADAY, Y. and E. TADMOR (1989), Analysis of the spectral viscosity method for periodic conservation laws, *SIAM J. Numer. Anal.* **26**, 854–870.

MAJDA, A., J. MCDONOUGH and S. OSHER (1978), The Fourier method for nonsmooth initial data, *Math. Comput.* **32**, 1041–1081.

MAVRIPLIS, C. (1989), Laguerre polynomials for infinite-domain spectral elements, *J. Comput. Phys.* **80**, 480–488.

MERCIER, B. (1982), Stabilité et convergence des méthodes spectrales polynômiales: Application à l'équation d'advection, *R.A.I.R.O. Anal. Numér.* **16**, 67–100.

MERCIER, B. (1989), *An Introduction to the Numerical Analysis of Spectral Methods* (Springer, Berlin).

MÉTIVET, B. (1987), Résolution spectrales des équations de Navier–Stokes par une méthode de sous-domaines courbes, Thesis, Université Pierre et Marie Curie, Paris.

MOCK, M.S. and P.D. LAX (1978), The computation of discontinuous solutions of linear hyperbolic equations, *Comm. Pure Appl. Math.* **31**, 423–430.

MONTIGNY-RANNOU, F. and Y. MORCHOISNE (1987), A spectral element method with staggered grid for incompressible Navier–Stokes equations, *Internat. J. Numer. Methods Fluids* **7**, 175–189.

MORCHOISNE, Y. (1983), Résolution des équations de Navier–Stokes par une méthode spectrale de sous-domaines, in: P. Lascaux, ed., *Comptes-rendus du 3ème Congrès International sur les Méthodes Numériques de l'Ingénieur* (Paris).

NEČAS, J. (1967), *Les Méthodes Directes en Théorie des Équations Elliptiques* (Masson, Paris).

NICOLAIDES, R.A. (1982), Existence, uniqueness and approximation for generalized saddle point problems, *SIAM J. Numer. Anal.* **19**, 349–357.

ORSZAG, S.A. (1972), Comparison of pseudospectral and spectral approximations, *Stud. Appl. Math.* **51**, 253–259.

ORSZAG, S.A. (1980), Spectral methods for problems in complex geometries, *J. Comput. Phys.* **37**, 70–92.

OSHER, S. (1985), Convergence of generalized MUSCL schemes, *SIAM J. Numer. Anal.* **22**, 947–961.

OULD KABER, S.M. (1991), Résolution par méthodes spectrales de problèmes hyperboliques non linéaires, Thesis, Université Pierre et Marie Curie, Paris.

OULD KABER, S.M. and H. VANDEVEN (1993), Reconstruction d'une fonction discontinue à partir de ses coefficients de Legendre, *C.R. Acad. Sci. Paris Série I* **317**, 527–532.

PASCIAK, J.E. (1980), Spectral and pseudospectral methods for advection equations, *Math. Comput.* **35**, 1081–1092.

PATERA, A.T. (1984), A spectral element method for fluid dynamics: Laminar flow in a channel expansion, *J. Comput. Phys.* **54**, 468–488.

PIRONNEAU, O. (1988), *Méthode des Éléments Finis pour les Fluides* (Masson, Paris), English translation: *Finite Element Methods for Fluids* (Wiley, Chichester, 1989).

QUARTERONI, A. (1987), Blending Fourier and Chebyshev interpolation, *J. Approx. Theory* **51**, 115–126.

RØNQUIST, E.M. (1988), Optimal spectral element methods for the unsteady 3-dimensional incompressible Navier–Stokes equations, Thesis, Massachusetts Institute of Technology, Cambridge.

SACCHI-LANDRIANI, G. and H. VANDEVEN (1989), Polynomial approximation of divergence-free functions, *Math. Comput.* **52**, 103–130.

SCHOCHET, S. (1989), The rate of convergence of spectral-viscosity methods for periodic scalar conservation laws, *SIAM J. Numer. Anal.* **27**, 1142–1159.

SCHWARTZ, L. (1966), *Théorie des Distributions* (Hermann, Paris).

SCOTT, L.R. and M. VOGELIUS (1985), Conforming finite element methods for incompressible and nearly incompressible continua, in: *Lectures in Applied Mathematics* **22** (Amer. Mathematical Soc., Providence, RI), 221–244.

SHU, C.-W. and S. OSHER (1989), Efficient implementation of essentially non-oscillatory shock capturing schemes II, *J. Comput. Phys.* **83**, 32–78.

SIDILKOVER, D. and G.E. KARNIADAKIS (1992), Non-oscillatory spectral element Chebyshev method for shock wave calculations, in: *Fifth International Symposium on Domain Decomposition Methods for Partial Differential Equations*, Norfolk, VA, 1991 (SIAM, Philadelphia, PA) 566–585.

SMOLLER, J. (1983), *Shock Waves and Reaction–Diffusion Equations* (Springer, New York).

SZEGÖ, G. (1978), *Orthogonal Polynomials* (Amer. Mathematical Soc., Providence, RI).

TADMOR, E. (1989), Convergence of spectral methods for nonlinear conservation laws, *SIAM J. Numer. Anal.* **26**, 30–44.

TEMAM, R. (1977), *Theory and Numerical Analysis of the Navier–Stokes Equations* (North-Holland, Amsterdam).

VANDEVEN, H. (1989a), Family of spectral filters for discontinuous problems, Thesis, École Polytechnique, Palaiseau.

VANDEVEN, H. (1989b), Compatibilité des espaces discrets pour l'approximation spectrale du problème de Stokes périodique/non périodique, *Modél. Math. Anal. Numér.* **23**, 649–688.

VANDEVEN, H. (1990), On the eigenvalues of second-order spectral differentiation operators, *Comput. Methods Appl. Mech. Engrg.* **80**, 313–318.

VANDEVEN, H. (1991), On the eigenvalues of second-order spectral differentiation operators, Internal Report of the Laboratoire d'Analyse Numérique, Université Pierre et Marie Curie, Paris.

VANDEVEN, H. (1993), A family of spectral filters for discontinuous problems, personal communication.

List of Symbols

Geometry

$\mathcal{O}$	Lipschitz-continuous open set in $\mathbb{R}^d$	219
$\partial\mathcal{O}$	Boundary of $\mathcal{O}$	219
Γ	Part of the boundary of $\mathcal{O}$	220
Λ	Interval $]-1,1[$	234
ζ	Generic point in Λ	234
Ω_d	Hypercube $]-1,1[^d$	220
$\boldsymbol{x}$	Generic point in Ω_d, with coordinates $x_1, \ldots$ and x_d	220
Γ_j	$(d-1)$-face of Ω	228
$\boldsymbol{n}_j$	Unit normal vector to Γ_j	228
Ω	Square $]-1,1[^2$	220
$\boldsymbol{x} = (x, y)$	Generic point in Ω	226
Γ_J	Edge of Ω	226
$\boldsymbol{a}_J$	Corner of Ω	226
$\boldsymbol{n}_J$	Unit normal vector Γ_J	226
$\boldsymbol{\tau}_J$	Unit tangential vector to Γ_J	226
$\mathbb{R}_+$	Half-line of positive real numbers	286

Standard functional spaces

$[X,Y]_\theta$	Interpolate space of index θ between Hilbert spaces X and Y	222
$\mathscr{C}^\infty(\overline{\mathcal{O}})$	Space of restrictions to $\overline{\mathcal{O}}$ of infinitely differentiable functions	220
$\mathscr{D}(\mathcal{O})$	Space of infinitely differentiable functions with a compact support in $\mathcal{O}$	220
$\mathscr{D}'(\mathcal{O})$	Space of distributions on $\mathcal{O}$	220
$L^2(\mathcal{O})$	Space of square integrable functions on $\mathcal{O}$	220
$H^s(\mathcal{O})$	Sobolev space of order s on $\mathcal{O}$	223
$H^s_0(\mathcal{O})$	Closure of $\mathscr{D}(\mathcal{O})$ in $H^s(\mathcal{O})$	224
$H^{-s}(\mathcal{O})$	Dual space of $H^s_0(\mathcal{O})$	224
$H^{s-1/2}(\Gamma)$	Space of traces over Γ of functions in $H^s(\mathcal{O})$	225

Weighted functional spaces

$\rho_\alpha(\zeta)\,\mathrm{d}\zeta$	Jacobi weighted measure on Λ	334
$\varpi_\alpha(\boldsymbol{x})\,\mathrm{d}\boldsymbol{x}$	Jacobi weighted measure on Ω_d	337
$L^2_\alpha(\Lambda)$	Space of square integrable functions on Λ for the weighted measure	334
$H^s_\alpha(\Lambda)$	Sobolev space of order s on Λ for the weighted measure	334
$H^s_{\alpha,0}(\Lambda)$	Closure of $\mathscr{D}(\Lambda)$ in $H^s_\alpha(\Lambda)$	334
$H^{-s}_\alpha(\Lambda)$	Dual space of $H^s_{\alpha,0}(\Lambda)$	334
$L^2_\alpha(\Omega_d)$	Space of square integrable functions on Ω_d for the weighted measure	337
$H^s_\alpha(\Omega_d)$	Sobolev space of order s on Ω for the weighted measure	337
$H^s_{\alpha,0}(\Omega_d)$	Closure of $\mathscr{D}(\Omega_d)$ in $H^s_\alpha(\Omega)$	338
$H^{-s}_\alpha(\Omega_d)$	Dual space of $H^s_{\alpha,0}(\Omega_d)$	338
$L^2_\diamond(\mathbb{R}_+)$	Space of square integrable functions on $\mathbb{R}_+$ for the measure $\mathrm{e}^{-\zeta}\,\mathrm{d}\zeta$	287
$H^s_\diamond(\mathbb{R}_+)$	Sobolev space of order s on $\mathbb{R}_+$ for the measure $\mathrm{e}^{-\zeta}\,\mathrm{d}\zeta$	287
$H^s_{\diamond\alpha}(\mathbb{R}_+)$	Sobolev space of order s on $\mathbb{R}_+$ for the measure $\zeta^\alpha\,\mathrm{e}^{-\zeta}\,\mathrm{d}\zeta$	288

Trace operators

T^Γ_m	Trace operator on Γ	225
R_m	Lifting operator of $T^{\partial\mathcal{O}}_m$	225

Differential operators

$\mathrm{d}^k/\mathrm{d}\zeta^k$	Derivative of order k	220
$\partial^{\boldsymbol{k}}$	Partial derivative of order k_j with respect to each x_j, $\boldsymbol{k} = (k_1, \dots, k_d)$	220
$\partial/\partial x_j$	Partial derivative with respect to each x_j	220
grad	Gradient operator	220
div	Divergence operator	373
curl	Scalar curl operator	379
curl	Vectorial curl operator	379
Δ	Laplace operator	229
Δ^2	Bilaplacian operator	230
$\partial/\partial n$	Normal derivative	225
$\partial/\partial n_J$	Normal derivative on Γ_J	226
$\partial/\partial \tau_J$	Tangential derivative on Γ_J	226
A	Sturm–Liouville operator for Legendre polynomials	235
A_α	Sturm–Liouville operator for Jacobi polynomials	349
$A_\diamond$	Sturm–Liouville operator for Laguerre polynomials	288

Inverse operators

T_{D}	Inverse of the Laplace operator with Dirichlet boundary conditions	233
T_{D4}	Inverse of the bilaplacian operator with Dirichlet boundary conditions	233
T_{N}	Inverse of the Laplace operator with Neumann boundary conditions	233
T_{D}^{α}	Inverse of the weighted Laplace operator with Dirichlet boundary conditions	344
T_{S}	Inverse of the Stokes operator	381

Polynomials

$\mathbb{P}_N(\Lambda)$	Space of polynomials with degree $\leqslant N$ on Λ	240
$\mathbb{P}_N^0(\Lambda)$	Space $\mathbb{P}_N(\Lambda) \cap H_0^1(\Lambda)$	261
$\mathbb{P}_N^{k,0}(\Lambda)$	Space $\mathbb{P}_N(\Lambda) \cap H_0^k(\Lambda)$	261
$\mathbb{P}_N(\mathbb{R}_+)$	Space of polynomials with degree $\leqslant N$ on $\mathbb{R}_+$	289
$\mathbb{P}_N(\Omega_d)$	Space of polynomials with degree $\leqslant N$ with respect to each variable on Ω_d	269
$\mathbb{P}_N^0(\Omega_d)$	Space $\mathbb{P}_N(\Omega_d) \cap H_0^1(\Omega_d)$	270
$\mathbb{P}_N^{k,0}(\Omega_d)$	Space $\mathbb{P}_N(\Omega_d) \cap H_0^k(\Omega_d)$	270
$\mathbb{P}_{n_1,\ldots,n_d}(\Omega_d)$	Space of polynomials with degree $\leqslant n_j$ with respect to x_j on Ω_d	419
$(L_n)_n$	Family of Legendre polynomials	234
$(J_n^{\alpha})_n$	Family of Jacobi polynomials	348
$(T_n)_n$	Family of Chebyshev polynomials	349
$(\mathcal{L}_n)_n$	Family of Laguerre polynomials	288

Quadrature formulas

ζ_j	Zeros of L_N, Gauss–Legendre nodes	242
ω_j	Weights in the Gauss–Legendre formula	242
ξ_j	Zeros of $(1-\zeta^2)L_N'$, Gauss–Lobatto–Legendre nodes	248
ρ_j	Weights in the Gauss–Lobatto–Legendre formula	249
ζ_j^m	Zeros of $\mathrm{d}^m L_N / \mathrm{d}\zeta^m$	241
ω_j^m	Weights in the related Gauss–Legendre formula	240
σ_j^m	Weights in the generalized Gauss–Legendre formula	241
$\sigma_{\pm}^{m,k}$	Weights in the generalized Gauss–Legendre formula	242
ζ_j^{α}	Zeros of J_N^{α}, Gauss–Jacobi nodes	351
ω_j^{α}	Weights in the Gauss–Jacobi formula	351
ξ_j^{α}	Zeros of $(1-\zeta^2)J_N^{\alpha\prime}$, Gauss–Lobatto–Jacobi nodes	354
ρ_j^{α}	Weights in the Gauss–Lobatto–Jacobi formula	354

$\zeta_j^{\alpha,m}$	Zeros of $\mathrm{d}^m J_N^\alpha / \mathrm{d}\zeta^m$	355
$\sigma_j^{\alpha,m}$	Weights in the generalized Gauss–Jacobi formula	355
$\sigma_\pm^{\alpha,m,k}$	Weights in the generalized Gauss–Jacobi formula	356
Σ_N	Gauss–Legendre grid	304
Ξ_N	Gauss–Lobatto–Legendre grid	306
Σ_N^m	Generalized Gauss–Legendre grid	307
Σ_N^α	Gauss–Jacobi grid	364
Ξ_N^α	Gauss–Lobatto–Jacobi grid	364
$\Sigma_N^{\alpha,2}$	Generalized Gauss–Jacobi grid	365

Polynomial projection operators

π_N	Orthogonal projection operator from $L^2(\Lambda)$ onto $\mathbb{P}_N(\Lambda)$	257
$\pi_N^{k,0}$	Orthogonal projection operator from $H_0^k(\Lambda)$ onto $\mathbb{P}_N^{k,0}(\Lambda)$	261
$\tilde{\pi}_N^k$	Projection operator from $H^k(\Lambda)$ onto $\mathbb{P}_N(\Lambda)$	264
$\tilde{\pi}_N^{t,p,0}$	Projection operator from $H^t(\Lambda) \cap H_0^p(\Lambda)$ onto $\mathbb{P}_N(\Lambda) \cap H_0^p(\Lambda)$	265
$\pi_N^\diamond$	Orthogonal projection operator from $L_\diamond^2(\mathbb{R}_+)$ onto $\mathbb{P}_N(\mathbb{R}_+)$	289
π_N^α	Orthogonal projection operator from $L_\alpha^2(\Lambda)$ onto $\mathbb{P}_N(\Lambda)$	356
$\pi_N^{\alpha,k,0}$	Orthogonal projection operator from $H_{\alpha,0}^k(\Lambda)$ onto $\mathbb{P}_N^{k,0}(\Lambda)$	357
$\tilde{\pi}_N^{\alpha,k}$	Projection operator from $H_\alpha^k(\Lambda)$ onto $\mathbb{P}_N(\Lambda)$	358
$\tilde{\pi}_N^{\alpha,t,p,0}$	Projection operator from $H_\alpha^t(\Lambda) \cap H_{\alpha,0}^p(\Lambda)$ onto $\mathbb{P}_N(\Lambda) \cap H_{\alpha,0}^p(\Lambda)$	358
Π_N	Orthogonal projection operator from $L^2(\Omega_d)$ onto $\mathbb{P}_N(\Omega_d)$	269
$\Pi_N^{k,0}$	Orthogonal projection operator from $H_0^k(\Omega_d)$ onto $\mathbb{P}_N^{k,0}(\Omega_d)$	270
Π_N^k	Orthogonal projection operator from $H^k(\Omega_d)$ onto $\mathbb{P}_N(\Omega_d)$	272
$\widetilde{\Pi}_N^{t,p,0}$	Projection operator from $H^t(\Omega_d) \cap H_0^p(\Omega_d)$ onto $\mathbb{P}_N(\Omega_d) \cap H_0^p(\Omega_d)$	273
Π_N^α	Orthogonal projection operator from $L_\alpha^2(\Omega_d)$ onto $\mathbb{P}_N(\Omega_d)$	359
$\Pi_N^{\alpha,k,0}$	Orthogonal projection operator from $H_{\alpha,0}^k(\Omega_d)$ onto $\mathbb{P}_N^{k,0}(\Omega_d)$	359
$\Pi_N^{\alpha,k}$	Orthogonal projection operator from $H_\alpha^k(\Omega_d)$ onto $\mathbb{P}_N(\Omega_d)$	359
$\widetilde{\Pi}_N^{\alpha,t,p,0}$	Projection operator from $H_\alpha^t(\Omega_d) \cap H_{\alpha,0}^p(\Omega_d)$ onto $\mathbb{P}_N(\Omega_d) \cap H_{\alpha,0}^p(\Omega_d)$	360

Polynomial interpolation operators

j_{N-1}	Lagrange interpolation operator on the ζ_j	294
i_N	Lagrange interpolation operator on the ξ_j	303
k_{N+m-1}^m	Hermite interpolation operator on the ζ_j^m	300
j_{N-1}^α	Lagrange interpolation operator on the ζ_j^α	360
i_N^α	Lagrange interpolation operator on the ξ_j^α	362
$k_{N+1}^{\alpha,2}$	Hermite interpolation operator on the $\zeta_j^{\alpha,2}$	363
$\mathcal{J}_{N-1}$	Lagrange interpolation operator on Σ_N	304

$\mathcal{I}_N$	Lagrange interpolation operator on Ξ_N	306
$\mathcal{K}_{N+m-1}$	Lagrange interpolation operator on Σ_N^m	307
$\mathcal{J}_{N-1}^{\alpha}$	Lagrange interpolation operator on Σ_N^{α}	364
$\mathcal{I}_N^{\alpha}$	Lagrange interpolation operator on Ξ_N^{α}	364
$\mathcal{K}_{N+1}^{\alpha,2}$	Hermite interpolation operator on $\Sigma_N^{\alpha,2}$	365

Bilinear forms

$\langle\cdot,\cdot\rangle$	Duality pairing between $H^{-m}(\mathcal{O})$ and $H_0^m(\mathcal{O})$	221
$\langle\cdot,\cdot\rangle_{\partial\mathcal{O}}$	Duality pairing between $(H^{1/2}(\partial\mathcal{O}))'$ and $H^{1/2}(\partial\mathcal{O})$	232
$[\cdot,\cdot]_{N,m}$	Discrete analogue of the $L^2(\Lambda)$ scalar product	300
$(\cdot,\cdot)_N$	Discrete analogue of the $L^2(\Omega_d)$ scalar product	309
$a(\cdot,\cdot)$	Bilinear form associated with the Laplace operator	274
$a_N(\cdot,\cdot)$	Discretization of $a(\cdot,\cdot)$	309
$d(\cdot,\cdot)$	Bilinear form associated with the bilaplacian operator	277
$(\cdot,\cdot)_{N-1,2}$	Discrete analogue of the $L^2(\Omega_d)$ scalar product	319
$d_{N-1,2}(\cdot,\cdot)$	Discretization of $d(\cdot,\cdot)$	319
$(\cdot,\cdot)_{N,m}$	Discrete analogue of the $L^2(\Omega_d)$ scalar product	326
$d_{N,m}(\cdot,\cdot)$	Discretization of $d(\cdot,\cdot)$	326
$b(\cdot,\cdot)$	Bilinear form associated with the divergence operator	379
$b_N(\cdot,\cdot)$	Discretization of $b(\cdot,\cdot)$	389
$(\cdot,\cdot)_{N\sim}$	Discrete analogue of the $L^2(\Omega_d)$ scalar product	427
$\tilde{a}_N(\cdot,\cdot)$	Discretization of $a(\cdot,\cdot)$	427
$\tilde{b}_N(\cdot,\cdot)$	Discretization of $b(\cdot,\cdot)$	427
$a_*(\cdot,\cdot)$	Scalar product in $L^2(\Omega)$	441
$a_{*N}(\cdot,\cdot)$	Discretization of $a_*(\cdot,\cdot)$	443
$b_*(\cdot,\cdot)$	Bilinear form associated with the Laplace operator	441
$b_{*N}(\cdot,\cdot)$	Discretization of $b_*(\cdot,\cdot)$	443
$\langle\cdot,\cdot\rangle_{\alpha}$	Duality pairing between $H_{\alpha}^{-m}(\Omega_d)$ and $H_{\alpha,0}^{m}(\Omega_d)$	338
$a_{\alpha}(\cdot,\cdot)$	Bilinear form associated with the Laplace operator	341
$(\cdot,\cdot)_{\alpha,N}$	Discrete analogue of the $L_{\alpha}^2(\Omega_d)$ scalar product	367
$a_{\alpha,N}(\cdot,\cdot)$	Discretization of $a_{\alpha}(\cdot,\cdot)$	367
$d_{\alpha}(\cdot,\cdot)$	Bilinear form associated with the bilaplacian operator	346
$(\cdot,\cdot)_{\alpha,N-1,2}$	Discrete analogue of the $L_{\alpha}^2(\Omega_d)$ scalar product	370
$d_{\alpha,N-1,2}(\cdot,\cdot)$	Discretization of $d_{\alpha}(\cdot,\cdot)$	370
$b_{\alpha i}(\cdot,\cdot)$	Bilinear forms associated with the divergence operator	382
$b_{\alpha i,N}(\cdot,\cdot)$	Discretization of $b_{\alpha i}(\cdot,\cdot)$	409

Navier–Stokes equations

ν	Viscosity	373
$\boldsymbol{u}$	Velocity	373
p	Pressure	373

ψ	Stream function	380
$\boldsymbol{\psi}$	Vector potential	380
ω	Vorticity	441
K	Space of divergence-free functions	380
K_N	Space of divergence-free polynomials	389
M_N	Space of discrete pressures	388
Z_N	Space of spurious modes for the pressure	389
D_N	Image of X_N by the divergence operator	389
X_N	Space of discrete velocities	426
M_N^α	Space of discrete pressures in the weighted case	409
$K_{\alpha i,N}$	Space of divergence-free and pseudo-divergence-free polynomials	409
$Z_{\alpha i,N}$	Space of spurious modes for the pressure and the test-functions	409

Hyperbolic systems

$[\bullet(s)]$	Jump of a function at a discontinuity point s	452
C_N	First Cesàro mean	451
$C_{N,j}$	Cesàro mean of order j	451
$\mathcal{K}_N$	Convolution kernel for π_N	454
$C_{N,\sigma}^{-1/2}$	Filtering operator for the Chebyshev expansion	457
$\mathcal{A}_N$	Cell averaging operator	459
$\mathcal{R}_N$	Reconstruction operator	460
B_N	Boundary term in the viscosity method	464
Q_N	Spectral viscosity term	464
(ε_N, m_N)	Parameters in the viscosity method	464

Subject Index

Aubin–Nitsche duality method, 274, 278, 313, 317, 321, 330
averaging operator, 451, 454, 457, 459
averaging
  see averaging operator
    cell averaging
averaging operator, 451, 454, 457, 459

Bessel function, 284
bilaplacian operator, 229, 381
boundary conditions
  see Dirichlet boundary conditions
    Neumann boundary conditions
Bramble–Hilbert lemma, 221
Bürgers' equation, 461, 467, 468

cell averaging, 449, 458–462
Cesàro mean, 450, 451, 457
Chebyshev polynomials, 294, 333, 349, 350
Chebyshev type methods, 333, 366, 374, 378, 408, 454, 457, 459
Christoffel–Darboux formula, 239, 243, 244, 350, 352
collocation method, 275, 293, 303, 308, 309, 314–316, 318, 320, 323, 324, 326, 327, 331–333, 365–367, 370, 371, 374, 403–405, 415, 424, 426, 432, 436, 464
compatibility condition, 226–228, 279, 315, 323, 340, 341, 366, 380, 381, 383, 385, 399, 404, 405, 420
condition number, 286
conservation laws, 449, 454, 458
  see hyperbolic systems
convex open set, 220, 233, 275, 280, 379, 442
Courant–Friedrichs–Lévy condition, 286

differential equations, 234, 245, 247, 249, 258, 281, 283, 288, 349, 351, 354
Dirichlet boundary conditions, 224, 229, 257, 273, 276, 281, 286, 293, 308, 315, 318, 323, 327, 329, 331, 333, 341, 358, 366, 373, 374, 378, 381, 383
divergence, 378, 379, 384, 385, 389, 393, 395, 407, 409, 411, 415–417, 424, 433
duality method
  see Aubin–Nitsche duality method
duality pairing, 220, 221, 228, 232, 334, 338, 375, 442

eigenfunction, 235, 281–286, 349, 356
eigenvalue, 244, 249, 281–286
eigenvector, 244
entropy, 458, 463, 467
equation
  see Bürgers' equation
    differential equations
    Euler's equations
    Navier–Stokes equations
    Poisson equation
error estimate, 274, 277, 278, 280, 283, 293, 311–313, 317, 321, 322, 324, 325, 327, 329, 330, 360, 368, 370, 371, 374, 386, 387, 392, 394, 395, 401, 406–408, 412, 414, 416, 417, 423, 424, 431, 432, 434, 437, 439, 440, 445, 447, 448
Euler's equations, 462, 467, 469
Euler's Gamma function, 348
exponential weight, 286

Fast Fourier Transform, 333, 366, 367, 370
filters, 374, 449, 454, 457, 460, 462, 468
finite differences, 374, 425, 458–468
finite elements, 260, 444
formula
  see Gauss formula
    Gauss–Lobato formula
    generalized Gauss formula
    induction formula
    Stirling's formula
    Stokes formula

Fourier series, 333, 452, 454, 457, 459
function
see Euler's Gamma function
stream function

Gagliardo–Nirenberg inequality, 455
Galerkin methods, 257, 274, 276, 277, 279, 281, 293, 309, 312, 314, 315, 318
Gamma function
see Euler's Gamma function
Gauss formula, 219, 240, 241, 243, 248, 249, 293, 294, 301, 304, 351, 427, 432
Gauss grid, 304, 305, 364, 425
Gauss–Lobatto formula, 240, 251, 293, 294, 309, 354, 364, 427
Gauss–Lobatto grid, 306, 364
generalized Gauss formula, 240, 241, 294, 300, 307, 319, 355, 363
generalized Gauss grid, 307, 365
Gibbs phenomenon, 449
grid
see Gauss grid
Gauss–Lobatto grid
generalized Gauss grid
staggered grids
Gram–Schmidt procedure, 234

Hardy's inequality, 294, 296, 302, 336, 342, 361, 382
Heaviside function, 450–452, 460
Hermite interpolation, 363, 365
Hermite polynomials, 286, 292
hyperbolic systems, 449, 458

incompressibility, 373, 375, 441
induction formula, 237–239, 288, 350
inequality
see Gagliardo–Nirenberg inequality
Hardy's inequality
inverse inequality
Poincaré–Friedrichs inequality
inflow boundary, 459, 464–466
inf–sup condition, 374–377, 379, 383–386, 388, 400, 401, 410–415, 417, 422, 425, 432, 440, 442, 443
interpolation
see interpolation at Gauss nodes
interpolation at Gauss–Lobatto nodes
interpolation at general Gauss nodes
interpolation between Hilbert spaces
Hermite interpolation
Lagrange interpolation
interpolation at Gauss nodes, 244, 294, 299, 304, 307, 333, 360, 361, 364
interpolation at Gauss–Lobatto nodes, 303, 306, 307, 315, 364
interpolation at general Gauss nodes, 293, 300, 304, 307, 308, 326, 363
interpolation between Hilbert spaces, 221, 255, 287, 290, 334, 337
inverse inequalities, 219, 252, 256, 282, 286, 290, 291, 303, 363, 399, 400, 443

Jacobi polynomials, 234, 333, 348–350
Jacobi weight, 333

Lagrange interpolation, 244, 293, 303, 304, 306, 307, 315, 328, 364, 449
Lagrange multiplier, 375, 465
Lagrange polynomials, 241, 242, 314, 331, 333, 390, 391, 403, 464
Laguerre polynomials, 286, 288
Laplace operator, 229, 343, 374
Lax–Milgram lemma, 219, 228–232, 263, 277, 309, 310, 312, 316, 321, 327, 329, 343, 368
Lebesgue measure, 220, 241
Legendre polynomials, 219, 233–235, 237, 238, 240, 241, 253, 258, 259, 349, 350, 450, 452
lifting operator, 225, 233, 316, 340, 341, 407, 424
Lipschitz-continuous open set, 219, 220, 225

mass matrix, 318
measure
see Lebegue measure
weighted measure
momentum, 441

Navier–Stokes equations, 373–375, 432, 433, 440, 441, 448
Neumann boundary conditions, 224, 229, 257, 279, 286, 293, 328, 332, 343
normal derivative, 225, 273, 321, 339

open set
see convex open set
Lipschitz-continuous open set
orthogonal polynomials
see Chebyshev polynomials
Hermite polynomials
Jacobi polynomials
Laguerre polynomials
Legendre polynomials
ultraspherical polynomials
orthogonal projection, 257, 261, 265, 266, 268, 270, 272, 274, 289, 356, 357, 359, 401, 414, 444
outflow boundary, 465

Padé approximant, 284
Poincaré–Friedrichs inequality, 221, 263, 271, 310, 312, 379, 397
Poisson equation, 468
polynomial interpolation
  see interpolation at ...
    Hermite interpolation
    Lagrange interpolation
pressure, 373–375, 378, 385, 388, 389, 391, 392, 395, 401, 404, 406, 409, 412, 415, 417, 423, 425, 426, 428, 432, 440, 442, 448, 462
projection
  see orthogonal projection

reconstruction operator, 459, 460, 462
regularity, 233, 273, 278, 344, 372, 381, 412, 434, 439
reiteration theorem, 222
Richardson acceleration, 453
Rodrigues' formula, 236, 251, 284, 349

saddle-point problem, 374, 375, 377
scalar product, 220, 222, 334, 389, 414
Sobolev imbedding, 224, 264, 294, 296, 433
spectral viscosity, 463, 464, 466–469
spurious modes, 374, 389, 390, 404, 409, 415, 424, 425, 428, 432
staggered grids, 374, 425, 426, 428
stiffness matrix, 286, 318
Stirling's formula, 247, 248, 350
Stokes formula, 378, 405
Stokes problem, 373–375, 377–381, 388, 394, 406, 408, 412, 415, 424, 425, 434, 439, 441, 442, 448
stream function, 348, 375, 380, 381, 393, 441, 444, 447
Sturm–Liouville operator, 258, 288, 349
symmetric bilinear form, 281, 320, 367, 368, 377
symmetric matrix, 244, 249, 318

tensorization, 257, 268, 269, 282, 287, 304, 333, 359, 366, 468
trace, 225, 226, 229, 233, 255, 273, 333, 335, 336, 338, 339, 341, 343, 367, 378, 404

ultraspherical polynomials, 348
unbounded domain, 286, 292

variational formulation, 219, 229–232, 257, 273, 277, 279, 309, 316, 319, 323, 328, 341, 343, 346, 347, 360, 368, 370, 371, 374, 379, 382, 388, 408, 427, 442
velocity, 373, 374, 378, 388, 394, 404, 407, 409, 415, 416, 425, 432, 433, 435, 440, 441, 444, 462
viscosity, 373, 389, 411, 465

Wallis integral, 236
weight
  see exponential weight
    Jacobi weight
weighted measure, 286, 333, 348, 349, 351, 354, 355

Numerical Analysis for Nonlinear and Bifurcation Problems

Gabriel Caloz

I.R.M.A.R., Université de Rennes I
35042 Rennes Cedex
France

Jacques Rappaz

D.M.A., École Polytechnique Fédérale de Lausanne
1015 Lausanne
Switzerland

HANDBOOK OF NUMERICAL ANALYSIS, VOL. V
Techniques of Scientific Computing (Part 2)
Edited by P.G. Ciarlet and J.L. Lions

Contents

PREFACE 491

CHAPTER I. Introduction 495

1. Preliminaries 495
2. Inverse and implicit function theorems 497

CHAPTER II. Model Examples 503

3. Semilinear problems 503
4. A discretization of the Navier–Stokes equations 517
5. A stationary heat problem with convection 523

CHAPTER III. Nonlinear Problems and Their Numerical Approximation 529

6. An abstract framework 529
7. Galerkin approximation of nonlinear problems 534
8. Application to a model problem 540
9. Estimates for an approximation of the Navier–Stokes problem 550
10. Error estimates for the stationary heat problem 553

CHAPTER IV. Parametrized Nonlinear Problems. Approximation of Regular Solutions 561

11. Orientation 561
12. Approximation of a parametrized solution set 563
13. Galerkin approximation of a parametrized solution set 569
14. Error estimates for a parameter dependent semilinear problem 572
15. Solution set containing a simple limit point 580
16. Galerkin approximation of simple limit points 585
17. Approximation of simple limit points. An example 593

CHAPTER V. Approximation of Simple Bifurcation Points 597

18. Lyapunov–Schmidt procedure. Bifurcation equations 597
19. The case with parameter and state spaces 601
20. Approximate bifurcation equations 610
21. Approximation of simple bifurcation points 614
22. A model example 623
23. Bibliographical comments 630

REFERENCES 633

SUBJECT INDEX 637

Preface

Computational applications generally involve nonlinear problems and often contain parameters. They may represent properties of the physical system they describe or quantities which can be varied. A basic problem in approximation consists in studying existence and convergence of approximated solutions for a given nonlinear problem, for instance when the parameters are fixed. Another problem is to represent the families or manifolds of solutions under variations of some parameters. Apart from a theoretical approach, such representations are computed and continuation methods are concerned with generating the solution manifolds. By varying one parameter, we can follow a path of solutions. Then to study the effects of change of parameters on a system, it is of prime interest to know the effects of numerical approximation on its behavior.

The goal of this article is to present a general framework in which approximations of nonlinear problems and approximations of solution manifolds can be studied. We will consider regular solutions, regular solution families, and singular solutions. Even though we will illustrate the general theory only with elementary finite element approximations of model boundary value problems, it can be applied to a much wider range of problems in connection with approximation methods. Our presentation is a remodelling of the one proposed by CROUZEIX and RAPPAZ [1989] taking its origin in DESCLOUX and RAPPAZ [1982] and in BREZZI, RAPPAZ and RAVIART [1980, 1981a,b].

The general problem we will handle and which covers a lot of applications is the following: find $x \in X$ such that

$$F(x) = 0$$

where X and Z are Banach spaces, $F : X \to Z$ is a smooth nonlinear mapping. Of particular interest is the case where the space X has the form $\mathbb{R}^m \times Y$, where $\mathbb{R}^m$ with $m \geqslant 1$ is the parameter space and the Banach space Y is the state space. We will work under the assumption that the derivative of F is a Fredholm operator of index $n \geqslant 0$. Both cases $n = 0$ and $n \geqslant 1$ with a surjective derivative are studied separately. Note that when n is positive, the family of solutions to $F(x) = 0$ is a differentiable manifold. The singular situation with a nonsurjective derivative is also studied.

In the general setting, the approximation schemes are written in the form

$$F_h(x) = 0$$

where h is a parameter in $(0, 1]$ and $F_h : X \to Z$ is an approximation of F. The family $\{F_h\}_{0<h\leqslant 1}$ converges to F as h tends to 0. Our work then is to compare the

solution set of the problem $F_h(x) = 0$ to the one of the problem $F(x) = 0$. The error estimates will generally be deduced from the inverse function theorem. Our program is the following.

In Sections 1 and 2, we present the functional analysis material we use to establish approximation results. It covers mainly variants of the inverse and of the implicit function theorems.

Sections 3, 4 and 5 present model examples which are used later to illustrate the general theory. They do not reflect the range of applications of the theory, but have been chosen because they are simple or familiar to the reader. In each case we discuss the exact problem and give some standard approximations of it.

Then we introduce a general framework to study the approximation of a nonlinear problem without parameter (or with fixed parameters). The originality of this approach is a generalization of the well-known inf–sup conditions to nonlinear problems. Furthermore, we are naturally led to discuss some a posteriori estimates in the nonlinear case, which are effective in an adaptive mesh procedure. The general case is studied in Section 6, Galerkin approximations are analyzed in Section 7 while examples are detailed in Sections 8 through 10.

Sections 11 through 17 are devoted to the study of the approximation of parametrized nonlinear problems. We consider exclusively regular solutions, that is points $(\lambda, u) \in \mathbb{R}^m \times Y \equiv X$ where the derivative $\mathrm{D}F(\lambda, u)$ is surjective, as it is described in Section 11. First we study the simplest case of regular solution families, the ones parametrized by λ. The approximation in an abstract setting is developed in Section 12, Galerkin approximations are studied in Section 13 and an example is presented in Section 14. In the remainder of the chapter devoted to the approximation of regular solutions, simple limit points and their Galerkin approximations are also discussed. By introducing a new equation we extend the system and transform the problem to apply the results of Section 12.

In Sections 18 through 23 we are concerned with singular solutions and their approximations. Section 18 is devoted to the classical Lyapunov–Schmidt decomposition procedure, which leads to a bifurcation equation. A situation frequently encountered in the applications is presented in Section 19 where the bifurcation equation is studied carefully. Then approximations are analyzed in Section 20 and we get estimates between the bifurcation equation and the approximate bifurcation equations. In Section 21 we compare thoroughly the bifurcation equation and the approximate bifurcation equations when these are defined in $\mathbb{R}^2$. A simple model example illustrates in Section 22 the general theory. In Section 23 we address bibliographical complements.

As mentioned previously the emphasis here is put on the analysis of the convergence and the error estimates of the approximate problems. The methods necessary to understand the approximation of a parametrized nonlinear problem are described in a simple way with a minimum of prerequisite. Our approach is based on a detailed analysis of the error estimates and gives a global insight on the perturbation of a solution set by an approximation procedure. We notice that the solution family of the exact problem or its approximation is under mild hypotheses a differentiable manifold. An approach based on modern differential geometry is proposed in FINK and RHEINBOLDT [1983a,b] or in RHEINBOLDT [1986] to study regular points. Although of great importance, we do not emphasize on the wide variety of algorithms to numerically solve the problem $F_h(x) = 0$, such as the predictor–corrector continuation method or the piecewise linear method, see KELLER [1977], KUEPPER, MITTELMANN and WEBER [1984], ALLGOWER and GEORG [1990] for instance and the literature therein.

CHAPTER I

Introduction

Among the tools commonly used to study parameter dependent problems are the inverse and the implicit function theorems, two basic results of the differential calculus in Banach spaces. It turns out that the approach of nonlinear problems based on contraction mappings is well-suited to study numerical perturbation.

We collect in the next two sections various notations and we recall the inverse and the implicit function theorems in a form convenient for our later use. Our presentation is voluntarily brief. We refer to the literature for further details.

1. Preliminaries

Let X and Z be two real Banach spaces with the norms $\|\bullet\|_X$ and $\|\bullet\|_Z$ respectively. The set $B(x,\delta) = \{y \in X\colon\ \|x-y\|_X < \delta\}$ is the open ball in X centered at x with radius δ, while $\overline{B}(x,\delta) = \{y \in X\colon\ \|y-x\|_X \leqslant \delta\}$ denotes the closed ball. The product space $X \times Z$ is equipped with the norm $\|(x,z)\|_{X\times Z} = \|x\|_X + \|z\|_Z$.

We denote by X' the dual space of the real Banach space X and by $\langle \bullet, \bullet \rangle_{X'X}$ the duality pairing between X' and X. The space of continuous linear mappings from X into Z is denoted by $\mathscr{L}(X;Z)$ and endowed with the norm

$$\|A\|_{X;Z} = \sup_{\substack{x\in X\\ \|x\|_X \leqslant 1}} \|Ax\|_Z$$

for all $A \in \mathscr{L}(X;Z)$, is a Banach space. The spaces of multilinear mappings are then defined by

$$\mathscr{L}^k(X;Z) = \mathscr{L}\big(X;\mathscr{L}^{k-1}(X;Z)\big), \quad k = 2,\dots.$$

Given an invertible mapping $A \in \mathscr{L}(X;Z)$ and a mapping $B \in \mathscr{L}(X;Z)$, the mapping $A + B \in \mathscr{L}(X;Z)$ is invertible if

$$\|A^{-1}B\|_{X;X} < 1. \tag{1.1}$$

Then the following bound holds

$$\|(A+B)^{-1}\|_{Z;X} \leqslant \frac{1}{1-\|A^{-1}B\|_{X;X}} \|A^{-1}\|_{Z;X}. \tag{1.2}$$

Let $\{A_n\}_{n\geqslant 1} \subset \mathscr{L}(X;Z)$ be a sequence of linear operators converging punctually to $A \in \mathscr{L}(X;Z)$, that is for all $x \in X$,

$$\lim_{n\to\infty} A_n x = Ax.$$

Let Y be a Banach space and $B \in \mathscr{L}(Y;X)$ be a compact operator. Then the sequence $\{A_n B\}_{n\geqslant 1} \subset \mathscr{L}(Y;Z)$ converges to AB in the operator norm, that is

$$\lim_{n\to\infty} \|A_n B - AB\|_{Y;Z} = 0. \tag{1.3}$$

A continuous mapping $G: X \to Z$ is Fréchet differentiable at a point $x \in X$ if there exists an operator $\mathrm{D}G(x) \in \mathscr{L}(X;Z)$ such that

$$\big\|G(y) - G(x) - \mathrm{D}G(x)(y-x)\big\|_Z = \mathrm{o}\big(\|x-y\|_X\big).$$

The mapping $\mathrm{D}G: X \to \mathscr{L}(X;Z)$ is the (Fréchet) derivative of G. The mapping G is said to be of class C^1 when the mapping $\mathrm{D}G$ is continuous on X. The mapping G is said to be of class C^2 when $\mathrm{D}G$ is of class C^1; the second derivative $\mathrm{D}^2G(x)$ at the point x belongs to $\mathscr{L}^2(X;Z)$. We can repeat this procedure in order to define the class C^p with $p \geqslant 2$. Let $G: X \to Z$ be a C^p mapping with $p \geqslant 1$. Then the Taylor expansion of G at the point $x \in X$ reads

$$\begin{aligned} G(y) = G(x) &+ \sum_{k=1}^{p-1} \frac{1}{k!} \mathrm{D}^k G(x)\underbrace{(y-x,\dots,y-x)}_{k \text{ times}} \\ &+ \frac{1}{(p-1)!}\int_0^1 (1-t)^{p-1}\mathrm{D}^p G\big(x+t(y-x)\big)(y-x,\dots,y-x)\,\mathrm{d}t. \end{aligned} \tag{1.4}$$

The literature in linear and nonlinear functional analysis is rich, for further details we refer for instance to BONIC [1969], BREZIS [1983], CARTAN [1967], DIEUDONNÉ [1968], DUNFORD and SCHWARTZ [1958].

Let us finally introduce some notation useful in the applications. In the examples presented in the next chapter, the appropriate function spaces are the Sobolev spaces $W^{m,p}(\Omega)$, where Ω is an open set in $\mathbb{R}^2$ and the numbers m, p are nonnegative real with $p \geqslant 1$. These Sobolev spaces are equipped with the norms $\|\bullet\|_{m,p,\Omega}$ and semi-norms $|\bullet|_{m,p,\Omega}$. The space $W_0^{m,p}(\Omega)$ is the closure of $\mathscr{D}(\Omega)$ in $W^{m,p}(\Omega)$ and $W^{-m,q}(\Omega)$ is the dual space of $W_0^{m,p}(\Omega)$ where $1/p + 1/q = 1$. For $p = 2$, we shall write $H^m(\Omega)$ and $H_0^m(\Omega)$ instead of $W^{m,2}(\Omega)$ and $W_0^{m,2}(\Omega)$ respectively, $\|\bullet\|_{m,\Omega}$ and $|\bullet|_{m,\Omega}$ instead of $\|\bullet\|_{m,2,\Omega}$ and $|\bullet|_{m,2,\Omega}$. Numerous presentations of the Sobolev spaces can be found, as for example in the books of ADAMS [1975], LIONS and MAGENES [1968], MAZ'YA [1985] or NEČAS [1967].

If v_1, v_2 are two functions on Ω, $v_1, v_2 : (x_1, x_2) \in \Omega \to v_1(x_1, x_2), v_2(x_1, x_2) \in \mathbb{R}$ and $\boldsymbol{v} = (v_1, v_2)$, then

$$\mathbf{grad}\, v_1 = \left(\frac{\partial v_1}{\partial x_1}, \frac{\partial v_1}{\partial x_2} \right), \qquad \operatorname{div} \boldsymbol{v} = \frac{\partial v_1}{\partial x_1} + \frac{\partial v_2}{\partial x_2}$$

and

$$\mathbf{grad}\, \boldsymbol{v} = \begin{pmatrix} \dfrac{\partial v_1}{\partial x_1} & \dfrac{\partial v_1}{\partial x_2} \\ \dfrac{\partial v_2}{\partial x_1} & \dfrac{\partial v_2}{\partial x_2} \end{pmatrix}, \qquad (\boldsymbol{v}\boldsymbol{\nabla})\boldsymbol{v} = \begin{pmatrix} v_1 \dfrac{\partial v_1}{\partial x_1} + v_2 \dfrac{\partial v_1}{\partial x_2} \\ v_1 \dfrac{\partial v_2}{\partial x_1} + v_2 \dfrac{\partial v_2}{\partial x_2} \end{pmatrix}.$$

2. Inverse and implicit function theorems

In our analysis, the inverse and implicit function theorems will play an essential role. We present here the results in a form convenient for our use. Let $G : X \to Z$ be a C^1 mapping and $v \in X$ be such that $\mathrm{D}G(v) \in \mathscr{L}(X; Z)$ is an isomorphism. We introduce the notations

$$\begin{aligned} \varepsilon &= \|G(v)\|_Z, \\ \gamma &= \|\mathrm{D}G(v)^{-1}\|_{Z;X}, \\ L(\alpha) &= \sup_{x \in \overline{B}(v,\alpha)} \|\mathrm{D}G(v) - \mathrm{D}G(x)\|_{X;Z}, \end{aligned} \tag{2.1}$$

and we are interested in the problem to find $x \in X$ such that

$$G(x) = 0. \tag{2.2}$$

THEOREM 2.1. *We assume that* $\mathrm{D}G(v) \in \mathscr{L}(X; Z)$ *is an isomorphism and that* $2\gamma L(2\gamma\varepsilon) \leqslant 1$. *Then problem* (2.2) *has a unique solution* u *in the ball* $\overline{B}(v, 2\gamma\varepsilon)$ *and* $\mathrm{D}G(u) \in \mathscr{L}(X; Z)$ *is invertible with*

$$\|\mathrm{D}G(u)^{-1}\|_{Z;X} \leqslant 2\gamma. \tag{2.3}$$

Moreover, for all $x \in \overline{B}(v, 2\gamma\varepsilon)$,

$$\|x - u\|_X \leqslant 2\gamma \|G(x)\|_Z. \tag{2.4}$$

PROOF. The proof follows BREZZI, RAPPAZ and RAVIART [1980] or CROUZEIX and RAPPAZ [1989]. Let $H : X \to X$ be the mapping defined by

$$H(x) = x - \mathrm{D}G(v)^{-1} G(x). \tag{2.5}$$

Clearly x is a fixed point of H if and only if x is a zero of the mapping G.

For any $x \in \overline{B}(v, 2\gamma\varepsilon)$ we can write

$$H(x) - v = \mathrm{D}G(v)^{-1}\big[\mathrm{D}G(v)(x-v) - \big(G(x) - G(v)\big)\big] - \mathrm{D}G(v)^{-1}G(v).$$

With the Taylor expansion (1.4) with $p = 1$ we get

$$\begin{aligned}\|H(x) - v\|_X &\leqslant \gamma\left[\varepsilon + \left\|\int_0^1 \big(\mathrm{D}G(v) - \mathrm{D}G(v + t(x-v))\big)(x-v)\,\mathrm{d}t\right\|_Z\right] \\ &\leqslant \gamma\big[\varepsilon + L(2\gamma\varepsilon)2\gamma\varepsilon\big] \leqslant 2\gamma\varepsilon.\end{aligned}$$

Thus H maps the closed ball $\overline{B}(v, 2\gamma\varepsilon)$ into itself.

Let now x, y be in $\overline{B}(v, 2\gamma\varepsilon)$, then

$$H(x) - H(y) = \mathrm{D}G(v)^{-1}\int_0^1 \big[\mathrm{D}G(v) - \mathrm{D}G\big(y + t(x-y)\big)\big](x-y)\,\mathrm{d}t$$

and

$$\big\|H(x) - H(y)\big\|_X \leqslant \gamma L(2\gamma\varepsilon)\|x - y\|_X \leqslant \tfrac{1}{2}\|x - y\|_X.$$

We have proved that H is a strict contraction from the ball $\overline{B}(v, 2\gamma\varepsilon)$ into itself. So H possesses a unique fixed point u in the ball $\overline{B}(v, 2\gamma\varepsilon)$. Since

$$\big\|\mathrm{D}G(v)^{-1}\big[\mathrm{D}G(u) - \mathrm{D}G(v)\big]\big\|_{X;X} \leqslant \gamma L(2\gamma\varepsilon) \leqslant \tfrac{1}{2}$$

and

$$\mathrm{D}G(u) = \mathrm{D}G(v) + \big(\mathrm{D}G(u) - \mathrm{D}G(v)\big),$$

by (1.1) and (1.2), $\mathrm{D}G(u)$ is invertible and (2.3) is true, i.e.,

$$\big\|\mathrm{D}G(u)^{-1}\big\|_{Z;X} \leqslant 2\gamma.$$

Finally to prove the estimate (2.4), we write for $\alpha > 0$ and $x \in \overline{B}(v, \alpha)$

$$\begin{aligned}u - x &= H(u) - x \\ &= \mathrm{D}G(v)^{-1}\left[-G(x) + \int_0^1 \big[\mathrm{D}G(v) - \mathrm{D}G\big(u + t(x-u)\big)\big](u-x)\,\mathrm{d}t\right]\end{aligned}$$

and

$$\|u - x\|_X \leqslant \gamma\big[\big\|G(x)\big\|_Z + L(\alpha)\|u - x\|_X\big].$$

Thus we finally get

$$\big(1-\gamma L(\alpha)\big)\|u-x\|_X \leqslant \gamma\big\|G(x)\big\|_Z. \qquad \square$$

REMARK 2.1. The uniqueness result in Theorem 2.1 can be improved in the following way. Under the hypotheses of the theorem and for all $\alpha \geqslant 2\gamma\varepsilon$ satisfying $\gamma L(\alpha) < 1$, problem (2.2) has a unique solution in the ball $\overline{B}(v,\alpha)$. Indeed if u_1 and u_2 are two different fixed points in $\overline{B}(v,\alpha)$ then

$$\|u_1-u_2\|_X = \big\|H(u_1)-H(u_2)\big\|_X \leqslant \gamma L(\alpha)\|u_1-u_2\|_X < \|u_1-u_2\|_X$$

and we have a contradiction.

REMARK 2.2. In Theorem 2.1 the hypothesis on the differentiability of G can be relaxed. If there exists an isomorphism $A \in \mathscr{L}(X;Z)$ and $v \in X$ such that $2\gamma L(2\gamma\varepsilon) \leqslant 1$, with

$$\varepsilon = \big\|G(v)\big\|_Z,$$
$$\gamma = \big\|A^{-1}\big\|_{Z;X},$$
$$L(\alpha) = \sup_{x,y\in\overline{B}(v,\alpha)} \frac{\|G(x)-G(y)-A(x-y)\|_Z}{\|x-y\|_X},$$

then problem (2.2) has a unique solution in $\overline{B}(v,2\gamma\varepsilon)$. Moreover, the estimate (2.4) still holds.

Note that if $\lim_{\alpha\to 0} L(\alpha) = 0$, then G is strongly differentiable at v and satisfies a Lipschitz condition in a neighborhood of v, see NIJENHUIS [1974] for details.

REMARK 2.3. Theorem 2.1 is the basic result to start an error analysis when we consider approximations of nonlinear problems. Usually u will be a solution of the approximate problem and $x = v$ the solution of the exact problem. The estimate (2.4) is the main relation to analyze the error between exact and approximated solutions.

The results stated in Theorem 2.1 will be sufficient for most applications. In fact under similar hypotheses we get the following variant of the inverse function theorem.

THEOREM 2.2. *For $v \in X$ and the function $G: X \to Z$ of class C^p, $p \geqslant 1$, we assume that $\mathrm{D}G(v) \in \mathscr{L}(X;Z)$ is an isomorphism and that α satisfies $2\gamma L(\alpha) \leqslant 1$, with $\gamma = \|\mathrm{D}G(v)^{-1}\|_{Z;X}$. Then there exists a C^p mapping $F: B(G(v),\alpha/2\gamma) \to B(v,\alpha)$ such that for all $z \in B(G(v),\alpha/2\gamma)$ we have*

$$G\big(F(z)\big) = z \tag{2.6}$$

and

$$\mathrm{D}F(z) = \big[\mathrm{D}G\big(F(z)\big)\big]^{-1}.$$

Moreover for all z_1, z_2 *in* $B(G(v), \alpha/2\gamma)$

$$\|F(z_1) - F(z_2)\|_X \leqslant 2\gamma\|z_1 - z_2\|_Z. \tag{2.7}$$

PROOF. We refer to BREZZI, RAPPAZ and RAVIART [1980] for a detailed proof. We will only sketch it here. For $z \in B(G(v), \alpha/2\gamma)$ we define the mapping $H : X \to X$ by

$$H(x) = x + \mathrm{D}G(v)^{-1}(z - G(x)).$$

As in the proof of Theorem 2.1, we can prove that for x, $y \in \overline{B}(v, \alpha)$

$$\|H(x) - H(y)\|_X \leqslant \tfrac{1}{2}\|x - y\|_X$$

and for $x \in \overline{B}(v, \alpha_1)$ with $\alpha_1 = 2\gamma\|z - G(v)\|_Z < \alpha$

$$\|H(x) - v\|_X \leqslant \|H(x) - H(v)\|_X + \|H(v) - v\|_X \leqslant \alpha_1.$$

Hence the mapping H maps the closed ball $\overline{B}(v, \alpha_1)$ into itself and is a contraction. So H has a unique fixed point in $\overline{B}(v, \alpha_1)$; that is for any $z \in B(G(v), \alpha/2\gamma)$ the equation $G(x) = z$ has a unique solution $x = F(z)$ in $\overline{B}(v, \alpha_1)$, which is unique in $B(v, \alpha)$ (in a similar way as in Remark 2.1). Whence the mapping

$$F : B(G(v), \alpha/2\gamma) \to B(v, \alpha)$$

is well defined. In a same way as we proved (2.3), we get for $x \in B(v, \alpha)$ that $\mathrm{D}G(x)$ is an isomorphism from X onto Z with $\|\mathrm{D}G(x)^{-1}\| \leqslant 2\gamma$. Then in a standard way we prove that $\mathrm{D}F(z) = [\mathrm{D}G(F(z))]^{-1}$ and that F is of class C^p. Finally the estimate (2.7) is a direct consequence of (2.6) and of a Taylor expansion of F. □

If we consider a C^1 mapping $G : \Lambda \times X \to Z$ where Λ is another Banach space and if (λ_0, x_0) is an element of $\Lambda \times X$, then $\mathrm{D}G(\lambda_0, x_0) \in \mathscr{L}(\Lambda \times X; Z)$, $\mathrm{D}_\lambda G(\lambda_0, x_0) \in \mathscr{L}(\Lambda; Z)$, and $\mathrm{D}_x G(\lambda_0, x_0) \in \mathscr{L}(X; Z)$ are respectively the total derivative of G, the derivative of G with respect to λ and the derivative of G with respect to x at the point (λ_0, x_0).

Finally we recall a version of the implicit function theorem. Let Λ, X, and Z be three Banach spaces and $G : \Lambda \times X \to Z$ be a C^p mapping with $p \geqslant 1$. For a given $(\lambda_0, x_0) \in \Lambda \times X$, we assume that $\mathrm{D}_x G(\lambda_0, x_0) \in \mathscr{L}(X; Z)$ is an isomorphism and we introduce the notations

$$\begin{aligned} \varepsilon &= \|G(\lambda_0, x_0)\|_Z, \\ \gamma_0 &= \|\mathrm{D}_\lambda G(\lambda_0, x_0)\|_{\Lambda;Z}, \\ \gamma_1 &= \|\mathrm{D}_x G(\lambda_0, x_0)^{-1}\|_{Z;X}, \\ L(\alpha) &= \sup_{(\lambda,x) \in \overline{B}((\lambda_0,x_0),\alpha)} \|\mathrm{D}G(\lambda_0, x_0) - \mathrm{D}G(\lambda, x)\|_{\Lambda \times X;Z}. \end{aligned} \tag{2.8}$$

THEOREM 2.3. *Let α satisfy $2\gamma L(\alpha) \leqslant 1$ with $\gamma = \max(\gamma_1, 1 + \gamma_0\gamma_1)$. If $\varepsilon < \alpha/4\gamma$, then there exists a unique C^p mapping*

$$g: B(\lambda_0, \alpha/4\gamma) \subset \Lambda \to B(x_0, \alpha) \subset X$$

satisfying for all $\lambda \in B(\lambda_0, \alpha/4\gamma)$

$$G(\lambda, g(\lambda)) = 0$$

and

$$\|g(\lambda) - x_0\|_X \leqslant 2\gamma(\varepsilon + \|\lambda - \lambda_0\|_\Lambda). \tag{2.9}$$

PROOF. We consider the mapping $F: \Lambda \times X \to \Lambda \times Z$ given by

$$F(\lambda, x) = (\lambda, G(\lambda, x)).$$

Clearly F is of class C^p and for $(\lambda, x) \in \Lambda \times X$ we have

$$\mathrm{D}F(\lambda_0, x_0)(\lambda, x) = (\lambda, \mathrm{D}_\lambda G(\lambda_0, x_0)\lambda + \mathrm{D}_x G(\lambda_0, x_0)x).$$

Since $\mathrm{D}_x G(\lambda_0, x_0)$ is an isomorphism from X onto Z, so is $\mathrm{D}F(\lambda_0, x_0)$ from $\Lambda \times X$ onto $\Lambda \times Z$ and for all $(\lambda, z) \in \Lambda \times Z$ we get

$$\|\mathrm{D}F(\lambda_0, x_0)^{-1}(\lambda, z)\|_{\Lambda \times X} \leqslant \max(1 + \gamma_0\gamma_1, \gamma_1)(\|\lambda\|_\Lambda + \|z\|_Z).$$

So we get the estimate

$$\|\mathrm{D}F(\lambda_0, x_0)^{-1}\|_{\Lambda \times Z; \Lambda \times X} \leqslant \gamma.$$

Moreover for all $(\lambda, x) \in \overline{B}((\lambda_0, x_0), \alpha)$, we have

$$\|\mathrm{D}F(\lambda, x) - \mathrm{D}F(\lambda_0, x_0)\|_{\Lambda \times X; \Lambda \times Z} \leqslant L(\alpha).$$

We are in a position to apply Theorem 2.2 with $G = F$ and $v = (\lambda_0, x_0)$. There exists a unique C^p mapping $\mathscr{F}: B(F(\lambda_0, x_0), \alpha/2\gamma) \to B((\lambda_0, x_0), \alpha)$ such that

$$F(\mathscr{F}(\lambda, z)) = (\lambda, z).$$

Since we assume $\varepsilon < \alpha/4\gamma$, then $(\lambda, 0)$ is in the ball $B((\lambda_0, G(\lambda_0, x_0)), \alpha/2\gamma)$ provided $\|\lambda - \lambda_0\|_\Lambda \leqslant \alpha/4\gamma$. Hence the mapping $g: B(\lambda_0, \alpha/4\gamma) \to B(x_0, \alpha)$ given by

$$g(\lambda) = P\mathscr{F}(\lambda, 0)$$

with P the projection operator from $\Lambda \times X$ onto X, is well defined on $B(\lambda_0, \alpha/4\gamma)$, of class C^p, and satisfies

$$G(\lambda, g(\lambda)) = 0 \quad \text{for} \quad \lambda \in B(\lambda_0, \alpha/4\gamma).$$

Finally we can use the estimate (2.7) to get (2.9) in the following way. For $\lambda \in B(\lambda_0, \alpha/4\gamma)$,

$$\begin{aligned} \|g(\lambda) - x_0\|_X &= \|P\mathscr{F}(\lambda, 0) - P\mathscr{F}(\lambda_0, G(\lambda_0, x_0))\|_X \\ &\leqslant 2\gamma\big(\|\lambda - \lambda_0\|_\Lambda + \|G(\lambda_0, x_0)\|_Z\big). \qquad \square \end{aligned}$$

REMARK 2.4. The hypothesis that G is C^1 can be relaxed in a same way it is done in Remark 2.2. For an operator $A \in \mathscr{L}(\Lambda; Z)$ and an isomorphism $B \in \mathscr{L}(X; Z)$, we introduce the notations

$$\begin{aligned} \gamma_0 &= \|A\|_{\Lambda;Z}, \\ \gamma_1 &= \|B^{-1}\|_{Z;X}, \\ L(\alpha) &= \sup_{\substack{(\lambda,x),(\lambda^*,x^*) \\ \in \overline{B}((\lambda_0,x_0),\alpha)}} \frac{\|G(\lambda, x) - G(\lambda^*, x^*) - A(\lambda - \lambda^*) - B(x - x^*)\|_Z}{\|\lambda - \lambda^*\|_\Lambda + \|x - x^*\|_X}. \end{aligned}$$

We assume that α is such that $2\gamma L(\alpha) \leqslant 1$ and $\varepsilon < \alpha/4\gamma$ with $\gamma = \max(\gamma_1, 1+\gamma_0\gamma_1)$. Then Theorem 2.3 holds with $p = 0$. This generalization has several applications, see GIRAULT and RAVIART [1982], RAPPAZ [1984], CROUZEIX and RAPPAZ [1989], BARRETT and ELLIOTT [1989], CALOZ [1987, 1991] for instance.

CHAPTER II

Model Examples

In this chapter we present examples of nonlinear problems and introduce standard finite element approximations for all of them. It is not our purpose to present many applications where parametrized equations arise. We prefer to choose model problems and discuss thoroughly their finite element approximations. We shall consider successively the case of semilinear second-order elliptic boundary value problem, the Navier–Stokes equations, and a stationary heat problem with convection.

3. Semilinear problems

A generic example to our theory is the case of semilinear boundary value problems. This class includes problems of the form

$$\begin{aligned} &-\Delta u + g(\lambda, u) = 0 \quad \text{in } \Omega, \\ &u = 0 \quad \text{on } \partial\Omega, \end{aligned} \tag{3.1}$$

where Ω is a bounded open set in $\mathbb{R}^2$ with smooth boundary $\partial\Omega$, Δ is the Laplacian operator, λ is a real parameter, and $g: \mathscr{D}(g) \subset \mathbb{R} \times H_0^1(\Omega) \to L^2(\Omega)$ is a given mapping. To study problem (3.1) in a convenient functional setting, we define the mapping $F: \mathscr{D}(g) \subset \mathbb{R} \times H_0^1(\Omega) \to H^{-1}(\Omega)$ by

$$\text{for all } (\lambda, v) \in \mathscr{D}(g), \quad F(\lambda, v) = -\Delta v + g(\lambda, v)$$

or equivalently for all $(\lambda, v) \in \mathscr{D}(g)$ and for all $w \in H_0^1(\Omega)$

$$\langle F(\lambda, v), w \rangle_{H^{-1}(\Omega)\, H_0^1(\Omega)} = \int_\Omega \mathbf{grad}\, v\, \mathbf{grad}\, w \, \mathrm{d}x + \int_\Omega g(\lambda, v) w \, \mathrm{d}x. \tag{3.2}$$

Then we look for solutions $(\lambda, u) \in \mathscr{D}(g)$ of the problem

$$F(\lambda, u) = 0. \tag{3.3}$$

The interesting point here is to study how the solutions u of (3.3) depend on λ and to develop numerical methods to compute them.

Given a function $f \in L^2(\Omega)$, we set $w \in H_0^1(\Omega)$ to be the unique solution to

$$\text{for all } v \in H_0^1(\Omega), \quad \int_\Omega \mathbf{grad}\, w\, \mathbf{grad}\, v \,\mathrm{d}x = \int_\Omega f v \,\mathrm{d}x. \tag{3.4}$$

Thus we have defined the continuous linear operator $T: f \in L^2(\Omega) \to Tf = w \in H_0^1(\Omega)$ which maps f onto the solution w of problem (3.4). It is useful but not always necessary for further development to assume that

$$T \in \mathscr{L}\big(L^2(\Omega); H_0^1(\Omega) \cap H^2(\Omega)\big), \tag{3.5}$$

which is a restriction on the regularity of $\partial\Omega$, see the standard elliptic regularity result in GILBARG and TRUDINGER [1977] or GRISVARD [1985] for instance. Then the operator T is compact from $L^2(\Omega)$ into $H_0^1(\Omega) \cap C^0(\overline{\Omega})$. Introducing the operator T, problem (3.3) can be written in the equivalent way

$$u + Tg(\lambda, u) = 0. \tag{3.6}$$

With some specific choices of the function g, the solution set to problem (3.3) or (3.6) can be described precisely. Several cases will be developed later.

Given a finite-dimensional space $V_h \subset H_0^1(\Omega)$, a Galerkin approximation to problem (3.3) consists in finding $(\lambda, u_h) \in (\mathbb{R} \times V_h) \cap \mathscr{D}(g)$ such that

$$\text{for all } v_h \in V_h, \quad \int_\Omega \mathbf{grad}\, u_h\, \mathbf{grad}\, v_h \,\mathrm{d}x + \int_\Omega g(\lambda, u_h) v_h \,\mathrm{d}x = 0. \tag{3.7}$$

Usually problem (3.7) is not solved as it is but with numerical quadrature rules. We introduce a projector P_h onto V_h and consider the problem of finding $(\lambda, u_h) \in (\mathbb{R} \times V_h) \cap \mathscr{D}(g)$ such that

$$\text{for all } v_h \in V_h, \quad \int_\Omega \mathbf{grad}\, u_h\, \mathbf{grad}\, v_h \,\mathrm{d}x + \int_\Omega P_h\big[g(\lambda, u_h) v_h\big] \,\mathrm{d}x = 0. \tag{3.8}$$

In order to write the approximate problem (3.7) in a way similar to (3.6), we set for a given $f \in L^2(\Omega)$, $w_h \in V_h$ to be the unique solution to

$$\text{for all } v_h \in V_h, \quad \int_\Omega \mathbf{grad}\, w_h\, \mathbf{grad}\, v_h \,\mathrm{d}x = \int_\Omega f v_h \,\mathrm{d}x.$$

So we have defined a mapping $T_h \in \mathscr{L}(L^2(\Omega); V_h)$ and (3.7) reads

$$u_h + T_h g(\lambda, u_h) = 0. \tag{3.9}$$

REMARK 3.1. Note that the two problems (3.6) and (3.9) have a structure identical to the general one studied by BREZZI, RAPPAZ and RAVIART [1980, 1981a,b]

$$u + Tg(\lambda, u) = 0, \qquad u_h + T_h g(\lambda, u_h) = 0,$$

where T is in $\mathscr{L}(W; V)$, W and V are two real Banach spaces, $g : \mathscr{D}(g) \subset \mathbb{R} \times V \to W$ is a regular mapping, $\{V_h\}_{0<h\leqslant 1}$ is a family of finite-dimensional subspaces of V and for $0 < h \leqslant 1$, T_h is a linear mapping in $\mathscr{L}(W; V_h)$. Under the main hypothesis

$$\lim_{h\to 0} \|T - T_h\|_{W;V} = 0$$

they have analyzed regular branches of solutions in BREZZI, RAPPAZ and RAVIART [1980], limit points [1981a], and simple bifurcation points [1981b]. Bifurcations at multiple eigenvalues have been studied in RAPPAZ and RAUGEL [1982].

Two model choices of a function $g : \mathscr{D}(g) \subset \mathbb{R} \times H_0^1(\Omega) \to L^2(\Omega)$ are presented now. First for a given function $f \in L^2(\Omega)$, we consider problem (3.1) with $g(\lambda, u) = u^3 - f$ and the problem reads: find $u \in H_0^1(\Omega)$ such that for all $v \in H_0^1(\Omega)$

$$\langle F(u), v\rangle_{H^{-1}(\Omega)\, H_0^1(\Omega)} \equiv \int_\Omega \mathbf{grad}\, u\, \mathbf{grad}\, v\, \mathrm{d}x + \int_\Omega (u^3 - f)v\, \mathrm{d}x = 0. \tag{3.10}$$

The mapping g is independent of λ and this example will illustrate our general approach applied to a problem not depending on a parameter. It is a classical matter to prove the existence of a solution to problem (3.10). The following result holds.

THEOREM 3.1. *There exists a unique solution $u \in H_0^1(\Omega)$ to the problem* (3.10) *and u is in $H^2(\Omega)$. Moreover the solution u is regular, in the sense that the derivative $\mathrm{D}F(u) \in \mathscr{L}(H_0^1(\Omega); H^{-1}(\Omega))$ is an isomorphism.*

PROOF. With the notation introduced in (3.6), our problem $F(u) = 0$ is equivalent to

$$u + T(u^3 - f) = 0. \tag{3.11}$$

Since the mapping $u \in H_0^1(\Omega) \to u^3 - f \in L^2(\Omega)$ is continuous and compact, the mapping $u \in H_0^1(\Omega) \to T(u^3 - f) \in H_0^1(\Omega)$ is compact. If the real number μ is in $[0, 1]$ and the function $w \in H_0^1(\Omega)$ satisfies

$$\text{for all } v \in H_0^1(\Omega), \quad \int_\Omega \mathbf{grad}\, w\, \mathbf{grad}\, v\, \mathrm{d}x + \mu \int_\Omega (w^3 - f)v\, \mathrm{d}x = 0,$$

by taking $v = w$ we get the estimate

$$|w|_{1,\Omega}^2 + \mu\|w\|_{0,4,\Omega}^4 \leqslant \mu\|f\|_{0,\Omega}\|w\|_{0,\Omega}$$

which implies

$$|w|_{1,\Omega}^2 \leqslant \|f\|_{0,\Omega}\|w\|_{0,\Omega}.$$

With the Poincaré inequality we obtain a bound for $|w|_{1,\Omega}$ independent of μ. As a consequence of the Leray–Schauder homotopy theorem, see LERAY and SCHAUDER ([1934], p. 64), we get the existence of a solution to (3.11). The regularity assumption implies that u is in $H^2(\Omega)$.

Let now u_1 and u_2 be two solutions to (3.11), then

$$\text{for all } v \in H_0^1(\Omega), \quad \int_\Omega \mathbf{grad}(u_1 - u_2)\,\mathbf{grad}\,v\,\mathrm{d}x + \int_\Omega (u_1^3 - u_2^3)v\,\mathrm{d}x = 0$$

or with $v = u_1 - u_2$

$$\int_\Omega \left|\mathbf{grad}(u_1 - u_2)\right|^2 \mathrm{d}x + \int_\Omega (u_1 - u_2)^2\left[\frac{u_1^2 + u_2^2}{2} + \frac{(u_1 + u_2)^2}{2}\right]\mathrm{d}x = 0,$$

which implies $u_1 = u_2$.

Let $u \in H_0^1(\Omega)$ be the solution to (3.11). For all $v, w \in H_0^1(\Omega)$ we have

$$\langle \mathrm{D}F(u)v, w\rangle_{H^{-1}(\Omega)\,H_0^1(\Omega)} = \int_\Omega \mathbf{grad}\,v\,\mathbf{grad}\,w\,\mathrm{d}x + 3\int_\Omega u^2 vw\,\mathrm{d}x \equiv b(v,w).$$

The bilinear form $b\colon H_0^1(\Omega) \times H_0^1(\Omega) \to \mathbb{R}$ is coercive on $H_0^1(\Omega)$. So the derivative $\mathrm{D}F(u) \in \mathscr{L}(H_0^1(\Omega); H^{-1}(\Omega))$ is an isomorphism. □

We now consider finite element approximations to problem (3.10). For ease of exposition we assume that Ω is a convex polygonal domain in $\mathbb{R}^2$, so it makes sense to consider $\{\mathscr{T}_h\}_{0<h\leqslant 1}$ to be a regular family of triangulations of $\overline{\Omega}$, in the sense defined in CIARLET ([1978], p. 124) or in CIARLET ([1991], p. 128). The maximum diameter of the triangles $T \in \mathscr{T}_h$ is equal to h. The approximation space and the test space are chosen to be

$$V_h = \{v \in C^0(\overline{\Omega})\colon\ v|_T \in \mathscr{P}_1(T),\ \text{for all } T \in \mathscr{T}_h\} \cap H_0^1(\Omega), \tag{3.12}$$

where $\mathscr{P}_1(T)$ represents the space of polynomials of degree $\leqslant 1$ defined on T. The corresponding finite element approximation to (3.10) reads

$$\text{for all } v_h \in V_h, \quad \int_\Omega \mathbf{grad}\,u_h\,\mathbf{grad}\,v_h\,\mathrm{d}x + \int_\Omega (u_h^3 - f)v_h\,\mathrm{d}x = 0. \tag{3.13}$$

Suppose that in the system (3.13), we would like to use numerical integration. For instance we could use the generalized trapezoidal quadrature rule which is exact for polynomials of degree up to 1 on each triangle. Then the problem reads: find $u_h \in V_h$ such that for all $v_h \in V_h$,

$$\int_\Omega \mathbf{grad}\, u_h\, \mathbf{grad}\, v_h \,\mathrm{d}x + \int_\Omega r_h\big[(u_h^3 - f)v_h\big]\,\mathrm{d}x = 0, \tag{3.14}$$

where the V_h-interpolation operator $r_h \in \mathscr{L}(C^0(\overline{\Omega}); V_h)$ is defined for $v \in C^0(\overline{\Omega})$ by

$$r_h v \in V_h, \qquad r_h v(x_i) = v(x_i) \quad \text{for all node } x_i \text{ of } \mathscr{T}_h.$$

Note that here the function f is supposed to be continuous in $\overline{\Omega}$. The main goal of the next chapter will be to develop a general formalism to study nonlinear problems. As an example we shall prove that problem (3.13) or (3.14) has a unique solution u_h in some neighborhood of the solution u to problem (3.10). In particular we will show that u_h converges to u in $H_0^1(\Omega)$ when h tends to zero and we will establish a priori and a posteriori error estimates for $\|u - u_h\|_{1,\Omega}$.

The second example of a function $g\colon \mathscr{D}(g) \subset \mathbb{R} \times H_0^1(\Omega) \to L^2(\Omega)$ is

$$g(\lambda, u) = -\lambda f(u),$$

where f is the Nemytskii operator of some function $\phi\colon \Omega \times \mathbb{R} \to \mathbb{R}$, that is for every function $u\colon \Omega \to \mathbb{R}$ we have

$$\text{for all } x \in \Omega, \quad f(u)(x) = \phi\big(x, u(x)\big).$$

The function $\phi\colon (x, z) \in \Omega \times \mathbb{R} \to \phi(x, z) \in \mathbb{R}$ is assumed to satisfy the following conditions:

$$\begin{gathered}\text{the function } \phi \text{ is continuously differentiable with respect to } z\\ \text{uniformly in } \Omega \text{ and for all } z \in \mathbb{R},\ \phi(\bullet, z) \text{ and } \frac{\partial \phi}{\partial z}(\bullet, z) \text{ are in } L^\infty(\Omega),\end{gathered} \tag{3.15}$$

$$\phi(\bullet, 0) > 0 \quad \text{in } \Omega, \tag{3.16}$$

$$\frac{\partial \phi}{\partial z}(x, z) > 0 \quad \text{for } x \in \Omega,\ z \geqslant 0. \tag{3.17}$$

Under these regularity and monotonicity hypotheses on ϕ, we shall study the problem: for $\lambda > 0$ find $u \in H_0^1(\Omega)$ nonnegative in Ω solution to

$$u - \lambda T f(u) = 0. \tag{3.18}$$

The hypothesis (3.15) implies that $f : H_0^1(\Omega) \cap C^0(\overline{\Omega}) \to L^2(\Omega)$ is well defined and C^1 in a neighborhood of a solution of (3.18).

We mention that the corresponding problems have physical motivations. For instance they concern the temperature distribution in an object heated by an electric current, in which case only the positive temperature u is of interest. It is known that a limit current exists beyond which steady state solutions do not exist, see JOSEPH [1965]. The choice $\lambda = 0$ represents the steady state problem with no current. The boundary condition is assumed to be homogeneous, only to subtract a harmonic function with a prescribed value on $\partial\Omega$.

To prove the existence of a value λ^* beyond which no solution exists, we shall use the following corollary of the maximum principle.

THEOREM 3.2. *Let ρ be a nonnegative, not identically zero function in $L^\infty(\Omega)$ and μ_1 be the least (or principal) eigenvalue of*

$$\begin{aligned} &-\Delta\zeta - \mu\rho\zeta = 0 \quad \text{in } \Omega, \\ &\zeta = 0 \quad \text{on } \partial\Omega. \end{aligned} \tag{3.19}$$

Then the eigenspace E_1 corresponding to μ_1 is one-dimensional and any nonzero function $\zeta \in E_1$ has a constant sign in Ω. Furthermore, let $\psi \in H_0^1(\Omega) \cap H^2(\Omega)$ satisfy

$$-\Delta\psi - \lambda\rho\psi \gneqq 0 \quad \text{in } \Omega. \tag{3.20}$$

Then ψ is positive in Ω if and only if $\lambda < \mu_1$.

PROOF. The theorem will be a consequence of the classical maximum principle. A function $v \in H_0^1(\Omega) \cap H^2(\Omega)$ with a nonpositive not identically zero Laplacian in Ω is strictly positive in Ω. The first point is a classical result.

The eigenvalue μ_1 is characterized by the Rayleigh quotient

$$\mu_1 = \min_{\substack{v \in H_0^1(\Omega) \\ v \neq 0}} \frac{|v|_{1,\Omega}^2}{\|\sqrt{\rho}v\|_{0,\Omega}^2},$$

and $E_1 = \{v \in H_0^1(\Omega)\colon |v|_{1,\Omega}^2 - \mu_1\|\sqrt{\rho}v\|_{0,\Omega}^2 = 0\}$. Now let ζ be a nonzero function in E_1. Then the function $\eta = |\zeta|$ is in $H_0^1(\Omega)$ and necessarily

$$|\eta|_{1,\Omega}^2 - \mu_1\|\sqrt{\rho\eta}\|_{0,\Omega}^2 = 0,$$

which means η is in E_1. Since $\zeta_+ \equiv \sup(\zeta, 0) = \frac{1}{2}(\zeta + \eta)$ and $\zeta_- \equiv -\inf(\zeta, 0) = \frac{1}{2}(\eta - \zeta)$, we conclude, E_1 being a vector space, that ζ_+ and ζ_- are also in E_1, that is ζ_+ and ζ_- are in $H_0^1(\Omega)$ and

$$-\Delta\zeta_+ = \mu_1\rho\zeta_+ \quad \text{in } \Omega, \qquad -\Delta\zeta_- = \mu_1\rho\zeta_- \quad \text{in } \Omega.$$

By the maximum principle and since $\zeta \not\equiv 0$, either $\zeta_+ > 0$ and then $\zeta_- \equiv 0$, or $\zeta_+ \equiv 0$ and then $\zeta_- > 0$.

Set $p = -\Delta\psi - \lambda\rho\psi$. Assume first that $\lambda < \mu_1$. Then let $\varphi \in H_0^1(\Omega)$ be the unique minimum of the functional

$$J(v) = \tfrac{1}{2} \int_\Omega \mathbf{grad}\, v\, \mathbf{grad}\, v \,\mathrm{d}x - \frac{\lambda}{2} \int_\Omega \rho v^2 \,\mathrm{d}x - \int_\Omega p v \,\mathrm{d}x$$

over $v \in H_0^1(\Omega)$. Such a function exists, is unique, and by standard elliptic regularity φ belongs to $H^2(\Omega)$, thus $\varphi = \psi$. Suppose that ψ is negative somewhere. Then we define the function $\eta = |\psi|$ which is in $H_0^1(\Omega)$. Since the function p is nonnegative, we get $J(\eta) \leqslant J(\psi)$, which contradicts the fact that ψ is the unique minimum of J. By the maximum principle and the relation (3.20), ψ cannot be zero in any open subset of Ω.

We assume now that ψ is positive and let ζ_1 be the positive eigenvector corresponding to μ_1 with $\|\zeta_1\|_{0,\Omega} = 1$. Then for v_1, $v_2 \in H_0^1(\Omega)$, we have

$$\int_\Omega \mathbf{grad}\,\psi\, \mathbf{grad}\, v_1 \,\mathrm{d}x - \lambda \int_\Omega \rho\psi v_1 \,\mathrm{d}x = \int_\Omega p v_1 \,\mathrm{d}x,$$

$$\int_\Omega \mathbf{grad}\,\zeta_1\, \mathbf{grad}\, v_2 \,\mathrm{d}x - \mu_1 \int_\Omega \rho\zeta_1 v_2 \,\mathrm{d}x = 0,$$

or setting $v_1 = \zeta_1$, $v_2 = \psi$ and subtracting

$$(-\lambda + \mu_1) \int_\Omega \rho\psi\zeta_1 \,\mathrm{d}x = \int_\Omega p\zeta_1 \,\mathrm{d}x,$$

so $-\lambda + \mu_1 > 0$. □

We first note that if ϕ satisfies (3.16), (3.17) then ϕ is positive for positive z and problem (3.18) can have a positive solution only for positive λ. The only solution of (3.18) with $\lambda = 0$ is $u \equiv 0$. With the implicit function Theorem 2.3, it is not difficult to check that for λ in a right neighborhood of 0, (3.18) has a positive solution. So let $\lambda^* > 0$ be given by

$$\lambda^* = \sup \{\lambda \geqslant 0\colon\ (3.18) \text{ has a positive solution}\}.$$

THEOREM 3.3. *We assume that ϕ satisfies* (3.15), (3.16), *and* (3.17). *Then for all $\lambda \in (0, \lambda^*)$, problem* (3.18) *has a minimal positive solution $u(\lambda)$, that is if v is a positive solution then $u(\lambda) \leqslant v$ in Ω. Moreover $u(\lambda)$ is an increasing function of λ, that is if $0 < \lambda_1 < \lambda_2 < \lambda^*$ then*

$$0 \leqslant u(\lambda_1) < u(\lambda_2) \quad \text{in } \Omega.$$

PROOF. Let $\lambda \in (0, \lambda^*)$ be fixed. There exists $\bar{\lambda}$, $\lambda \leqslant \bar{\lambda} \leqslant \lambda^*$, such that $(\bar{\lambda}, \bar{u})$ is a solution of (3.18) with $\bar{u} \geqslant 0$. We then define the sequences

$$\begin{aligned} \underline{u}_0 &\equiv 0, \qquad & \underline{u}_{i+1} &= \lambda T f(\underline{u}_i), \\ \bar{u}_0 &= \bar{u}, \qquad & \bar{u}_{i+1} &= \lambda T f(\bar{u}_i). \end{aligned}$$

We prove that the sequence $\{\underline{u}_i\}_{i\geqslant 0}$ is monotonous increasing, that is

$$\text{for all } x \in \Omega, \quad \underline{u}_{i+1}(x) > \underline{u}_i(x).$$

With the assumption (3.16), the function $\underline{u}_1 \in H_0^1(\Omega)$ satisfies

$$-\Delta \underline{u}_1 = \lambda \phi(\bullet, 0) > 0 \quad \text{in } \Omega,$$

and by the maximum principle $\underline{u}_1 > 0$. Suppose now we have $0 < \underline{u}_1 < \underline{u}_2 < \cdots < \underline{u}_i$. This implies that the difference $\underline{u}_{i+1} - \underline{u}_i$ is a solution to

$$\begin{aligned} &-\Delta(\underline{u}_{i+1} - \underline{u}_i) = \lambda\big[f(\underline{u}_i) - f(\underline{u}_{i-1})\big] \quad \text{in } \Omega, \\ &\underline{u}_{i+1} - \underline{u}_i = 0. \end{aligned}$$

So using (3.17) we get $\underline{u}_i < \underline{u}_{i+1}$. Finally with the same arguments we prove

$$0 \equiv \underline{u}_0 < \underline{u}_1 < \cdots < \underline{u}_n < \cdots < \bar{u}_n \cdots < \bar{u}_1 < \bar{u}_0.$$

The sequence $\{\underline{u}_i\}_{i\geqslant 0}$ is bounded in $L^\infty(\Omega)$; since T is compact as an operator in $\mathscr{L}(L^2(\Omega); H_0^1(\Omega) \cap C^0(\overline{\Omega}))$, there exists a subsequence $\{\underline{u}_{n_i}\}_{i\geqslant 0}$ converging to u in $H_0^1(\Omega) \cap C^0(\overline{\Omega})$, with $u \leqslant \bar{u}$. In fact the whole sequence converges to $u \geqslant 0$, which is a solution of (3.18). Clearly from our construction u is minimal and we set $u = u(\lambda)$. Moreover if $\lambda < \bar{\lambda} < \lambda^*$, then

$$u(\lambda)(x) < u(\bar{\lambda})(x) \quad \text{for all } x \in \Omega. \qquad \square$$

The method of proof used here is known as the monotonous iteration principle.

If we add some hypotheses on ϕ, then we can get more information on λ^*. Suppose for instance there exists a positive function $\varphi \in L^2(\Omega)$ such that

$$\text{for all } x \in \Omega \text{ and } z > 0, \quad \phi(x, z) < \varphi(x). \tag{3.21}$$

Then we can start the monotonous iteration scheme with

$$\underline{u}_0 = 0 \quad \text{and} \quad \bar{u}_0 = \lambda T \varphi.$$

Therefore we get the existence of a positive solution for all $\lambda > 0$ and we have $\lambda^* = \infty$. Suppose on the contrary that ϕ satisfies

$$\phi(x,z) > \varphi(x) + \rho(x)z \quad \text{for all } x \in \Omega \text{ and } z > 0 \tag{3.22}$$

with positive functions $\varphi, \rho \in L^\infty(\Omega)$. Then $\lambda^* \leqslant \mu_1$, where μ_1 is the least eigenvalue of (3.19). Indeed suppose that the problem (3.18) has a solution $u > 0$ for some $\lambda > \mu_1$. This implies

$$-\Delta u = \lambda\phi(x,u) > \lambda\rho u \quad \text{in } \Omega.$$

We get a contradiction from Theorem 3.2. So under the hypothesis (3.22), there exists a $\lambda^* < \infty$ beyond which problem (3.18) has no positive solution.

So far we have used essentially the monotonicity of ϕ. We assume furthermore

$$\text{for all } x \in \Omega, \quad \frac{\partial\phi}{\partial z}(x,\bullet) \quad \text{is strictly increasing in } \mathbb{R}_+. \tag{3.23}$$

Then we can say more on $u(\bullet)$ as a function of λ.

THEOREM 3.4. *We assume that ϕ satisfies* (3.15)–(3.17), *and* (3.23). *Then λ^* is bounded by μ_1, the principal characteristic value of*

$$\psi = \mu T\,\mathrm{D}f(0)\psi$$

and the mapping $\lambda \in [0,\lambda^*) \to u(\lambda) \in H_0^1(\Omega) \cap C^0(\overline{\Omega})$, where $u(\lambda)$ denotes the minimal positive solution, is C^1 and $(\lambda, u(\lambda))$ is a regular point of (3.18), *that is $I - \lambda T\,\mathrm{D}f(u(\lambda)) \in \mathscr{L}(H_0^1(\Omega) \cap C^0(\overline{\Omega}); H_0^1(\Omega) \cap C^0(\overline{\Omega}))$ is an isomorphism.*

PROOF. Since ϕ is convex in z, see (3.23), we note that

$$\phi(x,z) > \phi(x,0) + \frac{\partial}{\partial z}\phi(x,0)z \quad \text{for } x \in \Omega \text{ and } z > 0.$$

With the arguments developed in (3.22), we get that λ^* is bounded by μ_1.

The mapping $u : \lambda \in [0,\lambda^*) \to u(\lambda) \in H_0^1(\Omega) \cap C^0(\overline{\Omega})$ is monotone increasing, so the limits for fixed λ, $u(\lambda-)$ and $u(\lambda+)$ both exist, are positive, and are solution of (3.18). So

$$0 \leqslant u(\lambda-) \leqslant u(\lambda) \leqslant u(\lambda+).$$

Thus u is left continuous by the minimality of $u(\lambda)$, that is $u(\lambda-) = u(\lambda)$.

We introduce the operator $K(\lambda) : H_0^1(\Omega) \to H_0^1(\Omega)$ by

$$\text{for } w \in H_0^1(\Omega), \quad K(\lambda)w = \lambda T\,\mathrm{D}f\big(u(\lambda)\big)w,$$

and the set

$$A = \{\lambda \in [0, \lambda^*):\ \big(K(\lambda)w, w\big)_{1,\Omega} < (w, w)_{1,\Omega} \text{ for all } w \in H_0^1(\Omega),\ w \neq 0\}$$

with $(w, v)_{1,\Omega} = \int_\Omega \mathbf{grad}\, w\ \mathbf{grad}\, v\,\mathrm{d}x$. Since $\mathrm{D}f(u(\lambda))$ is an increasing function of λ and

$$\big(K(\lambda)w, w\big)_{1,\Omega} = \lambda \int_\Omega \mathrm{D}f\big(u(\lambda)\big)w^2\,\mathrm{d}x,$$

the set A is an interval containing zero. For $\lambda \in A$, $I - \lambda T\,\mathrm{D}f(u(\lambda))$ is an isomorphism on $H_0^1(\Omega)$ (or $H_0^1(\Omega) \cap C^0(\overline{\Omega})$). By using the implicit function theorem, it is a simple matter to prove that $u : A \to H_0^1(\Omega) \cap C^0(\overline{\Omega})$ is a C^1 function. Suppose that $A \neq [0, \lambda^*)$ and let $\bar\lambda = \sup A$, $\bar\lambda < \lambda^*$. Since u is continuous on A and left continuous at $\bar\lambda$, we have $\lim_{\lambda\to\bar\lambda,\ \lambda<\bar\lambda} u(\lambda) = u(\bar\lambda)$ and $\bar\lambda = \mu_1(\bar\lambda)$ is the principal characteristic value of $T\,\mathrm{D}f(u(\bar\lambda))$. Since the function $u(\bullet)$ is increasing, we have for $\lambda > \bar\lambda$

for all $x \in \Omega$,

$$\phi\big(x, u(\lambda)(x)\big) > \phi\big(x, u(\bar\lambda)(x)\big) + \frac{\partial}{\partial z}\phi\big(x, u(\bar\lambda)(x)\big)\big(u(\lambda)(x) - u(\bar\lambda)(x)\big).$$

So we obtain

$$\begin{aligned}
&-\Delta\big(u(\lambda) - u(\bar\lambda)\big) - \bar\lambda\frac{\partial}{\partial z}\phi\big(x, u(\bar\lambda)\big)\big(u(\lambda) - u(\bar\lambda)\big)\\
&\quad = (\lambda - \bar\lambda)f\big(u(\lambda)\big) + \bar\lambda\big(f\big(u(\lambda)\big) - f\big(u(\bar\lambda)\big)\big)\\
&\qquad - \bar\lambda\,\mathrm{D}f\big(u(\bar\lambda)\big)\big(u(\lambda) - u(\bar\lambda)\big) > 0.
\end{aligned}$$

We know that $u(\lambda) - u(\bar\lambda)$ is positive and we apply Theorem 3.2 to get $\bar\lambda < \mu_1(\bar\lambda)$ which is a contradiction. □

In Theorem 3.4, we prove that the minimal positive solutions of (3.18) form a regular solution path parametrized by λ, see Fig. 3.1.

The property $\mu_1(\lambda) > \lambda$ gives a stability result of the solution branch. We shall develop a more precise study in a left neighborhood of λ^* when illustrating the study of simple limit points.

REMARK 3.2. If we consider the case

$$\phi(x, z) = z^2 + 1,$$

the hypotheses (3.15), (3.16) and (3.23) are satisfied. It is a simple matter to modify Theorem 3.4 when the hypothesis (3.17) is relaxed to

$$\frac{\partial\phi}{\partial z}(x, z) > 0 \quad \text{for } x \in \Omega,\ z > 0.$$

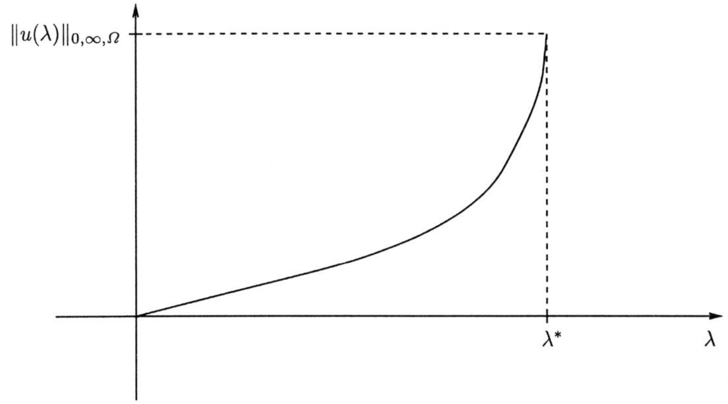

FIG. 3.1.

We can apply the above theory and λ^* is bounded. Moreover, one can prove that there exists $u^* \in H_0^1(\Omega) \cap C^0(\overline{\Omega})$ satisfying (3.18) with $\lambda = \lambda^*$, that is

$$u^* - \lambda^* T\big(u^{*2} + 1\big) = 0$$

and

$$\lim_{\lambda \to \lambda^*} u(\lambda) = u^*;$$

furthermore, (λ^*, u^*) is a singular point of the solution branch, in the sense that $DF(u^*)$ is not an isomorphism, where the mapping $F(v) = v - \lambda^* T(v^2 + 1)$.

Suppose we can get a bound in the $L^2(\Omega)$ norm of the right-hand side, that is

$$\big\|u^2(\lambda) + 1\big\|_{0,\Omega} \leqslant C \quad \text{for all } \lambda \in [0, \lambda^*), \tag{3.24}$$

this implies that the increasing function $u^2(\lambda) + 1$ converges in $L^2(\Omega)$ as λ tends to λ^*, so $u(\lambda) = \lambda T(u^2(\lambda) + 1)$ converges to u^* in $H_0^1(\Omega) \cap C^0(\overline{\Omega})$. Clearly u^* is a solution of (3.18) with $\lambda = \lambda^*$, and the operator $I - 2\lambda^* T(u^*.)$ is singular. To prove (3.24) it is sufficient to prove

$$\big\|u(\lambda)\big\|_{0,4,\Omega} \leqslant C \quad \text{for all } \lambda \in [0, \lambda^*) \tag{3.25}$$

since Ω is bounded.

For a given $\lambda \in [0, \lambda^*)$, we note $u = u(\lambda)$. We have

$$\text{for all } v \in H_0^1(\Omega), \quad \int_\Omega \mathbf{grad}\, u \,\mathbf{grad}\, v \,\mathrm{d}x = \lambda \int_\Omega \big(u^2 + 1\big) v \,\mathrm{d}x$$

and

$$\text{for all } w \in H_0^1(\Omega), \quad 2\lambda \int_\Omega u w^2 \, \mathrm{d}x \leqslant \int_\Omega |\,\mathbf{grad}\, w|^2 \, \mathrm{d}x.$$

We choose $w = u^{3/2}$ and $v = u^2$ to write successively

$$\begin{aligned} 2\lambda \int_\Omega u^4 \, \mathrm{d}x &\leqslant \int_\Omega \tfrac{9}{4} u \,\mathbf{grad}\, u \,\mathbf{grad}\, u \, \mathrm{d}x \\ &= \tfrac{9}{8} \int_\Omega \mathbf{grad}\, u^2 \,\mathbf{grad}\, u \, \mathrm{d}x = \tfrac{9}{8}\lambda \int_\Omega (u^2+1) u^2 \, \mathrm{d}x \end{aligned}$$

and so

$$\int_\Omega u^2 \big(u^2 - \tfrac{9}{7}\big) \, \mathrm{d}x \leqslant 0. \tag{3.26}$$

Let us define the two sets

$$A = \left\{ x \in \Omega\colon\ u(x) < \frac{3\sqrt{2}}{\sqrt{7}} \right\} \quad \text{and} \quad B = \left\{ x \in \Omega\colon\ u(x) \geqslant \frac{3\sqrt{2}}{\sqrt{7}} \right\}.$$

Then for $x \in B$, $u^2(x) - \frac{9}{7} \geqslant u^2(x)/2$, and so we have with (3.26)

$$\int_A u^2 \big(u^2 - \tfrac{9}{7}\big) \, \mathrm{d}x + \int_B \frac{u^4}{2} \, \mathrm{d}x \leqslant 0.$$

We are now in a position to estimate

$$\begin{aligned} \int_\Omega u^4 \, \mathrm{d}x &= \int_A u^4 \, \mathrm{d}x + \int_B u^4 \, \mathrm{d}x \\ &\leqslant \int_A \big[-u^4 + \tfrac{18}{7} u^2 \big] \, \mathrm{d}x \leqslant \int_A u^2 \big(\tfrac{18}{7} - u^2 \big) \, \mathrm{d}x \leqslant 4 \big(\tfrac{9}{7}\big)^2 |\Omega|. \end{aligned}$$

REMARK 3.3. If we consider the case

$$\phi(x, z) = \mathrm{e}^z,$$

the hypotheses (3.15)–(3.17) and (3.23) are satisfied and we can apply the above theory. Moreover, one can prove that there exists $u^* \in H_0^1(\Omega) \cap C^0(\overline{\Omega})$ satisfying (3.18), that is

$$u^* - \lambda^* T \mathrm{e}^{u^*} = 0, \qquad \lim_{\lambda \to \lambda^*} u(\lambda) = u^*;$$

furthermore (λ^*, u^*) is a singular point of the solution branch. The proof of this result is analogous to the one developed in Remark 3.2, see CROUZEIX and RAPPAZ [1989] for details or CRANDALL and RABINOWITZ [1973, 1975].

REMARK 3.4. We consider both cases when $\phi(x, z) = z^2 + 1$ or $\phi(x, z) = \mathrm{e}^z$. Then we know that the operator $I - \lambda^* T\,\mathrm{D}f(u^*)$ is singular with a kernel of dimension 1, see Theorem 3.2.

In fact since T is compact, $I - \lambda^* T\,\mathrm{D}f(u^*)$ is a self-adjoint Fredholm operator with index zero in $H_0^1(\Omega)$ and if $\varphi^* \in H_0^1(\Omega)$, $\varphi^* \neq 0$, is such that

$$\mathrm{Ker}\big(I - \lambda^* T\,\mathrm{D}f(u^*)\big) = \mathrm{span}\{\varphi^*\}$$

then the range of $(I - \lambda^* T\,\mathrm{D}f(u^*))$ is orthogonal to $\mathrm{span}\{\varphi^*\}$ in $H_0^1(\Omega)$ with respect to the scalar product $(\bullet, \bullet)_{1,\Omega}$ in $H_0^1(\Omega)$.

We now define the mapping

$$\mathscr{F} : (s, \lambda, u) \in \mathbb{R} \times \mathbb{R} \times H_0^1(\Omega) \cap C^0(\overline{\Omega}) \to \mathscr{F}(s, \lambda, u) \in \mathbb{R} \times H_0^1(\Omega) \cap C^0(\overline{\Omega})$$

by the relation

$$\mathscr{F}(s, \lambda, u) = \begin{pmatrix} s - (u - u^*, \varphi^*)_{1,\Omega} \\ u - \lambda T f(u) \end{pmatrix}.$$

We have $\mathscr{F}(0, \lambda^*, u^*) = 0$ and since $Tf(u^*)$ is a positive function and φ^* has a sign we can check that $\mathrm{D}_{\lambda u}\mathscr{F}(0, \lambda^*, u^*)$ is an isomorphism on $\mathbb{R} \times H_0^1(\Omega) \cap C^0(\overline{\Omega})$. As a consequence of the implicit function theorem, we can parametrize λ and u by s. There exists a mapping

$$(\lambda, u) : s \in (-\varepsilon, \varepsilon) \to \big(\lambda(s), u(s)\big) \in \mathbb{R} \times H_0^1(\Omega) \cap C^0(\overline{\Omega})$$

satisfying

$$\lambda(0) = \lambda^*, \qquad u(0) = u^*, \qquad u(s) \neq u^* \quad \text{when } s \neq 0,$$

and

$$u(s) - \lambda(s) T f\big(u(s)\big) = 0 \quad \text{for all } s \in (-\varepsilon, \varepsilon).$$

So we conclude that (λ^*, u^*) can be considered as a regular point of the mapping $\mathscr{F}$. A similar study is presented in a general setting in Section 15.

REMARK 3.5. There is an extensive literature concerning positive solutions of problems of the type (3.18). We mention the work of KELLER and COHEN [1967] where minimal positive solutions are studied with the monotonous iteration principle. In

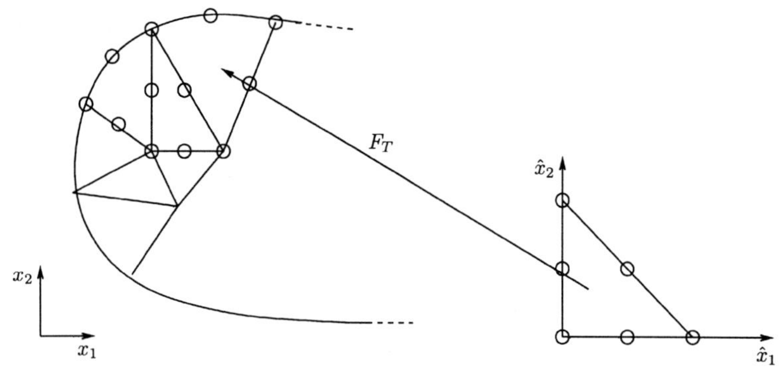

FIG. 3.2.

CRANDALL and RABINOWITZ [1973], the problem is studied by continuation and variational methods, with emphasis on the turning point λ^*. We can also refer to the review works of AMANN [1976], MIGNOT and PUEL [1980] and LIONS [1982].

We consider now finite element approximations of problem (3.18). We introduce first a family of isoparametric triangulations $\mathscr{T}_h$ with polynomials of degree 1 or 2, where $h = \max_{T\in\mathscr{T}_h} h_T$ (h_T is the diameter of T) and we denote by Ω_h the interior of $\overline{\Omega}_h = \bigcup_{T\in\mathscr{T}_h} T$, see CIARLET ([1978], p. 224) or CIARLET ([1991], p. 230) for further details.

Actually each element $T \in \mathscr{T}_h$ is the image $F_T(\widehat{T})$, where $(\widehat{T}, \widehat{\mathscr{P}}_k, \widehat{\Sigma}_k)$ is the Lagrange reference element of type (k) ($\widehat{\mathscr{P}}_k$ is the space of polynomials of degree $\leqslant k$ on $\widehat{T}$), and where the invertible mapping $F_T : \widehat{T} \to T$ has as its components polynomials of degree 1 or 2 and

$$F_T(\hat{a}_i) = a_i \quad \text{for all nodes } \hat{a}_i \in \widehat{\Sigma}_k,$$

see Fig. 3.2 (for $k = 2$).

The approximation space is given by

$$V_h = \big\{ v \in C^0(\mathbb{R}^2) : \ v(x) = 0 \text{ for all } x \notin \Omega_h, \\ v|_T \in P_T \text{ for all } T \in \mathscr{T}_h \big\}, \tag{3.27}$$

with $P_T = \{p : T \to \mathbb{R}\colon\ p \circ F_T \in \widehat{\mathscr{P}}_k\}$. The corresponding finite element approximation to (3.18) reads for all $v_h \in V_h$

$$a_h(u_h, v_h) = \int_{\Omega\cap\Omega_h} \lambda f(u_h) v_h \,\mathrm{d}x, \tag{3.28}$$

where

$$a_h(u_h, v_h) = \int_{\Omega_h} \mathbf{grad}\, u_h \,\mathbf{grad}\, v_h \,\mathrm{d}x.$$

Generally, problem (3.28) is not solved as it is, but with numerical quadrature rules, in which case we suppose that for $v_h \in V_h$, $f(v_h)$ is continuous. In the case $k = 1$, the elements $F_T(\widehat{T})$ are triangles and we consider the problem for all $v_h \in V_h$

$$a_h(u_h, v_h) = \sum_{T \in \mathscr{T}_h} \left[\frac{S_T}{3} \sum_{i=1}^{3} \big(f(u_h)v_h\big)(a_{i,T}) \right] \tag{3.29}$$

where S_T is the measure of T and $a_{1,T}$, $a_{2,T}$, $a_{3,T}$ are the vertices of $T \in \mathscr{T}_h$.

To study the case $k = 2$, we introduce for u, $v \in V_h$ and $T \in \mathscr{T}_h$

$$a_T(u, v) = \sum_{i=1}^{3} \tfrac{1}{6} \det\big(\mathrm{D}F_T(\widehat{m}_i)\big)\, \mathbf{grad}\, u(m_{i,T})\, \mathbf{grad}\, v(m_{i,T});$$

here $\widehat{m}_i$, $i = 1, 2, 3$, are the middles of the sides of $\widehat{T}$ and $m_{i,T}$ the images $F_T(\widehat{m}_i)$. Then problem (3.28) with numerical quadrature rules reads

for all $v_h \in V_h$,

$$\sum_{T \in \mathscr{T}_h} a_T(u_h, v_h) = \sum_{T \in \mathscr{T}_h} \sum_{i=1}^{3} \tfrac{1}{6} \det\big(\mathrm{D}F_T(\widehat{m}_i)\big)\big(f(u_h)v_h\big)(m_{i,T}). \tag{3.30}$$

As an illustration of our abstract theory developed in the next chapter, we shall analyze the problems (3.28), (3.29) and (3.30).

4. A discretization of the Navier–Stokes equations

Let Ω be a polygonal open domain in $\mathbb{R}^2$ with boundary $\partial\Omega$. The stationary incompressible Navier–Stokes problem consists in finding a velocity vector $\boldsymbol{u}: \Omega \to \mathbb{R}^2$ and a pressure $p: \Omega \to \mathbb{R}$ satisfying

$$-\nu\Delta\boldsymbol{u} + (\boldsymbol{u}\boldsymbol{\nabla})\boldsymbol{u} + \mathbf{grad}\, p = \boldsymbol{f} \quad \text{in } \Omega, \tag{4.1}$$

$$\operatorname{div}\boldsymbol{u} = 0 \quad \text{in } \Omega, \tag{4.2}$$

$$\boldsymbol{u} = 0 \quad \text{on } \partial\Omega, \tag{4.3}$$

where the positive number ν and the function $\boldsymbol{f} \in (L^2(\Omega))^2$ are given. We introduce the spaces $L_0^2(\Omega) = \{f \in L^2(\Omega)\colon \int_\Omega f\,\mathrm{d}x = 0\}$ and $X = (H_0^1(\Omega))^2 \times L_0^2(\Omega)$. A weak formulation of the problem reads: find $(\boldsymbol{u}, p) \in X$ such that

for all $\boldsymbol{v} \in \big(H_0^1(\Omega)\big)^2$,

$$\nu\int_\Omega \mathbf{grad}\,\boldsymbol{u}\,\mathbf{grad}\,\boldsymbol{v}\,\mathrm{d}x + \int_\Omega (\boldsymbol{u}\boldsymbol{\nabla})\boldsymbol{u}\boldsymbol{v}\,\mathrm{d}x - \int_\Omega p \operatorname{div}\boldsymbol{v}\,\mathrm{d}x = \int_\Omega \boldsymbol{f}\boldsymbol{v}\,\mathrm{d}x, \tag{4.4}$$

$$\text{for all } q \in L_0^2(\Omega), \quad \int_\Omega q \operatorname{div} \boldsymbol{u} \, \mathrm{d}x = 0. \tag{4.5}$$

Note that if $\boldsymbol{u} \in (H_0^1(\Omega))^2$, then $\boldsymbol{u} \in (L^p(\Omega))^2$ for all $p \in [1, \infty[$ and $(\boldsymbol{u}\boldsymbol{\nabla})\boldsymbol{u}$ is in $(L^q(\Omega))^2$ for all $q \in [1, 2[$. Since $\boldsymbol{v} \in (H_0^1(\Omega))^2$, the term $(\boldsymbol{u}\boldsymbol{\nabla})\boldsymbol{u}\boldsymbol{v}$ is in $L^1(\Omega)$.

We introduce the multilinear forms $a\colon (H_0^1(\Omega))^2 \times (H_0^1(\Omega))^2 \to \mathbb{R}$ given by

$$a(\boldsymbol{u}, \boldsymbol{v}) = \nu \int_\Omega \mathbf{grad}\, \boldsymbol{u}\, \mathbf{grad}\, \boldsymbol{v} \, \mathrm{d}x,$$

$c\colon (H_0^1(\Omega))^2 \times (H_0^1(\Omega))^2 \times (H_0^1(\Omega))^2 \to \mathbb{R}$ given by

$$c(\boldsymbol{w}; \boldsymbol{u}, \boldsymbol{v}) = \int_\Omega (\boldsymbol{w}\boldsymbol{\nabla})\boldsymbol{u}\boldsymbol{v} \, \mathrm{d}x,$$

and $b\colon (H_0^1(\Omega))^2 \times L_0^2(\Omega) \to \mathbb{R}$ given by

$$b(\boldsymbol{u}, p) = -\int_\Omega p \operatorname{div} \boldsymbol{u} \, \mathrm{d}x.$$

Then the problem (4.4)–(4.5) is equivalent to find $(\boldsymbol{u}, p) \in X$ such that

$$\text{for all } (\boldsymbol{v}, q) \in X,$$
$$a(\boldsymbol{u}, \boldsymbol{v}) + c(\boldsymbol{u}; \boldsymbol{u}, \boldsymbol{v}) + b(\boldsymbol{v}, p) - \int_\Omega \boldsymbol{f}\boldsymbol{v} \, \mathrm{d}x + b(\boldsymbol{u}, q) = 0. \tag{4.6}$$

The Navier–Stokes problem (4.6) has received a lot of attention and has been widely studied. It is well known that it has at least one solution. See TEMAM [1977], GIRAULT and RAVIART [1986] and the bibliography therein for further references.

Let $(\boldsymbol{u}, p) \in X$ be a solution of (4.6); we assume that $\boldsymbol{u} \in (H^2(\Omega))^2$. We also want to assume that the solution $(\boldsymbol{u}, p) \in X$ is regular, in the sense that there exists $\varepsilon > 0$ such that the only solution to (4.6) in the ball $B((\boldsymbol{u}, p), \varepsilon)$ is $(\boldsymbol{u}, p)$. Recall that on X we introduce the norm $\|(\boldsymbol{w}, r)\|_X = |\boldsymbol{w}|_{1,\Omega} + \|r\|_{0,\Omega}$. Referring to Theorem 2.1 the hypothesis is stated precisely in the following terms. The bilinear form $A\colon X \times X \to \mathbb{R}$ given by

$$\begin{aligned} A(\underline{w}, \underline{v}) &= A\big((\boldsymbol{w}, r), (\boldsymbol{v}, q)\big) \\ &= a(\boldsymbol{w}, \boldsymbol{v}) + b(\boldsymbol{w}, q) + b(\boldsymbol{v}, r) + c(\boldsymbol{u}; \boldsymbol{w}, \boldsymbol{v}) + c(\boldsymbol{w}; \boldsymbol{u}, \boldsymbol{v}) \end{aligned} \tag{4.7}$$

satisfies the so called inf–sup condition, see BABUŠKA and AZIZ [1972], that is

$$\begin{aligned} &\inf_{\substack{\underline{w} \in X \\ \|\underline{w}\|_X = 1}} \sup_{\substack{\underline{v} \in X \\ \|\underline{v}\|_X = 1}} A(\underline{w}, \underline{v}) > 0, \\ &\sup_{\substack{\underline{w} \in X \\ \|\underline{w}\|_X = 1}} A(\underline{w}, \underline{v}) > 0 \quad \text{for all } \underline{v} \in X, \ \underline{v} \neq 0. \end{aligned} \tag{4.8}$$

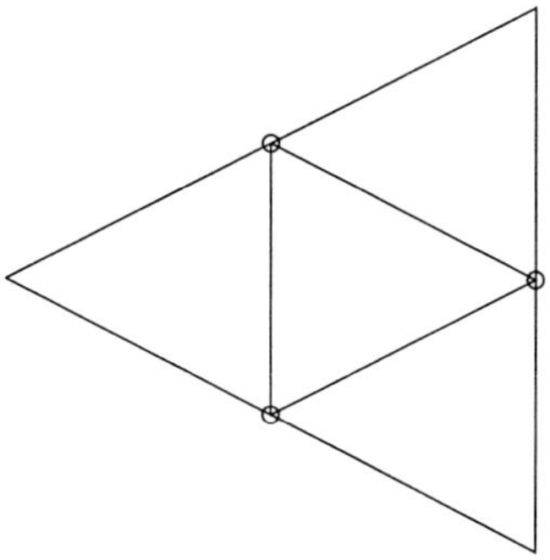

FIG. 4.1.

For simplicity we have assumed that Ω is polygonal, so we can consider $\mathscr{T}_h^c$ a regular triangulation of $\overline{\Omega}$ made up of triangles T of diameter inferior to h and $\mathscr{T}_h^f$ a finer triangulation obtained from $\mathscr{T}_h^c$ by dividing each triangle $T \in \mathscr{T}_h^c$ in four triangles as shown in Fig. 4.1.

The parameter h will tend to zero. We assume that the family $\{\mathscr{T}_h^c\}_{0<h\leqslant 1}$ is regular, see CIARLET ([1978], p. 124). To define the approximation spaces X_h to X we introduce the finite-dimensional spaces

$$V_h = \{v \in C^0(\overline{\Omega}) \colon v|_{T'} \in \mathscr{P}_1(T') \text{ for all } T' \in \mathscr{T}_h^f,\ v|_{\partial\Omega} = 0\}$$

and

$$W_h = \left\{w \in C^0(\overline{\Omega}) \colon w|_T \in \mathscr{P}_1(T) \text{ for all } T \in \mathscr{T}_h^c,\ \int_\Omega w \,\mathrm{d}x = 0\right\},$$

where $\mathscr{P}_1(T)$ is the space of polynomials of degree up to 1 on T. The velocity approximation space is then $V_h \times V_h$ while the pressure approximation space is W_h, so we set $X_h = V_h \times V_h \times W_h$. This is a variant of the Hood–Taylor method. A finite element approximation of (4.6), which is called the $(4\mathscr{P}_1 - \mathscr{P}_1)$ finite element approximation, consists in finding $(\boldsymbol{u}_h, p_h) \in X_h$ such that

$$\text{for all } (\boldsymbol{v}_h, q_h) \in X_h \quad a(\boldsymbol{u}_h, \boldsymbol{v}_h) + c(\boldsymbol{u}_h; \boldsymbol{u}_h, \boldsymbol{v}_h) + b(\boldsymbol{v}_h, p_h) - \int_\Omega \boldsymbol{f}\boldsymbol{v}_h \,\mathrm{d}x + b(\boldsymbol{u}_h, q_h) = 0. \tag{4.9}$$

On $\mathscr{T}_h^c$ we make the following hypothesis: there exists a set of interior nodes in Ω, $P_1, \ldots, P_l$, such that the finite element stars $S_1, \ldots, S_l$

$$S_i = \bigcup_{T \ni P_i} T$$

have the property

$$\text{for all } i \neq j, \quad \overset{\circ}{S}_i \cap \overset{\circ}{S}_j = \emptyset, \qquad \overline{\Omega} = \bigcup_{i=1}^{l} S_i, \tag{4.10}$$

where $\overset{\circ}{S}_i$ denotes the interior of S_i. Note that we always can realize such a triangulation only to refine some elements. The study of the $(4\mathscr{P}_1 - \mathscr{P}_1)$ finite element for the Stokes problem is well known, see GIRAULT and RAVIART ([1986], p. 181) for instance.

From the theory developed in the next chapter, we will be able to analyze the approximated Navier–Stokes problem (4.9) and get a priori and a posteriori error estimates. It is then essential to get the discrete analogue of the hypothesis (4.8). From the study of the $(4\mathscr{P}_1 - \mathscr{P}_1)$ finite element for the Stokes problem, we know that under the hypothesis (4.10) we have

$$\inf_{\substack{\underline{w}_h \in X_h \\ \|\underline{w}_h\|_X = 1}} \sup_{\substack{\underline{v}_h \in X_h \\ \|\underline{v}_h\|_X = 1}} S(\underline{w}_h, \underline{v}_h) \geqslant \beta > 0, \tag{4.11}$$

where the bilinear form corresponding to the Stokes problem $S : (H_0^1(\Omega))^2 \times L_0^2(\Omega) \to \mathbb{R}$ is given by

$$S(\underline{w}, \underline{v}) = S\big((\boldsymbol{w}, r), (\boldsymbol{v}, q)\big) = a(\boldsymbol{w}, \boldsymbol{v}) + b(\boldsymbol{w}, q) + b(\boldsymbol{v}, r). \tag{4.12}$$

THEOREM 4.1. *Let* $(\boldsymbol{u}, p) \in X$ *be a solution of* (4.6) *with* $\boldsymbol{u} \in (H^2(\Omega))^2$, *such that* (4.8) *is valid. Then under the assumptions* (4.10) *there exists* $\delta > 0$ *such that*

$$\inf_{\substack{\underline{w}_h \in X_h \\ \|\underline{w}_h\|_X = 1}} \sup_{\substack{\underline{v}_h \in X_h \\ \|\underline{v}_h\|_X = 1}} A(\underline{w}_h, \underline{v}_h) \geqslant \delta > 0. \tag{4.13}$$

PROOF. We define the linear operator $\mathscr{A} : X' \to X$ by: for $\underline{f} \in X'$, $\mathscr{A}\underline{f} \in X$ is the unique solution to

$$\text{for all } \underline{v} \in X, \quad A(\mathscr{A}\underline{f}, \underline{v}) = \langle \underline{f}, \underline{v} \rangle_{X'X},$$

where $\langle \bullet, \bullet \rangle_{X'X}$ is the duality pairing between X' and X. From the hypothesis (4.8), we know that $\mathscr{A}$ is an isomorphism from X' onto X. In the same way from the Stokes form (4.12), we define the linear operator $\mathscr{S} : X' \to X$ by: for $\underline{f} \in X'$, $\mathscr{S}\underline{f} \in X$ is the unique solution to

$$\text{for all } \underline{v} \in X, \quad S(\mathscr{S}\underline{f}, \underline{v}) = \langle \underline{f}, \underline{v} \rangle_{X'X},$$

which is an isomorphism. We also define the linear operator $N: X \to X'$ by for all $\underline{w} = (\boldsymbol{w}, r) \in X$, for all $\underline{v} = (\boldsymbol{v}, q) \in X$

$$\langle N\underline{w}, \underline{v}\rangle_{X'X} = \int_\Omega \big[(\boldsymbol{u}\boldsymbol{\nabla})\boldsymbol{w} + (\boldsymbol{w}\boldsymbol{\nabla})\boldsymbol{u}\big]\boldsymbol{v}\,\mathrm{d}x.$$

For $\underline{f} \in X'$ we have the relation for all $\underline{v} \in X$

$$A(\mathscr{A}\underline{f}, \underline{v}) = S(\mathscr{A}\underline{f}, \underline{v}) + \big\langle N(\mathscr{A}\underline{f}), \underline{v}\big\rangle_{X'X} = \langle \underline{f}, \underline{v}\rangle_{X'X}$$

or

$$\big\langle (\mathscr{S}^{-1} + N)(\mathscr{A}\underline{f}), \underline{v}\big\rangle_{X'X} = \langle \underline{f}, \underline{v}\rangle_{X'X}. \tag{4.14}$$

So we have proved that

$$\mathscr{S}^{-1} + N = \mathscr{A}^{-1} \quad \text{or} \quad I + \mathscr{S}N = \mathscr{S}\mathscr{A}^{-1} \tag{4.15}$$

where I is the identity in X.

The discrete analogue of $\mathscr{S}$ is denoted by $\mathscr{S}_h : X' \to X_h$ and is given by: for $\underline{f} \in X'$

$$\text{for all } \underline{v}_h \in X_h, \quad S(\mathscr{S}_h\underline{f}, \underline{v}_h) = \langle \underline{f}, \underline{v}_h\rangle_{X'X}.$$

As a consequence of hypothesis (4.10), we get

$$\lim_{h\to 0} \|\mathscr{S}_h\underline{f} - \mathscr{S}\underline{f}\|_X = 0 \quad \text{for all } \underline{f} \in X'. \tag{4.16}$$

The convergence property (4.16) and the compactness of $N: X \to X'$ imply

$$\lim_{h\to 0} \|\mathscr{S}N - \mathscr{S}_hN\|_{X;X} = 0, \tag{4.17}$$

see (1.3).

Writing

$$I + \mathscr{S}_hN - (I + \mathscr{S}N)\big[I + (I + \mathscr{S}N)^{-1}(\mathscr{S}_hN - \mathscr{S}N)\big]$$

we deduce from (4.15), (4.17) and (1.2) that there exists an $h_0 > 0$ such that $I + \mathscr{S}_hN$ is an isomorphism from X onto X for $h \leqslant h_0$ with a uniformly bounded inverse.

Let us prove now the discrete analogue of (4.15) or in other words

$$(I + \mathscr{S}_hN)|_{X_h} = \mathscr{S}_h\mathscr{A}^{-1}|_{X_h}. \tag{4.18}$$

We deduce from relation (4.14) that for $\underline{w} \in X$ and $\underline{v}_h \in X_h$,

$$S\big([\mathscr{S}_h\mathscr{A}^{-1} - (I + \mathscr{S}_h N)]\underline{w}, \underline{v}_h\big) = 0.$$

In particular with the inf–sup relation (4.11) we get

$$\text{for all } \underline{w}_h \in X_h, \quad [\mathscr{S}_h\mathscr{A}^{-1} - (I + \mathscr{S}_h N)]\underline{w}_h = 0.$$

Now since $I + \mathscr{S}_h N$ is an isomorphism from X onto X with a uniformly bounded inverse, we have

$$\text{for all } \underline{w}_h \in X_h, \quad \|\mathscr{S}_h\mathscr{A}^{-1}\underline{w}_h\|_X = \|(I + \mathscr{S}_h N)\underline{w}_h\|_X \geqslant \gamma\|\underline{w}_h\|_X, \tag{4.19}$$

with $\gamma > 0$, γ independent on h.

We are now in a position to prove (4.13). Let $\underline{w}_h \in X_h$ be such that $\|\underline{w}_h\|_X = 1$. Then by using (4.11) and (4.19) we get

$$\begin{aligned}
\sup_{\substack{\underline{v}_h \in X_h \\ \|\underline{v}_h\|_X = 1}} A(\underline{w}_h, \underline{v}_h) &= \sup_{\substack{\underline{v}_h \in X_h \\ \|\underline{v}_h\|_X = 1}} \langle \mathscr{A}^{-1}\underline{w}_h, \underline{v}_h \rangle_{X'X} \\
&= \sup_{\substack{\underline{v}_h \in X_h \\ \|\underline{v}_h\|_X = 1}} S\big(\mathscr{S}_h\mathscr{A}^{-1}\underline{w}_h, \underline{v}_h\big) \\
&\geqslant \beta\|\mathscr{S}_h\mathscr{A}^{-1}\underline{w}_h\|_X \geqslant \beta\gamma\|\underline{w}_h\|_X = \beta\gamma,
\end{aligned}$$

which proves (4.13) with $\delta = \beta\gamma$. □

REMARK 4.1. We could have formulated an abstract version of the above theorem. The generalization with X a reflexive Banach space, bilinear forms $A(\bullet, \bullet)$ and $S(\bullet, \bullet)$, a compact operator $N : X \to X'$, and $\{X_h\}_{0<h\leqslant 1}$ a family of finite-dimensional subspaces in X satisfying

$$\lim_{h\to 0} \inf_{u_h \in X_h} \|u - u_h\|_X = 0 \quad \text{for all } u \in X,$$

is straightforward.

In the next chapter we present how to use the above result to analyze the $(4\mathscr{P}_1 - \mathscr{P}_1)$ approximation of the Navier–Stokes problem.

5. A stationary heat problem with convection

The last example we develop is a convection diffusion type problem. Let $\Omega \subset \mathbb{R}^2$ be a bounded domain with a regular boundary $\partial\Omega$. Given the functions $f \in L^2(\Omega)$, $k \in L^\infty(\mathbb{R})$ and the vector field $\mathbf{c} \in (L^\infty(\Omega))^2$, we shall consider the problem to find $u \in H_0^1(\Omega)$ satisfying

$$-\operatorname{div}\big(k(u)\,\mathbf{grad}\,u\big) + \mathbf{c}\,\mathbf{grad}\,u = f \quad \text{in } \Omega. \tag{5.1}$$

To study it we introduce some further assumptions on the data. The function $k \in C^2(\mathbb{R})$ satisfies the hypotheses

$$\text{for all } s \in \mathbb{R}, \quad k(s) \geqslant \alpha > 0, \tag{5.2}$$

$$\text{for all } s \in \mathbb{R}, \quad \big|\mathrm{D}^l k(s)\big| \leqslant \gamma_l, \quad l = 0, 1, 2, \tag{5.3}$$

with positive real numbers α, γ_0, γ_1, and γ_2; here $\mathrm{D}^l k(s)$ stands for the lth derivative of k. The vector field $\mathbf{c}$ satisfies

$$\mathbf{c} \in \big(W^{1,\infty}(\Omega)\big)^2 \quad \text{and} \quad \operatorname{div}\mathbf{c} = 0. \tag{5.4}$$

In the approximation schemes, we shall keep the nonlinear term in the principal part of (5.1), although to analyze the exact problem, we reduce it to a semilinear problem. Under the assumption (5.2) the primitive K,

$$\text{for all } s \in \mathbb{R}, \quad K(s) = \int_0^s k(t)\,\mathrm{d}t,$$

of k is a strictly increasing function. Let G be the inverse function of K and g be the derivative of G. Then (5.1) can be written in the form

$$-\Delta K(u) + \mathbf{c}\,\mathbf{grad}\,u = f \quad \text{in } \Omega$$

or when setting $U = K(u) \in H_0^1(\Omega)$

$$-\Delta U + g(U)\mathbf{c}\,\mathbf{grad}\,U = f \quad \text{in } \Omega, \tag{5.5}$$

and the weak form reads: find $U \in H_0^1(\Omega)$ such that

$$\text{for all } \varphi \in H_0^1(\Omega),$$
$$\int_\Omega \mathbf{grad}\,U\,\mathbf{grad}\,\varphi\,\mathrm{d}x + \int_\Omega g(U)\mathbf{c}\,\mathbf{grad}\,U\varphi\,\mathrm{d}x = \int_\Omega f\varphi\,\mathrm{d}x. \tag{5.6}$$

THEOREM 5.1. *In addition to hypotheses* (5.2), (5.3) *and* (5.4) *we assume that* $f \in L^\infty(\Omega)$. *Problem* (5.6) *has a unique solution* U; *moreover, the solution* U *is in* $W^{2,p}(\Omega)$ *for all* $1 \leqslant p < \infty$.

PROOF. To prove the existence of a solution we introduce first the operator $T \in \mathscr{L}(H^{-1}(\Omega); H_0^1(\Omega))$, the inverse of the minus Laplacian operator with homogeneous boundary conditions defined in (3.4), and write (5.6) in the equivalent way:

$$U - T\big(f - g(U)\mathbf{c}\,\mathbf{grad}\,U\big) = 0. \tag{5.7}$$

The mapping $U \in H_0^1(\Omega) \to f - g(U)\mathbf{c}\,\mathbf{grad}\,U \in H^{-1}(\Omega)$ is continuous and compact. To apply the Leray–Schauder homotopy theorem, see LERAY and SCHAUDER [1934], we derive a a priori estimate for $V \in H_0^1(\Omega)$ with

$$V - \lambda T\big(f - g(V)\mathbf{c}\,\mathbf{grad}\,V\big) = 0,$$

which is independent of $\lambda \in [0,1]$. For $\varphi \in H_0^1(\Omega)$ we have

$$\int_\Omega \mathbf{grad}\,V\,\mathbf{grad}\,\varphi\,\mathrm{d}x - \lambda\int_\Omega \big(f - g(V)\mathbf{c}\,\mathbf{grad}\,V\big)\varphi\,\mathrm{d}x = 0$$

or with $\varphi = V$ and $\mathscr{G}(s) = \int_0^s g(t)t\,\mathrm{d}t$

$$\int_\Omega |\mathbf{grad}\,V|^2\,\mathrm{d}x - \lambda\int_\Omega fV\,\mathrm{d}x + \lambda\int_\Omega \mathbf{c}\,\mathbf{grad}\,\mathscr{G}(V)\,\mathrm{d}x = 0.$$

With hypothesis (5.4) we check that $\mathbf{c}\,\mathbf{grad}\,\mathscr{G}(V) = \operatorname{div}(\mathbf{c}\mathscr{G}(V))$ and since $\mathscr{G}(V)$ is in $H_0^1(\Omega)$ we get

$$\int_\Omega |\mathbf{grad}\,V|^2\,\mathrm{d}x = \lambda\int_\Omega fV\,\mathrm{d}x$$

and so

$$|V|_{1,\Omega}^2 \leqslant \|f\|_{0,\Omega}\|V\|_{0,\Omega} \leqslant C\|f\|_{0,\Omega}|V|_{1,\Omega}.$$

Thus we conclude the existence of a solution to (5.7).

Considering the relation (5.7) and the regularity of the domain, we get from the classical elliptic regularity result that the function U is in $H^2(\Omega)$, from which it follows that $f - g(U)\mathbf{c}\,\mathbf{grad}\,U \in L^p(\Omega)$, $1 \leqslant p < \infty$, and then $U \in W^{2,p}(\Omega)$.

Let us finally check the uniqueness result. Suppose that U and V are two solutions of (5.5). Then

$$-\Delta(U - V) + g(U)\mathbf{c}\,\mathbf{grad}\,U - g(V)\mathbf{c}\,\mathbf{grad}\,V = 0 \quad \text{in } \Omega. \tag{5.8}$$

Since G is the primitive of g with $G(0) = 0$ and since $\operatorname{div}\mathbf{c} = 0$, relation (5.8) can be written in the form

$$-\Delta(U - V) + \operatorname{div}\big(\mathbf{c}\big(G(U) - G(V)\big)\big) = 0 \quad \text{in } \Omega \tag{5.9}$$

or with $W = U - V$,

$$-\Delta W + \operatorname{div}\left(\mathbf{c}W \int_0^1 g\big(sU + (1-s)V\big)\,\mathrm{d}s\right) = 0 \quad \text{in } \Omega. \tag{5.10}$$

Our goal is to prove that $W \equiv 0$. The weak form of (5.10) reads with $\zeta = \int_0^1 g(sU + (1-s)V)\,\mathrm{d}s$,

$$\text{for all } \varphi \in H_0^1(\Omega), \quad \int_\Omega \mathbf{grad}\, W\ \mathbf{grad}\, \varphi \,\mathrm{d}x - \int_\Omega \zeta W \mathbf{c}\,\mathbf{grad}\,\varphi\,\mathrm{d}x = 0.$$

Given $\varepsilon > 0$, we choose $\varphi = W^+/(W^+ + \varepsilon)$ where W^+ stands for the positive part of W. Then we have since $\|\zeta\|_{0,\infty} \leqslant 1/\alpha$,

$$\begin{aligned}
\int_\Omega \left|\mathbf{grad}\log\left(1 + \frac{W^+}{\varepsilon}\right)\right|^2 \mathrm{d}x &= \int_\Omega \frac{1}{(\varepsilon + W^+)^2}\left|\mathbf{grad}\, W^+\right|^2 \mathrm{d}x \\
&= \int_\Omega \frac{1}{\varepsilon}\,\mathbf{grad}\, W\ \mathbf{grad}\,\frac{W^+}{\varepsilon + W^+}\,\mathrm{d}x \\
&= \int_\Omega \frac{1}{\varepsilon} W^+ \zeta \mathbf{c}\,\mathbf{grad}\,\frac{W^+}{\varepsilon + W^+}\,\mathrm{d}x \\
&= \int_\Omega \frac{W^+}{\varepsilon + W^+}\zeta \mathbf{c}\,\mathbf{grad}\log\left(1 + \frac{W^+}{\varepsilon}\right)\mathrm{d}x \\
&\leqslant \frac{\|\mathbf{c}\|_{0,\infty}}{\alpha}\int_\Omega \left|\mathbf{grad}\log\left(1 + \frac{W^+}{\varepsilon}\right)\right|\mathrm{d}x.
\end{aligned}$$

So with the Cauchy–Schwarz inequality we get

$$\int_\Omega \left|\mathbf{grad}\log\left(1 + \frac{W^+}{\varepsilon}\right)\right|^2 \mathrm{d}x \leqslant C.$$

Since $\log(1 + W^+/\varepsilon)$ is in $H_0^1(\Omega)$, it follows from the Poincaré inequality that

$$\int_\Omega \left|\log\left(1 + \frac{W^+}{\varepsilon}\right)\right|^2 \mathrm{d}x \leqslant C$$

with C independent of ε. Letting ε tend to 0, we get that W^+ must vanish in Ω. In the same way we prove that W^- must vanish in Ω. □

In effective computations problem (5.1) is most often discretized as it is, but not problem (5.6). For that reason we shall focus on the approximation of (5.1). The mapping $F : H_0^1(\Omega) \to H^{-1}(\Omega)$ is defined by: for $v \in H_0^1(\Omega)$,

$$F(v) = -\operatorname{div}\big(k(v)\,\mathbf{grad}\, v\big) + \mathbf{c}\,\mathbf{grad}\, v - f.$$

Let U be the unique solution of (5.6) given in Theorem 5.1. Then $u \equiv G(U)$ is the unique zero of F,

$$F(u) = 0, \tag{5.11}$$

and $u \in W^{2,p}(\Omega)$, $1 \leqslant p < \infty$.

Generally the mapping F is not C^1, which would be a sufficient condition to apply the approximation theory developed in the next section. In fact the appropriate spaces to work with are $W^{1,p}(\Omega)$ with $2 < p < \infty$, see POUSIN and RAPPAZ [1992]; the mapping F is then considered as $F: W_0^{1,p}(\Omega) \to W^{-1,p}(\Omega)$ and has desired differentiability properties.

THEOREM 5.2. *Under the hypotheses of Theorem 5.1, the mapping $F: W_0^{1,p}(\Omega) \to W^{-1,p}(\Omega)$ is of class C^1 when $2 < p < \infty$ and moreover the derivative $\mathrm{D}F(u) \in \mathscr{L}(W_0^{1,p}(\Omega); W^{-1,p}(\Omega))$ at the solution u is an isomorphism.*

PROOF. Let q be the conjugate of p, $1/p + 1/q = 1$. Then for all $v \in W_0^{1,p}(\Omega)$ and $\psi \in W_0^{1,q}(\Omega)$

$$\big\langle F(v), \psi \big\rangle_{W^{-1,p}(\Omega) W_0^{1,q}(\Omega)} = \int_\Omega k(v)\, \mathbf{grad}\, v\, \mathbf{grad}\, \psi \,\mathrm{d}x + \int_\Omega \mathbf{c}\, \mathbf{grad}\, v \psi \,\mathrm{d}x - \int_\Omega f \psi \,\mathrm{d}x,$$

$\langle \bullet, \bullet \rangle_{W^{-1,p}(\Omega) W_0^{1,q}(\Omega)}$ denoting the duality bracket. Then for all v, w in $W_0^{1,p}(\Omega)$, ψ in $W_0^{1,q}(\Omega)$

$$\big\langle \mathrm{D}F(v)w, \psi \big\rangle_{W^{-1,p}(\Omega) W_0^{1,q}(\Omega)} = \int_\Omega \mathbf{grad}\, \big(k(v)w\big)\, \mathbf{grad}\, \psi \,\mathrm{d}x + \int_\Omega \mathbf{c}\, \mathbf{grad}\, w \psi \,\mathrm{d}x \tag{5.12}$$

and with the regularity assumption on k in (5.2), (5.3), we easily verify that F is C^1.

It is known that the Laplacian operator with homogeneous boundary conditions is an isomorphism from $W_0^{1,p}(\Omega)$ onto $W^{-1,p}(\Omega)$, see for instance in SIMADER [1972] or in DAUGE [1992] or the regularity result in DAUTRAY and LIONS ([1987], p. 538) and the characterization of $W^{-1,p}(\Omega)$ given in GRISVARD ([1985], p. 17). Relation (5.12) with $v = u$ also reads

$$T\,\mathrm{D}F(u)w = k(u)w + T(\mathbf{c}\, \mathbf{grad}\, w) \tag{5.13}$$

where T is the inverse of the minus Laplacian operator. We introduce the mapping $R: \varphi \in W_0^{1,p}(\Omega) \to R\varphi = \varphi / k(u) \in W_0^{1,p}(\Omega)$, which is an isomorphism, in (5.13) with $w = R\varphi$ to get

$$T\,\mathrm{D}F(u)R\varphi = \varphi + T\big(\mathbf{c}\, \mathbf{grad}(R\varphi)\big). \tag{5.14}$$

Since the mapping $T \in \mathscr{L}(L^p(\Omega); W_0^{1,p}(\Omega))$ is compact, so the mapping

$$T\,\mathrm{D}F(u)R \in \mathscr{L}\big(W_0^{1,p}(\Omega); W_0^{1,p}(\Omega)\big)$$

is a Fredholm operator with index 0. To prove the theorem it is then sufficient to prove that $\mathrm{D}F(u) \in \mathscr{L}(W_0^{1,p}(\Omega); W^{-1,p}(\Omega))$ is injective.

Let $\varphi \in \mathrm{Ker}(\mathrm{D}F(u))$, then $\varphi \in W_0^{1,p}(\Omega)$ satisfies

$$-\Delta\big(k(u)\varphi\big) + \mathbf{c}\,\mathbf{grad}\,\varphi = 0 \quad \text{in } \Omega. \tag{5.15}$$

With the notation $\psi = k(u)\varphi$, $\boldsymbol{\alpha} = \mathbf{c}/k(u)$, (5.15) takes the form

$$-\Delta\psi + \mathrm{div}(\boldsymbol{\alpha}\psi) = 0 \quad \text{in } \Omega. \tag{5.16}$$

From the relations (5.15) and (5.16) we can conclude that $\psi \equiv \varphi \equiv 0$ following the proof of uniqueness in Theorem 5.1. □

We introduce now a finite element approximation of (5.11). For the sake of simplicity we suppose that Ω is a polygonal domain in $\mathbb{R}^2$. Given a regular family $\{\mathscr{T}_h\}_{0<h\leqslant 1}$ of triangulations of Ω, we associate to $\mathscr{T}_h$ the finite element space

$$V_h = \big\{\varphi \in C^0(\overline{\Omega}) \colon \varphi|_T \in \mathscr{P}_1 \text{ for all } T \in \mathscr{T}_h \text{ and } \varphi|_{\partial\Omega} = 0\big\},$$

where $\mathscr{P}_1$ denotes the space of polynomials of degree up to 1. A finite element approximation of (5.11) consists in finding $u_h \in V_h$ satisfying

$$\text{for all } v_h \in V_h,$$
$$\int_\Omega k(u_h)\,\mathbf{grad}\,u_h\,\mathbf{grad}\,v_h\,\mathrm{d}x + \int_\Omega \mathbf{c}\,\mathbf{grad}\,u_h v_h\,\mathrm{d}x = \int_\Omega f v_h\,\mathrm{d}x \tag{5.17}$$

or equivalently

$$\text{for all } v_h \in V_h, \quad \big\langle F(u_h), v_h\big\rangle_{W^{-1,p}(\Omega)W_0^{1,q}(\Omega)} = 0. \tag{5.18}$$

In the next chapter we will prove that (5.18) has a unique solution $u_h \in V_h$ and get a priori and a posteriori error estimates for $\|u - u_h\|_{1,p,\Omega}$.

CHAPTER III

Nonlinear Problems and Their Numerical Approximation

It is the purpose of this chapter to develop an approximation method for nonlinear problems not depending on a parameter. We follow here the presentation of POUSIN and RAPPAZ [1991, 1992].

A general framework is first established. Let X and Z be two real (or complex) Banach spaces, $F: X \to Z$ be a C^1 mapping, and $\{F_h\}_{0<h\leqslant 1}$ be a family of C^1 mappings defined from X to Z. Under equicontinuity, consistency, and stability hypotheses we prove the convergence of a solution u_h of $F_h(u_h) = 0$ to a solution u of $F(u) = 0$.

When Z is the dual space of a Banach space, a particular attention is paid to the Galerkin approximations of the problem $F(u) = 0$. Our approach is presented as a generalization of the linear theory. We get the analogue of the inf-sup conditions and both a priori and a posteriori error estimates.

Finally the abstract results are applied to the model problem of Section 3, to the approximation of the Navier–Stokes problem given in Section 4, and to the stationary heat problem with convection of Section 5.

6. An abstract framework

The two real (or complex) Banach spaces X and Z are endowed with the norms $\|\bullet\|_X$ and $\|\bullet\|_Z$ respectively. Given a family $\{F_h\}_{0<h\leqslant 1}$ of C^1 mappings, $F_h: X \to Z$, we assume there exists a $u \in X$ such that the three following hypotheses hold. The mappings $\mathrm{D}F_h$ are Lipschitzian at u, uniformly in a neighborhood of u, with constant L_h, that is there exists $\eta > 0$ and for all $h \in (0,1]$ there exists a constant L_h such that for all $v \subset \overline{B}(u,\eta)$,

$$\|\mathrm{D}F_h(u) - \mathrm{D}F_h(v)\|_{X;Z} \leqslant L_h\|u - v\|_X; \tag{6.1}$$

$$\lim_{h\to 0}(1 + L_h)\|F_h(u)\|_Z = 0; \tag{6.2}$$

$\mathrm{D}F_h(u)$ is an isomorphism from X onto Z with an inverse uniformly bounded in h, that is there exists C_1 such that

$$\text{for all } h \in (0,1], \quad \|DF_h(u)^{-1}\|_{Z;X} \leqslant C_1. \tag{6.3}$$

In the examples we shall develop, we will take $u \in X$ as a zero of a mapping $F: X \to Z$ and $\{F_h\}_{0<h\leqslant 1}$ as an approximation family of F. Most often in the applications L_h is bounded and the assumption (6.2) can be interpreted as a consistency hypothesis while (6.3) is a stability hypothesis. Note that with our presentation we can work with L_h unbounded but with $\|F_h(u)\|_Z$ tending to zero faster than L_h tends to ∞. Naturally the goal of this section is to prove that consistency, stability, and equicontinuity imply existence and convergence for the approximations.

THEOREM 6.1. *Let $\{F_h\}_{0<h\leqslant 1}$ be a family of C^1 mappings from X into Z and $u \in X$ satisfy hypotheses* (6.1), (6.2) *and* (6.3). *Then there exist $\delta_0 > 0$ and $h_0 > 0$ such that for all $h \leqslant h_0$, there is a unique u_h satisfying*

$$F_h(u_h) = 0 \quad \textit{and} \quad \|u - u_h\|_X \leqslant \frac{\delta_0}{1 + L_h}. \tag{6.4}$$

Moreover, for all $h \leqslant h_0$

$$\|u - u_h\|_X \leqslant 2\|DF_h(u)^{-1}\|_{Z;X}\|F_h(u)\|_Z. \tag{6.5}$$

PROOF. We set

$$\begin{aligned}
\varepsilon_h &= \|F_h(u)\|_Z, \\
\gamma_h &= \|DF_h(u)^{-1}\|_{Z;X}, \\
\widetilde{L}_h(\alpha) &= \sup_{v \in \overline{B}(u,\alpha)} \|DF_h(u) - DF_h(v)\|_{X;Z}.
\end{aligned}$$

From the stability (6.3), we get $\gamma_h \leqslant C_1$. If η is the radius of the ball $\overline{B}(u,\eta)$ in (6.1) we set $\delta_0 = \min(\eta, 1/(2C_1))$ and $\delta_h = \delta_0/(1 + L_h)$. From (6.1) we have for all $h \in (0,1]$

$$\widetilde{L}_h(\delta_h) \leqslant (2C_1)^{-1}.$$

Now from the consistency (6.2), there exists h_0, $0 < h_0 \leqslant 1$, such that for all $h \in (0, h_0]$,

$$2C_1\varepsilon_h \leqslant \delta_h.$$

Thus for $h \leqslant h_0$,

$$2\gamma_h\widetilde{L}_h(2C_1\varepsilon_h) \leqslant 2C_1\widetilde{L}_h(\delta_h) \leqslant 1.$$

We apply Theorem 2.1 with $G = F_h$. So there exists a unique $u_h \in \overline{B}(u, 2\gamma_h\varepsilon_h)$ with $F_h(u_h) = 0$. Moreover, the following estimate holds

$$\|u - u_h\|_X \leqslant 2\gamma_h \|F_h(u)\|_Z.$$

Applying Remark 2.1, we get the uniqueness of u_h in $\overline{B}(u, \delta_h)$. Finally Theorem 6.1 is proved. □

REMARK 6.1. In Theorem 6.1, if L_h is bounded with respect to h, we have proved the existence and uniqueness of a zero of F_h in a fixed neighborhood of u.

We consider now a C^1 mapping $F: X \to Z$, $u \in X$ a zero of F, and a family $\{F_h\}_{0<h\leqslant 1}$ of mappings satisfying (6.1), (6.2) and (6.3). To relate u and u_h we have the following result.

THEOREM 6.2. *If u is a zero of F such that $\mathrm{D}F(u) \in \mathscr{L}(X; Z)$ is an isomorphism from X onto Z and if the family $\{F_h\}_{0<h\leqslant 1}$ satisfies* (6.1), (6.2) *and* (6.3), *then there exists $0 < \bar{h}_0 \leqslant h_0$ such that for all $h \leqslant \bar{h}_0$, $\mathrm{D}F(u_h) \in \mathscr{L}(X; Z)$ is an isomorphism with a uniformly bounded inverse and*

$$\|u - u_h\|_X \leqslant 2\|\mathrm{D}F(u_h)^{-1}\|_{Z;X} \|F(u_h)\|_Z, \tag{6.6}$$

where h_0 and u_h are given in Theorem 6.1.

PROOF. To apply Theorem 2.1, we set

$$\begin{aligned} \varepsilon_h &= \|F(u_h)\|_Z, \\ \gamma_h &= \|\mathrm{D}F(u_h)^{-1}\|_{Z;X}, \\ \widetilde{L}_h(\alpha) &= \sup_{v \in \overline{B}(u_h,\alpha)} \|\mathrm{D}F(v) - \mathrm{D}F(u_h)\|_{X;Z}. \end{aligned}$$

Let h_0, δ_0, and u_h be given in Theorem 6.1. A consequence of the consistency (6.2) and the estimate (6.5) is

$$\lim_{h\to 0} \|u - u_h\|_X = 0.$$

Since $\mathrm{D}F(u)$ is an isomorphism from X onto Z, we can choose $h_1 \leqslant h_0$ such that $\mathrm{D}F(u_h)$ is an isomorphism with

$$\gamma_h = \|\mathrm{D}F(u_h)^{-1}\|_{Z;X} \leqslant 2\|\mathrm{D}F(u)^{-1}\|_{Z;X} \equiv \gamma.$$

There exist $0 < h_2 \leqslant h_1$ and $\delta > 0$ such that

$$\widetilde{L}_h(\delta) \leqslant \frac{1}{2\gamma} \quad \text{and} \quad \|u - u_h\|_X < \delta.$$

Now let $h_3 \leqslant h_2$ be such that for all $h \leqslant h_3$,

$$2\gamma\varepsilon_h \leqslant \delta.$$

We are in position to apply Theorem 2.1 with $G = F$, $v = u_h$, $0 < h \leqslant h_3$. We conclude that there exists a unique $w \in \overline{B}(u_h, 2\gamma_h\varepsilon_h)$ such that

$$F(w) = 0 \quad \text{and} \quad \|w - u_h\|_X \leqslant 2\big\|\mathrm{D}F(u_h)^{-1}\big\|_{Z,X}\big\|F(u_h)\big\|_Z.$$

Applying Remark 2.1, we get the uniqueness of w in $\overline{B}(u_h, \delta)$. Since $F(u) = 0$, we have $w = u$ (actually we must choose $\bar{h}_0 < h_3$) and (6.6) is proved. □

The estimates (6.5) and (6.6) have a different meaning. From the first estimate we can deduce an a priori estimate bounded by the consistency error $\|F_h(u)\|_Z$. On the other hand from the second estimate we get an a posteriori estimate bounded by the residual $\|F(u_h)\|_Z$.

THEOREM 6.3. *Let $F : X \to Z$ be a C^1 mapping and $\{F_h\}_{0<h\leqslant 1}$, $F_h : X \to Z$, be a family of C^2 mappings satisfying*

$$\begin{cases} \lim_{h\to 0} \|F(v) - F_h(v)\|_Z = 0 \quad \textit{for all } v \in X, \\ \lim_{h\to 0} \|\mathrm{D}F(v) - \mathrm{D}F_h(v)\|_{X;Z} = 0 \quad \textit{for all } v \in X, \\ \mathrm{D}^2F_h \textit{ is uniformly bounded with respect to } h \textit{ on all} \\ \quad \textit{bounded domain of } X. \end{cases} \tag{6.7}$$

Let $u \in X$ be such that

$$F(u) = 0, \tag{6.8}$$

$$\mathrm{D}F(u) \textit{ is an isomorphism from } X \textit{ onto } Z. \tag{6.9}$$

Then there exist $h_0 > 0$, $\delta_0 > 0$ such that for all $h \leqslant h_0$ there exists a unique $u_h \in X$ satisfying

$$F_h(u_h) = 0 \quad \textit{and} \quad \|u - u_h\|_X \leqslant \delta_0.$$

Moreover, for all $h \leqslant h_0$ we have

$$\|u - u_h\|_X \leqslant 2\big\|\mathrm{D}F_h(u)^{-1}\big\|_{Z;X}\big\|F_h(u)\big\|_Z, \tag{6.10}$$

$$\|u - u_h\|_X \leqslant 2\big\|\mathrm{D}F(u_h)^{-1}\big\|_{Z;X}\big\|F(u_h)\big\|_Z. \tag{6.11}$$

PROOF. It is easy to prove that hypotheses (6.7), (6.8) and (6.9) imply the ones of Theorems 6.1 and 6.2 with the constants L_h bounded with respect to h. □

REMARK 6.2. For simplicity, the function u in Theorem 6.1 is not supposed to depend on h. In Section 12, we assume a dependence on h. A convenient choice of a function depending on h may lead to error estimates in different norms, see Section 14.

In the next result we present a simple way to derive error estimates in different norms, see RAPPAZ [1983] for a variant of this technique.

We assume there exist two Banach spaces $\mathscr{X}$ and $\mathscr{Z}$, with the norms $\|\bullet\|_{\mathscr{X}}$ and $\|\bullet\|_{\mathscr{Z}}$, satisfying the following assumptions:

$$X \subset \mathscr{X} \quad \text{and} \quad Z \subset \mathscr{Z} \quad \text{with continuous injection;} \tag{6.12}$$

there exist $\varepsilon > 0$ and $\kappa > 0$ such that for all $v \in B_X(u, \varepsilon) = \{w \in X\colon \|w - u\|_X < \varepsilon\}$, the operators $\mathrm{D}F_h(v)$ and $\mathrm{D}F(v)$ can be extended as operators in $\mathscr{L}(\mathscr{X}; \mathscr{Z})$ and

$$\begin{cases} \text{for all } v \in B_X(u,\varepsilon), & \|\mathrm{D}F(v) - \mathrm{D}F(u)\|_{\mathscr{X};\mathscr{Z}} \leqslant \kappa\|v - u\|_X, \\ \text{for all } v \in B_X(u,\varepsilon), & \|\mathrm{D}F_h(v) - \mathrm{D}F_h(u)\|_{\mathscr{X};\mathscr{Z}} \leqslant \kappa\|v - u\|_X; \end{cases} \tag{6.13}$$

$$\lim_{h\to 0} \|\mathrm{D}F(u) - \mathrm{D}F_h(u)\|_{\mathscr{X};\mathscr{Z}} = 0; \tag{6.14}$$

$$\mathrm{D}F(u) \text{ is an isomorphism from } \mathscr{X} \text{ onto } \mathscr{Z}. \tag{6.15}$$

The following result holds.

THEOREM 6.4. *Under the assumptions of Theorem* 6.3 *and* (6.12)–(6.15), *there exists* $0 < h_1 \leqslant h_0$ *such that for all* $0 < h \leqslant h_1$, *we have*

$$\|u - u_h\|_{\mathscr{X}} \leqslant 2\|\mathrm{D}F_h(u)^{-1}\|_{\mathscr{Z};\mathscr{X}} \|F_h(u)\|_{\mathscr{Z}}, \tag{6.16}$$

$$\|u - u_h\|_{\mathscr{X}} \leqslant 2\|\mathrm{D}F(u_h)^{-1}\|_{\mathscr{Z};\mathscr{X}} \|F(u_h)\|_{\mathscr{Z}}. \tag{6.17}$$

PROOF. We first prove estimate (6.16). Noticing that $F_h(u_h) = 0$, we use the Taylor expansion to write

$$\begin{aligned} u - u_h = {}& \mathrm{D}F_h(u)^{-1}\left[\int_0^1 \big(\mathrm{D}F_h(u) - \mathrm{D}F_h\big(su + (1-s)u_h\big)\big)(u - u_h)\,\mathrm{d}s\right] \\ & + \mathrm{D}F_h(u)^{-1}F_h(u). \end{aligned} \tag{6.18}$$

Clearly, assumptions (6.14), (6.15), result (1.1), and estimate (6.10) imply the existence of $0 < \bar{h} \leqslant h_0$ and $\beta > 0$ such that for all $0 < h \leqslant \bar{h}$, the mapping $\mathrm{D}F_h(u)$ is an isomorphism from $\mathscr{X}$ onto $\mathscr{Z}$ with

$$\|\mathrm{D}F_h(u)^{-1}\|_{\mathscr{Z};\mathscr{X}} \leqslant \beta$$

and

$$\|u - u_h\|_X < \varepsilon.$$

With (6.18), we get

$$\|u - u_h\|_{\mathscr{X}} \leqslant \|\mathrm{D}F_h(u)^{-1}\|_{\mathscr{Z};\mathscr{X}} \times \{\kappa\|u - u_h\|_X \|u - u_h\|_{\mathscr{X}} + \|F_h(u)\|_{\mathscr{Z}}\}. \tag{6.19}$$

Let us choose now $0 < h_1 \leqslant \overline{h}$ such that for all $0 < h \leqslant h_1$ the following holds:

$$\beta\kappa\|u - u_h\|_X \leqslant \tfrac{1}{2}.$$

Then from estimate (6.19) we deduce (6.16).

To prove estimate (6.17), we proceed as before with the relation

$$u - u_h = \mathrm{D}F(u)^{-1}\left[\int_0^1 \left(\mathrm{D}F(u) - \mathrm{D}F\left(su + (1-s)u_h\right)\right)(u - u_h)\,\mathrm{d}s\right] - \mathrm{D}F(u)^{-1}F(u_h). \qquad \square$$

7. Galerkin approximation of nonlinear problems

In order to relate the results of the previous section with the ones relative to the Galerkin approximation of linear problems, we shall focus on a particular abstract framework. We suppose from now on that we have $Z = Y'$, Y' being the dual space of a real Banach space Y. We assume

$$\begin{aligned} &F: X \to Y' \text{ is a } C^1 \text{ mapping and } u \in X \text{ is such that } F(u) = 0 \\ &\text{and } \mathrm{D}F(u) \text{ is an isomorphism from } X \text{ onto } Y'. \end{aligned} \tag{7.1}$$

Note that

$$F(u) = 0 \Longleftrightarrow \text{for all } y \in Y,\ \langle F(u), y\rangle_{Y'Y} = 0.$$

Let now $\{X_h\}_{0<h\leqslant 1}$ be a family of finite-dimensional subspaces of X and $\{Y_h\}_{0<h\leqslant 1}$ be a family of finite-dimensional subspaces of Y. A Galerkin approximation of the problem to find $u \in X$ such that

$$F(u) = 0 \tag{7.2}$$

reads: find $u_h \in X_h$ such that

$$\text{for all } y_h \in Y_h,\quad \langle F(u_h), y_h\rangle_{Y'Y} = 0. \tag{7.3}$$

To introduce the analogue of the inf–sup conditions for the nonlinear case, we define the bilinear form $b: X \times Y \to \mathbb{R}$ by

$$\text{for all } x \in X,\ y \in Y, \quad b(x,y) = \langle \mathrm{D}F(u)x, y\rangle_{Y'Y}.$$

We assume there exists a constant $\beta_h > 0$ such that

$$\inf_{\substack{x \in X_h \\ \|x\|_X = 1}} \sup_{\substack{y \in Y_h \\ \|y\|_Y = 1}} b(x,y) \geqslant \beta_h \tag{7.4}$$

and

$$\dim X_h = \dim Y_h \tag{7.5}$$

for $0 < h \leqslant 1$. Finally we assume

$$\lim_{h \to 0} \inf_{x_h \in X_h} \beta_h^{-2} \|u - x_h\|_X = 0. \tag{7.6}$$

Remark that $\beta_h \leqslant \|\mathrm{D}F(u)\|_{X;Y'}$ and if β_h is bounded from below by $\beta > 0$ for all h, then (7.4) is a standard stability condition for linear problems and (7.6) is a standard approximation property of X by X_h.

We are in a position to state the major result of this section.

THEOREM 7.1. *Let us suppose that hypotheses* (7.1), (7.4), (7.5) *and* (7.6) *are satisfied. Moreover, we assume that* $\mathrm{D}F$ *is Lipschitzian at* u, *that is there exist* $\varepsilon_0 > 0$ *and* $L > 0$ *such that for all* $v \in X$ *with* $\|u - v\|_X \leqslant \varepsilon_0$,

$$\|\mathrm{D}F(u) - \mathrm{D}F(v)\|_{X;Y'} \leqslant L\|u - v\|_X.$$

Then there exist two constants $h_0 > 0$, $\delta_0 > 0$, *and for* $h \leqslant h_0$ *a unique solution* u_h *to problem* (7.3) *in the closed ball* $\overline{B}(u, \delta_h)$ *with* $\delta_h = \delta_0 \beta_h$. *Moreover, there exists a constant* C *independent of* h *such that*

$$\|u - u_h\|_X \leqslant C\beta_h^{-1} \inf_{x_h \in X_h} \|u - x_h\|_X \tag{7.7}$$

and

$$\|u - u_h\|_X \leqslant 2\|\mathrm{D}F(u_h)^{-1}\|_{Y';X} \|F(u_h)\|_{Y'}. \tag{7.8}$$

PROOF. The proof is based on Theorems 6.1 and 6.2. To define the mapping F_h, we introduce the two projectors $\Pi_{X_h} \in \mathscr{L}(X; X_h)$ and $\Pi_{Y_h} \in \mathscr{L}(Y; Y_h)$ by setting

$$\text{for all } y \in Y_h, \quad b(x - \Pi_{X_h} x, y) = 0, \tag{7.9}$$

$$\text{for all } x \in X_h, \quad b(x, y - \Pi_{Y_h} y) = 0. \tag{7.10}$$

Under hypotheses (7.4) and (7.5), both operators are well defined. Therefore we can construct a family $\{F_h\}_{0<h\leqslant 1}$ of mappings, $F_h : X \to Y'$, by

$$\text{for all } x \in X,\ y \in Y,\quad \langle F_h(x), y\rangle_{Y'Y} = \langle F(x), \Pi_{Y_h} y\rangle_{Y'Y} + b(x, y - \Pi_{Y_h} y). \tag{7.11}$$

Then we can analyze the problem to find $x \in X$ such that

$$F_h(x) = 0. \tag{7.12}$$

Problems (7.3) and (7.12) are equivalent. Clearly if $u_h \in X_h$ is a solution to (7.3) then it is a solution to (7.12). Conversely let $u_h \in X$ be a solution to (7.12). To prove that u_h is a solution to (7.3) it is sufficient to prove that $u_h \in X_h$. By taking $y = y_1 - \Pi_{Y_h} y_1$ with $y_1 \in Y$ in (7.11) we get

$$b(u_h, y_1 - \Pi_{Y_h} y_1) = 0$$

and so

$$\text{for all } y_1 \in Y,\quad b(u_h - \Pi_{X_h} u_h, y_1) = 0.$$

Since $\mathrm{D}F(u)$ is an isomorphism, we deduce that

$$u_h = \Pi_{X_h} u_h \in X_h.$$

In order to conclude, we will apply Theorems 6.1 and 6.2 to the family $\{F_h\}_{0<h\leqslant 1}$ defined in (7.11). So let us check that hypotheses (6.1)–(6.3) are fulfilled.

The Fréchet derivative of F_h at $v \in X$ is,

$$\text{for all } x \in X,\ y \in Y,\quad \langle \mathrm{D}F_h(v)x, y\rangle_{Y'Y} = \langle \mathrm{D}F(v)x, \Pi_{Y_h} y\rangle_{Y'Y} + b(x, y - \Pi_{Y_h} y).$$

It follows that for $x \in X$, $y \in Y$,

$$\langle (\mathrm{D}F_h(u) - \mathrm{D}F_h(v))x, y\rangle_{Y'Y} = \langle (\mathrm{D}F(u) - \mathrm{D}F(v))x, \Pi_{Y_h} y\rangle_{Y'Y}$$

and consequently for $v \in X$ with $\|u - v\|_X \leqslant \varepsilon_0$ we get

$$\|\mathrm{D}F_h(u) - \mathrm{D}F_h(v)\|_{X;Y'} \leqslant L\|u - v\|_X \|\Pi_{Y_h}\|_{Y;Y}. \tag{7.13}$$

With the notation $\|b\| = \|\mathrm{D}F(u)\|_{X;Y'}$, we easily check with (7.4) that

$$\|\Pi_{X_h}\|_{X;X} \leqslant \frac{\|b\|}{\beta_h}. \tag{7.14}$$

In order to bound the norm of Π_{Y_h} we write for $y \in Y$

$$\begin{aligned}\|\Pi_{Y_h} y\|_Y &= \sup_{\substack{\varphi \in Y' \\ \|\varphi\|_{Y'}=1}} \langle \varphi, \Pi_{Y_h} y\rangle_{Y'Y} = \sup_{\substack{\varphi \in Y' \\ \|\varphi\|_{Y'}=1}} b\big(\mathrm{D}F(u)^{-1}\varphi, \Pi_{Y_h} y\big) \\ &= \sup_{\substack{\varphi \in Y' \\ \|\varphi\|_{Y'}=1}} b\big(\Pi_{X_h} \mathrm{D}F(u)^{-1}\varphi, y\big).\end{aligned}$$

By using (7.14) we finally obtain

$$\|\Pi_{Y_h}\|_{Y;Y} \leqslant \frac{\|b\|^2}{\beta_h} \|\mathrm{D}F(u)^{-1}\|_{Y';X}. \tag{7.15}$$

We notice that $\beta_h \leqslant \|b\|$ and $\|\mathrm{D}F(u)^{-1}\|_{Y';X}^{-1} \leqslant \|b\|$, so inequalities (7.13) and (7.15) imply that hypothesis (6.1) holds with

$$L_h = (1 + L)\frac{\|b\|^2}{\beta_h} \|\mathrm{D}F(u)^{-1}\|_{Y';X} - 1.$$

Let us check hypothesis (6.2). By the definition of F_h we have

$$\begin{aligned}\|F_h(u)\|_{Y'} &= \sup_{\substack{y \in Y \\ \|y\|_Y=1}} \langle F_h(u), y\rangle_{Y'Y} = \sup_{\substack{y \in Y \\ \|y\|_Y=1}} b(u, y - \Pi_{Y_h} y) \\ &= \sup_{\substack{y \in Y \\ \|y\|_Y=1}} b(u - \Pi_{X_h} u, y) \leqslant \|b\| \|u - \Pi_{X_h} u\|_X. \end{aligned} \tag{7.16}$$

Using assumption (7.4), we get for $x_h \in X_h$,

$$\begin{aligned}\beta_h \|x_h - \Pi_{X_h} u\|_X &\leqslant \sup_{\substack{y \in Y_h \\ \|y\|_Y=1}} b(x_h - \Pi_{X_h} u, y) \\ &\leqslant \sup_{\substack{y \in Y_h \\ \|y\|_Y=1}} b(x_h - u, y) \leqslant \|b\| \|u - x_h\|_X.\end{aligned}$$

So we deduce that for $x_h \in X_h$,

$$\|u - \Pi_{X_h} u\|_X \leqslant \left(1 + \frac{\|b\|}{\beta_h}\right) \|u - x_h\|_X$$

and using (7.16) we have

$$\|F_h(u)\|_{Y'} \leqslant \frac{C}{\beta_h} \inf_{x_h \in X_h} \|u - x_h\|_X \tag{7.17}$$

where C is independent of h.

When taking into account that

$$L_h = (1 + L)\|b\|^2 \|\mathrm{D}F(u)^{-1}\|_{Y'X}/\beta_h - 1$$

and using hypothesis (7.6), we prove that hypothesis (6.2) is fulfilled. Going back to the definition of F_h we have $\mathrm{D}F_h(u) = \mathrm{D}F(u)$, which is an isomorphism from X onto Y' with a uniformly bounded inverse. Then we conclude that hypothesis (6.3) is satisfied.

We conclude with Theorems 6.1 and 6.2. The error estimate (7.7) is a direct consequence of (6.5), (7.17), and of the relation $\mathrm{D}F_h(u) = \mathrm{D}F(u)$. □

REMARK 7.1. Theorem 7.1 can be viewed as a generalization of the linear theory. If the mapping F is smooth enough and the solution regular in the sense of (7.1), then we have reduced the approximation study to a verification of an inf–sup condition, see (7.4), and of an approximability result, see (7.6). Then the error estimates are deduced via an interpolation result for the a priori estimate (7.7) and via a computation of the residual for the a posteriori estimate (7.8).

REMARK 7.2. The approach proposed by RHEINBOLDT [1986] to study the approximation of parameter dependent problems with projection methods is in some respect close to ours.

REMARK 7.3. Quite often in the applications the spaces X and Y are identical Hilbert spaces and there exists a Hilbert space W satisfying $X \subset W$ with continuous embedding and X dense in W. The space W takes the place of a pivot space in the embeddings

$$X \subset W \subset X'.$$

The scalar product in X (respectively W) is denoted by $(\bullet, \bullet)_X$ (respectively $(\bullet, \bullet)_W$).

We assume that $X_h = Y_h$, the hypotheses of Theorem 7.1 hold, and the mapping F is of class C^2. Then we can derive an estimate in the norm of W from the estimate in the norm of X using a standard duality argument.

Indeed we have

$$\|u - u_h\|_W = \sup_{\substack{g \in W \\ g \neq 0}} \frac{|(g, u - u_h)_W|}{\|g\|_W}.$$

For a given g in W, let w be the unique solution to

$$\text{for all } v \in X, \quad b(v, w) = (v, g)_W.$$

Then there exists a constant C independent of g such that

$$\|w\|_X \leqslant C\|g\|_W.$$

So for a given g in W we get

$$\left|(g, u-u_h)_W\right| = \left|b(u-u_h, w)\right| \leqslant \left|b(u-u_h, w-w_h)\right| + \left|b(u-u_h, w_h)\right|$$

for any function $w_h \in X_h$. We can estimate the term $|b(u-u_h, w_h)|$ in the following way:

$$\begin{aligned} b(u-u_h, w_h) &= \langle \mathrm{D}F(u)(u-u_h), w_h \rangle_{X'X} \\ &= \langle -F(u) + F(u_h) - \mathrm{D}F(u)(u_h - u), w_h \rangle_{X'X} \\ &= \left\langle \int_0^1 (1-t)\mathrm{D}^2F\big(u + t(u_h - u)\big)(u_h - u)^2, w_h \right\rangle_{X'X} \end{aligned}$$

and so

$$\left|b(u-u_h, w_h)\right| \leqslant C\|u-u_h\|_X^2 \|w_h\|_X.$$

For any g in W and $w_h \in X_h$, we get

$$\left|(u-u_h, g)_W\right| \leqslant C\left\{\|u-u_h\|_X \|w-w_h\|_X + \|u-u_h\|_X^2 \|w_h\|\right\}.$$

We have proved that for any $g \in W$ and $w_h \in X_h$

$$\begin{aligned} \left|(u-u_h, g)_W\right| \leqslant C\big\{&\|u-u_h\|_X\left[1 + \|u-u_h\|_X\right]\|w-w_h\|_X \\ &+ \|u-u_h\|_X^2 \|g\|_W\big\}. \end{aligned}$$

An illustration of the method is given in Theorem 8.2.

REMARK 7.4. When the mapping $F: X \to Y'$ has the following structure

$$F(x) = Lx + G(x)$$

where $L \in \mathscr{L}(X; Y')$ is an isomorphism and $G: X \to Y'$ is such that

$$\mathrm{D}G(x) \in \mathscr{L}(X; Y')$$

is compact, then we can use the simple bilinear form $b: X \times Y \to \mathbb{R}$ given by

$$b(x, y) = \langle Lx, y \rangle_{Y'Y}.$$

In Section 16, this idea is used to simplify the study of the approximation of simple limit points.

8. Application to a model problem

The general results of the last section are applied to the semilinear problem (3.10). We shall study both approximation schemes (3.13) and (3.14).

In this particular case, the notations are the following: $X = Y = H_0^1(\Omega)$ equipped with the norm

$$\|v\|_X = \|v\|_Y = |v|_{1,\Omega} = \left(\int_\Omega |\,\mathbf{grad}\, v|^2 \,\mathrm{d}x \right)^{1/2}$$

equivalent to $\|\bullet\|_{1,\Omega}$, $Z = Y' = H^{-1}(\Omega)$, and $F : X \to Y'$ is given by

$$\text{for all } u, v \in H_0^1(\Omega),$$
$$\langle F(u), v \rangle_{H^{-1}(\Omega) H_0^1(\Omega)} = \int_\Omega \mathbf{grad}\, u \,\mathbf{grad}\, v \,\mathrm{d}x + \int_\Omega u^3 v \,\mathrm{d}x - \int_\Omega f v \,\mathrm{d}x, \tag{8.1}$$

where f is an $L^2(\Omega)$ function. Theorem 3.1 implies there exists a unique solution $u \in H_0^1(\Omega)$ to the problem

$$F(u) = 0; \tag{8.2}$$

furthermore, u is in $H^2(\Omega)$ and the derivative $\mathrm{D}F(u) \in \mathscr{L}(H_0^1(\Omega); H^{-1}(\Omega))$ is an isomorphism, which means that (7.1) is satisfied; moreover, $\mathrm{D}F$ is Lipschitzian at u.

The approximation space X_h and the test space Y_h are both V_h given in (3.12), the space of continuous piecewise polynomials of degree $\leqslant 1$, so the hypothesis (7.5) is guaranteed. The bilinear form $b : H_0^1(\Omega) \times H_0^1(\Omega) \to \mathbb{R}$,

$$b(v, w) = \int_\Omega \mathbf{grad}\, v \,\mathbf{grad}\, w \,\mathrm{d}x + 3 \int_\Omega u^2 v w \,\mathrm{d}x,$$

is coercive on $H_0^1(\Omega)$ and this property ensures hypothesis (7.4) with $\beta_h = 1$. Since the family of triangulations $\{\mathscr{T}_h\}_{0<h\leqslant 1}$ is regular and β_h is equal to 1, assumption (7.6) is satisfied.

We can apply Theorem 7.1 to study the existence, the local uniqueness and the convergence of problem (3.13), that is find $u_h \in V_h$ such that

$$\text{for all } v_h \in V_h, \quad \langle F(u_h), v_h \rangle_{H^{-1}(\Omega) H_0^1(\Omega)} = 0. \tag{8.3}$$

A definition useful in the next theorem is the following. Let w be a function in $W^{1,1}(\Omega)$ (so that the trace operator can be defined), T_i, T_j, with $i < j$, be two adjacent triangles, and let T' denote the common side. If w_i and w_j denote the restrictions of w on T_i and T_j respectively, we set

$$[w] = w_i - w_j \quad \text{on } T',$$

$[w]$ is the jump of w through the edge T'. If T' is a side of T_i on the boundary of Ω, then $[w] = w_i$ on T'.

THEOREM 8.1. *There exist positive constants h_0, δ_0, C, and for $h \leqslant h_0$ a unique solution u_h to (8.3) in $\overline{B}(u, \delta_0)$. Moreover, the following a priori and a posteriori estimates hold for $h \leqslant h_0$:*

$$|u - u_h|_{1,\Omega} \leqslant Ch\|f\|_{0,\Omega} \tag{8.4}$$

and

$$|u - u_h|_{1,\Omega} \leqslant C\Big(\sum_{T\in\mathcal{T}_h} \eta(T)^2\Big)^{1/2}, \tag{8.5}$$

where $\eta(T)$ is the local estimator given by

$$\eta(T) = \left(h_T^2\| - \Delta u_h + u_h^3 - f\|_{0,T}^2 + \sum_{i=1}^{3} h_{t_i}\left\|\left[\frac{\partial u_h}{\partial n}\right]\right\|_{0,t_i}^2\right)^{1/2}$$

with h_T the diameter of $T \in \mathcal{T}_h$, t_i, $i = 1, 2, 3$, the edges of T of length h_{t_i}, and $[\bullet]$ the jump through an edge.

PROOF. The hypotheses of Theorem 7.1 were verified before with $\beta_h = 1$, so there exist positive constants h_0, δ_0, and for $h \leqslant h_0$ a unique solution u_h to (8.3) in $\overline{B}(u, \delta_0)$. From estimates (7.7) and (7.8), we shall deduce (8.4) and (8.5) respectively.

Since the solution u to (8.2) is in $H^2(\Omega)$ and $\beta_h = 1$, we deduce from (7.7) that

$$|u - u_h|_{1,\Omega} \leqslant Ch\|f\|_{0,\Omega}.$$

So we have proved (8.4).

The residual $\|F(u_h)\|_{-1,\Omega}$ is estimated following the method presented in BARANGER and EL AMRI [1991]. For $v \in H_0^1(\Omega)$ we have

$$\begin{aligned}\langle F(u_h), v\rangle_{H^{-1}(\Omega)H_0^1(\Omega)} &= \int_\Omega \mathbf{grad}\, u_h\, \mathbf{grad}\, v\, \mathrm{d}x + \int_\Omega (u_h^3 - f)v\, \mathrm{d}x \\ &= \int_\Omega \mathbf{grad}\, u_h\, \mathbf{grad}(v - \tilde{r}_h v)\, \mathrm{d}x \\ &\quad + \int_\Omega (u_h^3 - f)(v - \tilde{r}_h v)\, \mathrm{d}x,\end{aligned}$$

where $\tilde{r}_h \in \mathscr{L}(H_0^1(\Omega), V_h)$ is the Clément interpolation operator onto V_h, see CLÉMENT [1975]. So for all $v \in H_0^1(\Omega)$

$$\langle F(u_h), v\rangle_{H^{-1}(\Omega)H_0^1(\Omega)} = \sum_{T\in\mathscr{T}_h} \left\{ \int_T \mathbf{grad}\, u_h\, \mathbf{grad}(v - \tilde{r}_h v)\,\mathrm{d}x + \int_T \left(u_h^3 - f\right)(v - \tilde{r}_h v)\,\mathrm{d}x \right\}. \tag{8.6}$$

For all $T \in \mathscr{T}_h$,

$$\int_T \mathbf{grad}\, u_h\, \mathbf{grad}(v - \tilde{r}_h v)\,\mathrm{d}x = -\int_T \Delta u_h (v - \tilde{r}_h v)\,\mathrm{d}x + \int_{\partial T} \frac{\partial u_h}{\partial n_T}(v - \tilde{r}_h v)\,\mathrm{d}s,$$

where n_T is the outward unit normal along the sides of T. Since u_h is in the space $\mathscr{P}_1$ on T the above equality reads

$$\int_T \mathbf{grad}\, u_h\, \mathbf{grad}(v - \tilde{r}_h v)\,\mathrm{d}x = \int_{\partial T} \frac{\partial u_h}{\partial n_T}(v - \tilde{r}_h v)\,\mathrm{d}s.$$

Let S_h be the set of all sides of triangles in $\mathscr{T}_h$. For each $t \in S_h$ we choose the direction of a unit normal n and denote by h_t its length, then

$$\left| \sum_{T\in\mathscr{T}_h} \int_T \mathbf{grad}\, u_h\, \mathbf{grad}(v - \tilde{r}_h v)\,\mathrm{d}x \right| \leqslant \sum_{t\in S_h} \int_t \left| \left[\frac{\partial u_h}{\partial n}\right](v - \tilde{r}_h v)\right| \mathrm{d}s;$$

if t is a side on $\partial\Omega$, we naturally set to zero the value outside $\overline{\Omega}$. Then the Cauchy–Schwarz inequality gives

$$\left| \sum_{T\in\mathscr{T}_h} \int_T \mathbf{grad}\, u_h\, \mathbf{grad}(v - \tilde{r}_h v)\,\mathrm{d}x \right| \leqslant \left(\sum_{t\in S_h} h_t \left\| \left[\frac{\partial u_h}{\partial n}\right] \right\|_{0,t}^2 \right)^{1/2} \left(\sum_{t\in S_h} h_t^{-1} \|v - \tilde{r}_h v\|_{0,t}^2 \right)^{1/2}. \tag{8.7}$$

Using the technique of the reference element, see CIARLET [1978] or CIARLET [1991], it is a simple matter to prove the existence of a constant C such that for all $T \in \mathscr{T}_h$ with edges t_1, t_2 and t_3, and for all $w \in H^1(T)$ we have

$$\sum_{j=1}^{3} h_{t_j}^{-1} \|w\|_{0,t_j}^2 \leqslant C\left(h_T^{-2}\|w\|_{0,T}^2 + |w|_{1,T}^2\right).$$

So inequality (8.7) is transformed into

$$\left|\sum_{T\in\mathcal{T}_h}\int_T \mathbf{grad}\, u_h\, \mathbf{grad}(v-\tilde{r}_h v)\,\mathrm{d}x\right|$$

$$\leqslant C\left(\sum_{t\in S_h} h_t\left\|\left[\frac{\partial u_h}{\partial n}\right]\right\|_{0,t}^2\right)^{1/2}$$

$$\times\left(\sum_{T\in\mathcal{T}_h}\left(h_T^{-2}\|v-\tilde{r}_h v\|_{0,T}^2+|v-\tilde{r}_h v|_{1,T}^2\right)\right)^{1/2}. \tag{8.8}$$

With Clément's interpolation estimates, see CLÉMENT [1975], for $v\in H^1(\Omega)$,

$$\|v-\tilde{r}_h v\|_{0,T}\leqslant Ch_T\sum_{T'\in S_T}|v|_{1,T'}\quad\text{and}\quad |v-\tilde{r}_h v|_{1,T}\leqslant C\sum_{T'\in S_T}|v|_{1,T'},$$

where S_T is the set of triangles T' with $T'\cap T\neq\emptyset$, and with (8.8), relation (8.6) is transformed by using the Cauchy–Schwarz inequality into

$$\langle F(u_h),v\rangle_{H^{-1}(\Omega)H_0^1(\Omega)}\leqslant C\left[\left(\sum_{t\in S_h}h_t\left\|\left[\frac{\partial u_h}{\partial n}\right]\right\|_{0,t}^2\right)^{1/2}|v|_{1,\Omega}\right.$$

$$\left.+\left(\sum_{T\in\mathcal{T}_h}h_T^2\|u_h^3-f\|_{0,T}^2\right)^{1/2}|v|_{1,\Omega}\right]. \tag{8.9}$$

Finally estimate (8.5) is a consequence of (7.8), the fact that $\mathrm{D}F(u_h)$ for $h\leqslant h_0$ is an isomorphism from $H_0^1(\Omega)$ onto $H^{-1}(\Omega)$ with a uniformly bounded inverse because $\lim_{h\to 0}u_h=u$ and estimate (8.9). □

REMARK 8.1. Although the a posteriori estimate (8.5) is a crude one, it can justify an adaptive mesh refinement procedure, based on the minimization of the residual. The theory of a posteriori estimates for nonlinear problems has advanced considerably in recent years. Model problems in one space dimension are studied in BABUŠKA and RHEINBOLDT [1982] and in RHEINBOLDT [1981]. An other approach is presented in RHEINBOLDT [1985] which produces reliable a posteriori estimates. We refer to the review paper of VERFÜRTH [1993] for recent results on a posteriori estimates.

It is also possible to obtain a priori estimates in other norms than the energy norm $|\cdot|_{1,\Omega}$. For instance with duality arguments of Aubin–Nitsche, see Remark 7.3, we shall derive an error estimate in the $L^2(\Omega)$ norm.

THEOREM 8.2. *Let h_0, δ_0, and for all $h\leqslant h_0$, $u_h\in\overline{B}(u,\delta_0)$ be given in Theorem* 8.1. *There exists a constant C such that*

$$\|u-u_h\|_{0,\Omega}\leqslant Ch^2. \tag{8.10}$$

PROOF. Since the form $b: H_0^1(\Omega) \times H_0^1(\Omega) \to \mathbb{R}$, $b(v,w) = \langle \mathrm{D}F(u)v, w\rangle_{H^{-1}(\Omega)H_0^1(\Omega)}$, is coercive on $H_0^1(\Omega)$, there exists a unique solution w to the problem: find $w \in H_0^1(\Omega)$ satisfying

$$\text{for all } v \in H_0^1(\Omega), \quad b(v,w) = \int_\Omega (u - u_h)v \,\mathrm{d}x. \tag{8.11}$$

By the standard elliptic regularity, we have $w \in H_0^1(\Omega) \cap H^2(\Omega)$ and

$$\|w\|_{2,\Omega} \leqslant C\|u - u_h\|_{0,\Omega}; \tag{8.12}$$

for ease of exposition we have assumed that Ω is a convex polygonal domain in $\mathbb{R}^2$, see the definition of V_h in (3.12). If we choose $w_h \in V_h$ to be the solution of

$$\text{for all } v \in V_h, \quad b(v, w_h) = \int_\Omega (u - u_h)v \,\mathrm{d}x,$$

the following estimate holds:

$$\|w - w_h\|_{1,\Omega} \leqslant Ch\|w\|_{2,\Omega}$$

and with (8.12)

$$\|w - w_h\|_{1,\Omega} \leqslant Ch\|u - u_h\|_{0,\Omega}. \tag{8.13}$$

Starting from (8.11), we deduce from inequality (8.13)

$$\begin{aligned}\|u - u_h\|_{0,\Omega}^2 &= b(u - u_h, w)\\ &\leqslant Ch\|u - u_h\|_{0,\Omega}\|u - u_h\|_{1,\Omega} + \left|b(u - u_h, w_h)\right|.\end{aligned} \tag{8.14}$$

Since

$$F(u) = 0 \quad \text{and} \quad \text{for all } v_h \in V_h,\ \langle F(u_h), v_h\rangle_{H^{-1}(\Omega)H_0^1(\Omega)} = 0,$$

we get with the C^2 regularity of F and for $w_h \in V_h$,

$$\begin{aligned} &b(u - u_h, w_h)\\ &\quad = \left\langle \mathrm{D}F(u)(u - u_h), w_h\right\rangle_{H^{-1}(\Omega)H_0^1(\Omega)}\\ &\quad = \left\langle F(u) - F(u_h), w_h\right\rangle_{H^{-1}(\Omega)H_0^1(\Omega)}\\ &\qquad + \left\langle \int_0^1 (1-t)\mathrm{D}^2F\big(u + t(u_h - u)\big)(u_h - u)^2 \,\mathrm{d}t, w_h\right\rangle_{H^{-1}(\Omega)H_0^1(\Omega)}\end{aligned}$$

and so

$$\left|b(u-u_h,w_h)\right| \leqslant C\|u_h-u\|_{1,\Omega}^2\left[\|w_h-w\|_{1,\Omega}+\|w\|_{1,\Omega}\right]. \tag{8.15}$$

With (8.4) and (8.12)–(8.15) we finally get (8.10). □

To study the approximation scheme with numerical integration (3.14), we cannot use directly the result of Section 7 relative to Galerkin approximation. We have to adapt the formalism and use the general results of Section 6. The elliptic projection operator $\pi_h \in \mathscr{L}(H_0^1(\Omega);V_h)$ is defined for $v \in H_0^1(\Omega)$ by

$$\text{for all } v_h \in V_h, \quad \int_\Omega \mathbf{grad}(v-\pi_h v)\,\mathbf{grad}\,v_h\,\mathrm{d}x = 0.$$

We assume that f is in $C^0(\overline{\Omega})$ and then we construct the mapping $F_h : H_0^1(\Omega) \to H^{-1}(\Omega)$ by the formula, for $v, w \in H_0^1(\Omega)$,

$$\begin{aligned}\left\langle F_h(v),w\right\rangle_{H^{-1}(\Omega)H_0^1(\Omega)} &= \int_\Omega \mathbf{grad}\,v\,\mathbf{grad}\,w\,\mathrm{d}x \\ &\quad + \int_\Omega r_h\left[\left((\pi_h v)^3 - f\right)\pi_h w\right]\mathrm{d}x. \end{aligned} \tag{8.16}$$

It is a simple matter to check that problem (3.14) is equivalent to find $u_h \in H_0^1(\Omega)$ satisfying

$$\text{for all } v \in H_0^1(\Omega), \quad \left\langle F_h(u_h),v\right\rangle_{H^{-1}(\Omega)H_0^1(\Omega)} = 0. \tag{8.17}$$

Indeed let $u_h \in V_h$ be a solution to (3.14). Then clearly u_h is also a solution to (8.17). Conversely let $u_h \in H_0^1(\Omega)$ be a solution to (8.17). If the test function v in (8.17) has the form $w - \pi_h w$ with $w \in H_0^1(\Omega)$, then Eq. (8.17) reads

$$\int_\Omega \mathbf{grad}\,u_h\,\mathbf{grad}(w-\pi_h w)\,\mathrm{d}x = 0,$$

which implies

$$u_h = \pi_h u_h \in V_h.$$

If we take $v_h \in V_h$ to be a test function in (8.17) then we get (3.14). Thus a solution $u_h \in H_0^1(\Omega)$ of (8.17) is a solution to (3.14).

Before studying problem (8.17), we present a technical result.

THEOREM 8.3. *Let V_h be the space of continuous piecewise polynomials of degree $\leqslant 1$ given in* (3.12) *and $r_h \in \mathscr{L}(C^0(\overline{\Omega});V_h)$ be the Lagrange interpolation operator. Then* *for $M \in \mathbb{N}$, there exists a constant C such that for any $u_j \in V_h$, $j = 1, 2, \ldots, M$, we have*

$$\left| \int_\Omega \left(r_h \left(\prod_{j=1}^M u_j \right) - \left(\prod_{j=1}^M u_j \right) \right) \mathrm{d}x \right| \leqslant Ch \prod_{j=1}^M \|u_j\|_{1,\Omega}.$$

PROOF. Since

$$\left| \int_\Omega \left(r_h \left(\prod_{j=1}^M u_j \right) - \left(\prod_{j=1}^M u_j \right) \right) \mathrm{d}x \right| \leqslant \sum_{T \in \mathcal{T}_h} \int_T \left| r_h \left(\prod_{j=1}^M u_j \right) - \left(\prod_{j=1}^M u_j \right) \right| \mathrm{d}x,$$

we derive an estimate for each $T \in \mathcal{T}_h$. On an element T with the reference element technique, we get

$$\int_T \left| r_h \left(\prod_{j=1}^M u_j \right) - \left(\prod_{j=1}^M u_j \right) \right| \mathrm{d}x = |\det B| \int_{\widehat{T}} \left| \hat{r} \left(\prod_{j=1}^M \hat{u}_j \right) - \left(\prod_{j=1}^M \hat{u}_j \right) \right| \mathrm{d}\hat{x},$$

where $x \equiv F_T(\hat{x}) = B\hat{x} + b$ is the affine mapping from the reference element $\widehat{T}$ onto T, $\hat{r}$ the Lagrange $\widehat{\mathcal{P}_1}$-interpolant on $\widehat{T}$, and $\hat{u}_j(\hat{x}) = u_j(x)$; $\widehat{\mathcal{P}_l}$ denotes the space of polynomials of degree $\leqslant l$ on $\widehat{T}$. For any $\hat{p} \in \widehat{\mathcal{P}_0}$, we have

$$\int_T \left| r_h \left(\prod_{j=1}^M u_j \right) - \left(\prod_{j=1}^M u_j \right) \right| \mathrm{d}x = |\det B| \int_{\widehat{T}} \left| (I - \hat{r}) \left(\left(\prod_{j=1}^M \hat{u}_j \right) - \hat{p} \right) \right| \mathrm{d}\hat{x}.$$

Clearly the mapping $\hat{r} : \widehat{\mathcal{P}_M} \to \widehat{\mathcal{P}_1}$ is linear and continuous. So

$$\int_T \left| r_h \left(\prod_{j=1}^M u_j \right) - \left(\prod_{j=1}^M u_j \right) \right| \mathrm{d}x \leqslant C |\det B| \left\| \left(\prod_{j=1}^M \hat{u}_j \right) - \hat{p} \right\|_{1,1,\widehat{T}}$$

for any $\hat{p} \in \widehat{\mathcal{P}_0}$. Taking the infimum over $\hat{p}$ as in the Bramble–Hilbert Lemma and using the affine change of variables we get

$$\begin{aligned} \int_T \left| r_h \left(\prod_{j=1}^M u_j \right) - \left(\prod_{j=1}^M u_j \right) \right| \mathrm{d}x &\leqslant C |\det B| \left| \left(\prod_{j=1}^M \hat{u}_j \right) \right|_{1,1,\widehat{T}} \\ &\leqslant C \|B\| \left| \left(\prod_{j=1}^M u_j \right) \right|_{1,1,T}, \end{aligned}$$

where $\|\bullet\|$ is the matrix norm induced by the Euclidean vector norm. Noticing that

$$\|B\| \leqslant Ch_T$$

and that

$$\int_T \left|\mathbf{grad}\left(\prod_{j=1}^M u_j\right)\right| \mathrm{d}x \leqslant \sum_{\sigma\in\Delta}\int_T |\mathbf{grad}\, u_{\sigma_1}|\left|\prod_{l=2}^M u_{\sigma_l}\right| \mathrm{d}x$$
$$\leqslant \sum_{\sigma\in\Delta}\left(\|\mathbf{grad}\, u_{\sigma_1}\|_{0,T}\prod_{l=2}^M \|u_{\sigma_l}\|_{0,2(M-1),T}\right)$$

with Δ the set of circular permutations of $\{1,2,\ldots,M\}$, we write successively

$$\left|\int_\Omega \left(r_h\left(\prod_{j=1}^M u_j\right) - \left(\prod_{j=1}^M u_j\right)\right)\mathrm{d}x\right|$$
$$\leqslant Ch\sum_{T\in\mathcal{T}_h}\left(\sum_{\sigma\in\Delta}|u_{\sigma_1}|_{1,T}\prod_{l=2}^M\|u_{\sigma_l}\|_{0,2(M-1),T}\right)$$
$$\leqslant Ch\sum_{\sigma\in\Delta}|u_{\sigma_1}|_{1,\Omega}\prod_{l=2}^M\|u_{\sigma_l}\|_{0,2(M-1),\Omega}$$
$$\leqslant \widetilde{C}h\sum_{\sigma\in\Delta}\left(\prod_{j=1}^M\|u_{\sigma_j}\|_{1,\Omega}\right) = \widetilde{C}hM\prod_{j=1}^M\|u_j\|_{1,\Omega}$$

and Theorem 8.3 is proved. □

As a consequence of Theorem 6.1, we get the following result.

THEOREM 8.4. *In the framework developed above, assume that f is in $W^{1,p}(\Omega)$ for some $p > 2$. Then there exist positive constants h_0, δ_0, C, and for $h \leqslant h_0$ a unique solution $\tilde{u}_h$ to* (8.17) *(or equivalently to* (3.14)) *in $\overline{B}(u,\delta_0)$. Moreover, the following a priori estimate holds*

$$\|u - \tilde{u}_h\|_{1,\Omega} \leqslant Ch. \tag{8.18}$$

PROOF. The result will be a consequence of Theorem 6.1. Thus we shall check the hypotheses (6.1), (6.2) and (6.3). Note that for $v \in H_0^1(\Omega)$, the derivative $\mathrm{D}F_h(v) \in \mathcal{L}(H_0^1(\Omega); H^{-1}(\Omega))$ is given by

for all $w_1, w_2 \in H_0^1(\Omega)$,

$$\langle \mathrm{D}F_h(v)w_1, w_2\rangle = \int_\Omega \mathbf{grad}\, w_1\ \mathbf{grad}\, w_2\,\mathrm{d}x + 3\int_\Omega r_h\big[(\pi_h v)^2\pi_h w_1 \pi_h w_2\big]\,\mathrm{d}x,$$

with the notation $\langle\bullet,\bullet\rangle = \langle\bullet,\bullet\rangle_{H^{-1}(\Omega)H_0^1(\Omega)}$.

Let us first check (6.1) with $L_h =$ constant. By the definitions for $v \in H_0^1(\Omega)$ we have

$$\|\mathrm{D}F_h(u) - \mathrm{D}F_h(v)\|_{H_0^1(\Omega);H^{-1}(\Omega)} = \sup_{\substack{w_1 \in H_0^1(\Omega) \\ |w_1|_{1,\Omega} \leqslant 1}} \sup_{\substack{w_2 \in H_0^1(\Omega) \\ |w_2|_{1,\Omega} \leqslant 1}} \langle [\mathrm{D}F_h(u) - \mathrm{D}F_h(v)] w_1, w_2 \rangle .$$

Let v, w_1, w_2 in $H_0^1(\Omega)$, then following the technique used in the proof of Theorem 8.3, we get

$$\begin{aligned} &\langle [\mathrm{D}F_h(u) - \mathrm{D}F_h(v)] w_1, w_2 \rangle \\ &\quad = 3 \int_\Omega r_h \big[\big((\pi_h u)^2 - (\pi_h v)^2 \big) \pi_h w_1 \pi_h w_2 \big] \,\mathrm{d}x \\ &\quad \leqslant C \|\pi_h (u - v)\|_{0,4,\Omega} \|\pi_h (u + v)\|_{0,4,\Omega} \|\pi_h w_1\|_{0,4,\Omega} \|\pi_h w_2\|_{0,4,\Omega}. \end{aligned}$$

Since the $L^4(\Omega)$ norm is bounded by the $H^1(\Omega)$ norm and $\pi_h \in \mathscr{L}(H_0^1(\Omega); H_0^1(\Omega))$ is bounded with respect to h, we finally get for $v \in \overline{B}(u, 1)$

$$\|\mathrm{D}F_h(u) - \mathrm{D}F_h(v)\|_{H_0^1(\Omega);H^{-1}(\Omega)} \leqslant C |u - v|_{1,\Omega}, \tag{8.19}$$

which implies (6.1) with $L_h = C$ and $\eta = 1$.

To check hypothesis (6.3), we notice that, since $\mathrm{D}F(u) \in \mathscr{L}(H_0^1(\Omega); H^{-1}(\Omega))$ is an isomorphism, it is sufficient to check that $\lim_{h \to 0} \mathrm{D}F_h(u) = \mathrm{D}F(u)$. For $w_1, w_2 \in H_0^1(\Omega)$ we have

$$\begin{aligned} &\langle [\mathrm{D}F_h(u) - \mathrm{D}F(u)] w_1, w_2 \rangle \\ &\quad = 3 \int_\Omega r_h \big[(\pi_h u)^2 \pi_h w_1 \pi_h w_2 \big] \,\mathrm{d}x - 3 \int_\Omega (\pi_h u)^2 \pi_h w_1 \pi_h w_2 \,\mathrm{d}x \\ &\qquad + 3 \int_\Omega (\pi_h u)^2 \pi_h w_1 \pi_h w_2 \,\mathrm{d}x - 3 \int_\Omega u^2 w_1 w_2 \,\mathrm{d}x. \end{aligned}$$

Making use of Theorem 8.3 and of the fact that $\pi_h \in \mathscr{L}(H_0^1(\Omega); H_0^1(\Omega))$ is uniformly bounded with respect to h, we can prove that the first difference in the right-hand side of the last equation tends to zero. The second difference reads

$$\begin{aligned} &\int_\Omega (\pi_h u)^2 \pi_h w_1 \pi_h w_2 \,\mathrm{d}x - \int_\Omega u^2 w_1 w_2 \,\mathrm{d}x \\ &\quad = \int_\Omega (\pi_h u - u)(\pi_h u + u) w_1 w_2 \,\mathrm{d}x \\ &\qquad + \int_\Omega (\pi_h u)^2 (\pi_h w_1 - w_1) w_2 \,\mathrm{d}x + \int_\Omega (\pi_h u)^2 \pi_h w_1 (\pi_h w_2 - w_2) \,\mathrm{d}x \end{aligned}$$

$$\leqslant \|u-\pi_h u\|_{0,4,\Omega}\|u+\pi_h u\|_{0,4,\Omega}\|w_1\|_{0,4,\Omega}\|w_2\|_{0,4,\Omega}+\|\pi_h u\|^2_{0,4,\Omega}\|w_1 - \pi_h w_1\|_{0,4,\Omega}\|w_2\|_{0,4,\Omega}+\|\pi_h u\|^2_{0,4,\Omega}\|\pi_h w_1\|_{0,4,\Omega}\|\pi_h w_2 - w_2\|_{0,4,\Omega}.$$

The family of triangulations being regular, we can check that

$$\lim_{h\to 0}\ \sup_{\substack{w\in H_0^1(\Omega)\\ w\neq 0}} \frac{\|w-\pi_h w\|_{0,4,\Omega}}{|w|_{1,\Omega}} = 0.$$

So hypothesis (6.3) has been verified.

Let us finally check hypothesis (6.2). For $v \in H_0^1(\Omega)$ with $|v|_{1,\Omega} \leqslant 1$

$$\begin{aligned}\langle F_h(u), v\rangle &= \int_\Omega \mathbf{grad}\, u\, \mathbf{grad}\, v\, \mathrm{d}x + \int_\Omega r_h\big[\big((\pi_h u)^3 - f\big)\pi_h v\big]\, \mathrm{d}x\\ &= \int_\Omega \big\{r_h\big[(\pi_h u)^3\pi_h v\big] - (\pi_h u)^3\pi_h v\big\}\, \mathrm{d}x + \int_\Omega \big[(\pi_h u)^3 - u^3\big]\pi_h v\, \mathrm{d}x\\ &\quad - \int_\Omega \big\{r_h\big[f\pi_h v\big] - f\pi_h v\big\}\, \mathrm{d}x + \int_\Omega \mathbf{grad}(u-\pi_h u)\, \mathbf{grad}\, v\, \mathrm{d}x.\end{aligned}$$

We study successively all four terms in the last equality. We have

$$\left|\int_\Omega \mathbf{grad}(u-\pi_h u)\, \mathbf{grad}\, v\, \mathrm{d}x\right| \leqslant |u-\pi_h u|_{1,\Omega} \leqslant Ch. \tag{8.20}$$

For the third difference we notice that, with $v_h = \pi_h v$,

$$\begin{aligned}\int_\Omega \big[r_h(fv_h) - fv_h\big]\, \mathrm{d}x &= \int_\Omega \big[r_h(r_h f v_h) - fv_h\big]\, \mathrm{d}x\\ &= \int_\Omega \big[r_h(r_h f v_h) - r_h f v_h\big]\, \mathrm{d}x + \int_\Omega (r_h f - f)v_h\, \mathrm{d}x\end{aligned}$$

so with Theorem 8.3 we get the estimate

$$\begin{aligned}\left|\int_\Omega \big[r_h(fv_h) - fv_h\big]\, \mathrm{d}x\right| &\leqslant Ch\|r_h f\|_{1,\Omega}\|v_h\|_{1,\Omega} + \|f - r_h f\|_{0,\Omega}\|v_h\|_{0,\Omega}\\ &\leqslant Ch|v_h|_{1,\Omega}.\end{aligned} \tag{8.21}$$

The second difference can be estimated in the following way

$$\left|\int_\Omega \big[(\pi_h u)^3 - u^3\big]\pi_h v\, \mathrm{d}x\right| \leqslant Ch, \tag{8.22}$$

where C depends on u but is independent of h. As a consequence of Theorem 8.3, an estimate for the first difference reads

$$\left|\int_{\Omega} \{r_h[(\pi_h u)^3 \pi_h v] - (\pi_h u)^3 \pi_h v\} \, dx\right| \leqslant Ch. \tag{8.23}$$

As a consequence of (8.20) through (8.23), we get

$$\|F_h(u)\|_{-1,\Omega} \leqslant Ch, \tag{8.24}$$

where C is independent of h (but actually depends on u).

We apply Theorem 6.1 and estimate (8.24) with (6.5) to conclude. □

For a model problem we have developed a method to get a priori and a posteriori error estimates in the cases of Galerkin approximation without and with numerical integration. The interest of the technique lies in its ability to be adapted to other cases.

9. Estimates for an approximation of the Navier–Stokes problem

The abstract results of Section 7 are applied now to the commonly called $(4\mathscr{P}_1 - \mathscr{P}_1)$ approximation of the Navier–Stokes problem introduced in Section 4.

In this case we are using the following notations:

$$L_0^2(\Omega) = \left\{f \in L^2(\Omega); \int_{\Omega} f \, dx = 0\right\},$$

$$X = Y = \left(H_0^1(\Omega)\right)^2 \times L_0^2(\Omega),$$

and $F : X \to X'$ is defined for $\underline{u} = (\boldsymbol{u}, p) \in X$,

$$\begin{aligned}
&\text{for all } \underline{v} = (\boldsymbol{v}, q) \in X, \\
&\quad \langle F(\boldsymbol{u}, p), (\boldsymbol{v}, q)\rangle_{X'X} \\
&\qquad = \nu \int_{\Omega} \mathbf{grad}\, \boldsymbol{u}\, \mathbf{grad}\, \boldsymbol{v} \, dx + \int_{\Omega} (\boldsymbol{u}\boldsymbol{\nabla})\boldsymbol{u}\boldsymbol{v} \, dx - \int_{\Omega} p \operatorname{div} \boldsymbol{v} \, dx \\
&\qquad\quad - \int_{\Omega} \boldsymbol{f}\boldsymbol{v} \, dx - \int_{\Omega} q \operatorname{div} \boldsymbol{u} \, dx \\
&\qquad \equiv a(\boldsymbol{u}, \boldsymbol{v}) + c(\boldsymbol{u}; \boldsymbol{u}, \boldsymbol{v}) + b(\boldsymbol{v}, p) + b(\boldsymbol{u}, q) - \int_{\Omega} \boldsymbol{f}\boldsymbol{v} \, dx.
\end{aligned} \tag{9.1}$$

Let $\underline{u} = (\boldsymbol{u}, p)$ denote a regular solution to $F(\underline{v}) = 0$, in the sense that (4.8) is satisfied.

The approximation spaces and the test spaces are $X_h \equiv Y_h = V_h^2 \times W_h$, where the finite element spaces V_h and W_h are defined in Section 4. A Galerkin approximation of the problem $F(\underline{v}) = 0$ reads (see (4.9)): find $\underline{u}_h \in X_h$ such that

$$\text{for all } \underline{v}_h \in X_h, \quad \langle F(\underline{u}_h), \underline{v}_h \rangle_{X'X} = 0. \tag{9.2}$$

THEOREM 9.1. *Let $\underline{u} = (\boldsymbol{u}, p) \in X$ be a solution to $F(\underline{v}) = 0$ with $\boldsymbol{u} \in (H^2(\Omega))^2$. We assume that hypothesis* (4.8) *and assumption* (4.10) *on the triangulations are satisfied. Then there exist constants $h_0 > 0$, $\delta_0 > 0$, $C > 0$, and for $h \leqslant h_0$ a unique solution $\underline{u}_h = (\boldsymbol{u}_h, p_h) \in X_h$ to problem* (9.2) *satisfying $\|\underline{u} - \underline{u}_h\|_X \leqslant \delta_0$. Furthermore, the following estimates hold*

$$|\boldsymbol{u} - \boldsymbol{u}_h|_{1,\Omega} + \|p - p_h\|_{0,\Omega} \leqslant C \inf_{(\boldsymbol{v}_h, q_h) \in X_h} \big(|\boldsymbol{u} - \boldsymbol{v}_h|_{1,\Omega} + \|p - q_h\|_{0,\Omega}\big) \tag{9.3}$$

and

$$|\boldsymbol{u} - \boldsymbol{u}_h|_{1,\Omega} + \|p - p_h\|_{0,\Omega} \leqslant C\Big(\sum_{T \in \mathscr{T}_h^f} \eta(T)^2\Big)^{1/2}, \tag{9.4}$$

where for any triangle $T \in \mathscr{T}_h^f$ the estimator $\eta(T)$ is expressed by

$$\eta(T)^2 = h_T^2 \big\| -\nu \Delta \boldsymbol{u}_h + (\boldsymbol{u}_h \boldsymbol{\nabla})\boldsymbol{u}_h + \mathbf{grad}\, p_h - \boldsymbol{f} \big\|_{0,T}^2 + \sum_{i=1}^{3} h_{t_i} \left\| \left[\frac{\partial \boldsymbol{u}_h}{\partial n} \right] \right\|_{0,t_i}^2 + \|\operatorname{div} \boldsymbol{u}_h\|_{0,T}^2$$

with h_T the diameter of T, t_i, $i = 1, 2, 3$, the edges of T of length h_{t_i}, and $[\bullet]$ the jump through an edge.

PROOF. We will first check that we are in a position to apply Theorem 7.1. Since we assume that $\underline{u} = (\boldsymbol{u}, p) \in X$ satisfies (4.8), hypothesis (7.1) is satisfied. Assumption (7.4) is verified in Theorem 4.1 while (7.5) and (7.6) are easy to check with β_h independent of h. We can apply Theorem 7.1 with β_h constant.

Estimate (9.3) is immediately deduced from (7.7) while from (7.8) we shall deduce the estimate (9.4). In our case (7.8) reads

$$|\boldsymbol{u} - \boldsymbol{u}_h|_{1,\Omega} + \|p - p_h\|_{0,\Omega} \leqslant C \big\| F(\boldsymbol{u}_h, p_h) \big\|_{X'}. \tag{9.5}$$

For $\underline{v} = (\boldsymbol{v}, q) \in X$ and $\underline{v}_h = (\boldsymbol{v}_h, q_h) \in X_h$ with $\boldsymbol{v}_h = \tilde{r}_h \boldsymbol{v}$, where $\tilde{r}_h \in \mathscr{L}((H_0^1(\Omega))^2; V_h^2)$ is the Clément interpolation operator for discontinuous functions, see CLÉMENT [1975], we have

$$\begin{aligned}
&\langle F(\underline{u}_h), \underline{v}\rangle_{X'X} \\
&\quad = \langle F(\underline{u}_h), \underline{v} - \underline{v}_h\rangle_{X'X} = a(\boldsymbol{u}_h, \boldsymbol{v} - \boldsymbol{v}_h) + c(\boldsymbol{u}_h; \boldsymbol{u}_h, \boldsymbol{v} - \boldsymbol{v}_h) \\
&\qquad + b(\boldsymbol{v} - \boldsymbol{v}_h, p_h) + b(\boldsymbol{u}_h, q - q_h) - \int_\Omega \boldsymbol{f}(\boldsymbol{v} - \boldsymbol{v}_h)\,\mathrm{d}x \\
&\quad = \sum_{T \in \mathscr{T}_h^f} \bigg\{ \int_T \big(-\nu\Delta \boldsymbol{u}_h + (\boldsymbol{u}_h \boldsymbol{\nabla})\boldsymbol{u}_h + \mathbf{grad}\, p_h - \boldsymbol{f}\big)(\boldsymbol{v} - \boldsymbol{v}_h)\,\mathrm{d}x \\
&\qquad - \int_T \operatorname{div} \boldsymbol{u}_h (q - q_h)\,\mathrm{d}x + \nu \int_{\partial T} \frac{\partial \boldsymbol{u}_h}{\partial n}(\boldsymbol{v} - \boldsymbol{v}_h)\,\mathrm{d}s \bigg\}.
\end{aligned} \tag{9.6}$$

Let us bound each of these terms. Setting $\boldsymbol{w} = -\nu\Delta\boldsymbol{u}_h + (\boldsymbol{u}_h\boldsymbol{\nabla})\boldsymbol{u}_h + \mathbf{grad}\, p_h - \boldsymbol{f}$ on T, we get successively, since $\boldsymbol{v}_h = \tilde{r}_h \boldsymbol{v}$,

$$\int_T \boldsymbol{w}(\boldsymbol{v} - \boldsymbol{v}_h)\,\mathrm{d}x \leqslant \|\boldsymbol{w}\|_{0,T}\|\boldsymbol{v} - \boldsymbol{v}_h\|_{0,T} \leqslant C\|\boldsymbol{w}\|_{0,T} h \sum_{T' \in S_T} \|\boldsymbol{v}\|_{1,T'}$$

where S_T is the set of triangles T' in $\mathscr{T}_h^f$ with $T' \cap T \neq \emptyset$, and summing over all the elements

$$\begin{aligned}
\bigg| \sum_{T\in\mathscr{T}_h^f} \int_T \boldsymbol{w}(\boldsymbol{v} - \boldsymbol{v}_h)\,\mathrm{d}x \bigg| &\leqslant C \bigg(\sum_{T\in\mathscr{T}_h^f} h_T^2 \|\boldsymbol{w}\|_{0,T}^2 \bigg)^{1/2} \bigg(\sum_{T\in\mathscr{T}_h^f} \|\boldsymbol{v}\|_{1,T}^2 \bigg)^{1/2} \\
&\leqslant \widetilde{C} |\boldsymbol{v}|_{1,\Omega} \bigg(\sum_{T\in\mathscr{T}_h^f} h_T^2 \|\boldsymbol{w}\|_{0,T}^2 \bigg)^{1/2}.
\end{aligned} \tag{9.7}$$

We use the same techniques as the ones developed in the proof of Theorem 8.1 to estimate

$$\bigg| \sum_{T\in\mathscr{T}_h^f} \nu \int_{\partial T} \frac{\partial \boldsymbol{u}_h}{\partial n}(\boldsymbol{v} - \tilde{r}_h\boldsymbol{v})\,\mathrm{d}s \bigg| \leqslant C|\boldsymbol{v}|_{1,\Omega} \bigg(\sum_{t\in S_h} h_t \bigg\| \bigg[\frac{\partial \boldsymbol{u}_h}{\partial n} \bigg] \bigg\|_{0,t}^2 \bigg)^{1/2} \tag{9.8}$$

where S_h is the set of all sides t of triangles in $\mathscr{T}_h^f$. Since q is only a $L_0^2(\Omega)$ function, we take $q_h = 0$ and so

$$\bigg| \sum_{T\in\mathscr{T}_h^f} \int_T \operatorname{div} \boldsymbol{u}_h (q - q_h)\,\mathrm{d}x \bigg| \leqslant \|q\|_{0,\Omega} \bigg(\sum_{T\in\mathscr{T}_h^f} \|\operatorname{div} \boldsymbol{u}_h\|_{0,T}^2 \bigg)^{1/2}. \tag{9.9}$$

From estimates (9.7)–(9.9) we get the following bound for (9.6) noticing that $\Delta \boldsymbol{u}_h = 0$ on T,

$$\begin{aligned} &|\langle F(\underline{u}_h), \underline{v}\rangle_{X'X}| \\ &\leqslant C\Bigg\{ \sum_{T\in\mathcal{T}_h^f} \Bigg[h_T^2 \|(\boldsymbol{u}_h\boldsymbol{\nabla})\boldsymbol{u}_h + \mathbf{grad}\, p_h - \boldsymbol{f}\|_{0,T}^2 \\ &\quad + \sum_{i=1}^{3} h_{t_i} \Bigg\| \Bigg[\frac{\partial \boldsymbol{u}_h}{\partial n}\Bigg] \Bigg\|_{0,t_i}^2 + \|\operatorname{div} \boldsymbol{u}_h\|_{0,T}^2 \Bigg] \Bigg\}^{1/2} \|(\boldsymbol{v}, q)\|_X. \end{aligned} \tag{9.10}$$

Estimates (9.10) and (9.5) imply the error estimate (9.4). □

10. Error estimates for the stationary heat problem

The abstract results of Section 7 are applied now to the approximation (5.18) of the stationary heat problem with convection. Our framework is well-suited to study this effective Galerkin approximation.

We are using the following notations: $X = W_0^{1,p}(\Omega)$, $Y = W_0^{1,q}(\Omega)$ with p, q in $\mathbb{R}$, $p > 2$ and $1/p + 1/q = 1$, and $F: X \to Y'$ defined for $v \in W_0^{1,p}(\Omega)$ by for all $w \in W_0^{1,q}(\Omega)$

$$\begin{aligned} \langle F(v), w\rangle_{W^{-1,p}(\Omega)W_0^{1,q}(\Omega)} &= \int_\Omega k(v)\, \mathbf{grad}\, v\, \mathbf{grad}\, w \,\mathrm{d}x \\ &\quad + \int_\Omega \mathbf{c}\, \mathbf{grad}\, v w \,\mathrm{d}x - \int_\Omega f w \,\mathrm{d}x. \end{aligned} \tag{10.1}$$

When the domain Ω is regular, hypotheses (5.2)–(5.4) satisfied, and f in $L^\infty(\Omega)$, then the problem $F(v) = 0$ has a unique solution $u \in W_0^{1,p}(\Omega)$, which is in $W^{2,r}(\Omega)$ with $1 \leqslant r < \infty$ and $\mathrm{D}F(u) \in \mathcal{L}(W_0^{1,p}(\Omega); W^{-1,p}(\Omega))$ is an isomorphism, see Theorem 5.2.

For the sake of simplicity, we assume that Ω is a polygonal domain such that the Laplacian operator $\Delta: W_0^{1,p}(\Omega) \to W^{-1,p}(\Omega)$ is an isomorphism. Given a family $\{\mathcal{T}_h\}_{0<h\leqslant 1}$ of triangulations, h is the maximum of the diameter of $T \in \mathcal{T}_h$, the approximation spaces and the test spaces are the spaces V_h of continuous piecewise polynomials of degree 1. The family is assumed to be regular and quasi-uniform, see CIARLET ([1978], p. 132 and p. 140). We study the Galerkin approximation: find $u_h \in V_h$ satisfying

$$\text{for all } v_h \in V_h, \quad \langle F(u_h), v_h\rangle_{W^{-1,p}(\Omega)W_0^{1,q}(\Omega)} = 0. \tag{10.2}$$

REMARK 10.1. In Theorem 5.2, we have assumed that Ω is regular, so that the Laplacian operator with homogeneous boundary conditions is an isomorphism from

$W_0^{1,p}(\Omega)$ onto $W^{-1,p}(\Omega)$, see SIMADER [1972]. Here we assume Ω polygonal so that $\overline{\Omega} = \bigcup_{T\in\mathcal{T}_h} T$ and the Laplacian operator is an isomorphism, see DAUGE [1992]. Actually when Ω is regular, we should use a family of isoparametric triangulations of Ω.

To apply Theorem 7.1, we need to prove a discrete inf–sup condition for the bilinear form $b\colon W_0^{1,p}(\Omega) \times W_0^{1,q}(\Omega) \to \mathbb{R}$ given by: for $v \in W_0^{1,p}(\Omega)$, $w \in W_0^{1,q}(\Omega)$,

$$b(v,w) = \big\langle \mathrm{D}F(u)v, w\big\rangle_{W^{-1,p}(\Omega)W_0^{1,q}(\Omega)}.$$

THEOREM 10.1. *Under the hypotheses of Theorem* 5.1, *let* $u \in W_0^{1,p}(\Omega)$ *be the unique solution to* $F(v) = 0$ *and* $\mathrm{D}F(u) \in \mathcal{L}(W_0^{1,p}(\Omega); W^{-1,p}(\Omega))$ *be an isomorphism. We assume that the family of triangulations is regular and quasi-uniform. Then there exists* $\zeta > 0$ *independent of* h *such that*

$$\inf_{\substack{v_h\in V_h\\ |v_h|_{1,p,\Omega}=1}} \ \sup_{\substack{w_h\in V_h\\ |w_h|_{1,q,\Omega}=1}} b(v_h,w_h) \geqslant \zeta. \tag{10.3}$$

PROOF. Going back to the definition of $\mathrm{D}F(u)$ in (5.12), we have for $w \in W_0^{1,p}(\Omega)$, $\psi \in W_0^{1,q}(\Omega)$

$$\begin{aligned}\big\langle \mathrm{D}F(u)w, \psi\big\rangle_{W^{-1,p}(\Omega)W_0^{1,q}(\Omega)} &= \int_\Omega \mathbf{grad}\,\big(k(u)w\big)\,\mathbf{grad}\,\psi\,\mathrm{d}x\\ &\quad + \int_\Omega \mathbf{c}\,\mathbf{grad}\,w\psi\,\mathrm{d}x.\end{aligned} \tag{10.4}$$

Let $-T \in \mathcal{L}(W^{-1,p}(\Omega); W_0^{1,p}(\Omega))$ denote the inverse of the Laplacian operator with homogeneous boundary condition, which is an isomorphism. Then (10.4) reads for $w \in W_0^{1,p}(\Omega)$,

$$T\mathrm{D}F(u)w = k(u)w + T(\mathbf{c}\,\mathbf{grad}\,w). \tag{10.5}$$

If we introduce the mapping $R\colon \varphi \in W_0^{1,p}(\Omega) \to R\varphi = \varphi/k(u) \in W_0^{1,p}(\Omega)$ which is an isomorphism, then (10.5) reads with $w = R\varphi$,

$$T\mathrm{D}F(u)R\varphi = \varphi + T\big(\mathbf{c}\,\mathbf{grad}(R\varphi)\big). \tag{10.6}$$

We have proved in Theorem 5.2 that $T\mathrm{D}F(u)R \in \mathcal{L}(W_0^{1,p}(\Omega); W_0^{1,p}(\Omega))$ is an isomorphism.

Let $T_h \in \mathcal{L}(W^{-1,p}(\Omega); V_h)$ be the discrete analogue of T defined for $g \in W^{-1,p}(\Omega)$ by

$$\text{for all } v_h \in V_h, \quad \int_\Omega \mathbf{grad}(T_h g)\,\mathbf{grad}\,v_h\,\mathrm{d}x = \int_\Omega g v_h\,\mathrm{d}x.$$

We easily check that for $\varphi_h \in V_h$,

$$T_h \mathrm{D}F(u)R\varphi_h = \varphi_h + T_h\big(\mathbf{c}\,\mathbf{grad}(R\varphi_h)\big).$$

The family of triangulations is regular and quasi-uniform, so there exists a constant $C > 0$ such that for all $w \in W^{2,p}(\Omega)$

$$\|w - \pi_h w\|_{1,p,\Omega} \leqslant Ch\|w\|_{2,p,\Omega},$$

where $\pi_h \in \mathscr{L}(W^{1,p}(\Omega); V_h)$ is the elliptic projector, see RANNACHER and SCOTT [1982].

Since we have $T_h = \pi_h T$, we get for $\varphi_h \in V_h$,

$$T_h \mathrm{D}F(u)R\varphi_h = T\,\mathrm{D}F(u)R\varphi_h + (T_h - T)\big(\mathbf{c}\,\mathbf{grad}(R\varphi_h)\big)$$

and so there exists a constant γ_1 independent of h such that

$$\big|T_h \mathrm{D}F(u)R\varphi_h\big|_{1,p,\Omega} \geqslant \gamma_1|\varphi_h|_{1,p,\Omega} - \varepsilon_h\big\|\mathbf{c}\,\mathbf{grad}(R\varphi_h)\big\|_{0,p,\Omega}$$

with ε_h tending to zero for h tending to zero, independent of φ_h. So for h small enough

$$\big|T_h \mathrm{D}F(u)R\varphi_h\big|_{1,p,\Omega} \geqslant \frac{\gamma_1}{2}|\varphi_h|_{1,p,\Omega}. \tag{10.7}$$

We are working under the assumption that the Laplacian operator $\Delta\colon W_0^{1,p}(\Omega) \to W^{-1,p}(\Omega)$ is an isomorphism, where for all $v \in W_0^{1,p}(\Omega)$, for all $w \in W_0^{1,q}(\Omega)$,

$$\langle \Delta v, w\rangle_{W^{-1,p}(\Omega)W_0^{1,q}(\Omega)} = \int_\Omega \mathbf{grad}\,v\,\mathbf{grad}\,w\,\mathrm{d}x.$$

So we can check there exists a $\gamma_2 > 0$ with

$$\inf_{\substack{v\in W_0^{1,p}(\Omega)\\ |v|_{1,p,\Omega}=1}} \ \sup_{\substack{w\subset W_0^{1,q}(\Omega)\\ |w|_{1,q,\Omega}=1}} \int_\Omega \mathbf{grad}\,v\,\mathbf{grad}\,w\,\mathrm{d}x \geqslant \gamma_2 > 0.$$

The elliptic projector $\pi_h \in \mathscr{L}(W_0^{1,q}(\Omega); V_h)$ is stable in $W_0^{1,q}(\Omega)$, which means there exists a constant $C > 0$ such that for all $v \in W_0^{1,q}(\Omega)$,

$$|\pi_h v|_{1,q,\Omega} \leqslant C|v|_{1,q,\Omega},$$

see RANNACHER and SCOTT [1982]. Let v_h be in V_h with $|v_h|_{1,p,\Omega} = 1$. Then there exists a function $w \in W_0^{1,q}(\Omega)$ with $|w|_{1,q,\Omega} = 1$ such that

$$\frac{\gamma_2}{2} \leqslant \int_\Omega \mathbf{grad}\, v_h\, \mathbf{grad}\, w\, \mathrm{d}x = \int_\Omega \mathbf{grad}\, v_h\, \mathbf{grad}\, \pi_h w\, \mathrm{d}x.$$

So

$$\int_\Omega \mathbf{grad}\, v_h\, \mathbf{grad}\, \frac{\pi_h w}{|\pi_h w|_{1,q,\Omega}}\, \mathrm{d}x \geqslant \frac{\gamma_2}{2|\pi_h w|_{1,q,\Omega}}.$$

With the stability result of π_h, we conclude there exists a constant $\gamma_3 > 0$ independent of h such that

$$\inf_{\substack{v_h \in V_h \\ |v_h|_{1,p,\Omega}=1}} \sup_{\substack{w_h \in V_h \\ |w_h|_{1,q,\Omega}=1}} \int_\Omega \mathbf{grad}\, v_h\, \mathbf{grad}\, w_h\, \mathrm{d}x \geqslant \gamma_3. \tag{10.8}$$

Let the function φ_h be in V_h, then with estimates (10.7) and (10.8) we get

$$\begin{aligned}
\sup_{\substack{w_h \in V_h \\ |w_h|_{1,q,\Omega}=1}} b(R\varphi_h, w_h) &= \sup_{\substack{w_h \in V_h \\ |w_h|_{1,q,\Omega}=1}} \langle \mathrm{D}F(u)R\varphi_h, w_h \rangle_{W^{-1,p}(\Omega)W_0^{1,q}(\Omega)} \\
&= \sup_{\substack{w_h \in V_h \\ |w_h|_{1,q,\Omega}=1}} \int_\Omega \mathbf{grad}\, \big(T_h \mathrm{D}F(u)R\varphi_h\big)\, \mathbf{grad}\, w_h\, \mathrm{d}x \\
&\geqslant \gamma |\varphi_h|_{1,p,\Omega},
\end{aligned}$$

and so for all $\varphi_h \in V_h$,

$$\sup_{\substack{w_h \in V_h \\ |w_h|_{1,q,\Omega}=1}} b(R\varphi_h, w_h) \geqslant \gamma |\varphi_h|_{1,p,\Omega}. \tag{10.9}$$

Using hypotheses (5.2) and (5.3) on k, the regularity of the triangulations and the regularity of the solution $u \in W^{2,r}(\Omega)$, $1 \leqslant r < \infty$, we get by standard calculations on the reference element

$$\lim_{h \to 0} \max_{\substack{\chi_h \in V_h \\ |\chi_h|_{1,p,\Omega}=1}} \big\| R^{-1}\chi_h - r_h\big(R^{-1}\chi_h\big) \big\|_{1,p,\Omega} = 0, \tag{10.10}$$

where r_h is the Lagrange interpolation operator.

Finally estimates (10.9) and (10.10) imply there exists $\zeta > 0$ such that

$$\inf_{\substack{v_h \in V_h \\ |v_h|_{1,p,\Omega}=1}} \sup_{\substack{w_h \in V_h \\ |w_h|_{1,q,\Omega}=1}} b(v_h, w_h) \geqslant \zeta$$

and the proof is complete. □

We are in a position to prove the main result of this section.

THEOREM 10.2. *We assume the hypotheses of Theorem 5.1 and denote by* $u \in W_0^{1,p}(\Omega)$, $2 < p < \infty$, *the unique solution to the problem* $F(v) = 0$. *Furthermore, we assume that the family* $\{\mathcal{T}_h\}_{0<h\leqslant 1}$ *is regular and quasi-uniform. Then there exist positive constants* h_0, δ_0, C, *and for* $h \leqslant h_0$ *a unique solution* u_h *to problem* (10.2) *in the closed ball* $\overline{B}(u, \delta_0)$. *Moreover, the following estimates hold*

$$|u - u_h|_{1,p,\Omega} \leqslant Ch \tag{10.11}$$

and

$$|u - u_h|_{1,p,\Omega} \leqslant C\Big(\sum_{T\in\mathcal{T}_h} \eta(T)^p\Big)^{1/p}, \tag{10.12}$$

where for any $T \in \mathcal{T}_h$ *the estimator* $\eta(T)$ *is expressed by*

$$\eta(T) = h_T\big\| - \operatorname{div}\big(k(u_h)\,\mathbf{grad}\,u_h\big) + \mathbf{c}\,\mathbf{grad}\,u_h - f\big\|_{0,p,T} + h_T^{1-2/q}\sum_{i=1}^{3} h_{t_i}^{1/q}\Big\|\Big[k(u_h)\frac{\partial u_h}{\partial n}\Big]\Big\|_{0,p,t_i},$$

where h_T *is the diameter of the triangle* $T \in \mathcal{T}_h$, h_{t_i} *is the length of the edge* t_i *of the triangle* T, $i = 1, 2, 3$, *and* $[\bullet]$ *the jump across the considered edge (we adopt the convention of a zero function outside* $\overline{\Omega}$).

PROOF. Let us prove that we are in a position to apply Theorem 7.1. Hypothesis (7.1) is verified in Theorems 5.1 and 5.2. The inf–sup condition (7.4) is verified in Theorem 10.1 with $\beta_h = \beta$ independent of h while hypothesis (7.5) is satisfied. With the classical interpolation result, we can verify (7.6). With assumptions (5.2) and (5.3) on k, we easily check that DF is Lipschitzian at u.

The hypotheses of Theorem 7.1 are verified. The proof is complete once we have proved the following two error estimates:

$$\inf_{v_h\in V_h} |u - v_h|_{1,p,\Omega} \leqslant Ch \tag{10.13}$$

and

$$\|F(u_h)\|_{-1,p,\Omega} \leqslant C\Big(\sum_{T\in\mathcal{T}_h} \eta(T)^p\Big)^{1/p}. \tag{10.14}$$

Estimate (10.13) is a consequence of the interpolation result, for $v \in W^{2,p}(\Omega)$,

$$|v - r_h v|_{1,p,\Omega} \leqslant Ch\|v\|_{2,p,\Omega}.$$

Let us estimate now the term $\|F(u_h)\|_{-1,p,\Omega}$ with the same method as used in the proof of Theorem 8.1. For $w \in W_0^{1,q}(\Omega)$ and $w_h \in V_h$ we have

$$\begin{aligned}
&\langle F(u_h), w\rangle_{W^{-1,p}(\Omega)W_0^{1,q}(\Omega)} \\
&\quad = \langle F(u_h), w - w_h\rangle_{W^{-1,p}(\Omega)W_0^{1,q}(\Omega)} \\
&\quad = \sum_{T\in\mathcal{T}_h} \Big\{ \int_T k(u_h)\, \mathbf{grad}\, u_h\, \mathbf{grad}(w - w_h)\, \mathrm{d}x + \int_T \mathbf{c}\, \mathbf{grad}\, u_h\, (w - w_h)\, \mathrm{d}x \\
&\qquad - \int_T f(w - w_h)\, \mathrm{d}x \Big\} \\
&\quad = \sum_{T\in\mathcal{T}_h} \Big\{ \int_T \big[-\operatorname{div}\big(k(u_h)\, \mathbf{grad}\, u_h\big) + \mathbf{c}\, \mathbf{grad}\, u_h - f \big](w - w_h)\, \mathrm{d}x \\
&\qquad + \int_{\partial T} k(u_h) \frac{\partial u_h}{\partial n_T}(w - w_h)\, \mathrm{d}s \Big\},
\end{aligned}$$

where $\partial u_h/\partial n_T$ is the exterior normal derivative of u_h on the boundary ∂T of T. Using the Hölder inequality we get

$$\begin{aligned}
&\langle F(u_h), w\rangle_{W^{-1,p}(\Omega)W_0^{1,q}(\Omega)} \\
&\quad \leqslant \sum_{T\in\mathcal{T}_h} \Big\{ \big\| -\operatorname{div}\big(k(u_h)\, \mathbf{grad}\, u_h\big) + \mathbf{c}\, \mathbf{grad}\, u_h - f \big\|_{0,p,T} \|w - w_h\|_{0,q,T} \\
&\qquad + \sum_{i=1}^{3} \Big\| \Big[k(u_h) \frac{\partial u_h}{\partial n} \Big] \Big\|_{0,p,t_i} \|w - w_h\|_{0,q,t_i} \Big\}
\end{aligned} \tag{10.15}$$

with $[\partial u_h/\partial n]$ the jump of a normal derivative of u_h on the side t_i of T with the convention $u_h = 0$ outside $\overline{\Omega}$.

We choose in inequality (10.15) $w_h = \tilde{r}_h w$ where $\tilde{r}_h$ is the Clément interpolation operator, see CLÉMENT [1975]. Let $S_T = \{T' \in \mathcal{T}_h\colon\ T' \cap T \neq \emptyset\}$. Then for all functions $v \in W^{1,q}(\bigcup_{T'\in S_T} T')$ the following interpolation results hold

$$\|v - \tilde{r}_h v\|_{0,q,T} \leqslant C h_T \sum_{T'\in S_T} \|v\|_{1,q,T'}$$

and

$$\|v - \tilde{r}_h v\|_{1,q,T} \leqslant C \sum_{T'\in S_T} \|v\|_{1,q,T'}.$$

Using the technique of the reference element and the above interpolation result, it is a simple matter to prove there exists a constant C such that for all $v \in W^{1,q}(\bigcup_{T' \in S_T} T')$,

$$\|v - \tilde{r}_h v\|_{0,q,t_i} \leqslant C h_{t_i}^{1/q} h_T^{1-2/q} \Big(\sum_{T' \in S_T} \|v\|_{1,q,T'} \Big).$$

It follows that

$$\begin{aligned} &\langle F(u_h), w \rangle_{W^{-1,p}(\Omega) W_0^{1,q}(\Omega)} \\ &\quad \leqslant C \Big(\sum_{T \in \mathcal{T}_h} \Big\{ h_T \| - \operatorname{div}\big(k(u_h)\, \mathbf{grad}\, u_h\big) + \mathbf{c}\, \mathbf{grad}\, u_h - f \|_{0,p,T} \\ &\qquad + \sum_{i=1}^{3} h_{t_i}^{1/q} h_T^{1-2/q} \Big\| \Big[k(u_h) \frac{\partial u_h}{\partial n} \Big] \Big\|_{0,p,t_i} \Big\}^p \Big)^{1/p} |w|_{1,q,\Omega}, \end{aligned}$$

which completes the proof. □

REMARK 10.2. Actually problem (10.2) is not solved as it is. The integrals are not computed exactly but with numerical quadrature rules. For the sake of simplicity we have analyzed an approximate problem without numerical integration, so that we could apply Theorem 7.1.

REMARK 10.3. We can obtain a priori and a posteriori error estimates in the norm $|\bullet|_{1,\Omega}$ similar to estimates (10.11) and (10.12) with the method developed in Theorem 6.4. Here we have $X = W_0^{1,p}(\Omega)$ with $p > 2$, $Z = W^{-1,p}(\Omega)$. Then let $\mathscr{X} = H_0^1(\Omega)$ and $\mathscr{Z} = H^{-1}(\Omega)$. If q denotes the conjugate of p, the following inclusions hold

$$X \subset \mathscr{X} \subset W_0^{1,q}(\Omega) \subset L^2(\Omega) \subset Z \subset \mathscr{Z}.$$

Considering expression (5.12) for $DF(w)$ with $w \in W_0^{1,p}(\Omega)$, $2 < p < \infty$, we easily see that $DF(w) \in \mathscr{L}(W_0^{1,p}(\Omega); W^{-1,p}(\Omega))$ admits a continuous extension $DF(w) \in \mathscr{L}(H_0^1(\Omega); H^{-1}(\Omega))$ and the mapping

$$w \in W_0^{1,p}(\Omega) \to DF(w) \in \mathscr{L}\big(H_0^1(\Omega); H^{-1}(\Omega)\big)$$

is continuous.

Let us check now that the mapping $DF(u) \in \mathscr{L}(H_0^1(\Omega); H^{-1}(\Omega))$ is an isomorphism. We can check that the operator $TDF(u)R$ defined in the proof of Theorem 10.1, see (10.6), is a Fredholm operator with index zero from $H_0^1(\Omega)$ into itself. The density of $W_0^{1,p}(\Omega)$ into $H_0^1(\Omega)$ implies that the range of $TDF(u)R$ considered in $H_0^1(\Omega)$ is $H_0^1(\Omega)$ itself. It follows that $TDF(u)R$ is an isomorphism on $H_0^1(\Omega)$, and since $T \in \mathscr{L}(H^{-1}(\Omega); H_0^1(\Omega))$ is an isomorphism, $\mathrm{D}F(u) \in \mathscr{L}(H_0^1(\Omega); H^{-1}(\Omega))$ is also an isomorphism.

As a consequence of (10.11), we have

$$\lim_{h \to 0} |u - u_h|_{1,p,\Omega} = 0.$$

We proceed as in the proof of Theorem 6.4 to get, if h is small enough, say $h \leqslant h_0$,

$$|u - u_h|_{1,\Omega} \leqslant 2 \| \mathrm{D}F(u)^{-1} \|_{H^{-1}(\Omega); H_0^1(\Omega)} \| F(u_h) \|_{-1,\Omega}.$$

Such a posteriori estimates have been used in adaptive mesh refinement techniques applied to a two-dimensional regularized Stefan problem, see PICASSO [1992], and yield excellent results.

To derive a priori error estimates in the norm $| \bullet |_{1,\Omega}$, we proceed in the same way as above but with the mapping F_h defined in (7.11).

CHAPTER IV

Parametrized Nonlinear Problems. Approximation of Regular Solutions

Generally computational applications contain parameters. In Section 3 we have presented a parameter dependent problem, see problem (3.18), and we have proved the existence of a solution path. Our goal now is to study approximations of the problem $F(\lambda, u) = 0$, where X and Z are Banach spaces and F is a mapping defined on $\mathbb{R}^m \times X$ with values in Z. For a given value of the parameter λ the study can be reduced to the one of Sections 6 and 7, where no parameter does occur. But then we can study how the approximation varies with respect to the parameter, which is precisely the object of this chapter, when the solutions are regular.

First we introduce the notions of coordinate spaces and state spaces. Then the case where the parameter space $\mathbb{R}^m$ can be used as the coordinate space is analyzed both for general and Galerkin approximations. The results are applied to the example of Section 3. When the parameter space can no more be used as the coordinate space for the regular solution path, we transform the problem by adding equations. That situation is also analyzed and our results are illustrated with the example developed in Section 3.

11. Orientation

To present parameter dependent problems, it is adequate to work in the following general framework. Let X and Z be two real Banach spaces and $F: X \to Z$ be a C^p mapping with $p \geqslant 1$. A point $x \in X$ is called regular if

$$\begin{aligned} &\text{(i)} \quad \mathrm{D}F(x) \in \mathscr{L}(X; Z) \text{ has a finite-dimensional kernel,} \\ &\text{(ii)} \quad \mathrm{D}F(x) \in \mathscr{L}(X; Z) \text{ is surjective.} \end{aligned} \tag{11.1}$$

In this chapter we are interested to analyze and compute the solution set S of the problem

$$F(x) = 0 \tag{11.2}$$

in a neighborhood of the regular point $x_0 \in X$ satisfying $F(x_0) = 0$. Such a point x_0 is called a regular solution. Here we assume that $\dim \operatorname{Ker}(\mathrm{D}F(x_0)) \geqslant 1$. The case $\dim \operatorname{Ker}(\mathrm{D}F(x_0)) = 0$ has been studied in the previous chapter.

Since the dimension of $\operatorname{Ker}(\mathrm{D}F(x_0))$ is finite this subspace is direct, that is there exists a closed subspace $X_1 \subset X$ such that $X = \operatorname{Ker}(\mathrm{D}F(x_0)) \oplus X_1$. The point x_0 is regular, so the mapping $\mathrm{D}F(x_0)|_{X_1} : X_1 \to Z$ is an isomorphism, the inverse of which is called $A \in \mathscr{L}(Z; X_1)$. A precise description of the solution set S in a neighborhood of x_0 is the following.

THEOREM 11.1. *There exist a positive number ε, a neighborhood U of $x_0 \in X$, and a unique function $g : B(0,\varepsilon) \subset \operatorname{Ker}(\mathrm{D}F(x_0)) \to Z$ with*

$$g(0) = 0,$$
$$S \cap U = \{x(\lambda) = x_0 + \lambda + Ag(\lambda) \text{ for all } \lambda \in B(0,\varepsilon)\}.$$

Moreover, the mapping $x : \lambda \in B(0,\varepsilon) \to x(\lambda) = x_0 + \lambda + Ag(\lambda) \in S \cap U$ is a C^p diffeomorphism.

PROOF. The affine mapping $H : \operatorname{Ker}(\mathrm{D}F(x_0)) \times Z \to X$ is given by

$$H(\lambda, z) = x_0 + \lambda + Az.$$

Then we define the mapping $G : \operatorname{Ker}(\mathrm{D}F(x_0)) \times Z \to Z$ by

$$G(\lambda, z) = F(H(\lambda, z)),$$

which is of class C^p. We notice that $G(0,0) = 0$ and that the derivative $\mathrm{D}_z G(0,0) = \mathrm{D}F(x_0)A \in \mathscr{L}(Z; Z)$ is an isomorphism. We apply the implicit function Theorem 2.3. There exist positive constants ε_1, η_1, and a unique mapping $g : B(0,\varepsilon_1) \subset \operatorname{Ker}(\mathrm{D}F(x_0)) \to B(0,\eta_1) \subset Z$, which is of class C^p and such that

$$g(0) = 0,$$
$$\text{for all } \lambda \in B(0,\varepsilon_1), \quad F(x_0 + \lambda + Ag(\lambda)) = 0.$$

We set $U_1 = H(B(0,\varepsilon_1) \times B(0,\eta_1))$ and the uniqueness implies

$$S \cap U_1 = \{x = x_0 + \lambda + Ag(\lambda), \text{ for all } \lambda \in B(0,\varepsilon_1)\}.$$

To prove that the mapping $x(\lambda) = x_0 + \lambda + Ag(\lambda)$ is a C^p diffeomorphism, we remark that for all $\lambda \in B(0,\varepsilon_1)$, $F(x(\lambda)) = 0$ and $\mathrm{D}F(x(0))Dx(0) = 0$. So $\mathrm{D}x(0) \in \mathscr{L}(\operatorname{Ker}(\mathrm{D}F(x_0)); X)$ is an isomorphism onto $\operatorname{Ker}(\mathrm{D}F(x_0))$ and we define the mapping $K : B(0,\varepsilon_1) \times X_1 \to X$ by

$$K(\lambda, x_1) = x(\lambda) + x_1.$$

Then K is of class C^p and $\mathrm{D}K(0,0) \in \mathscr{L}(\mathrm{Ker}(\mathrm{D}F(x_0)) \times X_1, X)$ is an isomorphism. From the inverse function Theorem 2.2, we deduce that K is a C^p diffeomorphism from $B((0,0),\varepsilon) \subset \mathrm{Ker}(\mathrm{D}F(x_0)) \times X_1$ onto a neighborhood U of x_0 with $\varepsilon \leqslant \varepsilon_1$ and $U \subset U_1$. We can conclude since $K(\lambda, 0) = x(\lambda)$. □

With the decomposition $X = \mathrm{Ker}(\mathrm{D}F(x_0)) \oplus X_1$, we get a representation of the solution set close to x_0 in terms of $\lambda \in B(0,\varepsilon) \subset \mathrm{Ker}(\mathrm{D}F(x_0))$. So we say that $\mathrm{Ker}(\mathrm{D}F(x_0))$ is a coordinate space while X_1 is said to be a state space. Then Theorem 11.1 means that if $m = \dim \mathrm{Ker}(\mathrm{D}F(x_0))$, the mapping $x(\lambda)$ defines a local coordinate system of the m-dimensional C^p manifold S in a neighborhood of x_0. Clearly the choice of a coordinate space is not unique. A suitable decomposition of X could be any pair of closed subspaces T and X_1 such that

$$X = T \oplus X_1, \qquad \dim T = m, \qquad X_1 \cap \mathrm{Ker}\big(\mathrm{D}F(x_0)\big) = \{0\}. \tag{11.3}$$

Further developments of this analysis based on modern differential geometry can be found in FINK and RHEINBOLDT [1986, 1987], and in RHEINBOLDT [1986].

We have already noticed in the examples of Section 3 that some quantities are naturally identified as parameters. For example the space X could have the form $X = \mathbb{R}^m \times X_1$ with X_1 a Banach space and the identified parameters in $\mathbb{R}^m$.

In this chapter we shall focus on the following situation. Given a mapping $F : \mathbb{R}^m \times X \to Z$ of class C^p and a regular solution $(\lambda_0, u_0) \in \mathbb{R}^m \times X$ in the solution set $S = \{(\lambda, u)\colon\ F(\lambda, u) = 0\}$, we study approximations of S in a neighborhood of (λ_0, u_0). First we shall analyze the case when the parameter space $\mathbb{R}^m$ (in fact $\mathbb{R}^m \times \{0\}$), can be used as the coordinate space, see (11.3). In that situation the solution set can be parametrized by $\lambda \in \mathbb{R}^m$ and since (λ_0, u_0) is regular the operator $\mathrm{D}_u F(\lambda_0, u_0) \in \mathscr{L}(X; Z)$ is an isomorphism. The second case we shall analyze is when $\mathrm{D}_u F(\lambda_0, u_0)$ is no more an isomorphism from X onto Z. Then it is no more possible to describe the solution set of $F(\lambda, u) = 0$ in a neighborhood of (λ_0, u_0) as a regular function of the parameter $\lambda \in \mathbb{R}^m$. Then one needs either to choose a coordinate space (different from $\mathbb{R}^m$) satisfying (11.3) or transform the problem. We decide for the second method. By adding equations we transform the problem to apply the theory developed in the first case. Some further developments on local coordinate systems of S can be found in RHEINBOLDT ([1986], Chapter 4).

12. Approximation of a parametrized solution set

Let m be a positive integer, X and Z be two real Banach spaces. Given a C^1 mapping $F : \mathbb{R}^m \times X \to Z$, we consider the problem to find $(\lambda, x) \in \mathbb{R}^m \times X$ such that

$$F(\lambda, x) = 0. \tag{12.1}$$

We assume that

(λ_0, x_0) is a solution to (12.1) and $\mathrm{D}_x F(\lambda_0, x_0) \in \mathscr{L}(X; Z)$ is an isomorphism; (12.2) so (λ_0, x_0) is a regular solution to (12.1). We can apply the implicit function theorem and get the following result. There exist a neighborhood I of λ_0 in $\mathbb{R}^m$, a neighborhood U of x_0 in X, and a C^1 mapping $u: I \to U$ such that $\{(\lambda, u(\lambda)) \colon \lambda \in I\}$ is a regular solution set containing (λ_0, x_0). Moreover, if $\{(\lambda, u(\lambda)) \colon \lambda \in I\}$ and $\{(\lambda, \tilde{u}(\lambda)) \colon \lambda \in \widetilde{I}\}$ are two regular solution sets containing (λ_0, x_0), then $u(\lambda)$ and $\tilde{u}(\lambda)$ coincide on the connected component of $I \cap \widetilde{I}$ containing λ_0.

Note that in the case $m = 1$, the terminology solution branch is often used.

We introduce now a family $\{F_h\}_{0<h\leqslant 1}$ of C^1 mappings

$$F_h : (\lambda, x) \in \mathbb{R}^m \times X \to F_h(\lambda, x) \in Z,$$

which are approximations of F. Our goal is to study the existence and the convergence of solutions of the problem: find $(\lambda, x_h) \in \mathbb{R}^m \times X$ such that

$$F_h(\lambda, x_h) = 0 \tag{12.3}$$

in a neighborhood of the regular solution set $\{(\lambda, u(\lambda)) \colon \lambda \in I\}$.

For a fixed λ, we consider a solution $(\lambda, u) \in \mathbb{R} \times X$ of (12.1) and associate with it an element $\tilde{u}_h \in X$, $h \in (0, 1]$. In some cases it will be sufficient to choose $\tilde{u}_h = u$ but in others an appropriate choice of $\tilde{u}_h$ will lead to different error estimates.

Now for that fixed λ, the theory developed in Section 6 can be applied when $\tilde{u}_h = u$. Here we will work under analogous hypotheses of equicontinuity, consistency and stability. We assume that the mapping $\mathrm{D}_x F_h(\lambda, \bullet)$ is Lipschitzian at $\tilde{u}_h$ with constant L $(= L(\lambda))$, that is there exist $\eta > 0$ and L such that for all $v \in \overline{B}(\tilde{u}_h, \eta)$, for all $h \in (0, 1]$,

$$\left\|\mathrm{D}_x F_h(\lambda, \tilde{u}_h) - \mathrm{D}_x F_h(\lambda, v)\right\|_{X;Z} \leqslant L\|\tilde{u}_h - v\|_X. \tag{12.4}$$

Moreover, we assume that

$$\lim_{h\to 0} \left\|F_h(\lambda, \tilde{u}_h)\right\|_Z = 0 \tag{12.5}$$

and for all $h \in (0, 1]$, $\mathrm{D}_x F_h(\lambda, \tilde{u}_h)$ is an isomorphism from X onto Z with an inverse uniformly bounded in h, that is there exists C_1 $(= C_1(\lambda))$ such that for all $h \in (0, 1]$,

$$\left\|\mathrm{D}_x F_h(\lambda, \tilde{u}_h)^{-1}\right\|_{Z;X} \leqslant C_1. \tag{12.6}$$

THEOREM 12.1. *Let $\{F_h\}_{0<h\leqslant 1}$ be a family of C^1 mappings, $F_h : \mathbb{R}^m \times X \to Z$. For a given $\lambda \in \mathbb{R}^m$, we associate an element $\tilde{u}_h \in X$, $h \in (0, 1]$. We assume that hypotheses* (12.4), (12.5) *and* (12.6) *hold. Then there exist $h_0 > 0$, $\delta_0 > 0$, and for all $0 < h \leqslant h_0$ a unique $u_h \in X$ satisfying*

$$F_h(\lambda, u_h) = 0 \quad \textit{and} \quad \|u_h - \tilde{u}_h\|_X \leqslant \delta_0.$$

Moreover, the mapping $\mathrm{D}_x F_h(\lambda, u_h) \in \mathscr{L}(X; Z)$ *is invertible with the bound*

$$\|\mathrm{D}_x F_h(\lambda, u_h)^{-1}\|_{Z;X} \leqslant 2\|\mathrm{D}_x F_h(\lambda, \tilde{u}_h)^{-1}\|_{Z;X},$$

and

$$\|\tilde{u}_h - u_h\|_X \leqslant 2\|\mathrm{D}_x F_h(\lambda, \tilde{u}_h)^{-1}\|_{Z;X}\|F_h(\lambda, \tilde{u}_h)\|_Z. \tag{12.7}$$

PROOF. The proof goes along the same lines as the proof of Theorem 6.1. For the given $\lambda \in I$, we set

$$\varepsilon_h = \|F_h(\lambda, \tilde{u}_h)\|_Z,$$
$$\gamma_h = \|\mathrm{D}_x F_h(\lambda, \tilde{u}_h)^{-1}\|_{Z;X},$$
$$\widetilde{L}_h(\alpha) = \sup_{v \in \overline{B}(\tilde{u}_h, \alpha)} \|\mathrm{D}_x F_h(\lambda, \tilde{u}_h) - \mathrm{D}_x F_h(\lambda, v)\|_{X;Z}.$$

From the stability hypothesis (12.6), we get $\gamma_h \leqslant C_1$. We set $\delta_0 = \min(\eta, (2C_1 L)^{-1})$. Then from the Lipschitzian condition (12.4) we have

$$\widetilde{L}_h(\delta_0) \leqslant (2C_1)^{-1}.$$

Now from the consistency (12.5), there exists $h_0 \leqslant 1$ such that for all $h \in (0, h_0]$

$$2C_1\varepsilon_h \leqslant \delta_0.$$

Thus for $h \leqslant h_0$,

$$2\gamma_h \widetilde{L}_h(2\gamma_h\varepsilon_h) \leqslant 2C_1\widetilde{L}_h(\delta_0) \leqslant 1.$$

We apply Theorem 2.1 with $G = F_h(\lambda, \bullet)$, $v = \tilde{u}_h \in X$. For all $h \leqslant h_0$, there exists a unique $u_h \in \overline{B}(\tilde{u}_h, 2\gamma_h\varepsilon_h)$ satisfying $F_h(\lambda, u_h) = 0$, and $\mathrm{D}_x F_h(\lambda, u_h) \in \mathscr{L}(X; Z)$ is invertible with the bound

$$\|\mathrm{D}_x F_h(\lambda, u_h)^{-1}\|_{Z;X} \leqslant 2\gamma_h.$$

The following estimate holds

$$\|\tilde{u}_h - u_h\|_X \leqslant 2\gamma_h\|F_h(\lambda, \tilde{u}_h)\|_Z.$$

To get the uniqueness of u_h in $\overline{B}(\tilde{u}_h, \delta_0)$ for all $h \in (0, h_0]$, we use the argument developed in Remark 2.1. □

If we choose in Theorem 12.1, $\tilde{u}_h = u$ where (λ, u) is a solution of (12.1), then inequality (12.7) will lead to a priori error estimates for $\|u - u_h\|_X$. To get a posteriori error estimates, we use the following result.

THEOREM 12.2. *Let (λ, u) be a solution of problem (12.1). We assume that the family $\{F_h\}_{0<h\leqslant 1}$ of C^1 mappings satisfies the hypotheses (12.4), (12.5) and (12.6) with $\tilde{u}_h = u$. Then there exists $0 < \bar{h}_0 \leqslant h_0$ such that for all $h \leqslant \bar{h}_0$ the mapping $\mathrm{D}_x F(\lambda, u_h) \in \mathscr{L}(X; Z)$ is invertible with a uniformly bounded inverse and*

$$\|u - u_h\|_X \leqslant 2\big\|\mathrm{D}_x F(\lambda, u_h)^{-1}\big\|_{Z;X}\big\|F(\lambda, u_h)\big\|_Z, \tag{12.8}$$

where h_0 and u_h are given in Theorem 12.1.

PROOF. The proof is identical to that of Theorem 6.2. It is an application of Theorem 2.1 with $G = F(\lambda, \bullet)$ and $v = u$. □

Actually in Theorems 12.1 and 12.2 we have worked with the parameter $\lambda \in I$ fixed. We study now the dependence in λ of u_h.

The mappings $\mathrm{D}_x F_h(\lambda, \bullet)$ are assumed to be Lipschitzian at $\tilde{u}_h(\lambda)$, that is there exist $\eta > 0$, $L > 0$ such that for all $0 < h \leqslant 1$, $\lambda \in I$, and $v \in \overline{B}(\tilde{u}_h(\lambda), \eta)$,

$$\big\|\mathrm{D}_x F_h\big(\lambda, \tilde{u}_h(\lambda)\big) - \mathrm{D}_x F_h(\lambda, v)\big\|_{X;Z} \leqslant L\big\|\tilde{u}_h(\lambda) - v\big\|_X; \tag{12.9}$$

moreover, we assume the consistency condition

$$\lim_{h\to 0} \sup_{\lambda\in I} \big\|F_h\big(\lambda, \tilde{u}_h(\lambda)\big)\big\|_Z = 0 \tag{12.10}$$

and the stability condition: there exists a constant C such that for all $0 < h \leqslant 1$, $\lambda \in I$,

$$\big\|\mathrm{D}_x F_h\big(\lambda, \tilde{u}_h(\lambda)\big)^{-1}\big\|_{Z;X} \leqslant C. \tag{12.11}$$

THEOREM 12.3. *Let $\{(\lambda, u(\lambda))\colon\ \lambda \in I\}$ be a set of regular solutions of the problem (12.1). We assume that I is compact, the mappings $\tilde{u}_h : \lambda \in I \to \tilde{u}_h(\lambda) \in X$ are continuous, and the hypotheses (12.9)–(12.11) hold. Then there exist $h_0 > 0$, $\delta_0 > 0$, and for all $h \leqslant h_0$ a C^1 mapping $u_h : \lambda \in I \to u_h(\lambda) \in X$ such that*

$$\big(F_h(\lambda, v) = 0 \quad \text{and} \quad v \in \overline{B}\big(\tilde{u}_h(\lambda), \delta_0\big)\big) \Leftrightarrow v = u_h(\lambda).$$

Moreover, the mappings $\mathrm{D}_x F_h(\lambda, \tilde{u}_h(\lambda))$ and $\mathrm{D}_x F(\lambda, u_h(\lambda)) \in \mathscr{L}(X; Z)$ are isomorphisms with uniformly bounded inverse and for all $\lambda \in I$

$$\big\|u_h(\lambda) - \tilde{u}_h(\lambda)\big\|_X \leqslant 2\big\|\mathrm{D}_x F_h\big(\lambda, \tilde{u}_h(\lambda)\big)^{-1}\big\|_{Z;X}\big\|F_h\big(\lambda, \tilde{u}_h(\lambda)\big)\big\|_Z, \tag{12.12}$$

$$\big\|u_h(\lambda) - u(\lambda)\big\|_X \leqslant 2\big\|\mathrm{D}_x F\big(\lambda, u_h(\lambda)\big)^{-1}\big\|_{Z;X}\big\|F\big(\lambda, u_h(\lambda)\big)\big\|_Z. \tag{12.13}$$

PROOF. For each $\lambda \in I$ we can apply Theorem 12.1. So for each $\lambda \in I$, there exist $\delta_{0,\lambda}$, $h_{0,\lambda}$, and $u_{h,\lambda} \in X$ for $h \in (0, h_{0,\lambda}]$ satisfying

$$F_h(\lambda, u_{h,\lambda}) = 0, \qquad \|u_{h,\lambda} - \tilde{u}_h(\lambda)\|_X \leqslant \delta_{0,\lambda}$$

and

$$\|\tilde{u}_h(\lambda) - u_{h,\lambda}\|_X \leqslant 2\|\mathrm{D}_x F_h(\lambda, \tilde{u}_h(\lambda))^{-1}\|_{Z;X}\|F_h(\lambda, \tilde{u}_h(\lambda))\|_Z.$$

Furthermore, thanks to the uniformity in λ of assumptions (12.9), (12.10) and (12.11), it is immediate to check that $\delta_{0,\lambda}$ and $h_{0,\lambda}$ can be chosen independently of $\lambda \in I$. So for $h \leqslant h_0$, we can define the function $u_h : \lambda \in I \to u_h(\lambda) = u_{h,\lambda} \in X$.

To prove that the function $u_h(\bullet)$ is C^1 in I, we apply the implicit function Theorem 2.3 taking into account the uniqueness of $u_h(\lambda)$ in $\overline{B}(\tilde{u}_h(\lambda), \delta_0)$.

The estimate (12.13) is a consequence of Theorem 12.2 only to restrict h_0 if necessary. □

REMARK 12.1. Assumption (12.11) can be replaced by

$$\lim_{h\to 0} \sup_{\lambda\in I} \|\mathrm{D}_x F_h(\lambda, \tilde{u}_h(\lambda)) - \mathrm{D}_x F(\lambda, u(\lambda))\|_{X;Z} = 0. \tag{12.11'}$$

From the equality

$$\mathrm{D}_x F_h(\lambda, \tilde{u}_h(\lambda)) = [\mathrm{D}_x F_h(\lambda, \tilde{u}_h(\lambda)) - \mathrm{D}_x F(\lambda, u(\lambda))] + \mathrm{D}_x F(\lambda, u(\lambda))$$

and estimate (12.11′), we deduce from (1.1) that for h small enough the mapping $\mathrm{D}_x F_h(\lambda, \tilde{u}_h(\lambda)) \in \mathscr{L}(X; Z)$ is an isomorphism with an inverse uniformly bounded in h and λ.

If we assume that the mappings F_h are of class C^p with $p \geqslant 1$, then $u_h(\bullet)$ is also of class C^p. Then error estimates for the derivatives can also be deduced.

THEOREM 12.4. *If the assumptions of Theorem* 12.3 *hold and the mappings F_h are of class C^p with $p \geqslant 1$, then $u_h(\bullet)$ is of class C^p in I. Moreover, if the functions $\tilde{u}_h(\bullet)$ are of class C^p then there exist constants C_j, $j = 0, \ldots, p$, depending only on the quantities*

$$\max_{\lambda\in I} \max_{v\in\overline{B}(\tilde{u}_h(\lambda),\delta_0)} \|\mathrm{D}^k F_h(\lambda, v)\| \quad \textit{and} \quad \max_{\lambda\in I} \|\mathrm{D}^k \tilde{u}_h(\lambda)\|, \quad 0 \leqslant k \leqslant j,$$

such that for all $\lambda \in I$,

$$\|\mathrm{D}^j u_h(\lambda) - \mathrm{D}^j \tilde{u}_h(\lambda)\| \leqslant C_j \sum_{k=0}^{j} \left\| \frac{\mathrm{d}^k}{\mathrm{d}\lambda^k} F_h(\lambda, \tilde{u}_h(\lambda)) \right\| \tag{12.14}$$

for $j = 0, \dots, p-1$ *and*

$$\left\|\mathrm{D}^p u_h(\lambda)\right\| \leqslant C_p. \tag{12.15}$$

For simplicity $\|\bullet\|$ *denotes the norm in the different spaces.*

PROOF. From the implicit function theorem we deduce that u_h is of class C^p in I. We then have to prove estimates (12.14) and (12.15). The differentiation of the relation $F_h(\lambda, u_h(\lambda)) = 0$ for $\lambda \in I$ gives

$$\mathrm{D}u_h(\lambda) = -\mathrm{D}_x F_h\big(\lambda, u_h(\lambda)\big)^{-1}\mathrm{D}_\lambda F_h\big(\lambda, u_h(\lambda)\big),$$

which leads to the estimate

$$\left\|\mathrm{D}u_h(\lambda)\right\| \leqslant 2\left\|\mathrm{D}_x F_h\big(\lambda, \tilde{u}_h(\lambda)\big)^{-1}\right\|\left\|\mathrm{D}_\lambda F_h\big(\lambda, u_h(\lambda)\big)\right\|;$$

we notice that the estimate for $\|\mathrm{D}_x F_h(\lambda, u_h(\lambda))^{-1}\|$ is given in Theorem 12.1. This last estimate and (12.12) give estimates (12.14) and (12.15) with $p = 1$.

Assuming the result holds for l, $1 \leqslant l < p$, that is

$$\left\|\mathrm{D}^j u_h(\lambda) - \mathrm{D}^j \tilde{u}_h(\lambda)\right\| \leqslant C_j \sum_{k=0}^{j} \left\|\frac{\mathrm{d}^k}{\mathrm{d}\lambda^k} F_h\big(\lambda, \tilde{u}_h(\lambda)\big)\right\| \tag{12.16}$$

for $j = 0, \dots, l-1$ and

$$\left\|\mathrm{D}^l u_h(\lambda)\right\| \leqslant C_l', \tag{12.17}$$

we prove it for the index value $l+1$. We differentiate l times with respect to λ the function $F_h(\lambda, \tilde{u}_h(\lambda))$, so

$$\begin{aligned}\frac{\mathrm{d}^l}{\mathrm{d}\lambda^l} F_h\big(\lambda, \tilde{u}_h(\lambda)\big) = {} & \mathrm{D}_x F_h\big(\lambda, \tilde{u}_h(\lambda)\big)\mathrm{D}^l \tilde{u}_h(\lambda) \\ & + R_l\big(\lambda, \tilde{u}_h(\lambda), \dots, \mathrm{D}^{l-1}\tilde{u}_h(\lambda)\big)\end{aligned} \tag{12.18}$$

where R_l has the form

$$\begin{aligned} & R_1(\lambda, v) = \mathrm{D}_\lambda F_h(\lambda, v), \\ & R_l(\lambda, v, \dots, \mathrm{D}^{l-1}v) \\ & \quad = \sum_{i=2}^{l} \sum_{\substack{k_1+k_2+\dots+k_i=l \\ k_1,\dots,k_i \geqslant 1}} \alpha_{ik}\mathrm{D}^i F_h(\lambda, v)\big((\lambda, v)^{(k_1)}, \dots, (\lambda, v)^{(k_i)}\big)\end{aligned}$$

with

$$\alpha_{ik} \in \mathbb{R} \quad \text{and} \quad (\lambda, v)^{(k_j)} = \left(\mathrm{D}^{k_j}\lambda, \mathrm{D}^{k_j} v\right).$$

We derive now l times with respect to λ the relation $F_h(\lambda, u_h(\lambda)) = 0$ so

$$0 = \mathrm{D}_x F_h\left(\lambda, u_h(\lambda)\right)\mathrm{D}^l u_h(\lambda) + R_l\left(\lambda, u_h(\lambda), \dots, \mathrm{D}^{l-1} u_h(\lambda)\right). \tag{12.19}$$

From the two expressions (12.18) and (12.19), we get

$$\begin{aligned}
&\mathrm{D}^l \tilde{u}_h(\lambda) - \mathrm{D}^l u_h(\lambda) \\
&\quad = \mathrm{D}_x F_h\left(\lambda, u_h(\lambda)\right)^{-1}\left\{\frac{\mathrm{d}^l}{\mathrm{d}\lambda^l} F_h\left(\lambda, \tilde{u}_h(\lambda)\right) + \left[\mathrm{D}_x F_h\left(\lambda, u_h(\lambda)\right)\right.\right. \\
&\qquad \left. - \mathrm{D}_x F_h\left(\lambda, \tilde{u}_h(\lambda)\right)\right]\mathrm{D}^l \tilde{u}_h(\lambda) - R_l\left(\lambda, \tilde{u}_h(\lambda), \dots, \mathrm{D}^{l-1}\tilde{u}_h(\lambda)\right) \\
&\qquad \left. + R_l\left(\lambda, u_h(\lambda), \dots, \mathrm{D}^{l-1} u_h(\lambda)\right)\right\}.
\end{aligned} \tag{12.20}$$

The mapping R_l is Lipschitzian, so we deduce from (12.20) and (12.16) estimate (12.14) for l.

Estimate (12.15) for the index value $l+1$ is immediately deduced from the formula

$$\mathrm{D}^{l+1} u_h(\lambda) = -\mathrm{D}_x F_h\left(\lambda, u_h(\lambda)\right)^{-1} R_{l+1}\left(\lambda, u_h(\lambda), \dots, \mathrm{D}^l u_h(\lambda)\right). \qquad \square$$

REMARK 12.2. The method presented in Theorem 6.4 to derive error estimates in some other norms than the norm of X can be immediately generalized to parameter dependent problems.

13. Galerkin approximation of a parametrized solution set

The goal of the present section is to transcribe the results of the previous one in the particular case of Galerkin approximations. For simplicity we will consider problem (12.1) with $m = 1$.

Let X, Y be two real Banach spaces, and let Y' denote the dual space of Y. Given a C^1 mapping $F : \mathbb{R} \times X \to Y'$, we shall analyze Galerkin approximations of the problem

$$F(\lambda, x) = 0 \tag{13.1}$$

in a neighborhood of the regular solution $(\lambda_0, x_0) \in \mathbb{R} \times X$, such that (λ_0, x_0) satisfies

$$F(\lambda_0, x_0) = 0 \quad \text{and} \quad \mathrm{D}_x F(\lambda_0, x_0) \in \mathscr{L}(X; Y') \text{ is an isomorphism.} \tag{13.2}$$

As a consequence of hypothesis (13.2), we get the existence of a regular solution branch $\{(\lambda, u(\lambda))\colon\ \lambda \in I\}$ containing (λ_0, x_0), where the mapping $u : I \to X$ is of class C^1.

Let now $\{X_h\}_{0<h\leqslant 1}$ be a family of finite-dimensional subspaces of X and $\{Y_h\}_{0<h\leqslant 1}$ be a family of finite-dimensional subspaces of Y. A Galerkin approximation of the problem (13.1) consists in finding $\lambda \in \mathbb{R}$ and $x_h \in X_h$ such that

$$\text{for all } y_h \in Y_h, \quad \langle F(\lambda, x_h), y_h \rangle_{Y'Y} = 0. \tag{13.3}$$

For each $\lambda \in I$, we define the bilinear form $b : X \times Y \to \mathbb{R}$ by

$$\text{for all } x \in X,\ y \in Y, \quad b(x, y) = \langle \mathrm{D}_x F(\lambda, u(\lambda)) x, y \rangle_{Y'Y}. \tag{13.4}$$

Although we do not write it explicitly for notation simplicity, b is depending on λ. We assume that there exists a constant $\beta > 0$ such that for all $h \in (0, 1]$ and $\lambda \in I$

$$\inf_{\substack{x \in X_h \\ \|x\|_X = 1}} \ \sup_{\substack{y \in Y_h \\ \|y\|_Y = 1}} b(x, y) \geqslant \beta > 0 \tag{13.5}$$

and

$$\dim X_h = \dim Y_h. \tag{13.6}$$

Moreover, we assume that for all $\lambda \in I$

$$\lim_{h \to 0} \inf_{x_h \in X_h} \|u(\lambda) - x_h\|_X = 0. \tag{13.7}$$

Note that hypotheses (13.5) and (13.6) are standard stability conditions in the linear case while the (13.7) one is a standard approximation property of X by X_h.

The following theorem presents the convergence results in the case of Galerkin approximation.

THEOREM 13.1. *Let $F : \mathbb{R} \times X \to Y'$ be a C^1 mapping and $\{(\lambda, u(\lambda))\colon\ \lambda \in I \subset \mathbb{R},\ I$ compact interval$\}$ be a regular solution branch. We assume that there exist $L > 0$ and $\eta > 0$ such that for all $\lambda \in I$ and $v \in \overline{B}(u(\lambda), \eta)$*

$$\|\mathrm{D}_x F(\lambda, u(\lambda)) - \mathrm{D}_x F(\lambda, v)\|_{X;Y'} \leqslant L \|u(\lambda) - v\|_X.$$

Then under assumptions (13.5), (13.6) and (13.7), there exist positive constants h_0, δ_0, and for all $\lambda \in I$ and $h \in (0, h_0]$ a unique solution $(\lambda, u_h(\lambda)) \in \mathbb{R} \times X_h$ to problem (13.3) in $I \times \overline{B}(u(\lambda), \delta_0)$. The mapping $u_h : \lambda \in I \to X$ is of class C^1 and for $\lambda \in I$, $\mathrm{D}_x F(\lambda, u_h(\lambda)) \in \mathscr{L}(X, Y')$ is an isomorphism. Furthermore, there exists a constant $C > 0$ (independent of h and λ) such that for all $h \in (0, h_0]$ and $\lambda \in I$,

$$\|u(\lambda) - u_h(\lambda)\|_X \leqslant C \inf_{x_h \in X_h} \|u(\lambda) - x_h\|_X \tag{13.8}$$

and

$$\|u(\lambda) - u_h(\lambda)\|_X \leqslant 2\|\mathrm{D}_x F(\lambda, u_h(\lambda))^{-1}\|_{Y';X}\|F(\lambda, u_h(\lambda))\|_{Y'}. \tag{13.9}$$

PROOF. The proof is analogous to the proof of Theorem 7.1 and will be deduced from Theorems 12.1 and 12.2. To define the mapping $F_h : I \times X \to Y'$, it is useful to introduce for each $\lambda \in I$ the two projection operators $\Pi_{X_h} \in \mathscr{L}(X; X_h)$ and $\Pi_{Y_h} \in \mathscr{L}(Y; Y_h)$ (depending on λ) given by

$$\text{for all } y_h \in Y_h, \quad b(x - \Pi_{X_h} x, y_h) = 0 \tag{13.10}$$

and

$$\text{for all } x_h \in X_h, \quad b(x_h, y - \Pi_{Y_h} y) = 0. \tag{13.11}$$

Under assumptions (13.5) and (13.6) both operators Π_{X_h} and Π_{Y_h} are well defined. Then we construct the family $\{F_h\}_{0<h\leqslant 1}$ of mappings $F_h : I \times X \to Y'$ by: for $\lambda \in I$ and for $x \in X$, $y \in Y$,

$$\langle F_h(\lambda, x), y\rangle_{Y'Y} = \langle F(\lambda, x), \Pi_{Y_h} y\rangle_{Y'Y} + b(x, y - \Pi_{Y_h} y).$$

Then with the arguments developed in the proof of Theorem 7.1 it is a simple matter to check that problem (13.3) with $\lambda \in I$ is equivalent to find $(\lambda, x_h) \in I \times X_h$ such that

$$F_h(\lambda, x_h) = 0. \tag{13.12}$$

In the same way as in the proof of Theorem 7.1, there exist constants h_0, δ_0 and for $h \in (0, h_0]$ a unique solution $u_h \in \overline{B}(u(\lambda), \delta_0)$ to $F_h(\lambda, x_h) = 0$. Moreover, the estimates (13.8) and (13.9) are taken from (12.12) and (12.13).

Only to restrict h_0 if necessary, we can apply the implicit function Theorem 2.3 and the above uniqueness result to prove that $u_h : \lambda \in I \to u_h(\lambda) \in X$ is a C^1 mapping. □

REMARK 13.1. The duality argument used in Remark 7.3 is also valid here to get error estimates in some other norms than the norm of X. The generalization to the parameter dependent problem is immediate.

REMARK 13.2. If the mapping F is of class C^p with $p \geqslant 1$, then under the assumptions of Theorem 13.1, the conclusions of Theorem 12.4 are true with $\tilde{u}_h(\lambda) = u(\lambda)$ of class C^p. More precisely the function $u_h(\bullet)$ is of class C^p in I and there exist constants C_j, $j = 0, \ldots, p$, depending only on the quantities

$$\max_{\lambda\in I} \max_{v\in\overline{B}(u(\lambda),\delta_0)} \|\mathrm{D}^k F_h(\lambda, v)\| \quad \text{and} \quad \max_{\lambda\in I} \|\mathrm{D}^k u(\lambda)\|, \quad 0 \leqslant k \leqslant j,$$

such that for all $\lambda \in I$,

$$\left\|\mathrm{D}^j u_h(\lambda) - \mathrm{D}^j u(\lambda)\right\| \leqslant C_j \sum_{k=0}^{j} \left\| \frac{\mathrm{d}^k}{\mathrm{d}\lambda^k} F_h\big(\lambda, u(\lambda)\big) \right\| \tag{13.13}$$

for $j = 0, \ldots, p-1$ and

$$\left\|\mathrm{D}^p u_h(\lambda)\right\| \leqslant C_p. \tag{13.14}$$

For simplicity $\| \bullet \|$ denotes the norm in the different spaces.

14. Error estimates for a parameter dependent semilinear problem

An application of the abstract results obtained in the two previous sections to the parameter dependent problems developed in Section 3 is presented here. First we shall analyze a simple case of the general problem (3.18).

Let Ω be a polygonal convex open bounded set in $\mathbb{R}^2$ and let $f : H_0^1(\Omega) \to L^2(\Omega)$ be given by: for $v \in H_0^1(\Omega)$

$$\text{for all } x \in \Omega, \quad f(v)(x) = v^2(x) + 1. \tag{14.1}$$

The mapping $F : \mathbb{R} \times H_0^1(\Omega) \to H^{-1}(\Omega)$ is defined by: for all $(\lambda, v) \in \mathbb{R} \times H_0^1(\Omega)$, $w \in H_0^1(\Omega)$,

$$\big\langle F(\lambda, v), w \big\rangle_{H^{-1}(\Omega)\, H_0^1(\Omega)} = \int_\Omega \mathbf{grad}\, v \,\mathbf{grad}\, w \,\mathrm{d}x - \lambda \int_\Omega f(v) w \,\mathrm{d}x. \tag{14.2}$$

Since the mapping f is C^∞, so is F.

There exist a positive number λ^* and a C^0 mapping $u : [0, \lambda^*] \to u(\lambda) \in H_0^1(\Omega)$, which is C^∞ on $[0, \lambda^*)$, of minimal positive solutions to the problem

$$F(\lambda, v) = 0, \tag{14.3}$$

see Remark 3.2. Moreover, for each $\lambda \in [0, \lambda^*)$, the solution $(\lambda, u(\lambda))$ to (14.3) is regular, in fact $\mathrm{D}_v F(\lambda, u(\lambda)) \in \mathscr{L}(H_0^1(\Omega); H^{-1}(\Omega))$ given by: for all $w_1, w_2 \in H_0^1(\Omega)$

$$\big\langle \mathrm{D}_v F\big(\lambda, u(\lambda)\big) w_1, w_2 \big\rangle_{H^{-1}(\Omega)\, H_0^1(\Omega)} = \int_\Omega \mathbf{grad}\, w_1 \;\mathbf{grad}\, w_2 \,\mathrm{d}x - 2\lambda \int_\Omega u(\lambda) w_1 w_2 \,\mathrm{d}x,$$

is an isomorphism. This means that assumption (13.2) is satisfied for $\lambda \in [0, \lambda^*)$.

Given a regular family $\{\mathscr{T}_h\}_{0<h\leqslant 1}$ of triangulations of $\overline{\Omega}$, we consider the corresponding family $\{V_h\}_{0<h\leqslant 1}$ of finite element subspaces of degree 1,

$$V_h = \{v \in C^0(\overline{\Omega})\colon\ v|_T \in \mathscr{P}_1 \text{ for all } T \in \mathscr{T}_h\} \cap H_0^1(\Omega). \tag{14.4}$$

A Galerkin approximation to problem (14.3) consists in finding $\lambda \in \mathbb{R}$ and $v_h \in V_h$ such that

$$\text{for all } w_h \in V_h, \quad \langle F(\lambda, v_h), w_h\rangle_{H^{-1}(\Omega)\, H_0^1(\Omega)} = 0. \tag{14.5}$$

To study the discrete problem (14.5), we shall apply the results of Section 13, with $X = Y = H_0^1(\Omega)$, F given in (14.2), $X_h = Y_h = V_h$ and for $\lambda \in I = [0, \bar{\lambda}]$ with $0 < \bar{\lambda} < \lambda^*$, the bilinear form $b\colon H_0^1(\Omega) \times H_0^1(\Omega) \to \mathbb{R}$ given by

$$\text{for all } w_1, w_2 \in H_0^1(\Omega),$$
$$b(w_1, w_2) = \int_\Omega \mathbf{grad}\, w_1\, \mathbf{grad}\, w_2 \,\mathrm{d}x - 2\lambda \int_\Omega u(\lambda) w_1 w_2 \,\mathrm{d}x.$$

The set $\{(\lambda, u(\lambda))\colon\ \lambda \in I\}$ is a regular solution branch. The mapping F being of class C^k for any k, the derivative $\mathrm{D}_v F(\lambda, \bullet)$ is clearly Lipschitzian at $u(\lambda)$. To check the inf–sup condition, we notice that for $w_h \in V_h$,

$$\begin{aligned}
&\int_\Omega \mathbf{grad}\, w_h\, \mathbf{grad}\, w_h \,\mathrm{d}x - 2\lambda \int_\Omega u(\lambda) w_h w_h \,\mathrm{d}x \\
&\quad \geqslant \left(1 - \frac{\lambda}{\lambda^*}\right) \int_\Omega |\mathbf{grad}\, w_h|^2 \,\mathrm{d}x \\
&\qquad + \frac{\lambda}{\lambda^*}\left[\int_\Omega |\mathbf{grad}\, w_h|^2 \,\mathrm{d}x - 2\lambda^* \int_\Omega u(\lambda) w_h^2 \,\mathrm{d}x\right] \\
&\quad \geqslant \left(1 - \frac{\lambda}{\lambda^*}\right) \int_\Omega |\mathbf{grad}\, w_h|^2 \,\mathrm{d}x
\end{aligned}$$

since the expression inside the brackets is positive, see the proof of Theorem 3.4. This implies condition (13.5) with $\beta = (\lambda^* - \bar{\lambda})/\lambda^*$, while (13.6) is immediate with our choice of subspaces $X_h = Y_h = V_h$. Finally, assumption (13.7) is a standard polynomial interpolation result. We can apply Theorem 13.1 and get the following result.

THEOREM 14.1. *Let $F\colon \mathbb{R} \times H_0^1(\Omega) \to H^{-1}(\Omega)$ be the mapping defined in* (14.2). *The set $\{(\lambda, u(\lambda))\colon\ \lambda \in I = [0, \bar{\lambda}]\}$ with $\bar{\lambda} < \lambda^*$ denotes a set of minimal positive regular solutions to problem* (14.3). *Let the finite element subspaces V_h be given in* (14.4).

Then there exist positive constants h_0, δ_0, C, and for all $\lambda \in I$, $h \in (0, h_0]$, a solution $(\lambda, u_h(\lambda)) \in \mathbb{R} \times V_h$ to the Galerkin approximation (14.5); *for $\lambda \in I$, $u_h(\lambda)$*

is unique in the closed ball $\overline{B}(u(\lambda), \delta_0)$. *Moreover, the mapping* $u_h : \lambda \in I \to H_0^1(\Omega)$ *is of class* C^1 *and the following error estimates hold: for all* $h \in (0, h_0]$ *and* $\lambda \in I$

$$|u(\lambda) - u_h(\lambda)|_{1,\Omega} \leqslant Ch \tag{14.6}$$

and

$$|u(\lambda) - u_h(\lambda)|_{1,\Omega} \leqslant C\Big(\sum_{T\in\mathcal{T}_h} \eta(T)^2\Big)^{1/2}, \tag{14.7}$$

where the local estimator $\eta(T)$ *is*

$$\eta^2(T) = h_T^2 \|\Delta u_h(\lambda) + \lambda(u_h^2(\lambda) + 1)\|_{0,T}^2 + \sum_{i=1}^{3} h_{t_i} \left\| \left[\frac{\partial u_h(\lambda)}{\partial n} \right] \right\|_{0,t_i}^2;$$

here h_T *is the diameter of* T, t_i, $i = 1, 2, 3$, *are the edges of* T *of length* h_{t_i}, *and* $[\bullet]$ *denotes the jump through an edge.*

PROOF. The hypotheses of Theorem 13.1 were checked before with $\beta = 1 - \bar{\lambda}/\lambda^*$. From estimates (13.8) and (13.9), we can deduce (14.6) and (14.7) in the same way as in the proof of Theorem 8.1. Notice that the constant C is depending on β. □

The remainder of this section is devoted to study finite element approximations of problem (3.18) including numerical quadrature rules. In the previous example we have applied the theory of Galerkin approximations, now we will apply the general results of Section 12.

Let Ω be a regular open bounded set in $\mathbb{R}^2$ and f be the Nemytskii operator of some function $\phi : \Omega \times \mathbb{R} \to \mathbb{R}$, that is for a function $u : \Omega \to \mathbb{R}$ we have

$$\text{for all } x \in \Omega, \quad f(u)(x) = \phi(x, u(x)).$$

We assume that the function ϕ satisfies

$$\begin{aligned}&\text{the function } \phi \text{ is twice continuously differentiable with respect to the}\\ &\text{second argument uniformly in } \Omega \text{ and for all } z \in \mathbb{R},\ \phi(\bullet, z),\ \frac{\partial\phi}{\partial z}(\bullet, z),\\ &\text{and } \frac{\partial^2\phi}{\partial z^2}(\bullet, z) \text{ are in } L^\infty(\Omega)\end{aligned} \tag{14.8}$$

and assumptions (3.16), (3.17), (3.23). Remark that hypothesis (14.8) is stronger than (3.15), so that $f : H_0^1(\Omega) \cap C^0(\overline{\Omega}) \to L^2(\Omega)$ is a C^2 mapping.

In Section 3, we have studied the positive solutions of

$$u - \lambda T f(u) = 0, \tag{14.9}$$

with $-T$ the inverse of the Laplacian operator with homogeneous Dirichlet boundary conditions, see Theorem 3.4. There exist a $\lambda^* > 0$ and a C^2 mapping

$$\lambda \in [0, \lambda^*) \to u(\lambda) \in H_0^1(\Omega) \cap C^0(\overline{\Omega}),$$

where $u(\lambda)$ is the minimal positive solution of (14.9) with the parameter value λ. The derivative

$$I - \lambda T\, \mathrm{D}f\big(u(\lambda)\big) \in \mathscr{L}\big(H_0^1(\Omega) \cap C^0(\overline{\Omega}); H_0^1(\Omega) \cap C^0(\overline{\Omega})\big)$$

is an isomorphism.

Our goal is to study approximations of the regular solution set $\{(\lambda, u(\lambda)) \colon \lambda \in [0, \lambda^*)\}$ when we use the schemes (3.29) and (3.30) to numerically solve (14.9). Since we have introduced numerical quadrature rules to compute the integrals, we can no longer use the Galerkin approximation approach detailed in Section 13. We need to go back to the general results of Section 12. We are within the framework of Section 12 with $m = 1$, $X = Z = H_0^1(\Omega) \cap C^0(\overline{\Omega})$, and $F : \mathbb{R} \times X \to Z$ given by: for all $\lambda \in \mathbb{R}$, $v \in H_0^1(\Omega) \cap C^0(\overline{\Omega})$,

$$F(\lambda, v) = v - \lambda T f(v). \tag{14.10}$$

Both approximation schemes (3.29) and (3.30) can be written in the form

$$F_h(\lambda, u_h) = 0 \tag{14.11}$$

where $F_h : \mathbb{R} \times (H_0^1(\Omega) \cap C^0(\overline{\Omega})) \to H_0^1(\Omega) \cap C^0(\overline{\Omega})$ is given by: for all $\lambda \in \mathbb{R}$, $v \in H_0^1(\Omega) \cap C^0(\overline{\Omega})$

$$F_h(\lambda, v) = v - \lambda T_h f(v). \tag{14.12}$$

Here the operator $T_h \in \mathscr{L}(C^0(\overline{\Omega}); V_h)$ is defined by: for $g \in C^0(\overline{\Omega})$

$$\text{for all } v_h \in V_h, \quad a_h(T_h g, v_h) = (g, v_h)_h, \tag{14.13}$$

where

$$a_h(u_h, v_h) = \int_{\Omega_h} \mathbf{grad}(u_h)\, \mathbf{grad}\, v_h \, \mathrm{d}x, \tag{14.14}$$

$$(g, v_h)_h = \sum_{T \in \mathscr{T}_h} \left[\frac{S_T}{3} \sum_{i=1}^{3} (g v_h)(a_{i,T}) \right], \tag{14.15}$$

in the piecewise $\mathscr{P}_1$ approximation (3.29) and

$$a_h(u_h, v_h) = \sum_{T\in\mathscr{T}_h} \left[\sum_{i=1}^{3} \tfrac{1}{6} \det\big(\mathrm{D}F_T(\hat{m}_i)\big)(\mathbf{grad}\, u_h\, \mathbf{grad}\, v_h)(m_{i,T})\right], \tag{14.16}$$

$$(g, v_h)_h = \sum_{T\in\mathscr{T}_h} \sum_{i=1}^{3} \left[\tfrac{1}{6} \det\big(\mathrm{D}F_T(\hat{m}_i)\big)(g v_h)(m_{i,T})\right], \tag{14.17}$$

in the piecewise $\mathscr{P}_2$ approximation (3.30).

The main approximation properties of the operator T_h are summarized in the next theorem.

THEOREM 14.2. *Let T_h be the operator defined in* (14.13) *with* (14.14) *and* (14.15) *in the case $k = 1$ and with* (14.16) *and* (14.17) *in the case $k = 2$. Then there exists a constant C independent of h such that*

(i) *for all $f \in W^{1,q}(\Omega)$ with $q > 2$, the following estimate holds:*

$$|Tf - T_h f|_{1,\Omega} + \|Tf - T_h f\|_{0,\infty,\Omega} \leqslant Ch\|f\|_{1,q,\Omega}; \tag{14.18}$$

(ii) *for all $f \in W^{k+1,q}(\Omega)$ with $q > 1$, the following estimate holds:*

$$\|Tf - T_h f\|_{0,\Omega} \leqslant Ch^{k+1}\|f\|_{k+1,q,\Omega}; \tag{14.19}$$

(iii) *for $k = 2$ and $f \in W^{2,q}(\Omega)$ with $q > 1$, the following estimate holds:*

$$\|Tf - T_h f\|_{1,\Omega\cap\Omega_h} \leqslant Ch^2\|f\|_{2,q,\Omega}. \tag{14.20}$$

Moreover, for all $\varepsilon > 0$ there exists a constant $C(\varepsilon)$ such that for all $f \in W^{k+1,1}(\Omega)$,

$$\|Tf - T_h f\|_{0,\infty,\Omega} \leqslant C(\varepsilon)h^{k+1-\varepsilon}\|f\|_{k+1,1,\Omega}. \tag{14.21}$$

If $r_h : C^0(\overline{\Omega}) \to V_h$ denotes the corresponding Lagrange interpolation operator, then there exists a constant C such that for all $f \in C^0(\overline{\Omega})$ and $1 \leqslant p < \infty$,

$$\|T_h f\|_{1,p,\Omega} \leqslant C\|r_h f\|_{0,\Omega}. \tag{14.22}$$

These estimates are standard finite element estimates. For the proofs, see for instance CIARLET [1991], CROUZEIX and RAPPAZ [1989], and the paper of WAHLBIN [1978].

With each solution $(\lambda, u(\lambda))$ we associate the function $\tilde{u}_h(\lambda) = \lambda T_h f(u(\lambda)) \in H_0^1(\Omega) \cap C^0(\overline{\Omega})$. Since $(\lambda, u(\lambda))$ is a solution to (14.9), the following relation holds

$$\tilde{u}_h(\lambda) = u(\lambda) - \lambda T f\big(u(\lambda)\big) + \lambda T_h f\big(u(\lambda)\big) = u(\lambda) - \lambda(T - T_h) f\big(u(\lambda)\big),$$

which implies that $\tilde{u}_h(\lambda)$ is close to $u(\lambda)$ in the norm of $H_0^1(\Omega) \cap C^0(\overline{\Omega})$. This choice is well adapted to get error estimates.

From assumption (14.8), we deduce that F is of class C^2, so hypothesis (12.4) is clearly satisfied.

The estimate of $\|F_h(\lambda, \tilde{u}_h(\lambda))\|_{H_0^1(\Omega)\cap C^0(\overline{\Omega})}$ goes as follows. We set

$$\begin{aligned}\varepsilon_h &= \|F_h(\lambda, \tilde{u}_h(\lambda))\|_{H_0^1(\Omega)\cap C^0(\overline{\Omega})} = \|\tilde{u}_h(\lambda) - \lambda T_h f(\tilde{u}_h(\lambda))\|_{H_0^1(\Omega)\cap C^0(\overline{\Omega})} \\ &= \lambda \|T_h(f(u(\lambda)) - f(\tilde{u}_h(\lambda)))\|_{H_0^1(\Omega)\cap C^0(\overline{\Omega})}.\end{aligned}$$

As a consequence of (14.22), we get

$$\varepsilon_h \leqslant C\|r_h(f(u(\lambda)) - f(\tilde{u}_h(\lambda)))\|_{0,\Omega}.$$

Since for all $a, v \in C^0(\overline{\Omega})$,

$$\|r_h(av)\|_{0,\Omega} \leqslant C\|a\|_{0,\infty,\Omega}\|r_h v\|_{0,\Omega},$$

we have

$$\varepsilon_h \leqslant C\|r_h(u(\lambda) - \tilde{u}_h(\lambda))\|_{0,\Omega} = C\lambda\|(r_h T - T_h) f(u(\lambda))\|_{0,\Omega}.$$

From the classical properties of the Lagrange interpolation operator r_h and from the estimate (14.19), we get the estimate

$$\varepsilon_h \leqslant C h^{k+1}, \tag{14.23}$$

with $k = 1$ or 2 corresponding to both approximations (3.29) and (3.30) respectively. So assumption (12.5) is verified.

To check stability (12.6), let us first notice that $\mathrm{D}_v F(\lambda, u(\lambda)) \in \mathscr{L}(H_0^1(\Omega) \cap C^0(\overline{\Omega}); H_0^1(\Omega) \cap C^0(\overline{\Omega}))$ is an isomorphism. Using a compactness argument on T, it is a simple matter to check that $\mathrm{D}_v F(\lambda, u(\lambda)) \in \mathscr{L}(W_0^{1,p}(\Omega); W_0^{1,p}(\Omega))$, with $p > 2$, is an isomorphism. So there exists a constant C independent of h such that

$$\|\mathrm{D}_v F(\lambda, u(\lambda))^{-1}\|_{H_0^1(\Omega)\cap C^0(\overline{\Omega}); H_0^1(\Omega)\cap C^0(\overline{\Omega})} \leqslant C, \tag{14.24}$$

$$\|\mathrm{D}_v F(\lambda, u(\lambda))^{-1}\|_{W_0^{1,p}(\Omega); W_0^{1,p}(\Omega)} \leqslant C. \tag{14.25}$$

Let us notice that

$$\mathrm{D}_v F_h(\lambda, \tilde{u}_h(\lambda)) = \mathrm{D}_v F(\lambda, u(\lambda)) + \mathrm{D}_v F_h(\lambda, \tilde{u}_h(\lambda)) - \mathrm{D}_v F(\lambda, u(\lambda)).$$

So to prove the invertibility of $\mathrm{D}_v F_h(\lambda, \tilde{u}_h(\lambda)) \in \mathscr{L}(H_0^1(\Omega) \cap C^0(\overline{\Omega}); H_0^1(\Omega) \cap C^0(\overline{\Omega}))$, we will bound the operator

$$A_h(\lambda) = \mathrm{D}_v F(\lambda, u(\lambda))^{-1}\big[\mathrm{D}_v F_h(\lambda, \tilde{u}_h(\lambda)) - \mathrm{D}_v F(\lambda, u(\lambda))\big].$$

Let v be in $H_0^1(\Omega) \cap C^0(\overline{\Omega})$, then

$$\begin{aligned} A_h(\lambda)v &= \mathrm{D}_v F(\lambda, u(\lambda))^{-1}\big[\lambda T \mathrm{D}f(u(\lambda))v - \lambda T_h \mathrm{D}f(\tilde{u}_h(\lambda))v\big] \\ &= \mathrm{D}_v F(\lambda, u(\lambda))^{-1}\big[\lambda(T - T_h)\mathrm{D}f(u(\lambda))v \\ &\quad + \lambda T_h\big(\mathrm{D}f(u(\lambda)) - \mathrm{D}f(\tilde{u}_h(\lambda))\big)v\big]. \end{aligned} \tag{14.26}$$

Since the mapping $T_h \in \mathscr{L}(C^0(\overline{\Omega}); W^{1,p}(\Omega))$ is uniformly bounded, see (14.22), we get from (14.26) and (14.25),

$$\|A_h(\lambda)v\|_{1,p,\Omega} \leqslant C\|v\|_{0,\infty,\Omega} \leqslant C\|v\|_{H_0^1(\Omega)\cap C^0(\overline{\Omega})}. \tag{14.27}$$

On the other hand,

$$\begin{aligned} \|A_h(\lambda)v\|_{H_0^1(\Omega)\cap C^0(\overline{\Omega})} \leqslant C\Big[&\|(T - T_h)\mathrm{D}f(u(\lambda))v\|_{H_0^1(\Omega)\cap C^0(\overline{\Omega})} \\ &+ \|\big[\mathrm{D}f(u(\lambda)) - \mathrm{D}f(\tilde{u}_h(\lambda))\big]v\|_{0,\infty,\Omega}\Big], \end{aligned}$$

so with the regularity of f,

$$\begin{aligned} \|A_h(\lambda)v\|_{H_0^1(\Omega)\cap C^0(\overline{\Omega})} \leqslant C\Big[&\|(T - T_h)\mathrm{D}f(u(\lambda))v\|_{H_0^1(\Omega)\cap C^0(\overline{\Omega})} \\ &+ \|(T - T_h)f(u(\lambda))\|_{H_0^1(\Omega)\cap C^0(\overline{\Omega})}\|v\|_{1,p,\Omega}\Big]. \end{aligned}$$

Using the convergence property (14.18), we get

$$\|A_h(\lambda)v\|_{H_0^1(\Omega)\cap C^0(\overline{\Omega})} \leqslant Ch\|v\|_{1,p,\Omega}. \tag{14.28}$$

From estimates (14.27) and (14.28) we deduce

$$\|A_h^2(\lambda)v\|_{H_0^1(\Omega)\cap C^0(\overline{\Omega})} \leqslant Ch\|v\|_{H_0^1(\Omega)\cap C^0(\overline{\Omega})},$$

which implies

$$\|A_h^2(\lambda)\|_{H_0^1(\Omega)\cap C^0(\overline{\Omega}); H_0^1(\Omega)\cap C^0(\overline{\Omega})} \leqslant Ch.$$

It is not difficult to check that condition (1.1) can be relaxed to $\|(A^{-1}B)^2\|_{X;X} \leqslant l^2$, with $l \in (0,1)$, see for instance CROUZEIX and RAPPAZ [1989], to get that $A+B$ is an isomorphism. If h is chosen sufficiently small, then $\mathrm{D}_v F_h(\lambda, \tilde{u}_h(\lambda))$ is an isomorphism with a uniformly bounded inverse. So we have checked the stability assumption (12.6).

The mapping $\mathrm{D}_v F_h(\lambda, \bullet)$ is Lipschitzian since $T_h \in \mathscr{L}(C^0(\overline{\Omega}); H_0^1(\Omega) \cap C^0(\overline{\Omega}))$ is uniformly bounded and f is C^2, that is (12.4) is satisfied.

THEOREM 14.3. *Let* $\{(\lambda, u(\lambda)) \colon \lambda \in [0, \bar{\lambda}]\}$ *with* $\bar{\lambda} < \lambda^*$ *be a regular solution set to the problem* $F(\lambda, v) = 0$ *with* F *given in* (14.10) *and let* F_h *be the mapping defined in* (14.12). *There exist two constants* $h_0 > 0$, $\delta_0 > 0$ *and for* $h \leqslant h_0$ *a* C^2 *mapping* $u_h : \lambda \in [0, \bar{\lambda}] \to u_h(\lambda) \in V_h$ *such that for all* $\lambda \in [0, \bar{\lambda}]$,

$$\big(F_h(\lambda, v_h) = 0 \text{ and } \|v_h - \tilde{u}_h(\lambda)\|_{H_0^1(\Omega)\cap C^0(\overline{\Omega})} \leqslant \delta_0\big) \Leftrightarrow v_h = u_h(\lambda),$$

and the set $\{(\lambda, u_h(\lambda)) \colon \lambda \in [0, \bar{\lambda}]\}$ *is a regular solution set of problem* (14.11). *Moreover, the following a priori estimates hold:*

$$\|u(\lambda) - u_h(\lambda)\|_{0,\Omega} + h\|u(\lambda) - u_h(\lambda)\|_{1,\Omega} \leqslant C h^{k+1} \tag{14.29}$$

and

$$\|u(\lambda) - u_h(\lambda)\|_{0,\infty,\Omega} \leqslant C(\varepsilon) h^{k+1-\varepsilon}. \tag{14.30}$$

PROOF. The hypotheses of Theorem 12.3 are a consequence of the above developments together with the compactness of the interval $[0, \bar{\lambda}]$. There exist $h_0 > 0$, $\delta_0 > 0$, and for all $h \leqslant h_0$ a C^2 mapping $u_h : \lambda \in [0, \bar{\lambda}] \to u_h(\lambda) \in H_0^1(\Omega) \cap C^0(\overline{\Omega})$ such that

$$\big(F_h(\lambda, v_h) = 0 \text{ and } \|v_h - \tilde{u}_h(\lambda)\|_{H_0^1(\Omega)\cap C^0(\overline{\Omega})} \leqslant \delta_0\big) \Leftrightarrow v_h = u_h(\lambda)$$

and the mapping

$$\mathrm{D}_v F_h\big(\lambda, u_h(\lambda)\big) \in \mathscr{L}\big(H_0^1(\Omega) \cap C^0(\overline{\Omega}); H_0^1(\Omega) \cap C^0(\overline{\Omega})\big)$$

is invertible, with a uniformly bounded inverse.

Estimates (12.12) and (14.23) give

$$\|\tilde{u}_h(\lambda) - u_h(\lambda)\|_{H_0^1(\Omega)\cap C^0(\overline{\Omega})} \leqslant C h^{k+1}. \tag{14.31}$$

Since

$$u(\lambda) - \tilde{u}_h(\lambda) = \lambda(T - T_h) f\big(u(\lambda)\big),$$

we deduce estimates (14.29) and (14.30) from (14.31) and from Theorem 14.2. □

REMARK 14.1. In the same way we proved estimate (14.23), we can prove that

$$\left\| \frac{\mathrm{d}}{\mathrm{d}\lambda} F(\lambda, \tilde{u}_h(\lambda)) \right\|_{H_0^1(\Omega)\cap C^0(\overline{\Omega})} \leqslant Ch^{k+1}.$$

As a consequence of Theorem 12.4, there exist constants C and for $\varepsilon > 0$, $C(\varepsilon)$ such that for $h \leqslant h_0$

$$\|\mathrm{D}u(\lambda) - \mathrm{D}u_h(\lambda)\|_{0,\Omega} + h\|\mathrm{D}u(\lambda) - \mathrm{D}u_h(\lambda)\|_{1,\Omega} \leqslant Ch^{k+1}$$

and

$$\|\mathrm{D}u(\lambda) - \mathrm{D}u_h(\lambda)\|_{0,\infty,\Omega} \leqslant C(\varepsilon)h^{k+1-\varepsilon}.$$

REMARK 14.2. In Theorem 14.3, we have got a priori estimates by using the relation (12.12). Inequality (12.13) will give a posteriori estimates. We do not want to emphasize this point here.

15. Solution set containing a simple limit point

Let X and Z be two real Banach spaces and $F:\mathbb{R}\times X \to Z$ be a nonlinear mapping of class C^1. Here for simplicity we have considered the parameter space $\mathbb{R}^m$ with $m = 1$; our approach can be immediately generalized for any $m \in \mathbb{N}$.

We assume that $(\lambda_0, x_0) \in \mathbb{R}\times X$ is a regular solution to

$$F(\lambda, x) = 0, \tag{15.1}$$

but $\mathrm{D}_x F(\lambda_0, x_0) \in \mathscr{L}(X; Z)$ is no more an isomorphism. More precisely, the following assumption holds:

$$\begin{aligned}&(\lambda_0, x_0) \text{ is a solution to (15.1), } \mathrm{D}F(\lambda_0, x_0) \in \mathscr{L}(\mathbb{R}\times X; Z) \text{ is}\\ &\text{a Fredholm operator of index 1, } Z = \mathrm{Range}(\mathrm{D}F(\lambda_0, x_0)), \text{ and}\\ &\mathrm{D}_x F(\lambda_0, x_0) \in \mathscr{L}(X; Z) \text{ is not an isomorphism.}\end{aligned} \tag{15.2}$$

Assumption (15.2) means that the solution $(\lambda_0, x_0) \in \mathbb{R}\times X$ to (15.1) is such that $\mathrm{D}_x F(\lambda_0, x_0) \in \mathscr{L}(X; Z)$ is a Fredholm operator of index 0, $\dim \mathrm{Ker}(\mathrm{D}_x F(\lambda_0, x_0)) = 1$, and $\mathrm{D}_\lambda F(\lambda_0, x_0)$ does not belong to the range of $\mathrm{D}_x F(\lambda_0, x_0)$.

Let $\varphi \in X$, $\varphi \not\equiv 0$, satisfy $\mathrm{D}_x F(\lambda_0, x_0)\varphi = 0$; so $\mathrm{Ker}(\mathrm{D}_x F(\lambda_0, x_0)) = \mathrm{span}\{\varphi\}$.

Since $\mathrm{D}_x F(\lambda_0, x_0)$ is no more an isomorphism, it is not possible to describe the solution set of (15.1) in a neighborhood of (λ_0, x_0) as a solution branch parametrized by λ. To study it we introduce the mapping $\Phi:\mathbb{R}\times\mathbb{R}\times X \to \mathbb{R}\times Z$ given by

$$\Phi(s, \lambda, x) = \big(B(x - x_0) - s, F(\lambda, x)\big), \tag{15.3}$$

where B is a functional in X' with $B(\varphi) = 1$. We check that

$$\Phi(0, \lambda_0, x_0) = 0$$

and for $\delta \in \mathbb{R}$, $v \in X$,

$$\mathrm{D}_{\lambda x}\Phi(0, \lambda_0, x_0)(\delta, v) = \big(Bv, \mathrm{D}_\lambda F(\lambda_0, x_0)\delta + \mathrm{D}_x F(\lambda_0, x_0)v\big).$$

The derivative $\mathrm{D}_{\lambda x}\Phi(0, \lambda_0, x_0) \in \mathscr{L}(\mathbb{R}\times X; \mathbb{R}\times Z)$ is a Fredholm operator of index 0. Since dim $\mathrm{Ker}(\mathrm{D}_x F(\lambda_0, x_0))$ is finite, there exists a closed subspace $X_1 \subset X$ such that $X = \mathrm{span}\{\varphi\} \oplus X_1$. If $(\delta, v \equiv \alpha\varphi + v_1) \in \mathbb{R} \times X$ with $\alpha \in \mathbb{R}$ and $v_1 \in X_1$ satisfies

$$\mathrm{D}_{\lambda x}\Phi(0, \lambda_0, x_0)(\delta, v) = 0,$$

then $v_1 = 0$ and $\delta = 0$ since $\mathrm{D}F(\lambda_0, x_0) \in \mathscr{L}(X_1; Z)$ is an isomorphism. Since $B(\varphi) = 1$ so $\alpha = 0$. Finally $\mathrm{D}_{\lambda x}\Phi(0, \lambda_0, x_0)$ is injective and therefore

$$\mathrm{D}_{\lambda x}\Phi(0, \lambda_0, x_0) \in \mathscr{L}(\mathbb{R} \times X; \mathbb{R} \times Z)$$

is an isomorphism.

To study the solutions of

$$\Phi(s, \lambda, x) = 0 \tag{15.4}$$

in a neighborhood of $(0, \lambda_0, x_0)$, we can use the implicit function theorem to prove the following result.

THEOREM 15.1. *Under assumption* (15.2), *there exist a positive number* ε, *a neighborhood* W *of* $(\lambda_0, x_0) \in \mathbb{R} \times X$, *and a unique mapping* $(\lambda, u): s \in [-\varepsilon, \varepsilon] \to (\lambda(s), u(s)) \in \mathbb{R} \times X$ *of class* C^1 *such that for all* $s \in [-\varepsilon, \varepsilon]$,

$$F\big(\lambda(s), u(s)\big) = 0 \quad \textit{and} \quad s = B\big(u(s) - x_0\big), \tag{15.5}$$

$$\lambda(0) = \lambda_0 \quad \textit{and} \quad u(0) = x_0, \tag{15.6}$$

and if $(\lambda, u) \in W$ *is a solution to* (15.1), *then necessarily* $\lambda = \lambda(s)$, $u = u(s)$ *with* $s = B(u - x_0)$. *Moreover, for* $s \in [-\varepsilon, \varepsilon]$, *the mapping* $\mathrm{D}_{\lambda x}\Phi(s, \lambda(s), u(s)) \in \mathscr{L}(\mathbb{R} \times X; \mathbb{R} \times Z)$ *is an isomorphism.*

Clearly $(\lambda(s), u(s)) \neq (\lambda_0, x_0)$ for all $s \neq 0$ since $B(u(s) - x_0) = s$. So the mapping $s \in [-\varepsilon, \varepsilon] \to (\lambda(s), u(s)) \in \mathbb{R}\times X$ cannot be the constant function equal to (λ_0, x_0) and the set of solutions of (15.1) is not isolated. We can get more information on the variation of λ and u as a function of s. When differentiating with respect to s the identity $\Phi(s,\lambda(s),u(s))=0$, we get

$$B\left(\frac{\mathrm{d}}{\mathrm{d}s}u(0)\right)=1,$$

$$\mathrm{D}_\lambda F(\lambda_0,x_0)\frac{\mathrm{d}}{\mathrm{d}s}\lambda(0)+\mathrm{D}_x F(\lambda_0,x_0)\frac{\mathrm{d}}{\mathrm{d}s}u(0)=0,$$

and so $(\mathrm{d}\lambda/\mathrm{d}s)(0)=0$, $(\mathrm{d}u/\mathrm{d}s)(0)=\varphi$.

A pair $(\lambda_0,x_0)\in\mathbb{R}\times X$ satisfying to the condition (15.2) is called a simple limit point of (15.1).

If we assume that F is a mapping of class C^p with $p\geqslant 2$, then we can get a more precise description of the solution set $\{(\lambda(s),u(s))\colon s\in[-\varepsilon,\varepsilon]\}$. We differentiate twice with respect to s the identity $\Phi(s,\lambda(s),u(s))=0$ to get

$$\mathrm{D}_\lambda F(\lambda_0,x_0)\frac{\mathrm{d}^2}{\mathrm{d}s^2}\lambda(0)+\mathrm{D}^2_{xx}F(\lambda_0,x_0)(\varphi,\varphi)+\mathrm{D}_x F(\lambda_0,x_0)\frac{\mathrm{d}^2}{\mathrm{d}s^2}u(0)=0,$$

since $(\mathrm{d}\lambda/\mathrm{d}s)(0)=0$ and $(\mathrm{d}u/\mathrm{d}s)(0)=\varphi$, and

$$B\left(\frac{\mathrm{d}^2}{\mathrm{d}s^2}u(0)\right)=0.$$

Consequently $(\mathrm{d}^2\lambda/\mathrm{d}s^2)(0)$ is different from 0 if and only if

$$\mathrm{D}^2_{xx}F(\lambda_0,x_0)(\varphi,\varphi)\notin \mathrm{Range}\,\big(\mathrm{D}_x F(\lambda_0,x_0)\big). \tag{15.7}$$

If (λ_0,x_0) is a simple limit point of (15.1) satisfying to (15.7), then for s in a neighborhood of 0 the following development holds

$$\lambda(s)=\lambda_0+\frac{(\mathrm{d}^2\lambda/\mathrm{d}s^2)(0)}{2}s^2+\mathrm{o}(s^2).$$

In Fig. 15.1, we have represented the branch $\{(\lambda(s),B(u(s))\}$ in a neighborhood of 0.

A pair $(\lambda_0,x_0)\in\mathbb{R}\times X$ satisfying conditions (15.2) and (15.7) is a nondegenerate simple limit point.

In the degenerate case $(\mathrm{d}^2\lambda/\mathrm{d}s^2)(0)=0$, different cases can occur, see for instance Fig. 15.2.

REMARK 15.1. A nondegenerate simple limit point is in fact a junction point of two regular solution branches parametrized by λ. Let (λ_0,x_0) be a nondegenerate simple limit point and let $s\in[-\varepsilon,\varepsilon]\to(\lambda(s),u(s))\in\mathbb{R}\times X$ be the solution set given in Theorem 15.1. We assume here that F is of class C^p with $p\geqslant 2$. Then we will prove

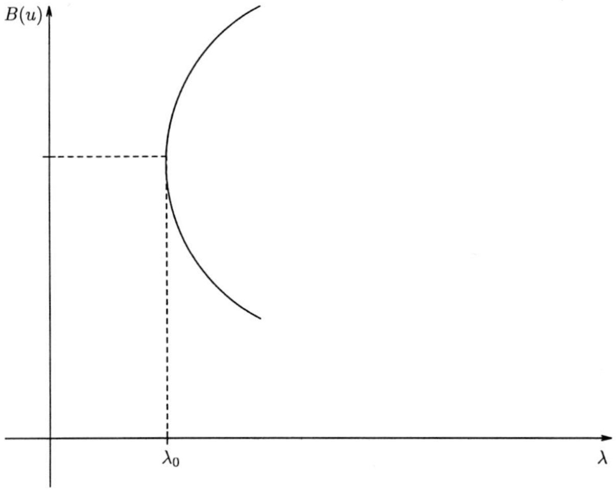

FIG. 15.1. $D^2\lambda(0) > 0$.

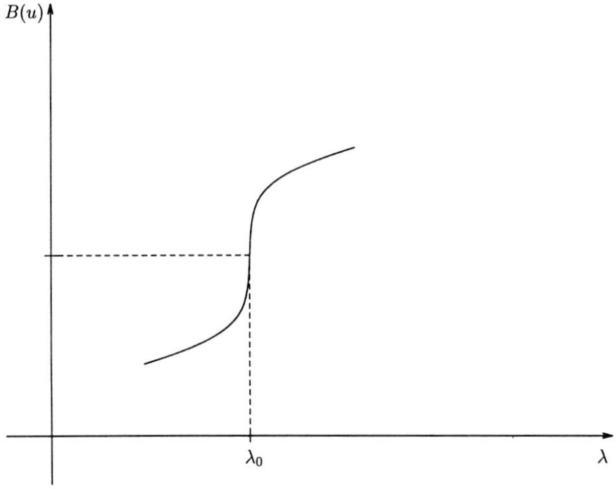

FIG. 15.2. $D^2\lambda(0) = 0$ and $D^3\lambda(0) \neq 0$.

that there exist positive numbers $0 < \varepsilon_1 \leqslant \varepsilon$ and C such that for all $s \in [-\varepsilon_1, \varepsilon_1]$, $s \neq 0$,

$$D_x F\big(\lambda(s), u(s)\big) \in \mathscr{L}(X; Z) \text{ is an isomorphism} \tag{15.8}$$

and

$$\big\| D_x F\big(\lambda(s), u(s)\big)^{-1} \big\|_{Z;X} \leqslant \frac{C}{|s|}. \tag{15.9}$$

Since $(\mathrm{d}^2\lambda/\mathrm{d}s^2)(0) \neq 0$, there exist ε_1, $0 < \varepsilon_1 \leqslant \varepsilon$, and a constant C_1 such that

$$\text{for all } s \in [-\varepsilon_1, \varepsilon_1], \qquad \left|\frac{\mathrm{d}}{\mathrm{d}s}\lambda(s)\right| \geqslant C_1|s|.$$

The derivative $\mathrm{D}_{\lambda x}\Phi(s, \lambda(s), u(s)) \in \mathscr{L}(\mathbb{R} \times X; \mathbb{R} \times Z)$ is an isomorphism with an inverse uniformly bounded with respect to $s \in [-\varepsilon_1, \varepsilon_1]$. For all $s \in [-\varepsilon_1, \varepsilon_1]$ and $f \in Z$, there exists a unique $(\delta, v) \in \mathbb{R} \times X$ with

$$\mathrm{D}_{\lambda x}\Phi\big(s, \lambda(s), u(s)\big)(\delta, v) = (0, f)$$

and

$$|\delta| + \|v\|_X \leqslant C_2\|f\|_Z,$$

where C_2 is independent of s. Given $f \in Z$ and $s \neq 0$, we use both relations

$$\mathrm{D}_\lambda F\big(\lambda(s), u(s)\big)\delta + \mathrm{D}_x F\big(\lambda(s), u(s)\big)v = f,$$

$$\mathrm{D}_\lambda F\big(\lambda(s), u(s)\big)\frac{\mathrm{d}}{\mathrm{d}s}\lambda(s) + \mathrm{D}_x F\big(\lambda(s), u(s)\big)\frac{\mathrm{d}}{\mathrm{d}s}u(s) = 0,$$

and set

$$w = v - \frac{\delta}{\frac{\mathrm{d}}{\mathrm{d}s}\lambda(s)}\frac{\mathrm{d}}{\mathrm{d}s}u(s),$$

to obtain

$$\mathrm{D}_x F\big(\lambda(s), u(s)\big)w = f$$

and

$$\|w\|_X \leqslant C_2\left(1 + \frac{1}{C_1|s|}\left\|\frac{\mathrm{d}}{\mathrm{d}s}u(s)\right\|_X\right)\|f\|_Z.$$

We have to check now that $\mathrm{D}_x F(\lambda(s), u(s))$ is injective with $s \in [-\varepsilon_1, \varepsilon_1]$, $s \neq 0$. Let $\psi \in X$ be such that

$$\mathrm{D}_x F\big(\lambda(s), u(s)\big)\psi = 0.$$

Then since $B((\mathrm{d}u/\mathrm{d}s)(s)) = 1$, we have

$$\mathrm{D}_{\lambda x}\Phi\big(s, \lambda(s), u(s)\big)\left(-B(\psi)\frac{\mathrm{d}}{\mathrm{d}s}\lambda(s), \psi - B(\psi)\frac{\mathrm{d}}{\mathrm{d}s}u(s)\right) = 0,$$

and consequently $\psi = B(\psi)(\mathrm{d}u/\mathrm{d}s)(s)$ and $B(\psi)(\mathrm{d}\lambda/\mathrm{d}s)(s) = 0$. Therefore $\psi = 0$.

16. Galerkin approximation of simple limit points

Let X, Y be two real Banach spaces, and let Y' denote the dual space of Y. Given a C^p mapping $F:\mathbb{R}\times X\to Y'$, with $p\geqslant 2$, we shall analyze Galerkin approximations of the problem

$$F(\lambda,x)=0 \tag{16.1}$$

in a neighborhood of a regular solution $(\lambda_0,x_0)\in\mathbb{R}\times X$ satisfying (15.2). The study of the exact problem has been done in Section 15.

To analyze the approximation of the simple limit point (λ_0,x_0), we consider the case where the mapping F has the following structure

$$F(\lambda,x)=Lx+G(\lambda,x) \tag{16.2}$$

where

$$L\in\mathscr{L}(X;Y')\quad\text{is an isomorphism} \tag{16.3}$$

and the C^p mapping $G:\mathbb{R}\times X\to Y'$ satisfies:

$$\text{for all } (\lambda,x)\in\mathbb{R}\times X,\quad \mathrm{D}_xG(\lambda,x)\in\mathscr{L}(X;Y') \text{ is compact.} \tag{16.4}$$

Then problem (16.1) reads: find $\lambda\in\mathbb{R}$, $x\in X$ such that

$$\text{for all } y\in Y,\quad \langle Lx,y\rangle_{Y'Y}+\big\langle G(\lambda,x),y\big\rangle_{Y'Y}=0. \tag{16.5}$$

Given a family $\{X_h\}_{0<h\leqslant 1}$ of finite-dimensional subspaces of X and a family $\{Y_h\}_{0<h\leqslant 1}$ of finite-dimensional subspaces of Y, a Galerkin approximation of (16.5) consists in finding $\lambda\in\mathbb{R}$ and $x_h\in X_h$ such that

$$\text{for all } y_h\in Y_h,\quad \langle Lx_h,y_h\rangle_{Y'Y}+\big\langle G(\lambda,x_h),y_h\big\rangle_{Y'Y}=0. \tag{16.6}$$

Let $b:X\times Y\to\mathbb{R}$ be the bilinear form defined by

$$\text{for all } (x,y)\in X\times Y,\quad b(x,y)=\langle Lx,y\rangle_{Y'Y}. \tag{16.7}$$

Then the stability assumption reads: for all $h\in(0,1]$,

$$\inf_{\substack{x\in X_h\\ \|x\|_X=1}}\ \sup_{\substack{y\in Y_h\\ \|y\|_Y=1}} b(x,y)\geqslant\beta>0 \tag{16.8}$$

and

$$\dim X_h=\dim Y_h. \tag{16.9}$$

The consistency assumption reads

$$\text{for all } x \in X, \quad \lim_{h \to 0} \inf_{x_h \in X_h} \|x - x_h\|_X = 0. \tag{16.10}$$

To study approximation (16.6) we proceed in a way similar to the one in Theorem 7.1. Let $\Pi_{X_h} \in \mathscr{L}(X; X_h)$ and $\Pi_{Y_h} \in \mathscr{L}(Y; Y_h)$ be the two projectors defined by

$$\text{for all } y_h \in Y_h, \quad b(x - \Pi_{X_h} x, y_h) = 0 \tag{16.11}$$

and

$$\text{for all } x_h \in X_h, \quad b(x_h, y - \Pi_{Y_h} y) = 0. \tag{16.12}$$

Then we construct the family $\{F_h\}_{0<h\leqslant 1}$ of mappings $F_h : \mathbb{R} \times X \to Y'$ by setting: for $\lambda \in \mathbb{R}$ and for $x \in X$, $y \in Y$,

$$\langle F_h(\lambda, x), y \rangle_{Y'Y} = \langle F(\lambda, x), \Pi_{Y_h} y \rangle_{Y'Y} + b(x, y - \Pi_{Y_h} y).$$

It is a simple matter (see the proof of Theorem 7.1) to check that problem (16.6) is equivalent to find $\lambda \in \mathbb{R}$ and $x_h \in X$ such that

$$F_h(\lambda, x_h) = 0. \tag{16.13}$$

To study problem (16.13) in some neighborhood of the simple limit point $(\lambda_0, x_0) \in \mathbb{R} \times X$, we define the mapping $\Phi_h : \mathbb{R} \times \mathbb{R} \times X \to \mathbb{R} \times Y'$ by

$$\Phi_h(s, \lambda, x) = \big(B\big(x - u(s)\big), F_h(\lambda, x)\big),$$

the discrete analogue of Φ given in (15.3). Here $(\lambda, u) : [-\varepsilon, \varepsilon] \to \mathbb{R} \times X$ is the exact solution set in a neighborhood of (λ_0, x_0). Clearly if $(s, \lambda_h, x_h) \in \mathbb{R} \times \mathbb{R} \times X$ satisfies

$$\Phi_h(s, \lambda_h, x_h) = 0 \tag{16.14}$$

then (λ_h, x_h) is a solution to (16.13) and consequently to (16.6). Conversely if $(\lambda_h, x_h) \in \mathbb{R} \times X$ is a solution to (16.6) or (16.13), then $(s \equiv B(x_h) - B(x_0), \lambda_h, x_h)$ is a solution to (16.14).

The study of problem (16.14) can be done as an application of the results of Section 12 with the mapping Φ_h and the solution set $\{(s, \lambda(s), u(s)) \colon s \in [-\varepsilon, \varepsilon]\}$. Let us check that Φ_h satisfies successively consistency (12.10), stability (12.11), and the Lipschitzian condition (12.9) at the solutions $(\lambda(s), u(s))$.

For a given $s \in [-\varepsilon, \varepsilon]$, the following estimate holds:

$$\begin{aligned}
&\big\|\Phi_h\big(s, \lambda(s), u(s)\big)\big\|_{\mathbb{R}\times Y'} \\
&\quad = \sup_{\substack{y\in Y\\ \|y\|_Y=1}} \big\langle F_h\big(\lambda(s), u(s)\big), y\big\rangle_{Y'Y} = \sup_{\substack{y\in Y\\ \|y\|_Y=1}} b\big(u(s), y - \Pi_{Y_h} y\big) \\
&\quad = \sup_{\substack{y\in Y\\ \|y\|_Y=1}} b\big(u(s) - \Pi_{X_h} u(s), y\big) \leqslant \|L\|_{X;Y'} \big\|u(s) - \Pi_{X_h} u(s)\big\|_X.
\end{aligned} \tag{16.15}$$

For $x \in X$ and $x_h \in X_h$, the inf–sup condition (16.8) implies

$$\begin{aligned}
\|x_h - \Pi_{X_h} x\|_X &\leqslant \frac{1}{\beta} \sup_{\substack{y_h\in Y_h\\ \|y_h\|_Y=1}} b(x_h - \Pi_{X_h} x, y_h) \\
&= \frac{1}{\beta} \sup_{\substack{y_h\in Y_h\\ \|y_h\|_Y=1}} b(x_h - x, y_h) \leqslant \|L\|_{X;Y'} \frac{1}{\beta} \|x_h - x\|_X.
\end{aligned}$$

This last inequality and the triangular inequality give

$$\|x - \Pi_{X_h} x\|_X \leqslant \left(1 + \frac{\|L\|_{X;Y'}}{\beta}\right) \inf_{x_h\in X_h} \|x - x_h\|_X. \tag{16.16}$$

Then we use estimate (16.15) with (16.16) and the consistency assumption (16.10) to check consistency (12.10).

Noticing that for $\delta \in \mathbb{R}$ and $v \in X$,

$$\begin{aligned}
&\mathrm{D}_{\lambda x}\Phi_h\big(s, \lambda(s), u(s)\big)(\delta, v) \\
&\quad = \big(B(v), \mathrm{D}_\lambda F_h\big(\lambda(s), u(s)\big)\delta + \mathrm{D}_x F_h\big(\lambda(s), u(s)\big)v\big) \\
&\quad = \mathrm{D}_{\lambda x}\Phi\big(s, \lambda(s), u(s)\big)(\delta, v) \\
&\qquad + \Big(0, \big(\mathrm{D}_\lambda F_h\big(\lambda(s), u(s)\big) - \mathrm{D}_\lambda F\big(\lambda(s), u(s)\big)\big)\delta \\
&\qquad + \big(\mathrm{D}_x F_h\big(\lambda(s), u(s)\big) - \mathrm{D}_x F\big(\lambda(s), u(s)\big)\big)v\Big)
\end{aligned}$$

and since $\mathrm{D}_{\lambda x}\Phi(s, \lambda(s), u(s)) \in \mathscr{L}(\mathbb{R} \times X, Y')$ is an isomorphism, the stability assumption (12.11) is checked, see (1.1), if we prove

$$\lim_{h\to 0} \max_{s\in[-\varepsilon,\varepsilon]} \big\|\mathrm{D}_\lambda F_h\big(\lambda(s), u(s)\big) - \mathrm{D}_\lambda F\big(\lambda(s), u(s)\big)\big\|_{Y'} = 0 \tag{16.17}$$

and

$$\lim_{h\to 0} \max_{s\in[-\varepsilon,\varepsilon]} \big\|\mathrm{D}_x F_h\big(\lambda(s), u(s)\big) - \mathrm{D}_x F\big(\lambda(s), u(s)\big)\big\|_{X;Y'} = 0. \tag{16.18}$$

We have successively

$$\begin{aligned}
&\big\|\mathrm{D}_\lambda F_h\big(\lambda(s),u(s)\big)-\mathrm{D}_\lambda F\big(\lambda(s),u(s)\big)\big\|_{Y'}\\
&\quad=\sup_{\substack{y\in Y\\ \|y\|_Y=1}}\big\langle \mathrm{D}_\lambda F_h\big(\lambda(s),u(s)\big)-\mathrm{D}_\lambda F\big(\lambda(s),u(s)\big),y\big\rangle_{Y'Y}\\
&\quad=\sup_{\substack{y\in Y\\ \|y\|_Y=1}}\big[\big\langle \mathrm{D}_\lambda F\big(\lambda(s),u(s)\big),\Pi_{Y_h}y-y\big\rangle_{Y'Y}\big]\\
&\quad=\sup_{\substack{y\in Y\\ \|y\|_Y=1}} b\big(L^{-1}\mathrm{D}_\lambda F\big(\lambda(s),u(s)\big),\Pi_{Y_h}y-y\big)\\
&\quad=\sup_{\substack{y\in Y\\ \|y\|_Y=1}} b\big((I-\Pi_{X_h})L^{-1}\mathrm{D}_\lambda F\big(\lambda(s),u(s)\big),y\big)\\
&\quad\leqslant \|L\|_{X;Y'}\big\|(I-\Pi_{X_h})L^{-1}\mathrm{D}_\lambda F\big(\lambda(s),u(s)\big)\big\|_X.
\end{aligned}$$

Then with estimates (16.16) and (16.10), we conclude that

$$\lim_{h\to 0}\max_{s\in[-\varepsilon,\varepsilon]}\big\|\mathrm{D}_\lambda F_h\big(\lambda(s),u(s)\big)-\mathrm{D}_\lambda F\big(\lambda(s),u(s)\big)\big\|_{Y'}=0.$$

Let us prove now relation (16.18). For $w\in X$ and $y\in Y$ given we have

$$\begin{aligned}
&\big\langle\big(\mathrm{D}_x F_h\big(\lambda(s),u(s)\big)-\mathrm{D}_x F\big(\lambda(s),u(s)\big)\big)w,y\big\rangle_{Y'Y}\\
&\quad=\big\langle \mathrm{D}_x F\big(\lambda(s),u(s)\big)w,\Pi_{Y_h}y-y\big\rangle_{Y'Y}+b(w,y-\Pi_{Y_h}y)\\
&\quad=\big\langle\big(L+\mathrm{D}_x G\big(\lambda(s),u(s)\big)\big)w,\Pi_{Y_h}y-y\big\rangle_{Y'Y}+\big\langle Lw,y-\Pi_{Y_h}y\big\rangle_{Y'Y}\\
&\quad=b\big(L^{-1}\mathrm{D}_x G\big(\lambda(s),u(s)\big)w,\Pi_{Y_h}y-y\big)\\
&\quad=-b\big((I-\Pi_{X_h})L^{-1}\mathrm{D}_x G\big(\lambda(s),u(s)\big)w,y\big).
\end{aligned}$$

Let us prove now that

$$\lim_{h\to 0}\big\|(I-\Pi_{X_h})L^{-1}\mathrm{D}_x G\big(\lambda(s),u(s)\big)\big\|_{X;X}=0.$$

From assumptions (16.3) and (16.4), we get that the operator $L^{-1}\mathrm{D}_x G(\lambda(s),u(s))\in\mathscr{L}(X;X)$ is compact. Then from consistency (16.10), estimate (16.16), and result (1.3), we get that for $s\in[-\varepsilon,\varepsilon]$ the above limit is zero. With the compactness of the interval $[-\varepsilon,\varepsilon]$ and the continuity of $s\in[-\varepsilon,\varepsilon]\to(\lambda(s),u(s))\in\mathbb{R}\times X$ we conclude that relation (16.18) is true.

Finally we verify the Lipschitzian condition (12.9) for $\mathrm{D}_{\lambda x}\Phi_h$. For $(\lambda,x)\in\mathbb{R}\times X$, $s\in[-\varepsilon,\varepsilon]$, and $(\delta,v)\in\mathbb{R}\times X$,

$$\begin{aligned}
&\big[\mathrm{D}_{\lambda x}\Phi_h\big(s,\lambda(s),u(s)\big)-\mathrm{D}_{\lambda x}\Phi_h(s,\lambda,x)\big](\delta,v)\\
&\quad=\big(0,\big[\mathrm{D}F_h\big(\lambda(s),u(s)\big)-\mathrm{D}F_h(\lambda,x)\big](\delta,v)\big).
\end{aligned}\tag{16.19}$$

Moreover,

$$\begin{aligned}
&\big\|\big(\mathrm{D}F_h(\lambda(s),u(s)) - \mathrm{D}F_h(\lambda,x)\big)(\delta,v)\big\|_{Y'} \\
&\quad = \sup_{\substack{y\in Y\\ \|y\|_Y=1}} \big\langle\big(\mathrm{D}F_h(\lambda(s),u(s)) - \mathrm{D}F_h(\lambda,x)\big)(\delta,v),y\big\rangle_{Y'Y} \\
&\quad = \sup_{\substack{y\in Y\\ \|y\|_Y=1}} \big\langle\big(\mathrm{D}F(\lambda(s),u(s)) - \mathrm{D}F(\lambda,x)\big)(\delta,v),\Pi_{Y_h}y\big\rangle_{Y'Y} \\
&\quad \leqslant \|\Pi_{Y_h}\|_{Y;Y}\big\|\mathrm{D}F(\lambda(s),u(s)) - \mathrm{D}F(\lambda,x)\big\|_{\mathbb{R}\times X;Y'}\big(|\delta| + \|v\|_X\big). \qquad (16.20)
\end{aligned}$$

With relation (16.19), estimate (16.20) and the fact that $\mathrm{D}F$ is Lipschitzian at $(\lambda(s),u(s))$, (F is assumed to be C^2), we conclude that (12.9) is verified for $\mathrm{D}_{\lambda x}\Phi_h$ at the points $(s,\lambda(s),u(s))$, $s\in[-\varepsilon,\varepsilon]$.

We can now apply Theorem 12.3.

THEOREM 16.1. *Let $F:\mathbb{R}\times X\to Y'$ satisfy assumptions* (16.2), (16.3) *and* (16.4) *with $p\geqslant 2$. Let $(\lambda_0,x_0)\in\mathbb{R}\times X$ be a simple limit point of problem* (16.1). *Under hypotheses* (16.8), (16.9) *and* (16.10), *there exist $h_0>0$, $\delta_0>0$, and for all $0<h\leqslant h_0$ a C^p mapping $(\lambda_h,u_h):s\in[-\varepsilon,\varepsilon]\to(\lambda_h(s),u_h(s))\in\mathbb{R}\times X$ such that*

$$\begin{aligned}
&\big(\Phi_h(s,\lambda,x_h) = 0 \text{ and } (\lambda,x_h)\in\overline{B}\big((\lambda(s),u(s)),\delta_0\big)\big) \\
&\quad \Leftrightarrow \big(\lambda=\lambda_h(s),\ x_h = u_h(s)\big).
\end{aligned}$$

Moreover, there exists a constant $C>0$ such that for all $s\in[-\varepsilon,\varepsilon]$,

$$\big|\lambda(s)-\lambda_h(s)\big| + \big\|u(s)-u_h(s)\big\|_X \leqslant C\inf_{x_h\in X_h}\big\|u(s)-x_h\big\|_X. \qquad (16.21)$$

PROOF. We have checked before that we can apply Theorem 12.3. Estimate (16.21) is a consequence of (12.12) and of relations (16.15) and (16.16). □

We assume now that the mapping F is of class C^p, $p\geqslant 3$, and that (λ_0,x_0) is a nondegenerate simple limit point, which means, see (15.7),

$$\mathrm{D}^2_{xx}F(\lambda_0,x_0)(\varphi,\varphi)\notin \text{Range}\,\big(\mathrm{D}_xF(\lambda_0,x_0)\big).$$

We have for $j=0,1,2$,

$$\begin{aligned}
\left\|\frac{\mathrm{d}^j}{\mathrm{d}s^j}\Phi_h\big(s,\lambda(s),u(s)\big)\right\|_{\mathbb{R}\times Y'} &= \left\|\frac{\mathrm{d}^j}{\mathrm{d}s^j}F_h\big(\lambda(s),u(s)\big)\right\|_{Y'} \\
&= \sup_{\substack{y\in Y\\ \|y\|_Y=1}}\left\langle\frac{\mathrm{d}^j}{\mathrm{d}s^j}F_h\big(\lambda(s),u(s)\big),y\right\rangle_{Y'Y}
\end{aligned}$$

$$= \sup_{\substack{y\in Y\\ \|y\|_Y=1}} b\left(\frac{\mathrm{d}^j}{\mathrm{d}s^j}u(s), y - \Pi_{Y_h}y\right)$$

$$= \sup_{\substack{y\in Y\\ \|y\|_Y=1}} b\left(\frac{\mathrm{d}^j}{\mathrm{d}s^j}u(s) - \Pi_{X_h}\frac{\mathrm{d}^j}{\mathrm{d}s^j}u(s), y\right).$$

So with this last relation and (16.16) we get for $j = 0, 1, 2,$

$$\lim_{h\to 0}\max_{s\in[-\varepsilon,\varepsilon]}\left\|\frac{\mathrm{d}^j}{\mathrm{d}s^j}\Phi_h\big(s, \lambda(s), u(s)\big)\right\|_{\mathbb{R}\times Y'} = 0. \tag{16.22}$$

Then with relation (16.22), Theorem 12.4 implies for $j = 0, 1, 2,$

$$\lim_{h\to 0}\max_{s\in[-\varepsilon,\varepsilon]}\left\{\left|\frac{\mathrm{d}^j}{\mathrm{d}s^j}\big(\lambda(s) - \lambda_h(s)\big)\right| + \left\|\frac{\mathrm{d}^j}{\mathrm{d}s^j}\big(u(s) - u_h(s)\big)\right\|_X\right\} = 0. \tag{16.23}$$

We have assumed that (λ_0, x_0) is a nondegenerate simple limit point, so

$$\lambda(0) = \lambda_0, \qquad \frac{\mathrm{d}}{\mathrm{d}s}\lambda(0) = 0, \qquad \frac{\mathrm{d}^2}{\mathrm{d}s^2}\lambda(0) \neq 0.$$

There exist ε_1 with $0 < \varepsilon_1 \leqslant \varepsilon$, h_1 with $0 < h_1 \leqslant h_0$, $C > 0$, and for $0 < h \leqslant h_1$ a unique $s_h \in [-\varepsilon_1, \varepsilon_1]$ such that

$$\frac{\mathrm{d}}{\mathrm{d}s}\lambda_h(s_h) = 0, \tag{16.24}$$

$$\left|\frac{\mathrm{d}^2}{\mathrm{d}s^2}\lambda_h(s)\right| \geqslant C \quad \text{for all } s \in [-\varepsilon_1, \varepsilon_1]. \tag{16.25}$$

So the point $(\lambda_h(s_h), u_h(s_h))$ is a nondegenerate simple limit point of problem (16.13). We know then that the solutions $(\lambda_h(s), u_h(s))$ form two regular solution branches parametrized by λ which meet at the point $(\lambda_h(s_h), u_h(s_h))$, see Remark 15.1.

Notice that estimate (16.21) does not give in general an error between (λ_0, x_0) and $(\lambda_h(s_h), u_h(s_h))$ where $(\lambda_h(s_h), u_h(s_h))$ is the nondegenerate simple limit point of the approximate problem. Let us try to compare $\lambda_0 \equiv \lambda(0)$ and $\lambda_{0h} \equiv \lambda_h(s_h)$. We write

$$\lambda_0 - \lambda_{0h} = \lambda_0 - \lambda_h(0) - \int_0^{s_h}\frac{\mathrm{d}}{\mathrm{d}s}\lambda_h(s)\,\mathrm{d}s.$$

Estimate (16.25) implies that $(\mathrm{d}^2\lambda_h/\mathrm{d}s^2)(s)$ does not change sign in $[0, s_h]$ and with (16.24)

$$|\lambda_0 - \lambda_{0h}| \leqslant |\lambda_0 - \lambda_h(0)| + |s_h|\left|\frac{\mathrm{d}}{\mathrm{d}s}\lambda_h(0)\right|.$$

With the two relations (16.24) and (16.25), we deduce that

$$|s_h| \leqslant \frac{1}{C}\left|\frac{\mathrm{d}}{\mathrm{d}s}\lambda_h(0)\right| \tag{16.26}$$

and then from the above inequality we get

$$|\lambda_0 - \lambda_{0h}| \leqslant |\lambda_0 - \lambda_h(0)| + \frac{1}{C}\left|\frac{\mathrm{d}}{\mathrm{d}s}\lambda(0) - \frac{\mathrm{d}}{\mathrm{d}s}\lambda_h(0)\right|^2. \tag{16.27}$$

From estimate (16.26) and the boundedness of $(\mathrm{d}u_h/\mathrm{d}s)(s)$, $s \in [-\varepsilon, \varepsilon]$, we get

$$\|u_h(0) - u_h(s_h)\|_X \leqslant \widetilde{C}\left|\frac{\mathrm{d}}{\mathrm{d}s}\lambda_h(0)\right|, \tag{16.28}$$

and then

$$\|u(0) - u_h(s_h)\|_X \leqslant \|u(0) - u_h(0)\|_X + \widetilde{C}\left|\frac{\mathrm{d}}{\mathrm{d}s}\lambda_h(0) - \frac{\mathrm{d}}{\mathrm{d}s}\lambda(0)\right|. \tag{16.29}$$

Estimates (16.27) and (16.29) measure the error between the nondegenerate simple limit point (λ_0, x_0) and its approximation $(\lambda_h(s_h), u_h(s_h))$.

It is often possible to improve the estimate for $|\lambda(0) - \lambda_h(0)|$ along the following lines. The Taylor expansion gives

$$\begin{aligned} F(\lambda_h(0), u_h(0)) = {} & F(\lambda_0, x_0) + \mathrm{D}_x F(\lambda_0, x_0)(u_h(0) - x_0) \\ & + \mathrm{D}_\lambda F(\lambda_0, x_0)(\lambda_h(0) - \lambda_0) \\ & + \mathrm{O}(|\lambda_h(0) - \lambda_0|^2 + \|u_h(0) - x_0\|_X^2). \end{aligned} \tag{16.30}$$

On the other hand, since $F_h(\lambda_h(0), u_h(0)) = 0$, $u_h(0)$ is a solution of (16.6) and

$$\text{for all } y \in Y, \quad \langle F(\lambda_h(0), u_h(0)), \Pi_{Y_h} y\rangle_{Y'Y} = 0. \tag{16.31}$$

We choose now $y_0 \in Y$, $\|y_0\| = 1$, such that

$$\text{for all } \psi \in X, \quad \langle \mathrm{D}_x F(\lambda_0, x_0)\psi, y_0\rangle_{Y'Y} = 0. \tag{16.32}$$

From relations (16.30)–(16.32) we deduce

$$\begin{aligned}&\big\langle D_\lambda F(\lambda_0, x_0), \Pi_{Y_h} y_0\big\rangle_{Y'Y}\big(\lambda_h(0) - \lambda_0\big)\\&\quad + \big\langle \mathrm{D}_x F\big(\lambda_0, x_0\big)\big(u_h(0) - x_0\big), \Pi_{Y_h} y_0 - y_0\big\rangle_{Y'Y}\\&\qquad = \mathrm{O}\big(\big|\lambda_h(0) - \lambda_0\big|^2 + \big\|u_h(0) - x_0\big\|_X^2\big).\end{aligned} \tag{16.33}$$

The point (λ_0, x_0) is a simple limit point, so $\mathrm{D}_\lambda F(\lambda_0, x_0) \notin \mathrm{Range}(\mathrm{D}_x F(\lambda_0, x_0))$ and then

$$\big\langle D_\lambda F(\lambda_0, x_0), y_0\big\rangle_{Y'Y} \neq 0. \tag{16.34}$$

The consistency assumption (16.10) and estimate (16.16) imply

$$\lim_{h\to 0} \big\langle D_\lambda F(\lambda_0, x_0), y_0 - \Pi_{Y_h} y_0\big\rangle_{Y'Y} = 0. \tag{16.35}$$

From relations (16.33)–(16.35) we deduce the existence of a constant $C > 0$ independent of h such that

$$\begin{aligned}\big|\lambda_0 - \lambda_h(0)\big| \leqslant C\big[\big\|u_h(0) - x_0\big\|_X + \|y_0 - \Pi_{Y_h} y_0\|_Y\big]\\ \times\big\|u_h(0) - x_0\big\|_X.\end{aligned} \tag{16.36}$$

With (16.36) estimate (16.27) can be improved in the following way

$$\begin{aligned}|\lambda_0 - \lambda_{0h}| \leqslant C\bigg(&\big\|u_h(0) - x_0\big\|_X^2 + \left|\frac{\mathrm{d}}{\mathrm{d}s}\lambda(0) - \frac{\mathrm{d}}{\mathrm{d}s}\lambda_h(0)\right|^2\\&+ \big\|u_h(0) - x_0\big\|_X \|y_0 - \Pi_{Y_h} y_0\|_Y\bigg).\end{aligned} \tag{16.37}$$

REMARK 16.1. So far we do not have presented a posteriori error estimates in the case of simple limit points. In fact under the assumptions of Theorem 16.1, we can prove there exists a constant $C > 0$ such that for all $s \in [-\varepsilon, \varepsilon]$,

$$\big|\lambda_h(s) - \lambda(s)\big| + \big\|u_h(s) - u(s)\big\|_X \leqslant C\big\|F\big(\lambda_h(s), u_h(s)\big)\big\|_{Y'}. \tag{16.38}$$

Indeed estimate (12.13) of Theorem 12.3 with the mapping Φ gives

$$\big|\lambda_h(s) - \lambda(s)\big| + \big\|u_h(s) - u(s)\big\|_X \leqslant C\big\|\Phi\big(s, \lambda_h(s), u_h(s)\big)\big\|_{\mathbb{R}\times Y'}.$$

Going back to the definition of Φ_h and Φ we deduce that

$$B\big(u_h(s) - x_0\big) - s = B\big(u_h(s) - u(s)\big) + B\big(u(s) - x_0\big) = 0$$

and then

$$\Phi\big(s, \lambda_h(s), u_h(s)\big) = \big(0, F\big(\lambda_h(s), u_h(s)\big)\big). \qquad \Box$$

REMARK 16.2. The goal of this section was to present a general but simple method to study approximations of simple limit points. We have not emphasized on what are the weakest assumptions on F.

Assumption (16.2) on the shape of F can be avoided, but then the study is different, see CALOZ [1994].

The function F does not need to be a C^2 function but $\mathrm{D}F$ needs to be Lipschitzian at $(\lambda(s), u(s))$ for $s \in [-\varepsilon, \varepsilon]$.

When we have applied the results of Section 12, we have decided to associate with each solution $(s, \lambda(s), u(s))$, $s \in [-\varepsilon, \varepsilon]$, the solution $(\lambda(s), u(s))$ itself instead of some approximations of it, say $(\tilde{\lambda}_h(s), \tilde{u}_h(s))$. Adding the appropriate assumptions on these approximations, see Sections 12 and 14, we can reproduce the approach of the present section to get estimates better suited to study the error in different norms. Another method would be to generalize Theorem 6.4.

17. Approximation of simple limit points. An example

The goal of this section is to apply the general results on the approximation of simple limit points in Section 16 to a simple and concrete example.

For simplicity we consider again problem (14.3). Let $\Omega \subset \mathbb{R}^2$ be a regular convex open bounded set and let $f : H_0^1(\Omega) \to L^2(\Omega)$ be given by: for all $v \in H_0^1(\Omega)$,

$$\text{for all } x \in \Omega, \quad f(v)(x) = v^2(x) + 1. \tag{17.1}$$

We define the mapping $F : \mathbb{R} \times H_0^1(\Omega) \to H^{-1}(\Omega)$ by: for all $(\lambda, v) \in \mathbb{R} \times H_0^1(\Omega)$, $w \in H_0^1(\Omega)$,

$$\big\langle F(\lambda, v), w \big\rangle_{H^{-1}(\Omega)\, H_0^1(\Omega)} = \int_\Omega \mathbf{grad}\, v\, \mathbf{grad}\, w\, \mathrm{d}x - \lambda \int_\Omega f(v) w\, \mathrm{d}x. \tag{17.2}$$

There exists a solution $(\lambda^*, u^*) \in \mathbb{R} \times H_0^1(\Omega)$, u^* positive in Ω, to the problem

$$F(\lambda, v) = 0, \tag{17.3}$$

such that $\mathrm{D}_v F(\lambda^*, u^*) \in \mathscr{L}(H_0^1(\Omega); H^{-1}(\Omega))$ is not an isomorphism, see Remark 3.2. We can more precisely study the mapping $\mathrm{D}_v F(\lambda^*, u^*)$. If $T \in \mathscr{L}(H^{-1}(\Omega); H_0^1(\Omega))$ is the inverse of $(-\Delta)$ with homogeneous Dirichlet boundary conditions, then for $w \in H_0^1(\Omega)$,

$$T\,\mathrm{D}_v F(\lambda^*, u^*) w = w - \lambda^* T\,\mathrm{D}f(u^*) w$$

and the mapping $T\,\mathrm{D}_v F(\lambda^*, u^*) \in \mathscr{L}(H_0^1(\Omega); H_0^1(\Omega))$ has a kernel of dimension 1, namely $\operatorname{span}\{\varphi\}$, where the function $\varphi \in H_0^1(\Omega)$, $\varphi > 0$ in Ω, is an eigenvector

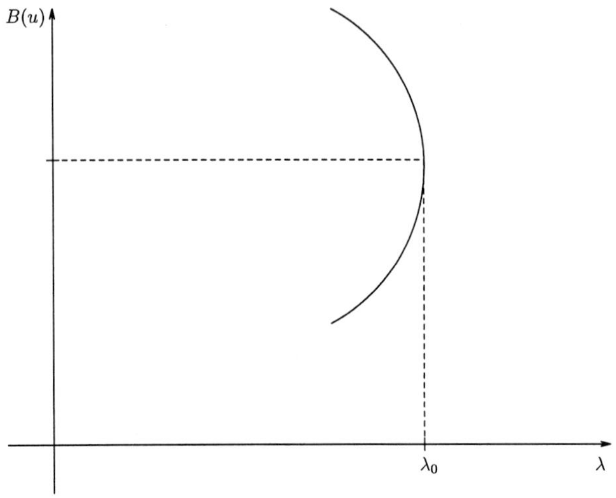

FIG. 17.1. (λ^*, u^*): Nondegenerate simple limit point.

corresponding to the least eigenvalue λ^* of the problem: find $\delta \in \mathbb{R}$ and $w \in H_0^1(\Omega)$, $w \neq 0$, such that

$$-\Delta w = \delta\, \mathrm{D}f(u^*)w \quad \text{in } \Omega$$

(see Remark 3.4, the proof of Theorem 3.4, and Theorem 3.2). Therefore the mapping $T\,\mathrm{D}_v F(\lambda^*, u^*)$ is a Fredholm operator of index 0. The range of $T\,\mathrm{D}_v F(\lambda^*, u^*)$ is orthogonal to $\varphi \in H_0^1(\Omega)$, with respect to the scalar product

$$(u, v)_{1,\Omega} = \int_\Omega \mathbf{grad}\, u\, \mathbf{grad}\, v\, \mathrm{d}x.$$

Since both u^* and φ are positive in Ω, we easily check that

$$T\,\mathrm{D}F(\lambda^*, u^*) \in \mathscr{L}\big(\mathbb{R} \times H_0^1(\Omega); H_0^1(\Omega)\big)$$

is a Fredholm operator of index 1 and $H_0^1(\Omega) = \mathrm{Range}(T\,\mathrm{D}F(\lambda^*, u^*))$. The operator $T \in \mathscr{L}(H^{-1}(\Omega); H_0^1(\Omega))$ is an isomorphism, so

$$\mathrm{D}F(\lambda^*, u^*) \in \mathscr{L}\big(\mathbb{R} \times H_0^1(\Omega); H^{-1}(\Omega)\big)$$

is a Fredholm operator of index 1.

The pair $(\lambda^*, u^*) \in \mathbb{R} \times H_0^1(\Omega)$ satisfies condition (15.2). It is a simple limit point of problem (17.3). Notice that this result has already been checked in Remark 3.4. It is a simple matter to prove that $\mathrm{D}^2_{vv} F(\lambda^*, u^*)(\varphi, \varphi)$ does not belong to the range of $\mathrm{D}_v F(\lambda^*, u^*)$, so relation (15.7) is satisfied and (λ^*, u^*) is a nondegenerate simple limit point.

Let $B \in H^{-1}(\Omega)$ with $B(\varphi) = 1$ denote a functional appearing in the definition of the mapping Φ in (15.3). For instance we could consider

$$B(v) = \frac{1}{\|\varphi\|_{0,\Omega}^2} \int_\Omega \varphi v \, \mathrm{d}x.$$

Then graphically we have the situation of Fig. 17.1. Here $(\lambda, u) : s \in [-\varepsilon, \varepsilon] \to (\lambda(s), u(s)) \in \mathbb{R} \times H_0^1(\Omega)$ denotes the C^∞ mapping such that for all $s \in [-\varepsilon, \varepsilon]$,

$$F\big(\lambda(s), u(s)\big) = 0 \quad \text{and} \quad B\big(u(s) - u^*\big) = s,$$
$$\lambda(0) = \lambda^* \quad \text{and} \quad u(0) = u^*.$$

Given a family $\{\mathcal{T}_h\}_{0<h\leqslant 1}$ of isoparametric triangulations of Ω, we consider the corresponding family $\{V_h\}_{0<h\leqslant 1}$ of finite element subspaces of degree $k = 1$ or 2,

$$V_h = \big\{v \in C^0(\mathbb{R}^2)\colon\ v(x) = 0 \text{ for all } x \notin \Omega_h,\ v|_T \in P_T \\ \text{for all } T \in \mathcal{T}_h\big\}, \tag{17.4}$$

see (3.27). A Galerkin approximation to problem (17.3) consists in finding $\lambda \in \mathbb{R}$ and $u_h \in V_h$ such that

$$\text{for all } v_h \in V_h, \quad \big\langle F(\lambda, u_h), v_h \big\rangle_{H^{-1}(\Omega)\, H_0^1(\Omega)} = 0. \tag{17.5}$$

We will study problem (17.5) in a neighborhood of the nondegenerate simple limit point (λ^*, u^*) by applying the results of Section 16 with $X = Y = H_0^1(\Omega)$.

It is not difficult to check that the mapping F in (17.2) has the structure (16.2) with the isomorphism $L = (-\Delta) : H_0^1(\Omega) \to H^{-1}(\Omega)$ and the C^∞ mapping $G(\lambda, \bullet) = -\lambda f(\bullet)$ such that $\lambda \mathrm{D} f(v) \in \mathscr{L}(H_0^1(\Omega); H^{-1}(\Omega))$ is compact for all $v \in H_0^1(\Omega)$.

The bilinear form $b : H_0^1(\Omega) \times H_0^1(\Omega) \to \mathbb{R}$ in our example is

$$\text{for all } v, w \in H_0^1(\Omega), \quad b(v, w) = \int_\Omega \mathbf{grad}\, v \ \mathbf{grad}\, w \, \mathrm{d}x.$$

Then problem (17.5) can be written equivalently in the form: find $\lambda \in \mathbb{R}$ and $u_h \in H_0^1(\Omega)$ such that for all $v \in H_0^1(\Omega)$,

$$\big\langle F_h(\lambda, u_h), v \big\rangle_{H^{-1}(\Omega)\, H_0^1(\Omega)} \\ \equiv \big\langle F(\lambda, u_h), \Pi_{V_h} v \big\rangle_{H^{-1}(\Omega)\, H_0^1(\Omega)} + b(u_h, v - \Pi_{V_h} v) = 0 \tag{17.6}$$

or equivalently

$$\int_\Omega \mathbf{grad}\, u_h \ \mathbf{grad}\, v \, \mathrm{d}x - \lambda \int_\Omega f(u_h) \Pi_{V_h} v \, \mathrm{d}x = 0,$$

with the elliptic projector $\Pi_{V_h} \in \mathscr{L}(H_0^1(\Omega); V_h)$. Then stability assumptions (16.8) and (16.9) are immediate. Furthermore, the approximability hypothesis (16.10) is a consequence of the classical polynomial interpolation results.

We can apply Theorem 16.1.

THEOREM 17.1. *There exist $h_0 > 0$, $\delta_0 > 0$, and for all $0 < h \leqslant h_0$ a C^∞ mapping $(\lambda_h, u_h) : s \in [-\varepsilon, \varepsilon] \to (\lambda_h(s), u_h(s)) \in \mathbb{R} \times H_0^1(\Omega)$ such that*

$$\Big((\lambda, u_h) \in \mathbb{R} \times V_h \text{ is a solution to } (17.5),\ B\big(u_h - u(s)\big) = 0, \text{ and}$$
$$(\lambda, u_h) \in \overline{B}\big((\lambda(s), u(s)), \delta_0\big)\Big)$$
$$\Leftrightarrow \big(\lambda = \lambda_h(s),\ u_h = u_h(s)\big). \tag{17.7}$$

Moreover, there exists a constant $C > 0$ such that for all $s \in [-\varepsilon, \varepsilon]$ and $0 < h \leqslant h_0$,

$$\big|\lambda(s) - \lambda_h(s)\big| + \big|u(s) - u_h(s)\big|_{1,\Omega} \leqslant Ch^k, \tag{17.8}$$

$k = 1$ or 2 depending on V_h.

There exists a $s_h \in [-\varepsilon, \varepsilon]$ such that the solution $(\lambda_h(s_h), u_h(s_h))$ is a nondegenerate simple limit point of problem (17.6) and the following estimates hold:

$$\big|\lambda^* - \lambda_h(0)\big| \leqslant Ch^{2k} \tag{17.9}$$

and

$$\big|\lambda^* - \lambda_h(s_h)\big| \leqslant Ch^{2k}. \tag{17.10}$$

PROOF. We have checked before that we can apply the results of the previous section. Estimate (17.8) is a consequence of (16.21). Estimates (17.9), (17.10) are got from (16.36), (16.37). Notice that an estimate for the term

$$\left|\frac{\mathrm{d}}{\mathrm{d}s}\lambda(0) - \frac{\mathrm{d}}{\mathrm{d}s}\lambda_h(0)\right|$$

is a consequence of Theorem 12.4 via estimate (12.14). □

CHAPTER V

Approximation of Simple Bifurcation Points

Let X, Z be two real Banach spaces and $F: X \to Z$ be a C^p mapping, $p \geqslant 1$. In this chapter we shall analyze approximations of the problem

$$F(x) = 0$$

in a neighborhood of a solution $x_0 \in X$ when $\mathrm{D}F(x_0)$ is a nonsurjective Fredholm operator of index greater than 1, that is to say

(i) $\mathrm{D}F(x_0) \in \mathscr{L}(X; Z)$ has a finite-dimensional kernel,

(ii) $\mathrm{Range}(\mathrm{D}F(x_0))$ is closed, different from Z,

(iii) $\dim \mathrm{Ker}(\mathrm{D}F(x_0)) - \dim(Z/\mathrm{Range}(\mathrm{D}F(x_0))) \geqslant 1$,

where $Z/\mathrm{Range}(\mathrm{D}F(x_0))$ denotes the quotient space of Z by $\mathrm{Range}(\mathrm{D}F(x_0))$. Such a solution x_0 is called singular. Compared to the case treated in the previous chapter, where the solution x_0 was supposed to be regular, that is satisfying assumption (11.1), the study is much more involved. Our goal is to present the method in a simple way and to complete it by bibliographical comments.

In Section 18 we develop a procedure to study the exact problem. This method is the standard Lyapunov–Schmidt decomposition, leading to the bifurcation equation. The common case when $X = \mathbb{R} \times \mathscr{X}$, $\mathbb{R}$ representing the parameter space and the Banach space $\mathscr{X}$ being the state space, is detailed in Section 19. Then approximations are analyzed in Section 20 where we get estimates between the bifurcation equation and the approximation bifurcation equations. In Section 21, we thoroughly compare the solutions of the bifurcation equation and those of the approximation bifurcation equations when these equations are defined in $\mathbb{R}^2$. A simple model example is presented in Section 22 to illustrate the general theory. Finally in Section 23 we address some further comments and bibliographical complements.

18. Lyapunov–Schmidt procedure. Bifurcation equations

Let X, Z be two real Banach spaces and $F: X \to Z$ be a C^p mapping, $p \geqslant 1$. A point $x \in X$ is called singular if

$$\mathrm{D}F(x) \in \mathscr{L}(X;Z) \text{ is a Fredholm operator of index } m,\ m \in \mathbb{N},\ m \geqslant 1,$$
$$\text{and } \mathrm{Range}(\mathrm{D}F(x_0)) \neq Z; \tag{18.1}$$

the index of the operator $\mathrm{D}F(x)$ is

$$\mathrm{index}\big(\mathrm{D}F(x)\big) = \dim \mathrm{Ker}\big(\mathrm{D}F(x)\big) - \dim \mathrm{Coker}\big(\mathrm{D}F(x)\big) = m,$$

where $\mathrm{Coker}(\mathrm{D}F(x))$ is the quotient space $Z/\mathrm{Range}(\mathrm{D}F(x))$, and the codimension of $\mathrm{Range}(\mathrm{D}F(x))$ (the dimension of the cokernel) is equal to $k \geqslant 1$.

In this section, we will present a method to analyze the solution set S of the problem

$$F(x) = 0, \tag{18.2}$$

in a neighborhood of the singular point $x_0 \in X$ satisfying $F(x_0) = 0$. Such a point x_0 is called a singular solution.

Let m be the index of $\mathrm{D}F(x_0)$ and k be the dimension of $\mathrm{Coker}(\mathrm{D}F(x_0))$.

Since the dimension of $\mathrm{Ker}(\mathrm{D}F(x_0))$ is finite this subspace is direct, that is there exists a closed subspace $X_1 \subset X$ such that

$$X = X_1 \oplus \mathrm{Ker}\big(\mathrm{D}F(x_0)\big).$$

The range of $\mathrm{D}F(x_0) \subset Z$ is of finite codimension k, so this subspace is direct. There exists a subspace $Z_1 \subset Z$, of dimension k, such that

$$Z = Z_1 \oplus \mathrm{Range}\big(\mathrm{D}F(x_0)\big).$$

Subsequently, in this chapter we shall use the following notations. A point $x \in X$ has a unique representation of the form

$$x = x_1 + x_2 \quad \text{with } x_1 \in X_1 \text{ and } x_2 \in \mathrm{Ker}\big(\mathrm{D}F(x_0)\big)$$

and $P_1 \in \mathscr{L}(X; X_1)$, $P_2 \in \mathscr{L}(X; \mathrm{Ker}(\mathrm{D}F(x_0)))$ denote the projectors associated with the decomposition of X, given by

$$P_1(x) = x_1, \qquad P_2(x) = x_2.$$

In a similar way a point $z \in Z$ has a unique representation of the form

$$z = z_1 + z_2 \quad \text{with } z_1 \in Z_1 \text{ and } z_2 \in \mathrm{Range}\big(\mathrm{D}F(x_0)\big)$$

and $Q_1 \in \mathscr{L}(Z; Z_1)$, $Q_2 \in \mathscr{L}(Z; \mathrm{Range}(\mathrm{D}F(x_0)))$ denote the projectors associated with the decomposition of Z, given by

$$Q_1(z) = z_1, \qquad Q_2(z) = z_2.$$

With these notations problem (18.2) can be written in the equivalent form

$$Q_1F(x) = 0 \quad \text{and} \quad Q_2F(x) = 0.$$

Since we study problem (18.2) in a neighborhood of the singular solution x_0, we write it in the following equivalent way: find $x_1 \in X_1$, $x_2 \in \operatorname{Ker}(DF(x_0))$ such that

$$Q_1F(x_0 + x_1 + x_2) = 0 \tag{18.3}$$

and

$$Q_2F(x_0 + x_1 + x_2) = 0. \tag{18.4}$$

Let us look first at Eq. (18.4). The mapping $\mathscr{F} : X_1 \times \operatorname{Ker}(DF(x_0)) \to \operatorname{Range}(DF(x_0))$ is defined for $(x_1, x_2) \in X_1 \times \operatorname{Ker}(DF(x_0))$ by

$$\mathscr{F}(x_1, x_2) = Q_2F(x_0 + x_1 + x_2). \tag{18.5}$$

We have $\mathscr{F}(0,0) = 0$ and the derivative $D_{x_1}\mathscr{F}(0,0) \in \mathscr{L}(X_1; \operatorname{Range}(DF(x_0))$ is equal to $Q_2DF(x_0)$, which is an isomorphism from X_1 onto $\operatorname{Range}(DF(x_0))$.

We apply the implicit function Theorem 2.3. There exist positive constants ε, η, and a mapping $g : B(0,\varepsilon) \subset \operatorname{Ker}(DF(x_0)) \to X_1$ which is of class C^p, such that for all $x_2 \in B(0,\varepsilon)$,

$$\big(\mathscr{F}(x_1, x_2) = 0 \text{ and } x_1 \in B(0,\eta) \subset X_1\big) \Longleftrightarrow x_1 = g(x_2).$$

In a neighborhood of x_0, the solution set to $Q_2F(x) = 0$ is the set

$$S_{\mathscr{F}} = \{x \in X\colon\ x = x_0 + g(x_2) + x_2,\ x_2 \in B(0,\varepsilon)\},$$

so

$$Q_2F\big(x_0 + g(x_2) + x_2\big) = 0 \quad \text{for all } x_2 \in B(0,\varepsilon). \tag{18.6}$$

To complete our study, we need to analyze Eq. (18.3) on the solution set $S_{\mathscr{F}}$. The mapping $\mathscr{G} : B(0,\varepsilon) \subset \operatorname{Ker}(DF(x_0)) \to Z_1$ is defined for $x_2 \in B(0,\varepsilon)$ by

$$\mathscr{G}(x_2) = Q_1F\big(x_0 + g(x_2) + x_2\big). \tag{18.7}$$

Notice that the mapping $\mathscr{G}$ can be identified to a mapping defined in $\mathbb{R}^{k+m}$ with values in $\mathbb{R}^k$. Indeed let $\zeta_1, \ldots, \zeta_k$ be a basis of Z_1 and $\varphi_1, \ldots, \varphi_{k+m}$ be a basis of

$\mathrm{Ker}(\mathrm{D}F(x_0))$. Then we can consider for a convenient $\delta > 0$ the function $\tilde{\mathscr{G}} : B(0,\delta) \subset \mathbb{R}^{k+m} \to \mathbb{R}^k$ given by: for $\boldsymbol{\alpha} = (\alpha_1, \ldots, \alpha_{k+m}) \in B(0,\delta)$

$$\tilde{\mathscr{G}}(\boldsymbol{\alpha}) = \left(\mathscr{G}_1\left(\sum_{j=1}^{k+m} \alpha_j \varphi_j\right), \ldots, \mathscr{G}_k\left(\sum_{j=1}^{k+m} \alpha_j \varphi_j\right)\right)$$

where

$$\mathscr{G}(x_2) = \sum_{l=1}^{k} \mathscr{G}_l(x_2)\zeta_l.$$

Therefore solving problem (18.2) in a neighborhood of the singular solution x_0 has been reduced to solving in a neighborhood of the origin the equation

$$\mathscr{G}(x_2) = 0 \tag{18.8}$$

or the finite nonlinear system of k equations with $k + m$ unknowns

$$\tilde{\mathscr{G}}(\boldsymbol{\alpha}) = 0. \tag{18.8$'$}$$

Equation (18.8) (or (18.8′)) is called the bifurcation equation of problem (18.2) at x_0. The method we have used to get the bifurcation equation is called the Lyapunov–Schmidt procedure.

REMARK 18.1. There is a different way to derive the bifurcation equation, as proposed in CROUZEIX and RAPPAZ [1989]. The idea is similar to the one we have used to study a solution set containing a simple limit point. It consists in transforming the problem to reduce it to a simpler one.

Let the mapping $\Phi : \mathbb{R}^k \times X \to Z$ be given by: for $(\boldsymbol{\mu}, x) \in \mathbb{R}^k \times X$

$$\Phi(\boldsymbol{\mu}, x) = F(x) + \sum_{j=1}^{k} \mu_j \zeta_j.$$

Recall that the elements $\zeta_1, \ldots, \zeta_k$ form a basis of Z_1. Then finding the solutions of problem (18.2) in a neighborhood of x_0 is equivalent to find the solutions $(\boldsymbol{\mu}, x) \in \mathbb{R}^k \times X$ of

$$\Phi(\boldsymbol{\mu}, x) = 0 \tag{18.9}$$

in a neighborhood of $(0, x_0)$ satisfying $\boldsymbol{\mu} = 0$. Then problem (18.9) is solved in two steps. First we look for the solutions of (18.9) in a neighborhood of $(0, x_0)$. Secondly we determine all the solutions $(\boldsymbol{\mu}, x)$ with $\boldsymbol{\mu} = 0$.

Remark that $\mathrm{D}\Phi(0, x_0) \in \mathscr{L}(\mathbb{R}^k \times X; Z)$ is a Fredholm operator with index $k + m$. Indeed

$$\operatorname{Ker}\big(\mathrm{D}\Phi(0, x_0)\big) = \big\{(0, x) \in \mathbb{R}^k \times X\colon\ x \in \operatorname{Ker}\big(\mathrm{D}F(x_0)\big)\big\}.$$

Moreover, the range of $\mathrm{D}\Phi(0, x_0)$ is equal to Z. We have a problem studied in Section 15 when $k = 1$. The case $k \geqslant 2$ can be treated in a similar way.

In the second step we look at the solutions of (18.9) with $\boldsymbol{\mu} = 0$, which is precisely the bifurcation equation.

REMARK 18.2. Since $F(x_0) = 0$, we easily check that $g(0) = 0$ and then $\mathscr{G}(0) = 0$. Moreover, when differentiating (18.7) with respect to x_2, we get $\mathrm{D}\mathscr{G}(0) = 0$. Hence problem (18.8) is singular. In the next section we will study the bifurcation equation more carefully.

19. The case with parameter and state spaces

In this section, we analyze more carefully the bifurcation equation deduced from the Lyapunov–Schmidt procedure when the C^p mapping F, $p \geqslant 2$, is

$$F\colon (\lambda, \kappa) \in \mathbb{R} \times \mathscr{X} \to F(\lambda, \kappa) \in Z$$

where $\mathbb{R}$ represents the parameter space while $\mathscr{X}$ is the state space.

Let $(\lambda_0, \kappa_0) \in \mathbb{R} \times \mathscr{X}$ be a singular solution of

$$F(\lambda, \kappa) = 0 \tag{19.1}$$

such that

$$\begin{aligned} &\text{(i)} \quad \dim \operatorname{Ker}(\mathrm{D}_\kappa F(\lambda_0, \kappa_0)) = 1, \\ &\text{(ii)} \quad \operatorname{codim}(\mathrm{D}_\kappa F(\lambda_0, \kappa_0)) = 1, \\ &\text{(iii)} \quad \mathrm{D}_\lambda F(\lambda_0, \kappa_0) = 0. \end{aligned} \tag{19.2}$$

The last assumption in (19.2) is satisfied when we study bifurcation from the trivial branch of solutions $\{(\lambda, 0)\colon\ \lambda \in \mathbb{R}\}$, when F satisfies $F(\lambda, 0) = 0$ for all $\lambda \in \mathbb{R}$. At the end of this section, we shall extend our results with weaker assumptions than (19.2).

Under assumptions (19.2), we can write the decomposition of both spaces X and Z in a more detailed way than in the previous section. Notice that the derivative $\mathrm{D}F(\lambda_0, \kappa_0) \in \mathscr{L}(\mathbb{R} \times \mathscr{X}; Z)$ satisfies: for all $(\delta, w) \in \mathbb{R} \times \mathscr{X}$

$$\mathrm{D}F(\lambda_0, \kappa_0)(\delta, w) = \mathrm{D}_\kappa F(\lambda_0, \kappa_0) w.$$

Since the kernel of $\mathrm{D}_\kappa F(\lambda_0, \kappa_0)$ is of dimension 1, there is a closed subspace $\mathscr{X}_1 \subset \mathscr{X}$ and $\xi \in \operatorname{Ker}(\mathrm{D}_\kappa F(\lambda_0, \kappa_0))$, $\xi \neq 0$, such that

$$\mathscr{X} = \mathscr{X}_1 \oplus \operatorname{Ker}\big(\mathrm{D}_\kappa F(\lambda_0, \kappa_0)\big) \equiv \mathscr{X}_1 \oplus \operatorname{span}\{\xi\}. \tag{19.3}$$

From assumption (19.2), we deduce that the kernel of $DF(\lambda_0, \kappa_0)$ is of dimension 2 and that

$$\operatorname{Ker}\bigl(DF(\lambda_0, \kappa_0)\bigr) = \mathbb{R} \times \operatorname{Ker}\bigl(D_\kappa F(\lambda_0, \kappa_0)\bigr).$$

Hence we can write a decomposition of X:

$$X = X_1 \oplus \operatorname{Ker}\bigl(DF(\lambda_0, \kappa_0)\bigr) \equiv X_1 \oplus \operatorname{span}\{\varphi_1, \varphi_2\}, \tag{19.4}$$

where $X_1 = \{0\} \times \mathscr{X}_1$, $\varphi_1 = (1, 0) \in \mathbb{R} \times \mathscr{X}$, and $\varphi_2 = (0, \xi) \in \mathbb{R} \times \mathscr{X}$.

Since the range of $DF(\lambda_0, \kappa_0)$ is equal to the range of $D_\kappa F(\lambda_0, \kappa_0)$ and the codimension of $D_\kappa F(\lambda_0, \kappa_0)$ is 1, the following decomposition of Z holds:

$$\begin{aligned} Z &= Z_1 \oplus \operatorname{Range}\bigl(D_\kappa F(\lambda_0, \kappa_0)\bigr) \\ &\equiv \operatorname{span}\{\zeta_1\} \oplus \operatorname{Range}\bigl(D_\kappa F(\lambda_0, \kappa_0)\bigr), \end{aligned} \tag{19.5}$$

where ζ_1 does not belong to the range of $D_\kappa F(\lambda_0, \kappa_0)$. Let P_1, P_2 denote the two projectors associated with decomposition (19.4) of X and Q_1, Q_2 associated with decomposition (19.5) of Z.

We can apply the Lyapunov–Schmidt procedure given in Section 18. Here we have $k = m = 1$. There exist positive constants ε, η, and a mapping

$$g : (\alpha_1, w) \in B(0, \varepsilon) \subset \mathbb{R} \times \operatorname{Ker}\bigl(D_\kappa F(\lambda_0, \kappa_0)\bigr) \to g(\alpha_1, w) \in \mathscr{X}_1$$

of class C^p such that for all $(\alpha_1, w) \in B(0, \varepsilon)$

$$\bigl(Q_2 F(\lambda_0 + \alpha_1, \kappa_0 + \kappa_1 + w) = 0 \text{ and } \kappa_1 \in B(0, \eta) \subset \mathscr{X}_1\bigr) \Longleftrightarrow \kappa_1 = g(\alpha_1, w).$$

In fact $w = \alpha_2 \xi$ and g can be identified as a function defined on $B(0, \delta) \subset \mathbb{R}^2$, δ small enough, by setting

$$g(\alpha_1, \alpha_2) = g(\alpha_1, \alpha_2 \xi).$$

Then g satisfies for all $(\alpha_1, \alpha_2) \in B(0, \delta)$

$$Q_2 F\bigl(\lambda_0 + \alpha_1, \kappa_0 + g(\alpha_1, \alpha_2) + \alpha_2 \xi\bigr) = 0. \tag{19.6}$$

In particular $g(0, 0) = 0$. Let the functional B in Z' satisfy

$$B(\zeta_1) = 1, \qquad B(z) = 0 \quad \text{for all } z \in \operatorname{Range}\bigl(DF(\lambda_0, \kappa_0)\bigr).$$

We can then write the bifurcation equation in the following way:

$$f(\alpha_1, \alpha_2) \equiv B\bigl(F(\lambda_0 + \alpha_1, \kappa_0 + g(\alpha_1, \alpha_2) + \alpha_2 \xi)\bigr) = 0, \tag{19.7}$$

where $f: B(0,\delta) \subset \mathbb{R}^2 \to \mathbb{R}$ is of class C^p. We easily check that

$$f(0,0)=0, \qquad \frac{\partial f}{\partial \alpha_1}(0,0)=0, \quad \text{and} \quad \frac{\partial f}{\partial \alpha_2}(0,0)=0,$$

which are relations we expect for a bifurcation equation, see Remark 18.2. To study the solutions of the bifurcation equation in a neighborhood of the origin, we need to compute the second derivative of f with respect to α_1 and α_2, the Hessian matrix $\mathrm{D}^2 f(0,0)$. When deriving Eq. (19.6) with respect to α_1 and α_2, at (0,0), we get with (19.2) (iii)

$$\mathrm{D}_\kappa F(\lambda_0,\kappa_0)\frac{\partial g}{\partial \alpha_1}(0,0)=0, \qquad \mathrm{D}_\kappa F(\lambda_0,\kappa_0)\left(\frac{\partial g}{\partial \alpha_2}(0,0)+\xi\right)=0,$$

which implies

$$\frac{\partial g}{\partial \alpha_1}(0,0)=0=\frac{\partial g}{\partial \alpha_2}(0,0). \tag{19.8}$$

Then a simple calculation gives

$$\begin{aligned}
\frac{\partial^2 f}{\partial \alpha_1^2}(0,0) &= B\big(\mathrm{D}^2_{\lambda\lambda}F(\lambda_0,\kappa_0)\big),\\
\frac{\partial^2 f}{\partial \alpha_1 \partial \alpha_2}(0,0) &= B\big(\mathrm{D}^2_{\lambda\kappa}F(\lambda_0,\kappa_0)\xi\big),\\
\frac{\partial^2 f}{\partial \alpha_2^2}(0,0) &= B\big(\mathrm{D}^2_{\kappa\kappa}F(\lambda_0,\kappa_0)(\xi,\xi)\big).
\end{aligned}$$

The determinant of the Hessian matrix $\mathrm{D}^2 f(0,0)$ is

$$\begin{aligned}
\det\big(\mathrm{D}^2 f(0,0)\big) &= B\big(\mathrm{D}^2_{\lambda\lambda}F(\lambda_0,\kappa_0)\big)B\big(\mathrm{D}^2_{\kappa\kappa}F(\lambda_0,\kappa_0)(\xi,\xi)\big)\\
&\quad - \big(B\big(\mathrm{D}^2_{\lambda\kappa}F(\lambda_0,\kappa_0)\xi\big)\big)^2.
\end{aligned} \tag{19.9}$$

The bifurcation equation is commonly studied with the Morse lemma. In fact the results are strongly depending on the sign of the determinant of $\mathrm{D}^2 f(0,0)$.

THEOREM 19.1 (Morse lemma). *Let H be a Hilbert space with the scalar product $(\cdot,\cdot)_H$ and the norm $\|\cdot\|_H$. Let $f: V \subset H \to \mathbb{R}$ be a mapping of class C^1 defined on a neighborhood V of the origin. We assume that $\mathrm{D}^2 f(0)$ exists. Let the continuous symmetric linear operator $S: H \to H$ be defined by: for all u, v in H*

$$(Su,v)_H = \mathrm{D}^2 f(0)(u,v).$$

We assume that $\mathrm{D}f(0)=0$ and that S is an isomorphism onto H.

Then there exist two neighborhoods $\mathcal{V}$, $\mathcal{W}$ of the origin in H, and a C^1 diffeomorphism $\Phi : \mathcal{V} \subset V \to \mathcal{W}$ satisfying

$$\Phi(0) = 0, \qquad \mathrm{D}\Phi(0) = I,$$

and for all $v \in \mathcal{V}$

$$f(v) = f(0) + \tfrac{1}{2}\mathrm{D}^2 f(0)\big(\Phi(v), \Phi(v)\big).$$

Moreover, if f is in $C^m(V;\mathbb{R})$ with $m \geqslant 1$ and if $\mathrm{D}^{m+1}f(0)$ exists, then Φ is of class C^m.

We can find proofs of the Morse lemma for instance in NIRENBERG [1974] or in CROUZEIX and RAPPAZ [1989]. We do not repeat the proof here.

We assume from now on that the determinant of the Hessian matrix $\mathrm{D}^2 f(0,0)$ is different from 0 and apply the Morse lemma to the bifurcation equation (19.7). There exist two neighborhoods $\mathcal{V}$, $\mathcal{W}$ of the origin in $\mathbb{R}^2$ and a C^1 diffeomorphism $\boldsymbol{\Phi} : \mathcal{V} \subset B(0,\delta) \subset \mathbb{R}^2 \to \mathcal{W} \subset \mathbb{R}^2$ satisfying

$$\boldsymbol{\Phi}(0,0) = 0, \qquad \mathrm{D}\boldsymbol{\Phi}(0,0) = I,$$

and for all $(\alpha_1, \alpha_2) \in \mathcal{V}$,

$$f(\alpha_1, \alpha_2) = \tfrac{1}{2}\big(\mathrm{D}^2 f(0,0)\boldsymbol{\Phi}(\alpha_1,\alpha_2), \boldsymbol{\Phi}(\alpha_1,\alpha_2)\big),$$

here $(\bullet, \bullet)$ is the scalar product in $\mathbb{R}^2$.

We analyze now the two different cases which can occur depending on the sign of the determinant of $\mathrm{D}^2 f(0,0)$. First let us assume that

$$\det\big(\mathrm{D}^2 f(0,0)\big) > 0, \tag{19.10}$$

which is the elliptic case. Since the matrix $\mathrm{D}^2 f(0,0)$ is symmetric, there exists a orthogonal matrix Q such that

$$Q^{\mathrm{t}} D Q = \mathrm{D}^2 f(0,0),$$

where the diagonal matrix D has the diagonal elements d_1 and d_2, d_i eigenvalue of $\mathrm{D}^2 f(0,0)$. Let the vector $\boldsymbol{\Psi}$ of components Ψ_1, Ψ_2, be

$$\boldsymbol{\Psi}(\alpha_1,\alpha_2) = Q\boldsymbol{\Phi}(\alpha_1,\alpha_2),$$

then for $(\alpha_1, \alpha_2) \in \mathcal{V}$,

$$f(\alpha_1,\alpha_2) = \frac{d_1}{2}\Psi_1^2(\alpha_1,\alpha_2) + \frac{d_2}{2}\Psi_2^2(\alpha_1,\alpha_2).$$

Then the bifurcation equation reads

$$d_1\Psi_1^2(\alpha_1,\alpha_2)+d_2\Psi_2^2(\alpha_1,\alpha_2)=0,$$

where d_1 and d_2 have the same sign. Since we have

$$\boldsymbol{\Psi}(0,0)=0 \quad \text{and} \quad \mathrm{D}\boldsymbol{\Psi}(0,0) \text{ regular,}$$

so in a neighborhood of (0,0), the bifurcation equation has the only solution (0,0) and the solution set

$$S=\{(\lambda,\kappa)\in\mathbb{R}\times\mathscr{X}\colon\ F(\lambda,\kappa)=0\}$$

is reduced to $\{(\lambda_0,\kappa_0)\}$ in a neighborhood of (λ_0,κ_0). Such a solution (λ_0,κ_0) to problem (19.1) is called an isolated solution.

We assume now that

$$\det\big(\mathrm{D}^2 f(0,0)\big)<0, \tag{19.11}$$

which is the hyperbolic case. Since the matrix $\mathrm{D}f(0,0)$ is regular and symmetric, there exists a regular matrix R such that

$$R^{\mathrm{t}}\begin{pmatrix}0 & 1/2\\ 1/2 & 0\end{pmatrix}R=\mathrm{D}^2 f(0,0).$$

Let the vector $\boldsymbol{\Psi}$ of components Ψ_1,Ψ_2, be

$$\boldsymbol{\Psi}(\alpha_1,\alpha_2)=R\boldsymbol{\Phi}(\alpha_1,\alpha_2),$$

then the bifurcation equation reads: for all $(\alpha_1,\alpha_2)\in\mathscr{V}$

$$f(\alpha_1,\alpha_2)\equiv\Psi_1(\alpha_1,\alpha_2)\Psi_2(\alpha_1,\alpha_2)=0. \tag{19.12}$$

Notice that

$$\boldsymbol{\Psi}(0,0)=0 \quad \text{and} \quad \mathrm{D}\boldsymbol{\Psi}(0,0) \text{ is regular.} \tag{19.13}$$

From (19.12), (19.13), and the Morse lemma, we deduce that the solution set of f

$$\mathscr{S}_f\big(\equiv\{(\alpha_1,\alpha_2)\in B(0,\delta)\colon\ f(\alpha_1,\alpha_2)=0\}\big)\cap\mathscr{V}$$

is C^{p-1} diffeomorphic to two parts of straight lines (in the coordinates Ψ_1,Ψ_2) which intersect at $(0,0)\in\mathbb{R}^2$. Consequently in a neighborhood of (λ_0,κ_0) the solution set S of (19.1) consists in two curves which intersect at the point (λ_0,κ_0).

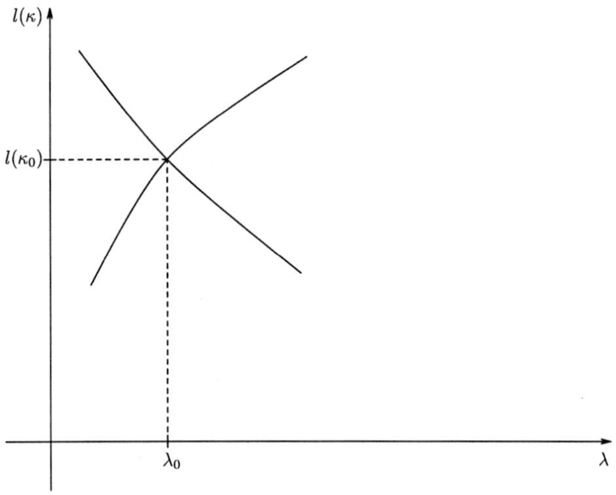

FIG. 19.1. Simple bifurcation point.

More precisely, let $\boldsymbol{\alpha}(\boldsymbol{\Psi})$ denote the inverse of the C^{p-1} diffeomorphism $\boldsymbol{\Psi}(\boldsymbol{\alpha})$ defined for $\boldsymbol{\Psi}$ in $R\mathscr{W}$, say for $|\Psi_1| + |\Psi_2| \leqslant \varepsilon$ for simplicity. Then

$$\mathscr{S}_f = \{(\alpha_1(s,0), \alpha_2(s,0)) \colon |s| \leqslant \varepsilon\} \cup \{(\alpha_1(0,s), \alpha_2(0,s)) \colon |s| \leqslant \varepsilon\}.$$

For the parameter value s, with $|s| \leqslant \varepsilon$, we set

$$\begin{aligned} \lambda_1(s) &= \lambda_0 + \alpha_1(s,0), \quad & \kappa_1(s) &= \kappa_0 + g(\alpha_1(s,0), \alpha_2(s,0)) + \alpha_2(s,0)\xi, \\ \lambda_2(s) &= \lambda_0 + \alpha_1(0,s), \quad & \kappa_2(s) &= \kappa_0 + g(\alpha_1(0,s), \alpha_2(0,s)) + \alpha_2(0,s)\xi. \end{aligned}$$

Clearly for $i = 1$ or 2 and $|s| \leqslant \varepsilon$,

$$\begin{aligned} &F(\lambda_i(s), \kappa_i(s)) = 0, \\ &\lambda_i(0) = \lambda_0, \ \kappa_i(0) = \kappa_0 \text{ and if } s \neq 0, \ (\lambda_i(s), \kappa_i(s)) \neq (\lambda_0, \kappa_0). \end{aligned}$$

Moreover, the two branches of solutions intersect at (λ_0, κ_0). If l is a linear form on $\mathscr{X}$, then in a neighborhood of (λ_0, κ_0) the solution set of (19.1) can be represented as in Fig. 19.1.

Such a solution (λ_0, κ_0) of (19.1) is called a simple bifurcation point.

Before studying bifurcation approximations of simple bifurcation points, we end up the section with some remarks.

REMARK 19.1. If the mapping $F : \mathbb{R} \times \mathscr{X} \to Z$ satisfies

$$F(\lambda, 0) = 0 \quad \text{for all } \lambda \in \mathbb{R}$$

and if $(\lambda_0, 0)$ satisfies (19.2) (i) and (ii), then we have

$$\mathrm{D}_\lambda F(\lambda_0, 0) = 0 \quad \text{and} \quad \mathrm{D}^2_{\lambda\lambda} F(\lambda_0, 0) = 0.$$

Clearly assumption (19.2) (iii) is satisfied. The determinant of the Hessian matrix $\mathrm{D}^2 f(0,0)$ given in (19.9) reads

$$\det\bigl(\mathrm{D}^2 f(0,0)\bigr) = -\bigl(B\bigl(\mathrm{D}^2_{\lambda\kappa} F(\lambda_0, 0)\xi\bigr)\bigr)^2.$$

So $(\lambda_0, 0)$ is a simple bifurcation point if and only if $B(\mathrm{D}^2_{\lambda\kappa} F(\lambda_0, 0)\xi) \neq 0$ or in other words

$$\mathrm{D}^2_{\lambda\kappa} F(\lambda_0, 0)\xi \notin \mathrm{Range}\bigl(\mathrm{D}_\kappa F(\lambda_0, 0)\bigr),$$

which is the Crandall–Rabinowitz condition, see CRANDALL and RABINOWITZ [1973]. Notice that approximations of simple bifurcation points on the trivial branch $\{(\lambda, 0)\colon\ \lambda \in \mathbb{R}\}$ are studied in CROUZEIX and RAPPAZ ([1989], Chapter 5), where they use a simpler approach than the Lyapunov–Schmidt procedure.

REMARK 19.2. If we replace assumption (19.2) (iii), namely $\mathrm{D}_\lambda F(\lambda_0, \kappa_0) = 0$, by the weaker one

$$\mathrm{D}_\lambda F(\lambda_0, \kappa_0) \in \mathrm{Range}\bigl(\mathrm{D}_\kappa F(\lambda_0, \kappa_0)\bigr), \tag{19.14}$$

then we can use the above development with minor changes and get a similar conclusion.

Indeed the kernel of $\mathrm{D}F(\lambda_0, \kappa_0)$ is of dimension 2 and

$$\mathrm{Ker}\bigl(\mathrm{D}F(\lambda_0, \kappa_0)\bigr) = \mathrm{span}\{\varphi_1, \varphi_2\}$$

with $\varphi_1 = (1, \eta) \in \mathbb{R} \times \mathscr{X}_1$, η the unique element satisfying

$$\mathrm{D}_\lambda F(\lambda_0, \kappa_0) + \mathrm{D}_\kappa F(\lambda_0, \kappa_0)\eta = 0,$$

and with $\varphi_2 = (0, \xi) \in \mathbb{R} \times \mathscr{X}$, $\xi \neq 0$ and

$$\mathrm{D}_\kappa F(\lambda_0, \kappa_0)\xi = 0.$$

With that choice of φ_1 and φ_2, decomposition (19.4) for X still remains valid, that is

$$X = X_1 \oplus \mathrm{span}\{\varphi_1, \varphi_2\} \tag{19.15}$$

where $X_1 = \{0\} \times \mathscr{X}_1$. Then Eq. (19.6) is modified in

$$Q_2 F\bigl(\lambda_0 + \alpha_1, \kappa_0 + g(\alpha_1, \alpha_2) + \alpha_1\eta + \alpha_2\xi\bigr) = 0 \tag{19.16}$$

and the bifurcation equation (19.7) in

$$f(\alpha_1,\alpha_2) \equiv B\big(F(\lambda_0+\alpha_1, \kappa_0+g(\alpha_1,\alpha_2)+\alpha_1\eta+\alpha_2\xi)\big)=0. \tag{19.17}$$

When deriving Eq. (19.16) with respect to α_1 and α_2, at $(0,0)$, we get

$$\mathrm{D}_\lambda F(\lambda_0,\kappa_0)+\mathrm{D}_\kappa F(\lambda_0,\kappa_0)\left(\frac{\partial g}{\partial \alpha_1}(0,0)+\eta\right)=0,$$
$$\mathrm{D}_\kappa F(\lambda_0,\kappa_0)\left(\frac{\partial g}{\partial \alpha_2}(0,0)+\xi\right)=0,$$

which implies

$$\frac{\partial g}{\partial \alpha_1}(0,0)=0=\frac{\partial g}{\partial \alpha_2}(0,0).$$

Then a simple calculation gives the following expressions for the elements of the Hessian matrix $\mathrm{D}^2 f(0,0)$,

$$\frac{\partial^2 f}{\partial \alpha_1^2}(0,0)=B\big(\mathrm{D}^2_{\lambda\lambda}F(\lambda_0,\kappa_0)+2\mathrm{D}^2_{\lambda\kappa}F(\lambda_0,\kappa_0)\eta+\mathrm{D}^2_{\kappa\kappa}F(\lambda_0,\kappa_0)(\eta,\eta)\big),$$
$$\frac{\partial^2 f}{\partial \alpha_1\partial\alpha_2}(0,0)=B\big(\mathrm{D}^2_{\lambda\kappa}F(\lambda_0,\kappa_0)\xi+\mathrm{D}^2_{\kappa\kappa}F(\lambda_0,\kappa_0)(\eta,\xi)\big),$$
$$\frac{\partial^2 f}{\partial \alpha_2^2}(0,0)=B\big(\mathrm{D}^2_{\kappa\kappa}F(\lambda_0,\kappa_0)(\xi,\xi)\big).$$

Then the study of the bifurcation equation goes along the same line as before, depending on the sign of the determinant of $\mathrm{D}^2 f(0,0)$.

REMARK 19.3. A similar development can be carried out if we replace the assumption (19.2) by

$$\begin{aligned}&\text{(i)}\quad \dim\operatorname{Ker}(\mathrm{D}_\kappa F(\lambda_0,\kappa_0))=2,\\&\text{(ii)}\quad \operatorname{codim}(\mathrm{D}_\kappa F(\lambda_0,\kappa_0))=2,\\&\text{(iii)}\quad D_\lambda F(\lambda_0,\kappa_0)\notin \operatorname{Range}(\mathrm{D}_\kappa F(\lambda_0,\kappa_0)).\end{aligned} \tag{19.18}$$

Indeed the kernel of $\mathrm{D}F(\lambda_0,\kappa_0)$ is of dimension 2 and

$$\operatorname{Ker}\big(\mathrm{D}F(\lambda_0,\kappa_0)\big)=\operatorname{span}\{\varphi_1,\varphi_2\}$$

with $\varphi_1=(0,\xi_1)\in\mathbb{R}\times\mathscr{X}$ and $\varphi_2=(0,\xi_2)\in\mathbb{R}\times\mathscr{X}$ such that

$$\operatorname{Ker}\big(\mathrm{D}_\kappa F(\lambda_0,\kappa_0)\big)=\operatorname{span}\{\xi_1,\xi_2\}.$$

There exists a closed subspace $\mathscr{X}_1 \subset \mathscr{X}$ such that

$$\mathscr{X} = \mathscr{X}_1 \oplus \operatorname{span}\{\xi_1, \xi_2\}$$

and we can write a decomposition of X,

$$X = X_1 \oplus \operatorname{span}\{\varphi_1, \varphi_2\} \tag{19.19}$$

with $X_1 = \mathbb{R} \times \mathscr{X}_1$. The codimension of $\mathrm{D}F(\lambda_0, \kappa_0)$ is 1, so the following decomposition of Z holds

$$Z = \operatorname{span}\{\zeta_1\} \oplus \operatorname{Range}\big(\mathrm{D}F(\lambda_0, \kappa_0)\big). \tag{19.20}$$

The mapping g can be identified to a function defined on $B(0,\delta) \subset \mathbb{R}^2$ with values in $\mathbb{R} \times \mathscr{X}_1$, that is

$$g : (\alpha_1, \alpha_2) \in B(0,\delta) \subset \mathbb{R}^2 \to \big(g_1(\alpha_1, \alpha_2), g_2(\alpha_1, \alpha_2)\big) \in \mathbb{R} \times \mathscr{X}_1,$$

and satisfies for all $(\alpha_1, \alpha_2) \in B(0,\delta)$

$$Q_2 F\big(\lambda_0 + g_1(\alpha_1, \alpha_2), \kappa_0 + g_2(\alpha_1, \alpha_2) + \alpha_1 \xi_1 + \alpha_2 \xi_2\big) = 0. \tag{19.21}$$

The bifurcation equation reads

$$\begin{aligned} &f(\alpha_1, \alpha_2) \\ &\quad \equiv B\big(F(\lambda_0 + g_1(\alpha_1, \alpha_2), \kappa_0 + g_2(\alpha_1, \alpha_2) + \alpha_1 \xi_1 + \alpha_2 \xi_2)\big) = 0 \end{aligned} \tag{19.22}$$

where $f : B(0,\delta) \subset \mathbb{R}^2 \to \mathbb{R}$ is of class C^p. When deriving Eq. (19.21) with respect to α_1 and α_2, at $(0,0)$, we get

$$\begin{aligned} &\mathrm{D}_\lambda F(\lambda_0, \kappa_0) \frac{\partial g_1}{\partial \alpha_1}(0,0) + \mathrm{D}_\kappa F(\lambda_0, \kappa_0) \left(\frac{\partial g_2}{\partial \alpha_1}(0,0) + \xi_1 \right) = 0, \\ &\mathrm{D}_\lambda F(\lambda_0, \kappa_0) \frac{\partial g_1}{\partial \alpha_2}(0,0) + \mathrm{D}_\kappa F(\lambda_0, \kappa_0) \left(\frac{\partial g_2}{\partial \alpha_2}(0,0) + \xi_2 \right) = 0. \end{aligned}$$

Since

$$\operatorname{Range}\big(\mathrm{D}F(\lambda_0, \kappa_0)\big) = \operatorname{span}\big\{\mathrm{D}_\lambda F(\lambda_0, \kappa_0)\big\} \oplus \operatorname{Range}\big(\mathrm{D}_\kappa F(\lambda_0, \kappa_0)\big)$$

and since ξ_1 and ξ_2 are in the kernel of $\mathrm{D}_\kappa F(\lambda_0, \kappa_0)$ we deduce that

$$\frac{\partial g_i}{\partial \alpha_j}(0,0) = 0, \quad 1 \leqslant i, j \leqslant 2.$$

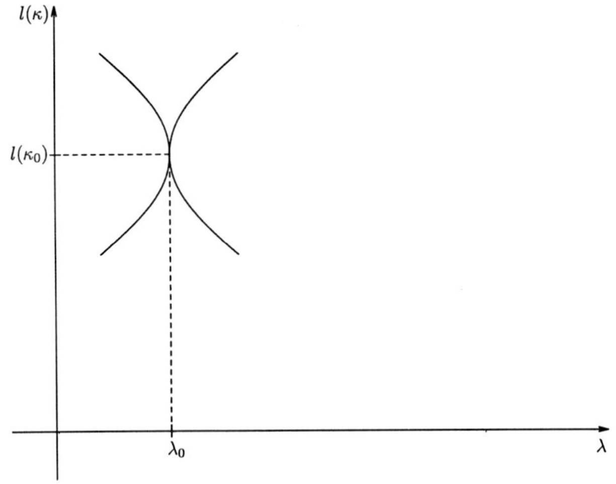

FIG. 19.2. Double nondegenerate limit point.

Then a simple calculation gives the following expressions

$$\frac{\partial^2 f}{\partial \alpha_1^2}(0,0) = B\big(\mathrm{D}^2_{\kappa\kappa}F(\lambda_0,\kappa_0)(\xi_1,\xi_1)\big),$$
$$\frac{\partial^2 f}{\partial \alpha_1 \partial \alpha_2}(0,0) = B\big(\mathrm{D}^2_{\kappa\kappa}F(\lambda_0,\kappa_0)(\xi_1,\xi_2)\big),$$
$$\frac{\partial^2 f}{\partial \alpha_2^2}(0,0) = B\big(\mathrm{D}^2_{\kappa\kappa}F(\lambda_0,\kappa_0)(\xi_2,\xi_2)\big).$$

The study of the bifurcation equation goes along the same line as before, depending on the sign of the determinant of $\mathrm{D}^2 f(0,0)$.

In the hyperbolic case, the solution set of $F(\lambda, x) = 0$ in a neighborhood of (λ_0, κ_0) is represented in Fig. 19.2, where l is a linear form on $\mathscr{X}$. Such a solution (λ_0, κ_0) of $F(\lambda, \kappa) = 0$ is called a double nondegenerate limit point.

20. Approximate bifurcation equations

Let X, Z be two real Banach spaces, $F: X \to Z$ be a C^p mapping, $p \geqslant 1$, and $x_0 \in X$ be a singular solution of the problem: find $x \in X$ such that

$$F(x) = 0. \tag{20.1}$$

We assume that the solution x_0 satisfies

$$\mathrm{D}F(x_0) \in \mathscr{L}(X; Z) \text{ is a nonsurjective Fredholm operator of index } m,\quad m \geqslant 1. \tag{20.2}$$

The study of the solution set S of (20.1) in a neighborhood of x_0 has been done in Sections 18 and 19. We will keep the notation introduced there. Let P_1, P_2 and Q_1, Q_2 be the projectors associated to the decomposition of X and Z,

$$X = X_1 \oplus \mathrm{Ker}\big(\mathrm{D}F(x_0)\big), \qquad Z = Z_1 \oplus \mathrm{Range}\big(\mathrm{D}F(x_0)\big).$$

We introduce now a family $\{F_h\}_{0<h\leqslant 1}$ of C^p mappings, $p \geqslant 1$,

$$F_h : x \in X \to F_h(x) \in Z$$

which are approximations of F. Our goal is to study the existence and the convergence of solutions of the problem: find $x \in X$ such that

$$F_h(x) = 0 \tag{20.3}$$

in a neighborhood of the singular solution x_0. We assume that the family $\{F_h\}_{0<h\leqslant 1}$ satisfies: there exists a real number $r > 0$ such that

$$\lim_{h\to 0} \sup_{x\in B(x_0,r)} \big\|\mathrm{D}^l F(x) - \mathrm{D}^l F_h(x)\big\| = 0, \tag{20.4}$$

for $l = 0, 1$.

To study problem (20.3) under assumption (20.4), we will apply the Lyapunov–Schmidt procedure to the mapping F_h with respect to the decomposition of X and Z given in the analysis of problem (20.1). Since we study problem (20.3) in a neighborhood of x_0, we write it in the following way: find $x_1 \in X_1$, $x_2 \in \mathrm{Ker}(\mathrm{D}F(x_0))$ such that

$$Q_1 F_h(x_0 + x_1 + x_2) = 0 \tag{20.5}$$

and

$$Q_2 F_h(x_0 + x_1 + x_2) = 0. \tag{20.6}$$

Corresponding to the mapping $\mathscr{F}$ in (18.5) for the exact problem, we define the mapping $\mathscr{F}_h : X_1 \times \mathrm{Ker}(\mathrm{D}F(x_0)) \to \mathrm{Range}(\mathrm{D}F(x_0))$ by: for $x_1 \in X_1$, $x_2 \in \mathrm{Ker}(\mathrm{D}F(x_0))$,

$$\mathscr{F}_h(x_1, x_2) = Q_2 F_h(x_0 + x_1 + x_2).$$

To study the solutions of (20.6) or of $\mathscr{F}_h(x_1, x_2) = 0$, we use Theorem 12.3, Remark 12.1, and Theorem 12.4, whose assumptions are not difficult to check with the hypothesis (20.4), since $\mathrm{D}_{x_1}\mathscr{F}(0,0) = Q_2\mathrm{D}F(x_0) \in \mathscr{L}(X_1; \mathrm{Range}(\mathrm{D}F(x_0)))$ is an isomorphism.

Let $g : B(0,\varepsilon) \subset \mathrm{Ker}(\mathrm{D}F(x_0)) \to X_1$ be the C^p mapping corresponding to the Lyapunov–Schmidt decomposition for F, satisfying (18.6). By Theorem 12.3, there exist positive constants $h_0 \leqslant 1$, η, and a C^p mapping $g_h : B(0,\varepsilon) \subset \operatorname{Ker}(\mathrm{D}F(x_0)) \to X_1$ such that for all $x_2 \in B(0,\varepsilon)$, $0 < h \leqslant h_0$,

$$\big(\mathscr{F}_h(x_1, x_2) = 0 \text{ and } x_1 \in B(0,\eta) \subset X_1\big) \Longleftrightarrow x_1 = g_h(x_2).$$

Moreover, the following estimates hold: for all $x_2 \in B(0,\varepsilon)$

$$\big\|g(x_2) - g_h(x_2)\big\|_X \leqslant C\big\|F_h\big(x_0 + g(x_2) + x_2\big)\big\|_Z, \tag{20.7}$$

$$\big\|\mathrm{D}g_h(x_2)\big\|_{X;X} \leqslant C. \tag{20.8}$$

Notice that we also have estimates for higher order derivatives, see (12.14) and (12.15), which are not needed here for our purpose.

To complete our analysis we need to study Eq. (20.5) on the solution set of $\mathscr{F}_h$ in $(B(0,\varepsilon) \times B(0,\eta))$. The approximate analogue of the mapping $\mathscr{G}$, is $\mathscr{G}_h : B(0,\varepsilon) \subset \operatorname{Ker}(\mathrm{D}F(x_0)) \to Z_1$ defined by: for $x_2 \in B(0,\varepsilon)$

$$\mathscr{G}_h(x_2) = Q_1 F_h\big(x_0 + g_h(x_2) + x_2\big). \tag{20.9}$$

Then the approximate bifurcation equation reads: find $x_2 \in B(0,\varepsilon)$ such that

$$\mathscr{G}_h(x_2) = 0. \tag{20.10}$$

Before undertaking with more details the study of (20.10), which is the object of Section 21 in the case developed in Section 19, we still need a comparison result between the solutions of the exact problem (20.1) and the ones of the approximate problems (20.3). For that purpose we define the sets

$$\mathscr{S}_{\mathscr{G}} = \big\{x_2 \in \mathscr{V}\colon\ \mathscr{G}(x_2) = 0\big\}, \qquad \mathscr{S}_{\mathscr{G}_h} = \big\{x_2 \in \mathscr{V}\colon\ \mathscr{G}_h(x_2) = 0\big\},$$

where

$$\mathscr{V} \subset B(0,\varepsilon) \subset \operatorname{Ker}\big(\mathrm{D}F(x_0)\big)$$

is a compact convex neighborhood in $\operatorname{Ker}(\mathrm{D}F(x_0))$ of 0; we also introduce

$$\mathscr{S} = \big\{x = x_0 + g(x_2) + x_2 \in X\colon\ x_2 \in \mathscr{S}_{\mathscr{G}}\big\},$$
$$\mathscr{S}_h = \big\{x = x_0 + g_h(x_2) + x_2 \in X\colon\ x_2 \in \mathscr{S}_{\mathscr{G}_h}\big\}.$$

Notice that for $x \in \mathscr{S}$, we have $F(x) = 0$ and for $x_h \in \mathscr{S}_h$, we have $F_h(x_h) = 0$. The distance we will use to compare the sets $\mathscr{S}_{\mathscr{G}}$, $\mathscr{S}_{\mathscr{G}_h}$ and $\mathscr{S}$, $\mathscr{S}_h$ is the Hausdorff metric.

Let Y be a vector space endowed with the metric d and $\mathscr{Y}$ denote the set of all nonempty bounded closed sets in Y. For $(A, B) \in \mathscr{Y} \times \mathscr{Y}$, we define

$$\rho(A, B) = \sup_{y \in A} d(y, B) \equiv \sup_{y \in A} \Big(\inf_{z \in B} d(y, z) \Big),$$

$$\delta(A, B) = \max \big(\rho(A, B), \rho(B, A)\big).$$

δ is a metric in $\mathscr{Y}$ called the Hausdorff metric.

THEOREM 20.1. *In the framework developed above, there exist positive constants C and h_0 such that for all $0 < h \leqslant h_0$,*

$$\delta(\mathscr{S}, \mathscr{S}_h) \leqslant C\Big[\delta(\mathscr{S}_{\mathscr{G}}, \mathscr{S}_{\mathscr{G}_h}) + \sup_{x \in \mathscr{S}} \big\| F_h(x) \big\|_Z\Big]. \tag{20.11}$$

PROOF. Let $x_2 \in \mathscr{S}_{\mathscr{G}}$, $x_{2h} \in \mathscr{S}_{\mathscr{G}_h}$, and

$$x = x_0 + g(x_2) + x_2, \qquad x_h = x_0 + g_h(x_{2h}) + x_{2h}.$$

By definition we have $x \in \mathscr{S}$ and $x_h \in \mathscr{S}_h$. Furthermore,

$$\begin{aligned} \|x - x_h\|_X \leqslant \|x_2 - x_{2h}\|_X &+ \big\|g(x_2) - g_h(x_2)\big\|_X \\ &+ \big\|g_h(x_2) - g_h(x_{2h})\big\|_X. \end{aligned} \tag{20.12}$$

We introduce in (20.12) both estimates (20.7) and (20.8), and deduce

$$\|x - x_h\|_X \leqslant C\big(\|x_2 - x_{2h}\|_X + \big\|F_h\big(x_0 + g(x_2) + x_2\big)\big\|_Z\big)$$

and

$$\|x - x_h\|_X \leqslant C\Big(\|x_2 - x_{2h}\|_X + \sup_{x \in \mathscr{S}} \big\|F_h(x)\big\|_Z\Big). \tag{20.13}$$

The infimum over $x_h \in \mathscr{S}_h$ in (20.13) leads to

$$\inf_{x_h \in \mathscr{S}_h} \|x - x_h\|_X \leqslant C\Big(\rho(\mathscr{S}_{\mathscr{G}}, \mathscr{S}_{\mathscr{G}_h}) + \sup_{x \in \mathscr{S}} \big\|F_h(x)\big\|_Z\Big)$$

and

$$\rho(\mathscr{S}, \mathscr{S}_h) \leqslant C\Big(\rho(\mathscr{S}_{\mathscr{G}}, \mathscr{S}_{\mathscr{G}_h}) + \sup_{x \in \mathscr{S}} \big\|F_h(x)\big\|_Z\Big). \tag{20.14}$$

In the same way we can prove

$$\rho(\mathscr{S}_h, \mathscr{S}) \leqslant C\Big(\rho(\mathscr{S}_{\mathscr{G}_h}, \mathscr{S}_{\mathscr{G}}) + \sup_{x \in \mathscr{S}} \big\|F_h(x)\big\|_Z\Big). \tag{20.15}$$

The two inequalities (20.14) and (20.15), and the definition of the Hausdorff metric imply estimate (20.11). □

To obtain some estimates with respect to h for $\delta(\mathscr{S}, \mathscr{S}_h)$, it remains to bound the quantity $\delta(\mathscr{S}_{\mathscr{G}}, \mathscr{S}_{\mathscr{G}_h})$ to some power of h with (20.11). In the next section we will study this quantity when the functions $\mathscr{G}$ and $\mathscr{G}_h$ are defined on a neighborhood of 0 in $\mathbb{R}^2$ with values in $\mathbb{R}$.

21. Approximation of simple bifurcation points

To complete our study we still need to present a precise comparison between the solution sets of the bifurcation equation $\mathscr{G}(x_2) = 0$ and of the approximate bifurcation equation $\mathscr{G}_h(x_2) = 0$. Then from estimate (20.11), we could derive error estimates.

For simplicity we analyze the case of a simple bifurcation point which is developed in Section 19. Let $F : \mathbb{R} \times \mathscr{X} \to Z$ be a C^p mapping, $p \geqslant 2$, and let $(\lambda_0, \kappa_0) \in \mathbb{R} \times \mathscr{X}$ be a singular solution to (19.1) satisfying to (19.2). Then the bifurcation equation reads

$$f(\alpha_1, \alpha_2) = 0, \tag{21.1}$$

where $f : B(0, \delta) \subset \mathbb{R}^2 \to \mathbb{R}$ is of class C^p, $p \geqslant 2$, and satisfies

$$f(0,0) = 0, \qquad \frac{\partial f}{\partial \alpha_1}(0,0) = 0 \quad \text{and} \quad \frac{\partial f}{\partial \alpha_2}(0,0) = 0.$$

The simple bifurcation case occurs under the assumption

$$\det\bigl(\mathrm{D}^2 f(0,0)\bigr) < 0. \tag{21.2}$$

In the following we will compare the solutions of approximate bifurcation equations with respect to the solution of the given bifurcation equation (21.1).

We can apply the theory of Section 20 to a family $\{F_h\}_{0<h\leqslant 1}$ of C^p mappings, $p \geqslant 2$,

$$F_h : (\lambda, \kappa) \in \mathbb{R} \times \mathscr{X} \to F_h(\lambda, \kappa) \in Z,$$

satisfying

$$\lim_{h\to 0} \sup_{(\lambda,\kappa)\in B((\lambda_0,\kappa_0),r)} \bigl\|\mathrm{D}^l F(\lambda, \kappa) - \mathrm{D}^l F_h(\lambda, \kappa)\bigr\| = 0, \tag{21.3}$$

for some $r > 0$, $l = 0, 1$. Then we get a family of bifurcation equations, say for $h \in (0, h_0]$,

$$f_h(\alpha_1, \alpha_2) = 0, \tag{21.4}$$

where $f_h : B(0,\delta) \subset \mathbb{R}^2 \to \mathbb{R}$ is of class C^p, only to restrict δ we keep the same as the one in the definition of f. With assumption (21.3) we deduce that

$$\lim_{h\to 0} \sup_{(\alpha_1,\alpha_2)\in B(0,\delta)} \left|f(\alpha_1,\alpha_2) - f_h(\alpha_1,\alpha_2)\right| = 0. \tag{21.5}$$

We define the two solution sets

$$\mathscr{S}_f = \{(\alpha_1,\alpha_2) \in B(0,\delta)\colon\ f(\alpha_1,\alpha_2) = 0\},$$
$$\mathscr{S}_{f_h} = \{(\alpha_1,\alpha_2) \in B(0,\delta)\colon\ f_h(\alpha_1,\alpha_2) = 0\}.$$

We now compare the behavior of $\mathscr{S}_{f_h}$ and of $\mathscr{S}_f$. Our goal is to get error estimates for the Hausdorff distance $\delta(\mathscr{S}_f, \mathscr{S}_{f_h})$ and to give qualitative results for $\mathscr{S}_{f_h}$.

THEOREM 21.1. *Under assumptions* (21.2) *and* (21.5), *there exists a constant* $C > 0$ *independent of* h *such that*

$$\delta(\mathscr{S}_f, \mathscr{S}_{f_h}) \leqslant C\Big(\sup_{(\alpha_1,\alpha_2)\in B(0,\delta)} \left|f(\alpha_1,\alpha_2) - f_h(\alpha_1,\alpha_2)\right|\Big)^{1/2}. \tag{21.6}$$

PROOF. For notation simplicity we set

$$\varepsilon_h = \sup_{(\alpha_1,\alpha_2)\in B(0,\delta)} \left|f(\alpha_1,\alpha_2) - f_h(\alpha_1,\alpha_2)\right|.$$

Applying the Morse lemma, we deduce that there exists a C^{p-1} diffeomorphism $\boldsymbol{\Psi}$ from $B(0,\delta)$, only to restrict δ, into $\mathbb{R}^2$ such that the bifurcation equation reads

$$f(\alpha_1,\alpha_2) \equiv \Psi_1(\alpha_1,\alpha_2)\Psi_2(\alpha_1,\alpha_2) = 0,$$

see (19.12). Consequently, without loss of generality, we can assume that the function f is, for $(\alpha_1,\alpha_2) \in B(0,\delta)$,

$$f(\alpha_1,\alpha_2) = \alpha_1\alpha_2.$$

Then $\rho(\mathscr{S}_{f_h}, \mathscr{S}_f)$ given by

$$\rho(\mathscr{S}_{f_h}, \mathscr{S}_f) = \sup_{(\alpha_1,\alpha_2)\in \mathscr{S}_{f_h}} \Big(\inf_{(\beta_1,\beta_2)\in \mathscr{S}_f} \left\|(\alpha_1,\alpha_2) - (\beta_1,\beta_2)\right\|\Big),$$

where $\|\bullet\|$ denotes the Euclidean norm in $\mathbb{R}^2$, can be bounded by

$$\rho(\mathscr{S}_{f_h}, \mathscr{S}_f) \leqslant \varepsilon_h. \tag{21.7}$$

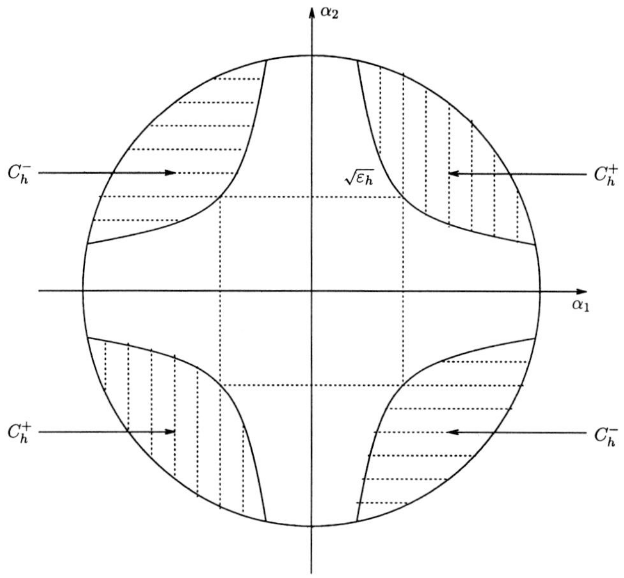

FIG. 21.1. Sets C_h^+ and C_h^-.

Let us estimate now $\rho(\mathscr{S}_f, \mathscr{S}_{f_h})$. We define the sets

$$C_h^+ = \{(\alpha_1, \alpha_2) \in B(0,\delta)\colon\ f(\alpha_1,\alpha_2) \geqslant \varepsilon_h\},$$
$$C_h^- = \{(\alpha_1, \alpha_2) \in B(0,\delta)\colon\ f(\alpha_1,\alpha_2) \leqslant -\varepsilon_h\}.$$

From assumption (21.5) we know that ε_h tends to 0 when h tends to 0, so for h small enough both sets C_h^+ and C_h^- are not empty. Let (α_1, α_2) be in $\mathscr{S}_f$. There exist $(\alpha_{1h}^+, \alpha_{2h}^+)$ in C_h^+ and $(\alpha_{1h}^-, \alpha_{2h}^-)$ in C_h^- satisfying

$$\|(\alpha_1,\alpha_2) - (\alpha_{1h}^+, \alpha_{2h}^+)\| \leqslant \sqrt{2\varepsilon_h}, \qquad \|(\alpha_1,\alpha_2) - (\alpha_{1h}^-, \alpha_{2h}^-)\| \leqslant \sqrt{2\varepsilon_h}.$$

By the definition of ε_h we have

$$f_h(\alpha_{1h}^+, \alpha_{2h}^+) \geqslant f(\alpha_{1h}^+, \alpha_{2h}^+) - \varepsilon_h \geqslant 0,$$
$$f_h(\alpha_{1h}^-, \alpha_{2h}^-) \leqslant f(\alpha_{1h}^-, \alpha_{2h}^-) + \varepsilon_h \leqslant 0.$$

So there exists $\theta \in [0,1]$ such that

$$f_h\big(\theta(\alpha_{1h}^+, \alpha_{2h}^+) + (1-\theta)(\alpha_{1h}^-, \alpha_{2h}^-)\big) = 0$$

and if by $(\alpha_{1h}^0, \alpha_{2h}^0)$ we denote that solution of f_h, we have

$$\|(\alpha_1,\alpha_2) - (\alpha_{1h}^0, \alpha_{2h}^0)\| \leqslant \sqrt{2\varepsilon_h}.$$

Then we deduce

$$\rho(\mathscr{S}_f, \mathscr{S}_{f_h}) \leqslant \sqrt{2\varepsilon_h}. \tag{21.8}$$

Finally from the two estimates (21.7) and (21.8) we can conclude. □

REMARK 21.1. If we do not add any assumptions to (21.2) and (21.5), then the estimate (21.6) is optimal. Consider $f(\alpha_1, \alpha_2) = \alpha_1\alpha_2$ and $f_h(\alpha_1, \alpha_2) = \alpha_1\alpha_2 + \varepsilon_h$, $\varepsilon_h > 0$ for $h \in (0, 1]$, then we have precisely

$$\delta(\mathscr{S}_f, \mathscr{S}_{f_h}) = \sqrt{2\varepsilon_h}.$$

REMARK 21.2. In the proof of Theorem 21.1, we have only used the continuity of f_h and both relations (21.2) and (21.5). In fact with assumption (21.3), we also have

$$\lim_{h\to 0} \sup_{(\alpha_1,\alpha_2)\in B(0,\delta)} \left\|\mathrm{D}f(\alpha_1, \alpha_2) - \mathrm{D}f_h(\alpha_1, \alpha_2)\right\| = 0. \tag{21.9}$$

Then we will derive in the following the estimate

$$\delta(\mathscr{S}_f, \mathscr{S}_{f_h}) \leqslant C\Big(\sqrt{f_h(0,0)} + \sup_{(\alpha_1,\alpha_2)\in B(0,\delta)} \left\|\mathrm{D}f(\alpha_1, \alpha_2) - \mathrm{D}f_h(\alpha_1, \alpha_2)\right\|\Big). \tag{21.10}$$

It is sometimes possible to improve the error analysis by using estimate (21.10) rather than (21.6).

To check estimate (21.10) we proceed like in the proof of Theorem 21.1. Notice first that with the Taylor expansion we get for $(\alpha_1, \alpha_2) \in B(0, \delta)$,

$$f(\alpha_1, \alpha_2) - f_h(\alpha_1, \alpha_2) = f(0,0) - f_h(0,0) + \int_0^1 \big(\mathrm{D}f\big(t(\alpha_1, \alpha_2)\big) - \mathrm{D}f_h\big(t(\alpha_1, \alpha_2)\big)\big)(\alpha_1, \alpha_2)\,\mathrm{d}t.$$

Since $f(0,0) = 0$, we deduce that

$$\left|f(\alpha_1, \alpha_2) - f_h(\alpha_1, \alpha_2)\right| \leqslant \left|f_h(0,0)\right| + \varepsilon_h^1\left\|(\alpha_1, \alpha_2)\right\|,$$

where ε_h^1 stands for

$$\varepsilon_h^1 - \sup_{(\alpha_1,\alpha_2)\in B(0,\delta)} \left\|\mathrm{D}f(\alpha_1, \alpha_2) - \mathrm{D}f_h(\alpha_1, \alpha_2)\right\|.$$

As in the proof of Theorem 21.1, we can assume that f is given by $f(\alpha_1, \alpha_2) = \alpha_1\alpha_2$ and satisfies

$$\left|f(\alpha_1, \alpha_2) - f_h(\alpha_1, \alpha_2)\right| \leqslant \left|f_h(0,0)\right| + C\varepsilon_h^1\big(|\alpha_1| + |\alpha_2|\big).$$

Then we have

$$\rho(\mathscr{S}_{f_h}, \mathscr{S}_f) \leqslant C\big(|f_h(0,0)| + \varepsilon_h^1\big). \tag{21.11}$$

To get an estimate for the distance $\rho(\mathscr{S}_f, \mathscr{S}_{f_h})$, we define the two sets

$$C_h^+ = \{(\alpha_1, \alpha_2) \in B(0,\delta)\colon\ f(\alpha_1, \alpha_2) \geqslant |f_h(0,0)| + C\varepsilon_h^1(|\alpha_1| + |\alpha_2|)\},$$
$$C_h^- = \{(\alpha_1, \alpha_2) \in B(0,\delta)\colon\ f(\alpha_1, \alpha_2) \leqslant -[|f_h(0,0)| + C\varepsilon_h^1(|\alpha_1| + |\alpha_2|)]\}.$$

From the two assumptions (21.5) and (21.9), we know that the sets C_h^+ and C_h^- are not empty. Using the same method as in the proof of Theorem 21.1, we get

$$\rho(\mathscr{S}_f, \mathscr{S}_{f_h}) \leqslant C\big(\sqrt{|f_h(0,0)|} + \varepsilon_h^1\big). \tag{21.12}$$

The two estimates (21.11) and (21.12) imply (21.10).

REMARK 21.3. We still can modify estimate (21.10) in the following way. We assume (21.2), (21.5), (21.9), and

$$\lim_{h\to 0} \sup_{(\alpha_1,\alpha_2)\in B(0,\delta)} \|D^2 f(\alpha_1,\alpha_2) - D^2 f_h(\alpha_1,\alpha_2)\| = 0. \tag{21.13}$$

Notice that assumption (21.13) is true if we assume (21.3) with $l = 2$. Then the following estimate holds

$$\delta(\mathscr{S}_f, \mathscr{S}_{f_h}) \leqslant C\Big(\sqrt{f_h(0,0)} + \sup_{(\alpha_1,\alpha_2)\in \mathscr{S}_f} \|Df(\alpha_1,\alpha_2) - Df_h(\alpha_1,\alpha_2)\|\Big). \tag{21.14}$$

Indeed let (α_1, α_2) be in $B(0,\delta)$ and (α_1^0, α_2^0) be in $\mathscr{S}_f$. From the Taylor expansion we deduce that

$$\begin{aligned} &f_h(\alpha_1,\alpha_2) - f(\alpha_1,\alpha_2) \\ &\quad = f_h(\alpha_1^0,\alpha_2^0) - f(\alpha_1^0,\alpha_2^0) + \big(Df_h(\alpha_1^0,\alpha_2^0) - Df(\alpha_1^0,\alpha_2^0)\big)(\alpha_1-\alpha_1^0, \alpha_2-\alpha_2^0) \\ &\qquad + \tfrac{1}{2}\big(D^2 f_h(\tilde\alpha_1,\tilde\alpha_2) - D^2 f(\tilde\alpha_1,\tilde\alpha_2)\big)\big((\alpha_1-\alpha_1^0,\alpha_2-\alpha_2^0),(\alpha_1-\alpha_1^0,\alpha_2-\alpha_2^0)\big) \end{aligned}$$

where $(\tilde\alpha_1, \tilde\alpha_2)$ belongs to the segment passing by (α_1, α_2) and (α_1^0, α_2^0). So

$$\begin{aligned} &|f_h(\alpha_1,\alpha_2) - f(\alpha_1,\alpha_2)| \\ &\quad \leqslant |f_h(\alpha_1^0,\alpha_2^0) - f(\alpha_1^0,\alpha_2^0)| \\ &\qquad + \varepsilon_h^1(\mathscr{S}_f)\|(\alpha_1-\alpha_1^0,\alpha_2-\alpha_2^0)\| + \tfrac{1}{2}\varepsilon_h^2\|(\alpha_1-\alpha_1^0,\alpha_2-\alpha_2^0)\|^2, \end{aligned}$$

where

$$\varepsilon_h^1(\mathscr{S}_f) = \sup_{(\alpha_1,\alpha_2)\in\mathscr{S}_f} \left\|\mathrm{D}f(\alpha_1,\alpha_2) - \mathrm{D}f_h(\alpha_1,\alpha_2)\right\|,$$

$$\varepsilon_h^2 = \sup_{(\alpha_1,\alpha_2)\in B(0,\delta)} \left\|\mathrm{D}^2 f(\alpha_1,\alpha_2) - \mathrm{D}^2 f_h(\alpha_1,\alpha_2)\right\|.$$

Noticing that

$$\begin{aligned}\left|f_h(\alpha_1^0,\alpha_2^0) - f(\alpha_1^0,\alpha_2^0)\right| &\leqslant \left|f_h(0,0) - f(0,0)\right| + C\varepsilon_h^1(\mathscr{S}_f)\left\|(\alpha_1^0,\alpha_2^0)\right\| \\ &\leqslant \left|f_h(0,0)\right| + C\varepsilon_h^1(\mathscr{S}_f)\left\|(\alpha_1^0,\alpha_2^0)\right\|,\end{aligned}$$

we get

$$\begin{aligned}\left|f_h(\alpha_1,\alpha_2) - f(\alpha_1,\alpha_2)\right| \leqslant{}& \left|f_h(0,0)\right| + C\varepsilon_h^1(\mathscr{S}_f)\big(\left\|(\alpha_1^0,\alpha_2^0)\right\| \\ &+ \left\|(\alpha_1-\alpha_1^0,\alpha_2-\alpha_2^0)\right\|\big) \\ &+ \tfrac{1}{2}\varepsilon_h^2\left\|(\alpha_1-\alpha_1^0,\alpha_2-\alpha_2^0)\right\|^2.\end{aligned}$$

We then follow the arguments in Remark 21.2 by reducing $f(\alpha_1,\alpha_2)$ to $\alpha_1\alpha_2$, to check (21.14).

The next result describes precisely the nature of the solution set $\mathscr{S}_{f_h}$.

THEOREM 21.2. *Let us assume that* (21.2) *holds and in addition that*

$$\lim_{h\to 0} \mathrm{D}f_h(0,0) = 0, \tag{21.15}$$

$$\lim_{h\to 0} \mathrm{D}^2 f_h(0,0) = \mathrm{D}^2 f(0,0), \tag{21.16}$$

and there exists a constant L such that for all $h \in (0,h_0]$ and $(\alpha_1,\alpha_2) \in B(0,\delta)$,

$$\left\|\mathrm{D}^2 f_h(\alpha_1,\alpha_2) - \mathrm{D}^2 f_h(0,0)\right\| \leqslant L\left\|(\alpha_1,\alpha_2)\right\|. \tag{21.17}$$

Then there exist two positive constants C and h_0, a neighborhood $\mathscr{V}$ of $(0,0)$, and for $h \leqslant h_0$ a point $(\alpha_{1h},\alpha_{2h}) \in \mathbb{R}^2$ and a C^1 diffeomorphism $\boldsymbol{\Phi}_h$ from $\mathscr{V}$ into $\mathbb{R}^2$ such that

$$\left\|(\alpha_{1h},\alpha_{2h})\right\| \leqslant C\left\|\mathrm{D}f_h(0,0)\right\|, \tag{21.18}$$

$$\mathrm{D}f_h(\alpha_{1h},\alpha_{2h}) = 0, \tag{21.19}$$

$$\boldsymbol{\Phi}_h(\alpha_{1h},\alpha_{2h}) = (\alpha_{1h},\alpha_{2h}), \qquad \mathrm{D}\boldsymbol{\Phi}_h(\alpha_{1h},\alpha_{2h}) = I,$$

$\boldsymbol{\Phi}_h(\mathscr{S}_{f_h} \cap \mathscr{V})$ *is a part of a hyperbola.*

PROOF. Under assumptions (21.15)–(21.17), we can apply Theorem 6.1 with $F_h = \mathrm{D}f_h$ and $L_h = L$ to prove the existence of $(\alpha_{1h}, \alpha_{2h}) \in B(0,\delta)$, for $h \in (0, h_0]$, only to restrict h_0 if necessary, such that both relations (21.18) and (21.19) hold. Moreover, for $h \leqslant h_0$ we have

$$\det\bigl(\mathrm{D}^2 f_h(\alpha_{1h}, \alpha_{2h})\bigr) < 0. \tag{21.20}$$

We are in a position to apply Theorem 19.1 to the mapping f_h at the point $(\alpha_{1h}, \alpha_{2h})$. There exists a C^{p-1} diffeomorphism $(\alpha_1, \alpha_2) \in \mathscr{V} \to \boldsymbol{\Phi}_h(\alpha_1, \alpha_2) \in \mathbb{R}^2$ such that

$$\begin{aligned} f_h(\alpha_1, \alpha_2) &= f_h(\alpha_{1h}, \alpha_{2h}) \\ &\quad + \tfrac{1}{2}\bigl(\mathrm{D}^2 f_h(\alpha_{1h}, \alpha_{2h})\bigl(\boldsymbol{\Phi}_h(\alpha_1, \alpha_2) - (\alpha_{1h}, \alpha_{2h}), \boldsymbol{\Phi}_h(\alpha_1, \alpha_2) - (\alpha_{1h}, \alpha_{2h})\bigr)\bigr) \end{aligned} \tag{21.21}$$

where $\mathscr{V}$ is a neighborhood of $(\alpha_{1h}, \alpha_{2h})$, which is a priori depending on h. Moreover, $\boldsymbol{\Phi}_h(\alpha_{1h}, \alpha_{2h}) = (\alpha_{1h}, \alpha_{2h})$ and $\mathrm{D}\boldsymbol{\Phi}_h(\alpha_{1h}, \alpha_{2h}) = I$. Consequently, the set $\mathscr{S}_{f_h}$ is diffeomorphic to a part of the hyperbola given in the $\boldsymbol{\Phi}$-plane by

$$\begin{aligned} &f_h(\alpha_{1h}, \alpha_{2h}) \\ &\quad + \tfrac{1}{2}\bigl(\mathrm{D}^2 f_h(\alpha_{1h}, \alpha_{2h})\bigl(\boldsymbol{\Phi} - (\alpha_{1h}, \alpha_{2h}), \boldsymbol{\Phi} - (\alpha_{1h}, \alpha_{2h})\bigr)\bigr) = 0. \end{aligned} \tag{21.22}$$

Going back to the proof of the Morse lemma, see the appendix in CROUZEIX and RAPPAZ [1989], it is not difficult to check that under assumptions (21.15)–(21.17) and with $p = 2$, we can choose $\mathscr{V}$ independently of h and containing the point $(0, 0)$, only to restrict h_0 if necessary. □

REMARK 21.4. With relation (21.21), we deduce that solving the bifurcation equation (21.4) is equivalent to solving equation (21.22). In the $\boldsymbol{\Phi}$-plane, the equation

$$f_h(\alpha_{1h}, \alpha_{2h}) + \tfrac{1}{2}\bigl(\mathrm{D}^2 f_h(\alpha_{1h}, \alpha_{2h})\bigl(\boldsymbol{\Phi} - (\alpha_{1h}, \alpha_{2h}), \boldsymbol{\Phi} - (\alpha_{1h}, \alpha_{2h})\bigr)\bigr) = 0$$

is the equation of a hyperbola centered at $(\alpha_{1h}, \alpha_{2h})$, since $\det(\mathrm{D}^2 f_h(\alpha_{1h}, \alpha_{2h})) < 0$. The hyperbola represents the solution set to (21.4) since $(\alpha_{1h}, \alpha_{2h})$ is invariant through $\boldsymbol{\Phi}_h$ and $\boldsymbol{\Phi}_h$ is tangent to the identity at $(\alpha_{1h}, \alpha_{2h})$. The point $(\alpha_{1h}, \alpha_{2h})$ is the center of the hyperbola and with the two relations (21.18) and (21.15) we know that $(\alpha_{1h}, \alpha_{2h})$ tends to 0.

The solution set $\mathscr{S}_{f_h}$ is described by Eq. (21.22) and we can distinguish two cases. If the value of $f_h(\alpha_{1h}, \alpha_{2h})$ is nonzero which is the common case,

$$f_h(\alpha_{1h}, \alpha_{2h}) \neq 0, \tag{21.23}$$

then $\mathscr{S}_{f_h}$ is diffeomorphic to a part of a nondegenerate hyperbola.

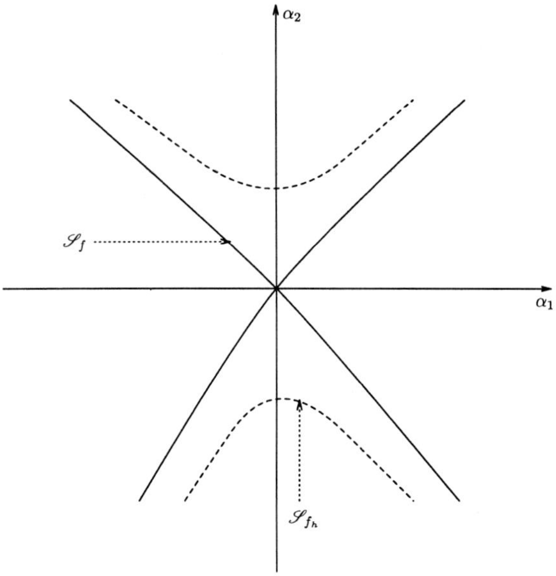

FIG. 21.2. Imperfect numerical bifurcation.

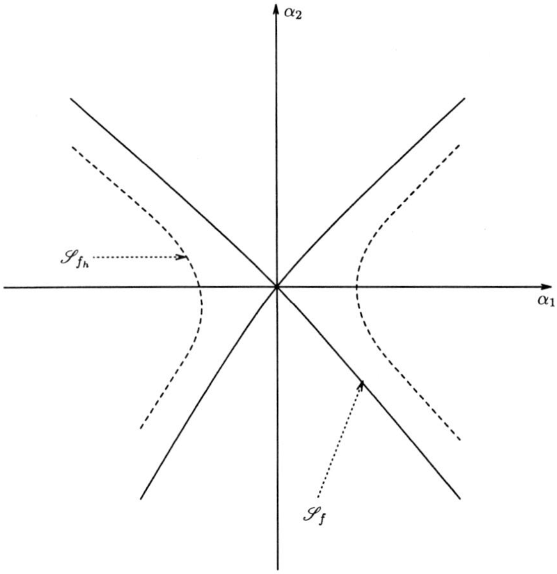

FIG. 21.3. Imperfect numerical bifurcation.

That situation is represented in Figs. 21.2 and 21.3, depending on the sign of $f_h(\alpha_{1h}, \alpha_{2h})$. We say that we have an imperfect numerical bifurcation.

The second case which can occur is

$$f_h(\alpha_{1h}, \alpha_{2h}) = 0. \tag{21.24}$$

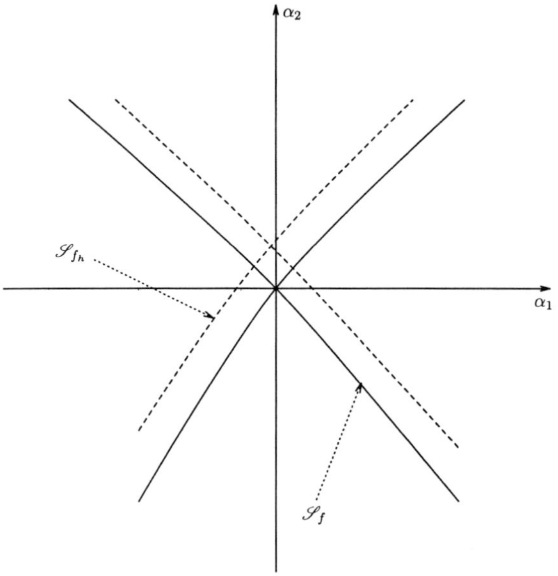

FIG. 21.4. Numerical bifurcation.

Then $\mathscr{S}_{f_h}$ is diffeomorphic to a part of a degenerate hyperbola.

That situation is represented in Fig. 21.4. We say that we have a numerical bifurcation. Condition (21.24) is satisfied for instance in the case of bifurcation from the trivial branch, when $F(\lambda, 0) = 0$ for all $\lambda \in \mathbb{R}$, see the example developed in the next section. It can also be satisfied when we have symmetries, see SATTINGER [1979].

REMARK 21.5. In the framework of Theorem 21.2, we assume furthermore that (21.5), (21.9) hold, and that $f_h(\alpha_{1h}, \alpha_{2h}) = 0$. Here $(\alpha_{1h}, \alpha_{2h})$ is the center of the hyperbola given in Theorem 21.2. Then from estimate (21.10) we get

$$\delta\big(\mathscr{S}_f, \mathscr{S}_{f_h}\big) \leqslant C \sup_{(\alpha_1,\alpha_2)\in B(0,\delta)} \big\|\mathrm{D}f(\alpha_1,\alpha_2) - \mathrm{D}f_h(\alpha_1,\alpha_2)\big\|. \tag{21.25}$$

Indeed we notice that

$$\big|f_h(0,0)\big| \leqslant C\big\|(\alpha_{1h},\alpha_{2h})\big\|^2$$

since $f_h(\alpha_{1h}, \alpha_{2h}) = 0$ and $\mathrm{D}f_h(\alpha_{1h}, \alpha_{2h}) = 0$, see (21.19). From estimate (21.18) we show

$$\begin{aligned}\big|f_h(0,0)\big| &\leqslant C\big\|\mathrm{D}f(0,0) - \mathrm{D}f_h(0,0)\big\|^2 \\ &\leqslant C\Big(\sup_{(\alpha_1,\alpha_2)\in B(0,\delta)} \big\|\mathrm{D}f(\alpha_1,\alpha_2) - \mathrm{D}f_h(\alpha_1,\alpha_2)\big\|\Big)^2.\end{aligned}$$

Estimate (21.25) is then a direct consequence of (21.10).

REMARK 21.6. In the framework of Theorem 21.2, we assume furthermore that (21.5), (21.9), (21.13) hold, and that $f_h(\alpha_{1h}, \alpha_{2h}) = 0$. Like in Remark 21.5, we can derive from estimate (21.14) the following one

$$\delta(\mathscr{S}_f, \mathscr{S}_{f_h}) \leqslant C \sup_{(\alpha_1,\alpha_2)\in\mathscr{S}_f} \big\|\mathrm{D}f(\alpha_1, \alpha_2) - \mathrm{D}f_h(\alpha_1, \alpha_2)\big\|. \tag{21.26}$$

REMARK 21.7. It is possible to generalize the results of Section 21 to the case where the mappings f, f_h are defined in $\mathbb{R}^3$, that is

$$f: B(0,\delta) \subset \mathbb{R}^3 \to \mathbb{R}, \qquad f_h: B(0,\delta) \subset \mathbb{R}^3 \to \mathbb{R}.$$

Then under analogous assumptions, we can prove that f is C^{p-1} diffeomorphic to a part of a cone and that f_h is C^{p-1} diffeomorphic to a part of a hyperboloid with one or two sheets.

22. A model example

To illustrate the general results in Sections 18 through 21, we will develop a simple model example putting emphasis on the methodology.

Let $\Omega \subset \mathbb{R}^2$ be a convex polygonal domain with the boundary $\partial\Omega$. We shall study the problem to find $\lambda \in \mathbb{R}$ and $u \in H_0^1(\Omega)$ satisfying

$$-\Delta u - \lambda u + u^2 = 0 \quad \text{in } \Omega \tag{22.1}$$

or in variational form, for all $v \in H_0^1(\Omega)$,

$$\int_\Omega \mathbf{grad}\, u\ \mathbf{grad}\, v \,\mathrm{d}x - \lambda \int_\Omega uv \,\mathrm{d}x + \int_\Omega u^2 v \,\mathrm{d}x = 0. \tag{22.2}$$

If the mapping $T \in \mathscr{L}(H^{-1}(\Omega); H_0^1(\Omega))$ denotes the inverse of the $(-\Delta)$ operator with homogeneous boundary conditions, then problem (22.2) reads: find $\lambda \in \mathbb{R}$ and $u \in H_0^1(\Omega)$ such that

$$u + T\big(u^2 - \lambda u\big) = 0, \tag{22.3}$$

see (3.6). Clearly the trivial branch $\{(\lambda, 0): \lambda \in \mathbb{R}\}$ belongs to the solution set of (22.3). Subsequently, we look for bifurcation from the trivial branch and study a finite element approximation of the problem.

We are in the framework of Section 19 with the following notations: $\mathscr{X} = H_0^1(\Omega)$, $X = \mathbb{R} \times H_0^1(\Omega)$, $Z = H_0^1(\Omega)$, and the mapping $F: \mathbb{R} \times H_0^1(\Omega) \to H_0^1(\Omega)$ given by, for $(\lambda, v) \in \mathbb{R} \times H_0^1(\Omega)$,

$$F(\lambda, v) = v + T\big(v^2 - \lambda v\big). \tag{22.4}$$

Clearly the mapping F is of class C^∞. For $\lambda \in \mathbb{R}$, the derivative $\mathrm{D}F(\lambda,0)$ is, for $(\delta,w) \in \mathbb{R} \times H_0^1(\Omega)$,

$$\mathrm{D}F(\lambda,0)(\delta,w) = w - \lambda T w. \tag{22.5}$$

We immediately deduce that

$$\operatorname{Ker}\big(\mathrm{D}F(\lambda,0)\big) = \mathbb{R} \times \operatorname{Ker}(I - \lambda T) \tag{22.6}$$

with I the identity operator in $H_0^1(\Omega)$ and T the restriction to $H_0^1(\Omega)$ of the above operator $T \in \mathscr{L}(H^{-1}(\Omega); H_0^1(\Omega))$. Let now λ_0 be the first eigenvalue of the eigen-problem: find $\lambda \in \mathbb{R}$, $\xi \in H_0^1(\Omega)$, $\xi \neq 0$ such that

$$-\Delta\xi = \lambda\xi \quad \text{in } \Omega.$$

It is well known that λ_0 is a simple positive eigenvalue and that a corresponding eigenvector ξ can be chosen positive and with a norm equal to 1, that is

$$\xi > 0 \text{ in } \Omega \quad \text{and} \quad \int_\Omega |\,\mathbf{grad}\,\xi|^2 \,\mathrm{d}x = 1.$$

Then

$$\operatorname{Ker}\big(\mathrm{D}F(\lambda_0,0)\big) = \mathbb{R} \times \operatorname{span}\{\xi\}, \tag{22.7}$$

which is of dimension 2. With the expression (22.5), we get

$$\operatorname{Range}\big(\mathrm{D}F(\lambda_0,0)\big) = \operatorname{Range}(I - \lambda_0 T). \tag{22.8}$$

Since the operator $T \in \mathscr{L}(H_0^1(\Omega); H_0^1(\Omega))$ is self-adjoint when $H_0^1(\Omega)$ is endowed with the scalar product, for v_1, $v_2 \in H_0^1(\Omega)$

$$(v_1, v_2)_{1,\Omega} = \int_\Omega \mathbf{grad}\, v_1 \;\mathbf{grad}\, v_2 \,\mathrm{d}x,$$

the range of $\mathrm{D}F(\lambda_0,0)$ is the orthogonal complement of ξ, that is

$$\operatorname{Range}\big(\mathrm{D}F(\lambda_0,0)\big) = \big\{v \in H_0^1(\Omega)\colon\ (v,\xi)_{1,\Omega} = 0\big\}, \tag{22.9}$$

which is of codimension 1.

The solution $(\lambda_0, 0)$ satisfies the following hypotheses:

$$\begin{aligned} &\text{(i)} \quad \operatorname{Ker}(\mathrm{D}_v F(\lambda_0,0)) = \operatorname{span}\{\xi\},\\ &\text{(ii)} \quad \text{the range of } \mathrm{D}_v F(\lambda_0,0) \text{ is closed, of codimension 1,}\\ &\text{(iii)} \quad \mathrm{D}_\lambda F(\lambda_0,0) = 0. \end{aligned} \tag{22.10}$$

Thus $(\lambda_0, 0)$ is a singular solution of (22.3) and assumptions (19.2) are verified. In our case a decomposition of both spaces $\mathbb{R} \times H_0^1(\Omega)$ and $H_0^1(\Omega)$ can be written explicitly. In (19.4) and (19.5), we choose

$$X_1 = \{0\} \times \mathscr{X}_1 \equiv \{0\} \times \text{Range}(I - \lambda_0 T),$$
$$Z_1 = \text{Ker}(I - \lambda_0 T) = \text{span}\{\xi\}.$$

Then the two projectors $Q_1 \in \mathscr{L}(H_0^1(\Omega); Z_1)$ and $Q_2 \in \mathscr{L}(H_0^1(\Omega); \text{Range}(I - \lambda_0 T))$ are the orthogonal projectors with respect to the scalar product $(\bullet, \bullet)_{1,\Omega}$ in $H_0^1(\Omega)$. The two projectors $P_1 \in \mathscr{L}(\mathbb{R} \times H_0^1(\Omega); X_1)$ and $P_2 \in \mathscr{L}(\mathbb{R} \times H_0^1(\Omega); \text{Ker}(DF(\lambda_0, 0)))$ can be written in the following way

$$P_1(\delta, v) = (0, Q_2 v), \qquad P_2(\delta, v) = (\delta, Q_1 v).$$

There exist positive constants γ, η, and a mapping $g : B(0, \gamma) \subset \mathbb{R}^2 \to \mathscr{X}_1$ of class C^∞ such that for all $(\alpha_1, \alpha_2) \in B(0, \gamma)$

$$\big(Q_2 F(\lambda_0 + \alpha_1, \kappa_1 + \alpha_2 \xi) = 0 \text{ and } \kappa_1 \in B(0, \eta) \subset \mathscr{X}_1\big) \iff \kappa_1 = g(\alpha_1, \alpha_2).$$

In particular for $(\alpha_1, \alpha_2) \in B(0, \gamma)$,

$$Q_2 F(\lambda_0 + \alpha_1, g(\alpha_1, \alpha_2) + \alpha_2 \xi) = 0, \tag{22.11}$$

and $g(\alpha_1, 0) = 0$ for $(\alpha_1, 0) \in B(0, \gamma)$.

If the functional $B \in H^{-1}(\Omega)$ satisfies

$$B(\xi) = 1 \quad \text{and} \quad B(z) = 0 \text{ for all } z \in \text{Range}(I - \lambda_0 T),$$

then clearly for all $z \in H_0^1(\Omega)$,

$$B(z) = (z, \xi)_{1,\Omega}. \tag{22.12}$$

The bifurcation equation reads, for $(\alpha_1, \alpha_2) \in B(0, \gamma)$,

$$f(\alpha_1, \alpha_2) = \big(F(\lambda_0 + \alpha_1, g(\alpha_1, \alpha_2) + \alpha_2 \xi), \xi\big)_{1,\Omega} = 0. \tag{22.13}$$

The function f can be written in a simpler form. With definition (22.4) of F and the relations

$$\big(g(\alpha_1, \alpha_2), \xi\big)_{1,\Omega} = 0 = \big(Tg(\alpha_1, \alpha_2), \xi\big)_{1,\Omega}$$

we get

$$\begin{aligned} f(\alpha_1,\alpha_2) &= \big(\xi, \alpha_2\xi + T\big((\alpha_2\xi + g(\alpha_1,\alpha_2))^2 - (\lambda_0+\alpha_1)\alpha_2\xi\big)\big)_{1,\Omega} \\ &= \big(\xi, T\big((\alpha_2\xi + g(\alpha_1,\alpha_2))^2 - \alpha_1\alpha_2\xi\big)\big)_{1,\Omega} \\ &= \int_\Omega \xi\big(\alpha_2\xi + g(\alpha_1,\alpha_2)\big)^2\,\mathrm{d}x - \frac{\alpha_1\alpha_2}{\lambda_0}. \end{aligned} \tag{22.14}$$

From relations (22.11) and (22.14), we check with some simple computations

$$g(0,0)=0 \quad \text{and} \quad f(0,0)=0, \tag{22.15}$$

$$\begin{cases} \dfrac{\partial g}{\partial\alpha_1}(0,0)=0, & \dfrac{\partial g}{\partial\alpha_2}(0,0)=0, \\ \dfrac{\partial f}{\partial\alpha_1}(0,0)=0, & \dfrac{\partial f}{\partial\alpha_2}(0,0)=0, \end{cases} \tag{22.16}$$

$$\begin{cases} \dfrac{\partial^2 g}{\partial\alpha_1^2}(0,0)=0, & \dfrac{\partial^2 g}{\partial\alpha_1\partial\alpha_2}(0,0)=0, \\ \dfrac{\partial^2 g}{\partial\alpha_2^2}(0,0)=-2(I-\lambda_0 T)^{-1}Q_2T\xi^2, \end{cases} \tag{22.17}$$

$$\begin{cases} \dfrac{\partial^2 f}{\partial\alpha_1^2}(0,0)=0, & \dfrac{\partial^2 f}{\partial\alpha_1\partial\alpha_2}(0,0)=-1/\lambda_0, \\ \dfrac{\partial^2 f}{\partial\alpha_2^2}(0,0)=2\displaystyle\int_\Omega \xi^3\,\mathrm{d}x; \end{cases} \tag{22.18}$$

notice that here the operator $I-\lambda_0T$ is an isomorphism onto $\mathscr{X}_1$. Since

$$\det\big(\mathrm{D}^2 f(0,0)\big) = -\left(\frac{1}{\lambda_0}\right)^2,$$

we are precisely in the case where $(\lambda_0,0)$ is a simple bifurcation point. Moreover, we notice that for all $(\alpha_1,0)\in B(0,\gamma)$,

$$f(\alpha_1,0)=0, \tag{22.19}$$

since $g(\alpha_1,0)=0$.

Once studied the solution set of (22.3) in a neighborhood of the singular solution $(\lambda_0,0)$, we turn our attention to finite element approximations of (22.3). For ease of exposition we have assumed that Ω is a convex polygonal domain in $\mathbb{R}^2$, so we can consider a regular family $\{\mathscr{T}_h\}_{0<h\leqslant 1}$ of triangulations of $\overline{\Omega}$. To each triangulation we associate the finite element subspace

$$V_h = \{v \in C^0(\overline{\Omega})\colon\ v|_T \in \mathscr{P}_1(T) \text{ for all } T \in \mathscr{T}_h\} \cap H_0^1(\Omega), \tag{22.20}$$

see (3.12). A finite element approximation of (22.2) consists in finding $\lambda \in \mathbb{R}$ and $u_h \in V_h$ such that for all $v_h \in V_h$

$$\int_\Omega \mathbf{grad}\, u_h\, \mathbf{grad}\, v_h \,\mathrm{d}x - \lambda \int_\Omega u_h v_h \,\mathrm{d}x + \int_\Omega u_h^2 v_h \,\mathrm{d}x = 0. \tag{22.21}$$

We define the mapping $T_h \in \mathscr{L}(H^{-1}(\Omega); V)$, the discrete analogue of T; given $f \in H^{-1}(\Omega)$, $T_h f \equiv w_h \in V_h$ is the unique solution of

$$\text{for all } v_h \in V_h, \quad \int_\Omega \mathbf{grad}\, w_h\, \mathbf{grad}\, v_h \,\mathrm{d}x = \langle f, v_h\rangle_{H^{-1}(\Omega)\, H_0^1(\Omega)}. \tag{22.22}$$

Then problem (22.21) can be written in the equivalent way: find $\lambda \in \mathbb{R}$ and $u_h \in H_0^1(\Omega)$ such that

$$u_h + T_h(u_h^2 - \lambda u_h) = 0. \tag{22.23}$$

We notice that if $u_h \in H_0^1(\Omega)$ satisfies (22.23), then u_h belongs to the range of T_h, which is precisely V_h, so u_h is a solution to (22.21). The discrete analogue of the mapping F in (22.4) is $F_h : \mathbb{R}\times H_0^1(\Omega) \to H_0^1(\Omega)$ defined by, for $(\lambda, v) \in \mathbb{R}\times H_0^1(\Omega)$

$$F_h(\lambda, v) = v + T_h(v^2 - \lambda v). \tag{22.24}$$

To study the approximate problem (22.23), we shall apply the results of Sections 20 and 21, with F defined in (22.4) and F_h in (22.24). We first check assumption (20.4). For $(\lambda, v) \in \mathbb{R} \times H_0^1(\Omega)$,

$$F_h(\lambda, v) - F(\lambda, v) = (T_h - T)(v^2 - \lambda v), \tag{22.25}$$

for $(\delta, w) \in \mathbb{R} \times H_0^1(\Omega)$

$$\begin{aligned}&\big(\mathrm{D}F_h(\lambda, v) - \mathrm{D}F(\lambda, v)\big)(\delta, w)\\ &\quad = -\delta(T_h - T)v + (T_h - T)(2vw - \lambda w),\end{aligned} \tag{22.26}$$

and for $(\delta_1, w_1), (\delta_2, w_2) \in \mathbb{R} \times H_0^1(\Omega)$,

$$\begin{aligned}&\big(\mathrm{D}^2F_h(\lambda, v) - \mathrm{D}^2F(\lambda, v)\big)\big((\delta_1, w_1), (\delta_2, w_2)\big)\\ &\quad = -(T_h - T)(\delta_1 w_2 + \delta_2 w_1) + 2(T_h - T)(w_1 w_2).\end{aligned} \tag{22.27}$$

Moreover, for $l > 2$,

$$\mathrm{D}^l F(\lambda, v) = \mathrm{D}^l F_h(\lambda, v) = 0. \tag{22.28}$$

From the classical finite element estimates, for all $f \in L^2(\Omega)$ we have

$$\left|(T - T_h)f\right|_{1,\Omega} \leqslant Ch\|f\|_{0,\Omega}, \tag{22.29}$$

where T and T_h are given in (3.4) and (22.22) and so

$$\lim_{h\to 0} \|T - T_h\|_{L^2(\Omega);H_0^1(\Omega)} = 0. \tag{22.30}$$

Then from the formulae (22.25)–(22.28) and the continuity of the embedding $H_0^1(\Omega) \subset L^p(\Omega)$, we get

$$\lim_{h\to 0} \sup_{(\lambda,w)\in B} \left\|\mathrm{D}^l F_h(\lambda, w) - \mathrm{D}^l F(\lambda, w)\right\| = 0 \tag{22.31}$$

for all $l = 0, 1, 2, \ldots$ and $B \subset H_0^1(\Omega)$ bounded. Assumption (20.4) is verified and we can apply the theory developed in Section 20 at the singular solution $(\lambda_0, 0)$.

There exist positive constants $h_0 \leqslant 1$, η, and a C^∞ mapping $g_h : B(0,\gamma) \subset \mathbb{R}^2 \to \mathscr{X}_1$ such that for all $(\alpha_1, \alpha_2) \in B(0,\gamma)$, $0 < h \leqslant h_0$,

$$\big(Q_2 F_h(\lambda_0 + \alpha_1, \kappa_1 + \alpha_2\xi) = 0 \text{ and } \kappa_1 \in B(0,\eta) \subset \mathscr{X}_1\big) \Longleftrightarrow \kappa_1 = g_h(\alpha_1, \alpha_2).$$

In particular for $(\alpha_1, \alpha_2) \in B(0,\gamma)$,

$$Q_2 F_h\big(\lambda_0 + \alpha_1, g_h(\alpha_1, \alpha_2) + \alpha_2\xi\big) = 0, \tag{22.32}$$

and $g_h(\alpha_1, 0) = 0$ for $(\alpha_1, 0) \in B(0,\gamma)$. Moreover, we can derive the following error estimates, see (20.7), (20.8), (12.14), (12.15), (22.25)–(22.29),

$$\sup_{(\alpha_1,\alpha_2)\in B(0,\gamma)} \left|\mathrm{D}^l g(\alpha_1, \alpha_2) - \mathrm{D}^l g_h(\alpha_1, \alpha_2)\right|_{1,\Omega} \leqslant C_l h, \tag{22.33}$$

where the constants C_l, $0 \leqslant l < \infty$, are independent of h.

The approximate bifurcation equation reads, for $(\alpha_1, \alpha_2) \in B(0,\gamma)$,

$$f_h(\alpha_1, \alpha_2) \equiv \big(F_h\big(\lambda_0 + \alpha_1, g_h(\alpha_1, \alpha_2) + \alpha_2\xi\big), \xi\big)_{1,\Omega} = 0. \tag{22.34}$$

We can compute with definitions (22.13) of f and (22.34) of f_h, the difference $f - f_h$,

$$\begin{aligned} &f(\alpha_1, \alpha_2) - f_h(\alpha_1, \alpha_2) \\ &\quad = \big(T\big(g^2(\alpha_1, \alpha_2) - g_h^2(\alpha_1, \alpha_2) + (2\alpha_2\xi - \alpha_1)\big(g(\alpha_1, \alpha_2) - g_h(\alpha_1, \alpha_2)\big)\big), \xi\big)_{1,\Omega} \\ &\qquad + \big((T - T_h)\big(\big(g_h(\alpha_1, \alpha_2) + \alpha_2\xi\big)^2 - (\lambda_0 + \alpha_1)\big(g_h(\alpha_1, \alpha_2) + \alpha_2\xi\big)\big), \xi\big)_{1,\Omega}. \end{aligned}$$

With estimates (22.29) and (22.33) we deduce the following:

$$\sup_{(\alpha_1,\alpha_2)\in B(0,\gamma)} \left|\mathrm{D}^l f(\alpha_1,\alpha_2) - \mathrm{D}^l f_h(\alpha_1,\alpha_2)\right| \leqslant C_l' h, \tag{22.35}$$

where the constants C_l', $0 \leqslant l < \infty$, are independent of h. Since for all $(\alpha_1, 0) \in B(0,\gamma)$ we have $g_h(\alpha_1, 0) = 0$, it is not difficult to check with the definition of f_h in (22.34) that for all $(\alpha_1, 0) \in B(0,\gamma)$,

$$f_h(\alpha_1, 0) = 0, \tag{22.36}$$

which is the analogue of (22.19).

To study the approximate bifurcation equation, we are exactly in the framework presented in Section 21. The functions f, f_h are of class C^∞ defined on $B(0,\gamma) \subset \mathbb{R}^2$ with values in $\mathbb{R}$, see (22.13) and (22.34). We have checked that

$$\det\left(\mathrm{D}^2 f(0,0)\right) = -(1/\lambda_0)^2 < 0,$$

so $(\lambda_0, 0)$ is a simple bifurcation point. Moreover, the difference $\mathrm{D}^l f - \mathrm{D}^l f_h$ is estimated in (22.35).

We can apply Theorem 21.2. For h small enough and only to restrict γ, there exists a C^∞ diffeomorphism from $\mathscr{S}_{f_h} = \{(\alpha_1,\alpha_2) \in B(0,\gamma) \subset \mathbb{R}^2\colon\ f_h(\alpha_1,\alpha_2) = 0\}$ onto a part of a hyperbola centred at $(\alpha_{1h}, \alpha_{2h})$. Going back to the definition of f_h, we can check that the center of the hyperbola is in fact $(\alpha_{1h}, 0)$ with

$$f_h(\alpha_{1h}, 0) = 0 \quad \text{and} \quad \mathrm{D} f_h(\alpha_{1h}, 0) = 0,$$

which means that the hyperbola is degenerate and we are in the case of Fig. 21.4.

From estimates (21.18) and (22.35), we have

$$|\alpha_{1h}| \leqslant Ch. \tag{22.37}$$

Estimates (21.25) and (22.35) lead to

$$\delta(\mathscr{S}_f, \mathscr{S}_{f_h}) \leqslant Ch, \tag{22.38}$$

and estimates (20.11), (22.29) and (22.38) imply

$$\delta(\mathscr{S}, \mathscr{S}_h) \leqslant Ch, \tag{22.39}$$

where $\mathscr{S}$ and $\mathscr{S}_h$ are the solution sets for F and F_h introduced in Section 20.

REMARK 22.1. In fact we can check that there exists a $\xi_h \in V_h$, $\xi_h \neq 0$ such that

$$\text{for all } v_h \in V_h, \quad \int_\Omega \mathbf{grad}\, \xi_h\, \mathbf{grad}\, v_h \,\mathrm{d}x = (\lambda_0 + \alpha_{1h}) \int_\Omega \xi_h v_h \,\mathrm{d}x.$$

It is known from the theory of approximation of eigenproblems, see BABUŠKA and OSBORN [1991], that the following estimate holds:

$$|\alpha_{1h}| \leqslant Ch^2.$$

Estimate (22.37) is not optimal. Optimal estimates are derived with a different method in CROUZEIX and RAPPAZ [1989].

REMARK 22.2. If we take polynomials of degree $k > 1$ on each triangle in the definition (22.20) of V_h, then we cannot improve estimate (22.33) but estimate (22.39) is modified into

$$\delta(\mathscr{S}, \mathscr{S}_h) \leqslant Ch^k.$$

23. Bibliographical comments

The brief comments below are intended only to outline the wealthy literature which is related to our work. A lot of additional results could have been included or quoted.

In Sections 18 through 22 we describe the different steps to study approximations of singular solutions. Only the main features of the method are developed and the example of Section 22 is elementary but it illustrates the general theory. An example of a simple bifurcation point not on the trivial branch is studied into detail in CAUSSIGNAC, DESCLOUX and RAPPAZ [1987] and an example of bifurcation for a problem on an unbounded domain is presented in DESCLOUX and RAPPAZ [1982], see also CROUZEIX and RAPPAZ [1989].

Bifurcation phenomena often occur in parameter dependent nonlinear problems. In KELLER and ANTMAN [1969] different examples are presented. General results on bifurcation problems can be found in the books of CHOW and HALE [1982], GOLUBITSKY and SCHAEFFER [1985], GOLUBITSKY, STEWART and SCHAEFFER [1988], IOOSS and JOSEPH [1981], NIRENBERG [1974], and SATTINGER [1973, 1979], and in the bibliographies therein.

The approximation of parameter dependent problems has been widely studied in the literature. For a general approach we can mention the works of BREZZI, RAPPAZ and RAVIART [1980, 1981a,b], CROUZEIX and RAPPAZ [1989], FUJI and YAMAGUTI [1980], KELLER [1977], KIKUCHI [1976], RABIER [1985], RHEINBOLDT [1986] and SIMPSON [1972].

Further developments concerning turning point can be found for instance in GRIEWANK and REDDIEN [1984], KIKUCHI [1979], and PAUMIER [1981], and concerning bifurcation can be found in RAPPAZ and RAUGEL [1982] for bifurcation from multiple eigenvalues, in BERNARDI [1982] and DESCLOUX [1984] for Hopf bifurcation. A generalization in the direction of nondifferentiable problems was mentioned in Remark 2.4.

When the exact problem and its approximation are covariant with respect to representations of a finite group, see SATTINGER [1979], GOLUBITSKY and SCHAEFFER

[1985], GOLUBITSKY, STEWART and SCHAEFFER [1988], then the study can be sharpened, see CROUZEIX and RAUGEL [1988], DELLNITZ and WERNER [1989], RAUGEL [1984] for instance.

The applications of the general approximation theory are numerous and quite diverse. For instance we can find an account on the analysis of discretization of the Navier–Stokes equations in GIRAULT and RAVIART [1986]. Approximations of the von Kármán equations, see CIARLET and RABIER [1980], can be found in BREZZI, RAPPAZ and RAVIART [1980, 1981b], KESAVAN [1979, 1980], REINHARDT [1982].

In our work we have developed a general method based on the inverse and the implicit function theorems to study approximations of regular solutions of nonlinear problems and approximations of regular and singular solutions of parametrized nonlinear problems. There are other methods to study nonlinear problems, which are not even mentioned here. For instance the ones based on the maximum principle or on monotonicity are well-suited to variational inequalities, see for instance BARRETT and ELLIOTT [1991], GLOWINSKI [1984], CONRAD and CORTEY-DUMONT [1987], ŽENÍŠEK [1990].

There is a wide variety of algorithms to solve nonlinear approximate problems. Some fundamental results are due to Keller, see KELLER [1970], KEENER and KELLER [1974] and KELLER [1987]. For an overview we refer to the conference proceedings edited by KUEPPER, MITTELMANN and WEBER [1984], KUEPPER, SEYDEL and TROGER [1987], MITTELMANN and WEBER [1980]. There is an extensive bibliography on the topic of numerical continuation methods in the book of ALLGOWER and GEORG [1990].

References

ADAMS, R. (1975), *Sobolev Spaces* (Academic Press, New York).

ALLGOWER, E.L. and K. GEORG (1990), *Numerical Continuation Methods. An Introduction*, Springer Series in Computational Mathematics (Springer, Berlin).

AMANN, H. (1976), Fixed point equations and nonlinear eigenvalue problems in ordered Banach space, *SIAM Rev.* **18**, 620–709.

BABUŠKA, I. and A.K. AZIZ (1972), Survey lectures on the mathematical foundations of the finite element method, in: A.K. AZIZ, ed., *The Mathematical Foundations of the Finite Element Method with Application to Partial Differential Equations* (Academic Press, New York) 3–359.

BABUŠKA, I. and J. OSBORN (1991), Eigenvalue problems, in: *Handbook of Numerical Analysis* **II** (North-Holland, Amsterdam) 641–787.

BABUŠKA, I. and W.C. RHEINBOLDT (1982), Computational error estimates and adaptive processes for some nonlinear structural problems, *Comput. Methods Appl. Mech. Engrg.* **34**, 895–937.

BARANGER, J. and H. EL AMRI (1991), Estimateur a posteriori d'erreur pour le calcul adaptatif d'écoulements quasi-newtoniens, *RAIRO Modél. Math. Anal. Numér.* **25**, 31–48.

BARRETT, J.W. and C.M. ELLIOTT (1989), Finite element approximation of a plasma equilibrium problem, *IMA J. Numer. Anal.* **9**, 443–464.

BARRETT, J.W. and C.M. ELLIOTT (1991), Finite element approximation of a free boundary problem arising in the theory of liquid drops and plasma physics, *RAIRO Modél. Math. Anal. Numér.* **25**, 213–252.

BERNARDI, C. (1982), Approximation of Hopf bifurcation, *Numer. Math.* **39**, 14–37.

BONIC, R.A. (1969), *Linear Functional Analysis*, Notes on Mathematics and its Applications (Gordon and Breach, New York).

BREZIS, H. (1983), *Analyse Fonctionnelle. Théorie et Applications*, Collection Mathématiques Appliquées pour la Maîtrise (Masson, Paris).

BREZZI, F., RAPPAZ, J. and P.A. RAVIART (1980), Finite dimensional approximation of nonlinear problems. Branches of nonsingular solutions, *Numer. Math.* **36**, 1–36.

BREZZI, F., RAPPAZ, J. and P.A. RAVIART (1981a), Finite dimensional approximation of nonlinear problems. Limit points, *Numer. Math.* **37**, 1–28.

BREZZI, F., RAPPAZ, J. and P.A. RAVIART (1981b), Finite dimensional approximation of nonlinear problems. Simple bifurcation points, *Numer. Math.* **38**, 1–30.

CALOZ, G. (1987), Simulation numérique des équilibres d'un plasma dans un tokamak: Modélisation et études mathématiques, Thèse 650, École Polytechnique Fédérale de Lausanne.

CALOZ, G. (1991), Approximation by finite element method of the model plasma problem, *RAIRO Modél. Math. Anal. Numér.* **25**, 213–252.

CALOZ, G. (1994), Stability on a regular branch of solutions, Tech. Rpt. 94, I.R.M.A.R., Université de Rennes I.

CARTAN, H. (1967), *Cours de Calcul Différentiel*, Collection Méthodes (Hermann, Paris).

CAUSSIGNAC, PH., DESCLOUX, J. and J. RAPPAZ (1987), Study of an elliptic problem with nonlinear boundary conditions, *Math. Methods Appl. Sci.* **9**, 261–275.

CHOW, S.N. and J. HALE (1982), *Methods of Bifurcation Theory*, Grundlehren **251** (Springer, Berlin).

CIARLET, P.G. (1978), *The Finite Element Method for Elliptic Problems*, Studies in Mathematics and its Applications (North-Holland, Amsterdam).

CIARLET, P.G. (1991), Basic error estimates for elliptic problems, in: *Handbook of Numerical Analysis* **II** (North-Holland, Amsterdam) 17–351.

CIARLET, P.G. and P. RABIER (1980), *Les Équations de von Kármán*, Lecture Notes in Mathematics **826** (Springer, Berlin).

CLÉMENT, P. (1975), Approximation by finite element functions using local regularization, *RAIRO Anal. Numér.* **2**, 77–84.

CONRAD, F. and P. CORTEY-DUMONT (1987), Nonlinear eigenvalue problems in elliptic variational inequalities: Some results for the maximal branch, *Numer. Funct. Anal. Optim.* **9**, 1059–1114.

CRANDALL, M.G. and P.H. RABINOWITZ (1973), Bifurcation, perturbation of simple eigenvalues and linearized stability, *Arch. Rational Mech. Anal.* **52**, 161–180.

CRANDALL, M.G. and P.H. RABINOWITZ (1975), Some continuation and variational methods for positive solutions of nonlinear eigenvalue problems, *Arch. Rational Mech. Anal.* **58**, 241–269.

CROUZEIX, M. and J. RAPPAZ (1989), *On Numerical Approximation in Bifurcation Theory*, RMA **13** (Masson, Paris).

CROUZEIX, M. and G. RAUGEL (1988), Invariance under the dihedral group and application to bifurcation problems, *Nonlinear Anal.* **12**, 75–99.

DAUGE, M. (1992), Neumann and mixed problems on curvilinear polyhedra, *Integral Equations Operator Theory* **15**, 227–261.

DAUTRAY, R. and J.L. LIONS (1987), *Analyse Mathématique et Calcul Numérique pour les Sciences et les Techniques* (Masson, Paris).

DELLNITZ, M. and B. WERNER (1989), Computational methods for bifurcation problems with symmetries, *J. Comput. Appl. Math.* **26**, 97–123.

DESCLOUX, J. (1984), On Hopf and subharmonic bifurcations, in: *Numerical Methods for Bifurcation Problems* (Birkhauser, Basel) 145–161.

DESCLOUX, J. and J. RAPPAZ (1982), Approximation of solution branches of nonlinear equations, *RAIRO Anal. Numér.* **16**, 319–349.

DIEUDONNÉ, J. (1968), *Éléments d'Analyse* **I** (Gauthier-Villars, Paris).

DUNFORD, N. and J.T. SCHWARTZ (1958), *Linear Operators* (Interscience, New York).

FINK, J.P. and W.C. RHEINBOLDT (1983a), On the discretization error of parametrized nonlinear equations, *SIAM J. Numer. Anal.* **20**, 732–746.

FINK, J.P. and W.C. RHEINBOLDT (1983b), Solution manifolds and submanifolds of parametrized equations and their discretization errors, Tech. Rpt. ICMA-83-59, Institute for Computational Mathematics and Applications, University of Pittsburgh (1984), *Numer. Math.* **45**, 323–343.

FINK, J.P. and W.C. RHEINBOLDT (1986), Folds on the solution manifold of a parametrized equation, *SIAM J. Numer. Anal.* **23**, 693–706.

FINK, J.P. and W.C. RHEINBOLDT (1987), A geometric framework for the numerical study of singular points, *SIAM J. Numer. Anal.* **24**, 618–633.

FUJI, H. and M. YAMAGUTI (1980), Structure of singularities and its numerical realization in nonlinear elasticity, *J. Math. Kyoto Univ.* **20**, 489–590.

GILBARG, D. and N.S. TRUDINGER (1977), *Elliptic Partial Differential Equations of Second Order* (Springer, Berlin).

GIRAULT, V. and P.A. RAVIART (1982), An analysis of upwind schemes for the Navier–Stokes equations, *SIAM J. Numer. Anal.* **19**, 312–333.

GIRAULT, V. and P.A. RAVIART (1986), *Finite Element Methods for Navier–Stokes Equations* (Springer, Berlin).

GLOWINSKI, R. (1984), *Numerical Methods for Nonlinear Variational Problems*, Springer Series in Computational Physics (Springer, Berlin).

GOLUBITSKY, M. and D.G. SCHAEFFER (1985), *Singularities and Groups in Bifurcation Theory* **1** (Springer, Berlin).

GOLUBITSKY, M., STEWART, I. and D.G. SCHAEFFER (1988), *Singularities and Groups in Bifurcation Theory* **2** (Springer, Berlin).

GRIEWANK, A. and G.W. REDDIEN (1984), Characterization and computation of generalized turning points, *SIAM J. Numer. Anal.* **21**, 176–185.

GRISVARD, P. (1985), *Elliptic Problems in Nonsmooth Domains* (Pitman, Boston).

IOOSS, G. and D.D. JOSEPH (1981), *Elementary Stability and Bifurcation Theory* (Springer, Berlin).

JOSEPH, D.D. (1965), Non-linear heat generation and stability of the temperature distribution in conduction solids, *Internat. J. Heat Mass Transfer* **8**, 281–288.

KEENER, J.P. and H.B. KELLER (1974), Perturbed bifurcation theory, *Arch. Rational Mech. Anal.* **50**, 159–175.

KELLER, H.B. (1970), Nonlinear bifurcation, *J. Differential Equations* **7**, 417–434.

KELLER, H.B. (1977), Numerical solution of bifurcation and nonlinear eigenvalue problems, in: P.H. RABINOWITZ, ed., *Application of Bifurcation Theory* (Academic Press, New York) 359–384.

KELLER, H.B. (1987), *Lectures on Numerical Methods in Bifurcation Problems* (Springer, Berlin).

KELLER, H.B. and S. ANTMAN, eds. (1969), *Bifurcation Theory and Nonlinear Eigenvalue Problems* (Benjamin, New York).

KELLER, H.B. and D.S. COHEN (1967), Some positone problems suggested by nonlinear heat generation, *J. Math. Mech.* **16**, 1361–1376.

KESAVAN, S. (1979), La méthode de Kikuchi appliquée aux équations de von Kármán, *Numer. Math.* **32**, 209–232.

KESAVAN, S. (1980), Une méthode d'éléments finis mixtes pour les équations de von Kármán, *RAIRO Anal. Numér.* **14**, 149–173.

KIKUCHI, F. (1976), An iterative finite element scheme for bifurcation analysis of semilinear elliptic equations, Report 542, Inst. Space Aero. Sc., Tokyo University.

KIKUCHI, F. (1979), Finite element approximations to bifurcation problems of turning point type, *Theoret. Appl. Mech.* **27**, 99–144.

KUEPPER, T., MITTELMANN, H.D. and H. WEBER (1984), *Numerical Methods for Bifurcation Problems*, ISNM **70** (Birkhauser, Basel).

KUEPPER, T., SEYDEL, R. and H. TROGER (1987), *Bifurcation: Analysis, Algorithms, Applications*, ISNM **79** (Birkhauser, Basel).

LERAY, J. and J. SCHAUDER (1934), Topologie et équations fonctionnelles, *Ann. Sci. École Norm. Sup.* **51**, 45–78.

LIONS, J.L. and E. MAGENES (1968), *Problèmes aux Limites non Homogènes et Applications*, Travaux et Recherches Mathématiques (Dunod, Paris).

LIONS, P.L. (1982), On the existence of positive solutions of semilinear elliptic equations, *SIAM Rev.* **24**, 441–467.

MAZ'YA, V.G. (1985), *Sobolev Spaces* (Springer, Berlin).

MIGNOT, F. and J.P. PUEL (1980), Sur une classe de problèmes nonlinéaires avec non linéarité positive, croissante, convexe, *Comm. P.D.E.* **5**, 791–836.

MITTELMANN, H.D. and H. WEBER (1980), *Bifurcation Problems and their Numerical Solution*, ISNM **54** (Birkhauser, Basel).

NEČAS, J. (1967), *Les Méthodes Directes en Théorie des Équations Elliptiques* (Masson, Paris).

NIJENHUIS, A. (1974), Strong derivatives and inverse mappings, *Amer. Math. Monthly* **81**, 969–980.

NIRENBERG, L. (1974), *Topics in Nonlinear Functional Analysis*, Courant Inst. Math. Sci. New York Univ.

PAUMIER, J.C. (1981), Une méthode numérique pour le calcul de points de retournement. Application à un problème aux limites non-linéaires, *Numer. Math.* **37**, 433–452.

PICASSO, M. (1992), Simulation numérique des traitements de surface par laser, Thèse 1011, École Polytechnique Fédérale de Lausanne.

POUSIN, J. and J. RAPPAZ (1991), Consistance, stabilité, erreurs à priori et à posteriori pour des problèmes non linéaires, *C. R. Acad. Sci. Paris, Sér. I* **312**, 699–703.

POUSIN, J. and J. RAPPAZ (1992), Consistency, stability, a priori and a posteriori errors for Petrov–Galerkin methods applied to nonlinear problems, Tech. Rpt. 04-92, D.M.A. École Polytechnique Fédérale de Lausanne; *Numer. Math.* **69**, 213–231.

RABIER, P. (1985), *Topics in One-parameter Nonlinear Problems*, Tata Institute Lecture Notes (Springer, Berlin).

RANNACHER, R. and R. SCOTT (1982), Some optimal error estimates for piecewise linear finite element approximations, *Math. Comp.* **38**, 437–445.

RAPPAZ, J. (1983), Estimations d'erreur dans différentes normes pour l'approximation de problèmes de bifurcation, *C. R. Acad. Sci. Paris Sér. I* **296**, 179–182.

RAPPAZ, J. (1984), Approximation of a nondifferentiable nonlinear problem related to MHD equilibria, *Numer. Math.* **45**, 117–133.

RAPPAZ, J. and G. RAUGEL (1982), Approximation of double bifurcation points for nonlinear eigenvalue problems, in: J. WHITEMAN, ed., *MAFELAP* 1981 (Academic Press, New York).

RAUGEL, G. (1984), Approximation numérique de problèmes non linéaires, Thèse Université de Rennes I.

REINHARDT, L. (1982), On the numerical analysis of the von Kármán equations, *Numer. Math.* **39**, 371–404.

RHEINBOLDT, W.C. (1981), Numerical analysis of continuation methods for nonlinear structural problems, *Comput. & Structures* **13**, 103–116.

RHEINBOLDT, W.C. (1985), Error estimates for nonlinear finite element computations, *Comput. & Structures* **20**, 91–98.

RHEINBOLDT, W.C. (1986), *Numerical Analysis of Parametrized Nonlinear Equations* (Wiley–Interscience, New York).

SATTINGER, D.H. (1973), *Topics in Stability and Bifurcation Theory*, Lecture Notes in Mathematics **309** (Springer, Berlin).

SATTINGER, D.H. (1979), *Group Theoretic Methods in Bifurcation Theory*, Lecture Notes in Mathematics **762** (Springer, Berlin).

SIMADER, C.G. (1972), *On Dirichlet's Boundary Value Problem*, Lecture Notes in Mathematics **268** (Springer, Berlin).

SIMPSON, R.B. (1972), Existence and error estimates for solutions of a discrete analog of nonlinear eigenvalue problems, *Math. Comp.* **26**, 359–375.

TEMAM, R. (1977), *Theory and Numerical Analysis of the Navier–Stokes Equations* (North-Holland, Amsterdam).

VERFÜRTH, R. (1993), A review of a posteriori error estimation and adaptive mesh-refinement techniques, Tech. Rpt., Institut für Angewandte Mathematik Universität Zürich.

WAHLBIN, L.R. (1978), Maximum norm error estimates in the finite element method with isoparametric quadratic elements and numerical integration, *RAIRO Anal. Numér.* **12**, 173–202.

ŽENÍŠEK, A. (1990), *Nonlinear Elliptic and Evolution Problems and their Finite Element Approximations*, Computational Mathematics and Applications (Academic Press, London).

Subject Index

a posteriori estimate, 532, 541, 543, 550, 559
a priori estimate, 532, 541, 543, 550, 559
approximate bifurcation equation, 610, 612, 614, 628

bifurcation equation, 597, 600–602, 605, 608, 609, 614, 620, 625

Clément's interpolation, 543, 551, 558
codimension, 598
consistency, 564, 566, 586
convection diffusion problem, 523, 553
coordinate space, 563
Crandall–Rabinowitz condition, 607

direct subspace, 562, 598
double nondegenerate limit point, 610
duality argument, 538, 543

elliptic projector, 555
equicontinuity, 564
estimator, 541, 551, 557

finite element approximation, 506, 516, 519, 527, 574, 627
Fréchet derivative, 496
Fredholm operator, 491

Galerkin approximation, 504, 534, 550, 551, 553, 569, 573, 585, 595

Hausdorff metric, 612, 613, 615
homotopy theorem, 506, 524
Hood–Taylor method, 519

implicit function theorem, 497, 500
index, 598
inf–sup condition, 518, 535, 554
inverse function theorem, 497, 499

Lyapunov–Schmidt procedure, 597, 600, 602, 611

maximum principle, 508
minimal positive solution, 509, 511, 512, 515, 575
monotonous iteration principle, 510
Morse lemma, 603

Navier–Stokes problem, 517, 550
Nemytskii operator, 507
numerical bifurcation, 622
  imperfect, 621
numerical integration, 545

quadrature rule, 517, 574

Rayleigh quotient, 508
regular point, 561
regular solution, 518, 562
regular solution set, 564, 575
residual, 532, 541

semilinear problem, 503, 540
simple bifurcation point, 606, 607, 614, 626, 630
simple limit point, 580, 582, 585, 593
  nondegenerate, 582, 590, 594
singular point, 597
singular solution, 598, 625
Sobolev spaces, 496
stability, 564, 566, 570, 585
state space, 563

Taylor expansion, 496
turning point, 630

Wavelets and Fast Numerical Algorithms

Y. Meyer

CEREMADE
Université de Paris-Dauphine
Place du Maréchal de Lattre de Tassigny
75775 Paris Cedex 16
France

HANDBOOK OF NUMERICAL ANALYSIS, VOL. V
Techniques of Scientific Computing (Part 2)
Edited by P.G. Ciarlet and J.L. Lions

Contents

CHAPTER I. Preface, Panorama and Heuristics 643

1. What is wrong with Fourier analysis? 643
2. Grossmann–Morlet wavelets 645
3. Scales versus frequencies 647
4. Wavelets and Littlewood–Paley analysis 649
5. Wavelets and paradifferential operators 651
6. The Haar system 652
7. The Daubechies wavelets 653
8. Back to pointwise multiplication 654
9. Donoho's optimal denoising 655
10. Wavelets and multipole algorithm 657

CHAPTER II. Wavelets and Fast Numerical Algorithms 659

11. Introduction 659
12. From Littlewood–Paley analysis to orthonormal wavelets 663
13. Quadrature mirror filters 670
14. Quadrature mirror filters, pyramidal algorithms in image processing and multigrid algorithms 675
15. Multiresolution analysis 677
16. How to extract wavelets from a multiresolution analysis 680
17. A. Cohen's criterion 683
18. Regularity of the Daubechies wavelets 684
19. Convergence of the cascade algorithm 686
20. Wavelets on the interval 688
21. n-dimensional wavelets 692
22. Where do wavelets come from? 692
23. Wavelets provide unconditional bases for most of the functional spaces 693
24. Definition of singular integral operators 696
25. The non-standard representation of an operator 701
26. Characterization of general Calderón–Zygmund operators 702
27. Paraproducts 704
28. Quality of the approximation of general Calderón–Zygmund operators by banded matrices 705
29. The case of finite matrices 706

REFERENCES 709

SUBJECT INDEX 713

CHAPTER I

Preface, Panorama and Heuristics

J.L. Lions was eager to have a substantial part of this book devoted to wavelets. I have however to warn the reader that wavelet based algorithms in numerical analysis are hardly off the ground. Once this remark is made, I would also like to stress that an important application already exists. This application is a wavelet based version of the celebrated multipole algorithm designed by V. Rokhlin. It applies to large classes of singular integral operators and will be expounded in full details, together with some improvements obtained by Yang Qi Xiang.

Roughly speaking Rokhlin's algorithm (as well as the related algorithms discovered by Beylkin and Coifman) is a fast algorithm which applies to the evaluation of integral operators. It can be compared to the "panel clustering algorithms" discovered by W. Hackbush. These algorithms have been successfully tested on non-academical situations but we still do not know when one be preferred to the other.

Before entering into the technicalities of the Beylkin–Coifman–Rokhlin algorithm, let us provide the reader some heuristics about wavelets as well as some panorama over applications of wavelets to other scientific issues. Therefore this preface will be rather large and divided into ten sections.

1. What is wrong with Fourier analysis?

Wavelet analysis is a substitute for the standard Fourier analysis. One of the main drawbacks of a standard Fourier analysis is the following crucial fact: *Being given a function $f(t)$ or being given its Fourier transform $\hat{f}(\xi)$ does not mean at all being given the same information.*

Information about the geometrical features of a function $f(t)$ (or a signal, an image, ...) cannot in general be extracted from its Fourier transform $\hat{f}(\xi)$ and this information loss may even happen in cases when this Fourier transform is given with full precision.

By geometrical features we mean the following. In the case of an image *edges* are relevant geometrical features. In the case of a signal people are interested in the *sizes, heights and location of peaks, jump discontinuities, bursts, ...* . In the case of a function f, one would like to have some access to its *pointwise behavior*, *local scaling exponents* or *multifractal structure* which are completely hidden when our information is only provided by the Fourier transform of f.

These remarks which are based on a long training might be objected both by the pure mathematician or by the engineer.

The pure mathematician would say: take an inverse Fourier transform and you will recover f together with all the information you need about it.

The engineer would say: apply a *Fast Fourier Transform* (FFT) and f will be fully recovered within a few seconds.

The numerical analyst is used to the kind of difficulties we are alluding to. He knows that some problems which, from the viewpoint of a pure mathematician, have a unique solution are however ill-conditioned from the viewpoint of the numerical analyst. *Let me use again this metaphor and say that the issue of extracting geometrical information on a function f from the knowledge of its Fourier transform can be viewed as being an ill-conditioned problem.*

Here is a fascinating example, provided by the great Riemann himself. He considered the trigonometric series $\sum_1^\infty n^{-2}\sin(n^2 t)$ which converges uniformly to a 2π-periodic continuous function $f(t)$. In this example what is given is the Fourier transform of f and the access to $f(t)$ is indirect since it amounts summing a series.

Riemann guessed that $f(t)$ was nowhere differentiable. *For more than one century nobody could prove or disprove Riemann's conjecture.*

Let us stress again that people had a full knowledge of the Fourier transform of the Riemann's function $f(t)$ and no information about the function itself!

Serge Lang routinely suggested this "conjecture of Riemann" to his students and, to the general surprise, Joseph L. Gerver, one of Lang's undergraduate students, resolved the problem in the opposite direction by proving that $f'(\pi) = -\frac{1}{2}$. A year later (1970) Gerver proved that $f'(t_0) = -\frac{1}{2}$ whenever $t_0 = [(2m+1)/(2n+1)]\pi$, $m, n = 0, 1, 2, \ldots$. This completely solved the problem raised by Riemann since it was already known by G.H. Hardy (1916) that $f'(t)$ does not exist at other points.

Gerver's proof relies on messy computations. Today quite simple proofs exist. They are based upon the following idea. *Since the information given by a Fourier series is not pointwise behavior oriented, one should expand Riemann's function $f(t)$ into another series of functions with the hope that a pointwise behavior would be transparent in this new representation. Wavelets do the job.* When a function is expanded into a wavelet series, one is given a straightforward access to this pointwise behavior. This spectacular fact was proved at the same time by S. Jaffard, M. Holschneider and P. Tchamitchian. As it is often the case in science, the problem does not disappear entirely. The wavelet series expansion of our functions, our signals and our images should be accessed.

Returning to the special case of the Riemann's function, one plays with Poisson summation formula together with complex methods to obtain its wavelet series expansion. Using a correctly tuned wavelet, Holschneider proved that the wavelet coefficients are given by the celebrated Jacobi θ function. Since this Jacobi function is fully understood, Gerver's theorem follows from the general relation between pointwise behavior and size of wavelet coefficients.

This proof is detailed in MEYER [1992].

It seems that we are lost inside the realm of pure mathematics. Indeed the opposite is true. Physicists want efficient algorithms for describing the multifractal structure of signals and images. Let us be more precise and consider the case of a continuous function f of the real variable x_0. If f is not differentiable at x_0, let us define its Hölder exponent $\alpha(x_0)$ by

$$\alpha(x_0) = \liminf_{x \to x_0} \frac{\log|f(x) - f(x_0)|}{\log|x - x_0|}$$

with obvious modifications when $f'(x_0)$ exists ($f(x) - f(x_0)$ is then replaced by the error term in the Taylor expansion at x_0).

The spectrum of singularities of f is a function $D(\alpha)$ of the local Hölder exponent α: $D(\alpha)$ is the Hausdorff dimension of the set of all points x_0 for which $\alpha(x_0) = \alpha$. Physicists want to have access to that function $D(\alpha)$ through fast and robust algorithms which can even be used on noisy experimental signals. In the case of the fully developed turbulence signal, a theory predicts that $D(\alpha)$ is a concave function of α which reveals the multifractal structure of turbulence. This application motivated A. Arneodo and his team to design a wavelet based algorithm for computing $D(\alpha)$.

In the case of the Riemann's function, α ranges from $\frac{1}{2}$ to $\frac{3}{2}$. More precisely $\alpha(x_0) = \frac{3}{2}$ if and only if $x_0 = (m/n)\pi$ where m and n are both odd integers and $\alpha(x_0) = \frac{1}{2}$ when $x_0 = (m/n)\pi$ where either m or n is even. J.J. Duistermaat proved that $\alpha(x) = \frac{3}{4}$ almost everywhere. The spectrum of singularities of the Riemann's function has been unveiled by S. Jaffard (JAFFARD [1996]).

A crucial issue can be addressed: Is there an optimal data representation? MARR [1982] raised this problem in his famous book and gave the following answer.

Any particular representation makes certain information explicit at the expense of information that is pushed into the background and may be quite hard to recover.

This important observation can be viewed as a generalization of Heisenberg's uncertainty principle.

2. Grossmann–Morlet wavelets

The goal of wavelet analysis is to be used as an alternative to the standard Fourier analysis. Wavelet analysis has the property that the local behavior of a function f at a given point x_0 is given by simple and directly accessed size conditions on the wavelet coefficients which, in a natural way are attached to x_0.

A mother wavelet is any function ψ of the real variable t, belonging to $L^2(\mathbb{R})$ and fulfilling

$$\int_0^\infty |\widehat{\psi}(\xi)|^2 \frac{d\xi}{\xi} = \int_0^\infty |\widehat{\psi}(-\xi)|^2 \frac{d\xi}{\xi} = 1 \tag{2.1}$$

where $\widehat{\psi}$ denotes the Fourier transform of ψ.

A sufficient condition is the following: $\psi(t)$ is real valued, for some positive δ the integral

$$\int_{-\infty}^{\infty} (1+|t|)^{1+\delta} |\psi(t)|^2 \, dt$$

is finite and

$$\int_{-\infty}^{\infty} \psi(t) \, dt = 0$$

(localization and cancellation). Then the normalization requirement can always be fixed by multiplying $\psi(t)$ by a suitable constant.

There are infinitely many choices for the mother wavelets. Some lead to an interesting analysis which is essentially equivalent to the Littlewood–Paley theory. Other choices do not provide valuable information.

Once the mother wavelet $\psi(t)$ is fixed, the other wavelets in the family are defined by

$$\psi_{(a,b)}(t) = \frac{1}{\sqrt{a}} \psi\left(\frac{t-b}{a}\right), \quad a > 0, \ -\infty < b < \infty. \tag{2.2}$$

For analyzing any L^2 function $f(t)$ one computes its wavelet coefficients

$$W(a,b) = \langle f, \psi_{(a,b)} \rangle = \int_{-\infty}^{\infty} f(t) \overline{\psi}_{(a,b)}(t) \, dt. \tag{2.3}$$

One recovers $f(t)$ by the inversion formula

$$f(t) = \int_0^{\infty} \int_{-\infty}^{\infty} W(a,b) \psi_{(a,b)}(t) \, db \frac{da}{a^2}. \tag{2.4}$$

Everything works as if our wavelets would constitute an *orthonormal basis* for $L^2(\mathbb{R})$. This remark will lead to the construction of orthonormal wavelet bases (Haar (1909), Strömberg (1981), Daubechies (1987); see STRÖMBERG [1981] and DAUBECHIES [1992]).

Identity (2.4) was already known and used by A. Calderón in the sixties and was discovered again by J. Morlet in the eighties.

Both A. Calderón and J. Morlet wanted to replace Fourier expansions by algorithms which are more robust when f belongs to some functional spaces that model large classes of signals or images (of course, A. Calderón was not interested in signal processing while J. Morlet was waving hands instead of dealing with functional spaces).

Our wavelet $\psi_{(a,b)}$ is localized around b, with a spread given by a. The wavelet coefficient $W(a,b)$ is intended to measure the fluctuations of f over the interval

$[b-a, b+a]$. However, for a given position b and for a given interval length $2a$, $W(a,b)$ heavily depends on the choice of the mother wavelet ψ.

This remark has to be taken very seriously. It implies that extracting a diagnostic from too few wavelet coefficients is impossible.

One wavelet coefficient does not provide any information. For measuring the fluctuation of f over the interval $[b-a, b+a]$, all the wavelet coefficients $W(a', b')$ for which $0 < a' < a$ and $|b'-b| \leqslant 2a$ should be used (I am assuming that the mother wavelet ψ is supported by the interval $[-1, 1]$).

3. Scales versus frequencies

A folk wisdom in image processing and numerical analysis is that small scales correspond to high frequencies. It can be said that wavelet analysis provides a precise formulation of this folk belief.

Let us be more specific.

As compared to the waves used in Fourier analysis, wavelets have a sharp localization with respect to the time/space variable but a poor frequency localization.

For improving this frequency localization one could impose the following condition

the Fourier transform $\widehat{\psi}$ of ψ is supported by the union of the two intervals $[-\omega_1, -\omega_0]$ and $[\omega_0, \omega_1]$ where $\omega_1 > \omega_0 > 0$. (3.1)

For improving the localization of the mother wavelet, one could add a second condition like

$$\psi(t) = \mathrm{O}(t^{-q}) \quad \text{for every } q \text{ as } |t| \text{ tends to infinity.} \tag{3.2}$$

If both (3.1) and (3.2) are imposed, ψ belongs to the Schwartz class $\mathcal{S}(\mathbb{R})$.

Weaker conditions often replace (3.1) and (3.2) since (3.1) cannot be satisfied when ψ is compactly supported. It is required that, for some large integer N, we have

$$|\widehat{\psi}(\xi)| \leqslant C_N |\xi|^N \quad \text{when } |\xi| \leqslant 1, \tag{3.3}$$

$$|\widehat{\psi}(\xi)| \leqslant C'_N |\xi|^{-N} \quad \text{when } |\xi| \geqslant 1 \tag{3.4}$$

and that ψ would be compactly supported.

The weaker form of frequency localization will be satisfied in the case of Daubechies wavelets.

In both cases, the information given by the wavelet coefficients $W(a,b)$ of a function $f(t)$ is a slight variant on the information given by a Littlewood–Paley analysis.

This remark is crucial since it permits the new-born wavelet analysis to inherit all the scientific wisdom of the Littlewood–Paley analysis. Littlewood–Paley analysis will be defined in the next section.

For the sake of simplicity, let us restrict our attention to the case where ψ fulfills (3.1) and (3.2).

The Fourier transform of $\psi_{(a,b)}$ is $\sqrt{a}\,e^{-ib\xi}\widehat{\psi}(a\xi)$ and is therefore supported by the intervals $[-\omega_1/a, -\omega_0/a]$ and $[\omega_0/a, \omega_1/a]$.

The folk belief that the inverse of a scale is a frequency applies to all our wavelets $\psi_{(a,b)}$.

Let us follow this line of thought.

When a is quite small, the corresponding frequency localization of $\psi_{(a,b)}$ is still poor. What remains acceptable is $\Delta\xi/\xi$ and the wavelet transformation is exactly what people in signal processing call "Q-constant filters" since the quality (Q) of a filter is measured by its normalized spread $(\Delta\xi/\xi)$ in the frequency domain.

These remarks lead to the following conclusion: As a tool in signal or image processing, Fourier analysis stands at one extreme (no localization in time, perfect localization in frequency). Wavelet analysis stands at the other extreme (excellent time resolution, poor frequency resolution).

We cannot have both a sharp time resolution together with a sharp frequency resolution (Heisenberg's uncertainty principle). They are several trade-offs between Fourier analysis and wavelet analysis. In acoustics people are not using a wavelet analysis since they need a better frequency resolution. They cannot use a standard Fourier analysis since they want local algorithms. They are using a windowed Fourier transform providing a spectrogram.

In image processing, wavelet analysis has been widely used much before the wavelet phraseology became popular. Let me give two examples.

The first one is the pyramidal algorithm discovered by P. Burt and E. Adelson. It amounts to *filtering* an image by a sequence of low-pass filters tuned over a large range of scales and down-sampling the outputs of this filter bank. The scales are of the form 2^{-j} and the down-sampling does not violate Shannon's sampling theorem. Therefore an increasing sequence $\Gamma_j = 2^{-j}\mathbb{Z}^2$ of finer and finer grids is naturally entering the game. The fundamental discovery of P. Burt and E. Adelson is that such algorithms can be given a hierarchical structure whenever the filters are carefully designed. More precisely in order to compute the pixels corresponding to the grid Γ_j, it suffices to average the pixels which correspond to the finer grid Γ_{j+1} and it is not necessary to return to the full image. Moreover, the algorithm providing these averages uses finite length filters and does not depend on j.

If one is incorporating quadrature mirror filters, Burt and Adelson's pyramidal algorithm provides a discrete version of the wavelet transformation. This beautiful remark is due to S. Mallat and will be fully developed in Section 14.

Here is a second example where wavelet algorithms were used before wavelet theory was built.

D. Marr tried to explain the unbelievable efficiency of human vision in solving the problem of edge detection. He showed that the human vision system does not use a Fourier type algorithm to process visual information to the brain. He used very convincing arguments which backed the following theory.

The edge detection algorithm used in human vision consists of

(i) a smoothing of the given image with a large collection of Gaussian filters tuned to cover many distinct scales;

(ii) a differentiation applied to the outputs of this filterbank.

Since he wanted his algorithm to be fully isotropic, he used the Laplace operator.

Today we know that D. Marr was performing a 2D wavelet analysis in which the mother wavelet ψ is the Laplacian of a 2D Gaussian function.

A more detailed account on Marr's ideas can be found in his famous book or in MEYER [1992].

4. Wavelets and Littlewood–Paley analysis

Littlewood–Paley analysis has a long history. It is either used to analyze functions or to analyze operators. The efficiency of wavelet analysis for analyzing functions and compressing operators is deep-rooted inside the corresponding performances of Littlewood–Paley analysis.

What is today called Littlewood–Paley analysis stems from the problem of estimating the L^p-norm of a 2π-periodic function given by its Fourier expansion.

Since the relationship between Littlewood–Paley analysis and wavelet analysis is so important, we would like to provide the reader a detailed account on this problem.

Let f be a 2π-periodic L^2-function given by its Fourier expansion $\sum_{-\infty}^{\infty} c_n \, \mathrm{e}^{int}$. Here we mean that we have a *direct* access to the Fourier coefficients of $f(t)$ and an *indirect* access to the function $f(t)$ itself.

If one addresses the problem of computing the mean value of $|f|^2$, the answer is obvious and given by

$$\sum_{-\infty}^{\infty} |c_n|^2 = \frac{1}{2\pi} \int_0^{2\pi} |f(t)|^2 \, \mathrm{d}t.$$

If sharper information about the localization of the energy is needed, it is convenient to compare the L^4-norm of f, defined as

$$\left(\frac{1}{2\pi} \int_0^{2\pi} |f(t)|^4 \, \mathrm{d}t \right)^{1/4} = \|f\|_4$$

to the L^2-norm.

If $\|f\|_4$ is much larger than $\|f\|_2$, it indicates that most of the energy is localized on a set with small measure.

In principle, the knowledge of the Fourier coefficients (c_n) should provide a full information about this L^4-norm. But it is a well-known fact that it is impossible to estimate our L^4-norm in terms of the magnitudes $(|c_n|)$ of the Fourier coefficients c_n. The phases φ_n (defined by $c_n = \mathrm{e}^{\mathrm{i}\varphi_n} |c_n|$) are playing a crucial role, too intricate to be deciphered by human beings.

Computing an L^4-norm is not an academical problem. In quantum field theory, one needs to estimate

$$\|(I-\Delta)^{-1/2}u(x)\|_4 \quad \text{where } \Delta = \frac{\partial^2}{\partial x_1^2}+\frac{\partial^2}{\partial x_2^2}+\frac{\partial^2}{\partial x_3^2}$$

and $u(x)$ is a random function of three variables. This problem led to some algorithms which are almost identical to the solution given by orthonormal wavelet expansions MEYER [1990].

These remarks lead to the so-called Littlewood–Paley decomposition. In the thirties, Littlewood and Paley proposed the following algorithm for computing (or estimating) L^p-norms of 2π-periodic functions given by their Fourier expansions $\sum_{-\infty}^{\infty} c_n\, e^{int}$. Let us define the dyadic blocks $\Delta_j(f)$ by

$$\Delta_j(f) = \sum_{2^j \leqslant |n| < 2^{j+1}} c_n\, e^{int}, \quad j \in \mathbb{N}.$$

Then for $1 < p < \infty$ there exist two constants α_p and β_p such that

$$\alpha_p\|f\|_p \leqslant |c_0| + \left\| \left(\sum_0^\infty |\Delta_j(f)|^2 \right)^{1/2} \right\|_p \leqslant \beta_p\|f\|_p.$$

In other terms, the L^p-norms are provided by stable and robust algorithms when Fourier expansions are given up and replaced by the representation of a function given by the so-called Littlewood–Paley decomposition

$$f = c_0 + \sum_0^\infty \Delta_j(f).$$

It seems from this introduction to Littlewood–Paley analysis that it is a by-product of a Fourier expansion. We will see in a moment that the Littlewood–Paley expansion can be obtained by another algorithm.

In several dimensions, the estimation of an L^p norm by a Littlewood–Paley decomposition is no longer valid. A counter-example was found by C. Fefferman, see MEYER [1990].

One should replace the "hard blocks" $\Delta_j(f)$ by similar blocks defined by smoother cut-off functions. More precisely, let θ belong to $C_0^\infty(\mathbb{R}^n)$ and satisfy the following two conditions: $\theta(\xi) = 0$ if $|\xi| \leqslant \frac{1}{2}$ and

$$\sum_{-\infty}^{\infty} |\theta(2^{-j}\xi)|^2 = 1$$

for $\xi \neq 0$. The case where $\theta(\xi) = 1$ when $\frac{1}{2} \leqslant |\xi| < 1$ is an example where the two last conditions are satisfied but not the regularity condition. When $n = 1$, this choice will give the "hard blocks" $\Delta_j(f)$.

When $n \geqslant 2$, the smooth blocks $\Delta_j(f)$ are defined by

$$\Delta_j(f)(x) = \sum_k c_k \theta(k 2^{-j}) \, \mathrm{e}^{\mathrm{i}k \bullet x}$$

when $f(x) = \sum_k c_k \, \mathrm{e}^{\mathrm{i}k \bullet x}$.

The mother wavelet ψ will show up as the inverse Fourier transform of θ which now assumed to be both real-valued and radial.

Then we have the elementary but fundamental identity

$$\Delta_j(f)(x) = 2^{nj/2} \langle f, \psi_{(2^{-j}, x)} \rangle$$

where, as in the case of Grossmann–Morlet's wavelets,

$$\psi_{(a,b)}(x) = a^{-n/2} \psi\left(\frac{x-b}{a}\right).$$

Wavelet analysis and Littlewood–Paley analysis coincide!

5. Wavelets and paradifferential operators

The preceding remarks pave the way to a huge scientific program which is hardly off the ground.

Let us stress again the efficiency of Littlewood–Paley analysis to deal with non-linear problems. This is not accidental since estimating an L^p-norm is already a non-linear problem.

But today a variety of non-linear (or multilinear) problems have been successfully attacked using Littlewood–Paley analysis.

Let us mention two examples.

The first one is the improved version of the div–curl lemma which has been proposed by P.L. Lions. This version is the following theorem. If $E_1(x), \ldots, E_n(x)$ together with $B_1(x), \ldots, B_n(x)$ belong to $L^2(\mathbb{R}^n)$ and if $\operatorname{div} E(x) = 0$, $\operatorname{curl} B(x) = 0$ where $E = (E_1, \ldots, E_n)$, $B = (B_1, \ldots, B_n)$, then $E_1(x)B_1(x) + \cdots + E_n(x)B_n(x)$ belongs to the Hardy space $\mathcal{H}^1(\mathbb{R}^n)$ in its real variable version which was proposed by Stein and Weiss. The second example is the discovery and the proof by J.Y. Chemin of the smooth evolution of vortex sheets in 2D fluid dynamics. This discovery is striking since numerical simulations suggested the opposite.

A general methodology was developed by J.M. Bony and was given the name of paradifferential calculus. This calculus uses the Littlewood–Paley decomposition exactly the same way as the standard pseudo-differential calculus is using the Fourier transformation.

The paradifferential calculus provides the simplest proof to P.L. Lions' div–curl lemma. The same calculus was used by J.Y. Chemin in his first proof of his theorem.

Wavelet algorithms are playing with respect to the Littlewood–Paley analysis exactly the same crucial role as an FFT with respect to the Fourier transformation.

It means that wavelets pave the way to eventual applications of the paradifferential calculus to numerical analysis.

The algebra of paradifferential operators contains the subalgebra of classical pseudo-differential operators. These classical pseudo-differential operators are well compressed in an orthonormal wavelet basis, as it will be later shown in full details.

But paradifferential operators provide algorithms which are out of reach when one is using standard pseudo-differential operators. Paradifferential operators provide a sharp bilinear analysis of the pointwise product between any two rough (non-smooth) functions. If one of these two functions happens to be smooth, pseudo-differential operators would have done the job.

That sharp bilinear analysis of pointwise multiplication is suited to the div–curl lemma and to many problems in homogeneization.

One would also like to multiply two functions given by their wavelet expansions through a fast algorithm running on the wavelet coefficients.

If one needs to multiply the two corresponding functions, one should return to the "physical space" since we do not know how to perform the pointwise product at the coefficients level.

This limitation prevents wavelets to be successfully applied to non-linear PDEs.

Since the paradifferential calculus was designed to circumvent this difficulty and since there are so many connections between this calculus and wavelet based algorithms, it is very likely that G. Beylkin will succeed in his program. Beylkin's program is to feed wavelet algorithms with the best of the paradifferential calculus.

It seems that the ground is fully prepared for spectacular applications of wavelets to fluid dynamics, turbulence, ...

S. Dobyinsky executed the following experiment. She compared two proofs of the "improved div–curl lemma". The first one was based on the paradifferential calculus and the second one was a wavelet proof. The good news is that these two proofs are quite similar. The bad news is that the wavelet based proof is more technical and less transparent.

6. The Haar system

The Haar system is important since it was the first example of an orthonormal wavelet basis. It is also important from the viewpoint of J.M. Bony's paradifferential operators. The corresponding theory, once adapted to the Haar system, provides an operator algebra known as "martingale transforms". This remark can be interpreted as a justification for J.M. Bony's theory.

Let us remind the reader about A. Haar's motivation. A. Haar knew that the Fourier series of a continuous function might diverge at a given point. Was it possible to construct an orthonormal basis for $L^2[0, 1]$ with the property that the expansion of any continuous function would converge uniformly to that function? Such an orthonormal basis is nowadays called a Schauder basis.

The solution given by A. Haar is the simplest one but has some limitations if one modifies the functional setting, as it will be explained below.

A. Haar started his construction with a function $h(x)$ of the real variable x defined by $h(x) = 1$ if $0 \leqslant x \leqslant \frac{1}{2}$, $h(x) = -1$ if $\frac{1}{2} \leqslant x < 1$ and $h(x) = 0$ elsewhere. When $n \geqslant 1$ is written $n = 2^j + k$, $j \geqslant 0$, $0 \leqslant k \leqslant 2^j - 1$, $h_n(x) = 2^{j/2}h(2^j x - k)$. Let us observe that $h_n(x)$ is supported by $[0, 1)$.

Finally $h_0(x) = 1$, by definition, and HAAR [1909] proved that the sequence $\{h_0, h_1, \ldots\}$ is an orthonormal basis for $L^2[0,1]$. This basis is a Schauder basis for $C[0,1]$. The approximation to a continuous function given by the Haar system is quite familiar. It consists in averaging our continuous function $f(x)$ on each dyadic interval $[k2^{-j}, (k+1)2^{-j})$ and approximating f by the step function which is defined by these averages.

It is clear that it is not what we would like to do if $f(x)$ happens to be a C^1 function since the quality of the approximation would remain rather poor.

If one is interested in constructing an orthonormal basis for $L^2(\mathbb{R})$ which is, at the same time, a Schauder basis for $C_0(\mathbb{R})$, it suffices to define $h_{j,k}(x) = 2^{j/2}h(2^j x - k)$ for $j \in \mathbb{Z}$, $k \in \mathbb{Z}$. This family has the required property and is also called the Haar system. We have denoted by $C_0(\mathbb{R})$ the space of continuous functions on the line which tend to 0 at infinity.

7. The Daubechies wavelets

The Haar system is the prototype of an orthonormal wavelet basis. However, the corresponding mother wavelet (i.e. $h(x)$) only satisfies (3.3) and (3.4) for $N = 1$. The frequency resolution of $h(x)$ is too poor. People in signal processing would say that "hard blocks create numerical artifacts". I. Daubechies' wavelets lie somewhere in between the Haar system (one extreme) and a Littlewood–Paley analysis for which (3.1) and (3.2) are satisfied (the other extreme). The construction depends on an integer which is usually denoted by N. When $N = 1$, Daubechies' wavelets are identical to the Haar system (on the line). As N gets larger, the regularity of the mother wavelet $\psi(x) = \psi_N(x)$ increases as much as the size of its support.

For each integer N, the collection $2^{j/2}\,\psi(2^j - k)$, $j \in \mathbb{Z}$, $k \in \mathbb{Z}$, is an orthonormal basis for $L^2(\mathbb{R})$.

The function ψ is not given by some explicit analytical form but the coefficients of the wavelet expansion can be reached through a fast pyramidal algorithm. This fast algorithm is quite similar to the pyramidal algorithms created by Burt and Adelson. It is also a special case of the "quadrature mirror filters" designed by people in signal processing.

Such quadrature mirror filters happen to be quite effective in image compression when the filters are correctly specified. Before Daubechies' construction, these correct filters were obtained by try and error methods. Afterwards it was observed that these empirical filters were the ones leading to smooth wavelets in Daubechies' work.

8. Back to pointwise multiplication

I cannot resist describing a nice result obtained in collaboration with S. Dobyinsky. Let us assume that ψ is a Daubechies (mother) wavelet with some regularity. The function $h(x)$ of the Haar system does not work but the next one in Daubechies' construction will already do the job. Pick any two functions f and g in $L^2(\mathbb{R})$. Expand f and g into the orthonormal wavelet basis $\psi_{(j,k)}$. One obtains

$$f(x) = \sum_{-\infty}^{\infty}\sum_{-\infty}^{\infty} \alpha(j,k)\psi_{(j,k)}(x)$$

$$g(x) = \sum_{-\infty}^{\infty}\sum_{-\infty}^{\infty} \beta(j,k)\psi_{(j,k)}(x).$$

Then the pointwise product $f(x)g(x)$ is the sum $v(x) + w(x)$ where

$$v(x) = \sum_{-\infty}^{\infty}\sum_{-\infty}^{\infty} \alpha(j,k)\beta(j,k)\psi_{j,k}^2(x).$$

Since the Daubechies wavelets are real valued, one has

$$\int_{-\infty}^{\infty} v(x)\,\mathrm{d}x = \sum_{-\infty}^{\infty}\sum_{-\infty}^{\infty} \alpha(j,k)\beta(j,k) = \int_{-\infty}^{\infty} f(x)g(x)\,\mathrm{d}x.$$

It implies that $\int_{-\infty}^{\infty} w(x)\,\mathrm{d}x = 0$. But this "error term" $w(x)$ has a stronger cancellation property since it belongs to the same Hardy space $\mathcal{H}^1(\mathbb{R})$ that plays such a crucial role in the div–curl lemma.

S. Dobyinsky was able to use the n-dimensional version of the preceding remark in order to give a very interesting proof of the div–curl lemma, DOBYINSKY [1995].

From a function analyst's viewpoint, $w(x)$ might be viewed as an error term since the space $\mathcal{H}^1(\mathbb{R})$ is much smaller than $L^1(\mathbb{R})$. For a numerical analyst there are no reasons for stating that $w(x)$ is smaller than $v(x)$.

In order to prove the superiority of orthonormal wavelet bases to any other orthogonal system, let us perform the same experiment on the trigonometric system. Let us start with two functions f and g belonging to $L^2(0, 2\pi)$ and write

$$f(x) = \sum_{0}^{\infty} (a_n \cos nx + b_n \sin nx),$$

$$g(x) = \sum_{0}^{\infty} (\alpha_n \cos nx + \beta_n \sin nx).$$

We then write $f(x)g(x) = v(x) + w(x)$ where

$$v(x) = \sum_{0}^{\infty} \left(a_n \alpha_n \cos^2 nx + b_n \beta_n \sin^2 nx\right)$$

and where $\int_0^{2\pi} w(x)\,\mathrm{d}x = 0$.

In that case nothing better can be said about $w(x)$ which can be any L^1 function with a zero mean.

9. Donoho's optimal denoising

David Donoho designed a remarkable denoising algorithm which is aimed to extract a signal or an image from noisy and indirect data. This algorithm is optimal (in a sense to be made precise) over any linear or non-linear procedure. Applications range from extragalactic astronomy to medical imaging. This algorithm has been fully tested and the corresponding software is available.

A priori knowledge on the signal is needed as well as some assumption on the noise and on the ill-posed problem which is being solved.

Let us begin with the signal. The a priori assumption allows a certain number of bumps with different widths, different heights and different locations. Their cardinality should not be too large.

Similarly functions with bounded variations are welcomed. What is certainly excluded are signals with long high frequency components since the algorithm would hardly be able to distinguish these components from the noise. Jump discontinuities are fine as well as edges in the 2D case. The denoising algorithm is aimed at the same time to provide a sharp edge detector and to yield the signal back inside any region where the signal or the image are smooth.

For modelling the type of signals or images he has in mind, Donoho introduces Besov spaces $B_p^{s,q}$ where $n/p \leqslant s$ (n being the dimension). The a priori knowledge is expressed as some bound on the Besov norm of our signal (or image). Therefore strong textures are not admitted and will be wiped out by the denoising algorithms. However, the restitution of edges will be optimal.

There are two versions of Donoho's algorithm. In the first case, we are just dealing with noisy data without the extra difficulty coming from the ill-posed problem. The second version will be described later on.

In the first case we consider an image defined by a grey-level denoted by $f(x, y)$, $0 \leqslant x \leqslant 1$, $0 \leqslant y \leqslant 1$. This function $f(x, y)$ is assumed to belong to a given ball of some Besov space. But neither the size of the ball nor the exponents s, p or q will be present in the algorithm.

This image is corrupted by a white noise with energy σ. Our data are samples of this corrupted image over a fine grid.

We want to build an estimator $\hat{f}$ for f. This estimator should belong to the given ball of the given Besov space or at worst to the double ball. The error is measured by the expectation of $\|\hat{f} - f\|_2^2$ and we want the worst error to be as small as possible.

Worst means that we compute the supremum over the Besov ball and small means that we are interested to the rate at which the worst expected error tends to 0 as σ tends to 0.

Donoho's estimator will be described in the discrete case. Let us assume that the data are $f(t_i) + \sigma z_i$, $1 \leqslant i \leqslant N$, $t_i = i/N$, z_i being i.i.d. $\mathrm{N}(0, 1)$.

In that case Donoho's estimator consists in (1) computing the empirical wavelet coefficients $\alpha_{j,k}$, (2) applying a smooth non-linear thresholding $\hat{\alpha}_{j,k} = \theta(\alpha_{j,k})$ to these empirical coefficients and (3) defining $\hat{f}$ by $\sum\sum \hat{\alpha}_{j,k}\,\psi_{j,k}$.

Let us define a threshold τ by $\tau = \sigma\sqrt{(2\log N)/N}$. Then

$$\begin{aligned}
&\theta(u) = u - \tau && \text{if } u \geqslant \tau,\\
&\theta(u) = 0 && \text{if } -\tau \leqslant u \leqslant \tau,\\
&\theta(u) = u + \tau && \text{if } u \leqslant -\tau.
\end{aligned}$$

Donoho is using a wavelet analysis which is not exactly the one which was described above. They differ in two aspects. Firstly, samples are given on a finite interval and not on the whole real line. Secondly, we are not given a continuum of data but rather a finite collection. It means that the real line is now replaced by the grid $\{0, 1/N, 2/N, \ldots, (n-1)/N\}$ where N is large and similarly the plane is replaced by the corresponding grid in the 2D case.

In a forthcoming paper, A. Cohen, I. Daubechies and P. Vial found the correct way to adapt the construction of orthonormal wavelet bases to this situation in such a way that Donoho's algorithm can run.

Donoho designed a much more involved version of his denoising algorithm. This version applies to ill-posed problems in the presence of noise. If f denotes the signal or the image we want to recover or estimate, the data we collect are of the form $T(f) + r$ where T is a compact operator and r is the noise. In image processing T models the blurring caused by the experimental measurement. An example is provided by images provided by the Hubble spatial telescope.

This problem does not reduce to the first one since inverting T is forbidden: the unbounded inverse operator would drastically increase the noise and f would be buried inside this huge noise.

A traditional algorithm which is routinely applied to this type of ill-posed problems is the so-called Tychonov regularization. Let us take a few moments for describing Tychonov regularization. We will then analyze the drawbacks of this algorithm and indicate when and why Donoho is superior.

For performing a Tychonov regularization, one should first compute the eigenvectors $e_0, e_1, \ldots, e_n, \ldots$ and the eigenvalues $\lambda_0, \lambda_1, \ldots, \lambda_n, \ldots$ of the selfadjoint compact operator T^*T. Then if our noisy and indirect data are denoted by y, one should expand $T^*(y)$ into the orthonormal basis provided by the eigenvectors $e_0, \ldots, e_n, \ldots$. Since $y = T(f) + r$, one would like to estimate f by the series $\alpha_0\lambda_0^{-1}e_0 + \cdots + \alpha_n\lambda_n^{-1}e_n + \cdots$. But λ_n tends to 0 as n tends to infinity and in most of the cases, this series is awfully divergent.

Tychonov regularization amounts to using a summation procedure. A weight is introduced and our divergent series is replaced by

$$\alpha_0 \lambda_0^{-1} w_0 e_0 + \cdots + \alpha_n \lambda_n^{-1} w_n e_n + \cdots.$$

This weight somehow measures how much we trust the empirical coefficients $\alpha_0, \alpha_1, \ldots$. The simplest case is when $w_n = 1$ for $0 \leqslant n \leqslant N$ and $w_n = 0$ when $n > N$ but in many cases smoother weights are needed.

The integer N should be adjusted to the noise level, to the growth of λ_n^{-1} and the a priori knowledge we might have on the signal (which in some cases might be reflected by the decay of the empirical coefficients α_n).

Tychonov regularization has a severe drawback. The orthonormal basis $e_0, e_1, \ldots$ is perfectly adapted to the operator T. However, in many cases, this basis is the trigonometrical system and we are returning to the fundamental issue of expressing some a priori information on the signal by a property of its Fourier coefficients.

A precise location of the edges of an image is out of reach unless a quite large number of terms are kept in the Fourier expansion. But these high frequency contributions which are so crucial for a sharp edge detection are systematically killed by a Tychonov regularization.

Tychonov regularization is inadequate when the relevant information is of geometrical nature.

What Donoho was looking for is an orthonormal basis which would be at the same time operator oriented and object oriented. The price to be paid is that such a basis would be suboptimal in both cases but that is much better than a basis which is optimal for describing the operator and poorly adapted to the characterization of the geometrical features of our signal (image, ...).

Donoho is assuming that T^*T belongs to the general class of operators which are well compressed in an orthonormal wavelet basis. This class of operators will be detailed in this chapter on wavelets.

The operator T itself might be much more involved. Fourier integral operators or the Radon transformation have the above-mentioned property.

Then Donoho proves that optimal recovery is provided by a wavelet based algorithm.

Applications are important and Donoho's algorithm paves the way to some more sophisticated algorithms like the optimal denoising of the Hubble spatial telescope blurred images. This optimal denoising was designed by A. Lannes and S. Roques (Observatoire Midi Pyrénées, Toulouse).

10. Wavelets and multipole algorithm

Wavelets provide fast algorithms for computing $Y = AX$ when A is a large $N \times N$ matrix of a certain type and X is an arbitrary vector.

The fast Fourier transform applies to the case when A is of convolution type (some periodicity is also needed). For using wavelets, we need another assumption about A. Roughly speaking large entries should be located on a few smooth lines and the remaining entries $a(i, j)$ should smoothly depend on the location of (i, j) in the plane when we move away from the few lines carrying the singularity of the matrix.

A typical and trivial example is given by $a(i,j) = 1/(i-j)$ if $1 \leqslant i,j \leqslant N$ with the convention that $a(i,i) = 0$.

Applying orthonormal wavelets for compressing such matrices means using a unitary change of coordinates. If U is this unitary matrix, UAU^{-1} becomes a sparse matrix when A satisfies the above mentioned properties. This observation paves the way to the so-called Beylkin–Coifman–Rokhlin algorithms which will be expounded in this chapter. Since I wanted to yield some heuristics in this preface, let me indicate the connection between Beylkin–Coifman–Rokhlin algorithms and multipole algorithms.

Multipole algorithms were designed by V. Rokhlin in order to evaluate the potential created by some huge cloud of charges. The locations of the charges are $x_1, \ldots, x_N$ and the corresponding masses are $m_1, \ldots, m_N$. We are working in three dimensions. The algorithm does not rely on some specific geometrical properties of the locations.

A brute force computation amounts to applying a huge $N \times N$ matrix A (whose entries are $a(i,j) = |x_i - x_j|^{-1}$) to a huge vector $(m_1, \ldots, m_N)$. This brute force calculation requires N^2 operations. The goal of a multipole algorithm is to replace N^2 by $C(\varepsilon)N$ where the constant $C(\varepsilon)$ depends on the required precision.

The multipole algorithm consists in a hierarchical organization of the space. This organization permits to regroup our charges into clusters. When the potential generated by one specific cluster Γ is computed far away from Γ, this cluster Γ can be replaced by a multipole centered at the center of the cluster. The order of the multipole depends on the required accuracy. If one iterates these regroupings, at the first stage charges are becoming clusters, then clusters are also regrouped into superclusters and so on. Then our N^2 computation is replaced by a $C(\varepsilon)N$ computation where ε is the required precision. Once ε is given, the order of the multipole can be fixed as well as the domain in which the approximation of a cluster by a multipole is valid.

R. Coifman made the beautiful observation that the multipole algorithm could be rephrased in the wavelet terminology. More generally wavelet methods provide efficient algorithms for computing integral or singular integral operators of the type $\int K(x,y)f(y)\,dy$ whenever the kernel $K(x,y)$ fulfills some natural conditions which are satisfied in the case of the coulombian potential.

These general conditions are also satisfied for any (classical) pseudo-differential operator. This remark applies to the special case of the Hilbert transform and solves a problem raised by A. Zygmund, as it will be explained below (Chapter II).

For ending this panorama, let us stress that an algorithm which in principle is optimal has to be experimented a very long time and carefully fixed before becoming number one in practice. For instance, if one is using Daubechies wavelets of length 10, the mere evaluation of one wavelet coefficient will in three dimensions require 10^3 operations. That means that the number N of charges has to be much larger than 10^3 in order that the multipole algorithm in its wavelet based version would be optimal.

Donoho's denoising algorithm heavily relies on the Beylkin–Coifman–Rokhlin algorithm. Moreover, one has to invert a banded matrix which is roughly speaking the matrix of a T wavelet basis. This bounded matrix is playing the role of the diagonal matrix which appears in Tychonov's regularization.

CHAPTER II

Wavelets and Fast Numerical Algorithms

11. Introduction

A. Zygmund used to raise the following problem: *is it possible to prove the boundedness of the Hilbert transformation without the Plancherel theorem?* Let us remind the reader how the Hilbert transformation is defined and then indicate the deep relevance of Zygmund's problem.

The Hilbert transform of f is the convolution $f * (1/\pi)$ p.v. $(1/x)$ and is denoted by $\tilde{f}(x)$. In other words

$$\tilde{f}(x) = \lim_{\varepsilon \downarrow 0} \frac{1}{\pi} \int_{|x-y| \geqslant \varepsilon} \frac{f(y)}{x-y} \, \mathrm{d}y. \tag{11.1}$$

This singular integral exists when f satisfies a Hölder condition at x and belongs to $L^2(\mathbb{R})$.

A trivial proof of

$$\int_{-\infty}^{\infty} \left|\tilde{f}(x)\right|^2 \mathrm{d}x = \int_{-\infty}^{\infty} \left|f(x)\right|^2 \mathrm{d}x$$

is the following. The Fourier transform (in the distributional sense) of $1/\pi$ p.v. $(1/x)$ is $-\mathrm{i} \operatorname{sign} \xi$ and $|-\mathrm{i} \operatorname{sign} \xi| = 1$. Therefore Plancherel's theorem yields $\|H(f)\|_2 = \|f\|_2$.

A. Zygmund was interested in another proof for the following reason. Together with A. Calderón he wanted to prove L^2-estimates for operators which are no longer given by convolutions but have a singular kernel with size (and smoothness) estimates similar to the Hilbert transformation.

Calderón's operators are

$$Tf(x) = \lim_{\varepsilon \downarrow 0} \int_{|x-y| \geqslant \varepsilon} K(x,y) f(y) \, \mathrm{d}y, \tag{11.2}$$

where $K(x,y)$ satisfies the following conditions:

$$|K(x,y)| \leqslant |x-y|^{-1}, \tag{11.3}$$

$$K(x,y) = -K(y,x) \tag{11.4}$$

and

$$\left|\frac{\partial}{\partial x} K(x,y)\right| \leqslant |x-y|^{-2}. \tag{11.5}$$

Then the singular integral (11.2) certainly exists in the distributional sense. It means that whenever f and g are two testing functions,

$$(Tf, g) = \lim_{\varepsilon \downarrow 0} \iint_{|x-y| \geqslant \varepsilon} K(x,y) f(y) g(x) \, \mathrm{d}y \, \mathrm{d}x$$

exists. The crucial cancellation is coming from (11.4). A. Calderón was convinced that such operators are always bounded on $L^2(\mathbb{R})$ and A. Zygmund thought that a proof of the boundedness of the Hilbert transform that does not use the Plancherel theorem would extend to these new operators.

This program happened to fail since the author of this paper produced a counterexample. Indeed it will be later explained that (11.3), (11.4) and (11.5) do not imply the required L^2-estimate. If we add the necessary and sufficient condition for such an estimate (which is $T(1) \in$ BMO and will be later developed), then the proof will not follow an abstract Hilbert space organization. Indeed the problem has deep real-variable roots and the fact that the John and Nirenberg space BMO plays a crucial role shows how far we are from abstract arguments.

Before moving further, let us observe that the singular integral operators defined by (11.2) will be incorporated inside the famous method of rotations (discovered by A. Calderón and A. Zygmund) and will generate the n-dimensional operators A. Calderón's was interested in. One among Calderón's motivations was the solution of elliptic problems in non-smooth domains with Lipschitzian boundaries by the method of the double layer potential. Then the "main term" which appears is given by

$$Tf(x) = \text{p.v.} \int_{\mathbb{R}^n} K(x,y) f(y) \, \mathrm{d}y, \tag{11.6}$$

where

$$|K(x,y)| \leqslant C|x-y|^{-n}, \qquad \left|\frac{\partial}{\partial x_j} K(x,y)\right| \leqslant C|x-y|^{-n-1} \tag{11.7}$$
$$1 \leqslant j \leqslant n \text{ and } K(y,x) = -K(x,y).$$

Coming back to A. Zygmund's program, two substitutes for the Plancherel theorem have been found. These new methods are much more flexible and can be adapted to settings where no group structure is available. However, using these new tools needs some practice.

The first tool is *Cotlar's lemma* (which was discovered in its general formulation by M. Cotlar and E. Stein independently).

The second tool is the *wavelet approach* based upon the so-called Schur lemma.

Let us begin with Cotlar's lemma.

It is an abstract tool. Let H be a Hilbert space and $T_j : H \to H$, $j = 0, \pm 1, \pm 2, \ldots$ be bounded operators. We want to estimate the operator norm of $\sum_{-\infty}^{\infty} T_j$ by something much better than $\sum_{-\infty}^{\infty} \|T_j\|$.

Let us assume that, for each j and each k, one has

$$\|T_j^* T_k\| \leqslant \omega(j-k) \quad \textit{and} \quad \|T_j T_k^*\| \leqslant \omega(j-k), \tag{11.8}$$

where

$$\sum_{-\infty}^{\infty} \sqrt{\omega(j)} < \infty. \tag{11.9}$$

Then, for each $x \in H$, the series $\sum_{-\infty}^{\infty} T_j(x)$ is norm-convergent and

$$\left\| \sum_{-\infty}^{\infty} T_j \right\| \leqslant \sum_{-\infty}^{\infty} \sqrt{\omega(j)}. \tag{11.10}$$

When one is using Cotlar's lemma, one is given an operator T and not a series $\sum_{-\infty}^{\infty} T_j$ of bounded operators. For instance, in order to apply Cotlar's lemma to the Hilbert transformation H, one has to break H into pieces T_j in order to get (11.8) and (11.9). This analysis of the operator is based upon some understanding of the operator. In other terms one needs a serious training for using Cotlar's lemma.

The second method for proving that an operator is bounded on some Hilbert space is the celebrated Schur lemma.

Let H be a Hilbert space and $(e_j)_{j \in J}$ be an orthonormal basis. The scalar product between two vectors x, y in H will be denoted by $\langle x, y \rangle$.

If we are given a possibly unbounded operator T (defined on a dense subspace $V \subset H$, such that $e_j \in V$ for every $j \in J$), we consider the entries $\alpha(i,j) = \langle T(e_i), e_j \rangle$ of the matrix M of T in our orthonormal basis.

Assume we are given a positive weight $\omega(j)$, $j \in J$, such that

$$\sum_{j \in J} |\alpha(i,j)| \omega(j) \leqslant \omega(i) \quad \textit{for each } i \in I \tag{11.11}$$

and

$$\sum_{i\in I} |\alpha(i,j)|\omega(i) \leqslant \omega(j) \quad \textit{for each } j \in I. \tag{11.12}$$

Then $\|T\| \leqslant 1$.

The proof is almost trivial and left to the reader. What is crucial is the flexibility in the choice of $\omega(j)$. We do not impose any growth condition on the sequence $\omega(j)$ which might be unbounded ... .

It is clear that Schur's lemma is as subtle as Cotlar's lemma. In the everyday life of the working analyst one is given an operator $T: V \to H$ where V is dense in H. In order to apply Schur's lemma one has to find a convenient orthonormal basis for H (a difficult task) as well as a weight $\omega(j)$ (an easier task).

Let us return to the problem raised by A. Zygmund and treat the case of the Hilbert transform. One can use Cotlar's lemma for proving the L^2-boundedness of the Hilbert transform. The operators T_j will be defined by $T_j f(x) = (f * K_j)(x)$ where $K_j(x) = 0$ if $|x| \leqslant 2^{-j-1}$ or if $|x| > 2^{-j}$ and $K_j(x) = 1/\pi x$ if $2^{-j-1} < |x| \leqslant 2^{-j}$. Observe that $K_j(x) = 2^j\theta(2^j x)$, θ being supported by $\frac{1}{2} < |x| \leqslant 2$ and that $\int \theta(x)\,dx = 0$. It is easy to check that (11.11) and (11.12) are satisfied with $\omega(j) = C2^{-|j|}$.

That is the first answer to Zygmund's problem. Unfortunately, this approach does not work for the general case of singular kernels satisfying (11.3), (11.4) and (11.5). Today we know that this approach could not work since these operators might be unbounded.

A second answer to A. Zygmund problem is to use Schur's lemma with the Haar basis. This answer was found by COIFMAN, JONES and SEMMES [1989], see also SEMMES [1989], and the Haar basis is the prototype for a wavelet basis.

Let us start with a function $h(x)$ which is defined by $h(x) = 1$ on $[0, \frac{1}{2})$, $h(x) = -1$ on $[\frac{1}{2}, 1)$ and $h(x) = 0$ elsewhere. Then $2^{j/2}h(2^j x - k)$, $j \in \mathbb{Z}$, $k \in \mathbb{Z}$, is an orthonormal basis for $L^2(\mathbb{R})$.

This basis was discovered by HAAR [1909]. It is convenient to denote by $I = I(j,k)$ the dyadic interval $[k2^{-j}, (k+1)2^{-j}[$, by $\mathcal{I}$ the collection of all such dyadic intervals and to observe that the support of $2^{j/2}h(2^j x - k)$ is precisely $I = I(j,k)$. For that reason we will label this function with I and write h_I instead of $h_{j,k}$.

The second answer to Zygmund's problem is given by the Haar basis. Indeed the entries $\alpha(I, I') = \langle Hh_I, h_{I'}\rangle$ of the Hilbert transformation satisfy Schur's criterion with the weight $\omega(I) = |I|^{1/2}$ (SEMMES [1989]).

In Sections 12 to 19, we will construct orthonormal wavelet bases. They improve the properties of the Haar basis and the function $h(x)$ is replaced by other functions $\psi(x)$ (the so-called Daubechies wavelets) which are smoother and have more vanishing moments. These properties are needed for improving the properties of the matrix $\beta(I, I') = \langle H\psi_I, \psi_{I'}\rangle$.

A. Zygmund was defined by one of my colleagues as the mathematician who spent fifty years in his life studying the properties of the Hilbert transform. Here, as in Zygmund's work, the Hilbert transform paves the way to the study of a larger class of singular integral operators which can be characterized by size conditions on the corresponding entries in a wavelet basis. Here we have to use wavelets with some regularity.

It is striking to observe that either Cotlar's lemma or Schur's lemma work efficiently for a certain class of Calderón–Zygmund operators which, in some sense, are close to being convolution operators and which are named pseudo-convolution for that reason. A pointwise multiplication with a function $m(x) \in L^\infty$ does not belong to that subclass.

However, under some circumstances the wavelet approach can be tailored to more general Calderón–Zygmund operators. These problems will be studied in Sections 24 to 28.

The last section will be concerned with the numerical algorithms in the situation of matrices. The issue will be to efficiently compress such matrices by a unitary change of variables.

Wavelet analysis could be defined as harmonic analysis invigorated by signal processing. Many concepts which have strong roots in signal processing are now playing a key role in wavelet analysis. Among the most important concepts are *pyramidal algorithms* and *quadrature mirror filters.*

That explains why we have chosen a leisure route and have tried to provide the reader with heuristics and intuitive approaches to wavelet analysis.

In doing it, we have omitted other roads leading to wavelets. Indeed it is true that wavelets were already there in K. Wilson's work on renormalization (K. Wilson was awarded the Nobel prize in physics for this work).

The interested reader is referred to BATTLE [1994].

An other approach to wavelets is provided by the so-called subdivision algorithms which are used in CAGD. The reader is referred to MEYER [1990] for a detailed exposition.

12. From Littlewood–Paley analysis to orthonormal wavelets

Loosely speaking a wavelet $\psi(x)$ is a function of the real variable x enjoying the four following properties

$\psi(x)$ is smooth (12.1)

$\psi(x)$ is localized around 0 (12.2)

$\psi(x)$ is oscillating as much as possible (12.3)

there is a fast algorithm for decomposing any function f in $L^2(\mathbb{R})$ into a linear combination of translates and dilates of $\psi(x)$. (12.4)

Let us be more precise. The ideal case for (12.1) and (12.2) would be a C^∞ function with compact support. For (12.3) the ideal case would be that the Fourier transform of ψ vanishes around 0 and be compactly supported.

Finally (12.4) means that any function in $L^2(\mathbb{R})$ can be written as a sum or integral of $(1/\sqrt{a})\,\psi((x-b)/a)$ where $a>0$ and $b\in\mathbb{R}$. These wavelets are generated by the "mother wavelet" ψ and will be denoted by $\psi_{(a,b)}$.

A *wavelet analysis* is an algorithm for decomposing any function in $L^2(\mathbb{R})$ as such a linear combination of wavelets $\psi_{(a,b)}$. We expect the coefficients of this decomposition to provide some useful information about the function $f(x)$.

Let us try to be more specific.

We define a wavelet as being a real-valued function of the real variable x with the three following properties

$$\int_{-\infty}^{\infty}(1+x^2)\left|\psi(x)\right|^2\,\mathrm{d}x<\infty, \tag{12.5}$$

$$\int_{-\infty}^{\infty}\psi(x)\,\mathrm{d}x=0 \tag{12.6}$$

and the Fourier transform $\hat{\psi}$ of ψ should satisfy

$$\int_0^{\infty}\left|\widehat{\psi}(\xi)\right|^2\frac{\mathrm{d}\xi}{\xi}=\int_0^{\infty}\left|\widehat{\psi}(-\xi)\right|^2\frac{\mathrm{d}\xi}{\xi}=1. \tag{12.7}$$

This third condition is just a normalization. The two integrals are the same since ψ is real-valued and are convergent by (12.5) and (12.6).

We next define the wavelets $\psi_{(a,b)}$, $a>0$, $b\in\mathbb{R}$, generated by the mother wavelet ψ:

$$\psi_{(a,b)}(x)=\frac{1}{\sqrt{a}}\psi\left(\frac{x-b}{a}\right). \tag{12.8}$$

The wavelet transform of $f\in L^2(\mathbb{R})$ is defined by

$$W(a,b)=\langle f,\psi_{(a,b)}\rangle \tag{12.9}$$

where $\langle u,v\rangle=\int u(x)\overline{v(x)}\,\mathrm{d}x$.

Finally Calderón's reproducing identity is

$$f(x)=\int_0^{\infty}\int_{-\infty}^{\infty}W(a,b)\psi_{(a,b)}(x)\,\mathrm{d}b\,\frac{\mathrm{d}a}{a^2}. \tag{12.10}$$

A function ψ fulfilling (12.5) and (12.6) is called a molecule in G. Weiss' terminology. A decomposition of $f(x)$ given by (12.10) is then called an atomic (or molecular) decomposition. It should be noticed that one might replace $W(a,b)$ by some other coefficients $\gamma(a,b)$ and still recover $f(x)$. The coefficients $W(a,b)$ are just the most convenient ones. Similarly one might use another wavelet $\widetilde{\psi}$ for the recovery of f. The relation between ψ and $\widetilde{\psi}$ would be given by

$$\int_0^\infty \widehat{\widetilde{\psi}}(\xi)\overline{\widehat{\psi}}(\xi)\,\frac{\mathrm{d}\xi}{\xi} = \int_0^\infty \widehat{\widetilde{\psi}}(-\xi)\overline{\widehat{\psi}}(-\xi)\frac{\mathrm{d}\xi}{\xi} = 1. \tag{12.11}$$

We systematically use $\hat{f}$ for denoting the Fourier transform of f.

This identity has an n-dimensional analogue. A wavelet is defined by the three conditions

$$\int (1+|x|)^{n+1}|f(x)|^2\,\mathrm{d}x < \infty, \qquad \int f(x)\,\mathrm{d}x = 0 \quad \text{and}$$

$$\int_0^\infty |\widehat{\psi}(t\xi)|^2\,\frac{\mathrm{d}t}{t} = 1 \quad \text{if } \xi \neq 0.$$

Then

$$\psi_{(a,b)}(x) = a^{-n/2}\psi\left(\frac{x-b}{a}\right)$$

are the wavelets generated by the mother wavelet $\psi(x)$. The wavelet coefficients of $f \in L^2(\mathbb{R})$ are $W(a,b) = \langle f, \psi_{(a,b)}\rangle$ and one recovers $f(x)$ by Calderón's identity

$$f(x) = \int_0^\infty \int_{\mathbb{R}^n} W(a,b)\,\psi_{(a,b)}(x)\,\mathrm{d}b\,\frac{\mathrm{d}a}{a^{n+1}}. \tag{12.12}$$

Let us give a well-known example of a wavelet analysis. Starting with a function f in $L^2(\mathbb{R}^n)$, one considers its harmonic extension $u(x,t)$ to the upper half-space. Either the full gradient $\nabla u(x,t)$ or the last component $\partial u/\partial t$ is evaluated. We consider the second option and for some homogeneity consideration, we form $v(x,t) = -t\partial u(x,t)/\partial t$. Then many properties of the function $f(x)$ can be characterized by size properties on $v(x,t)$. An excellent reference is STEIN [1970].

It is interesting to observe that $v(x,t)$ is a wavelet coefficient of the function $f(x)$. Indeed

$$u(x,t) = c_n \int \frac{t}{(|x-y|^2+t^2)^{(n+1)/2}} f(y)\,\mathrm{d}y$$

which implies

$$v(x,t) = \frac{1}{t^n}\int \psi\left(\frac{x-y}{t}\right) f(y)\,\mathrm{d}y = t^{-n/2}\langle f, \psi_{(t,x)}\rangle$$

where the wavelet ψ is given by

$$\psi(x) = \frac{n - |x|^2}{(|x|^2 + 1)^{(n+3)/2}}.$$

We have $\int \psi(x)\,\mathrm{d}x = 0$ but $x\,\psi(x)$ does not belong to $L^1(\mathbb{R}^n)$ which limits the applications. This is why one should use $v_m(x,t) = (t\,\partial/\partial t)^m\,u(x,t)$ for larger values of m in some applications. Observe that $v_m(x,t) = t^{-n/2}\langle f, \psi^{(m)}_{(t,x)}\rangle$ for an other wavelet $\psi^{(m)}$.

A large part of STEIN [1970] can be incorporated inside the wavelet analysis. It remains to indicate why this generalization is interesting.

Before giving an answer, let us consider a second viewpoint on wavelet analysis.

We consider two functions φ and ψ belonging to the Schwartz class $\mathcal{S}(\mathbb{R}^n)$ and satisfying the following condition:

there exists a positive number $\delta > 0$, $\delta < 1$, such that

$$\widehat{\varphi}(\xi) = 1 \quad \text{if } |\xi| \leqslant 1 - \delta, \tag{12.13}$$

$$\widehat{\varphi}(\xi) = 0 \quad \text{if } |\xi| > 1 + \delta, \tag{12.14}$$

$$\left|\widehat{\psi}(\xi)\right|^2 + \left|\widehat{\varphi}(\xi)\right|^2 = \left|\widehat{\varphi}(\xi/2)\right|^2 \quad \text{for every } \xi. \tag{12.15}$$

We denote by φ_j and ψ_j the functions $2^{nj}\varphi(2^j x)$ and $2^{nj}\psi(2^j x)$ and by S_j and Δ_j the convolution operators with φ_j and ψ_j.

Then (12.13) and (12.15) imply

$$I = S_0 S_0^* + \sum_0^\infty \Delta_j \Delta_j^*. \tag{12.16}$$

This identity is obviously related to a wavelet analysis. Indeed we have

$$\Delta_j^* f(x) = \int f(y)\,\overline{\psi_j(y - x)}\,\mathrm{d}y.$$

This scalar product can be rewritten $2^{nj/2}\langle f, \psi_{2^{-j},x}\rangle$ where

$$\psi_{(a,b)}(x) = \frac{1}{a^{n/2}}\psi\left(\frac{x - b}{a}\right)$$

and (12.16) is the discrete analogue of Calderón's identity. The integral with respect to all scales $a > 0$ is replaced by a sum over all dyadic scales $a = 2^j$.

The decomposition of a function by the identity (12.16) is called a *Littlewood–Paley decomposition.*

Let us say a few words about Littlewood–Paley analysis. It was created by J. Littlewood and R.E.A.C. Paley in the early thirties. The goal was to estimate the L^p-norm of a Fourier series

$$f(\theta) = \sum_0^\infty (a_k \cos\theta + b_k \sin\theta)$$

when $1 < p < \infty$ and $p \neq 2$. The case $p = 2$ is omitted, being trivial.

What Littlewood and Paley found is a brand of analysis which lies half the way between the "physical world" and the "Fourier world". Indeed full information on the Fourier coefficients of $f(\theta)$ is useless for computing the L^p-norm. One breaks the Fourier series of f into dyadic blocks

$$\Delta_j f(\theta) = \sum_{2^j \leqslant k < 2^{j+1}} (a_k \cos k\theta + b_k \sin k\theta)$$

and one has the following:

$$c_p \|f\|_p \leqslant \|g\|_p \leqslant c'_p \|f\|_p, \tag{12.17}$$

where $c'_p \geqslant c_p > 0$ are two constants and

$$g(\theta) = \left(|a_0|^2 + \sum_0^\infty |\Delta_j f(\theta)|^2 \right)^{1/2}. \tag{12.18}$$

Here

$$\|f\|_p = \left(\frac{1}{2\pi} \int_0^{2\pi} |f(\theta)|^p \, d\theta \right)^{1/p}.$$

This theorem had a tremendous impact on real analysis. Soon after, Marcinkiewicz used (12.17) in order to prove his celebrated multiplier theorem. If a sequence μ_k, $k \in \mathbb{Z}$, of complex numbers satisfies the conditions

$$\sup_k |\mu_k| \leqslant C_0 \quad \text{and} \quad \sup_{j \geqslant 0} \sum_{2^j}^{2^{j+1}} |\mu_{k+1} - \mu_k| \leqslant C_1, \tag{12.19}$$

then (μ_k) is a multiplier for the Fourier coefficients of L^p-functions when $1 < p < \infty$.

The case of the Hilbert transform is given by $\mu_k = -\mathrm{i} \operatorname{sign} k$ and (12.19) is obviously satisfied. Extending these results to the n-dimensional setting was not at all easy. For instance, one cannot define the dyadic blocks by the condition $2^j \leqslant |k| < 2^{j+1}$ where $k = (k_1, \ldots, k_n) \in \mathbb{Z}^n$. The reason is that the characteristic function of the ball is not a multiplier for $L^p(\mathbb{R}^n)$ when $p \neq 2$. Therefore the analogue of the Littlewood–Paley function $g(\theta)$ is

$$g(x) = \left(\left|S_0(f)\right|^2 + \sum_0^\infty \left|\Delta_j(f)\right|^2 \right)^{1/2}, \tag{12.20}$$

where S_0 and Δ_j are defined by (12.13), (12.14) and (12.15). With this definition, we still have (12.17).

The Littlewood–Paley analysis of a function is an extremely versatile tool. It happens to be useful even in non-linear PDEs where it is incorporated in the celebrated paraproducts (BONY [1983]; MEYER [1990]).

We now arrive to the last and most precise definition of a wavelet. We first treat the one-dimensional case.

A wavelet $\psi(x) \in L^2(\mathbb{R})$ is a function satisfying the three following properties:

localization: $|\psi(x)| \leqslant C_m(1+|x|)^{-m}$ for every integer m; (12.21)

smoothness: there exists an integer $r \geqslant 0$ such that

$$\left|\left(\frac{\mathrm{d}}{\mathrm{d}x}\right)^q \psi(x)\right| \leqslant C_m\big(1+|x|\big)^{-m}$$

for $0 \leqslant q \leqslant r$ and every integer m; (12.22)

orthogonality: the collection $\psi_{j,k}(x)$, $j \in \mathbb{Z}$, $k \in \mathbb{Z}$, defined by

$\psi_{j,k}(x) = 2^{j/2}\,\psi(2^j x - k)$ is an orthonormal basis for $L^2(\mathbb{R})$. (12.23)

The *cancellation* properties of $\psi(x)$ are hidden inside (12.23): we necessarily have $\int_{-\infty}^{\infty} x^q \psi(x)\,\mathrm{d}x = 0$ for $0 \leqslant q \leqslant r$. Then the wavelet coefficients of $f(x)$ are precisely $\langle f, \psi_{j,k}\rangle = \alpha(j,k)$. One has

$$\alpha(j,k) = 2^{-j/2}\,\Delta_j^* f\big(k2^{-j}\big), \tag{12.24}$$

where $\psi_j(x) = 2^j\psi(2^j x)$ and Δ_j is the convolution product with $\psi_j(x)$.

It means that the wavelet coefficients of $f(x)$ are obtained by the two following operations: apply the Littlewood–Paley adjoint operator Δ_j^* and then sample with a sampling rate which is about the inverse of the cut-off frequency of Δ_j. Moreover the relation between the Littlewood–Paley identity

$$f = \sum_{-\infty}^{\infty} \Delta_j \Delta_j^*(f)$$

and the identity

$$f(x) = \sum_{-\infty}^{\infty} \sum_{-\infty}^{\infty} \alpha(j,k)\,\psi_{j,k}(x)$$

is clear.

What seems missing is the lowpass filter S_0 and the function $\varphi(x)$. It will be shown in Section 22 that this function always exist for any orthonormal wavelet basis fulfilling (12.21), (12.22) and (12.23).

From this description it might seem that orthonormal wavelets offer a mere rephrasing of the celebrated Littlewood–Paley analysis.

This is not true and wavelets bring a substantial improvement on Littlewood–Paley analysis. They are building the bridge between Littlewood–Paley analysis and the theory of unconditional bases for Banach spaces.

For the reader's convenience we give the definition of a Schauder basis and of an unconditional basis for a Banach space E.

Let E be a Banach space. A *Schauder basis* for E is a collection $e_0, e_1, \dots, e_k, \dots$ of vectors of E with the following properties:

$$\text{for each } x \in E \text{ there exists a scalar sequence } \alpha_0, \alpha_1, \dots, \alpha_k, \dots$$
$$\text{such that } \big\|x - (\alpha_0 e_0 + \cdots + \alpha_m e_m)\big\|_E \to 0 \text{ as } m \text{ tends to infinity;} \tag{12.25}$$
$$\text{the sequence } \alpha_0, \alpha_1, \dots, \alpha_k, \dots \text{ fulfilling this requirement}$$
$$\text{is unique.} \tag{12.26}$$

This definition allows to state that E is isomorphic to a sequence space. But we do not know how to compute the norm $\|x\|_E$ when $(\alpha_k)_{k\in\mathbb{N}}$ is given. We say that $(e_k)_{k\in\mathbb{N}}$ is an *unconditional basis* whenever it is a Schauder basis and there is a constant C such that

$$\text{for every } m \geqslant 0, \text{ every sequence } \alpha_0, \dots, \alpha_m$$
$$\text{and every sequence } \beta_0, \dots, \beta_m, \text{the conditions } |\alpha_k| \leqslant |\beta_k| \ (1 \leqslant k \leqslant m)$$
$$\text{imply } \|\alpha_0 e_0 + \cdots + \alpha_m e_m\|_E \leqslant C\|\beta_0 e_0 + \cdots + \beta_m e_m\|_E. \tag{12.27}$$

In other terms, the series $\sum_0^\infty \alpha_k e_k$ represents a vector x in E if and only if $\sum_0^\infty |\alpha_k| e_k$ represents a vector y in E (in the sense given by (12.25).

For instance, the trigonometric system $\mathrm{e}^{\mathrm{i}kx}$, $k \in \mathbb{Z}$, is certainly not an unconditional basis for $L^p(0, 2\pi)$ when $p \neq 2$. This is the reason why the Littlewood–Paley theory was created.

But the Haar system for $L^2[0,1]$ is indeed an unconditional basis for $L^p[0,1]$ as it was proved by Marcinkiewicz in 1938 $(1 < p < \infty)$.

The relation between this fact and the Littlewood–Paley theory could not be clarified in the thirties. Today wavelets are building the bridge between the Haar system and the Littlewood–Paley theory.

It might seem that we are lost in the abstract world of pure mathematics. It has been quite a shock when D. Donoho found that the best denoising of an image or a signal for which an a priori information is given is provided by the so-called wavelet shrinkage.

At least this statement is true when the noise is so-called white noise and when the a priori information knowledge is of geometric nature (like the shape of objects, etc).

Then Donoho needed orthonormal bases which would be at the same time unconditional bases for a full range of functional spaces. White noise is fully decorrelated in any orthonormal basis and the a priori knowledge is expressed as a size condition on wavelet coefficients.

Applications range from Extragalactic Astronomy to Molecular Spectroscopy, Medical Imaging and Computer Vision.

13. Quadrature mirror filters

Where do orthonormal wavelets come from? It is an impossible task to solve the problem as it is stated. In other terms the definition given by (12.21), (12.22) and (12.23) does not provide an algorithm for constructing all orthonormal wavelet bases.

An algorithm for building some orthogonal bases was discovered in a joint work by S. Mallat and the author. It is based upon *multiresolution analysis* which will be described below (Section 15). But then the problem is moved a little bit further. Where does multiresolution analysis come from? The only previously known examples are based upon splines associated with refinements of meshes.

S. Mallat suggested that every multiresolution analysis could be constructed from a pair of two quadrature mirror filters. This insight has been crucial for the subsequent research. Indeed quadrature mirror filters were discovered about ten years before wavelets became fashionable. Quadrature mirror filters are deeply related to another well-known technique in signal processing, the so-called subband coding.

Mallat's program revealed the existing connections between signal processing, image processing, numerical analysis and pure mathematics.

Some experts in image processing first claimed that wavelets could not improve their work since it had already reached its top level. This is not true and a counterexample is provided by the striking discoveries made by D. Donoho on denoising algorithms. This work is by no means an empirical approach but relies on some deep properties of wavelet bases.

The purpose of Sections 13 to 19 is to unveil the relations between (1) quadrature mirror filters, (2) multiresolution analysis and (3) orthonormal wavelets.

The deep results we are going to describe in these sections are due to Pascal Auscher, Ingrid Daubechies, Pierre-Gilles Lemarié-Rieusset, Stéphane Mallat.

Let us begin the definition of *quadrature mirror filters*. The discussion will be easier if quadrature mirror filters are compared to the so-called ideal filters which take advantage of *Shannon's sampling theorem.*

Shannon's theorem is the following well-known observation. Let $[a, b]$ be any interval of length $l = b - a$. Let us fix $p \in (1, \infty)$ and denote by V the translation invariant closed subspace of $L^p(\mathbb{R})$ consisting of all f whose Fourier transform is supported by $[a, b]$. Here this Fourier transform is defined by

$$\hat{f}(\xi) = \int e^{-ix\xi} f(x)\, dx.$$

Then we consider the grid $\Gamma = (2\pi/l)\mathbb{Z}$ and Shannon's theorem is the following statement.

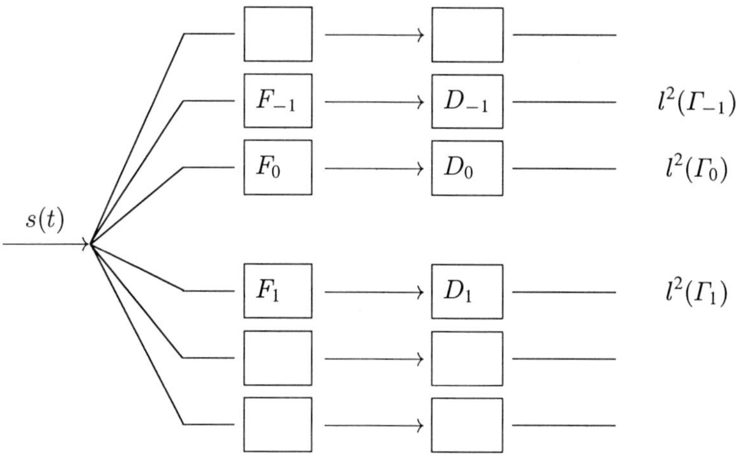

FIG. 13.1.

The restriction of any $f \in V$ to the grid Γ belongs to $l^p(\Gamma)$ and this restriction operator is an isomorphism D between V and $l^p(\Gamma)$.

This is no longer true when $p = 1$ and $p = \infty$ (MEYER [1990], Chapter I).

What we have in mind is the following application of Shannon's theorem. Let us consider an increasing sequence a_j of real numbers such that $\lim_{j\to+\infty} a_j = +\infty$ and $\lim_{j\to-\infty} a_j = -\infty$. Let us denote by $W_j \subset L^2(\mathbb{R})$ the translation invariant closed subspace defined by the condition that the Fourier transform $\hat{f}$ of any $f \in W_j$ should be supported by $[a_j, a_{j+1}]$.

Then $L^2(\mathbb{R}) = \bigoplus_{-\infty}^{\infty} W_j$ and each W_j is isomorphic to $l^2(\Gamma_j)$ where Γ_j is the grid $(2\pi)/l_j\mathbb{Z}$, $l_j = a_{j+1} - a_j$. This restriction operator is denoted by D_j (and is called a decimation operator in signal processing). Figure 13.1 illustrates the resulting isomorphism between $L^2(\mathbb{R})$ and $\bigoplus_{-\infty}^{\infty} l^2(\Gamma_j)$.

The ideal filters F_j are the projection operators from $L^2(\mathbb{R})$ onto W_j. Each F_j is a convolution with

$$\frac{1}{\pi}\frac{\sin l_j x}{l_j x}\,e^{i\omega_j x} \qquad \text{where} \quad \omega_j = \tfrac{1}{2}(a_j + a_{j+1}).$$

This function

$$\frac{1}{\pi}\frac{\sin l_j x}{l_j x}\,e^{i\omega_j x}$$

has a poor decay at infinity and for this reason, the ideal filters F_j are unrealistic.

Our goal is to construct another orthogonal decomposition $L^2(\mathbb{R}) = \bigoplus_{-\infty}^{\infty} W_j$ and to define the corresponding decimation operators $D_j : W_j \to l^2(\Gamma_j)$ in such a way that *the projection operators* $F_j : L^2(\mathbb{R}) \to W_j$ *are obtained*

through finite length filters; (13.1)

W_j *is invariant by translations by* $\tau \in \Gamma_j$. (13.2)

If (13.2) was replaced by a full translation invariance, we would we back to the case of ideal filters which implies unrealistic filters.

Our goal is to define the "channels" W_j, $j \in \mathbb{Z}$, the corresponding projection operators $F_j : L^2(\mathbb{R}) \to W_j$ and the corresponding decimation operators $D_j : W_j \to l^2(\Gamma_j)$. Before giving the solution to this problem which was found by Galand, we would like to indicate that any orthonormal wavelet basis with compact supports provides a solution.

Returning to the definition given in Section 12, let us assume that $\psi(x)$ has a compact support. Let us define W_j as the closed span of $2^{j/2}\,\psi(2^j x - k)$, $k \in \mathbb{Z}$.

Then the projection operator $F_j : L^2(\mathbb{R}) \to W_j$ is defined by

$$(F_j s)(t) = \sum_k \langle s, \psi_{j,k} \rangle \psi_{j,k}(t), \quad \psi_{j,k}(t) = 2^{j/2}\psi(2^j t - k)$$

while $D_j F_j(s)$ is simply the sequence $\langle s, \psi_{j,k} \rangle$, $k \in \mathbb{Z}$.

For giving the solution found by Croisier, Esteban and Galand, we need to fix some definitions and some notations. A filter is by definition a continuous translation invariant operator T acting on $l^2(\mathbb{Z})$. Then T is a convolution operator with a sequence $(\omega_k)_{k\in\mathbb{Z}}$ fulfilling the following condition: the transfer function $m(\theta) = \sum_{-\infty}^{\infty} \omega_k e^{ik\theta}$ belongs to $L^\infty(\mathbb{R})$. The operator norm of T is $\|m(\theta)\|_\infty$.

It will be convenient to write $H = l^2(\mathbb{Z})$ and to denote by H_0 the Hilbert space $l^2(2\mathbb{Z})$ which is still unrelated to H.

An elementary but useful observation is the following. If $T : H \to H_0$ is a continuous linear operator which commutes with even translations, then $T = DF$ where $F : H \to H$ commutes with all translations and $D : H \to H_0$ is the decimation operator defined by $D[(x_k)] = (x_{2k})$. Let T_0 and $T_1 : H \to H_0$ be two such operators (commuting with even translations).

DEFINITION 13.1. With the above mentioned notation, T_0 and T_1 are called *quadrature mirror filters* if, for every $x \in H$,

$$\|x\|^2 = \|T_0(x)\|^2 + \|T_1(x)\|^2. \tag{13.3}$$

This condition is obviously equivalent to

$$T_0^* T_0 + T_1^* T_1 = I \tag{13.4}$$

which is named perfect reconstruction by the signal processing community.

With a little more work, the following theorem can be proved.

THEOREM 13.1. *Let $T_0 : H \to H_0$ and $T_1 : H \to H_0$ be two bounded operators commuting with even translations and let $q_0(\theta)$ and $q_1(\theta)$ be the transfer functions of the corresponding filters F_0 and F_1.*

Then the following three properties are equivalent:

T_0 and T_1 are quadrature mirror filters; (13.5)

the operator $T = (T_0, T_1) : H \to H_0 \times H_1$ is a unitary isomorphism; (13.6)

the matrix

$$M(\theta) = \frac{1}{\sqrt{2}} \begin{pmatrix} q_0(\theta) & q_1(\theta) \\ q_0(\theta + \pi) & q_1(\theta + \pi) \end{pmatrix}$$

is unitary for almost every $\theta \in \mathbb{R}$. (13.7)

Here are some examples.

A trivial one is given by $T_0[(x_k)] = (x_{2k})$ and $T_1[(x_k)] = (x_{2k+1})$. In other terms, T_0 is the decimation operator D while $T_1 = DS$ where S is the shift by one. In that case $q_0(\theta) = 1$ and $q_1(\theta) = \mathrm{e}^{-\mathrm{i}\theta}$.

A less trivial example is given by $T_0[(x_k)] = (1/\sqrt{2})(x_{2k} + x_{2k+1})$ and $T_1[(x_k)] = (1/\sqrt{2})(x_{2k} - x_{2k+1})$ leading to $q_0(\theta) = (1/\sqrt{2})\,(1 + \mathrm{e}^{-\mathrm{i}\theta})$ and $q_1(\theta) = (1/\sqrt{2})\,(1 - \mathrm{e}^{-\mathrm{i}\theta})$.

This second example is the prototype of something we have in mind. In the applications, F_0 should behave like a discrete integration while F_1 should be similar to a discrete differentiation.

A second approach to quadrature mirror filters will be crucial. Let us begin with the obvious embedding $2\mathbb{Z} \subset \mathbb{Z}$ and try to mimic this embedding by a partial isometry $J : l^2(2\mathbb{Z}) \to l^2(\mathbb{Z})$ commuting with even translations.

Starting from such an operator J, there always exists a pair $T = (T_0, T_1)$ of quadrature mirror filters such that $T_0^* = J$.

For proving this remark, let us represent a sequence $(x_{2k}) \in l^2(\mathbb{Z})$ by the corresponding series

$$f(2\theta) = \sum_{-\infty}^{\infty} x_{2k}\, \mathrm{e}^{2k\,\mathrm{i}\theta}.$$

Then

$$\sum_{-\infty}^{\infty} |x_{2k}|^2 = \frac{1}{2\pi} \int_0^{2\pi} |f(\theta)|^2\, \mathrm{d}\theta.$$

Via this rephrasing J becomes an operator $\widehat{J}$ mapping π-periodic L^2 functions into 2π-periodic L^2 functions. Moreover, $\widehat{J}(m(2\theta) f(2\theta)) = m(2\theta) g(\theta)$ whenever $\widehat{J}(f(2\theta)) = g(\theta)$. This reflects the fact that J commutes with even translations.

Therefore, $\widehat{J}(f(2\theta)) = m_0(\theta) f(2\theta)$ where $m_0(\theta) = \widehat{J}(1)$. The operator $\widehat{J}$ is a partial isometry if and only if $|m_0(\theta)|^2 + |m_0(\theta+\pi)|^2 = 1$.

It is then easy to build F_0 and F_1. Indeed,

$$q_0(\theta) = \sqrt{2m_0(\theta)} \quad \text{and} \quad q_1(\theta) = \sqrt{2}\,\mathrm{e}^{\mathrm{i}\theta}\,\overline{m_0(\theta+\pi)}$$

certainly fulfill condition (13.7) in the theorem. It is easily checked that $F_0^* = J$.

A functional setting in which this remark is useful is the following.

Let E and $F \subset E$ be two closed subspaces of $L^2(\mathbb{R})$. We assume that E is invariant by translation by 1 while F is assumed to be invariant by translation by 2. Furthermore, E is assumed to be equipped with an orthonormal basis $(e_k)_{k\in\mathbb{Z}}$ which is the orbit of e_0 by the $\mathbb{Z}$-action on E. Similarly, F is equipped with an orthonormal basis (f_{2k}) which is the orbit of f_0 by the $2\mathbb{Z}$-action on F.

With these assumptions we denote by $J : l^2(2\mathbb{Z}) \to l^2(\mathbb{Z})$ the partial isometry defined by

$$J[(\alpha_{2k})] = (\beta_k)$$

whenever $\sum \alpha_{2k} f_{2k} = \sum \beta_k e_k$.

This operator J obviously commutes with even translations. Therefore J is associated to a pair (T_0, T_1) of two quadrature mirror filters. Moreover, $T_0[(\beta_k)] = (\alpha_{2k})$ if and only if $\sum \alpha_{2k} f_{2k}$ is the orthogonal projection of $\sum \beta_k e_k$ onto F.

We denote by G the orthogonal complement of F inside E. Then the theory of quadrature mirror filters amounts to building an orthonormal basis $(g_{2k})_{k\in\mathbb{Z}}$ for G which is also an orbit by the $2\mathbb{Z}$-action. Moreover, $T_1[(\beta_k)] = (\gamma_{2k})$ if and only if $\sum \gamma_{2k}\, g_{2k}$ is the orthogonal projection of $\sum \beta_k e_k$ onto G.

In order to organize hierarchies of quadrature mirror filters the following convention should be used. If $q \in \mathbb{N}$ is any integer and if $T = (T_0, T_1)$ is a pair of two quadrature mirror filters, then the following diagram provides the definition of

$$T = (T_0, T_1) : l^2(2^{-q}\mathbb{Z}) \to l^2(2^{-q+1}\mathbb{Z}) \times l^2(2^{-q+1}\mathbb{Z}):$$

$$\begin{array}{ccc} l^2(\mathbb{Z}) & \xrightarrow{\quad T \quad} & l^2(2\mathbb{Z}) \times l^2(2\mathbb{Z}) \\ \vdots & & \vdots \\ l^2(2^{-q}\mathbb{Z}) & \xrightarrow[\quad T^{(q)} \quad]{} & l^2(2^{-q+1}\mathbb{Z}) \times l^2(2^{-q+1}\mathbb{Z}) \end{array}$$

The identification between $l^2(\mathbb{Z})$ and $l^2(2^{-q}\mathbb{Z})$ is consistent with the identification between $l^2(2\mathbb{Z})$ and $l^2(2^{q-+1}\mathbb{Z})$. In both cases, the sequence (x_k) is indexed by $k2^{-q}$.

By abuse of notation, this operator $T^{(q)} = (T_0^{(q)}, T_1^{(q)})$ will be simply denoted $T = (T_0, T_1)$ and the value of q will be obvious from the context.

Soon after quadrature mirror filters were designed, hierarchical algorithms were considered. One starts with a signal sampled on a fine grid $2^{-q}\mathbb{Z}$. As usual, this sampling S_q is given by a filtering followed by an ordinary sampling. The filtering is obtained by a convolution with $2^q G(2^q t)$ where G is a correctly designed function. The correct choice of G will be given in Section 19.

We will study a string of the following type:

$$s(t) \xrightarrow{S_q} l^2(2^{-q}\mathbb{Z}) \xrightarrow{T_0} \cdots \xrightarrow{T_0} \cdots \xrightarrow{T_0} l^2\left(\tfrac{1}{2}\mathbb{Z}\right) \xrightarrow{T_0} l^2(\mathbb{Z})$$

and ask ourselves the following condition: is this algorithm tending to a limit as q tends to infinity? More precisely, we will compute, for each $m \in \mathbb{Z}$,

$$\lim_{q \to +\infty} T_0^q \, S_q(s)[m] \tag{13.8}$$

for $s(t) \in L^2(\mathbb{R})$.

The answer will depend on G as well as on T_0 and will be given in Section 19.

14. Quadrature mirror filters, pyramidal algorithms in image processing and multigrid algorithms

As the title indicates, quadrature mirror filters, pyramidal algorithms and multigrid algorithms offer striking similarities. Further on, the main differences will be stressed.

Let us begin with multigrid algorithms. One starts with a fine grid $h\mathbb{Z}^n$, $h = 2^{-m}$, which will be denoted by Γ_m. We then consider a ladder Γ_q $(0 \leqslant q \leqslant m-1)$ of coarser and coarser grids defined by $\Gamma_q = 2^{-q}\mathbb{Z}^n$.

This ladder is illustrated by the following scheme

$$\Gamma_m \supset \Gamma_{m-1} \supset \cdots \supset \Gamma_0 = \mathbb{Z}^n.$$

In a multigrid algorithm, a computation which is too expensive to be performed on the grid Γ_m is lifted up to the grid Γ_{m-1} (which contains four times fewer points). If necessary the computation is further lifted to a coarser grid and so on until we reach the coarsest grid.

The operator which is used to move from one grid to the next is not the trivial restriction to Γ_{q-1} of a sequence $X \in l^2(\Gamma_q)$. This trivial restriction will be named the decimation operator (and denoted by D).

Before decimating, we should filter $X \in l^2(\Gamma_q)$ by a low pass filter (denoted by F_0). The resulting operator is $T_0 = DF_0$. Similarly, once the computation is performed on the coarse grid Γ_{q-1}, one should come back to the fine grid Γ_q. This operation could be done using a trivial interpolation. This trivial interpolation is denoted by E and $Y = E(x)$ is defined by $Y = X$ and Γ_{q-1} and $Y = 0$ elsewhere. This wrong decision should be filtered with a second low pass filter (denoted by $\widetilde{F_0}$).

We are now in a good position for defining the trend and the fluctuation of a given sequence $X \in l^2(\Gamma_m)$ and for proving that X is the sum of the trend and the fluctuation (around the trend).

The trend could be defined by $T_0(X)$ where $T_0 = DF_0$ is the lifting operator. But this sequence belongs to $l^2(\Gamma_{m-1})$ and cannot be subtracted from X.

Therefore, the trend is defined as $\widetilde{T}_0 T_0(X)$ where $\widetilde{T}_0 : l^2(\Gamma_{m-1}) \to l^2(\Gamma_m)$ is the operator $\widetilde{F}_0 E$. One should observe that $\widetilde{T}_0 T_0$ is no longer a convolution operator acting on $l^2(\Gamma_m)$.

The fluctuation (around the trend) is defined by $X - \widetilde{T}_0 T_0(X)$.

Finally any vector $X \in l^2(\Gamma_m)$ is represented by a pair (Y, Z) where $Y = T_0(X)$ and $Z = X - \widetilde{T}_0 T_0 X$.

The same idea has been introduced by Burt and Adelson in image processing. It is then called a pyramidal algorithm. An image is represented by a vector $X \in l^2(h\mathbb{Z}^2)$ where $h > 0$ is small and $h\mathbb{Z}^2$ is the fine grid. The pyramid algorithm will exploit the fact that "most of the time" the gray level in neighboring pixels remains the same. In some other places there are some abrupt discontinuities which are either revealing edges or are artifacts due to the noise.

One should at the same time compress the image, eliminate the noise and enhance the edges.

This is achieved by a correctly designed pyramidal algorithm. When, at some scale, the gray level is constant on neighboring pixels, the fluctuations can be ignored and the corresponding portion of our image can be downsampled. This downsampling is a constitutive piece of the pyramidal algorithm and is obtained by the mapping T_0. Our iterates and obtain the "herring bone" scheme

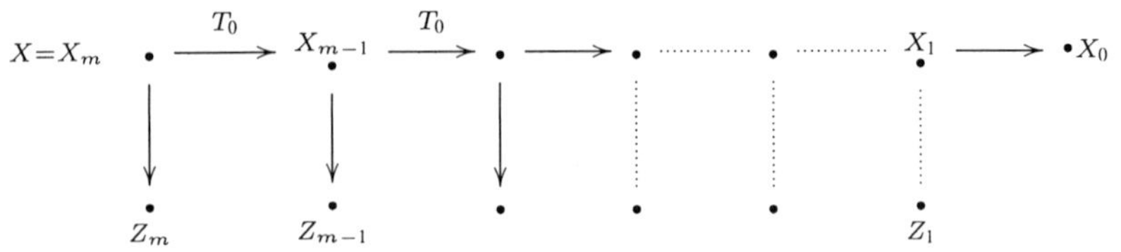

The Burt and Adelson algorithm consists in coding $X = X_m \in l^2(\Gamma_m)$ by $X_0 \in l^2(\Gamma_0)$ together with the succession of fluctuations calculated through dyadic scales, i.e. $Z_m, \ldots, Z_1$. As it was already mentioned most of these fluctuations will vanish and the final trend X_0 contains 4^m times less points than the original image. Indeed X_0 is downsampled by a factor of 4^m.

Let us compute the total amount of information that has been consumed if X_m vanishes outside the unit square $0 \leqslant x < 1$, $0 \leqslant y < 1$. The amount of information contained in X_m is 4^m. The trend contains 4^{m-1} data and the fluctuation is given on the fine grid with 4^m data. In such a way that 4^m data become, at the next step, $4^{m-1}+4^{m-2}$ data and finally we obtain at the end of the scheme $4^m+4^{m-1}+\cdots+4+1$ data (all the terms, but the last one, correspond to fluctuations at distinct scales and the last one corresponds to the trend at the coarsest scale).

Finally 4^m data are transformed into $\frac{4}{3}4^m$ data and this algorithm is too expensive in principle.

As was told before, this calculation is unfair since most of the fluctuations vanish most of the time.

The reason of this lack of efficiency is the following. The fluctuations belong to very specific subspaces of $l^2(\Gamma_q)$ and this spatial geometry should be used for coding the fluctuation.

Let us assume that $\widetilde{T}_0 : l^2(\Gamma_{q-1}) \to l^2(\Gamma_q)$ *is the adjoint of* $T_0 : l^2(\Gamma_q) \to l^2(\Gamma_{q-1})$ *and that* $\widetilde{T}_0$ *is a partial isometry.*

Then the fluctuation $Z = X - T_0^* T_0(X)$ will be orthogonal to the trend $Y = T_0^* T_0(X)$. Let us denote by $V \subset l^2(\Gamma_m)$ the range of $T_0^* : l^2(\Gamma_{m-1}) \to l^2(\Gamma_m)$. Then V is closed in $l^2(\Gamma_m)$ and T_0, once restricted to V, becomes an isomorphism V and $l^2(\Gamma_{m-1})$. The operator $T_0^* T_0$ is the orthogonal projection from $l^2(\Gamma_m)$ onto V and the fluctuations belong to the orthonormal complement W of V in $l^2(\Gamma_m)$. Returning to the case where $X \in l^2(\Gamma_m)$ is supported by the unit square, the fluctuation Z belongs to a subspace W with dimension $\frac{3}{4} 4^m$. This factor $\frac{3}{4}$ compensates the loss with 4/3 in the general Burt and Adelson scheme.

For being more efficient, one should have a direct access to some convenient orthonormal basis of W. But this is precisely what quadrature mirror filters yield.

15. Multiresolution analysis

Before giving a precise definition, let us start with two examples. The first one is the standard Littlewood–Paley analysis in its most crude definition.

Starting with $f \in L^2(\mathbb{R})$, $S_j f$ and $\Delta_j f$, $j \in \mathbb{N}$, are defined by $\mathcal{F}(S_j f)(\xi) = \hat{f}(\xi)\, \mathbf{1}_{\{|\xi| \leqslant 2^j\}}$ and $\mathcal{F}(\Delta_j f)(\xi) = \hat{f}(\xi)\, \mathbf{1}_{\{2^j \leqslant |\xi| \leqslant 2^{j+1}\}}$ where $\mathcal{F}f = \hat{f}$ is the Fourier transform of f.

The Littlewood–Paley decomposition of f is given by

$$f = S_0(f) + \sum_0^\infty \Delta_j(f). \tag{15.1}$$

We denote by $V_0 \subset L^2(\mathbb{R})$ and $W_j \subset L^2(\mathbb{R})$ the closed subspaces of $L^2(\mathbb{R})$ which are the ranges of S_0 and Δ_j and (15.1) can be rewritten

$$L^2(\mathbb{R}) = V_0 \oplus W_0 \oplus W_1 \oplus \cdots \oplus W_j \oplus \cdots. \tag{15.2}$$

The drawbacks of this crude Littlewood–Paley analysis are the following ones: (1) the Hölder classes $C^\alpha(\mathbb{R})$ are not characterized by size properties of $\Delta_j(f)$, (2) when the dimension n exceeds 1, this Littlewood–Paley analysis is not convenient for $L^p(\mathbb{R}^n)$.

That is the reason why the crude Littlewood Paley analysis was replaced by the soft one.

The crude partitioning of the identity

$$1 = \mathbf{1}_{\{|\xi| \leqslant 1\}} + \sum_0^\infty \mathbf{1}_{\{2^j \leqslant |\xi| \leqslant 2^{j+1}\}}$$

is replaced by a soft partitioning of the identity given by

$$1 = \widehat{\varphi}(\xi) + \sum_{0}^{\infty} \widehat{\psi}(2^{-j}\xi)$$

where $\widehat{\varphi}$ and $\widehat{\psi}$ belong to $C_0^\infty(\mathbb{R}^n)$, $\widehat{\varphi}(\xi) = 1$ in some neighborhood of 0 and $\widehat{\psi}(\xi) = \widehat{\varphi}(\xi/2) - \widehat{\varphi}(\xi)$.

The operators S_0 and Δ_j are now defined by convolution with φ and ψ_j (where $\psi_j(x) = 2^{nj}\psi(2^j x)$). Then this new definition is adapted to the study of most of the functional spaces.

The drawback of this new definition is the fact that the ranges of the operators S_0 and Δ_j are no longer closed. Moreover, these ranges are no longer orthogonal.

The goal of a multiresolution analysis is to offer a substitute for the Littlewood–Paley analysis. More precisely, we are looking for an orthogonal decomposition

$$L^2(\mathbb{R}^n) = V_0 \oplus W_0 \oplus \cdots \oplus W_j \oplus \cdots,$$

where V_0 and W_j are closed and where the usual functional spaces can be studied through simple size properties of the dyadic blocks f_j. These dyadic blocks f_j are defined as the orthogonal projections of f onto W_j.

We now arrive to the definition of a multiresolution analysis.

DEFINITION 15.1. A multiresolution analysis of $L^2(\mathbb{R})$ is, by definition, an increasing sequence V_j, $j \in \mathbb{Z}$, of closed linear subspaces of $L^2(\mathbb{R}^n)$ with the following properties:

$$\bigcap_{-\infty}^{\infty} V_j = \{0\}, \qquad \bigcup_{-\infty}^{\infty} V_j \text{ is dense in } L^2(\mathbb{R}^n); \tag{15.3}$$

for every $j \in \mathbb{Z}$ and every $f \in L^2(\mathbb{R}^n)$, we have

$$f(x) \in V_j \Longleftrightarrow f(2x) \in V_{j+1}; \tag{15.4}$$

there exists a function g in V_0 such that

$$g(x-k),\ k \in \mathbb{Z}, \text{ is a Riesz basis for } V_0. \tag{15.5}$$

This last condition means that functions $f \in V_0$ admit a simple and powerful expansion given by $\sum \alpha_k\, g(x-k)$ where $(\sum_{-\infty}^{\infty} |\alpha_k|^2)^{1/2}$ and $\|f\|_2$ are equivalent norms on V_0. This would have been impossible if V_0 were not closed.

DEFINITION 15.2. Let $r \in \mathbb{N}$ and V_j, $j \in \mathbb{Z}$, be a multiresolution analysis of $L^2(\mathbb{R}^n)$. Then this multiresolution analysis is r regular if there is a choice for the function g in (15.5) such that

$$\left|\partial^\alpha g(x)\right| \leqslant C_m\big(1 + |x|\big)^{-m} \tag{15.6}$$

for every $x \in \mathbb{R}^n$, every $m \in \mathbb{N}$ and every $\alpha = (\alpha_1, \ldots, \alpha_n)$ such that $|\alpha| = \alpha_1 + \cdots + \alpha_n \leqslant r$.

Let us provide the reader with some examples.

A first series of examples is given by embedded subspaces of spline functions associated to refinements of meshes. More precisely, $g(x) = \xi * \cdots * \xi$ (r convolution products and $r+1$ terms) where ξ is the indicator of $[0, 1)$. Then $g(x-k)$, $k \in \mathbb{Z}$, is a Riesz basis for V_0 and the corresponding V_j are constructed through (15.4).

A second class of examples is closer to the Littlewood–Paley analysis.

Let φ be an even function in the $\mathcal{S}(\mathbb{R})$ class of Schwartz with the following properties: $\widehat{\varphi}(\xi) = 1$ on $[-\frac{2}{3}\pi, \frac{2}{3}\pi]$, $\widehat{\varphi}(\xi) = 0$ outside $[-\frac{4}{3}\pi, \frac{4}{3}\pi]$ and

$$\sum_{-\infty}^{\infty} \left|\widehat{\varphi}(\xi + 2k\pi)\right|^2 = 1.$$

This last condition reduces to $|\widehat{\varphi}(\xi)|^2 + |\widehat{\varphi}(2\pi - \xi)|^2 = 1$ whenever $|\xi - \pi| \leqslant \frac{1}{3}\pi$.

Then $\varphi(x-k)$, $k \in \mathbb{Z}$, is an orthonormal sequence in $L^2(\mathbb{R})$ and we denote by V_0 the closed linear span of this sequence. It is easily checked that the corresponding V_j, $j \in \mathbb{Z}$, as defined by (15.4), form a multiresolution analysis.

With this example in mind, we arrive at the notion of frequency channels.

DEFINITION 15.3. Let V_j, $j \in \mathbb{Z}$, be a multiresolution analysis of $L^2(\mathbb{R})$. Let $W_j \subset V_{j+1}$ be the orthogonal complementary of $V_j : V_{j+1} = V_j \oplus W_j$.

Then these W_j, $j \in \mathbb{Z}$, are the frequency channels associated with the multiresolution analysis.

Let us return to the multiresolution analysis which is close to the Littlewood–Paley decomposition. In that case any band-limited function f will belong to V_j when $\frac{2}{3}\pi 2^j \geqslant T$ ($\hat{f}$ being supported by $[-T, T]$). It follows that

$$f \in W_j \Longrightarrow \text{support } \hat{f}(\xi) \subset \left\{\frac{2\pi}{3} 2^j \leqslant |\xi| \leqslant \frac{8\pi}{3} 2^j\right\}.$$

In other terms, if $f \in W_j$, we have frequency information: the order of magnitude of the frequencies of f is 2^j. At the same time

$$f(x) \in W_j \Longleftrightarrow f(2^{-j}x) \in W_0$$

and it will be shown that the functions in W_0 can be sampled over the grid $\mathbb{Z}$.

It means that the functions in W_j are correctly sampled over the grid $2^{-j}\mathbb{Z}$.

In other terms the frequency channel W_j is also providing a scale information: the fundamental scale associated with W_j is 2^{-j}.

Up to a normalization factor this scale 2^{-j} and the corresponding "average frequency" $2\pi \bullet 2^j$ of functions in W_j are related by the Heisenberg's uncertainty principle.

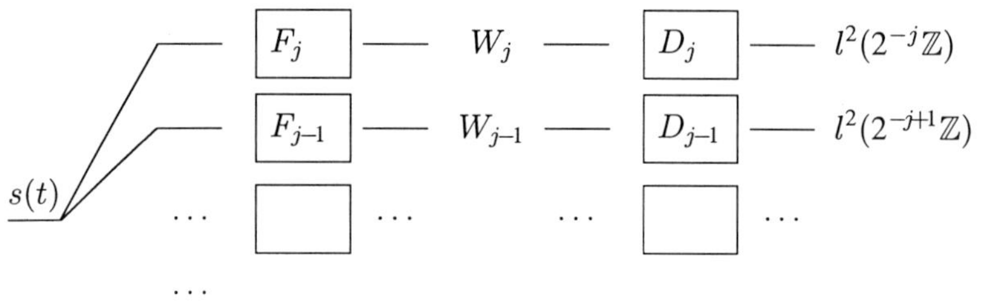

FIG. 15.1.

We now return to the fundamental diagram and ask for the filters F_j and the decimation operators D_j associated with the frequency channels W_j.

The answer is given by orthonormal wavelet bases.

THEOREM 15.1. *Let $(V_j)_{j\in\mathbb{Z}}$ be a multiresolution analysis of $L^2(\mathbb{R})$. Then there a function $\varphi \in V_0$ (the so-called sampling function) and there is a function $\psi \in W_0$ (the so-called wavelet) with the following properties:*

$$\varphi(x-k),\ k \in \mathbb{Z},\ \textit{is an orthonormal basis of } V_0, \tag{15.7}$$

$$\psi(x-k), k \in \mathbb{Z},\ \textit{is an orthonormal basis of } W_0, \tag{15.8}$$

$$\textit{if } g(x) \textit{ fulfills } (15.6) \textit{ so do } \varphi \textit{ and } \psi. \tag{15.9}$$

This theorem provides the filters F_j. Indeed, let us denote by $\widetilde{\psi}(x)$ the function $\overline{\psi}(-x)$. For any j, we write $\widetilde{\psi}_j(x) = 2^j\,\widetilde{\psi}(2^j x)$.

Finally, F_j denotes the convolution operator with the function $\widetilde{\psi}_j$. The decimation operator D_j is the restriction to the grid $2^{-j}\mathbb{Z}$. Then $D_jF_j(s)$ precisely yields the wavelet coefficients of s.

However, the frequency channel is not the range of F_j: this range is not generally closed. In order to build W_j from F_j, we use the sampling D_jF_j and interpolate this sampling $\alpha(j,k)$ using the wavelets $2^{j/2}\psi(2^jt-k)$. One obtains $\sum_k \alpha(j,k)2^{j/2}\psi(2^jt-k)$.

16. How to extract wavelets from a multiresolution analysis

Our task in this section is to construct a function $\psi(x)$ in W_0 such that $\psi(x-k)$, $k \in \mathbb{Z}$, be an orthonormal basis for W_0. The definition of W_0 is given in Section 15.

In this case then (for every j), $2^{j/2}\,\psi(2^jx-k)$, $k \in \mathbb{Z}$, will be an orthonormal basis for W_j.

But $L^2(\mathbb{R}) = \bigoplus_{-\infty}^{\infty} W_j$, the sum being direct orthogonal. Therefore, the full collection $\{2^{j/2}\psi(2^jx-k)\colon\ j \in \mathbb{Z},\ k \in \mathbb{Z}\}$ is an orthonormal basis for $L^2(\mathbb{R})$.

For constructing ψ let us return to the decomposition $V_1 = V_0 \oplus W_0$. We are given an orthonormal basis $\{\sqrt{2}\,\varphi(2x-k)\colon\ k \in \mathbb{Z}\}$ for V_1, an orthonormal basis $\{\varphi(x-k)\colon\ k \in \mathbb{Z}\}$ for V_0 and an orthonormal basis $\{\psi(x-k)\colon\ k \in \mathbb{Z}\}$ for W_0. The first one will be denoted by e_k, $k \in \mathbb{Z}$, the second one f_{2k}, $k \in \mathbb{Z}$, and the third one g_{2k}, $k \in \mathbb{Z}$. The indices give the localization of the corresponding functions if the unit scale is $\frac{1}{2}$. Since $V_1 = V_0 \oplus W_0$, we know that every vector $\sum_{-\infty}^{\infty} \alpha_k e_k$, $\sum |\alpha_k|^2 < \infty$, can be uniquely written

$$\sum_{-\infty}^{\infty} \alpha_k e_k = \sum_{-\infty}^{\infty} \beta_{2k} f_{2k} + \sum_{-\infty}^{\infty} \gamma_{2k} g_{2k}. \tag{16.1}$$

We analyze the mapping $T: l^2(\mathbb{Z}) \to l^2(2\mathbb{Z}) \times l^2(2\mathbb{Z})$ defined by $T[(\alpha_k)] = [(\beta_{2k}), (\gamma_{2k})]$. This mapping commutes with even translations and is a unitary isomorphism. Therefore, $T = (T_0, T_1)$ where T_0 and T_1 are two quadrature mirror filters.

From the above theory of quadrature mirror filters, one singles out the operator J which corresponds to the embedding $V_0 \subset V_1$. Writing $\varphi(x) = 2\sum \alpha_k \varphi(2x+k)$ and passing to the Fourier transform, one obtains $\widehat{\varphi}(\xi) = m_0(\xi/2)\widehat{\varphi}(\xi/2)$ or

$$\widehat{\varphi}(2\xi) = m_0(\xi)\widehat{\varphi}(\xi). \tag{16.2}$$

Similarly, $\psi(x) = 2\sum \beta_k \varphi(2x+k)$ will lead to

$$\widehat{\psi}(2\xi) = m_1(\xi)\widehat{\varphi}(\xi). \tag{16.3}$$

The matrix

$$U(\xi) = \begin{pmatrix} m_0(\xi) & m_1(\xi) \\ m_0(\xi+\pi) & m_1(\xi+\pi) \end{pmatrix} \tag{16.4}$$

is unitary and a convenient choice of m_1 is given by

$$m_1(\xi) = \mathrm{e}^{-\mathrm{i}\xi}\, \overline{m_0(\xi+\pi)}. \tag{16.5}$$

If the multiresolution analysis (V_j) is r-regular, then the 2π-periodic function $m_0(\xi)$ will be indefinitely differentiable. Therefore ψ which is defined by (16.3) will satisfy (15.6).

We have just proved Theorem 15.1 which will be restated.

If a multiresolution analysis (V_j) is r-regular then there exists a wavelet ψ satisfying (15.6) and such that $2^{j/2}\psi(2^j x - k)$, $j \in \mathbb{Z}$, $k \in \mathbb{Z}$, is an orthogonal basis for $L^2(\mathbb{R})$.

The other problems to be solved are the following ones.

(A) *Is it always true that any orthonormal basis $2^{j/2}\psi(2^j x - k)$, $j \in \mathbb{Z}$, $k \in \mathbb{Z}$, of $L^2(\mathbb{R})$ should come from a multiresolution analysis?*

(B) *Is it always possible to invert the preceding construction and to build every multiresolution analysis from a pair of two quadrature mirror filters?*

The reason why these problems are important is the following: if we stick to the classical examples of multiresolution analysis, we will never end up with wavelets with compact support.

We begin with the following observation.

LEMMA. *If $\varphi(x)$ is a compactly supported function in $L^2(\mathbb{R})$, if φ is a solution to the two-scales equation*

$$\varphi(x) = 2\sum \alpha_k\, \varphi(2x+k) \tag{16.6}$$

and if finally $\varphi(x-k)$ is an orthogonal sequence, then $\alpha_k = 0$ for $|k| \geqslant T$.

Indeed (16.6) implies $\alpha_k = \int \varphi(x)\overline{\varphi(2x+k)}\,\mathrm{d}x = 0$ for $|k| \geqslant T$.

The second observation is that for any such function φ, $\varphi(x+m)$, $m \in \mathbb{N}$, is also a solution where α_k is replaced by α_{k+m}. This observation reduces the problem to the case where (16.6) reads

$$\varphi(x) = 2\sum_{0}^{N} \alpha_k \varphi(2x+k). \tag{16.7}$$

For any r-regular multiresolution analysis, the scaling function φ satisfies the so-called Fix–Strang conditions (FIX and STRANG [1969])

$$\widehat{\varphi}(0) = 1 \quad \text{and} \quad \widehat{\varphi}(2k\pi) = 0,\ k \in \mathbb{Z},\ k \neq 0. \tag{16.8}$$

More precisely, $\partial^\alpha \widehat{\varphi}(2k\pi) = 0$, $k \in \mathbb{Z}$, $k \neq 0$, for every multi-index α such that $|\alpha| \leqslant r$.

These remarks imply that the only way for constructing a multiresolution analysis for which φ has a compact support is by the following recipe:

$$\text{start with a trigonometric sum } m_0(\xi) = \sum_{0}^{N} \alpha_k\, \mathrm{e}^{\mathrm{i}k\xi} \tag{16.9}$$

fulfilling $m_0(0) = 1$ and $|m_0(\xi)|^2 + |m_0(\xi+\pi)|^2 = 1$,

$$\text{define } \varphi \text{ by } \widehat{\varphi}(\xi) = m_0(\xi/2)m_0(\xi/4)\cdots m_0(\xi/2^j)\cdots, \tag{16.10}$$

$$\text{define } \psi \text{ by } \widehat{\psi}(\xi) = m_1(\xi/2)\widehat{\varphi}(\xi/2), \tag{16.11}$$

where $m_1(\xi) = \mathrm{e}^{-\mathrm{i}\xi}\,\overline{m_0(\xi+\pi)}$.

This construction raises two problems. The first one concerns the converse statement: is it true that (16.9), (16.10), (16.11) necessarily lead to orthonormal wavelets? Is this recipe the only way for constructing such bases? In other terms is it true that every orthonormal wavelet basis with compact supports originates from a multiresolution analysis where φ is compactly supported?

A series of papers by A. Cohen takes care of the first problem and a positive solution to the second one was recently found by P.G. Lemarié-Rieusset.

17. A. Cohen's criterion

The main difficulty in the above mentioned construction is to show that $\varphi(x-k),\ k \in \mathbb{Z}$, is an orthonormal sequence. It is not always the case and a classical counter-example is provided by $m_0(\xi) = \frac{1}{2}(1 + \mathrm{e}^{-3\mathrm{i}\xi})$. Then the function $\varphi(x)$ which is obtained by (16.10) is $\frac{1}{3}$ over $[0, 3)$ and 0 elsewhere. In that case $\varphi(x-k),\ k \in \mathbb{Z}$, is certainly not an orthonormal sequence.

A. Cohen's criterion (see COHEN [1995] or DAUBECHIES [1992]) is a necessary and sufficient condition on the trigonometric sum $m_0(\xi) = \alpha_0 + \cdots + \alpha_N \mathrm{e}^{\mathrm{i}N\xi}$ ensuring that $\varphi(x-k),\ k \in \mathbb{Z}$, is an orthonormal sequence yielding a multiresolution analysis. It will remain to relate the regularity r of the corresponding wavelet ψ to some properties of $m_0(\xi)$. This second problem has also been solved by A. Cohen in a joint work with J.P. Conze (see COHEN and CONZE [1992]).

A. Cohen's criterion can be given two equivalent forms. The first one is

$$\lim_{j\to+\infty} m_0(\omega)m_0(2\omega)\cdots m_0(2^j\omega) = 0 \quad \text{if } 0 < |\omega| \leqslant \pi. \tag{17.1}$$

Let us observe that $m_0(\xi) = \frac{1}{2}(1+\mathrm{e}^{-3\mathrm{i}\xi})$ does not fulfill (17.1) since $m_0(2^j\omega_0) = 1$ for $\omega_0 = \frac{2}{3}\pi$ and every $j \in \mathbb{N}$.

This condition has a very nice intuitive meaning.

Since F_0 has finite length, the corresponding operator $T_0 = DF_0$ acts from $l^\infty(\mathbb{Z})$ into $l^\infty(2\mathbb{Z})$. The normalization is distinct from the one used for $l^2(\mathbb{Z})$ if we impose the condition $T_0(\mathbf{1}) = \mathbf{1}$ (here $\mathbf{1}$ denotes the sequence $(\ldots, 1, 1, 1, \ldots)$). We should replace T_0 by $(1/\sqrt{2})T_0$. But this is irrelevant for what will be discussed now.

Let us iterate the operator T_0 which gives

$$l^\infty(\mathbb{Z}) \underset{T_0}{\longrightarrow} l^\infty(2\mathbb{Z}) \underset{T_0}{\longrightarrow} \cdots \underset{T_0}{\longrightarrow} l^\infty(2^{q-1}\mathbb{Z}) \underset{T_0}{\longrightarrow} l^\infty(2^q\mathbb{Z}) \tag{17.2}$$

and ask for the limit of $T_0^q[\mathrm{e}^{\mathrm{i}\omega k})]$.

At the first step, we have

$$T_0\big[(\mathrm{e}^{\mathrm{i}\omega k})\big] = \frac{1}{\sqrt{2}}\, q_0(\omega)\,\mathrm{e}^{2\mathrm{i}\omega k} = \overline{m_0(\omega)}\,\mathrm{e}^{2\mathrm{i}\omega k}.$$

Iterating, one obtains $\overline{m_0(\omega)}\,\overline{m_0(2\omega)}\,\mathrm{e}^{4\mathrm{i}\omega k}$, etc.

Then Cohen's criterion is the following statement: when one iterates T_0, the limiting action is an average of the given sequence.

This fact is obviously necessary if we want T_0 to be associated with a multiresolution analysis where φ is smooth and compactly supported.

Indeed let us start with

$$f(x) = \sum_{-\infty}^{\infty} \mathrm{e}^{\mathrm{i}\omega k}\, \varphi(x-k) \in L^\infty(\mathbb{R}).$$

Letting T_0^q act on $f(x)$ means projecting $f(x)$ on $V_{(-q)}$. An orthonormal basis for V_{-q} is provided by $2^{-q/2}\varphi(2^{-q}x-k)$, $k \in \mathbb{Z}$, and the scalar product between a given function $f(x) \in L^\infty(\mathbb{R})$ and $2^{-q}\varphi(2^{-q}x-k)$ is an average of f, around $k2^q$, at the scale 2^q. A last observation is that the average of $f(x)$ is the same as the average of the sequence $\mathrm{e}^{\mathrm{i}\omega k}$, $k \in \mathbb{Z}$.

A second and equivalent form of Cohen's criterion is given by the following theorem.

THEOREM 17.1. *If* $m_0(\xi) = \alpha_0 + \alpha_1 \mathrm{e}^{i\xi} + \cdots + \alpha_N \mathrm{e}^{\mathrm{i}N\xi}$ *fulfills* $m_0(0) = 1$ *and* $|m_0(\xi)|^2 + |m_0(\xi+\pi)|^2 = 1$, *then there exists* ω *such that* $0 < |\omega| < \pi$ *and*

$$\lim_{j\to+\infty} \left|m_0(\omega)\cdots m_0\left(2^j\omega\right)\right| > 0$$

if and only if there exist two relatively prime odd numbers $m \geqslant 1$ *and* $n \geqslant 3$ *such that* $|m_0(\pi 2^q m/n)| = 1$ *for all* $q \geqslant 1$.

Returning to the counter-example given by $m_0(\xi) = \frac{1}{2}(1+\mathrm{e}^{-3\mathrm{i}\xi})$ we have $|m_0(\xi)| = |\cos 3\xi/2| = 1$ if and only if $\xi = 2k\pi/3$. Here $m=1$, $n=3$.

A simpler sufficient condition (DAUBECHIES [1992, p. 186]) is $m_0(\xi) \neq 0$ for $-\frac{1}{3}\pi \leqslant \xi \leqslant \frac{1}{3}\pi$.

This sufficient condition applies to the particular case of the Daubechies wavelets which will be studied in the next section.

18. Regularity of the Daubechies wavelets

Let us first define the Daubechies wavelets. For any $N \geqslant 1$, we define the trigonometric sum

$$g_N(\xi) = \sum_{-(2N-1)}^{2N-1} \gamma_k\, \mathrm{e}^{\mathrm{i}k\xi}$$

by

$$g_N(\xi) = c_N \int_\xi^\pi (\sin t)^{2N-1}\, \mathrm{d}t$$

where $c_N > 0$ is chosen such that $g_N(0) = 1$.

We then have $g_N(\xi) + g_N(\xi + \pi) = 1$ and $0 \leqslant g_N(\xi) \leqslant 1$. A well-known result yields a trigonometric sum

$$m_0(\xi) = \alpha_0 + \alpha_1 \mathrm{e}^{-\mathrm{i}\xi} + \cdots + \alpha_{2N-1} \mathrm{e}^{-\mathrm{i}(2N-1)\xi}$$

with real coefficients such that $m_0(0) = 1$ and $|m_0(\xi)|^2 = g_N(\xi)$. However, $m_0(\xi)$ is not unique.

From $m_0(\xi)$ the scaling function φ is constructed via

$$\widehat{\varphi}(\xi) = m_0(\xi/2) \cdots m_0(\xi/2^j) \cdots. \tag{18.1}$$

Let us observe that $|\widehat{\varphi}(\xi)|^2 = g_N(\xi/2)\, g_N(\xi/4) \cdots$ is uniquely defined.

It is easily proved that

$$m_0(\xi) = \left(\frac{1 + \mathrm{e}^{-\mathrm{i}\xi}}{2}\right)^N \mu_N(\xi)$$

and that

$$\left|\mu_N(\xi)\right|^2 = P_N\left(\sin^2 \xi/2\right)$$

where

$$P_N(x) = \sum_0^{N-1} \binom{N-1+n}{n} x^n$$

is the solution to

$$x^N P_N(1-x) + (1-x)^N P_N(x) = 1.$$

The regularity of the scaling function φ will be measured by the decay of its Fourier transform at infinity. It is proved in DAUBECHIES [1992, p. 225], that

$$|\widehat{\varphi}(\xi)| \leqslant C\left(1 + |\xi|\right)^{-\alpha}, \tag{18.2}$$

where

$$\alpha = \log_2 \frac{1}{|m_0(2\pi/3)|}$$

this estimate being optimal.

The optimality is obvious. Indeed

$$\widehat{\varphi}(2^q\xi) = m_0(2^{q-1}\xi) \cdots m_0(\xi)\widehat{\varphi}(\xi)$$

which implies, if $\xi = \frac{2}{3}\pi$,

$$|\widehat{\varphi}(2^q\xi)| = \left|m_0\left(\frac{2\pi}{3}\right)\right|^q \left|\widehat{\varphi}\left(\frac{2\pi}{3}\right)\right|.$$

Since $m_0(\xi)$ does not vanish on $[-\frac{1}{3}\pi, \frac{1}{3}\pi]$, we have $|\widehat{\varphi}(\frac{2}{3}\pi)| > 0$ and the estimate given by (18.2) is sharp.

It remains to estimate α. It is proved in DAUBECHIES [1992, p. 226], that

$$\frac{1}{\sqrt{3}} N^{-1/4} \left(\frac{\sqrt{3}}{2}\right)^N \leqslant \left|m_0\left(\frac{2\pi}{3}\right)\right| \leqslant \left(\frac{\sqrt{3}}{2}\right)^N$$

which implies $\alpha = N \log_2(2/\sqrt{3}) + \mathrm{O}(\log N)$.

This answers the problem of computing the asymptotic smoothness of φ when N is large. It should be observed that (18.2) implies $\varphi \in C^{\alpha-1}$ and that conversely $\varphi \in C^{\alpha}$ implies (18.2).

Therefore we have

THEOREM 18.1. *For large N, the Daubechies scaling function φ and the corresponding wavelet belong to C^r for $r = N \log_2(2/\sqrt{3}) + \mathrm{O}(\log N)$.*

19. Convergence of the cascade algorithm

The scaling function φ is the unique solution (normalized by $\int \varphi(x)\,\mathrm{d}x = 1$) to the two-scales equation

$$\varphi(x) = 2 \sum_{0}^{2N-1} \alpha_k \varphi(2x + k).$$

One solution for calculating $\varphi(x)$ is to use the Fourier transformation. One obtains (18.1).

However, a more natural approach would have been to follow a canonical fixed point algorithm.

One starts with a compactly supported L^∞ function $f_0(x)$ such that $\int f_0(x)\,\mathrm{d}x = 1$. We would like to know whether the sequence f_j, $j \in \mathbb{N}$, defined by

$$f_{j+1}(x) = 2 \sum_{0}^{2N-1} \alpha_k f_j(2x + k) \tag{19.1}$$

converges to $\varphi(x)$ in $L^2(\mathbb{R})$.

It is obvious that

$$f_j(x) = \sum_k \alpha_{j,k} f_0(2^j x + k) \tag{19.2}$$

and one should compute these coefficients $\alpha_{j,k}$. We follow DURAND [1996] and observe that these $\alpha_{j,k}$ are unique and that $f_0 = \varphi$ implies $f_j = \varphi$ for every $j \in \mathbb{N}$.

Therefore,

$$\varphi(x) = \sum_k \alpha_{j,k}\varphi(2^j x + k)$$

and

$$\alpha_{j,k} = 2^j \int \varphi(y)\varphi(2^j y + k)\,\mathrm{d}y$$

since $2^{j/2}\varphi(2^j x + k),\ k \in \mathbb{Z}$, is an orthonormal basis for V_j.

Finally,

$$f_j(x) = \int K_j(x,y)\varphi(y)\,\mathrm{d}y = T_j(\varphi)(x)$$

where

$$K_j(x,y) = 2^j K(2^j x, 2^j y)$$

and

$$K(x,y) = \sum_k f_0(x+k)\varphi(y+k).$$

Therefore $K(x,y) = 0$ whenever $|x-y| \geqslant T$, $|K(x,y)| \leqslant C$ and $K(x+1, y+1) = K(x,y)$.

There exists a general necessary and sufficient condition implying that, for such kernels $K(x,y)$, $2^j K(2^j x, 2)$ of operators T_j defined by the kernels $2^j K(2^j x, 2^j y)$ would converge to the identity. This condition is $\int K(x,y)\,\mathrm{d}y = 1$ and is equivalent to $\|T_j(f) - f\|_2 \to 0$ $(j \to +\infty)$ for any $f \in L^2(\mathbb{R})$ but also to $\|T_j(f) - f\|_2 \to 0$ for *one* particular function $f \in L^2(\mathbb{R})$ (we except the zero function).

We can conclude the following:

THEOREM 19.1. *The sequence f_j converges in L^2 to φ if and only if f_0 satisfies the Fix and Strang condition, i.e. if $\sum_{-\infty}^{\infty} f_0(x-k) = 1$ identically.*

Let us now assume that both the scaling function and f_0 belong to the ordinary space C_0^m of compactly supported m times continuously differentiable functions. We then would like $f_j(x)$ to converge to φ with respect to the C^m-norm.

It is easily seen that this condition is equivalent to the following one:

$$\int K(x,y)y^q\,\mathrm{d}y = x^q + P_{q-1}(x) \tag{19.3}$$

for $0 \leqslant q \leqslant m$, $P_{q-1}(x)$ being a polynomial of degree less than q.

Since

$$K(x,y)=\sum_k f_0(x+k)\varphi(y+k),$$

this condition is easily seen to be equivalent to

$$\sum_k (x+k)^q f_0(x+k)=Q_{q-1}(x), \tag{19.4}$$

where $Q_{q-1}(x)$ is a polynomial of degree less than q. Finally, (19.4) is equivalent to the generalized Fix and Strang conditions which read

$$\left(\left(\frac{\mathrm{d}}{\mathrm{d}\xi}\right)^q \hat{f}_0\right)(2k\pi)=0 \quad \text{for } 0\leqslant q\leqslant m \text{ and } |k|\geqslant 1. \tag{19.5}$$

The last problem of interest is the speed of convergence of f_j to φ, measured in L^∞-norm when both f_j and φ belong to C^m. The reader is referred to DURAND [1996].

The dual problem to the convergence of the cascade algorithm is the convergence of the pyramidal algorithm. The latter problem was raised in Section 14. The function G plays the role of the function f_0 in such a way that the solution to the open problems in Section 14 is also given by the Fix and Strang conditions.

20. Wavelets on the interval

If we except the case of the Haar system, it is not at all clear how wavelets can be fixed in such a way that they would provide an orthonormal basis for $L^2[0,1]$. We want this orthonormal basis to be an unconditional basis for the spaces $C^s[0,1]$ or $H^s[0,1]$ consisting of restrictions to $[0,1]$ of functions in $C^s(\mathbb{R})$ or $H^s(\mathbb{R})$. For achieving this goal, we want our wavelets to be orthonormal to polynomials of degree $\geqslant N$. The difference with the Haar system is the following. Let us fix an integer $N\geqslant 1$ and consider the corresponding Daubechies wavelet ψ supported by $[0,2N+1]$ with a vanishing integral and N vanishing moments. Let us consider all the wavelets $2^{j/2}\psi(2^jx-k)$, $j\geqslant j_0$, $k\in\mathbb{Z}$, which are supported by a sub-interval of $[0,1]$. Then the closed linear span of these wavelets has an infinite codimension in $L^2[0,1]$.

The problem will be to build the "missing edge wavelets". The theorem we want to prove is the following.

THEOREM 20.1. *There are $N+1$ "edge wavelets" associated with 0 and denoted by $\psi^{(0)},\dots,\psi^{(N)}$, together with $N+1$ "edge wavelets" associated with 1 and denoted by $\widetilde{\psi}^{(0)},\dots,\widetilde{\psi}^{(N)}$ such that, if $2^{j_0}\geqslant 4N+2$,*

the functions $2^{j/2}\psi^{(q)}(2^jx)$, $0\leqslant q\leqslant N$, $j\geqslant j_0$,

together with $2^{j/2}\widetilde{\psi}^{(q)}(2^j(x-1))$, $0\leqslant q\leqslant N$, $j\geqslant j_0$,

and with the inner wavelets $2^{j/2}\psi(2^jx-k)$,

$$1 \leqslant k \leqslant 2^j - (2N+2),\ j \geqslant j_0,\ \textit{are an orthonormal family} \tag{20.1}$$

the closed linear span of this orthogonal family has

a finite codimension in $L^2[0,1]$; (20.2)

the edge wavelets as well as the inner wavelets are

orthogonal in $L^2[0,1]$ *to all polynomials of degree* $\leqslant N$. (20.3)

As usual wavelets are coming from a multiresolution analysis and before constructing the frequency channel W_j one should build V_j.

Before giving the detailed construction for the interval $[0,1]$, we consider the case of the half-line $[0,\infty)$ and of the corresponding Hilbert space $L^2[0,\infty)$. We will take advantage from the invariance by dilation.

We denote by $V_j = V_j(\mathbb{R})$ the multiresolution analysis associated with the Daubechies wavelets. For defining the corresponding spaces $V_j([0,\infty))$ it will suffice to consider the case when $j = 0$ since the general case will then be deduced by an obvious dilation.

Then $V_0([0,\infty))$ is not the space of restrictions to $[0,\infty)$ of functions in V_0. Such a choice would lead to the "inner scaling functions" $\varphi(x-k),\ k \geqslant 1$, together with the restrictions to $[0,\infty)$ of $\varphi(x), \varphi(x+1), \ldots, \varphi(x+2N)$. Instead we keep the same "inner functions" and will add to them $N+1$ more functions in V_0 (instead of $2N+1$). For choosing these $N+1$ functions, it should be noticed that

$$\sum_{n\in\mathbb{Z}} (x-n)^k \varphi(x-n) = m_k, \quad 0 \leqslant k \leqslant N, \tag{20.4}$$

where

$$m_k = \int_0^{2N+1} t^k \varphi(t)\, dt. \tag{20.5}$$

An easy consequence of the identity (20.4) is the identity

$$\sum_{n\in\mathbb{Z}} n^k \varphi(x-n) = p_k(x), \quad 0 \leqslant k \leqslant N, \tag{20.6}$$

where $p_k(x)$ is a polynomial whose leading order term is exactly x^k. We consider $N+1$ "edge scaling functions" on $[0,\infty)$ which are denoted by $\theta_0(x), \ldots, \theta_N(x)$ and defined by

$$\theta_0(x) = \varphi(x) + \varphi(x+1) + \cdots + \varphi(x+2N)$$
$$\theta_2(x) = \varphi(x+1) + \cdots + 2N\varphi(x+2N)$$
$$\vdots$$

$$\theta_k(x) = \varphi(x+k) + \cdots + \binom{l}{k}\varphi(x+l) + \cdots + \binom{2N}{k}\varphi(x+2N) \tag{20.7}$$

$$\vdots$$

$$\theta_N(x) = \varphi(x+N) + \cdots + \binom{2N}{N}\varphi(x+2N).$$

Let us denote by E the vector space consisting in all sums

$$a_0 + a_1 x + \cdots + a_N x^N + \sum_1^\infty \gamma_n \varphi(x-n), \quad \gamma_n = \mathrm{O}(n^N),$$

restricted to $[0,\infty)$. Since

$$a_0 + a_1 x + \cdots + a_N x^N = \sum_{n\in\mathbb{Z}} P(n)\varphi(x-n),$$

where P is a polynomial of degree N, then we also have

$$\begin{aligned} a_0 &+ a_1 x + \cdots + a_N x^N \\ &= \alpha_0\theta_0(x) + \cdots + \alpha_N\theta_N(x) + \sum_{n\geqslant 1} P(n)\varphi(x-n) \end{aligned}$$

on $[0,\infty)$. We have taken advantage of the fact that $\varphi(x+m) = 0$ on $[0,\infty)$ when $m \geqslant 2N-1$.

In other terms, the "edge scaling functions" are the $N+1$ missing entities which complement the "inner scaling functions in order to generate $1, x, \ldots, x^N$.

From the definition of E and the properties of scaling functions, it is obvious that $f(x) \in E$ implies $f(x/2) \in E$.

We then define $V_0^{[0,\infty)}$ by $E \cap L^2(0,\infty)$. Then the collection $\{\theta_0, \theta_1, \ldots, \theta_N, \varphi(x-1), \varphi(x-2), \ldots\}$ is a Riesz basis of $V_0^{[0,\infty)}$. Since the tail of this Riesz basis is already orthonormal, it is an easy job to replace $\theta_0, \theta_1, \ldots, \theta_N$ by $\theta_0^\sharp, \ldots, \theta_N^\sharp$ in order that $\{\theta_0^\sharp, \ldots, \theta_N^\sharp, \varphi(x-1), \varphi(x-2), \ldots\}$ be an orthonormal basis for $V_0^{[0,\infty)}$.

Let us now consider the ladder of spaces $V_j^{[0,\infty)}$ which are obtained from $V_0^{[0,\infty)}$ by dyadic dilations.

In order to construct wavelets on the half line, we define $W_j^{[0,\infty)}$ as the orthogonal complement of $V_j^{[0,\infty)}$ inside $V_{j+1}^{[0,\infty)}$. "Inner wavelets" are defined as being $2^{j/2}\psi(2^j x - k)$, $k \geqslant 1$. Together with these inner wavelets one considers $N+1$ "edge wavelets" $\psi^{(0)} \in W_0^{[0,\infty)}, \ldots, \psi^{(N)} \in W_0^{[0,\infty)}$ which are the orthogonal projections on $W_0^{[0,\infty)}$ of $\theta_0(2x), \ldots, \theta_N(2x)$. These "edge wavelets" are already orthogonal to the "inner wavelets" $\psi(x-1), \psi(x-2), \ldots$ and one checks that the "edge wavelets" together with the "inner wavelets" form a Riesz basis of $W_0^{[0,\infty)}$. It suffices to orthonormalize the edge wavelets to obtain an orthonormal basis for $W_0^{[0,\infty)}$.

Everything which has been done for $[0,\infty)$ is worked out for $(-\infty,0]$ and the corresponding edge scaling functions are denoted by $\eta_0(x),\ldots,\eta_N(x)$.

If $T \geqslant 4N+2$ is an integer, we are ready for defining $V_0^{[0,T]}$. A basis for this space is given by the $N+1$ "edge scaling functions" $\theta_0(x),\ldots,\theta_N(x)$ associated with 0, the $N+1$ "edge scaling functions" $\eta_0(x-T),\ldots,\eta_N(x-T)$ associated with T together with the "inner scaling functions" $\varphi(x-1),\ldots,\varphi(x-T+2N+2)$.

The dimension of $V_0^{[0,T]}$ is therefore T. When $T=2^j$ and $j \geqslant j_0$, one defines $V_j^{[0,1]}$ by an obvious rescaling.

The orthogonal complement of $V_j^{[0,1]}$ inside $V_{j+1}^{[0,1]}$ is $W_j^{[0,1]}$. Then a basis for $W_j^{[0,1]}$ is given by

$$\text{the } N+1 \text{ “edge wavelets” } 2^{j/2}\psi^{(0)}(2^jx),\ldots,2^{j/2}\psi^{(N)}(2^jx) \tag{20.8}$$

which are associated with 0;

$$\begin{gathered}\text{the} \quad N+1 \text{ “edge wavelets”} \\ 2^{j/2}\widetilde{\psi}^{(0)}(2^j(x-1)),\ldots,2^{j/2}\widetilde{\psi}^{(N)}(2^j(x-1))\end{gathered} \tag{20.9}$$

which are associated with 1;

$$\begin{gathered}\text{the } 2^j-(2N+2) \text{ “inner wavelets” } 2^{j/2}\psi(2^jx-k), \\ 1 \leqslant k \leqslant 2^j-(2N+2).\end{gathered} \tag{20.10}$$

Once more it is a trivial task to orthonormalize each bunch of edge wavelets which amounts to replacing $\psi^{(q)}$ by $\Psi^{(q)}$ and $\widetilde{\psi}^{(q)}$ by $\widetilde{\Psi}^{(q)}$.

Theorem 20.1 is therefore proved.

In most of the applications, one deals with the discrete analogues of the above algorithms. Then one identifies $V_j^{[0,1]}$ to $l^2\{0,2^{-j},\ldots,(2^j-1)2^{-j}\}$. This identification associates the point $(k+N)/2^j$ with the inner scaling function $2^{j/2}\varphi(2^jx-k)$. The points $0,2^{-j},\ldots,N2^{-j}$ are reserved to the edge scaling functions associated with 0 while the points $(2^j-N-1)2^{-j},\ldots,(2^j-1)2^{-j}$ are associated with 1.

Once this identification is performed, we identify $W_j^{[0,1]}$ with $l^2\{0,2^{-j},\ldots,(2^j-1)2^{-j}\}$.

Then the orthonormal decomposition $V_{j+1}^{[0,1]} = V_j^{[0,1]} \oplus W_j^{[0,1]}$ becomes a unitary mapping between $l^2(F_{j+1})$ and $l^2(F_j)\oplus l^2(F_j)$ where $F_j=\{0,2^{-j},\ldots,(2^j-1)2^{-j}\}$.

One might like to have a direct access to this unitary mapping. It is a simple modification on quadrature mirror filters. Indeed if we except what is happening on the end points $0,2^{-j},\ldots,N2^{-j}$ or $(2^j-n-1)2^{-j},\ldots,(2^j-1)2^{-j}$, the filters to be used are the same which would be used on the line. The coefficients of the end points are treated with distinct coefficients which do not depend on the scale (2^{-j}). The reader is referred to COHEN [1993].

21. n-dimensional wavelets

In order to construct n-dimensional wavelets from the one-dimensional case, the simplest way is the tensor product method. This method would not work without a scaling function φ associated with the wavelet ψ. For the sake of simplicity, let us stick to the 2-dimensional case. We obtain the following theorem

THEOREM 21.1. *Let us define* $\psi^{(1)}, \psi^{(2)}$ *and* $\psi^{(3)}$ *in* $L^2(\mathbb{R}^2)$ *by*

$$\psi^{(1)}(x, y) = \psi(x)\varphi(y), \qquad \psi^{(2)}(x, y) = \varphi(x)\psi(y),$$
$$\psi^{(3)}(x, y) = \psi(x)\psi(y),$$

where φ *and* ψ *are respectively the Daubechies scaling function and the Daubechies wavelet (the index* N *being dropped). Then the family* $\{2^j\psi^{(1)}(2^jx - k, 2^jy - l),\ 2^j\psi^{(2)}(2^jx - k, 2^jy - l),\ 2^j\psi^{(3)}(2^jx - k, 2^jy - l);\ j \in \mathbb{Z},\ k \in \mathbb{Z},\ l \in \mathbb{Z}\}$ *is an orthonormal basis for* $L^2(\mathbb{R}^2)$.

The proof runs as follows. We denote by V_j, $j \in \mathbb{Z}$, the multiresolution analysis which generates φ and ψ. We consider the family $\varphi(x - k)\varphi(y - l)$, $k \in \mathbb{Z}$, $l \in \mathbb{Z}$, which is an orthonormal sequence in $L^2(\mathbb{R}^2)$ and denote by $\mathcal{V}_0$ the closed linear span of this family. Then it is easily checked that the corresponding sequence $\mathcal{V}_j$ defined by $\mathcal{V}_j = \overline{V_j \otimes V_j}$ is a multiresolution analysis for $L^2(\mathbb{R}^2)$ ($\mathcal{V}_j$ is the closure of the algebraic tensor product).

Finally $V_{j+1} = V_j \oplus W_j$ implies

$$\begin{aligned}\mathcal{V}_{j+1} &= (V_j \oplus W_j) \otimes (V_j \oplus W_j)\\ &= [V_j \otimes V_j] \oplus [V_j \otimes W_j] \oplus [W_j \otimes V_j] \oplus [W_j \otimes W_j].\end{aligned}$$

We have reached the three closed subspaces which form the orthogonal complement of $\mathcal{V}_j$ inside $\mathcal{V}_{j+1}$ and the theorem is proved.

A similar construction can be worked out on the unit square $[0, 1] \times [0, 1]$. One replaces V_j by the corresponding multiresolution analysis adapted to the interval and the details are left to the reader.

22. Where do wavelets come from?

P.G. Lemarié-Rieusset proved the following fundamental theorem.

THEOREM 22.1. *Let* ψ *be a compactly supported function in* $C^r(\mathbb{R})$, $r > 0$. *Let us assume that* $2^{j/2}\psi(2^jx - k)$, $j \in \mathbb{Z}$, $k \in \mathbb{Z}$, *is an orthonormal basis for* $L^2(\mathbb{R})$.

Then there exists a compactly supported function φ *in* $C^r(\mathbb{R})$ *with the following properties:*

$$\varphi(x - k),\ k \in \mathbb{Z},\ \text{is an orthonormal sequence;} \tag{22.1}$$

if V_j *denotes the closed linear span of the sequence*

$2^{j/2}\varphi(2^jx-k)$, $k\in\mathbb{Z}$, then $(V_j)_{j\in\mathbb{Z}}$

is a multiresolution analysis for $L^2(\mathbb{R})$; (22.2)

if W_j denotes the orthogonal complement of V_j

inside V_{j+1}, then $2^{j/2}\psi(2^jx-k)$, $k\in\mathbb{Z}$,

is an orthonormal basis for W_j. (22.3)

Extensions of this result to more general settings (ψ does not have a compact support but satisfies some decay conditions at infinity or the 2D case) can be found in AUSCHER [1995].

Theorem 22.1 means that in the one-dimensional case, wavelet orthonormal bases are always constructed from a multiresolution analysis. This is not a purely algebraic fact (it is not true if the only hypothesis on ψ is $\psi\in L^2(\mathbb{R})$ and $2^{j/2}\psi(2^jx-k)$, $j\in\mathbb{Z}$, $k\in\mathbb{Z}$, is an orthonormal basis for $L^2(\mathbb{R})$).

It is not true in two dimensions (MEYER [1990, p. 301], and AUSCHER [1995]).

23. Wavelets provide unconditional bases for most of the functional spaces

Let E be a Banach space of functions defined on $\mathbb{R}^n$. It means that $\mathcal{S}(\mathbb{R}^n)\subset E\subset\mathcal{S}'(\mathbb{R}^n)$ and that the two embeddings are continuous ones.

A Schauder basis is a sequence $e_0,e_1,\dots,e_n,\dots$ of elements of E such that every $x\in E$ admits a unique expansion as a series $\sum_0^\infty \alpha_k e_k$ such that

$$\left\|x-\sum_0^m \alpha_k e_k\right\|\to 0\quad (m\to+\infty).$$

A Schauder basis is an unconditional basis if there exists a constant C such that for any $m\geqslant 0$, any two sequences $(\alpha_0,\dots,\alpha_m)$ and $(\beta_0,\dots,\beta_m)$ the conditions $|\alpha_0|\leqslant|\beta_0|,\dots,|\alpha_k|\leqslant|\beta_k|,\dots,|\alpha_m|\leqslant|\beta_m|$ imply

$$\left\|\sum_0^m \alpha_k e_k\right\|\leqslant C\left\|\sum_0^m \beta_k e_k\right\|.$$

Let us give a second definition. If E has a Schauder basis, it means that E can be identified to a space of sequences $(\alpha_k)_{k\in\mathbb{N}}$. But every sequence does not correspond to a vector $x\in E$. We still need a criterion for deciding upon a given sequence if it originates from a vector $x\in E$ or not.

If E has an unconditional basis, it means that this criterion is a size condition on the coefficients α_k, $k\in\mathbb{N}$. Let us consider an example. Let E be the Banach space $L^p[0,2\pi]$ for a given $p\in(1,\infty)$. Consider the trigonometric system $(1/\sqrt{2\pi})\,\mathrm{e}^{\mathrm{i}kx}$, $k=0,\pm1,\pm2,\pm3,\dots$ which is an orthonormal basis for $L^2(-\pi,\pi]$.

Then $(1/\sqrt{2\pi})\,\mathrm{e}^{ikx}$, $k \in \mathbb{Z}$, is still a Schauder basis for $L^p[0, 2\pi]$ when $1 < p < \infty$. It means that

$$\lim_{m \to +\infty} \left\| f(x) - \sum_{-m}^{m} c_k e^{ikx} \right\|_p = 0,$$

where c_k are the Fourier coefficients of $f(x)$.

However, this basis is never an unconditional basis for $L^p[0, 2\pi]$ if we except the case $p = 2$.

It is interesting to observe that this problem of constructing Schauder bases or unconditional bases has paved the way to wavelet bases. Indeed A. Haar built the so-called Haar basis for constructing a Schauder basis for the space $\mathcal{C}[0, 1]$ of continuous functions on the interval $[0, 1]$. Indeed it was proved by Dubois-Reymond that the trigonometric system fails to have this property.

In 1938, Marcinkiewicz proved that the Haar system is an unconditional basis for $L^p[0, 1]$ when $1 < p < \infty$. It means that the Haar basis suffices to our needs as long as we do not need to differentiate or integrate.

A wavelet basis (1) shares with the Haar system the above mentioned properties, (2) keeps these properties after differentiation or integration. It implies that a wavelet basis is a universal unconditional basis for most of the functional spaces if we except these spaces which are constructed from L^1 or L^∞.

These considerations seem to belong to the abstract realm of pure mathematics. But D. Donoho proved that this discussion is crucial for optimal denoising of images or signals. Let us explain why.

Let us consider those geometrical images which can be modelled by smooth functions on some finite domains with smooth boundaries with jump discontinuities on the boundaries. We define such an image by the corresponding values of the gray level on a "fine grid". We assume that this image is delimited by the unit square $0 \leqslant x, y \leqslant 1$ and that the "fine grid" is defined by $x = k/N$, $y = l/N$, $0 \leqslant k \leqslant N$, $0 \leqslant l < N$.

This image is corrupted by a white noise with variance σ^2. This white noise in its discrete version is given by $\sigma g_{k,l}(\omega)$, $0 \leqslant k < N$, $0 \leqslant l < N$, where $g_{k,l}(\omega)$ are i.i.d. normal Gaussian variables.

The goal of denoising is for a given signal, to find an optimal orthogonal transformation which improves the conditioning of our image. This means that a few coefficients should stay above the zero level and that a reconstruction based upon these coefficients would have an optimal quality.

If such an image is corrupted by white noise, this white noise will be spread over all the coefficients (but the amount of noise will be small for each coefficient). The denoising will be achieved through a thresholding (this is a crude option and a more sophisticated one will be given in a moment). Then, at the same time, the noise and the smallest coefficients are wiped out. We are left with a few large coefficients which will provide an optimal recovery.

The difficulty consists in finding this optimal orthogonal transformation. Options are given in (MEYER [1992, 1993]) and range from pure Fourier analysis to wavelet packets or other sophisticated windowed Fourier analysis.

Let us return to the special case of geometrical images without textures and let us compare Fourier techniques to standard orthonormal wavelets.

The Fourier coefficients are defined by

$$c(p,q) = \frac{1}{N^2}\sum_{0}^{N-1}\sum_{0}^{N-1} f\left(\frac{k}{N},\frac{l}{N}\right)\exp 2\pi\mathrm{i}\left(\frac{pk}{N}+\frac{ql}{N}\right), \tag{23.1}$$

where $0 \leqslant p \leqslant N-1$, $0 \leqslant q \leqslant N-1$.

If N is even, we may as well suppose that $-N/2 \leqslant p \leqslant N/2-1$ and $-N/2 \leqslant q \leqslant N/2-1$ which will be our option. Let us observe that the operator defined by $f \mapsto c(p,q)$ becomes unitary if

$$\|f\|_2 = \left(\sum_{0}^{N-1}\sum_{0}^{N-1} N^{-2}\left|f\left(\frac{k}{N},\frac{l}{N}\right)\right|^2\right)^{1/2}$$

and

$$\|c\|_2 = \left(\sum_{p}\sum_{q}\left|c(p,q)\right|^2\right)^{1/2}.$$

If the values of f are corrupted by a white noise $\sigma g_{k,l}(\omega)$, then the corresponding Fourier coefficients $c(p,q)$ are corrupted by $(\sigma/N)\hat{g}_{p,q}(\omega)$ where $\hat{g}_{p,q}$ denote other i.i.d. normal Gaussian random variables.

On the other hand the Fourier coefficients of our simple geometrical images are $\mathrm{O}((|p|+|q|)^{-3/2})$. If one tries to wipe out the white noise by a thresholding at the level $2(\sigma/N)\sqrt{\log N}$, the error (measured with the L^2-norm) will be of the order of magnitude of $(\sigma/N)^{1/3}(\log N)^{1/6}$. Let us use N^2 orthonormal wavelets instead. Let us forget the few scaling functions which are needed to construct the orthonormal basis of $L^2([0,1]\times[0,1])$.

Then there are two cases. If the wavelet does not hit the boundary of one of the regions, then what is measured is the scalar product between the wavelet and the white noise. The result is $\mathrm{O}((\sigma/N)\sqrt{\log N})$.

On the other hand, if the wavelet hits the boundary of one region, the scalar product between the wavelet and the corresponding indicator function will be large (of the order of magnitude of $c2^{-j}$ where c is some positive constant). This $c2^{-j}$ is corrupted by white noise which is $\mathrm{O}((\sigma/N)\sqrt{\log N})$ but it means that most of the coefficients are much above the noise level.

The further discussion depends upon the value of σ. It is clear that the total noise energy cannot exceed the corresponding signal energy. We therefore assume that σ is much smaller than c. Then the error, as measured with the L^2 norm, is $\mathrm{O}((\sigma/N^{1/2})(\log N)^{1/2})$. It is a striking improvement upon a Fourier based method.

D. Donoho carefully designed a (near) optimal recovery algorithm. All the wavelet coefficients of an image which has been corrupted by white noise are shifted towards

0 by an amount of $2(\sigma/N)\sqrt{\log N}$ (and every coefficient with an absolute value less than $2(\sigma/N)\sqrt{\log N}$ is replaced by 0). This denoising technique is the one which we used and which beats the Fourier method. It is called "wavelet shrinkage" and depends upon the fact that one does not alter a function too much in some Besov space by diminishing the wavelet coefficients.

This denoising algorithm also relies upon the fact that within our special class of images, the wavelet coefficients are well-conditioned (either they are "rather large" or "rather small") as opposed to the Fourier coefficients which are ill-conditioned.

If the interesting information contained in our image concerned its texture, it is clear that the wavelet approach would not work. A texture is generally ill-conditioned in any wavelet basis.

At the proof level, our "natural images without textures" are becoming functions of two variables which are bounded in suitably designed Besov spaces and one uses in a crucial way the property of wavelets to yield unconditional bases for such Besov spaces.

24. Definition of singular integral operators

As it was stressed in the introduction, there are several classes of Calderón–Zygmund singular integral operators. The first one consists in convolution operators with kernels of the type $S(x) = \text{p.v.}\, \Omega(x)/|x|^n$ where $\Omega(x)$ is homogeneous of degree 0 in $\mathbb{R}^n \setminus \{0\}$, smooth on the unit sphere and a vanishing integral on the unit sphere.

In the second class, we consider convolution operators with a larger class of kernels $S(x)$. These kernels will satisfy the following estimates:

$$|S(x)| \leqslant C_0|x|^{-n}, \qquad |S(x') - S(x)| \leqslant C_1|x' - x|^{\gamma}|x - y|^{-n-\gamma} \tag{24.1}$$

whenever $|x' - x| \leqslant \frac{1}{2}|x - y|$. Here C_0 and C_1 are two constants and $\gamma \in (0, 1)$ is a fixed exponent.

Then the necessary and sufficient condition for the L^2-continuity is the existence of a constant C such that for any $\varepsilon > 0$ and any $R > \varepsilon$, one has

$$\left| \int_{\varepsilon \leqslant |x| \leqslant R} S(x)\, \mathrm{d}x \right| \leqslant C. \tag{24.2}$$

The third class will be unveiled later on and the last class is the broadest. It consists in operators T which cannot be defined by a singular integral $\int K(x, y) f(y)\, \mathrm{d}y$ since this integral cannot be given an a priori meaning.

In order to avoid this problem of giving a meaning to this singular integral, one treats the local and the global behavior of $Tf(x)$ separately.

(A). *Local information*: one has $Tf(x) = \int K(x, y) f(y)\, \mathrm{d}y$ for all x which do not belong to the support of f and the kernel $K(x, y)$ satisfies the following estimates:

$$|K(x, y)| \leqslant C_0|x - y|^{-n}, \tag{24.3}$$

$$\left|K(x',y)-K(x,y)\right| \leqslant C_1|x'-x|^{\gamma}|x-y|^{-n-\gamma} \tag{24.4}$$

whenever $|x'-x| \leqslant \frac{1}{2}|x-y|$ and

$$\left|K(x,y')-K(x,y)\right| \leqslant C_1|y'-y|^{\gamma}|x-y|^{-n-\gamma} \tag{24.5}$$

whenever $|y'-y| \leqslant \frac{1}{2}|x-y|$ (C_0 and C_1 are two constants and $\gamma \in (0,1]$ is a fixed exponent).

(B). *Global information*: one has $(Tf,g) = (S, g \otimes f)$ where $S \in \mathcal{D}'(\mathbb{R}^n \times \mathbb{R}^n)$ is a given distribution and $(g \otimes f)(x,y) = g(x)f(y)$.

What is aimed to is to find a necessary and sufficient condition on this distribution $S(x,y)$ for the L^2-continuity of T.

Let us observe that T is fully determined by S and that the restriction of the distribution $S(x,y)$ to the open subset of $\mathbb{R}^n \times \mathbb{R}^n$ defined by $y \neq x$ is precisely $K(x,y)$.

The two main problems concerning singular integral operators and wavelets are (1) the *comparison issue* and (2) the *relation between the operator norm of* $T: L^2(\mathbb{R}^n) \to L^2(\mathbb{R}^n)$ *and the matrix of* T *in the given orthonormal wavelet basis.*

Let us give some examples of singular integral examples. Inside the first class, one finds the celebrated Hilbert transform (in the one dimensional case) and the Riesz transforms R_j, $1 \leqslant j \leqslant n$. The Riesz transforms are defined by $R_j = \partial_j(-\Delta)^{-1/2}$ where $\partial_j = -\mathrm{i}\partial/\partial x_j$. The distributional kernel of R_j is c_n p.v. $(x_j - y_j)/|x-y|^{n+1}$ which obviously satisfies (24.4) and (24.5).

Among the second class, one finds the well-known Marcinkiewicz operator $M_\gamma = (-\Delta)^{\mathrm{i}\gamma/2}$, $\gamma \in \mathbb{R}$. It is a convolution operator with a distribution S_γ. The restriction of S_γ to $x \neq 0$ is given by $c(n,\gamma)|x|^{-n-\mathrm{i}\gamma}$. Conditions (24.4) and (24.5) are clearly satisfied.

The third class is illustrated by the double layer potential. The distributional kernel of this operator is

$$c_n \text{ p.v. } \frac{A(x)-A(y)-(x-y)\bullet\nabla A(y)}{[|x-y|^2+(A(x)-A(y))^2]^{(n+1)/2}},$$

where $x \in \mathbb{R}^n$, $y \in \mathbb{R}^n$ and $A: \mathbb{R}^n \to \mathbb{R}$ is a Lipschitz function.

Another example of an operator belonging to this third class is the pointwise multiplication with an L^∞ function $m(x)$.

From this example, one observes a striking difference between operators of the third class and operators in the first one.

Let us return to the Hilbert transform. If $(\psi_{j,k})$ is an orthonormal wavelet basis, we have $H(\psi_{j,k}) = \widetilde{\psi}_{j,k}$ where $\widetilde{\psi} = H(\psi)$ and $\widetilde{\psi}_{j,k}$ is an other orthonormal wavelet basis. This is due to the three properties characterizing the Hilbert transformation H: it commutes with translations, with positive dilation and H is a unitary operator.

If we consider the case of the Riesz transformations R_j, $1 \leqslant j \leqslant n$, we still have information which is reminiscent of the case of the Hilbert transform. We need a definition.

DEFINITION. A *smooth molecule* is a function $f(x)$ enjoying the following properties:

$$|f(x)| \leqslant C_0\big(1+|x|\big)^{-n-\alpha} \tag{24.6}$$

for some positive exponent α and some constant C_0,

$$\big|f(x')-f(x)\big| \leqslant C_1|x'-x|^{\beta} \tag{24.7}$$

for some positive exponent β and some constant C_1,

$$\int_{\mathbb{R}^N} f(x)\,\mathrm{d}x = 0. \tag{24.8}$$

If $f(x)$ is a smooth molecule, so are $R_1(f), \ldots, R_n(f)$ but $m(x)f(x)$ is no longer a smooth molecule. This observation motivates the following definition.

DEFINITION. A Calderón–Zygmund operator T is a *pseudo-convolution* if, for any smooth molecule $f(x)$, both $Tf(x)$ and $T^*f(x)$ have a vanishing integral.

We would like to stress that operators in the first and the second class are pseudo-convolution.

Here is an example of a pseudo-convolution which is not a true convolution operator. We consider an orthonormal wavelet basis and define an operator by $T(\psi_{j,k}) = \lambda_{j,k}\,\psi_{j,k}$ where the scalars $\lambda_{j,k}$ satisfy $|\lambda_{j,k}| \leqslant 1$.

Then it is easy to observe that T is a pseudo-convolution but T does not commute with translations (not even with integral translations).

The first main theorem concerning pseudo-convolution is the following statement.

THEOREM 24.1. *If T is a pseudo-convolution and if $f(x)$ is a smooth molecule, then $T(f)$ is also a smooth molecule.*

The converse statement is true if some invariance under the affine group action is imposed.

Let us define the operators R_b, $b \in \mathbb{R}^n$, and U_a, $a > 0$, by $(R_bf)(x) = f(x-b)$ and $U_af(x) = a^{-n/2}f(x/a)$. Let us now assume that for every smooth molecule $f(x)$ the family $g_{a,b}(x) = (R_b^{-1}U_a^{-1}TU_aR_b)[f](x)$ satisfies (24.6), (24.7) and (24.8) uniformly with respect to a and b.

Then T is a pseudo-convolution.

Another important characterization of pseudo-convolutions is the fact that the corresponding matrices in an orthonormal wavelet basis are "almost diagonal". We have to give a precise meaning to "almost diagonal" and this meaning is better understood if a distance is introduced on (dyadic) intervals. If $\lambda = [k2^{-j}, (k+1)2^{-j}) \in \Lambda$ where

Λ is the collection of all dyadic intervals, we will label by λ the wavelet $\psi_{j,k}$. Then the condition on the entries of T in some orthonormal wavelet basis will be a decay property for $|\langle T\psi_\lambda, \psi_{\lambda'}\rangle|$ in terms of the distance between λ and λ'.

Let us begin with defining a distance $d(I, I')$ between two arbitrary intervals $I = [a, b]$, $I' = [a', b']$. This distance will be invariant through simultaneous translations (I and I' are replaced by $I + x_0$ and $I' + x_0$) and through simultaneous dilations (I and I' are replaced by $\lambda I = [\lambda a, \lambda b]$ and $\lambda I' = [\lambda a', \lambda b']$, $\lambda > 0$).

The first option for defining $d(I, I')$ is with associate with each $I = [a, b]$, the complex number $z_I = x + iy$ where $x = (a + b)/2$ and $y = (b - a)/2$. Then $d(I, I')$ will be the hyperbolic instance between z_I and $z_{I'}$.

This distance can be easily computed. Keeping the above notations there exist two positive constants $C' > C > 0$ such that $C \log(y'/y) \leqslant d(I, I') \leqslant C' \log(y'/y)$ if $0 < y \leqslant y'$ and if $|x - x'| \leqslant y + y'$.

If $|x - x'| > y + y'$, one has

$$C \log \frac{|x - x'| + y'}{y} \leqslant d(I, I') \leqslant C' \log \frac{|x - x'| + y'}{y}.$$

A second approach to the distance is now given. For any integer $m \geqslant 1$, let us denote by mI the interval $[x - my, x + my]$ when $I = [x - y, x + y]$. Then if I and I' are any two intervals, we define

$$m(I, I') = \inf\{m \geqslant 1\colon\ mI \supset I' \text{ and } mI' \supset I\}.$$

Finally, we have, for two positive constants C and C',

$$C \log m(I, I') \leqslant d(I, I') \leqslant C' \log m(I, I').$$

If I is any interval, the condition $d(I, I') \leqslant C_0$ (for some constant C_0) means that I and I' have comparable lengths and that the ordinary distance between I and I' is bounded by some fixed multiple of that length.

If $\psi_{(j,k)}$ is an orthonormal wavelet basis, we associate with the pair $\lambda = (j, k)$ the corresponding dyadic interval

$$I = I(j, k) = [k2^{-j}, (k + 1)2^{-j}).$$

We then decide that the distance $d(\lambda, \lambda')$ is given by $d(I, I')$.

We now arrive at the characterization of those Calderón–Zygmund operators which are pseudo-convolutions.

Let us now assume that the mother wavelet $\psi(x)$ is smooth, compactly supported and has a large number of vanishing moments. The required smoothness and number of vanishing moments should exceed the exponent γ which appears in (24.4) and (24.5).

One might argue that γ is always less than one in (24.4) and (24.5). Indeed if $\gamma > 1$, $\gamma = m + r$ where $0 < r \leqslant 1$ and $m \in \mathbb{N}$ and (24.4) should be replaced by

$$\left|\partial_x^\alpha K(x', y) - \partial_x^\alpha K(x, y)\right| \leqslant C|x' - x|^r |x - y|^{-n-\gamma}$$

whenever

$$|x' - x| \leqslant \tfrac{1}{2}|x - y|$$

and a similar condition in y.

We then have

THEOREM 24.2. *Let T be a pseudo-convolution. Then if $\lambda = (j, k)$, $\lambda' = (j', k')$, $x = \left(k + \frac{1}{2}\right)2^{-j}$, $x' = \left(k' + \frac{1}{2}\right)2^{-j'}$, $l = 2^{-j}$, $l' = 2^{-j'}$ and $0 < l \leqslant l'$, we have*

$$\left|\langle T\psi_\lambda, \psi_{\lambda'}\rangle\right| \leqslant C\left(\frac{l}{l'}\right)^{(1/2+\gamma)} \left(\frac{l'}{|x - x'| + l'}\right)^{1+\gamma}. \tag{24.9}$$

Conversely, any matrix $M = (m_{\lambda,\lambda'})$ with entries satisfying (24.9) *is the matrix of a pseudo-convolution T with a kernel $K(x, y)$ satisfying* (24.4) *and* (24.5) *but where γ should be replaced by any $\gamma' < \gamma$.*

In contrast with this statement, general Calderón–Zygmund operators T cannot be characterized by size estimates on the corresponding entries $\langle T\psi_\lambda, \psi_{\lambda'}\rangle$. More precisely, we have

THEOREM 24.3. *Let T be a Calderón–Zygmund operator. Let us assume that every operator L satisfying*

$$\left|\langle L\psi_\lambda, \psi_{\lambda'}\rangle\right| \leqslant \left|\langle T\psi_\lambda, \psi_{\lambda'}\rangle\right| \tag{24.10}$$

is a Calderón–Zygmund operator.

Then T is a pseudo-convolution.

In other terms, if one excepts the special case of pseudo-convolutions, general Calderón–Zygmund operators cannot be characterized by size conditions on the entries $\langle T\psi_\lambda, \psi_{\lambda'}\rangle$.

Another and related problem is the approximation of T given by the following procedure. For $m \geqslant 1$, T_m is defined by $\langle T_m\psi_\lambda, \psi_{\lambda'}\rangle = \langle T\psi_\lambda, \psi_{\lambda'}\rangle$ if $d(\lambda, \lambda') \leqslant m$ and $\langle T_m\psi_\lambda, \psi_{\lambda'}\rangle = 0$ if $d(\lambda, \lambda') > m$.

We want to know if $\|T - T_m\| \to 0$ as $m \to +\infty$ where $\| \bullet \|$ denotes the operator norm. This is certainly true if T is a pseudo-convolution but is not the case if T is a general Calderón–Zygmund operator.

These remarks show that the standard matrix representation (as given by $\langle T\psi_\lambda, \psi_{\lambda'}\rangle$) is inadequate for analyzing general Calderón–Zygmund operators.

25. The non-standard representation of an operator

As was indicated above, it is a dead end to analyze general Calderón–Zygmund operators using orthonormal wavelet bases. However, this difficulty disappears if one gives up the operator and instead considers the distributional kernel $S(x,y)$ of T. The idea is to analyze $S(x,y)$ as if it were a function of $(x,y) \in \mathbb{R}^n \times \mathbb{R}^n$.

In order to keep the notation as simple as possible we consider the one-dimensional case.

Let us assume that $(V_j)_{j\in\mathbb{Z}}$ is a multiresolution analysis of $L^2(\mathbb{R})$ where the corresponding scaling function $\varphi(x)$ is compactly supported and belongs to $C^r(\mathbb{R})$.

Let us denote by $P_j : L^2(\mathbb{R}) \to V_j$ the orthonormal projection and write $Q_j = P_{j+1} - P_j$. Then the non-standard analysis of T is given by

$$T = \sum_{-\infty}^{\infty} \left(P_{j+1}TP_{j+1} - P_jTP_j\right), \tag{25.1}$$

a telescopic series which converges to T under general conditions we would not like to make more precise at this level.

Furthermore, $P_{j+1}TP_{j+1} - P_jTP_j = Q_jTP_j + P_jTQ_j + Q_jTQ_j$ which leads to

$$T = A + B + C, \tag{25.2}$$

where

$$A = \sum_{-\infty}^{\infty} Q_jTP_j, \tag{25.3}$$

$$B = \sum_{-\infty}^{\infty} P_jTQ_j, \tag{25.4}$$

and

$$C = \sum_{-\infty}^{\infty} Q_jTQ_j. \tag{25.5}$$

A simple and useful observation concerns the kernels $A_j(x,y)$ of Q_jTP_j, $B_j(x,y)$ of P_jTQ_j and $C_j(x,y)$ of Q_jTQ_j. Indeed

$$A_j(x,y) = \sum_k \sum_l \alpha_{j,k,l}\psi_{j,l}(x)\overline{\varphi_{j,k}(y)}, \tag{25.6}$$

where

$$\alpha_{j,k,l} = \langle T\varphi_{j,k}, \psi_{j,l}\rangle = \langle S, \psi_{j,l} \otimes \varphi_{j,k}\rangle$$

$$B_j(x,y) = \sum_k \sum_l \beta_{j,k,l}\varphi_{j,l}(x)\overline{\psi_{j,k}(y)} \tag{25.7}$$

where

$$\beta_{j,k,l} = \langle T\psi_{j,k}, \varphi_{j,l}\rangle = \langle S, \varphi_{j,l} \otimes \psi_{j,k}\rangle$$

and finally

$$C_j(x,y) = \sum_k \sum_l \gamma_{j,k,l}\psi_{j,l}(x)\overline{\psi_{j,k}(y)}, \tag{25.8}$$

where

$$\gamma_{j,k,l} = \langle T\psi_{j,k}, \psi_{j,l}\rangle = \langle S, \psi_{j,l} \otimes \psi_{j,k}\rangle.$$

It is clear that

$$S(x,y) = \sum_{-\infty}^{\infty} A_j(x,y) + \sum_{-\infty}^{\infty} B_j(x,y) + \sum_{-\infty}^{\infty} C_j(x,y)$$

is a 2D wavelet decomposition of the function $S(x,y)$. Up to now, everything is formal and we do not care about convergence. This attitude could be incorrect since there are situations where this convergence issue exists. Let us assume for instance that $S(x,y) = 1$ identically. The corresponding operator T maps L^1 into constant functions (inside L^∞) by $T(f) = \int_{-\infty}^{\infty} f(x)\,\mathrm{d}x$. This operator is identified to 0 by the above algorithm. What is great in the so-called non-standard algorithm (defined by (25.3), (25.4) and (25.5)) is that interactions between distinct scales seem to have disappeared. That cannot be the case and existing couplings between distinct scales should be hidden somewhere. Indeed each function $\varphi_{j,k}$ can be further decomposed into larger scales wavelets $\psi_{j',k'}$ where $j' \leqslant j-1$.

However, the non-standard algorithm is adapted to general Calderón–Zygmund operators which can be characterized in terms of the entries $\alpha_{j,k,l}$, $\beta_{j,k,l}$, $\gamma_{j,k,l}$.

26. Characterization of general Calderón–Zygmund operators

One more definition is needed.

DEFINITION. A continuous linear operator $T : \mathcal{D}(\mathbb{R}) \to \mathcal{D}'(\mathbb{R})$ has the *weak boundedness property* if there exists a constant C such that, for each interval I and each pair (u, v) of two functions in $\mathcal{D}(\mathbb{R})$ supported by I, one has

$$\bigl|\langle T(u), v\rangle\bigr| \leqslant C|I|^3\|u'\|_\infty\|v'\|_\infty \tag{26.1}$$

(u' and v' are the derivatives of u and v).

If T happens to be bounded on $L^2(\mathbb{R})$, this is satisfied since $\|u\|_2 \leqslant |I|^{3/2}\|u'\|_\infty$ and this explains the normalization factor $|I|^3$ in (26.1).

The first theorem characterizes the operators fulfilling (24.3), (24.4), (24.5) and (26.1).

THEOREM 26.1. *A continuous linear operator satisfies* (24.3), (24.4), (24.5) *and* (26.1) *if and only if the corresponding entries* $\alpha_{j,k,l}$, $\beta_{j,k,l}$ *and* $\gamma_{j,k,l}$ *satisfy*

$$|\alpha_{j,k,l}| + |\beta_{j,k,l}| + |\gamma_{j,k,l}| \leqslant C\big(1 + |k - l|\big)^{-1-\gamma}. \tag{26.2}$$

Here we only need wavelets of class C^1 (the Haar system does not suffice).

However, this theorem can be further generalized. If $\gamma > 1$, we write $\gamma = m + \rho$, $0 < \rho \leqslant 1$, and generalize (24.4) by imposing the condition

$$\begin{aligned} &\big|\partial_x^m K(x', y) - \partial_x^m K(x, y)\big| \leqslant C|x' - x|^\rho |x - y|^{-1-\gamma} \\ &\quad \text{if } |x - x'| \leqslant \tfrac{1}{2}|x - y|. \end{aligned} \tag{26.3}$$

A similar condition will replace (24.5).

If we have at our disposal wavelets of class C^{m+1} (with compact support), then the theorem can be generalized in the obvious way. Conditions (26.2) characterize those operators which satisfy (26.1), (26.3) and the symmetric condition.

The second theorem characterizes these operators T which are bounded on $L^2(\mathbb{R})$.

A dyadic interval $[k2^{-j}, (k+1)2^{-j}[$ will be denoted by $I(j,k)$, $j \in \mathbb{Z}$, $k \in \mathbb{Z}$ and a sequence $\alpha(j,k)$ satisfies the Carleson condition if there exists a constant C such that for every dyadic interval I one has

$$\sum_{\{I(j,k)\subset I\}} |\alpha(j,k)|^2 \leqslant C|I|. \tag{26.4}$$

Then the characterization of general Calderón–Zygmund operators is given by the following condition.

THEOREM 26.2. *Keeping the same notations as in Theorem* 16.1, T *is a general Calderón–Zygmund operator if and only if* (26.2) *is satisfied and moreover,*

$$\tilde{\alpha}_{j,l} = 2^{-j/2}\sum_k \alpha_{j,k,l} \quad \textit{and} \quad \tilde{\beta}_{j,k} = 2^{-j/2}\sum_l \beta_{j,k,l}$$

both satisfy the Carleson condition (26.4).

An example is given by $\alpha_{(j,k,k)} = 1$ and $\alpha_{(j,k,l)} = 0$ for $|k - l| \geqslant 1$. Then $\tilde{\alpha}_{j,l} = 2^{-j/2}$ and for each j,

$$\sum_{\{I(j,k)\subset I\}} |\tilde{\alpha}_{j,l}|^2 = |I|.$$

Then the sum over j is infinite and the corresponding operator is unbounded. If $\alpha_{(j,k,k)} = 1$ for $j_0 \leqslant j \leqslant j_1$, $k \in \mathbb{Z}$, and $\alpha_{(j,k,l)} = 0$ otherwise, then the operator norm of the corresponding operator A will be of the order of magnitude of $\sqrt{j_1 - j_0}$, a remark which will be important in the last section.

A generalization of this example leads to a class of operators which is so remarkable that an entire section will be devoted to their properties.

27. Paraproducts

Let us assume that $\psi_{j,k}(x)$ is the orthogonal basis of $L^2(\mathbb{R})$ consisting of Daubechies wavelets. A natural way for defining an operator on $L^2(\mathbb{R})$ is to pick a bounded sequence $f_{j,k}$, $j \in \mathbb{Z}$, $k \in \mathbb{Z}$, in $L^2(\mathbb{R})$ and to impose $T(\psi_{j,k}) = f_{j,k}$. However, this sequence $(f_{j,k})_{(j,k)\in\mathbb{Z}\times\mathbb{Z}}$ should satisfy some remarkable properties if we want T to be bounded on $L^2(\mathbb{R})$.

At this level of generality, it is an hopeless problem to find a non tautological necessary and sufficient condition on a bounded sequence $(f_{j,k})$ insuring that T be bounded on $L^2(\mathbb{R})$.

This difficulty disappears if there exist some constants $\mu_{j,k}$ such that $f_{j,k} = \mu_{j,k}\,\varphi_{j,k}$ where $\varphi_{j,k}(x) = 2^{j/2}\,\varphi(2^j x - k)$ and where φ is the "father function" associated with ψ.

Indeed we have

THEOREM 27.1. *With the above notations, T is bounded on $L^2(\mathbb{R})$ if and only if $2^{-j/2}\,\mu_{j,k}$ satisfies the Carleson condition* (26.4).

The distributional kernel of T is

$$S(x,y) = \sum_j \sum_k 2^j \mu_{j,k}\varphi(2^j x - k)\overline{\psi(2^j y - k)}$$

and the adjoint of T is defined by the distributional kernel

$$S^*(x,y) = \sum_j \sum_k 2^j \overline{\mu}_{j,k}\psi(2^j x - k)\overline{\varphi}(2^j y - k).$$

Let us recall the definition of the space BMO($\mathbb{R}$) which was created by John and Nirenberg.

A function $f(x)$ belongs to BMO($\mathbb{R}$) if $f(x)$ belongs to $L^1_{\mathrm{loc}}(\mathbb{R})$ and if there exists a constant C such that for any interval I one can find a number $\gamma = \gamma(I,f)$ such that $(1/|I|)\int_I |f(x) - \gamma|\,\mathrm{d}x \leqslant C$.

A candidate for γ is always given by $(1/|I|)\int_I f(t)\,\mathrm{d}t$.

Then a sequence $\alpha(j,k)$, $j \in \mathbb{Z}$, $k \in \mathbb{Z}$, satisfies the Carleson condition (26.4) if and only if $\sum\sum \alpha(j,k)\psi_{j,k}(x)$ belongs to BMO($\mathbb{R}$). This statement excludes the Haar system and our mother wavelet ψ will always be assumed to belong to some Hölder space $C^\alpha(\mathbb{R})$ for some $\alpha > 0$.

We denote by $\beta(x)$ this BMO-function and write $g = \pi(\beta, f)$ instead of

$$g(x) = \sum_j \sum_k \langle \beta, \psi_{j,k} \rangle \langle f, \psi_{j,k} \rangle 2^{j/2} \varphi_{j,k}(x).$$

Now the adjoint operator of the (linear) operator $f \to \pi(\beta, f)$ is defined by

$$g \to \sum_j \sum_k 2^{j/2} \langle g, \varphi_{j,k} \rangle\rangle \beta, \psi_{j,k} \rangle \psi_{j,k}(x) = \pi_*(\beta, g).$$

Here we are assuming that g be real (for the sake of simplicity).

This adjoint operator is called the paraproduct between β and g.

Returning to the non-standard algorithm, paraproducts and their adjoint correspond to the case where $(\beta_{j,k,l})$ is diagonal or to the case where $(\alpha_{j,k,l})$ is diagonal.

Finally one can decompose the general matrix $\alpha_{j,k,l}$ into a sum $\sigma_{j,k} + \tilde{\alpha}_{j,k,l}$ where $\sigma_{j,k}$ is diagonal and $\tilde{\alpha}_{j,k,l}$ satisfies $\sum_k \tilde{\alpha}_{j,k,l} = 0$.

Similarly one can decompose the general matrix $\beta_{j,k,l}$ into a sum $\tau_{j,k} + \tilde{\beta}_{j,k,l}$ where $\tau_{j,k}$ is also diagonal and $\sum_l \tilde{\beta}_{j,k,l} = 0$.

This yields the celebrated David-Journé splitting of a Calderón–Zygmund operator $T = L + M + R$ where L is the paraproduct with a BMO-function $\beta(x)$, where M is the adjoint of such a paraproduct and R is a pseudo-convolution.

Unlike L and M, this operator R is well compressed in our wavelet basis.

28. Quality of the approximation of general Calderón–Zygmund operators by banded matrices

Returning to Theorem 26.2, we now assume that $\sum_k \alpha_{j,k,l} = 0$ and $\sum_l \beta_{j,k,l} = 0$. We know that

$$|\alpha_{j,k,l}| + |\beta_{j,k,l}| + |\gamma_{j,k,l}| \leqslant C\big(1 + |k - l|\big)^{-1-\gamma}$$

and would like to replace these three matrices by banded matrices where all entries for which $|k - l| > R$ are replaced by 0.

But this destroys the cancellation condition $\sum_k \alpha_{j,k,l} = 0$ and $\sum_l \beta_{j,k,l} = 0$. Something even worse might happen. By replacing small entries by 0, one might replace a bounded operator by an unbounded one.

Therefore one should be more clever. We define $\alpha^\sharp_{j,k,l}$ and $\beta^\sharp_{j,k,l}$ by $\alpha^\sharp_{j,k,l} = \alpha_{j,k,l}$ if $|k - l| \geqslant R + 1$, $\alpha^\sharp_{j,k,l} = 0$ if $1 \leqslant |k - l| \leqslant R$ and $\alpha^\sharp_{j,k,k} = -\sum_{|l-k| \geqslant R+1} \alpha_{j,k,l}$.

A similar operation is performed on $\beta_{j,k,l}$. This clever thresholding defines an operator T_R.

Yang Qi Xiang [1996] proved the following.

THEOREM 28.1. *There is a constant C such that $\|T_R\| \leqslant CR^{-\gamma}\sqrt{\log R}$ and this estimation is optimal. More precisely, there exists a Calderón–Zygmund operator T and a positive constant c for which for each $R \geqslant 1$, one has $\|T_R\| \geqslant cR^{-\gamma}\sqrt{\log R}$.*

The proof of Theorem 27.1 is a clever application of Schur's lemma. Let us detail the construction of the counter-example. We define $\alpha_{j,k,l} = (1 + |k - l|)^{-1-\gamma}$ if $|k - l| \geqslant 1$ and fix $\alpha_{j,k,k}$ such that $\sum_l \alpha_{j,k,l} = 0$. Both $\beta_{j,k,l}$ and $\gamma_{j,k,l}$ are zero.

For computing the operator norm of T_R, we use the following lemmata.

LEMMA 28.1. *If $S(x, y)$ is the distributional kernel of a Calderón–Zygmund operator T, if $f \in C_0^\infty(\mathbb{R})$, then uniformly in $m > 0$, the product $f((x - y)/m)S(x, y)$ defines a Calderón–Zygmund operator T_m satisfying*

$$\|T_m\|_{L^2,L^2} \leqslant \|\hat{f}\|_1 \|T\|_{L^2,L^2}.$$

LEMMA 28.2. *If T is a Calderón–Zygmund operator, then uniformly in $j \in \mathbb{Z}$, P_jTP_j is a Calderón–Zygmund operator.*

Let us return to Theorem 28.1. We consider $P_jTP_j - P_0TP_0$ which is $A_1 + A_2 + \cdots + A_{j-1}$ in the notation of Section 25. Let $A_1(x, y), \ldots, A_{j-1}(x, y)$ be the corresponding distributional kernels. Then

$$A_q(x, y) = 2^q \sum_k \sum_l \alpha_{q,k,l} \varphi(2^q x - k) \overline{\psi}(2^q y - l)$$

vanishes if

$$2^{-q}(2N - 1) \leqslant |x - y| \leqslant 2^{-q}\big(R - (2N - 1)\big).$$

We then relate R and j by $2(2N - 1) = 2^{-j}(R - (2N - 1))$ and apply Lemma 28.1 to $P_jTP_j - P_0TP_0$. Then m and f are fixed in such a way that

$$f\left(\frac{x - y}{m}\right) A_q(x, y) = 2^q \sum_k \alpha_{q,k,k} \varphi(2^q x - k) \overline{\psi}(2^q y - k)$$

for $0 \leqslant q \leqslant j$.

Finally we assume that $|\alpha_{q,k,k}| \geqslant 1$ which is compatible with the other assumptions. If the operator with kernel $f((x - y)/m)(A_0(x, y) + \cdots + A_j(x, y))$ was bounded on $L^2(\mathbb{R})$, it would be of paraproduct type. The operator norm of this operator is computed with Theorem 26.2 and exceeds $c\sqrt{j}$.

29. The case of finite matrices

Let us consider the case of a finite matrix $a(k, l)$, $1 \leqslant k \leqslant N$, $1 \leqslant l \leqslant N$, fulfilling the following conditions

$$\big|a(k, k)\big| \leqslant C_0, \qquad \big|a(k, l)\big| \leqslant \frac{C_0}{|k - l|}, \tag{29.1}$$

$$\left|a(k+1,l) - a(k,l)\right| \leqslant \frac{C_1}{|k-l|^2} \tag{29.2}$$

and

$$\left|a(k,l+1) - a(k,l)\right| \leqslant \frac{C_1}{|k-l|^2}. \tag{29.3}$$

We add the weak boundedness property which reads

$$\sum_{u\leqslant k\leqslant v} \left| \sum_{u\leqslant l\leqslant v} a(k,l)\right| \leqslant C_2(v-u). \tag{29.4}$$

We want to compute the product TX in $\mathrm{O}(N)$ operations when

$$T = \big((a(k,l))\big)_{1\leqslant k,l\leqslant N}$$

satisfies (29.2), (29.3) and (29.4) and X is an arbitrary vector.

Let us assume that $N = 2^q$. The general case can be easily reduced to this special situation.

One then uses the discrete version of the wavelets over the interval. We use the non-standard decomposition and consider

$$A = \big((a_{j,k,l})\big) = \big((T\varphi_{j,k}, \psi_{j,l})\big),$$

$$B = \big((b_{j,k,l})\big) = \big((T\psi_{j,k}, \varphi_{j,l})\big)$$

and

$$C = \big((c_{j,k,l})\big) = \big((T\psi_{j,k}, \psi_{j,l})\big)$$

where $0 \leqslant j \leqslant q$, $0 \leqslant k \leqslant N2^{-j}$ and $0 \leqslant l \leqslant N2^{-j}$.

We replace each one of these three matrices by the corresponding band matrix A_m, B_m and C_m where $a^{(m)}_{j,k,l} = a_{j,k,l}$ if $|k-l| < m$ and 0 if not, and where $b^{(m)}_{j,k,l}$ and $c^{(m)}_{j,k,l}$ are defined similarly.

Then $A - A_m = R_m + D_m$ where D_m is diagonal and $\sum_k r^{(m)}_{j,k,l} = 0$. The entries of D_m are uniformly $\mathrm{O}(m^{-1})$.

This diagonal matrix D_m corresponds to a paraproduct in the non-standard representation. Therefore $\|D_m\| \leqslant Cm^{-1}\sqrt{\log N}$ (by Theorem 27.1). On the other hand $\|R_m\| \leqslant Cm^{-1}\sqrt{\log m}$. It means that the error, measured with the operator norm will not exceed $Cm^{-1}(\sqrt{\log N} + \sqrt{\log m})$.

If instead of (29.2) and (29.3), one has the corresponding estimates on the second differences (with $C_1|k-l|^{-2}$ replaced by $C_1|k-l|^{-3}$), then this error would not exceed $Cm^{-2}(\sqrt{\log N} + \sqrt{\log m})$.

Let us show that the computation of TX is a $\mathrm{O}(mN)$ algorithm if we accept an error which is $\mathrm{O}(\varepsilon\|X\|)$ where $\varepsilon \leqslant Cm^{-1}(\sqrt{\log N} + \sqrt{\log m})$ in the first case, $\varepsilon \leqslant Cm^{-2}(\sqrt{\log N} + \sqrt{\log m})$ in the second one, etc.

Indeed the product by a band matrix with width m is a $\mathrm{O}(mN)$ operation. The extra cost might come from the computation of the entries $\langle X, \varphi_{j,k}\rangle$ or $\langle X, \psi_{j,k}\rangle$ which are needed to feed our band matrix. But these wavelet coefficients are calculated from a pyramidal algorithm which is as fast as an FFT is (and even faster).

We would like to conclude in observing the strong similarities between this algorithm and the multipole algorithms which have been created by V. Rokhlin (see GREENGARD and ROKHLIN [1987]). Other references are HACKBUSH [1985, 1991], HACKBUSH and NOWAK [1989].

References

ALPERT, B.K. (1990), Sparse representation of smooth linear operators, Dept. of Computer Science, Yale University.

ALPERT, B.K. (1992a), Wavelets and other bases for fast numerical linear algebra, in: Charles K. CHUI, ed., *Wavelets, a Tutorial in Theory and Applications* (Academic Press, New York).

ALPERT, B.K. (1992b), Construction of simple multiscale bases for fast matrix operations, in: Mary Beth RUSKAI, ed., *Wavelets and Their Applications* (Jones and Bartlett).

AUSCHER, P. (1992a), Wavelets with boundary conditions on the interval, in: C. CHUI, ed., *Wavelets – A Tutorial in Theory and Applications* (Academic Press, Orlando, FL).

AUSCHER, P. (1992b), Il n'existe pas de bases d'ondelettes régulières dans l'espace de Hardy $\mathbb{H}^2(\mathbb{R})$, *C.R. Acad. Sci. Paris* **315**(1), 769–772.

AUSCHER, P. (1994), Remarks on the local Fourier bases, in: J. BENEDETTO and M.W. FRAZIER, eds., *Wavelets: Mathematics and Applications* (CRC Press).

AUSCHER, P. (1995), Solution of two problems on wavelets, *J. Geom. Anal.* **5**(2).

BATTLE, G. (1994), Wavelet refinement of the Wilson recursion formula, in: Larry L. SCHUMAKER and Glen WEBB, eds., *Recent Advances in Wavelet Analysis* (Academic Press, New York).

BEYLKIN, G. (1991), *Wavelets, Multiresolution Analysis and Fast Numerical Algorithms*, INRIA Lecture Notes.

BEYLKIN, G., R. COIFMAN and V. ROKHLIN (1991a), Fast wavelet transforms and fast algorithms, in: Y. MEYER, ed., *Wavelets and Applications*, RMA **20** (Masson, Paris).

BEYLKIN, G., R. COIFMAN and V. ROKHLIN (1991b), Fast wavelet transforms and numerical algorithms, I, *Comm. Pure Appl. Math.* **44**, 141–183.

BEYLKIN, G., R. COIFMAN and V. ROKHLIN (1992), Wavelets in numerical analysis, in: Mary Beth RUSKAI, ed., *Wavelets and Their Applications* (Jones and Bartlett).

BONY, J.M. (1983), Propagation et interaction des singularités pour les solutions des équations aux dérivées partielles non-linéaires, *Proc. ICM*, Warszawa, 1143–1147.

BRADIE, B., R. COIFMAN and A. GROSSMAN (1993), Fast numerical computations of oscillatory integrals related to acoustic scattering, *Appl. Comp. Harmonic Anal.* **1**(1), 94–99.

BURT, P. and E. ADELSON (1983), The Laplacian pyramid as a compact image code, *IEEE Trans. Comm.* **31**, 482–540.

COHEN, A. (1995), *Wavelets and Multiscale Signal Processing*, Appl. Math. & Math. Comp. **11** (Chapman and Hall, London).

COHEN, A. and J.P. CONZE (1992), Régularité des bases d'ondelettes et mesures ergodiques, *Rev. Mat. Iberoamericana* **8**(3), 351–365.

COHEN, A., I. DAUBECHIES and P. VIAL (1993), Wavelets on the interval and fast wavelets transforms, *Appl. Comp. Harmonic Anal.* **1**(1), 54–81.

COIFMAN, R. (1990), Adapted multiresolution analysis, computation, signal processing and operator theory, *Proc. ICM Kyoto*, 879–887.

COIFMAN, R., P. JONES and S. SEMMES (1989), Two elementary proofs of the L^2 boundedness of the Cauchy integral on Lipschitz curves. *J. Amer. Math. Soc.* **2**, 553–564.

DAUBECHIES, I. (1992), *Ten Lectures on Wavelets*, CBMS-NSF, Regional Conference Series in Applied Mathematics (SIAM, Philadelphia, PA).

DESLAURIERS, D. and S. DUBUC (1987), Interpolation dyadique, *Fractals Dimensions Non-entières Appl.*, 44–55.

DE VORE, R.A., B. JAWERTH and B.J. LUCIER (1992), Image compression through wavelet transform coding, *IEEE-IT* **38**(2), 719–746.

DOBYINSKY, S. (1995), La version ondelettes du théorème du Jacobien, *Rev. Mat. Iberoamericana* **11**(2), 309–333.

DONOHO, D. (1993), Unconditional bases are optimal bases for data compression and for statistical estimation, *Appl. Comp. Harmonic Anal.* **1**(1), 100–115.

DONOHO, D. (1995), Nonlinear solution of linear inverse problems by wavelet–vaguelette decomposition, *Appl. Comp. Harmonic Anal.* **2**(2), 101–126.

DONOHO, D. and I. JOHNSTONE (1996), Ideal spatial adaptation by wavelet shrinkage, Preprint.

DURAND, S. (1996), Etude de la vitesse de convergence de l'algorithme en cascade dans la construction des ondelettes d'Ingrid Daubechies, *Rev. Mat. Iberoamericana* **12**(2).

FIX, G. and G. STRANG (1969), Fourier analysis of the finite element method in Ritz–Galerkin theory, *Stud. Appl. Math.* **48**, 265–273.

GREENGARD, L. and V. ROKHLIN (1987), Rapid evaluation of potential fields in three dimensions, in: *Vortex Methods Proceedings, Los Angeles*, Lecture Notes in Mathematics (Springer, Berlin).

HACKBUSH, W. (1985), *Multigrid Methods and Applications* (Springer, Berlin).

HACKBUSH, W. (1991), The solution of large systems..., *Rend. Sem. Mat. Univ. Pol. Torino*, Fascicolo Speciale 1991 "Numerical Methods".

HACKBUSH, W. and Z.P. NOWAK (1989), On fast matrix multiplication in the boundary element method by panel clustering, *Numer. Math.* **54**, 463–491.

HERVÉ, L. (1992), Construction et régularité des fonctions d'échelle, IRMAR, Campus de Beaulieu, 35042 Rennes.

JAFFARD, S. (1996), The spectrum of singularities of the Riemann function, *Rev. Mat. Iberoamericana*, to appear.

KAHANE, J.P. and P.G. LEMARIÉ-RIEUSSET (1996), *Fourier Series and Wavelets* (Gordon and Breach Publishing Group, London).

LEMARIÉ-RIEUSSET, P.G. (1992a), Analyses multi-résolutions non orthogonales, commutation entre projecteurs et dérivation et ondelettes vecteurs à divergence nulle, *Rev. Mat. Iberoamericana* **8**(2), 221–237.

LEMARIÉ-RIEUSSET, P.G. (1992b), Sur l'existence des analyses multi-résolutions en théorie des ondelettes, *Rev. Mat. Iberoamericana* **8**(3), 457–474.

MARR, D. (1982), *Vision* (Freeman, New York).

MEYER, Y. (1990), *Ondelettes et Opérateurs*, Tomes I, II & III (Hermann, Paris).

MEYER, Y. (1992), *Les Ondelettes, Algorithmes et Applications* (Armand Colin).

MEYER, Y. (1993a), *Ondelettes et Algorithmes Concurrents* (Hermann, Paris).

MEYER, Y. (1993b), *Wavelets and Operators* (Cambridge University Press, Cambridge).

MEYER, Y. and F. PAIVA, Convergence de l'algorithme de Mallat, *J. Anal. Math.*, volume dedicated to Sjolem Mandelbrojt, to appear.

RISLER, J.J. (1991), *Méthodes Mathématiques pour la CAO*, RMA **18** (Masson, Paris).

ROKHLIN, V. (1985), Rapid solution of integral equations of classical potential theory, *J. Comput. Phys.* **60**(2) (1985).

SEMMES, S. (1989), Nonlinear Fourier analysis, *Bull. Amer. Math. Soc.* **20**(1), 1–18.

STEIN, E. (1970), *Singular Integrals and Differentiability Properties of Functions* (Princeton Univ. Press, Princeton, NJ).

STRANG, G. and T. NGUYEN (1996), *Wavelets and Filterbanks* (Wellesley–Cambridge Press).

STRÖMBERG, J.O. (1983), A modified Franklin system and high order spline systems on $\mathbb{R}^n$ as unconditional bases for Hardy spaces, in: W. BECKNER, ed., *Conference on Harmonic Analysis in honor of A. Zygmund* (Wadeworth International Group, Belmont, CA).

VETTERLI, M. and J. KOVAČEVIĆ (1995), *Wavelets and Subband Coding* (Prentice–Hall, Englewood Cliffs, NJ).

VILLEMOES, L. (1992), Energy moments in time and frequency for two-scale difference equations solution and wavelets, *SIAM J. Math. Anal.* **23**.

VILLEMOES, L., Wavelet analysis of two-scale difference equations, *SIAM J. Math. Anal.*, to appear.

WICKERHAUSER, M.V. (1994), *Adapted Wavelet Analysis, From Theory to Software, with diskette* (AK Peters Ltd., Wellesley, MA).
YANG QI XIANG (1996), Fast algorithms for Calderón–Zygmund singular integral operators, *Appl. Comp. Harmonic Anal.* **3**(2), 120–126.
YSERENTANT, H. (1991), Hierarchical bases, *Proc. ICIAM* **91**, 256–276.

Subject Index

Computer Aided Geometric Design

Jean-Jacques Risler

Laboratoire d'Analyse Numérique
Université Pierre et Marie Curie
Tour 55–65, 5ème étage
4 place Jussieu
75252 Paris Cédex 05
France

HANDBOOK OF NUMERICAL ANALYSIS, VOL. V
Techniques of Scientific Computing (Part 2)
Edited by P.G. Ciarlet and J.L. Lions

Contents

Introduction

This article is an introduction to Computer Aided Geometric Design, henceforth abbreviated as "CAGD", from a mathematical point of view. The main part (namely the first two chapters) presents the traditional modeling using B-spline functions, in the spirit of RISLER [1991].

The modeling of curves and surfaces is based on "nonuniform rational B-splines", henceforth abbreviated as "NURBS", which we now briefly describe. The classical spline functions are that serves as functions of class $\mathcal{C}^2$ and piecewise polynomial of degree 3, with junctions at a finite number of points t_i. They have a characterization that serves as a justification of their use in CAGD: among all the functions f of class $\mathcal{C}^2$ interpolating a finite number of points $y_i = f(t_i)$, the spline functions minimize the integral $\int |f''|^2$. Their name "spline" comes in fact from this property. The space of (nonuniform) spline functions is a generalization of these "splines": they are piecewise polynomials of some degree k, with junctions of class $\mathcal{C}^{r_i}$, $1 \leqslant r_i \leqslant k+1$, at t_i. The splines are "uniform" when the sequence (t_i) is a (finite or not) uniform sequence of points (e.g., $t_i = i$).

The *B-spline functions* constitute a privileged basis of this space called the *B-spline basis*. This basis has some remarkable properties allowing in particular the applications of simple and numerically stable algorithms. In particular, the B-spline functions have a "small" support, and they constitute a partition of unity.

The modelization of curves and surfaces using B-splines goes back to Bézier and De Casteljau; the main idea is to create curves controlled by the coefficients of the spline functions in the B-spline basis instead of interpolating points: those coefficients are the coordinates of the "control points", and the curve does not in general pass through the control points.

In other words, let $B_0(t), \ldots, B_{n-1}(t)$, $a \leqslant t \leqslant b$, be the set of B-spline functions which constitute a B-spline basis and let $P_0, \ldots, P_{n-1}$ be points in $\mathbb{R}^s$. The corresponding *B-spline curve* is by definition the curve $S(t) = \sum_{i=0}^{n-1} P_i B_i(t)$. The points P_i form a polygonal line called the *control polygon* of the curve.

A special case is that of *Bézier curves*, first introduced by Bézier (see BÉZIER [1972]): The space of spline functions is the space of polynomials of degree $\leqslant k$ on $[0, 1]$; the corresponding B-spline basis is then called the *Bernstein basis*, and a B-spline curve $S(t) = \sum_{i=0}^{k} P_i B_i(t)$ is called a *Bézier curve of degree k*. Note that a B-spline curve (or a Bézier curve) does not in general pass through the points P_i, but is *controlled* by those points: moving one point P_i changes locally the shape of the curve in the direction of the motion of P_i, but with an "inertia" due to the other control points: the spline curve will always remain "tight".

A slight generalization is obtained by the introduction of denominators for the B-spline functions: this allows the exact representation of any arc of conic, which is in general impossible with piecewise polynomial functions. This gives the "Non Uniform Rational B-splines" (NURBS). Note that the interpolation is relatively easy, using the B-spline basis. Moreover, the fact that spline functions are piecewise polynomials of low degree (or quotient of two polynomials of low degree) give the hope that techniques of Algebraic Geometry can be used in geometric problems involving spline functions; this aspect is nowadays one of the main branches of research in the domain.

The surfaces are usually modeled using *tensor products* of spline functions: the space of tensor product splines is generated by the products $B_i(u)B_j(v)$, where $(B_i(u), 0 \leqslant i \leqslant n-1)$, (resp. $(B_j(v))$) form a B-spline basis in one variable. Tensor products are easy to handle, because all the algorithms for one variable splines can be used separately in u and v. A set of points $P_{ij} \in \mathbb{R}^s$, $0 \leqslant i \leqslant n-1$, $0 \leqslant j \leqslant n-1$, determine a *tensor product patch* given by $S(u, v)$: $\sum P_{ij} B_i(u) B_j(v)$. All those properties give a way of modeling well adapted for *geometric design*; it is this approach that we will describe in this article.

In the first chapter, the B-splines in one variable are presented, and their classical applications for the modeling of curves is emphasized. Note that our definition of B-splines does not uses the technique of "divided differences". The Bézier curves are introduced as a particular case of B-spline curves.

The second chapter is devoted to surfaces, and three ways of modeling surfaces are presented: tensor products of B-splines, triangular Bézier patches, and polyhedral splines. Each of these three techniques has its own interest; the first one because it allows to use algorithms for one variable spline curves (or one variable Bézier curves) and because it is the most popular nowadays to model surfaces, the second one because it allows the modeling of triangular patches with very simple algorithms, and the third one because polyhedral splines are in some sense direct generalization of NURBS in one variable, with the same advantages in relation to the triangular Bézier surface patches as the NURBS in relation to Bézier curves.

The third chapter is devoted to some practical aspects of Algebraic Geometry (mainly real Algebraic Geometry). With the development of symbolic computation, made possible by high-speed computers, it seems that the techniques of Algebraic Geometry will be more and more available for graphics and CAGD. This fact is asserted by numerous research papers on this field. This chapter presents some techniques of Algebraic Geometry (elimination theory for instance) to potential users, a priori not specialists in the field. Although there is no introduction to Differential Geometry, it is clear that this matter is also important for CAGD (see the appendix by Böhm in FARIN [1988]).

CHAPTER I

Spline Curves

In this chapter, we give the main properties of the B-spline basis and B-spline curves, following closely the spirit of RISLER [1991]. B-spline functions are introduced by their induction relation, following an idea of DE BOOR and HÖLLIG [1987], and their basic properties are proved in Sections 1–4. Bernstein polynomials are then introduced as a special case of B-splines (Section 6), and spline curves (and therefore Bézier curves) are studied in Sections 7–10, Section 11 being devoted to the junction between curves.

Two sections present a different point of view about B-splines: Section 5 presents the divided differences (which is the usual way of introducing B-splines, as in SHUMAKER [1981] for instance), and Section 12 introduces the "blossoming", which gives an alternative way for proving the properties of B-splines, as in SEIDEL [1989].

1. B-splines: Fundamental properties

We give the inductive definition of B-spline functions, and quote the immediate consequences.

NOTATION. Let us take in $\mathbb{R}$ a sequence $t_0, \ldots, t_m$ of points called *knots* such that for all i, $t_i \leqslant t_{i+1}$. If there are r successive t_i's equal to τ, one says that τ is a node *of order* r or of *multiplicity* r. Moreover, for $1 \leqslant j \leqslant m+1-i$, one sets:

$$\omega_{i,j}(x) = \begin{cases} \dfrac{x - t_i}{t_{i+j} - t_i} & \text{if } t_i < t_{i+j}, \\ 0 & \text{otherwise.} \end{cases}$$

For the definition of B-splines, we follow the way of DE BOOR and HÖLLIG [1987].

DEFINITION (*Definition of B-splines*). With the above notation, set $t = (t_0, \ldots, t_m)$. For $x \in \mathbb{R}$, k positive integer, $0 \leqslant i \leqslant m-k-1$, the functions $B_{i,k,t}(x)$ (also denoted $B_{i,k}(x)$ when the sequence t is fixed) are defined by induction on k in the following way

$$B_{i,0}(x) = \begin{cases} 1 & \text{if } t_i \leqslant x < t_{i+1}, \\ 0 & \text{otherwise,} \end{cases} \tag{1.1}$$

$$B_{i,k}(x) = \omega_{i,k}(x)B_{i,k-1}(x) + \bigl(1 - \omega_{i+1,k}(x)\bigr)B_{i+1,k-1}(x) \quad \text{for } k \geqslant 1.$$

By definition of $\omega_{i,k}(x)$, one has

$$B_{i,k}(x) = \frac{x - t_i}{t_{i+k} - t_i}B_{i,k-1}(x) + \frac{t_{i+k+1} - x}{t_{i+k+1} - t_{i+1}}B_{i+1,k-1}(x)$$

if $t_i < t_{i+k}$ and $t_{i+1} < t_{i+k+1}$. Note that if $t_{i+1} = t_{i+k+1}$, $B_{i+1,k-1}(x) \equiv 0$ (see Remark 1.1).

This relation is known as the "De Boor–Cox algorithm", after DE BOOR [1972] and COX [1972].

REMARKS 1.1. (1) One may define in the same way B-splines for an *infinite* sequence of knots t_i ($t_i \leqslant t_{i+1}$), since the definition of each $B_{i,k}$ uses only a finite numbers of knots (see Theorem 1.1 below: the spline $B_{i,k}(x)$ has $[t_i, t_{i+k+1}]$ for support, and its definition uses only those t_j such that $i \leqslant j \leqslant i + k + 1$).

(2) If, for an index i, $t_i = t_{i+k+1}$ (and so t_i is a node of multiplicity $\geqslant k + 2$), one has $B_{i,k} \equiv 0$; the converse is also true: see Theorem 1.1 below.

THEOREM 1.1. *With the above notation, one has the following properties*:

(i) $B_{i,k}(x)$ *is a piecewise polynomial of degree* k.

(ii) $B_{i,k}(x) = 0$ *for* $x \notin [t_i, t_{i+k+1}[$.

(iii) $B_{i,k}(x) > 0$ *for* $x \in]t_i, t_{i+k+1}[$; $B_{i,k}(t_i) = 0$ *unless* $t_i = t_{i+1} = \cdots = t_{i+k} < t_{i+k+1}$ (*node of order* $k + 1$), *and then* $B_{i,k}(t_i) = 1$.

(iv) *Let* $[a, b]$ *be an interval such that* $t_k \leqslant a$, $t_{m-k} \geqslant b$. *Then* $\sum_{i=0}^{m-k-1} B_{i,k}(x) = 1$ *for all* $x \in [a, b[$.

(v) *Let* $x \in]t_i, t_{i+k+1}[$. *Then* $B_{i,k,t}(x) = 1$ *if and only if* $t_{i+1} = \cdots = t_{i+k} = x$.

(vi) $B_{i,k}(x)$ *is right-continuous* (*and even right-infinitely differentiable*), *for all* $x \in \mathbb{R}$ (*recall that* $B_{i,k}(x) = 0$ *for* x *outside* $[t_0, t_m]$).

PROOF. Properties (i)–(iv) and (vi) are clear for $k = 0$; one deduces immediately (i)–(iii) and (vi) for $B_{i,k}$ by induction on k.

Let us prove property (iv) by induction on k. Let $x \in [a, b]$. Then there exists j, $j \geqslant k$, $j \leqslant m - k - 1$, such that $x \in [t_j, t_{j+1}[$.

If $x = t_j$ and $B_{j,k}(x) = 1$, the property is clear (see (iii)).

In the other cases, one has

$$\sum_{i=0}^{m-k-1} B_{i,k}(x) = \sum_{i=j-k}^{j} B_{i,k}(x)$$

by (ii), and

$$\sum_{j-k}^{j} B_{i,k}(x) = \sum_{j-k}^{j} \omega_{i,k}B_{i,k-1}(x) + \sum_{j-k}^{j}(1 - \omega_{i+1,k})B_{i+1,k-1}(x)$$

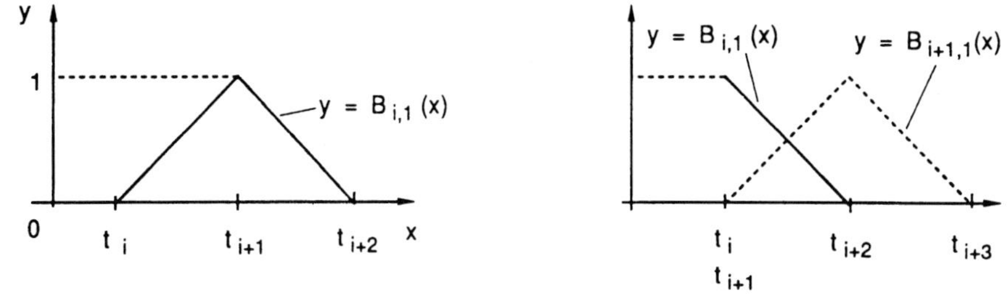

FIG. 1.1. B-splines of degree 1. Left: case of simple knots; right: case of a double knot.

by the definition of the B-splines, which implies, grouping terms together,

$$\sum_{i=j-k}^{j} B_{i,k}(x) = \omega_{j-k} B_{j-k,k-1}(x) + \sum_{i=j+1-k}^{j} B_{i,k-1}(x) + (1-\omega_{j+1,k}) B_{j+1,k-1}(x).$$

But $B_{j-k,k-1}(x) = 0$ and $B_{j+1,k-1}(x) = 0$ because $x \in [t_j, t_{j+1}[$ (see (ii)) and

$$\sum_{i=j+1-k}^{j} B_{i,k-1}(x) = 1$$

by the induction hypothesis, which implies (iv).

(v) is an easy consequence of (iv) and (iii). □

REMARK 1.2. One already sees in this proposition some remarkable properties of the B-splines: (iv) shows that they constitute a *partition of unity* on a convenient interval, and (ii) shows that each $B_{i,k}(x)$ has a "*small support*".

REMARK 1.3. If $t_{m-k} = \cdots = t_m = b$, property (iv) is valid only on the interval $[a, b[$, because for all i, $1 \leqslant i \leqslant m-k-1$, one has $B_{i,k}(b) = 0$ (see (vi) above: B-splines are right-continuous). If we want a formula valid on $[a, b]$, we have to set $B_{m-k-1,k}(b) = 1$, which makes the B-spline $B_{m-k-1,k}(x)$ *left-continuous* at b.

We will systematically use this convention, which is necessary only if there is a knot of multiplicity $k+1$ at b.

EXAMPLES. (1) $k = 1$, see Fig. 1.1.

(2) $k = 2$ (i.e., piecewise quadratic functions), see Fig. 1.2.

The curve $y = B_{i,2}(x)$ is built from three arcs of parabolas and two half-lines with junction C^1 at the knots t_i, t_{i+1}, t_{i+2} and t_{i+3}.

(3) Let us represent the 9 elements of the B-spline basis of $\mathcal{P}_{3,t}$ when $n = 9$, i.e., with 5 knots (simple) in $]a, b[$, and $t_0 = \cdots = t_3 = a$ and $t_9 = \cdots = t_{12} = b$.

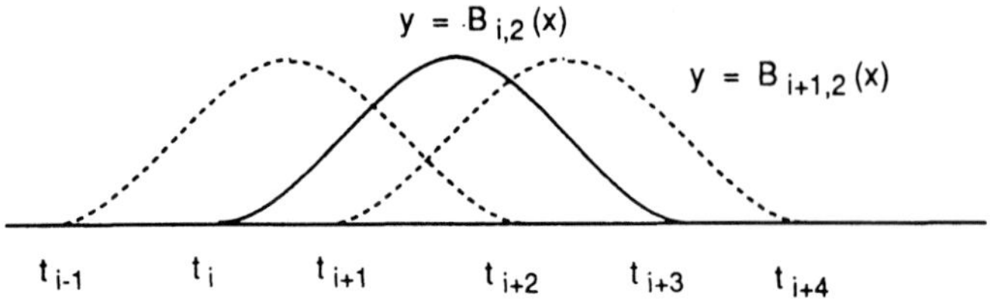

FIG. 1.2. B-splines of degree 2: Case of simple knots.

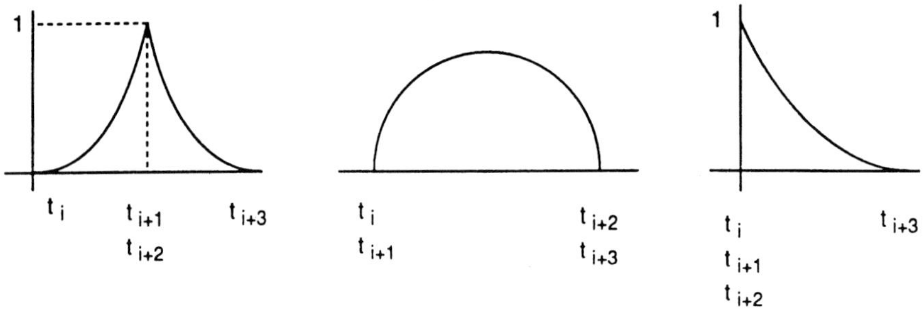

FIG. 1.3. $y = B_{i,2}(x)$ in the case of multiple knots.

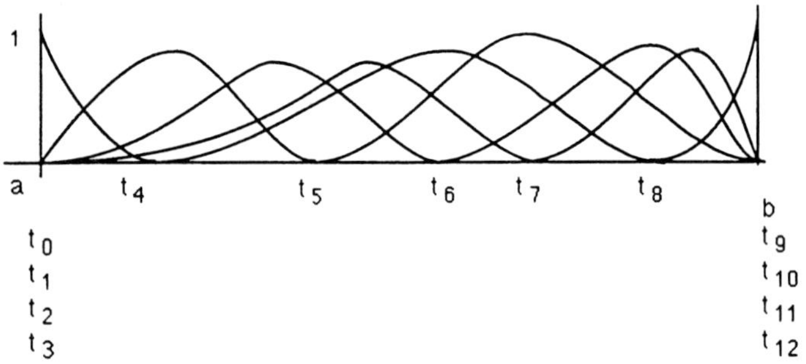

FIG. 1.4.

REMARK 1.4. We will give later (Section 10) the expressions for the B-splines $B_{i,3}(x)$ in the case of a uniform sequence of knots.

THEOREM 1.2. *For all $k \geqslant 0$ and all $x \in \mathbb{R}$, $B_{i,k}$ is right-differentiable and one has*

$$B'_{i,k}(x) = k\left[\frac{B_{i,k-1}(x)}{t_{i+k} - t_i} - \frac{B_{i+1,k-1}(x)}{t_{i+k+1} - t_{i+1}}\right],$$

where, by convention, one replaces an expression whose denominator is equal to zero by 0.

PROOF. By induction on k using (1.1), the case $k = 0$ being trivial. □

2. The B-spline basis

DEFINITION. Let t_i, $0 \leqslant i \leqslant m$, be a sequence of points in $\mathbb{R}$ such that $t_i \leqslant t_{i+1}$ and $[a,b]$ be an interval such that $t_k \leqslant a$, $t_{m-k} \geqslant b$, where k is an integer $\geqslant 0$. One denotes by $\mathcal{P}_{k,t}([a,b])$ (or simply $\mathcal{P}_{k,t}$) the vector space of piecewise polynomial functions on $[a,b]$, of degree $\leqslant k$, with $\mathcal{C}^{k-p_j}$-continuity at t_j, if t_j is a knot of multiplicity p_j.

By assumption, a $\mathcal{C}^{k-p_j}$-continuity with $k - p_j < 0$ gives no condition at t_j.

If all the knots are of multiplicity $\leqslant k+1$, we set $n = m - k$. We then have $\dim \mathcal{P}_{k,t} = n$, as is easily checked.

THEOREM 2.1. *With the above notation, assume that all the knots are of multiplicity $\leqslant k+1$; then the functions $B_{i,k,t}$, $0 \leqslant i \leqslant n-1$, constitute a basis for the space $\mathcal{P}_{k,t}$.*

PROOF. Recall first that by the hypothesis on the sequence (t_i), there is no knot of multiplicity $\geqslant k+2$, so that none of the $B_{i,k,t}$'s is identically zero.

One has $n = \dim \mathcal{P}_{k,t}$; it is then enough to prove that, denoting by S the space generated by the $B_{i,k,t}$'s (in the space of the functions on $[a,b]$), we have $\mathcal{P}_{k,t} \subset S$. It is easy to see that a basis of $\mathcal{P}_{k,t}$ is given by

$$\begin{cases} (x-a)^\nu, & 0 \leqslant \nu \leqslant k, \\ ((x-\tau_i)_+)^\nu, & 1 \leqslant i \leqslant l-1,\ r_i \leqslant \nu \leqslant k; \end{cases}$$

where each t_j is equal to some τ_i and where there are $(k+1-r_i)$ nodes t_j equal to τ_i, $0 \leqslant r_i \leqslant k$. Let us prove that without any hypothesis on the multiplicity of the knots, one has for $t \in [a,b]$,

$$(x-t)^k = \sum_{i=0}^{n-1} \psi_{i,k}(t) B_{i,k}(x), \tag{2.1}$$

and

$$\big((x-t_j)_+\big)^k = \sum_{i \geqslant j} \psi_{i,k}(t_j) B_{i,k}(x) \quad \text{for } 0 \leqslant j \leqslant n-1, \tag{2.2}$$

with $\psi_{i,k}(t) = (t_{i+1}-t)\cdots(t_{i+k}-t)$. The function $\psi_{i,k}$ is defined for $k \geqslant 1$, and is of degree k in t; one sets $\psi_{i,0}(t) = 1$ for all i. Formula (2.2) is a consequence of (2.1) because for $x \leqslant t_j$ and $i \geqslant j$, $B_{i,k}(x) = 0$, and then the two terms of (2.2) are zero. For $x \geqslant t_j$ and $i < j$, either $\psi_{i,k}(t_j) = 0$ (for $i \geqslant j-k$), or $B_{i,k}(x) = 0$ (for $i < j-k-1$), and so

$$\sum_{i=0}^{j-1} \psi_{i,k}(t_j) B_{i,k}(x) = 0;$$

we now only have to apply (2.1). Formula (2.1) is proven by induction on k, the case $k = 0$ being clear. For the general case, we replace in the expression

$$\sum_{i=0}^{n-1} \psi_{i,k}(t) B_{i,k}(x)$$

the function $B_{i,k}(x)$ by its expression in terms of $B_{i,k-1}$ and $B_{i+1;k-1}$ (see (1.1)). We then apply the induction hypothesis which says that

$$\sum_{i=1}^{n-1} \psi_{i,k-1}(t) B_{i,k-1}(x) = (x-t)^{k-1}.$$

Recall that we have to prove that $\mathcal{P}_{k,t} \subset S$. We have $(x-t)^k \in S$ by (2.1), and differentiating formula (2.1) with respect to t, we see that $(x-t)^{k-s} \in S$, $0 \leqslant s \leqslant k$. Doing the same for $(x-t_j)_+^k$ proves the theorem. □

3. Basic algorithms for B-splines

We give the three basic algorithms for B-spline functions, namely the evaluation (or De Casteljau–De Boor–Cox) algorithm, the differentiation algorithm, and the knot insertion (or Böhm) algorithm. We then prove the variation diminishing property for B-splines functions.

With the same notation as in the previous section, let

$$S(x) = \sum_{i=0}^{n-1} a_i B_{i,k}(x)$$

be an element of $\mathcal{P}_{k,t}$; we will call $S(x)$ a *spline function*. We will use the notation of Sections 1 and 2: the B-splines are computed with a knot sequence $t_0, \dots, t_m$ and defined over all $\mathbb{R}$, and the algorithms described are independent of the chosen interval $[a,b]$ (as long as $t_k \leqslant a$ and $t_{m-k} \geqslant b$); in the algorithms described below, we have set $n = m - k$. One generally assumes that there is no knot of multiplicity $\geqslant k+2$, which implies that the functions $B_{i,k,t}$, $0 \leqslant i \leqslant n-1$, form a basis of $\mathcal{P}_{k,t}$ (by Theorem 2.1). However, this hypothesis is not in general satisfied for the spaces $\mathcal{P}_{k',t}$, $k' < k$ (which are considered in the inductions if there are multiple knots; for instance, one often has $t_0 = \dots = t_k = a$ and $t_n = \dots = t_{n+k} = b$).

We will describe three algorithms for the function $S(x)$: the "De Boor–Cox" or "De Casteljau" algorithm, allowing the evaluation of $S(x)$ at a point $x \in [a,b]$; the algorithm giving the coefficients (in relation to the $B_{i,k-1}$'s) of the derivative $S'(x)$; and finally the "knot insertion" algorithm, which computes the coefficients of $S(x)$ expanded in the basis $B_{i,k,t'}$, $0 \leqslant i \leqslant n$, the knot sequence t' being obtained from the sequence t by adding a new knot $\hat{t}$.

Evaluation algorithm. See, e.g., DE BOOR [1972], or BÖHM, FARIN and KALMAN ([1984], p. 16). This algorithm comes from the following theorem:

THEOREM 3.1. *Let* $x \geqslant t_k$, $S(x) = \sum_{i=0}^{n-1} a_i B_{i,k}(x)$. *We then have*

$$S(x) = \sum a_i^0 B_{i,k}(x) = \sum a_i^1(x) B_{i,k-1}(x) = \cdots = \sum a_i^k(x) B_{i,0}(x),$$

with

$$a_i^0 = a_i,$$
$$a_i^{r+1}(x) = \omega_{i,k-r}(x) a_i^r(x) + \big(1 - \omega_{i,k-r}(x)\big) a_{i-1}^r(x).$$

Recall that

$$\omega_{i,k}(x) = \begin{cases} \dfrac{x - t_i}{t_{i+k} - t_i} & \text{if } t_i < t_{i+k}, \\ 0 & \text{otherwise.} \end{cases}$$

PROOF. It is a direct application of the inductive definition of the B-splines (1.1). □

REMARKS 3.1. (1) The algorithm is also valid for $x < t_k$, with the convention that the B-splines of index $\leqslant 0$ are identically zero.

(2) If one has to evaluate $S(x)$ for $x \in [t_j, t_{j+1}[$, one then sets $S(x) = a_j^k(x)$, as $B_{j,0}(x)$ is the characteristic function of the interval $[t_j, t_{j+1}[$. For the computation of $a_j^k(x)$, it is enough to evaluate a_i^r for $j - k + r \leqslant i \leqslant j$, the other $B_{i,k-r}(x)$ being identically zero, as $x \in [t_j, t_{j+1}[$ (see Theorem 1.1). We therefore have

$$S(x) = \sum_{i=j-k}^{j} a_i^0(x) B_{i,k}(x) = \sum_{i=j-k+1}^{j} a_i^1 B_{i,k-1}(x) = \cdots = a_j^k(x).$$

The algorithm is then "triangular", each element being a convex linear combination of the two ones above it:

$$\begin{array}{ccccccccccc}
a_{j-k} & & a_{j-k+1} & & \cdots & \cdots & \cdots & & a_{j-1} & & a_j \\
& a_{j-k+1}^1 & & \cdots & \cdots & \cdots & \cdots & \cdots & & a_j^1 & \\
& & \ddots & & & & & & & & \\
& & & & a_{j-1}^{k-1} & & a_j^{k-1} & & & & \\
& & & & & a_j^k & & & & &
\end{array}$$

(3) This algorithm is quite expensive: it requires the computation of $k(k+1)/2$ linear convex combinations, and for each one, it uses two multiplications, one division, and one subtraction (for the computation of $1 - \omega_{i,k-r}(x)$). Nevertheless, it has several advantages:

– It is numerically stable.

– The computation of the $\omega_{i,k-r}(x)$'s is often very simple (for instance if x and the knots t_i are integers).
– It is well adapted to the tracing of spline curves (see the next chapter).

(4) If one has to evaluate the function $S(x)$ at several points of $[t_i, t_{i+1}[$, one generally proceeds in another way:

– One computes once and for all the polynomial expression of S between t_i and t_{i+1}:

$$S(x) = \sum_{j=0}^{k} \frac{\alpha_j}{j!}(x - t_i)^j$$

with $\alpha_j = \mathrm{D}^j(S)(t_i)$ (the right derivative evaluated with the differentiation algorithm, described below).
– One evaluates $S(x)$ by the "Hörner rule" (see SCHUMAKER ([1981], p. 106)) which requires only $2k$ operations. Let us recall the Hörner rule: Let P be a polynomial of degree d: $P = a_d X^d + \cdots + a_1 X + a_0$; the program to evaluate P at x amounts to writing $P(x)$ in the following way

$$P(x) = x(\cdots(x(xa_d + a_{d-1}) + a_{d-2}) + \cdots + a_1) + a_0,$$

setting $P_d = a_d$, $P_i = a_i + xP_{i+1}$, $d - 1 \geqslant i \geqslant 0$, and $P = P_0$.

Differentiation algorithm. It comes from the following:

THEOREM 3.2. *Let* $S(x) = \sum_{i=0}^{n-1} a_i B_{i,k}(x)$. *Then the right derivative* $\mathrm{D}(S(x))$ *is given by the following formula*

$$\mathrm{D}(S(x)) = \sum_{i=1}^{n-1} b_i B_{i,k-1}(x), \tag{3.1}$$

with

$$b_i = \begin{cases} k\dfrac{(a_i - a_{i-1})}{(t_{i+k} - t_i)} & \text{if } t_i < t_{i+k}, \\ 0 & \text{otherwise.} \end{cases}$$

PROOF. It is enough to replace in $\mathrm{D}(S(x))$, $\mathrm{D}(B_{i,k}(x))$ by its expression (Theorem 1.2). □

REMARK 3.2. With the previous notation, if $S(x) \in \mathcal{P}_{k,t}$, then $\mathrm{D}(S(x)) \in \mathcal{P}_{k-1,t}$, and one always has $B_{0,k-1}|[a,b] \equiv 0$, which explains why in $\mathrm{D}(S(x))$ the sum goes from 1 to $n-1$. For efficient simultaneous calculation of derivatives, see BÖHM, FARIN and KAHMANN [1984], or FIOROT [1987].

Knot insertion algorithm. (BÖHM [1980].) Let $S(x) = \sum_{i=0}^{n-1} a_i B_{i,k}(x)$ be a spline function defined with a knot sequence $(t_i)_{0 \leqslant i \leqslant m=n+k}$ and let us add a new knot $\hat{t}$ to the sequence t_i in such a way that $\hat{t} \leqslant t_{n-1}$ ($\hat{t}$ may be equal to one of the t_i's). To the new sequence so obtained (denoted t') there corresponds a set of B-splines $(\widehat{B}_{i,k})$, $0 \leqslant i \leqslant n$, and one has $\mathcal{P}_{k,t} \subset \mathcal{P}_{k,t'}$.

THEOREM 3.3. *One has*

$$\sum_{i=0}^{n-1} a_i B_{i,k}(x) = \sum_{i=0}^{n} \hat{a}_i \widehat{B}_{i,k}(x), \tag{3.2}$$

with

$$\hat{a}_i = \begin{cases} a_i & \textit{if } t_{i+k} \leqslant \hat{t}, \\ \omega_{i,k}(\hat{t})a_i + (1 - \omega_{i,k}(\hat{t}))a_{i-1} & \textit{if } t_i < \hat{t} < t_{i+k}, \\ a_{i-1} & \textit{if } \hat{t} \leqslant t_i. \end{cases}$$

PROOF. The proof is simple, but a little tedious, by induction on k. It consists in applying the first step of the evaluation algorithm to the two terms, and the induction hypothesis; see RISLER [1991, p. 28–29]. □

As an application, let us prove the "variation diminishing property" for spline functions. This proof comes from DE BOOR and HÖLLIG [1987].

DEFINITION. (1) Let $\boldsymbol{a} = (a_0, \ldots, a_{n-1})$ be a sequence of real numbers; the *variation* of $\boldsymbol{a}$, noted $V(\boldsymbol{a})$, is by definition the *number of sign changes* of the sequence $\boldsymbol{a}$, i.e., the number of indices i such that $a_i \bullet a_{i+r} < 0$, with $a_{i+1} = \cdots = a_{i+r-1} = 0$. For instance, for the sequence $\boldsymbol{a} = (-1, 0, 0, 2, 0, 4, 0, 0, -5, 6)$, one has $V(\boldsymbol{a}) = 3$.

(2) Let $f : [a, b] \to \mathbb{R}$ be a function; set $V(f)$ for the variation of f, or the number of change of signs of f, defined as $\sup V(f(x_1), \ldots, f(x_r))$, the sup being taken over all sequences of points $x = (x_1, \ldots, x_r)$ $(x_1 < \cdots < x_r)$ in $[a, b]$.

With the previous notation, we have:

THEOREM 3.4 (Variation diminishing property for splines functions). *Let*

$$S(x) = \sum_{i=0}^{n-1} a_i B_{i,k}(x)$$

be a spline function defined over an interval $[a, b]$ *such that* $t_k \leqslant a$ *and* $t_n \geqslant b$. *Then* $V(S) \leqslant V(\boldsymbol{a})$.

With the notation of Theorem 3.3 (knot insertion algorithm), let us add a new knot $\hat{t}$ to the sequence t_i, and let $\hat{\boldsymbol{a}} = (\hat{a}_i)_{1 \leqslant i \leqslant n+1}$ be the sequence of coefficients of the function $S(x)$ in the basis $\widehat{B}_{i,k}$. Let us prove that

$$V(\hat{\boldsymbol{a}}) \leqslant V(\boldsymbol{a}).$$

Each $\hat{a}_i$ is a linear convex combination of a_{i-1} and a_i (see Theorem 3.3); then if we insert $\hat{a}_j$ between a_{j-1} and a_j, the variation of the sequence remains constant. For in passing from the sequence $(\boldsymbol{a})$ to the sequence $(\hat{\boldsymbol{a}})$, we may begin by adding, for each index i, $\hat{a}_i$ between a_{i-1} and a_i, which does not change the variation, and remove the a_j's after, which may only lower the variation. Now, let $x_1, \ldots, x_r$, $x_i < x_{i+1}$, be r points in $[a, b]$; we have to show that $V(S(x_1), \ldots, S(x_r)) \leqslant V(\boldsymbol{a})$. Let us add knots equal to x_1 to the sequence t_i, until we obtain a knot of multiplicity k at x_1 (if x_1 is not equal to a knot t_i, we add k knots equal to x_1; we may in fact always assume that this is the case). We obtain in this way a new sequence $(\tilde{a}_i^1)_{0 \leqslant i \leqslant n+k-1}$ of coefficients for the function $S(x)$, corresponding to B-splines $\widetilde{B}_i^1$, and such that $V(\tilde{a}_i^1) \leqslant V(a_i)$, as follows from above. Then there is a knot of multiplicity k at x_1: therefore, there exists an index j such that $\widetilde{B}_j^1(x_1) = 1$ (see Theorem 1.1(v)), and so $S(x_1)$ is an element of the sequence $(\tilde{a}_i^1)$; proceeding in the same way for all the x_i's, one gets a sequence $(\tilde{a}_i^r)$ such that $V(\tilde{a}_i^r) \leqslant V(a_i)$, and containing $S(x_i)$, $1 \leqslant i \leqslant r$; removing then the superfluous $\tilde{a}_i^r$, which may only diminish the variation, we obtain $V(S(x_1), \ldots, S(x_r)) \leqslant V(\boldsymbol{a})$.

REMARK 3.3. Theorem 3.4 implies that if we assume that the roots of $S(x)$ are simple, then the number of zeros of $S(x)$ in $[a, b]$ is bounded by $V(a_i)$. Therefore, Theorem 3.4 may be considered as a generalization to spline functions of *Descartes' lemma* (or *Descartes' rule of signs*) which applies to real polynomials.

Let us recall this lemma:

THEOREM 3.5. *Let* $P = \sum_{i=0}^{r} a_i X^i \in \mathbb{R}[X]$. *If* $Z_+(P)$ *designates the number of roots* > 0 *of* P *(counted with multiplicity), then* $Z_+(P) \leqslant V(\boldsymbol{a})$.

The proof is easy by induction on r, and it uses Rolle's lemma (see BENEDETTI and RISLER ([1990], p. 14)). In fact, there exists a proof of Theorem 3.4 that uses a generalization of Rolle's lemma to spline functions: see SCHUMAKER [1981].

4. Approximation by a spline function

Let $g(x)$ be a function of class C^2 defined on $[a, b]$, $(t_i)_{0 \leqslant i \leqslant n+k}$ a sequence of knots as in the previous section, i.e., such that $t_k \leqslant a$ and $t_{n+1} \geqslant b$, allowing us to define B-splines $B_{i,k}$ of degree k ($k \geqslant 1$, $0 \leqslant i \leqslant n-1$). Set $t_i^* = (t_{i+1} + \cdots + t_{i+k})/k$ and $Sg(x) = \sum_{i=0}^{n-1} g(t_i^*) B_{i,k}(x)$. Then $Sg(x)$ is a spline function defined on $[a, b]$, which will be the function approximating g, and will be denoted Sg.

Let us first give a property of the operator S (called Schoenberg's operator, see DE BOOR [1978]), which motivates the choice of the sequence t^*.

THEOREM 4.1. *The operator* S *"reproduces lines", i.e., one has* $Sl = l$ *for any polynomial* $l(X)$ *of degree* $\leqslant 1$.

PROOF. We have to show that $S1 = 1$ and $SX = X$; the first equality comes from Theorem 1.1(iv); for the second, proving $(SX)' = 1$ is enough, which is easy using the differentiation algorithm above. □

THEOREM 4.2. *Set* $|t| = \sup_{k \leqslant i \leqslant n-1}(t_{i+1} - t_i)$. *Then for* $k \geqslant 1$, *there exists a constant* $C(k)$ $(C(k) = k^2/2)$ *such that*

$$\|g - Sg\| \leqslant C(k)|t^2| \|g''\|,$$

the norm being the uniform norm on $[a, b]$, *namely,* $\|f\| = \sup_{x \in [a,b]} |f(x)|$.

PROOF. The proof uses Taylor's formula (see RISLER ([1991], p. 32), or DE BOOR [1978]). □

The curve $S\gamma : y = Sg(x)$ has an additional property, linked to the "variation diminishing property" of the spline functions (Theorem 3.4): it is *regularizing*, that is, for any line D, $\mathrm{card}(D \cap S\gamma) \leqslant \mathrm{card}(D \cap \gamma)$ (see Theorem 4.3 below). Here "$\mathrm{card}(A)$" means the number of elements of the finite set A.

THEOREM 4.3. *Let* γ *be a curve of class* C^1 *defined on* $[a, b]$ *by the equation* $y = g(x)$. *Then for any line* D *transversal to* $S\gamma$, *one has*

$$\mathrm{card}(S\gamma \cap D) \leqslant \mathrm{card}(\gamma \cap D).$$

PROOF. The transformation S reproduces lines, i.e., $Sl = l$ for any affine function $l(x)$. Let $y = l(x)$ be the equation of D; since D intersects $S\gamma$ transversally, one has $\mathrm{card}(S\gamma \cap D) \leqslant V(Sg - l)$, V being the variation in the interval $[a, b]$. However,

$$\begin{aligned} V(Sg - l) &= V\big(S(g - l)\big) \quad (\text{as } Sl = l) \\ &= V\left(\sum_{i=0}^{n-1}(g - l)(t_i^*)B_i(x)\right) \\ &\leqslant V\big((g - l)(t_i^*)\big) \quad (\text{by Theorem 3.4}) \\ &\leqslant \mathrm{card}(\gamma \cap D). \qquad \square \end{aligned}$$

5. Divided differences

B-splines are classically defined with "divided differences".

We will recall their definition and some properties, for the sake of completeness, and in order to help the reader to make connection with classical books on B-splines. Moreover, this definition will allow us to generalize B-splines to several variables.

One may look for instance at DE BOOR [1978] for more details.

NOTATION. If $t_0 < \cdots < t_r$ are points of $\mathbb{R}$ and $g_0, \ldots, g_r$ denote functions defined on $[t_0, t_r]$, one denotes by

$$D\begin{pmatrix} t_0, \ldots, t_r \\ g_0, \ldots, g_r \end{pmatrix}$$

the determinant of the matrix $(g_j(t_i))$ which is of order $(r+1)$, where rows and columns are numbered from 0 to r.

If we assume only that $t_0 \leqslant \cdots \leqslant t_r$, one denotes in the same way the determinant of the matrix $(D^{d_i} g_j(t_i))$, where $d_i = \sup\{j\colon\ t_i = t_{i-1} = \cdots = t_{i-j}\}$, i.e., the determinant of the matrix

$$\begin{pmatrix} g_0(t_0) & g_1(t_0) & \cdots & \cdots \\ g_0'(t_0) & \cdots & \cdots & \cdots \\ \vdots & & & \\ g_0^{(l-1)}(t_0) & \cdots & \cdots & \cdots \\ g_0(t_l) & \cdots & \cdots & \cdots \\ \vdots & & & \end{pmatrix}.$$

DEFINITION. Let $t_0 \leqslant \cdots \leqslant t_r$ be points of $\mathbb{R}$ and let f be a sufficiently differentiable function. One defines the rth *divided difference* of f, denoted $[t_0, \ldots, t_r]f$, in one of the two following equivalent ways:

(i) $[t_0, \ldots, t_r]f$ is the coefficient of X^r in the unique polynomial P of degree $\leqslant r$ such that $P(t_i) = f(t_i)$ for $0 \leqslant i \leqslant r$ (and $P^{(j)}(t_i) = f^{(j)}(t_i)$, $0 \leqslant j \leqslant l_i - 1$, if there are l_i points t_j equal to t_i).

(ii)

$$[t_0, \ldots, t_r]f = \frac{D\binom{t_0, \ldots\ldots\ldots, t_r}{1, \ldots, X^{r-1}, f}}{D\binom{t_0, \ldots, t_r}{1, \ldots, X^r}} \tag{5.1}$$

(see notation above).

For the proof, see, for example, SCHUMAKER [1981].

THEOREM 5.1. *Let k be an integer $\geqslant 1$, $(t_i)_{0 \leqslant i \leqslant n+k}$ be a sequence of points such that $t_i \leqslant t_{i+1}$. We then have*

$$B_{i,k,t}(x) = (t_{i+k+1} - t_i)[t_i, \ldots, t_{i+k+1}](\bullet - x)_+^k,$$

the notation $[t_i, \ldots, t_{i+k+1}](\bullet - x)_+^k$ meaning that one applies the divided difference operator $[t_i, \ldots, t_{i+k+1}]$ to the function $y \mapsto (y-x)_+^k$.

For the proof, see DE BOOR [1978].

Using induction on the integer k, one can prove easily the following formula, which can be generalized to several variables. Formula (5.2) is known as the *Hermite–Genocchi formula*.

Let

$$S_k = \left\{ (\tau_0, \ldots, \tau_k),\ \tau_i \geqslant 0,\ \sum_{i=0}^{k} \tau_i = 1 \right\}$$

be the k-dimensional simplex (embedded in $\mathbb{R}^{k+1}$). We then have, for any C^k function g,

$$[t_0, \ldots, t_k]g = \int_{S_k} g^{(k)}(t_0\tau_0 + \cdots + t_k\tau_k)\, \mathrm{d}\tau_1 \cdots \mathrm{d}\tau_k, \tag{5.2}$$

where the notation $\int_{S_k}$ means that we have set $\tau_0 = 1 - \tau_1 - \cdots - \tau_k$ and that we integrate over the domain $\tau_i \geqslant 0$ $(1 \leqslant i \leqslant k)$ and $\sum_{i=1}^{k} \tau_i \leqslant 1$.

THEOREM 5.2. *Let g be a C^1 function on $\mathbb{R}$ with compact support. One then has the formula*

$$\int_{-\infty}^{+\infty} g(x)B_{i,k,t}(x)\, \mathrm{d}x$$
$$= k!(t_{i+k+1} - t_i) \int_{S_{k+1}} g(t_i\tau_0 + \cdots + t_{i+k+1}\tau_{k+1})\, \mathrm{d}\tau_1 \cdots \mathrm{d}\tau_{k+1}.$$

PROOF. Let G be a function sufficiently differentiable, such that $G^{(k+1)} = g$. We then have

$$\int_{S_{k+1}} g(t_i\tau_0 + \cdots + t_{i+k+1}\tau_{k+1})\, \mathrm{d}\tau_1 \cdots \mathrm{d}\tau_{k+1} = [t_i, \ldots, t_{i+k+1}]G$$

by (5.2). It is enough to prove that

$$\int_{-\infty}^{+\infty} g(x)B_{i,k,t}(x)\, \mathrm{d}x = k!(t_{i+k+1} - t_i)[t_i, \ldots, t_{i+k+1}]G.$$

This can be done by induction on k. □

The formula of Theorem 5.2 may be generalized to the case where x is a variable of $\mathbb{R}^s$, which defines $B_{i,k}(x)$ as a "distribution" on $\mathbb{R}^s$, generalizing the one-variable B-splines. These "B-splines" in several variables will be defined and studied in Section 20.

6. Bernstein polynomials

We will introduce the Bernstein polynomials as special cases of B-splines; Bézier curves will then appear as special spline curves, and will be treated specifically only as examples.

Let us fix k, and set $[a, b] = [0, 1]$; we will look at the B-splines $B_{i,k,t}$, the knot sequence $t = (t_i)$ being defined in the following way:

$$\begin{cases} t_0 = \cdots = t_k = 0, \\ t_{k+1} = \cdots = t_{2k+1} = 1. \end{cases}$$

The functions $B_{i,k}$, $0 \leqslant i \leqslant k$, are then polynomials of degree k on [0,1], verifying (from Theorem 1.1):

$$B_{i,k}(X) = XB_{i,k-1}(X) + (1 - X)B_{i+1,k-1}(X).$$

The polynomials $B_{i,k}$ then make a basis of the space of polynomials (on $[0, 1]$) of degree $\leqslant k$ (Theorem 2.1), called *the Bernstein basis.*

NOTATION. The Bernstein polynomials $B_{i,k}$ are traditionally denoted B_i^k, which makes a distinction with the general B-splines. We will use this notation B_i^k when we consider Bernstein polynomials.

THEOREM 6.1. *We have*

$$B_i^k(X) = \binom{k}{i}(1 - X)^{k-i}X^i \quad (0 \leqslant i \leqslant k).$$

PROOF. The induction relation of B-splines is

$$B_{i,k,t}(X) = XB_{i,k-1,t}(X) + (1 - X)B_{i+1,k-1,t}(X), \tag{6.1}$$

where $B_{i,k}$ is the B-spline of degree k defined with the knot sequence

$$\begin{cases} t_0 = \cdots = t_k = 0, \\ t_{k+1} = \cdots = t_{2k+1} = 1. \end{cases} \tag{t}$$

On the other hand, the Bernstein polynomial B_j^{k-1} is by definition equal to $B_{j,k-1,t'}$, the spline $B_{j,k-1,t'}$ being now evaluated with the knot sequence

$$\begin{cases} t'_0 = \cdots = t'_{k-1} = 0, \\ t'_k = \cdots = t'_{2k-1} = 1. \end{cases} \tag{t'}$$

We then have $B_{j,k-1,t} = B_{j-1,k-1,t'} = B_{j-1}^{k-1}$, $1 \leqslant j \leqslant k$, the spline $B_{0,k-1,t}$ being identically zero (as there is a knot of multiplicity $k + 1$ at 0, for the sequence (t)); relation (6.1) then becomes (for Bernstein polynomials):

$$B_i^k(X) = XB_{i-1}^{k-1}(X) + (1 - X)B_i^{k-1}(X),$$

and so

$$B_i^k(X) = \binom{k}{i}(1 - X)^{k-i}X^i,$$

after the formula: $\binom{k}{i} = \binom{k-1}{i} + \binom{k-1}{i-1}$. □

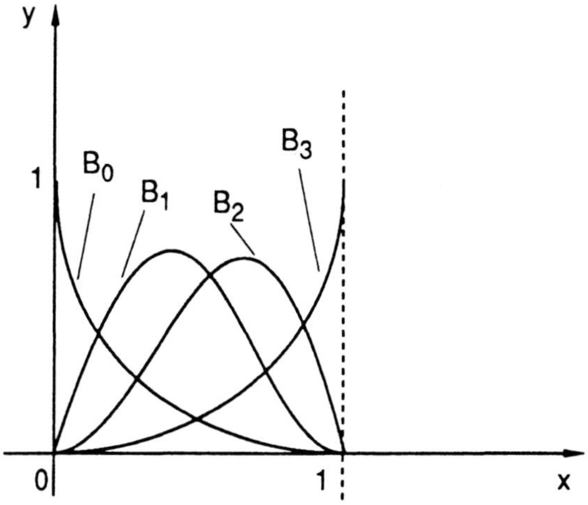

FIG. 6.1. Bernstein polynomials of degree 3.

Let us now state several properties of the Bernstein basis, similar to Theorem 1.1 for B-splines.

THEOREM 6.2.

(i) *The functions* $B_i^k(X)$, $0 \leqslant i \leqslant k$, *constitute a basis of the space* $\mathcal{P}_k$ *(polynomials of degree* $\leqslant k$*).*

(ii) $B_i^k(X) \geqslant 0$, $0 \leqslant X \leqslant 1$.

(iii) $\sum_{i=0}^k B_i^k(X) = 1$.

(iv) *If one sets* $f(X) = \sum_{i=0}^n a_i B_i^k(X)$, *one has* $V(f) \leqslant V(a_i)$.

(v) $(B_i^k)'(X) = k(B_{i-1}^{k-1}(X) - B_i^{k-1}(X))$.

(vi) *One has* $B_i^k(X) = B_{k-i}^k(1 - X)$, $0 \leqslant i \leqslant k$, *for* $X \in [0, 1]$.

PROOF. The proof is the same as for B-splines (Theorem 1.1). $\square$

EXAMPLE. The elements B_i^3, $0 \leqslant i \leqslant 3$, of the Bernstein basis of degree 3 are given by:

$$B_0^3(X) = (1 - X)^3, \qquad B_1^3(X) = 3(1 - X)^2 X,$$
$$B_2^3(X) = 3(1 - X)X^2, \qquad B_3^3(X) = X^3.$$

They are represented in Fig. 6.1.

REMARK 6.1. In the case of an interval $[a, b]$, with

$$\begin{cases} t_0 = \cdots = t_k = a, \\ t_{k+1} = \cdots = t_{2k+1} = b, \end{cases}$$

the corresponding $B_{i,k,t}$'s, $0 \leqslant i \leqslant k$, constitute the *Bernstein basis* for the interval $[a, b]$; if B_i^k, $0 \leqslant i \leqslant k$, is the Bernstein basis on $[0, 1]$, we then have

$$B_{i,k,t}(X) = B_i^k\left(\frac{X - a}{b - a}\right), \quad a \leqslant X \leqslant b,$$

as is easily seen.

7. B-spline curves

We give the definition and several examples of B-spline curves.

We will denote by t the variable of a parametric curve. Apart from this change, the notation remains the same as in Section 1. Let $P_0, \ldots, P_{n-1}$ be n points in $\mathbb{R}^s$.

DEFINITION. (i) The B-spline curve (one also says *spline curve*) associated with the polygon $P_0, \ldots, P_{n-1}$, is the parametric curve γ defined by

$$S(t) = \sum_{i=0}^{n-1} P_i B_{i,k}(t), \quad a \leqslant t \leqslant b. \tag{7.1}$$

The polygon $P_0, \ldots, P_{n-1}$ is called the *control polygon* of the curve γ.

(ii) In the special case of Bernstein polynomials $B_i^k(t)$, the curve

$$B(t) = \sum_{i=0}^{k} P_i B_i^k(t), \quad 0 \leqslant t \leqslant 1,$$

is called the *Bézier curve* associated with the polygon $(P_0, \ldots, P_k)$; in this case, the number of P_i's is equal to $k + 1$, if we look at Bernstein polynomials of degree k.

The properties of B-spline functions studied in Section 1 give the following.

THEOREM 7.1.

(i) *The curve γ does not in general pass through the points P_i; one has however $S(a) = P_0$ and $S(b) = P_{n-1}$ when $t_0 = \cdots = t_k = a$, $t_n = \cdots = t_{n+k} = b$, and in this case, γ is tangent at P_0 and P_{n-1} to the sides of the control polygon.*

(ii) *The curve γ is in the convex hull of the points $P_0, \ldots, P_{n-1}$. More precisely, if $t_i \leqslant t < t_{i+1}$, $S(t)$ is in the convex hull of $(P_{i-k}, \ldots, P_i)$. We say that the curve γ has the convex hull property.*

(iii) *When the knots t_i, $k+1 \leqslant i \leqslant n-1$, are simple, the curve γ is a C^{k-1}-curve, and is formed with n parametrized polynomial arcs of degree $\leqslant k$.*

(iv) *The curve γ is invariant by affine maps of $\mathbb{R}^s$: if h is such a map, the points $h(P_i)$ are the control points of the curve $h(S(t))$.*

(v) *The operator $S(t)$ is regularizing, that is, if $\mathbf{P}$ denotes the control polygon of γ, and if H is a transversal hyperplane to γ, one has*

$$\operatorname{card}(H \cap \gamma) \leqslant \operatorname{card}(H \cap \mathbf{P}).$$

REMARK 7.1. The B-spline curves have a *local* behavior, in the following sense:

(1) If X is a point on the curve with parameter value t_0, the position of X depends only on at most $k+1$ points P_i, since if $t_j \leqslant t_0 < t_{j+1}$, we have

$$S(t_0) = \sum_{i=0}^{n-1} P_i B_{i,k}(t_0) = \sum_{i=j-k}^{j} P_i B_{i,k}(t_0).$$

(2) In the same way, each point P_j "has an influence" on the curve γ for the values of t such that $B_{j,k}(t) \neq 0$, i.e., for $t_j \leqslant t < t_{j+k+1}$. Thus, if one moves the point P_j, only a small part of the curve $S(t)$ is modified.

EXAMPLES (*see* SABLONNIERE [1987]). (1) Let us look at a quadratic curve ($k=2$) corresponding to the knot sequence in Figs. 7.1 and 7.2. The reader will check that the curve is tangent to the sides P_1P_2, P_2P_3 and P_3P_4 at their middle, which are the points $S(t_3)$, $S(t_4)$ and $S(t_5)$ respectively (this is special for the case $k=2$).

(2) If we want a *closed* $\mathcal{C}^1$ curve, we may take a uniform sequence of knots (setting for instance $t_i = i$, $i \in \mathbb{Z}$). We then have $B_i(t) = B_{i+r}(t+r)$, $r \in \mathbb{Z}$ (i.e., the B-spline basis elements are integer translates of each other: see Fig. 7.3).

Let $S(t) = \sum_{i=-\infty}^{+\infty} P_i B_{i,2}(t)$, setting $P_{6+i} = P_i$; we then have $S(t-6) = S(t)$ and the curve is defined on any interval of length 6 (Fig. 7.4).

(3) Let us draw in Fig. 7.5 two Bézier curves of degree 3, for two different shapes of the control polygon.

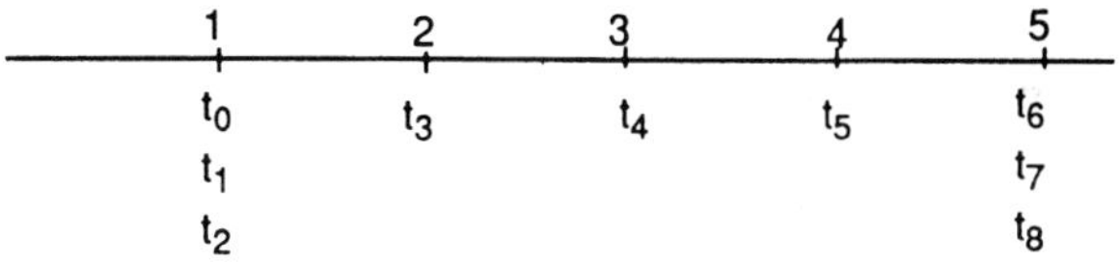

FIG. 7.1.

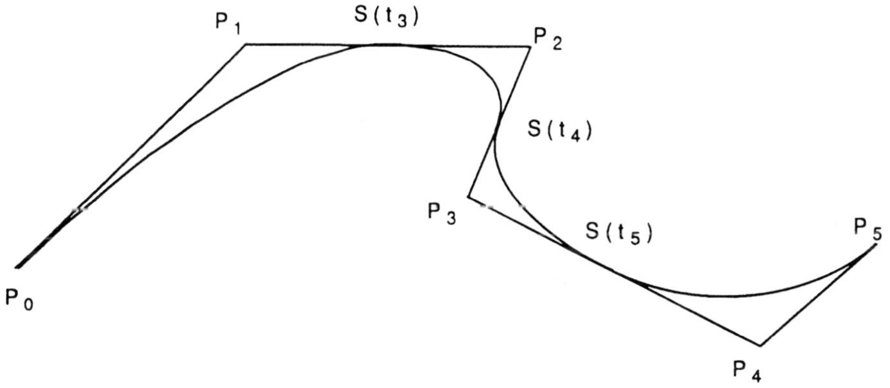

FIG. 7.2. A quadratic B-spline curve.

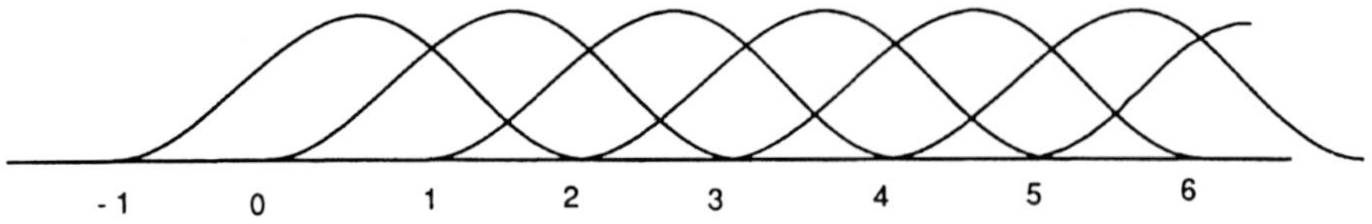

FIG. 7.3. Uniform B-splines.

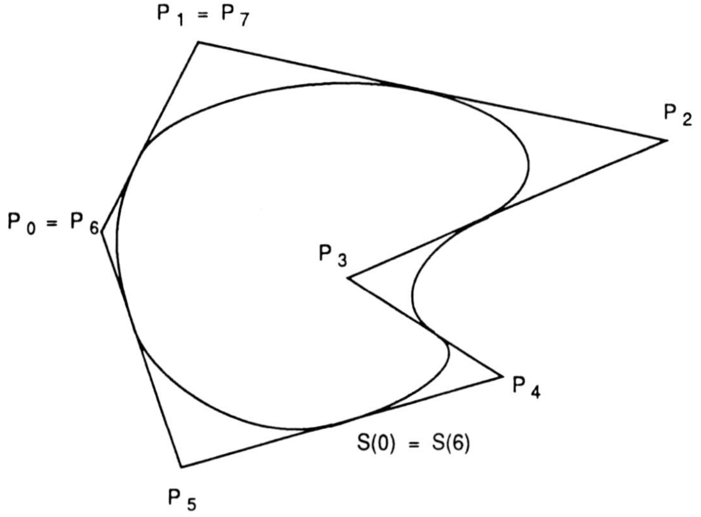

FIG. 7.4. Closed spline curve of degree 2.

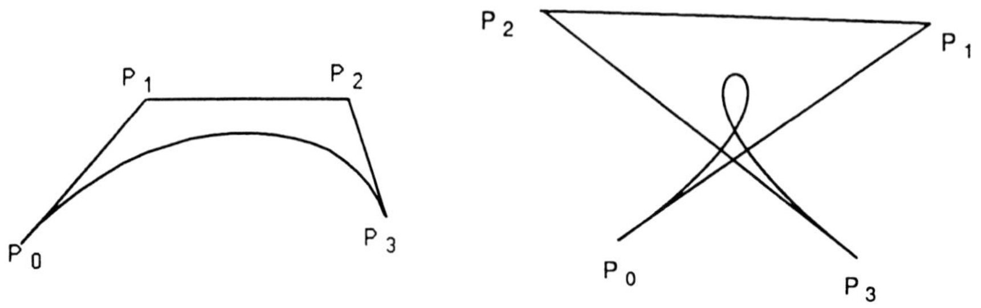

FIG. 7.5. Bézier curves of degree 3.

8. Algorithms for spline curves

We give applications of the algorithms of Section 3 to spline curves. In particular, the knot insertion algorithm allows for the subdivision of spline (or Bézier) curves.

Evaluation algorithm. This algorithm is also called "De Boor–Cox" in the B-spline case, or "De Casteljau" in the Bézier curve case. It is a simple adaptation of the evaluation algorithm for spline curves (Theorem 3.1). Set

$$S(t) = \sum_{i=0}^{n-1} P_i B_{i,k}(t),$$

and assume $t_j \leqslant t < t_{j+1}$. Then if we set

$$\begin{aligned} P_i^0 &= P_i, \quad j-k \leqslant i \leqslant j, \\ P_i^{r+1}(t) &= \omega_{i,k-r}(t) P_i^r + \big(1 - \omega_{i,k-r}(t)\big) P_{i-1}^r \\ &= \frac{(t - t_i) P_i^r(t) + (t_{i+k-r} - t) P_{i-1}^r}{t_{i+k-r} - t_i} \end{aligned}$$

for $0 \leqslant r \leqslant k-1$ and $j-k+r+1 \leqslant i \leqslant j$, we have,

$$P_j^k(t) = S(t).$$

REMARK 8.1. In the case of a Bézier curve, the algorithm is simpler, as the $\omega_{i,k}(t)$'s do not depend on k. If $B(t) = \sum_{i=0}^k P_i B_i^k(t)$ is a Bézier curve (with control points $P_i \in \mathbb{R}^s$), the evaluation algorithm (or De Casteljau algorithm) is the following:

$$\begin{aligned} P_i^0(t) &= P_i, \quad 0 \leqslant i \leqslant k-1, \\ P_i^{r+1}(t) &= (1-t) P_{i-1}^r + t P_i^r, \quad 0 \leqslant r \leqslant k-1, \quad r+1 \leqslant i \leqslant k, \\ P_k^k(t) &= B(t). \end{aligned}$$

Differentiation algorithm. The algorithm is explained in Theorem 3.2.

Let $S(t) = \sum_{i=0}^{n-1} P_i B_{i,k}(t)$ be a spline function, with $t_j \leqslant t < t_{j+1}$. We then have

$$\mathrm{D}^r S(t) = \sum_{i=j-k+r}^{j} P_i^r B_{i,k-r}(t)$$

with

$$P_i^0 = P_i, \quad j-k \leqslant i \leqslant j,$$

$$P_i^r = \begin{cases} (k-r+1) \dfrac{P_i^{r-1} - P_{i-1}^{r-1}}{t_{i+k-r+1} - t_i} & \text{if } t_i < t_{i+k-r+1},\ j-k+r \leqslant i \leqslant j, \\ 0 & \text{otherwise.} \end{cases}$$

To evaluate the derivative $\mathrm{D}^r S(t)$ at a point t, one may then use the above evaluation algorithm to compute $B_{i,k-r}(t)$; this allows us for instance to find the values of the successive derivatives at a knot t_i, and hence to find the polynomial expression of the curve $S(t)$ between t_i and t_{i+1}, with Taylor's formula.

Let us look at the case of Bézier curves (see FIOROT ([1987], p. 14)): let $B(t) = \sum_{i=0}^{k} P_i B_i^k(t)$ be a Bézier curve of degree k.

Set

$$\Delta P_i = P_{i+1} - P_i,$$

$$\Delta^2 P_i = \Delta(\Delta P_i) = P_{i+2} - 2P_{i+1} - P_i,$$

$$\Delta^k P_i = \Delta(\Delta^{k-1} P_i) = \sum_{j=0}^{k} (-1)^{k-j} \binom{k}{j} P_{i+j}.$$

THEOREM 8.1. *We have*

$$\mathrm{D}^r B(t) = \frac{k!}{(k-r)!} \sum_{i=0}^{k-r} B_i^{k-r} \Delta^r P_i.$$

PROOF. Immediate by induction on r, using Theorem 6.2(v). □

If we now want to evaluate $\mathrm{D}^r B$ at t, we may use De Casteljau's algorithm.

Note that this algorithm commutes with the Δ operation, as the latter is linear ($\Delta(\lambda P_{i-1} + \mu P_i) = \lambda \Delta P_{i-1} + \mu \Delta P_i$). The evaluation algorithm of B and its derivatives $\mathrm{D}^r B$ at a point t then becomes:

(1) Evaluate $B(t)$ with the evaluation algorithm, which gives a triangular array $T^{(0)}$ of points P_i^r, $0 \leqslant r \leqslant k$, $r \leqslant i \leqslant k$.

(2) Pass from array $T^{(l)}$ to array $T^{(l+1)}$ by difference (operation Δ).

(3) $\mathrm{D}^r B(t)$ is given by the last element of the array $T^{(r)}$.

Knot insertion algorithm. The algorithm is described in Theorem 3.3. Recall that if

$$S(t) = \sum_{i=0}^{n-1} P_i B_{i,k}(t)$$

and if we add a new knot $\hat{t}$, we have

$$S(t) = \sum_{i=0}^{n} \widehat{P}_i \widehat{B}_{i,k}(t),$$

with

$$\widehat{P}_i = \begin{cases} P_i & \text{if } t_{i+k} \leqslant \hat{t}, \\ \omega_{i,k}(\hat{t}) P_i + \big(1 - \omega_{i,k}(\hat{t})\big) P_{i-1} & \text{if } t_i < \hat{t} < t_{i+k}, \\ P_{i-1} & \text{if } \hat{t} \leqslant t_i. \end{cases} \tag{8.1}$$

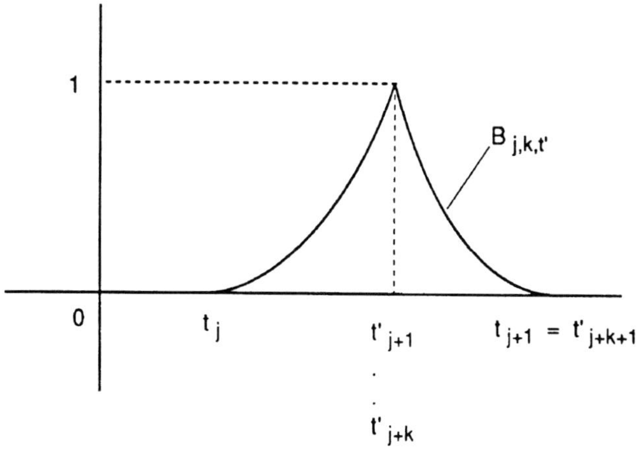

FIG. 8.1. After the insertion of k knots.

We see that these formulas are the same as those of the first step of the evaluation algorithm (if for instance we have $t_j < \hat{t} < t_{j+1}$, the condition $t_i < \hat{t} < t_{i+k}$ is equivalent to $j-k+1 \leqslant i \leqslant j$, which corresponds to the evaluation algorithm). We see again that, if we successively insert k knots at the same place τ (with $t_j < \tau < t_{j+1}$), and if at the rth time we get the points P_i^r ($P_i^0 = P_i$), then $S(\tau) = P_j^k$. Let us look at the situation after the insertion of k knots at τ, and assume $t_j < \tau < t_{j+1}$. If we set $(t_i')_{0 \leqslant i \leqslant n+2k}$ for the new knot sequence, the support of the B-spline $B_{j,k,t'}$ is $[t_j, t_{j+1}]$ (as then $t_{j+1} = t'_{j+k+1}$), and one has $B_{j,k,t'}(\tau) = 1$ (cf. Theorem 1.1 and Fig. 8.1).

We may then write, for a new sequence (P_i') of control points:

$$S(t) = \sum_{i=0}^{n+k-1} P_i' B_{i,k,t'}(t),$$

with $S(\tau) = P_j'$.

Let us add now a $(k+1)$th knot $\hat{t}$ equal to τ. Formulas (8.1) then give

$$\widehat{P}_i = P_i' \quad \text{for } i \leqslant j, \text{ and}$$
$$\widehat{P}_i = P_{i-1}' \quad \text{for } i > j.$$

In other words, the point P_j' has been *doubled* and the B-spline $B_{j,k,t'}$ "divided in two" (one half $\widehat{B}_{j,k}$ with support $[t_j, \tau]$ and the other $\widehat{B}_{j+1,k}$ with support $[\tau, t_{j+1}] = [\tau, t'_{j+k+1}]$). The supports of all the B-splines $\widehat{B}_{i,k}$ are then contained either in $[a, \tau]$, or in $[\tau, b]$, and the spline curve $S(t)$ splits into two spline curves (of the same degree k), one defined on $[a, \tau]$ and the other on $[\tau, b]$. These two curves are defined in the following way (setting $B_{i,k,t'}(t) = \widehat{B}_{i,k}(t)$),

$$S_1(t) = \sum_{i=0}^{j} P_i^k \widehat{B}_{i,k}(t) \quad \text{for } a \leqslant t \leqslant \tau, \tag{8.2}$$

and

$$S_2(t) = \sum_{i=j+1}^{n+k-1} P_i^k \widehat{B}_{i,k}(t) \quad \text{for } \tau \leqslant t \leqslant b. \tag{8.3}$$

REMARK 8.2. One has in fact $\widehat{B}_{j,k}(\tau) = 0$ and $\widehat{B}_{j+1,k}(\tau) = 1$, as the B-splines are always-right continuous. In the formula (8.2) above, one generally sets $\widehat{B}_{j,k}(\tau) = 1$ so that $S_1(t)$ is continuous at τ (giving $S(t) = S_1(t) + S_2(t)$ except for $t = \tau$). This preceding fact allows to subdivide the spline curves; we will begin by studying in detail the case of Bézier curves, for which the algorithm is especially simple.

Subdivision algorithm for a Bézier curve. Set for instance $\tau = \frac{1}{2}$. If

$$B(t) = \sum_{i=0}^{k} P_i B_i^k(t)$$

is a Bézier curve of degree k, the De Casteljau algorithm for the evaluation of $B(\frac{1}{2})$ is then

$$\begin{aligned} P_i^0 &= P_i, \quad 0 \leqslant i \leqslant k, \\ P_i^{j+1} &= \tfrac{1}{2}(P_{i-1}^j + P_i^j), \quad j+1 \leqslant i \leqslant k, \\ P_k^k &= B(\tfrac{1}{2}). \end{aligned}$$

After once duplicating the point P_k^k, and applying what has been done for the splines above, we see that the curve splits into two curves:

$$B_1(t) = \sum_{i=0}^{k} P_i^i \widehat{B}_i^k(t) \quad \text{for } 0 \leqslant t \leqslant \tfrac{1}{2},$$

with $\widehat{B}_i^k(t) = B_i^k(2t)$ (see Remark 6.1) and

$$B_2(t) = \sum_{i=0}^{k} P_i^i \widehat{B}_i^{\prime k}(t) \quad \text{for } \tfrac{1}{2} \leqslant t \leqslant 1,$$

now with $\widehat{B}_i^{\prime k}(t) = B_i^k(2t - 1)$ (see Remark 6.1).

EXAMPLE ($k = 3$). The algorithm may be represented by a triangular array:

$$\begin{array}{ccccccc} P_0 & & P_1 & & P_2 & & P_3 \\ & P_1^1 & & P_2^1 & & P_3^1 & \\ & & P_2^2 & & P_3^2 & & \\ & & & P_3^3 & & & \end{array}$$

We may now iterate the process, applying the same algorithm to $B_1(t)$ for $t = \frac{1}{4}$, and to $B_2(t)$ for $t = \frac{3}{4}$. Continuing in this way, one obtains at the nth step 2^n Bézier curves whose union is the initial curve, and 2^n control polygons Π_n^i whose total number of vertices is $2^n k + 1$. Among those vertices, there are $2^n + 1$ points of the initial curve: the points corresponding to $t = p/2^n$ (p integer, $0 \leqslant p \leqslant 2^n$), see Fig. 8.2.

THEOREM 8.2. *The sequence of polygons Π_n^k converges uniformly towards the Bézier curve $B(t)$, with convergence rate $\lambda/2^n$ in the case where $\tau = \frac{1}{2}$; the constant λ is a constant independent of n.*

The proof is easy, see RISLER ([1991], p. 54).

REMARK 8.3. The subdivision algorithm gives the vertices of the polygon Π_n^i; after a certain step, and taking in account the resolution of the screen, the polygon Π_n^i is "indistinguishable" from the curve. This allows a quick drawing (as the convergence is exponential) of a Bézier curve $B(t)$.

Subdivision algorithm for uniform splines. Let $B_k(X)$ be the spline of degree k, built with integer knots, and with support $[0, \ldots, k+1]$. Its translates $B_k(X - j)$ satisfy $\sum_{j \in \mathbb{Z}} B_k(X - j) = 1$, as follows from Theorem 1.1 applied to a sufficiently large interval $[a, b]$ containing X. The $B_k(X - j)$'s are by definition the *uniform B-splines* corresponding to the knot sequence $\tau = \mathbb{Z}$.

Let now $\tau_p = \{j/p,\ j \in \mathbb{Z}\}$, $p \geqslant 2$, be a finer uniform knot sequence: the corresponding B-splines of degree k are the $B_k(pX - j)$'s, and we may expand $B_k(X)$ with respect to the basis $B_k(pX - j)$ as:

$$B_k(X) = \sum_j a_p^k(j/p) B_k(pX - j).$$

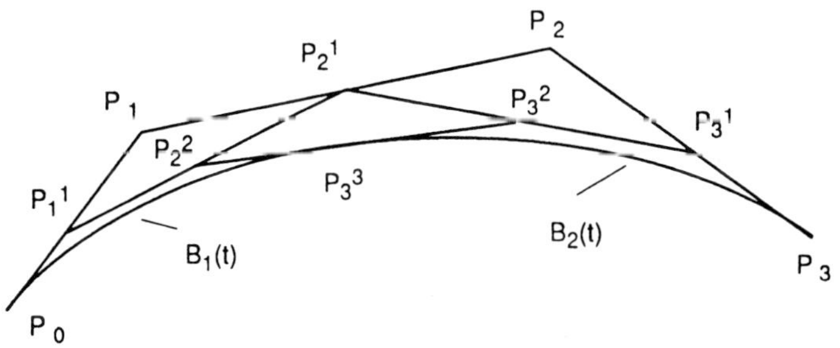

FIG. 8.2. Subdivision of a Bézier curve of degree 3.

REMARK 8.4. It is easy to see (see RISLER ([1991], p. 60)) that we have

$$B_k(X) = \sum_j a_p^k(j/p) B_k(pX - j),$$

setting $a_p^k(j/p) = 0$ for $j \notin [0, (k+1)(p+1)]$.

Let now

$$B(t) = \sum_{i \in \mathbb{Z}} P_i B_k(t - i)$$

be a spline curve with control polygon (P_i); $B(t)$ is well defined, as the sum is finite for each t. It is often convenient to consider an infinite sequence of P_i's; one may return to the notation of Section 1, looking at functions on $[a, b] = [k, n]$ for instance, and to the knots $(t_i)_{0 \leqslant i \leqslant n+k}$. With the above notation and conventions (see Remark 8.4), we have:

$$B(t) = \sum_{j \in \mathbb{Z}} Q_p^k(j/p) B_k(pt - j),$$

where the new control polygon $Q_p^k(j/p)$ satisfies the following relations

$$Q_p^k(j/p) = \sum_{i \in \mathbb{Z}} a_p^k(j/p - i) P_i.$$

The subdivision algorithm then follows from the following theorem whose proof is omitted.

THEOREM 8.3. *The numbers* $a_p^k(j/p)$ *are the coefficients of the development of the rational fraction*

$$\frac{1}{p^k} \left(\frac{Z^p - 1}{Z - 1} \right)^{k+1};$$

in other words, one has

$$\frac{1}{p^k} \left(\frac{Z^p - 1}{Z - 1} \right)^{k+1} = \sum_{j=0}^{(k+1)(p-1)} a_p^k(j/p) Z^j. \tag{8.4}$$

For the proof, see RISLER ([1991], p. 61).

As an easy consequence, we obtain:

THEOREM 8.4 (Subdivision algorithm). *With the above notation, the following induction formula holds for the control polygons*:

$$Q_p^{k+1}(j/p) = \frac{1}{p}\sum_{l=0}^{p-1} Q_p^k(j/p - l/p) \quad (j \in \mathbb{Z}).$$

In particular, we obtain for $p = 2$,

$$Q_2^{k+1}(j) = \tfrac{1}{2}\big(Q_2^k(j) + Q_2^k(j - \tfrac{1}{2})\big),$$
$$Q_2^{k+1}(j + \tfrac{1}{2}) = \tfrac{1}{2}\big(Q_2^k(j + \tfrac{1}{2}) + Q_2^k(j)\big),$$

which allows a very easy construction of the control polygon $Q_p^k(j/p)$, starting from the relations $Q_p^0(j) = \cdots = Q_p^0(j + (p-1)/p) = P_j$ $(j \in \mathbb{Z})$, which follow from the fact that $a_p^0(j/p) = 1$ for $0 \leqslant j \leqslant p-1$, and 0 otherwise.

9. Interpolation

Let $P_0, \ldots, P_{n-1}$ be n points in $\mathbb{R}^s$; we look for a spline curve $B(t)$ of degree k passing through the points P_i. To this end, let us take a knot sequence $(t_i)_{0 \leqslant i \leqslant n+k}$ (as always, we assume that $t_i < t_{i+k+1}$, $0 \leqslant i \leqslant n-1$). We then try to find control points Q_j and points $u_i \in \mathbb{R}$, $u_i \leqslant u_{i+1}$, such that the curve $B(t) = \sum_{j=0}^{n-1} B_{j,k}(t) Q_j$ satisfies $B(u_i) = P_i$, $0 \leqslant i \leqslant n-1$; we then have to solve the linear system:

$$\sum_{j=0}^{n-1} B_{j,k}(u_i) Q_j = P_i, \quad 0 \leqslant i \leqslant n-1.$$

THEOREM 9.1. *The matrix $N = (B_{j,k}(u_i))$ is invertible if and only if all the diagonal elements are nonzero, i.e., if $B_{j,k}(u_j) \neq 0$, $0 \leqslant j \leqslant n-1$, or else if $t_i < u_i < t_{i+k+1}$, $0 \leqslant i \leqslant n-1$.*

REMARKS 9.1. (1) One may possibly have $u_i = t_i$ and $Q_i = P_i$ when t_i is a knot of order $k+1$ ($t_i = t_{i+1} = \cdots = t_{i+k} < t_{i+k+1}$; see Theorem 1.1).

(2) One generally chooses $u_i = t_i^* = (t_{i+1} + \cdots + t_{i+k})/k$ (see Section 4), which implies that the condition of Theorem 9.1 is fulfilled, except when t_{i+1} is a knot of order $k+1$ ($t_i < t_{i+1} = \cdots = t_{i+k+1}$). Moreover, the choice $u_i = t_i^*$ implies that the spline curve will have a smaller number of oscillations than the polygon (P_i) (cf. Theorem 4.3), i.e., that this oscillation will be "minimal" among all the interpolating curves.

(3) In some cases, the values of the parameters u_i (such that $B(u_i) = P_i$) are a priori fixed, but we have to choose the values of t_j; we then generally set (see SABLONNIERE ([1987], p. 8)):

$$t_0 = \cdots = t_k = u_0,$$

$$t_{i+k} = \frac{u_i + \cdots + u_{i+k-1}}{k}, \quad 1 \leqslant i \leqslant n-k-1,$$

$$t_n = \cdots = t_{n+k} = u_{n-1}.$$

PROOF of Theorem 9.1. *Necessity.* Assume that there exists i such that $B_i(u_i) = 0$ (with $0 \leqslant i \leqslant n-1$); assume for instance that $u_i \leqslant t_i$ (and $t_i < t_{i+k}$ if $u_i = t_i$, see Section 1). Then each of the $(i+1)$ first rows of the matrix N contains at most i nonzero terms (namely $B_0(u_r), \ldots, B_{i-1}(u_r)$ for the rth row ($r \leqslant i$), as $B_j(u_r) = 0$ for $j \geqslant i$), and then these $(i+1)$ rows are dependent, which implies that the matrix N is singular.

Sufficiency. Let

$$\sum_{j=0}^{n-1} \lambda_j B_j(u_i) = 0, \quad 0 \leqslant i \leqslant n-1,$$

be a relation between the columns of N. Set $B(t) = \sum_{j=0}^{n-1} \lambda_j B_j(t)$: we then have by hypothesis $B(u_i) = 0$, $0 \leqslant i \leqslant n-1$. We have to prove that, with the hypothesis $B_i(u_i) \neq 0$, $0 \leqslant i \leqslant n-1$, this implies $\lambda_j = 0$, $0 \leqslant j \leqslant n-1$. This is a consequence of the variation diminishing property of B-splines (see Section 3), which implies a property similar to the Descartes' rule of signs. For details, see RISLER ([1991], p. 68–69). □

REMARKS 9.2. (1) The same proof implies that under the assumptions of Theorem 9.1, all the *principal minors* of the matrix N (i.e., all the determinants of the matrix of the form $(B_j(u_i))$ $(i = i_1, \ldots, i_r,\ j = i_1, \ldots, i_r)$) are nonzero.

(2) The matrix N is "sparse". More precisely, it is a "band" matrix, which has at most $k+1$ nonzero elements in each row: the $(i+1)$th row of the matrix N is the row $B_j(u_i)$, $0 \leqslant j \leqslant n-1$; if $t_l \leqslant u_i \leqslant t_{l+1}$, only the B-splines $B_{l-k}, \ldots, B_l$ are possibly nonzero at u_i.

THEOREM 9.2. *Assume that the matrix $N = (B_j(u_i))$ satisfies the conditions of Theorem 9.1 (i.e., $|N| \neq 0$). One may then write $N = LU$, where L is an inferior triangular matrix, and U a superior triangular one.*

PROOF. This results from the fact that the principal minors of N are all nonzero: see CIARLET ([1982], p. 83–84). □

REMARKS 9.3. (1) The existence of the factorization $N = LU$ is equivalent to the fact that the resolution of the linear system $NX = Y$ can be made without *pivoting*, which makes the resolution much easier, especially when N is a band matrix; see CIARLET [1982], or DE BOOR [1978].

(2) Once the factorization $N = LU$ is computed, the solution of any linear system of the form $NX = Y$ is very fast: one has to solve successively the two following triangular systems:

$$LZ = Y,$$

$$UX = Z.$$

(3) If the points P_i, $0 \leqslant i \leqslant n-1$, are given and $[a,b] = [0,1]$, and if we want to determine a spline curve $B(t)$ of degree k, and points $\theta_i \in [0,1]$ such that $B(\theta_i) = P_i$, $0 \leqslant i \leqslant n-1$, the simplest method consists in setting

$$L_i = |P_0P_1| + \cdots + |P_{i-1}P_i|, \qquad 1 \leqslant i \leqslant n-1,$$

and taking $\theta_i = L_i/L_n$. (In other words, we take for θ_i the curvilinear abscissa of the piecewise linear spline passing through the points P_i.) One then deduces the value of the knots t_i, using Remark 9.1. For cubic spline interpolation, see BÖHM, FARIN and KAHMANN ([1984], p. 22). For the choice of the points θ_i, see EPSTEIN [1976].

10. Matrix representation

A useful way to represent uniform B-splines curves, or Bézier curves, is the matrix representation; one can then use matrix algebra to compute rotations, change of reference frame, etc.

Let us look at a spline curve $B(t) = \sum_{i=0}^{n-1} P_i B_{i,k}(t)$ of degree k; by restricting $B(t)$ to the interval $[t_i, t_{i+1}[$ ($t_i < t_{i+1}$), we can identify it with a polynomial in t of degree $\leqslant k$; we will always assume in this section that the interval $[t_i, t_{i+1}[$ is in fact $[0,1[$; this can be obtained with an affine change of variable. We may then write (on $[0,1\,[$):

$$B(t) = \sum_{j=i-k}^{i} B_{j,k}P_j = \begin{pmatrix} B_{i-k}(t) & \cdots & B_i(t) \end{pmatrix} \begin{pmatrix} P_{i-k} \\ \vdots \\ P_i \end{pmatrix},$$

and each $B_{i,k}(t)$ is a polynomial in t, which may be written as its Taylor series at 0 (see the differentiation algorithm, Section 8); we may therefore write:

$$\begin{pmatrix} B_{i-k}(t) & \cdots & B_i(t) \end{pmatrix} = \begin{pmatrix} 1 & t & \cdots & t^k \end{pmatrix} M,$$

M being a $(k+1) \times (k+1)$ matrix; $B(t)$ is thus represented (on $[t_i, t_{i+1}[$) in the following matrix form:

$$B(t) = \begin{pmatrix} 1 & t & \cdots & t^k \end{pmatrix} M \begin{pmatrix} P_{i-k} \\ \vdots \\ P_i \end{pmatrix}, \quad 0 \leqslant t < 1.$$

We will give some examples of matrix representation, in the (mostly used) case where $k = 3$. Sometimes, the uniform B-splines are defined directly in matrix form. See for instance FAUX and PRATT [1979] and BARSKY, BARTELS and BEATTY [1987].

Uniform B-splines with integer knots. These B-splines are defined in Section 8.

Let us begin by computing the expression of $B_3(t)$ on each interval $[t_i, t_{i+1}[$, $0 \leqslant i \leqslant 3$, $B_3(t)$ being the B-spline of degree 3 with knots at $0, 1, 2, 3, 4$. To do that, it is enough to apply Definition 1.1, viz.,

$$\begin{aligned}
B_3(t) = B_{0,3}(t) &= \tfrac{1}{3}\big(tB_{0,2}(t) + (4-t)B_{1,2}(t)\big),\\
B_{0,2}(t) &= \tfrac{1}{2}\big(tB_{0,1}(t) + (3-t)B_{1,1}(t)\big),\\
B_{1,2}(t) &= \tfrac{1}{2}\big((t-1)B_{1,1}(t) + (4-t)B_{2,1}(t)\big),\\
B_{0,1}(t) &= tB_{0,0}(t) + (2-t)B_{1,0}(t),\\
B_{1,1}(t) &= (t-1)B_{1,0}(t) + (3-t)B_{2,0}(t),\\
B_{2,1}(t) &= (t-2)B_{2,0}(t) + (4-t)B_{3,0}(t),
\end{aligned}$$

$$B_{i,0}(t) = \begin{cases} 1 & \text{if } t_i \leqslant t < t_{i+1},\\ 0 & \text{otherwise,}\end{cases}$$

which gives

$$B_3(t) = \begin{cases} \frac{1}{6}t^3 & \text{for } 0 \leqslant t < 1,\\ \frac{1}{6}\big(-3t^3 + 12t^2 - 12t + 4\big) & \text{for } 1 \leqslant t < 2,\\ \frac{1}{6}\big(3t^3 - 24t^2 + 60t - 44\big) & \text{for } 2 \leqslant t < 3,\\ \frac{1}{6}(4-t)^3 = \frac{1}{6}\big(-t^3 + 12t^2 - 48t + 64\big) & \text{for } 3 \leqslant t < 4.\end{cases} \tag{10.1}$$

Let us now look at a uniform spline curve $\sum_i B_3(t-i)P_i$ defined on $\mathbb{R}$. On the interval $[0, 1[$, it reduces to

$$\begin{aligned}
&B_3(t+3)P_{-3} + B_3(t+2)P_{-2} + B_3(t+1)P_{-1} + B_3(t)P_0\\
&\quad = \tfrac{1}{6}\big[(1-t)^3P_{-3} + \big(3(t+2)^3 - 24(t+2)^2 + 60(t+2) - 44\big)P_2\\
&\qquad + \big(-3(t+1)^3 + 12(t+1)^2 - 12(t+1) + 4\big)P_{-1} + t^3P_0\big],\\
&\quad = \tfrac{1}{6}\big[\big(-t^3 + 3t^2 - 3t + 1\big)P_{-3} + \big(3t^3 - 6t^2 + 4\big)P_{-2}\\
&\qquad + \big(-5t^3 + 3t^2 + 3t + 1\big)P_{-1} + t^3P_0\big],
\end{aligned}$$

which gives the following matrix representation:

$$B(t) = \frac{1}{6}\begin{pmatrix}1 & t & t^2 & t^3\end{pmatrix}\begin{pmatrix}1 & 4 & 1 & 0\\ -3 & 0 & 3 & 0\\ 3 & -6 & 3 & 0\\ -1 & 3 & -3 & 1\end{pmatrix}\begin{pmatrix}P_{-3}\\ P_{-2}\\ P_{-1}\\ P_0\end{pmatrix} \tag{10.2}$$

for $0 \leqslant t < 1$.

The portion of the curve $B(t)$ corresponding to $i \leqslant t < i+1$ is written with the same matrix, as:

$$B(\theta) = \frac{1}{6}\begin{pmatrix}1 & \theta & \theta^2 & \theta^3\end{pmatrix}\begin{pmatrix}1 & 4 & 1 & 0\\ -3 & 0 & 3 & 0\\ 3 & -6 & 3 & 0\\ -1 & 3 & -3 & 1\end{pmatrix}\begin{pmatrix}P_{i-3}\\ P_{i-2}\\ P_{i-1}\\ P_i\end{pmatrix}, \quad 0 \leqslant \theta < 1.$$

Bézier curves. See Section 7. Let P_i $(0 \leqslant i \leqslant 3)$ be points in $\mathbb{R}^s$; we have:

$$f(t) = P_0 B_0^3(t) + P_1 B_1^3(t) + P_2 B_2^3(t) + P_3 B_3^3(t),$$

where the $B_i^3(t)$'s are the Bernstein polynomials of degree 3 (see Section 6); whence

$$f(t) = \frac{1}{6}\begin{pmatrix}1 & t & t^2 & t^3\end{pmatrix}\begin{pmatrix}1 & 0 & 0 & 0\\ -3 & 3 & 0 & 0\\ 3 & -6 & 3 & 0\\ -1 & 3 & -3 & 1\end{pmatrix}\begin{pmatrix}P_0\\ P_1\\ P_2\\ P_3\end{pmatrix}, \quad 0 \leqslant t \leqslant 1. \tag{10.3}$$

11. Junction between two curves

After studying the C^r junction between curves, the concept of geometric continuity is introduced. As an example of application, the β-splines are considered, after BARSKY [1988].

Case of spline curves. Let us consider two spline curves of degree k in $\mathbb{R}^s$:

$$S(t) = \sum_{i=0}^{n-1} B_{i,k}(t) P_i$$

defined on $[a, b]$ with knots $t = (t_0, \ldots, t_{n+k})$, and

$$S'(t) = \sum_{j=0}^{n'-1} B_{j,k} P'_j$$

defined on $[a', b']$ with knots $t' = (t'_0, \ldots, t'_{n'+k})$. If we want to join S and S', it is most convenient to assume that $t'_0 = t_n, \ \ldots, \ t'_k = t_{n+k}$ (recall that we have $t_i \geqslant b$ for $n \leqslant i \leqslant n+k$ and $t'_j \leqslant a'$ for $0 \leqslant j \leqslant k$). If one sets

$$Q_i = P_i, \quad 0 \leqslant i \leqslant n-1,$$
$$Q_{n+j} = P'_j, \quad 0 \leqslant j \leqslant n'-1,$$

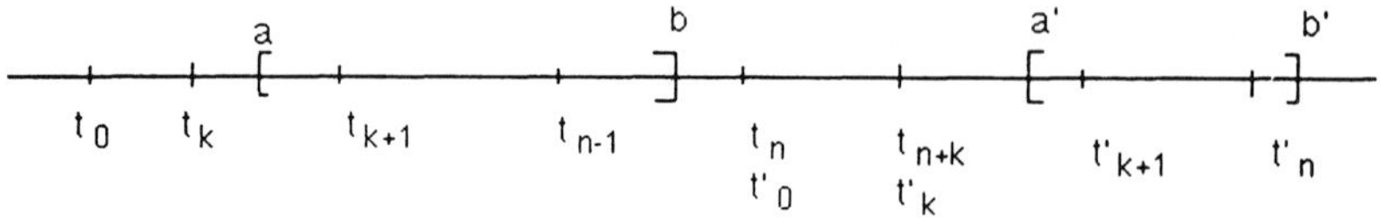

FIG. 11.1.

the curve

$$\Gamma(t) = \sum_{i=0}^{n+n'-1} B_{i,k}(t) Q_i$$

is a spline curve which coincides with $S(t)$ for $t \in [a, b[$, and with $S'(t)$ for $t \in [a', b']$ (see Fig. 11.1).

REMARKS 11.1. (1) The junction is clearly simpler to make in the case of spline curves with uniform knots.

(2) If we require $b = a'$, we have to set $t_n = \cdots = t_{n+k} = b$ and $t'_0 = \cdots = t'_k = a' = b$; for instance this will be the case for Bézier curves, which we will study below; if moreover we want a continuous junction between the two curves, we then have to assume that $P_{n-1} = P'_0$. The junction will then in general be only of class $\mathcal{C}^0$ at P_{n-1}, as the spline curve $\Gamma(t)$ will have at this point a knot of order $k + 1$.

We will study the case of a differentiable junction in the case of Bézier curves, leaving to the reader the analogous case of spline curves (for which the question arises only when one wants $a' = b$ and $S(b) = S'(a')$).

Case of Bézier curves. Let $B(t) = \sum_{i=0}^{k} B_i^k(t) P_i$ $(0 \leqslant t \leqslant 1)$ be a Bézier curve of degree k, $B_i^k(t)$ being the ith Bernstein polynomial (see Section 6). Set

$$\Delta P_i = P_{i+1} - P_i,$$

$$\Delta^r P_i = \Delta(\Delta^{r-1} P_i) = \sum_{j=0}^{r} (-1)^{r-j} \binom{r}{j} P_{i+j}.$$

If D denotes the derivation operator, we then have (see Theorem 8.1),

$$\mathrm{D}B(t) = k \sum_{i=0}^{k-1} \Delta(P_i) B_i^k(t),$$

$$\mathrm{D}^r B(t) = \frac{k!}{(k-r)!} \sum_{i=0}^{k-r} \Delta^k(P_i) B_i^{k-r}(t).$$

If now $B'(t) = \sum_{i=0}^{k} B_i^k(t-1) P'_i$ is another Bézier curve, of the same degree k, defined for $1 \leqslant t \leqslant 2$, we can then join the curve B' to the curve B at the point $t = 1$ under the following conditions:

C^0-*continuity*: $P_k = P'_0$.
C^1-*continuity*: the same condition, and $\Delta P_{k-1} = \Delta P'_0$, or $P_k - P_{k-1} = P'_1 - P'_0$.
C^2-*continuity*: the same conditions, and $\Delta^2 P_{k-2} = \Delta^2 P'_0$, or

$$P_k - 2P_{k-1} + P_{k-2} = P'_2 - 2P'_1 + P'_0,$$

or else

$$P'_2 - P_{k-2} = 2(P'_1 - P_{k-1}) \quad \text{since } P_k = P'_0.$$

C^p-*continuity*: $\Delta^i P_{k-i} = \Delta^i P'_0$ $(0 \leqslant i \leqslant p)$.

EXAMPLES. (1) C^2-continuity between two Bézier curves, see Fig. 11.2; we must have:

$$|P_{k-1}P_k| = |P_k P'_1|,$$
$$|P_{k-2}P'_2| = 2|P_{k-1}P'_1|.$$

(2) A closed Bézier curve of degree 5 and of class C^2, see Fig. 11.3: $P_5 = P_0$, $|P_1P_0| = |P_0P_4|$, $|P_2P_3| = 2|P_1P_4|$.

REMARKS 11.2. (1) The condition of C^2-continuity implies in particular that the points $P_{k-2}, P_{k-1}, P_k = P'_0, P'_1$ and P'_2 are coplanar.
(2) If $P_k = P'_0$ and if P_{k-1}, P_k and P'_1 are on a straight line without the condition $|P_{k-1}P_k| = |P_k P'_1|$, there is a discontinuity in the length of the tangent vector (at P_k),

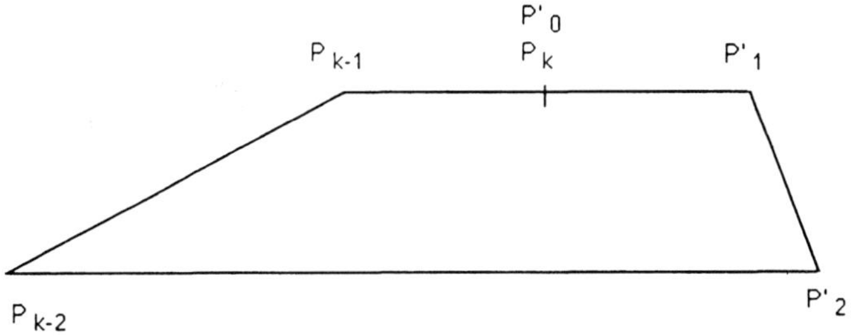

FIG. 11.2. Condition for C^2-continuity at $P'_0 = P_k$.

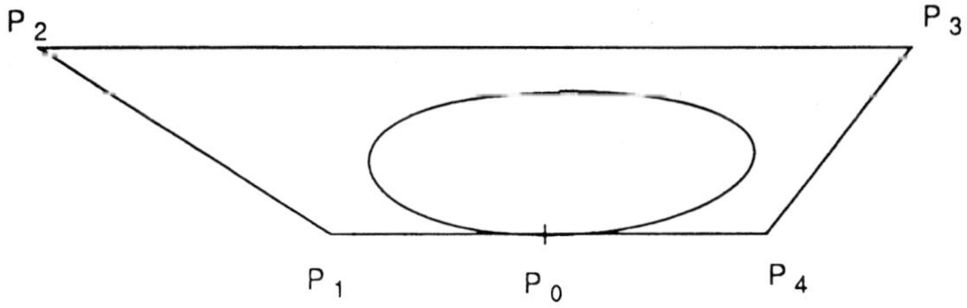

FIG. 11.3. A closed Bézier curve of degree 5.

but not in its direction. In other words, there will be a junction with C^1-continuity after a (for instance linear) change of parameters for one of the two curves (which does not change the image of the curve in $\mathbb{R}^s$). The discontinuity of the tangent is thus "invisible" on the image of the curve.

One then says that there is *visual continuity* of the tangent, or that the junction at P_k is *visually* C^1. One sometimes also says that there is *geometric continuity* or G_1-*continuity* at the point P_k (see "Geometric continuity" below).

There are similar notions for more differentiable junctions, for instance the continuity of the curvature (see below).

Degree elevation in the case of Bézier curves. If the two Bézier curves $B^k(t)$ and $B^{k'}(t)$ are not of the same degree, with for instance $k < k'$, one can turn to the case $k = k'$ by "elevating the degree" of $B^k(t)$ in the following way. Assuming by induction that $k' = k+1$, one may consider the curve $B^k(t) = \sum_{i=0}^{k} P_i B_i^k(t)$ as a special curve of degree $\leqslant k+1$, viz.,

$$B^k(t) = \sum_{i=0}^{k+1} Q_i B_i^{k+1}(t).$$

THEOREM 11.1. *With the above notation, we have:*

$$Q_0 = P_0,$$
$$Q_{k+1} = P_k,$$
$$Q_i = \frac{iP_{i-1} + (k-i+1)P_i}{k+1}, \quad 1 \leqslant i \leqslant k.$$

PROOF. It suffices to remark that we can write

$$B_i^k(t) = (1-t)B_i^k(t) + tB_i^k(t) = \left(\frac{k+1-i}{k+1}\right) B_i^{k+1}(t) + \left(\frac{i+1}{k+1}\right) B_{i+1}^{k+1}(t)$$

(see Section 6), and to use this relation in the expression

$$B^k(t) = \sum_{i=0}^{k} B_i^k(t) P_i. \qquad \square$$

Curvature of a Bézier curve. In the same spirit as above, let us point out how the curvature of a Bézier curve is related to its control polygon (for more details, see FAUX and PRATT [1979]). Recall that if $B(t)$ is a curve in $\mathbb{R}^s$ of class C^2, its curvature at the point $B(t_0)$ is given by the formula:

$$K(t_0) = \frac{|B'(t_0) \wedge B''(t_0)|}{|B'(t_0)|^3},$$

the symbol "$\wedge$" denoting the vector product of two vectors.

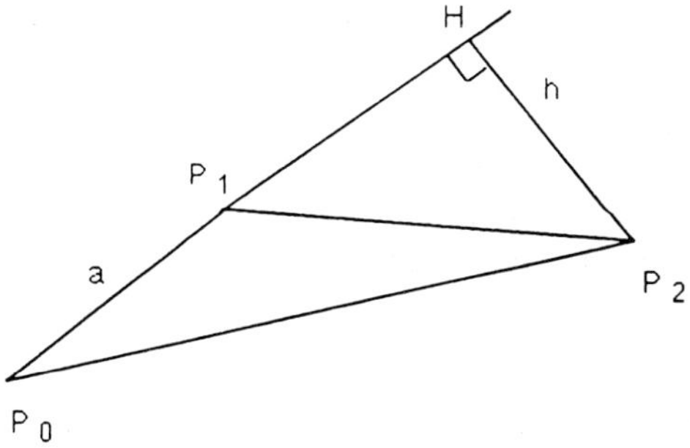

FIG. 11.4.

In the case of a Bézier curve of degree k: $B(t) = \sum_{i=0}^{k} B_i^k(t)P_i$, we therefore have:

$$K(0) = \frac{k-1}{k}\frac{|(P_1 - P_0) \wedge (P_2 - P_1)|}{|P_1 - P_0|^3},$$

$$K(1) = \frac{k-1}{k}\frac{|(P_{k-1} - P_k) \wedge (P_{k-2} - P_{k-1})|}{|P_{k-1} - P_k|^3},$$

since for instance (see Theorem 8.1):

$$\begin{aligned} B'(0) &= k(P_1 - P_0), \\ B''(0) &= k(k-1)[P_0 - 2P_1 + P_2] \\ &= k(k-1)\big[(P_2 - P_1) - (P_1 - P_0)\big]. \end{aligned}$$

Figure 11.4 shows a geometric interpretation of the curvature:

$$K(0) = \frac{2(k-1)}{k} \times \left(\frac{h}{a^2}\right),$$

with $a = |P_1 - P_0|$, $h = |P_2 - H|$.

Geometric continuity.

DEFINITION. Let r be an integer $\geqslant 0$ and $S_1(t)$, $0 \leqslant t \leqslant 1$, and $S_2(t)$, $1 \leqslant t \leqslant 2$, be two parametric curves of class C^r in $\mathbb{R}^s$ such that $S_1(1) = S_2(1)$ and $S_1'(1) \neq 0$ or $S_2'(1) \neq 0$; if i is an integer such that $1 \leqslant i \leqslant r$, one says that there is a *G_i-junction* (or *G_i-continuity*) at the point $S_1(1) = S_2(1)$ if there exists a change of parameter of class C^i $t = \phi(u)$, $0 \leqslant u \leqslant 1$, $\phi(1) - 1$, $\phi'(1) > 0$ such that the curve $S_1(\phi(u))$ has a C^i-junction with $S_2(t)$ at $u = t = 1$.

Let us describe in more details the conditions for G_i-continuity, $0 \leqslant i \leqslant 2$. The notation is the same as in the definition above.

(i) The G_0-junction condition means that $S_1(1) = S_2(1)$: it is thus equivalent to the C^0-junction condition.

(ii) There is G_1-junction if and only if there is G_0-junction and if there exists a constant $\beta_1 > 0$ such that $\beta_1 S_1'(1) = S_2'(1)$ (one has $\beta_1 = \phi'(1)$ with the above notation). The G_1-continuity means then that the tangents to the two curves (at the point corresponding to $t = 1$) have the same direction and the same orientation.

The C^1-continuity case corresponds to the case $\beta_1 = 1$.

(iii) There is G_2-continuity if and only if there is G_1-continuity, and if there exists a constant β_2 such that

$$S_2''(1) = \beta_2 S_1'(1) + \beta_1^2 S_1''(1). \tag{11.1}$$

Let us prove this last formula; we compute the derivatives of $T(u) = S_1(\phi(u))$ at $u = 1$:

$$\begin{aligned} T'(1) &= S_1'(1)\phi'(1) = \beta_1 S_1'(1), \\ T''(1) &= \beta_1^2 S_1''(1) + \phi''(1) S_1'(1). \end{aligned}$$

Setting $\beta_2 = \phi''(1)$, we find formula (11.1) above; the C^2-case corresponds to $\beta_1 = 1$, $\beta_2 = 0$.

REMARKS 11.3. (1) The G_2-continuity has the same geometric meaning as the C^2-continuity: the curves $S_1(t)$ and $S_2(t)$ have the same image in $\mathbb{R}^s$. They are only covered with different rates. The advantage of the G_2-condition is that it introduces two additional parameters, and hence more flexibility.

(2) Traditionally, β_1 is called the parameter of *bias* and β_2 the parameter of *tension*.

As an example, we will describe the uniform "β-splines" of degree three, following BARSKY and BEATTY [1983].

Let a uniform knot sequence be given (for instance $t_i = i$), along with two constants β_1 and β_2, with $\beta_1 > 0$. We want to generalize the B-splines of degree 3, replacing them by piecewise polynomial functions $S_i(t)$, with G_2 junctions at the knots t_i, the parameters of bias and tension being respectively equal to β_1 and β_2 at each t_i. Since the knot sequence is uniform, we have only to define the function $S_0(t)$, whose support will be $[0, 4]$. It is equivalent (and often more convenient), to compute the matrix

$$M = (a_{i,j})_{0 \leqslant i \leqslant 3,\ -3 \leqslant j \leqslant 0}$$

such that if we set $S(t) = \sum_j P_j S_j(t)$, we have

$$S(t) = (1 \quad t \quad t^2 \quad t^3) M \begin{pmatrix} P_{-3} \\ P_{-2} \\ P_{-1} \\ P_0 \end{pmatrix}.$$

So

$$S(t) = \sum_j P_j S_j(t) = \sum_{j=-3}^{0} P_j S_0(t-j),$$

and we set

$$S_0(t-j) = S_j(t) = a_{0,j} + a_{1,j}t + a_{2,j}t^2 + a_{3,j}t^3.$$

Recall from Section 10 that in the B-spline case ($\beta_1 = 1,\ \beta_2 = 0$), we have

$$M = \frac{1}{6}\begin{pmatrix} 1 & 4 & 1 & 0 \\ -3 & 0 & 3 & 0 \\ 3 & -6 & 3 & 0 \\ -1 & 3 & -3 & 1 \end{pmatrix}.$$

We then need a system of 16 linear equations to determine the $a_{i,j}$'s. Let us describe these equations. Continuity of $S_0(t)$ at the points $0, 1, 2, 3, 4$ yields the following five equations:

$$a_{0,0} = 0 \quad \text{(continuity at 0)},$$

$$a_{0,0} + a_{1,0} + a_{2,0} + a_{3,0} = a_{0,-1} \quad \text{(continuity at 1)},$$

$$a_{0,-1} + a_{1,-1} + a_{2,-1} + a_{3,-1} = a_{0,-2},$$

$$a_{0,-2} + a_{1,-2} + a_{2,-2} + a_{3,-2} = a_{0,-3},$$

$$a_{0,-3} + a_{1,-3} + a_{2,-3} + a_{3,-3} = 0.$$

Then the G_1- and G_2-continuity conditions at the points $0, 1, 2, 3, 4$ respectively give the two following systems:

$$\begin{aligned} &a_{1,0} = 0, \\ &\beta_1(a_{1,0} + 2a_{2,0} + 3a_{3,0}) = a_{1,-1}, \\ &\beta_1(a_{1,-1} + 2a_{2,-1} + 3a_{3,-1}) = a_{1,-2}, \\ &\beta_1(a_{1,-2} + 2a_{2,-2} + 3a_{3,-2}) = a_{1,-3}, \\ &a_{1,-3} + 2a_{1,-3} + 3a_{3,-3} = 0, \end{aligned} \qquad \begin{aligned} &a_{2,0} = 0, \\ &\beta_1^2 a_{2,0} + \beta_2 a_{1,0} = a_{2,-1}, \\ &\beta_1^2 a_{2,-1} + \beta_2 a_{1,-1} = a_{2,-2}, \\ &\beta_1^2 a_{2,-2} + \beta_2 a_{1,-2} = a_{2,-3}, \\ &\beta_1^2 a_{2,-3} + \beta_2 a_{1,-3} = 0. \end{aligned}$$

This yields a total of 15 equations in 16 unknowns; we add the equation $\sum_{j=-3}^{0} S_j(0) = 1$, i.e., $a_{0,-3} + a_{0,-2} + a_{0,-1} + a_{0,0} = 1$ to get a system whose determinant is

$$\Delta = \beta_2 + 2\beta_1^3 + 4\beta_1^2 + 4\beta_1 + 2$$

which is nonzero if for instance $\beta_1 > 0$ and $\beta_2 \geqslant 0$. We then find a unique solution for the $a_{i,j}$'s, hence we obtain a set of basis functions generalizing the uniform B-splines of degree three, which we call the β-splines (cf. BARSKY [1988]). The matrix $(a_{i,j})$ is given by

$$M = \frac{1}{\Delta}\begin{pmatrix} 2\beta_1^2 & \beta_2 + 4\beta_1^2 + 4\beta_1 & 2 & 0 \\ -6\beta_1^3 & 6\beta_1^3 - 6\beta_1 & 6\beta_1 & 0 \\ 6\beta_1^3 & -(3\beta_2 + 6\beta_1^3 + 6\beta_1^2) & 3\beta_2 + 6\beta_1^2 & 0 \\ -2\beta_1^3 & 2\beta_2 + 2\beta_1^3 + 2\beta_1^2 + 2\beta_1 & -(2\beta_2 + 2\beta_1^2 + 2\beta_1 + 2) & 2 \end{pmatrix}.$$

It is easy to verify that $\sum S_j(t) = 1$, which implies the convex hull property for the curves constructed with these base functions (see Theorem 7.1).

Let us look at the influence of the tension parameter β_2; when β_2 tends to $+\infty$ and $\beta_1 = 1$, the base functions become

$$\begin{aligned} &S_{-3}(t) = 0, \\ &S_{-2}(t) = 1 - (3t^2 - 2t^3), \\ &S_{-1}(t) = 3t^2 - 2t^3, \\ &S_0(t) = 0, \end{aligned} \qquad (0 \leqslant t \leqslant 1).$$

The curve $S(t) = \sum S_j(t)P_j$ then has the expression $(1-\alpha)P_{-2} + \alpha P_{-1}$ for $0 \leqslant t \leqslant 1$, with $\alpha = 3t^2 - 2t^3$. Its image is the segment $P_{-2}P_{-1}$. We see that when $\beta_2 \to +\infty$, the curve $S(t)$ tends to coincide with the control polygon (always staying G_2-continuous), which explains the name "tension" given to the parameter β_2.

The bias parameter is not so clear to interpret; it controls the shape of the curve at the junction points between two polynomial pieces. If β_1 is large, the curve is closer to the tangent at the "right side" of the knot than at the "left side".

12. Blossoming

Another way of introducing B-spline functions has been proposed by Ramshaw (see RAMSHAW [1987]). This technique is called blossoming, and is based on the work of DE CASTELJAU [1985]; it yields very elegant proofs.

We will make a short exposition of the main properties of blossoming in one variable, following the way of SEIDEL [1989]. Similar properties are also true for tensor product surfaces: see SEIDEL [1991]. One characteristic property of blossoming is that it allows the computation of the control points of a Bézier or B-spline curve from its "blossom" expression. This is interesting from a theoretical point of view, but not so much from a practical one, because one is generally interested by the opposite direction, namely the definition of the curve by the control points.

The blossoming is the operation which consists in replacing a polynomial $F(X)$ of degree d by a d-multi-affine symmetric function $f(u_1, \ldots, u_d)$ such that $f(X, \ldots, X) = F(X)$, according to the following definition:

DEFINITION. A function $f(u_1, \ldots, u_d) : \mathbb{R}^d \to \mathbb{R}$ is *multi-affine* (or *d-affine*) if f is affine with respect to each of the variables u_i separately, i.e., if it satisfies

$$f\Big(\sum_j a_j v_j\Big) = \sum_j a_j f(v_j)$$

for $u_i = \sum_j a_j v_j$, $\sum_j a_j = 1$.

THEOREM 12.1. *Let $d \geqslant 1$ be an integer. If F is a polynomial in X of degree $\leqslant d$, there exists a unique symmetric d-affine function $f : \mathbb{R}^d \to \mathbb{R}$ satisfying $f(u, \ldots, u) = F(u)$. This function is called the d-blossom (or simply the blossom) of F.*

PROOF. For the existence of f, we may assume by linearity that $F(u) = u^k$, with $k \leqslant d$. We then set

$$f(u_1, \ldots, u_d) = \frac{k!(d-k)!}{d!} \sum_{1 \leqslant i_1 < \cdots < i_k \leqslant d} u_{i_1} \cdots u_{i_k}.$$

Let us now prove the uniqueness. Let a_j be the coefficient of the monomial $u_1 \cdots u_j$ in some blossom of F (for instance the one constructed above). Because of the symmetry, the coefficient of X^j in $F(X)$ is $(d!/j!(d-j)!)a_j$, and hence:

$$F^{(j)}(0) = \frac{d!}{(d-j)!} a_j \tag{12.1}$$

which proves the uniqueness of a_j. We then prove (by induction on j) the following formula:

$$f(0, \ldots, 0, \underbrace{1, \ldots, 1}_{j}) = a_j + j a_{j-1} + \cdots + \binom{j}{k} a_k + \cdots + a_0, \tag{12.2}$$

which proves the uniqueness of f, since f is clearly determined by the numbers $f(0, \ldots, 0, 1, \ldots, 1)$. □

EXAMPLE. The blossom of the polynomial

$$F(X) = a_0 + a_1 X + a_2 X^2 + a_3 X^3$$

is given by

$$f(u_1, u_2, u_3) = a_1 + \frac{a_1}{3}(1 + u_2 + u_3) + \frac{a_2}{3}(u_1 u_2 + u_2 u_3 + u_1 u_3) + a_3 u_3.$$

THEOREM 12.2 (see SEIDEL ([1989], p. 24–25)). *Consider a polynomial F of degree d over the parameter interval $[s, t]$, and let f be its blossom. If we set*

$$b_i = f(\underbrace{s, \ldots, s}_{d-i} \underbrace{t, \ldots, t}_{i}),$$

then

$$F(X) = \sum_{i=0}^{d} b_i B_i^d(X)$$

is the Bézier representation of F, and b_i are the Bézier points.

The proof is simple, and uses the De Casteljau algorithm: see SEIDEL ([1989], p. 25).

We may now apply the same idea to each polynomial piece of a B-spline basis element $B_{i,k}$ of degree k. Assume $t_j < t_{j+1}$. Let $b_{i,j,k}$ be the blossom of the restriction of $B_{i,k}$ to the interval $[t_j, t_{j+1}[$.

THEOREM 12.3 (see SEIDEL ([1989], p. 26)). *We have the following induction formula, (similar to the De Boor–Cox algorithm* 1.1):

$$b_{i,j,0} = \delta_{i,j},$$
$$b_{i,j,k}(u_1, \ldots, u_k) = \frac{u_k - t_j}{t_{j+k} - t_j} b_{i,j,k-1}(u_1, \ldots, u_{k-1}) + \frac{t_{j+k+1} - u_k}{t_{j+k+1} - t_{j+1}} b_{i+1,j,k-1}(u_1, \ldots, u_{k-1}).$$

The proof is immediate, using (1.1) and induction on k.

THEOREM 12.4. *Let $S(t) = \sum P_i B_{i,k}(t)$ be a spline curve. Let $S_j(t)$ be the restriction of $S(t)$ to the interval $[t_j, t_{j+1}[$, and let $s_j(u_1, \ldots, u_k)$ be its blossom. Assume $t_j < t_{j+1}$ and $j - p \leqslant l \leqslant j$. Then the control point P_l of the curve $S(t)$ is given by the formula*

$$P_l = s_j(t_{l+1}, \ldots, t_{l+k}).$$

PROOF. On the interval $[t_j, t_{j+1}[$, we have

$$s_j(u_1, \ldots, u_k) = \sum P_i B_{i,j,k}(u_1, \ldots, u_k),$$

where $B_{i,j,k}$ is the blossom of $B_{i,k}$ on $[t_j, t_{j+1}[$. But it is easy to see from Theorem 12.3 that

$$B_{i,j,k}(u_{l+1}, \ldots, u_{l+k}) = \delta_{i,l}. \qquad \square$$

All the properties of Sections 1–4 can now be proved using blossoming: see SEIDEL [1989]. For a generalization to the tensor product case, see SEIDEL [1991].

CHAPTER II

Spline Surfaces

This chapter is devoted to surfaces. Three main ways of modeling surface patches are considered: the *tensor products* (Sections 13–15), the *triangular Bézier patches* (Sections 16–18), with a special emphasis on *rational Bézier patches* and *base points* (Section 19), and *polyhedral splines* (Section 20), which can be viewed as a direct generalization of B-spline functions in one variable. Finally, Section 21 is devoted to *box splines*, which are an important and useful particular case of polyhedral splines.

13. Tensor products

For parametric surface patches $S(u, v)$, tensor products allow to use one-variable algorithms for spline (or Bézier) curves separately in u (for fixed v), or in v (for fixed u).

Let us consider two knot sequences: $u = (u_i)_{0 \leqslant i \leqslant m+k}$, $v = (v_j)_{0 \leqslant j \leqslant n+l}$ satisfying

$$\begin{aligned} &u_0 \leqslant \cdots \leqslant u_k \leqslant a, \quad u_i \leqslant u_{i+1}, \\ &b \leqslant u_m \leqslant \cdots \leqslant u_{m+k}, \quad \text{and} \\ &v_0 \leqslant \cdots \leqslant v_l \leqslant c, \quad v_j \leqslant v_{j+1}, \\ &d \leqslant v_n \leqslant \cdots \leqslant v_{n+l}, \end{aligned}$$

and let $B_{i,k}(u)$ (resp. $B_{j,l}(v)$) be the corresponding B-splines.

DEFINITION. With the above notation, let $P_{i,j}$, $0 \leqslant i \leqslant m-1$, $0 \leqslant j \leqslant n-1$, be points in $\mathbb{R}^s$; one then defines a "*B-spline patch*" as the following application of $[a, b] \times [c, d]$ into $\mathbb{R}^s$:

$$S(u, v) = \sum_{i=0}^{m-1} \sum_{j=0}^{n-1} P_{i,j} B_{i,k}(u) B_{j,l}(v).$$

REMARKS 13.1. (1) We may write

$$S(u, v) = \sum_{i=0}^{m-1} P_i(v) B_{i,k}(u), \quad \text{with}$$

$$P_i(v) = \sum_{j=0}^{n-1} P_{ij} B_{j,l}(v).$$

The image by S of the line segment $v = v_0$ is therefore a spline curve of degree k, with the set of points $P_i(v_0)$, $0 \leqslant i \leqslant m-1$, as control polygon. We have a similar result for the image by S of the segments $u =$ constant.

(2) Each coordinate of the function $S(u, v)$ is of degree $l + k$ with respect to u and v; for instance, in the most common case $l = k = 3$, the coordinates of $S(u, v)$ are of degree 6.

(3) The linear subspace of functions on $[a, b] \times [c, d]$ spanned by the $B_{i,k}(u)B_{j,l}(v)$'s, $0 \leqslant i \leqslant m-1$, $0 \leqslant j \leqslant n-1$, is called the *tensor product* of the spaces $\mathcal{P}_{k,u}$ and $\mathcal{P}_{l,v}$.

(4) In the case where

$$\begin{aligned} u_0 &= \cdots = u_k = a, \\ u_m &= \cdots = u_{m+k} = b, \\ v_0 &= \cdots = v_l = c, \\ v_n &= \cdots = v_{n+l} = d, \end{aligned}$$

the boundary of the surface $S(u, v)$ is a union of spline curves which have portions of the boundary of the polyhedron (P_{ij}) as control polygon. In particular, the surface passes through the points $P_{0,0}$, $P_{0,n-1}$, $P_{m-1,n-1}$, $P_{m-1,0}$.

Evaluation algorithm. The advantage of tensor products is that one can use the algorithms developed for the case of one variable. We will illustrate this observation with the evaluation algorithm. Let

$$S(u, v) = \sum_{i=0}^{m-1} \sum_{j=0}^{n-1} P_{i,j} B_{i,k}(u) B_{j,l}(v)$$

be a B-spline patch, which we want to evaluate at the point (u_0, v_0); let us write as above:

$$S(u, v) = \sum_{i=0}^{m-1} P_i(v) B_{i,k}(u),$$

with

$$P_i(v) = \sum_{j=0}^{n-1} P_{i,j} B_{j,l}(v).$$

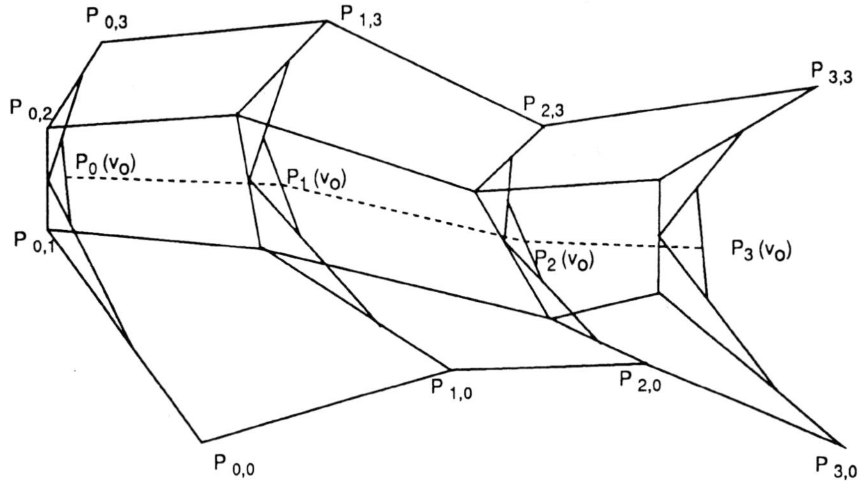

FIG. 13.1. Computation of $P_i(v_0)$ $(v_3 < v_0 < v_4)$.

We begin by evaluating $P_i(v_0)$, $0 \leqslant i \leqslant m-1$, by the evaluation algorithm (Section 8) also called "De Boor–Cox" (for each coordinate of $P_i(v_0)$). We then evaluate $S(u_0, v_0)$ using once more this algorithm (for each coordinate of S). Assuming

$$(u_0, v_0) \in [u_r, u_{r+1}] \times [v_s, v_{s+1}],$$

we obtain the following algorithm (see Section 8).

Evaluation of $P_i(v_0)$. One sets

$$P^0_{i,j}(v_0) = P_{i,j}, \quad r-k \leqslant i \leqslant r,\ s-l \leqslant j \leqslant s,$$

$$P^{p+1}_{i,j}(v_0) = \frac{(v_0 - v)P^p_{i,j}(v_0) + (v_{j+l-p} - v_0)P^p_{i,j-1}(v_0)}{v_{j+l-p} - v_j},$$

$$\text{for } r-k \leqslant i \leqslant r,\ 0 \leqslant p \leqslant l-1,\ s-l+p+1 \leqslant j \leqslant s.$$

Then we have $P_i(v_0) = P^l_{i,s}(v_0)$.

Evaluation of $S(u_0, v_0)$ *viewed as a spline curve in* u *with* $(P_i(v_0))$ *as its control polygon*. One sets:

$$P^0_i(u_0, v_0) = P^l_{i,\varepsilon}(v_0) \quad \text{for } r-k+1 \leqslant i \leqslant r,$$

$$P^{q+1}_i(u_0, v_0) = \frac{(u_0 - u_i)P^q_i(u_0, v_0) + (u_{i+k-q} - u_0)P^q_{i-1}(u_0, v_0)}{u_{i+k-q} - u_i},$$

$$\text{for } 0 \leqslant q \leqslant k-1 \text{ and } r-k+q+1 \leqslant i \leqslant r.$$

We have finally $S^r(u_0, v_0) = P^k_r(u_0, v_0)$.

There is clearly a similar algorithm obtained by exchanging the roles of u and v.

EXAMPLE. In the bicubic case ($k = l = 3$), the algorithm is illustrated in Fig. 13.1.

14. Particular case of Bézier surfaces

Let us take $(u, v) \in [0, 1]^2$, and consider

$$B(u,v) = \sum_{i=0}^{k}\sum_{j=0}^{l} P_{i,j} B_i^k(u) B_j^l(v),$$

where $B_i^k(u)$ is the ith Bernstein polynomial of degree k (Section 6). The image of $B(u, v)$ is generally called a *Bézier patch.*

Computation of the derivatives. We have

$$\frac{\partial^{r+s}}{\partial u^r \partial v^s} B(u,v) = \frac{k!}{(k-r)!}\frac{l!}{(l-s)!}\sum_{i=0}^{k-r}\sum_{j=0}^{l-s} \Delta^{rs} P_{i,j} B_i^{k-r}(u) B_j^{l-s}(v), \qquad (14.1)$$

with

$$\Delta^{rs} P_{i,j} = \Delta_1^r \big(\Delta_2^s P_{i,j}\big) = \Delta_2^s \big(\Delta_1^r P_{i,j}\big),$$

where

$$\Delta_1 P_{i,j} = P_{i+1,j} - P_{i,j},$$
$$\Delta_2 P_{i,j} = P_{i,j+1} - P_{i,j}$$

(see Theorem 8.1).

This computation allows to "join" several Bézier patches. Let us for instance consider two Bézier patches of the same degree (k, l): $B(u, v)$, $(u, v) \in [0, 1]^2$, with $(P_{i,j})$ as control polygon, and $B'(s, t)$, $(s, t) \in [0, 1]^2$, with $P'_{i,j}$ as control polygon. Assume that the junction is made along the curves $B(1, v)$ and $B'(0, t)$, given by

$$B(1,v) = \sum_{j=0}^{l} P_{k,j} B_j^l(v),$$

$$B'(0,t) = \sum_{j=0}^{l} P'_{0,j} B_j^l(t).$$

If we want C^0-continuity between these two patches, the two curves have to coincide, i.e., we must have

$$P_{k,j} = P'_{0,j}, \quad 0 \leqslant j \leqslant l. \tag{14.2}$$

Let us compute the first partial derivatives of B and B':

$$\begin{cases} \dfrac{\partial B}{\partial u}(1,v) = D_{1,0}B(1,v) = k\displaystyle\sum_{j=0}^{l}(P_{k,j} - P_{k-1,j})B_j^l(v), \\ \dfrac{\partial B}{\partial v}(1,v) = D_{0,1}B(1,v) = l\displaystyle\sum_{j=0}^{l-1}(P_{k,j+1} - P_{k,j})B_j^{l-1}(v), \end{cases} \tag{14.3}$$

$$\begin{cases} \dfrac{\partial B'}{\partial s}(0,t) = k\displaystyle\sum_{j=0}^{l}(P'_{1,j} - P'_{0,j})B_j^l(t), \\ \dfrac{\partial B'}{\partial t}(0,t) = l\displaystyle\sum_{j=0}^{l-1}(P'_{0,j+1} - P'_{0,j})B_j^{l-1}(t); \end{cases} \tag{14.4}$$

note that $(\partial B/\partial v)(1,v) = (\partial B'/\partial t)(0,v)$, $v \in [0,1]$, if (14.2) is fulfilled.

If we want C^1-continuity, it is necessary that $P_{k-1,j} + P'_{1,j} = 2P_{k,j}$, $0 \leqslant j \leqslant l$. In other words, the vertices $P_{k,j} = P'_{0,j}$ have to be at the middle of the segments $P_{k-1,j}P'_{1,j}$. This means that for each j, the Bézier curves (in u) with control polygons $P_{i,j}$, $0 \leqslant i \leqslant k$, and $P'_{i,j}$, $0 \leqslant i \leqslant k$, have C^1-continuity at $P_{k,j} = P'_{0,j}$ (see Section 11).

REMARKS 14.1. (1) In the same manner, we can give conditions for C^r-continuity $(r > 1)$ for the two Bézier patches. It is for instance easy to see that, for C^2-continuity along the curves $B(1,v)$ and $B'(0,t)$, the following condition must be fulfilled in addition to (14.3) and (14.4):

$$P'_{2,j} - P_{k-2,j} = 2\big(P'_{1,j} - P_{k-1,j}\big), \quad 0 \leqslant j \leqslant l, \tag{14.5}$$

(see Section 11). More generally, we immediately see that two patches $B(u,v)$ and $B'(s,t)$ have a C^r-continuity along the curves $B(1,v)$ and $B'(0,t)$ if and only if, for $0 \leqslant j \leqslant l$, the Bézier curves with control polygons $P_{i,j}$ and $P'_{i,j}$, $0 \leqslant i \leqslant k$, have a C^r-continuity at $P_{k,j} = P'_{0,j}$.

(2) The C^r-continuity condition may be interpreted as follows. Set $s = u-1$, $t = v$; then the surface $\widetilde{B}(u,v)$, $0 \leqslant u \leqslant 2$, $0 \leqslant v \leqslant 1$, defined by $\widetilde{B}(u,v) = B(u,v)$ for $0 \leqslant u \leqslant 1$ and $\widetilde{B}(u,v) = B'(u,v)$ for $1 \leqslant u \leqslant 2$ is of class C^r.

(3) If the patches B and B' are parametrized by $(u_1,v_1) \in [a,b] \times [c,d]$ and $(s_1,t_1) \in [b,b'] \times [d,d']$, and if we want C^r-continuity with respect to the parameters

(u_1, v_1) and (s_1, t_1), it suffices to use the preceding case, with

$$u = \frac{u_1 - a}{b - a}, \qquad v = \frac{v_1 - c}{d - c},$$

$$s = \frac{s_1 - b}{b' - b}, \qquad t = \frac{t_1 - d}{d' - d}.$$

The $\mathcal{C}^1$-continuity is then expressed by the relations

$$(b' - b)P_{k-1,j} + (b - a)P'_{1,j} = (b' - a)P_{k,j}, \quad 0 \leqslant j \leqslant l.$$

The points $P_{k-1,j}$, $P_{k,j} = P'_{0,j}$ and $P'_{1,j}$ are again on the same line, but $P_{k,j}$ is no longer necessarily at the middle of $[P_{k-1,j}, P'_{1,j}]$.

(4) The condition $\lambda P_{k-1,j} + \mu P'_{1,j} = (\lambda + \mu)P_{k,j}$ (see (2) above) means that the two Bézier patches have the same tangent plane at the point $P_{k,j} = P'_{0,j}$. More generally, one may define G_i-continuity ($i \geqslant 0$) between two parametric surfaces, in a similar way as for curves (see Section 11), as follows.

DEFINITION. Two parametric surfaces $S_1(u, v)$, $0 \leqslant u \leqslant 1$, $0 \leqslant v \leqslant 1$, and $S_2(s, t)$, $0 \leqslant s \leqslant 1$, $0 \leqslant t \leqslant 1$, have *$G_i$-continuity* ($i \geqslant 0$) along the curves $S_1(1, v)$ and $S_2(0, t)$ (we have $S_1(1, t) = S_2(0, t)$, $0 \leqslant t \leqslant 1$, by assumption), if there exists a change of parameters:

$$(s', t') \mapsto (s, t) = (\psi_1(s', t'), \psi_2(s', t')),$$

with $\psi_1(0, t') = 0$, $\psi_2(0, t') = t'$, such that the surfaces $S_1(u, v)$ and $S_2(\psi_1(s', t'), \psi(s', t'))$ have $\mathcal{C}^i$-continuity along the curves $S_1(1, v)$ and $S_2(\psi_1(0, t'), \psi_2(0, t')) = S_2(0, t')$.

As with curves, the condition of G_0-continuity is the same as that for $\mathcal{C}^0$-continuity. Let us look at the G_1-continuity condition. (See also RISLER ([1991], p. 102).)

THEOREM 14.1. *Assume that there is a $\mathcal{C}^0$-junction between the two surfaces $S_1(u, v)$ and $S_2(s, t)$ along the curve $\gamma : S_1(0, v) = S_2(1, t)$.*

There is G_1-continuity along this curve if and only if S_1 and S_2 have the same tangent plane at all the points of the common curve γ.

Let us apply the above result to the case of two Bézier patches $B(u, v)$ and $B'(u, v)$, joining them along the curve $B(1, v) = B'(0, v)$ (one then has $P_{k,j} = P'_{0,j}$, $0 \leqslant j \leqslant l$).

The tangent plane continuity along the common curve is expressed by the following relation (see Theorem 14.1):

$$\alpha(v)k \sum_{j=0}^{l} (P_{k,j} - P_{k-1,j}) B_j^l(v)$$

$$+\beta(v)k\sum_{j=0}^{l}(P'_{1,j}-P'_{0,j})B_j^l(v)+\gamma(v)l\sum_{j=0}^{l-1}(P_{k,j+1}-P_{k,j})B_j^{l-1}(v)=0,$$

$\alpha(v)$, $\beta(v)$ and $\gamma(v)$ being arbitrary functions of v. It is natural to assume that α, β and γ are polynomials in v; the simplest choice is to take α and β constants, and γ linear in v; one then sets

$$\begin{aligned}
&\alpha(v)=\alpha,\\
&\beta(v)=\beta,\\
&\gamma(v)=\gamma_0(1-v)+\gamma_1(v).
\end{aligned}$$

Since

$$\begin{aligned}
(1-v)B_j^{l-1}(v)&=\frac{l-j}{l}B_j^l(v), \quad 0\leqslant j\leqslant l-1,\\
vB_j^{l-1}(v)&=\frac{j+1}{l}B_{j+1}^l(v), \quad 0\leqslant j\leqslant l-1,
\end{aligned}$$

(see Theorem 6.1), one finds the following relations:

$$\begin{aligned}
&P_{k,j}=P'_{0,j},\\
&\alpha k(P_{k,j}-P_{k-1,j})+\beta k(P'_{1,j}-P'_{0,j})\\
&\quad+\gamma_0(l-j)(P_{k,j+1}-P_{k,j})+\gamma_1(j)(P_{k,j}-P_{k,j-1})=0, \quad 0\leqslant j\leqslant l,
\end{aligned}$$

which give sufficient conditions for the G_1-continuity of two Bézier patches along their common boundary $B(1,v)=B'(0,v)$.

For more details about these matters, the reader may look at SCHMITT and DU [1987] or DU [1988].

Matrix representation. Let us consider a bicubic patch (i.e., $k=l=3$). Recall (Section 10) that a cubic Bézier curve is represented in the following way:

$$f(t)=\begin{pmatrix}1 & t & t^2 & t^3\end{pmatrix}\begin{pmatrix}1 & 0 & 0 & 0\\ -3 & 3 & 0 & 0\\ 3 & -6 & 3 & 0\\ -1 & 3 & -3 & 1\end{pmatrix}\begin{pmatrix}P_0\\ P_1\\ P_2\\ P_3\end{pmatrix}. \tag{14.6}$$

THEOREM 14.2. *Let*

$$P(u,v)=\sum_{i=0}^{3}\sum_{j=0}^{3}P_{i,j}B_i^3(u)B_j^3(v)$$

be a Bézier patch, with $P_{i,j} \in \mathbb{R}^s$*; one may then write*

$$P(u,v) = U\ M\ P\ {}^{t}M\ {}^{t}V,$$

with $U = (1 \quad u \quad u^2 \quad u^3)$, $V = (1 \quad v \quad v^2 \quad v^3)$, $P = (P_{i,j})_{0 \leqslant i \leqslant 3,\ 0 \leqslant j \leqslant 3}$, *and*

$$M = \begin{pmatrix} 1 & 0 & 0 & 0 \\ -3 & 3 & 0 & 0 \\ 3 & -6 & 3 & 0 \\ -1 & 3 & -3 & 1 \end{pmatrix}.$$

PROOF. One may write

$$P(u,v) = \bigl(B_0(u) \ \ldots \ B_3(u)\bigr) P \begin{pmatrix} B_0(v) \\ \vdots \\ B_3(v) \end{pmatrix};$$

it then suffices to apply the formula

$$\bigl(B_0(u) \ \ldots \ B_3(u)\bigr) = (1 \quad u \quad u^2 \quad u^3) M,$$

and a similar formula obtained by exchanging u and v, and then by transposing. □

15. Interpolation and approximation

Interpolation of a set of points. Let $Q_{\lambda,\mu}$, $0 \leqslant \lambda \leqslant m-1$, $0 \leqslant \mu \leqslant n-1$, be a set of points in $\mathbb{R}^s$. The interpolation problem consists in finding a spline surface

$$S(u,v) = \sum_{i=0}^{m-1} \sum_{j=0}^{n-1} P_{i,j} B_{i,k}(u) B_{j,l}(v),$$

and parameters $(\alpha_\lambda, \beta_\mu)$ such that

$$S(\alpha_\lambda, \beta_\mu) = Q_{\lambda,\mu}, \quad 0 \leqslant \lambda \leqslant m-1, \quad 0 \leqslant \mu \leqslant n-1. \tag{15.1}$$

As in Section 9, we will assume that $B_{\lambda,k}(\alpha_\lambda) B_{\mu,l}(\beta_\mu) \neq 0$ for all pairs (λ, μ); this is equivalent, by Theorem 9.1, to assuming that the matrices

$$A = (a_{\lambda,i}) = \bigl(B_{i,k}(\alpha_\lambda)\bigr), \quad (\lambda, i) \in [0, m-1] \times [0, m-1],$$
$$B = (b_{\mu,j}) = \bigl(B_{j,l}(\beta_\mu)\bigr), \quad (\mu, j) \in [0, n-1] \times [0, n-1],$$

are *invertible*. Relation (15.1) then becomes: $Q = AP\,{}^{t}B$, where Q is the matrix $(Q_{i,j})$ and P the matrix $(P_{i,j})$ (which we have to determine). To solve this equation, set $U = P\,{}^{t}B$, and compute U, solving $AU = Q$ (if the points $Q_{\lambda,\mu}$ are in $\mathbb{R}^3$, this implies that we have to solve $3n$ linear systems with matrix A: for that, see Remarks 9.1). One then solves the system $B\,{}^{\mathrm{t}}P = {}^{\mathrm{t}}U$, in order to determine the matrix P, which is equivalent to solving $3m$ scalar linear systems with the same matrix B.

Approximation of a surface (or "quasi-interpolation"). If we now want to *approximate* a parametric surface generated by the function $f(u,v)$ defined on $[a,b]\times[c,d]$, we may, as in Section 4, set

$$S_f(u,v)=\sum_{i=0}^{m-1}\sum_{j=0}^{n-1} f(\theta_i,\eta_j)B_{i,k}(u)B_{j,l}(v),$$

with

$$\theta_i=\frac{u_{i+1}+\cdots+u_{i+k}}{k},\qquad \eta_j=\frac{v_{j+1}+\cdots+v_{j+l}}{l}.$$

With this choice of θ_i and η_j, this spline surface will have the property *"to soften the oscillations"* of the surface defined by f: this comes from the similar fact in one variable (see Theorem 7.1): the operator S_f is regularizing for each curve "$u=$ constant" or "$v=$ constant".

Boolean sum of two operators. To approximate the surface, one may also take the *boolean sum* of the two operators (with the above notation):

$$S_m g(u)=\sum_{i=0}^{m-1} g(\theta_i)B_{i,k}(u),$$

$$S_n h(v)=\sum_{j=0}^{n-1} h(\eta_j)B_{j,l}(v),$$

and then set

$$\begin{aligned}(S_m\oplus S_n)&f(u,v)\\ &=S_m f(\bullet,v)+S_n f(u,\bullet)-S_f(u,v)\\ &=\sum_{i=0}^{m-1} f(\theta_i,v)B_{i,k}(u)+\sum_{j=0}^{n-1} f(u,\eta_j)B_{j,l}(v)\\ &\quad-\sum_{i=0}^{m-1}\sum_{j=0}^{n-1} f(\theta_i,\eta_j)B_{i,k}(u)B_{j,l}(v).\end{aligned}$$

As $S_m l=l$ (and $S_n l=l$) for any linear form l (see Theorem 4.1), we see that

$$(S_m\oplus S_n)l(u)v^j=l(u)v^j,\quad 0\leqslant j\leqslant n-1,$$
$$(S_m\oplus S_n)u^i l(v)=u^i l(v),\quad 0\leqslant i\leqslant m-1,$$

for any linear form l; in particular, $(S_m \oplus S_n)P(u,v) = P(u,v)$ for any polynomial P of degree $\leqslant 3$. This operator then gives a very good approximation of the surface, but the operation needs the knowledge of the "isoparametric curves" $f(u,\eta_j)$ and $f(\theta_i,v)$.

Coons patches. See COONS [1967]. One may also consider the linear operator

$$r_1 g(u) = \frac{1}{b-a}\big[(b-u)g(a) + (u-a)g(b)\big]$$

which, for a function g of one variable defined on an interval $[a,b]$, interpolates $g(a)$ and $g(b)$. Set in the same way

$$r_2 h(v) = \frac{1}{d-c}\big[(d-v)h(c) + (v-c)h(d)\big]$$

which interpolates $h(c)$ and $h(d)$ for a function h defined on an interval $[c,d]$.

For a function $f(u,v)$, $(u,v) \in [a,b] \times [c,d]$, the direct sum $(r_1 \oplus r_2)f(u,v)$ becomes

$$\begin{aligned}
&\frac{1}{b-a}[(b-u)f(a,v) + (u-a)f(b,v)] + \frac{1}{d-c}[(d-v)f(u,c) + (v-c)f(u,d)] \\
&\quad - \frac{1}{(b-a)(d-c)}\big[(b-u)(d-v)f(a,c) + (b-u)(v-c)f(a,d) \\
&\quad + (u-a)(d-v)f(b,c) + (u-a)(v-c)f(b,d)\big].
\end{aligned}$$

This surface, called a *"Coons patch"*, interpolates the four curves of the boundary of the parametric surface defined by $f(u,v)$, namely the curves $f(u,c)$, $f(u,d)$, $f(a,v)$, $f(b,v)$, $(u,v) \in [a,b] \times [c,d]$. Its construction needs in fact only the knowledge of these four curves.

16. Bernstein polynomials

Tensor products are often used because they are easy to compute, and because one can apply to them the algorithms of the one variable case; nevertheless, they have several disadvantages:

They are well adapted for rectangular portions of surfaces, but not for triangular ones for instance.

There are two privileged families of curves on a spline surface defined as a tensor product: the ones defined by $u =$ constant, and by $v =$ constant; for these curves, which are one variable splines, one easily gets many properties; for instance, it is easy to compute curvature, we have the regularity property (Theorem 7.1), etc. On the other hand, it is very difficult to study other curves, for instance $u+v =$ constant.

Even in the bicubic case, the most often used, each coordinate is of *total degree 6* in u and v, which implies that the intrinsic equation $f(X,Y,Z) = 0$ of a bicubic patch is

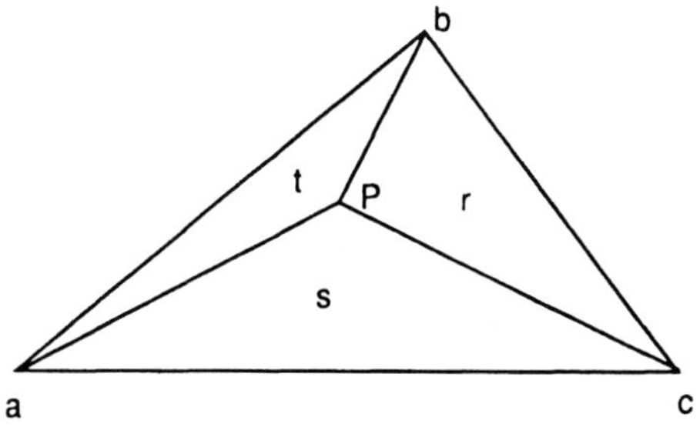

FIG. 16.1. Barycentric coordinates.

a portion of an algebraic surface of degree 18. The intersection of two such surfaces is then an algebraic curve of degree 324, hardly accessible to computer algebra systems.

The above considerations lead us to try to directly generalize spline curves, and Bézier curves, to the case of surfaces. We begin with the case of Bézier patches (Sections 16–19), and then study the case of polyhedral splines in Sections 20–21.

Let $\triangle \subset \mathbb{R}^2$ be a triangle with vertices a, b, c and area 1, and let (r, s, t) be the barycentric coordinates of a point $P \in \mathbb{R}^2$ with respect to (a, b, c).

Recall that

– $r + s + t = 1$.

– P is in the interior of the triangle $\triangle$ if and only if $r > 0,\ s > 0,\ t > 0$.

– r, s and t are respectively the areas of the triangles Pbc, Pac and Pab (Fig. 16.1).

DEFINITION. Let us fix a triangle $\triangle$ and an integer n; the *Bernstein polynomials* of degree n are defined in the following way:

$$B^n_{i,j,k} = \frac{n!}{i!j!k!} r^i s^j t^k \quad \text{for } i + j + k = n. \tag{16.1}$$

Let us quote the main properties of the $B_{i,j,k}$'s.

(1) There are $(n+1)(n+2)/2$ Bernstein polynomials of degree $\leqslant n$.

(2) The $B^n_{i,j,k}(r, s, 1-r-s)$'s form a *basis* of the space of polynomials in r and s of degree $\leqslant n$; this basis is called the *Bernstein basis*. For the proof, it is enough, after (1), to show that the $B^n_{i,j,k}(r, s, 1-r-s)$'s are linearly independent, which is clear, as the lowest degree term in the expression $r^i s^j (1-r-s)^k$ is $r^i s^j$.

(3) One has the following induction formula

$$B^n_{i,j,k} = rB^{n-1}_{i-1,j,k} + sB^{n-1}_{i,j-1,k} + tB^{n-1}_{i,j,k-1}. \tag{16.2}$$

In fact, the coefficient of the term $r^i s^j t^k$ on the right is

$$\frac{(n-1)!}{(i-1)!(j-1)!(k-1)!}\left[\frac{1}{jk} + \frac{1}{ik} + \frac{1}{ij}\right],$$

which is equal to $n!/i!j!k!$, since $i + j + k = n$.

(4) One has

$$\sum_{i+j+k=n} B^n_{i,j,k}(r, s, 1-r-s) = 1. \tag{16.3}$$

In fact, $\sum_{i+j+k=n} B^n_{i,j,k}(r,s,t)$ is equal to the expansion of $(r+s+t)^n$; we then have only to set $t = 1-r-s$.

(5) If $P \in \mathbb{R}^2$ has barycentric coordinates r, s, t with respect to the triangle $\triangle$, one has $B^n_{i,j,k}(r,s,t) > 0$ in the interior of $\triangle$.

(6) *Degree elevation.* Any polynomial of degree n (in r, s, t) may be viewed as a polynomial of degree $\leqslant n+1$, and thus may be expanded in the basis $B^{n+1}_{i,j,k}$. One has in particular $B^n_{i,j,k} = (r+s+t)B^n_{i,j,k}(r,s,t)$, because $r+s+t=1$, and furthermore

$$B^n_{i,j,k} = \frac{1}{n+1}\left[(i+1)B^{n+1}_{i+1,j,k} + (j+1)B^{n+1}_{i,j+1,k} + (k+1)B^{n+1}_{i,j,k+1}\right], \tag{16.4}$$

as one immediately infers from (16.1).

17. Bézier triangular patches

As in the case of curves (see (7.1)), let us fix an integer n, a (closed) triangle $\triangle \subset \mathbb{R}^2$, and let us consider a "triangular" network $(P_{i,j,k})(i+j+k=n)$ of points in $\mathbb{R}^s$.

DEFINITION. A *Bézier triangular patch* is by definition the parametric surface generated by the map $B : \triangle \to \mathbb{R}^s$, defined as follows:

$$B(r,s,t) = \sum_{i+j+k=n} P_{i,j,k} B^n_{i,j,k}(r,s,t). \tag{17.1}$$

EXAMPLE. Case $n = 3$. The basis $B^n_{i,j,k}(r,s,t)$ is made from 10 elements, which we may represent in the following triangular array:

$$\begin{array}{ccccccc} & & & s^3 & & & \\ & & 3s^2t & & 3rs^2 & & \\ & 3st^2 & & 6rst & & 3r^2s & \\ t^3 & & 3rt^2 & & 3r^2t & & r^3 \end{array}$$

Let us draw the points $P_{i,j,k}$ in $\mathbb{R}^3$. To be sure that the patch will not have "self-intersection", one may for instance assume, as in Fig. 17.1, that if π denotes a linear projection $\mathbb{R}^3 \to \mathbb{R}^2$, $\pi(P_{i,j,k})$ is the point of barycentric coordinates $(i/n, j/n, k/n)$.

Let us now quote the main properties of the patch $B(r,s,t)$.

(1) The patch $B(r,s,t) = \sum_{i+j+k=n} P_{i,j,k} B^n_{i,j,k}(r,s,t)$ passes through the points $P_{n,0,0}$, $P_{0,n,0}$ and $P_{0,0,n}$; one has in fact

$$B^n_{n,0,0}(1,0,0) = B^n_{0,n,0}(0,1,0) = B^n_{0,0,n}(0,0,1) = 1.$$

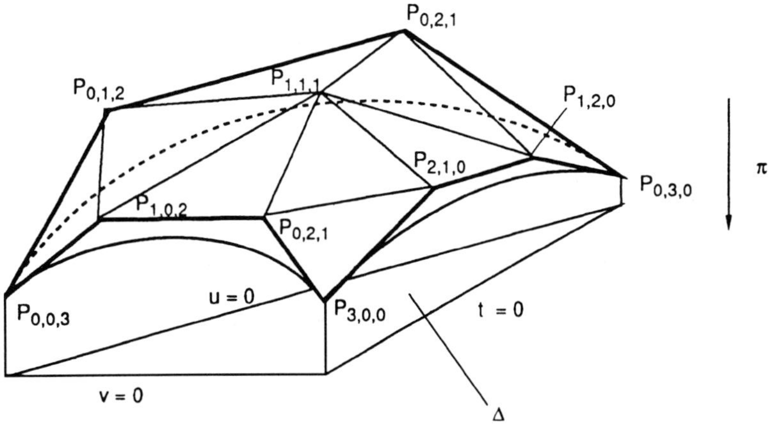

FIG. 17.1. A Bézier patch of degree 3.

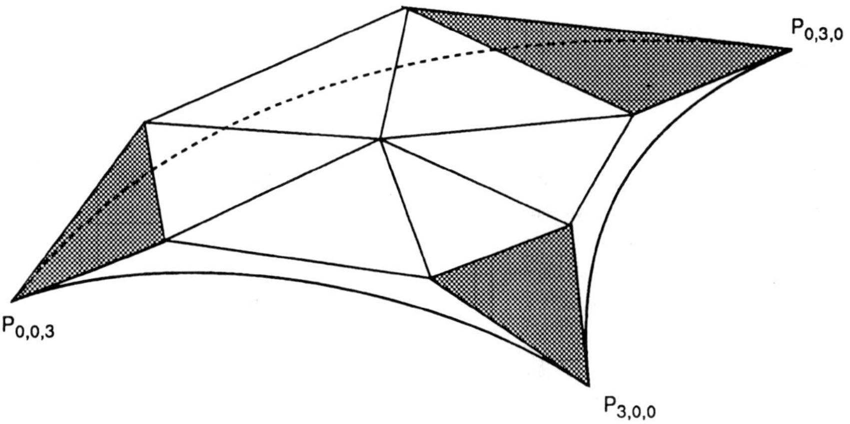

FIG. 17.2. The tangent planes are shaded.

(2) The boundary of the patch is formed with three Bézier curves, which are the images of the sides of the triangle $\triangle$ (these sides are defined by $r = 0$, $s = 0$, and $t = 0$). For instance, the curve image of the side $r = 0$ is the Bézier curve of degree n with control polygon $(P_{0,0,n}, P_{0,1,n-1}, \ldots, P_{0,n,0})$; the $B^n_{0,j,l}(0, s, 1-s)$'s are in fact the Bernstein polynomials of degree n in s, as is immediately seen.

(3) The patch is included in the convex hull of the points $P_{i,j,k}$. This results from the relation $\sum_{i+j+k=n} B^n_{i,j,k} = 1$ (see (16.3)).

(4) The tangent planes to the patch $B(r, s, t)$ at the points $P_{n,0,0}$, $P_{0,n,0}$, $P_{0,0,n}$ are generated by the two neighbouring points of the network; for instance, the tangent plane to the patch at the point $P_{n,0,0}$ is the plane spanned by the points $(P_{n,0,0},\ P_{n-1,1,0},\ P_{n_1,0,1})$. (See Fig. 17.2.) This results immediately follow from the fact that the vectors $\overrightarrow{P_{n,0,0}P_{n-1,1,0}}$ and $\overrightarrow{P_{n,0,0}P_{n-1,0,1}}$ are tangent to the two Bézier curves of the boundary of the patch, passing through the point $P_{n,0,0}$.

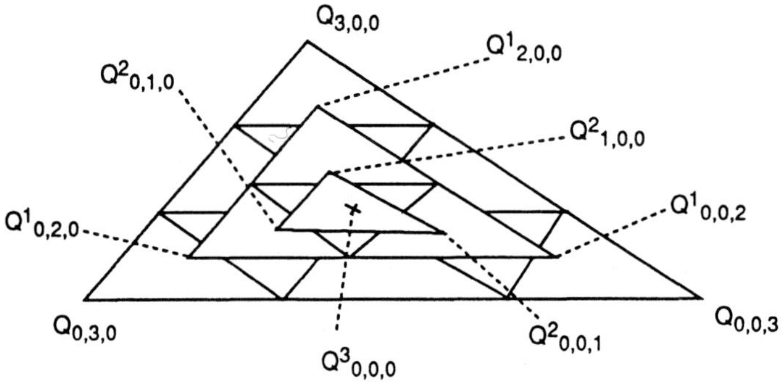

FIG. 17.3.

(5) The image under B of any straight line of the r, s, t plane, is a parametric curve of degree n. (Note that this property does not hold for tensor products.)

Let us now discuss some algorithms used for computing Bézier patches.

Evaluation algorithm (*or "De Casteljau algorithm"*). It simply uses the induction formula (16.2).

Let $B(r,s,t) = \sum P_{i,j,k}B^n_{i,j,k}(r,s,t)$ be a triangular Bézier patch; we want to evaluate $B(r,s,t)$ for fixed values of r, s, t.

– Set $P^0_{i,j,k} = P_{i,j,k}$, $i+j+k=n$.

– For $1 \leqslant l \leqslant n$, solve

$$P^l_{i,j,k} = rP^{l-1}_{i+1,j,k} + sP^{l-1}_{i,j+1,k} + tP^{l-1}_{i,j,k+1} \quad \text{for } i+j+k = n-l.$$

– We then have $B(r,s,t) = P^n_{0,0,0}(r,s,t)$.

Let us illustrate this algorithm in the case of a patch in $\mathbb{R}^3$ of degree 3, whose image under the projection π is the triangle Δ: let $P_{i,j,k}$, $i+j+k=3$, be points in $\mathbb{R}^3$ such that $\pi(P_{i,j,k}) = Q_{i,j,k} = (i/3, j/3, k/3) \in \Delta$, and set $Q^l_{i,j,k} = \pi(P^l_{i,j,k})$, $0 \leqslant l \leqslant 3$.

Figure 17.3 shows this evaluation algorithm; one has $B(r,s,t) = Q^3_{0,0,0}$.

REMARK 17.1. The De Casteljau algorithm above, which has a rather high number of operations, may give more information than the simple evaluation of $B(r,s,t)$. For instance, the intermediate points computed by the De Casteljau algorithm will be useful for the computation of the derivatives; in particular, the tangent plane to the Bézier patch at the point $B(r,s,t)$ is the plane containing the points $P^{n-1}_{1,0,0}$, $P^{n-1}_{0,1,0}$ and $P^{n-1}_{0,0,1}$. Moreover, the De Casteljau algorithm also gives, as for Bézier curves, a subdivision algorithm of the Bézier patch (this time in three patches; see RISLER ([1991], p. 114)).

Degree elevation. Let us look at a patch

$$B(r,s,t) = \sum_{i+j+k=n} P_{i,j,k}B^n_{i,j,k}.$$

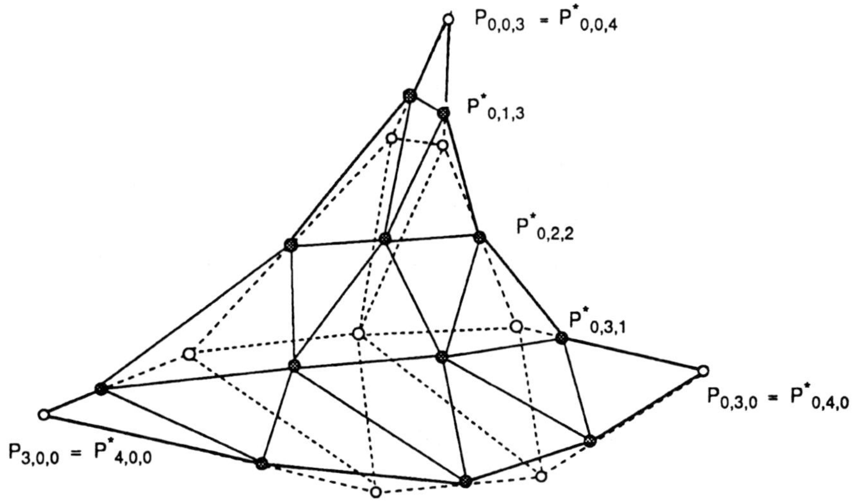

FIG. 17.4. Degree elevation from 3 to 4.

Applying (16.4), we obtain

$$B(r,s,t) = \sum_{i+j+k=n+1} P^*_{i,j,k} B^{n+1}_{i,j,k},$$

with a new set of control points $(P^*_{i,j,k})_{i+j+k=n+1}$ defined by

$$P^*_{i,j,k} = \frac{1}{n+1}(iP_{i-1,j,k} + jP_{i,j-1,k} + kP_{i,j,k-1}).$$

As in the case of curves, one may easily prove, that if this operation is iterated, the sequence of networks obtained in this fashion converges towards the patch. (See Fig. 17.4.)

18. Junction between Bézier patches

To model surfaces, it is necessary to have several triangular patches at a junction.

C^1-junction of two triangular patches. Let $\triangle$ and $\triangle'$ be two triangles, with barycentric coordinates (r,s,t) and (r',s',t') respectively, and two Bézier patches of the same degree n, $B(r,s,t)$ and $B'(r',s',t')$ (one may assume that the two patches have the same degree because of the degree elevation formula (16.4)). We may for instance make the junction of the two patches along the curves $r = 0$ and $r' = 0$. We will assume that $\triangle$ has the side $r = 0$ common with the side $r' = 0$ of $\triangle'$, $\triangle$ and $\triangle'$ being on each side of it (Fig. 18.1).

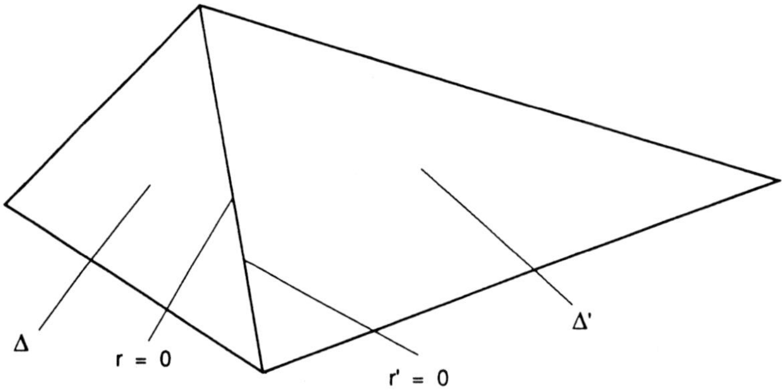

FIG. 18.1.

The parameters are then related by a relation of the following form:

$$\begin{aligned} r' &= -\alpha r, \\ s' &= s + \beta r, \quad \alpha > 0,\ 1 + \alpha = \beta + \gamma, \\ t' &= t + \gamma r, \end{aligned} \tag{18.1}$$

and in order to have a $\mathcal{C}^0$-continuity, the curves image of $r = 0$ and $r' = 0$ must coincide at B and B', which means that $P_{0,j,k} = P'_{0,j,k}$ $(j + k = n)$.

To have a $\mathcal{C}^1$-continuity, the first derivatives of B and B' must coincide along the common curve Γ, the image under B or B' of the segment $r = r' = 0$. For fixed j, let g_j (resp. g'_j) be the affine map which sends $\triangle$ (resp. $\triangle'$) onto the triangle $T_j = (P_{0,j,k}, P_{0,j+1,k-1}, P_{1,j,k-1})$ (resp. onto the triangle $T'_j = (P_{0,j,k}, P_{0,j+1,k-1}, P'_{1,j,k-1})$).

Let us denote by $\overrightarrow{X}$ the vector $\overrightarrow{X_0X_1}$. We then have, denoting by $\mathrm{D}_{\overrightarrow{X}}$ the differentiation operator in the direction of the vector $\overrightarrow{X}$,

$$\mathrm{D}_{\overrightarrow{X}} B = n \sum_{i+j+k=n-1} B^{n-1}_{i,j,k}(\triangle_{r_0} P_{i+1,j,k} + \triangle_{s_0} P_{i,j+1,k} + \triangle_{t_0} P_{i,j,k+1}).$$

The barycentric coordinates of $\overrightarrow{X}$ are $\triangle_{r_0}$, $\triangle_{s_0}$ and $\triangle_{t_0}$, hence we have

$$g_j(\overrightarrow{X}) = P_{0,j,k}\triangle_{r_0} + P_{0,j+1,k-1}\triangle_{s_0} + P_{1,j,k-1}\triangle_{t_0}.$$

Now, if x is a point of the segment $r = 0$ in $\triangle$, it follows that

$$\mathrm{D}_{\overrightarrow{X}} B(x) = n \sum_{j+k=n-1} B^{n-1}_{0,j,k}(x) g_j(\overrightarrow{X}).$$

In the same way, we have

$$D_{\overrightarrow{X}} B'(x) = n \sum_{j+k=n-1} B^{n-1}_{0,j,k}(x) g'_j(\overrightarrow{X}).$$

The C^1-continuity condition along Γ between the patches B and B' is therefore the identity of the maps g_j and g'_j, $0 \leqslant j \leqslant n$, which means that $T_j \cup T'_j$ is the image of $\triangle \cup \triangle'$ under a unique affine map. This implies in particular that the triangles T_j and T'_j are *coplanar*.

19. Base points of rational Bézier patches

In this section, we follow WARREN [1990], who explained the behavior of a triangular rational Bézier patch at a base point, and used this notion for devising multisided patches.

Let r, s, t be barycentric coordinates in a triangle $\triangle \subset \mathbb{R}^2$, and $B^n_{i,j,k}(r,s,t)$ be the corresponding Bernstein basis (see (16.1)). Let also P_{ijk} $(i+j+k=n)$ be control points in $\mathbb{R}^s$, and let w_{ijk} $(i+j+k=n)$ be real numbers not all zero, called *weights*. As for the case of curves, one defines the rational triangular Bézier patches as follows.

DEFINITION. The *triangular rational Bézier patch* associated with the points P_{ijk}'s and with the weights w_{ijk}'s is the image of the triangle $\triangle$ under the map

$$R(r,s,t) = \frac{B(r,s,t)}{T(r,s,t)},$$

with

$$B(r,s,t) = \sum_{i+j+k=n} P_{ijk} w_{ijk} B^n_{i,j,k}(r,s,t),$$

$$T(r,s,t) = \sum_{i+j+k=n} w_{ijk} B^n_{i,j,k}(r,s,t).$$

For simplicity, we will assume that $w_{ijk} \geqslant 0$, which is indeed a common assumption in this matter.

DEFINITION. A *base point* of the rational Bézier patch $R(r,s,t)$ is a point (r_0, s_0, t_0) of $\triangle$ such that $B(r_0,s_0,t_0) = T(r_0,s_0,t_0) = 0$.

REMARKS 19.1. (1) If $T(r_0,s_0,t_0) = 0$, but $B(r_0,s_0,t_0) \neq 0$, the image of the point (r_0,s_0,t_0) is defined if one allows points "at infinity", i.e., if one assumes that the patch lies in the *projective space*.

(2) The map $R(r,s,t)$ is not defined at a base point. In this section, we will look at the behavior of the patch near a base point; however, we will not treat the general case, but rather some special ones, which may be useful for design purposes, and which give a good insight into the general situation.

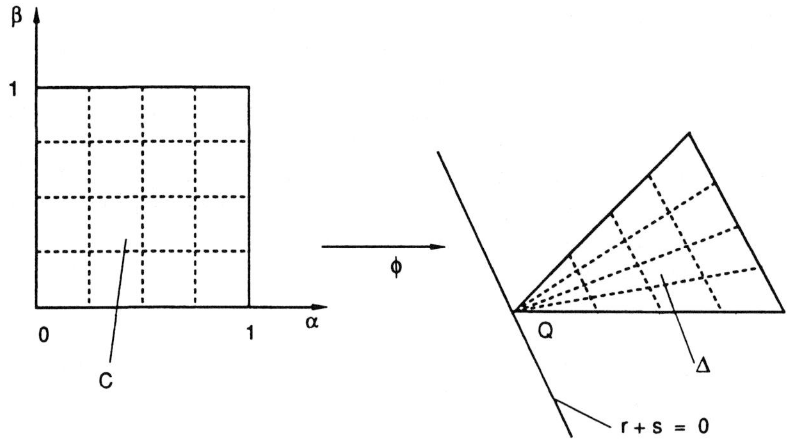

FIG. 19.1. Blowing-up of Q.

(3) For example, if one has $w_{00n} = 0$, the point $(0, 0, 1)$ of $\triangle$ is a base point. In fact, one then has $B(0, 0, 1) = T(0, 0, 1) = 0$, since

$$B^n_{i,j,k}(r, s, t) = \frac{n!}{i!j!k!} r^i s^j t^k,$$

and therefore $B^n_{i,j,k}(0, 0, 1) = 0$ if $i > 0$ or $j > 0$.

Let us now define the "blowing-up" of a point $Q \in \triangle$. For simplicity, we will assume that Q is one of the three vertices of $\triangle$, say the point $(0, 0, 1)$. This is the most useful case for possible applications, as the blowing up of Q will let us interpret any triangular patch as a rectangular one (see below). For the general definition of blowing up, see FULTON [1969] or BENEDETTI and RISLER ([1990], p. 150).

DEFINITION. Let $Q = (0, 0, 1) \in \triangle$. The *blowing-up* of Q is by definition the map ϕ of the square $C = [0, 1] \times [0, 1]$ with Cartesian coordinates (α, β), onto $\triangle$ (with barycentric coordinates (r, s, t)) defined by

$$r = \alpha(1 - \beta), \qquad s = \alpha\beta, \qquad t = 1 - \alpha.$$

REMARKS 19.2. (1) The map ϕ is invertible on $\triangle \setminus \{Q\}$. In fact, the inverse mapping ψ of ϕ is defined by the following formulas

$$\alpha = r + s = 1 - t, \qquad \beta = \frac{s}{r + s}.$$

The map ψ is in fact defined on all the plane of $\triangle$, except on the line $r + s = 0$, which intersects $\triangle$ only at the point Q.

Let us illustrate the map $\phi : C \to \triangle$ by Fig. 19.1. The image of the side $\alpha = 0$ of C is the point Q, and the images of the horizontal lines are the lines passing through

Q. The map ϕ induces a bijection of $C \setminus \{\alpha = 0\}$ onto $\triangle \setminus \{Q\}$. This map is called the *"blowing-up"* of Q, since the map ψ, defined on $\triangle \setminus \{Q\}$, "blows up" the point Q into the segment $\{\alpha = 0\}$, $0 \leqslant \beta \leqslant 1$ of C.

(2) Let $R(r,s,t): \triangle \to \mathbb{R}^m$ be a rational Bézier patch of degree n. The map $R \circ \phi: C \to \mathbb{R}^m$, which has the same image as ϕ, defines the patch as a rational Bézier tensor product patch of bidegree (n,n) defined over the unit square C. This patch has the same control polygon as R, but with modified weights, and with the image of a side degenerated to a point.

Following WARREN [1990], let us illustrate this fact in the case $n = 2$. One then has

$$\begin{aligned} B &= w_{002}P_{002}t^2 + 2w_{101}P_{101}rt + w_{200}P_{200}r^2 + 2w_{110}P_{110}rs + w_{020}P_{020}s^2 \\ &\quad + 2w_{011}st, \\ T &= w_{002}t^2 + 2w_{101}rt + 2w_{011}st + w_{200}r^2 + 2w_{110}rs + w_{020}s^2, \end{aligned}$$

where P_{ijk} $(i+j+k=2)$ is the set of control points in $\mathbb{R}^m$. Applying the transformation ϕ yields the following tensor product rational quadratic surface:

$$\begin{aligned} B \circ \phi(\alpha,\beta) &= w_{002}P_{002}(1-\alpha)^2 + \big(w_{101}P_{101}(1-\beta) + w_{011}P_{011}\beta\big)\big(2\alpha(1-\alpha)\big) \\ &\quad + \big(w_{200}P_{200}(1-\beta)^2 + w_{110}P_{110}2(1-\beta)\beta + w_{020}P_{020}\beta^2\big)(\alpha)^2, \\ T \circ \phi(\alpha,\beta) &= w_{002}(1-\alpha)^2 + \big(w_{101}(1-\beta) + w_{011}\beta\big)\big(2\alpha(1-\alpha)\big) \\ &\quad + \big(w_{200}(1-\beta)^2 + w_{110}2(1-\beta)\beta + w_{020}\beta^2\big)(\alpha)^2. \end{aligned}$$

The constant terms, as in $w_{002}(1-\alpha)^2$, or the terms of degree one, can be raised to quadratics to bring the patch into standard product form.

Let us now apply this transformation, assuming Q is a base point (i.e., $w_{002} = 0$ in this example). Then both $B \circ \phi$ and $T \circ \phi$ become divisible by α. Set

$$\widetilde{B}(\alpha,\beta) = \frac{B \circ \phi}{\alpha}, \qquad \widetilde{T}(\alpha,\beta) = \frac{T \circ \phi}{\alpha}.$$

Then the expression $\widetilde{B}(\alpha,\beta)/\widetilde{T}(\alpha,\beta)$ is equal to $B \circ \phi(\alpha,\beta)/T \circ \phi(\alpha,\beta)$ for $\alpha \neq 0$, and if w_{101} and w_{011} are nonzero, it is defined for $\alpha = 0$. In the case of degree n, this defines a patch, which is a tensor product of bidegree $(n-1,n)$ coinciding with the previous patch on $C \setminus \{\alpha = 0\}$, but which has four sides, the image of the segment $\alpha = 0$ being the line connecting $P_{1,0,n-1}$ and $P_{0,1,n-1}$: the triangular surface (with base point Q) is in fact four-sided. More generally, if all weights w_{ijk} with $i+j<m$ vanish, $B \circ \phi$ and $T \circ \phi$ become divisible by α^m, and, after division of $B \circ \phi$ and $T \circ \phi$ by α^m, the triangular rational Bézier patch becomes four-sided, the fourth side (the image of $\alpha = 0$) being the rational Bézier curve of degree m with control points P_{ijk}, $i+j=m$, $i+j+k=n$, as is immediately seen from the expressions of $B \circ \phi$ and $T \circ \phi$ (see (19.1) for the case $n = 2$).

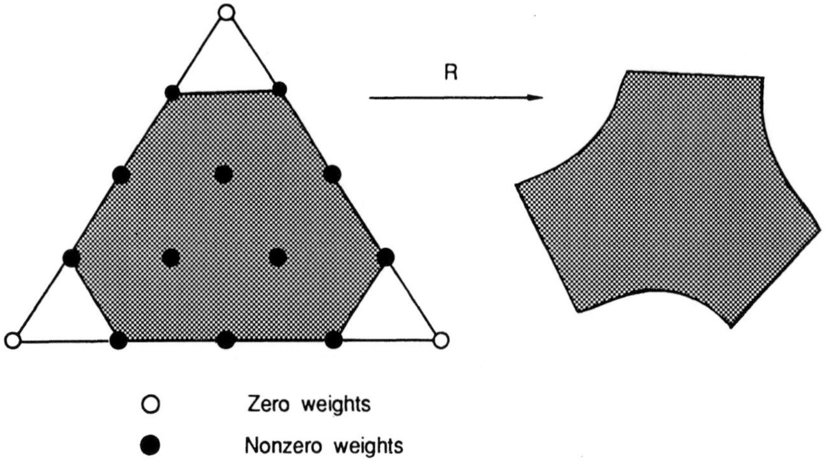

FIG. 19.2. Creation of a six-sided patch.

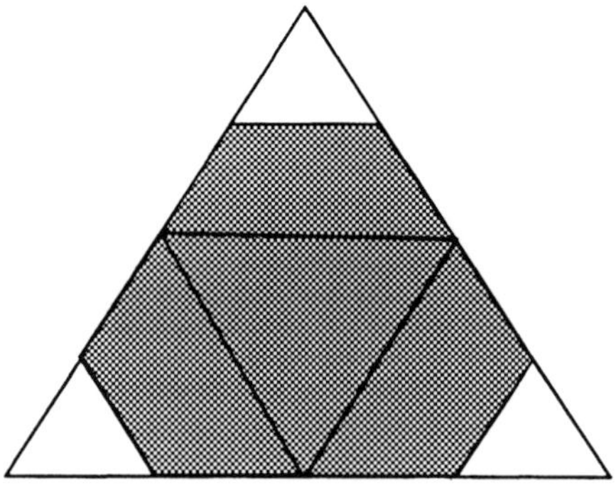

FIG. 19.3.

Let us see now how this property can be used to create multi-sided patches (for more details, the reader may refer to WARREN [1990]). The technique described above (i.e., setting weights to zero to create a four-sided patch) may be generalized to create five- and six-sided patches, by treating each vertex of the triangular domain independently. Figure 19.2 illustrates the creation of a six-sided patch, with degree $n = 4$; three portions of the boundary are formed with segments, and the others by rational quadratic Bézier curves.

From a computational point of view, one has to treat the three vertices of $\triangle$ independently. A solution, proposed in WARREN [1990], is to subdivide $\triangle$ into four triangles (and the patch into four patches), as shown in Fig. 19.3.

For another technique for constructing multi-sided patches, see LOOP and DE ROSE [1990].

We will not study base points in the general case (for instance in the case where the weights are not assumed to be $\geqslant 0$). One needs some machinery from algebraic geometry (see FULTON [1969]) to prove that the situation is similar to the one described above: after a sequence of blowing-ups, the base points disappear, replaced by a union of lines in a projective space.

For more facts about base points, see WARREN [1990] or RISLER ([1991], p. 125).

20. Polyhedral splines

We will define polyhedral splines and study their fundamental properties; the reader may refer to DAHMEN and MICCHELLI [1983] for a relatively complete study of the subject. The two following sections are based on RISLER [1991]; for another presentation, see SEIDEL [1991].

In view of the Hermite–Genocchi formula (Section 5), one natural way of generalizing one-variable B-splines is to consider "simplicial" splines (see (20.1) below), or more generally "polyhedral" splines (see (20.2)).

Let $X_0, \ldots, X_n$ be $n+1$ points in $\mathbb{R}^s$, and let $[X_0 \cdots X_n]$ be their convex hull; we will assume $\mathrm{Vol}[X_0 \cdots X_n] > 0$, which implies $n \geqslant s$ (Vol means here the s-dimensional volume in $\mathbb{R}^s$).

DEFINITION. Let $S_n \subset \mathbb{R}^{n+1}$ the standard simplex of dimension n. We define the *B-spline associated with* $X_0, \ldots, X_n$, denoted $B(x|X_0, \ldots, X_n)$, as the unique function on $\mathbb{R}^s$ satisfying

$$\int_{\mathbb{R}^s} f(x) B(x|X_0, \ldots, X_n) = n! \int_{S_n} f(t_0 X_0 + \cdots + t_n X_n)\, \mathrm{d}t_1 \cdots \mathrm{d}t_n \qquad (20.1)$$

for any function f of class $\mathcal{C}^\infty$ with compact support on $\mathbb{R}^s$. Formula (20.1) introduces B as a *distribution*. We will see below that, with the hypotheses we have made, B is in fact a piecewise polynomial function on $\mathbb{R}^s$.

The notation

$$\int_{S_n} f(t_0 X_0 + \cdots + t_n X_n)\, \mathrm{d}t_1 \cdots \mathrm{d}t_n$$

means that we set $t_0 = 1 - t_1 - \cdots - t_n$, and that we integrate on the projection S'_n of S_n in $\mathbb{R}^n$: see Fig. 20.1. The projection $S'_n \subset \mathbb{R}^n$ is defined by $t_i \geqslant 0$, $\sum_{i=1}^n t_i \leqslant 1$; this fixes in particular the orientation of S_n. We may moreover replace t_0 by any t_j, $0 \leqslant j \leqslant n$, and multiply the result by $(-1)^j$, in such a way that the orientation is preserved.

The preceding definition suggests a generalization to the case where the simplex S_n (or S'_n) is replaced by any polyhedron.

DEFINITION. Let $P \subset \mathbb{R}^n$ be a compact polyhedron, let $\pi : \mathbb{R}^n \to \mathbb{R}^s$ be an affine map. One defines the *polyhedral spline* $B_P(x)$ associated with P and to π, by the formula

$$\int_{\mathbb{R}^s} f(x) B_P(x) = \int_P f \circ \pi\, \mathrm{d}t_1 \wedge \cdots \wedge \mathrm{d}t_n \qquad (20.2)$$

for any function f of class $\mathcal{C}^\infty$ with compact support on $\mathbb{R}^s$.

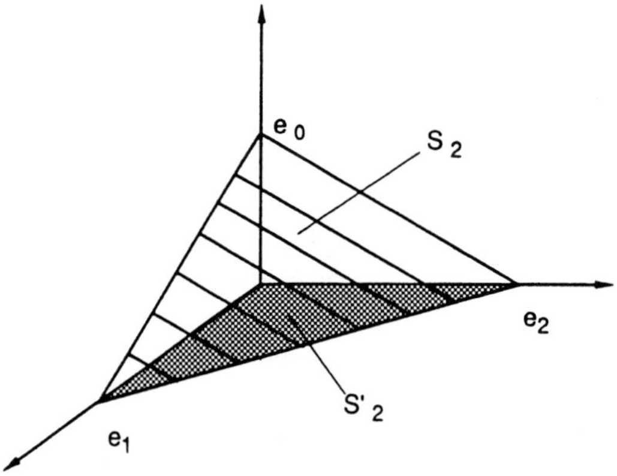

FIG. 20.1.

Note that the B-spline $B_P(x)$ is defined by the above formula only as a distribution; we will see below under which conditions $B_P(x)$ is in fact a function. The B-spline defined in (20.1) then corresponds (up to the factor $n!$) to the case where P is the projection S'_n of S_n in $\mathbb{R}^n$. We thus have $B(x|X_0,\dots,X_n) = n!\,B_{S'_n}(x)$. The factor $n!$ here is indeed commonly introduced.

The preceding definition easily implies the following result:

THEOREM 20.1. *The support of $B_P(x)$ is equal to $\pi(P)$. In particular, the support of $B(x|X_0,\dots,X_n)$ is the convex hull $[X_0\cdots X_n]$ of the X_i's.*

If π is surjective, we have $B_P(x) = \mathrm{Vol}_{n-s}(\pi^{-1}(x)\cap P)$; this proves in particular that $B_P(x)$ is a function if $\mathrm{Vol}_{n-s}(\pi^{-1}(x)\cap P)$ is defined, i.e., as soon as $\pi(P)$ generates $\mathbb{R}^s$, as is the case in formula (20.1).

If $n = s$, one has

$$B(x|X_0,\dots,X_n) = \frac{\chi_{[X_0\cdots X_n]}(x)}{\mathrm{Vol}([X_0\cdots X_n])},$$

where χ_A denotes the characteristic function of a set A.

Let us now state two fundamental properties of the polyhedral splines: the *differentiation formula*, and the *induction formula*. For simplicity, we will restrict ourselves to the case of B-splines (a polyhedral spline is in fact always a sum of B-splines, obtained by a triangulation of P, i.e., by a decomposition of P into a union of simplices; note however that this decomposition is not canonical, the triangulation of P being nonunique).

Let us denote by $\mathrm{D}_{X_i-X_j}$ the derivative in the direction of the vector $X_i - X_j$, and by $(X_0,\dots,\widehat{X_j},\dots,X_n)$ the sequence $(X_0,\dots,X_n)$ in which the point X_j has been omitted.

THEOREM 20.2. *With the above notation, we have*

$$\begin{aligned} &\mathrm{D}_{X_i-X_j}B(x|X_0,\dots,X_n)\\ &\quad = n\big(B(x|X_0,\dots,\widehat{X_i},\dots,X_n) - B(x|X_0,\dots,\widehat{X_j},\dots X_n)\big). \end{aligned}$$

For the proof, see RISLER ([1991], p. 131), or DAHMEN and MICHELLI [1983].

As an immediate consequence, we have the following formula: let $Z = \sum_{i=0}^{n} \mu_i X_i$, with $\sum_{i=0}^{n} \mu_i = 0$. Then

$$\mathrm{D}_Z B(x|X_0,\dots,X_n) = n\sum_{j=0}^{n} \mu_j B\big(x|X_0,\dots,\widehat{X_j},\dots,X_n\big).$$

THEOREM 20.3. *The function $B(x|X_0,\dots,X_n)$ is a piecewise polynomial of degree $\leqslant n-s$, the pieces being defined by the partition of $[X_0\cdots X_n]$ generated by the mutual intersections of sets of the form $[X_{i_0}\cdots X_{i_s}]$ $(0 \leqslant i_j < i_{j+1} \leqslant n)$.*

PROOF. We use induction on n, the case $n = s$ resulting from the fact that

$$B(x|X_0,\dots,X_s) = \frac{\chi_{[X_0\dots X_s]}(x)}{\mathrm{Vol}([X_0\cdots X_s])}, \tag{20.3}$$

and the iteration from n to $n+1$ being an immediate consequence of Theorem 20.2. □

THEOREM 20.4. *The B-spline function $B(x|X_0,\dots,X_n)$ is a function on $\mathbb{R}^s$ of class C^{d-2}, if d is the smallest of the cardinalities of the subsets $Y \subset X = (X_0,\dots,X_n)$ such that $X \setminus Y$ does not affinely span $\mathbb{R}^s$.*

Again this theorem is an easy consequence of Theorem 20.2.

REMARKS 20.1. (1) In the case $s = 1$, we recognize the properties of the B-spline functions of degree $n-1$ in one variable.

(2) We see that when the X_i's are in a *general position*, $B(x|X_0,\dots,X_n)$ is of class C^{n-s-1} (a general position meaning here that all the subsets of $(X_0,\dots,X_n)$ with $s+1$ elements generate $\mathbb{R}^s$ in the affine sense).

Let us now state the induction relation on the B-splines, which generalizes to polyhedral splines the definition of the one-variable B-splines. For describing a correct algorithm, we have to take care of the support of B-splines, as in dimension 1, where $B_{i,0}(t)$ is the characteristic function $\chi([t_i,t_{i+1}[)$ (and not $\chi([t_i,t_{i+1}])$. A solution has been given by SEIDEL ([1991], p. 8.29) for $s = 2$, who defined the "half-open convex hull" of three points $x_0, x_1, x_2 \in \mathbb{R}^2$.

DEFINITION. Let $\triangle = [x_0, x_1, x_2]$ be the convex hull of $\{x_0, x_1, x_2\}$, $\overset{\circ}{\triangle}$ its interior, e_1 the first unitary vector. The *half-open convex hull* $[x_0, x_1, x_2)$ is the set of points

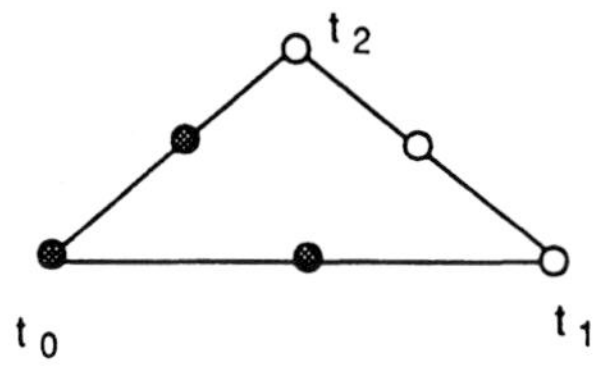

FIG. 20.2.

$u \in \triangle$ such that there exists a vector η with positive slope, and a number $\varepsilon > 0$ such that the set

$$\{u + se_1 + t\eta = 0,\ \ s, t, s + t < \varepsilon\}$$

is completely contained in $\mathring{\triangle}$.

EXAMPLE. In Fig. 20.2, the black points belong to the half-open convex hull, while the white points do not.

Assuming $s = 2$, and defining $B(x|X_i, X_j, X_k)$ as

$$\frac{\chi([X_i, X_j, X_k])}{\mathrm{Vol}([X_i, X_j, X_k])},$$

we have the following:

THEOREM 20.5. *Let x be a point in $[X_0 \cdots X_n]$ of the form $x = \sum_{j=0}^n \lambda_j X_j$, with $\sum_{j=0}^n \lambda_j = 1$. Then*

$$B(x|X_0, \dots, X_n) = \frac{n}{n-s} \sum_{j=0}^{n} \lambda_j B\big(x|X_0, \dots, \widehat{X_j}, \dots, X_n\big).$$

In a similar way as in the one-variable case (Section 6), the Bernstein polynomials introduced in Section 16 may be viewed as a special case of B-spline functions: let $X_0, \dots, X_s$ in $\mathbb{R}^s$ be given, each X_i being of multiplicity $(m_i + 1)$, i.e., counted $m_i + 1$ times, with $\sum_{i=0}^s (m_i + 1) = n + 1$. Then, if $\lambda_i,\ 0 \leqslant i \leqslant s$, are such that

$$X = \sum_{i=0}^{s} \lambda_i X_i, \quad \sum \lambda_i = 1,$$

we have, with self-explanatory notation

$$B\big(x|(X_0)^{m_0+1}, \dots, (X_s)^{m_s+1}\big) = n! \frac{\lambda_0^{m_0}}{m_0!} \cdots \frac{\lambda_s^{m_s}}{m_s!} \frac{\chi_{[X_0 \cdots X_s]}(x)}{\mathrm{Vol}[X_0 \cdots X_s]}. \tag{20.4}$$

In other words, we recognize, up to a constant factor, the generalization to an arbitrary dimension s of Bernstein polynomials of degree $\sum_{i=0}^{s} m_i = n - s$ as defined in (16.1). Note that the λ_i's are the barycentric coordinates of x with respect to the X_i's, and so are uniquely determined.

21. Box splines

We will consider a particular case of polyhedral splines, when the polyhedron P is the cube $[0,1[^n \subset \mathbb{R}^n$. We may then look at such a spline as a generalization of the one-dimensional spline $B_n(x)$ (defined on the interval $[0, n+1]$: see Section 8, and in particular we will consider the space generated by the translates (by vectors in the lattice $\mathbb{Z}^n$) of the hypercube P. These splines are then called *Box splines*.

Let X be a given set of $n+1$ points $X_0, X_1, \dots, X_n$ in $\mathbb{R}^s$, satisfying the following assumptions:

(i) $X_0 = 0$, and the X_i's have integer coordinates, several X_i's being possibly equal.

(ii) The X_i's generate $\mathbb{R}^s$ (in this section, we will identify the points X_i, $i \geqslant 1$, with the vectors $(X_i - X_0)$ of $\mathbb{R}^s$).

DEFINITION. Let $(e_1, \dots, e_n)$ be the canonical basis of $\mathbb{R}^n$, $\pi : \mathbb{R}^n \to \mathbb{R}^s$ be the linear map defined by $\pi(e_i) = X_i$ $(1 \leqslant i \leqslant n)$, $P = [0,1[^n \subset \mathbb{R}^n$. Then the spline $B_P(x)$ associated with these data (see (20.2)) is called the *box spline* associated with X and is denoted $M(x|X)$.

To mark a difference with the B-splines of Section 20, we have avoided the classical notation $B(x|X)$. We then have, by definition,

$$\int_{\mathbb{R}^s} f(x) M(x|X) = \int_{[0,1[^n} (f \circ \pi)\, \mathrm{d}t_1 \wedge \cdots \wedge \mathrm{d}t_n \tag{21.1}$$

for all f that are C^∞ with compact support.

EXAMPLE. $n = 3$, $s = 2$. In Fig. 21.1, the edges of the projection of P are the vectors X_1, X_2 and X_3, and the edges represented in dotted lines; $\pi(P)$ is the support of $B_P(x) = M(x|X)$.

REMARKS 21.1. (1) Let us denote by X the set of X_i's, $1 \leqslant i \leqslant n$; if X does not generate $\mathbb{R}^s$, $M(x|X)$ is, as in Section 20, a distribution on $\mathbb{R}^s$, with support $[X_0, \dots, X_n]$. However, when we write the expression $M(x|X)$, we will always assume either that X generates $\mathbb{R}^s$, or that $x \notin [X_0 \cdots X_n]$, in which case we will set $M(x|X) = 0$.

(2) If X does not generate $\mathbb{R}^s$, the distribution $M(x|X)$ on $\mathbb{R}^s$ may also be interpreted as a box-spline function on the linear subspace H of $\mathbb{R}^s$ generated by X, by setting

$$\int_H f(x) M(x|X) = \int_{[0,1[^n} (f \circ \pi)\, \mathrm{d}t_1 \wedge \cdots \wedge \mathrm{d}t_n$$

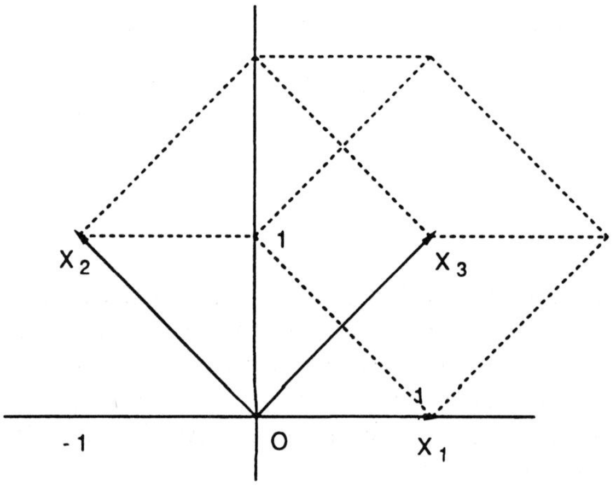

FIG. 21.1. Support of a piecewise-linear Box spline.

for any function f on H of class $\mathcal{C}^\infty$ with compact support.

Let us now look at those properties of $M(x|X)$ that are similar to those of Theorems 20.2, 20.3 and 20.4. Let f be a function on $\mathbb{R}^s$. Set

$$\Delta_{X_i} f = f(x + X_i) - f(x),$$
$$\nabla_{X_i} f = f(x) - f(x - X_i).$$

THEOREM 21.1. *If* D_{X_i} *denotes, for* $i > 0$, *the differentiation operator in the direction of the vector* X_i, *and if* $X \setminus X_i$ *generates* $\mathbb{R}^s$, *then the derivative* $\mathrm{D}_{X_i} M(x|X)$ *exists, and further we have*

$$\mathrm{D}_{X_i} M(x|X) = \nabla_{X_i} M(x|X \setminus X_i). \tag{21.2}$$

As a consequence, we have that the function $M(x|X)$ is piecewise polynomial of degree $\leqslant n - s$, and of class $\mathcal{C}^{d-1}$, if $(d+1)$ is the smallest of the cardinalities of the subsets $Y \subset X = (X_1, \ldots, X_n)$ such that $X \setminus Y$ does not generate $\mathbb{R}^s$ as a vector space. In other words, for any subset $Y \subset X$ with cardinality $\leqslant d$, the set $X \setminus Y$ generates $\mathbb{R}^s$.

We also have an induction relation, as in Theorem 20.5, considering the definition of the half-convex hull (Section 20):

THEOREM 21.2. *Let* x *be a vector of* $\mathbb{R}^2$. *We may then write* $x = \sum_{i=1}^{n} t_i(X_i - X_0)$, *in a non-unique way if* $n > 2$. *We then have*

$$M(x|X) = \frac{1}{n-s} \sum_{i=1}^{n} \big[t_i M(x|X \setminus X_i) + (1 - t_i) M(x - X_i|X \setminus X_i)\big]. \tag{21.3}$$

PROOF. Use Remark 21.1. □

We will now consider the vector space S_X generated by the functions $M(x-\alpha|X)$, $\alpha \in \mathbb{Z}^s$, understood in the following sense: S_X is the set of functions

$$\sum_{\alpha\in\mathbb{Z}^s} \lambda_\alpha M(x-\alpha|X), \qquad \lambda_\alpha \in \mathbb{R}.$$

We thus allow infinite sums, which are meaningful, as the family of the $M(x|X)$'s is locally finite. This space is a possible generalization to the case of s variables of the space generated by the (one variable) uniform B-splines with integer knots. Two questions arise in a natural way, namely: If $\mathbb{P}_r$ denotes the space of polynomials of degree $\leqslant r$ on $\mathbb{R}^s$, determine the integers r such that $\mathbb{P}_r \subset S_X$, and study the linear independence of the functions $M(x-\alpha|X)$.

Let us look at the first question, which is crucial, since it determines the possible order of approximation of the functions on $\mathbb{R}^s$ by elements of S_X.

THEOREM 21.3. *Let d be the integer defined just after Theorem* 21.1; *we then have* $\mathbb{P}_d \subset S_X$.

For a proof of Theorem 21.3, and for the second question, see DAHMEN and MICHELLI [1983], or RISLER [1991].

REMARK. In the "generic" case, where any subset of X with s elements generates $\mathbb{R}^s$, we have $d = n - s$, as for the one-dimensional splines. For instance, in the case treated in Example 21.1, we have $d = 1$.

CHAPTER III

Effective Algebraic Geometry

In this chapter, some basic facts of *algebraic geometry* are stated, which are of practical interest for CAGD, possibly with the additional help of *computer algebra systems*. A complete overview of the matter should clearly contain other subjects, such as some aspects of *differential geometry* (as in the appendix by Böhm in FARIN [1988]), or some elementary facts of *singularity theory*.

Sections 22–24 are devoted to some aspects of *elimination theory*: Elimination is requested for any geometric operation on piecewise algebraic patches, such as ray tracing, intersection of patches, etc. In Section 22, the "computation", or "isolation" of the real roots of a one variable polynomial is studied. This is the final step of any geometric algorithm about algebraic or semi-algebraic sets. Section 23 is devoted to the classical theory of *discriminant* of one polynomial, and of the *resultant* of two polynomials. In Section 24, the *Groëbner basis technique* is presented. This technique is nowadays very popular among computer scientists.

Section 25 is devoted to the geometric and algebraic properties of *algebraic curves*; Section 26 presents the important particular case of *conics*, and Section 27 presents the *Steiner surface patches*. Steiner patches give an example of possible "co-existence" of a parametric representation, together with an intrinsic equation, for modeling surfaces. Finally, elementary facts of real *semi-algebraic geometry* are presented in Section 28.

22. Roots of one-variable polynomials

Let $P \in \mathbb{R}[X]$. Three natural questions arise about the roots of P:
- Determine the exact number of real roots of P.
- Isolate the roots of P, i.e., find a family of disjoint intervals $[a_i, b_i]$ of $\mathbb{R}$, such that each interval contains a single root of P.
- Find an approximate value for each root. This problem needs first the resolution of the second problem above.

In this section, we will study the first two questions without going into deep details; the third one requires tools from approximation theory that are beyond the scope of this article.

For more details, the reader may consult BENEDETTI and RISLER [1990].

Number of real roots of a polynomial.

DEFINITION (*see* (3.1)). Let $a = (a_0, \dots, a_n)$ be a sequence of real numbers. Then the *variation* (or number of change of signs) of a, denoted $V(a)$, is the number of pairs $(i, i+k)$, $k \geqslant 1$, such that
- $a_i a_{i+k} < 0$,
- $a_{i+r} = 0$ for $0 < r < k$.

DEFINITION. Let $K \subset \mathbb{R}$ be a field, and P and Q be two elements of $K[X]$. Let R be the remainder of the Euclidean division algorithm of $P'Q$ by P, P' being the derivative of P. The *Sturm sequence* of (P, Q), denoted $S(P, Q)$, is the sequence of polynomials S_i, $0 \leqslant i \leqslant t$, defined by:

$$S_0 = P,$$
$$S_1 = R,$$
$$S_{i-1} = S_i A_i - S_{i+1}, \quad \deg(S_{i+1}) < \deg(S_i), \ 1 \leqslant i \leqslant t-1,$$
$$S_t = (P, R),$$

where (P, R) denotes the greatest common divisor (abbreviated as GCD) of P and R. In other words, the Sturm sequence is, except for the signs, the sequence of remainders of the Euclidean algorithm. Note that if $Q = 1$, we have $R = P'$, and the above notion of Sturm sequence is the classical one.

THEOREM 22.1. *Let $[a, b] \subset \mathbb{R}$ be an interval such that $P(a)P(b) \neq 0$. If $S = S(P, Q)$ denotes the Sturm sequence of P and Q, we have*

$$V\big(S(a)\big) - V\big(S(b)\big) = Z_{Q>0}^{[a,b]}(P) - Z_{Q<0}^{[a,b]}(P),$$

where $Z_{Q>0}^{[a,b]}(P)$ denotes the number of roots α of P in the interval $[a, b]$ such that $Q(\alpha) > 0$, and $Z_{Q<0}^{[a,b]}(P)$ denotes the number of roots β in $[a, b]$ such that $Q(\beta) < 0$.

For a proof, see RISLER ([1991], Prop. 6.13).

Let us indicate two useful corollaries. The notation is the same as in Theorem 22.1.

THEOREM 22.2. *Let $S_1 = S(P, 1)$ be the Sturm sequence of the polynomials P and 1. Then $V(S_1(a)) - V(S_1(b))$ is equal to the number of roots of P in the interval $[a, b]$.*

The proof is immediate, from Theorem 22.1.

THEOREM 22.3. *Let $S(+\infty)$ be the sequence of leading coefficients of the sequence of polynomials $S = (S_i)$, $0 \leqslant i \leqslant t$, and $S(-\infty)$ be the sequence of leading coefficients of the sequence $(-1)^{d_i} S_i$, where d_i is the degree of S_i. We then have*

$$V\big(S(-\infty)\big) - V\big(S(+\infty)\big) = Z_{Q>0}(P) - Z_{Q<0}(P),$$

where $Z_{Q>0}(P)$*, resp.* $Z_{Q<0}(P)$*, denotes the number of roots* α *of* P *such that* $Q(\alpha) > 0$*, resp.* $Q(\alpha) < 0$.

PROOF. This theorem follows from the following simple fact: there exists a number $A > 0$ such that for any $M > A$, we have

$$V\big(S(M)\big) = V\big(S(+\infty)\big),$$
$$V\big(S(-M)\big) = V\big(S(-\infty)\big),$$

and any root α of one of the polynomials S_i satisfies $|\alpha| < M$.

If $P = a_0 + a_1 X + \cdots + a_n X^n$ with $a_n \neq 0$, one may for instance take

$$A = 1 + \max\left(\frac{|a_0|}{|a_n|}, \ldots, \frac{|a_{n-1}|}{|a_n|}\right)$$

(see BENEDETTI and RISLER ([1990], p. 19) or MIGNOTTE [1982]); this result is due to Cauchy.

For a complete proof of Theorem 22.3, see RISLER ([1991], p. 177). □

REMARK 22.1. It is possible to multiply (or divide) each term of the Sturm sequence by a positive element of K, without modifying the variation at a given point a. This observation is useful when the polynomials P and Q are in $A[X]$, where A is a ring with quotient field K. We may then obtain, using the pseudo-division (see the definition below), a sequence equivalent to the Sturm sequence, but with all the polynomials in $A[X]$ (generally, we have $A = \mathbb{Z}$).

DEFINITION. Let

$$P = a_p X^p + \cdots + a_1 X + a_0, \quad a_p \neq 0,$$
$$Q = b_q X^q + \cdots + b_1 X + b_0, \quad b_q \neq 0,$$

be two polynomials in $A[X]$ such that $p \geqslant q$. The *pseudo-division* of P by Q is the Euclidean division of $(b_q)^{p-q+1} P$ by Q. The remainder of this division is called *pseudo-remainder*.

We immediately see that, if we make the division:

$$(b_q)^{p-q+1} P = BQ + R$$

with $\deg R < q$, then B and R have coefficients in A. Note that it is not necessarily the case for the division of P by Q.

If P and Q have coefficients in $\mathbb{Z}$, and if we perform the Euclidean algorithm using pseudo-divisions (to compute for instance the GCD of P and Q, or the Sturm sequence), the size of the coefficients of the successive remainders grow in general exponentially in terms of the degrees of P and Q; following DAVENPORT, SIRET and TOURNIER ([1987], p. 80–81), let us describe the "pseudo-remainder sequence algorithm" of Collin and Loos (see LOOS [1982]), which performs the Euclidean algorithm with a linear growing of coefficients.

Let P_1 and P_2 be two polynomials with integer coefficients. We define a sequence P_i of polynomials, with integer coefficients, which will be the sequence of successive remainders in the Euclidean algorithm, up to multiplication by a non zero constant, and for which the size of coefficients grows linearly in terms of the degrees of P_1 and P_2.

Let

$$d_i = \deg P_i, \qquad d_i - d_{i+1} = \delta_i,$$

and let p_i be the leading coefficient of P_i ($P_i = p_i X^{d_i} + \cdots$). Then, as remarked after the preceding definition, the remainder of the division of $p_i^{\delta_{i-1}+1} P_{i-1}$ by P_i is a polynomial with integer coefficients. Set $\beta_{i+1} P_{i+1}$ for this polynomial, β_{i+1} being an integer we have to compute. Because of these coefficients β_i, the growth of the coefficients of the polynomials P_i will not be too fast. In the classical Euclidean algorithm, we take $\beta_i = 1$. Set

$$\beta_3 = (-1)^{\delta_1+1},$$
$$\beta_i = -p_{i-2}\psi_i^{\delta_{i-2}},$$

where the ψ_i's are defined by

$$\psi_3 = -1,$$
$$\psi_i = (-p_{i-1})^{\delta_{i-3}}\psi_{i-1}^{1-\delta_{i-3}}.$$

Then it is proved for instance in LOOS [1982] that the P_i's have integer coefficients, and that the growth of coefficients is linear.

If this algorithm is used for the computation of Sturm sequences, special care must be exercised when computing the signs of the successive β_i's; see GONZALEZ, LOMBARDI, RECIO and ROY [1990] for details on this matter. This algorithm is related to the calculation of the "subresultants" of P_1 and P_2; see Section 23 below.

Isolation of roots. We will describe some methods, and the related algorithms, that let us isolate the (real) roots of a polynomial $P \in \mathbb{Z}[X]$. For more details, see COLLINS and LOOS [1982]. Isolating roots means finding a sequence $[a_i, b_i]$ of intervals with rational end-points, such that each interval $[a_i, b_i]$ contains a single root of P.

THEOREM 22.4 (minimal separation). *There exists a number m (depending on the degree n of P and of an appropriate bound on the size of the coefficients) such that if α_i and α_j are two distinct roots of P, then $|\alpha_i - \alpha_j| \geqslant m$.*

It can be proved that $m \geqslant \exp^{-cp\log(p)}$, where c is a constant > 0, and p is the degree of P: see for instance BENEDETTI and RISLER ([1990], p. 21).

Kronecker's method consists of the following steps:

(i) Let M be a bound for the $|\alpha_i|$'s, where the α_i's are the roots of P. One can take $M = \sup_{0 \leqslant i \leqslant p-1} |a_i/a_p|$ as above, or

$$M = 2 \sup \left(\left| \frac{a_{p-1}}{a_p} \right|, \left| \frac{a_{p-2}}{a_p} \right|^{1/2}, \ldots, \left| \frac{a_0}{a_p} \right|^{1/p} \right)$$

which is in fact a "better" choice (see KNUTH [1981]). We then divide the interval $[-M, +M]$ in $2M/m$ intervals (a_i, a_{i+1}), $1 \leqslant i \leqslant 2M/m$, of length m.

(ii) We evaluate $P(a_i)$ $(1 \leqslant i \leqslant 2M/m)$. If $P(a_i) = 0$, we find a root of P.

(iii) If $P(a_i) \neq 0$ and $P(a_{i+1}) \neq 0$, we compute $P(a_i)P(a_{i+1})$.

(iv) If $P(a_i)P(a_{i+1}) < 0$, there is a real root of P in $]a_i, a_{i+1}[$ (and only one, since $|a_i a_{i+1}| \leqslant m$), and if $P(a_i)P(a_{i+1}) > 0$, there is no root of P in $]a_i, a_{i+1}[$.

Another method, called Sturm's method, consists in using the properties of Sturm sequences described above: Let P be a square-free polynomial.

(i) We construct a Sturm sequence $S(P, P')$ by the Euclidean algorithm (using well chosen pseudo-divisions: see above). If $x \in \mathbb{R}$, let $w(x)$ denote the variation of the sequence $S(P, P')$ at x.

(ii) Set $x_0 = -M$, $x_1 = M$, $w_1 = w(x_0) - w(x_1)$ and $w := w_1$. If $w = 0$, P has no real root, and if $w = 1$, P has exactly one real root in $[-M, +M]$ (and then also in $\mathbb{R}$), and the algorithm ends.

(iii) If $w_1 > 1$, set $x_2 = (x_0 + x_1)/2$. We compute $w_2 = w(x_0) - w(x_1)$, and $w_3 = w(x_2) - w(x_1)$, and we come back to (ii), by setting successively $w := w_2$ and $w := w_3$. The algorithm ends when all the computed w_i's are equal to 0 or 1. This indeed happens because of Theorem 22.4.

The most costly part of this algorithm is the computation of the Sturm sequence $S(P, P')$.

23. Resultants and discriminants

The resultant is an algebraic tool allowing the elimination of a variable between two algebraic equations. More specifically, it gives the condition on the coefficients under which two polynomials $P(X)$ and $Q(X)$ have a common root (real or complex).

The resultant gives then a way of algebraically defining the intersection of two algebraic surfaces, or the intersection of a curve and a surface for instance. Another elimination technique, of a more algorithmic nature, is that of *Groëbner bases*, also called sometimes *standard bases*, which will be studied in Section 25.

In this section, we define resultants, and give some elementary properties. For other approaches, see CANNY and ROJAS [1991], STURMFELS [1993] or HOFFMANN [1992].

Let

$$P = a_0 + a_1 X + \cdots + a_p X^p$$

and

$$Q = b_0 + b_1 X + \cdots + b_q X^q$$

be two polynomials with coefficients in a commutative ring A ($A = \mathbb{Z}$ for instance). However, we will assume in general that the a_i's and the b_j's are independent variables, i.e., that $A = \mathbb{Z}[a_0, \ldots, a_p, b_0, \ldots, b_q]$.

DEFINITION. The *resultant* $R(P,Q)$ of P and Q is the determinant of the "*Sylvester matrix*" $M(P,Q)$ defined by

$$M(P,Q) = \begin{pmatrix} a_0 & \cdots & a_{q-1} & \cdots & a_{p-1} & a_p & & \\ & \ddots & & & & & \ddots & \\ & & a_0 & \cdots & \cdots & & & a_p \\ b_0 & \cdots & \cdots & b_q & & & & \\ & \ddots & & & & & & \\ & & & & b_0 & & \cdots & b_q \end{pmatrix}. \tag{23.1}$$

This matrix is a square matrix of order $p+q$: it has q rows of a_i's and p of b_j's. Note that $R(P,Q)$ is an element of A.

REMARK 23.1. If d is an integer, let $\mathcal{P}_d(A)$ denote the set of polynomials of degree $\leqslant d$ with coefficients in A. The degree of the polynomial P is denoted by $d(P)$. If j is an integer such that $1 \leqslant j \leqslant \inf(p,q)$, let Ψ_j be the linear map $\mathcal{P}_{q-1-j} \times \mathcal{P}_{p-1-j} \to \mathcal{P}_{p+q-1-j}$ defined by $\Psi_j(U,V) = UP + VQ$. Considering the canonical bases of the spaces of polynomials (with the natural order on the monomials, i.e., by increasing degrees), we see that the matrix $M(P,Q)$ is the matrix of the linear map Ψ_0.

THEOREM 23.1. (i) *Assume A is factorial (which is the case if A is a field, or the ring $\mathbb{Z}[a_0, \ldots, a_p, b_0, \ldots, b_q]$). Then $R(P,Q) = 0$ if and only if $a_p = b_q = 0$, or if the polynomials P and Q have a common non trivial factor in $A[X]$.*

(ii) *We have $R(P,Q) = UP + VQ$ with $U \in A[X]$ of degree $\leqslant q-1$, and $V \in A[X]$ of degree $\leqslant p-1$.*

PROOF. If $a_p = b_q = 0$, it is clear that $R(P,Q) = 0$. If P and Q have a common factor S in $A[X]$, we may write $P = SP_1$ and $Q = SQ_1$, and $Q_1P - P_1Q = 0$, with $d(Q_1) \leqslant q-1$ and $d(P_1) \leqslant p-1$. The map Ψ_0 is then not injective, which implies $R(P,Q) = 0$.

Conversely, assume $R(P,Q) = 0$, and $a_pb_q \neq 0$. As the map Ψ_0 is not injective, there exist polynomials P_1 of degree $\leqslant p-1$ and Q_1 of degree $\leqslant q-1$ such that $Q_1P = P_1Q$. This equality implies that P cannot be prime with respect to Q (otherwise, P would divide P_1 by the "Gauss Lemma", which is absurd), and so P and Q have a common factor. This proves (i).

Let us multiply the ith column of the matrix $M(P,Q)$ by X^{i-1}. We obtain a matrix M', whose determinant is equal to $X^{\alpha}R(P,Q)$, with $\alpha = (p+q-1)(p+q)/2$. But we may also compute the determinant of M' by adding all the columns to the first. The first column then becomes

$$P(X), XP(X), \ldots, X^{q-1}P(X), Q(X), \ldots, X^{p-1}Q(X).$$

If we now compute the determinant by expanding with respect to the first column, we obtain $X^\alpha(UP + VQ)$, with $d(U) \leqslant q - 1$ and $d(V) \leqslant p - 1$, whence $R(P, Q) = UP + VQ$. Hence (ii) is proved. □

We assume $a_p b_q \neq 0$, and that we may write

$$P = a_p(X - \alpha_1) \cdots (X - \alpha_p),$$
$$Q = b_q(X - \beta_1) \cdots (X - \beta_q),$$

with α_i and β_j in a field K (it is always possible in a convenient extension K of the fractions field of A). We then have:

THEOREM 23.2. *Let $a_p b_q \neq 0$. Using the above notation, we have*

$$R(P, Q) = a_p^q b_q^p \prod (\alpha_i - \beta_j), \quad 1 \leqslant i \leqslant p, \ 1 \leqslant j \leqslant q, \tag{23.2}$$

$$R(P, Q) = a_p^q \prod_{i=1}^{p} Q(\alpha_i) = (-1)^{pq} b_q^p \prod_{j=1}^{q} P(\beta_j). \tag{23.3}$$

PROOF. See any good textbook in algebra, LANG [1971] for instance. □

As an example of application, let us describe how the computation of the resultant may help for the computation of the intersection of two plane curves: roughly speaking, the roots of the resultant of the two equations (viewed as polynomials in Y), are the projections on the X-axis of the intersection points of the two curves. Let K be the field $\mathbb{C}$ of complex numbers, or the field $\mathbb{R}$ of real numbers, and let P and Q be two polynomials in $K[X, Y]$ (in fact in $\mathbb{Z}[X, Y]$), defining two plane curves C_1 $(P(X, Y) = 0)$ and C_2 $(Q(X, Y) = 0)$. Set $R(X) = R_Y(P, Q)$, P and Q being viewed as polynomials in Y, with coefficients in $\mathbb{Z}[X]$. An algorithm to compute the intersection of C_1 and C_2 proceeds along the following lines: *Isolate the real roots of* $R(X)$. Each root of $R(X)$ is the projection on the X-axis of a point of intersection of C_1 and C_2 (however, this point may be with complex Y-coordinate, or "at infinity"). Then,

For each real root α of $R(X)$, compute (or isolate) the real roots of $P(\alpha, Y)$.

Finally,

For each real root β of $P(\alpha, Y)$, test if $Q(\alpha, \beta) = 0$.

This algorithm requires computations with *real algebraic numbers*. See RIOBOO [1991] for this matter. Let $P = a_0 + a_1 X + \cdots + a_p X^p \in A[X]$, A being equal either to $\mathbb{Z}[a_0, \ldots, a_p]$, or to a factorial ring which we do not specify. Let $P' =$

$a_1 + \cdots + pa_pX^{p-1}$ be the derivative polynomial of P. It is then clear from its definition that the resultant $R(P, P')$ is divisible by a_p.

DEFINITION. The *discriminant* $D(P)$ is defined as

$$D(P) = \begin{cases} 0 & \text{if } a_p = 0, \\ (1/a_p)R(P, P') & \text{if } a_p \neq 0. \end{cases} \tag{23.4}$$

EXAMPLES. (1) Set $P = aX^2 + bX + c,\ a \neq 0$. Then

$$R(P, P') = \begin{vmatrix} c & b & a \\ b & 2a & 0 \\ 0 & b & 2a \end{vmatrix} = a\big(4ac - b^2\big)$$

and then $D(P) = 4ac - b^2$.

(2) Set $P = X^3 + pX + q$. It is then easy to verify that $D(P) = 4p^3 - 27q^2$.

THEOREM 23.3. *Let* α_i $(1 \leqslant i \leqslant p)$ *be the roots of* P *in a field containing* A. *Then* $D(P) = (a_p)^{2p-2} \prod_{i \neq j}(\alpha_i - \alpha_j)$.

PROOF. It is an immediate consequence of Theorem 23.2. We have

$$R(P, P') = (a_p)^{p-1} \prod_{i=1}^{p} P'(\alpha_i)$$

by Theorem 23.2, whence the expression of Theorem 23.3 follows by evaluating $P'(X)$, and replacing X by α_i. □

For two polynomials $P = a_0 + a_1X + \cdots + a_pX^p$ and $Q = b_0 + b_1X + \cdots + b_qX^q$, the vanishing of the resultant $R(P, Q)$ means that P and Q have a common factor (see Theorem 23.1). If the coefficients are in the field $\mathbb{C}$ of complex numbers (for instance if they are integers), this is equivalent to the existence of at least one common complex root for P and Q.

One may ask if it is possible to specify the exact number of common roots (counted with multiplicities), i.e., the degree of the common factor. The answer is given by the notion of subresultant. We will use it for the theory of semi-algebraic sets, in particular for the cylindric algebraic decomposition algorithm (see Section 28 below). Let $M(P, Q)$ be the Sylvester matrix of P and Q.

DEFINITION. Let i be such that $0 \leqslant i \leqslant \inf(p, q)$. The ith *subresultant* $r_i(P, Q)$ is the determinant of the $(p + q - 2i) \times (p + q - 2i)$ submatrix of $M(P, Q)$, obtained in the following way: delete from $M(P, Q)$ the last i rows (they consist of coefficients of P) and the rows numbered $(q - i + 1)$ to q (i.e., the last i rows made of coefficients of Q), the first i columns, and the last i columns.

Note that $r_0(P, Q)$ is the usual resultant as defined above. We then have the following.

THEOREM 23.4. *The following assertions are equivalent:*
(i) *P and Q have exactly j complex common roots (counted with multiplicities)*;
(ii) $r_0(P, Q) = \cdots = r_{j-1}(P, Q) = 0,\ r_j(P, Q) \neq 0$.

For a proof, see BENEDETTI and RISLER ([1990], p. 32 or 294), COSTE [1988] or SCHWARTZ and SHARIR ([1983], p. 324).

24. Groëbner basis

A thorough algebraic presentation of *Groëbner basis* is made in LEJEUNE [1985]. See also DAVENPORT, SIRET and TOURNIER ([1987], p. 105–112), and BUCHBERGER [1976, 1985].

The resultant defined in Section 23 above is used for eliminating one variable between two equations. Namely, if $P = 0$, $Q = 0$ are two equations involving a variable X, the polynomial $R(P, Q) = 0$, which depends on the coefficients of P and Q, is independent of X: the variable X has been "eliminated". The equation $R(P, Q) = 0$ then gives a necessary and sufficient condition for P and Q to have a common factor (if their coefficients are in a factorial ring A), or to have a common root in $\mathbb{C}$ (if their coefficients are in $\mathbb{C}$). When several variables have to be eliminated, different technics exist (see HOFFMANN [1992]). We describe one of them, very popular nowadays among computer scientists, called "Groëbner basis", sometimes also called "standard basis".

We begin by some basic definitions. Let $P_1, \ldots P_k$ be polynomials in

$$A = K[X_1, \ldots, X_n],$$

where K is a field (usually $\mathbb{Q}, \mathbb{R}$ or $\mathbb{C}$).

DEFINITION. The ideal

$$I = I(P_1, \ldots, P_k)$$

generated by $P_1, \ldots, P_k$ is the set of all polynomials of the form $Q_1P_1 + \cdots + Q_kP_k$ for any $Q_1, \ldots, Q_k$ in A. The set $(P_1, \ldots, P_k)$ is called a *set of generators* or a *basis* for I.

Ideals are the basic objects in algebraic geometry. For instance, if we look at a set of polynomial equations

$$\begin{cases} P_1(X_1, \ldots, X_n) = 0, \\ \quad\vdots \\ P_k(X_1, \ldots, X_n) = 0, \end{cases}$$

with $P_i \in K[X_1, \dots, X_n]$, the sets of common solutions (in K^n) of these equations depends only on the ideal $I = I(P_1, \dots, P_k)$, and is denoted $V(I)$. The set of generators of an ideal I is not unique, and a Groëbner basis is a set of generators $(Q_1, \dots, Q_p)$ of I which is algorithmically computable from any other basis $(P_1, \dots, P_k)$, and which is convenient for several purposes, that of elimination for instance. We need a *total order* on the set M of unitary monomials (i.e., with coefficient 1) of $A = K[X_1, \dots, X_n]$. This order must be *multiplicative*, i.e., it should verify the two following properties:

$$m > 1 \text{ for every nonconstant monomial } m, \tag{24.1}$$

$$m > n \Rightarrow pm > pn \quad \forall p \in M. \tag{24.2}$$

Examples of the most usual multiplicative orders are given by the following examples (see BILLERA and ROSE [1989]).

EXAMPLES. (1) *The lexicographic ordering.* We say that $X_1^{a_1} \cdots X_n^{a_n} > X_1^{b_1} \cdots X_n^{b_n}$ if and only if for the smallest index i such that $a_i - b_i \neq 0$, then $a_i - b_i > 0$. For instance, if $n = 3$, we have $X_1 > X_2 > X_3 > 1$, and

$$X_1^2 > X_1X_2 > X_1X_3 > X_1 > X_2^2 > X_2X_3 > X_2 > X_3^2 > X_3 > 1.$$

(2) *The graduated lexicographic ordering.* We say that $X_1^{a_1} \cdots X_n^{a_n} > X_1^{b_1} \cdots X_n^{b_n}$ if and only if $a_1 + \cdots + a_n > b_1 + \cdots + b_n$, or $a_1 + \cdots + a_n = b_1 + \cdots + b_n$ and for the smallest index i such that $a_i - b_i$ is nonzero, then $a_i - b_i > 0$. For instance, if $n = 3$, we have $X_1 > X_2 > X_3 > 1$ and

$$X_1^2 > X_1X_2 > X_1X_3 > X_2^2 > X_2X_3 > X_3^2 > X_1 > X_2 > X_3 > 1.$$

(3) *The graduated reverse lexicographic ordering.* We say that $X_1^{a_1} \cdots X_n^{a_n} > X_1^{b_1} \cdots X_n^{b_n}$ if and only if $a_1 + \cdots + a_n > b_1 + \cdots + b_n$, or $a_1 + \cdots + a_n = b_1 + \cdots + b_n$ and for the smallest index i such that $a_i - b_i$ is nonzero, then $a_i - b_i < 0$. For instance, if $n = 3$, we have $X_1 > X_2 > X_3 > 1$, and

$$X_1^2 > X_1X_2 > X_2^2 > X_1X_3 > X_2X_3 > X_3^2 > X_1 > X_2 > X_3 > 1.$$

Definition of a Groëbner basis. Let us choose a multiplicative order on the monomials, and set $X^\alpha = X_1^{\alpha_1} \cdots X_n^{\alpha_n}$. We first give a preliminary definition.

DEFINITION. Let $P \in A$, $P = \sum_{1 \leqslant i \leqslant k} \lambda_i X^{\alpha_i}$, with $\lambda_i \in K$ and $X^{\alpha_i} > X^{\alpha_{i+1}}$. Then $\lambda_1 X^{\alpha_1}$ is the *initial term* of P, denoted $\text{in}(P)$, and X^{α_1} is the *principal monomial* of P, denoted $m(P)$. If $I \subset A$ is an ideal, $\text{in}(I)$ will design the ideal generated by all the $\text{in}(P)$ for $P \in I$.

EXAMPLE. If $f(X,Y,Z,W) = 5X + 3Z^2 - YW + 2Y$, where $X > Y > Z > w$, then $\text{in}(f) = X$ under the lexicographic order, $\text{in}(f) = YW$ under graduated lexicographic order, or graduate reverse lexicographic order.

We give the definition of a Groëbner basis for an ideal $I \subset A$.

DEFINITION. Let $I \subset A$ be an ideal; a finite set $(P_1, \ldots, P_k)$, $P_i \in I$, is a *Groëbner basis* for I if the set of monomials $(\text{in}(P_1), \ldots, \text{in}(P_k))$ is a minimal set of generators for the ideal $\text{in}(I)$.

EXAMPLE. Let us take $A = K[X_1, X_2]$, and choose the lexicographic order. Then, if $I = (X_1^3, X_1^2 X_2 - X_2^3)$, a Groëbner basis for I is given by $(X_1^3, X_1^2 X_2 - X_2^3, X_1 X_2^3, X_2^5)$.

The reader can check this fact, using the properties of Groëbner basis quoted below.

Let us give some elementary properties of a Groëbner basis; we choose a multiplicative order on the monomials of A.

(1) A Groëbner basis of an ideal I is a set of generators of I. The converse is not true (see the example above).

(2) Any ideal has a finite Groëbner basis.

(3) A Groëbner basis for an ideal I is not unique, but if $(P_1, \ldots, P_k)$ is a Groëbner basis, the set of exponents of $\text{in}(P_1), \ldots, \text{in}(P_k)$ does not depend on the Groëbner basis, but only on the ideal I. It is called sometimes the *frontier*, or the *stair* of I.

DEFINITION. Let $(B) = (Q_1, \ldots, Q_s)$ be a finite set of polynomials. A polynomial P is *B-reduced* in relation to (B) if $\text{in}(P)$ is not a multiple of any monomial $\text{in}(Q_i)$ for $Q_i \in (B)$.

In other words, P is B-reduced if, for any polynomial R of the form $R = P - HQ_i$, the principal monomial $m(R)$ of R is not smaller than $m(P)$. If P is not B-reduced, there exist $H \in A$ and $Q_i \in (B)$ such that if $P_1 = P - HQ_i$, then $m(P_1) < m(P)$. If P_1 is again not B-reduced, we can repeat this procedure, and find a polynomial P_2 such that $m(P_2) < m(P_1)$. This process ends in a finite number of steps, with a polynomial $\widetilde{P}$ that is B-reduced, because of the following theorem.

THEOREM 24.1 (see LEJEUNE ([1985], p. 38)). *Let us choose a total multiplicative order on the monomials. There is no infinite sequence of monomials*

$$m_0 > m_1 > \cdots > m_i > \cdots.$$

The polynomial $\widetilde{P}$ is called a *reduction* (or a *B-reduction*) of P.

THEOREM 24.2 (see DAVENPORT, SIRET and TOURNIER ([1987], p. 105)). *Let (B) be a system of generators of an ideal I. Then (B) is a Groëbner basis of I if and only if any B-reduction of an element of I is* 0.

Let us now describe the basic algorithm of the theory, namely the Buchberger algorithm (see BUCHBERGER [1970]).

As usual, we choose a multiplicative order on the set of monomials of A.

DEFINITION. Let P and Q be two non zero polynomials, h the least common multiple of $\mathrm{in}(P)$ and $\mathrm{in}(Q)$. Then the polynomial

$$S(P,Q) = \frac{h}{\mathrm{in}(P)}P - \frac{h}{\mathrm{in}(Q)}Q \tag{24.3}$$

is called the *S-polynomial* of P and Q.

The S-polynomial is the basic tool in the Buchberger algorithm described below, because the initial terms of $(h/\mathrm{in}(P))P$ and $(h/\mathrm{in}(Q))Q$ are equal, and therefore disappear in $S(P,Q)$. We can now give a more precise version of Theorem 24.2.

THEOREM 24.3. *Let (B) be a basis of I. Then (B) is a Groëbner basis of I if and only if for all $P_i, P_j \in (B)$, the B-reduction of $S(P_i, P_j)$ is* 0.

From this, we may now define the Buchberger algorithm, which proceeds along the following line: take a basis $(B) = (P_1, \ldots, P_k)$ of I, and compute all the S-polynomials $S(P_i, P_j)$. If the reduction of all these polynomials are 0, then (B) is a Groëbner basis. If not, add to (B) a nonzero reduction $\widetilde{S}$ of a $S(P_i, P_j)$, and repeat this procedure with the new set $(B)\cup\{\widetilde{S}\}$. Buchberger proves in BUCHBERGER [1970] that this process is finite, and gives an algorithm of construction of a Groëbner basis, starting from a basis of the ideal I.

Let us now give some applications of Groëbner bases. Groëbner bases are very useful for actual computations in algebraic geometry. Let us give some examples. We choose a multiplicative order on the monomials of $A = K[X_1, \ldots, X_n]$, and let $(B) = (Q_1, \ldots, Q_s)$ be a Groëbner basis of an ideal $I = (P_1, \ldots, P_k)$. Then $1 \in I$ if and only if $Q_j = 1$ for some j (this is clear from the definition of Groëbner basis and from the one of multiplicative order). Note that if $K = \mathbb{C}$, this property is equivalent to the fact that the system of equations

$$\begin{cases} P_1(X_1, \ldots, X_n) = 0, \\ \qquad\vdots \\ P_k(X_1, \ldots, X_n) = 0, \end{cases}$$

has no solutions in $\mathbb{C}^n$, by *Hilbert's Nulstellensatz* (see, e.g., BENEDETTI and RISLER ([1990], p. 120)). We next have the membership property:

THEOREM 24.4. *Let $P \in A$. Then $P \in I$ if and only if its B-reduction is* 0 (*see Theorem* 24.1 *above*).

This gives an algorithm to decide whether $P \in I$, or not. Let us now choose the lexicographic ordering on the monomials of $K[X_1, \ldots, X_n]$. Then we have the following elimination result:

THEOREM 24.5 (see LEJEUNE ([1985], p. 51)). *Let $I \subset K[X_1, \dots, X_n]$ be an ideal, and define $J = I \cap K[X_t, \dots, X_n]$, for some t, $1 \leqslant t \leqslant n$. Then, if $(B) = (Q_1, \dots, Q_s)$ is a Groëbner basis for I, the set $(B) \cap K[X_t, \dots, X_n]$ is a Groëbner basis for J, again for the lexicographic ordering.*

REMARKS 24.1. (1) If $J = (0)$, then the set of $Q_i \in (B)$ such that $Q_i \in K[X_t, \dots, X_n]$ is empty.

(2) Assume $t = n$, and $I \cap K[X_n] \neq 0$. $I \cap K[X_n]$ is then an ideal of $K[X_n]$ generated by a polynomial $Q(X_n)$, which can be viewed as the result of the elimination process of of the variables $(X_1, \dots, X_{n-1})$ among any set $(P_1, \dots, P_k) \subset I$. In fact, assume $K = \mathbb{C}$. Then any solution $(x_1, \dots, x_n)$ of the system

$$P_1 = \cdots = P_k = 0 \tag{S}$$

satisfies $Q(x_n) = 0$, because $Q \in I$, hence, $Q = \sum_{i=1}^{k} H_i P_i$ with $H_i \in \mathbb{C}[X_1, \dots, X_n]$. Further, the set of solutions of the system (S) is non empty because $1 \notin I$ (by Hilbert's Nulstellensatz). This means that the projection on the X_n-axis of the set of solutions of (S) is non empty, and contained in the set of roots of $Q(X_n)$. The polynomial $Q(X_n)$ is therefore a generalization of the resultant $R(P, Q)$ introduced above (see Section 23). Moreover, it can be proved (see LEJEUNE ([1985], p. 55)), that for each root of $Q(X)$, there is at least one solution of (S) of the form $(x_1, \dots, x_{n-1}, \alpha)$. The computation of the roots of $Q(X)$ is then the starting point of an algorithm for solving the system (S).

25. Rational curves

Bézier curves and spline curves are (piecewise) rational curves of a special type, since they are piecewise curves parametrized by polynomials. In this section, we study elementary properties of rational algebraic curves from the point of view of algebraic geometry, and then we define rational spline and Bézier curves.

We begin with the following observation: it is impossible to represent an arc of a conic $C \subset \mathbb{R}^2$ (except for a parabola) in the parametric form

$$\begin{cases} X = g(t), \\ Y = h(t), \end{cases} \tag{25.1}$$

where g and h are polynomials in t; this follows from the fact that the degree of $g(t)$ and $h(t)$ must be $\leqslant 2$, since C cuts any line in at most two (real or complex) points, and from the following result.

THEOREM 25.1. *The parametric curve*

$$\begin{cases} X = a_2 t^2 + a_1 t + a_0, \\ Y = b_2 t^2 + b_1 t + b_0, \end{cases}$$

$0 \leqslant t \leqslant 1$, *is an arc of parabola.*

PROOF. Eliminating t between these two equations (see Section 23), we find the relation

$$a_2T^2 - a_1\alpha T + (a_0 - X)\alpha^2 = 0$$

with

$$T = b_2(a_0 - X) - a_2(b_0 - Y),$$
$$\alpha = b_2a_1 - b_1a_2,$$

which is the equation of a parabola, because it is a curve with only one asymptotic direction, namely $T = 0$. □

The circle $X^2 + Y^2 = 1$ can however be represented in parametric form with rational fractions, as

$$X = \frac{1-t^2}{1+t^2}, \qquad Y = \frac{2t}{1+t^2}. \tag{25.2}$$

This observation motivates the introduction of rational fractions in the definition of spline functions. We will begin by defining those curves in $\mathbb{R}^2$ that have a parametrization by rational fractions.

$\mathbb{R}[X,Y]$ denotes as usual the polynomial ring in two variables with coefficients in $\mathbb{R}$.

DEFINITION. An algebraic curve $\Gamma \subset \mathbb{R}^2$ (defined by an equation $P(X,Y) = 0$, with $P \in \mathbb{R}[X,Y]$) is called *rational* if it admits a rational parametrization of the form

$$\begin{cases} X = g(t), \\ Y = h(t), \end{cases} \tag{25.3}$$

where g and h are rational fractions.

REMARKS 25.1. (1) If we want all the points of Γ in the form (25.3), it is sometimes necessary to admit the value "∞" for t (which gives for instance the point $(-1,0)$ of the circle $X^2 + Y^2 = 1$ parametrized by (25.2)). Also, some values of t (for which for instance the denominator is zero) may correspond to "points at infinity" of Γ. We will not study further these notions of elementary algebraic geometry (which motivate the introduction of projective spaces). In fact, they are very easy and intuitive in the cases considered here.

(2) One can make similar considerations for (not necessarily plane) curves in $\mathbb{R}^s$: we have considered here only the case where $s = 2$. The definition of a general algebraic curve is otherwise more involved.

DEFINITION. Let $\Gamma \subset \mathbb{R}^2$ be an algebraic curve of degree d, i.e., one that is defined by an equation $P(X,Y) = 0$, with $P \in \mathbb{R}[X,Y]$ of degree d. One says that the origin $0 \in \Gamma$ is a *point of order* $d-1$, if the equation $P(X,Y) = 0$ of Γ is of the form

$$P_{d-1}(X,Y) + P_d(X,Y) = 0, \tag{25.4}$$

where P_{d-1} (resp. P_d) is *homogeneous* of degree $d-1$ (resp. d); if $d > 1$, one says that 0 is *singular*. A point $Q \in \Gamma$ is of *order* $(d-1)$, if, after a linear change of coordinates which sends Q to 0, the origin becomes a point of order $(d-1)$.

THEOREM 25.2. (i) *Any nonempty conic* $C \subset \mathbb{R}^2$ *without singular point* (*i.e., that is not the union of two distinct* (*or otherwise*) *lines*) *is a rational curve.*

(ii) *If* $\Gamma \subset \mathbb{R}^2$ *is a curve of degree* d *that has a point* Q *of order* $(d-1)$ *and that does not contain a straight line passing by* Q, *then* Γ *is rational.*

PROOF. We see that (i) is a particular case of (ii), taking for Q any point of C, as then $d = 2$. For the proof of (ii), set $Y = tX$; Eq. (25.4) then becomes

$$X^{d-1}\left[P_{d-1}(1,t) + XP_d(1,t)\right] = 0.$$

We find in this way the solutions $(X = 0,\ Y = 0)$ which correspond to the origin "counted $(d-1)$ times", and

$$\begin{cases} X = -\dfrac{P_{d-1}(1,t)}{P_d(1,t)}, \\[2ex] Y = tX = -\dfrac{tP_{d-1}(1,t)}{P_d(1,t)}, \end{cases}$$

which gives a rational parametrization of Γ. □

For completeness, we recall (without going into details, see for instance FULTON [1969]) that one can define the *genus* g of an algebraic irreducible curve Γ (i.e., defined by an equation $P(X,Y) = 0$ where P is an irreducible polynomial over the field $\mathbb{C}$ of complex numbers). The genus g is an integer that has in particular the following properties: If Γ has no singular point, ("real or complex, at finite or infinite distance"), and P is of degree d, then $g = (d-1)(d-2)/2$. If P is of degree d and Γ has k singular ordinary quadratic points (real or complex, at finite or infinite distance), $g = (d-1)(d-2)/2 - k$.

THEOREM 25.3. *Let* $\Gamma \subset \mathbb{R}^2$ *be an irreducible algebraic curve; then* Γ *is rational if and only if its genus is* 0.

We see for instance that the only rational curves without singular points are the straight lines and the conics, and that the cubics are rational only if they have a singular point (which, for this particular degree, can only be a cusp or an ordinary quadratic point).

We now introduce rational splines. We always keep the same notation; let $B_{i,k,t}$, $0 \leqslant i \leqslant n-1$, denote the B-splines of degree k on $[a,b]$, corresponding to a knot sequence $t = (t_0, \ldots, t_{n+k})$.

DEFINITION (*see* FIOROT *and* JEANNIN [1989]). Let $P_0, \ldots, P_{n-1}$ be points in $\mathbb{R}^s$, and let $w_0, \ldots, w_{n-1}$ be n real numbers not all zero. The *rational B-spline* curve associated with this data is the parametric curve Γ defined by

$$S(t) = \frac{\sum_{i=0}^{n-1} P_i w_i B_i(t)}{\sum_{i=0}^{n-1} w_i B_i(t)}. \tag{25.5}$$

The w_i's are the *weights* associated with the curve Γ; as is usually done, we shall assume in the sequel that the range of t is an interval $[a,b]$ such that $t_k \leqslant a$, $t_n \geqslant b$.

Let us describe some properties of the rational spline curves: One may write

$$S(t) = \sum_{i=0}^{n-1} P_i \psi_i(t), \quad \text{where } \psi_i(t) = \frac{w_i B_i(t)}{\sum_{i=0}^{n-1} w_i B_i(t)}$$

is a rational fraction; Γ is then a rational curve as defined above (see (25.3)). If $w_i = 1$, $0 \leqslant i \leqslant n-1$, we find the usual notion of spline curve, when the range of t is restricted to $[a,b]$, $t_k \leqslant a$, $t_n \geqslant b$, as we then have the relation $\sum_{i=0}^{n-1} B_{i,k}(t) = 1$ for all $t \in [a,b]$ (see Theorem 1.1). If $w_i > 0$, $0 \leqslant i \leqslant n-1$, the denominator of $S(t)$ is not zero in $[a,b]$; this is the most frequent case. We have $\sum_{i=0}^{n-1} \psi_i(t) = 1$, $t \in [a,b[$, and then, as for polynomial splines in the case where the w_i's are > 0, the curve Γ is situated in the convex hull of the points P_i. As we have

$$\psi_i(t) = \frac{w_i B_i(t)}{\sum_{i=0}^{n-1} w_i B_i(t)},$$

the same properties hold for the support for the functions ψ_i as those for the polynomial splines (see Theorem 1.1).

Let us take for instance $s = 3$, i.e., $P_i \in \mathbb{R}^3$. Let $\widetilde{\Gamma} \subset \mathbb{R}^4 = \mathbb{R}^3 \times \mathbb{R}$ be the curve defined by the parametrization

$$\widetilde{S}(t) = \begin{cases} \displaystyle\sum_{i=0}^{n-1} w_i P_i B_i(t), \\ \displaystyle\sum_{i=0}^{n-1} w_i B_i(t). \end{cases}$$

The curve $\widetilde{\Gamma}$ is therefore the B-spline in $\mathbb{R}^4$ having for the vertices of its control polygon the set of points

$$Q_i = (w_i P_i, w_i) \in \mathbb{R}^3 \times \mathbb{R}, \quad 0 \leqslant i \leqslant n-1.$$

Let X_1, X_2, X_3, X_4 be coordinates on $\mathbb{R}^4$; the curve Γ is then the image of $\widetilde{\Gamma}$ by the map

$$\pi : (X_1, X_2, X_3, X_4) \in \mathbb{R}^4 \setminus \{X_4 = 0\} \to (X_1/X_4, X_2/X_4, X_3/X_4) \in \mathbb{R}^3.$$

We then may apply to $\widetilde{\Gamma}$ all the algorithms of the preceding sections, which will yield algorithms for Γ using the projection π.

When the curve $\widetilde{\Gamma}$ passes through the origin of $\mathbb{R}^4$ (or, more generally, if the denominator and the numerator of one of the coordinates of $S(t)$ are simultaneously zero for one value of t), the corresponding point of Γ is indeterminate, it is then called a *base point*. This happens for instance for the rational Bézier curves (see below) when $w_0 = 0$, for $t = 0$. This notion is studied in Section 19 for rational Bézier patches.

Let us now examine the special case of Bézier curves. Recall that the ith Bernstein polynomial of degree k on $[0, 1]$ is defined by $B_i^k(t) = \binom{k}{i} t^i (1-t)^{k-i}$, $0 \leqslant i \leqslant k$; in the same manner, $B_i^k((t-a)/(b-a))$ is the ith Bernstein polynomial on $[a, b]$ (see Theorem 6.1 and Remark 6.1). A rational Bézier curve is thus defined as above by

$$B(t) = \frac{\sum_{i=0}^{k} w_i P_i B_i^k(t)}{\sum_{i=0}^{k} w_i B_i^k(t)}, \quad 0 \leqslant t \leqslant 1.$$

We may then, without modifying the curve, make a *homographic* transformation on the parameter: $t = (a\theta + b)/(c\theta + d)$ (with $ad - bc \neq 0$); the term $t^i(1-t)^{k-i}$ is then transformed in $(a\theta + b)^i((c-a)\theta + d - b)^{k-i}/(c\theta + d)^k$. This may sometimes simplify the expression for $B(t)$.

26. Rational representations of conics

This section follows LEE [1987]. Let us look at Bézier curves of degree two; we have

$$B_0^2(t) = (1-t)^2, \qquad B_1^2(t) = 2t(1-t), \qquad B_2^2(t) = t^2.$$

A *rational Bézier curve of degree two* is thus of the form

$$\gamma(t) = \frac{w_0 P_0 (1-t)^2 + 2 w_1 P_1 t(1-t) + w_2 P_2 (t^2)}{w_0 (1-t)^2 + 2 w_1 t(1-t) + w_2 t^2}, \tag{26.1}$$

where the real numbers w_0, w_1 and w_2 are not all zero (see Fig. 26.1)

We list below some properties of the rational Bézier curves of degree two:

(1) Assume $w_i > 0$, $0 \leqslant i \leqslant 2$. Then

$$\gamma(0) = P_0, \qquad \gamma(1) = P_2,$$

$$\gamma'(0) = \frac{2w_1}{w_0}(P_1 - P_0), \qquad \gamma'(1) = \frac{2w_1}{w_0}(P_2 - P_1),$$

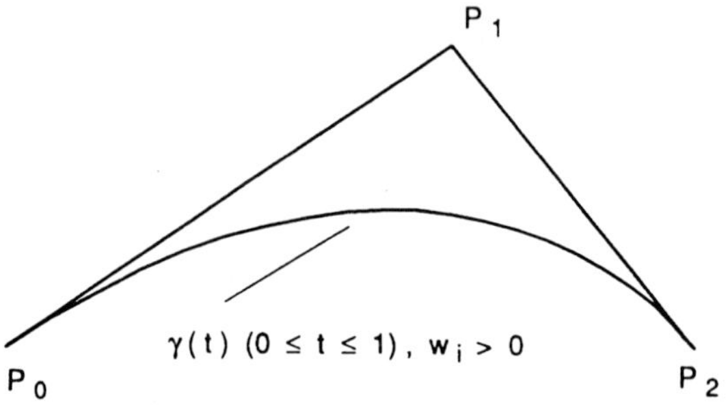

FIG. 26.1. A rational Bézier curve of degree two.

and $\gamma(t)$ is an arc of *conic* C; more precisely, let

$$\gamma(t) = P_0\psi_0(t) + P_1\psi_1(t) + P_2\psi_2(t),$$

and let us consider the coordinate system with origin P_1, and with unit vectors $\vec{u} = P_0 - P_1$, and $\vec{v} = P_2 - P_1$. We may write

$$\gamma(t) = \psi_0(t)(P_0 - P_1) + \psi_2(t)(P_2 - P_1) + P_1$$

(since $\sum_{i=0}^{2}\psi_i(t) = 1$), and then in this system the coordinates of $\gamma(t)$ become

$$\alpha = \psi_0(t), \qquad \beta = \psi_2(t).$$

Set $w(t) = w_0(1-t)^2 + 2w_1t(1-t) + w_2t^2$; we have

$$\alpha\beta = \frac{w_0w_2}{(w(t))^2}B_0(t)B_2(t) = \frac{w_0w_2}{(w(t))^2}\frac{(B_1^2(t))^2}{4} = \frac{w_0w_2}{4w_1^2}(\psi_1(t))^2,$$

which gives the intrinsic equation of C, again in the same system of coordinates:

$$\alpha\beta = \frac{w_0w_2}{4w_1^2}(1-\alpha-\beta)^2, \tag{26.2}$$

which is of degree two, and which depends only on the value of $k = w_0w_2/4w_1^2$.

(2) If we change w_1 into $-w_1$, the intrinsic equation does not change; we then find, for $0 \leqslant t \leqslant 1$, another arc $\tilde{\gamma}(t)$ of the same conic (which is no longer in the convex hull of $P_0P_1P_2$). If we set

$$\tilde{w}(t) = w_0(1-t)^2 - 2w_1t(1-t) + w_2t^2,$$

we have

$$\tilde{\gamma}(t)\tilde{w}(t) - \gamma(t)w(t) = 2w_1P_1B_1(t),$$

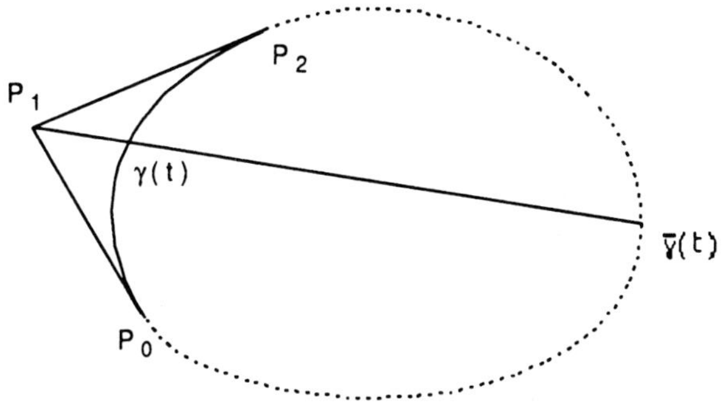

FIG. 26.2. The two arcs of a conic in Bézier form.

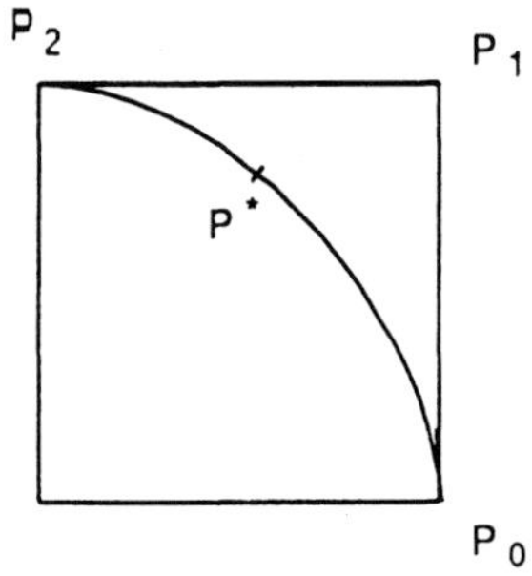

FIG. 26.3. An arc of a circle.

whence

$$\tilde{\gamma}(t) - P_1 = \frac{w(t)}{\tilde{w}(t)}\big(\gamma(t) - P_1\big).$$

Therefore the points $\gamma(t)$ and $\tilde{\gamma}(t)$ are on the line containing P_1, and we deduce that the conic C of Eq. (26.2) is the union of the arcs $\gamma(t)$ and $\tilde{\gamma}(t)$, $0 \leqslant t \leqslant 1$ (see Fig. 26.2).

The denominator $\tilde{w}(t)$ of $\tilde{\gamma}(t)$ may have a zero. Its discriminant is equal to $w_1\sqrt{1-4k}$, with $k = w_0w_2/4w_1^2$, which implies that if $4k > 1$, C is an ellipse, if $4k = 1$, C is a parabola, and if $4k < 1$, C is an hyperbola. We see from the expression of $\gamma(t)$ that $\tilde{\gamma}(t) = \gamma(t/(2t-1))$.

(3) Given points P_0, P_1 and P_2, the conic C and its parametrization are determined by the data of an additional point P^* and of a value t^* $(0 < t^* < 1)$ such that $\gamma(t^*) = P^*$; this allows to compute $k = w_0w_2/4w_1^2$ from Eq. (26.2).

EXAMPLE. Suppose that

$$P_0 = (1,0), \qquad P_1 = (1,1), \qquad P_2 = (0,1), \qquad P^* = \left(\tfrac{3}{5}, \tfrac{4}{5}\right), \qquad t^* = \tfrac{1}{2}.$$

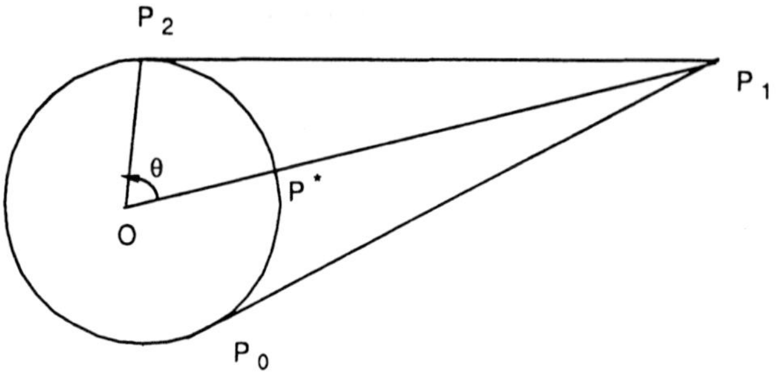

FIG. 26.4. A circle in rational Bézier form.

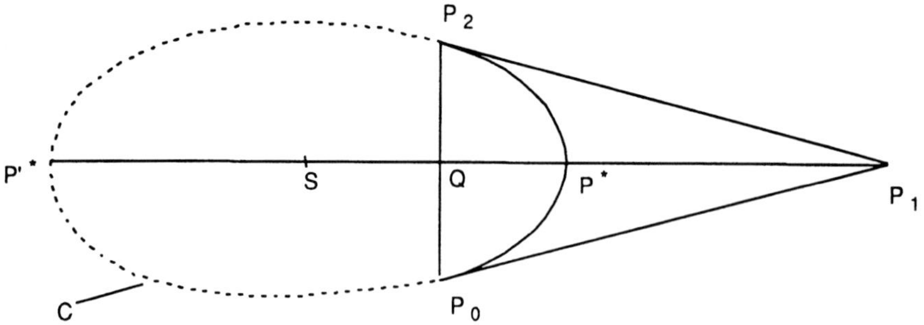

FIG. 26.5. Center of an ellipse.

Set $w_0 = 1$: we then find, putting $t = t^*$ in (25.1) that $w_1 = 1,\ w_2 = 2$, whence

$$X(t) = \frac{1 - t^2}{1 + t^2}, \qquad Y(t) = \frac{2t}{1 + t^2}.$$

It is thus nothing but the usual parametrization of the circle S^1 (see Fig. 26.3).

EXAMPLE. Let C be the circle of center 0 and let the points P_0, P_1, P_2 be arranged as in Fig. 26.4. Let $t = \frac{1}{2}$ and let P^* be the middle point of the arc P_0P_2; we then find the parametrization

$$\gamma(t) = \frac{P_0(1-t)^2 + 2\cos\theta P_1 t(1-t) + P_2 t^2}{(1-t)^2 + 2\cos\theta t(1-t) + t^2}.$$

(4) Let us now compute the (symmetry) center S of the conic (in the case where $k \neq \frac{1}{4}$, i.e., for an ellipse or a hyperbola).

Let Q be the middle of P_0P_2; the center of C is on the line containing P_1 and Q, at the middle of the segment $P^*P'^*$, which is the intersection of P_1Q with the conic; we then find immediately (with $\gamma(\frac{1}{2}) = P^*$), that

$$S = \big(2k(P_0 + P_2) - P_1\big)\frac{1}{4k - 1}.$$

27. Steiner patches

In specific cases, special techniques of elimination can be used. We give here two such examples, well adapted to CAGD, following ANDERSON and SEDERBERG [1985]. For a complete study of this question, see RONGA and VUST [1990].

Let us first look at a *rational curve* Γ of degree $\leqslant 3$, i.e., that is defined by the following equations:

$$X = \frac{P(t)}{Q(t)}, \qquad Y = \frac{S(t)}{Q(t)}, \tag{27.1}$$

where P, Q, S are polynomials of degree $\leqslant 3$ and $0 \leqslant t \leqslant 1$. Two questions naturally arise: find the "intrinsic equation" $\phi(X,Y) = 0$ of Γ by eliminating t, and find the "inversion formula" $t = \phi(X,Y)$, which gives the value of the parameter t at a point $(X,Y) \in \Gamma$.

THEOREM 27.1. *The equation* $\phi(X,Y) = 0$ *is of degree* $\leqslant 3$.

PROOF. Look at the intersection of Γ with a line $aX + bY + c = 0$; we find no more than three points, if Γ does not contain the line. □

Now Γ is a parametric curve of degree $\leqslant 3$; assume it is of degree 3, which is the general case. It then has a singular point by Theorem 25.3. The idea to solve the two above problems is to find first the singular point S, and then to reparametrize Γ using a line passing through S, with the same parameter t for the line and for the point of Γ. Let (X_0, Y_0) be the coordinates of S; the equation of a line that passes through S is $Y - Y_0 = \theta(X - X_0)$, and each value of t gives a point of Γ, hence a value of θ which is thus a function of t. Conversely, for each value of θ, the line $Y - Y_0 = \theta(X - X_0)$ intersects Γ at one point, apart from S, which gives in general one value of t: the correspondence between t and θ is therefore a *homographic* one (see Theorem 25.2):

$$\theta = \frac{ct + d}{at + b}.$$

We then obtain the equation

$$\frac{Y - Y_0}{X - X_0} = \frac{ct + d}{at + b}. \tag{27.2}$$

Replacing X and Y by their values from (27.1), and setting

$$\begin{cases} a_1 = aY_0 - cX_0, \\ b_1 = bY_0 - dX_0, \end{cases} \tag{27.3}$$

we find the relation $(ct + d)P(t) - (at + b)S(t) + (a_1 t + b_1)Q(t) \equiv 0$. This relation gives five linear equations by cancelling the coefficients of t^4, t^3, t^2, t and the constant coefficient, in five unknowns a, b, c, a_1, b_1 (setting for instance $d = 1$). We assume that this system is nonsingular, (this is the general case), and thus we can compute the coordinates (X_0, Y_0) of the singular point from (27.3). Then (27.2) gives the inversion formula $t = \psi(X, Y)$.

REMARK 27.1. The above computation proves that X_0 and Y_0 are rational numbers, if P and Q have integer coefficients.

Now, from the first equation of (27.1), we find the intrinsic equation $\phi(X, Y) = 0$. We next study *Steiner patches* which are by definition *rational surfaces of degree* 2:

$$X = \frac{P(u,v)}{Q(u,v)}, \qquad Y = \frac{R(u,v)}{Q(u,v)}, \qquad Z = \frac{S(u,v)}{Q(u,v)}, \tag{27.4}$$

with $P, Q, R, S \in \mathbb{Z}[u, v]$ of degree 2, and $0 \leqslant u \leqslant 1, 0 \leqslant v \leqslant 1$.

The process of implicitization and inversion is similar to the one for rational plane curves of degree 3, because there is a "triple point" on the surface (note that this triple point is not necessarily in the image of the square $[0, 1] \times [0, 1]$).

THEOREM 27.2. *The degree of the implicit equation of a Steiner patch is in general* 4 (*and always* $\leqslant 4$).

The proof is similar to the one of Theorem 27.1; a line intersects the surface at at most four (complex) points.

As in ANDERSON and SEDERBERG [1985] or RONGA and VUST [1990], we base our method of implicitization on the one for the curves of degree 3 described above. For a theoretical justification of the method, see RONGA and VUST [1990].

Let Σ be the Steiner surface (closure of the image of $\mathbb{R}^2$ by the map defined by the system (27.4)), and let (X_0, Y_0, Z_0) be the coordinates of the triple point S. We consider a "general" plane passing through S:

$$\begin{aligned}(X - X_0)(au + bv + c) + (Y - Y_0)(du + ev + f)\\ + (Z - Z_0)(gu + hv + k) = 0\end{aligned} \tag{27.5}$$

whose coefficients are linear functions of the parameters u and v. A line, intersection of two such planes, will in general intersect Σ at the point S, and at another point. Let us assume that it is the point of Σ with parameters u and v. Then Eq. (27.2) becomes an identity, and setting

$$\begin{cases} a_1 = aX_0 + dY_0 + gZ_0, \\ b_1 = bX_0 + eY_0 + hZ_0, \\ c_0 = cX_0 + fY_0 + kZ_0, \end{cases} \tag{27.6}$$

and replacing X, Y, Z by their values given in (27.4) gives a system of 10 linear equations (obtained by cancelling all the monomials in u, v in the relation (27.5)) in 12 unknowns, namely $(X_0, Y_0, Z_0, a, b, c, d, e, f, g, h, i)$. But the nine last ones are defined up to multiplication by a nonzero constant, therefore there are in fact only 11 unknowns, and in general two independent solutions. From these solutions, we see that X_0, Y_0, Z_0 (computed from system (27.6)) are necessarily rational numbers, and we can compute u and v as functions of X, Y, Z: these are the inversion equations. Now, the intrinsic equation is given for instance by the equation $Q(u, v)X = P(u, v)$, once u, v are replaced by their values in functions of X, Y, Z.

28. Semi-algebraic geometry

Semi-algebraic sets (defined by equations and inequations with integer coefficients) permit to modelize any usual shape in $\mathbb{R}^n$, and they can be described algorithmically, at least in theory. For this reason, they are useful in graphics and robotics, for instance for the "piano mover's problem" described below, however with algorithms of very high complexity.

DEFINITION. A set $X \subset \mathbb{R}^n$ is *algebraic*, if there exist polynomials $P_1, \ldots, P_k$ in $\mathbb{R}[X_1, \ldots, X_n]$ such that $X = \{x \in \mathbb{R}^n \mid P_i(x) = 0,\ 1 \leqslant i \leqslant k\}$.

REMARKS 28.1. (1) Any real algebraic set defined by equations $P_i(X) = 0,\ 1 \leqslant i \leqslant k$, is the set of zeros of a unique polynomial P, namely $P = \sum_{i=1}^{k} P_i^2$.

(2) The algebraic subsets of $\mathbb{R}$ are the finite sets.

(3) Examples of real algebraic sets are the real algebraic curves, the real algebraic surfaces, etc. … . Note however that a connected component of an algebraic set is not in general an algebraic set (consider for instance the hyperbola $XY - 1 = 0$), and the projection of an algebraic set is not necessarily algebraic. For instance, the unit circle in $\mathbb{R}^2$ projects on the x coordinates onto the interval $[-1, 1]$, which is not algebraic (see (2)). This remark motivates (among other things) the definition below.

(4) Any real algebraic set has a finite number of connected components.

More precisely, it can be proved (see BENEDETTI and RISLER [1990]), that if $A \subset \mathbb{R}^n$ is the set of zeros of $P \in \mathbb{R}[X_1, \ldots, X_n]$, the degree of P being d, then the number of connected components of A is $\leqslant d^n$.

DEFINITION. A set $A \subset \mathbb{R}^n$ is *semi-algebraic* if it can be written as

$$A = \bigcup_{i=1}^{s} \bigcap_{j=1}^{r_i} \{x \in \mathbb{R}^n\colon\ P_{i,j}(x)\ s_{ij}\ 0\}, \tag{28.1}$$

where, for each $i = 1, \ldots, s$ and each $j = 1, \ldots, r_i$, $s_{ij} \in \{>, =, <\}$, and $P_{i,j} \in \mathbb{R}[X_1, \ldots, X_n]$. In other words, the set of semi-algebraic sets of $\mathbb{R}^n$ is the smallest family of subsets of $\mathbb{R}^n$ that contains all the sets of the form

$$\{x \in \mathbb{R}^n\colon\ P(x) \geqslant 0\}, \quad P(X) \in \mathbb{R}[X],$$

and that is closed under the elementary set operations (finite unions, finite intersections, and complementarity).

REMARKS 28.2. (1) Any real algebraic subset of $\mathbb{R}^n$ is semi-algebraic.

(2) The semi-algebraic subsets of $\mathbb{R}$ are the finite unions of intervals (and points).

(3) The following assertions can be proved (see BENEDETTI and RISLER ([1990], p. 60–61)): Any connected component of a semi-algebraic set is semi-algebraic, the interior and the closure of a semi-algebraic set are semi-algebraic, the linear projection of a semi-algebraic set from $\mathbb{R}^n$ onto $\mathbb{R}^{n-k}$ is a semi-algebraic set.

(4) Any polyhedron is a semi-algebraic set.

(5) Any semi-algebraic set has a finite number of connected components, which are themselves semi-algebraic sets; to see this, we can use for instance Remark 28.1(4), and the property that any semi-algebraic set is the projection of an algebraic set.

(6) The fact that any polyhedron is semi-algebraic implies that the semi-algebraic sets let us model "shapes" in $\mathbb{R}^3$. They have moreover some remarkable properties of "stability", by projection, closure, etc., and are in principle well adapted for the techniques of symbolic computation, for which the methods sketched in this section are used. They are currently the object of intense study in "theoretical robotics", but not yet much in CAGD. However, they probably will be more and more used in CAGD, graphics, etc.

We give a description of the cylindric algebraic decomposition algorithm (CAD for short) of Collins (see COLLINS [1975]). This algorithm gives a decomposition of any semi-algebraic set A defined by polynomials with integer coefficients in a finite numbers of "cells", i.e., connected sets homeomorphic to open balls, which we will be able to "compute". In particular, this algorithm gives answers to questions of the form: Is A nonempty, connected, compact, etc.? The problem is that this algorithm has a high "complexity": it is doubly-exponential in the number of variables. Let us first give two definitions.

DEFINITION. Let $(P) = (P_1, \ldots, P_k)$ be a set of polynomials in $\mathbb{Z}[X_1, \ldots, X_n]$. An *invariant sign-partition* of $\mathbb{R}^n$ (for the set (P)) is a finite partition $\mathbb{R}^n = \bigcup A_i$ in semi-algebraic sets A_i, such that the sign of each P_j is constant on each A_i (the sign being by definition > 0, < 0, or $= 0$).

DEFINITION. Let the system of coordinates $(X_1, \ldots, X_n)$ be fixed. We will identify $\mathbb{R}^{n-1}$ with the hyperplane $\{X_n = 0\}$. Then a *cylindric algebraic decomposition* D of $\mathbb{R}^n$ is defined recursively as follows: it is a partition $\mathbb{R}^n = \bigcup A_i$ such that: Each A_i is semi-algebraic, connected and homeomorphic to an open ball. We call such an A_i a "*cell*". In each cell A_i, there is a computed point $P_i \in A_i$, which is called a "*testing point*". There exists a cylindric algebraic decomposition D' of $\mathbb{R}^{n-1}$: $\mathbb{R}^{n-1} = F_1 \cup \cdots \cup F_m$ for the system of coordinates $(X') = (X_1, \ldots, X_{n-1})$. For each element A_i of D, there exists an element F_j of D' such that A_i is of one of the following forms:

$$\{(x)\colon\ (x') \in F_j \text{ and } x_n < f_k(x')\},$$

$$\{(x)\colon\ (x') \in F_j \text{ and } x_n = f_k(x')\},$$

$$\{(x)\colon\ (x') \in F_j \text{ and } f_k(x') < x_n \text{ and } x_n < f_{k'}(x')\},$$

$$\{(x)\colon (x') \in F_j \text{ and } x_n > f_k(x')\},$$

where the f_k are solutions of polynomials equations in X_n with coefficients in $\mathbb{Z}[X_1, \ldots, X_{n-1}]$.

Let $(P) = (P_1, \ldots, P_k)$ be a set of polynomials, $P_i \in \mathbb{R}[X_1, \ldots, X_n]$. A cylindric algebraic decomposition of $\mathbb{R}^n$, $\mathbb{R}^n = \bigcup E_i$ is *adapted* to (P) if it is sign invariant for the set (P).

In other words, for each $F_j \in D'$, the "cylinder" $\pi^{-1}(F_j) \subset \mathbb{R}^n$ is decomposed into a finite number of semi-algebraic subsets, bounded by the real roots of a finite number of polynomials in X_n with coefficients in $\mathbb{Z}[X_1, \ldots, X_{n-1}]$. These roots include the roots of the P_i's.

EXAMPLES. (1) A CAD of $\mathbb{R}$ is simply a finite partition of $\mathbb{R}$ in open intervals and points. If $P \in \mathbb{R}[X]$, the CAD is adapted to P if all the real roots of P are among the cells of dimension 0. Recall that each cell of dimension 0 is a point

(2) Let $P = X^2 + Y^2 - 1 \in \mathbb{R}[X, Y]$. A CAD of $\mathbb{R}^2$ adapted to P is made of a CAD of $\mathbb{R}$ adapted to $X^2 - 1$ with five cells:

$$(-\infty, -1], \quad \{-1\}, \quad]-1, 1[, \quad \{1\}, \quad]1, +\infty),$$

which are labeled successively $(0, 1, 2, 3, 4)$, and of decomposition of each cylinder of $\mathbb{R}^2$ (above a cell of $\mathbb{R}$) that depends on the roots of $P(X, Y)$, viewed as a polynomial in Y. For instance, above the cell (2) of $\mathbb{R}$, there are five cells of $\mathbb{R}^2$, labeled (2,0), (2,1), (2,2), (2,3), (2,4), limited by the two roots of $P(X, Y)$ when $X \in]-1, 1[: Y_1 = -\sqrt{1 - X^2}$, and $Y_2 = +\sqrt{1 - X^2}$.

The knowledge of a CAD adapted to (P) is very useful for any problem involving semi-algebraic sets defined with equations and inequalities involving elements of (P). Note that any such nonempty semi-algebraic set is a union of cells. In particular, a CAD of $\mathbb{R}^n$ adapted to (P) computes a decomposition of the *projection* of any semi-algebraic set $A \subset \mathbb{R}^n$ defined by a formula involving elements of (P) into $\mathbb{R}^{n-1}$. This should be useful for robotics (see below the "Piano Mover's Problem").

THEOREM 28.1 (COLLINS [1975]). *There exists an algorithm which computes a CAD of $\mathbb{R}^n$ adapted to a family (P) of polynomials with integer coefficients. The complexity of this algorithm is of the form $d^{2^{O(n)}}$ (where d is a bound for the degree of the polynomials in (P)), i.e., it is doubly exponential with respect to n.*

Let us describe the principle of this algorithm. For details of implementation using language "axiom", see RIOBOO [1991]. To begin with, we describe the basic operation of the algorithm, which will consist, given a list $(P) = (P_1, \ldots, P_k)$ of polynomials in $\mathbb{Z}[X_1, \ldots, X_n]$, in computing a new list $(Q) = (Q_1, \ldots, Q_r)$ of polynomials in $\mathbb{Z}[X_1, \ldots, X_{n-1}]$, in such a way that a CAD of $\mathbb{R}^{n-1}$ adapted for (Q) will serve as a "base" for a CAD of $\mathbb{R}^n$ adapted to (P). For this reason, one sets $(Q) = \mathrm{Proj}(P)$ (Proj stands for "projection"). Let us take for $\mathrm{Proj}(P)$ the following set of polynomials in the variables $X_1, \ldots, X_{n-1}$: All the coefficients of the polynomials P_i of (P) (viewed as polynomials in X_n with coefficients in $\mathbb{Z}[X_1, \ldots, X_{n-1}]$), all the subresultants $r_i(P_j, P_l)$ for $P_j, P_l \in (P)$ of degree $\geqslant 1$ (see Section 23), and all the subresultants $r_i(P_j, P_j')$ (for $P_j \in (P)$), viewed as a polynomial in X_n of degree $\geqslant 2$ (P_j' is the derivative of P_j).

EXAMPLE. If $(P) = \{X_1^2 + X_2^2 - 1\} \in \mathbb{R}[X_1, X_2]$, then $\operatorname{Proj}(P) = \{X_1^2 - 1\}$.

Now, we have the following theorem, which makes the CAD work (see COSTE ([1988], p. 14) or BENEDETTI and RISLER ([1990], p. 54)).

THEOREM 28.2. *Let $(P) = (P_1, \ldots, P_r)$, and $C \subset \mathbb{R}^{n-1}$ be a semi-algebraic connected subset which is* Proj(P)*-invariant. Then the total number of real roots of $P_1, \ldots, P_r$ (after among these polynomials those which vanish on C have been eliminated) is constant on C, and these real roots are given by continuous semi-algebraic functions $\xi_1 < \cdots < \xi_l : C \to \mathbb{R}$.*

Let us now describe the CAD algorithm (after COSTE [1988]). Given a list of polynomials in one variable X_1, count, and isolate in intervals with rational bounds, all the real roots of these polynomials, using for instance Sturm's method described in Section 22. The cells of this decomposition $\mathcal{C}_1$ are the roots and the intervals between these roots. The roots are characterized by the polynomial that they cancel, and the interval with rational bounds that isolates them. These bounds of intervals may also serve as testing points for the intervals between the roots, although there may exist more convenient choices. Given a list $(P) = (P_1, \ldots, P_r)$ of polynomials in $(X_1, \ldots, X_n)$ with $n > 1$, compute $\operatorname{Proj}(P)$, and call the algorithm with as input this list of polynomials in $(X_1, \ldots, X_{n-1})$. One gets $\mathcal{C}_{n-1}$, which is a partition of $\mathbb{R}^{n-1}$ into $\operatorname{Proj}(P)$-invariant cells, together with a testing point a_C for each $C \in \mathcal{C}_{n-1}$. To such a cell C, one may apply Theorem 28.2, in order to partition the cylinder $C \times \mathbb{R}$ into (P)-invariant cells. To know how many cells there are in the cylinder, and to get a testing point for each of them, one has to look for the real roots of $P_i(a_C, X_n)$. This follows again from Sturm's theorem, but one has to take into account the fact that the coefficients may be algebraic numbers.

REMARK 28.3. The knowledge of a CAD is useful to decide questions of the kind: is a given semi-algebraic set A (defined by a formula involving polynomials $P_1, \ldots, P_r$) empty or not?

If we want to know more precise facts about A, for instance if it is connected, we have to know more about the CAD, namely the adjacency relations between cells, i.e., which are the indices i and j such that $C_i \subset \overline{C}_j$. This is a delicate matter; see COSTE ([1988], p. 18–22), and SCHWARTZ and SHARIR ([1983], p. 339–349) for an analysis of this question.

Let us give the principle of an application of the preceding considerations to a problem of robotics: the "Piano mover's problem" (see SCHWARTZ and SHARIR [1983] for details). Let $K \subset \mathbb{R}^3$ be a compact semi-algebraic set (the "piano"), and let $F \subset \mathbb{R}^3$ be a closed semi-algebraic set (the set of "walls" that the piano is not allowed to “touch”). For simplicity, we will identify the set D of isometries of $\mathbb{R}^3$ (i.e., maps made up by a translation followed by a rotation) with $\mathbb{R}^6$. Set $\mathbb{RP}^3$ for the real projective space of dimension 3. Then the set D is topologically the product $\mathbb{R}^3 \times \mathbb{RP}^3$; taking a chart of $\mathbb{RP}^3$ gives for instance the above identification; see SCHWARTZ and SHARIR ([1983], p. 301) for details.

The piano mover’s problem can now be described as follows: given two positions K_1 and K_2 of the piano K in $\mathbb{R}^3 \setminus F$, decide if there is a motion of K in $\mathbb{R}^3 \setminus F$ from K_1 to K_2, and if yes, describe such a motion (as a semi-algebraic set in D). To solve this problem, we look at the following diagram

$$\begin{array}{ccc} \mathbb{R}^3 \times D & \xrightarrow{\ \phi\ } & \mathbb{R}^3 \\ \Big\downarrow{\scriptstyle \pi} & & \\ D & & \end{array}$$

where π is the second projection, and ϕ is defined by $\phi(x, d) = d(x)$.

Let $\widetilde{D} \subset D$ be the set of “allowed positions”, i.e., deplacements d such that $d(K) \cap F = \emptyset$. We have the following easy proved relation:

$$D \setminus \widetilde{D} = \pi\big(\phi^{-1}(F) \cap (K \times D)\big). \tag{28.2}$$

As $\phi^{-1}(F) \cap (K \times D)$ is clearly a semi-algebraic set, we see that $D \setminus \widetilde{D}$, which is the projection of a semi-algebraic set, is also semi-algebraic. It follows that $\widetilde{D}$ is a semi-algebraic set in $\mathbb{R}^6$. To describe $\widetilde{D}$, and decide whether there exists a path joining two given positions of K in $\widetilde{D}$, we then use the CAD algorithm described above, in $\mathbb{R}^9 \simeq \mathbb{R}^3 \times D$, adapted to the polynomials describing the semi-algebraic set $\phi^{-1}(F) \cap (K \times D)$.

References

ANDERSON, D.C. and T.W. SEDERBERG (1985), Steiner surface patches, *IEEE Computer Graphics Appl.* **5**(5), 23–36.

BARSKY, B.A. (1988), *Computer Graphics and Geometric Modeling Using Beta-Splines* (Springer, Berlin).

BARTELS, R.H., J.C. BEATTY and B.A. BARSKY (1987), *An Introduction to Splines for Use in Computer Graphics and Geometric Modeling* (Morgan Kaufmann, Palo Alto).

BEATTY, J.C. and B.A. BARSKY (1983), Local control of bias and tension in beta-splines, *Comput. Graph.* **17**(3), 193–216.

BENEDETTI, R. and J.-J. RISLER (1990), *Real Algebraic and Semi-Algebraic Sets* (Hermann, Paris).

BÉZIER, P. (1972), *Numerical Control, Mathematics and Applications* (Wiley, London).

BILLERA, L.J. and L. ROSE (1989), Gröbner basis methods for multivariate splines, in: T. LYCHE and L.L. SCHUMAKER, eds., *Math. Methods in CAGD* (Academic Press, New York) 93–104.

BÖHM, W. (1980), Inserting new knots into B-spline curves, *Comput. Aided Design* **12**, 199–201.

BÖHM, W., G. FARIN and J. KAHMANN (1984), A survey of curve and surface methods in CAGD, *Comput. Aided Geom. Design* **1**, 1–60.

BUCHBERGER, B. (1970), Ein algorithmisches Kriterium für die Lösbarkeit eines Algebraischen Gleichungssystem, *Aequationes Math.* **4**, 374–383.

BUCHBERGER, B. (1976), Theoretical basis for the reduction of polynomials to canonical forms, *SIGSAM Bull.* **39**, 19–29.

BUCHBERGER, B. (1985), Gröbner bases: An algorithmic method in polynomial theory, in: N.K. BOSE, ed., *Multidimensional Systems Theory* (Reidel, Dordrecht) 184–232.

CANNY, J. and D. MANOCHA (1992), Algorithm for implicitizing rational parametric surfaces, *Comput. Aided Geom. Design* **9**, 25–53.

CANNY, J. and J.M. ROJAS (1991), An optimal condition for determining the exact number of roots of a polynomial system, in: *Proceedings ISSAC 91* (ACM Press, New York) 96–102.

CIARLET, P.G. (1982), *Introduction à l'Analyse Numérique Matricielle et à l'Optimisation* (Masson, Paris). English translation: *Introduction to Numerical Linear Algebra and Optimisation* (Cambridge Univ. Press, Cambridge, 1989).

COLLINS, G.E. (1975), Quantifier elimination for real closed fields by cylindrical algebraic decomposition, in: *Proc. 2nd Conf. Automata Science Theory and Formal Language*, Lect. Notes in Comp. Science **33** (Springer, Berlin) 134–183.

COLLINS, G.E. and R. LOOS (1982), Real zeros of polynomials, *Symbol. Algebraic Comput.* (Computing Supplementum) **4**, 83–94.

COONS, S.A. (1967), Surfaces for computed aided design of space forms, MIT Project MAC-TR-41.

COSTE, M. (1988), Effective semi-algebraic geometry, in: J.D. BOISSONNAT and J.P. LAUMOND, eds., *Geometry and Robotics*, Lecture Notes in Comp. Science **391** (Springer, Berlin) 1–27.

COX, M.G. (1972), The numerical evaluation of B-splines, *J. Inst. Math. Appl.* **10**, 134–149.

DAHMEN, W. and CH.A. MICHELLI (1983), Recent progress in multivariate splines, in: C.K. CHUI, L.L. SCHUMAKER and J.D. WARD, eds., *Approx. Theory* **IV** (Academic Press, New York) 27–121.

DAVENPORT, J., Y. SIRET and E. TOURNIER (1987), *Calcul Formel* (Masson, Paris).

DE BOOR, C. (1972), On calculating with B-splines, *J. Approx. Theory* **8**, 19–45.

DE BOOR, C. (1978), *A Practical Guide to Splines* (Springer, New York).

DE BOOR, C. and K. HÖLLIG (1987), B-splines without divided differences, in: G. FARIN, ed., *Geometric Modeling* (SIAM, Philadelphia, PA) 21–27.

DE CASTELJAU, P. (1985), *Formes à Pôles* (Hermès, Paris).

DU, W.H. (1988), Etude sur la représentation de surfaces complexes, Thèse, ENST Paris, groupe image.

EPSTEIN (1976), *On the influence of parametrization in parametric interpolation*, SIAM J. Numer. Anal. **13**(2), 261–268.

FARIN, G. (1988), *Curves and Surfaces for CAGD* (Academic Press, New York).

FAUX, I. and M. PRATT (1979), *Computational Geometry for Design and Manufacture* (Ellis Horwood, Chichester, UK).

FIOROT, J.-C. (1987), Courbes Bézier, in: A. LE MÉHAUTÉ and P. SABLONNIÈRE, eds., *Courbes et Surfaces Bézier/B-splines* (INSA, Rennes).

FIOROT, J.-C. and P. JEANNIN (1989), *Courbes et Surfaces Rationnelles*, RMA **12** (Masson, Paris).

FULTON, W. (1969), *Algebraic Curves* (Benjamin, New York).

GONZALEZ, L., H. LOMBARDI, T. RECIO and M.-F. ROY (1990), Spécialisation de la suite de Sturm et sous-résultants, *RAIRO Inform. Théor. Appl.* **24**(6), 561–588.

HOFFMANN, C.M. (1992), Implicit curves and surfaces in CAGD, Techn. Report CSD-TR-92-002, Purdue University, West Lafayette, IN 47907, USA.

KNUTH, D.E. (1981), *The Art of Computer Programming* **II**, *Semi-Numerical Algorithms* (Addison-Wesley, New York, 2nd ed.).

LANG, S. (1971), *Algebra* (Addison-Wesley, New York).

LEE, T.Y. (1987), The rational Bézier representation for conics, in: G. FARIN, ed., *Geometric Modeling* (SIAM, Philadelphia, PA) 3–19.

LEJEUNE, M. (1985), Effectivité de calculs polynomiaux, Preprint, Institut Fournier, Univ. de Grenoble I.

LOOP, C. and T. DE ROSE (1990), Generalized B-spline surfaces of arbitrary topology, *Comput. Graph.* **24**(4), 347–356.

LOOS, R. (1982), Polynomial remainder sequences, in: B. BUCHBERGER, G.F. COLLINS and R. LOOS, eds., *Symbol. Algebraic Comput.* (Computing Supplementum 4) (Springer, Berlin) 115–137.

MIGNOTTE, M. (1982), Some useful bounds, in: B. BUCHBERGER, G.F. COLLINS and R. LOOS, eds., *Symbol. and Algebraic Comput.* (Computing Supplementum 4) (Springer, Berlin) 259–263.

RAMSHAW, L. (1987), Bézier and B-splines as multiaffine maps, in: *Theorical Foundations of Computer Graphics and CAD* (Springer, New York) 757–776.

RENNES (1987), *Courbes et Surfaces Bézier-B-splines*, P. SABLONNIÈRE and A. LE MÉHAUTÉ, eds. (INSA, Rennes).

RIOBOO, R. (1991), Quelques aspects du calcul exact avec les nombres réels, PhD Thesis, LITP, Univ. Paris 6.

RISLER, J.-J. (1991), *Méthodes Mathématiques pour la CAO* (Masson, Paris). English translation: *Mathematical Methods for CAD* (Cambridge Univ. Press, Cambridge).

RONGA, F. and T. VUST (1990), Les equations des surfaces de Steiner, Preprint, Université de Genève, Suisse.

SABLONNIÈRE, P. (1982), Bases de Bernstein et approximants splines, PhD thesis, University of Lille, France.

SABLONNIÈRE, P. (1987), Courbes splines, in: A. LE MÉHAUTÉ and P. SABLONNIÈRE, eds., *Courbes et Surfaces Bézier/B-splines* (INSA, Rennes).

SCHMITT, J. and W.H. DU (1987), Bézier patches with local shape control parameters, in: *Eurographics '87*, Amsterdam.

SCHUMAKER, L.L. (1981), *Spline Functions: Basic Theory* (Wiley, New York).

SCHWARTZ, J.-T. and M. SHARIR (1983), On the piano Mover's problem 2, *Adv. Appl. Math.* **4**, 296–351.

SEIDEL, H.-P. (1989), A new multiaffine approach to B-splines, *CAGD* **6**, 23–32.

SEIDEL, H.-P. (1991), Polar forms and triangular B-spline surfaces, *SIIGRAPH '91*.

STURMFELS, B. (1993), Sparse elimination theory, in: D. EISENBUD and L. ROBBIANO, eds., *Computational Algebraic Geometry and Commutative Algebra Proc. Cortona* (Cambridge Univ. Press, Cambridge) 377–396.

WARREN, J. (1990), The effect of base points on rational Bézier surfaces, Rice Techn. Report TR-90-122.

Subject Index